Biosphere 2000

Protecting Our Global Environment

Donald G. Kaufman
Miami University

Cecilia M. Franz
Miami University

HarperCollinsCollegePublishers

Executive Editor: Glyn Davies
Developmental Editor: Kathleen Dolan
Project Editor: David Nickol
Art Director: Teresa J. Delgado
Art Coordinator: Kathy Skultety
Text Design: A Good Thing, Inc.
Cover Design: Teresa J. Delgado
Cover Photo: Envision © Peter Dombrovskis
Photo Researcher: Lynn Mooney
Production Administrator: Jeffrey Taub
Compositor: Black Dot Graphics
Printer and Binder: Von Hoffmann Press, Inc.
Cover Printer: The Lehigh Press, Inc.

Biosphere 2000: Protecting Our Global Environment

Library of Congress Cataloging-in-Publication Data

Kaufman, Donald G.
    Biosphere 2000: protecting our global environment / Donald G.
  Kaufman, Cecilia M. Franz.
      p. cm.
    Includes bibliographical references and index.
    ISBN 0-06-043576-3
    1. Human ecology. 2. Environmental protection. 3. Conservation
  of natural resources. I. Franz, Cecilia. II. Title.
GF41.K38 1993
363.7—dc20   92-19689
                                              CIP

92 93 94 95   9 8 7 6 5 4 3 2 1

"A society whose youth believe only in the Now is deceiving itself. It denies man's basic and oldest characteristic, that he is a creation of memory, a bridge into the future, a time binder."

Loren Eiseley

Dedication

This book is dedicated to all of the students who helped to create it and who believed in it—and in the power of youth and love—to make a difference in the world. It is especially dedicated to the memory of Laura Luedeke, who was among the group of students who originally conceived of the book. Long before the book was completed, indeed, while it was still taking shape in our minds, she envisioned it, believed in it, and strove to realize it. Our fondest wish is that her belief and dedication—like the belief and dedication of all those very special students—may encourage others to act as time binders, building solid bridges to bring us all into the 21st century.

D.G.K.

C.M.F.

Contents in Brief

Contents in Detail

Preface

Living on Round River

"One of the marvels of early Wisconsin," wrote the ecologist Aldo Leopold, "was the Round River, a river that flowed into itself, and thus sped around and around in a never-ending circuit. Paul Bunyan discovered it, and the Bunyan saga tells how he floated many a log down its restless waters.

"No one has suspected Paul of speaking in parables, yet in this instance he did. Wisconsin not only *had* a round river, Wisconsin *is* one. The current is the stream of energy which flows out of the soil into plants, thence into animals, thence back into the soil in a never-ending circuit of life."

Implicit in Leopold's writing is the idea that the entire earth is also a round river. And we—the humans who live on the river—are bound up with it, part and parcel of the environment that we affect and are affected by.

Themes

This book is an environmental primer, a foray into the workings and the wonder of the earth and the problems that beset it. It is written for use in introductory environmental science college courses or in general ecology courses with an environmental emphasis, but it contains a message and information for all individuals and communities. In all the topics we discuss, we emphasize four major themes.

First, despite the scope and gravity of environmental problems, we find reason to hope. This book is not a death knell for a doomed planet, nor is it an apologia for past failures. Inarguably, humanity faces many complex environmental problems, but focusing on the difficulty of resolving problems can only paralyze us. We have attempted to examine the many environmental problems that beset us, the dimensions of those problems, and their varied and interrelated causes. While we acknowledge the difficulty of the present situation, we have taken a positive approach. For example, we consider the opportunity solid wastes present as well as the problem; we look at solid wastes as "unrealized resources" that society can and should begin to use

more fully. Moreover, notice that this text contains no unit or section on pollution per se. Rather, Chapter 6, Ecosystem Degradation, presents a general discussion of the topic, including the types and sources of pollution and their effects and associated problems. Specific information on pollution is then integrated into appropriate chapters, especially those on air, water, and soil. Finally, throughout the text we offer lists of practical things—called "What You Can Do"—that individuals can do to help protect and preserve resources and natural systems. These suggestions enable students, teachers, classes, and communities to become active participants in the greatest challenge humans have ever faced: the struggle to preserve our global environment.

A second major theme of the book is that environmental problems, although complex, can be solved through the use of an interdisciplinary problem-solving model. This flexible and effective model, introduced in Chapter 2, is based on similar models developed for use in the social sciences, and it has been applied successfully to many environmental problems.

A third important theme of the text is that many problems can be avoided altogether through environmentally sound resource management. Such management is proactive and cost-effective: by preventing environmental problems, we avoid the necessity of costly remedies. Of course, "management" is a tricky idea; experience has shown that we can't manage natural systems in the sense that we can do whatever we like. Further, many environmentalists argue that effective management is impossible until we first learn to manage our own species, primarily by controlling population growth and conserving resources.

The fourth major theme of the text is that all human societies must develop a stewardship or land ethic of resource use. Aldo Leopold, whom many consider the spiritual founder of the modern environmental movement in this country, was among the first to merge the scientific lessons of ecology with the philosophical awareness of ethics. Leopold explained his land ethic in his influential writings, particularly *A Sand County Almanac* and *Sketches Here and There*. Leopold encouraged us to view "the land" as a community of organisms intimately bound up with its physical environment and sparked by a current of energy flowing through its living parts. His works sparked the imagination of generations of scientists and nonscientists alike, who came to realize that treating nature as if it were disposable property would eventually lead to the demise of this community and hence to the collapse of modern civilization. Leopold maintained that each human is "but a plain member and citizen of the land-community," entitled to a share of resources but charged with the responsibility to act as a steward, or caretaker, of the land-community.

How This Book Began

During the fall semester in 1983 one of us (Donald Kaufman) taught an environmental science class for honors students at Miami University, Oxford, Ohio. At the end of the semester the students agreed to do additional research on specific natural systems or resources with the long-term goal of incorporating their research into a textbook for college freshmen or sophomores taking their first, and possibly only, environmental science course. While there are many books targeted for these students, none had been developed with the assistance of undergraduate nonmajors. The students believed that a book developed *by* students would be helpful to their peers. This was the origin of the book development team. One of the first things we did was send a questionnaire to university professors nationwide who teach environmental science or resource management courses. We asked them to suggest resources or natural systems that best illustrate environmental principles. Each student selected a topic from these suggestions, developed a comprehensive research plan, and, for the next three years, pursued that plan in extensive library and field investigations. One student traveled to Boston to meet with a leading authority on whales, another headed to Marathon County, Wisconsin, to study soil and land use, and yet another ventured to the Everglades of Florida.

These students, and others who later joined the project, focused on the environmental problems associated with each resource and how it had been managed in the past. They summarized and evaluated various management strategies and made suggestions for future management. After years of effort, they completed extensive reports. These reports formed the basis for the **Environmental Science in Action** sections that supplement Chapters 3–24.

These essays give concreteness to the environmental principles discussed.

It's impressive enough to realize that these students—18 and 19 years old when the project began—took on this work simply because they thought it important and worthwhile. What's even more impressive is that they took the initiative to seek and secure the financial support needed to fund their research. A grant from the George Gund Foundation of Cleveland, Ohio, supported student researchers during the summer months. Additional grants from Miami University's Fund for Excellence financed student travel and provided students with support as they worked to refine their preliminary reports. (See pages E-1 through E-3 for a list of these students and what they are doing now.)

Our project had its detractors. A fellow professor maintained that undergraduates were not capable of either the intense research or the long-term commitment necessary to accomplish such an ambitious goal. Clearly, the students proved him wrong; more importantly, they have proven to themselves that individuals working toward a common goal can achieve it, despite the difficulty. And so, in honor of the students whose work and enthusiasm gave birth to this text, we are donating all royalties from the first edition to support future undergraduate research. The newly established **Miami University Foundation Global Heritage Endowment** will award grants on a competitive basis to enable undergraduate students nationwide to undertake environmental projects. We invite students or their teachers to write to us for details.

Organization

The Prologue invites readers to examine their worldview. Simply put, a worldview is how we look at the world and how we define our place in it. Examining our worldview is an important first step in the process of becoming environmentally aware. Such an examination helps us to understand the values, beliefs, and attitudes that shape our actions toward the environment, how we use resources, and how we respond to environmental problems.

The body of the text is divided into seven major units. Unit I presents an overview of environmental problems and the state of the biosphere (Chapter 1, Where Are We Now?) and shows how problem-solving techniques and environmentally sound management can help us to achieve a sustainable future (Chapter 2, And Where Do We Want To Go?).

Unit II examines how the biosphere works. Chapters 3 to 5 present ecological principles of ecosystem structure, function, development, and equilibrium.

Chapter 6 looks at the degradation of ecosystems, with emphasis on degradation caused by human activities. Chapter 7 discusses the application of ecological principles to restoring and preserving ecosystems.

Units III through VI examine the ways in which humans affect the environment. Each unit looks at resources that have common qualities or are linked in a significant way. Unit III considers the environmental imperative to balance the human population, food, and energy resources. The three chapters on energy explore the environmental issues related to energy consumption, society's use of conventional fuels (oil, coal, and natural gas), and alternative sources of energy.

Protecting the environmental triad of air, water, and soil is examined in Unit IV.

Resources and products critical to industrial societies—minerals, nuclear resources, toxic and hazardous substances, and "unrealized resources"— are discussed in Unit V.

Unit VI is concerned with public lands, wilderness, biological resources, and cultural resources. Clean air, clean water, and fertile soil can help to feed our bodies, but wildlands, other living creatures, and human artifacts feed our minds and spirits.

Finally, Unit VII, An Environmental Legacy: Shaping Human Impacts on the Biosphere, looks at the cultural systems that shape human attitudes and behavior toward the natural world. Although the past impact of cultural systems, such as religion, ethics, economics, and politics, is included, the emphasis is on the present and future. We attempt to show how these powerful factors can be and are being used to modify and enhance human interactions with the biosphere.

Special Features

Biosphere 2000 has many special features. No other environmental science text asks its readers to examine their worldview before they undertake an examination of the environment, and no other text offers proven models for solving environmental problems and for managing resources in an environmentally sound manner.

Biosphere 2000 also includes three chapters that are unique to environmental science texts: applied ecology (Chapter 7), cultural resources (Chapter 24), and environmental education (Chapter 28). Just as an understanding of ecological concepts is crucial to environmental awareness, a knowledge of the diverse social forces that affect environmental

decisions is also vital. Chapters on religion and ethics (Chapter 25), politics and economics (Chapter 26), and law and dispute resolution (Chapter 27) illustrate how attitudes and behaviors are shaped by these important disciplines and how these disciplines can be used to help or hinder environmentally sound management.

Special sections of the text, entitled **Focus On**, give readers a closer look at a specific resource, illustrate an ecological principle, or delve more deeply into a particular environmental issue. Many are written by guest essayists and thus offer a variety of points of views.

Chapters 3 through 24 (excluding Chapter 9) include special chapter supplements called **Environmental Science in Action**. Based on the reports of our undergraduate book development team, the supplements emphasize the environmental problems associated with a particular system or resource and discuss the ways in which the system or resource is or could be managed.

To make these special supplements and the text itself even more accurate and up-to-date, we drew on the research and experience of graduates of the master's program of Miami University's Institute of Environmental Sciences (IES). The graduates work in both the public and private sectors, and many have gone on to earn a doctoral degree in a specialized environmental field. Many of these professionals reviewed and augmented the chapter supplements and text. They and the academic reviewers of the manuscript are acknowledged below (see "Reviewers").

Other special features of the text include:

- Chapter objectives
- Key terms
- Discussion questions
- Bibliography for further reading and research
- Lists of environmental organizations and publications
- Glossary
- Photographs supplied by the National Geographic Society, World Wildlife Fund, and numerous individuals. Because of the not-for-profit nature of the text, these organizations and individuals have allowed us to use their photographs at a nominal cost. Our thanks to them for their generous support of this project.
- Use of many real-life, current examples to illustrate principles throughout the text. We have selected numerous examples from regions throughout North America. We also provide esamples from other continents in order to provide students with a global perspective.
- **What You Can Do** sections containing specific

suggestions on ways individuals can become involved in environmental issues, help to preserve natural systems, and protect resources. We invite students and teachers to write to us with additional suggestions for future editions of *Biosphere 2000*.

Supplementary Materials

The supplementary materials accompanying *Biosphere 2000* are:

- *Instructor's Resource Manual and Test Bank* by Lisa Breidenstein, Donald G. Kaufman, John Hurley (Eastern Kentucky University), LuAnne Clark (Lansing Community College), and William Rogers (Winthrop University) includes for each chapter an overview, lecture outline, student objectives, teaching tips, and an F.Y.I. section. The test bank features over 3000 multiple-choice, true/false, sentence-completion, and critical-thinking questions. The test bank is also available on HarperCollins Testmaster software for IBM and Macintosh computers.
- *Student Resource Guide* by Cheryl Puterbaugh and Donald G. Kaufman contains, for each chapter, a chapter overview and outline, learning objectives, key terms, suggested activities, and approximately 20 review questions.
- *Finding Our Niche: The Human Role in Healing the Earth* by Karla Armbruster and Donald G. Kaufman presents expanded versions of the Environmental Science in Action supplements featured in the text. Each of the 21 cases describes a real-world application of environmental management.
- *Transparency Acetates:* Over 120 full-color acetates of art from the text are available to adopters.
- *The Student Environmental Action Guide:* Published in conjunction with the Earth Works Group, this handbook lists 25 practical things students can do to help improve the environment, complete with background statistics, examples, organizing tips, resources, and contacts.
- *A Short Guide to Writing About Biology:* by Jan Pechenik of Tufts University: This popular brief guide teaches biology students how to read more thoughtfully and think more carefully, and then translate these skills into effective writing in the science area.

Field Testing

The textbook and its supplementary materials were used by Donald G. Kaufman in his environmental science courses prior to publication. Our sincere thanks to the more than 300 students who evaluated the manuscript and offered so many valuable suggestions for improvements. Thanks, too, to the panel of undergraduates, selected from those courses, who evaluated later drafts of the manuscript. Their insight and commitment have resulted in a far better text.

A Note to Teachers

There is no one right way to teach environmental science and no one right sequence of topics to be covered. Some instructors may want to start with the Prologue; others may wish to use it at the end of their course as a capstone experience. Some may want to delay introducing our environmental problem-solving model (Chapter 2) until they feel students will be more comfortable with or receptive to it. Some instructors may choose to start with human population dynamics and examine the consequences of human impact on the earth before they examine ecological concepts or specific environmental or resource management topics. We believe that the contents of this text can accommodate the creativity and teaching needs of all instructors. We hope to receive feedback from many of you. In that way, our own classes will improve and we will be able to incorporate your experiences in subsequent editions. (Please see our *Instructors Resource Manual* for suggestions about teaching and learning techniques.)

A Note About the Authors

Any book reflects the concerns and background of its authors, and so it is with this text. Its major characteristics—a positive approach, the use of a problem-solving model, an emphasis on the interdisciplinary nature of environmental science—derive from its co-authors. Donald Kaufman has taught environmental science courses for more than twenty years and is a former deputy director for research and a current affiliate of the Institute of Environmental Sciences at Miami University. A practicing environmental scientist, he has used the problem-solving method in many situations: to establish several recycling centers, to design a county park, to manage a wildlife preserve, to develop an interpretive center for a nature preserve, and to convert an abandoned gravel pit into an apartment-lake complex, to name just a few. Because he is also an ecologist, the text reflects his background in biology.

The interdisciplinary nature of environmental science is illustrated by the fact that the other co-author, Cecilia Franz, is a writer with a literature background who specializes in the communication of scientific and technical materials. In addition to her work at Miami University, she is involved in writing projects for the Center for Reproduction of Endangered Wildlife (CREW) at the Cincinnati Zoo and Botanical Garden. We feel that nonscience majors will appreciate and respond to the influence of social sciences and humanities she brings to *Biosphere 2000*.

Acknowledgments

Since its inception, this textbook has benefitted from the help and participation of students, educators, and environmental professionals. (Profiles of the students who researched and developed the Environmental Science in Action chapter supplements appear at the back of the book on pages E-1 through E-3.) Any errors that may remain are, of course, the responsibility of the authors.

There are many people at HarperCollins who helped to guide and shape this book throughout the long development and production process, and we owe our thanks to the editors and artists whose work and talents have made this a better book. Several people must be singled out for their efforts. First and foremost, we want to thank LeeAnne Fisher, who first approached us about the idea of writing an environmental science textbook with HarperCollins (then Harper & Row) and who remained, throughout the ensuing years, both a personal friend and our staunchest supporter. Kathleen Dolan helped to develop the manuscript, guided it through review and revision stages, and continued to check on the book as it passed through the production phase. Barbara Conover provided immensely helpful suggestions at a critical juncture, thereby transforming the manuscript into a cohesive whole. Judith Kahn's careful, "leave no stones unturned" approach to copyediting made the manuscript tighter and more exacting. Lynn Mooney, photo researcher, found wonderful images to accompany the text; we are thankful for both her perceptive nature and her persistence. David Nickol saw the manuscript through the production stages with a calm, low-key approach that helped to sooth otherwise jangled nerves. Finally, Glyn Davies skillfully oversaw the entire process, providing human and financial resources when needed.

We have also benefited from the assistance of many people here at home. Our sincere thanks go to Dr. Gene Willeke, Director of Miami's Institute of Environmental Sciences; Dr. Richard Nault, Director of the University Honors Program; and Dr. Robert Sherman, former Chair of the Zoology Department, Miami University, for their continuing and unfailing support. And on the subject of support, a special thanks to our families, whose encouragement and patience over the last few years made it possible for us to see this book through to its completion.

We tapped the special talents of certain friends and associates to produce this textbook. Susan Friedmann, a freelance artist living and working in Cincinnati, Ohio, created many of the original illustrations that grace the book. Ken Constant, a computer graphics specialist with a Master's degree in environmental science, supplied computer images that provided the model for many of the text's graphics. The maps that accompany the Environmental Science in Action pieces were created by Laura Taylor, a graduate of the School of Public and Environmental Affairs (SPEA), Indiana University. The expertise, enthusiasm, and hard work of library scientist Nancy Moeckel made our trips to Miami's Brill Science Library far more productive. Mary Songhen, a former English professor at Miami, provided valuable advice early in the project as the students researched and developed the Environmental Science in Action supplements; she helped us again years later in proofreading the text galleys.

We owe a special debt of gratitude to several individuals who helped to research and draft selected chapters that correspond to their particular interests or fields. Wright Gwyn, Director of the Forest Park, Ohio, Environmental Awareness and Recycling Program, worked diligently on the energy chapters and on issues related to nuclear resources (Chapters 11–13 and Chapter 18). Greg McNelly, an associate with the Washington D.C.-based Clean Sites, Inc., a nonprofit group that assists in the cleanup of hazardous waste sites nationwide, showed considerable persistence in helping us to present a difficult topic, toxic and hazardous substances (Chapter 19), in understandable language. Andy Jones, currently a Master's degree candidate in Miami University's botany program, translated his enthusiasm for recycling into a comprehensive look at the management options available to us for unrealized resources (Chapter 20). Finally, Melinda Thiessen, formerly manager of the technical communications department at the Sacramento, California, office of Radian Corporation, an environmental consulting firm, eagerly tackled the topic of the social forces that shape environmental issues (Chapters 25–28). Her interest in dispute resolution, which served her so well during the research of these chapters, has since led her to Trinity College in Dublin, Ireland, where she is currently working on a Master of Philosophy Degree in Peace Studies.

Last but not least, we wish to acknowledge the many friends who willingly did the often mundane but very important tasks necessary in putting together a textbook. Those people whose assistance was particularly helpful and unfailing include Cheryl Puterbaugh, Greg McNelly, Lisa Breidenstein, Wright Gwyn, Andy Jones, Pam Marsh, Bobbie Stringer, Jim Tish, Eric Lazo, Kate Grady, Chris Brueske, Chris Carter, Laura Taylor, Karla Armbruster, and finally, Lisa Taylor, who came through in the dark winter months of 1991–1992 and put us over the top. We owe them our deepest gratitude.

Donald G. Kaufman
Cecilia M. Franz

Reviewers

We would like to thank the following academic and technical reviewers for their suggestions and advice throughout the development of this text:

Clark E. Adams, Texas A&M University

Doug Ammon, Clean Sites, Inc., Alexandria, Virginia

Valerie A. Anderson, California State University, Sacramento

Susan Arentsen, U.S. EPA, Cincinnati, Ohio

Lisa Bardwell, University of Michigan, Ann Arbor

Jeff Binkley, Roy F. Weston, Inc. (hazardous waste remediation), Okemos, Michigan

Dale J. Blahna, Northeastern Illinois University

Anne Brataas, Pioneer Press, St. Paul, Minnesota

W.M. Brock, Great Basin National Park, Baker, Nevada

LuAnne Clark, Lansing Community College

Paul Doscher, Society for the Protection of New Hampshire Trees, Concord, New Hampshire

Robert Dulli, National Geographic Society, Washington, D.C.

Steve Edwards, Miami Oxford Recycling Enterprise, Oxford, Ohio

Hardy Eshbaugh, Miami University,

Eric Fitch, Director, Coastal Zone and Natural Resource Studies, University of West Florida

Lloyd Fitzpatrick, University of North Texas

Herman S. Forest, State University of New York, Geneseo

Dwain Freels, FFA Director, Oxford, Ohio

Gerald Gaffney, Southern Illinois University

Greg Githens, O.H. Materials Company (hazardous and toxic waste management), Findlay, Ohio

Wright Gwyn, Director of the Environmental Awareness and Recycling Program for the cities of Forest Park and Green Hills, Ohio

Elmer B. Hadley, University of Illinois, Chicago

John Harley, Eastern Kentucky University

Anne Heise, Washtenaw Community College

Clyde Hibbs, Professor Emeritus, Ball State University

William Hinger, Environmental Specialist, Huntington Mortgage Company, Columbus, Ohio

Harry L. Holloway, Jr., University of North Dakota

Verne B. Howe, Southern Vermont College

Mario Huerta, Environmental Education Specialist, Foundation Mexicana para la Educacion Ambiental, Tepotzotlan, Mexico

Hugo H. John, University of Connecticut, Storrs

Mary Mederios Kent, Population Reference Bureau, Inc., Washington, D.C.

Robert Kleinhenz, Ross Incineration Services, Inc., Grafton, Ohio

Barbara Knuth, Cornell University, Ithaca, New York

Kirk E. La Gory, Argonne National Laboratory, Argonne, Illinois

Terry A. Larson, Texas A&M University

Rebecca Lorato, Lansing Community College

Anne Lynn, U.S. Dept. of Agriculture Soil Conservation Service, College Park, Maryland

Margaret Lynn, Ohio Department of Health, Columbus, Ohio

John Lytle, Director of Air Pollution Services, Burgess & Niple, Ltd., Columbus, Ohio

Margaret Martin, Planergy, Inc., Houston, Texas

Lisa McDaniel, Environ-Protection Systems, Chantilly, Virginia

Greg McNelly, International Technology Corporation, Cincinnati, Ohio

Rick Melberth, SUNY–Plattsburgh

Mary Beth Metzler, Technical Communications, Rensselaer Polytechnic Institute

Dennis Miller, Oak Ridge National Laboratory, Oak Ridge, Tennessee

James Millette, McCrone Environmental Services, Inc., Norcross, Georgia

Graham Mitchell, Ohio EPA, Dayton, Ohio

Nancy Moeckel, Natural Resources Specialist, Brill Science Library, Miami University

Earnie Montgomery, Tulsa Junior College

Joseph Moore, California State University, Northridge

Mary Ann Moore, Wellesley College

Steven Oberjohn, Westinghouse Environmental Management Company, Cincinnati, Ohio

Kay Phillips, Miami University

Anantha Prasad, University of Illinois, Urbana

Joseph Priest, Miami University

Ernest Puber, Northeastern University

Claire Puchy, Oregon Department of Fish and Wildlife, Portland, Oregon

Carl H. Reidel, University of Vermont

Joe Rogers, Environmental Control Technology Corporation, Ann Arbor, Michigan

William Rogers, Winthrop College

Judith Salisbury, New Alchemy Institute, East Falmouth, Massachusetts

F.A. Schoolmaster, North Texas State University

Philbin Scott, McDonnell Douglas, Mesa, Arizona

Clark Sorensen, Manager of Information Systems and Services, Indiana University, Bloomington, Indiana

Melinda Thiessen, Environmental Services Department, Radian Corporation, Davis, California

Jim Thomas, Hanford Education Action League (HEAL), Spokane, Washington

John Thompson, Central States Education Center, Urbana, Illinois

Catherine Tompson, Baltimore Zoo, Baltimore, Maryland

Don Van Meter, Ball State University

Barbara Venneman-Fedders, New River Gorge National River, Beckley, West Virginia

John D. Vitek, Oklahoma State University

Mary C. Wade, College of Lake County

Robert Whyte, Lake County Health Department, Waukegan, Illinois

John Wilson, Aullwood Audubon Center and Farm, Dayton, Ohio

Arthur L. Youngman, Wichita State University

Prologue
Discovering Our Worldview

"Every form of refuge has its price."*
The Eagles

On December 21, 1968, Apollo 8 broke free of the earth's gravitational field and sped outward into the blackness of space. Its mission: to orbit the moon, enabling the crew to photograph and study our planet's only natural satellite. The mission was important to the lunar program of the United States because it would supply information about the lunar surface and pave the way for a future moon landing. But the craft's journey turned out to have a greater, unexpected significance. As Apollo 8 emerged from the dark side of the moon at the end of its fourth orbit, the crew—Frank Borman, James A. Lovell, Jr., and William A. Anders—rotated the craft. As they did so, they became the first human beings to witness an earthrise (Figure P-1). Anders, recalling the sight, said, "We saw the beautiful orb of the planet coming up over this relatively stark, inhospitable lunar horizon and it brought back to

me that indeed even though our flight was focused on the moon, it was really the earth that was most important to us." That realization resonated in the hearts of millions who saw the photographs and video footage the Apollo 8 crew sent back to a waiting earth. Those images forever altered humankind's view of our home planet. For the first time we gazed at the earth from the vantage point of outer space, and what we saw was a strikingly beautiful oasis of life dwarfed by the black void of space through which it traveled.

The heavens have always been an object of wonder, mystery, and speculation. As human cultures arose, people developed different explanations of the origin of the earth and the universe and of the human place in the cosmos. These accounts helped to form their worldview, a way of looking at reality. The earliest worldviews

FIGURE P-1: Photo of earthrise taken by Apollo 8 crew. The most important result of humankind's ventures into space is that we have a clearer, more accurate view of Planet Earth. The physical reality of the planet—its significance in comparison to the universe—and the possibility that it is the only planet capable of supporting life as we know it have profoundly altered the way we think about Earth.

(a)

(c)

(b)

(d)

FIGURE P-2: Diverse worldviews. (a) The Senecas, a native American tribe, believed that the earth formed when a woman dropped from the sky, fell to the sea, and was helped by a turtle who rose to the surface and allowed her to ride upon its back. (b) The god Shiva symbolizes the ancient Hindu belief in a cyclic universe. The halo of fire surrounding the god represents his dominion over the cycle of creation, destruction, and rebirth. (c) According to Egyptian mythology, Sky is a feminine god who arches over the male Earth. (d) Ge, or Gaia, is the ancient Greek earth goddess, whose abundance nurtures life.

FIGURE P-3: How different worldviews shape attitudes and behavior toward nature and resources. (a) Michaelangelo's painting on the ceiling of the Sistine Chapel depicts a fundamental Christian belief, that God created humankind in His image. God's touch imbues humanity with the soul that sets humans apart from all other creatures. (b) Developed by the philosopher Lao-tzu, the yin and yang represent complementary opposites—earth/sky, positive/negative, male/female, light/dark—that together form an unbroken whole and symbolize the harmonious balance of nature. (c) A leading scientific explanation for the origin of the universe is that it resulted when an incomprehensibly small, dense mass (or superatom) exploded, a concept popularly known as the big bang.

took voice in a rich diversity of creation myths, legends, and tales; they took shape in drawings, paintings, and sculpture (Figure P-2). Ultimately, worldviews take life in the beliefs, attitudes, and values of people. Worldviews are reflected in endeavors of the human spirit, like religion, philosophy, art, music, and literature (Figure P-3).

What do worldviews have to do with environmental science? As it turns out, quite a bit. Because a worldview includes basic assumptions about the self, others, nature, space, and time, and about the relationships between the self and everything else, it shapes our attitudes toward nature and influences how we use resources.

More than two decades have passed since the world watched its first earthrise. Successive space missions have sharpened our first images of Planet Earth. In 1972 astronauts took the first picture of the full earth, a planet of deep blues, verdant greens, and variegated browns, cloaked in swirling clouds of pure white. Pictures from the space shuttles and human-made satellites have further clarified our image of Earth. What we are learning is that this planet we call home is an unbelievably complex and possibly unique entity. We are learning, too, just how much humans have changed and continue to change the earth. As our physical understanding of the earth becomes more sophisticated, our worldview is also changing. Slowly—perhaps too slowly—we are realizing that the earth is not a sophisticated mechanical "spaceship," an analogy popular during the exciting era of the nation's space program two decades ago, nor are we the drivers of this craft. Many people feel that we need a new way of looking at our planet, a worldview that accurately accounts for the complexity of earth and the interdependence of its life forms.

Take some time to think about your own worldview. Jot down some notes in your text or a notebook. As you read, think about the consequences of your beliefs, attitudes, and values. What is the price of the refuge you take in Earth? What sort of earth do you want to leave for your children, for all future generations, to inherit? What will it take to make that legacy possible?

> *And don't you know the life that lives within the silent hills is just as rich and beautiful, and just as unfulfilled as man with all his intellect, his reason, and his choice. Oh who's to say the nightingale has any less a voice?*
> John Denver, "Children of the Universe"

The Biosphere and Environmental Science

Where Are We Now?

An Overview of Environmental Problems

Right now, at this moment in history and for the long haul into the next century and centuries beyond, no other issue is more relevant to the physical quality of life for the human species than the condition of our environment and status of our natural resources—air, water, soil, minerals, scenic beauty, wilderness, wildlife habitat, forests, rivers, lakes, and oceans. These resources determine the physical condition of our lives and dramatically influence the human condition. . . .

Gaylord Nelson

To the extent that the world surrenders its richness and diversity, it surrenders its poetry . . . as its options (no matter how absurd or unlikely) diminish, so do its chances for the future.

Tom Robbins

Learning Objectives

When you finish reading this chapter, you should be able to:

1. Describe the biosphere and explain how scientists believe it evolved.

2. Summarize the current state of the biosphere and identify the three root causes of environmental problems.

3. Explain how environmental problems and the human activities that cause them are related to cultural attitudes, values, and beliefs.

4. Distinguish between a biocentric and an anthropocentric worldview.

We invite you to try an experiment. Fill a clean jar with water from a pond, lake, marsh, or ocean tidal pool, seal the jar so that it is airtight, and place it on a windowsill where it receives indirect sunlight. Observe the jar for a period of a few weeks or months. What you will see is a self-sustaining *microcosm*, a world within a world (Figure 1-1). Nutrients within the water will nourish algae and aquatic plants, which in turn will provide food for microscopic animals. If you were lucky enough to include snails, water fleas, or other animals along with the water, you will find that they thrive in the world inside your jar. You need not add anything. Simply take care that the jar is not exposed to direct sunlight for long periods, or your world within a world will become too hot! If you wish, you can keep your microcosm going for years. A colleague, Bob Hays, started a microcosm in 1964; even today, the Robert Hays Memorial Microcosm, as it is affectionately known, continues to thrive on a windowsill in our lab.

Another experiment in the creation of worlds within worlds is currently taking place about 35 miles north of Tucson, Arizona, near the small com-

FIGURE 1-1: Microcosms in jars. The world within a microcosm can spark greater interest in our own world.

FIGURE 1-2: Artist's rendering of Biosphere II, longitudinal section. The structure encloses five natural and two cultural simulated earth environments: tropical rain forest, savanna, marsh, ocean, desert, intensive agriculture, and human habitat. The latter includes apartments, laboratories, libraries, and an amphitheater. Nearly 150 crops, including staples such as rice, wheat, barley, and vegetables and other crops such as coffee, grapes, and herbs, are grown in the agricultural zone. Domestic animals such as chickens and pygmy goats are also raised in the agricultural zone. Various birds, insects, reptiles, and small mammals inhabit the varied natural environments. In all, more than 3,800 species of macroscopic plants and animals inhabit Biosphere II.

munity of Oracle Junction. There, a futuristic structure of glass and steel rises abruptly from the surrounding scrub of the Sonoran desert. It is Biosphere II, a completely enclosed model space colony. Covering more than three acres, Biosphere II has been touted as a world unto itself, a $30 million "miniplanet." Powered by solar energy and, *in theory*, containing everything needed to sustain life, it is modeled after Biosphere I, the Planet Earth.

In 1991 eight researchers entered Biosphere II, where they are to remain for two years. Sealed off from the earth, dependent entirely upon an artificial environment, they are conducting numerous ongoing experiments. Within the structure, which contains seven simulated earth environments, all water and wastes are to be recycled and the air and water purified (Figure 1-2). Biosphere II, however, cannot compare with the complexity of Biosphere I. Moreover, the project has been criticized by some people, both scientists and lay observers, who charge that a carbon dioxide recovery system was secretly installed to augment Biosphere II's air recycling system, that outside air was quietly let in, and that a large supply of food was hidden inside the structure. The creators of Biosphere II deny the charges. Ultimately, it will be left to the scientific community to determine the credibility of the Biosphere II venture. Whatever the outcome, it is likely that researchers worldwide will continue to try to develop a self-sustaining system that could be used as a model for future space stations or bases on the moon, Mars, and other sites, enabling humans to travel and live in space far from earthbound supplies.

To travel and live in space: the very phrase calls to mind hundreds of scenes and images from comic

books, science fiction novels, television shows, and movies. Humans have had a long and abiding interest in the heavens, an infatuation with the notion of exploring the universe. The notion that other worlds and other life forms may exist was first suggested by the human imagination and later encouraged by scientific observation. Thus, we spend large amounts of money and energy each year to venture out into the vastness of space.

Not everyone is enthralled with the idea of space travel. Indeed, many people argue that it is ironic to spend billions of dollars to research and construct a safe environment for space travel and habitation, while the majority of earthbound inhabitants live in poverty and environmental squalor. They contend that we already have a planet home of beauty, diversity, and mystery and that that home is in critical need of our care. It would be far better, they argue, to spend the money now allocated to space research on combating the many serious environmental problems the earth faces. Otherwise, space habitation may someday become a necessity rather than a luxury.

We certainly are not suggesting an end to space exploration. Despite the danger, cost, and uncertainty, the human race should continue to venture out into the galaxy and beyond. Discovering life in any form would excite and stimulate our uniquely human curiosity about our origin and our place in the cosmos. More importantly, as researchers gain the knowledge needed to construct an artificial life-support environment that simulates the earth, they are likely to gain unprecedented insights about how our planet functions. Technological innovations that result from such research could improve the quality of life for the earth's inhabitants. It is important,

FIGURE 1-3: Planet Earth, as seen from Apollo 17.

however, that when we reach out to other worlds, we do so confident that the earth can sustain us upon our return. We must not be forced to explore the universe in search of a new home because we have made the earth inhospitable, even uninhabitable. For if we do not solve the environmental and related social problems that beset us on earth—pollution, toxic contamination, resource depletion, prejudice, poverty, hunger—those problems will surely accompany us to other worlds.

Solving the earth's environmental problems requires an understanding of how and why these problems arise. That, in a nutshell, is the purpose of this book. This first chapter presents an overview of the environmental problems that are explored in greater detail throughout the text and briefly discusses why those problems occur. A good grasp of environmental problems, however, is impossible without an understanding of the biosphere and how life on earth evolved.

What Is the Biosphere?

Viewed from a great distance, earth presents a startling and beautiful contrast to the vast blackness of space (Figure 1-3). Teeming with living things, and bejeweled with colors, it is home to millions of different species of organisms intricately woven into a complex tapestry we call life. That life exists on earth is due to the **biosphere**, the thin layer of air (atmosphere), water (hydrosphere), and soil and rock (lithosphere) that surrounds the planet and

contains the conditions to support life. The biosphere includes any place on, above, within, or below the earth's surface where life can be found, such as areas around thermal vents in the ocean deeps, underground caverns, the deepest penetration of roots into soil, and the highest reaches of avian flight. However vast it might seem, the biosphere is relatively small compared to the earth's total mass. Further, this life zone is unevenly distributed across the earth; some places are too hot, too cold, or too dry to support life.

The biosphere is made up of living and nonliving components. Nonliving physical environments comprise the **abiota**, or abiotic component, of the biosphere. Living organisms collectively comprise the **biota**, or biotic component. The millions of different species (including us, *Homo sapiens*) that depend upon the biosphere for their existence are at the same time an integral part of the biosphere. They influence the cycling of chemicals and water between the abiotic and biotic components and thus help to regulate the global environment. Because humans are part of the biota, the biosphere encompasses human, or cultural, systems such as farmlands and cities as well as natural systems such as marshes and forests.

To see how the biota and abiota are intricately linked, let's look more closely at our atmosphere. It is composed of about 78 percent nitrogen, 21 percent oxygen, 0.03 percent carbon dioxide, and trace gases. These proportions fluctuate somewhat, but over time they remain relatively constant. Incoming solar radiation is reflected off the surface of the earth and radiated back into space as heat energy. Carbon dioxide, along with the trace gases, traps some of the reflected radiation and the trapped heat warms the earth, much as the glass of a greenhouse traps the reradiated energy from sunlight and thereby warms the interior of the structure. Earth's proportion of greenhouse gases enables life to flourish.

Life on earth also helps to maintain the composition of the atmosphere. Water, nitrogen, and trace atmospheric elements such as sulfur and phosphorus all continually cycle through the biotic and abiotic components of the biosphere. For example, green plants take in carbon dioxide from the atmosphere and give off oxygen. Animals, in turn, take in oxygen and give off carbon dioxide. Thus, the biota help to maintain the composition of the atmosphere at relatively constant levels.

We cannot leave this topic without noting that human activities increase the amount of carbon dioxide in the atmosphere. The combustion of oil, coal, natural gas, and wood (all of which contain carbon) adds carbon dioxide to the atmosphere, and the clearing of forests reduces the amount of carbon dioxide taken from the atmosphere by trees for use

(a)

(c)

(b)

(d)

FIGURE 1-4: The fabulous diversity of life on earth. Among earth's estimated 30 million species are (a) winter mushrooms, seen growing on a red elm tree, (b) foamflowers, seen here in Wisconsin's Nicolet National Forest, (c) millipedes, like this one on a moss-covered rock in the Great Smoky Mountains of Tennessee, and (d) mountain gorillas of Central Africa.

in photosynthesis. These activities increase the proportion of greenhouse gases in the atmosphere, thus intensifying the greenhouse effect and contributing to global warming.

Life is the predominant characteristic of the biosphere. In all its diversity—from the familiar to the exotic, the microscopic to the gigantic, the fiercely beautiful to the beautifully fierce—life is the hallmark of Biosphere I (Figure 1-4). The earth may be the only planet in the galaxy, even in the universe, capable of supporting life as we know it. But this statement begs the question, how *do* we know life? Indeed, what is life? Scientists, philosophers, and theologians have been trying to answer that question for ages. So far, no one has come up with a definition that satisfies everyone. Scientists have, however, compiled a list of characteristics that describe living organisms.

First and foremost all organisms *live at the expense of their environment*. From the simplest one-celled bacteria (which, in one of life's delightfully ironic

twists, are far from simple) to the most complex mammals, all organisms must extract materials and energy from the biosphere in order to live. Consequently, all organisms modify their environment. All living things *have a cellular structure*. The cell is the smallest unit of life that has the structures and chemical mechanisms needed to conduct the activities associated with living. Many living organisms *exhibit movement*. Movement is one of the most reliable indicators of life (not surprising when you remember that someone or something which is being hunted will often "play possum" to confuse its predator). Living things also *show growth*, gaining mass over time. A plant manufactures the molecules that form the raw material for added mass. In contrast, an animal must acquire the raw materials needed for growth from its environment. Living organisms also *reproduce*. Not all individuals of a given species may reproduce, but living species as a whole must be capable of reproduction or the species would die out as its members died. Living

FIGURE 1-5: Response of living things to stimuli. All living things respond to stimuli in their environment. A scent on the wind has alerted this deer.

things *respond to stimuli*. Response allows an organism to react to changes in the environment, such as the movement of the sun or a scent on the wind (Figure 1-5). Finally, living things *evolve and adapt*. Just as an individual must be able to respond to changes in its immediate environment, species evolve, or change over generations, as they adapt to long-term changes in the environment. Charles Darwin, the English naturalist who first proposed the theory of evolution by natural selection, called it "descent with modification."

Describing characteristics of living organisms is quite different from defining life. Keep in mind that none of the characteristics listed above alone defines life; they must be considered as a sort of tableau which distinguishes the living from the nonliving. Moreover, some of these characteristics do not apply to all organisms, and others apply to nonliving things as well.

How Did the Biosphere Develop?

To understand how the biosphere developed, we must first understand the milieu in which it developed, that is, we need some inkling of the origins of the universe and the earth. Of course, no one can explain with absolute certainty the origins of the universe. The most widely accepted scientific theory, developed by theoretical physicists using powerful mathematical tools, holds that the universe came into being as a result of what is popularly known as the big bang. According to the **big bang theory,** the universe—all matter, energy, and space—arose from an infinitely dense, infinitely hot point called a singularity, roughly the size of a speck of dust. About 13 to 20 billion years ago, the singularity exploded and space began to expand. Only energy had existed in the singularity, where the

temperature reached billions of degrees Celsius, but as the expanding universe began to cool, particles and then matter formed. Hydrogen and helium formed during the first 500,000 years after time zero, the moment of explosion. In the young expanding universe, gravity caused the newly formed matter to clump together in large masses, the precursors of galaxies. As matter rushed outward into the void, some of the galaxies fragmented, forming stars. Nuclear reactions within the cores of the stars gave off heat and light and created the heavier elements that comprise the planets and our bodies. Scientists theorize that, about 5 to 10 billion years ago, our solar system developed from gas and dust on the outer edge of the galaxy we call the Milky Way.

The big bang theory may be difficult to imagine, but astronomic observations lend credence to the concept. In the 1920s, the American astronomer Edwin Hubble observed that the universe is expanding, an observation that would be expected if the universe had indeed arisen from some sort of explosion. Moreover, astronomers have noted the presence of cosmic background radiation—radiation that seems to originate equally from all directions in the cosmos. The characteristics of this type of radiation are consistent with what would be expected if the universe had arisen from a small, dense area that underwent a big bang.

Scientists theorize that the earth formed about 4.6 billion years ago, condensing out of interstellar gas and dust. Hot, molten, and volcanic, the planet was a lifeless sphere veiled by a thin layer of hot gases. A heavy cloud cover, miles thick, cloaked it in darkness. The only light came from cracks in the planet's molten surface, revealing the fiery heat below, and the almost continual lightning that split the dark from above. Water vapor that condensed and fell to the earth's scorching surface was immediately vaporized and returned to the atmosphere. The desolate, surreal setting gave little evidence that upon this stage the great drama of life would soon begin. Yet life on earth seems to have developed soon after the surface of the young planet cooled, somewhere between 3.5 to 4 billion years ago. And it began, scientists believe, as a result of chemical and physical processes on the planet's surface and in its atmosphere.

The early atmosphere is believed to have consisted largely of hydrogen, far different from the present atmosphere. Hydrogen tends to combine with many other elements, among them nitrogen, oxygen, and carbon. Consequently, the primitive atmosphere was likely characterized by ammonia (hydrogen and nitrogen), water (hydrogen and oxygen), and methane (hydrogen and carbon).

As the planet's surface cooled, water pooled in valleys and other low places. But the surface was still too warm for the water to remain liquid for long. It evaporated, cooled in the atmosphere, and fell to the earth once more. The rains that pelted the planet's surface washed minerals and salts from the exposed rocks. The mineral-laden waters filled cracks and crevices; streams and rivers formed, eventually giving way to oceans as the waters filled the earth's deepest valleys and canyons. The hydrogen-rich molecules that had formed in the atmosphere became dissolved in the primitive planet's seas. Ultraviolet radiation and lightning from above and the heat trapped within the earth supplied energy that broke apart the hydrogen-rich molecules, enabling them to recombine into new and more complex molecules. Thus, the oceans became a sort of "primordial soup," a mixture of molecules and substances of increasing complexity. And it was from this primordial soup, scientists theorize, that life on earth originated.

In the harsh environment of the young planet, evolution proceeded slowly. (Table 1-1 summarizes what are believed to have been the major events in evolution.) Evolution occurs primarily as a result of organisms' adaptations to changes in environmental conditions. The factors that drive evolution are mutation and natural selection. A **mutation** is a random change within the genetic material of an individual that can be passed on to that individual's offspring. Most mutations are harmful or result in

changes that are useless to the organism. Some mutations, however, cause the organism to differ in a way that allows it (or its offspring) to adapt to a change in environmental conditions. Because the individual is better suited to the environment than others in the population, it is more likely to survive and reproduce, a phenomenon known as **natural selection.** To be advantageous, natural selection must improve an organism's chances for successful reproduction. The flip side of natural selection is that other organisms which are less suited to environmental conditions are eventually eliminated. Over time natural selection may give rise to the formation of new species, a process called **speciation.**

Once the evolutionary process took hold, the planet's watery surface teemed with a rich diversity of life forms. The proliferation of certain life forms that generated oxygen slowly changed the atmosphere to an oxygen-rich environment. This change, about 600 million years ago, signaled the start of the Cambrian period. Many species, particularly bacteria which had thrived in the low-oxygen environment, could not tolerate the increased oxygen atmosphere and died out. But it is during the Cambrian period that evolution really took off, and a great many new marine species appeared. Approximately 425 million years ago atmospheric oxygen reached its present level. About the same time, the land mass became increasingly drier as the great

▶ Table 1-1
Geologic Timetable

Era and Duration	Period	Epoch	Years Before Present (In Millions)	Principal Events of the Era
CENOZOIC 65 Million Years	Quaternary	Recent Pleistocene	1	Age of man; end of last ice age; warmer climate. First human societies; large scale extinctions of plant and animal species; repeated glaciation.
	Tertiary	Pliocene	11	Appearance of man; volcanic activity; decline of forests; grasslands spreading.
		Miocene	25	Appearance of anthropoid apes; rapid evolution of mammals. Formation of Sierra Mountains.
		Oligocene	36	Appearance of most modern genera of mammals and monocotyledons; warmer climate.
		Eocene	54	Appearance of hoofed mammals and carnivores; heavy erosion of mountains.
		Paleocene	65	First placental mammals.
	Cretaceous		135	Appearance of monocots; oak and maple forests; first modern mammals; beginning of extinction of dinosaurs. Formation of Andes, Alps, Himalayas, and Rocky Mountains.

▶ Table 1-1
Geologic Timetable (continued)

Era and Duration	Period	Epoch	Years Before Present (In Millions)	Principal Events of the Era
MESOZOIC 160 Million Years	Jurassic		181	Appearance of birds and mammals; rapid evolution of dinosaurs; first flowering plants; shallow seas over much of Europe and North America.
	Triassic		220	Appearance of dinosaurs; gymnosperms dominant; extinction of seed ferns; continents rising to reveal deserts.
	Permian		280	Widespread extinction of animals and plants; cooler, drier climates; widespread glaciation; mountains rising; atmospheric carbon dioxide and oxygen reduced.
	Pennsylvanian		310	Appearance of reptiles; amphibians dominant; insects common. Gymnosperms appear; vast forests; life abundant. Climates mild; low lying land; extensive swamps; formation of enormous coal deposits.
	Mississippian		355	Many sharks and amphibians; large trees and seed ferns; climate warm and humid.
	Devonian		405	Appearance of seed plants; ascendance of bony fishes; first amphibians; higher, drier lands; glaciations.
PALEOZOIC 360 Million Years	Silurian		425	Atmospheric oxygen reaches second critical level. Explosive evolution of many forms of life over the land; first land plants and animals. Great continental seas; continents increasingly dry.
	Ordovician		500	Appearance of vertebrates, but invertebrates and algae dominant. Land largely submerged. Warm climates worldwide.
	Cambrian		600	Atmospheric oxygen reaches first critical level. Explosive evolution of life in the oceans; first abundant marine fossils formed; trilobites dominant; appearance of most phyla of invertebrates. Lowlying lands; climates mild.
PRE-CAMBRIAN 3500 Million Years			2700	Life confined to shallow pools, fossil formation extremely rare. Volcanic activity, mountain building, erosion, and glaciation. Photosynthetic life.

Source: Robert A. Wallace, *Biology: The World of Life*, 6th ed. (New York: HarperCollins, 1992), 272–273.

oceans began to shrink. The first land plants and animals evolved. Seed plants, bony fishes, and the first amphibians appeared soon after. Some 310 million years ago reptiles and flowering plants known as gymnosperms appeared. Vast forests and extensive swamps covered the land surfaces. Dinosaurs appeared about 220 million years ago, and before long (40 million years later) birds and mammals appeared. The end of the Cretaceous period, about 65 million years ago, marked a period of mass extinctions during which many species, including the great dinosaurs, disappeared. Some scientists believe that prehistoric glacial or meteoric activity may have altered environmental conditions and caused these mass extinctions. About 25 million years ago, mammals began to evolve more rapidly; the first humanlike apes appeared. *Homo sapiens* came on the scene only about 3 million years ago, scarcely more than a blink of the eye in geologic time.

To get a better understanding of just how young our species is, in geologic terms, let's compress the 15-billion year history of the universe into a single 24-hour day. Starting at midnight, atoms form in the first four seconds. Stars and galaxies appear by about 5 A.M. Not until 6 P.M., a full thirteen hours later, do our sun and solar system form. Life on earth appears about 8 P.M., but the first vertebrates (animals with backbones) do not appear until about 10:30 P.M. Dinosaurs roam the earth from approximately 11:35 to 11:56 P.M. Finally, ten seconds before midnight, *Homo sapiens* appears. The Age of Exploration to our nuclear age (fifteenth century to the present) occurs in the last thousandth of a second.

It has been said that the only constant we can count on is change. Individuals change. Countries and nations change. As this abbreviated discussion of evolutionary history illustrates, the earth's environment and its life forms have changed, and changed dramatically, over the millennia. In the last 450 years, a new agent of change has begun to exert its influence on the planet. Since the beginning of the Age of Exploration, the activities of a single species, *Homo sapiens*, have accelerated the rate of environmental change and species extinctions to levels not seen on earth since the end of the Age of Dinosaurs. To understand how our species is altering the planet, let's take a close look at conditions on Biosphere I.

What Is the State of Biosphere I?

In 1988 *Time* Magazine broke with tradition. In place of its Man of the Year award, given to the person who most influenced events in the world that year, the magazine named Endangered Earth the Planet of the Year. The reasons behind *Time's* unusual action were obvious. In 1988 the United States experienced one of the worst heat waves ever recorded, with temperatures across the nation soaring into the 100s for days on end. At the same time many regions of the nation suffered through a severe three-month drought. Stunted crops withered in the fields; the country's grain harvest was reduced by almost a third. The drought also affected natural systems. With awe and horror, the public watched as fires raged through rain-parched Yellowstone National Park and many western forests. Elsewhere, devastating hurricanes tore through the Caribbean

and a catastrophic flood ravaged Bangladesh.

In addition to the natural catastrophes of 1988, many human-caused environmental calamities grabbed the public's attention. Along the U.S. coastline, particularly in the East, sunbathers found beaches despoiled by garbage, raw sewage, and medical wastes. On the beaches of the Mediterranean and the North seas, Europeans encountered similar kinds of pollution. Scientists watched a growing hole in the ozone layer. The hole had appeared over the Antarctic each fall for the previous several years, and scientists had linked its appearance to chemicals produced and released to the atmosphere by humans. Ozone, O_3, a naturally occurring gas in the upper atmosphere, shields the planet from ultraviolet (UV) radiation, which is harmful to life. The destruction of the ozone layer, scientists warned, meant an increase in the amount of ultraviolet radiation that would reach the earth, with possible adverse health effects.

The extraordinary events of 1988 spawned a global environmental awakening. Many people believed that even the year's weather-related catastrophes— the heat wave, drought, flood, and hurricane—were a result of human activities. Since the early 1980s scientists had warned that carbon dioxide and other gases, when released to the atmosphere through the combustion of fossil fuels and wood and other activities, could cause a rise in the earth's temperature by preventing solar radiation from being radiated back to space as heat. This global warming, some experts cautioned, might be enough to alter the earth's climate.

Whether or not the natural catastrophes of 1988 were the result of global warming cannot be proven. Variations in temperature and rainfall can be caused by a number of natural phenomena, including ocean currents and increased solar radiation brought on by sunspots and solar flares. Nevertheless, the public perception was that human actions were responsible for the year's environmental woes. By naming Earth the Planet of the Year, *Time* held up a mirror to the world, and what we saw in the mirror was a small and fragile planet. The vision in that mirror was frightening and sobering, for it reflected what scientists refer to as the three root causes of environmental problems.

What Are the Three Root Causes of Environmental Problems?

Environmental scientists agree that the three root causes of environmental problems are population growth, abuse of resources and natural systems, and pollution. They exert the most disruptive pressures on the biosphere, altering environmental conditions in such a way that one or more members of the biota are adversely affected (Figure 1-6). The three root causes are interrelated; consequently, environmental problems are complex and are difficult to solve.

Population Growth

Sometime between July 1986 and July 1987 the human population reached 5 billion. Person number 5 billion (5B) is an unwitting participant in an already crowded race; by the time 5B is 11 or 12 years old, another billion contestants will have joined the race. When 5B is 40 years old, as many as 5 billion more racers may swell the ranks of the contestants. If 5B was born in an affluent country, he or she can reasonably expect to live past the age of 70, barring any worldwide catastrophe. At that time 5B could be just one of 15 billion racers—triple the number when 5B was born!

What is this race and what are the racers trying to win? It is the population growth race, and the contestants are trying to win some share of the world's space, food, water, and shelter. How much they are likely to win depends on where they were born, to whom, and the educational opportunities they

FIGURE 1-7: A crowd in the United States. As the human population grows—it is expected to double within the next forty years to reach 10 billion people—it exerts ever-increasing pressure on the biosphere.

enjoy. But the racers have no say in these matters, and they cannot decline the invitation to race. Birth automatically enters each one of us (Figure 1-7).

The race can only become progressively more intense because population growth is exponential, that is, it grows by doubling. Unlike arithmetic growth, which is additive ($2 + 2 = 4$, $4 + 2 = 6$, $6 + 2 = 8$), exponential growth is compounding, or geometric. To illustrate the difference, imagine that you

POPULATION	RESOURCE ABUSE	POLLUTION
Africa Still the World's "Sick Child"	Cost Is Too Steep for Atomic Power	TOXIC FOOD ADDITIVES A GROWING CONCERN
Urban Areas Draw More Population	Native Americans Hold Key to Nature's Resource Needs	Vast Water Contamination From Chernobyl
BANKS BALK AT LENDING TO THIRD WORLD	Protection of Groundwater Urged	Silo for Hazardous Materials Worries Neighbors
World Population Hits "High Five"	CHEAP OIL : BOOM MAY BACKFIRE	Pollution Heating Up North Pole
Baby Boomers Failing to Meet Parents' Financial Achievements	Campaign and Boycott Aim to Save Tropical Rain Forests	FARMERS ASK FOR LITTER HELP
MORE PEOPLE ARE A GOOD THING	Environmental Concerns Go Beyond Vatican's Borders	Sharp drop in Antarctica's Ozone Worries Scientists
The Population Bomb Ticks On	Castoff Plastics Endanger Animals	THE RHINE: RIVER OF ROMANCE TURNS INTO RIVER OF RUBBISH

FIGURE 1-6: Headlines reflecting disruptive pressures on the biosphere. These headlines illustrate the link between population growth, resource abuse, and pollution, the three root causes of environmental problems.

are offered a job that will last thirty days. You will be paid 1 cent the first day, with a promise of an arithmetic increase of 1 cent on each succeeding day. On the thirtieth day you would receive just 30 cents, for a grand total of $4.65 for the month. But if your salary were to increase exponentially each day, the amount you receive on each succeeding day will be twice the amount you received the previous day (1 cent, 2, 4, 8, 16, and so on). On the thirtieth day you would receive over $5 million! Which job would you take?!

Exponential doubling may be a good thing where your salary is concerned, but with regard to population growth, it presents serious problems. As Figure 1-8 illustrates, exponential growth, when graphed, takes the shape of the letter J. When human populations were small, the doubling caused by exponential growth did not produce a dramatic increase in total population numbers. In fact, the human population did not reach 1 billion until about 1800 and 130 years passed before it doubled to 2 billion. But from 1930 to 1975, just 45 years, it doubled again, to 4 billion. Clearly, once the population reached a critical size, the doubling effect became significant. At current growth rates the global population will double again in just over 40 years, from 5 billion in 1987 to 10 billion about 2030. Our population growth, which began so slowly, has speeded up dramatically now that we have rounded the J curve. It will be difficult to slow the pace of this growth (and as we shall see later in Chapters 8 and 9, virtually impossible within the next few generations), with the consequence that contestants in the race for the necessities of life face a continual and increasing struggle.

Numbers tell only part of the story. What does the global population mean in terms of how people live? According to the United Nations, of the more than 5 billion people who now inhabit the earth, only one-fifth have adequate food, housing, and safe drinking water. Four-fifths do not. Each day, approximately 109,600 people die from the effects of starvation. Starvation-related illnesses kill an estimated 40 million people each year. Most of the severely stressed are in the developing (also called the less-developed or less-industrialized) countries, but even in some developed (industrialized) nations many people do not have enough food to eat and cannot afford a warm, dry place in which to live.

Abuse of Resources and Natural Systems

A **resource** is anything that serves a need; it is useful and available at a particular cost. All organisms consume resources (another way of saying that all living things live at the expense of their environment). They produce wastes, which then become resources

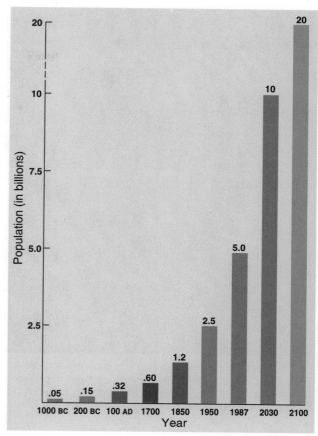

FIGURE 1-8: J-curve of human population growth. For most of human history, our population grew slowly. Within the last 150 years, however, the doubling effect of exponential growth has created a dramatic increase in population. Note: the bars do not represent exact population levels at particular times in history, but rather they illustrate the general increase in human numbers.

for other organisms. Some resources are **renewable.** They can be replaced by the environment, and as long as they are not used up faster than they can be restored, supplies are not depleted. Forests, solar energy, and the famed blue crabs of Chesapeake Bay are renewable resources. In contrast, **nonrenewable** resources exist in finite supply or are replaced by the environment so slowly (in human terms) that, for all practical purposes, the supply might as well be finite. With use, the supply of a nonrenewable resource is depleted. Coal, oil, and other fossil fuels are nonrenewable resources.

All organisms exploit the environment to the best of their ability. Humans are one of the most successful of all species at doing so. Technology has enabled us to extract and use resources more efficiently and has allowed us to overcome or minimize the effects of such things as extreme climate, other predators, and certain diseases that might otherwise keep our population at lower levels. The technologies we have developed, the fact that we tend to satisfy wants rather than needs, and our sheer numbers

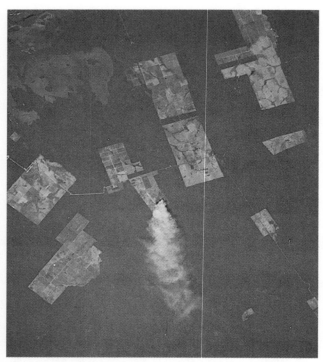

FIGURE 1-9: Clearings and roads in a tropical rain forest in Goias State, Brazil. Rain forest is being destroyed to provide pastureland and agricultural land. The smoke in the photo is rising from a fire used to clear vegetation.

have resulted in an environmental impact greater than that exerted by any other single species. To understand the extent of this impact, let's look at a concept scientists call the **net primary productivity (NPP)**. The net primary productivity is the total amount of solar energy fixed biologically through the process of photosynthesis minus the amount of energy that plants use for their own needs. Humans appropriate an estimated 40 percent of the potential NPP on land, chiefly for food but also for fiber, lumber, and fuel. What are the implications of this for all other species?

Human demand for renewable resources appears to be accelerating faster than the biosphere can renew them. We are, so to speak, living off the capital of our environment rather than the interest. As the World Resources Institute, in its 1990–1991 report on the global environment, warns, "The earth cannot produce its bounty if we continue to damage its productive capacity." For example, some experts estimate that human activities have transformed 10 percent of the land surface of the planet from forests and rangelands into desert, and another 25 percent is at risk. Existing agricultural land lost through erosion alone is estimated at 14.8 to 17 million acres (6 to 7 million hectares) annually; an additional 3.7 million acres (1.5 million hectares) are lost to other factors related to human activities. Though the total world fish catch is up, the catch

per person is declining, a reflection of the rapid increase in the human population. Moreover, pollution and overharvesting are taking their toll on some fisheries, and the catch from 4 of the world's 16 major fisheries has exceeded the limit that scientists believe can be sustained over time.

The current pace of tropical deforestation is frightening (Figure 1-9). The World Resources Institute estimates that almost half of the original 3.7 to 3.9 billion acres (1.5 to 1.6 billion hectares) of tropical mature forests have been cleared to accommodate other uses. Satellite observations suggest that the rate of tropical deforestation worldwide is 40.5 to 50.4 million acres (16.4 to 20.4 million hectares) a year. These forests will likely continue to disappear at accelerating rates to support the short-term economic needs of growing populations. Their destruction may well have dire consequences. Tropical rain forests house about half of all the species on earth, and they help to regulate the global climate. Moreover, like all forests, they prevent erosion and recycle carbon dioxide and other gases produced by our activities.

A debate rages regarding the actual cause of resource abuse and depletion. Some people contend that population growth is consuming resources faster than the earth's living systems can replace them. Others argue that resources are available to support the world's people, if we can find the means to distribute the resources more evenly and to stop their wasteful use. They point out that 20 percent of the world's people living in developed nations (especially western Europe, the United States, Canada, Australia, the former Soviet republics, and Japan) use nearly 80 percent of the resources consumed each year. The other 80 percent of the world's people—4 billion human beings—consume just 20 percent of the resources. The United States alone, with just 5 percent of the world's people, uses 35 percent of the world's raw materials and about 35 percent of the world's energy. Further, the United States wastes about 50 percent of the materials and energy it uses (Figure 1-10).

Resource use is expected to rise sharply in the next 20 to 30 years. During that same period the less developed countries will try to become more prosperous. What are the environmental implications if countries with a standard of living barely above subsistence want a standard of living similar to that of western European nations, Japan, Canada, Australia, or the United States?

Pollution

Most of us find it easy to identify and describe pollution. Newspapers, radio, and television draw our attention to dramatic pollution events, such as the

1991 oil slick in the Persian Gulf, the 1989 Exxon *Valdez* oil spill in Alaska's Prince William Sound, the 1986 disaster at the Soviet Union's Chernobyl nuclear plant, and the 1984 release of deadly methyl isocyanate from Union Carbide of India's pesticide production facility in Bhopal, India. Worldwide, there are countless constant reminders of the less dramatic, but more insidious continual and increasing pollution of our environment—dead fish on stream banks, litter in national parks, decaying bridges and buildings, leaking landfills, and dying forests and lakes.

In the United States an increasing number of people are experiencing the effects of pollution firsthand. Train and truck accidents resulting in the release of toxic pollutants force us to evacuate our homes. We become ill after drinking contaminated water or breathing contaminated air. We can no longer swim at favorite beaches because of sewage contamination (Figure 1-11). We restrict fish, shellfish, and meat consumption because of the presence of harmful chemicals, cancer-causing substances, and hormone residues. We are exposed to nuclear

FIGURE 1-11: Hypodermic needle and other waste on beach. Contamination by sewage and medical wastes has forced the closing of many beaches along the East Coast of the United States in recent years.

contaminants released to the air and water from uranium-processing plants in areas such as Hanford, Washington, and Fernald, Ohio. Such exposures have been linked to fatal and debilitating conditions such as liver, kidney, and nerve damage, lung and bone marrow cancers, birth defects in the unborn, stillbirths, spontaneous abortions, and reduced fertility.

But serious pollution problems do not occur only in the more developed, affluent countries. Around the globe, humans have seemingly launched a direct assault on air, water, and soil. Poor, densely populated, less-developed countries often suffer from pollution. Long-term contamination of water supplies by human sewage is common. Open garbage and solid waste dumps degrade water and air (Figure 1-12). Industrial plants are often unconcerned with controlling air pollution, since the processes that produce the pollution are associated with jobs and

FIGURE 1.10: In One Day . . . .

▶ In one day, people in the United States

1. Throw out 200,000 tons of edible food.
2. Use 450 billion gallons of water for their homes, factories, and farms.
3. Use 313 million gallons of fuel—enough to drain 26 tractor-trailer trucks every minute.
4. Use 6.8 billion gallons of water to flush their toilets.
5. Throw 1 million bushels of litter out the windows of their cars and trucks.
6. Add 10,000 minks to their closets and coat racks.
7. Eat 170 million eggs and more than 4 million pounds of bacon.
8. Spend $200 million on advertising.
9. Use 72 million pounds of wheat flour.
10. Saw up and pound together 100 million board feet of wood. Flooring mills alone turn out 15 acres of hardwood floor.
11. Use 57 billion kilowatt hours of energy.
12. Wrest 18 million tons of raw material from its natural state. If all Americans shared the job equally, they would each have to dig up, saw down, or gather together more than 150 pounds of stuff every day.
13. Use 250,000 tons of steel.
14. Use 187,000 tons of paper.
15. Eat 1.7 million pounds of canned tuna, 6.5 million gallons of popcorn, more than 400 bushels of bananas, 1,465 miles of licorice twists, 95 tons of sardines, 1.2 million bushels of potatoes, and 47 million hot dogs. In fact, Americans eat 815 billion calories of food each day, roughly 200 billion more than they need to maintain a moderate level of activity.

FIGURE 1-12: Copsa Mica, Romania, 1990. Carbon snowfall blankets the town and this shepherd's flock. In many Eastern European countries, open dumps and severe industrial pollution are the legacy of years of little or no environmental regulation.

economic development. In Mexico City 500 million tons of industrial pollutants are released to the air each year. Industrial pollution is perhaps nowhere more severe than in eastern Europe, where the environmental movement has only recently begun to gain strength.

People have become increasingly aware that many pollution problems are global in nature and can be solved only through cooperation among nations. Three such global problems are the increase in greenhouse gases such as carbon dioxide, methane, and nitrous oxide; the decrease in atmospheric ozone; and the acidification of soils, forests, and lakes. Too often political haggling, economic concerns, and social barriers hamper progress in addressing global problems. Consider, for example, that despite years of study, acid precipitation related to the use of high-sulfur coal and gasoline continues to damage lakes and forests in the eastern United States and Canada, Germany, eastern Europe, and Scandinavia.

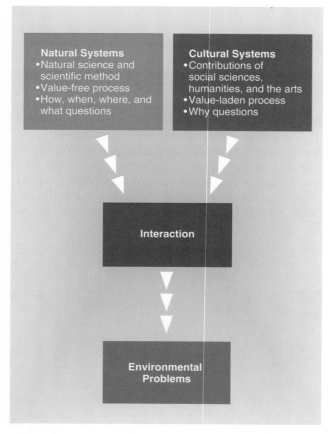

FIGURE 1-13: Interaction of natural and cultural systems. Environmental problems stem from the interaction of natural and cultural systems. Natural systems are studied through the natural sciences and the scientific method, a value-free process that can help us answer such questions as "how," "when," "where," and "what." Cultural systems are studied through the social sciences, humanities, and the arts. These fields use a value-laden process to answer the question "why." Solving environmental problems requires the use of both the natural sciences and the social sciences, the humanities, and the arts.

While pollution is often easy to identify, it is less easy to define. A **pollutant** is a substance that adversely affects the physical, chemical, or biological quality of the earth's environment or that accumulates in the cells or tissues of living organisms in amounts that threaten the health or survival of those organisms.

Pollution can come from natural or cultural sources. It is an inevitable consequence of the biological, chemical, and physical processes of the earth. Volcanoes, fires caused by lightning, products from the decomposition of materials in swamps and soils, sulfurous gas seeps, and other natural processes all contribute pollutants to the earth's soil, water, and air. Further, every living organism produces wastes. Usually, these waste products become a resource for other organisms, keeping the chemical and physical consistency of the earth's air, water, and atmosphere relatively constant over time.

Historically, as long as human population numbers remained small, human pollutants could be adequately processed by the environment. But as population numbers increased, human wastes began to overwhelm natural systems. As societies became technologically sophisticated, humans began producing large numbers and volumes of new substances each year. It is difficult, and sometimes impossible, for the environment to absorb and process these substances.

To complicate the situation even further, consumption of resources per person in the developed world has dramatically increased. A prevailing attitude of "consumerism," leading to resource abuse, results in ever-increasing amounts of pollution released to the environment. Consequently, pollution has become an increasingly significant source of environmental problems.

Why Do Uncontrolled Population Growth, Resource Abuse, and Pollution Occur?

The problems of population growth, resource abuse, and pollution occur because human attitudes, values, and beliefs often lead to behaviors that are incompatible with natural systems. Because environmental problems arise from the interaction of natural and cultural systems, they cannot be solved without addressing the underlying cultural factors that cause them (Figure 1-13).

Attitudes, values, and beliefs comprise a person's, or a society's, **worldview**—a way of perceiving reality. An important part of any worldview is a person's beliefs about the place of humans in the natural world. A worldview is reflected in and transmitted through culture. Culturally transmitted

FIGURE 1-14: Astronomical clock of Wells Cathedral, England. The astronomical clock reflects the belief, prevalent in the seventeenth and eighteenth centuries, that the universe acts like clockwork.

beliefs and values shape attitudes toward nature and the environment, which in turn lead to behaviors that can cause or minimize environmental problems.

The dominant worldview in the United States has been the worldview of the Europeans who colonized North America beginning in the fifteenth century. This Western worldview is a product of religious beliefs, science, democratic ideals, and economic concepts. Let's take a closer look at how this worldview developed and how it affects the way Western societies have traditionally viewed nature.

Christianity, with its roots in Judaism, was a major factor in the development of the Western worldview. Judaism and Christianity proposed a single God who created the universe but was separate from and outside of His creation. A basic Christian belief was that God gave humans dominion over creation, with the freedom to use the environment as they saw fit. Another important Judeo-Christian belief predicted that God would bring a cataclysmic end to the earth sometime in the future. Some people interpreted this to mean that the earth was only a temporary way station on the soul's journey to the afterlife. Because these beliefs tended to devalue the natural world, they fostered attitudes and behaviors that had a negative effect on the environment.

The rise of science was a second major factor in the development of the Western worldview. Francis Bacon, a seventeenth-century English philosopher and essayist, believed that God's kingdom would be reestablished on earth when humans, via science, achieved dominion over nature. According to Bacon,

God had created a "clockwork" universe that operated by certain rules and patterns, and science could arrive at the ultimate truth by discovering these patterns (Figure 1-14). Scientifically derived knowledge translated into new technologies that enabled humans to exploit resources more fully.

The eighteenth century brought the Enlightenment, when profound changes in economic, political, and social orders further influenced the Western worldview. Capitalism, an aggressive economic system based on the accumulation of wealth, gained wide acceptance. The political revolutions that established democracies in North America and France affirmed the rights of individuals to determine their own destinies by making and enforcing laws, owning property, and taking the initiative to develop the resources associated with private property.

About this time Europe shifted from a largely agrarian (farming) society to an industrial society. As mechanical inventions such as seed sowers enabled farmers to increase their crop yields, they were able to raise enough food to feed many more people. Consequently, fewer individuals were needed to farm the land. Many people found work in factories, as newly emerging industries developed and flourished. The Industrial Revolution started in England at the end of the eighteenth century; its success depended, to a great extent, upon the availability of a large labor force. Industrialization brought with it urbanization, as people began to cluster in the areas where industries were located. People physically separated from daily contact with the land began to lose the knowledge of nature, natural events, and natural phenomena—a knowledge that can best be called a **sense of the earth**—that had been handed down through generations.

Industrialization also brought with it the accelerated use of resources and pollution. Steam engines drove the Industrial Revolution, and coal fueled the steam engines. More factories and more engines meant more coal. Burning more coal, in turn, meant increased air pollution, especially immediately surrounding the industrial centers. The declining quality of life in urban areas was also due, in no small part, to mere numbers. With so many people concentrated in one area, sewage and wastes poured into rivers and streams, and burning refuse in open dumps polluted the air. Contagious diseases tended to spread easily under such poor conditions.

The dominant Western worldview, with its emphasis on the exploitation of nature and resources, the accumulation of wealth, faith in science and technology, and belief in the inherent rights of the individual, was a powerful and aggressive force. The Europeans who migrated to the New World found a wild and beautiful land of seemingly

Focus On:
Native American Beliefs

We did not think of the great open plains, the beautiful rolling hills, and winding streams with tangled growth, as "wild." Only to the white man was nature a "wilderness" and only to him was the land infested with "wild" animals and "savage" people. To us it was tame. Earth was bountiful and we were surrounded with the blessings of the Great Mystery.

Luther Standing Bear, Oglala band of Sioux

My people, the Blackfeet Indians, have always had a sense of reverence for nature that made them want to move through the world carefully, leaving as little mark behind them as possible. My mother once told me: "A person should never walk so fast that the wind cannot blow away his footprints."

Jamake Highwater, Blackfeet

Did you know that trees talk? Well, they do. They talk to each other, and they'll talk to you if you listen. Trouble is, white people don't listen. They never learned to listen to the Indians so I don't suppose they'll listen to other voices in nature. But I have learned a lot from trees: sometimes about the weather, sometimes about animals,

sometimes about the Great Spirit.

Walking Buffalo, Stoney tribe, Canada

In *Ani Yonwiyah*, the language of my people, there is a word for land: *Eloheh*. This same word also means history, culture and religion. This is because we Cherokees cannot separate our place on earth from our lives on it, nor from our vision and our meaning as a people. From childhood we are taught that the animals and even the trees and plants that we share a place with are our brothers and sisters.

So when we speak of land, we are not speaking of property, territory or even a piece of ground upon which our houses sit and our crops are grown. We are speaking of something truly sacred.

Jimmie Durham, Cherokee

Sell the earth? Why not sell the air, the clouds, the great sea?

Tecumseh, Shawnee

Every part of this soil is sacred in the estimation of my people. Every hillside, every valley, every plain and grove has been hallowed by some sad or happy event in days long vanished. Even the rocks, which seem to be dumb and dead

as they swelter in the sun along the silent shore, thrill with memories of stirring events connected with the lives of my people, and the very dust upon which you now stand responds more lovingly to [our] footsteps than to yours, because it is rich with the blood of our ancestors and our bare feet are conscious of the sympathetic touch. . . .

And when the last Red Man shall have perished, and the memory of my tribe shall have become a myth among the White Men, these shores will swarm with the invisible dead of my tribe, and when your children's children think themselves alone in the field, the store, the shop, upon the highway, or in the silence of the pathless woods, they will not be alone. In all the earth there is no place dedicated to solitude. At night when the streets of your cities and villages are silent and you think them deserted, they will throng with the returning hosts that once filled them and still love this beautiful land. The White Man will never be alone.

Let him be just and deal kindly with my people, for the dead are not powerless. Dead, did I say? There is no death, only a change of worlds.

Translation by Dr. Henry Smith of a speech attributed to Chief Sealth of the Duwamish tribe, 1853

unlimited natural resources, peopled by natives who had a far different worldview. (See Focus On: Native American Beliefs.) For several hundred years these settlers enjoyed the confluence of a large resource base, the technological means to maximize the use of this base, and the social acceptance of maximum use of resources. The result was what has been called the **frontier mentality,** a mindset that encouraged the aggressive exploitation of nature. If you doubt this, simply take a look at an old history book. Very likely, it will talk of the European settlement of North America as a struggle to "carve a new country out of an untamed wilderness inhabited by savages." By the way, native Americans had no name in their languages for "wilderness." To them, it was simply called home.

What Is the Environmental Revolution?

Not everyone in Europe and the United States accepted the dominant Western worldview. For example, the fourteenth-century Roman Catholic monk St. Francis of Assisi and other Christians believed that humans had a responsibility to care for the earth. Humans were not given a mandate to use resources wantonly, but instead were to be stewards of nature. Moreover, humans were part of nature; created out of the dust of the earth, they would return to dust after death. According to this interpretation, humans were the link between God and nature, part of a system created by God and pronounced by God as good. Sadly, the concept of humans as stewards of nature was not widely held. Starting in the nineteenth century, some Americans began to question society's attitudes and behaviors toward the environment. About 1864 George Perkins Marsh (a lawyer, diplomat, and scholar) warned of the destructive effects of the dominant cultural beliefs and practices on the environment. He was echoed by such visionaries as the late nineteenth-century naturalist and writer John Muir and twentieth-century scientists such as Aldo Leopold and Rachel Carson. Leopold and Carson both spoke of human responsibility for the earth. Leopold insisted that what society needed was an ethic of resource use, a **land ethic** that would guide our behavior toward the natural world. (See Focus On: Aldo Leopold.) Given our society's faith in science, the warnings issued by Leopold and Carson helped to crystallize an environmental revolution in the 1960s. That revolution, which had its roots in the work of Marsh, Muir, and others, continues today.

What Social and Environmental Factors Have Contributed to the Environmental Revolution?

Numerous factors helped to spark the environmental revolution of the 1960s. The bald eagle, our national symbol, like the coal miners' canary, was warning us of impending disaster. Eagles, like many other species, were suffering from accumulations of toxic substances in their tissues and were in danger of losing their ability to reproduce successfully. And, like many other species, eagles were in danger of becoming extinct. The extinction of plant and animal species, silt-choked streams, algae-laden lakes, foul drinking water, eye-stinging air, and dangerous chemical dumps shocked us into unprecedented action—and into a new understanding of nature and the human relationship to the environment.

Many environmentalists used the analogy of Spaceship Earth during this period, an era when our space program was successful and popular. This analogy was valuable because it enabled people to understand that, like a spaceship, the planet was essentially a closed system, with a finite resource base and a limited ability to recycle pollution. However, by placing humans at the helm of the spaceship, the analogy perpetuated the idea that the natural world was under our control. Although the idea of piloting a spaceship included accepting responsibility for that role, it also fostered the belief that if we could only develop the proper technologies and learn to manipulate them carefully, our species, and the spaceship itself, would survive. Like Bacon's clockwork universe, the concept that the earth was a machine, albeit a sophisticated one, implied that we could control it if we gained the necessary knowledge. Moreover, the spaceship analogy did not account for the earth's ability to maintain the conditions necessary for life, a form of self-regulation and self-repair that machines are incapable of performing.

Nevertheless, the emerging environmental movement accomplished a great deal. Congress passed several important pieces of legislation designed to reduce pollution, protect endangered species and wilderness, and promote sound resource management. Environmental awareness and protection were ideas whose time had come, as heralded by the first Earth Day and environmental teach-in in April 1970. Environmentalists joined scientists, politicians, citizen groups, government officials, students, conservation groups, and educators in an unprecedented national resolve to enforce these new actions. The late 1960s and 1970s were years of optimism and idealism. However, as a nation we did not significantly change our understanding of nature and of the human place in the environment. For the most part, we did not regain our sense of the earth. Consequently, the shifting winds of politics and economics soon dealt a blow to the environmental movement. Inflation and an economic downturn began to drive the monetary cost of a clean environment ever higher. "Jobs or environmental controls" became a standard argument in many hard-pressed regions of the country.

When Ronald Reagan became president in 1980, the idealism of the 1960s and 1970s was replaced by an emphasis on individual well-being, especially economic well-being. Economic goals took precedence in environmental and resource management decisions. Federal coal leases were sometimes sold below market price, and timber and grazing rights were also sold at low prices in a rush to reduce the resource holdings of the federal government. Throughout the 1980s the government attempted to reduce federal spending and to reduce the federal deficit, but to no avail. Instead, spending and the deficit increased. Defense spending made up the largest share of that increase.

Throughout this period the American people, in poll after poll, affirmed their desire to continue efforts to assure a clean, safe environment for themselves and their children. More people became members of environmental and conservation groups than ever before. Americans, determined to maintain a quality life, did not—and do not now—wish to sacrifice environmental quality; they continue to equate a quality life with high standards for environmental quality. Sensing this, both George Bush and Michael Dukakis—for the first time in American politics—made the environment a major presidential

Focus On: Aldo Leopold

J. Baird Callicott, Department of Philosophy,
University of Wisconsin

Aldo Leopold has been characterized as a "prophet" and his *A Sand County Almanac*, the "bible" of the modern environmental movement. Yet there has been remarkably little critical exploration of his "land ethic"—how he evolved it, what ideas lie behind it, and how it has informed our present attitudes and values about nature.

Aldo Leopold was born in Burlington, Iowa, in 1887 to a family of aristocratic German origin. As a boy, he was introduced to the outdoors and a rigorous code of sportsmanship by his father, Carl. He attended Lawrenceville Academy, a prep school in New Jersey, and the Yale Forest School. Upon graduation in 1909 he began his U.S. Forest Service career in the Southwest.

Steeped in the utilitarian conservation philosophy of Gifford Pinchot, Chief of the United States Forest Service (USFS), Leopold put into practice the then new "multiple use" concept of forest management. He sought to maximize the "recreational resource" potential of the national forests by clearing the range of predators—principally the timber wolf and grizzly bear.

A near fatal bout with nephritis forced him out of the field. He transferred to the USFS forest products laboratory in Madison, Wisconsin, in the mid-1920s. There he pursued his interest in game conservation as an avocation—just as the science of ecology came of age in the work of Frederick Clements, Charles Elton, and Arthur Tansley. Gradually, Leopold came to realize that predator removal and game stocking were artificial, myopic, and ultimately self-destructive. Renewable, sustainable game populations required appropriate habitat.

The intellectual stimulation and spiritual elevation of a professorship at the University of Wisconsin in "game management" further broadened his understanding. Practical habitat management demanded extensive knowledge of ecological relationships. Ecological research eventually led Leopold to the perception of "game" species as but members of larger biotic communities.

Thus, what began as a means—habitat improvement to the end of sport and meat production—became an end in itself—maintenance of the integrity, stability, and beauty of ecosystems.

Aldo Leopold's pivotal place in the contemporary environmental consciousness turns in part upon his own odyssey of discovery and conversion. He donned the mantle of Pinchot—he was a no-nonsense professional conservationist, a member, in fact, of the conservation establishment, and a passionate sportsman. But he spoke with the voice of Muir—he confessed his love, respect, and admiration for wild things, and he insisted upon the "philosophical value" of meadow mice no less than deer, plants no less than animals.

The intrinsic value Muir found in nature rested essentially upon theology—either an orthodox doctrine of creation or a transcendental doctrine of immanence. Muir, in other words, sometimes argued that all creatures, great and small, have a right to a share of life because God created them and invested them with an intrinsic value not essentially different from our own. Or, following Emerson and Thoreau, he more often argued that a divine presence animated all natural things, rocks no less than "spirit beams" (sunshine) and rattlesnakes as well as redwoods.

Such arguments are too easily dismissed as not only soft-hearted, but soft-headed. Therefore, Leopold marshalled the full intellectual resources of modern biology, which were unavailable to John Muir's generation of environmental advocates, and argued for the intrinsic value and "biotic rights" of non-

campaign issue in 1988, and Bush's promise to be "the environmental president" helped him win the election. By the start of the new decade environmental awareness and concern were at an all-time high, both in the United States and abroad. Earth Day 1990 was celebrated worldwide as the global community expressed its support for environmental action (Figure 1-15).

How Has the Environmental Revolution Caused Us to Reassess Our Relationship with Nature?

The environmental movement, now almost three decades old, has caused many people to reassess their relationship with nature. Many have become

FIGURE 1-15: Earth Day festivities in New York City. Worldwide celebrations on Earth Day, 1990, gave testament to the increased global interest in environmental issues.

human natural entities and nature as a whole in purely scientific terms. Drawing directly upon Charles Darwin's natural history of morals in *The Descent of Man*, Leopold explained that ethics evolved and grew in scope and complexity in tandem with the evolution and growth of human society. With each step in the expansion of human society—from the savage clan to the family of man—we may observe a corresponding extension in the scope of ethics. The next, and perhaps final, step in this sequence is the emergence of the land ethic. Ecology "enlarges the boundaries of the community to include soils, waters, plants, and animals, or collectively: the land." The land ethic is simply the natural response of intelligent social primates to this expansive ecological representation.

Science makes no claim to unimpeachable truth, but an ethic expressed exclusively in evolutionary and ecological terms cannot be dismissed as mere speculation or wishful thinking. The land ethic, thus, is persuasive and authoritative.

The land ethic reflects, moreover, the holistic cast of ecology. Although fellow-members of the biotic community command our respect, their several interests are subordinate to the health and integrity of the various ecosystems to which they belong. Hence, soils may be plowed, trees felled, animals trapped and hunted, as long as the overall integrity, stability, and beauty of biotic communities remain intact. This goal may not be readily or easily attained, but it does represent a practicable ideal of environmental health.

Leopold's land ethic thus perfectly underpinned the agenda of the new ecologically informed environmentalism which emerged in the 1960s and became a powerful social force in the 1970s. Two types of environmental legislation characterize the political achievement of environmentalists during the last twenty years. The Clean Air, the Clean Water, and other anti-pollution acts, designed to safeguard human health, were a response in large measure to an environmental alarm first sounded loudly and clearly by Rachel Carson. In contrast, legislation such as the Wilderness Preservation Act, the Endangered Species Act, and the Marine Mammals Protection Act were intended to directly benefit nature by preserving and protecting wilderness, endangered species, marine mammals, wildflowers, and wildlife. Legislation of this type was inspired by the "ecological conscience" first elegantly expressed in the language of hard science by Aldo Leopold.

Less visible than these impressive legislative and public policy legacies of Aldo Leopold is his seminal contribution to a new field of philosophy—environmental philosophy and ethics. Yet that contribution may ultimately prove to be the more momentous. Great and powerful ideas often have humble beginnings. (Indeed, that is the rule not the exception.) Leopold's land ethic and its contemporary philosophical amplifications have as a goal nothing less than the transformation of human attitudes and values toward nature. Only upon such a shift in the ground sea of human consciousness can human behavior in relation to nature be lastingly reformed. The success of the philosophical enterprise of which *A Sand County Almanac* is the well spring and ongoing source of inspiration cannot now be foreseen. Future historians of Western ideas may record Leopold as the Descartes of the twentieth century or they may relegate him to a footnote as the father of a philosophy that never flew. Which it will be is up to us.

interested in the worldviews of other cultures, particularly those of native Americans (Figure 1-16). We are learning that there is no one correct worldview that is compatible with healthy attitudes and behaviors toward the environment. Christianity, which promotes a stewardship ethic, can lead to an environmentally benign life-style as easily as can the Baha'i faith, which espouses the oneness of all living things.

Another way in which we are reassessing our relationship with nature is to examine the language we use when talking about nature, for example, the words *environment* and *nature*. When many people speak of their environment, they are simply referring to their physical surroundings. However, the term **environment** actually refers to a system of interdependent living and nonliving components in a

FIGURE 1-16: Native American art, dance, and rituals, which reflect the worldview of the continent's original inhabitants, are enjoying increased popularity as people reassess their relationship with nature.

given area, over a given period of time, and including all physical, chemical, and biological interactions. All organisms living in a particular area, including humans, and anything that influences or is influenced by those organisms are part of the environment.

When many people speak of nature, they are referring to something apart from humanity and often something to which humans are superior. However, **nature** is the sum of all living things interacting with the earth's physical and chemical components as a complete system. It is an integrated whole, a self-sufficient system in which life is sustained in dynamic equilibrium. The belief that humans are not part of this whole is called an **anthropocentric,** or human-centered, view.

Some advocates of the anthropocentric view define nature as that part of the world that is devoid of human influence. If humans are not part of nature, they reason, anything produced or shaped by humans, such as buildings, statues, and farms, is not natural either. Others believe that humans were "natural" at one time, but became "unnatural" at a certain point in their development. Some trace the human split with nature to the moment when Adam and Eve were banished from the Garden of Eden. Others believe the split occurred with the development of culture—the use of tools, the mastery of fire, and the development of language—which allowed humans to modify their world significantly. According to this reasoning, the greater the modifying power, or destructive power, of human artifacts, the less natural humans became. Bows and arrows, stone tools, and adobe huts were natural; rifles, bulldozers, and condominiums are not. Still another anthropocentric view holds that humans' instincts and basic drives are natural, but their social activities and inventions are not. Hence, humans are a part of nature, but human culture is not. This belief places humans in the unique position of being both a part of and apart from nature.

In contrast, a **biocentric** view holds that humans are as much a part of nature as anything else on earth. Humans are subject to all natural laws. Although human inventiveness appears to have circumvented many natural laws, it is only a matter of time before humans must confront the consequences of their activities.

Because the dominant Western worldview teaches us as children that we are superior to nature, we often grow up acting on the belief that aggression toward nature is acceptable. If we believe resources and natural systems are objects with no inherent value beyond their usefulness to the human race, it becomes easy to exploit and abuse those resources and systems. If we do not see ourselves as part of nature and as part of the forces affecting the environment, it is hard to see ourselves as part of the problem. Thus, changing the way we see the human relationship with nature—and hence the environment—is the first step to solving and minimizing environmental problems.

Summary

The biosphere is the thin layer of air, water, soil, and rock that surrounds the earth and contains the conditions to support life. The physical environments comprise the abiotic component of the biosphere, and living organisms comprise the biotic component. Over the millennia the earth's living things adapted to and continue to adapt to changes in the nonliving environment, and, in turn, living things influenced and continue to influence the nonliving. Thus, change has been a constant characteristic of the planet throughout its history. Recently (in geologic time), however, a new agent of change has altered the planet in unprecedented ways.

Modern culture has enabled humans to exert an influence far greater than that of any other single species in the history of the planet. We have been able to modify our environment significantly through technology and social institutions. Arrogance and ignorance, however, have blinded the human race to the intricacies of nature, for while we are able to modify certain aspects of our existence, we cannot control natural processes. *Our earth environment sets limits on all creatures.* Many people believe that we are approaching those limits; as evidence, they point to worldwide environmental degradation and problems such as global warming. The three root causes of environmental problems are population growth, abuse of resources and natural systems, and pollution.

Environmental problems arise from the interaction of natural and cultural systems. Accordingly, to solve environmental problems, individuals and societies must address the underlying cultural factors—attitudes, values, and beliefs—that cause the problems. Part of that process must be an examination of one's worldview, or way of looking at reality, which includes beliefs about the relationship of humans with the natural world. The dominant Western worldview, which encourages the aggressive exploitation of nature and resources, can be traced to religious, social, economic, and political developments throughout Western history. Other worldviews, which encourage less aggressive behaviors toward nature, began to gain greater popularity in Western societies in the 1960s and 1970s with the advent of the environmental revolution.

The decade of the 1990s promises to be one of renewed commitment to environmental action. Humans must develop and encourage aspirations within the limits imposed by the environment. This does not mean going back to the horse and buggy. It does mean minimizing human impact on the earth by managing population growth, using resources wisely, and controlling pollution. Ecological balance is the ultimate relationship humans must learn to maintain with nature. *No matter how far removed we become from direct contact with nature, we remain a part of it.* If our species is to survive and prosper, this contact must be nurtured.

Key Terms

abiota
anthropocentric view
big bang theory
biocentric view
biosphere
biota
environment
evolution
frontier mentality
land ethic

mutation
nature
net primary productivity
nonrenewable resource
pollutant
renewable resource
resource
sense of the earth
worldview

Discussion Questions

1. How might Western culture have differed if the dominant worldview had been one of stewardship?
2. How are the three root causes of environmental problems interrelated?
3. Could the development of Biosphere II have major implications for understanding Biosphere I? Explain.
4. What major developments brought about the environmental revolution?
5. Predict the implications for the biosphere (including humans) if the human population doubles in your lifetime and resource use continues to increase.

And Where Do We Want to Go?

Achieving a Sustainable Future

The unleashed power of the atom has changed everything except our way of thinking . . . we shall need an essentially new way of thinking if mankind is to survive.

Albert Einstein

We have entered an era characterized by syndromes of global change that stem from the interdependence between human development and the environment. As we attempt to move from merely causing these syndromes to managing them consciously, two central questions must be addressed: What kind of planet do we want? What kind of planet can we get?

What kind of planet we want is ultimately a question of values.

William C. Clark

Learning Objectives

When you finish reading this chapter, you should be able to:

1. Define a sustaining and sustainable earth society.

2. Describe the difference between the natural science of ecology and environmental science.

3. Identify three different methods for solving environmental problems.

4. List the five steps of the problem-solving model for environmental science and tell what each step entails.

5. Describe the major characteristics of environmentally sound management and explain why it is preferable to environmental problem solving.

To nurture and be nurtured, beyond mere survival needs, is a distinct characteristic of earth's more complex life forms, and nurturing reaches a peak of complexity in humans (Figure 2-1). For humans, to nurture means to care for over long periods of time and to sustain through adversity. If we adopt a nurturing attitude toward the earth, it would be difficult, if not impossible, to adopt an aggressive or destructive attitude toward the earth's resources and the earth's other inhabitants, human and nonhuman. Respect for all life forms, as well as for the chemical and physical components of the earth, can enable us to nurture the earth which nurtures life. Such reciprocal nurturing is the essence of a sustainable earth.

This chapter describes the characteristics of a sustainable and sustaining earth society and explores some ways in which we can act as environmental stewards in order to achieve such a society. To develop a sense of stewardship, we need a basic understanding of the nature of science, ecology, and environmental science. From that understanding, we can move to a discussion of how to solve environmental problems and how to minimize their occurrence through environmentally sound management.

What Is a Sustainable and Sustaining Earth Society?

Imagine a world in which smog-shrouded cities and acid rain–ravaged forests are only a distant memory. Imagine a world in which human societies cooperate with rather than abuse other cultures, in which the people of Ethiopia, the Amazonian rain forests, and other environmentally threatened areas are no longer driven from their homes. Imagine a world in which a stable human population seeks to preserve and nurture the diversity of both the earth's biota and its human cultures.

FIGURE 2-1: The nurturing of young in diverse life forms. Caring for their young are this (a) humpback whale, (b) family from the Solomon Islands, (c) wolf spider, and (d) great blue heron.

Such a world is what we call a sustainable and sustaining earth society: **sustainable** because human behavior, in harmony with natural systems, acts to maintain the health and integrity of the environment; **sustaining** because this harmonious and healthy natural world continues to nurture and support the rich diversity of life. Despite economic, cultural, and political barriers, a sustainable and sustaining earth is not a pipe dream. It is a vision of what can be. Realizing this vision, however, requires a significant change in the way we as a global community use resources.

For any society, its population size, available resource base, and culture (dominant beliefs, attitudes, and values) determine how that society uses resources and its resulting **standard of living.** In general, for a society:

$$\text{Culture} \times \frac{\text{Resource base}}{\text{Population}} = \text{Standard of living}$$

Keep in mind that this is a general model for the average standard of living; in any society there are always individuals whose standard of living is higher than average and those whose standard is lower.

If a society encourages consumption and a materialistic lifestyle and is able to secure the resources it wants, it will use a greater proportion of resources per capita than other societies. If another society has a large population and a relatively small resource base, it will have a much lower standard of living than that of the first society. Large populations and cultural attitudes that encourage consumption may stress the resource base and eventually result in a lower standard of living (either the lack of material necessities or a degraded environment or both). The great disparity in standards of living throughout the world is due in large measure to variations in population size coupled with inequitable distribution of resources and the lack of efficient methods to use the available resources.

To achieve a sustainable and sustaining earth society, we must incorporate into our culture an ethic of resource use, an ethic that redefines the human role in nature from conquerors to stewards of the biosphere. Once a stewardship ethic becomes an integral part of our culture, we will be better able to change our way of living to one compatible with a sustainable and sustaining earth. The health of our biosphere and our survival as a species depend on our willingness to accept responsibility for stewardship of the earth.

Change at a societal level results from the cumulative effect of change in individual life-styles. It is the responsibility of each individual to decide for himself or herself how to balance consumption and environmental quality. We must ask ourselves if we need an extensive wardrobe, two cars, large land holdings, and fax machines to be happy. Or, are we satisfied with enough clothes to keep us warm, a bicycle for transportation, a single radio, television, or telephone? Are we concerned about throwaway plastic utensils and excessive packaging of fast foods, or do we believe that waste is okay as long as the burgers are fast and cheap? Would we be satisfied with one meal per day if we're lucky, water if we can find it, and a cardboard shack for shelter if it hasn't been blown down or seized by someone stronger (Figure 2-2)?

How Can We Act as Environmental Stewards?

There are many ways in which we can act as stewards of nature. As individuals, we can learn about and care for the area in which we live. Direct, intimate contact with nature can help us to regain our sense of the earth and to appreciate the interdependence of humans and the environment. We can study and enjoy nature in cities, parks, and vacant lots. We can learn to value environments, such as farms and urban communities, that are managed in ways that are not destructive to the biosphere.

Contact with wild nature can also help us to become more effective environmental stewards. For many people, wilderness is emotionally and spiritually enriching. Wilderness provides a conceptual resource for the human mind, inspiring both scientific inquiry and poetic description. Indeed, in the scientist and the poet alike, a sense of the earth can lead to wonder, serenity, fear, awe, and discovery.

Many experts agree that individual action is necessary, but inadequate alone to address the planet's serious environmental problems. In the essay "Managing Planet Earth," William C. Clark writes: "It is as a global species that we are transforming the planet. It is only as a global species—pooling our knowledge, coordinating our actions, and sharing what the planet has to offer—that we have any prospect for managing the planet's transformation along path-

FIGURE 2-2: Homeless people in an alleyway in Calcutta, India.

ways of sustainable development." Those of us living in affluent countries can learn to "live more lightly upon the earth" by reducing the wasteful consumption of resources; promoting the recycling, recovery, and reuse of materials; using technologies that are appropriate to both natural and cultural systems; and conserving resources through energy-efficient processes.

Environmental stewardship requires that nations work together to solve environmental problems. A sustainable and sustaining earth depends upon achieving a balance between human population and the earth's finite resource base. Governments that effectively promote voluntary control of population growth can avert stringent, state-enforced policies or natural disasters that severely reduce or control population size. Through population control, nations could reduce total resource demand while providing a better quality of living for their people. Other steps that nations can take include using appropriate technologies, increasing energy efficiency, reusing and recycling materials, and avoiding policies that promise short-term economic gains at the expense of the resource base.

Finally, we can practice stewardship by becoming knowledgeable about our local, national, and global environment. Traditionally, humans have sought to understand the world around them through science and the process of scientific inquiry.

What Is Science?

Science is a way of knowing about the natural world. Science is not static or mysterious. It is open to anyone who wants to participate in a vigorous, exciting, and ongoing search for unbiased explanations for natural phenomena—from the world within the cell to the far reaches of the cosmos.

Scientists can provide information and data that may help us to tackle many health and environmental problems. But science cannot be hurried or put on a schedule. True, we desperately want to understand the mechanisms that cause acidified streams and lakes, produce cancer in our bodies, prevent the spread of acquired immune deficiency syndrome, increase protein content in soybeans, or safely control insect pest damage. But we can't guarantee that we will understand these mechanisms within any specified period of time. We *can* be reasonably sure, however, that without adequate support for scientific research, answers will not be found.

Scientific knowledge relating to environmental problems can be attained from many fields, including physics, chemistry, biology, and ecology. Scientific fields all share the belief that the acquisition of knowledge must be based on the process of observation, hypothesis development, and experimentation known as **scientific inquiry**.

Scientific Inquiry

The following brief account of the investigations of zoologists Windsor Watson and Carl Royce-Malgrem at the University of New Hampshire can help us to understand scientific inquiry, its limitations, and the role scientists play in helping us to understand and solve environmental problems.

Observation: Atlantic salmon are declining in numbers in acidified streams.

Hypothesis: Because observation and experimentation have shown that salmon are directed to spawning streams by odors imprinted on their brains as young fish, Watson and Royce-Malgrem hypothesize that acidified water interferes with the salmon's ability to detect odors, possibly preventing salmon from "recognizing" the proper home stream in which to spawn and thereby preventing spawning runs from the ocean.

Experiment: One group of Atlantic salmon are placed in water with acidity levels now found in many New England streams and subjected to odors that previously attracted them. A second group, known as the control group, is placed in nonacidified water and subjected to the same odors.

Findings: The salmon in the acidified water avoid or become indifferent to the odors.

Significance: The findings could explain the low numbers of salmon in acidified streams.

Scientific inquiry is a rigorous method of discovery that helps us to understand the biological and physical phenomena associated with natural systems. By observing the natural world, developing hypotheses, and conducting experiments, scientists attempt to explain natural phenomena. Colleagues then verify or dispute those explanations through repeated testing. When an explanation is widely confirmed and accepted, it becomes part of the basis of shared scientific knowledge. Other investigators can then build on this explanation as they seek to understand related, unexplained phenomena or to discover new phenomena.

The essence of scientific inquiry is the posing of questions and the search for precise answers to those questions. The inquiry is limited to questions that can be answered objectively through verifiable observation and experimentation. In the above example, the question is whether or not the hypothesis that acidification affects salmon's spawning behavior is true. Based on observation and experimentation, scientists draw conclusions that explain a phenomenon or establish a causal link between two events, such as acidified streams and nonspawning salmon.

Scientists must be careful that when they design their experiments and interpret data from their research, they are free of bias or coercion. Two hallmarks of scientific inquiry help to reduce bias and coercion: the controlled experiment and the complete reporting of methods and results to other scientists who may want to repeat the experiment in order to validate the results.

Controlled experiments are designed to compare two situations differing in only one variable. If the comparison yields different results, the difference is attributed to that variable. In the salmon experiment the only apparent difference between the conditions of the control and the experimental groups was the acidification of the experimental group's water.

If the results of the experiment support the hypothesis and the same results are obtained when the experiment is repeatedly conducted under the same conditions, the hypothesis is accepted by the scientific community as an explanation for the original observation. If the experimental results do not support the hypothesis, it is disregarded or reformed for further testing. A new look at the observation might lead to new questions and new hypothesis formation, and the process continues. (As of this writing, the results of the salmon experiment must be further validated before the conclusions of Watson and Royce-Malgrem are generally accepted by the scientific community.)

The more data collected to support the hypothesis, the more acceptable the hypothesis becomes. Experiments and data collection most often take place over several years; major scientific research requires several decades or more of experimentation and validation of hypotheses.

If, over long periods of time, many verifiable facts support the hypothesis, it might then be considered a theory. Theories unify many related facts or observations. Theories that withstand repeated testing over time become principles or laws. For example, four commonly held biological principles are: (1) all organisms and living systems evolve; (2) all organisms are made of cells; (3) all life is a competition for the capture and use of energy; (4) organisms share the common characteristics of metabolism, growth, movement, responsiveness (irritability), reproduction, and adaptability.

Because we refer to a statement as a principle or law doesn't mean that all observations support it or that all scientists agree about all aspects related to it. Nor does it mean that our understanding of it might not change as new information is discovered. The natural world is exciting precisely because there are so many interesting unanswered and, as yet, unposed questions to explore.

Limitations of Scientific Inquiry

Scientific inquiry is an important and useful tool in acquiring knowledge, but it does have limitations. First, it isn't always possible to conduct controlled experiments. The system under observation might be too large or complex or the time frame may be too long. For example, scientists have observed that the carbon dioxide content in the atmosphere is increasing. They suspect the increase is due to increased fossil fuel combustion and the widespread clearing and burning of forests. Further, they suspect that increasing levels of carbon dioxide will amplify the warming effect of the atmosphere, perhaps raising the average temperature of the planet by several degrees. Since scientists do not have an unindustrialized, uncleared "spare" earth to serve as a control group in experiments, they base their conclusions about the greenhouse effect on mathematical models. By constructing mathematical models of the atmosphere, scientists simulate its effects on the earth.

A second, related limitation of science is that it tends to be reductionist. In order to isolate variables and determine their effects, scientists must often look at ever-smaller units in laboratory settings. But determining the effect of a single variable in isolation from other related variables does not afford an accurate picture of how nature works.

A third limitation of science concerns the way scientists approach their work. The objective, value-free approach that is the hallmark of good science means that science cannot determine when or if something is good or bad or beautiful or worth preserving; neither can it answer questions about the spiritual or supernatural. Those conclusions and answers are not subject to scientific proof. Instead, they arise out of value judgments. Consequently, science cannot take the place of other ways of knowing, such as philosophy, religion, art, and poetry.

Role of Scientists and Nonscientists

Both scientists and nonscientists have a role to play in the effort to solve environmental problems. Most, perhaps all, environmental problems can ultimately be solved only through value judgments. For example, scientific research led to the development of insecticides, and scientists can now predict the possible biological side effects of insecticide use. But science cannot choose for us between preventing human death from insect-borne diseases like malaria and the possible toxic side effects of insecticide accumulation in food or groundwater. We all must participate, individually and collectively, in such decisions. Of course, scientists help us to make informed decisions by helping us to understand natural systems, such as the connection between acidified streams and poor spawning ability in salmon. But ultimately, the society as a whole must make the hard choices needed to solve the acid precipitation problem as it relates to increased acidity in streams.

Because scientists are part of the natural world they seek to explore and explain, honesty and objectivity must be part of their moral fiber. Although some scientists may disagree, we believe that scientists have a responsibility to try to prevent the misuse of their discoveries. For example, to understand the consequences of stopping all genetic engineering research, the public must first understand the basics of genetics and genetic engineering. One strain of bacteria can be induced to produce insulin for use by diabetics, another to produce crop plants resistant to frost, and a third to be resistant to antibiotics, making it a candidate for use in germ warfare. All three are scientifically possible, but each situation presents a different set of ethical and moral judgments for us to make.

What Is the Natural Science of Ecology?

Although many branches of science help us understand the physical, chemical, and biological processes of our environment, the natural science of ecology concentrates on the way those processes interact as systems. The holistic perspective of ecology is especially appropriate for studying environmental damage to living systems.

(a)

(d)

(b)

(e)

FIGURE 2-3: The tremendous variety of the earth's ecosystems. (a) The rain forest of eastern Brazil; (b) a pond in southeastern Indiana, United States; (c) a high-altitude meadow in western Canada; (d) the Kofa National Wildlife Refuge, Arizona; and (e) Lake Superior in winter.

Since environmental problems usually arise from a disturbance of natural systems, ecology is critically important to understanding, solving, and preventing environmental problems.

As a discipline, ecology grew out of scientific interest in the natural history of plants and animals and from the pioneering work of plant geographers. In the 1940s, many biologists urged that a course in ecology be offered for all majors in the biological sciences. The first widely used college textbook for ecology students, Eugene Odum's *Fundamentals of Ecology*, was published in 1953 and focused attention on understanding the dynamics of the biosphere through the study of its subunits, or ecosystems. **Ecosystems** are self-sustaining communities of organisms interacting with one another and with the physical environment within a given geographic area (Figure 2-3).

(c)

The term *ecology*, first proposed by Ernst Haeckel in 1866, comes from the Greek roots *oikos*, meaning "home," and *logos*, meaning "the study of." Literally, it refers to the study of the home. **Ecology** is the scientific study of the structure, function, and behavior of the natural systems that comprise the biosphere. Ecologists study the relationships of organisms with each other and with their physical environment.

Ecologists attempt to understand ecosystems based on research from the physical and biological sciences. They study nature as a functioning system instead of as a collection of distinct, unrelated parts. Understanding how the parts function together to form a forest, field, lake, or ocean helps ecologists to predict the effects of human-induced stress on the environment. Ecological research can help us to understand, for example, how nutrients and energy are transferred from organism to organism, how climate is modified by forest destruction, and how one species affects the population dynamics of another.

What Is Environmental Science?

While ecology can help us understand environmental problems such as acid rain or global warming, it cannot explain why human society has created these problems and what obstacles stand in the way of solving them. Both natural and cultural aspects of environmental issues are addressed by another field, environmental science (Table 2-1).

The discipline of **environmental science** attempts to understand and to solve the problems caused by the interaction of natural and cultural systems. It is an interdisciplinary field that draws on both natural science and a variety of disciplines pertaining to cultural systems (Figure 2-4). Environmental science can help us to propose and implement future-directed, goal-centered solutions to environmental problems and to practice sound environmental management that minimizes the degradation of natural systems.

Understanding how the biosphere makes all life possible and protecting the biosphere's integrity from the degrading effects of human activity are two distinctly different endeavors. Environmental science requires motivated people from all fields who are committed to solving environmental problems, people who believe those problems can be solved and who are not afraid to be at the center of debate on human values, ethics, and the implications of social policy.

Relation of Humanities and Social Sciences to Environmental Science

The arts and humanities, social and behavioral sciences, and communication and education shape our worldview. Consequently, a knowledge of these disciplines can help us to understand and modify our worldview and understand our behaviors and activities within social systems. They also help to explain the human dimensions of environmental problems: why those problems develop and how they can be solved within our cultural systems.

The arts and humanities enable us to define, record, interpret, and teach each other what it means to be human. Accordingly, they can encourage or discourage the practice of environmental stewardship. Through philosophy, literature, ethics, esthetics, art, music, and architecture, we seek to understand and express the value we place on resources, the natural world, and the relationships between natural and cultural systems. Moreover, we often learn about others' beliefs and values through their works. The knowledge we gain enables us to share in the thought and experiences of the artist and so it broadens our perspective.

Diverse behavioral and social sciences such as anthropology, history, psychology, sociology, theology, and ethnobotany (the study of how particular cultures use plants) enable us to study human origins and motivations. They contribute to our understanding of other societies, civilizations, and cultures. Other social sciences, such as political science and economics, provide us with information we need to understand local, national, and world events.

Social and behavioral sciences also enable us to understand how a society or culture perceives risks and whether or not it encourages conspicuous consumption or the conservation of natural resources.

▶ Table 2-1
Comparison of Ecology and Environmental Science

Ecology	Environmental Science
1. Uses the scientific method, which is grounded in experimentation.	1. Uses the problem-solving method, which is grounded in social process.
2. Defines natural system structure, function, and behavior, which may or may not have direct application to a particular environmental problem.	2. Defines a process for solving environmental problems.
3. Uses controlled laboratory experimentation or field testing of hypotheses.	3. Uses scientific research when appropriate to solving environmental problems.
4. Proposes hypotheses based on past observations.	4. Proposes solutions that are future-directed.
5. Follows a deliberate, thorough process, with theories or generalizations taking years to formulate and test adequately.	5. Follows a deliberate, thorough process, but, out of necessity, must sometimes proceed before all the facts are in.

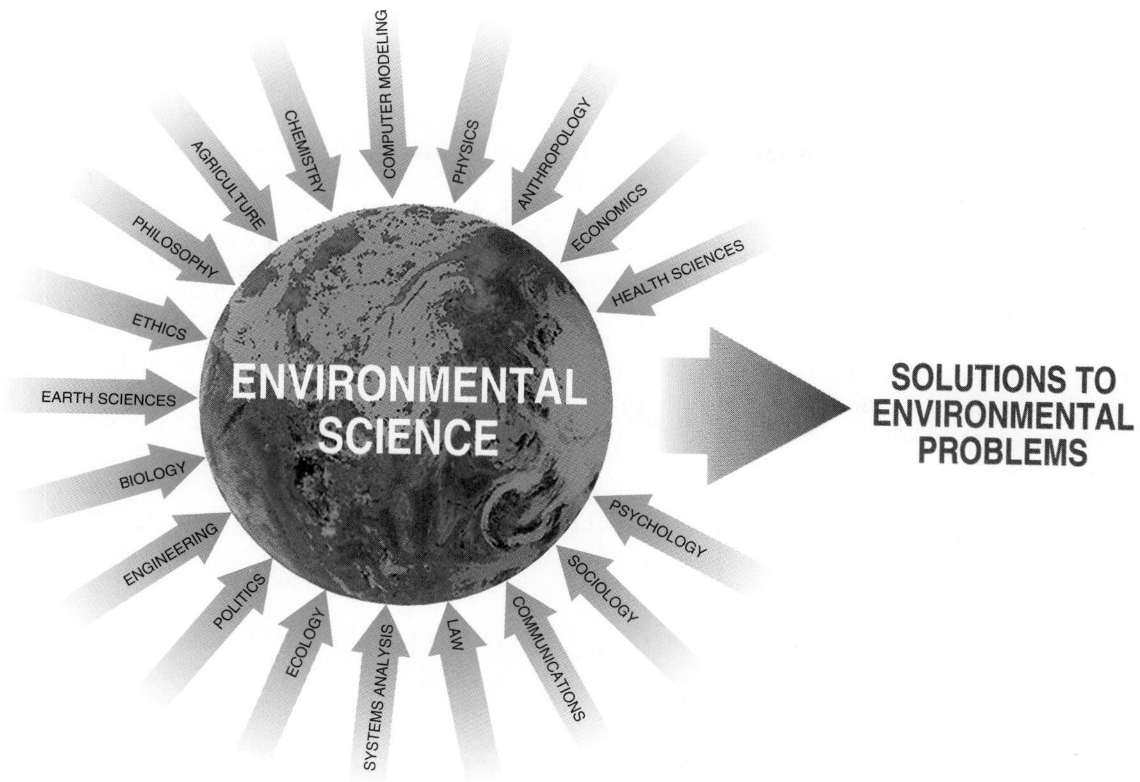

ENVIRONMENTAL
SCIENCE

COMPUTER MODELING
CHEMISTRY
PHYSICS
AGRICULTURE
ANTHROPOLOGY
PHILOSOPHY
ECONOMICS
ETHICS
HEALTH SCIENCES
EARTH SCIENCES
BIOLOGY
ENGINEERING
PSYCHOLOGY
POLITICS
SOCIOLOGY
ECOLOGY
LAW
COMMUNICATIONS
SYSTEMS ANALYSIS

SOLUTIONS TO
ENVIRONMENTAL
PROBLEMS

FIGURE 2-4: Fields contributing to environmental science. The search for solutions to environmental problems requires the participation of individuals from the natural sciences, social and behavioral sciences, and arts and humanities.

For example, in Japan, owning older or second-hand objects carries a social stigma; items are rarely recycled. New possessions are considered inherently better than old ones, regardless of the condition of the older object (Figure 2-5).

The disciplines of communication and education help us to acquire, organize, and transmit information and to mold opinion. They are shapers of public perception and attitudes and are powerful agents of change.

What Methods Can We Use to Solve Environmental Problems?

Perhaps the most difficult aspect of solving environmental problems is finding adequate solutions that are economically, politically, and ethically acceptable. A variety of ways to find such solutions are used today, including environmental problem solving, environmental activism, and litigation.

Environmental Problem Solving

Environmental science is driven by problem-solving models originally developed by research in the social

sciences. These models differ from the scientific method in that they start with a human-caused problem and take into account the human values

FIGURE 2-5: "Trash" in a Japanese landfill. In Japan, material such as this Indonesian plywood, which could be reused, or machines and appliances that are in perfect working order are sometimes discarded because of cultural preferences for new, unused items.

pertinent to the problem. The scientific method and environmental problem solving are both useful tools. However, they are applicable in different situations (Table 2-2).

The five-step environmental problem-solving model that we use can be adapted by a variety of groups to almost any situation. Because environmental problems are complex, involve many different disciplines in their solutions, and are affected by human value judgments, the problem-solving model is team-oriented, interdisciplinary, and goal-directed. Since no two environmental problems are alike, teams are assembled to fit the specific needs of the problem to be solved. The model is predictive,

allowing users to forecast the implications of possible solutions. Moreover, it involves reassessment and modification of proposed solutions as new information is gathered, additional parties become involved, or improved technologies become available. The model seeks reliable data and invites public participation.

The environmental problem-solving model can be used by dedicated nonspecialists and environmental professionals alike. For instance, a diverse group of undergraduate students at Miami University's Hamilton, Ohio, campus used the model to establish a recycling center in 1978.

The five steps in our environmental problem-solving model are: (1) identify and diagnose the problem, (2) set goals and objectives, (3) design and conduct a study, (4) propose alternative solutions, and (5) implement, monitor, and reevaluate the chosen solution.

Step 1: Identify and Diagnose the Problem. This step is critical, for problem solvers must correctly identify the problem and its boundaries before they can propose and implement appropriate, effective solutions. Identifying and diagnosing an environmental problem correctly usually involves considerable discussion among team members and is often more difficult than it initially appears to be. One of the most difficult tasks is to separate the actual problem from its symptoms.

Suppose, for instance, that in a hypothetical New England city, buildings, monuments, and statues are beginning to deteriorate. Discolored leaves signal that the trees are dying. Residents are reporting a variety of health problems, such as stinging eyes and breathing difficulties. All of these conditions are symptoms of air pollution. City leaders might adopt measures to alleviate the symptoms: repair the buildings, replant trees, and urge residents to remain indoors during specific periods. These measures, however, would not alleviate the problem itself and so the symptoms would be likely to reappear. An environmental problem-solving team, given the responsibility of finding a solution to the problem, must first attempt to identify, with as much certainty as possible, the sources of the city's air pollution. Using previous studies and findings, the team might identify five significant sources of local air pollution: carbon monoxide and nitrous oxide emissions from automobiles; ozone emissions from electric transformers; particulates from the city's coal-fired electric generating facility; sulfur oxide emissions from a local factory, the city's major employer; and sulfur oxide emissions from sources that are geographically far removed from the city.

Defining problem boundaries is an essential part of identifying a problem. The team members might

Table 2-2
Comparison of Scientific Method and Environmental Problem Solving

Scientific Method	Environmental Problem Solving
1. Involves observation, hypothesis formation, and experimentation within a specific field; not used to solve an environmental problem directly.	1. Involves problem identification, goal formation, objectives formation, data collection, and analysis from many fields applied to environmental problems.
2. Attempts to be objective, value-free, bias-free.	2. Considers human values, so cannot be value-free.
3. Involves an anthropocentric view.	3. Involves a biocentric view.
4. Involves continual questioning and testing to establish validity of hypotheses.	4. Involves continual evaluation and monitoring of situations to improve solutions.
5. Requires experimentation that has adequate controls and can be replicated by other scientists.	5. May not call for direct experimentation, but the results of experiments often form the basis for solutions.
6. Considers experimental design and testing of hypotheses to be the most critical steps of inquiry.	6. Considers problem identification to be the most critical step in finding solution, but other steps to be more important at different times in implementing solution.
7. Is interested in knowledge for its own sake; or, in the case of applied science, is interested in applications of knowledge which may or may not solve environmental problems.	7. Is interested in finding the best solution to actual environmental problems within a particular social setting.

ask: Is the problem confined to our city or to our entire air pollution control region? Does it occur throughout New England or in several parts of the United States and Canada? Is the problem confined to one industry? Does it occur only during specific seasons or specific weather patterns?

The answers to these questions, coupled with a thorough investigation of the problem and its probable causes, will enable team members to describe the problem as completely as possible. A complete description of the problem is necessary to ensure that the solution the team chooses to implement is appropriate and effective.

Step 2: Set Goals and Objectives. A goal is a clear statement of direction. It answers the question "Where do we want to go?" Objectives are developed to achieve a goal. They answer the question, "How can we get there?" That is, objectives outline, in measurable terms, a plan to attain a specific goal. The residents of the New England city might decide that their goal is "To achieve clean air for our city within three years." The problem-solving team would then set objectives carefully stating, in quantifiable terms, all the aspects of environmental quality that collectively form the standard for clean air in the city. Since they identified five major pollution sources, the team would set objectives for each one. One objective might call for the reduction of sulfur oxide emissions of 20 tons per year in each of the next three years.

Setting goals and objectives is often a difficult process, because "clean" or "safe" or "healthy" does not have the same quantitative meaning for everyone. Individual perceptions of environmental quality are influenced by emotions, knowledge, and values. Employees of the factory might initially fear that they will lose their jobs if the plant managers are forced to install costly pollution control equipment. Residents living near the factory might be greatly concerned about the risk of emphysema associated with exposure to sulfur oxide emissions. Their concern might cause them to doubt both the plant managers' assessment of the situation and their efforts to solve the problem, no matter how sincere those efforts are. The plant managers, in turn, might question the intent of local environmental groups and refuse to accept any data or studies that appear to link pollutant emissions with environmental degradation and increased health risks.

Fear and mistrust, while sometimes warranted, more often distort individual perceptions of the problem and thus adversely affect efforts to develop a common, clear goal and objectives. For example, the plant managers might initially maintain that no reduction in sulfur oxide emissions is needed, while citizens groups maintain that all sulfur oxide emissions should be eliminated. However, if the health risks and long-term effects of the emissions and the costs of pollution abatement are properly calculated and communicated to all parties, then individual and group perceptions about the goal might change. When people are well informed about the consequences of their actions, their perceptions of problems tend to be based on their values rather than their emotions. Increased knowledge might convince the employees, community residents, and plant managers to modify their positions and support a reduction in emissions.

Thus, in the final analysis, individuals will define "clean" or "safe" or "healthy" according to the value they place on employment versus health risks versus environmental degradation in both the short and long term. Economic security, acceptable personal risk, and acceptable risk to the environment are values that are individually and collectively arrived at. These values are incorporated into both qualitative and quantitative goals and objectives.

Step 3: Design and Conduct a Study. To design a study properly, problem solvers determine what information they need to support each objective and how they will collect and analyze that information. Typically, the environmental problem-solving model does not require team members to conduct long-term experiments to collect such data; instead, they use previous research. If the necessary data do not exist, they might design short-term experiments to generate the needed data within a predetermined time frame. In these cases, competent specialists conduct the studies, analyze the data, and relate the findings to both the objectives and goal.

To determine what information they need to acquire, problem solvers develop key questions for each objective. Regarding sulfur oxide emissions, they might ask:

1. How many tons of sulfur oxide are released into the air over the city each day, month, and year?
2. What sources release the greatest amounts of sulfur oxide?
3. How do various weather conditions affect the level of sulfur oxide in the city's air?
4. How do various levels of sulfur oxide emissions affect buildings, trees, and human lungs?
5. What processes or technologies exist to reduce sulfur oxide emissions?
6. What do these processes or technologies cost?
7. What happens to the sulfur oxide when it is removed from exhaust gases?
8. How important is the reduction of sulfur oxide to the community?

Team members carefully document the collection and analysis of data. Often, a thorough data search will uncover new areas of inquiry and might suggest new objectives. After careful data analysis, team members begin to propose alternative solutions to the problem.

Step 4: Propose Alternative Solutions. It is tempting to proceed too rapidly to this step. The great fictional detective Sherlock Holmes, with his brilliant deductions and solutions to problems, devoted most of his time to the steps leading up to his solutions. That's why they were so brilliant! If environmental problem solvers are careful in the steps leading up to their solutions, those solutions will be more effective and easier to implement.

Team members propose several alternative solutions to the problem. Each alternative will ordinarily have positive and negative points. For example, one solution might cost a lot initially, but have a short pay-back time. A second solution might be quick to implement, but be very expensive. A third solution might be healthful for people in the community, but create a problem for people in a neighboring community. Although cost, speed of implementation, and health would all be important criteria for the team, they have to decide which factors are more important to them than others and which solution offers the best combination of factors. In this way, their decisions will ultimately be based on value judgments.

Suppose, for instance, that one solution includes measures to curb pollution from automobile exhaust and the local electric generating facility. The measures could be instituted rather quickly but might not result in a sufficient reduction in sulfur oxide emissions. A second solution targets significant sources of pollution from outside the area, such as sulfur-emitting coal-fired power plants in the Midwest. The team might propose pooling their resources with other municipalities in the region to launch an educational and political campaign to put pressure on geographically removed sources to reduce sulfur emissions. If such a campaign was successful, it would produce the desired results in terms of cleaner air for the city. The disadvantage of this solution, however, is that the outcome of the campaign would be out of the control of the people in the city and surrounding region; those far removed from the area would have the final say in determining local air quality.

A third solution proposed by the team might call for the local factory to install a particular piece of equipment that removes sulfur oxide from emissions. This solution could have several consequences: higher consumer costs, layoffs for some employees as plant managers attempt to offset part of the cost of the new equipment, increased expenditures by the factory to dispose of the sulfur oxide after it is removed, and possible loss of tax revenues to local schools if the plant closes because of increased expenses. Nevertheless, the problem-solving team agrees that this solution is the best because it is the most cost-effective. Per dollar spent, it is, in their opinion, the most effective way to reduce emissions, thus preventing environmental degradation and the loss of plant and animal species and decreasing the incidence of emphysema among the workers and community residents. If, as a group, the people affected by the team's decision share these concerns, they will probably be willing to absorb the cost of the solution.

Step 5: Implement, Monitor, and Reevaluate the Chosen Solution. Finding an effective and long-term solution to a problem must be viewed as a dynamic, ongoing process. Unfortunately, it is human nature to propose a solution, implement it, and then turn our attention elsewhere, confident that the problem has been solved and we no longer need be concerned. If only it were that easy! It is critical that the situation is monitored to ensure that the chosen solution has its intended effect. We may discover that the solution didn't work well enough to meet the objectives. Or, we may find that the solution, while solving the original problem, has created an entirely new problem. In such cases we could return to step 2, 3, or 4. New information and insights could lead us to modify our solution or reexamine our goal or objectives.

At times, we do not understand natural and cultural systems well enough to be certain our solution will work, and yet we must try to effect some change before it is too late. For example, an irrefutable link between coal-fired power plants in the Midwest and acid precipitation in the East is not entirely established. Similarly, a link between acid precipitation and lake, forest, and soil deterioration in the Adirondack Mountains of New York State has not been established beyond a shadow of a scientific doubt. In environmental problem solving, we are sometimes compelled to take action before the natural system is irreversibly damaged while we wait for irrefutable scientific evidence of cause and effect. In other words, environmental problem solvers must sometimes treat the patient before the condition becomes terminal.

Environmental Activism

Environmental activists are a diverse group: parents who picket city hall to complain about pollution and related health risks in their neighborhood; the millions of members of mainstream environmental groups like the National Wildlife Federation, World Wildlife Fund, Sierra Club, and Audubon Society; the members of more radical environmental groups that take specific, direct action to prevent or halt environmental degradation. Though they are a diverse lot, environmental activists are alike in that they are passionate about their particular environmental cause.

Different activist groups hold different philosophies. Some believe that only lawful actions, such as protests and boycotts, are justified, while others are convinced that the defense of the environment takes priority over the law. Groups such as Greenpeace, Sea Shepherd, and Earth First! act directly to protect the environment from harm. Greenpeace members typically participate in nonviolent efforts, such as placing themselves between whales and whaling ships to try to prevent harpoons from being fired into the whales (Figure 2-6). Earth First! members often take more extreme measures, such as disabling bulldozers and other development equipment, to protect wilderness habitats.

Activists alert the public to situations that are harmful to the environment. Their efforts often generate media and public attention and thus may produce results. In 1989–1990 activists boycotted tuna because the nets used to catch the fish trapped

FIGURE 2-6: Greenpeace members aboard the *Rainbow Warrior II.*

and killed dolphins as well. The boycott was so successful that in 1990 Starkist and two other major tuna companies agreed to market only tuna that was "dolphin-safe." However, sometimes environmental activism is difficult to apply to complex problems, such as global warming, that require widespread changes in life-styles and improvements in technology.

Litigation

Litigation, or the initiation of lawsuits, is an adversarial process sometimes used to address environmental problems. The attorney on each side of the legal contest tries to provide evidence that supports his or her client's claims and refutes the arguments of the opposing side. The attorneys must abide by rigid rules for presenting or excluding information offered as evidence. A judgment is rendered between competing "truths." This system can be effective in stopping certain actions, such as violations of the Clean Air Act or Clean Water Act. It is less effective in initiating actions, such as a community effort to clean up a local waste dump.

The legal process often stops with the assessment of blame and is usually time consuming and expensive. In many instances there are no clear winners; all parties to the case, particularly the environment, can and do lose. Thus, although the adversarial system can be helpful, protracted and costly litigation is not an ideal answer to most environmental problems. Alternatives to litigation include negotiation, environmental dispute resolution, and the cooperation of attorneys interested in the environment (see Chapter 27, Law and Dispute Resolution, for a more detailed look at this subject).

How Can We Minimize Environmental Problems?

Environmental problem solving, activism, and litigation share one major drawback: they react to a specific problem after it has happened. Some degree of damage to the environment or human health typically occurs before the problem is recognized. To secure a sustainable and sustaining planet, we must minimize or avoid environmental degradation in the first place. The most effective way to do this is through future-oriented environmental management based on a stewardship ethic.

A stewardship ethic acknowledges that, unlike all other living creatures on earth, humans exercise options beyond instinct. We learn, modify, and transmit knowledge; we develop explanations for our origins in both biological and spiritual terms; we record the past and plan for the future. We are capable of much cultural variation: We have diverse value systems, ideologies, and cultures; we belong to local, national, and international organizations; we have many forms of governments. And, we alter our environments to satisfy our needs. Our human nature, in fact, is reflected in our individual and institutional propensity to try to control our environment through management. It is time to acknowledge our management responsibilities.

Environmentally sound management (ESM), introduced to the public during the environmental movement of the 1960s and 1970s, is based on the premise that any resource, natural or cultural, can be managed in environmentally sound ways. It attempts to produce the least environmentally disruptive decisions through a management plan which minimizes or prevents environmental degradation. ESM can also foster sustainable economic development and contribute to security throughout the world. Resources, of course, are the lifeblood of economic development. Development is sustainable only when resources are efficiently used in the present and conserved and maintained so that they can be used in the future.

Environmental management should not be based on arrogance or conceit. We cannot control nature. Thousands of years of failed attempts should make that clear to us. Indeed, as we have discovered, effective environmental management is based on understanding natural processes and ensuring that human activities do not violate these processes. When we "manage" within the constraints imposed by the earth's systems of checks and balances, we can attain a sustainable and sustaining environment. In the final analysis, we must learn to manage ourselves. Perhaps when we learn to harmonize our actions and cultural systems within the bounds imposed by nature, we will become, as the physician and writer Lewis Thomas hopes, "the collective mind for the earth's global support system."

How Can We Implement Environmentally Sound Management?

Environmentally sound management is based on: (1) a stewardship ethic, (2) a biocentric world view, (3) an understanding of natural system dynamics, (4) environmental education, (5) interdisciplinary planning, (6) data from sound natural and social system research, (7) sociocultural considerations, (8) knowledge of political systems, (9) sound economic analysis, (10) maximum public participation (Figure 2-7).

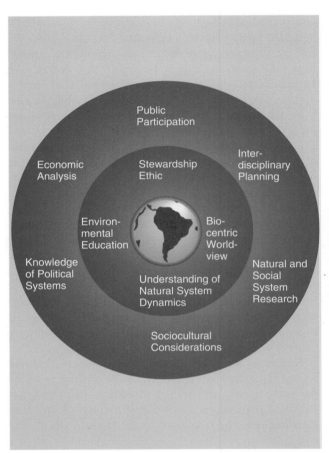

FIGURE 2-7: Environmentally sound resource management model. The resource management model is similar to problem-solving models, but because it includes an ethic of resource use, it is proactive and future-oriented. The model can be used to manage resources in an environmentally sound manner in order to avoid environmental problems.

You may have noticed that these elements seem familiar, and so they are. The environmental problem-solving method also incorporates natural system knowledge, interdisciplinary planning, sound natural and social system research, sociocultural considerations, knowledge of political systems, sound economic analysis, public participation, and environmental education. Environmentally sound management differs from problem solving in that it incorporates a stewardship ethic that allows it to be proactive rather than reactive. Thus, ESM is directed at the future. It can be used for any definable resource at the local, regional, national, or international level, and it invites the participation of individuals, institutions, and society at large.

In order to manage resources in a future-directed, environmentally sound manner, we must develop, implement, and monitor management plans to guide future activities. Those practicing ESM develop management plans in three parts. Part 1 is a comprehensive description of the resource that defines all physical, biological, and social boundaries and includes estimates of availability. Part 2 presents a thorough historical sketch of previous management practices, including an analysis of environmental problems. Part 3 establishes the management goal and objectives and plans for accomplishing them.

An ESM team uses their stated objectives to guide them in collecting data, proposing and evaluating alternative strategies, and selecting and implementing the best strategy. Further, they monitor results and, when necessary, redefine management strategies in order to attain their goal. They continually ask key questions:

1. What research is available?
2. What new research needs to be conducted?
3. How will the management plan affect natural systems?
4. Which experts should be consulted?
5. How can the findings be integrated into an interdisciplinary plan?
6. What are the costs of the plan?
7. What are the social, political, and educational impacts of the plan?
8. Has the public been adequately involved in the planning process?
9. What educational materials need to be developed for a better understanding by the public of the management strategies?
10. Does the management plan conform to a stewardship ethic of resource use?

The ESM model can be used to prepare management plans for any definable resource: coal, petroleum, minerals, food crops, historic sites, whales, lakes, wilderness areas, local parks, and wetlands, to name a few. It requires trust, compromise, and communication among the interdisciplinary team that develops the plan, the resource managers who implement it, and the consumers or users of the resource. The management plan must be continually monitored to ensure that it remains an effective and useful strategy.

We can preserve critical habitats, minimize intrusion into wilderness, protect wildlife, restore rivers, lakes, estuaries, and wetlands, build up soil fertility, and preserve genetic diversity *if* we are willing to place a high priority on environmentally sound management of these valuable resources. Most importantly, we can attain these goals as part of an overall mission to achieve sustained economic and social development. Environmental, social, and economic well-being are inseparable. Indeed, economic and social well-being are ultimately dependent on environmental well-being. In the long run, safeguarding and conserving natural systems and resources is the most effective way to ensure economic well-being.

Summary

A sustaining and sustainable earth society is the goal to which we must dedicate ourselves if we are to preserve the planetary conditions we now enjoy. A stewardship ethic to guide the use of resources is critical to achieving a sustainable society. There are many different things we can do to develop this ethic. One of the most important is to experience nature first-hand. Doing so will better enable us to live in harmony with natural systems as individuals, as societies, and as a species.

Science is an important way of learning about the natural world. The natural science of ecology views natural systems holistically, studying how all the individual parts interact with each other, and thus it can help us understand how culturally created stresses affect those systems.

Environmental science differs from the natural sciences because it takes into account human values and culture. Environmental problems contain social, political, and economic dimensions, and so environmental scientists must draw upon knowledge in numerous and diverse areas. There are no easy solutions to environmental problems. Our challenge in solving them is to understand the underlying scientific principles and the consequences of interrupting natural processes through the long-term effects of our social actions. This focus on basic scientific concepts, coupled with an understanding of social processes, forms a basis on which to analyze, propose solutions to, and, finally, manage environmental problems.

Environmental problems can be solved through the careful, systematic use of problem-solving methods. Our five-step model can be followed by anyone interested in working for a better environment: students (with any major or profession in mind), business leaders, city planners, architects, homemakers, ministers, politicians, labor leaders. The quality of the environment depends on the decisions of its citizens. When people understand the functioning of natural systems, the limitations of natural processes, and how humans culturally modify those processes, and when they are given opportunities to solve environmental problems, they will make informed decisions that lead to a better environment.

Environmental problems can be minimized or avoided through environmentally sound management. Environmentally sound management acknowledges that all forms of life on earth have value, opposes uncontrolled resource exploitation, promotes the wise use of resources, and minimizes waste and environmental damage. Since no one can predict what future generations will deem of value, the most important thing we can do is ensure a maximum of choice for future generations.

Key Terms

ecology
ecosystems
environmental science
environmentally sound management
science
scientific inquiry
standard of living
sustainable
sustaining

Discussion Questions

1. What are some activities that individuals can engage in to act as environmental stewards? Communities? Nations?

2. Suggest ways that respect for nature could be reflected in how we define a quality standard of living.

3. Using China, Mexico, the United States, and Kenya, describe how the standard of living in each of these nations is the result of its population size, resource base, and culture.

4. For the same countries as in question 3, describe the severity of and possible solutions for their environmental problems.

5. A major employer in your community must find a way to handle or dispose of its industrial waste or cease operations. One proposed solution is to place a new landfill near your home. How could this situation be resolved through environmental problem solving, activism, and litigation?

How the Biosphere Works

Ecological Principles and Application

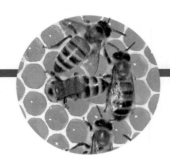

Ecosystem Structure

Economic deficits may dominate our headlines, but ecological deficits will dominate our future.

Lester Brown

Some of today's most intellectually challenging problems are found in the area of environmental sciences, to which the science of ecology has much to contribute.

Robert Leo Smith

Learning Objectives

When you finish reading this chapter, you should be able to:

1. Define ecosystem and list the two major components of any ecosystem.

2. Describe the roles played by producers, consumers, and decomposers in the ecosystem and the interactions among them.

3. Describe how the biota and abiota influence ecosystem structure and each other.

4. Identify three important limiting factors and give an example of how each might act to regulate the structure of an ecosystem.

We have what may seem to be a strange request to ask of you. Think, for a moment, of the languages you studied during high school or college—Spanish, French, Russian, or Greek. Before you could hope to read and speak Spanish or Greek, you had to learn the letters and words that are the building blocks of language. Only then could you begin to learn the rules that govern how the letters and words are formed into sentences to communicate ideas. As your understanding grew, you were able to express your own ideas and to read complete passages written in a language that at one time had seemed so alien.

You are probably wondering what this has to do with environmental science. In a real sense, nature, too, has its own language. Ecosystem structure is an alphabet, a collection of letters and words that can be combined in many different ways. In this chapter we will take a close look at the structural components—the alphabet and words—all ecosystems share.

Knowing letters and words alone does not allow one to speak a language, and so it is with nature. Simply knowing the structure of ecosystems will not allow us to understand how they function, change, and respond to human activities—in short, how to speak the language of nature. To do so, we must learn how the structural components of an ecosystem function as part of a greater whole. That topic is the focus of the next chapter.

Language, of course, is a cultural artifact, something created by humans for humans. As the makers of language, we establish the rules—grammar—that enable us to communicate with one another. Natural systems, too, have rules that govern how natural systems function and respond to change. But, obviously, natural systems are not a human creation, and so we do not always understand—and sometimes are not even aware of—their laws. To understand our environment, then, we must learn to decipher the intricate, beautiful language of the biosphere. Ecology, the discipline concerned with understanding natural systems, is still in its infancy, but already it shows great promise in enabling us to speak the language of the biosphere.

Why Do We Study Ecology?

In Chapter 2 we defined ecology as the scientific study of organisms in their natural surroundings. Ecologists attempt to understand how living things interact with their environment and with one another and how the earth's living systems maintain the integrity of the biosphere. An understanding of natural systems is essential to solving environmental problems and developing environmentally sound management plans. For example, ecological research on the flow of nutrients in ecosystems has helped us to understand how excessive amounts of nutrients added to soil result in the nutrient enrichment or pollution of waters; how pesticides applied to crops and forests affect other living organisms; and how radionuclides released from nuclear power plants or weapons production facilities enter the environment and become incorporated into plant and animal tissues.

Clearly, ecological research provides information that can serve as the basis for managing resources and the landscape in an environmentally sound manner. Unfortunately, ecological considerations have often been ignored in land development plans, for example, the design and construction of highways, subdivisions, and marinas and the siting and size requirements of parks, forests, and green spaces. Several attempts have been made to bring ecology into the planning process. Arguably, the most significant of these is the National Environmental Policy Act (NEPA), which requires that an environmental impact statement (EIS) be prepared for any project involving federal aid. The purpose of the EIS is to assess the project's effect on the natural system. Since NEPA's passage in 1969, many of the EISs prepared have had no bearing on the final project, but others have forced the planners to modify practices that would have adversely affected the environment. In order to ensure that ecological considerations are incorporated into the planning process—for projects that do not receive federal aid as well as for those that do—all those involved in making environmental decisions, from planners and engineers to sociologists and politicians, must develop an ecological awareness. The foundation of an ecological awareness is basic ecological knowledge.

What Are the Levels of Ecological Study?

Ecology is studied at many levels: individual, population, community, ecosystem, biome, and biosphere (Figure 3-1). An individual is a single member of a species. A **species** includes all organisms of a particular kind that are capable of producing viable offspring (i.e., individuals which can themselves produce offspring). For species that reproduce sexually, the individual members reproduce by mating, or interbreeding.

A **population** is a distinct group of individuals of a species that live and interact in the same geographic area, for example, largemouth bass in a particular

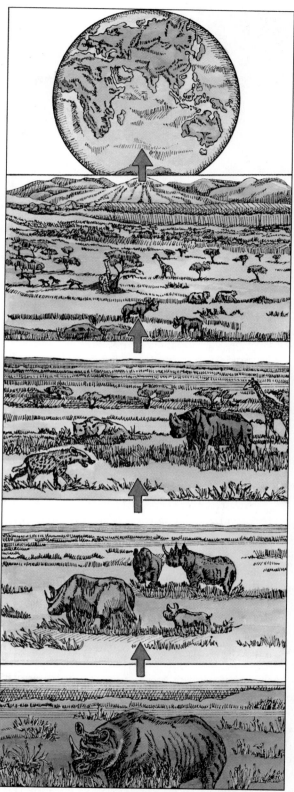

FIGURE 3-1: Levels of ecological study. The study of ecology spans the continuum from the individual organism (like this black rhinoceros), to populations, communities, ecosystems, and finally, the biosphere. Information at all these levels contributes to our understanding of the "ecos," our home.

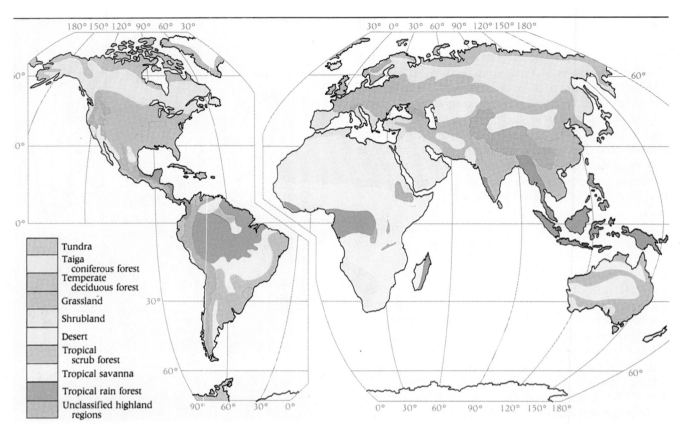

FIGURE 3-2: World map showing major biomes of the earth. Biomes are identified chiefly by their dominant plant life, which is a product of specific climatic conditions such as temperature, precipitation, and availability of light. These variables are affected by latitude and altitude.

pond or Canada goldenrod in one field. Populations have measurable group characteristics such as birth and death rates or seed dispersal and germination rates. Both individual organisms and populations have a place within the physical environment commonly called **habitat,** the place where the organism or population lives. Forests, streams, and soil are just a few examples of habitats where organisms or populations are found.

A **community** includes all of the populations of organisms that live and interact with one another in a given area at a given time. A community and its interactions with the physical environment comprise an **ecosystem.** The Great Lakes, wetlands, and estuaries are examples of aquatic ecosystems. Examples of terrestrial (land) ecosystems are oak-hickory forests, alpine meadows, and high mountain deserts.

Many ecosystems taken together in a large terrestrial area of the earth are referred to as a **biome** (Figure 3-2). Biomes are identified and classified according to their dominant vegetation type and its associated microbes and animals (Figure 3-3). Vegetation type is largely the product of climatic conditions, that is, long-term weather patterns, including temperature, precipitation, and availability of light.

The eastern United States, for example, is characterized by temperate deciduous forest, which is dominated by hardwood trees. Beech and maple might dominate in one area, oak and hickory in another, and tulip poplar in a third, but the entire area, collectively, is one biome. The union of all terrestrial and aquatic ecosystems is known as the **biosphere.**

In addition to the formal study of ecology, we can learn about ecological processes, the environment in which we live, and the impact of human activities in many ways. Your participation in this class is an important step in your becoming ecologically and environmentally aware. But perhaps the best way to learn about the ecosystems around you is to study them first-hand. Explore the fields, woods, parks, vacant lots, rivers, streams, or lakes in your area. Identify the kinds of plants and animals that live there. Perhaps one particular plant or animal will capture your interest, and you might want to study it more closely. By observing a plant, you can learn a great deal—how it produces seeds, what feeds on those seeds, when it blooms, and what soils it grows best in. By observing an animal, you can learn where it lives, what it eats, what animals prey on it, when it breeds, and how it raises its young.

FIGURE 3-3: Biomes in different parts of the earth. (a) The vast tundra, as seen here in Denali National Park, Alaska, is dominated by lichens, grasses, and dwarfed trees. (b) The taiga, a northern coniferous or boreal forest composed of great forests of spruce, hemlock, and fir, is found exclusively in the Northern Hemisphere. It is seen here in Winema National Forest in Oregon. (c) Plants adapted to extremely dry conditions dominate in the world's desert biomes. The Saguaro cactus, shown here in Organ Pipe Cactus National Monument, Arizona, is a familiar figure in the desert of the American Southwest. (d) The Caribbean National Forest, Puerto Rico, illustrates the lushness of the tropical rain forest, the earth's most diverse biome in both plant and animal life. (e) The tropical savanna biome of the African continent borders tropical rain forests. It is a grassland biome, but also contains some trees as shown here in Amboseli National Park, Kenya. (f) The temperate deciduous forest biome, which dominates the eastern half of the United States, is also found in parts of Europe, Asia, Australia, and South America.

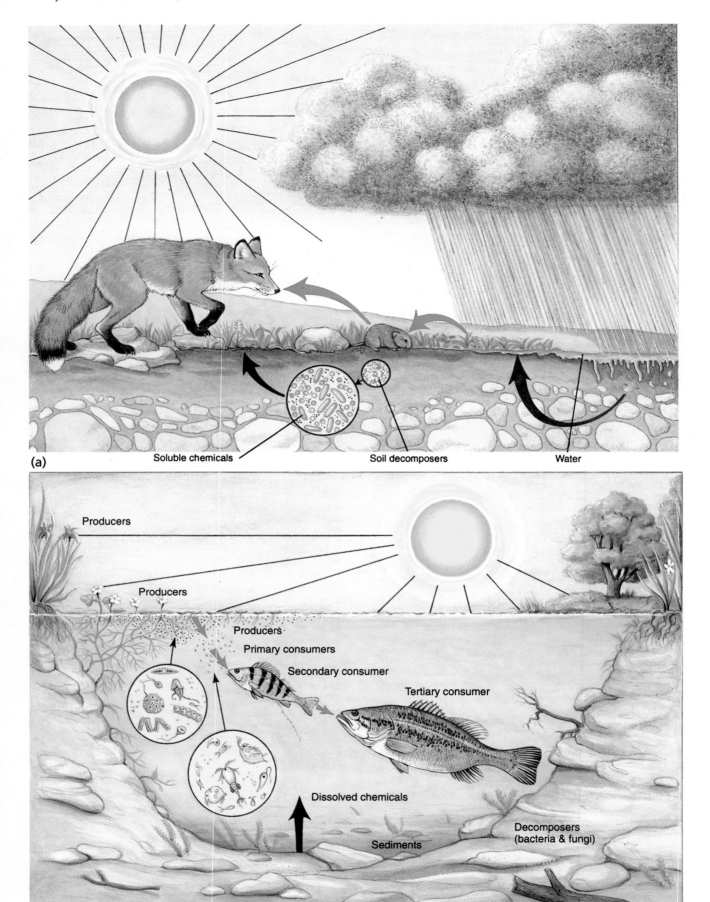

(a)

Soluble chemicals

Soil decomposers

Water

Producers

Producers

Producers

Primary consumers

Secondary consumer

Tertiary consumer

Dissolved chemicals

Sediments

Decomposers
(bacteria & fungi)

(b)

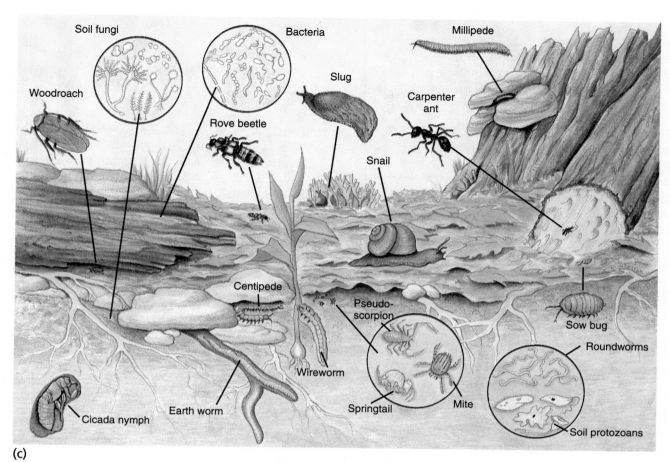

(c)

FIGURE 3-4: Examples of terrestrial and aquatic ecosystems. Ecosystems vary in size and in location, as these drawings illustrate. But whether it is a field (a), pond (b), or rotten log (c), every ecosystem is a self-sustaining community of organisms interacting with the physical environment within a defined geographic space. Blue arrows represent the one-way flow of energy; black arrows represent the cycling of materials.

What Is an Ecosystem?

Even though the biosphere can be viewed as a single large system, it is typically studied in smaller, but globally interrelated units called ecological systems or ecosystems. As we discussed in Chapter 2, an **ecosystem** is a self-sustaining, self-regulating community of organisms interacting with the physical environment within a defined geographic space. *Ecos* refers to the assemblages of life forms in a particular environment, and *system* refers to the processes necessary to maintain the integrity of the environment through an intricate set of checks and balances. Ecosystems occur in both terrestrial and aquatic habitats (Figure 3-4).

Ecosystems are not static. Even mature ecosystems like hardwood, broad-leaf forests constantly undergo internal changes as a result of changes in their external environments. For example, lack of rainfall can make older trees more susceptible to disease. Diseased trees may die, allowing younger, healthier trees to become dominant. Heavy winds can topple trees, opening up the forest floor to increased sunlight and temperature, which promote, for a time, the growth and success of species adapted to these changed conditions.

Sometimes, seemingly minor changes in one component of an ecosystem, such as spraying a field or forest with an insecticide, can have far-reaching effects on another component of the system. The insecticide might destroy nontarget insects more successfully than it does the insect pest, killing pollinators like honeybees, thus reducing pollination success in flowers and nectar production for the beehive. In addition, insecticides can kill predators of the targeted insect, predators that previously helped to control the target insect population.

Ecosystems can usually compensate for the stresses caused by external changes. But many ecologists believe that both individual ecosystems and the biosphere have thresholds beyond which catastrophic, potentially permanent change occurs. For example, at the ecosystem level, thresholds may be exceeded when a lake becomes acidified, a stream becomes overloaded with sewage, or a forest is cut and fragmented. Perhaps, if the external stress is eliminated, the system may recover, given enough time. It is uncertain, however, whether or not the system will return to its pre-stress state. Stratospheric ozone depletion, the buildup of greenhouse gases, and the rapid loss of biological diversity are some examples of such potentially disastrous biospheric changes.

What Are the Components of an Ecosystem?

Ecosystems are composed of a nonliving or abiotic component, the abiota, and a living or biotic component, the biota. The abiota and biota interact to provide the materials and energy necessary for organisms to survive.

Abiota

The abiotic component of ecosystems includes: energy, matter (nutrients and chemicals), and physical factors such as temperature, humidity, moisture, light, wind, and available space.

Energy. The ability to do work—to move matter from place to place or to change matter from one form to another—is known as **energy.** We use energy for many purposes: to build shelters and warm or cool them; to process and transport food; to keep the cells of our bodies active and functioning properly.

Energy reaches the earth in a continuous but unevenly distributed fashion as sunlight. Less than 1 percent (0.023 percent) of the total energy reaching the earth's atmosphere each day is actually captured through photosynthesis by living things; the rest is reflected by the earth's cloud cover and never reaches the biosphere or is radiated by the earth's surface back into space as heat. Huge amounts of energy, produced when the planet formed, are stored deep below the earth's surface. Stored energy makes its way to the surface through volcanoes, deep sea vents, and terrestrial cracks or fissures.

Energy cannot be recycled; when it is used, it is changed to another form and eventually radiated into space as heat. Consequently, the earth must be supplied with a constant flow of energy in order to support life. The internal energy of the earth accounts for only a small percentage of its energy; the vast majority is supplied by the sun. Thus, the earth is an open system for energy, continuously receiving and using energy from the sun and radiating waste heat into space.

The **first law of energy,** or **first law of thermodynamics,** states that during a physical or chemical change energy is neither created nor destroyed. However, it may be changed in form and it may be moved from place to place. The **second law of energy,** or **second law of thermodynamics,** states that with each change in form, some energy is degraded to a less useful form and given off to the surroundings, usually as low-quality heat (Figure 3-5). Thus, in the process of doing work, high-quality energy is converted to low-quality energy. For example, the high-quality energy available as gasoline in an automobile is converted to both mechanical energy used to move the car and heat. The energy stored in the sugars we eat is converted to chemical energy; this, in turn, is converted to mechanical energy that moves our muscles and creates heat. With each transfer of energy, heat is given off to the surroundings, eventually dissipating to the external environment, through our atmosphere to space, and throughout the universe.

Energy constantly changes or flows from a high-quality, concentrated form to a low-quality, dispersed, and less useful form. This tendency to dispersal or randomness is called **entropy.** Life slows down, but does not stop the process of entropy. Living organisms temporarily concentrate energy in their tissues and thus, for a time, create a more ordered system. Perhaps the best example of this is the capture and storage of the sun's energy by green plants, algae, and some bacteria. Inevitably, however, the plants die, they lose their leaves, and the system tends toward entropy.

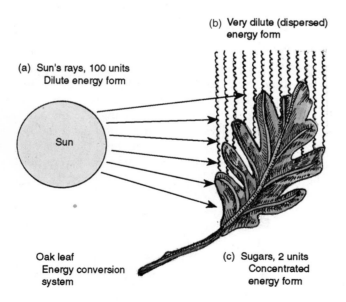

(a) Sun's rays, 100 units
Dilute energy form

Sun

(b) Very dilute (dispersed) energy form

(c) Sugars, 2 units
Concentrated energy form

Oak leaf
Energy conversion system

FIGURE 3-5: The two laws of thermodynamics. The conversion of solar energy (a) to food (c) through the process of photosynthesis illustrates the first law of thermodynamics. The second law of thermodynamics maintains that (c) is always less than (a) because heat (b) is lost during the conversion.

Matter. Matter is anything that has mass (weight) and takes up space. Matter is the "stuff of life"; everything on earth—animal, vegetable, and mineral—is composed of matter. Although meteors and meteorites sometimes enter the earth's atmosphere, adding matter to the biosphere, the earth is essentially a closed system for matter. Most of the matter that will be incorporated into objects in future generations is already present and has been present since the planet came into being.

All matter is composed of elements. **Elements** are substances that cannot be changed to simpler substances by chemical means. Each element has been given a name and a letter symbol. Some familiar elements are oxygen (O), carbon (C), nitrogen (N), sulfur (S), and hydrogen (H).

There are 92 naturally occurring elements and 15 synthetic ones; each has special characteristics that make it unique from all others. What makes an element unique is its atomic structure. All elements (and hence all matter) are composed of atoms. **Atoms** are the smallest parts of elements. They are composed of subatomic parts called protons, neutrons, and electrons. An atom can be divided into its subatomic parts, but when this happens, it loses its unique characteristics.

Each atom has a center or nucleus which contains a set number of protons and neutrons. Protons carry a positive charge; neutrons carry no charge. Negatively charged electrons revolve around the nucleus in special patterns called orbits. Each orbit represents an energy level that is maintained by the force of attraction between the electron in orbit and the nucleus.

All atoms of a particular element have the same number of electrons and protons, but they sometimes have different numbers of neutrons, which change the mass or weight of the atom. These different forms of the same atom are called **isotopes.** Isotopes may exhibit different properties. For example, carbon 12 is the normal, stable form of the element carbon. Its isotope, carbon 14, is radioactive. The nucleus of the carbon 14 atom is unstable and eventually decomposes to form atoms of other elements. As it does, it emits radiation.

Molecules are formed when two or more atoms combine. Some elements, such as oxygen (O_2), nitrogen (N_2) and hydrogen (H_2), are found in nature as molecules. Molecules composed of two or more different elements are known as **compounds.** In chemical notation, O, written by itself, represents one atom of oxygen; 2 O represents two atoms; and O_2 represents two atoms of oxygen that are joined to form a molecule of oxygen. Water is a compound formed of two hydrogen atoms and one oxygen atom, designated by the symbol H_2O. Glucose sugar is a compound formed of 6 carbon atoms, 12 hydrogen atoms, and 6 oxygen atoms; its symbol is $C_6H_{12}O_6$. Most matter exists as compounds held together by the forces of attraction in the chemical bonds between their constituent atoms.

Organic compounds all contain atoms of carbon. These may be combined with other carbon atoms or with atoms of one or more other elements. Hydrocarbons such as methane (CH_4), chlorinated hydrocarbons such as DDT ($C_{14}H_9Cl_5$), and simple sugars such as glucose ($C_6H_{12}O_6$) are representative of the millions of organic compounds. All other compounds are inorganic.

The **law of the conservation of matter** states that during a physical or chemical change, matter is neither created nor destroyed. However, its form may be changed, and it can be moved from place to place. Ecosystems function within the law of the conservation of matter by using processes that constantly recycle matter.

Carbon, oxygen, hydrogen, nitrogen, phosphorus, and sulfur are **macronutrients,** chemicals needed by living organisms in large quantities for the construction of proteins, fats, and carbohydrates. These six macronutrients are the major constituents of the complex organic compounds found in all living organisms. Along with numerous **micronutrients**—substances needed in trace amounts, such as copper, zinc, selenium, and lithium—macronutrients are regulated by cycles so that they remain available in the physical environment. The chemicals and water that form the complex compounds found in living organisms continually cycle between the abiota and the biota.

Physical Factors. In addition to energy and matter, physical factors are important to consider when studying the abiotic component of the ecosystem. Physical factors include temperature, precipitation, humidity, wind, light, shade, fire, salinity, and available space. They do not remain constant, but vary over space and time. Figure 3-6 illustrates the importance of physical factors.

The abiotic component of an ecosystem—the available energy; the type, amount, and distribution of nutrients; and physical factors—helps to determine which organisms will comprise the biotic component of that system.

Biota

The biota, or living organisms of an ecosystem, are grouped into two broad categories, autotrophs and heterotrophs, based on their nutritional needs and feeding type.

Autotrophs. Producers or **autotrophs** are self-nourishing organisms (*auto,* "self"; *troph,* "nourishment").

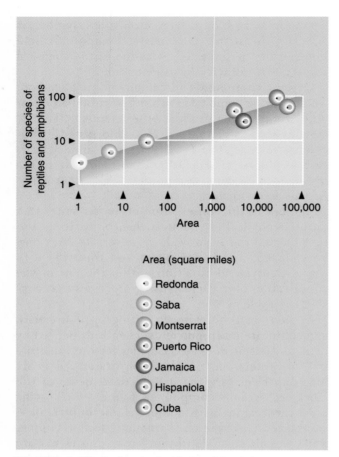

FIGURE 3-6: Effect of a physical factor, island size, on number and type of resident species. The size of an island affects the number and kind of species it can contain. As the chart shows, the number of species of reptiles and amphibians increases with increasing size.

They need water, nutrients, and a source of energy; they can produce the compounds necessary for their survival. Most producers, such as green plants and algae, are known as **phototrophs;** they use energy from the sun (*photo*, "light") to convert relatively simple chemicals (carbon dioxide, water, and nutrients) into complex chemicals—carbohydrates (sugars, starches), lipids (oils, waxes), and proteins. A second group of producers, known as **chemotrophs,** convert the energy found in inorganic chemical compounds into energy. One type of chemotroph is the bacteria which live in and around the thermal vents of the deep trenches of the oceans. These special bacteria use the energy of the compound hydrogen sulfide (H_2S) for their nutritional needs. They are the producers that supply the food energy for the consumers of that special ecosystem.

Heterotrophs. Consumers or **heterotrophs** (*hetero*, "other, different") eat by engulfing or predigesting the cells, tissues, or waste products of other organisms. All heterotrophic organisms obtain the energy-

rich chemicals they need either directly or indirectly from autotrophs and thus indirectly from the sun. Because heterotrophs cannot make their own food, they live at the expense of other organisms. They can be broadly categorized as macroconsumers or microconsumers.

Macroconsumers. **Macroconsumers** feed by ingesting or engulfing particles, parts, or entire bodies of other organisms, either living or dead. They include herbivores, carnivores, omnivores, scavengers, and detrivores. **Herbivores** or **primary consumers** (grasshoppers, mice, deer, for example) eat green plants directly. Other consumers (meadowlarks, black rat snakes, bobcats) feed indirectly on plants by eating herbivores. Because they eat other animals, they are referred to as *carnivores* or **secondary consumers.** Consumers that eat both plants and animals (black bears, Norway rats, humans) are **omnivores.** Carnivores that eat secondary consumers (hawks, large-mouth bass) are **tertiary consumers.**

Many heterotrophs consume dead organic material. Those that consume the entire dead organism are known as *scavengers*. Two familiar examples are vultures and hyenas. Consumers that ingest fragments of dead or decaying tissues or organic wastes are called **detrivores** or **detritus feeders** (bluecrabs, dung beetles, earthworms, shrimp).

Microconsumers. Like detrivores, **microconsumers** or **decomposers** feed on the waste products of living organisms or the tissues of dead organisms. They differ in that they digest materials outside of their cells and bodies, through the external activities of enzymes, and then absorb the predigested materials into their cells. Decomposers live on or within their food source. The result of the activity of the decomposers is what we call rot or decay. Eventually, decomposers reduce complex molecules to simple molecules and return them to the physical environment for reuse by producers. Microconsumers include some bacteria, some protozoans, and fungi (for example, yeasts and molds).

Decomposers play the major role in reducing complex organic matter to inorganic matter and returning nutrients to the physical environment in a form that can be used by producers. The importance of decomposers is often overlooked. We are only now beginning to get a clearer picture of how they perform their vital functions. Consider what it would be like if leaves were not decomposed after they fell from trees. How quickly nutrients in the soil would be depleted if decomposers did not continually recycle them after the death of producers and consumers! Because organisms continually remove necessary chemicals from the environment, decomposers form the vital link in the cycle that returns those chemicals to the soil, enabling material to proceed from death to life in ecosystems.

What Determines the Structure of Ecosystems?

Ecosystem structure is a product of both the abiotic environment and the biotic community.

Abiotic Regulators

Abiotic regulators, called **limiting factors,** determine ecosystem structure. Temperature, light, precipitation, and available phosphorus, oxygen, and carbon are limiting factors. For example, difference in average rainfall separates the major plant communities into forests, grasslands, or deserts. Typically, regions with more than 40 inches (100 centimeters) of precipitation per year are forests, 10 to 30 inches (25 to 75 centimeters) are grasslands, and less than 10 inches (25 centimeters) are deserts. Temperature is another important limiting factor. If an area usually gets 40 inches (100 centimeters) of rainfall per year and is hot, it will sustain a tropical savanna; if the area is temperate, it will sustain a deciduous forest (beeches, maples, oaks). Soil types within a forest are also a limiting factor. In a temperate climate with adequate rainfall, oaks and hickories are more successful on low-nutrient soils, while beeches and maples are more successful on high-nutrient soils. Abiotic factors form a complex set of interactions that limit or control the activities of organisms, pop-

ulations, and communities. Ecologists do not yet understand all the ways in which these factors interact.

In an aquatic ecosystem, oxygen, sunlight, and nutrients (phosphorus and nitrogen) are the most significant limiting factors. Generally, the availability of phosphorus (as the compound phosphate) in lakes and streams limits the growth of aquatic plants and algae. Increasing the amount of phosphates increases the growth of plants and algae. Nutrient enrichment of a lake, stream, or estuary can set in motion a mix of physical, chemical, and biological changes that collectively are known as **eutrophication,** the natural aging of a lake. The high input of nutrients may be the result of natural erosion or may be related to human activities, as in the case of runoff from agricultural fields that have been treated with fertilizers. When human activities lead to nutrient enrichment of a body of water, ecologists say that the aquatic system has undergone *cultural eutrophication.* Environmental Science in Action: Lake Erie (pages 53–56) illustrates how cultural eutrophication has dramatically altered Lake Erie.

Living organisms, populations, and communities have a *range of tolerances* for each of the limiting factors. This is known as the law of tolerances. Tolerances range along a continuum from the maximum amount or degree that can sustain life to the optimum or best amount for sustaining life to the minimum amount that can sustain life (Figure 3-7). Any change that approaches or exceeds the limits of tolerance, either the maximum or the minimum, becomes a limiting factor. For example, laboratory

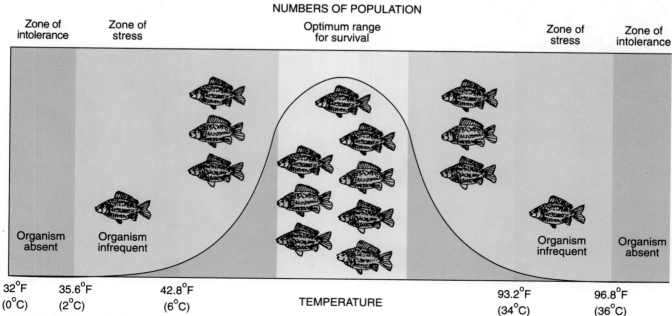

FIGURE 3-7: Law of tolerances. Goldfish can live at temperatures ranging from 35.6°F (2°C) to 96.8°F (36°C) but cannot tolerate temperatures above or below this range.

trials have demonstrated that speckled trout prefer water temperatures of 57° to 66° F (14° to 19° C) although they can tolerate temperatures as high as 77° F (25° C) for short periods of time. This finding is corroborated by field studies, which show that speckled trout are not found in streams where the temperature exceeds 75° F (24° C) for an extended period of time.

Some organisms have wide ranges of tolerances and others have narrow ones. Aquatic insects such as mayflies, fish such as hogsuckers, and flatworms such as planaria all have narrow tolerances for oxygen that limit their success in aquatic habitats. When the concentration of oxygen in streams or lakes is reduced, these species disappear quickly and are replaced by species more tolerant of lower oxygen levels, such as mosquito larvae, carp, or sludge worms. Species less tolerant of pollution can be used as indicators of the quality of a particular ecosystem. Scientists look for the presence of such organisms when determining the ecological health of both aquatic and terrestrial ecosystems.

Limiting factors affect the distribution and success of living things. No organism, population, or natural community is distributed evenly about the earth. Instead, each occupies a particular environment or habitat. Some organisms are distributed throughout large areas, while others live in very specific habitats (Figure 3-8). Some organisms are found only in the leaf litter of deciduous forests, under rocks in fast moving streams, or in the minute cracks of rocks in the Antarctic. Some organisms are found on the forest floor, but not in the forest canopy; some survive at great ocean depths, but not in shallow waters. Some organisms are successful in the wetter, cooler conditions on the northern or western slopes of mountains in the Northern hemisphere, while others are successful in the drier, warmer conditions of the southern or eastern slopes of the same mountain range.

Limiting factors have important implications for humans. For example, the success of a crop is sometimes limited by micronutrients, such as iron for soybeans or molybdenum for clover, even if macronutrients (phosphorus, nitrogen, and potassium) are applied in large quantities for that particular crop. We manipulate limiting factors when we try to maintain optimum conditions for growth to ensure or increase our harvest from gardens, farms, and ponds. (Text continues on page 56.)

(a)　　　　　　　　　　　　　　　　　　(b)

FIGURE 3-8: Habitat and distribution of species. (a) Some species are widely distributed. The common ragweed is found throughout North America from Mexico to Canada. (b) Other species, such as the giant, blood-red tube worm, are found in very specific habitats, such as near the thermal vents on the ocean floor.

Environmental Science in Action: Lake Erie
Lisa Bentley

Lake Erie is part of the 2,000-mile-long Great Lakes system, which contains nearly 20 percent of the world's and 95 percent of the United States' fresh surface water. More than 36 million people in Canada and the United States make the Great Lakes region their home.

Lake Erie is an important regional, national, and international resource. It is the second smallest of the Great Lakes in terms of surface area, is the shallowest, and has the smallest water volume. To manage this tremendous resource, to repair past damages, and to maintain its future health, we must closely examine Lake Erie's physical, biological, and social characteristics, and we must also understand the environmental problems associated with each.

Describing the Resource

Physical Boundaries

The Great Lakes are glacial in origin, formed by the gouging and scraping action of the ice sheets that once covered much of Canada and the upper Midwest of the United States. A vital link in the Great Lakes system, Erie accepts water from Lakes Superior, Michigan, and Huron via the Detroit River and discharges it into Lake Ontario via the Niagara River (Figure 3-9). The Lake Erie drainage basin has a relatively small ratio of land to water, about 2.5:1 (most lakes have a land to water ratio of around 6:1). Hence, the lake is recharged largely through precipitation. Consecutive years of higher-than-average rainfall, such as the early to mid 1980s, contribute significantly to higher lake levels. High waters erode shorelines and cause flooding in low-lying areas, damaging homes and businesses built too near the shore or on flood plains.

Erie is composed of three distinct basins; it is almost like three lakes in one. The shallow western basin, with an average depth of 24 feet, lies west of Cedar Point, Ohio, near Sandusky. The average depth of the central basin, which stretches between Cedar Point and Erie, Pennsylvania, is 61 feet. East of Erie lies the deep eastern basin, with an average depth of 120 feet and a maximum depth of 210 feet.

During the summer, as the sun warms a lake's surface, the water tends to form layers. Warm surface waters are less dense than cool, deeper waters. Because it contains a large phototrophic base, the sunny, oxygen-rich upper layer supports most of the life in the lake. The deepest layer receives less sunlight and supports less life.

Layering rarely occurs in the shallow western basin, where winds and wave action keep the water well mixed. In the central basin, where the bottom layer is narrow, layering poses serious problems. The limited oxygen supply is depleted by bacteria decomposing dead organic material, mainly algae. When the amount of dissolved oxygen in the water reaches zero, the water can no longer support life. In summer, up to 90 percent of the bottom layer of the central basin can become devoid of oxygen. Because the bottom layer of the deeper eastern basin is thicker and colder, it maintains its oxygen content throughout the summer.

In addition to sunlight and dissolved oxygen, the amount of life that a lake supports depends on nutrients available in the water. Scientists generally agree that the nutrient that acts as the most

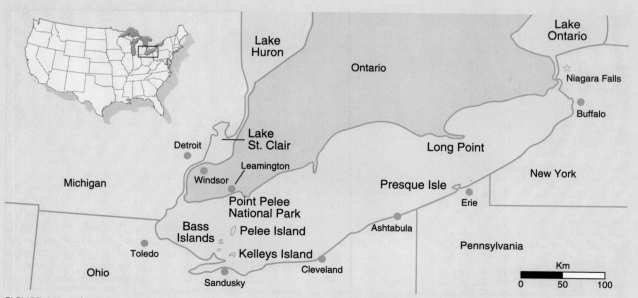

FIGURE 3-9: Lake Erie. Approximately 13 million people live in the Erie drainage basin, and 39 percent of the Canadian shoreline and 45 percent of the United States shoreline is devoted to residential use.

important limiting factor in Lake Erie is phosphorus.

Biological Boundaries

Up to a point, the more nutrients in the lake, the more biologically productive it becomes. Biological productivity is desirable, since phototrophs provide the food base in an aquatic ecosystem. However, the overproduction of producers can cause problems during the summer in the western and central basins. As the algae and green plants die, they sink to the bottom of the lake, where they are decomposed by bacteria. The decomposition of large amounts of organic material depletes the oxygen supply in the water, and, eventually, the bottom waters become devoid of oxygen. One indication of high biological productivity is eutrophication, the natural aging of a lake due to high inputs of nutrients, a process which is ongoing in Lake Erie.

When the first Europeans arrived in the Great Lakes basin in the sixteenth century, they found a diverse biotic community. Smallmouth and largemouth bass, muskellunge, northern pike, and channel catfish favored a near-shore habitat. Blue pike, lake whitefish, walleye, cisco, sauger, freshwater drum, lake trout, lake sturgeon, and white bass flourished in open waters. Mayflies were an important source of food for the large fish.

Today, the composition of the biotic community has changed dramatically. Although Lake Erie is still biologically productive, it is less diverse. Gone are the blue pike, lake whitefish, lake sturgeon, and cisco. The mayfly, which survives in vastly reduced numbers, is no longer the important food source it once was. Currently, walleye and yellow perch are the prized catch in both sport and commercial fishing. Other numerous fish include white bass, smelt, alewife, sheepshead, carp, and rock bass.

The changes in the biota of Lake Erie, including changes in species composition, community structure, and the declining health of organisms, are the result of human activity. Introduced species have contributed to changes in species composition. Toxins present in the lake and its tributaries accumulate in the tissues of organisms, eventually reaching a dangerous level of contamination. Tumors and genetic deformities are two signs of toxic contamination sometimes seen in fish and birds.

The adverse effects of human activity are also seen in habitats. Lake Erie's once extensive wetlands have been greatly diminished, a victim of the plow and urban development. Condominiums and marinas crowd its southern shore. The press of a burgeoning human population in the region means less room for wildlife.

Even so, Lake Erie is still a place of great beauty and diversity. Limestone islands dot the waters of the shallow western basin. Migrating birds and waterfowl, as well as the monarch butterfly, are regular visitors to the unusual sand spits that reach into the lake at Presque Isle, Pennsylvania, and Long Point and Point Pelee, Ontario. Quiet wetlands belie a teeming diversity of life (Figure 3-10). And, Erie's waters roar over the falls at Niagara, attracting visitors from around the world.

Social Boundaries

The Great Lakes opened the interior of the continent to European penetration and settlement. Drawn by the area's abundant resources, including fish, game, and lumber, settlers cleared forested land and drained wetlands for agricultural use. Today, 67 percent of the land in the Erie basin is used for agricultural purposes.

Lumbering became a booming enterprise in the 1870s. Later, iron ore from the deposits near Lake Superior was shipped to Lake Erie ports, and Appalachian coal was sent overland, giving birth to a steel industry. Erie's shores remain home to a heavy concentration of industry. Manufacturing industries, including auto production, steel production, glass manufacturing, and shipbuilding, use nearly 6 billion gallons of Lake Erie water each day. Municipalities use lake water for a variety of purposes, including drinking and disposal of sewage and other wastes.

Shoreline development, including marinas, condominiums, and urban revitalization projects such as the Cleveland Lakefront State Park, has a major economic impact on the lake. Tourism and recreational uses of the lake, including fishing, swim-

FIGURE 3-10: An egret in the Lake Erie wetlands. Wetlands, ecosystems of great diversity and beauty, are increasingly subjected to human pressures.

ming, sunbathing, and boating, are also significant. In recent years the improvement in the lake's appearance and the resurgence of sportfishing, particularly for walleye, have led to a rising demand for public access to the lake.

Sportfishing in Erie is a multimillion dollar industry. Of the Great Lakes, Erie is second only to Lake Michigan in sportfishing, and its western basin is known as the Walleye capital of the world. The increasing importance of the sportfishing industry affects the lives of a growing number of residents who offer charter boat services, sell sportfishing equipment, and provide accommodations for sportspeople and tourists. Hence, there is increased pressure to manage the fishery in the interest of sportfishers and recreational users.

As the area's economy becomes more service-oriented, commercial fishers often find their livelihood and life-style threatened. Competition from the sportfishing industry (particularly in the United States), the decline of valuable species, size limitations, and bans on certain species due to toxic contamination all contribute to the difficulty of commercial fishing. The commercial fishery is economically more vital in Canada than it is in the

United States. The Lake Erie fishery represents over two-thirds of Canada's total Great Lakes commercial harvest. Despite its decline, commercial fishing remains a relatively important industry. The primary targets of commercial fishing are perch, white bass, smelt, and walleye (called pickerel in Canada).

Looking Back

The extensive use of Lake Erie has brought many environmental problems. By the middle of this century, agricultural runoff, detergent-laden waste water, and insufficiently treated sewage carried phosphorus into Lake Erie, and the increased nutrient load stimulated the growth of autotrophs. Municipal sewage also contributed high levels of bacteria, posing a health threat to swimmers. By the late 1960s green plants and algae had flourished to such a degree that seasonal algal blooms, primarily of *Cladophora*, spread over entire portions of the central and western basins. Mats of green algae washed ashore, fouling beaches. Newspaper headlines announced "Lake Erie Is Dead," when actually the lake was more alive than ever. It was undergoing cultural eutrophication, aging caused by a high influx of nutrients due to human activities.

For years, the same industries that brought growth and prosperity to the area had also discharged dangerous pollutants such as polychlorinated biphenyls (PCBs) into tributary or lake waters. PCBs are organic chemicals that do not degrade easily. Although the discharge of PCBs has been banned, the chemicals are still present in the ecosystem. They have accumulated in bottom sediments, particularly near harbors and industrial sites.

The diverse and often competing concerns of residents, the fishing industry, the shipping industry, and politicians have hampered lake management. As early as 1909, with the signing of the Boundary Waters Treaty, Canada and the United States created the International Joint Commission (IJC) to approve all projects regarding the use of, construction on the shores of, and diversion of boundary waters. The IJC's power is advisory; it does not have the authority to enforce its recommendations. Leg-

islative efforts have been made even more difficult by the need to coordinate various jurisdictional authorities. A Canadian province, four state governments, the United States and Canadian federal governments, and numerous local authorities on both sides of the lake must cooperate in order to develop management policy. Such cooperation has often been hard to obtain. However, in some areas progress has been made.

The growing public demand for cleanup action resulted in the 1972 Great Lakes Water Quality Agreement (GLWQA), signed by representatives of Canada and the United States, which expressed "the determination of each country to restore and enhance the water quality of the largest freshwater system in the world." This precedent-setting agreement established objectives and procedures for monitoring the progress of water quality programs. It was renewed and enlarged with the adoption of the 1978 Great Lakes Water Quality Agreement, which heralded a new era in the management of the lakes. For the first time, management was officially approached on an ecosystemwide basis.

Another step toward cooperative, basinwide management took place in 1982, with the establishment of the Council of Great Lakes Governors (CGLG). In 1985 the CGLG signed the Great Lakes Charter, in which the council members agreed to consult on and cooperatively manage the region's water resources.

Like water quality, the management of the fisheries is also an international concern. Overfishing began to take its toll as early as the turn of the century. By the mid-1950s, the cumulative effect of industrial waste, municipal sewage, agricultural and urban runoff, drained wetlands, and channeled rivers, coupled with the pressures exerted by modern, highly efficient fishing fleets, depleted stocks of many species. In the late summer of 1953 calm, warm weather and a dense algal bloom depleted the oxygen content of the water in the lake's western basin, and the mayfly population crashed. Because the mayfly population did not fully recover, the populations of many popular and valuable fish species followed suit. In 1939 there were

40 commercial fishers working at South Bass Island in the western basin; in 1957 there were none. The 1955 catch of blue pike was 19.7 million pounds; by the decade's end the blue pike fishery had collapsed and the species is now officially listed as extinct in Lake Erie. The whitefish and cisco fisheries suffered similar fates.

In response, capital expenditures were made for pollution control, especially the redesign of sewage treatment systems. Over a billion dollars was spent by the Cleveland Metropolitan Sewage Treatment District alone. Agricultural techniques that reduce erosion and the need for chemical fertilization were introduced. Greater protection was given to wetlands, thus preserving vital spawning grounds. The overharvesting of fish populations was curtailed.

In 1955 the Great Lakes Fishery Commission (GLFC), a binational advisory body, was formed. The twofold mission of the GLFC was to maintain the maximum productivity of fish stocks and eradicate the sea lamprey, a fish parasite accidentally introduced into Great Lakes waters through the St. Lawrence Seaway and Welland Canal. A chemical was discovered that killed lamprey larvae in the streams where the species breeds without harming other fish. The lampricide brought the lamprey's population under control, although it is still considered a problem. Recently, much attention has focused on two other introduced species, first discovered in the Great Lakes in the early 1980s. Researchers are trying to determine what effect, if any, *Bythotrephes cederstroemi*, a microscopic crustacean, and *Dreissena polymorpha*, the zebra mussel, will have on Great Lakes fisheries. Both of these exotic species are thought to have entered the Great Lakes system in the bilge water of ocean-going vessels.

Looking Ahead

Protecting and preserving the unique Lake Erie ecosystem will require a comprehensive management effort that includes measures to ensure water quality, preserve vital habitats, and safeguard the lake's fish populations. Specific goals spelled out in the Great Lakes

Water Quality Agreement include restoring year-round oxygen levels in the bottom waters of Lake Erie's central basin, substantially reducing the present levels of algal biomass to below that of a nuisance condition in Lake Erie, decreasing erosion rates throughout the Lake Erie basin, controlling the entry of toxic substances into the lakes, developing programs and technologies necessary to eliminate or reduce the discharge of pollutants into the lake, and restoring and maintaining the health of wetlands so that they are able to perform vital functions.

Maintaining water quality will help to preserve vital habitats and protect the Lake Erie fishery. Other measures that can help to manage the fishery effectively include standardizing fishing limits and size restrictions in Canadian and United States waters. For example, gill net fishing, a method some people believe leads to overharvesting, is banned in United States waters but not in Canadian waters. Fish obviously do not acknowledge the international border; we should be equally unconcerned with it if we are to maintain a thriving fishery and a healthy, balanced ecosystem.

Placing quotas on both sport and commercial fishing will help to eliminate overfishing of desirable species, such as walleye and yellow perch. Currently, quotas are in effect for the western basin walleye fishery. We can help to maintain balance in the ecosystem by encouraging increased harvesting of less sought-after species, such as carp, gizzard shad, and freshwater drum, which compete with desirable ones.

Because the interrelationships and interdependencies among lake biota are so complex, and our knowledge of them is incomplete, we must protect the health of all organisms in the ecosystem, including forage fish and vegetation. Further, we must protect critical habitats, particularly spawning grounds. Knowledge of the natural system and data derived from long-term research are crucial.

Ultimately, the health of the Lake Erie ecosystem depends upon effective and comprehensive management. Such management would best be accomplished by a consolidated governing body, empowered by the Canadian and United States governments to enforce regulations. This is a challenging prospect con-sidering the legal, political, and bureaucratic difficulties, but the advantages are numerous. Duplication of research, monitoring, and surveillance efforts would be avoided. Interdisciplinary planning, in which specialists from many fields work together to devise management plans, would be far easier to achieve within such a framework.

In addition to political and scientific action, public participation is necessary. Citizen groups can and do make a difference throughout the Lake Erie basin. At Ashtabula, Ohio, public participation has made a real difference in the development of plans to clean up the Ashtabula River and the city's Lake Erie harbor. Such participation has been duplicated at many areas of concern in both the United States and Canada. At Erie, Pennsylvania, citizens have led the effort to clean up the city's bay and harbor and to protect the unique habitat at Presque Isle. At Point Pelee, Ontario, conservationists, concerned residents, and sportspeople are taking an active role in the effort to protect and manage Point Pelee National Park.

Biotic Regulators

Ecologists are beginning to understand the effect that organisms can have on ecosystem structure. Beavers are an excellent example of a biotic regulator. Through the construction of dams, beavers significantly alter the ecosystem structure of their habitats. Beaver dams alter stream channels and slow down the flow of water, creating and maintaining wetlands, thereby encouraging the growth of types of plants and animals that are successful in flooded areas and discouraging the growth of those that are successful in drier areas. By altering their habitats, beavers significantly influence the diversity of the biotic community. For this reason, they are known as a **keystone species,** a species whose activities determine the structure of the community.

The rabbits of southern England are a second example of a keystone species. A drastic reduction in the rabbit population through disease allowed a thick growth of meadow grass. The heavy grass caused the local extinction of open-ground ants and, in turn, the extinction of the large blue butterfly, whose caterpillar fed on the ants. The loss of one keystone species, the rabbit, caused the successful growth of a species of plant, the meadow grass, throughout the ecosystem and the local extinction of two animal species, the ant and the butterfly.

The structure of ecosystems can also be influenced by the interactions of predator and prey. **Prey,** living organisms that serve as food for other organisms, are often inhibited or eliminated by **predators,** organisms that obtain their food by eating other living organisms. For example, if the population of bass, a predator, in a pond increases, the population of prey species such as bluegills may decrease. The resultant drop in the bluegill population may then lead to an increase in the small animals (zooplankton) that the bluegills eat. The increase in zooplankton, in turn, may cause a decrease in the population of producers, or phytoplankton, in the pond.

Summary

Solving environmental problems and developing environmentally sound management plans require an understanding of natural systems. Ecology, the study of our planet home, can provide that understanding. Ecologists study the natural world at many levels: individual species, population, community, ecosystem, biome, and biosphere.

An individual is a single organism, which is a member

of a species. A species includes all organisms that are capable of breeding to produce viable, fertile offspring. Individuals of a particular species that live in the same geographic area comprise populations; populations have measurable group characteristics such as birth rates, death rates, seed dispersal rates, and germination rates. The place where the individual organism or population lives is its habitat. All of the populations of organisms that live and interact with one another in a given area at a given time are collectively known as a community. A community and its interactions with the physical environment comprise an ecosystem. A biome is a grouping of many ecosystems in a specific, large geographic area identified by a dominant vegetation type. The union of all terrestrial and aquatic ecosystems—and the largest system of life-physical interactions on earth—is called the biosphere.

Ecosystems are life-perpetuating systems having structural components called the biota (the community) and abiota (its physical surroundings). Energy, matter, and physical factors like temperature and rainfall constitute the abiota. According to the first law of energy, or first law of thermodynamics, energy can neither be created nor destroyed but may be changed in form and may be moved from place to place. The second law of energy, also known as the second law of thermodynamics, states that with each change in form, some energy is degraded to a less useful form and given off to the surroundings, usually as heat. Consequently, in the process of doing work, high-quality energy is converted to low-quality energy. Entropy refers to the tendency of natural systems toward dispersal or randomness.

Matter is anything that has mass (weight) and takes up space. Elements, substances that cannot be changed to simpler substances by chemical means, comprise all matter. Atoms, the smallest parts of elements, are composed of subatomic parts called protons, neutrons, and electrons. Molecules are formed when two or more atoms combine. Compounds are molecules composed of two or more different elements. Compounds containing atoms of carbon are known as organic compounds. According to the law of the conservation of matter, matter is neither created nor destroyed but its form may be changed and it can be moved from place to place.

The biota is composed of producers (autotrophs) and consumers and decomposers (heterotrophs) linked together in food chains and food webs. Autotrophs that convert the energy of the sun into chemical energy, which is then locked in the chemical bonds of starches and sugars, are called phototrophs. Autotrophs that use the energy found in inorganic chemical compounds in order to produce starches and sugars are known as chemotrophs. Consumers are categorized as macroconsumers and microconsumers. Macroconsumers include herbivores, or primary consumers, and carnivores, which may be secondary or tertiary consumers. At each level in a food chain, microconsumers, or decomposers, break down dead and decaying plant and animal matter, returning the nutrients to the environment to be used once again.

Both abiotic and biotic factors affect the structure of an ecosystem. Abiotic factors, known as limiting factors, include precipitation, temperature, and nutrient levels. According to the law of tolerances, living organisms, populations, and communities have a range of tolerances for each of the limiting factors that operate in a specific ecosystem. Examples of biotic factors include predator-prey interactions and keystone species, species whose activities determine the structure of the community.

Key Terms

atom	limiting factor
autotroph	macroconsumer
biome	macronutrient
carnivore	matter
chemotroph	microconsumer
community	micronutrient
compound	molecule
decomposer	omnivore
detritus feeder	organic compound
detrivore	phototroph
ecosystem	population
element	predator
energy	prey
eutrophication	primary consumer
habitat	range of tolerances
herbivore	scavenger
heterotroph	secondary consumer
isotope	species
keystone species	tertiary consumer

Discussion Questions

1. A bumper sticker reads "Nature bats last." Explain what that saying means in ecological terms; be sure to consider its implications in terms of the structure of the biotic community. Explain what it means in terms of environmental problem solving.

2. Because they are internally ordered, organisms, populations, and living systems tend to slow down the disorder or randomness of the physical earth, thus slowing, but not stopping, the process of entropy. The order found in an organism is paid for by creating disorder in its immediate environment. Humans, it seems, are an exception to this rule. Our societies have grown more complex and theoretically more ordered. Can you explain this seeming paradox?

3. According to the law of the conservation of matter, matter can neither be created nor destroyed. Many people argue that humans ignore this law. What might be some consequences of our failure to heed the conservation of matter law?

4. Using Lake Erie to illustrate your answer, explain how limiting factors affect the structure of an ecosystem. Discuss the concept of the law of tolerances in terms of the crash of the commercial fishery in Lake Erie in the 1950s.

5. Can you think of ways that nonhuman members of the biotic community alter or affect the structure of the ecosystems found in your area?

Ecosystem Function

We see, then, that chains of plants and animals are not merely "food chains," but chains of dependency for a maze of services and competitions, of piracies and cooperations.

Aldo Leopold

Only the sun gives without taking.

Anonymous

Learning Objectives

When you finish reading this chapter, you should be able to:

1. Identify the two functional processes that bind together the structural components of an ecosystem.

2. Explain how green plants and algae capture and use the sun's energy.

3. Trace the flow of energy through the biotic community.

4. Describe how energy flow affects the structure of an ecosystem.

5. Explain, in general, how materials cycle in an ecosystem.

While the biota and abiota are the structural components of an ecosystem, it is process—energy flow and materials cycling—that links them together as a functional unit. All ecosystems—aquatic or terrestrial, frigid or tropical, desert or rain forest—are dependent upon the flow of energy and the cycling of materials through the community of living organisms. In this chapter we will learn how phototrophs capture and use the sun's energy, how energy flows through the biotic community and affects its structure, and how materials cycle in an ecosystem.

How Do Autotrophs Capture and Use Energy?

There are two types of autotrophs—phototrophs and chemotrophs. Although the processes by which they capture and use energy are different, both are producers.

Through the process of **photosynthesis** (*photo*, "light"; *synthesis*, "to put together"), phototrophs—green plants, algae, and cyanobacteria (or blue-green algae)—capture and convert the sun's light energy to chemical energy stored primarily as carbohydrates (sugars and starches) or lipids (oils). Photosynthesis is roughly 1 to 3 percent efficient at converting light energy to chemical energy, that is, 100 units of light energy produce 1 to 3 units of chemical energy.

Annually, about 34 percent of the energy input from the sun is reflected directly back to space by clouds, dust, water, and chemicals in the atmosphere (Figure 4-1). Approximately 1.1 to 1.5 million kilocalories per square meter per year ($kcal/m^2/yr$) reach the earth's surface. Of the 66 percent that is absorbed by the biosphere, 42 percent heats the land and warms the atmosphere, 23 percent helps regulate the water cycle through evaporation, and about 1 percent generates wind currents. Only 0.023 percent of the sunlight reaching the earth is actually captured by phototrophs, yet that small fraction results in the huge amount—hundreds of billions of tons—of living matter or **biomass** that exists on the planet.

Phototrophs contain chlorophyll, the green pigment in plants, algae, and cyanobacteria responsible for absorbing the light energy required for photosynthesis. Using carbon dioxide (CO_2), water (H_2O),

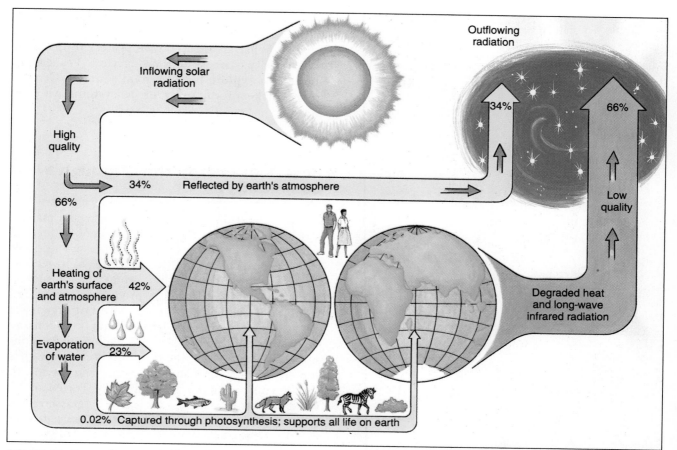

FIGURE 4-1: Energy flows to and from the earth. About one-third of incoming solar radiation is reflected by the earth's atmosphere back into space. Of the two-thirds that is not, 42 percent heats the earth's surface and atmosphere, 23 percent causes the evaporation of water (from oceans, lakes, rivers, and plant leaves), and less than 1 percent is captured by phototrophs. Notice that outflowing radiation (the amount immediately reflected by the earth's atmosphere plus the amount of degraded heat energy and long-wave infrared radiation) equals incoming radiation.

and light energy, they produce carbohydrates that store energy. Essentially, light energy is transferred to the carbon bonds that form carbohydrates. In the process, oxygen is given off.

Producers can then convert carbohydrates into lipids and complex carbohydrates such as starches. They store the complex carbohydrates in their tissues to be used later to meet energy needs. Stored energy in seeds, roots, or sap enables plants to call on reserves during germination, after winter, or during prolonged periods of cloudy days.

The energy produced by phototrophs is the most important driving force for the earth's biotic communities. Terrestrial plants (trees, grasses, herbs, shrubs), aquatic plants and algae, and some bacteria are phototrophs. Another important group of phototrophs are **phytoplankton,** or microscopic single-celled organisms. These organisms capture and store most of the energy in aquatic habitats.

A small percentage of organisms depend on the energy stored by chemotrophs, autotrophs that produce energy from chemicals available in their envi-

ronment. Chemotrophs are represented primarily by species of bacteria that live in and around deep thermal vents in the oceans, at the mud-water interface in high mountain lakes during winter, or in wetlands. These organisms use hydrogen sulfide (H_2S) as an energy source; they convert the energy in the chemical bonds of the hydrogen sulfide to make and store carbohydrates and, in doing so, give off sulfur compounds into the water. Unlike photosynthesis, **chemosynthesis** takes place without sunlight or chlorophyll.

The total amount of energy produced by autotrophs over a given period of time is called **gross primary productivity (GPP).** This is energy that can be used by the producers themselves. The release of energy from fuel molecules is known as **respiration.** Respiration enables all living organisms, including autotrophs, to carry out life processes such as growth, reproduction, and movement. Only organisms with chlorophyll can accomplish photosynthesis, thereby capturing the energy needed for life, but *all* living things respire. **Net primary productivity**

▶ Table 4-1 Estimated Gross Primary Production of the Biosphere and Its Distribution Among Major Ecosystems

Ecosystem	Area (millions of km²)	Gross Primary Productivity (kcal m² yr)	Total Gross Production (10¹⁶ kcal/yr)
MARINE			
Open ocean	326.0	1,000	32.6
Coastal zones	34.0	2,000	6.8
Upwelling zones	0.4	6,000	0.2
Estuaries and reefs	2.0	20,000	4.0
Subtotal	362.4	—	43.6
TERRESTRIAL			
Deserts and tundras	40.0	200	0.8
Grasslands and pastures	42.0	2,500	10.5
Dry forests	9.4	2,500	2.4
Northern coniferous forests	10.0	3,000	3.0
Cultivated lands with little or no energy subsidy	10.0	3,000	3.0
Moist temperate forests	4.9	8,000	3.9
Fuel-subsidized (mechanized) agriculture	4.0	12,000	4.8
Wet tropical and subtropical (broad-leaved ever-green) forests	14.7	20,000	29.0
Subtotal	135.0	—	57.4
Total for biosphere (round figures, not including ice caps)	500.0	2,000	100.0

Source: From *Fundamentals of Ecology*, 3rd Edition, by Eugene P. Odum. Copyright © 1971 by W. B. Saunders Company. Reprinted by permission of Holt, Rinehart and Winston, CBS College Publishing.

(NPP) is the amount of energy available for storage after the autotroph's own respiratory energy needs (R) are met: GPP – R = NPP. Measuring NPP can tell us how much energy is available for primary consumers.

One glance at Table 4-1 reveals some startling insights. The ecosystems that are most efficient at producing energy are the most threatened by human encroachment. Look at the energy produced by tropical rain forests, wetlands, swamps, estuaries, and coastal zones. Compare these ecosystems to cultivated areas and to areas occupied by cities, coastal zone developments, and drained swamps and wetlands. Can you understand why it is vitally important in terms of energy production to maintain the integrity of certain ecosystems? Contrast the energy production of open oceans, the continental shelf, swamps and marshes, estuaries, and tropical rain forests with croplands and cities. (See Environmental Science in Action: The Chesapeake Bay, pages 67–70.) What areas should we protect to maintain the integrity of ecosystem production? What areas should we explore for new food sources? Can we improve the productivity of ecosystems we now exploit?

How Does Energy Flow Through a Community?

To understand the movement of energy and materials through an ecosystem, we must look at feeding relationships. In every ecosystem successive levels of consumers depend upon the organisms at lower levels. These various levels—the producers and successive steps removed from the producers—are called **trophic levels.** Thus, the community of organisms forms a **food chain.** The first trophic level consists of the producers; the second level primary consumers, or herbivores; the third level secondary consumers, or carnivores and omnivores; and the fourth level tertiary consumers (Figure 4-2). For example, weedy plants (producers) are eaten by grasshoppers (primary consumers), which in turn are eaten by meadowlarks (secondary consumers), which in turn are eaten by Cooper's hawks (tertiary consumers); (Figure 4-3). Decomposers (microconsumers) operate at each trophic level in both aquatic and terrestrial ecosystems, converting dead organic matter into its constituent compounds. Unlike other food chains, detritus food chains begin with dead or decaying materials (Figure 4-4).

Food chains are simplified illustrations of the path that energy and materials follow in an ecosystem. But feeding relationships are much more complicated than food chains imply. Interlocking chains, woven into complex associations called **food webs,** more accurately define the feeding relationships and movements of energy and materials (Figure 4-5). Food webs start with producers (P) of many kinds, consumed by many species of consumers at several trophic levels (C_1, C_2, C_3), and culminate with decomposers (D) working at all levels. **Detritus food webs** are based on decomposing plant and animal material or animal waste products. They include several levels of consumers, perhaps as many as four in an aquatic habitat. It is important to remember, however, that the energy derived from the detritus originated with living producers.

Food webs account for particular assemblages of organisms and the feeding relationships among them. They also account for the flow of energy and the movement of materials through ecosystems, which are explored in the following sections.

How Does Energy Flow Affect the Structure of an Ecosystem?

Energy, which flows through an ecosystem according to the laws of thermodynamics, affects the system's structure because it determines trophic relationships. In general, each successive trophic level contains less energy, less biomass, and fewer numbers of organisms, resulting in pyramidal relationships for energy, biomass, and numbers.

The **pyramid of energy** depicts the production, use, and transfer of energy from one trophic level to another (Figure 4-6a). Energy transfers from one trophic level to another are not efficient and vary widely from species to species. As a general rule, 90 percent of the available energy is lost as heat when members of one trophic level are consumed by members of another. This is not a precise figure. In some cases the energy efficiency might be less than 10 percent, in others as high as 30 percent. On average, however, only 10 percent of the available energy is actually transferred at each step. Known as the 10 percent rule, this phenomenon helps us to understand why food chains are generally short and why pyramidal relationships exist.

The 10 percent rule has significant implications for human populations. Humans are omnivorous, capable of eating both plant and animal tissues. The

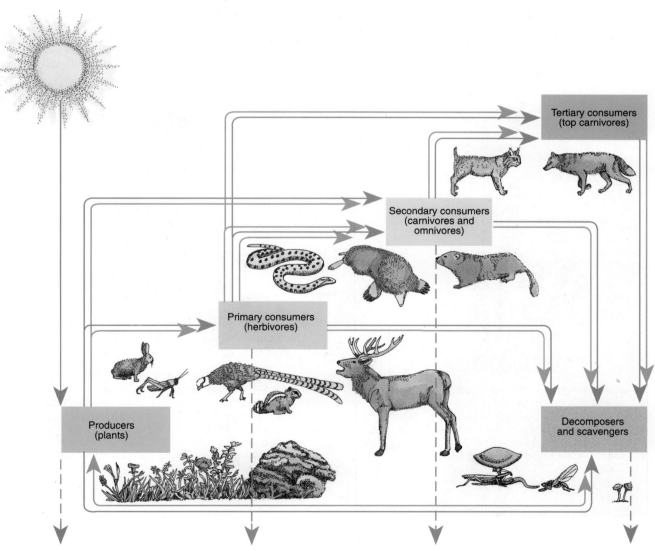

FIGURE 4-2: Energy flow through the components of an ecosystem. The energy of sunlight is captured by phototrophs through the process of photosynthesis. It then flows, as the energy in chemical bonds in food, through successive trophic levels (solid blue lines). Ultimately, the energy passes to decomposers. At each transfer some energy is lost as heat, a product of the respiratory activities that support all organisms (dashed blue lines). The red lines represent the cycling of materials through the ecosystem. Note the two-way red arrow between decomposers and producers.

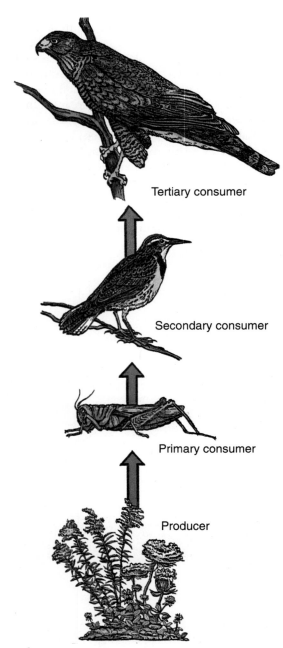

FIGURE 4-3: A typical food chain. Queen Anne's lace, yarrow, and clover capture and store the sun's energy, forming the basis of this food chain. Herbivorous grasshoppers (primary consumers) feed on the plants, and carnivorous meadowlarks (secondary consumers) feed on the grasshoppers. Carnivorous Cooper's hawks (tertiary consumers) feed on meadowlarks.

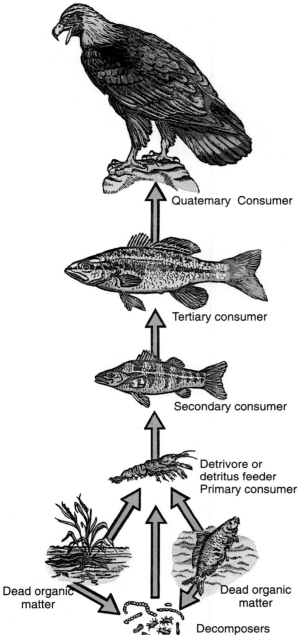

FIGURE 4-4: A typical detritus food chain. Detritus food chains are unique because they begin with dead matter. Bacteria decompose dead organic matter (both plant and animal) which is then eaten by detritus feeders, such as crawdads. Smaller fish that feed upon the crawdad are in turn fed upon by larger fishes. Top carnivores, such as eagles, consume large fish.

trophic level at which they consume the majority of their calories, then, depends upon the availability of food. Meat production is an expensive and energy-inefficient process; for this reason, people who cannot afford a diet heavy in meat or meat products tend to eat foods derived from plants, thus acting as primary consumers. Consumption at lower trophic levels allows consumers to "get more (energy and biomass) for their money." In many areas of Asia, by far the planet's most populous region, people consume most of their calories at the primary consumer level, supplementing their diet with protein-rich seafood or livestock such as chickens and hogs. Theirs is an energy-efficient and economic diet. The meat-rich diet of the United States and other Western nations is far less energy-efficient (Figure 4-7).

The **pyramid of biomass** depicts the total amount of living material at each trophic level (Figure 4-6b). Although total biomass tends to become smaller with each level, the size of each individual organism tends to become larger. Measuring biomass at each level enables us to estimate the total amount of living material at any one time. Together, estimates of the rate of energy produced and the amount of biomass enable us to better understand the potential for production of energy and materials (food and fiber) of any particular ecosystem.

The **pyramid of numbers** depicts the relative abundance of organisms at each trophic level (Figure 4-6c). For example, herbivores at the first consumer level, such as grasshoppers, mice, daphnia (a type of microscopic animal), and antelopes, are usually found in large numbers, while carnivores at higher levels, such as wolves, lions, bobcats, owls, and eagles, are found in smaller numbers. Carnivores at the top of the food chain are dependent upon huge numbers of organisms at the trophic levels below them. At the same time, they regulate the populations of organisms at lower trophic levels. What

would be the result if all top carnivores were eliminated from an ecosystem?

In some ecosystems the pyramid of numbers can become partly inverted (Figure 4-6d). In a temperate forest community during the summer, for instance, most of the producers are large trees and shrubs. The number of consumers they support—insects, birds, small mammals—far exceeds their own number.

It is important to remember that major differences in net primary productivity are apparent among ecosystems. In specific ecosystems and for specific crops (like fish or timber), we can determine the amount of food and fiber available for harvest that does not endanger the system's ability to sustain itself. Ecologists also try to determine how efficient our crops and animals are at providing energy and materials for our use. Small changes in efficiency can mean substantial increases in the production of food and fiber over an entire crop or herd. In farming or ranching, for example, all conditions—rainfall, temperature, soil conditions, seeds, fertilizers, and the mechanics of farming—must be optimal to realize the best yield.

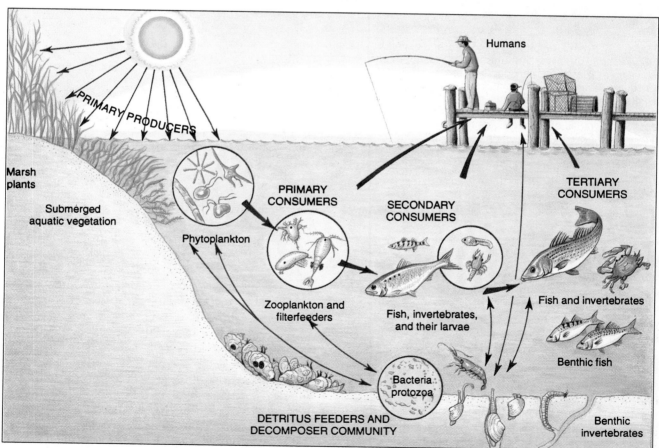

FIGURE 4-5: A simplified food web for the Chesapeake Bay. Arrows indicate the movement of energy and materials through the biota of the Chesapeake Bay.

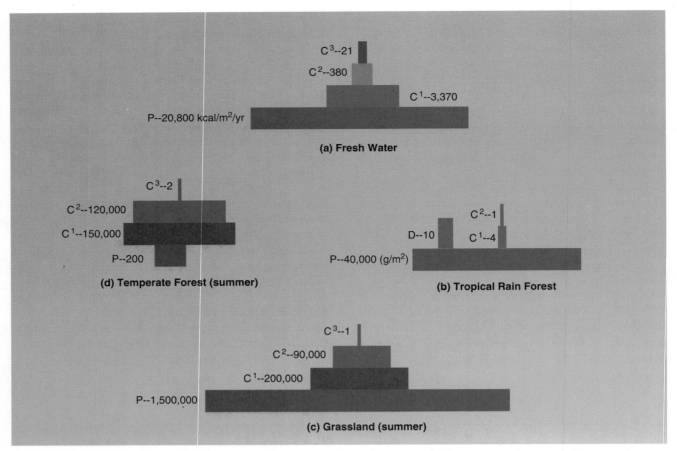

FIGURE 4-6: Pyramids of energy, numbers, and biomass. (a) Pyramid of energy. A freshwater aquatic community at Silver Springs, Florida, illustrates the inefficiency of energy transfers among trophic levels. Energy flow is expressed as kilocalories per square meter per year. (b) Pyramid of biomass. Measurements of biomass (grams per square meter) in a Panamanian rain forest reveal a large difference between the producer and consumer levels: 40,000 grams per square meter supports a total consumer mass of just 5 grams per square meter. D represents the decomposer biomass, which we would expect to be large in a community where nutrients are recycled rapidly. (c) Pyramid of numbers. The number of organisms at each trophic level of a grassland community results in a stepped pyramid. (d) A temperate forest in summer yields a different pyramid of numbers. The pyramid is partly inverted because the number of producers—large trees and shrubs—is much smaller than the number of consumers they support.

How Do Materials Cycle Through an Ecosystem?

Unlike energy, materials such as water, oxygen, carbon, and nutrients are used over and over again. Materials cycle through ecosystems by the workings of many processes. We call these processes **biogeochemical cycles** because they involve living organisms and geologic and chemical factors.

Materials cycle from the air, water, and soil through food webs and back to the air, water, and soil. Throughout the earth's history, materials have continued to cycle in this way. Nutrients and gases are released into soil, water, and air when microbes decompose once-living tissue to simpler molecules; the molecules are subsequently altered by chemical and physical changes until they are in a form that

can once again be used by living organisms. Although natural "sidetracks" of materials do occur—the formation of coal beds, for example, or the loss of phosphorus to the deep sediments of the oceans—these take place very slowly, over millions of years of time. Moreover, processes such as the movement of tectonic plates, eruption of volcanoes, and the upwelling of deep ocean currents eventually recycle even these materials. Thus, the chemicals making up our bodies and the pages of this book might once have been part of a tree fern in an ancient swamp, a dinosaur's bone, or a rock in the Blue Ridge Mountains. Perhaps the nutrients in the food or the oxygen in the air that once nourished Confucius or Socrates or Joan of Arc now nourish one of us. Perhaps some of these nutrients will be sidetracked for millions of years in one of

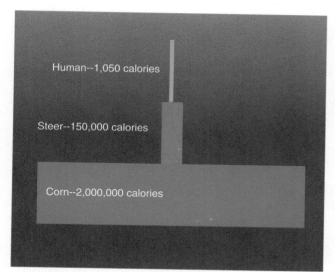

FIGURE 4-7: Humans as secondary consumers. When humans consume most of their food at the secondary consumer level, the transfer of energy is less efficient than it is when they consume at the primary level because there is a considerable loss of energy from the producers (corn) to the primary consumers (steer). For each 1,000 calories stored by beef-eating humans, the producers will have stored nearly 2 million calories.

humankind's many dumps or nonbiodegradable synthetic substances!

Biogeochemical cycles are grouped into three categories: hydrologic, gaseous, and sedimentary.

Hydrologic Cycle

The **hydrologic cycle** includes all of the biospheric processes that cause water to be moved from the hydrosphere through the atmosphere and lithosphere and back to the hydrosphere (Figure 4-8). Water enters the atmosphere through solar-driven vaporization from lakes, rivers, oceans (evaporation), and the leaves of plants (transpiration). It cools and condenses, forms clouds, and returns to the earth as some form of precipitation (rain, snow, hail, etc.) to begin the cycle once more. The earth's total water supply is not added to or subtracted from over time. The quality and availability of water at any one time or place, however, are dependent upon several factors: the uses to which humans put it, the condition and average temperature of the earth's surface where it falls, and the continued ability of the soil, forests, wetlands, and lakes to absorb, clean, and help store it.

Gaseous Cycles

Gaseous cycles take place primarily in the atmosphere. The most significant of these are the carbon, oxygen, and nitrogen cycles.

Carbon and Oxygen Cycles. Each year phototrophs produce billions of tons of organic matter. These producers, primarily terrestrial plants and aquatic algae, remove carbon, in the form of carbon dioxide

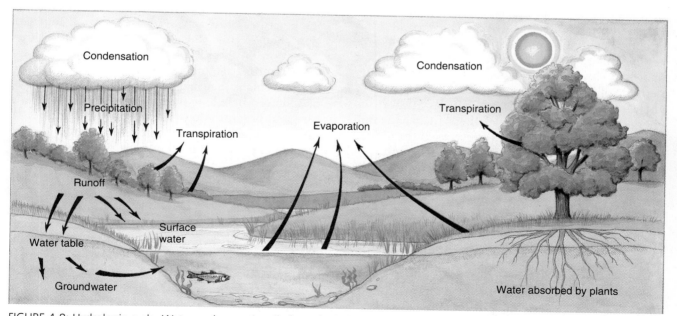

FIGURE 4-8: Hydrologic cycle. Water cycles continually from the hydrosphere through the atmosphere and lithosphere and back to the hydrosphere.

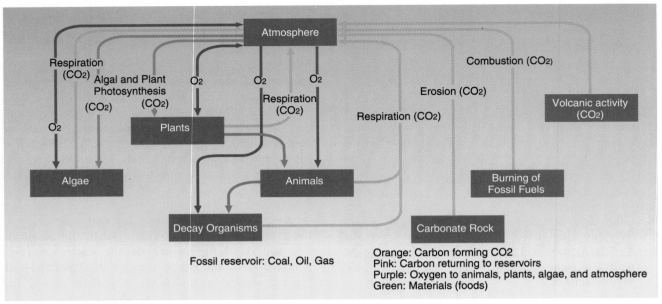

FIGURE 4-9: Carbon and oxygen cycles. The cycling of carbon begins when carbon dioxide (CO_2) enters plants and algae during photosynthesis. Through respiration, phototrophs release some carbon back to the atmosphere (and to the soil) as carbon dioxide. Phototrophs give off oxygen as a product of photosynthesis, which is then used by organisms during respiration. When animals eat plants, respiration returns more carbon dioxide back to the atmosphere. Volcanic activity and erosion from carbonate rock are other natural sources of carbon dioxide. The burning of fossil fuels by humans is a significant source of carbon dioxide.

(CO_2), from the atmosphere and hydrosphere to produce sugars and other complex organic molecules. In doing so, they produce oxygen (O_2), which is released to the environment. Simultaneously, respiration by living organisms converts organic matter back to carbon dioxide and water (Figure 4-9).

About 300 million years ago a surplus of organic matter began to accumulate in the sediments of oceans, seas, and swamps faster than consumers or decomposers could use it. According to the most widely accepted theory, sediments then buried the organic material. Physical processes of great pressure and heat slowly converted this detritus into huge deposits of coal, petroleum, and natural gas, materials we now consume in large quantities to meet our energy needs. Combustion of these fuels adds millions of tons of carbon dioxide to the atmosphere each year. An approximately equal amount is released to the atmosphere annually through the cutting, clearing, and burning of major portions of our forests, both temperate and tropical. Extensive forests act as carbon sinks, that is, they absorb and store excess carbon dioxide, which is used by trees and plants to produce carbohydrates during photosynthesis. The oceans also act as a carbon sink. Large amounts of carbon dioxide dissolve in them, where it may combine with calcium and magnesium to form limestone or dolomite precipitates.

Over the last 60 million years the earth has experienced oscillations in the balance of oxygen and carbon dioxide in the atmosphere. During the last century, however, human activity has resulted in an increase of atmospheric carbon dioxide. The increase seems to be accelerating and could signal significant warming of the earth's climate, since carbon dioxide (and other gases) retains heat in the atmosphere through the greenhouse effect. Changes of just a few degrees in average temperature could melt a portion of the polar ice caps, raising ocean levels. Ocean levels would also rise as the warmer temperatures caused the waters to expand. Temperature changes might also affect the amount of precipitation. Both precipitation and temperature are major limiting factors in an ecosystem. Some scientists speculate that increases in carbon dioxide will be followed by a long period of significant temperature and precipitation changes that will accelerate unless or until phototrophs absorb enough carbon dioxide to reduce its level in the atmosphere.

Nitrogen Cycle. Nitrogen is the fourth most common element in living tissues (after oxygen, carbon, and hydrogen). A chief constituent of proteins, nucleic acids, and chlorophyll, among other important molecules, it is essential to life. Proteins are used as structural components in cells and as part of the enzymes that catalyze cellular processes. Though nitrogen (N_2) is the most common gaseous constituent of the atmosphere (78 percent), plants and animals cannot use it directly. Plants can use nitrogen only when it is in the form of inorganic compounds, principally ammonium (NH_4^+) or

Environmental Science in Action:
The Chesapeake Bay
James Johnson

The Chesapeake Bay is the nation's largest and most productive estuary, a semienclosed coastal body of water composed of fresh and saline (salt) water. The "Queen of Estuaries" provides a bounty of resources. Indeed, the bay has shaped the character of the region and its people since the advent of European settlement over 300 years ago. Thus, when the bay's declining productivity and deteriorating water quality became widely recognized in the 1970s, many people became seriously concerned about restoring the estuary's ecological health.

Describing the Resource
Physical Boundaries

The main stem of the Chesapeake Bay begins near the Susquehanna flats in Maryland. From there, it stretches 190 miles to meet the Atlantic Ocean at tidewater Virginia, creating a drainage basin of over 64,000 square miles in 6 states (Figure 4-10). More than 150 rivers, streams, and creeks contribute fresh water to the bay. Fifty of these are considered major tributaries, and of these, 8—the Susquehanna, Patuxent, Potomac, Rappahannock, York, James, and Choptank rivers and the West Chesapeake drainage area (the Gunpowder, Patapsco, and Back rivers)—contribute about 90 percent of the inflowing fresh water. Consequently, the land use and management practices within these major river basins determine the volume and chemical characteristics of the fresh water discharged to the bay, and each has its own chemical composition and water quality problems. For example, the conditions on the heavily industrialized Elizabeth and Patapsco rivers are very different from those on the Choptank and Rappahannock rivers.

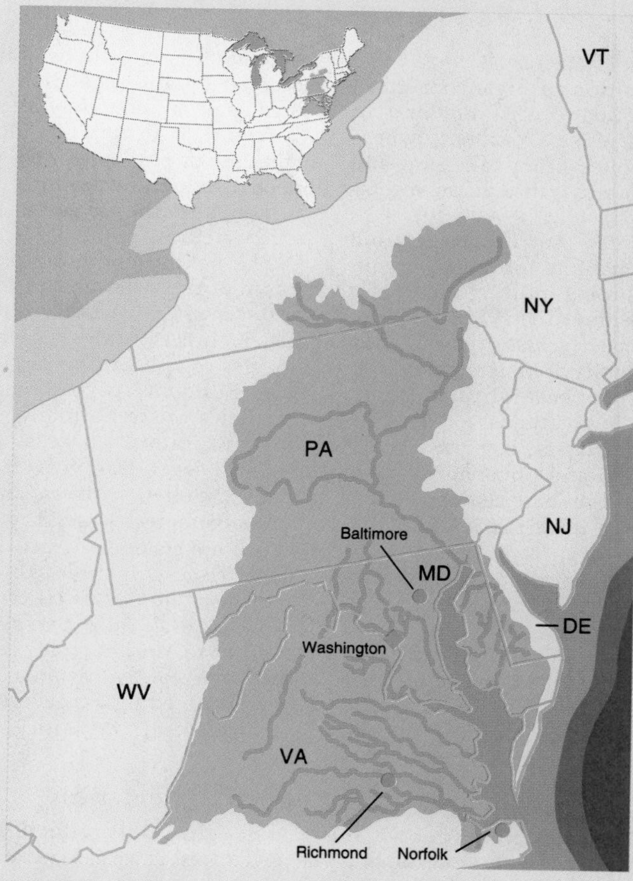

Circulation and salinity characteristics of the Chesapeake are also influenced by daily ocean tides, which vary in height and the degree to which they penetrate the estuary. (**Salinity** is a measure of the concentration of dissolved salts in the water.) Ocean and river current action and the resulting interaction of salt and fresh water are important physical processes in the Chesapeake. Because salt water is denser than fresh water, it flows beneath the fresh water. Mixing patterns also depend on the physical dimensions of an estuary. Comparatively shallow estuaries, such as the Chesapeake Bay, have a two-layer flow with a vertical mixing zone in which fresh water is mixed downward and salt water upward. A layer of fresh water flows seaward above the vertical mixing area, while an incoming layer of salt water flows below the mixing area.

Salinity varies throughout the Chesapeake. It is highest at the mouth of the bay, where marine waters enter the estuary, and gradually decreases toward the northern end of the main stem. Salinity levels also vary vertically and horizontally. Deeper waters and waters on the eastern side of the bay are more saline.

The varying chemical composition of estuarine waters makes for diverse habitats. If an estuary is ecologically healthy, it can be one of the most productive ecosystems in

FIGURE 4-10: Chesapeake Bay drainage basin. With its extensive drainage basin, the biologically productive "Queen of Estuaries" is truly a vital regional resource.

terms of its energy flow and the numbers and variety of its biota (look again at Table 4-1).

Biological Boundaries

Five interacting biological communities comprise the Chesapeake Bay ecosystem: marshes or wetlands, plankton, submerged aquatic vegetation (SAV) or bay grasses, benthic or bottom-dwelling organisms, and nekton or swimmers.

Wetlands are vegetated areas that act as a transitional zone between land and water. They are kept moist by runoff, groundwater, adjacent streams, and bay tides. Marsh wetlands are dominated by grasses, which filter nutrients from inflowing water, thus decreasing nutrient loading to the major body of water they border, in this case the bay. Waterfowl, furbearers and other animals, and the young of commercially important fish and shellfish depend on wetlands for food and shelter. The wetlands also provide a stopping ground for birds along the Atlantic Coast flyway. The draining and dredging of wetlands to provide land for agricultural and residential use pose a major threat to this vital biotic community.

Plankton, tiny floating organisms that drift with the water's movement, include phytoplankton (microscopic plants), zooplankton (microscopic animals), bacteria, and jellyfish larvae. Phytoplankton are producers that occupy the first trophic level in the bay ecosystem. Increased nutrient loading to the bay increases the amount of phytoplankton, causing algal blooms that pose a threat to both submerged aquatic vegetation and benthic communities.

There are numerous species of submerged aquatic vegetation. Growing in shallow waters where the light reaches the bottom, they must remain moist with their leaves at or below the water's surface. Bay grasses are a primary food source for a variety of organisms including herbivores, such as ducks and Canada geese. They are particularly important links in detritus food chains; many species of invertebrates feed on decaying grasses. The invertebrates are an important food source for small blue crabs and fish, such as striped bass and perch; these in turn provide food for wad-

FIGURE 4-11: Blackwater National Wildlife Refuge, Chesapeake Bay drainage basin. The beauty, biological richness, and natural diversity of the Chesapeake continues to attract residents and visitors alike to its shores.

ing birds, such as herons. The decomposition of organisms at all trophic levels returns nutrients to the physical environment, where they are once again taken up and used by submerged aquatic vegetation.

Submerged aquatic vegetation provides habitat for many organisms, including molting blue crabs and spawning fish. In addition, because bay grasses slow down water velocity, particulate matter settles at the base of their stems, resulting in clearer water. Finally, like marsh grasses, bay grasses take up nitrogen and phosphorus, thus buffering the bay against excess quantities of these nutrients.

Despite its importance, submerged aquatic vegetation is threatened throughout the Chesapeake. In the spring of 1984 it occupied only an estimated 24 percent of its total potential habitat. The increase in the amount of phytoplankton in the bay has been blamed for the decline in submerged aquatic vegetation, since phytoplankton block sunlight needed by the bay grasses.

Benthic communities consist of organisms that live on or in the bottom of the bay outside of the marsh and grass beds. They are composed primarily of such commercially important organisms as oysters,

blue crabs, and clams. Like submerged aquatic vegetation, benthic organisms are adversely affected by nutrient loading and increasing amounts of phytoplankton, because decomposition of dead phytoplankton depletes the oxygen in the bottom waters.

The nekton community is composed of the swimmers of the bay—fish, certain crustaceans, squid, and other invertebrates. Some 200 species of fish can be found in the bay. Resident species tend to be smaller and are normally found in shallow waters where submerged aquatic vegetation provides cover. They include killifishes, anchovies, and silversides. Migratory species, which are generally larger than resident fishes, include those that spawn in the bay or its tributaries (yellow and white perch, striped bass, shad, and herring) and those that spawn in the ocean and use the bay for feeding (croakers, drum, menhaden, and weakfish).

Social Boundaries

A healthy and productive Chesapeake Bay is vital to the region's economy. For the past 50 years, the average oyster catch has been about 27 million pounds (12.2 million kilograms) of meat per year. The bay is also the largest producer of blue crabs in the world, with yearly harvests of approximately 55 million pounds (24.9 million kilograms). The value of the finfish and shellfish harvests is approximately $1 billion annually. In the United States only two other areas outproduce the Chesapeake—the Atlantic and Pacific oceans.

As a recreational resource, the bay is unmatched on all of the East Coast. There are 1,750 miles (2,816 kilometers) of navigable shoreline, and the value of the sportfishing industry is nearly $300 million annually. Swimming and boating are supported by numerous beaches and safe harbors. In the past, much of the bay's charm was its clear blue water and the vast wetlands that surrounded it and provided a haven for a rich diversity of wildlife (Figure 4-11).

Land uses in the Chesapeake basin are diverse. The rich soils of the coastal plain support agriculture, large areas of the piedmont are blanketed by forests, and the eastern shore hosts poultry-, seafood-, and vegetable-processing industries. Located along the major tributaries are industrial facilities for steelmaking, chemical production, leather tanning, plastics and resin manufacturing, paper manufacturing, and shipbuilding. The economic importance of Chesapeake Bay ports is staggering. During 1979 more than 90 million tons of cargo, worth nearly $24 billion, were shipped via the bay.

Given the beauty and abundant resources of the Chesapeake Bay drainage basin, it is little wonder that the area has experienced a tremendous growth in population. For example, between 1959 and 1980 the population grew by 4.2 million. Its rising population has contributed significantly to the region's problems. Since the mid-twentieth century the rate at which land has been converted to residential use has increased by 182 percent. Developed land comes at the expense of wetland, forest, pasture, and cropland, and often exacerbates soil erosion and siltation. Population growth and urban development have brought increased municipal wastewater discharge and have concentrated industry, thus increasing pollution in certain areas.

Today, the Chesapeake is much

less productive than in earlier years. A 1987 baywide survey using aerial photography of submerged aquatic vegetation beds showed that bay grasses covered over 49,000 acres (19,800 hectares) of the Chesapeake bottom, an increase over 1984 levels, but still far below the 100,000 to 300,000 acres (40,486 to 121,457 hectares) estimated for the 1960s. The commercial harvests of striped bass, American shad, white perch, and oyster have declined over the past several decades. The oyster population, in fact, is at its lowest level ever, an estimated 1 percent of historical levels. The reasons for the oyster's decline are complex and interrelated: outbreaks of parasitic infection, poor reproduction, reduced survival of larvae and young oysters, loss of habitat due to sedimentation and anoxia, and commercial overharvesting. The history of human use and resulting environmental problems have contributed significantly to the dramatic drop in the bay's productivity.

Looking Back

As the population of early settlements in the bay area grew, forests were cleared to provide more farmland. As agriculture was conducted on an ever greater scale, siltation due to soil erosion began to be a major problem. By the early twentieth century urbanization intensified the decline of local water quality. Large cities dumped sewage and industrial waste into the bay's watershed. Soon, this type of pollution replaced siltation as the principal water quality problem in the bay.

Water quality continued to deteriorate through the middle of this century. Problems included siltation, algal blooms, concentrated toxins and organic compounds in the sediments and water, decreased dissolved oxygen, and a general decline of submerged aquatic vegetation.

Until recently, point source pollution—pollution that can be traced to an identifiable source—received most of the blame for the bay's degraded condition. Over 2,750 municipal sewage treatment plants and industrial facilities discharge directly into the bay. Some of the

toxic substances found on the bottom of the bay in high concentrations are PCBs, kepone, and DDT. Metals such as cadmium, chromium, lead, and zinc are found in the bay's tributary river systems.

Municipal treatment plants are also the source of a different and perhaps more far-reaching problem. Discharge from sewage plants contains phosphorus and nitrogen. Increased levels of these nutrients in the bay encourage the growth of phytoplankton, especially blue-green algae, with resulting problems for both bay grass and benthic organisms. The areas having the heaviest nutrient enrichment are the Patuxent, Potomac, and James rivers in the east and the north and central main bay.

Although point source pollution has been blamed for much of the bay's problems in the past, it is becoming increasingly apparent that nonpoint source pollution is also quite detrimental to the ecosystem. Nonpoint source pollution has no identifiable source; it includes such sources as agricultural and urban runoff. Such sources have a significant impact on the Chesapeake because the bay's drainage basin is so large. Farms and cities hundreds of miles away add to the bay's pollution problems. Nonpoint sources contribute 39 percent of the bay's excess phosphorus and 67 percent of its excess nitrogen.

Throughout the bay's drainage basin, programs to drain wetlands, which act as filters for inflowing water, increase nutrient loading and the flow of pesticides to the bay. Wetland loss is associated with the decline of the Chesapeake Bay. In Maryland agricultural drainage is the primary culprit; in Virginia channelization projects, especially for agricultural purposes, are the principal cause. Residential development, industrial projects, expansion and development of marinas, and dredge-and-fill activities also play a role.

In response to the disturbing state of the bay, the Chesapeake Bay Program (CBP) was formed in 1975. The CBP was to coordinate the efforts of many diversified groups in a single cooperative effort. It had four broad objectives: (1) describe historical trends, (2) determine the current state of the bay by evaluating ongoing research, (3) project future conditions of the bay, and (4) identify alternative strategies for managing bay resources.

CBP participants identified the most critical management areas. The decline in submerged aquatic vegetation, eutrophication, and toxic accumulation in the food chain were given high priority. Medium priority issues included dredging and disposal of spoils, shellfish bed closures, and fishing modifications. Wetland alteration, shoreline erosion, and effect on water quality of boating and shipping were given low priority.

The bay study was completed in 1981, after six years and $30 million. This comprehensive research effort characterized the bay as an integrated ecosystem. It acknowledged the differing needs and problems of various sections of the bay and recommended that control strategies be targeted by geographic area. Perhaps most importantly, the CBP recognized the need for a baywide management authority to coordinate the activities of the various federal and state planning and regulatory agencies.

In December 1983 the governors of Virginia, Pennsylvania, and Maryland, the mayor of the District of Columbia, the EPA administrator, and the director of the Chesapeake Bay Commission issued the Chesapeake Bay Agreement of 1983. The agreement established the Chesapeake Executive Council to oversee the implementation of programs designed to improve and protect the water quality and living resources of the bay.

While government and institutional programs are essential to the protection and improvement of the Chesapeake Bay ecosystem, the efforts of individuals and citizen groups have also proven invaluable. Since 1985 trained citizen monitors for the Alliance for Chesapeake Bay (a federation of citizen groups, businesses, scientists, and individuals) have been studying numerous near-shore sites along tidal tributaries of the bay, such as the James and Patuxent rivers. They collect data on water quality factors such as the amount of dissolved oxygen in the water, salinity, pH, clarity, and temperature. Their work, which supplements a baywide monitoring program begun in 1984, has yielded some important insights. For example, the shallower near-shore habitats have greater swings in the levels of dissolved oxygen than do the deeper waters of the rivers and streams (which are closely watched as part of the baywide monitoring program). Species in these habitats are thus subjected to greater stress than those in deeper waters.

Looking Ahead

Over 12 million people live in the Chesapeake Bay drainage basin. By the year 2000 that figure is expected to grow to 16 million. The growing human population is certain to place increasing stress on the ecosystem. A comprehensive, flexible management plan is necessary if the restoration plan developed for the Chesapeake is to achieve its purpose: to improve and protect the water quality and living resources of the estuarine system; to restore and maintain the bay's ecological integrity, productivity, and beneficial uses; and to protect human health.

Participants in the 1983 agreement established baywide goals and objectives for controlling nutrients and toxins, protecting and restoring the bay's living resources, addressing other related matters, and supporting a cooperative management approach. The goals are:

1. Reduce point and nonpoint source nutrient loadings.
2. Reduce or control point and nonpoint sources of toxic materials.
3. Provide for the restoration and protection of the living resources, their habitats, and ecological relationships.
4. Develop and manage related environmental programs with a concern for their impact on the bay.
5. Support and enhance a cooperative approach toward bay management at all levels of government.

Although the framework established by the 1983 agreement provides for flexible and cooperative management of the bay, management of the Chesapeake will continue to be a challenging task, because many questions remain. What specific nutrient reduction levels need to be met to assure water quality and protect living resources? Which key living resources need special protection? What specific conditions are required in each area of the bay to protect and restore those living resources? Continued research should provide the answers to these and other questions. As more information becomes available, and our knowledge of this intricate ecosystem grows, management of the Chesapeake Bay should improve.

---

nitrates (NO_3^-). Animals obtain the nitrogen they need by eating plant or animal tissues. Because plants cannot use gaseous nitrogen directly, and the conversion of nitrogen to ammonium or nitrates is slow and complicated, nitrogen is often a limiting factor for plants, especially terrestrial plants.

Nitrogen is converted to a usable form in three major ways (Figure 4-12). First, some kinds of bacteria, either free-living in the soil or living in swellings (known as nodules) in plant roots, convert free nitrogen to nitrates or to ammonia (NH_3), which is further converted to usable ammonium. This process is called **nitrogen fixation**. The nitrogen-fixing bacterium *Rhizobium* lives within the root nodules of legumes (such as peas, beans, alfalfa, locust trees, and redbud trees). It converts nitrogen to usable forms for the plant; the legume, in turn, provides the bacterium with a place to live and a food source, sugars.

Decomposition also makes nitrogen available to plants in a usable form. Some decomposers reduce tissues and waste products to less complex forms, eventually converting proteins and amino acids to ammonium. Other decomposers convert ammonium to nitrates and nitrites (salts that contain NO_2^-), forms that can be used by plants to produce organic molecules.

Nitrogen is also converted to a usable form by lightning. Lightning transforms nitrogen in the atmosphere to usable nitrates which then enter the soil in rain.

Nitrogen enters the biosphere in several ways.

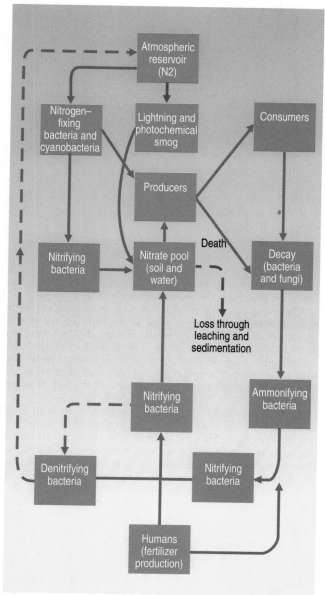

FIGURE 4-12: Nitrogen cycle. The blue arrows represent the pathway by which nitrogen becomes available to living organisms. The purple arrows indicate the movement of nitrogen, in the form of nitrates, through the biota. Nitrogen fixation by bacteria and cyanobacteria removes nitrogen from the atmosphere and converts it to a form that can be used by producers. The nitrogen then passes through successive trophic levels as proteins, nucleic acids, and other organic molecules. Nitrogen is also converted to a usable form through the decay of organic matter and the action of lightning and photochemical fog.

Some special bacteria living in nutrient-rich environments, such as estuaries, lakes, bogs, and the ocean floor, are capable of slowly producing nitrogen gas, which then enters the atmosphere. Nitrogen also enters the atmosphere from volcanoes and exposed nitrate deposits. Industrial and agricultural activities also contribute nitrogen to the biosphere. Emissions from factories, electric power plants, and automobiles contribute millions of tons of nitrous oxides (produced when nitrogen is oxidized during combustion) to the atmosphere each year. Intensively grown corn crops often get extra nitrogen in the form of ammonia fertilizer early in the planting cycle to ensure good germination of seeds and early health of the plant. Additionally, many new seed varieties need extra nitrogen to enhance their production of amino acids like lysine; high-lysine grain is then fed to animals to help increase their protein production.

Sedimentary Cycles

Sedimentary cycles involve those materials that move primarily from the land to the oceans and back to the land again (although the cycles may include a gaseous phase). Phosphorus, sulfur, and other nutrients, such as potassium, calcium, and magnesium, follow essentially the same pathway.

Phosphorus Cycle. Phosphorus is a major component of genetic material (DNA and RNA), energy molecules (ATP), and cellular membranes. It is also the major structural component of the shells, bones, and teeth of animals.

Phosphorus cannot be absorbed in its elemental form, but instead enters the environment in several ways (Figure 4-13). Decomposers convert phosphorus into phosphate (PO_4), a form producers can use. Wind and water erode phosphate from phosphate-rich rocks. In terrestrial systems the cycling of phosphate is usually very efficient. Even though small amounts are slowly lost to streams, rivers, and eventually to the oceans, most phosphate, because it is in a readily usable form, continues to cycle in the ecosystem. The phosphate that does eventually make its way to the oceans is lost to the biosphere for long periods of time. Sea bird droppings, known as guano, deposited on the land, and ocean floor uplifting help counteract this loss, but these are slow processes.

The implications for interrupting phosphorus cycling are particularly serious because phosphates are an important limiting factor in aquatic habitats. When aquatic plants or algae get extra phosphate from soil runoff, phosphate detergents, or sewage, a dense bloom often results and the water becomes choked with plant life. Lake Erie and Lake Washing-

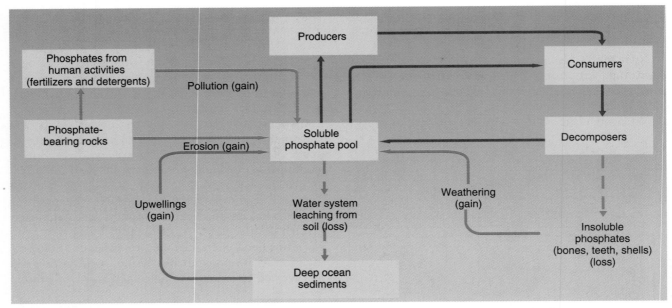

FIGURE 4-13: Phosphorus cycle. Phosphorus, in the form of usable phosphates, is found in soil, water, and aquatic living systems. Phosphates pass to consumers through trophic levels. While decomposers make most of the phosphates available again, a portion remains locked for long periods of time in bones, teeth, and shells and in the deep sediments of the oceans.

ton (near Seattle), among others, have been adversely affected by increased phosphate loads.

Sulfur Cycle. Sulfur is a constituent of amino acids and proteins. It helps to activate enzymes, it is important in energy metabolism, and it becomes a part of some vitamins. Like phosphorus, sulfur cannot be absorbed by plants in its elemental form. Instead, it is absorbed as sulfates (SO_4), which are produced from the weathering of rocks and soils. Sulfates released to the atmosphere contribute to the formation of sulfuric acid (H_2SO_4) and ammonium sulfide (NH_4S). Sulfur also enters the atmosphere as hydrogen sulfide (H_2S), which is produced by volcanoes, hot springs, decomposers, and industrial sources. Human-made sources such as power plants and industrial activities now account for the greatest release of sulfur into the atmosphere.

Summary

The biotic and abiotic components of the biosphere are inseparable, bound together by a complex and delicately balanced web of biological and physical processes that regulate the flow of energy and the cycling of materials. Ecosystems constantly receive energy from the sun; phototrophs produce complex organic chemicals from inorganic chemicals and sunlight, while chemotrophs produce organic chemicals from inorganic materials and thermal energy. Less than 1 percent of the energy that reaches the

biosphere is actually used by phototrophs, yet it is the lifeblood that fuels the earth's biomass.

The gross primary productivity of an ecosystem—the total amount of energy produced by autotrophs over a given period of time—represents energy that can be used as food for the producers themselves. After the autotroph's own respiratory energy needs are met, the amount of energy available for storage, and thus available to other organisms in the community, is the net primary productivity.

Food chains and food webs represent the feeding relationships and the movement of energy and materials among the organisms of the biotic community. Producers occupy the first trophic level in a food chain or web; primary consumers or herbivores, which feed directly on the producers, occupy the second. Secondary consumers or carnivores occupy the third trophic level, followed by tertiary consumers or top carnivores. An organism may operate at more than one level. Omnivores, for example, consume both plants and animals. At each trophic level decomposers break down dead and decaying organic matter, thereby returning nutrients to the ecosystem to be used once again. Feeding relationships based on decomposing plant and animal material or animal wastes are known as detritus food webs. They may include as many as four consumer trophic levels in an aquatic habitat.

Pyramidal relationships are used to depict the flow of energy and production of biomass in an ecosystem. The pyramid of energy represents the production, use, and transfer of energy from one trophic level to another. Generally, only about 10 percent of the available energy is transferred to the next successive trophic level; the rest is lost to the environment as low-quality heat. The pyramid of biomass represents the total amount of living material at each trophic level, and the pyramid of numbers depicts the relative abundance of organisms at each trophic level.

In general, lower levels contain a greater amount of biomass and a greater number of organisms than successive levels.

Unlike energy, materials cycle through ecosystems and are used over and over again by the biotic community. Because the processes by which materials cycle involve living organisms as well as geologic and chemical processes, they are known as biogeochemical cycles. The hydrologic cycle describes the movement of water between the air, seas, and land. Gaseous cycles include the carbon and oxygen cycle and the nitrogen cycle. Major sedimentary cycles include phosphorus and sulfur.

Key Terms

biogeochemical cycle
biomass
carbon cycle
detritus food web
food chain
food web
gaseous cycle
gross primary productivity
hydrologic cycle
net primary productivity
nitrogen cycle
nitrogen fixation

oxygen cycle
phosphorus cycle
photosynthesis
phytoplankton
pyramid of biomass
pyramid of energy
pyramid of numbers
respiration
sedimentary cycle
sulfur cycle
trophic level

Discussion Questions

1. Construct a diagram to illustrate an ecosystem; then explain how it works. Be sure to include energy, biomass, and number relationships.

2. Author and ecologist Barry Commoner once wrote, "There is no such thing as a free lunch in nature." What does this mean in ecological terms?

3. What would be the environmental, social, and economic effects if people in the United States consumed most of their food at a lower trophic level?

4. Explain the role of decomposers in the cycling of materials through the ecosystem.

5. Explain why food chains are seldom longer than three or four links.

6. How would a decline in submerged aquatic vegetation in the Chesapeake Bay affect net primary productivity? What effects would this have on the detritus food web? What effects would this have on human food consumption?

5

Ecosystem Development and Dynamic Equilibrium

A thing is right when it tends to preserve the integrity, stability and beauty of the biotic community. It is wrong when it tends otherwise.

Aldo Leopold

So almost every corner of the planet, from the highest to the lowest, the warmest to the coldest, above water and below, has acquired its population of interdependent plants and animals. It is the nature of these adaptations that has enabled living organisms to spread so widely through our varied planet.

David Attenborough

Learning Objectives

When you finish reading this chapter, you should be able to:

1. Explain how ecosystems develop over time through the process of succession.

2. Distinguish between primary and secondary succession.

3. Describe what ecologists mean by dynamic equilibrium.

4. Explain how feedbacks, species interactions, and population dynamics influence dynamic equilibrium.

Nature is not static; ecosystems change. In this chapter we will examine the development of ecosystems through a process called succession and also

examine the concept of dynamic equilibrium—the idea that ecosystems maintain stability by constantly responding to internal and external change. Chapter 6 looks more closely at how human activities degrade and alter ecosystems, and Chapter 7 focuses on human efforts to restore degraded ecosystems.

What Causes Ecosystems to Change?

Individuals, species, populations, communities, ecosystems, and the biosphere itself are subject to continual environmental change: differences in amount of rainfall, daily and seasonal temperature and wind fluctuations, differences in amount of sunlight. Natural systems can usually compensate for such variability without drastic alteration in the biota. Some changes, however, can dramatically alter ecosystems, either reducing the number or type of species present or making communities less productive. Such stresses include fire, flood, drought, hurricane, volcanic eruption, and human activities.

It may be a long time after a major stress before the community of organisms present at the time of the disturbance again occupies that habitat. On Mount St. Helens, for example, the site of volcanic activity in 1980, the forest ecosystem was destroyed. Although both plant and animal life have returned, it will be hundreds of years before Mount Saint Helens once again resembles its previously forested state (Figure 5-1). Similarly, Hurricane Hugo's assault on South Carolina's coast in September 1989 had a devastating effect on the Francis Marion National Forest. Over 100,000 acres (40,469 hectares) of the forest were leveled. Living in the forest were approximately 500 breeding pairs of the endangered red-cockaded woodpecker. Since the woodpecker prefers to nest in cavities of live trees—trees that are no longer available—the species

may not survive during the time that it takes the forest to recover.

Human activities can also stress ecosystems. Acid
precipitation, toxic wastes, deforestation, and the
draining and filling of wetlands are just some of the
many ways in which humans disturb and degrade
ecosystems.

How Do Ecosystems Develop?

Ecological succession is the process by which an
ecosystem matures; it is the gradual, sequential, and
somewhat predictable change in the composition of
the biotic community. While the term succession is
commonly accepted, it is more accurate to think of
the maturation process as **ecosystem development,**
which takes into account the accompanying modifications in the physical environment (such as microclimate and soil type) brought about by the actions
of living organisms (Table 5-1).

In immature or developing ecosystems, when
there is not much biomass to support, most of the
energy captured through photosynthesis goes into
growth or production (P) of new biomass. In maturing ecosystems less energy is available for production
because increasing amounts of energy are required to
maintain the biota through respiration (R). As an
ecosystem develops, the species composition of the
community changes until the **climax community,**
the association of organisms best adapted to the
physical conditions of a defined geographic area, is
reached. In a climax community, production equals
respiration; there are little net production and no
further increase in biomass (Table 5-2).

Climax communities are usually dominated by a
few abundant plant species, which give the community its name, such as the beech-maple forest or the
tall-grass prairie. In a beech-maple forest beech and
maple trees might account for 80 percent of the
number of trees. The other 20 percent might be
composed of 10 to 12 other varieties, such as tulip
poplar, black cherry, sweet gum, ash, oak, and hickory. Even though the climax community is named
for its dominant plant species, it contains a unique
assemblage of animals, microbes, and fungi.

Succession occurs in both terrestrial and aquatic
habitats and has been observed in all habitats that
support living organisms. There are two types of
succession, primary and secondary. **Primary succession** is the development of a new ecosystem in an
area previously devoid of organisms. The most common examples of primary succession are communities that develop on bare rock, after glaciers recede,
after wind-blown sand stabilizes, and after volcanic
islands form.

(a)

(b)

(c)

FIGURE 5-1: Mount St. Helens before eruption in 1980,
immediately after eruption, and in 1990. (a) Before it erupted Mount St. Helens was characterized by a forest ecosystem. (b) The volcanic eruption drastically altered the site. (c)
Although little more than a decade has passed, the site has
begun to recover as vegetation from nearby areas recolo-
nizes the stressed area.

▶ Table 5-1 Comparison of Development Stages and Mature Stages in Ecosystems

Ecosystem Attribute	Developmental Stages	Mature Stages
Gross primary productivity	Increasing	Stabilized at moderate level
Biomass	Low	High
Production-respiration	Unbalanced	Balanced
Growth and maintenance	Low	High
Use of primary production	Mostly via linear grazing food chains	Mostly via detritus food webs
Diversity	Low	High
Stability (resistance to stress)	Low	High

Secondary succession is the change that occurs after an ecosystem has been disturbed, often by human activity. Although some organisms are still present, the ecosystem is set back to an earlier successional stage. Typically, secondary succession occurs more rapidly than primary succession, chiefly because the soil is usually already in place, eliminating the long process of soil building. Examples of secondary succession are the new growth in abandoned plowed fields or gardens, forests that have been burned or altered by storms, and lowlands that have been affected by floods or hurricanes.

Primary Succession from Glacial Till or Bare Rock to Climax Forest

Figure 5-2 illustrates primary succession from glacial till or bare rock to a temperate, deciduous forest in North America after the Wisconsin glacier receded northward some 12,000 years ago.

Stage I: Lichen Pioneer Community. Almost immediately after the glacier receded, resistant spores of algae and fungi invaded the glacial till, that is, the lifeless deposits of sand, clay, gravel, and boulders left by the glacier. The invading organisms came from already established communities that had not been affected by glacial activity. These organisms were capable of withstanding the periodic dry and harsh conditions that occurred for long periods after the glaciers receded. Lichens, a hardy association of algae and fungi, began to colonize exposed rock surfaces. Slowly, the organic detritus of these **pioneer organisms** or **opportunistic species** mixed with bits of weathered rock and minerals to form soil. This primitive soil held moisture and nutrients well enough to support the second stage, mosses.

Stage II: Moss Community. As mosses became more successful, they crowded out the lichens. They, too, were autotrophs able to resist harsh conditions.

Acids formed by the mosses, weathering, and alternate freezing and thawing of water in small cracks in the rock combined to break down the rock into small particles, which slowly accumulated with the decaying organic matter to further build the soil.

The dense mats of moss attracted insects and other animals, which added more organic matter to the soil. Just as the lichens changed the environment and made it more suitable for the mosses which replaced them, mosses changed the environment and made it more suitable for other, larger plants. As the soil accumulated and matured, the communities that helped to form it were preparing it for the next community of organisms.

Stage III: Herbaceous Community. By stage III the soil accumulated enough nutrients, organic matter, and moisture to support larger herbaceous plants such as goldenrods, asters, milkweed, primrose, ragweed, and wild carrot. These attracted more insects and burrowing and grazing animals such as snails, moles, voles, and mice. We are describing a location in which the temperature and moisture conditions were right for the development of a temperate deciduous forest; in such a location the dominant organisms at this stage were herbaceous plants. In a different physical location, with drier climatic conditions, prairie grasses might dominate.

Stage IV: Shrub Community. Soil formation went on for several thousands of years before the soil was able to support a shrub community. As the glacier receded, the climate was changing, becoming warmer and more moist. Because more soil, organic matter, and moisture were available, taller, woody shrubs became established and outcompeted the herbaceous plants of the previous stage. The community began to assume a pronounced vertical structure. Sumac, black locust, wild rose, blackberry, and mulberry grew and increasingly shaded the soil. The seeds of these plants were often first

▶ Table 5-2
Production and Respiration as (kcal/m²/yr) in Growing and Climax Ecosystems

	Alfalfa Field (U.S.A)	Young Pine Plantations (England)	Medium-Aged Oak-Pine Forest (N.Y.)	Large Flowing Spring (Silver Springs, Fla.)	Mature Rain Forest (Puerto Rico)	Coastal Sound (Long Island, N.Y.)
Gross primary production	24,400	12,200	11,500	20,800	45,000	5,700
Autotrophic respiration	9,200	4,700	6,400	12,000	32,000	3,200
Net primary production	15,200	7,500	5,000	8,800	13,000	2,500
Heterotrophic respiration	800	4,600	3,000	6,800	13,000	2,500
Net community production	14,400	2,900	2,000	2,000	Little or none	Little or none

Source: Adapted from *Fundamentals of Ecology*, 3rd Edition, by Eugene P. Odum. Copyright © 1971 by W. B. Saunders Company. Reprinted by permission of Holt, Rinehart and Winston, CBS College Publishing.

brought into the community in bird droppings. Deer, opossum, and red fox were successful in the shrub community.

Stage V: Tree Community. The physical factors of light, temperature, and moisture continued to change. As they did, the soil was further enriched and was able to support sapling trees. After thousands of years, there was enough organic matter in the soil to support the first tree stages: red maple,

willow, and cottonwood in wet areas; locust, tulip poplar, oak, and hickory in drier areas. Shade-tolerant varieties of plants developed on the forest floor. Early-blooming spring wildflowers were prevalent. Squirrels, raccoons, and chipmunks inhabited the early forest community.

Stage VI: Climax Forest or Equilibrium Community. The soil matured and, in the absence of climatic change, remained, and will remain, essentially con-

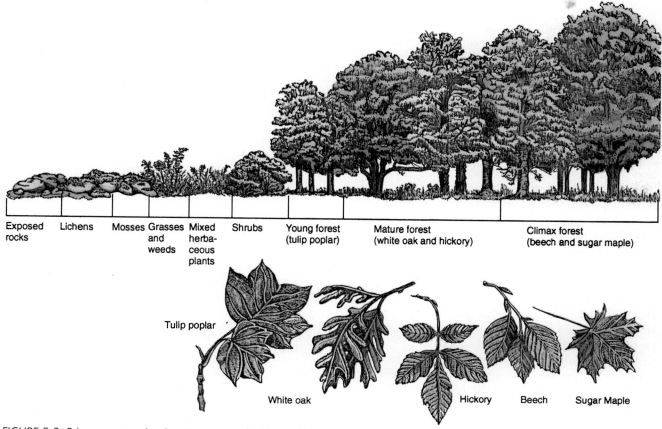

Exposed rocks | Lichens | Mosses | Grasses and weeds | Mixed herbaceous plants | Shrubs | Young forest (tulip poplar) | Mature forest (white oak and hickory) | Climax forest (beech and sugar maple)

Tulip poplar   White oak   Hickory   Beech   Sugar Maple

FIGURE 5-2: Primary succession in a temperate deciduous biome.

Focus On: Fields of Gold
Richard B. Fischer, Professor, Cornell University

It is late summer and the sun shines high above fields of gold—goldenrod fields (Figure 5-3). Although there are over a hundred species of goldenrod in the United States, they are most numerous in the Northeast. The apparent monotony of the goldenrod field is deceiving, for it teems with a diversity of life. If you are fortunate enough to live near a goldenrod field, you will find that it is a microcosm, a world within a world, which offers hours of interesting, and often surprising, observation.

A variety of insects feed on the goldenrod's foliage, pollen, nectar, flower, and sap. Look closely at the plant's leaves, and you will likely find evidence of *Trirhabda canadenis* and *Trirhabda virgata*, two species of beetle that feed primarily on goldenrod foliage. The adults of many species of beetles feed on the plant's pollen, while their young feed elsewhere. For example, the larvae of the locust borer (*Megacyllene robiniae*) tunnel in living black locust trees, but the adults are generally found on goldenrod blossoms. The eggs of short-horned grasshoppers nourish the larvae of the adult black blister beetle, which feeds on goldenrod pollen. The polistes wasp and the honeybee, two members of the order Hymenoptera, feed on the plant's pollen as well.

The pollen eaters illustrate an important point about goldenrod. This plant, with its heavy pollen, is insect pollinated rather than wind pollinated. Little of the heavy pollen gets into the air. But because goldenrod is in full bloom at the same time as the common ragweed, many people mistakenly assume that its pollen causes hayfever.

Goldenrod bloom in late summer, when most other wildflowers have gone to seed. Thus, for pollen-loving insects such as honeybees, goldenrod are the last big source of nectar before the first frosts of autumn. Many other species in the order Hymenoptera, including bumblebees, carpenter bees, solitary bees, and assorted wasps, also sip goldenrod nectar, as do flower flies, or hover flies, which resemble small yellow jackets but are true flies belonging to the family Syrphidae. Goldenrod nectar is also an important food source for the migrating monarch butterfly, providing this beautiful insect with energy for its journey southward to the forests of Mexico and Latin America.

Several insects eat the entire goldenrod blossom, consuming both nectar and pollen. These are the adult Japanese beetle and the adult *Acullia angustipennis*, a moth whose colorful larva also feeds on goldenrod flowers. If you look closely at

FIGURE 5-3: A goldenrod field in full bloom. Goldenrod fields provide important nesting habitat for goldfinches; cover for quail, pheasants, sparrows, shrew, mice, rabbits, and deer; and food for a wide diversity of insects.

the stems of goldenrod, you might notice small, active creatures feeding on the plant's sap. Among the insect families that take advantage of this liquid food source are those belonging to the orders Hemiptera and Homoptera. Like other sap suckers, they have piercing and sucking mouthparts.

With so many herbivores feeding on various parts of this special

stant. In general, conditions in the temperate location favored the seed germination, growth, and maintenance of large trees such as sugar maple and beech. However, microclimates within the climax forest community, characterized by different moisture, light, and temperature conditions, were favorable to different species, such as oak and hickory.

Rainfall of over 40 inches per year (102 centimeters) is needed to maintain a climax tree community, which in turn acts as a sponge to hold water. In a location where the average annual rainfall is just 10 to 20 inches (25.4 to 50.8 centimeters), the climax community might be a grassland prairie.

A climax community is dynamic. Some areas are set back by disease, fire, storms, or insect damage. These stressed areas eventually return to the climax stage.

Human activities can also interrupt the process of succession. Today, most of the original forests that

once blanketed millions of acres of land in states like Ohio are gone, having been replaced by towns, cities, roads, and millions of acres of farmland. Worldwide, natural communities that took thousands of years to form and mature are covered by roads, buildings, golf courses, cattle ranches, farms, and other less complex ecosystems. These simplified systems are not as effective at maintaining soil fertility, retaining groundwater, and anchoring soil.

Secondary Succession from Old Field to Climax Forest

When an agricultural field or garden is abandoned and left to natural processes, succession begins to operate. Figure 5-4 illustrates secondary succession from old field to climax forest in a temperate deciduous biome.

plant, it is little wonder that the goldenrod also hosts many organisms at higher trophic levels which feed on the herbivores. Most of these predators are insects. The ambush bug, resembling a discolored bloom, hides among the flowers, lying in wait for pollen eaters or nectar sippers. It grasps its prey with its powerful forelegs and injects a proteolytic enzyme into the victim. Though the ambush bug is smaller than a Japanese beetle, it can capture insects as large as honeybees. The flower heads host other carnivores as well, including the assassin bug, damsel bug, and stink bug.

Many spiders feed on goldenrod herbivores. The banded argiope, the black and yellow argiope, and the harvestman ("daddy long-legs") prey on sap-feeding aphids. A very unusual arachnid is the crab spider. Hiding among the blossoms, it ambushes herbivorous insects ranging in size from small flies to large bumblebees.

While examining a goldenrod, you might notice strange swellings on the plant's roots, stems, leaves, or flowers. These are galls, the products of an odd assemblage of flies and moths. Although gall making involves many plant families, the goldenrod family bears more galls than any other. A female fly or moth of a particular species deposits an egg on the growing tissue of a particular species of goldenrod. The resulting larva secretes a minute amount of liquid which radically alters the tissue's genetic expression. A gall often results, sheltering the larva. Because different species of goldenrod produce different types of galls, if you can identify the gall, you can more easily identify the goldenrod itself. The most conspicuous gall belongs to the Canada goldenrod. It is a round swelling on the plant's stem made by the fly *Eurosta solidaginis*. By late summer, when the galls are fully developed, you might cut open a gall to discover the plump, cream-colored maggot. The maggot overwinters in its protective case, emerging as an adult in May. But not all gall-enclosed maggots survive. If when you open the gall, you find a small, brown object resembling a seed, you will have found the pupa of a minute, highly specialized wasp that preys only on the maggot of the gall-making fly. The pupal case that ensconces the wasp is produced by the fly in response to the larval predator's attack! If you find a small hole drilled into the gall, you won't find anything after cutting it open. The hole is the mark of the downy woodpecker. During the winter months maggots are a high-energy food source for these woodpeckers.

Not surprisingly, plant ecologists find goldenrod a rich and interesting object of study. They recognize one phase of old field or secondary succession as the goldenrod phase. Researchers have long been interested in the disproportionate length of time for which an area will remain in the goldenrod phase. Abandoned pastures may exist as goldenrod fields for fifty years or more! Research conducted by Jack McCormick and associates in Waterloo Mills, Pennsylvania, revealed the key to the mystery of the goldenrod field's longevity. Certain goldenrods and asters release chemicals that suppress the germination and seedling growth of other plants. Apparently, goldenrod inhibit the next stage of succession. In doing so, they help to maintain habitat diversity. The goldenrod field you study and enjoy today may continue to intrigue observers for many years.

Adapted from Richard B. Fischer, "Goldenrod: An Ecological Goldmine," *The American Biology Teacher* 47 (October 1985): 7.

FIGURE 5-4: Secondary succession in a temperate deciduous biome.

FIGURE 5-5: A hedgerow between bean and corn fields. An ecotone or edge community forms where two distinct habitats, such as forest and grassland, meet. A hedgerow between adjacent fields can also be considered an ecotone. Species diversity within the ecotone is greater than that within the adjacent habitats.

Stage I: Annual Weed Community. Invasion of an abandoned corn field usually begins before the last corn cob has been picked! In fact, the seeds of opportunistic pioneer plants, such as pigweed, ragweed, and wild mustard, are present in the soil in what scientists call a weedy seed bank. Once those species are no longer kept under control by human efforts (weeding and the application of herbicides, for example), they begin to take over the field.

These early plants are generally dominated in numbers and in species by annuals, which have a one-year life cycle and generally have small, windblown seeds.

Stage II: Perennial Weed Community. The annuals may dominate for a year or two. The next community, dominated by perennials like goldenrod, are better able to tolerate the amount of resources that are available in the soil after the annuals have taken what they need. Goldenrods inhibit other communities from getting started, perhaps by chemical suppression of other species, and may dominate the community for decades. (See Focus On: Fields of Gold.)

Stage III: Shrub or Young Tree Community. Eventually, shrubs and trees enter the old field with the goldenrod perennials. Perhaps disease, predation, or shade on the field edges sufficiently change the abiotic characteristics to allow the larger woody plants to get a start. Honeysuckle, multiflora rose, black raspberry, eastern red cedar, ash, and locust seem to be most successful at this stage.

Stage IV: Young Forest Community, Stage V: Climax Forest Community. For stages IV and V of secondary succession the community tends to follow the changes described for stages V and VI of primary succession. It takes hundreds of years to establish a forest after succession begins in an old field.

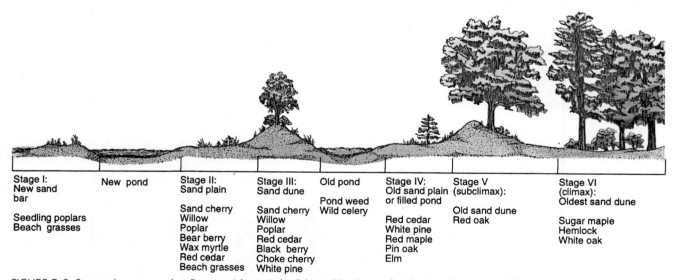

Stage I: New sand bar Seedling poplars Beach grasses	New pond	Stage II: Sand plain Sand cherry Willow Poplar Bear berry Wax myrtle Red cedar Beach grasses	Stage III: Sand dune Sand cherry Willow Poplar Red cedar Black berry Choke cherry White pine	Old pond Pond weed Wild celery	Stage IV: Old sand plain or filled pond Red cedar White pine Red maple Pin oak Elm	Stage V (subclimax): Old sand dune Red oak	Stage VI (climax): Oldest sand dune Sugar maple Hemlock White oak

FIGURE 5-6: Succession on sand at Presque Isle on Lake Erie at Erie, Pennsylvania. On the new sand bars that form near the water's edge, beach grasses and seedling poplars take root; they are the pioneer organisms that begin the process of succession. The sand plain represents the second stage of succession; there we might find sand cherry, willow, poplars, bear berry, wax myrtle, and red cedar. By the third stage, the sand dune, the trees have matured, joined by blackberry, choke cherry, and white pine. Older dunes may host other tree species as well, including red maple, pin oak, elm, and red oak. The climax community is represented by sugar maple, hemlock, and white oak. Note that as succession proceeds, vegetation gradually invades and fills in the oldest ponds—those farthest from the beach.

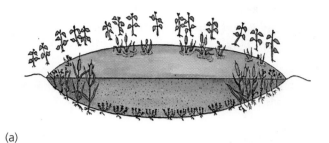

(a)

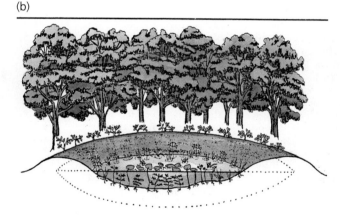

(b)

(c)

FIGURE 5-7: Succession in beach ponds on Presque Isle on Lake Erie at Erie, Pennsylvania. Careful observation reveals how the ponds have changed, proceeding toward dry land. (a) Young ponds are colonized by bulrush, chara, and cattail. (b) Older ponds, about 50 years of age, feature cottonwood, bayberry, and willow. (c) Ponds 100 years old and older are characterized by myriophyllum, yellow water lilly, pondweed, and cottonwood.

Ecotone. The stages of succession are not discrete; rather, one blends into another. Another type of blending occurs where different communities, such as forest and grassland, meet and compete for resources and space. Each community has an area in which its species composition dominates. Where the different communities meet, a zone of transition and intense competition, called an **ecotone** or edge community, is created (Figure 5-5). Ecotones usually contain species from both communities as well as some transitional species that appear to be most successful in the ecotone. They are interesting areas of greater diversity than the adjacent habitats and are characterized by intense competition between community types undergoing succession or between climax communities. To encourage diversity, wildlife managers often create and maintain numerous ecotones in nature preserves.

Succession in Ponds, Lakes, and Wetlands

Succession also occurs in aquatic habitats. The maturation or natural aging of a lake or pond is called eutrophication. An aquatic ecosystem that undergoes eutrophication becomes enriched with nutrients, which encourages the growth of aquatic plants. In deep lakes eutrophication is usually slow. However, it is rapid in farm or beach ponds or shallow lakes that receive high amounts of nutrients and sediments from the land that surrounds them. In these cases the aquatic system proceeds to dry land fairly quickly. It may pass through marsh, swamp, or bog-like conditions before dry land succession commences. Presque Isle, a recurring sand spit that juts out into Lake Erie at Erie, Pennsylvania, affords an excellent opportunity to study pond succession (Figures 5-6 and 5-7).

Usually, an aquatic ecosystem undergoing succession proceeds to the climax terrestrial community of that particular area. But if a pond or lake is in a wetland habitat, the climax stage may be a marsh or swamp community—as long as water dynamics remain constant. The climax community of the Everglades, which is sometimes referred to as a "great river of grass," is characterized by saw-grass sedges, cypress swamps, and hardwood hammocks, among others. Environmental Science in Action: The Everglades (pages 88–90) focuses on this unique ecosystem and the ongoing effort to preserve it.

What Is Dynamic Equilibrium?

Humans tend to think that nature, when left undisturbed, reaches and maintains a perfect harmony, that is, a perfect balance or state of static existence. Perhaps this is because our perceptions are limited by our inability to transcend the human time scale. We have difficulty imagining that some things occur over millions of years and that things that appear to have remained the same throughout human history are in fact in a slow process of development. But the fossil record provides clear evidence that ecosystems, biomes, and the biosphere itself have changed dramatically over the ages (Figure 5-8). Thus, while

Focus On: The Concept of the Balance of Nature
Frank N. Egerton, Department of History, University of Wisconsin, Parkside

The "balance of nature" is the oldest theoretical idea relating to that part of science we now call ecology. It was never explicitly formulated, so we should not call it a theory. However, it was a general assumption which functioned like a theory. The explicit scientific theories of the ancient Greeks, such as the geocentric universe, the four elements (earth, water, air, and fire), and Galen's understanding of the movement of bodily fluids, would all be refuted during the scientific revolution of the 16th and 17th centuries. They were refuted because a broader and more precise scientific knowledge made them seem implausible. Progress in ecological knowledge developed more slowly than in astronomy, physics, chemistry, and physiology. Furthermore, a general assumption is harder to refute than is an explicit theory. Thus, the balance of nature idea survived down to the present.

The earliest natural philosophers attempted to describe nature using sensory perceptions rather than myths. Later, Herodotus (about 500–about 425 B.C.), the "father of history," was undoubtedly influenced by the ideas of natural philosophers, but he had not lost entirely a reliance upon supernatural explanations. He reported an account from Arabia about why timid animals, like the hare, were

not all eaten up by fierce animals, like the lion: the timid kinds produced more young than did the fierce kinds. He believed a superintending Providence had created the different species with different reproductive capabilities so that predators would never exterminate their prey. Stories such as this do not add up to an ecological theory, but since nature appeared to remain constant from year to year, they seemed to illustrate a balance in nature.

The Athenian philosopher Plato (427–347 B.C.) wanted to teach about nature as well as about ethics, but he felt skeptical about what we can learn from sensory perceptions. In his writings he therefore had various philosophers tell stories about nature which he called myths. These myths were close to being scientific theories, except that he did not expect they could actually be proven. His creation myth drew upon insights such as those from Herodotus and suggested that each kind of animal was endowed with the means to assist its escape from enemies, its protection from weather, and its finding of sufficient food.

Another Greek philosopher, Aristotle (384–322 B.C.), had greater confidence than Plato in the evidence of sensory perception. Aristotle and his followers collected reports from farmers, hunters, and

fishermen about both wild and domestic animals. Like Herodotus, the Aristotelians knew that hares produce more offspring than lions do, but they sought a naturalistic cause rather than appealing to divine superintendence. They found it in physiology. Since lions must maintain large bodies, they cannot afford to divert large amounts of resources to producing many young. Smaller animals, such as hares, can produce more offspring since they do not have to grow so large. However, this explanation was offered not as a refutation of Herodotus, but as a different perspective.

The Romans admired Greek learning, yet made no advances beyond Greek science. Nevertheless, they did attempt to develop personal philosophies that drew upon Greek learning. Cicero (106–43 B.C.), in his book *On the Nature of God*, drew upon Greek learning to illustrate God's wisdom and concern. Examples of the apparent balance of nature were especially useful. Later, his book fit the needs of Christian theology about as well as it had Roman paganism, and Christians drew upon his book in medieval and later times.

By the 17th century, however, an interest in science was so widespread in western Europe that it was bound to lead to new observa-

the concept of the balance of nature pervades western culture, we know that it is not scientifically accurate. Even in climax communities, which represent the mature stage of the ecosystem, change is a constant. In fact, an ecosystem can persist through time only by constantly reacting to changes. Thus, a "stable" ecosystem is one that maintains what scientists call a **dynamic equilibrium** or a **dynamic steady state.** (See Focus On: The Concept of the Balance of Nature.)

What Factors Contribute to Dynamic Equilibrium?

Ecosystems, and ultimately the biosphere, maintain a dynamic equilibrium through a number of factors,

FIGURE 5-8: Trilobite fossils. Much of North America was once covered by an inland sea. Fossils like these trilobites found in Criner Hills, Oklahoma, provide evidence that our world has changed dramatically over eons.

tions and ideas concerning the balance of nature. A London merchant, John Graunt (1620–74), decided to study the mortality records of London and published the first book on statistics in 1662. He discovered that the sex ratio remains constant in humans and that about the same numbers died from violence and from the various diseases from year to year (excepting years of epidemics). It soon became obvious that the same must be true for animals. Another new area for scientific investigation concerned the nature of fossils. At the end of the century few naturalists doubted that these were the remains of plants and animals turned to stone. But this conclusion led to a puzzle: fossils seldom closely resembled living species. An obvious possibility was that they were remains of extinct species, but this idea seemingly implied that God had blundered and created some species unable to cope with their environment or foes. John Ray (1627–1705), a leading clergyman-naturalist in England, suggested that fossils might represent species now living in unexplored regions.

That possibility remained plausible for another century. As late as 1803 Thomas Jefferson asked Lewis and Clark, on their western expedition in the Mississippi valley, to be on the lookout for mammoths,

known as fossils. These elephants, too large to be overlooked if alive, were not seen. However, another possible interpretation was developing in Europe. Carl Linnaeus and Comte de Buffon, leading naturalists in the 18th century, both suspected that species must somehow change. Linnaeus, in 1749, established an explicit account of how the balance of nature seems to work. His model was a static one, which reflected the prevailing Newtonian view of nature. Linnaeus's ideas about changing species were not well enough developed to lead him to the view of nature as balanced, but changing.

At the beginning of the 19th century, the French naturalist Jean-Baptiste Lamarck (1744–1829) argued that the fossil species had not become extinct, but had gradually changed into the rather different species living today. Thus, an evolving balance of nature was an implicit aspect of Lamarck's unsuccessful theory of evolution.

An evolving or dynamic balance of nature was also implied in Charles Darwin's successful theory of evolution by natural selection. Darwin's theory states that all species produce more offspring than can survive, that these offspring are somewhat variable, and that the ones best adapted to their environment are the ones that sur-

vive to reproduce. Thus, a kind of balance is maintained, but with evolving rather than static species, and with some species also losing out in the struggle and becoming extinct. Darwin had a fine sensitivity to ecological matters. However, he raised far more questions than he could answer.

Since Darwin's time, naturalists have wondered how meaningful it is to imagine that a balance exists in nature when we know that environments do change, and that some species become extinct. Nevertheless, some of us still find the balance of nature concept useful. The difficulty arises when we try to elevate this convenient assumption to the level of a scientific theory, as in the recent "Gaia" hypothesis of James E. Lovelock. According to Lovelock, since organisms modify their own environment by their life functions, they actively try to maintain a stable environment for themselves. While there could be some truth to this, if one tries to make this generalization into a precise scientific theory, one finds it is like taking the measurements of a cloud. The balance of nature concept must remain what it has always been: a vague approximation of what we see.

including feedback, species interaction, and population dynamics. Diversity may also enable ecosystems to maintain equilibrium. These factors help to maintain equilibrium by resisting change or by restoring structure and function after a stress. **Inertia** refers to the ability of an ecosystem to resist change; **resiliency** refers to its capacity to return to a state of dynamic equilibrium after a stress.

Feedback

All life forms maintain an internal dynamic equilibrium. Maintaining equilibrium keeps organisms healthy in the face of constant external change. And so it is with ecosystems. In an ecosystem that is in equilibrium, organisms balance each other's inputs and outputs; the long-term average of the population of each species remains at a level that the envi-

ronment can support; and nutrients are continually recycled by biogeochemical processes. As energy flows and materials cycle through the ecosystem in countless ways, equilibrium is maintained by many checks and balances called feedbacks.

Both negative and positive feedbacks operate in living systems. **Positive feedback** continues a particular trend; **negative feedback** reverses it. In general, negative feedbacks are stabilizing, helping a mature system maintain equilibrium (inertia) or restore equilibrium (resiliency) when it is being disturbed. Positive feedbacks, in perpetuating changes, are destabilizing.

Feedbacks in nature are more complicated than those in human inventions (Figure 5-9). Consider, for example, the role decomposers play in reestablishing a forest community that has been stressed by fire (resiliency). Initially, conditions are not as favor-

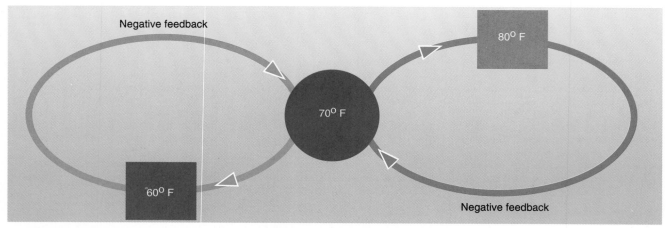

FIGURE 5-9: A simple negative feedback mechanism. When a thermostat to control room temperature is set at 70°F (21°C), if the room becomes warmer or cooler, negative feedback operates to return the room to 70°.

able as they were before the fire to bacterial populations, which thrive in the dark, moist environments of mature forests; consequently, organic matter is not decomposed as readily and tends to build up. As the forest matures, conditions become more favorable to bacteria and fungi, and their populations begin to thrive, an example of positive feedback. In other words, there is a positive feedback between the environment and the populations, causing the populations of decomposers to increase. Bacteria and fungi consume organic matter, which allows them to reproduce more rapidly, recycling more nutrients to plants and animals. Subsequently, in the mature forest the reproduction of microbial populations tends to be balanced by the production of organic matter by plants and animals. Negative feedbacks are at work, causing the microbial populations to stabilize.

Species Interactions

Species interactions are an important influence on the composition of populations and the structure of communities. Community structure, in turn, influences the inertia and resiliency of the ecosystem. That is why ecologists spend so much time trying to understand species interactions—besides, they are fun to learn about!

Species interact in numerous ways. Some of these are well known. The weasel stalks the rabbit, the butterfly pollinates the flower (Figure 5-10a), the flea sucks blood from the rat. Other species interactions are less familiar. The oxpecker, a bird native to East Africa, feeds on lice and other small insects found on rhinos. In coral reefs the cleaner wrasse fish cleans the teeth or gills of other fishes, receiving a free meal for its trouble (Figure 5-10b). The familiar acorn is home to and food for an almost bewildering array of herbivores, predators, parasites, and

decomposers. Weevils, ants, filbert worm moths, acorn moths, sap beetles, tachinid fly, braconid wasp, earthworms, termites, sow bugs, fungus gnat larvae, wire worms, slugs, millipedes, centipedes, and snails are just some of the animals that live in or use acorns for food. In addition, more than 80

(a)

(b)

FIGURE 5-10: Species interactions. (a) Butterflies pollinate the swamp milkweed. (b) Cleaner wrasse fish of the coral reef ecosystem cleans other fishes, such as the coral trout shown here.

species of North American birds and mammals eat acorns. It's a wonder that enough acorns are left to become oak trees!

Three significant types of species interactions are competition, cooperation, and predation.

Competition. Competition occurs when two or more individuals vie for resources (food and water), sunlight, space, or mates. Individuals that can out-compete others survive, reproduce, and pass on their attributes—a restatement of Darwin's theory of evolution by natural selection. (Evolution takes place at the population level, but the process goes on within the context of the ecosystem.) In this way populations adapt over time to changes in the environment. Understanding competitive interactions helps us to understand how populations react to change, thus affecting the structure of communities.

Basically, two types of competitive interaction take place: intraspecific and interspecific. **Intraspecific competition** is competition between members of the same species. When population densities (numbers per unit area) are low and the available nutrients, sunlight, space, and mates are adequate for all individuals in the population, little if any competition takes place. Competition is keenest when population densities are high.

Several examples illustrate the effects of intraspecific competition. Experiments have shown that frog tadpoles reared at high densities suffer higher mortality, experience slower growth, remain tadpoles longer, and metamorphose at smaller sizes than tadpoles reared at low densities. When plant densities are high and resources (sunlight, water, and soil nutrients) become limited, plants often change their reproductive strategies, producing fewer seeds and vegetative offspring. In animal populations competition often leads to the establishment of social

FIGURE 5-12: Showy orchis. The showy orchis, or showy lady's slipper, occupies a special niche within the forest community.

FIGURE 5-11: Eastern timber wolf pups. A wolf pack illustrates intraspecific competition and cooperation. These pups will learn social behaviors from older members of the pack. A high position in the social order gives the individual an advantage in the competition for food, mates, and territory.

hierarchies and defended territories and the development of behaviors that affect the individual's success within the population (Figure 5-11).

Interspecific competition is competition between members of different species. The greater the similarity in their needs and the smaller the supply of resources, the keener the competition. Interspecific competition is most intense when different species occupy similar ecological niches. A **niche** defines the functional role of an organism within its community. The habitat of a species (where it lives) is analogous to its address, while its niche is analogous to its occupation. Niche is the complete ecological description of an individual species, including all the physical, chemical, and biological factors that the species needs to survive and fulfill its role in the community. Niche includes the physical space occupied by the organism; its interactions with other species; its trophic level position; its place, time, and method of reproduction; its physical requirements (temperature, moisture, salinity, light, and others); its methods of protection and food gathering.

A biotic community can be thought of as a single unit composed of the niches of its various organisms. Some organisms, like the showy orchis, occupy a specialist niche within the community. This beautiful, rare orchid favors the rich, wet woods of beech-maple forests. It is found at the lower edges of steep slopes where the soil is deep and water abundant at flowering time in May and June (Figure 5-12). Other organisms, like the Canada thistle, occupy a generalist niche. It can be found in many different habitats.

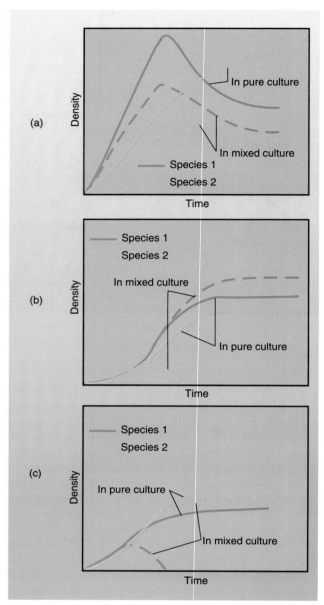

FIGURE 5-13: Effects of interspecific competition (a) Researchers at the University College of North Wales conducted experiments with clover that showed that the closely related species *Trifolium repens* (species 1) and *T. fragiferum* (species 2) are able to coexist in mixed culture, although at lower leaf densities than when each exists in pure culture. Their results indicate that competing species can coexist despite crowding and competition for limited resources as long as there are differences in some significant factor, for example, the timing of growth, a requirement for nutrients, water, or light, or sensitivity to grazing or toxic substances. (b) In the 1940s and 1950s, investigators at the University of Chicago carried out a series of competition experiments with laboratory cultures of flour beetles. They found that *Tribolium castaneum* (species 1) excludes *T. confusium* (species 2) under hot and wet climatic conditions (93°F or 34°C, 70 percent relative humidity), even though both species can thrive under these conditions separately. (c) The opposite is true under cool and dry conditions (75°F or 24°C, 30 percent relative humidity). *T. confusium* (species 2) excludes *T. castaneum* (species 1), even though both do well at these conditions in pure culture.

Rarely does interspecific competition lead to the exclusion of a species. More commonly, the competing species eventually coexist at reduced densities by partitioning the available resources (Figure 5-13a) or the less dominant species is forced to use other space or resources. In the tropical forests of New Guinea as many as four species of pigeons, differing primarily in weight, rely on a particular fruit tree as their sole food source. The successively smaller and lighter birds have adapted to feed on the fruit that grows on successively smaller branches outward from the trunk.

When interspecific competition leads to the exclusion of one of the competing species, the **competitive exclusion principle** is said to operate. This principle has been demonstrated in laboratory experiments with two closely related species of protozoans, fruit flies, mice, flour beetles, and annual plants (Figure 5-13b, c). It is more difficult to observe in the field, since ecologists may not have all the information they need, particularly about species' life-history requirements, to prove conclusively that the exclusion of one species is due to competition with another.

Cooperation. Intraspecific cooperation is best exemplified by the social structure of bees and termites and the behaviors of herds (for example, deer, bison, musk ox) and packs (for example, wolves and hyenas). Cooperative behaviors among members of the beehive, termite hill, herd, and pack aid the survival of the entire group.

Interspecific cooperation also occurs. **Symbiosis,** defined as the intimate association of two dissimilar species, regardless of the benefits or lack of them to both species, can take one of several forms: mutualism, commensalism, and parasitism.

Mutualism is an association of two species in which both benefit. Lichens, for example, are mutualistic associations of fungi and algae. Carbohydrates

FIGURE 5-14: Remora and manta ray. The remora fish and manta ray illustrate commensalism: The remora benefits from its association with the manta, but the manta neither benefits nor is harmed.

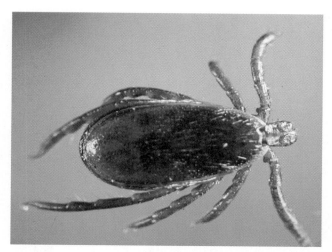

FIGURE 5-15: Deer tick. The deer tick is an ectoparasite, attaching itself to its host by means of biting mouth parts. Deer ticks can cause Lyme disease in humans.

produced by the algae are used by both the algae and the fungi, and the fungi provide moisture and protection for the algal cells. Mutualism is also illustrated by the association between termites and protozoa. While termites can chew through wood, they cannot digest the cellulose in it. Digestion of the cellulose is accomplished by protozoa that live in the guts of termites.

Commensalism is an association of two species in which one benefits and the other neither benefits nor is harmed. Remora fish attach themselves to rays with suckers and benefit from the sharks' feeding activities, but neither help nor harm the shark (Figure 5-14). Water collected in bromeliad plants of the South American rain forests provides a place for tree frogs to breed, but the plants neither benefit nor are harmed by the frogs' activities.

Parasitism is an association of two species in which one benefits and the other is harmed. The organism that benefits from the activity is called the parasite; the other organism is the host. The host

acts as the environment for the parasite, which derives food, shelter, and protection from the host. Parasites, such as the deer tick, that live outside the host are called ectoparasites (Figure 5-15); parasites, such as tapeworms and flukes, that live inside the host's body cavity, organs, or blood are called endoparasites. Typically, the parasite does not kill its host, but the host may weaken and die from complications associated with the parasite's presence.

Predation. Perhaps the most widely recognized form of species interaction is the predator-prey relationship, in which one species eats another. Predators are generally larger than their prey, live apart from their prey, and consume all or part of their prey. Prey species have evolved behavioral, reproductive, and physical and chemical characteristics to help them avoid being eaten. Predators, in turn, have evolved mechanisms to enhance capture of prey (Figure 5-16). Predators may be herbivores, omnivores, or carnivores, and prey may be producers, herbivores, omnivores, or carnivores.

Nonhuman Population Dynamics

Species interactions help ecologists to understand trophic level interactions, food webs, community structure, and the complex interactions between the biotic and abiotic components of ecosystems that help to regulate nonhuman population dynamics. In Chapter 3 we learned that a population is a group of organisms of the same species that occupies a particular habitat and has definable group characteristics. Seed dispersal and germination are characteristics of plant populations. Birth and death rates are major characteristics of animal populations.

If it were possible to construct a utopian environment, that is, an environment with no limitations, a population could experience unlimited growth. As long as the environment remained utopian, the population could continue to grow in an unrestricted

(a)

(b)

FIGURE 5-16: Camouflaged predators. Camouflage is an evolved mechanism that enhances the ability of the Arctic fox (a) and walkingstick (b) to capture prey.

Environmental Science in Action:
The Everglades
Cheryl Puterbaugh

Florida's Everglades illustrates what happens to a wetlands ecosystem when humans severely disturb it. The state's attractive climate has given rise to increased population, urbanization, and agriculture. The inland wetlands, the coastal ridge, and the coastal wetlands of southern Florida make up a vital, interconnected system that now exists precariously on the overbuilt, overdrained, and overpopulated peninsula. Future management must achieve a balance between human needs and the needs of this valuable, life-supporting ecosystem.

Describing the Resource

Physical Boundaries

The Everglades is in South Florida, in the southernmost part of a drainage basin that begins with the numerous lakes south of Orlando in central Florida. These lakes drain southward into the Kissimmee River, formerly a 98-mile (158-kilometer) stream that meandered through thousands of acres of marshland (Figure 5-21). The river has been replaced by a 52-mile (84-kilometer) canal that rapidly transports water into Lake Okeechobee. The water then flows through canals from Lake Okeechobee's southern rim, making its way through rich agricultural land. At the northern border of the Everglades, the water begins its slow movement through the grasses and hardwood hammocks to Florida Bay at the tip of the peninsula.

In its natural and healthy state the Everglades performs many essential ecosystem services. The wetlands stabilize the shoreline and reduce coastal storm damage. Wetland vegetation absorbs wave and storm energy, slows water currents, and prevents erosion. Wetlands slow down and store flood waters, lessening the extent of flood damage. They also continually recharge underground aquifers, maintaining the supply of fresh water. Groundwater pressure limits the intrusion of salt water, preventing it from contaminating South Florida's

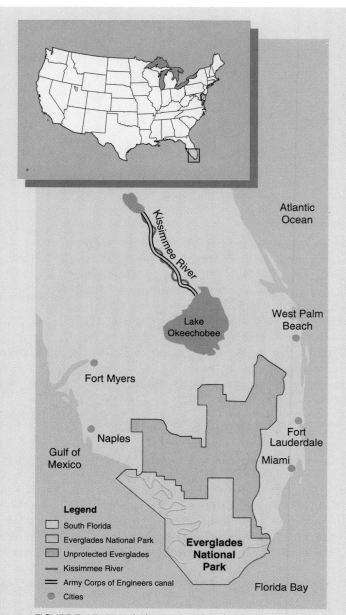

FIGURE 5-17: Everglades National Park, Florida. South Florida has been the site of tremendous change since the start of the twentieth century. The region's "river of grass"—the Everglades—now flanked by growing urban areas to the east and west and mostly agricultural land to the north, is a natural system under siege.

aquifers, the source of drinking water for much of the population. The ecosystem also maintains water quality. Wetlands are natural water treatment facilities; nutrients are recycled and pollutants settle out.

The Everglades also help to moderate the regional climate. Mean annual rainfall in South Florida is 60 inches (152.4 centimeters);

88

however, this may vary from 35 inches (89 centimeters) in dry years to 120 inches (305 centimeters) in wet years. This great variation in annual rainfall is accounted for by South Florida's rainfall cycles: periods of increasingly drier years followed by increasingly wetter years. Factors specific to South Florida account for at least two-thirds of the region's rainfall. For example, the amount of evaporation and transpiration that occurs over the Everglades, with its abundance of shallow water and vegetation, is considerable. The moisture from the rising wetlands air accounts for about 35 percent of the total rainfall. As the wetlands near the coast are "reclaimed" for urban development, the reduced amount of moisture and resulting increase in the heat of the ground mean less rain falls to recharge the aquifers.

A final ecosystem service provided by the Everglades is the maintenance of the region's soil. The organic muck of the wetland was formed by thousands of years of decaying vegetation, and unless it stays moist, it oxidizes (combines with oxygen) and disappears. When water is scarce, soil loss occurs. The most extensive soil loss has occurred in the Everglades agricultural area south of Lake Okeechobee, which produces about $700 million in farm products annually. Because farming requires the land to be drained, the loss of muck soil to oxidation occurs at the rate of 1 inch per year. The dried organic matter of the drained land is also extremely susceptible to fires, which have increasingly been a problem throughout South Florida. Fires burn beneath the surface of the ground and are difficult to put out.

Biological Boundaries

The Florida Everglades serves as habitat for migratory waterfowl, wading birds, and freshwater fish and as spawning and nursery grounds for marine fish and shellfish. The Everglades–Florida Bay area is home for hundreds of species of wildlife. In this ecosystem are

fresh- and saltwater fish, which support a vital fishing industry; many species of reptiles, amphibians, and mammals; and over 200 species of birds. Many of these animals, including the Florida panther, the American crocodile, the brown pelican, the Everglades snail kite, and the southern bald eagle, are endangered.

Grassy marshes, shrub thickets, and hardwood hammocks are the primary habitats in the Everglades. Species that rely on one of these wetlands habitats for nesting often rely on another for feeding or other needs. For example, surrounding marshes provide fire protection and food supply to the inhabitants of hammocks, while hammocks are sources of shelter during floods for the marsh species. The dynamic equilibrium of the wetland ecosystem is the result of such interrelationships.

The saw-grass prairies and marshes are covered with water 5 to 10 months each year. The inland marshes are dominated by Muhly grass and a grasslike sedge called saw grass and are important feeding, mating, and nesting grounds for many migratory waterfowl and wading birds such as herons and egrets. Coastal marshes provide spawning and nursery grounds for fish and shellfish as well as a supply of nutrients to offshore waters.

Wax myrtle thickets serve as nesting and roosting grounds for many wading birds and are also important habitats for white-tailed deer, bobcats, alligators, and the Florida indigo snake. Hardwood hammock forests occur in places that are elevated enough to be covered with water a month or less each year; they support a rich and diverse biotic community (Figure 5-22).

Social Boundaries

The fertile, peaty muck soil of the Everglades is viewed as desirable farmland, but before farmers will use it to grow crops or raise livestock, it must be drained of its water and cleared of its native plants. Millions of dollars in sugarcane, vegetables, and cattle are pro-

FIGURE 5-18: Hardwood hammock forest of the Everglades, Florida. Dotting the Everglades landscape are hardwood hammocks home to diverse species such as mahogany trees, royal palms, saw palmettos, the endangered Florida panther, the gopher tortoise, and the colorful liguus tree snail.

duced here annually.

Because of Florida's mild climate and proximity to the ocean, huge population centers have grown up on the state's eastern coastal ridge. As this population increases, demand for more space pushes back the wetlands. Before 1900, 75 percent of South Florida was wetland; presently, about half of that has been converted to agricultural or urban use.

The Everglades ecosystem also holds great recreational, esthetic, and scientific value. Tourism is one of Florida's most important industries, and hunting, fishing, and bird-watching are only a few of the activities offered in the wetlands. In addition, the wetlands are an invaluable site for ecological studies. The diverse and productive ecosystem provides many opportunities for research, which is especially important because of the potentially serious environmental problems facing the area.

Looking Back

Throughout the nineteenth century various efforts by both private interests and the government were made to control water movement in South Florida. By the mid-1920s,

440 miles of canals, 47 miles of levees, and 16 locks and dams had been constructed to improve navigation of inland channels, control floods, and convert marshland to agricultural use. After severe hurricanes in 1926 and 1928, demands for increased flood protection generated increased construction, and the southern rim of Lake Okeechobee was replaced by a dike that completely stopped the normal flow. Subsequent years of construction have resulted in almost 1,500 miles of levees and canals, along with pumping stations, dams, and tide and flood gates, which are maintained and operated by the U.S. Army Corps of Engineers and the South Florida Water Management District (SFWMD).

The Kissimmee River was the water source for perhaps as many as 200,000 acres (80,938 hectares) of river marsh and associated wetlands. Millions of migratory and wading birds depended on this wetland habitat. In 1971 the Corps of Engineers completed the 52-mile long, 200-foot wide, and 30-foot deep canal that replaced the river. In addition to flood protection for the urban area south of Orlando, the canal was to provide better flood protection for the farmers and ranchers of the lower basin, improve navigability, and expand the recreational uses of the river.

With the construction of the Kissimmee canal, water no longer took eleven days to reach Lake Okeechobee; it rushed through in two. Nitrogen and phosphorus runoff from the newly created pastures and farmland along the canal and wastewater from the Orlando area moved quickly downstream, and little of the pollutants were filtered out. The old oxbows of the meandering river slowly dried up. As the wetlands along the river basin disappeared, so did the wildlife, especially bald eagles, ducks, and coots.

Water is the lifeblood of the Everglades. Because the water supply is threatened by mismanagement, many things that depend on water for survival are threatened as well.

From 1980 to 1982 South Florida suffered its worst drought in history. With the increase in dry land in certain controlled areas called Water Conservation Areas, the white-tailed deer population boomed to an estimated 5,500. In the spring of 1982 heavy rains began to raise the water levels. Flood water was channeled into the controlled areas until only a few hundred acres of dry land were left. Under these conditions the deer population could not survive. The Game and Fresh Water Fish Commission called for an emergency out-of-season deer hunt in July 1982 to thin the herd and increase the chances of survival for the remaining deer. A two-day hunt was held and 723 deer were taken. The rest of the hunt was canceled in favor of a rescue attempt; 19 deer were captured in a two-day period, but only 6 survived. The commission estimated that 1,200 deer may have died in one area as a result of high water in 1982.

Occasionally, during the rainy season the Water Conservation Areas cannot contain all the water drained from surrounding land. In such cases some of the excess water is pumped into the Everglades National Park. This often floods the park at the worst possible time, when it already has enough water from rainfall and when the alligators and wading birds are nesting. After torrential rains in 1983, floodwaters into Everglades National Park released so much water that it destroyed the nests of wading birds. Some progress has been made; now three of the four floodgates are kept open continuously to allow a more natural flow.

In damaged and drained lands it is easy for exotic (nonnative) plant species to move in and interrupt the normal pattern of succession. The Brazilian pepper, the Australian pine, and the Melaleuca are three exotic species that are threatening native life in the Everglades. The eucalyptus-like Melaleuca was introduced in 1906 in an effort to attract commercial woodcutters. But inside its soft, thick bark the hardwood interior is runny with water and difficult to saw, which makes it unattractive to the timber industry. The tree uses three times as much water as other trees and has taken over about 60 square miles (155 square kilometers) of wetland. Its leaves are filled with eucalyptol, an oily, flammable substance that smokes and sparks when on fire. Its insulating bark protects the tree from fire.

Looking Ahead

During his term as governor from 1979 to 1987 Robert Graham took a special interest in the situation and organized Save Our Everglades, a group that has proven to be important in the effort to restore this vital ecosystem. (See Focus On: Marjory Stoneman Douglas.) The first goal of Save Our Everglades is the restoration of the Kissimmee River to its natural course. The first phase of the program, the diversion of canal water back to its original course, was begun in July 1984. Approximately 1,300 acres (526 hectares) will be restored. Later phases, including land acquisition and dike removal, will result in the restoration of 28,000 acres (11,331 hectares) of wetland.

Hydrologists from the South Florida Water Management District say that a reflooding of the valley would be better for everyone except the property owners along the river, who will be losing flood protection. Towns near the upper lakes will not be threatened, because the canal never provided the expected flood protection in that area. Restoration will reduce the need for emergency releases of water—sudden dumpings that cause damage in the coastal estuaries by drowning oyster and clam beds and depositing silt.

The population of Florida has increased in the last 40 years from 2 million to over 10 million and is expected to double between 1975 and 2000. The balance among the population, water, and wetlands is becoming more and more important. Steps must be taken now to prevent further degradation of what used to be thought of as "forbidding" swamplands. Without those prime wildlife breeding grounds and water sources, South Florida might not look so appealing.

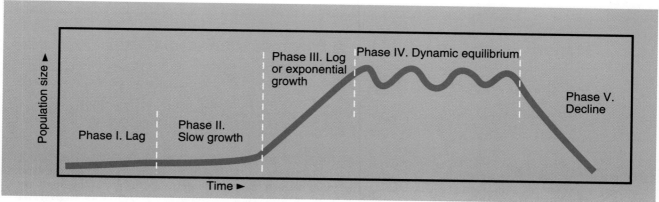

FIGURE 5-19: Five common phases of population growth.

manner. In just 24 hours one yeast cell could give rise to several thousand cells, and one *Escherichia coli* bacterium could give rise to 40 septillion bacteria; in a single year two houseflies could produce 6 trillion individuals! But environments are not utopian; all populations fluctuate over time and have an upper limit (set by the number of organisms) their environment can support. In controlled laboratory experiments, populations exhibit similar growth patterns. From studies of bacteria, fruit flies, mice, and rats that were provided with unlimited food and water, scientists have identified five common phases of population growth: I, lag phase; II, slow growth phase; III, log or exponential growth phase; IV, dynamic equilibrium phase; and V, decline (Figure 5-17). (Caution must be exercised when making conclusions about population growth based on these experiments, because only single populations were studied, environmental conditions were carefully controlled, space was restricted, and waste products were not removed.)

The experimental populations remained in phase I, lag, for some period of time. During this time organisms were adjusting to their new environments and few were mature enough to reproduce, so the population did not grow. In phase II the population grew slowly over time as more individuals began to reproduce. In phase III, log or exponential growth, many organisms were capable of reproducing, and the population grew rapidly over time until it reached phase IV, dynamic equilibrium. During phase IV the population fluctuated around a particular level as reproduction basically equaled mortality. In all the systems studied, the population went through phase V, decline, eventually reaching zero. If the size of the experimental world was increased, that is, if the organisms were provided with additional space, population numbers increased, but once again all stages were observed.

The unopposed, or utopian, reproduction represented in phase III illustrates a positive feedback system. Such a system cannot be maintained in a world

of limited resources, as studies of naturally occurring populations have shown. These populations exhibit S-curve, or oscillating, growth patterns (Figure 5-18). It is as if phase IV of our experimental population pattern were magnified for careful scrutiny. All natural populations fluctuate over time around the number of individuals the environment can best support. In other words, all known nonhuman populations thus far studied have an upper

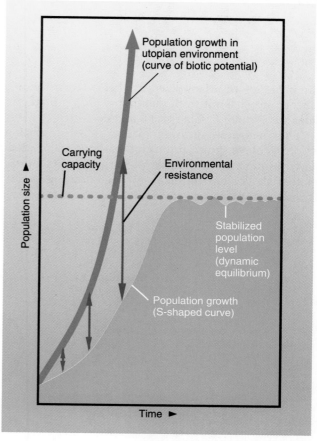

FIGURE 5-20: S-curve of population dynamics. The diagram illustrates the relationship of biotic potential, environmental resistance, and carrying capacity. The population in dynamic equilibrium oscillates around the level that can best be supported by the physical environment.

Focus On:
Marjory Stoneman Douglas—A Friend of the Everglades

The Save Our Everglades movement is the result of a lot of hard work by people concerned about the welfare of the threatened wetlands. One of the many dedicated individuals who have contributed their time to the cause is Marjory Stoneman Douglas.

Douglas is an environmental activist who has lived in Florida since 1915, when she moved to Miami, then a town of only 5,000, and became a reporter for the *Herald*. The history of the Everglades is not a textbook exercise to her; she has lived it. She survived the disastrous hurricane of 1928. She witnessed the building of the canals, levees, and dikes that now control the water flow. She understood from the beginning the terrible destruction occurring to a productive and vital ecosystem. In her 1947 book *The Everglades: River of Grass*, she wrote a poetic account of the mystery and beauty of the

Everglades. She later turned to political activism to save this threatened ecosystem.

In 1970 Douglas founded Friends of the Everglades, an organization that has fought to preserve North America's only subtropical wetland. Through private lobbying and public hearings the group has had many successes. Among other things they blocked the construction of a jetport in the east Everglades, forced two drainage canals to be plugged, and were instrumental in the plan to restore the Kissimmee River.

Douglas believes that the Army Corps of Engineers, which built the disastrous Kissimmee canal, "doesn't understand a complex natural system like the Everglades." She believes that the Corps approach is too mechanical and narrow. She talks about the people who have profited from the construction—developers, cattle ranchers, and

large agricultural interests. The desire for immediate economic gain has simply played too large a part in the area's management. For example, to grow sugarcane in the area, the land must be first drained and then irrigated. Mrs. Douglas finds it ridiculous to try to make the land something it isn't. "Why don't we import our sugar from Caribbean countries where growing conditions are right for it?" she asks. "Why don't we produce sugar from beets grown in the natural farmland of the Midwest?"

As president of Friends of the Everglades, Douglas is a formidable warrior in the battle for the environment. This centenarian describes herself as "a tough old woman," and she takes advantage of everything she can—including her age—because, as she says, her cause is just.

limit to growth. Many scientists believe that there is also an upper limit to human population growth, an idea we explore in Chapters 8 and 9.

The maximum growth rate that a population could achieve, given unlimited resources and ideal environmental conditions, is known as its **biotic potential**. The environment, through limiting factors known as **environmental resistance**, exerts a con-

trolling influence on population size. Those organisms that are able to adapt to changing environmental conditions survive and reproduce. Over time populations are regulated by the interaction between the biotic potential of the population (which acts as positive feedback) and environmental resistance (which acts as negative feedback). This dynamic equilibrium is referred to as the **carrying capacity**,

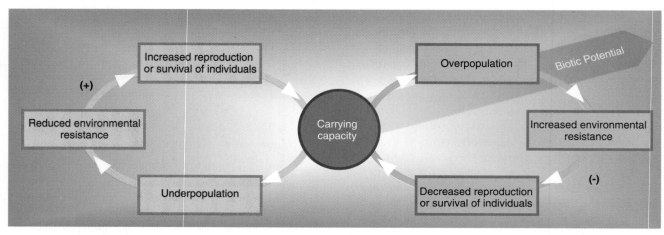

FIGURE 5-21: Population growth as the result of the interplay between biotic potential and environmental resistance. Carrying capacity represents a balance between biotic potential (positive feedback) and environmental resistance (negative feedback). Negative feedback tends to reduce the population if it exceeds the carrying capacity of the environment, and positive feedback enables a population to recover after a stress.

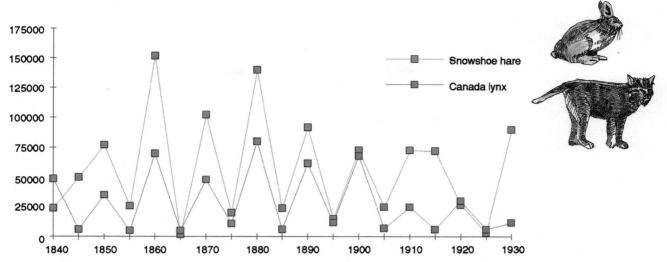

FIGURE 5-22: Snowshoe hare and lynx populations, northern North America. The graph shows the 9- to 10-year cyclic fluctuations that occur in the lynx and snowshoe hare populations.

that is, the population size that can best be supported by the environment over time (Figure 5-19).

Environmental resistance can be separated into two categories: density-dependent (biotic) factors and density-independent (abiotic) factors. **Density-dependent factors** are biotic; their effect is greater when the population density is high. They include a wide range of intra- and inter-species interactions. Some density-dependent factors are competition, predation, disease, spontaneous abortion, embryo absorption, infanticide, cannibalism, starvation, and stress. **Density-independent factors** tend to set upper limits on the population. They are abiotic and include a wide range of interactions between the physical environment and the population. Density-independent factors include climate, weather, and the availability of nutrients, water, space, den or nest sites, and seed germination sites.

For any given nonhuman species, it is likely that a combination of density-dependent and density-independent factors operate simultaneously to regulate populations. However, some factors may be more important than others for certain populations. For lynx-hare populations, the prime regulator might be predation by the lynx, a density-dependent factor, or it might be the amount of food available for the hare, a density-independent factor (Figure 5-20). Rainfall and temperature (density-independent factors) are important regulators of insect populations. Certain rodent populations (mice and rats) seem to be regulated by available space (density-independent) and the stress caused by overcrowding (density-dependent). Possibly, stress-induced hormonal changes cause abnormal social behaviors and decrease the rodents' resistance to disease, thus lowering birth rates and increasing death rates. Because

reproduction would then be less successful, population numbers would decline to a level where the stress caused by overcrowding would no longer be a major limiting factor.

A caution is in order. The results of population dynamics studies on nonhuman organisms should not be directly applied to humans. Conclusions cannot be made about the effect of environmental resistance on human populations in the absence of reproducible studies using human populations.

Species Diversity

Does species diversity contribute to dynamic equilibrium in ecosystems? Ecologists disagree. Until recently, many believed that the greater the ecosystem's diversity of species, the more easily it would recover from a disturbance. So long as the disturbance was not too severe, this assertion seemed reasonable. Recently, however, this belief has been disputed. Discoveries about the inability of the most diverse of all ecosystems, tropical rain forests, to recover from deforestation, suggest that ecosystems vary in their response to stress and that factors other than diversity—factors not yet identified or understood—may contribute significantly to dynamic equilibrium. Accordingly, most ecologists now believe that there is no simple relationship between diversity and dynamic equilibrium. Because each ecosystem has evolved to fit a particular physical environment, less complex systems may be just as stable and capable of recovering from disturbance as complex systems.

Perhaps the best position we can take at this time is to recognize that we do not understand the complexities of ecosystem relationships well. In terms of management that means that until we do, we should preserve the diversity in all ecosystems. Diversity and the arguments for its preservation are explored in greater detail in Chapter 23.

Summary

All ecosystems undergo succession, the gradual, sequential, and somewhat predictable changes in the composition of the biotic community. Succession, or ecosystem development, occurs in both terrestrial and aquatic habitats. Primary succession is the development of a new community in an area previously devoid of organisms; secondary succession is the development that occurs after an ecosystem has been disturbed. Successional stages are not discrete; they tend to blend into one another. A similar sort of blending occurs where different types of communities meet and compete for resources and space. The transitional zone between the communities is known as an ecotone, or edge community.

Ecosystems continually react to change and disturbance, thereby maintaining a dynamic equilibrium or dynamic steady state. Ecosystems maintain a dynamic equilibrium either by resisting change, known as inertia, or by restoring structure and function after a disturbance, known as resiliency.

Numerous factors contribute to ecosystem stability. Of particular importance are feedback, species interaction, and population dynamics. Feedback can be positive or negative; positive feedback continues a particular trend, and negative feedback reverses it.

Predator-prey relationships are perhaps the most familiar type of species interaction, but competition and cooperation are also forms of interaction. Intraspecific competition takes place between individuals of the same species; interspecific competition occurs between individuals of different species. Different species with similar requirements sometimes compete to the exclusion of one of them, a phenomenon known as the competitive exclusion principle. Interspecific competition may also result in coexistence of the two species at reduced densities. Organisms with a wide range of tolerances enabling them to compete successfully in different areas are said to occupy a generalist niche in the community. Organisms with a narrow range of tolerances allowing them to compete successfully in only specific conditions occupy a specialist niche.

Cooperation can take place between individuals of the same species or between individuals of different species. Intraspecific cooperation is exemplified by insect societies and animal herds and packs. Symbiosis is the intimate association of two different species, regardless of the benefits or lack of them. Mutualism, commensalism, and parasitism are forms of symbiosis. Mutualism is an association of two species in which both benefit. Commensalism is the association of two species in which one benefits and the other neither benefits nor is harmed. Parasitism is the association of two species in which one benefits (the parasite) and the other is harmed (the host).

Population dynamics also contribute to ecosystem stability. In laboratory studies scientists have found that populations of organisms commonly exhibit five phases of growth: lag, slow growth, log or exponential growth, stability or dynamic equilibrium, and decline. In contrast, natural populations fluctuate over time around the carrying capacity of the environment, the number of individuals the environment or habitat can best support. This number is lower than the population size that would result if the population achieved its biotic potential, that is, the maximum growth rate possible given unlimited resources and ideal environmental conditions.

Limiting factors, collectively known as environmental resistance, prevent a population from realizing its biotic potential. Environmental resistance factors may be density-dependent or density-independent. Density-dependent factors are biotic; their effect is greater when the population density is high. Density-independent factors are abiotic; generally, they set upper limits on the population.

Key Terms

biotic potential	interspecific competition
carrying capacity	interspecific cooperation
climax community	intraspecific competition
commensalism	intraspecific cooperation
competitive exclusion principle	mutualism
cooperation	negative feedback
density-dependent factor	niche
density-independent factor	parasitism
dynamic equilibrium	pioneer organism
ecological succession	positive feedback
ecosystem development	primary succession
ecotone	opportunistic species
environmental resistance	resiliency
inertia	secondary succession
	symbiosis

Discussion Questions

1. Differentiate between primary and secondary succession. Give examples of successional stages. What maintains succession at its climax stage?

2. What is the concept of the balance of nature? Can the concept be tested? Why or why not? How does the concept relate to dynamic equilibrium?

3. Discuss how feedback, species interaction, and population dynamics help to maintain an ecosystem in dynamic equilibrium.

4. What might happen to population dynamics at each trophic level in an ecosystem if the top carnivores are eliminated?

5. How will restoring the Kissimmee River to its former free-flowing state affect the Everglades? In your opinion, should this be done?

6

Ecosystem Degradation

We stand guard over works of art, but species representing the work of aeons are stolen from under our noses.

Aldo Leopold

The worst thing that can happen—will happen—is not energy depletion, economic collapse, limited nuclear war or conquest by a totalitarian government. As terrible as these catastrophes would be for us, they can be repaired within a few generations. The one process ongoing in the 1990's that will take millions of years to correct is the loss of genetic and species diversity by the destruction of natural habitats. This is the folly our descendants are least likely to forgive us.

E. O. Wilson

Learning Objectives

When you finish reading this chapter, you should be able to:

1. List the five D's of ecosystem degradation.
2. Describe what is meant by ecosystem damage and define pollutant.
3. Explain how human activities cause disruption of natural systems.
4. Identify the primary cause of ecosystem destruction.
5. Define desertification and identify three human activities that contribute to it.
6. Define deforestation and identify its primary causes.

When human activities alter environmental conditions in such a way that they exceed the range of tolerances for one or more organisms in the biotic community, the ecosystem becomes degraded. It loses some capacity to support the diversity of life forms that are best suited to its particular physical environment.

Although ecosystem degradation is complex, it is helpful to classify it into five broad categories which we call the five D's of natural system degradation: Damage, Disruption, Destruction, Desertification, and Deforestation. Although desertification and deforestation are subsets of the first three and deforestation can be a cause of desertification, we feel that each mechanism of degradation is serious enough to warrant special attention.

What Is Ecosystem Damage?

Ecosystem damage is an adverse alteration of a natural system's integrity, diversity, or productivity. Pollution is the major cause of ecosystem damage. A **pollutant** is a substance or form of energy, such as heat, that adversely alters the physical, chemical, or biological quality of natural systems or that accumulates in living organisms in amounts that threaten their health or survival (Figure 6-1).

Five major factors determine the damage done by pollutants: the quantity of the pollutant discharged to the environment, the persistence of the pollutant, the way in which the pollutant enters the environment, the effect of the pollutant, and the time it takes the ecosystem to remove the pollutant. Although we will examine these five major dimensions of pollution separately, they are clearly interrelated.

Quantity of the Pollutant

Some waste products, such as fertilizers from farm fields, human sewage, animal waste from feed lots, waste from food-processing plants, carbon monoxide from automobiles, and solid waste (refuse) are pro-

FIGURE 6-1: Catfish with tumor. Pollutants accumulate in the bottom sediments of rivers and lakes, where they may adversely affect aquatic life. Toxic pollutants may cause tumors and malignancies in organisms that feed or live in the bottom sediments. This catfish, a bottom-feeder, was taken from Ohio's heavily contaminated Black River.

duced in such large quantities that natural systems cannot efficiently or readily process them. Americans generate, on average, about 1 ton (0.91 metric tons) of solid waste each year, about 5 pounds (2.25 kilograms) each day. A city the size of New York produces approximately 9 million tons (8.2 million metric tons) of solid waste a year and 1 to 1.2 billion gallons (3.79 to 4.55 billion liters) of raw sewage a day!

Consider the case of the New York Bight, a bay formed by the curve of the Atlantic coastline from Montauk Point, at the end of Long Island, New York, to Cape May, New Jersey (Figure 6-2). From 1938 to 1987 New York City dumped sewage sludge (slightly refined sewage) into an area of the Bight known as the 12-mile site (twelve miles from the city). Eight other sewage authorities from Long Island and New Jersey also dumped sludge into the 12-mile site. In the early 1980s acid wastes and harbor-dredged sediments were also dumped into the bight. Raw sewage and industrial waste dumped into the Hudson River eventually ended up in the bight as well.

One sign that the bight is severely stressed is the magnitude of the occurrence of phytoplankton blooms. Phytoplankton thrive on the nitrogen and phosphorus content of the sewage and multiply so rapidly in the spring that zooplankton cannot consume them quickly enough. When the phytoplankton die in the summer, they sink to the ocean floor, where bacteria, decomposing the organic overload, deplete the oxygen supply. The lack of oxygen substantially reduces the populations of bottom-

dwelling fauna such as crabs, lobsters, snails, clams, and worms.

The cumulative effect of these massive amounts of waste was to turn the bight into an "organic soup" laced with high concentrations of toxic metals, including silver, lead, zinc, and mercury. Polychlorinated biphenyls (PCBs), known to be carcinogenic, are also present in high concentrations. The sewage and industrial wastes damage the bight in several ways. Bacteria, viruses, and trace metals accumulate in the tissues of clams, oysters, and mussels, presenting a danger to animals that feed on them. The tissues of striped bass, bluefish, white perch, and eels have high PCB levels. Other effects that may be related to pollution include abnormalities in the Atlantic mackerel's eggs and larvae, red blood cell mutations in windowpane flounder and red hake, and black gill disease in rock crabs and lobster.

As of December 1987 all dumping at the 12-mile site ceased. The Environmental Protection Agency then designated a deepwater municipal sludge dump site, 106 miles out into the ocean, off the continental shelf. Over 8 million wet tons of sewage sludge were dumped at the site annually. Preliminary study by researchers at Woods Hole Oceanographic Institute indicated that the sludge was not dispersing as expected, but had accumulated on the ocean floor. In 1988 Congress passed the Ocean Dumping Ban Act, which mandated that all ocean dumping of sewage sludge and industrial waste end as of December 31, 1991. New York City was unable to meet this deadline, but planned to end dumping by June 1992, by which time more than 32 million wet tons of sewage sludge would have been dumped at the 106 site—the largest known human-made stress of a deep-ocean environment.

Persistence of the Pollutant

Materials that can be broken down and rendered harmless by living systems are called **biodegradable**. For example, if processing methods reduce sewage so that the concentration of materials in the water leaving the treatment facility does not alter the population of organisms present in the river or stream, then the river or stream system is biodegrading, or recycling, the sewage. A substance is **nonbiodegradable** when it enters a system in a form unusable by the organisms present in that system. As the case of the New York Bight illustrates, even biodegradable substances, in large amounts, can adversely affect the productivity and species diversity of ecosystems.

A pollutant that accumulates in natural systems over time is called a **persistent pollutant**. Persistent pollutants include those that are nonbiodegradable and those that are only slowly biodegradable. Although some insects and bacteria begin to break

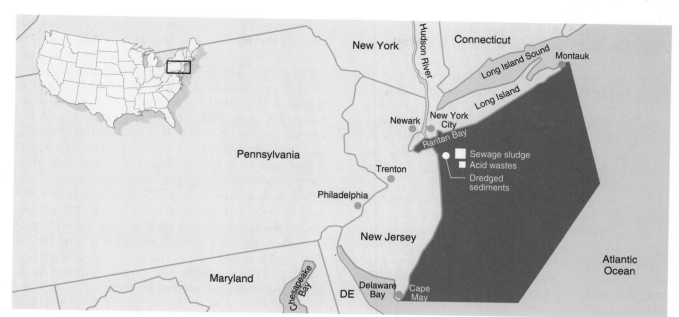

FIGURE 6-2: The New York Bight.

down the insecticide DDT almost as soon as it is released to the environment, it may persist in natural systems for 20 to 25 years. Moreover, its breakdown products (called metabolites), DDD and DDE, may persist in the environment for as long as 150 years. Plastics and other synthetic chemicals take hundreds, perhaps thousands, of years to degrade, and some radioactive substances persist in dangerous forms for hundreds of thousands of years. Archaeologists of the future may find artifacts of our culture that are too dangerous to handle. Some of our present-day pollutants will persist for longer periods than human cultures have been active on the earth!

How the Pollutant Enters the Environment

Point sources are identifiable, specific sources from which pollutants are released into the environment. Pipes that discharge sewage into streams and smokestacks that emit smoke and fumes into the air are point sources of pollution (Figure 6-3). Past regulatory efforts focused on eliminating or reducing point sources because they are usually easy to identify.

Nonpoint sources are not localized, but add pollutants from a wide area, such as surrounding agricultural and urban land. Nonpoint sources are much harder to identify than point sources and consequently, much harder to control. Puget Sound, framed by the Olympic and Cascade mountains in the State of Washington, is an ecosystem under siege by nonpoint-source pollution. Runoff from farm fields, pastures, parking lots, and streets contains

large amounts of silts mixed with fertilizers, herbicides, insecticides, animal wastes, oils, fuel additives, solvents, and cleaners. The runoff enters streams and rivers that feed into the sound. Like Chesapeake Bay and San Francisco Bay, which are

FIGURE 6-3: Steel mill in Birmingham, Alabama, 1970. Point sources, like the smokestacks of this steel mill, were the focus of efforts in the 1970s to control air pollution. Scenes like this one have virtually disappeared in the United States.

also greatly affected by nonpoint-source pollution, Puget Sound does not flush, or naturally cleanse itself, very rapidly.

Certain contaminants, called **cross-media pollutants,** can enter an ecosystem by moving from one type of environment to another. They can move from the air into soils or streams, seep from landfills into groundwater, or evaporate from surface water into the air. These pollutants can change their chemical character as they react with the chemicals in the medium into which they move.

Acid precipitation is the best known cross-media pollutant. However, synthetic compounds and their by-products, such as solvents and cleaners, might eventually become our most troublesome cross-media pollutants because they are so widely used. Much of the pollution affecting New Bedford Harbor (also known as Buzzards Bay) near Cape Cod in Massachusetts has entered the ecosystem as cross-media pollutants. Researchers at Woods Hole Oceanographic Institute report that, since 1940, 2 million pounds (900,000 kilograms) of PCBs have entered the harbor system from leaky landfills, illegal hazardous waste dumps, and unregulated industrial discharges to the air.

Effect of the Pollutant

Acute effects occur immediately upon or shortly after the introduction of a pollutant, and they are readily detected. For example, in September 1985, 50,000 gallons (189,500 liters) of sulfuric acid flowed into the Medina River from a train that derailed near San Antonio, Texas. The Texas Parks and Wildlife Department determined that 15,000 fish were killed in a 22-mile corridor downstream from the derailment site. In assessing the damage, the State of Texas billed the polluter for the cost of replacing all of the different species of fish, the cost of investigation and cleanup after the spill, and the lost recreation time to fishers.

Chronic, or long-term, pollution has a more insidious effect on the environment. Often changes are not noticed for several years or decades after the introduction of the pollutant. This delay, known as lag time, makes it particularly difficult to predict what effect the pollutant may have at the ecosystem level.

Tributyltin (TBT) illustrates the difficulties posed by chronic pollution. TBT is an effective antifouling compound, a substance that prevents the growth of aquatic organisms on boat hulls and dock pilings. TBT is used in paints to protect ships and docks from damage by organisms such as algae and barnacles. It is also possibly the most toxic compound ever deliberately introduced into aquatic ecosystems.

The first TBT paints became commercially available in the mid-1960s, and TBT compounds can now be found in a variety of plastics, oils, and paints. In the late 1970s researchers began to discover TBT compounds in marine sediments and oyster shells. TBT-caused deformities and mortalities in nontarget organisms, such as oysters, led both the French and British to restrict its use in the 1980s. By the late 1980s TBT was detected in some Great Lakes marina waters and in Lake St. Clair, near Detroit, Michigan.

What should be done? TBT compounds can save millions of dollars in maintenance costs on boats and docks. Some researchers are searching for safe levels of TBT in paints. Others, calling for a total ban on TBT marine products, point out their most serious drawback: the possibility that target organisms might become immune to present levels of TBT in paints. This would require producing paints that have higher TBT concentrations; such paints could pose an even greater threat to nontarget organisms.

The effects of a pollutant are sometimes the result of interactions with other substances. As pollutants enter and move through ecosystems, they may interact and form new combinations of chemicals that are more harmful than the separate components. Such pollutants are said to have a **synergistic effect,** that is, their combined effect is greater or more harmful than the sum of their individual effects. Air contains a mix of thousands of chemicals. Some are believed to be safe; others are known to be toxic. For example, when hydrocarbons and nitrous oxides mix in the atmosphere and are exposed to sunlight, interactions called **synergisms** can occur that produce photochemical smog containing new, more harmful compounds. Synergistic interactions may occur in aquatic or terrestrial habitats as well. For example, chlorine gas, which can be dangerous at certain levels, is used in small amounts to purify water. It may react with organic substances sometimes found in water to form chloramines, substances that may be toxic or carcinogenic. Synergistic interactions can also occur in decomposing landfills, resulting in new, potentially dangerous compounds that may then appear in groundwater.

Bioaccumulation is the storage of chemicals in an organism in higher concentrations than are normally found in the environment. For example, nitrogen and phosphorus are usually found in lower concentrations in aquatic environments than are needed by phytoplankton. In order to obtain sufficient amounts for growth and reproduction, phytoplankton actively transport these nutrients through their cell membranes. As they do, persistent pollutants, such as DDT and PCBs, also enter through the cell membranes. Like other persistent chemicals, they are lipid soluble. Because they dissolve in lipids, they tend to accumulate in lipid materials of the phyto-plankton. If they were water soluble, the chemicals would be flushed from the cells and excreted. The persistent chemicals may be present in the environment of the phytoplankton in small quantities, but biological activity concentrates them and they accumulate in the organisms. Another example of bioaccumulation is the amount of TBT found in Pacific oysters. Concentrations of TBT 2,000 to 20,000 times higher than found in the environment have been found in the oysters.

Biomagnification is the accumulation of chemicals in organisms in increasingly higher concentrations at successive trophic levels. Consumers at higher trophic levels ingest a significant number of individuals at lower levels, along with the persistent pollutants stored in their tissues. Biomagnification occurs in both aquatic and terrestrial habitats (Figure 6-4). Carnivores at the top of food chains, such as large fish and fish-eating birds (gulls, eagles, pelicans) in aquatic habitats or robins in terrestrial habitats, may accumulate poisons in concentrations high

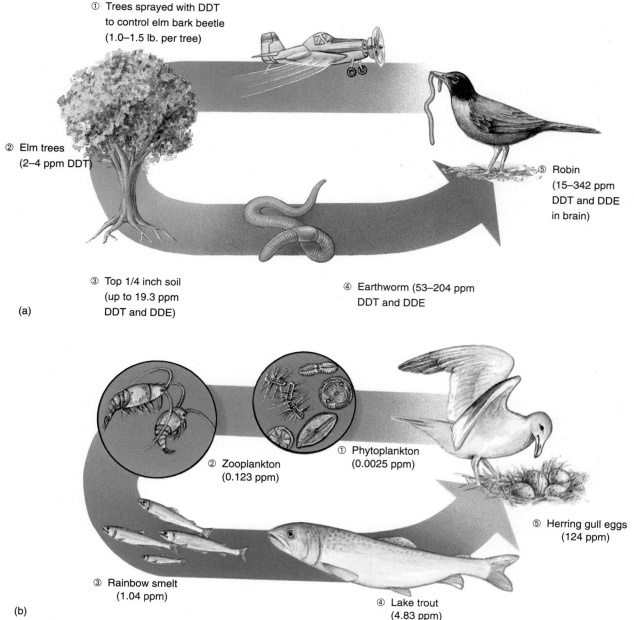

① Trees sprayed with DDT to control elm bark beetle (1.0–1.5 lb. per tree)

② Elm trees (2–4 ppm DDT)

③ Top 1/4 inch soil (up to 19.3 ppm DDT and DDE)

④ Earthworm (53–204 ppm DDT and DDE)

⑤ Robin (15–342 ppm DDT and DDE in brain)

(a)

② Zooplankton (0.123 ppm)

① Phytoplankton (0.0025 ppm)

③ Rainbow smelt (1.04 ppm)

④ Lake trout (4.83 ppm)

⑤ Herring gull eggs (124 ppm)

(b)

FIGURE 6-4: Biomagnification in terrestrial and aquatic habitats. Arrows show the movement of the substances through the food chain. (a) DDT was sprayed on trees to control the elm bark beetle, the insect responsible for the spread of Dutch elm disease. The insecticide entered the soil, where earthworms accumulated it in their tissues. The concentration of DDT was biomagnified in the tissues of earthworm-eating robins. (b) PCBs dumped into the Great Lakes move through the food chain. The concentration of PCBs increases with each step in the chain, and the highest levels are reached in the eggs of fish-eating birds such as herring gulls.

enough to prevent their eggs from hatching or to cause deformities or death. The concentrations of poisons in predators (nontarget organisms) can be a million times higher than the concentration of the chemical in the water or soil.

Synergism, bioaccumulation, and biomagnification make it more difficult to estimate or assess the effect of a particular pollutant and its long-range cost—both the cost to the environment and the cost of cleanup or removal.

Time It Takes the Ecosystem to Remove the Pollutant

After a pollutant enters the environment, it takes an ecosystem a certain length of time to remove it, depending on the amount of the pollutant and its persistence. Some are removed or recycled immediately, some take decades, while still others could take hundreds, thousands, or tens of thousands of years to be removed. No single human lifetime is long enough to witness a reversal in the degradation of natural systems caused by these long-term pollutants. We must have the vision to realize that the costs incurred for pollution control now will reap benefits in the future.

What Is Ecosystem Disruption?

An ecosystem disruption is a rapid change in the species composition of a community that can be traced directly to a specific human activity. (Although natural disruptions can and do occur, this discussion is limited to culturally induced disruptions.) Changes in species composition have been linked to the use of persistent pesticides, the introduction of new species (usually accidentally) into an area, construction such as dams or highways, and overexploitation of resources.

In 1955 the World Health Organization (WHO), in a massive campaign to reduce the incidence of malaria on North Borneo, sprayed dieldrin (similar to DDT) to kill disease-carrying mosquitoes. The campaign was a success; the dieldrin killed most of the mosquitoes, and malaria was nearly eliminated from the island. But the pesticide also killed many other insects. Small lizards that fed on the dead insects died as well. Soon, cats that fed on the lizards began dying. Before long the rodent population had increased significantly, since the population of their natural predator, the cats, was sharply reduced. Rats began to overrun the villages, and the people faced a new threat: sylvatic plague, carried by fleas on the rats. The situation was brought under control when the British air force, on instructions from the WHO, parachuted cats into Borneo.

The episode caused the villagers other troubles as well. The population of a particular type of caterpillar, which either avoided or was unaffected by the dieldrin, rose sharply when the populations of wasps and other insects, which were its natural predators, decreased. Caterpillar larvae ate their way through one of their favorite foods—the leaves that formed the thatched roofs of the villagers' homes.

In many areas where DDT has been sprayed extensively to control mosquitoes, populations of DDT-resistant mosquitoes have developed. How does this happen? Diversity or variation in genetic structure normally exists in living organisms. Some mosquitoes were naturally resistant to the levels of DDT sprayed to control their population numbers. Resistant mosquitoes were more likely to survive and breed with other resistant mosquitoes and produce entire populations of mosquitoes resistant to a certain level of DDT. When levels of DDT were increased in the sprays, the same thing happened again (an example of a positive feedback), until it became too dangerous to increase the insecticide level any further. In some parts of the world "super" mosquitoes now exist that are resistant to high levels of many types of insecticides. In fact, after years of using chemicals in an attempt to control pests, not one pest has been eradicated. Instead, resistant populations of many human, animal, and plant pests or disease-causing microorganisms now exist. These stronger pest strains and disease organisms have caused tremendous human misery throughout the world. Rats, cockroaches, fire ants, fleas, staphylococcus bacteria, streptococcus bacteria, and the bacteria that cause syphilis and gonorrhea all show increasing resistance to our control efforts.

When a species is introduced to an area in which it is not native, it rarely has natural predators that can keep its population under control. Because the Japanese beetle, accidentally introduced to the United States in 1911, had no natural predators, its population grew wildly. It continues to cause extensive damage to vegetation throughout many parts of the United States. But it is just one of a long list of nonnative species that have become, or may become, established in this country. Four excellent examples of the potential disruption to local ecosystems are the entry into Latin America and the southern region of Texas of Africanized bees, popularly (but erroneously) called "killer" bees; the accidental establishment of an eastern blue crab colony in San Francisco Bay; the escape from gardens of the plant kudzu in the southeastern United States; and the spread of the plant purple loosestrife in the wetlands and marshes of the Great Lakes and midwestern states (Figure 6-5). In each case the introduced species competes more successfully than native species that occupy the same niche, resulting in the displacement of the native species.

What Is Ecosystem Destruction?

Human activities can and do disrupt the species composition of ecosystems, but more often human activities destroy ecosystems. *Ecosystem destruction is the replacement of a natural system by a human system.* The functions performed by the natural system, however, are not replaced.

Urbanization, transportation (paving and the land required for freeways), and agriculture are the major causes of destruction. As the human population increases, we continue to take habitat away from other species at an accelerating rate. Habitat loss is the primary cause of species extinction—extinction that is conservatively estimated to be as high as 1,000 plant and animal species each year, or nearly 3 extinctions each day.

Each year farmland is lost to housing developments and other nonagricultural uses or becomes less productive due to erosion. At present rates nearly 20 percent of the world's arable land will be unavailable for productive use by the year 2000. As prime cropland is lost to urban and industrial uses, areas that are less suitable for agriculture, for example, wetlands and forests, are being lost to the plow. Over half of the original wetlands in the contiguous 48 states have been drained and developed since colonial times. The Congressional Office of Technology Assessment reported in 1986 that four-fifths of the 550,000 acres (222,581 hectares) of wetlands lost in the United States annually are converted for agricultural use. According to the Fish and Wildlife Service, more than 11 million acres (about 4.5 million hectares) of freshwater wetlands have been converted to croplands and urban projects since the 1950s. Agricultural and urban conversion simplifies ecosystems and makes them highly susceptible to disturbances caused by insect pests, pollution, or soil erosion. The draining of coastal wetlands (saltwater marshes, mangrove forests, and estuaries) seriously affects marine species that use these areas as nurseries, which in turn damages the fishing industry. One encouraging sign is that the rate of wetland loss has slowed in recent years because of increased regulations at both state and federal levels.

Coral reefs, especially those located near development projects in the Caribbean Sea, the Philippines, and Japan, are under constant threat of destruction. Reefs that took millions of years to form and grow are being destroyed, often in a matter of seconds by dynamiting. Major causes of ongoing destruction are dredging, sedimentation, oil spills, sewage, toxic pollutants, warm water emitted from power plants, and the collection of corals for sale to tourists, collectors, and jewelers. Reefs are not isolated ecosystems; they are important to the overall stability and productivity of marine systems. Reefs are connected to the grasslands and mangrove swamps along the

FIGURE 6-5: Beekeeper surrounded by Africanized bees. Popularly called "killer bees," African bees were introduced into Brazil to breed with local bees. The Africanized bees have spread northward through South and Central America and have been sighted in southern Texas.

Human constructions sometimes impede or prevent the migrations and movements of species. Dams block fish that are migrating to spawning grounds, oil and gas exploration interferes with the seasonal migrations of caribou and elk, cooling water intakes for electric-generating power plants suck in millions of planktonic larval fish and shrimp, and highway constructions interfere with animals' movements and complicate courtship or breeding behaviors.

Overexploiting resources alters the species composition of the community and can seriously hamper the functioning of ecosystems. Overgrazing, for example, promotes the spread of undesirable weed species over range grasses. Environmental Science in Action: The Blue, Gray, and Humpback Whales (pages 102–104) examines the possibility that several species of whales may soon become extinct because of human activity. Driving a species to extinction disrupts natural systems. Forests that are logged are often replaced with pine plantations or tree farms, stands consisting of one or two commercially valuable species. Overharvested areas no longer benefit from the basic functions that forests perform, including air and water cleansing, local climate moderation, and erosion control. This is especially true when **riparian** zones, vegetation zones near streams and rivers, are not maintained during timber harvests or when steep slopes are harvested, promoting rapid erosion.

When species disappear or when drastic population changes occur within a community, other species, and the ecosystem in general, are likely to be affected as well. Hence, ecosystem disruption and decreasing diversity form a vicious cycle.

Environmental Science in Action:
The Blue, Gray, and Humpback Whales
Pam Marsh and Amy Franz

The great whales are some of the most magnificent creatures on the planet. We know relatively little about them, yet we do know they are capable of communication and possess a large and complex brain. We also know they are important members of the biotic community. Recent work shows that gray whale carcasses in the deep sea act as "stepping stones," allowing organisms from hydrothermal vent communities to colonize new vent sites, which are often widely spaced. We know, too, that whales are ancient inhabitants of earth, having first lived on land and then having evolved to adapt to life in the ocean approximately 60 million years ago, long before the appearance of humans. And we know their chief predator is *Homo sapiens* (orcas, or killer whales, are the only other animal to prey on whales). If we fail to preserve these ancient and impressive creatures, whose fate we hold in our hands, can there be hope for other less spectacular species in the oceans or upon the land?

Describing the Resource

Two groups comprise the biological order Cetacea, toothed whales (Odontoceti) and baleen whales (Mysticeti). There are approximately 65 species of toothed whales, including sperm whales, porpoises, and dolphins. All are fast-swimming hunters, using teeth and jaws to capture and grasp prey, typically fish and squid. The baleen whales are all relatively large and include six major species. In order of increasing size, these are the gray, humpback, right, bowhead, fin, and blue (Table 6-1). These whales have horny plates of baleen, or whale-bone, hanging from their upper jaw. They use the baleen plates to strain food, particularly tiny shrimplike crustaceans known as krill and small fish, from the water. Baleen whales have been hunted more extensively than toothed whales, since their great size yields larger quantities of blubber and meat.

Biological Boundaries

Blue and humpback whales are found in all oceans of the world, and gray whales, known as California gray whales, are found only in the North Pacific. The blue whale takes its name from its characteristic slate blue color. It is also known as the sulphur bottom whale, a reference to the yellowish film of diatoms that develops on its underside. Its heart is the size of a compact car. The largest blue whale ever recorded was 110 feet 2½ inches long, over one-third the length of a football field. The gray whale is actually black. Its mottled gray appearance is caused by white parasitic barnacles that attach to its skin. The humpback takes its name from the way in which it dives, rounding and exposing its back and dorsal fin. Its flippers are flexible and very large, often one-third the size of its body. The humpback uses them for balance and propulsion. Courting pairs use their flippers to embrace and caress one another, and cows often gently pat or caress their calves with their flippers (Figure 6-6).

Blue whales feed almost exclusively on krill, but occasionally feed on somewhat larger organisms, such as small fish. Gray whales feed on krill, plankton, and small, bottom-dwelling organisms. The diet of humpbacks consists of krill and small fish, such as sand lances, anchovies, and sardines.

Blues and humpbacks typically feed in several ways. The whale may swim slowly with its mouth open, straining huge numbers of krill from the water, or it may take large gulps of water and then expel the water, leaving behind krill and other organisms. The whale then uses its tongue to scrape the food from the baleen. Humpbacks have devised another method for obtaining food. A group of seven or more individuals will dive in unison. Then, they swim toward the surface in a spiral fashion, forming a bubble net that concentrates a group of small fish or krill. Mouths open, the humpbacks rise through the concentrated prey.

The gray, unlike other baleen whales, feeds on bottom-dwelling organisms. Using its short, stiff baleen plates, the gray plows through bottom sediments, sifting crustaceans, mollusks, bristle worms, and other animals.

Physical Boundaries

Because whales are warm-blooded mammals, they feed in colder waters only during the summer months and migrate to warmer waters with the coming of autumn. There, they breed and give birth to their young.

Each autumn gray whales leave their summer feeding grounds in the Bering and Chukchi seas and migrate southward. In the warm waters of sheltered bays and lagoons of Baja California they court and mate, and cows give birth to the young who were conceived the previous breeding season. From late January to April the whales again migrate, heading northward to their Arctic feeding grounds. They feed

▶ Table 6-1 Characteristics of the Blue, Gray, and Humpback Whales

Species	Size (feet)	Weight (tons)	Where Found	Estimated Current Number
Blue	80–85	100	All oceans	11,700
Gray	39–45	35	North Pacific	17,000
Humpback	50–55	60	All oceans	10,275

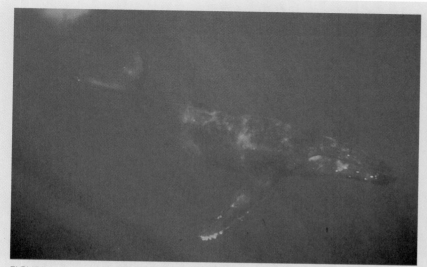

FIGURE 6-6: Humpback whale. The humpback, perhaps the most easily recognizable of the great whales, is a "flagship species" for conservation. Because it is so widely known and appreciated, the humpback draws public attention to the plight of endangered wildlife.

in colder water during the summer because of the incredible productivity of the Arctic waters during the long summer days.

Because blue and humpback whales are found in all oceans, the migratory routes they follow depend on the stock to which they belong. A stock is a group of whales of the same species that live in the same general location. Stocks are composed of pods, or groups of whales that migrate together. Blue whale pods typically include two or three individuals, gray pods as many as six, and humpback pods three or four.

The routes followed by migrating whales are important to animal management for several reasons. Their strict migrational patterns enable hunters to determine easily the times and places certain whales can be found and harvested. Also, because they migrate so extensively, it is difficult to manage a whale stock. One country may protect the stock within its national waters, but it cannot protect the whales after they have migrated elsewhere. Clearly, managing whales requires protecting them throughout their range. If whale populations are to be restored to healthy levels, inter-national cooperation is necessary to protect them from hunters, water pollution, and interferences, such as whale-watching groups, during calving season.

Social Boundaries

Human social systems exert both direct and indirect pressures on whale populations. Historically, direct pressures, in the form of whale harvesting and exploitation, have been of greater significance. In the future, indirect pressures, including habitat degradation and the depletion of their food sources, may pose the chief threat to the great whales' survival.

Although many species of whales, including the blue, gray, and humpback, are protected, illegal harvesting continues. Traditionally, whales were hunted to provide a variety of products, including oil (used to make margarine and soap, as cooking oil, as a lubricant, and to tan high-quality leather), baleen and bone (used in such products as umbrellas, corset stays, fishing rods, nets, brushes, and scrimshaw, the etching of pictures into bone) and meat and bone (in addition to human consumption of meat by the Japanese, Northern Europeans, and Aleutian Eskimos, other uses include dogfood, fertilizer, and livestock feed).

Today, cheaper, often synthetic, products are now used in place of the oil products once derived from whales. The market for baleen is limited, though it is still used occasionally in brushes and, in the Far East, in novelty products such as shoehorns and tea trays. And the market for meat is small.

Japan harvests more whales than other remaining whaling nations (chiefly Iceland), though its operations are restricted to the minke whale, according to the Institute of Cetacean Research in Tokyo. Its persistent whaling operations can be traced to its cultural heritage. Whale meat was once an important element in the Japanese diet, although statistics show that now less than 6 percent of the population of Japan eats whale meat. The Japanese also use the oil, whalebone, skin, and various internal organs to produce a variety of objects, including sandals, shoelaces, and cosmetics.

The direct exploitation of whales has slowed as other products are increasingly used in place of those obtained from whales. Indirect pressures are becoming more significant on whale populations. One of the most important indirect pressures is water pollution, which is certain to become of greater concern as long as the practice of ocean dumping of wastes continues. Little is known about the effects on the ocean biota of dumping wastes, including hazardous or toxic wastes. Another potential threat is the depletion of their food source through the harvesting of krill and anchovies for human consumption. The former Soviet Union, which employed large harvesting and processing ships, experimented with making krill palatable to humans.

Looking Back

Until the mid-1800s whalers used small wooden boats, hand-held harpoons, and nets to pursue slow-swimming whales and those that stayed close to shore—the right, bowhead, sperm, humpback, and gray whales. In the late 1800s the harpoon gun made whaling easier and enabled humans to hunt faster-swimming species. Whaling became even more profitable with the introduction, in 1925, of the pelagic (ocean) factory ship, which allowed whalers to process and store their catch aboard ship. Because they could remain at sea for longer periods of time, hunters took increased numbers of whales.

Few attempts were made to regulate whaling until 1930, when Norway formed the Bureau of International Whaling Statistics in order to investigate stocks and populations. In 1931 the International Convention for the Regulation of Whaling was established, making it illegal to hunt right whales or females of any species with calves. The convention proved ineffective, so a new regulatory system, the International Whaling Convention, was established in 1937. It prohibited the hunting of gray whales, defined a whaling season for the Southern Hemisphere, and set catch size limits for certain species, including the humpback. In 1938 specific humpback stocks were protected from whaling.

In 1946 the International Whaling Commission (IWC) replaced the International Whaling Convention. The IWC has since been the basis for the international regulation of all whale species. About 90 percent of the world's catch of whales is taken by member nations. The original objective of the IWC was to regulate whale stocks in the best interest of the whaling industry, which realized that its continued profitability depended upon healthy stocks.

Public opposition in the United States led to the cessation of this country's whaling operations in the early 1970s. Riding the wave of popular sentiment, Congress passed several pieces of legislation concerning whales and other endangered animals.

The United States was not alone in its enthusiasm for conservation during the environmental movement of the 1970s. Many other countries and organizations began to advocate the protection of endangered species. The Convention on International Trade in Endangered Species of Wild Fauna and Flora (CITES) took effect in 1975. The result of nearly ten years of effort by the International Union for the Conservation of Nature and Natural Resources, CITES established rules for wildlife trade. Unfortunately, a compliance loophole allows a country to enter a "reservation" on a species, thus notifying other countries that it does not intend to comply with the trade restrictions on that species. For example, Japan submitted a reservation on the fin whale.

In 1982 the IWC approved, by a vote of 25 to 7, a 10-year moratorium on commercial whaling, effective 1986. Japan, Norway, Iceland, Peru, Brazil, the Soviet Union, and South Korea cast dissenting votes and filed formal objections. (Peru withdrew its objection the following year, and the Soviet Union and Norway later ceased commercial whaling.) Since its adoption, however, the moratorium has been the focus of intense debate. Although the IWC voted in 1990 to keep the moratorium in place for at least another year, its long-term success is in doubt because the main whaling nations—Japan and Iceland—have not agreed to comply with it. Moreover, as with CITES, a country can enter a reservation on a particular species. IWC members may also continue to harvest whales for scientific research, as do Japan, Norway, and the former Soviet Union, a loophole that some whale biologists, conservation groups, and environmentalists fear is being exploited for nonscientific purposes.

Allowing countries to file objections and enter reservations is a major weakness of the IWC; it undermines many of the commission's efforts because countries can ignore any regulation with which

they disagree. Because the commission is an international agency, it lacks the authority to enforce recommendations upon a sovereign nation without that nation's consent.

Looking Ahead

Many scientists maintain that the IWC has developed effective and sound management plans for whale stocks. Others, particularly scientists from whaling nations, argue that further research, especially the estimation of population numbers, is needed before effective plans can be proposed. But even the best management plan cannot safeguard a resource that has been depleted beyond its ability to recover. Unless the moratorium is successful and whale stocks begin to rebound, the question of their management may be moot by the end of this century.

Because economic incentives are the basis of whaling operations, economic disincentives may be the most effective means to end whaling. Without economic pressure applied by nonwhaling countries, whaling nations will continue to ignore the IWC moratorium. Even if enough economic pressure is applied to whaling nations to ensure the strict observation of the 10-year moratorium, there is no guarantee that all whale stocks will be able to recover. However, the moratorium offers all stocks the best chance to recover from the pressure of whaling.

Before we can safeguard whale stocks from indirect pressures, such as habitat degradation, intensive research must be conducted to determine the effects of various human activities. There is some suspicion, for example, that the Exxon *Valdez* oil spill may have affected the gray whale populations. We know relatively little about the whales, especially about their migratory routes and calving areas, so it is difficult to determine which areas need protection and how large an area is required to safeguard the routes and calving grounds.

shore, the rivers of the land, and the open ocean. For example, pollution in the form of pesticides, heavy metals, and silts from the Orinoco and Amazon rivers end up in the Caribbean Sea and Atlantic Ocean, respectively, and can affect fish and coral organisms. In addition, fish that spawn in the marshes and mangroves live as adults in the coral reefs. At present, although 105 countries have coral reefs, only 12 countries have established reef reserves, protecting a total of just 40 reefs.

What Is Desertification?

Desertification is the expansion into or creation of desertlike conditions in areas where such conditions do not naturally occur through overgrazing, overcultivation, deforestation, and poor irrigation practices. It affects primarily agricultural lands and grasslands and is increasing at an alarming rate.

Desertification is a problem worldwide. According to the Worldwatch Institute, some 15 million acres (over 6 million hectares), an area the size of West Virginia, become desertified annually (Figure

6-7). The United Nations Environment Program estimates that over one-third of the earth's land surface is threatened by desertification. In the western and southwestern United States, the threat of desertification comes mainly from intense grazing by cattle and sheep and improper irrigation of crops. Long-term irrigation increases the salinity and alkalinity of soils. Fresh water contains low levels of salts and minerals, which are left behind in the soil when the irrigation water evaporates.

Intense grazing of grasslands is not necessarily harmful. Wandering herds of animals, such as the American buffalo before the European settlement of the American West, usually grazed an area intensely and then moved on, allowing the grassland to recover. Some studies have shown that short, intense periods of grazing are actually beneficial. However, constant pressure by cattle and sheep, as now occurs on marginal lands throughout much of the southwestern United States, has contributed greatly to the spread of deserts in these areas.

In some nations desertification results from the use of marginally productive lands. The landless poor are forced by political, economic, and ethnic

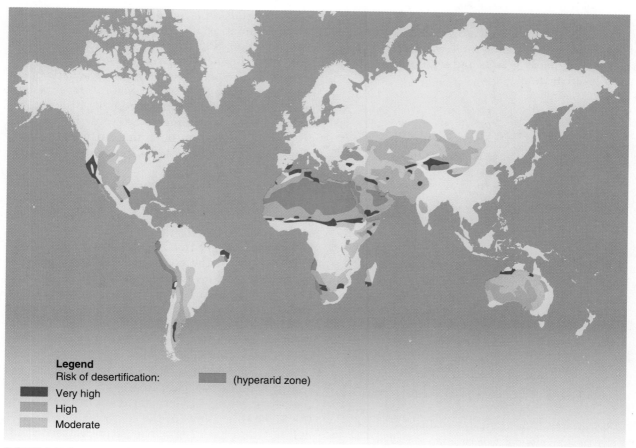

Legend
Risk of desertification:
- Very high
- High
- Moderate
- (hyperarid zone)

FIGURE 6-7: Deserts and areas at risk of desertification.

pressures to cultivate and graze herds on such lands. These areas, although semiarid or arid, are not dry (hyperarid) desert areas. They are subjected to periodic droughts, and the ecosystems there have evolved mechanisms to cope with periodic drying. Native grasses and the animals that graze on them are in balance. However, each year 29.6 million acres (12 million hectares) of marginal land used intensively for crops and rangeland deteriorate and become desertlike. Topsoil erodes because of overplowing and overgrazing. During the past 50 years 160.6 million acres (65 million hectares) in the southern fringes of the Sahara Desert alone have been converted to desert (Figure 6-8).

For the desperate people trying to live in these marginal areas, famine and misery are the rule. Their plight was brought to the world's attention by the Sahel disaster in the 1970s and the Ethiopian disaster in the 1980s. In both instances periods of better-than-average rainfall were followed by severe overcultivation and overgrazing on marginal lands. The overused lands were then subjected to normal, periodic but severe drought, and the grasslands were destroyed and new deserts formed.

Deserts expand slowly, but once expansion occurs, rehabilitation is costly and usually impossible. Instead, desertification can be halted through preventive measures: encouraging the full use of good cropland, using nondestructive grazing techniques, planting hardy grasses and windbreak trees, and using less irrigation.

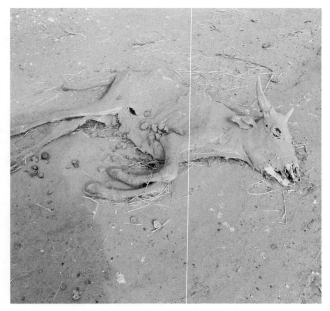

FIGURE 6-8: Drought-ravaged Senegal in the mid-1980s. Desertification threatens once productive semi-arid regions, and wildlife, livestock, and humans suffer.

What Is Deforestation?

Deforestation is the cutting down and clearing away of forests. According to the Global 2000 Report to the President, a government study commissioned by former President Jimmy Carter to predict the state of the biosphere in the year 2000, deforestation is the most serious problem confronting the global environment. Worldwide, the primary cause of deforestation is conversion to agricultural use. Other causes include conversion to pastureland; demand for fuel, timber, and paper products; and construction of roadways. In areas like rural India, Nepal, Haiti, and much of Africa and Asia, where natural gas, oil, and electricity are not available or are too expensive to use, women and children spend most of the day searching for fuel wood. Because wood and charcoal for cooking and heating are becoming scarce and expensive, people cut or gather any available trees, branches, and fallen leaves.

All forests, from the tropical rain forest to the temperate woodland to the boreal coniferous forest, perform many beneficial functions. Often called the lungs of the planet, forests help to regulate the carbon, nitrogen, and oxygen cycles. They help to regulate temperature and rainfall. The root systems of trees hold soil in place, preventing it from eroding and accumulating in streams and lakes, a process called **siltation**. Siltation impedes the flow of water. Moreover, the increased amount of sediments suspended in the water adversely affects aquatic organisms: the muddy or turbid water prevents light penetration to phototrophs, reducing primary productivity, and the sediments can settle on fish eggs, preventing them from obtaining sufficient oxygen. Thus, forests help to ensure a steady supply of clean air and water.

Deforested hillsides are vulnerable to the erosive action of rains and spring thaws. Rain and melting snow rushing down hillsides erode the soil and can cause flashfloods in the valleys below. Hence, forests also prevent flooding. Moreover, forest habitats are vital to the continued existence of many species. Forests are huge storehouses of biomass. They produce biomass quickly and maintain a great diversity of organisms; they also produce humus, thus building up the soil. Many forests are cathedral-like, serene, majestic, and beautiful; most of these are also very old (Figure 6-9).

While it is true that forest death and decline as a result of pollution and disease are showing up in temperate forests in Germany, eastern Europe, and the higher elevations of the eastern United States, deforestation has been particularly controversial in the rain forests of tropical regions and in the ancient forests of the Pacific Northwest of the United States.

FIGURE 6-9: Old-growth forest in the Pacific Northwest of the United States. Deforestation threatens the majestic old-growth forests of Douglas fir and hemlock found in the Pacific Northwest and Alaska.

Tropical Rain Forests

It is difficult to determine the extent of deforestation worldwide, and figures vary widely. Many are based on a 1980 study by the Food and Agricultural Organization, which estimated that 28 million acres (11.4 million hectares) were deforested each year. However, in a 1990 report the World Resources Institute indicated that the loss might be much higher—as much as 50.4 million acres (20.4 million hectares) annually. Researchers estimate that from 50 to 100 acres (20 to 40.5 hectares) are lost each minute. Such activity could reduce our tropical forests by half by the year 2050 (Figure 6-10). Tropical forests help to moderate the climate of the entire planet. Scientists cannot be sure how or to what degree the earth's weather patterns and overall temperatures will be affected if we continue to destroy them.

The regions of the vast rain forests of the Amazon, Congo, and Southeast Asia, which lie between the Tropic of Cancer and the Tropic of Capricorn,

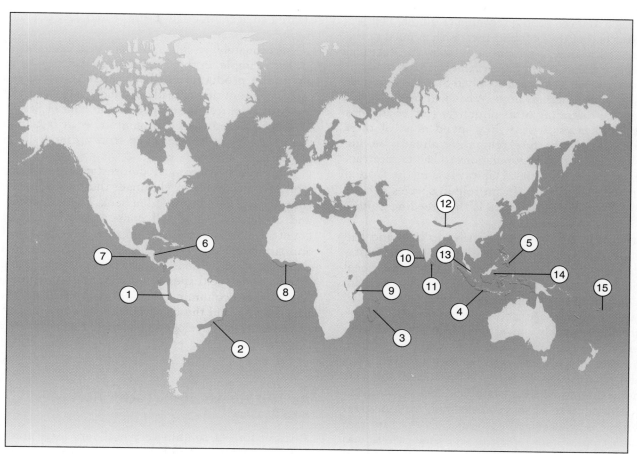

FIGURE 6-10: Deforestation in tropical rain forests. Tropical regions at greatest risk from deforestation are the Andes (1), the Atlantic forest of Brazil (2), Madagascar (3), Indonesia (4), and the Philippines (5). Other areas at risk are Central American lowland forests (6), Central American highland forests (7), Upper Guinean forests (8), the Eastern Arc Mountains of Tanzania (9), the Western Ghats of India (10), the Sinharaja Forest of Sri Lanka (11), the eastern Himalayas (12), peninsular Malaysia (13), northern Borneo (14), and Melanesia (15).

contain some of the least industrialized countries with some of the highest birth rates. Over 2 billion people live in these regions, in countries in which population could double in 30 to 35 years. The need to clothe, feed, house, and provide fuel for growing populations and the economic need to export wood products and beef to developed countries contribute to massive deforestation.

Researchers estimate that tropical forests contain from 3 to 30 million species of plants, animals, and microorganisms—few of which have been discovered, classified, named, or studied. Over 200 different tree species can be found in a single hectare (about 2.5 acres) of the Amazon rain forest. Compare this to the 10 or so usually found in a temperate forest. That same hectare of tropical forest is home to a bewildering array of insects, spiders, birds, and mammals. Half of all known species (and probably a much higher proportion of all species, known and unknown) live in tropical rain forests, yet these forests account for just 7 percent of the earth's land mass. Through deforestation, we face a huge loss of species diversity before many of these species have even been discovered (Figure 6-11).

Over half of the 650 bird species that breed in the United States migrate to and from the tropical forests of Central and South America. For many of them the time spent in the United States is minimal, yet we claim them as ours. What will be the result when the migrating birds no longer find conditions suitable in areas where they spend most of their adult lives? Some researchers are already linking declines in "our" bird populations to habitat destruction in tropical forests. Habitat loss in the United States may play a significant role in species decline as well, for North America is the breeding ground for many of these birds.

Ironically, although conversion to cropland and pastureland is the major cause of tropical deforestation, many of the soils do not respond to traditional agriculture and cattle ranching. Most of the soils are shallow because 75 percent of the nutrients are tied up in the lush foliage, above ground. Leaves or trees that fall are quickly converted to nutrients and recycled back to the foliage. Once an area is deforested, the exposed soil may be productive for a few years and then lose its fertility. Also, when the tree canopy is removed to create a field, the heavy tropical rains tend to wash away the soil, even in areas where it is relatively thick. Much of the rain that falls in tropical rain forests is generated by evaporation and by transpiration from the foliage itself; if large tracts of forest are removed, rainfall in the area may decrease markedly (Figure 6-12). The clearing of large tracts is preceded by the development of logging roads, which penetrate deep into the rain

FIGURE 6-11: Tropical deforestation. Conversion to cropland or pastureland is one of the leading causes of deforestation in Brazil and other tropical countries.

forest, allowing easier access to forest lands. The land comes under intense pressure as trees are cleared to accommodate farming and cattle ranching.

Unfortunately, raising cattle has little local benefit. For example, in the Central American tropics, as per capita production of beef cattle goes up, per capita consumption actually declines. Why? Most of the beef produced is shipped to the United States, where it is used as cheap hamburger in fast food chains, mixes in hot dogs and frozen dinners, and pet food.

The causes of tropical deforestation—and the strategies necessary to eliminate those causes—are complex and could easily fill a book. Political, social, and economic factors all come into play. Many countries, for example, rely on their export of beef and timber to raise cash to pay for their national debt. But debt burdens are often due, in large part, to oil imports and spending for military purposes. Short-term economic gains, which benefit only a wealthy few, and the disproportionate consumption of resources by developed countries, have also been implicated.

Ancient Forests of the Pacific Northwest and Alaska

Ancient forests contain very old, very large trees, including western hemlock, Sitka spruce, Douglas fir, western red cedar, and coastal redwood, to name a few. They range from 500 to 1,000 years old. Some grow more than 300 feet (90 meters) tall and are more than 10 feet (3 meters) in diameter. These

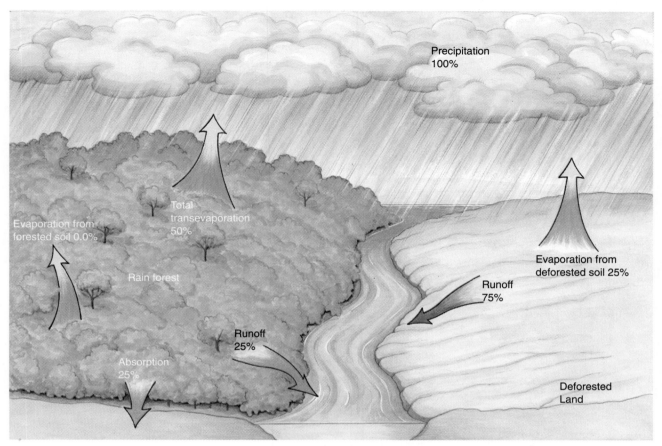

FIGURE 6-12: The cycle of evaporation, transpiration, and precipitation in the tropical rain forest. In an intact forest ecosystem, about half of the precipitation that falls either evaporates from, or is taken up and then transpired by, the vegetation. One-quarter of the precipitation is absorbed by the soil, and another quarter runs off the land into nearby surface waters. When the land is deforested, three-quarters of the precipitation ends up as runoff, carrying sediments from the denuded land into streams, rivers, and oceans. The remainder evaporates from the deforested soil.

majestic forests are home to unique communities of plants and animals.

Once vast, the ancient or old-growth forests of the Pacific Northwest are coming under increasing pressure. Before the arrival of European settlers, old-growth trees covered about 60 percent of the forested areas between the Cascade Mountains and the Pacific Ocean. They stretched 2,000 miles from Alaska to California, covering most of the 100 miles between the mountains and the sea. Today, only about 10 percent of the old-growth forest remains.

Is it necessary to cut ancient forests for timber? Representatives of the forest products industry say yes. Conservationists say no. The U.S. Forest Service and the Bureau of Land Management, both charged with the management of public lands and resources, have developed a plan to allow extensive cutting of most of the remaining old-growth stands. In the mid-1980s, the Forest Service approved a plan to cut two-thirds of the remaining old growth in Alaska's Tongass National Forest! The cost to the taxpay-

er was $253 million, most of which went into building access roads. The return on that "investment" was $2.9 million in actual tree sales. Most of this timber was shipped to Japan.

Conservationists argue that the federal government is subsidizing the forest products industry's destruction of trees that are actually not needed for forest products. The present market does not warrant the huge expenditure to harvest these trees. The forest products industry and the Forest Service maintain that the trees will die and rot if left alone. They claim that the depressed lumber industry and state economies in the Northwest can use the jobs provided by lumbering.

The debate between environmental protection and economics in the Pacific Northwest has developed into a confrontation of monumental proportions. The focal point of the confrontation is the northern spotted owl, a secretive species currently listed as threatened by the U.S. Fish and Wildlife Service and thus protected under the Endangered

FIGURE 6-13: Northern spotted owl. The rare northern spotted owl is officially listed as a threatened species by the U.S. Fish and Wildlife Service. Research has shown that a breeding pair needs 2,000 acres to be successful. If enough breeding pairs can be identified, larger areas can be set aside to protect them and the 160 other species that live only in the unique environments provided by the old-growth forests.

Species Act (Figure 6-13). If the Fish and Wildlife Service upgrades the spotted owl's status to endangered, the Forest Service might have to set aside all areas in which the owl is found. Whether or not the owl receives this increased protection may be left up to the "God Committee," so called because its members hold the power of life or death for endangered species. The committee has the authority to resolve "irreconcilable conflicts" by granting exemptions to the Endangered Species Act if "no reasonable and prudent alternatives exist" or if the project (in this case, lumbering) is of national or regional significance. The committee is composed of the secretaries of Agriculture, Interior, and the Army, the Chairperson of the Council of Economic Advisors, the heads of the Environmental Protection Agency and the National Oceanic and Atmospheric Administration, and a representative of the state affected by the situation (in this case, Oregon).

Summary

Ecosystems become degraded when human activities alter environmental conditions in such a way that they exceed the range of tolerances for one or more organisms in the biotic community. A degraded ecosystem loses some capacity to support the diversity of life forms that are best suited to its particular physical environment. The five D's of ecosystem degradation are damage, disruption, destruction, desertification, and deforestation.

Damage occurs when the integrity of natural systems is altered. Pollution is the most common form of environmental damage. A pollutant is a substance or form of energy, such as heat, that adversely alters the physical, chemical, or biological quality of natural systems or that accumulates in living organisms in amounts that threaten their health or survival.

Five major factors must be considered when determining how a specific pollutant damages natural systems: the quantity of pollutants discharged to the environment, the persistence of the pollutant, how the pollutant enters the environment, the effect of the pollutant, and the time it takes the natural system to remove the pollutant. Some otherwise harmless substances may be discharged to the environment in such large amounts that natural systems are unable to process, or biodegrade, them. Other pollutants may be nonbiodegradable, that is, they cannot be broken down by natural systems, no matter how small the amount of the pollutant. Persistent pollutants accumulate in the environment over time; they include nonbiodegradable pollutants as well as those that degrade slowly. Pollutants enter the environment through point or nonpoint sources. Cross-media pollutants, such as acid precipitation, move from one medium into another. Acute effects occur immediately upon or shortly after the introduction of the pollutant to the system, and they are readily detected. Chronic, or long-term, effects may not be noticed for several years or decades after the introduction of the pollutant, rendering it particularly difficult to predict what effect the pollutant may have on the ecosystem. The interaction of two or more pollutants may have a synergistic effect, that is, their combined effect is greater than the sum of their individual effects. A pollutant that bioaccumulates becomes stored in an organism in concentrations higher than those normally found in the environment. A pollutant that biomagnifies appears in increasingly higher concentrations in organisms at successive trophic levels. Biomagnification occurs as consumers at higher trophic levels ingest a significant number of lower-level individuals that have accumulated persistent pollutants; the ingested, stored pollutants are added to the persistent pollutants accumulated in the fatty tissues of the higher-level consumers.

A rapid change in the species composition of a community that can be traced directly to a specific human activity is known as a disruption. The use of persistent pesticides, the introduction of new species into an area, construction, and overexploitation of resources have all been linked to changes in species composition.

While human activities can and do disrupt the species composition of ecosystems, a more common form of natural system degradation is ecosystem destruction. A natural system is destroyed when it is replaced by a human system. The major causes of destruction are urbanization, transportation, and agriculture. Perhaps the most serious consequences of natural system destruction are habitat loss (the primary cause of species extinction) and the loss of the functions and services provided by natural systems, functions that human systems cannot duplicate.

Desertification and deforestation, two specific types of natural system degradation, are increasing at alarming rates. Overgrazing, overcultivation, deforestation, and poor irrigation practices impoverish semiarid lands, encouraging desertification—the spread of desertlike conditions in areas where such conditions do not naturally occur. Agricultural lands and grasslands are most susceptible to desertification. Deforestation, the cutting down and clearing away of forests, is believed by many people to be the most serious problem confronting the global environment. The primary cause of deforestation worldwide is conversion to agricultural land. The demand for pastureland, fuel, timber, and paper products and the construction of roadways also lead to deforestation.

Key Terms

acute effect

bioaccumulation

biodegradable

biomagnification

chronic effect

cross media pollutant

nonbiodegradable

nonpoint source

point source

pollutant

persistent pollutant

siltation

synergistic effect

Discussion Questions

1. Differentiate among the five D's of environmental degradation and give an example of each.

2. What factors determine how a pollutant damages an ecosystem? How can soil erosion into a stream, nutrient loading to a lake, and the discharge of a toxic synthetic chemical to the atmosphere all be considered pollutants?

3. Scan each issue of your local newspaper for one week. Identify several cases of pollution or potential pollution in your area. Relate each case to the factors that determine how pollutants affect ecosystems.

4. Differentiate between bioaccumulation and biomagnification. How is the concept of trophic levels related to biomagnification?

5. Identify three nonnative animal or plant species that are commonly found in your area. How did their introduction alter the biotic community?

6. Suggest ways in which habitat loss might be reduced or minimized. Include suggestions that are appropriate for your own area and suggestions that are appropriate for such species-rich areas as the tropical rain forests and coral reefs.

7. Some people argue that the decline or extinction of a single species is not an example of ecosystem degradation. What do you think? How might the biotic community of the open ocean be altered if the blue, gray, and humpback whales become extinct?

Applying Ecological Principles

The time has come for science to busy itself with the earth itself. The first step is to reconstruct a sample of what we had to start with.

Aldo Leopold

Restoration ecology is an emerging environmental profession with an important mission. For the first time, ecologists are being called upon not only to enhance knowledge about the structure and process of biotic communities, but also to take an active part in managing them. . . . The road from the discipline of ecology to the profession of restoration ecology carries with it the responsibility to balance detached science with public service.

Thomas Bonnicksen

Learning Objectives

When you finish reading this chapter, you should be able to:

1. Define applied ecology and explain how it differs from ecology.

2. Identify six subdisciplines of applied ecology and give examples to illustrate each.

3. Explain how applied ecologists use computer models in their work. List the advantages and disadvantages of computer modeling.

Environmental management is one of humankind's most difficult tasks and one that we cannot fail to attend to from now on. We cannot continue to degrade our environment without destroying our societies and possibly our species in the process. To manage natural systems intelligently, we must learn how natural systems function, and we must learn how to manage human activities (population growth, resource abuse, and pollution) within the limitations imposed by those systems.

In this chapter we explore some exciting new approaches to managing ecosystems through the application of knowledge about natural systems. Keep in mind, however, that the full potential of applied ecology can be realized only within the context of the social setting. We are slowly coming to realize that there are limits to nature's resiliency.

Any effective plan for environmentally sound management must integrate a knowledge of ecosystems with an awareness of human institutions and cultural considerations. Social factors such as politics, economics, ethics, and education have a profound effect on whether and how environmental problems will be addressed. (These factors are considered throughout the text, particularly in Chapters 25 to 28.)

What Is Applied Ecology?

Applied ecology is a scientific discipline that attempts to predict the ecological consequences of human activities and recommend ways to limit damage to, and to restore, ecosystems. The challenge for applied ecologists is to identify and lessen the effects of human activities. To do so they must understand how natural systems respond to, recover from, and adapt to disturbances. Their research increases our understanding of the natural world and provides a basis for action on environmental problems.

Applied ecology can be divided loosely into six subdisciplines: disturbance ecology, restoration ecology, landscape ecology, agroecology (agricultural ecology) conservation (or preservation) ecology, and ecological toxicology (ecotoxicology).

Disturbance Ecology

All the subdisciplines of applied ecology are concerned in some way with the effect of disturbances on natural systems. **Disturbance ecology,** however,

FIGURE 7-1: Clear-cut forest, Olympic Peninsula, Washington. A disturbance ecologist tries to determine the impact of major stresses such as clear cutting on a forest community.

narrows its area of concern to assessing the impact of particular stresses on particular organisms, populations, and ecosystems. Disturbance ecologists might try to predict the effects of the construction of an oil pipeline on a migratory species, determine the effects of climate warming on fish populations, assess the effects of clear-cutting on a forest, or evaluate the effects of pesticides on agricultural fields (Figure 7-1). They might be asked to provide research findings about the impacts of proposed projects such as housing developments, harbors, recreation areas, and refineries (Figure 7-2). Their research often contributes to the development of environmental impact statements, which are required by federal law for all projects receiving federal funding and by many states for projects covered by state funding.

Restoration Ecology

The three major goals of **restoration ecology** are: (1) to repair biotic communities after a disturbance, to reestablish them on the original site if the communities have been destroyed, or to establish them on other sites if the original sites can no longer be used; (2) to maintain the present diversity of species and

ecosystems by finding ways to preserve biotic communities or to protect them from human disturbances so that they can evolve naturally; and (3) to increase knowledge of biotic communities and restoration techniques that can be applied to restore or manage other systems. Restoration ecology can be used to preserve biotic diversity or to restore or maintain a particular species or ecosystem. It is a practical, hands-on approach to ecology. Under the guidance of a restoration ecologist, people from all walks of life and with varying skills can contribute to projects (Figure 7-3). A variety of ecosystems have been the focus of restoration projects, including Florida's phosphate mines, coal strip mines in Pennsylvania and Illinois, freshwater and marine reefs, grasslands and prairies, and forests. (See Environmental Science in Action, pages 118–120, for a report on the restoration of a prairie community.)

Restoration Techniques. Reefs are an important habitat for many aquatic organisms, both plant and animal. Restoration ecologists have constructed artificial reefs in Lake Erie and along the Atlantic and Pacific coasts to provide new habitat for freshwater and marine organisms. Using materials such as old car frames, bathroom fixtures, and abandoned freighters, restoration ecologists create artificial reefs to replace natural reefs destroyed by pollution or siltation. Oil-drilling rigs make especially good reefs because they extend upward in the water column, thus accommodating a variety of fish and shellfish that live at different water depths.

As with all restoration projects, the builders of artificial reefs must have a good understanding of the natural system they are trying to restore or replicate. For stability, artificial reefs should be built in

FIGURE 7-2: Waterfront condominium development at Presque Isle, Lake Erie. A disturbance ecologist should be consulted before potentially disruptive projects, such as this condominium, are undertaken. Note the attempts made to prevent wave action from damaging the property.

FIGURE 7-3: Workers helping to restore a prairie community in DuPage County, Illinois. A science background is not necessary for people who want to help restore ecosystems, such as this prairie. Under the guidance of a restoration ecologist, committed and hardworking volunteers can play a significant role in restoration efforts.

areas with a firm bottom and moderate currents. The depth at which the reef is placed affects the species that will inhabit it; in general, reefs sunk in deeper waters will attract larger fishes and more types of game fish. Japanese researchers have identified several shapes and sizes of artificial reef that seem to work well. They build the reefs in modular form and sink them in specific locations to provide just the right habitat for commercially desirable fish. To date, the Japanese have committed over $1 billion for basic and applied research on artificial reef construction.

In the United States over 400 artificial reef ecosystems provide habitat for barnacles, mussels, shrimp, sponges, corals, algae, small forage fish, and larger game fish (Figure 7-4). States that have constructed artificial reefs, such as Texas, Georgia, and Florida, have recorded increased numbers of fish and increased numbers of species, thus attracting commercial and sport fishing and realizing an economic return on their investments. Economics is also the driving force in California, where the commercial shellfish industry is interested in using sunken oil platforms to accommodate mussels that grow on the structures' support columns.

A note of caution: Although the technology exists to create artificial reefs, it would be impractical, if not impossible, to try to replace all of the natural reefs currently being destroyed with artificial reefs. Moreover, it has not been scientifically demonstrated that artificial reefs can mimic the variety and complexity of life provided by natural reefs.

The use of fire is another management technique. All stress on ecosystems is not detrimental. Some disturbances may harm or even destroy individual plants and animals, but not permanently disturb the ecosystems; in fact, they may be part of the natural processes that preserve the system. Historically, prairies were periodically burned as a result of lightning or the action of native hunters. In the central and western United States intensive grazing by large herds of migrating buffalo and elk also periodically stressed prairies. Both fires and intermittent, intense grazing patterns removed old vegetation and promoted the new growth of grasses. Prairies prospered because of these periodic stresses.

With the drastic reduction in grazing herds of bison and elk and the control of fire, the grasslands of the American West have suffered. To help restore and maintain grassland ecosystems, trained managers deliberately set prescribed fires. **Prescribed burning** is the use of fire, under appropriate conditions, to achieve specific biological objectives. Managers at the Samuel H. Ordway, Jr., Memorial Prairie in South Dakota periodically burn the prairie to help maintain soil fertility and a variety of prairie grasses (Figure 7-5). Scientists use carefully controlled fires to maintain certain ecosystems in a disturbed or preclimax successional stage or to prevent the encroachment of forest into grassland areas. In Yosemite National Park, for instance, the U.S. Park Service sets prescribed fires, cuts trees, and pulls up conifer seedlings in order to restore and maintain the meadows. Were it not for their efforts, plant succession would transform the meadows—which

FIGURE 7-4: Squirrel fish at home near an artificial reef, Hawaii. Studies conducted in waters off Florida, the Virgin Islands, and Hawaii have shown that artificial reefs do not just concentrate marine life, as some scientists once believed, but actually increase the biota by accommodating greater numbers of mature organisms and enabling greater numbers of young and juveniles to survive. Like natural systems, artificial reefs serve as breeding grounds and provide protection from predators and strong underwater currents.

are highly valued by society for their esthetic qualities—into dense woodlands.

The maintenance of Yosemite's meadows illustrates two important points about restoration ecology. First, there is some debate over whether or not it is appropriate for humans to intervene by preventing successional changes or restoring a system to an earlier successional change and maintaining it there. Societal values determine which communities or ecosystems are maintained. Prairies, grasslands, and meadows are valued for scientific, esthetic, and often, historic and cultural reasons. For example, because prairies are an important component of the American experience and the national psyche, efforts are often directed at restoring and maintaining them. Secondly, for the purposes of restoration ecology, it matters little whether or not the techniques used to maintain an ecosystem are those that operated historically. In other words, it is not important whether the agent that starts a fire is lightning or the action of native American hunters or the action of Park Service rangers. The essential thing is that the desired biotic community is maintained.

Fire is also used as a management tool in forest ecosystems. In northern Michigan forest managers use fire to preserve habitat for the endangered Kirtland's warbler, which nests only in young jack pines. Because the jack pine cone needs the heat of fire to open and release its seeds, controlled fires are set to

FIGURE 7-6: Kirtland's warbler. The jack pine forests of Michigan provide habitat for the endangered Kirtland's warbler.

keep parts of the forest in an early successional stage characterized by the pine. Were it not for the fires, the area would be transformed through succession to hardwood forests. Maintaining the jack pine stage of succession helps to prevent the extinction of this species (Figure 7-6).

Intentionally set fires must be carefully monitored to prevent them from spreading out of control. In most natural and intentionally set forest fires, the trees are not seriously damaged or burned to the ground. Typically, the undergrowth is cleared, but the larger trees are merely scorched. In areas where the undergrowth or understory of trees is so thick that managers deem fires to be hazardous, they use logging to restore the forest ecosystem. For example, in Yosemite National Park a dense understory of pole-size white pines gained a foothold in the Mariposa Grove of giant sequoias. This understory was so flammable that any intentionally set fire might easily rage out of control. To restore the integrity of the sequoia grove, managers logged the white pines.

Restoration ecologists are conducting research to identify other techniques and management strategies that might be used to restore and maintain damaged climax forest ecosystems. In the southern Appalachian and northern Adirondack mountains, monitoring projects are currently underway that are designed to help explain the causes of changes in tree growth rates, twig and limb die-back, and leaf discoloration. These studies examine the effects of disease, tree age, climatic patterns, toxic fog, and acid precipitation on forest decline. Throughout the natural forests of North America, studies are being conducted to determine the effects of roads, riparian areas, and insect pests on forest productivity and

FIGURE 7-5: Intentionally set fire in the Samuel H. Ordway, Jr., Memorial Preserve, South Dakota. By setting controlled fires, managers at the preserve help to maintain the conditions necessary for the prairie to thrive.

diversity. These studies could result in restoration techniques that might then be applied in forests that have been seriously degraded.

Landscape Ecology

A view of the earth from the air reveals the landscape as it truly is: an intricate patchwork quilt comprised of individual ecosystems sewn together by threads we call structure, function, and change. A **landscape** is a system of varied ecosystems, often interspersed with human constructions. We live in landscapes, moving from field to woodlot to stream to farmstead to town, unaware of the ways in which **patches,** or subunits of the landscape, are connected (Figure 7-7). Traditionally, ecologists have tried to understand individual ecosystems. This work is necessary and important, but it is also essential to understand how the ecosystem patches function as a landscape. Focusing on the quilt rather than the patches is the task of landscape ecologists.

Landscape ecologists look at a geographic area as a unit. They study the distribution of ecosystems and the movement of plants, animals, nutrients, and energy among those ecosystems. They examine how combinations of ecosystems—such as woods, fields, swamps, roads, ponds, parks, and towns—are structured, how they function and interact, and how they change. Of primary importance are the relations between human society and its living space, both the natural and managed components of the landscape. Landscape ecologists attempt to solve problems associated with intensive human use in rural areas, urban areas, and parklands. In order to do this, they may apply principles from ecology, geography,

FIGURE 7-8: A satellite image of Oregon. Ecologists use the information provided by remote sensing to improve management of the landscape.

wildlife management, engineering, landscape architecture, cultural anthropology, landscape planning, and resource management.

In their search for understanding of the landscape as a whole, landscape ecologists ask various questions. How do changes in landscape use affect the natural populations of the landscape? How do changes in populations of organisms affect the movement of nutrients and energy among the units of the landscape? Do changes in landscape enhance or retard the movement of pest species? An important new tool that can provide the data needed to answer these and other questions is remote sensing, the use of satellite imagery to take a new look at the landscape, track the movement of animals, note the changes in vegetation types, and determine human alterations of the landscape (Figure 7-8).

It is often instructive to study landscapes where humans have lived harmoniously with nature for long periods of time. One such area is the landscape surrounding Matsalu Bay, on the east coast of the Baltic Sea in Estonia. Since before recorded history, humans have lived in the area, occupied chiefly with hay making and cattle raising. The region is highly stable; pasturelands and cattle coexist with wildlife, particularly birds, in a landscape officially known as the Matsalu State Nature Reserve.

To understand the benefits of landscape ecology, consider the rangelands of the western United States. These lands are not homogeneous areas. Stream, riparian, grassland, and canyon ecosystems are several of the more obvious landscape patches. The patches differ in their biota, the ways in which they are used, and the disturbances to which they are exposed. Cattle graze in the grasslands, removing biomass and its associated nutrients from that

FIGURE 7-7: Aerial view revealing the intricate patchwork quilt of the landscape. The landscape is a geographic unit made up of many interacting subunits: farmland, abandoned fields, forests, ponds, and streams.

ecosystem, but often defecate and urinate near streams and watering holes. The nutrients then end up in the stream, polluting the waters. Moreover, the biotic community of the stream bank may be severely damaged by the trampling of the herd as it waters. Clearly, the grassland cannot be managed without concern for the stream and riparian ecosystems, since the cattle move among these various landscape patches. Landscape ecologists would attempt to mitigate the disturbances to the various ecosystems by developing management plans for the landscape as a whole.

Agroecology

Agroecology, or agricultural ecology, can be defined as the study of purely ecological phenomena within the crop field or **agroecosystem**. Although agroecosystems are not as complex or diverse as natural ecosystems, the same ecological factors operate in both. Thus, agroecology considers energy flow, nutrient cycles, soil organisms, successional changes, predator-prey relationships, and crop-weed competition. Unlike natural ecosystems, however, agroecosystems also require subsidies of energy, water, and nutrients. Further, they are modified by pesticides, mechanical technology, and plant and animal breeding developments. Because of these human interventions, agroecology incorporates a sensitivity to social and environmental concerns. It does not focus solely on production as does agricultural science, but on developing means of production that are ecologically sustainable. Agroecology has its roots in the agricultural sciences, ecology (particularly research on tropical and temperate ecosystems), the environmental movement, studies of the agricultural practices and knowledge of indigenous peoples (especially in the humid tropics), and studies of rural development.

Agroecologists are conducting research to answer the following questions. Can sewage sludge be used to improve soil productivity? Do strips of unplowed land around crop fields encourage spiders and other insect predators to move into crops to control pests? Do polycrops, such as combinations of corn, beans and squash, increase biomass without the need for extra fertilizers or herbicides? Do crop residues left on fields increase fertility by promoting recycling of nutrients? Does increasing crop diversity reduce pest damage?

Typically, agroecologists are concerned with managing agroecosystems over the long term. They emphasize protecting the health of the soil—the biological capital on which the health of the land and, ultimately, the health of people, depend. One technique that illustrates this concern is alley cropping, a multiple-use system in which crop species

FIGURE 7-9: Alley cropping, Costa Rica. Young trees grow among maize and yucca. The trees, *Leucaena leucophala*, fix nitrogen, which benefits the other crops.

are grown in alleys about 6.5 to 13 feet wide (2 to 4 meters) formed by leguminous shrubs or tree species, which fix nitrogen (Figure 7-9). Tree and shrub prunings provide nutrient-rich green manure for the crops. They are also put on the fields during the fallow period to suppress weeds and shade the soil, thus helping to maintain soil moisture. Prunings can also be used as animal feed or firewood. Research conducted in Nigeria has shown that alley cropping of maize with *Leucaena leucophala* resulted in substantial increases in crop production over control crops.

Conservation Ecology

Conservation or **preservation ecology** refers to the application of ecological principles in order to conserve species and communities. Sometimes, the term *conservation* is used interchangeably with *preservation*, but the terms have subtle differences in meaning. **Conservation** implies wise use, the careful and planned management of an area (such as a forest or prairie) or resource (such as petroleum or a particular species) in order to provide for its continued use. The National Forests are managed so as to provide recreational and esthetic opportunities as well as harvests of certain resources like timber. **Preservation** can be thought of as a strict form of conservation in which the use of an area or resource is limited to nonconsumptive activities. The area or resource is left essentially unchanged by humans; human activity is restricted in order to allow natural processes to run their course. Preserving a wilderness area would restrict human use to minimal impact camping; roads and structures would be prohibited. Preserving the biological diversity of an

Environmental Science in Action:
Prairie Restoration at Fermi National Accelerator Laboratory
Karla Armbruster

The prairie is a familiar image in American history and imagination—an "inland sea" of waving grass stretching to the horizon, grazed by herds of buffalo and traversed by wagon trains heading west. It symbolizes the freedom and openness, the frontier spirit, that we prize so highly. But now 99 percent of the great prairie that once covered most of the Midwest is gone, paved over by cities and supplanted by farms that have flourished on its rich, deep soil. Applied ecologists have begun to restore examples of the prairie community, and their efforts hold out the hope that this magnificent landscape may once more become a familiar sight. The largest and one of the oldest prairie restoration projects is taking place on the grounds of Fermi National Accelerator Laboratory (Fermilab), a nuclear physics research institution near Chicago, Illinois.

Describing the Resource
Physical Boundaries

In 1972 the Department of Energy (DOE) commissioned the construction of Fermi National Accelerator Laboratory. While scientists inside the lab delve into the nature of atomic particles, attempting to break down the proton into its smallest parts, other experimental work is occurring on the surrounding land outside. A dedicated team, including Robert Betz of Northeastern Illinois University, Fermilab staff, and volunteers, are working to build a prairie ecosystem out of its constituent parts—soil and prairie plant species—in order to preserve the community and all its complex interrelationships (Figure 7-10).

Biological Boundaries

The prairie that once blanketed the Midwest was dominated by grassland plants classified as grasses, forbs (wildflowers), or legumes; the vast prairie also included savanna and marsh species. The colorful forbs were the most eye-catching: goldenrod, purple gentians, blue

asters, white false indigo, yellow black-eyed Susans, and pink blazing stars. Although less spectacular than the forbs, the grasses dominated the prairie in numbers. Tall grasses such as big bluestem and Indian grass can grow to a height of 8 feet (even 12 feet under certain conditions). In the savanna areas tree species were primarily oak, hickory, and white ash. Marshy areas had their unique species, too, such as prairie sedge and water parsnip.

The plants supported a huge array of insects, some of which depended for survival on just one or two plant species. For example, the baptisia dusky-wing butterfly fed only on wild and false indigos (genus *Baptisia* of leguminous plants). Species of worms particular to the prairie helped maintain its unique soil characteristics. About 300 species of birds, many of them migratory, were at home on the prairie. These included bobolinks, eastern meadowlarks, savanna sparrows, grasshopper sparrows, Henslow's sparrows, and upland sandpipers. Common mammals were the meadow vole, groundhog, mink, beaver, muskrat, rabbit, coyote, and deer. Many of these species have successfully adapted to human alteration of the prairie, but others, such as the bison, wolf, elk and bobcat, have completely disappeared from the Illinois area.

As Betz and his colleagues attempted to create a truly prairie-sized Illinois prairie, they discovered, largely by trial and error, that a new prairie must be established in stages. They first planted approximately 70 prairie species on the Fermilab site, but only about 25 plants—what they call the "prairie matrix"—flourished. Not surprisingly, these same plants are the last to disappear when a piece of prairie is invaded by "weedy" species, many of them Eurasian in origin, brought to the Midwest by settlers. Without the benefit of fire, prairie plants gradually lose their hold to species like Hungarian brome, ragweed, and Queen Ann's lace.

The Fermilab project has been one of the innovators in the use of

fire to aid in prairie restoration. Without burning, constant weeding is required to keep weedy species and trees from invading the prairie. Betz and his colleague Ray Schulenberg were largely convinced to try fire by historical accounts of prairie fires in the area. Though most natural fires occurred in autumn, they found that burning in the spring, after the weedy species have begun to germinate but before the prairie plants get started, works best. After the first fire, the prairie matrix begins to outcompete the nonnative species.

Eventually, the matrix plants alter the soil composition and chemistry enough to enable a second wave of less hardy prairie plants to grow. These include sky-blue, smooth and heath asters, cream wild indigo, blazing stars, mountain mint, and prairie dropseed. Sometimes, seeds of these second-wave plants, sowed ten or fifteen years before, will suddenly sprout, indicating that something important has changed in the soil; exactly what has changed is unknown and is an important topic for research. After the first- and second-wave plants have modified the soil to almost presettlement conditions (estimates for this process range between 50 and 100 years), the prairie will be ready for the most delicate, rare species, such as milkwort gentians.

Social Boundaries

The prairie is a vital part of our national heritage. The bounty of its rich soil allowed the European settlers' farms to flourish, and the animals that inhabited it helped to feed and clothe the settlers. How quickly the 40,000 square miles of the Illinois prairie shrank to 4 or 5 square miles is a testament to the settlers' drive to move westward and to make use of their new-found land. The current drive to reconstruct prairies and other ecosystems where they once flourished represents an equally adventurous spirit, but this one is a spirit of preservation and stewardship.

In addition to its cultural importance, the prairie ecosystem is valu-

118

able for scientific reasons. Many of the plants and animals that live only on the prairie would become extinct if their habitat disappeared. These threatened species hold an untold wealth of genetic knowledge which we might use to create better crops or produce new medicines.

Looking Back

In the early 1970s Betz and Schulenberg learned that Robert Wilson, director of the newly created Fermilab, was interested in ways to landscape the thousands of acres owned by the lab. Betz describes Wilson as a man of vision who was already interested in the idea of prairie restoration. Wilson had even purchased a herd of bison to raise on the lab grounds. Their meeting was "like a positron and an electron hitting together," resulting in a huge burst of energy—energy that would fuel the largest prairie restoration project in the country. Once the restoration project received official backing from the Nature Conservancy, the DOE approved the plan. Fermilab set up a prairie committee made up of Fermilab staff to manage the project and allocated equipment and the time of its roads and grounds staff for prairie work.

The first step occurred in the fall of 1974. Almost 100 volunteers hand-collected prairie seed from local remnants, resulting in 400 pounds of seeds from over 70 species. The next spring those seeds were planted on 9.6 plowed and disked acres. Although weeds dominated in the first year, the hardier prairie plants eventually gained a foothold. By the mid-1980s virtually all the area then available for cultivation (385 acres) was planted with prairie species. In recent years additional acreage has been sown with the seeds of prairie plants. A 90-acre interpretive trail, named for Margaret Pearson, Fermilab's former public information officer and a fervent supporter of the prairie project, is open to the public and includes signs that explain the prairie ecosystem and the prairie restoration process.

FIGURE 7-10: Fermilab prairie. The restoration project on the land surrounding the Fermi National Accelerator Laboratory near Chicago, Illinois, is attempting to re-create a portion of our nation's once-vast midwestern grasslands.

Since no one had ever attempted a restoration on the scale of the Fermilab project, Betz and his colleagues had to experiment with ways of planting and managing the prairie. Hand harvesting and planting was simply impractical for a prairie that would eventually cover hundreds, perhaps thousands, of acres. They gradually developed successful harvesting and planting techniques which use farm equipment such as plows and combines—ironically, the same equipment that destroyed the original prairie. Their methods and techniques for establishing prairie plants over large areas can now serve as models for other restoration projects.

The planting process begins in the fall, when the area to be seeded is mowed and plowed. Exposing weed roots to the freezing temperatures helps to discourage their growth the next year. The following May the area is disked and harrowed, which levels and dislodges

roots, and then "cultipacked" to tamp down the soil slightly in order to provide a firm surface for the seeds. Once the seeds are spread, the cultipacker runs over the area to press the seeds lightly into the soil. Given two weeks of good weather, the crew can prepare and plant approximately 70 acres with prairie seed.

Seeds are planted using an all-terrain spreader (normally used to spread fertilizer), which can hold 1,000 pounds of seed. The crew uses a combine to harvest 5,000 to 10,000 pounds of seed from older plots on the Fermilab prairie each autumn. Volunteers still hand-collect 100 to 200 pounds from remnant prairies to increase genetic variability, adding new species and supplementing those already at Fermilab in small numbers.

While still overpopulated with grasses like big bluestem and infested with weeds like white sweet clover, the Fermilab site boasts 125 species of prairie and marsh plants. In the older tracts second-wave species are beginning to thrive, although many exist only in small numbers, and in some areas along service drives, big bluestem is even beginning to invade without the help of lab staff!

Of course, establishing the plant species is just the beginning of restoration; a prairie is also made up of many species of animals. Fermilab lacks many prairie species, especially invertebrates, but some animals are making an impressive comeback. Coyote, deer, mink, and weasel can be found on the site, and the frog population seems to be growing. As of 1989, a total of 224 species of birds had been sighted, including 2 from the federal list of endangered species (the peregrine falcon and the piping plover) and 15 from the state list. The endangered sandhill crane is actually breeding on the site, and the long- and short-eared owls, also endangered in Illinois, spend time there. Some rare trumpeter swans have been brought in to live in a pond on the lab grounds.

Looking Ahead

The Fermilab prairie has flourished under the dedicated direction of Dr. Betz and with the generous support of the lab's directors and staff. It promises to become a close reproduction of the original Illinois prairie and to provide other restoration projects with proven methods of establishing a prairie community on a large scale. It also holds the potential for endless research projects, from simple inventories of animal and plant species to investigations into the complex biochemical processes that regulate the ecosystem.

In the next ten years Betz and the Fermilab committee plan to work toward a number of goals. They will add new plots of prairie plants every year. Having plots planted in chronological order provides a wonderful opportunity for researchers to compare various stages of succession. The researchers plan to maintain their effort to reintroduce second-wave and even less tolerant third-wave species to the older plots. They also plan to enrich marsh areas with original dominant species such as great bulrush and swamp dock. Their goal is eventually to link all the separate plots into one large, contiguous prairie landscape, including marsh and savanna, so that the original complex interaction of ecosystems can occur.

Another major stage in the restoration will be the reintroduction of animal species. Although many species have successfully reintroduced themselves, the researchers are interested in establish-ing the prairie chicken and Franklin's ground squirrel at some point. Insects present perhaps the greatest challenge. Most invertebrate species cannot travel far enough from prairie remnants to reach the Fermilab prairie, and so a comprehensive program will have to be undertaken to establish populations of prairie earthworms, yellow-winged grasshoppers, and countless others.

Will the Fermilab prairie and other restored prairies ever be identical to the original ecosystem? Opinions vary, although the Fermilab project has accomplished in 20 years what it might have taken nature 40 or 50 to do. It may be impossible to eradicate all the weedy species, and an untold number of original species may already be extinct. However, the hope is that if the majority of prairie species and a few nonnative species can interact as a self-sustaining ecosystem, they will act as a reasonable substitute for the native prairie. Although they can never replace the original prairies, restored prairies will have their own special symbolism. They will represent the attempt by our society not only to halt the degradation of the environment, but actually to repair it.

area, such as an old-growth forest, would prohibit all but nonconsumptive human uses, such as scientific study, hiking, bird-watching, or photography; collecting plants or animals would be prohibited. In some cases, even many nonconsumptive uses would be prohibited. Preserving a fragile bog or marsh could mean prohibiting all use except scientific studies and strictly governing the activities of the scientists.

Conservation ecology attempts to prevent the elimination of communities and the extinction of species. In this way it is similar to restoration projects like the one designed to preserve habitat for Kirtland's warblers. But while there is overlap between conservation ecology and restoration ecology, there are differences, especially in focus and techniques. Restorationists use intrusive management tools such as fire and logging to duplicate the effects of natural forces that humans have displaced. Conservation ecologists focus on protecting habitat and mitigating the harmful effects of human activities on biotic communities and use less intrusive management techniques.

Conservation ecologists use a knowledge of population dynamics and specific habitats to improve the carrying capacity of the environment for both game (white-tailed deer and ruffed grouse, for example) and nongame species (bluebirds, tree frogs, and turtles, for example). Carrying capacity can be improved by reducing environmental resistance. For example, density-independent factors can be improved by creating den, nest, or breeding sites for animals as diverse as minks, bluebirds, and great horned owls. Trout streams can be improved by decreasing siltation, increasing oxygen content, and reducing toxic pollutants. These steps have the effect of improving the water quality for the insects which the trout feed on, which in turn increases the trout population.

Through the efforts of conservation ecologists, the endangered Knowlton's cactus has been reintroduced on federally owned land in New Mexico, river otters have been reintroduced into the Grand River of northern Ohio, red wolves have been reestablished in North Carolina, and elephant seals, sea lions, harbor seals, and sea birds like puffins and murres are once again seen in the Farallon Islands National Wildlife Refuge off the coast of northern California.

Ecological Toxicology

Ecological toxicology, or **ecotoxicology,** blends the principles and practices from ecology and **toxicology,** the study of the effect of toxic materials on individual organisms. Ecotoxicologists study the effect of **toxicants** (or **toxins**), chemicals that can cause serious illness or death, on population dynamics, community structure, and, ultimately, ecosystems. They attempt to understand, monitor, and predict the consequences of a wide variety of pollutants. Doing so enables them to offer suggestions for mitigating pollution effects.

Ecotoxicologists look for indicator species to help determine when pollutant levels are unsafe. Some aquatic species, such as mayflies and riffle beetles, are more sensitive than others to certain pollutants and succumb more readily to the effects of pollution, and other species, such as clams and mussels, accumulate toxic materials in their tissues at sublethal levels. These species can be monitored to track pollution movement and buildup in aquatic systems. Preliminary work suggests that earthworms may play a similar role in terrestrial systems. Studies conducted to assess the impact of sewage sludge treatments on old-field communities revealed that earthworms concentrate cadmium, copper, and zinc in their tissues at levels that exceed those found in the soil. Cadmium levels even exceed the concentrations found in the sludge. Thus, researchers suggest that earthworms could be used to monitor the effects of sludge disposal on terrestrial communities.

Ecotoxicological studies allow both structural characteristics, such as the changes in the density of species, and functional characteristics, such as gross primary productivity, to be studied in a natural setting. Interactions can be observed that could not be anticipated or simulated in the laboratory. For example, to determine the effect of heavy metals in a stream, researchers fill trays with substrate—the material (rocks, pebbles, and sand) found in the stream bed—and anchor the trays in the waterway (Figure 7-11). After a period of colonization—when animals of the stream take up residence on the new

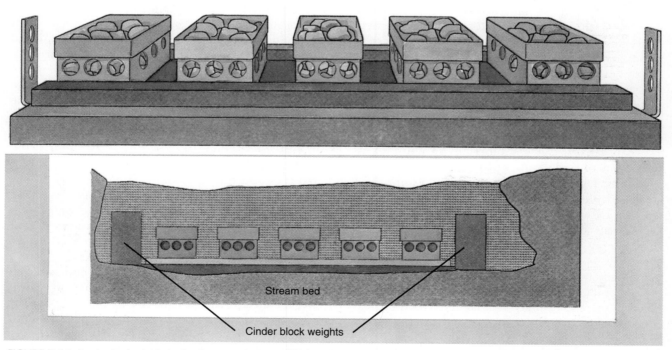

Stream bed

Cinder block weights

FIGURE 7-11: Ecotoxicological study of a stream. Trays filled with substrate similar to that found on the stream bed are lowered into the stream and anchored by cinder-block weights. After invertebrates and other animals colonize the trays, researchers can subject the biotic community to various pollutants.

substrate—the community is exposed to various concentrations of pollutants (such as copper) right in the stream. Alternatively, the trays can be moved to experimental, artificial streams for testing. These recolonized substrate boxes can also be placed above and below pollution sites to determine the effect of pollution on community structure at different sites in the stream. Potential sites include a power plant, a sewage treatment plant, or a manufacturing plant suspected of discharging copper or chromium into the stream.

Most of the toxicological research done to date has worked with individual species to try to determine in a laboratory setting the lethal levels of various pollutants so that safe levels may be given for the release of chemicals into the environment. Unfortunately, the high priority for testing individual chemicals on individual species has slowed research into the ecological effects of pollutants. Ecotoxicological research is very complex, and it requires significant expenditures of time, energy, and money.

How Can Computer Models Help Applied Ecologists?

Based on observations and on theory about ecosystem dynamics, applied ecologists develop **computer models,** sophisticated mathematical equations that help them to understand how ecosystems respond to stress and to predict the effects of management strategies. Powerful, high-speed computers are as important to the work of applied ecologists as the electron microscope is to the work of cell biologists. Massive amounts of data can be gathered and processed to simulate species interactions, energy flow, material cycling, and effects of stress and population changes.

Because natural ecosystems are so complex, it takes years of careful observation and data collection to develop a good model. The first models simulated small ecosystems, such as small streams and thermal springs. Later models simulated larger ecosystems, such as salt marshes, lakes, and prairies, and became increasingly complex. For example, Richard Weigart's comprehensive model of a salt marsh ecosystem in Georgia has undergone eight major revisions since its development in 1973. This model has become increasingly sophisticated in its ability to explain salt marsh food webs and the movement of carbon into and out of the system. Many subset models explore the ecology of particular species, such as shrimp, eel grass, or top minnows. Other subset models are currently being developed to

explore the relationships between various nutrients (for example, carbon-nitrogen ratios).

Computer models can also help us to understand succession and how climax communities are maintained in a dynamic steady state. Larry Kapustka, of the U.S. Environmental Protection Agency research laboratory in Corvallis, Oregon, developed a sophisticated computer model of a tall-grass prairie in Kansas. After many years of refining this model, Kapustka is beginning to understand the nitrogen cycle and the role of fire and grazing animals in perpetuating the tall-grass prairie climax community.

Other researchers are using computer models to assess the effects on lakes of introduced, nonnative species and to track pollution through aquatic systems. A computer model of the Chesapeake Bay provides information about nutrient loads to the bay, which is then used to simulate the effects of pollution on the bay itself. Initial results showed that significant nutrient loads come from states within the bay's watershed that are not part of the regional cleanup effort (the Chesapeake Bay Program)—New York, West Virginia, and Delaware. Computer models of Puget Sound and Boston Harbor revealed that tides and currents play a significant role in the movement of pollutants. Pollutants are not flushed through these semiclosed systems as quickly as scientists had believed. Rather, they tend to wash out and wash back in. Consequently, living organisms inside the sound and harbor accumulate greater amounts of the toxins in their tissues than do organisms in the open ocean.

Some of the most exciting and potentially most important computer models are those that attempt to simulate the planet's weather systems in order to determine the potential effects of global warming. General circulation models (GCM), as they are known, are based on mathematical equations so complex that they can be solved only by a few of the most advanced supercomputers. The equations take into account such factors as the balance of radiation to and from the earth; patterns of air circulation, evaporation and rainfall; ice cover; vegetative cover; and sea-surface temperature. Given a change in one variable—say, an increase in incoming solar radiation or an increase in atmospheric greenhouse gases—supercomputers calculate and map the resultant changes in global and regional climate over weeks, years, even centuries (Figure 7-12).

A computer model is, in effect, a hypothesis. It is the researcher's attempt to simulate nature's complexity and explain how nature functions. Even the most sophisticated model remains a hypothesis until observation and monitoring verify that it is accurately predicting the effects of a specific change. While computer modeling is a useful tool, we currently have no model capable of replicating all that occurs

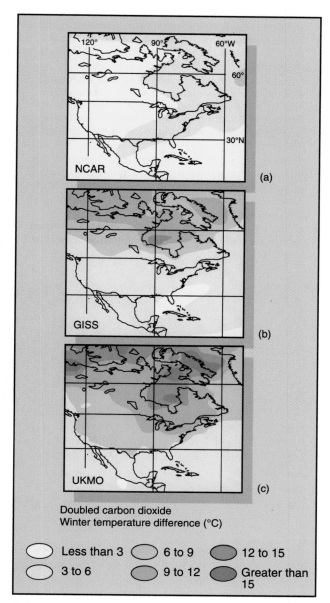

FIGURE 7-12: Computer models of potential global warming caused by a doubling of atmospheric carbon dioxide. The models were developed by (a) National Center for Atmospheric Research, (b) NASA/Goddard Institute for Space Studies, and (c) United Kingdom Meteorological Office. The differences in predicted effects illustrate the difficulties in attempting to model complicated natural systems.

in a single tree, let alone an ecosystem or the biosphere. In fact, scientists are currently unable to predict how most ecosystems will respond to most disturbances. They lack the necessary data from research to compare stressed and unstressed systems. As computers are capable of processing and storing more and more data, and as researchers continually refine the models, computer simulations should represent actual ecosystems more accurately and predict more precisely the consequences of both natural and culturally induced change.

Summary

Six subdisciplines of applied ecology are disturbance ecology, restoration ecology, landscape ecology, agroecology, conservation ecology, and ecotoxicology. Disturbance ecology is concerned with assessing the impact of specific stresses on specific organisms, populations, and ecosystems and predicting how they will affect natural systems. Disturbance ecology is primarily concerned with understanding the causes and effects of change in natural systems and with planning and designing appropriate projects to mitigate the adverse effects of human activities.

Although closely related to disturbance ecology, restoration ecology is a distinct subdiscipline that is goal-oriented and driven largely by the values of society. Its three major goals are to repair or reestablish biotic communities on their original site or to establish them on other sites if the original sites can no longer be used; to maintain the present diversity of species and ecosystems; and to increase knowledge of biotic communities and restoration techniques.

Landscape ecology attempts to understand how ecosystems interact to form a larger landscape. It studies the distribution of ecosystems and the movement of plants, animals, nutrients, and energy among those ecosystems. It is especially interested in the relations between human society and its living space, both the natural and managed components of the landscape.

Agroecology is concerned with developing methods of food production that are ecologically sustainable. Agroecologists must consider all the ecological factors that operate in agroecosystems, including internal energy flow, nutrient cycles, successional changes, predator-prey relationships, and crop-weed competition. However, because agroecosystems require subsidies of energy, water, and nutrients and are modified by pesticides, mechanical technology, and plant and animal breeding developments, agroecology often incorporates ideas and practices that reflect a sensitivity to social and environmental concerns.

Conservation ecology is the application of ecological principles and knowledge in order to conserve species and communities. Although there is some overlap between restoration and conservation ecology, they differ in focus and techniques. Conservation ecologists focus on protecting habitat (rather than restoring it) and mitigating the harmful effects of human activities on biotic communities. They rely on a knowledge of population dynamics and specific habitats to improve the carrying capacity of the environment for game and nongame species.

Ecological toxicology is a merger of ecology and toxicology, the study of the effect of toxic materials on individual organisms. Ecotoxicologists study the effect of toxicants on population dynamics, community structure, and, ultimately, ecosystems. Their goal is to understand, monitor, and predict the consequences of a wide variety of pollutants in order to offer suggestions for mitigating pollution effects.

The various subdisciplines of applied ecology share some common techniques and tools. One of the most important of these are computer models, sophisticated mathematical equations that enable applied ecologists to understand how ecosystems respond to stress and to predict the effects of management strategies. Because of the complexity of natural systems, it takes years of careful observation and data collection to develop an accurate computer model. Models have been developed to simulate both terrestrial and aquatic ecosystems and to predict the likely effects of global warming.

Key Terms

agroecology

agroecosystem

applied ecology

computer model

conservation

conservation ecology

disturbance ecology

ecological toxicology

landscape ecology

patches

preservation

restoration ecology

toxicant

Discussion Questions

1. How does applied ecology differ from ecology?

2. Give three examples of how applied ecology can mitigate the effects of the five D's of ecosystem degradation.

3. Suggest reasons why restoration ecologists might use controlled fires to stress forest or grassland ecosystems. Do you think such intentionally set fires are appropriate?

4. What ecosystems are found in the area where you live? What environmental problems are caused by the different ways in which these ecosystems are managed? What sorts of information could landscape ecology provide that would improve management in the area?

5. Drawing on your knowledge of ecosystem structure, function, change, and dynamic equilibrium, identify some of the factors and relationships that would have to be taken into account in order to develop an accurate model of an ecosystem.

6. What lessons have you learned about prairie restoration that would help you in a restoration project in your area?

An Environmental Imperative

Balancing Population, Food, and Energy

Chapter

8

Human Population Dynamics

Barring revolutionary advances in technology, life for most people on earth in the year 2000 will be more precarious than it is now—unless the nations of the world act decisively to alter current trends.

Global 2000 Report to the President, 1980

At the start of this critical decade, the choice must be made to act decisively to slow population growth, attack poverty and protect the environment. The alternative is to hand our children a poisoned inheritance.

Dr. Nafis Sadik

Learning Objectives

When you finish reading this chapter, you should be able to:

1. Explain the basic arguments of each side in the Great Population Debate.

2. Define demography and discuss the demographic statistics pertaining to population size and growth (how populations are measured and what information those statistics reveal).

3. Describe the relationship between carrying capacity and population growth.

4. Discuss how demographic statistics are indicators of the quality of life for a particular nation.

5. Compare population growth and quality of life in the more-developed countries and the less-developed countries.

Many students recognize the serious overpopulation in some less-developed countries, such as Kenya and Mexico, the rapid population growth in critical regions of the world, particularly the tropics, and the sheer numbers in the world's most highly populated countries, China and India. Others point to excessive consumption by people in stable or slow-growing developed countries; they maintain that many environmental problems would be nonexistent if resources were more equitably distributed (Figure 8-1). Our discussions with students either directly or indirectly relate environmental issues to problems caused by some aspect of human population dynamics. And the debate that always rages in these discussions is the same debate that rages among ecologists, economists, philosophers, and politicians. (See Box 8-1: The Great Population Debate.) Most people agree that the human population will grow (and for much of the world, grow rapidly) in the coming decades. The debate, then, is not over whether the population will grow; rather, it is essentially over whether it is desirable, for humanity and the biosphere as a whole, to allow our species to continue to grow unchecked. In simple terms, the topic of the debate is "Is human population growth a problem?" The answer to that question is critical, for it determines whether we—as individuals, communities, and nations—take deliberate action to limit the growth of our species.

Chapters 8 and 9 are designed to help you decide where you stand in the great population debate. A first step is to determine whether or not the human population can be considered a resource.

Describing the Human Resource: Biological Boundaries

Can the Human Population Be Considered a Resource?

Earlier, we defined a resource as anything that has value. By that definition, many people would say that the human population is the ultimate resource. They point to the arts, science, and technology as evidence that humans are the most powerful and impressive creatures on the planet. The other side of the coin, however, is that human accomplishments are also responsible for many of the ills that threaten

FIGURE 8-1: Uneven distribution of space among the world's people. Crowds throng the streets and alleys of Bombay (photo at left), while a solitary rider takes in the scenery in Matterhorn Canyon (photo at right).

the biosphere. Acid rain, ozone-destroying chlorofluorocarbons, and toxic wastes are just a few of the dangerous "products" of human activity. If our actions can prove so devastating to the biosphere, can humans really be considered a resource? And if the answer to that question is yes, can the adverse effects of human activities be minimized and the positive effects maximized?

In both theory and practice, we place values on the human population—on human life and human freedoms. One way in which these values are expressed is the way we label various groups of people. For example, comparisons are made between the "first world" (also known as the "North," "industrialized nations," "more-developed countries" (MDCs), "developed countries"), the "second world" or former eastern-European communist countries, and the "third world" (also known as the "South," "nonindustrialized nations," "less-developed countries" (LDCs), "developing countries"). With the fall of communism in many eastern European countries and the breakup of the Soviet Union, the geopolitical second world has all but disappeared. In an economic and ecological sense, however, the second world remains.

Although these terms have been widely used, not only are they, being labels, inaccurate, but they also reflect and perpetuate attitudes toward other groups. When the term third world was first coined, it referred to the "Group of 77" poorer countries that formed the majority in the United Nations General Assembly. Although the group, which now numbers 100 or so, form a majority, these countries lack the clout of a majority bloc because the ethnic, cultural, and ideologic diversity among them is often divisive. China, a communist nation, is considered part of the third world, yet China, with the world's largest population, is very different from Burkina Faso, a small African nation that is also a third world country. China's people are adequately fed, and they have access to health care and other social services (Figure 8-2). Citizens of Burkina Faso experience hunger and inadequate care. India, a democratic nation, is also part of the third world. It has a rising gross national product and exports grain, but it is home to more starving people than any nation on earth. Why are these three countries, all in very different situations, treated as if they have the same problems? What is it that these countries have in common that has earned them the label of third world? The answer is that they have not become industrialized in the manner of western Europe, Canada, the United States, Australia, New Zealand, and Japan.

The problem with labels is threefold. First, the label *third world* makes it less likely that we will pro-

FIGURE 8-2: A woman bringing produce to market in Dali, Yunnan Province, southwest China.

Box 8-1:
The Great Population Debate

Whether in a classroom, a scholarly journal, or the United Nations, passions run high when people answer the question "Is human population growth a problem?" The answers depend largely on beliefs and values. The beliefs and values of the different groups depicted below dictate their positions in the Great Population Debate. Groups with different ideologies may take the same position on the population issue, but for different reasons, of course!

Where do you stand in the Great Debate?

"There Is No Population Problem"
Cornucopians

Some people argue that people are the world's ultimate resource, and the more of them, the better. They adopt a cornucopian view of population which sees continued growth as a positive factor that can recreate the mythical "horn of plenty." Cornucopians contend that a growing population can actually help a country to become more prosperous, since additional people mean both more consumers and more producers. Moreover, they steadfastly maintain that there is no such thing as a "population problem," because human inventiveness will solve any problems—environmental,

economic, social, or otherwise—that may beset continued population growth. Cornucopians believe that a free market society should be the driving force to accomplish population stability. In general, this is the "unofficial" policy of the United States. Cornucopians emphasize that education is critical to understand the relationship between free enterprise, development, and population growth.

The economists Julian Simon and the late Herman Kahn are probably the most well-known proponents of this view. In a 1981 interview in *Forbes* Magazine, Simon, author of *The Ultimate Resource*, asserted, "I have no reason to think that population growth is not a good thing. . . . Population growth is a moral and material triumph. That's the flip side of the thing—what an enormous success for the human species. If the flies took over the world, the biologists would say, 'What a successful species!' But we get more and more people, and they say, 'What a failing species!' What a strange twist!"

Marxists

Marxists argue that poverty is the result of the unequal distribution of resources rather than unchecked population growth. They see the problem as one of power—the haves versus the have-nots. Consequently,

Marxists do not acknowledge a population problem. Instead, they maintain that efforts should focus on distributing resources to all peoples more evenly. In China, for example, the stated goal of family planning efforts is to produce healthier, happier children; one way to do this is to limit the number of children born so that resources can be shared among a smaller population. Such a situation is good for the children, good for families, and good for China.

"There Is a Population Problem"
Malthusians

Malthusians are proponents of the beliefs credited to Thomas Malthus, an eighteenth-century parson, who believed that humans tend to over-reproduce out of an innate drive to procreate. Because population increases geometrically, but food increases only arithmetically, he reasoned, population growth is greater than the earth's power to sustain it. War, famine, and pestilence act as negative feedbacks that help limit population growth. According to Malthus, these factors affect the poor in greater proportion than the rich, since the wealthy can avoid service in armies, can afford an adequate supply of food, and can afford medical attention.

vide a country with effective help, because effective aid must be tailored to the needs of the country and not based on a generalized concept of a third world nation. Second, the third world label makes it easier to reason that these people are accustomed to a particular way of life, that it is somehow not as bad for third world people to go hungry or to remain illiterate as it is for people in the first world. Third, and perhaps most significantly, using the words *first*, *second*, and *third* implies that the path the capitalist, industrialized nations have chosen is the best path.

The term third world carries a tone of superiority and condescension, a hint of pity for the nations that have not followed the path of industrialization. On

the surface, pity might seem justifiable. After all, the more-developed countries enjoy a standard of living that is far higher than other countries; we have access to more resources, live longer, and have luxuries and amenities unknown, even unimaginable, to most of the world's people. But where do those resources come from? At whose expense do we enjoy such high standards of living? Is our standard of living sustainable? Can it be enjoyed by all the world's people, or is it sustainable for us only if most of the world's population go hungry, remain impoverished, and suffer sickness and ill health? If humans are indeed a resource, then isn't it logical to assume that all humans, whether they live in

Neomalthusians

Malthus was convinced that food resources are the limiting factor on the size of the human population. Some contemporary thinkers argue that although food may not be as important as Malthus believed, his ideas about the link between poverty and family size are correct. They claim that the "Malthusian checks" (war, famine, pestilence) are more common in poorer nations. Those who espouse this adapted Malthusian view are known as neomalthusians.

Neomalthusians concede that the cornucopians have fashioned a powerful image—that of each child as a potential gift to humanity. But, they counter, how can we expect to find answers to the many problems facing the human population among the starving masses in Ethiopia, Bangladesh, and India? Can those who have little food and water, no home, and a bleak future fulfill their human potential? Neomalthusians see continued population growth as a threat to environmental quality, political stability, international relations, and economic development. They contend that even if the rate of population growth could be decreased immediately, it would take many years to slow and halt the total increase in the absolute numbers of human beings on earth. In the meantime, the human population would con-

tinue to stress and damage the earth's life support systems, with dire consequences for the biosphere and its human and nonhuman inhabitants.

Advocates of Zero Population Growth

Some neomalthusians translate their beliefs about the consequences of population growth into recommendations for public policy. Because they believe that poverty is the result of high fertility, they recommend that birth control and mandatory family planning programs be introduced to achieve zero population growth (ZPG) as soon as possible. They predict that, unless zero population growth is reached, natural checks on population will cause widespread misery and death. They point to Africa in general and Ethiopia in particular as examples of what the future holds for humans.

ZPG, Inc., is a group started by Drs. Paul and Anne Ehrlich, authors of *The Population Bomb* and *The Population Explosion*. The Ehrlichs' goal is to convince couples to have no more than two children through an extensive educational program to help people understand the relationships among population growth, poverty, environmental degradation, and standard of living. In an excerpt from *The Population Explo-*

sion, the Ehrlichs contend that "Even some environmentalists are taken in by the frequent assertion that 'There is no population problem, only a problem of distribution.' . . . Unfortunately, an important truth, that maldistribution is a cause of hunger now, has been used as a way to avoid a more important truth—that overpopulation is critical today and may well make the distribution question moot tomorrow."

"The 'Population Problem' Is a Complex Issue"

Many people contend that most of the problems associated with human population growth (hunger, landlessness, poverty, pollution) involve distribution of resources and resource consumption patterns, not simply population numbers and growth rate. People in the United States, Japan, Canada, and western Europe all consume far more resources than do people in less developed countries. In "The Grim Payback of Greed," Alan Durning, senior researcher at the Worldwatch Institute in Washington, D.C., writes "Overconsumption by the world's fortunate is an environmental problem unmatched in severity by anything but perhaps population growth."

Colombia, Haiti, Zaire, Germany, the United States, or Japan, are valuable and worthy of a quality life?

As you read Chapters 8 and 9, you must decide for yourself whether or not the human population can be considered a resource. We believe that humans are indeed a resource, and that is the position we have taken in writing this chapter and text. Moreover, we believe that it is necessary to study and understand human population dynamics before studying any other resource, simply because humans have such a significant impact on the biosphere and on all other resources. A powerful tool that can assist us in that study is the discipline known as demography.

What Is Demography?

All of us are born, age, and die. Many of us marry, have children and grandchildren. Most of us live in more than one location during our lifetimes; some of us migrate from one country to another. These are personal events, indeed, decisions about bearing children and raising a family are among the most intensely personal ones a human can make. But personal decisions are important on a societal level as well, for the sum of thousands and millions of individual acts has a tremendous impact on societies and the environment.

Demography is the scientific study of the sum of

our individual acts as they affect measurements of the population. Demographers study population statistics, also called **vital statistics.** They seek to understand human population dynamics by studying how populations change, for example, growth rates, birth rates (natality), death rates (mortality), fertility rates, and immigration rates. Population can be studied at many levels: all the individuals on earth; the distinct populations occupying continents or regions of the world (Europe or North America, the Northern or Southern hemisphere); the populations of distinct political nations (Germany or Mexico); the populations of regions, states, or republics within countries or continents (Bavaria, Germany, or Sonora, Mexico); the residents of cities (Munich or Mexico City); smaller units, such as towns, counties, and townships. What level is studied depends on what you are looking for.

Demographers are interested in questions such as: How long do the people of a particular society live? How many children do parents want to have? How many people are younger than 15 and older than 65? How often do people move and why? (One important question demographers do not ask is: How many people can a specific environment support?)

Why Do We Study Demography?

Demographers try to understand the causes and consequences of changes in populations. The information they find can help us to plan for health care needs, retirement needs, school facilities, transportation needs, and the allocation of food and materials. In other words, understanding population demographics helps us to make social, political, and resource use decisions for individuals, local regions, the country, and international efforts, such as extending aid to other countries. It also helps us to understand the environmental and cultural consequences of our decisions. Finally, demographics can help us to implement the decisions necessary for the quality of life we want for our children and our children's children.

To understand the significance of demographic statistics, we must have some understanding of the culture they describe. Birth, death, fertility, and migration are influenced by a culture's dominant beliefs and value systems. For example, in many less-developed countries where the population is primarily Catholic, a taboo against artificial means of birth control helps to keep birth rates higher and family size larger than in countries where religious plurality is the rule. In cultures where parents show a strong preference for boys over girls, parents may continue to have children in order to have the boy(s) they want, they may decide to abort a female fetus (when

FIGURE 8-3: Diverse peoples of the world. The human population is a wonderfully varied resource. Unfortunately, our differences—skin color, physical attributes, culture, language—are often divisive. Increasingly, however, we are coming to realize that the fate of one people is dependent on the fate of other peoples. Nowhere is this realization more evident than in demographic studies.

the sex of the child is known), or they may resort to infanticide after the child's birth.

Throughout this chapter we must not lose sight of the fact that impersonal numbers, statistics, and percentages can obscure the lives and aspirations of the people they represent. Relating population statistics to our personal decisions about birth, death, marriage, and migration helps us realize that each statistic represents a person whose life, goals, and dreams are just as real as our own. It helps us to remember that the real population story is not numbers; it is people (Figure 8-3).

How Are Populations Measured?

Number of People

In 1991 the world population reached an estimated 5.38 billion. Over 4 billion people live in the developing world and over 1 billion in the developed world. The most populous countries are China with

▶ Table 8-1
Basic Demographic Data, Selected Countries

	Population Estimate, Mid-1991 (millions)	Crude Birth Rate	Crude Death Rate	Annual Natural Increase (%)	Population Doubling Time at Current Rate (years)	Population Projection, 2010 (millions)	Population Projection, 2020 (millions)
World	5,384	27	9	1.7	40	7,189	8,645
More-developed countries	1,219	14	9	0.5	137	1,345	1,412
Less-developed countries	4,165	30	9	2.1	33	5,844	7,234
Less-developed countries, excluding China	3,014	34	10	2.3	30	4,424	5,643
Afghanistan	16.6	48	22	2.6	27	32.7	44.5
Australia	17.5	15	7	0.8	88	18.7	21.9
Bangladesh	116.6	37	13	2.4	28	176.6	226.4
Bolivia	7.5	38	12	2.6	27	11.4	14.3
China	1,151.3	21	7	1.4	48	1,420.3	1,590.8
Denmark	5.1	12	12	0.0	1,732	5.1	4.8
El Salvador	5.4	35	8	2.8	25	7.6	9.4
Haiti	6.3	45	16	2.9	24	9.4	12.3
Hungary	10.4	12	14	−0.2	—	10.5	10.1
India	859.2	31	10	2.0	34	1,157.8	1,365.5
Italy	57.7	10	9	0.1	1,155	55.9	52.3
Japan	123.8	10	7	0.3	210	135.8	134.6
Kenya	25.2	46	7	3.8	18	45.5	63.2
Mexico	85.7	29	6	2.3	30	119.5	143.3
Peru	22.0	31	8	2.3	30	31.0	37.4
Sierra Leone	4.3	48	22	2.7	26	7.0	10.0
Sweden	8.5	13	11	0.2	367	8.3	8.0
Thailand	58.8	20	7	1.3	53	70.7	78.1
Turkey	58.5	30	8	2.2	32	83.4	102.7
United States	252.8	17	9	0.8	88	299.0	333.7
Former Soviet Union	292.0	18	10	0.8	91	333.0	363.0
Yemen	10.1	51	16	3.5	20	19.0	29.9
Zaire	37.8	46	14	3.1	22	67.5	101.1

Note: Figures for growth rate are sometimes rounded and thus do not exactly equal (birth rate – death rate).
Source: Population Reference Bureau, *1991 World Population Data Sheet.*

1.15 billion, India 859 million, the former Soviet Union 292 million, and the United States 252.8 million. Table 8-1 presents demographic data for selected nations.

Demographers calculate the change in a population over a period of time (usually one year, from mid-July to mid-July). World population growth, expressed as an absolute number, is determined by subtracting the total number of deaths worldwide from the total number of live births. In any given year, as long as live births outnumber deaths, world population will grow in absolute numbers.

For a country or region, calculation of the change in absolute numbers differs somewhat from the formula for world population because demographers must consider **migration**, movement from one coun-

try or region to another for the purpose of establishing a new residence. Migration into a country or region is called **immigration**; migration out of a country or region is **emigration**. (In countries with substantial illegal immigration, accurate immigration data may not be available.) Assuming that the necessary data are available, the actual increase in absolute numbers for a specific country is determined by the formula: (births + immigration) – (deaths + emigration).

Growth Rate

Though population size and growth in absolute numbers are important, demographers are also interested in another measure of population growth, the rate at which a population is increasing or decreasing. Knowing how fast a population is growing (or shrinking) allows demographers to predict future population size, and that prediction allows others to predict the population's effect on the economic, political, and social structure of the society.

To determine how fast a population is growing, demographers must have information about birth and death rates. The **crude birth rate** (CBR) is the number of live births per 1,000 people. It is calculated as follows: (number of live births) ÷ (total population at mid-year) × 1,000. The **crude death rate** (CDR), the number of deaths per 1,000 people, is: (number of deaths) ÷ (total population at mid-year) × 1,000. The growth rate is: crude birth rate − crude death rate, with the difference expressed as a percent (that is, as a fraction of 100).

For example, in 1991 the number of live births for the world population was 27 per 1,000 people and the number of deaths was 9 per 1,000. Subtracting deaths from births, we find 18 per 1,000 (or 18/1,000) people were added. When expressed in decimal form, 18/1,000 is 0.018. To express this figure as a percent, we multiply by 100: 0.018 × 100 = 1.8 percent. Thus, in 1991 world population grew at a rate of 1.8 percent. (A shorthand way to calculate growth rate is to subtract the crude death rate from the crude birth rate and then divide the difference by 10 (27 − 9 = 18; 18/10 = 1.8).

Growing at a rate of 1.8 percent, the population adds almost 90 million people annually (about equal to the population of Mexico). If this growth rate remains constant, world population could reach over 10 billion by about 2025. Most projections of world population, however, are lower (Table 8-2). They are based on the assumption that the growth rate will decline until the population stabilizes at approximately 12 billion around the start of the twenty-second century. These projections are partly borne out by demographic trends over the past several decades. The annual growth rate dropped from a high of 2.1 percent in the mid-1960s to 1.7 percent for most of the 1980s and early 1990s.

Some people point to the decrease in the world growth rate as evidence that there is no population problem, but that argument is misleading. While the decrease is encouraging, we must not lose sight of the fact that a positive growth rate (of any amount) translates into an annual increase in absolute numbers. In other words, while the growth rate has slowed in the past several decades, it is still positive, and thus the population, in absolute numbers, continues to increase. The confusion over the meaning of falling growth rates versus increasing absolute numbers is complicated by the fact that demographers use rates per thousand for birth and death rates and rates per hundred (percent) for growth.

The important thing to remember is that as long as births outnumber deaths, the growth rate is positive, and absolute numbers increase. When births are equal to deaths, the growth rate is zero, and absolute numbers remain the same. This is the condition known as **zero population growth**, or **ZPG**. (ZPG also assumes zero net migration, that is, immigrants equal emigrants.) In 1991 only Denmark had a growth rate of 0.0 percent and thus a stable population. When deaths outnumber births, the growth rate is less than 0.0 (a negative growth rate), and absolute numbers decline. In 1991 Hungary had a negative growth rate (−0.2 percent) and thus a declining population. These figures, however, do not account for the effects of migration, which may either exaggerate or offset the natural rate of increase or decrease.

▶ Table 8-2
World Population: Number of Years to Add Each Billion

Billion Reached	Year	Years to Add
First	1800	All of human history
Second	1930	130
Third	1960	30
Fourth	1975	15
Fifth	1987	12
PROJECTED		
Sixth	1998	11
Seventh	2009	11
Eighth	2020	11
Ninth	2033	13
Tenth	2046	13
Eleventh	2066	20
Twelfth	About 2100	34

Source: Population Reference Bureau, *1991 World Population Data Sheet.*

Regions of Highest Population Increase

Some 85 to 90 percent of the people born in the coming decades will live in less-developed countries (LDCs), where life is undoubtedly harshest. Seventy percent of projected world population growth by the year 2025 will occur in just 20 countries (Figure 8-4). Growth rates are highest in Africa at 2.9 percent, western Asia 2.8 percent, Central America 2.5 percent, and southern Asia 2.3 percent (Figure 8-5).

Demographers generally attribute rapid population growth in LDCs to a complex mix of social factors, among them improved medical care, which has lowered death rates, and lack of access to sex education, family planning services, and safe and reliable means of birth control. Another factor closely related to high birth rates is a high incidence of infant

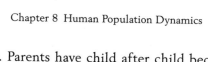

Country	Population (Millions)	Change from 1989 to 2025
Bangladesh		119.4
Brazil		95.4
China		357.1
Egypt		39.9
Ethiopia		65.6
India		592.2
Indonesia		82.7
Iran		65.6
Kenya		52.5
Mexico		61.5
Nigeria		188.3
Pakistan		144.4
Philippines		49.0
South Africa		28.0
Sudan		34.4
Tanzania		57.5
Turkey		34.0
Uganda		36.8
Vietnam		50.8
Zaire		63.5
		Total: 2,218.6

■ 1960 ■ 1989 ■ 2025

FIGURE 8-4: Past and projected population growth in 20 less-developed countries. Seventy percent of the projected growth in world population by 2025 will take place in these countries. (Source: "The Growing Human Population," by Nathan Keyfitz. Copyright © 1989 by Scientific American, Inc. All rights reserved.)

mortality. Parents have child after child because of the reality that many will die as infants or youngsters. In the absence of social programs, insurance, and other measures to care for the aged, ill, and injured, children represent security. In many parts of the world sons are seen as particularly desirable, because sons are expected by custom to care for aged parents. In rural areas children help their mothers with planting and harvest, child care, and all household activities. In urban areas they help support the family; children may work as peddlers (selling small trinkets and goods), scavengers (scouring dumps for cast-off items that can be used by the family or sold), or prostitutes. Bangkok, Thailand, for example, is notorious for its high number of child prostitutes.

Population Doubling Time

Less-developed countries have the fastest **doubling times**, the number of years it will take a population to double, assuming that the current growth rate remains constant. Doubling times are calculated by the rule of 70: 70 divided by the current annual growth rate yields the time it will take to double population, if the growth rate remains constant (Figure 8-6). Thus, at 1.8 percent the world population would double in just under 40 years (70/1.8), reaching 10 billion at about 2030. The doubling time for India, with a growth rate of 2.0 percent, is 35 years (70/2.0). The United States at 0.8 percent natural rate of increase would double in 87.5 years (70/0.8). However, because immigration raises the United States' growth rate to an estimated 1.0 percent, the actual doubling time is 70 years (70/1.0).

There are two things to keep in mind concerning doubling times. First, the effect on resources and the environment depends on the absolute size of the initial population. The doubling of a population of 1,000, for instance, will have less impact than the doubling of a population of 1,000,000.

Second, the calculation of doubling times assumes that the current growth rate will remain constant in the future. Because in fact it usually does not remain constant, the doubling time may not accurately predict the real growth of a population. Nevertheless, the doubling time is a valuable measure because it allows us to visualize what future environmental and social conditions would be like if present growth rates were to continue. Let's consider Kenya as a case in point. With 3.8 percent annual growth, Kenya's population would be 2 times as large in about 18 years, 4 times as large in 36 years, and 16 times as large in 72 years, growing from 25.2 million to a staggering 403.2 million! However, most demographers, economists, and ecologists agree that that growth rate could not be sustained. The stress

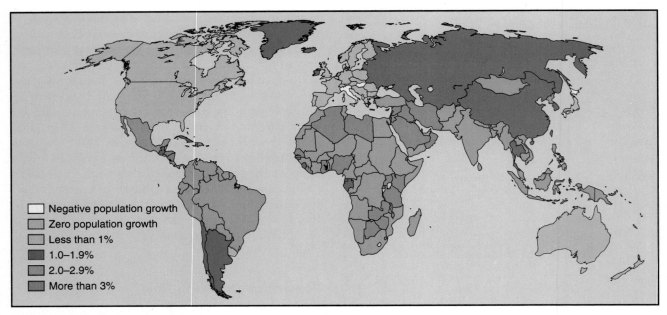

FIGURE 8-5: World population growth rates.

exerted by such unchecked growth would almost surely overwhelm both the natural resource base (which, according to many indicators, is already stressed) and the country's social, economic, and political systems.

Factors Affecting Growth Rates

Among the factors affecting growth rate are migration, fertility, and age distribution. The importance of any one factor in determining growth varies among countries. Migration, for instance, is one reason why the United States is one of the fastest growing nations in the industrialized world.

Migration. The *actual* rate of increase for a country or region differs from the *natural* rate of increase because the actual rate takes migration into account. Immigration can have a significant effect on the rate of population growth. For example, as we have said, in 1991 the natural rate of increase for the United States was 0.8 percent, but the actual rate of increase was over 1 percent. During the 1980s Asians (a diverse group that includes Pacific Islanders) constituted the fastest growing minority group in the United States. This group grew from 3.8 million to 6.9 million between 1980 and 1989, an increase of 80 percent. Most of that increase (2.4 million people) was due to immigration. Hispanics

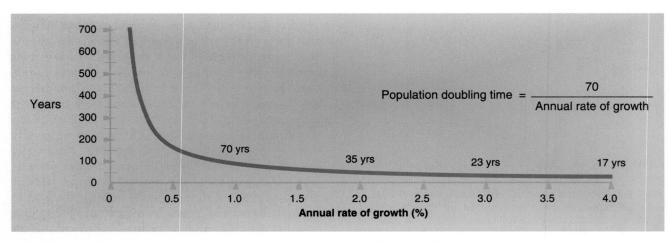

FIGURE 8-6: Doubling time and annual growth rate. A population's doubling time decreases dramatically as its growth rate rises. Even a small increase in growth rate results in a much shorter doubling time.

(primarily individuals whose cultural heritage can be traced to a Spanish-speaking country in Latin America) are another fast growing group. Between 1980 and 1988, the Hispanic population in the United States rose 34 percent, compared to a 7 percent increase in the nation's overall population. The U.S. Census Bureau estimates that during that period a net total of over 2 million Hispanic immigrants (including undocumented immigrants) entered the country (Figure 8-7).

Population of regions within a country can also be significantly affected by migration. The economic recession in the United States in the late 1970s and early 1980s resulted in significant emigration from the Midwest, home of the nation's hard-hit automobile and steel industries, to the Southwest. This movement hurt the Midwest's economy in several ways. Valuable and able workers left for better opportunities elsewhere, the area became less attractive to new industry, and the tax base was diminished in many localities as old industries died out and were not replaced with new ones. The mass migration posed many problems for the Southwest as well. The infrastructure of most communities (roads, schools, hospitals, and so forth) was not equipped to cope with demand as populations increased rapidly. Local and state governments were not always prepared to control or plan development, with the result that the environment was often degraded as many people came to live in what is essentially desert or semidesert land. Water short-

FIGURE 8-7: Festival of Cinco de Mayo, San Francisco, 1990. Immigrants can have a significant impact on the nation's politics, culture, and economics. While the population of the United States as a whole is not growing very rapidly, the population of Spanish-speaking people is, particularly in certain cities and regions of the United States, including Los Angeles, Miami, Denver, and San Antonio.

ages, increasing pollution, and crowding were just a few of the results of the large influx of people.

Fertility. For many countries and regions of the world, crude birth and death rates are the only demographic data available. Crude rates are so called because they do not include information about age and sex. Consequently, they do not enable us to make accurate predictions about the future dynamics of the population. For example, men, children, and the aged are included in the calculation for crude birth rate. But only women bear babies, and women of certain ages, specifically 15 to 49, are more likely to bear children than others. Thus, measures of **fertility**, or the actual bearing of offspring, are more accurate indicators of the potential for future population growth. The most important of these are the general fertility rate, age-specific fertility rates, and the total fertility rate.

The **general fertility rate** is the number of live births per 1,000 women of childbearing age per year. Ages 15 to 49 are considered by many demographers (including the Population Reference Bureau, the nonprofit educational organization cited throughout this text) to be the childbearing years. Other demographers list 15 to 44 as the childbearing years.

A more helpful indicator of potential future growth is the **age-specific fertility rate**, the number of live births per 1,000 women of a specific age group per year.

The **total fertility rate (TFR)** is the average number of children a woman will bear during her life, based on the current age-specific fertility rate, assuming that current birth rates remain constant throughout the woman's lifetime. A TFR of 2.1 children per woman is generally considered to be the replacement level fertility for the more-developed countries (MDCs). The **replacement fertility** is the fertility rate needed to ensure that each set of parents is "replaced" by their offspring. A fertility rate of 2.1 births per woman enables each woman to replace herself and her mate and allows for some mortality, that is, children who fail to survive to adulthood and reproduce. In many LDCs the average replacement fertility is 2.5, a reflection of the greater risk of mortality faced by children in those countries.

In 1991 the world total fertility rate was 3.4. The TFR for the developed world was just 1.9, below replacement level. In some western European countries—notably Italy, Austria, Germany, Luxembourg, the Netherlands, Belgium, Austria, Denmark, and Spain—the TFR ranged from 1.3 to 1.8. The populations of these countries are or soon will be declining. In contrast, the TFR for the developing world in 1991 was 3.9. Excluding China, which has

FIGURE 8-8: A young family of the nomadic Tuareg tribe settling around the outskirts of Timbuktu, Mali. In 1991, Mali's TFR was 7.1. At current growth rates, the country's population will double in 23 years. High birth rates in Africa are linked to high rates of infant and childhood mortality, economic insecurity, cultural beliefs that link male virility with number of offspring, low education levels and a lack of access to family planning services and safe, reliable means of birth control.

pursued an aggressive population control program for a number of years (see Chapter 9), the TFR for the developing world was 4.4 (Figure 8-8). The African nation of Rwanda had the world's highest total fertility rate, 8.1.

In the United States, the "baby boom" period (about 1946 to 1963) peaked in 1957 at a TFR of about 3.8 births per woman. By 1976 the TFR dropped to 1.7. It remained below the replacement level throughout the 1980s, then rose to 2.1 by 1991, a reflection of the large numbers of women in the childbearing years (both younger baby boomers and the children of older baby boomers).

Population Momentum. Consider two hypothetical populations. Population A has a small number of women ages 15 to 49, but a high fertility rate; women of childbearing age have, on average, five or six children. The population will grow rapidly in proportion to its size, but it will not add, in absolute terms, large numbers of people. In contrast, population B has large numbers of women in the fertile childbearing years, but a low fertility rate; women, on average, have just one or two children. The effect of these conditions is a large number of children, simply because there are so many women of reproductive age. While population B will grow slowly in relation to its size, it will add many more people than population A. Population B is undergoing population momentum.

Population momentum occurs when there are large numbers of children living as fertility rates begin to drop. Each successive generation that enters the childbearing years is larger than the preceding generation and thus bears more children overall, even if each set of new parents has fewer children than did parents of the previous generation. Because the number of children born is greater than the number of deaths among older generations, the population grows. Even when fertility drops below the replacement level, population momentum can cause the population to grow for as many as a hundred years. During that time span the population can undergo tremendous growth in absolute numbers. Theoretically, in less-developed countries where momentum is highest, the population might double as many as three times before reduced fertility eventually stabilizes population size.

Age Distribution. The age distribution of a population is the number of individuals of each sex and age (from birth through old age). It is graphically represented by an age structure histogram, or **population profile,** which is constructed from census data (Figure 8-9). (A census is the periodic collection of demographic information by a government about its citizens.) Horizontal bars are used to depict the numbers of males and females of each age group. Typically, each age group is plotted at 5-year intervals. Every 5 years, the bars are moved up one position. New bars are added to the bottom to represent new births and the uppermost bar is removed as the elderly die. Other upper bars are usually slightly reduced to reflect mortality related to aging.

Population profiles for LDCs generally have a pyramidal shape because about 40 percent of the population is below 15 years of age and only 4 percent or so is age 65 or older. High growth rates in LDCs are the result of high fertility coupled with reduced infant and child mortality rates. More people survive to their childbearing years, producing an even larger next generation and a larger and overall younger population. The populations of older age groups, where most deaths occur, remain small in

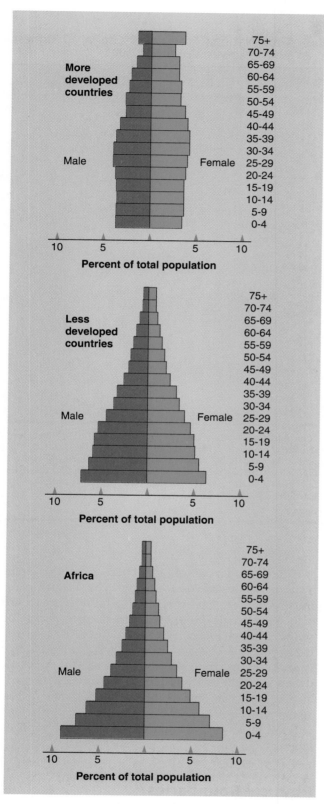

FIGURE 8-9: 1990 population profiles for more-developed countries, less-developed countries, and Africa. As the profiles show, the populations of LDCs will continue to expand, while those of MDCs will stabilize or have already stabilized. Africa, with the highest growth rate of any continent or region, has the youngest age structure and the greatest potential for future growth.

relation to younger groups. Even if the fertility rate drops, population momentum acts to expand the base of the pyramid as each new bar is added. If the fertility rate does not drop, that expansion can be substantial.

The population profile for MDCs differs greatly from that for LDCs. In 1991, 21 percent of the population was 15 or younger, and 12 percent was 65 or older. The number of people in the 29 to 45 age group is slightly larger because of the temporary high birth rates (the "baby boom") after World War II. Population growth in MDCs will stabilize and begin to decrease as the 29- to 45-year-olds move into old age categories. This should happen in 20 to 30 years as long as birth and death rates remain equal or current low fertility rates are maintained.

Age Distribution and Future Growth. Age distribution is an important indicator of a nation's future growth. For example, in 1991, about 50 percent of Kenya's population was under 15; about 2 percent was 65 or older. In contrast, 22 percent of the U.S. population was under age 15, and 12 percent was 65 or older. Kenya's young age distribution, coupled with a TFR of 6.7, means that there is a great potential for future growth because the majority of its population has not yet reached reproductive age. The older age structure of the U.S. population, coupled with a TFR of 2.1, means continued growth is not likely. Though the U.S. population will continue to grow slowly, thanks to the built-in momentum caused by the baby boom, it should stabilize and begin to decline after the boomers and their children have passed through their fertile years. This projection is based on the assumption that the U.S. total fertility rate will remain below replacement level. If that assumption proves correct, the U.S. population profile will eventually have the shape of (roughly) a straight column, indicating that there are about the same number of people in each age group. Sweden is one country with a columnar population profile. Its TFR has hovered around replacement level, 2.1, for a number of years. Each generation produces just enough children to replace itself.

Age Distribution and Dependency Load. Age distribution determines a nation's **dependency load,** the number of dependents—those under 15 or over 65—in the population. People in these age groups are generally assumed to be too young or too old to be in the formal work force and to contribute to the country's gross national product (Table 8-3). Worldwide, Kenya has the highest percentage of young people in the population, 50 percent. At least 30 percent of the population of every African country is under age 15, and in many the proportion is 45 percent.

▶ Table 8-3
Percentage of Population Under Age 15 and
Over Age 64, 10 Regions, 1991

Region	Population Under Age 15 (%)	Population Over Age 64(%)
World	33	6
More-developed countries	21	12
Less-developed countries	36	4
Less-developed countries, excluding China	39	4
Africa	45	3
Asia	33	5
Europe	20	13
Latin America	36	5
North America	22	12
Oceania	27	9

Source: Population Reference Bureau, *1991 World Population Data Sheet.*

▶ Table 8-4
Gross National Product per Capita, 10 Highest
and 10 Lowest Countries, 1991

Country	GNP ($ per capita)
10 HIGHEST	
Switzerland	30,270
Luxembourg	24,860
Japan	23,730
Finland	22,060
Norway	21,850
Sweden	21,710
Iceland	21,240
United States	21,100
Denmark	20,510
Canada	19,020
10 LOWEST	
Chad	190
Guinea-Bissau	180
Malawi	180
Bangladesh	180
Somalia	170
Laos	170
Nepal	170
Ethiopia	120
Tanzania	120
Mozambique	80

Note: Although Japan is now first in per capita GNP *among major industrial countries,* Switzerland is the world's richest nation, with a per capita GNP of $30,270. Most nations have per capita GNPs less than one-tenth that of Switzerland. About 50 countries fall below a "global poverty line" of $600 annual per capita GNP, established by the World Bank.
Source: World Bank.

Population Growth and Economic Development

Population growth affects both GNP and per capita GNP. In 1991 Mozambique, a fast-growing country (2.7 percent), had an annual per capita income of $80, the world's lowest. In contrast, the slow-growing population of Switzerland (0.3 percent) enjoys the world's highest per capita income, $30,270—over 378 times that of the average Mozambican's. In general, per capita income in fast-growth countries is far lower than per capita income in slow- or no-growth countries.

There is a direct relationship between population growth and economic development. For every 1 percent increase in population growth, the gross national product (GNP), an indicator of economic health, must increase 3 percent in order for the nation's economy to grow. This is a tall order for governments whose predominantly young populations are increasing at high rates. A growth rate of 3.5 percent, for instance, means that GNP must increase 10.5 percent a year in order to provide basic services (schools, housing, hospitals or medical clinics, and roads) and employment opportunities to all citizens. It is virtually impossible for real, sustainable economic development to occur when the government is overwhelmed by the difficult task of providing health care, education services, and employment opportunities for a rapidly growing population. Yet in the absence of sustainable economic development, illiteracy and unemployment become more widespread, strengthening the cycle of poverty (Table 8-4).

A large proportion of older citizens can also strain a nation's network of social services. Health care, nursing homes, and even transportation systems are affected by a sizable increase in the percentage of senior citizens. In the next several decades the growing proportion of older citizens in the United States is expected to strain the country's already overburdened Social Security system. If fertility remains relatively low, the responsibility for the care of the aging baby boomers (as a sector of the population) will fall on an ever-shrinking proportion of younger laborers.

Physical Boundaries

What Can We Learn from Studying Nonhuman Population Dynamics?

As we learned in Chapter 5, nonhuman populations are subject to numerous checks and balances. Further, environments are not utopian; all natural populations have a maximum number that their environment can support. In other words, all nonhuman populations studied thus far have an upper limit to growth. This upper limit, the **maximum population size,** is determined by the **biological carrying capacity** of the environment, that is, the maximum number of a particular species that the environment can support year after year without degradation. Typically, populations do not exist for extended periods at the maximum population size. Rather, they fluctuate over time around the number of individuals the environment can best support, known as the **optimum population size.** Populations remain at or near the optimum size when a balance is struck between the biotic potential of the species (positive feedback) and environmental resistance (negative feedback).

Why Do Biologists Calculate the Carrying Capacity of an Environment?

Biologists attempt to calculate the carrying capacity of an ecosystem, such as a grassland or lake, to determine how many organisms of a particular species can be sustained over a period of time. To do so, they must have a good understanding of the resources necessary for the survival of the species.

Numerous factors can raise or lower environmental carrying capacity: variations in climate or weather patterns, the relative availability of food in a particular year, and the relative number of predators, to name but a few. Consequently, estimates of carrying capacity often vary. Wildlife biologists try to keep the population of the species they are managing far enough below the carrying capacity to avoid the phenomenon of population explosion and crash. When some check on population is reduced (for example, when predators are eliminated), the population will increase rapidly for a period of time until it reaches a peak above the carrying capacity. This is a population explosion. The high numbers stress and degrade the environment, and eventually, the population declines (usually quickly) because the habitat cannot support it. This is the population crash. In some cases the explosion-crash phenomenon so seriously degrades the environment that it cannot regain its former productivity for a very long time, with the consequence that its carrying capacity is, in effect, permanently reduced.

Carrying Capacity as Applied to Human Populations

The concept of carrying capacity was developed by population biologists in their study of nonhuman population dynamics. Thus, it can be applied to humans by analogy only and with several considerations. First, as for nonhuman populations, there is no standard, universally agreed upon equation to calculate the carrying capacity for human populations.

Second, human populations are typically classified according to geopolitical units, and such cultural classifications are meaningless in ecological terms. Rivers, for example, often cross national borders. The amount of water available to people in one country (for drinking water, sanitation, agriculture, and industrial use) sometimes depends on the actions of peoples in other nations upstream. Diversion projects and dams may reduce a river's flow so much that it has a detrimental effect on the carrying capacity of the ecosystem downstream (see the discussion of the consequences of large-scale hydroelectric projects in Chapters 13 and 15).

Third, unlike nonhuman populations, humans are able to raise the carrying capacity of an environment in several ways. Technologies exploit otherwise unavailable resources. Mining processes, for instance, have been so refined that ores containing smaller and smaller percentages of the desired mineral can be used in production. Moreover, humans can bring in resources from outside their immediate environment. They can exploit resources in ecosystems in which they do not live in order to sustain themselves in their own environment. The Middle East, for example, supplies about one-half of the oil needed to meet U.S. demand.

Fourth, it seems reasonable to measure the carrying capacity for nonhuman populations in terms of the resources needed for survival, but most people agree that this strategy is not acceptable for people. Consequently, the most important factor that distinguishes the calculation of carrying capacity for humans from that for nonhumans is **quality of life** or **standard of living.** The carrying capacity of the environment will differ according to the quality of life a population aspires to.

Determining carrying capacity for the biosphere or selected countries is a complex and inexact science, but some attempts have been made. *Limits to Growth* (1972) presented the results of a global modeling study that attempted to determine the earth's carrying capacity. Using systems analysis, computer projections, and a wealth of data, researchers determined that the planet could support a population of 6 billion at a European standard of living. In Chapter 9, we discuss how China determined its carrying capacity based on available fresh water and food.

Standards of living vary widely from nation to nation and from region to region. Twenty percent of the world's population (living in MDCs) uses about 80 percent of the earth's resources (both energy resources and materials). The remaining 80 percent of the population uses just 20 percent. According to Paul Ehrlich, each birth of a child in the United States exerts 200 times the load on the world's natu-

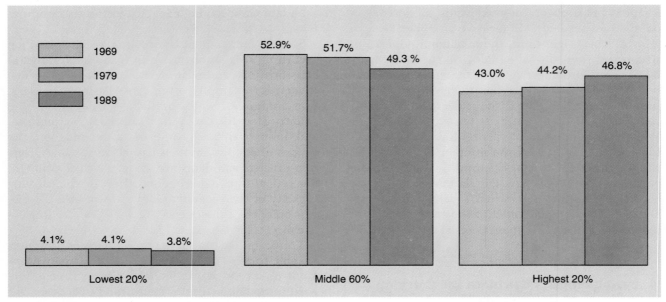

FIGURE 8-10: Poverty and its attendant ills—homelessness, hunger, and despair—are very real in the United States. Over the past three decades America's wealthy have received an increasing proportion of the nation's aggregate household income, while the poor's meager share has dwindled.

ral resources as does the birth of a child in Bangladesh. Thus, if the entire human population were to use resources at the rate of that found in the MDCs, resources would be depleted much more quickly and far more pollution would be produced. Disparities in resource use and standard of living can also be seen between different segments of a single society. In the United States, for example, the wealthiest 20 percent of the population receives almost 47 percent of the national income; the poorest 20 percent receives less than 4 percent (Figure 8-10).

When a population exceeds the carrying capacity, the environment is eventually degraded, and environmental resistance increases. Certain factors increasingly come into play: hunger, starvation, disease, competition for resources, lack of space. The amount of resources per capita decreases. There is simply less to go around—less drinkable water, less arable land, less food, less suitable living space. Consequently, the quality of life decreases. Unless resources are brought into the area from outside, the population will eventually be reduced to a size that can be supported by the environment.

In most nations where environmental degradation has reduced the carrying capacity, food and medical aid have been brought in to supplement (or replace) local resources. Some experts believe that this aid is misguided, for it falsely "props up" the carrying capacity of the land. The population continues to grow, and each year more and more people compete for fewer and fewer resources. Environmental stress increases, and the land's carrying capacity is further reduced, exacerbating and prolonging the misery of the entire population. When the situation in an area becomes intolerable, residents often flee (usually to cities or aid camps) in search of food. They become **environmental refugees,** people forced to abandon their homes because the land can no longer support them. In the absence of continued and accelerated relief aid, a population crash is inevitable.

Determining carrying capacity is a difficult and imprecise process, but can nations afford *not* to make the attempt? Environmental Science in Action: Ethiopia (pages 144–147) presents an example of the complex relationship between a human population and its environment, and the consequences that arise when the carrying capacity of the land is reduced.

Social Boundaries

What Do Demographic Statistics Tell About Quality of Life?

Demographic statistics reveal important information about the quality of life of the majority of people in a society. Four demographic factors are particularly telling: population density, urbanization, life expectancy, and infant mortality rate (Table 8-5).

Population Density

Humans are not uniformly distributed over the earth. Not only the size of a population, but also its

▶ Table 8-5
Demographic Statistics

	Infant Mortality Rate	Total Fertility Rate	Population Under Age 15/over 65 (%)	Life Expectancy at Birth (years)	Urban Population (%)	Married Women Using Contraception, Total/Modern/ Methods (%)	Per Capita GNP, 1987 (U.S. dollars)
World	68	3.4	33/6	65	43	56/48	3,760
More-developed countries	14	1.9	21/12	74	73	71/55	16,990
Less-developed countries	75	3.9	36/4	62	34	52/46	750
Less-developed countries, excluding China	85	4.4	39/4	59	37	44/36	910
Afghanistan	182.0	7.1	43/3	41	18	—	—
Australia	7.7	1.8	22/11	76	85	67/47	14,440
Bangladesh	120.0	4.9	44/3	53	14	33/26	180
Bolivia	93.0	4.9	41/4	56	50	30/12	600
China	33.0	2.3	27/6	69	26	71/70	360
Denmark	8.4	1.6	17/16	75	85	63/59	20,510
El Salvador	55.0	4.6	44/4	62	43	47/45	1,040
Haiti	107.0	6.4	45/4	53	28	10/9	400
Hungary	15.7	1.8	21/13	70	59	73/62	2,560
India	91.0	3.9	36/4	57	27	49/42	350
Italy	10.1	1.3	19/13	74	72	78/32	10,420
Japan	4.5	1.5	18/12	79	77	64/60	23,730
Kenya	62.0	6.7	50/2	63	22	27/18	380
Mexico	43.0	3.8	39/4	70	71	53/45	1,990
Peru	76.0	4.0	39/4	65	69	46/23	1,090
Sierra Leone	147.0	6.5	44/3	42	30	—	200
Sweden	5.8	2.1	18/18	78	83	78/71	21,710
Thailand	39.0	2.2	35/4	66	18	68/65	1,170
Turkey	62.0	3.7	38/4	64	60	77/49	1,360
United States	9.1	2.1	22/12	75	74	74/69	21,100
Former Soviet Union	23.0	2.3	26/9	70	66	—	—
Yemen	121.0	7.4	49/3	50	25	—	640
Zaire	83.0	6.1	46/3	52	40	—	260

Source: Population Reference Bureau, *1991 World Population Data Sheet.*

density, how closely people are grouped, are important demographic factors, especially with respect to the degree of resource use within defined geophysical or political boundaries. Consider, for example, that over 116 million Bangladeshis crowd into an area about the size of Arkansas, which has a population of about 2.5 million (Figure 8-11). India has more than three times the population of the United States in about one-third of the space. Though China is slightly larger than the United States in terms of surface area, the mountain and desert areas of the country are sparsely inhabited, and its population of over 1 billion occupies a space smaller than that of the continental United States. Because all of these densely populated countries have a smaller resource base than that of the United States, their people have a much smaller per capita resource base. In general, density in developing nations is more than double the density in developed nations.

Often, countries that are densely populated perceive the problem as one of an inadequate resource base rather than too large a population. When this perception is widely accepted, population pressures may be manifested in the desire to annex adjacent lands, colonize distant lands, acquire resources from outside the country, or encourage emigration. History is replete with examples of expansionist policies associated with population pressures: the Roman Empire, western European colonization of the Americas and Africa from the 1600s to the 1800s, Adolf Hitler's Germany, and the Japanese Empire of the 1930s and 1940s.

FIGURE 8-11: Bangladeshi victims of the April 1991 cyclone await handouts of rice in the town of Pekua. The storm killed more than 125,000 people and left many others homeless. Its high population density renders Bangladesh susceptible to environmental problems and strains the nation's ability to respond to natural catastrophes.

Urbanization

One indicator of increasing population density is **urbanization,** a rise in the number and size of cities. In 1970 about 37 percent of the world's population lived in cities. By the turn of the century nearly 50 percent will inhabit the planet's sprawling cities, and by 2025 over 60 percent will be urban dwellers (Figure 8-12).

Currently, the vast majority (about 75 percent) of people in MDCs live in urban areas, while the majority of people in most LDCs still live in rural areas. This situation is rapidly changing, however. In 1970 only about one-fourth of the population of LDCs was urban; in 1987 that proportion had risen to one-third; by 2025 well over one-half will live in cities. By 2000 eight of the ten most populous cities in the world will be in LDCs. Mexico City, at over 25 million, will head the list, followed by Sao Paulo,

Brazil; Tokyo, Japan; Calcutta and Bombay, India; New York, United States; Seoul, South Korea; Shanghai, China; Rio de Janeiro, Brazil; and Delhi, India.

Urban areas place a strain on the environment far greater than the absolute numbers of people. The concentration of people means that wastes and pollution are also concentrated, often to such a degree that natural processes are unable to cleanse the environment of contaminants. Moreover, while rural dwellers are typically engaged in food production, urbanites are strictly consumers.

Several factors are responsible for the rapid growth of cities in the poorer nations. In many regions a relatively small percentage of the population owns most of the land; wealthy landowners are able to buy out small landowners or charge such high prices to rent their land that peasants are unable to make a living farming the land. In desperation they move to the cities in search of work. This situation is particularly true in Latin America, where an estimated 75 percent of the population will be urban dwellers by the year 2000. In Africa the growth in urban populations is due in large measure to the fact that environmental degradation and droughts have reduced soil fertility, making it difficult for farmers to earn a living from the land. Thus, throughout the developing world, more and more people are leaving the countryside in search of opportunity and a better way of life in the city. In general, the poor in urban areas are better off than the rural poor—at least they are able to find or scavenge food, make a living (however meager), and find some kind of shelter in one of the many squatter's areas or shantytowns that invariably ring cities in the developing world. Shantytowns are typically crowded and dirty, with nonexistent or inadequate sewage facilities and contaminated water supplies.

The influx of people to cities in the developing nations places a significant strain on the financial resources of the governments. Poor nations simply do not have the money or the infrastructure to provide adequate housing, education, transportation, and health and sanitary services to their expanding populations; they are overwhelmed by a ceaseless tide of humanity. Saddled with heavy debt burdens, they must allocate a large portion of their financial resources to pay the interest on overdue foreign loans. To earn needed cash in the short run, they are often forced to take actions that deplete the countries' natural resource bases, consequently lowering the carrying capacity.

Governments in the industrialized world are also struggling to meet the needs of their urban poor. High rates of unemployment, increasing numbers of high school dropouts, and violent crime are all too common in the inner cities of the United States. In

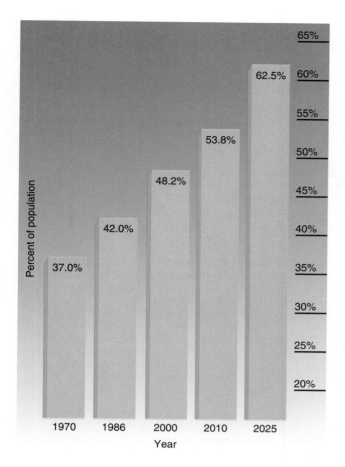

FIGURE 8-12: World urbanization trends, 1970–2025.

addition, the nation's urban areas appear to be experiencing a resurgence of racial unrest and ethnic violence as the economic gap widens between America's wealthy and her poor. Certainly, there are many differences between cities in LDCs and those in the United States, but sadly, the similarities are more important. Crowding, crime, pollution, sickness, economic deprivation, and desperation are the common experiences of the urban poor, whether those poor live in a shantytown in Mexico City, makeshift shacks on a Calcutta sidewalk, or the slums of Los Angeles or Detroit.

Life Expectancy

Life expectancy at birth, the average number of years a newborn can be expected to live, is a good indicator of a nation's standard of living. For both men and women, life expectancy is higher in the developed world, about 74 years, than in the developing world, about 62 years (see Table 8-5).

Europe and North America have the world's highest life expectancies, an average of 75 years. In 1991 life expectancy was highest in Iceland, Sweden, and Switzerland, all at 78 years. African nations have

among the world's lowest life expectancies, an average of 53 years. Within Africa, however, regional and national differences are striking. Contrast western Africa (49 years), eastern Africa (52 years), and middle Africa (50 years) with northern Africa (59 years) and southern Africa (63 years). A child born in 1991 in the French territory of Réunion, a small island in the Indian Ocean, can be expected to live to age 72, while life expectancy for a newborn in Ethiopia is just 46 years.

Disparities are evident in Asia and Latin America as well. Life expectancy in eastern Asia (which includes Japan and China) in 1991 was 70; in southern Asia, the figure was 57. Southern Asia includes the war-torn nation of Afghanistan, with the world's lowest life expectancy (41 years), but it also includes Sri Lanka, with a life expectancy of 70.

The Eastern Hemisphere does not have sole dominion over low life expectancies. The Caribbean nation of Haiti, plagued by political turmoil and corruption, violence, an epidemic of Acquired Immune Deficiency Syndrome (AIDS), poverty, and unemployment, has a life expectancy of 53 years, low when compared with the neighboring nations of Dominica, Netherlands Antilles, Barbados, Cuba, Jamaica, and Martinique, all with a life expectancy of 76 to 77 years.

What factors contribute to low life expectancy? Among the most visible are war and hunger. Ongoing strife is a factor in the lower life expectancy of both Angolans (46 years) and Afghans. The Ethiopian civil war exacerbated the problems caused by chronic famine; life expectancy in 1991 was 46 years. Like Ethiopia, the West African nation of Sierra Leone (life expectancy, 42 years) has been beset by famine and food shortages. By far, however, the most significant factors that keep average life expectancy for the developing world so far below that of the developed world are infant and childhood mortality.

Infant and Childhood Mortality

The **infant mortality rate** (IMR) is the annual number of infants under age 1 who die per 1,000 live births. It is widely considered the single best indicator of a society's quality of life. The **childhood mortality rate** is the annual number of children between the ages of 1 and 5 who die per 1,000 live births. Both rates are sometimes expressed as percentages.

In 1991 the average infant mortality rate worldwide was 68/1,000, or 6.8 percent (see Table 8-5). The rate for MDCs was 14/1000 (1.4 percent), while the rate for LDCs was 75/1000 (7.5 percent). The rate for LDCs excluding China was 85/1000 (8.5 percent). Worldwide, among countries with a population of 1 million or more, Japanese infants

(Text continues on page 148.)

Environmental Science in Action: Ethiopia

Tracy Linerode

During 1984 and 1985 catastrophic famine in Ethiopia dominated the news. Persistent drought, coupled with unchecked population growth, had exacerbated the problems of an already fragile agricultural system. But the tragedy in Ethiopia was more than the result of drought and population growth; complex, interrelated factors set the stage for the disaster that unfolded in this nation on the Horn of Africa: a history of oppressive governments, decades of civil war, and long-standing ethnic hostilities. Although countries and organizations all over the world rushed food and supplies to Ethiopia, the conditions that led to the famine persist and will continue to do so until a stabilizing federal government, acceptable to Ethiopia's various ethnic groups, makes a consistent, concerted effort to solve the problems that make the nation especially vulnerable to food shortages.

Describing the Resource

Physical Boundaries

Much of north and central Ethiopia consists of high plateaus and mountains, with a strip of tropical lowland extending along the Sudan border (Figure 8-13). The Great Rift valley, one of the most extensive faults on the earth's surface, splits the region into western and eastern highlands. The western highlands are characterized by massive mountains ranging from 8,000 feet (2,400 meters) to 12,000 feet (3,600 meters) above sea level, while the eastern highlands are lower, with more plateau regions. A less elevated geographic area encompasses the Rift valley and the lowland plains in the south.

Although Ethiopia lies just north of the equator, the temperature varies with elevation. The lowlands and most of the Rift valley are hot all year round. In contrast, the highest terrain experiences temperatures between 32°F (0°C) and 60.8°F (16°C). The climate of the highland plateaus, the country's most agriculturally productive region, is milder, with temperatures between 60.8°F (16°C) and 84.2°F (29°C) annually.

Although the southwestern region of Ethiopia receives rainfall throughout the year, precipitation follows a seasonal pattern in most of the highlands. The rainy season usually begins with light rains in May and June, followed by a short period of hot, dry weather. Between June and September torrential downpours are common. If a rainy season is missed, no precipitation can be expected for at least six months. When the rain does fall, it comes down with such force that any soil not protected by plant cover and secured by root systems is easily washed away, especially in the mountainous highlands.

In many areas the soil contains an extremely high percentage of clay and is thus easily waterlogged during rainy spells. Farmers must either wait for the soil to dry out, shortening the growing period, or plant water-tolerant crops, which usually yield less than other varieties. As a result, they tend to plant crops on slopes, which drain faster than flat fields (these are often left for grazing). Unfortunately, cultivating the slopes makes them more vulnerable to erosion. Already, half of the farmland in Ethiopia, 35 million acres (14.16 million hectares) has been severely degraded by erosion, and 4.9 million acres (1.98 million hectares) are now devoid of soil. Some soil experts predict that if the current rate of erosion continues, the amount of acreage lost to agriculture will increase fivefold by the year 2010 (Figure 8-14).

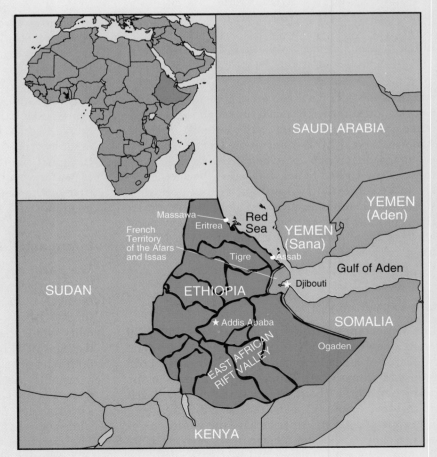

FIGURE 8-13: Ethiopia.

144

Biological Boundaries

Ethiopia is the historic home of a wide variety of crops. Some plant geneticists believe that varieties of millet, coffee, safflower, barley, and emmer (a type of hard red wheat) were first domesticated in the region. Ethiopian farmers also grow legumes such as chick-peas and lentils, oilseeds such as linseed and sesame, cotton, and sugarcane. Many farmers also raise livestock, primarily cattle. In fact, Ethiopia produces 25 percent of all the cattle raised on the African continent. Consequently, almost half of the arable land is used for grazing.

Agricultural production is the heart of Ethiopia's economy. Over 80 percent of the population is involved in farming. Agricultural products, especially coffee, make up 90 percent of Ethiopia's exports and account for 50 percent of its GNP. Even so, two-thirds of the country's farmers produce only enough to stay alive. As a consequence, Ethiopia imports about 15 percent of its food requirements even in nonfamine years.

Many observers see Ethiopia's rapidly growing population as the chief cause of its famines. Others maintain that Ethiopia has a potential carrying capacity much greater than its current population and should be able to feed itself for many years even with a rapidly growing population. They contend that the country's political system and agricultural practices have greatly limited its capacity to produce food, and thus a growing population causes food demand to outstrip production by a greater and greater margin. Consequently, Ethiopia must import more food every year.

Ethiopia's rapid population growth is due to a birth rate roughly two and a half times the death rate. Between 1984 and 1991 the population grew from 46 million to over 53 million, and this number would surely be higher if not for the soaring death rates during the 1984–1985 famine. Although the growth rate slowed to 2.9 percent by 1991 (in large measure because of deaths and emigration during the famine years), Ethiopia's population could double in just 24 years and is projected to swell to over 140 million by 2025.

Two major factors are responsible for high birth rates. First, poor or nonexistent health services and contaminated water contribute to an infant mortality rate of 130/ 1000, among the world's highest. Multiple offspring ensure that parents will have enough children to help them work their farms. Second, most Ethiopian men consider fathering many children to be a sign of virility.

Social Boundaries

Ethiopia is a nation of great cultural diversity, with 70 ethnic groups and at least as many languages. To understand present-day Ethiopia, one must first understand the historical roots of the nation's ethnic diversity, which is the source of its current dividedness and strife.

Ethiopia traces its history to the ancient kingdom of Aksum, which came to be known as Abyssinia. Encompassing most of present-day northern Ethiopia, the empire was ruled by a dynasty thought to be descended from the legendary Menelik I, son of King Solomon and the Queen of Sheba. During the late nineteenth century, Emperor Menelik II conquered provinces to the east, south, and west, extending the boundaries of the empire almost to their present-day limits. In the first half of the twentieth century Ethiopia gained the Ogaden area along the Somali border through treaties with Britain, France, and Italy. Finally, in 1952, at the recommendation of the United Nations, Britain granted Ethiopia the rule of Eritrea, the area to the north of Ethiopia along the Red Sea.

Cultural diversity has proved especially divisive in Ethiopia. The Amharas, an ethnic group from the nation's original central provinces, are at the top of the country's class structure. Predominantly Coptic Christians, members of the Ethiopian Orthodox church, they have long dominated the central government

FIGURE 8-14: A civil war and widespread environmental degradation have made refugees of many Ethiopians. Deforestation is a main cause of environmental degradation. Less than 12 percent of the land now sustains forest, compared to 40 percent in the late 1800s. Ethiopia's rapidly growing population has required more and more fuel and farmland, and cutting down forests provides both. Unrestrained clearing of forests has almost exhausted the supply of firewood and has left vast stretches of land unprotected against erosion.

in Addis Ababa, the nation's capital. The Amhara consider themselves to be the most sophisticated of the Ethiopians and therefore the best suited to rule the nation. The Oromos, Ethiopia's largest ethnic group with 40 percent of the population, resent the Amhara and the political power they wield. Though now dispersed throughout Ethiopia, their traditional homeland is in the southern provinces, and they feel stronger ties with the people of Somalia than with the Amhara. The Eritreans, who remain concentrated in their far northwestern province, have never accepted rule by Ethiopia. When the Eritrean parliament voted for full unification with Ethiopia in 1962, amid rumors of bribery and intimidation by the Amharan-dominated central government, Eritreans unhappy with the decision began a civil war that would last for three decades. Tigreans, also from the northern provinces, are ancient rivals of the Amhara; the animosity between the two groups dates back to the time when Aksum, located in the heart of Tigre, was the capital of the ancient Ethiopian kingdom.

Looking Back

Under Emperor Haile Selassie, who ruled Ethiopia from 1930 to 1974, farmers comprised 90 percent of the labor force, but few owned their own land. In a legacy of a feudal system that began in the twelfth century, landlords held more than 2 million acres (809,389 hectares) of the country's arable land and received up to three-quarters of what was produced on it. The peasants who farmed the land were obliged to give the landowner and managers money, free labor, provisions, and pack animals. Wealthy land-owners encouraged the planting of cash crops like coffee, cotton, and carnations rather than food. Consequently, farmers had no incentive to produce high yields or care for the land. When pressed to increase yields, they abandoned crop rotation, farmed marginal land, or deforested areas. Livestock were allowed to overgraze fields. The vicious cycle of destructive agricultural practices and population growth made Ethiopia more vulnerable to the effects of drought.

As the gap between rich and poor in Ethiopia widened, a severe drought struck the country in 1972. Twenty million people were perpetually hungry, and 8 percent of the population actually starved to death. So many people had to share the few available water sources that the sources became disease ridden; fewer than 3 percent of the population had access to uncontaminated water. Food aid was grossly mismanaged by the imperial Ethiopian government, which used available motor vehicles to control the dissatisfied population rather than to transport food. The Ethiopian Orthodox church and wealthy landlords sold food to relief groups for as much as a 300 percent profit.

Rising discontent fueled a revolution in 1974, when the government was taken over by Soviet-backed Marxists. Within a few years Lieutenant Colonel Mengistu Halie

Mariam was firmly in place as the country's leader. The new regime nationalized land, restricting the size of individual family farms to about 25 acres (10 hectares), but it did little to help farmers become self-sufficient. Nearly all of the funds spent on agriculture went directly to collectivized state farms; even so, these produced just 6 percent of the country's output. By far, the largest percentage of the government's money was spent battling various internal rebel groups, particularly in the north. To help feed the army and populous cities, food prices were set approximately 70 percent below market level. World Bank monies were used to increase the production of cash crops. Unable to own land or profit from their labors, destitute peasant farmers had even less incentive than before to increase yields or take care of the land.

The Famines of the Mid and Late 1980s

The famine of 1984–1985 was technically brought about by a drought, which was especially severe in the northern provinces. Up to 1 million people starved to death. In 1984 the government began a "resettlement" program to move farmers from the drought-stricken north to the more fertile southern provinces. Farmers were tricked into government camps by promises of grain or cattle and then abducted and forcibly moved. Thousands were separated from their families, kept for days with little food or water, dumped in areas with no shelter, source of water, or agricultural equipment, and forced to perform slave labor for the government. An estimated 10 to 20 percent of the 500,000 people resettled during 1984–1985 died in the process. Worldwide, many people suspected that Mengistu was trying to remove potential troublemakers from the rebellious north rather than help poor farmers.

Another practice criticized as inhumane was villagization. Claiming that it was easier to give farmers health, educational, and social services if they lived in centralized

villages, the government forced farmers in the southern provinces to pack up their belongings and move. Soldiers burned the homes of those who refused to go. Many victims fled the country to Somalia. The lack of services provided to the "villages" confirmed many groups' suspicions that the government was only trying to quell unrest by the Muslim Oromos.

Foreign aid to minimize the suffering inflicted by the drought was plentiful. By January 1985 the Ethiopian government had received $312.7 million and 560,907 tons (509,916 metric tons) of food from nongovernment international organizations alone. The United States, the largest donor, supplied 383,000 tons (348,530 metric tons) of food and $37.5 million in nonfood aid. Unfortunately, politics, rather than need, influenced aid distribution. The government refused to permit relief workers to provide food or money to anyone in the rebel provinces, especially Eritrea, even though most of those starving were located there. Government and army personnel were accused of confiscating equipment from relief workers, beating volunteers, and pillaging relief camps suspected of harboring rebels. What grain and livestock peasant farmers managed to obtain was confiscated by the government in order to keep the army and urban centers content.

In 1987 another drought threatened to devastate the country, but a full-scale crisis was averted by substantial foreign aid. When the rains failed again in 1989, international aid provided 650,000 tons (591,500 metric tons) of food and relief supplies. However, Mengistu refused to allow any of the aid to enter the rebel territory, and he announced that the government would abandon proposed development plans for 1989–1990 in order to use those funds for the suppression of the rebels.

The Revolution of 1991

Mengistu's government was finally overthrown, in May of 1991, by its main rivals—the Ethiopian People's Revolutionary Democratic Front

(EPRDF), the Oromo Liberation Front (OLF), and the Eritrean People's Liberation Front (EPLF). The EPRDF, a four-group umbrella organization dominated by the Tigre People's Liberation Front, took over Addis Ababa, with the sanction of the United States, in order to restore order and calm in the capital. Under terms worked out by an international group, the rebels promised to hold multiparty elections within a year. Meanwhile, the Eritrean People's Liberation Front announced that it would administer the province until a vote on Eritrea's status is held, a vote that is expected by most political observers to result in secession from Ethiopia. The United States announced that it supported the Eritreans' right to self-determination. Many non-Eritreans oppose the province's independence, however, because without Eritrea's Red Sea coast, Ethiopia would be landlocked. Food aid and supplies, so critical in famine years, enter the country chiefly through the Eritrean ports of Massawa and Assab. The Eritreans have promised not to interfere with the flow of goods through their territory, but in a country rife with historic ethnic divisions such promises are looked on with suspicion.

Looking Ahead

Ethiopia's future remains cloudy. The country set 1994 as a target date for self-sufficiency in food, but they will achieve this goal only if the political situation stabilizes, future relief efforts help to create self-sufficiency and promote sustainable agricultural methods, and population growth slows.

Political Situation

The prospects for assembling a stable, enduring central government in Ethiopia appear slim. In the absence of their common enemy Mengistu, the military allies that helped to overthrow him may turn to fighting among themselves. The EPRDF, dominated by the Tigreans, wants a unified country, but has pledged not to oppose Eritrean independence. Meanwhile, many Oromos

are also demanding a vote on either autonomy or independence for the southern provinces, a demand that has heightened tensions between them and the Tigreans. Thus, which provinces will form a new Ethiopia is unclear. What kind of government will eventually be established is also unclear. Although the Eritreans and the Tigreans claim to favor a market economy and political pluralism, both groups are former Marxists who have little experience with democracy.

Relief Efforts

In 1992 famine and starvation still threatened some 7 million Ethiopians. In an attempt to break the cycle of drought, unsustainable agriculture, and famine, aid donors are taking steps to help Ethiopians become more self-sufficient. Chief among these steps are providing farmers with drought-resistant seeds and basic farming implements and providing villages with better water supplies and health services to prevent the spread of disease during droughts. Encouraging the production of native grains such as teff can also help to alleviate hunger. Teff can survive at higher altitudes than corn or sorghum, and it needs less rain than wheat or barley, the other principal high-altitude crops. Ethiopians in both rural and urban areas eat bread made from teff as a staple of their diet.

Restoring soil fertility will also help Ethiopia's beleaguered food production system. The acacia tree, which fixes atmospheric nitrogen, could be planted on and around agricultural fields. In addition to fixing nitrogen, acacias would generate a 50 to 100 percent increase in soil organic matter, increase the capacity of the soil to hold water, increase microbiological activity in the soil, decrease wind erosion, and produce fodder for livestock.

Another way to rebuild soil is to encourage composting, irrigation, and no-till methods of agriculture. Some researchers in Ethiopia are experimenting with methods based on native practices uniquely adapted to Ethiopia's soil and climate. They have adapted a traditional plow so that it builds soil-catching

terraces as the farmer plows; these terraces can increase barley yields by 50 percent in two growing seasons. Another plow creates raised "broad" beds that allow water to drain more quickly. Still another tool facilitates the no-till technique with a blade that runs beneath the surface of the field and cuts the roots of weeds without disturbing the soil.

Population Growth

Even if political and agricultural reform could enable Ethiopia to support its population, rapid growth must end sooner or later. Many experts point out that farmers will remain opposed to family planning until the standard of living rises all over the country. People must feel sure that the children that they do have will live and that they have some economic security other than their offspring before they will see the merits of limiting family growth.

Education is a crucial factor in effective family planning. In a male-dominated culture like that of Ethiopia, men's understanding of the importance of family planning may be even more crucial than women's. International groups propose establishing regional offices to sponsor education programs and provide community health workers for each province, ensuring that couples who want help with family planning have access to it.

Because so many of Ethiopia's interrelated problems grow out of its history, politics, and culture, we who live in the United States and other countries can have little direct influence over the path that Ethiopia takes. However, we have some influence through the type of aid we are providing. Are we teaching Ethiopian farmers to be self-sufficient and conscious of soil conservation? Are we insisting that a stable government make appropriate reforms before we provide aid? Such small incentives may help the Ethiopian people to implement the changes needed to protect their environment, safeguard the integrity and self-sufficiency of the peasant farmers, and ultimately ensure the health and welfare of the nation.

had the best chance of surviving their first year; about 4.5 children per 1,000 died before their first birthday. Prospects were grimmest for Afghan infants; 182 per 1,000 (18 percent) never saw their first birthday. Many African nations also had IMRs well over 100/1,000, with the continent as a whole averaging 102/1000. In the Western Hemisphere only Haiti, at 107/1,000, had a 1991 IMR of 100/1,000 or greater, although Bolivia, at 93/1,000, was close.

The African nations of Rwanda and Kenya provide evidence that infant mortality rates can have a significant effect on a nation's growth rate. In 1991 the total fertility rate for Rwanda was 8.1, that is, a Rwandan woman could be expected to have 8.1 children over the course of her childbearing years. For Kenya the TFR was 6.7. Yet Kenya grew at a rate of 3.8 percent in 1989, while Rwanda's growth rate was 3.4 percent. What caused Kenya's faster growth? Two factors are especially important. First, Kenya's population is three times greater than Rwanda's, and thus there are more women of childbearing age. Secondly, Rwanda's IMR (117/1,000) is almost twice as high as Kenya's (62/1,000).

Though infant mortality rates are uniformly higher in LDCs, they vary widely among countries in both the developed and developing world. The infant mortality rate in the United States, for example, is twice as high as that in Japan. In fact, the United States, with a 1991 IMR of 9.1/1,000, ranks behind most other industrial nations. This poor ranking is a cause for great concern. There have been improvements in recent years—for most of the 1980s the nation's IMR remained above 10/1,000, and its recent decline is the first time the IMR has ever been below 10/1,000—but these overall rates must be viewed cautiously. The IMR for the poor and some minority groups in the United States is much higher than the national average, hovering at around 20/1,000 for blacks and Hispanics. Moreover, many people argue that the average rate would be even lower if the United States had a national health care system or better care (especially prenatal care) for its poor. The U.S. situation underscores the fact that, as Dr. Marsden Wagner of the World Health Organization has said, "Infant mortality is not a health problem; it is a social problem."

It seems almost bizarre that an IMR of 1 percent or 2 percent causes such controversy in the developed world while almost 10 percent of infants routinely die in developing nations each year. Why do so many infants and children in the developing nations die? Naturally, famine and malnutrition contribute to the high infant and childhood mortality rates. In 1985, in some Ethiopian villages in the war-torn region of Eritrea, 60 percent of children died before their first birthday. But famine, while unde-

niably a relentless killer, is not the most common cause of infant and childhood mortality. More children die each year because they are afflicted with diarrhea (which can cause life-threatening dehydration when left untreated), are not immunized against common childhood diseases (particularly measles), or do not receive treatment for curable illnesses and infections (such as malaria, whooping cough, pneumonia, and tetanus).

Often, the culprit in children's illnesses is disease-infested water (Figure 8-15). Sadly, parents generally have no choice but to continue to use contaminated water. Particularly in areas where fuel (usually wood) is in short supply, parents feel they cannot "waste" the fuel by boiling water that *looks* clean. Thus, even if children with water-borne diseases receive adequate medical attention and recover, they will pick up parasites and bacteria again and again.

Many children also die because they are improperly weaned. They may be weaned, or withdrawn, from mother's milk too early, if the mother does not sufficiently space the birth of her young, and not be able to get sufficient nourishment from a meager adult diet. In many areas in Africa, however, children are weaned too late; their only source of nour-

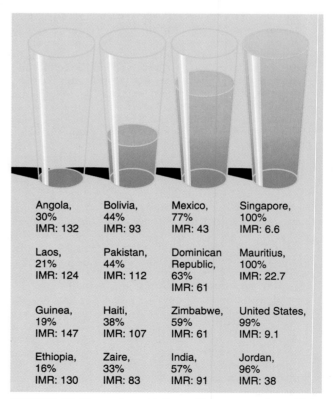

Angola, 30% IMR: 132	Bolivia, 44% IMR: 93	Mexico, 77% IMR: 43	Singapore, 100% IMR: 6.6
Laos, 21% IMR: 124	Pakistan, 44% IMR: 112	Dominican Republic, 63% IMR: 61	Mauritius, 100% IMR: 22.7
Guinea, 19% IMR: 147	Haiti, 38% IMR: 107	Zimbabwe, 59% IMR: 61	United States, 99% IMR: 9.1
Ethiopia, 16% IMR: 130	Zaire, 33% IMR: 83	India, 57% IMR: 91	Jordan, 96% IMR: 38

FIGURE 8-15: Share of population with access to clean drinking water and infant mortality rates, selected countries, 1990. Access to clean water is related to the infant mortality rate. In areas where people have access to clean water, even in the developing world, infant mortality is lower than in areas where they do not.

ishment may be breast milk until they are 12 to 18 months of age. When they are finally weaned, the children are given adult food, which may be difficult for their systems to digest.

Another cause of high infant mortality rates is the poor health of the mother and inadequate prenatal care. Low birth weight (like malnutrition) is one of the most highly reliable indicators of poor infant health. Low birth weight can be caused by a number of things, including the young age of the mother, poor maternal health, and lack of prenatal care, all of which are linked to the low status of women.

The key to saving young lives is to educate and empower women so that they can better care for their children. According to the United Nations' Children's Fund (UNICEF), one-half to two-thirds of the estimated 14 to 17 million children who die each year from the combined effects of poor nutrition, diarrhea, and disease (measles, pneumonia, whooping cough, and tetanus) could be saved through simple measures. Oral rehydration therapy (ORT), for instance, can save the life of a child dangerously ill from diarrhea-induced dehydration; ORT salts cost just about ten cents a pack. But the mother who does not know why her child is dying, who does not have access to medical care, and who cannot afford to pay for care even if it is available, will not be able to prevent her baby's death.

What Is the Relationship Between Women's Status and Population Growth?

Women's status is a significant, though not often recognized, factor in population growth. In nations where women's status is low, and women have few avenues open to them, motherhood becomes the only option and birth rates rise. As we have seen, high birth rates contribute to high population growth rates; a rising population stresses limited resources.

Women's status is reflected in such factors as access to education, access to adequate health care, legal rights, employment opportunities outside the home, wage earnings, marriage age, and number of children. In 1988, the Washington D.C.-based Population Crisis Committee, a private, nonprofit organization founded in 1965, published a report called *Country Rankings of the Status of Women: Poor, Powerless, and Pregnant*. Researchers used data compiled by a variety of sources, including the United Nations Educational, Social, and Cultural Organization (UNESCO), the International Labour Office, the U.S. Bureau of the Census, the Population Council, and the U.N. Statistical Office. They compared 99 countries, with a total of 92 percent of the world's

female population (2.3 billion women), to rate the countries in terms of the status of women in five sectors: health, marriage and children, education, employment, and social equality. Some of the criteria considered are listed in Table 8-6. No country received an overall rating of excellent. Just seven countries, representing only 6 percent of women, received a rating of very good. Over 60 percent of women lived in societies where women's status is poor, very poor, or extremely poor. Sweden earned the top ranking, Bangladesh the lowest. The United States ranked third. North America and northern Europe dominated the top-ranked countries, while Africa, the Middle East, and southern Asia earned the lowest marks.

The study clearly shows that the "gender gap" is worldwide. In Sweden, for instance, a professional female earns 90 percent of what a male does; in the United States the figure is 68 percent. The gender gap is far more severe in developing countries. In those societies that were in the bottom half of the Population Crisis Committee's ranking (51 out of the 99 surveyed), women and girls live under conditions that pose a threat to their health, limit or deny them choices regarding childbearing, restrict educational and economic opportunities, and do not grant them equal rights and freedoms with men. In over one-fifth of the nations studied, there are 25 to 50 percent more literate men than women, and in five nations (Libya, Benin, Syria, Tanzania, and Turkey) the gap is particularly wide. Worldwide, over 500,000 women per year die because of inadequate reproductive health care. Moreover, for every woman who dies, 10 to 15 are handicapped. Serious complications to the birthing process, such as hemorrhage and infection, afflict about 25 million women. In general, childbirth poses the greatest risk to the very young (under 19), women over 35, and those in poor health. In over one-fourth of the countries studied, 10 percent of women will die in their childbearing years (defined as between 15 and 45). In nine countries—Afghanistan, Benin, Cameroon, Malawi, Mali, Mozambique, Nepal, Nigeria, and the former North Yemen—20 percent will die. In contrast, in those countries with the lowest rates of adult female mortality, only 1 percent of women will die in their childbearing years.

In all countries in which one-third or more of women marry as adolescents (between ages 15 and 19), overall rankings for the countries fall into the lowest two categories (very poor and extremely poor). Early childbearing increases the risk of complications associated with pregnancy; adolescent mothers are twice as likely to die in childbirth as are women in their twenties. Early marriage is also linked to family size; in countries where most adolescent girls marry, women bear three times as many

▶ Table 8-6
Women's Status in Highest and Lowest Ranked Countries

In Sweden . . . *(Population: 8.4 million; Area: 173,730 square miles)*	In Bangladesh . . . *(Population: 109.5 million; Area: 55,598 square miles)*
Female life expectancy is 81 years.	Female life expectancy is 49 years.
Women live an average of seven years longer than men.	Women live an average of two years less than men.
One in 167 girls dies before her fifth birthday.	One in 5 girls dies before her fifth birthday.
One in 53 15-year-olds will not survive her childbearing years (with 1 percent of these deaths related to pregnancy and childbirth).	One in 6 15-year-olds will not survive her childbearing years (with about one-third of these deaths related to pregnancy and childbirth).
Fewer than 1 percent of 15- to 19-year-old women have already been married.	Almost 70 percent of 15- to 19-year-old women have already been married.
Women bear one to two children on average.	Women bear five to six children on average.
Over three-fourths of married women use contraception.	One-fourth of married women use contraception.
Almost all school-aged girls are in school.	One in 3 school-aged girls is in school.
Female university enrollment is 37 percent of women aged 20 to 24.	Female university enrollment is less than 2 percent of women aged 20 to 24.
About one-half of the secondary school teachers are women.	About one-tenth of the secondary school teachers are women.
Three in 5 women are in the paid labor force.	One in 15 women is in the paid labor force.
Two in 5 women are professionals.	Three in 1,000 women are professionals.
Women and men have similar literacy rates.	Some 24 percent more women than men are illiterate.
About 50 percent of the paid work force is female.	About 14 percent of the paid work force is female.
In 1988 women held 113 seats in Sweden's 349-member parliament.	In 1988 women held 4 seats in Bangladesh's 302-member parliament, out of 30 reserved for them.

Source: Population Crisis Committee

children as do women in countries where late marriage is more common. In many countries, age at marriage, contraceptive use, and total fertility are also linked with educational attainment and paid employment in the modern sector. For instance, at least 50 percent of all adult women in Canada and Finland, where family planning is common, have paid jobs in the formal work sector; 15 percent of them hold professional or managerial positions. In contrast, only 1 percent or less of adult women (age 15 and above) in Mali and Afghanistan—where total fertility levels are high and contraceptive use is low—are in the formal paid work force. In the few Asian countries where contraceptive use is high—Taiwan, China, Hong Kong, Singapore, and South Korea—total fertility rates are low, 1 to 3 children per woman. These countries also have low levels of adolescent marriages.

The study's findings include many other sobering statistics, but numbers can be numbing. Let's look instead at some of the practices faced by women in countries where the status of females is low. In many of the countries that received a rating of poor or below, women and girls eat only after men and

| 4:45 A.M.
Wake up, wash, and eat | 5:00 A.M.–5:30 A.M.
Walk to fields | 5:30 A.M.–3:00 P.M.
Work in fields | 3:00 P.M.–4:00 P.M.
Collect firewood | 4:00 P.M.–5:30 P.M.
Pound and grind corn | 5:30 P.M.–6:30 P.M.
Collect water | 6:30 P.M.–8:30 P.M.
Cook for family and eat | 8:30 P.M.–9:30 P.M.
Wash children and dishes | 9:30 P.M.
Go to bed |

FIGURE 8-16: Typical workday for a rural African woman. African women perform as much as 75 percent of all agricultural work in addition to their domestic work.

boys have eaten (a practice that undermines the health of girls and their mothers when so little food is available to begin with) in spite of the fact that women often work longer hours and harder than men (Figure 8-16). In some Islamic nations female heirs may legally inherit only half as much as male heirs. Women are not legally considered autonomous individuals in the most conservative Islamic countries; they must have a designated male "guardian," typically a father, brother, or husband. In many of these countries, very conservative interpretations of Islam forbid a woman to divorce her husband, but a man may easily divorce his wife (in Egypt, as well as some other nations, he need not even notify her), and men generally retain custody of the children. In Brazil battered women have little legal recourse, and husbands who murder their wives "to protect their honor" frequently do so with little fear of legal reprisal. And in Africa clitoridectomies and other extreme genital operations (commonly referred to as female circumcisions) for young girls remain common. These procedures are deemed necessary to keep women chaste and faithful, and in most conservative regions a woman is not considered marriageable unless she has undergone such an operation.

Summary

The scientific study of the sum of individual population acts is called demography. Demographers study how populations change over time in order to understand the causes and consequences of human population dynamics. Basic demographic data include the crude birth rate and crude death rate, the number of births or deaths, respectively, per 1,000 people. The crude birth (death) rate is equal to the number of births (deaths) divided by the total population at mid-year multiplied by 1,000. In 1991 the world population of 5.38 billion included over 4 billion people in the developing world and over 1 billion in the developed world. The world growth rate, determined by subtracting the crude death rate from the crude birth rate, was 1.8 percent in 1991.

At current growth rates world population could soar to over 20 billion by the end of the twenty-first century. However, it is expected to stabilize at approximately 12 billion around the start of the twenty-second century. Falling world annual growth rates over the past several decades—from a high in the mid-1960s of 2.1 percent to 1.7 percent for most of the 1980s and early 1990s—are responsible for that projection. When births are equal to deaths, the growth rate will be zero and no growth in actual numbers will occur, a condition known as zero population growth, or ZPG.

Most of the people born in the coming decades will live in less-developed countries (LDCs). Rapid population growth in LDCs is caused by falling death rates due to improved medical care; lack of access to family planning services and safe and reliable means of birth control; and a high incidence of infant mortality, which compels parents to have more children to account for those who die

young. LDCs all have the fastest doubling times, the number of years until a population doubles. Doubling times are calculated by using the rule of 70: assuming that the current growth rate remains constant, 70 divided by a nation's current growth rate yields the time it will take for a country to double its population. At a rate of 1.8 percent, the world population would double in just under 40 years.

Growth rates are affected by migration, fertility, and the population's age distribution. Migration is movement into (immigration) or out of (emigration) a country or region. The general fertility rate is the number of live births per 1,000 women of childbearing age for any one year. The age-specific fertility rate is the number of live births per 1,000 women of a specific age group for any one year. The total fertility rate is the average number of children a woman will bear during her life, based on the current age-specific fertility rate and assuming that the current birth rate remains constant throughout a woman's lifetime. The replacement fertility—the fertility rate needed to ensure that each set of parents is "replaced" by their offspring—is 2.1 for more-developed countries (MDCs) and 2.5 for LDCs. Even in nations where the fertility rate is less than the replacement level, the population may grow because of population momentum, the continued growth of a population because of the large number of children it contains.

The age distribution of a population—the number of individuals in each sex and age category (from birth through old age)—can be represented by an age structure histogram, or population profile. Profiles for LDCs have a pyramidal shape because about 40 percent of the population is below age 15 and about 4 percent is age 65 or older. The profile for MDCs is more rectangular, since each age group is closer in size to other age groups. Populations that have a large proportion of young people (under 15) or old people (over 65) are said to have a high dependency load, that is, a high number of people who are not in the formal work force. High dependency loads can strain a nation's infrastructure and system of social services.

Calculating the carrying capacity of environments for human populations requires factoring in the quality of life a population aspires to. Although determining cultural carrying capacity is difficult and imprecise, many people argue that it is absolutely necessary. The quality of life for a particular country can be derived from studying population density, urbanization, life expectancy, and infant mortality rate. Population density, how closely people are grouped, gives us some idea of the pressure exerted on the natural resource base of a geographic or political area. Population density in LDCs is more than twice that in MDCs.

Urbanization, a rise in the number and size of cities, indicates that population density is increasing. The greatest increases in urbanization will occur in LDCs, where well over one-half of the population will live in cities by the year 2025.

Life expectancy at birth is the average number of years a newborn can be expected to live. Life expectancy in MDCs is about 74 years, in LDCs about 62 years. Factors that contribute to low life expectancy include war, political oppression, hunger, and infant and childhood mortality.

The infant mortality rate (IMR) is the annual number of children under age 1 who die per 1,000 live births; the childhood mortality rate is the annual number of children between the ages of 1 and 5 who die per 1,000 live births. The IMR is widely considered the single best indicator of a society's quality of life. In 1991 the average IMR worldwide was 68/1000 (6.8 percent); the rate for MDCs was just 14/1000 (1.4 percent), while the rate for LDCs was 75/1000 (7.5 percent). Japan had the lowest IMR (4.5/1000), and Afghanistan had the highest (182/1000). IMRs, whether in LDCs or among the poor in MDCs, could be lowered if all people were given access to clean water, sanitation, and adequate health care, and if mothers were given the education and health services needed to care for their children.

Improving women's status could help to lower population growth by opening new avenues to them beyond motherhood. Women's status is reflected in access to education, access to adequate health care, legal rights, employment opportunities outside the home, wage earnings, marriage age, and number of children.

Key Terms

age-specific fertility rate	growth rate
childhood mortality rate	immigration
crude birth rate	infant mortality rate
crude death rate	life expectancy
cultural carrying capacity	migration
demography	population momentum
dependency load	population profile
doubling time	replacement fertility
emigration	total fertility rate
environmental refugee	urbanization
fertility	vital statistics
general fertility rate	zero population growth

Discussion Questions

1. Do you think that the human population should be considered a resource? Why or why not?

2. Do you think that continued growth of human population is a problem? Why or why not?

3. Using Ethiopia to illustrate your answer, explain how the biotic potential of the human species and environmental resistance factors might interact to raise or lower the carrying capacity of the land.

4. Suggest ways that the carrying capacity of the earth might be increased for humans. Suggest ways that the quality of life (or standard of living) for all humans could be increased. Can both of these objectives be met simultaneously? Why or why not?

5. Referring to the demographic statistics in Tables 8-1 and 8-5, compare population dynamics and the quality of life in more-developed and less-developed countries.

6. Some people argue that the status of women is not relevant to population growth and is not an appropriate topic for environmental science. Do you agree or disagree and why?

7. A 20-year-old college student (a citizen of the United States) in 1993 will be 54 years old in 2027, the year when the human population is expected to reach 10 billion, twice its present size. Speculate on what life will be like for that person. Consider housing, job opportunities, cost of living, availability of resources, and recreational activities such as wilderness camping or travel. Also speculate on what kind of world that person will inhabit. For example, are relations with other countries likely to be stronger and more secure or weaker and more precarious than they are now?

Chapter

9

Managing Human Population Growth

It was the best of times. It was the worst of times.

Charles Dickens

People with deeply personal commitments to a particular vision of the future are perfectly justified in struggling toward their goal, even if the outcome now seems remote and improbable. In life, as in any game whose outcome depends on both luck and skill, the rational response to bad odds is to try harder.

Marvin Harris

Learning Objectives

When you finish reading this chapter, you should be able to:

1. Briefly recount how the human population has grown historically.

2. Describe the demographic transition, identify the factors needed for the transition to occur, and explain what is meant by the demographic trap.

3. Define population policy and explain how policies can be used to encourage or discourage growth.

4. Summarize the arguments given in favor of controlling or limiting population growth.

5. Discuss the difference between family planning and birth control; give examples of each.

Chapter 8 focused on human demographics: birth, death, and growth rates; migration; fertility rates; infant mortality rates; dependency ratios; and so forth. We examined the biological, physical, and social boundaries of the human population in order to better understand what many people call "the ultimate resource." In this chapter, we look at how human populations have been managed historically,

the policies used to influence growth and manage populations, the advantages of limiting population growth, and the technologies used to control births.

History of Management of Human Population Growth

How Has the Human Population Grown Historically?

Population control—for some people the words bring to mind a futuristic, Orwellian society. But the idea of limiting population growth is not new. Anthropologists and historians seeking to unravel the tale of our species have shown that population control, in various forms, has taken place throughout human history.

The fossil record indicates that hominid (human-like) species existed as much as 3 million or more years ago. Some 40,000 to 50,000 years ago, our own species, *Homo sapiens*, made its appearance. For most of its tenure on earth, *Homo sapiens* has existed by gathering wild plants and hunting wild animals. The early hunter-gatherers were nomadic and probably lived in small bands of several dozen people. Dominated by the forces of nature, they possessed a strong sense of the earth—the location and growth of plants, the habits of animals, and seasonal weather patterns.

It is popularly believed that life for these tribal peoples was harsh, a struggle against the elements, predation, and disease, with high death rates keeping populations relatively low. However, anthropologists offer a different view of our early ancestors, maintaining that life for them was far less brutish than is widely believed (Figure 9-1). They contend that early hunter-gatherers were relatively healthy peoples. The mainstay of their diet was meat, the large mammals, including reindeer, mammoth, horses, bison and wild cattle, that roamed lush, grassy plains. They practiced intentional population con-

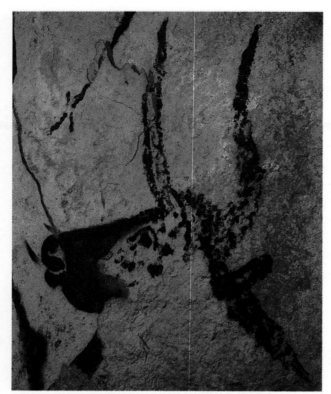

FIGURE 9-1: Cave drawing in Lascaux, France. Beautiful paintings dating from 20,000 B.C. on the walls of this cave suggest that life was not as brutish for our ancestors as was once believed.

trol through a variety of means, including abstinence from sexual intercourse, birth spacing, and infanticide. The global human population grew slowly. Up until about 10,000 years ago, when people first began to practice agriculture, the earth probably supported about 5 million people (a number equivalent to the 1991 population of Finland).

Anthropologists believe that humans developed agriculture out of necessity rather than out of any innate desire to establish permanent dwellings. About 10,000 to 12,000 years ago, the last ice age came to an end. As the glaciers receded northward, the climate began to warm significantly and forests of evergreens and birches invaded the grassy plains. Loss of habitat, along with human predation, led to the extinction of numerous species of large animals, including the wooly mammoth, steppe bison, and giant elk. Accordingly, both the diet and life-style of prehistoric peoples underwent a significant change. For the first time, humans began to cultivate their own food, relying more on plants than they had previously. For many populations the nomadic life-style became a thing of the past, and as it disappeared, so did many of the reasons for limiting births. Having numerous young children did not present the same practical difficulties as it had for nomadic women. In fact, with the onset of agriculture, labor—to

maintain homes and gardens, collect firewood and edible plants, and perform other necessary chores—became a strong impetus for having children. The rise of agriculture, then, led to increased human population pressures and increased impact on the environment of human activities.

The rise of cities dramatically increased the effect of humans upon the biosphere. Food produced in the countryside was consumed in the city. Food wastes were no longer returned to the soil, and the soil became less productive. Also, the concentration of population in the cities meant that human wastes were concentrated, sometimes in amounts too great for the local river and/or stream to decompose effectively. The problems resulting from urbanization grew over the centuries. The urban poor often lived in horrendous circumstances, suffering from the ill effects of crowding, poverty, and hunger.

In thirteenth- and fourteenth-century England, as well as other medieval societies, infanticide was a means of managing population. "Overlaying" was common—the instance of a nursing mother who falls asleep and (presumably accidentally) rolls over onto her baby, suffocating it. Beginning in 1348, the Black Death, or bubonic plague, greatly reduced the populations of Europe and Asia, perhaps by as much as one-half. Crowding in cities contributed to the spread and severity of the plague. Within several hundred years, however, populations had rebounded. By the sixteenth century, infanticide was again common. Overlaying, drugging a baby with gin or opiates, and outright starvation were used to kill unwanted children. Many children were left at "foundling" hospitals, which sprang up in the eighteenth century to handle the huge number of abandoned babies. Children left at such institutions rarely survived their first year of life.

During the early phases of the Industrial Revolution (around the late 1700s), children came to be viewed as valuable sources of cheap labor (for employers) and income (for parents). For the first time, the effects of **exponential** growth become apparent (Figure 9-2).

The population in those areas we now refer to as the developed world (Great Britain, Europe, the United States, and Canada) began to grow more rapidly. By the 1820s, human population reached 1 billion for the first time. Medical advances and improvements in sanitation and hygiene helped to control diseases and epidemics. These advances in death control ushered in the period of rapid growth in the global human population, as births began to outstrip deaths.

By 1900 another shift in human population growth occurred. The birth rate in the industrialized world dropped markedly, slowing the growth

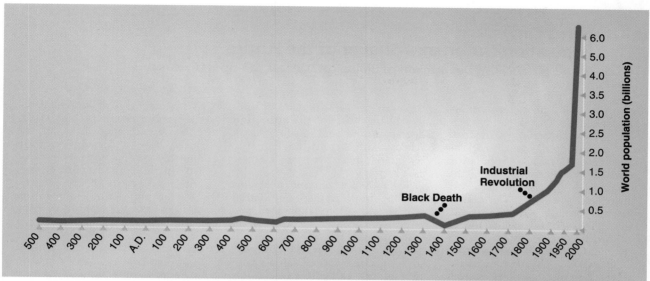

FIGURE 9-2: J-curve growth in human population. The human population is increasing exponentially, with potentially catastrophic consequences for civilization and the biosphere.

rate. Three factors contributed to this lowered growth. First, the Industrial Revolution led to a rise in living standards for many people, and lower growth rates typically accompany rises in the standard of living. Second, safe and inexpensive means of birth control (such as condoms made of sheep gut, vaginal douches, and vaginal plugs) were introduced. Third, an increase in the cost of child rearing (with the introduction of child labor laws and mandatory education statutes) meant that having more children yielded fewer material benefits. Thus, while the human population continued to grow, it grew far more slowly for most industrial nations.

What Are the Demographic Transition and the Demographic Trap?

The population path followed by the industrial nations is known as the **demographic transition**, that is, the movement of the population of a nation from high growth to low growth as it moves through stages of economic development, historically, through industrialization (Figure 9-3). It consists of four stages. In stage 1, birth and death rates are both high. Birth rates are actually fairly constant, but death rates fluctuate as seasonal and cyclic factors

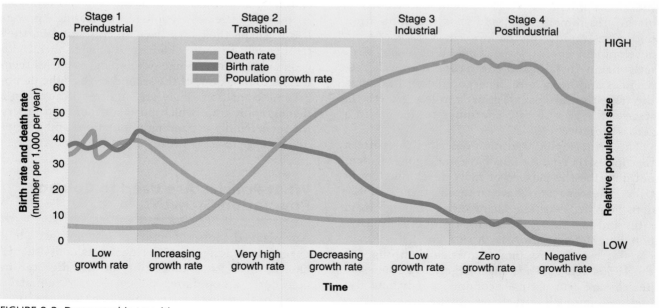

FIGURE 9-3: Demographic transition.

Focus On:
Population Outbreak—Shaper of the Future

Anne H. Ehrlich, Department of Biological Sciences, Stanford University

Nothing will more profoundly shape people's lives in the twenty-first century than the dynamics of the human population during the next few decades. Population growth continues, as it has since the middle of this century, at unprecedented rates. Most of the consequences of this outbreak, some foreseeable, some not, will be borne by future generations.

Demographers project another doubling of the global population—to over 10 billion—before growth ends. In a world where roughly one-fifth of the population is destitute and some 15 million children die each year unnecessarily because of poverty and hunger, while resources are increasingly depleted and environments are rapidly deteriorating everywhere, the prospect of another doubling is chilling. Recent calculations based on measurements of the planet's overall biological productivity show that humanity today uses directly or is coopting about 25 percent of that basic productivity on land—the total food supply of all terrestrial animals. Another 13 percent of the land's potential produc-

tivity has been lost because of land-use changes. Desertification and degradation of land in sub-Saharan Africa, which contributed to the recent famines there, are a tragic case in point.

Social and economic problems arising from rapid population growth are often recognized and cited as reasons to slow growth. Unfortunately, the absolute size of the population, current or anticipated, is rarely considered in a context of the carrying capacity for human life of the planet's life-support systems. As a result, social scientists tend to focus on the problems associated with rates of increase and ignore resource pressures exerted by populations because of their size.

Even if a decision were made to end population growth by reducing the number of births, however, it could not be done quickly because of a built-in momentum due to the predominance of young people in an expanding population. Even after reaching replacement reproduction, stopping population growth takes roughly a lifetime—about 70 years.

Of course, growth will end sooner if reproduction falls substantially below replacement level. In the United States and many European countries reproduction is below that level. In the United States growth continues because of immigration and the population momentum resulting from the 1945–1965 baby boom. Without immigration or a change in fertility, however, growth would end within a few decades. In some European countries growth has stopped and populations are gradually shrinking.

These demographic changes, however, have taken place in the absence of any policy changes. Most developed nations, including the United States, lack any coherent official population policy (except to regulate immigration). In contrast, the majority of developing nations have population policies. Some of these are limited to a national family planning program, which simply provides individuals with information and the means for birth control. Other countries have much stronger policies. Their leaders have realized that rapid population

temporarily raise or lower the carrying capacity of the environment. In stage 2, death rates fall but birth rates remain high, and thus the population undergoes rapid growth. In stage 3, as economic development improves the standard of living, birth rates begin to fall, and the nation's growth rate declines and nears zero. The population has made the transition from high growth to low growth. In stage 4 the growth rate continues to decline to a zero or negative rate.

The demographic transition describes the pattern that industrialized nations have undergone. It does not describe the pattern of most developing countries. These nations have entered stage 2 of the demographic transition, characterized by high birth rates and low death rates with resultant explosive population growth. In Asia, Africa, and Latin America, death rates dropped dramatically throughout the 1950s and 1960s, thanks largely to advances in health care. For a time economic conditions also improved in much of the developing world, but during the 1970s and 1980s economic development did not keep pace with rapid growth. Unchecked growth

and the environmental deterioration it brings are causing a downward spiral in the standard of living. As poverty increases, people feel increasingly powerless to shape their own lives. In this situation they tend to rely on more children to provide them with economic security. Some developing countries seem unable to break out of the second stage of the demographic transition. They are, in effect, caught in a **demographic trap,** which portends increased misery for their rapidly expanding populations and continued damage to their already degraded resource base.

What Policies Are Used to Control Population Growth?

Any planned course of action or inaction taken by a government designed to influence its constituents' choices or decisions on fertility or migration can be considered a **population policy.** Over 90 countries (most of them in the developing world) have official policies on population growth or size. (See Focus On: Population Outbreak—Shaper of the Future.)

growth is a major hindrance to successful development, because all the economic gains must go to provide for the additional people, and improvement of each individual's lot becomes increasingly difficult. In these cases, the family planning program is tied to a government policy specifically favoring smaller families.

Perhaps the best example of a successful, strong population policy is that of the People's Republic of China. Early on, China's leaders made improving their people's well-being their highest priority. This led them to establish a family planning program aimed at reducing population growth, which was seen to be consuming all their economic gains. The goal of the program, which was integrated into the health system, evidently was to maximize the survival of each child born while minimizing the number of births.

Then, unlike any other nation, China conducted a serious assessment of its resource base in order to determine how many people the nation could support over the long term. It concluded that only about 650 to 700 million people could be supported. Unfortunately, by then there were over a billion Chinese!

This dismaying discovery led to establishment of the "one-child family" program, which encouraged Chinese couples to pledge themselves to only one child in return for special privileges. It was hoped that half of today's parents would join the program and that the population would reach a peak size of 1.3 billion, then begin a slow decline. Sadly there were abuses, and the traditional preference for sons led to instances of female infanticide. Moreover, the Chinese realized that a generation of single children would lead to a heavy burden of old people later; each couple would be responsible for four elderly parents as well as their own children.

A way out of this dilemma was to lengthen generation times. China's policy already prescribed relatively late marriage and reproduction. If each couple's second birth were 8 to 10 years after the first, the population would stop growing and begin declining almost as soon as it would if half of all couples had only one child—with much less social disruption. In addition, with such a long delay, many couples would end up without a second child.

China's policies are of enormous significance, in part because that nation includes some 22 percent of the world's people and its slower growth accounts for much of the recent worldwide slowdown of population growth. Unfortunately, China's example has not been followed. The concept of formulating population policies to conform with a nation's long-term carrying capacity has yet to be adopted elsewhere.

For the other four-fifths of humanity, every day that passes without a global campaign to end and reverse population growth humanely increases the odds that the human outbreak will be ended nature's way: by rising death rates. Nuclear war, famine, AIDS, or some other epidemic will eventually cause a population crash—the usual result of an outbreak. Time is running out, but it is not too late to make the choice to build a sustainable future.

India, for example, has an official population policy that relies heavily on family planning, understanding human sexuality, and providing prospective parents with birth control.

The United States has no formal population policy. There is no legislation that limits or manages population growth, advocates attempts to determine ideal population size or cultural carrying capacity, supports family planning programs, or promotes education about population issues or human sexuality. Many people, however, contend that the government has unofficially adopted certain policies and promoted certain beliefs that affect population growth both at home and abroad. Consider the following: an income tax structure that provides deductions for all children in a family; a strong movement to limit legal abortions in the United States and to limit or stop funds to international family planning groups in countries with legalized abortion; increased benefits for each child born into a welfare family; a strong belief that the economy is based on continued population growth; and a strong belief that family size should be decided on by the family. According to many, the government's stance constitutes an unofficial **pronatalist policy** that encourages natality or births.

Other countries have official pronatalist policies. For example, France adopted pronatalist policies in the 1930s. All parents received extra payments and services to help with raising their children. In 1976 France initiated a more aggressive program to encourage larger families. The government agreed to pay an extra maternal salary (for up to three years) to mothers who gave birth to additional children past the first two. The former East Germany gave interest-free housing loans to new parents, then forgave part of the debt each time an additional child was born. In some cases pronatalist policies have been extreme. In an effort to encourage larger families, Romania banned abortion, the nation's leading birth control method, in 1966. No woman under the age of 45 with fewer than five children could legally terminate a pregnancy. Women were forced to submit to monthly "pregnancy checkups" at their place of work; these were supervised by a special branch of the secret police force. The "pregnancy police," as they were known, monitored pregnant women; any sudden termination of pregnancy had to be explained. A woman found guilty of having had an abortion could be sentenced to death. Married women who did not conceive were kept under surveillance; childless couples had to produce medical evidence of infertility in order to avoid a special tax (which was also levied on single adults over the age of 25). After the fall of the Communist regime in 1989 and the execution of ousted leader Nicolae Ceausescu, one of the first actions taken by Romania's provisional government was to repeal the ban on abortion.

Antinatalist policies are designed to prevent or discourage increases in fertility or to lower existing fertility rates. The People's Republic of China has strict laws on the marriage age for women that prevent marriage until a woman is in her early twenties, thus shortening her fertility years. Up until the late 1980s parents in Singapore faced reduced housing subsidies when children were born; families could be removed from low-cost housing if they had more than two children. Other policies sometimes used to discourage births were payments to couples who delayed having children, bonuses paid to women who reached a certain age without having children, and tax structures that penalized parents who chose to have large families. Singapore's antinatalist policy worked so well that by 1990 the government reversed itself and adopted a pronatalist policy! Figure 9-4, a reprint of a publication of the Singapore Ministry of Health, lists some of the measures instituted to encourage larger families.

Pronatalist or antinatalist policies are developed as a result of the way population changes are perceived: Is continued growth or lack of growth considered to be a problem? Are present growth rates too high or too low? Is the population size too large for the resource base or too small to defend itself from real (or perceived) aggression? Could standards of living be improved if the population were smaller? How we answer these questions and what we choose to do about them are very much a matter of values. We cannot determine in a completely value-free way whether fertility rates are too high or too low, if the population size is too small or too large, if certain birth control methods are acceptable while others are not, or when or if population growth is a threat to the environment and to the economy.

What Are the Arguments in Favor of Controlling Population Growth?

Those who argue that population growth should be controlled or limited fall into two broad categories: those who oppose any growth and those who oppose

rapid growth. The first group maintains that any growth in population will ultimately lead to environmental degradation; degradation will eventually increase environmental resistance, resulting in worsened living conditions and a subsequent lowering of the carrying capacity. In addition, they point out that violent conflict and population growth are strongly interrelated. Often, resource scarcity compels populations toward war. Population pressures and conflicts over resources were at least partially responsible for World War I and World War II. Many links have been established between antisocial behavior and increasing population density in an area. Abundant evidence exists to link increased violence and mental disorders to the crowded conditions that occur in and around large cities.

Those who oppose any population growth point out that countries with slow growth rates like the United States, Japan, and Canada exact a greater toll on the global biosphere because of their high rate of per capita resource consumption. To prevent irreversible damage to the biosphere, opponents of population growth maintain that it is essential to develop policies, educational programs, and, if need be, laws to limit the number of births to replacement levels in all countries. They advocate stabilizing population size at the culturally determined carrying capacity for a particular state, region, or country. If the carrying capacity is below present population levels, they recommend that steps be taken to lower the fertility rate below replacement level.

The second group distinguishes between slow and rapid growth. They see rapid growth, especially in developing countries, as having a negative impact on economic development and living standards. They agree that effective improvements in environmental conditions cannot be made unless or until rapid population growth is brought under control. But there is widespread disagreement over how to proceed. Should strong measures be taken to reduce population growth, on the assumption that economic development will improve? Or should developed countries provide poorer nations with economic and social aid, in the hope that improved living conditions will convince people that children are not their only means of economic security and thereby encourage them to have fewer children? Those who advocate providing social aid for the developing world argue that the development process by itself does not lower birth rates and help bring rapid population growth under control. Rather, they maintain that social development has a greater effect on population numbers than does economic development. Programs that improve people's lives through better education (especially for women), nutrition, health, and sanitation are more effective in the long run

Social Policies Related to Family Formation

建立家庭的
社会政策

1 Income Tax Relief

effective from year of assessment, 1988

a. Normal Child Relief

1rst child	$750
2nd child	$750
3rd child	$750
4th child	$300
5th child	$300

b. Enhanced Child Relief

1rst child $750 + 5%
of mother's earned income

2nd child $750 + 10%
of mother's earned income

3rd child $750 + 15%
of mother's earned income

4th child $750 + 15%
of mother's earned income

c. Tax Rebate

The parents of a 3rd born on or after 1 Jan. '87 will be eligible for tax rebates of $20,000 and 15% of the mother's earned income. The rebates must be absorbed within five consecutive years from the date of birth of the 3rd child. For further details, please contact the Inland Revenue Department.

2 Child Care Subsidy

A subsidy of $100 per child per month will be granted to the first three children attending approved childcare centers.

3 Public Housing Scheme

Priority in housing allocation will be given to families, owning 3-room or larger HDB flats, who wish to upgrade their flats upon the birth of the third child.

FIGURE 9-4: Pronatalist policy—a publication of the Singapore Ministry of Health. The Singapore government has developed a number of incentives to encourage larger families, among them tax incentives and advantages in housing and child care.

than relying on improvements in GNP to help lower fertility.

China, Cuba, Thailand, and the Indian state of Kerala are all moving from high to low population growth. Their governments have made birth control technologies available through strong, centralized family planning efforts (efforts partially funded through international family planning organizations). The programs include economic incentives to lower family size. In each instance, a decline in the growth rate was achieved while economic conditions improved at local levels.

What Is Family Planning?

Family planning is a term used to describe a wide variety of measures that enable parents to control the number of children they have and the spacing of their children's births. The basic goal of family planning is not to limit births; it is to enable couples to have healthy children, to care for their children, and to have the number of children they want (Figure 9-5). Family planning includes education in human sexuality, hygiene, prenatal and postnatal care, and preconception and postconception birth control measures.

Most LDCs have some form of family planning program (even if it is as simple as spacing births or improved health care), but it may or may not be implemented because of a lack of financial resources (Figure 9-6). In only a few instances (China and India under Indira Gandhi's first term) has a government imposed mandatory family planning.

In 1974, at the First World Conference on Population in Bucharest, the United States advocated that LDCs should develop family planning and contraception programs, based on the argument that unchecked population growth posed a threat to development and improved standards of living. LDCs generally were pessimistic about this approach, asserting that development was the best cure for population problems. Some openly opposed family planning efforts, which they felt were being forced on their countries out of racism. They argued that the wealthy MDCs were trying to suppress poor countries in order to continue dominating them economically. They maintained that economic development was more important than any other problem facing their countries, and that once economic development was in place, fertility and growth rates would drop. During the ten years between the Bucharest conference and the Second World Conference on Population, held in Mexico City in 1984, economic and environmental conditions in many LDCs worsened. Moreover, global population rose by about 800 million in that short period! At the Mexico City conference, delegates from LDCs seemed to change their opinions about the relationship between rapid population growth and development. Many expressed the belief that population planning is essential for improving both economic development and the health and welfare of women and children.

Just as the developing world appeared ready to take full advantage of assistance for family planning, the United States, under the Reagan administration, did an about-face, declaring at the Mexico City conference that population growth is not a problem, that there is no relationship between population growth and economic development, and that free market economics are the answer to development problems. In 1984 the administration denied funds to the International Planned Parenthood Federation (IPPF), and in 1986 it withheld funds from the UN Fund for Population Activities (UNFPA). The IPPF and the UNFPA are the largest nongovernmental organization and the largest multilateral organization, respectively, providing family planning assistance to LDCs. In what has become known as the Mexico City policy, the administration prohibited funding for organizations involved in abortion-related activities or for countries where family planning activities were deemed coercive. The administration pointed out that family planning aid could be channeled through organizations other than the IPPF and the UNFPA. Critics of the policy, however, argued that these two organizations had

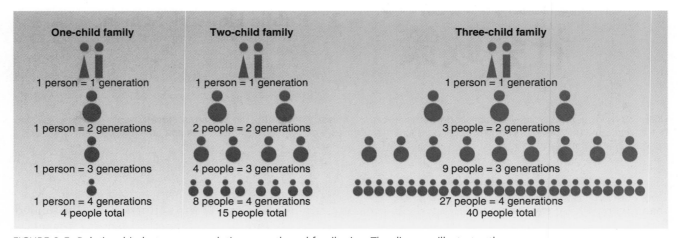

FIGURE 9-5: Relationship between population growth and family size. The diagram illustrates the consequences of family size across generations. If each generation has just one or two children above the replacement number of two, the result is a significant population increase. Family planning can help parents to have the number of children—that is, the family size—they desire. It does not limit family size or prescribe a certain size, however.

FIGURE 9-6: Promoting family planning in western Africa. A group of children gather in a marketplace. Behind them is a billboard promoting the government's population policy.

the most experience and the best infrastructures to help developing countries with their family planning programs. By cutting off aid to these organizations, the administration seriously damaged family planning efforts in the developing world.

The Reagan administration's policy, later upheld by the Bush administration, was a strong statement against family planning programs it found objectionable. It is true that some countries, notably Bangladesh and China, have imposed controversial limits on family size and have used debatable means to achieve these limits. In Bangladesh, for instance, both men and women are offered an economic incentive—a payment equal to about two weeks' wages, a piece of clothing, and wheat—for undergoing sterilization. Although the sterilization program is supposedly voluntary, many people argue that the hungry and poor have little real choice; they point out that the number of sterilizations tends to fluctuate according to the availability of food. Considered coercive by some and necessary by others, China's family planning strategies have aroused strong feelings on both sides. Let's examine the Chinese program more closely.

China's Program

China presents us with perhaps our best-known example of a population control program. But the Chinese wouldn't call it that. Because China follows a strong Marxist ideology, the Chinese do not officially acknowledge population growth as a problem. Instead, they maintain that the program was designed to "maximize the health and well-being of each child born." According to official government policy, births are to be controlled because healthy, happy children are good for the state, and what is good for the state is good for all of the people (Figure 9-7). Even though China's delegation to the 1974 Bucharest population conference formally came out against population limitations, their country had already instituted one of the strongest family planning efforts that had ever been attempted.

Why did the Chinese feel it was necessary to initiate this program?

The first census taken of the Chinese population revealed that in 1953 China had a population of over 580 million people. By the late 1970s that figure had reached over 1 billion people. Fully 20 percent of the world's people were trying to live on 7 percent of the world's arable land. After years of terrible famine (perhaps as many as 30 million people died), poverty, and political turmoil, the government became convinced that China could adequately support no more than 650 to 700 million people. The availability of fresh water and food was especially critical in the determination of carrying capacity. Because the population already exceeded the carrying capacity by more than 300 million people, a stringent family planning effort was needed to stop population growth and eventually reduce population size. The government estimated that, assuming growth rates continued to decline, it would take about a hundred years to reduce the population to 700 million. Because of population momentum, even if the growth rate fell to replacement level, China's population was predicted to peak between 1.2 and 1.5 billion people. China became the first nation to have as an official goal the end of population growth and the subsequent lowering of absolute numbers by a significant amount. From 1969 to 1979 China achieved a transition from high to low birth rates. Let's examine how they did it.

In the late 1960s a family planning program was begun that reached into every village across China. This program made birth control, including birth control pills, accessible to all. An extensive health care system was initiated that for the first time paid attention to China's rural poor. This important development would not have been possible without the major political change that took place in China. It was a change that stressed family planning as a

FIGURE 9-7: The goal of Chinese family planning—a healthy, happy child. A baby enjoys his day sunning on a sidewalk in Beijing, China, in the summer of 1990.

way to increase the standard of living for all, gave women increased access to jobs and education, redistributed wealth to the masses, and provided for social security in the form of old age pensions and food cooperatives.

The 1970s became known as the "later, longer, fewer" years. A mass educational effort, employing slogans, posters, and radio and television programs, encouraged people to have fewer children. The age for legal marriage for women was increased from as young as possible to the early twenties; contraceptives, abortions, and sterilizations were provided free of charge; and supplementary payments, longer maternity leaves, free education and health care for children, preference in housing, and retirement incomes were granted to couples who conformed to the goals of the program. Couples who did not conform lost benefits, paid heavy fines, and risked the loss of job promotions.

In 1979 China's leaders felt that population growth was not falling fast enough and the emphasis switched to one child per family. Thus began the most restrictive family planning program ever attempted. The outside world deemed it intensely coercive. Intense pressure was brought to bear on women who became "unofficially" pregnant. Few women were able to resist this pressure; many had abortions or were sterilized.

In the 1980s China began to change its guaranteed policies toward employment and old age. It began to encourage individual initiatives over collective agriculture. As personal and family incomes increased, families (particularly rural families) began to have extra children, especially boys. The government relaxed its policy on marriage age. Birth rates soon rose. In 1987 China's birth rate was 21 per

1,000, up from 18 per 1,000 the year before. In 1991 China's population of over 1.15 billion was growing at a rate of 1.4 percent and was projected to reach 1.59 billion by 2025.

Whether China can maintain its family planning program in the face of mounting pressure against it, both internally and abroad, remains to be seen. The Chinese government feels that its program is working. It maintains that the drop in growth is the result of education, not coercion, and that in the long run it will help to prevent widespread misery by offering a higher standard of living to a smaller population. The Chinese people, the government says, have agreed to put aside individual desires for the good of the society, at least temporarily. Evidence suggests, however, that the program is not working as well in the rural areas, where children are needed to help with chores and food production, as it is in the cities.

Though it is difficult for many of us to accept China's example, it is important to realize that family planning programs can arise from many different ideologies. Common elements of many programs are improving education, health care, sanitation, and the status of women (through education and employment opportunities). Critical to the success of these programs are education about reproduction and birth control and widespread access to effective and appropriate methods of birth control.

What Methods Are Used to Control Births?

Throughout most of human history, men and women had few options for controlling births. They relied on natural forms of contraception. Breast-feeding, for example, stimulates the production of hormones that *may* reduce the likelihood of ovulation; if a woman breast-feeds for two years or so, she *may* be able to delay pregnancy. The rhythm, or periodic abstinence, method is based on a woman's menstrual cycle. A couple abstains from intercourse during the time when the woman ovulates. Unfortunately, like breast-feeding, rhythm has a high failure rate; it is about 76 percent effective at preventing pregnancy. Moreover, for it to be as effective as possible, a couple must have a fairly sophisticated knowledge of the body processes on which it is based, and it requires a degree of cooperation between husband and wife that is unusual in more traditional cultures. Although rhythm is not reliable, it is popular among certain groups (worldwide 10 to 15 million people practice rhythm). Many Catholics throughout the developing world consider rhythm the only acceptable form of birth control, but this does not hold true for Catholics in more-developed countries, especially the United States. Religious beliefs do not always override the desire for control over family size. In Catholic Latin America, sterilization and the pill, though prohibited by the church, are the most popular methods of birth control, and Italy has one of the highest abortion rates in the world.

To meet the demand for reliable and safe birth control, researchers have developed a wide variety of modern contraceptive methods. These can be broadly categorized into two groups, those which prevent conception and those which act after conception has occurred. The effectiveness of all methods depends heavily on factors beyond the control of science: the availability of medical services and reliable sources of birth control devices, adequate education in the use of the method, religious beliefs, social conventions (especially regarding the status of women), and the laws of each particular country.

Preconception Birth Control Methods

Preconception birth control methods include barrier methods, hormonal methods, and sterilization. Barrier methods include the condom, vaginal sponge, diaphragm, and spermicides.

Condoms, generally made of latex and thus commonly called rubbers, are from 70 to 90 percent effective depending on their quality and how they are used. Using a condom in combination with a spermicide, a foam or fluid that kills sperm, increases its effectiveness to about 95 percent. The condom has become increasingly popular in recent years, primarily because it is the only contraceptive that effectively prevents the spread of sexually transmitted diseases such as syphilis, gonorrhea, and especially Acquired Immune Deficiency Syndrome (AIDS).

Barrier methods for female use include spermicides, the vaginal sponge, and the diaphragm. These prevent passage of the sperm into the uterus and thus prevent fertilization. Spermicides are 75 percent to 82 percent effective, with foams more effective than creams, jellies, and suppositories. The sponge and diaphragm are typically used with spermicides, which increase their effectiveness to 83 percent (sponge) and 87 percent (diaphragm).

Barrier methods are generally not as effective as hormonal contraceptives, which interrupt the reproductive cycle, preventing the cyclic maturation and release of eggs. The pill, an oral contraceptive, is the third most popular birth control method in the world. This contraceptive can be 99 percent effective if taken consistently. Women with inconsistent access to the pill and little education as to the way it works often have significantly lower rates of success; 4 to 10 percent of users worldwide become pregnant

because of improper use. For these reasons, the pill is generally not considered a good choice for women in LDCs, where supplies may be erratic. Moreover, many women in LDCs hesitate to take the pill because of its side effects (which pose a greater health threat in the absence of routine medical care).

Other hormonal methods, such as injections or implants, may be more efficient than the pill because they are less expensive and need to be administered much less frequently. Depo-Provera, an injectable hormonal contraceptive that has been in use for about a decade, is effective for up to three months. Norplant, which has been in use for a much briefer period of time, is a hormonal contraceptive that is implanted under the skin of the upper arm. Norplant is effective for up to five years, but must be removed by medical personnel after the duration of effectiveness expires. There are numerous side effects associated with these contraceptives: menstrual disorders, headaches, weight gain, depression, loss of libido, abdominal disorders, and delayed return in fertility after use of the method has been stopped. Some preliminary studies suggest that Depo-Provera may also be linked to an increased long-term risk of cervical cancer.

Sterilization, the most reliable method of birth control, is the most popular method throughout the world. Men and women can be sterilized: vasectomy is the cutting or tying of the tubes by which sperm leave the body, and tubal ligation is the tying of the oviducts by which eggs reach the uterus. Sterilization is becoming increasingly common in many LDCs. In China and India, for example, large numbers of women have undergone sterilization for contraceptive purposes.

Sterilization procedures, which are usually irreversible, are sometimes used in family planning programs that limit, rather than extend, the reproductive freedom of the individual. Among the countries and territories where abuses have been reported, Puerto Rico has been singled out by women's groups as having the most abusive sterilization program. The government began promoting sterilization in the mid-1940s, and by 1965 Puerto Rico had the highest sterilization rate in the world. Fully one-third of all women who had ever been married were sterilized, 40 percent of them before the age of 25. Many women were not told that the procedure was irreversible, and some women were sterilized unknowingly, while under anesthesia for other procedures.

Research on reversible sterilization is currently underway and appears promising. If reversible procedures are developed, they would be as effective as permanent sterilization, but more acceptable and less threatening.

Postconception Birth Control Methods

Postconception birth control methods include the intrauterine device, RU-486 pill, and abortion. An intrauterine device (IUD) is a small plastic or metal object inserted into the uterus that prevents implantation of the fertilized egg. Next to sterilization, the IUD is the most popular form of contraceptive in use worldwide. Chinese women account for three-quarters of all IUD users. The IUD is 95 percent effective; when used with a spermicide, it can be 98 percent effective. Some people prefer the IUD to the pill because it does not have to be administered every day. However, there is a possibility of side effects, including pelvic inflammatory disease and infertility. Moreover, skilled medical care is needed to insert the IUD properly and to conduct periodic checkups. In LDCs where medical services are inadequate, such a method is not a popular or practical option.

The RU-486 pill interrupts pregnancy. RU-486 either prevents implantation of the embryo or induces sloughing of the lining of the uterus (and thus the embryo) after implantation has occurred. It is used in France, where it was developed, but at present it is not available in the United States.

Abortion also interrupts pregnancy. It is an elective surgical procedure in which the lining of the uterus is removed along with the developing embryo. According to the Washington, D.C.-based Worldwatch Institute, about 55 million pregnancies worldwide end in abortion each year. About one-half of these are illegal, carried out mainly in LDCs. Only half of the world's people live in countries where abortions are freely available. Because many women prohibited from legal abortion attempt to induce abortion themselves (using knitting needles, wire coat hangers, and poisons) or go to unskilled practitioners, the risk of death from an illegal abortion is 30 times greater than from a legal abortion. The World Health Organization attributes roughly 200,000 women's deaths per year to illegal abortions.

In the United States the right to a legal abortion is a controversial and extremely divisive issue (Figure 9-8). The controversy revolves around the conflict between religious and moral beliefs about the status of the fetus—Is it a living human being? At what point does it gain human status?—and the right of women to choose whether or not they will bear children.

Some people argue that the right to an abortion ought to be legally ensured; they maintain that the option to terminate a pregnancy is essential to a woman's reproductive freedom. In several European countries where abortions are legal and where con-

Figure 9-8: Pro-choice and antiabortion demonstrators, Los Angeles, 1989. Abortion issues are complicated by the arguments and tactics of some people on both sides of the issue. The bombing of abortion clinics and harassment of female patients run directly counter to the pro-life movement's stance of love and compassion for fellow human beings. And certain pro-choice arguments—particularly, that women alone can and should make the decision regarding the fate of their unborn children—effectively eliminate male participation in what is clearly the concern of two people.

traception and family planning are widely accessible, the number of abortions has dropped. It has been shown that where abortions are illegal, the number of abortions remains about the same; what changes is the incidence of mortality for the mothers.

Many who are personally opposed to abortion are uneasy about intervening (or allowing the state to intervene) in a matter that is so personal and private. They feel that abortion is a matter between a woman and her conscience, and they are reluctant to impose their own values on anyone else. But for many antiabortionists the sanctity of human life supersedes a woman's right to determine whether or not she will bear her unborn child. For those who believe that life begins at conception, it is a moral imperative to try to prevent what they perceive as murder.

Sadly, using abortion as a means of birth control can lead to abuses. Women in the former Soviet Union, for example, had an average of 4 or 5 abortions throughout their childbearing years, a reflection of the fact that contraceptives were difficult to obtain. In certain Asian cultures, "son preference" is so strong that abortion may be used to end a pregnancy if the fetus is discovered to be female. In Bombay, India, researchers found that of 8,000 fetuses aborted after amniocentesis (a test designed to discover genetic defects, but sometimes used to ascertain the sex of a fetus), all but one were female. There is a terrible irony in this situation: abortion, hailed by many as a means of liberating women, is used in some cases to prevent the birth of females.

Given the controversial nature of abortion and the abuses it can invite, many people agree that preventing pregnancy is preferable to terminating a pregnancy through abortion.

Contraceptive Use Worldwide

Worldwide, contraceptive use varies widely; people in the industrial nations enjoy easy access to birth control methods, while those in the developing nations, in general, do not (Table 9-1). About 56 percent of married women worldwide used some form of contraception in 1991; 46 percent used modern contraceptives. In MDCs 71 percent used some form, with 55 percent using modern methods. In LDCs 44 percent used some form, with 36 percent using modern methods. The U.S. Agency for International Development (AID) estimates that, in order to stabilize populations in developing countries, 80 percent of women of childbearing age should use birth control.

Contraceptive use also varies widely within the United States. Teenagers (especially unwed teens) and poor women are the groups least likely to use birth control. Teenage pregnancy is a particularly serious problem in the United States; the Population Reference Bureau reports that more than a million teenage girls become pregnant annually (Figure 9-9). About 800,000 of these teens are unmarried and approximately 30,000 are under the age of 15. Less than half of these pregnancies result in birth; about 40 percent end in abortion, and another 13 percent end in stillbirth or miscarriage. Some researchers believe that teen pregnancy is a reasonable choice for poor women in the United States, given the fact that they are likely to have family support available and the probability that their health will deteriorate throughout their twenties.

A number of serious problems are associated with teen pregnancies. The teenagers who become pregnant often do not have adequate diets and do not get the prenatal or postnatal care they need. The rate of low birth weight babies born to teen mothers is higher than that of babies born to older mothers, and the survival rate for babies born to teens is lower. Moreover, maternal age has some effect upon

▶ Table 9-1
Estimated Use of Effective Birth Control Methods, 1986 (millions of users)[a]

	China	Developing World Except China	Total Developing World[c]	Developed World[c]	Total World
Female sterilization	53	45	98	15	113
Male sterilization	17	18	35	8	43
Oral contraceptives	9	28	37	27	64
IUDs	59	13	72	11	83
Condoms	5	12	17	28	45
Other effective methods[b]	3	8	11	13	24
Total number of contraceptive users	146	124	270	102	372
Total number of couples	200	463	663	197	860
Contraceptive users as percent of couples	73%	27%	41%	52%	43%
Annual incidence of abortion	12	16	28	26	54

[a] Data presented refer to number (millions) of couples of reproductive age in union. In most countries an additional 5 to 20 percent of couples in union may use less effective or traditional methods, and a widely variable percentage may also rely on abortion, either alone or as a backup for other methods of birth control. These prevalence figures therefore underestimate the total number of couples practicing some form of fertility control.

[b] The category "other effective methods" includes injectable contraceptives (worldwide use estimated at 6.5 million), other steroidal methods, and barrier methods other than the condom, such as the diaphragm and spermicides.

[c] Contraceptive prevalence levels for most developed countries are above 65 percent. However, prevalence levels for sterilization, oral contraceptives, and IUDs are unusually low in Japan, the former Soviet Union, and some eastern European countries, and therefore bring down the average significantly. Prevalence figures do not include women sterilized for noncontraceptive purposes.

Source: World Contraceptive Use, wall chart, United Nations, 1987. Issues in Contraceptive Development, table, p. 3, Briefing Paper No. 15, Population Crisis Committee, May 1985. General Office of the State Family Planning Commission, China, 1987. Induced Abortion: A World Review, 1986, Christopher Tietze, Stanley K. Henshaw, The Alan Guttmacher Institute. As cited in Population Crisis Committee, Population Briefing Paper No. 19, October, 1987.

Figure 9-9: A pregnant teenager talks with a counselor. Teen pregnancy is a serious problem in the United States.

the emotional, social, and intellectual development of the child—the younger the mother, the more likely the child will not be as bright or adjust as well as other children while growing up. Teen parents (mothers and fathers) do not complete as many years of schooling, on average, as those who delay parenthood. This consequence of teen parenthood may haunt the young parents throughout their life, since lack of education decreases occupational options and earnings potential. Society loses, too, when teen pregnancy rates are high: teen childbearing translates into greater demand for public health and welfare services and thus greater public expenditures. Citizens who are less educated and less well-adjusted cannot achieve their full human potential, and thus may not contribute as significantly to society.

Various factors contribute to high teen pregnancy rates. Premarital sex among teens has become common; according to a study by researchers at Johns Hopkins University, more than one-quarter of all 15-year-olds in the United States reported having intercourse, as did about one-third of all 16-year-olds and almost one-half of all 17-year-olds. In addition, contraceptive use among teens remains low. Only about one-third of all sexually active teenage girls (between the ages of 15 and 19) use birth control.

The low rate of contraceptive use by teenagers in the United States explains, in part, why the rates of teen pregnancy and abortion are much higher in this country than in all other developed nations—at least double rates in Canada, England, and France and seven times higher than rates in the Netherlands. Teens in these nations are as sexually active as teens in the United States, and socioeconomic conditions in these nations are similar to those in this country.

However, in other MDCs sex education and free or low cost contraceptive services are universally available.

Future Management of Human Population Growth

Most people agree that it is a worthwhile goal to provide all people, whatever the size of the global population, with the opportunity for a life of quality and dignity. That may appear to be an impossible task. Population growth may seem to be a problem the individual can do little about, but there are things that each of us can do, from encouraging smaller families to lessening our own impact, and that of our families, on the environment (see What You Can Do: Population Growth).

At the societal level, a fundamental population policy change is required. We must begin to consciously "manage" our species in relation to our natural environment. Management efforts should include protecting human health and the environment, preventing resource abuse through conservation, and preserving living systems.

Protect Human Health and the Environment

A number of steps can be taken to manage our species to protect human health and the environment. Of these, the most important is that every country establish a formal policy to guide population growth. The cornerstone of a population policy should be an effort to determine the cultural carrying capacity—the population size that can be sustained by the environment and resources of a country at a reasonably high standard of living for all people. Once the carrying capacity is determined, measures can be developed to help the country reach that standard. Two elements are especially important: efforts to reduce the population's impact on the environment and a strong family planning program.

The family planning program can be tailored to meet the needs of a particular society. For instance, in predominantly Catholic areas of the developing world, where people may resist the use of artificial means of birth control, educational efforts can be aimed at helping women to use the rhythm method. Even though rhythm is not one of the most reliable contraceptive methods, it can be fairly effective if used correctly. Moreover, the strict use of the rhythm method will be more effective than even the most advanced artificial contraceptive, if the artificial method goes largely unused because it is unacceptable, too expensive, or not understood. Other measures that can be used to encourage couples to

Cherish fewer children. Support relatives and friends who decide to have just one or two children or none. Avoid pressuring your children to bear children. And don't believe the stereotype that says single children and single adults are unhappy—it isn't true!

Spread the love around. If you've got a strong parental urge, consider adopting children rather than having your own. Make enriching the lives of other people's children a part of your life.

Onlies are OK. If you decide to have children, consider having only one or, at most, two. Each child born in the United States has an enormous impact on the environment due to our heavy consumption of water, energy, and goods.

Buy local produce or grow your own. When developing nations use scarce cropland to grow food for export, they deprive their populations of that land. Feed your family sustainably produced foods from your area—even start your own garden.

Mandate equal opportunity for women. Where women have better educational and economic opportunities, the birth rate has declined. Consider working for the passage of the Equal Rights Amendment, to ensure equal opportunities for men and women alike.

Make contraceptives available globally. During the next two decades three billion young people will enter their reproductive years. Currently, only about half of fertile women have access to contraception. Encourage your congressional representatives to support expanded family planning programs in the United States and abroad.

Work to eliminate the need for abortion. Outlawing abortion does not improve family planning or make for happy, wanted children; rather, it leads to dangerous illegal abortions, increased mortality rates for women, and unwanted children. Most people—both pro-life and pro-choice—would agree that eliminating the need for abortion is a worthier goal. Support family planning programs and efforts to make

contraceptives widely available. Work to encourage an attitude of respect for human life—the unborn, the poor, the homeless, the parentless—so that everyone can enjoy a life of dignity.

Limit development. Because of population growth, the world's farmers have to feed about 90 million more people with 24 million fewer tons of topsoil each year. Use your vote to promote land-use policies that preserve open space and farming, not only as a means of production, but as a way of life.

Sponsor a foster child in a developing country. For a small monthly fee, reputable organizations such as Childreach (formerly Foster Parents Plan) link caring people in the United States with needy children and their families overseas. The programs help families and communities become self-sufficient. Write Childreach, 155 Plan Way, Warwick, RI, 02886 or call 1-800-556-7918 toll-free (in RI, 401-738-5600)

Source: Adapted from a series on personal ecology by Monte Paulsen, editor and publisher of *Casco Bay Weekly*, in Portland, Maine.

---

have smaller families are lowering the infant mortality rate (by improving health care and sanitation systems), increasing educational and employment opportunities for women, enacting laws to raise the marriage age, and providing economic incentives to couples with only one or two children.

Prevent Resource Abuse Through Conservation

Most LDCs want to implement family planning programs. However, governments must be highly motivated to set these programs in place, and they need the support of international agencies to do so. The United States and other MDCs should increase funding for family planning efforts in the developing world. Reversing the Mexico City policy will allow U.S. funds to go to the two agencies with the most experience and best track record in providing family planning assistance to developing nations—the UN

Fund for Population Activities and the International Planned Parenthood Federation.

It is also important that the governments of LDCs resist depleting their countries' natural resource bases to realize short-term gain or to meet interest payments on foreign debts. This can only be done, obviously, with the cooperation of MDCs to which these debts are owed (see Chapter 26, Economics and Politics).

The single most important thing that slow-growth MDCs can do to prevent resource abuse is to reduce resource consumption. Creative thinking can offer alternatives that will keep the standard of living high while lessening the impact of the population.

Preserve Living Systems

Maintaining population size slightly below the carrying capacity and preventing resource abuse will help to preserve living systems and protect the natural resource base. A sustainable population should be at zero population growth if it is at or near the carrying capacity of its environment. When needed, programs can be designed to lower the growth rate until it reaches zero.

Summary

For most of its tenure on Earth, beginning some 40,000 or 50,000 years ago, *Homo sapiens* existed by gathering wild plants and hunting wild animals. Up until about 10,000 years ago the earth probably supported about 5 million people. With the rise of agriculture populations began to grow more rapidly and the environmental impact of human activities increased. By the beginning of the nineteenth century the human population had reached 1 billion. Advances in death control ushered in a period of rapid growth. Birth rates in the western world began to fall with the onset of the Industrial Revolution and subsequent rising standard of living, the introduction of safe and reliable means of birth control, and an increase in the cost of child rearing. However, the population continued to grow (though more slowly) because there were so many more people in total. Growth rates for most industrial nations are presently either low or zero.

The population path followed by the industrial nations, the demographic transition, describes the movement of a nation from high growth to low growth. It consists of four stages. In stage 1, birth and death rates are both high. In stage 2, death rates fall, but birth rates remain high, and thus the population undergoes rapid growth. In stage 3, birth rates begin to fall, and the growth rate declines until it eventually nears zero. In stage 4, the growth rate is at or below zero.

The human population reached 2.5 billion at about the mid-century mark and has since doubled. Much of the growth in the past fifty years occurred in LDCs; falling death rates and constant or slightly rising birth and fertility rates are responsible for this growth. Rapid growth and the environmental deterioration it causes are fueling a downward spiral in the standard of living. Some developing nations are caught in a demographic trap, unable to break out of stage 2 of the demographic transition.

Any planned course of action taken by a government designed to influence its constituents' choices or decisions on fertility or migration can be considered a population policy. A pronatalist policy encourages natality or births; an antinatalist policy discourages births. Pronatalist or antinatalist policies are developed as a result of the way population changes are perceived. Family planning refers to a wide variety of measures that enable parents to control the number of children they have and the spacing of their children's births. The goal of family planning is not to limit births, but to enable couples to have healthy children and to have the number of children they want.

China became the first nation to have as an official goal the end of population growth and the subsequent lowering of absolute numbers by a significant amount. From 1969 to 1979 China achieved a transition from high to low birth rates by implementing the strongest family planning measures ever attempted. These included free birth control, higher marriage ages for women, economic incentives to have fewer children, education, and media campaigns. The "one child per family" policy was effective. After China relaxed its family planning policies in the 1980s, birth rates rose slightly.

Birth control can be achieved through various preconception and postconception methods. Natural forms of contraception, such as rhythm, have a high failure rate. Other forms of birth control include barrier methods (the condom, spermicides, vaginal sponge, and diaphragm) and hormonal contraceptives. Sterilization, the most reliable method of birth control, is the most popular method of contraception in the world. It is usually irreversible. The IUD is the most popular form of reversible contraceptive in use worldwide. The pill, an oral hormonal contraceptive, is the third most popular birth control method in the world. Other hormonal methods, such as injections or implants, may be more efficient than the pill because they are less expensive and need to be administered much less frequently. Abortion continues to be widely used worldwide as a form of birth control. The abortion controversy revolves around the conflict between religious and moral beliefs about the status of the fetus.

Worldwide, contraceptive use varies widely. People in the industrial nations enjoy easy access to birth control methods, while those in the developing nations, in general, do not. Contraceptive use also varies widely within the United States. Teenagers (especially unwed teens) and poor women are the groups least likely to use birth control. Teenage pregnancy is a particularly serious problem with more than a million teenage girls becoming pregnant annually. The rates of teen pregnancy and abortion are much higher in the U.S. than in all other developed nations.

Key Terms

abortion

antinatalist policy

demographic transition

demographic trap

family planning

population policy

pronatalist policy

Discussion Questions

1. Describe the demographic transition. Which stage do you think the United States is in and why? Which stage do you think Kenya is in and why?

2. Explain the difference between pronatalist and antinatalist policies. What factors cause a country to adopt one or the other? Do you think the United States should adopt an official population control policy? Why or why not? If you answer yes, what policy should it adopt?

3. What are the goals of family planning? How is family planning different from ZPG policies?

4. List some important features of China's "one child per family" policy. Do you think such a policy would work in India, the United States, and Mexico? Why or why not?

5. List at least five methods of birth control. Which ones are most effective in LDCs and in MDCs? Explain.

Food Resources, Hunger, and Poverty

Hunger is real; scarcity is not.

Institute for Food and Development Policy

One evening a gentleman came to our house and told us there was a Hindu family with many children which had not eaten in several days. He asked us if we could do something. So I took some rice and went to them. When I got there I saw the hunger in the shallow eyes of the children, real hunger. The mother took the rice from my hand and divided it in two and left the room. She said simply, "Next door they are hungry, also."

Mother Teresa

Learning Objectives

When you finish reading this chapter, you should be able to:

1. Identify the critical components of a healthy diet.

2. Identify the major foods relied upon by the global human population and summarize the current status of food production.

3. Describe the various manifestations of hunger and explain how hunger affects human health.

4. Discuss how food consumption patterns vary worldwide.

5. Describe the relationship among hunger, poverty, and environmental degradation.

6. List various reasons why hunger continues to be a problem.

7. Briefly describe how food production patterns have changed in the past fifty years in both the developed and the developing worlds, and identify problems with modern agriculture in both MDCs and LDCs.

8. Explain the role of gene banks, biotechnology, and aquaculture in food production.

We obtain our food from our environment. Whether we harvest from cultivated fields and pastures or wild forests, lakes, and oceans; whether we glean from bush, vine, and shrub or collect from fish and fowl, we depend upon the intricately interwoven fabric of the biosphere for sustenance and nourishment. Food production, processing, storage, and distribution are subject to the laws of nature as well as to the whims of humans.

In this chapter we look at food and its relation to the human animal: the biological, physical, and social boundaries of the food resource. Environmental problems related to food production worldwide are also examined. We try to discover what hunger is and why it continues unabated in our world, and how we might effectively manage our food resources to alleviate and eventually eliminate hunger. Finally, we explore the relationship between hunger, poverty, and environmental degradation.

Describing Food Resources: Biological Boundaries

What Are the Components of a Healthy Diet?

The human body needs food to supply the energy needed to do work, to grow and develop, and to regulate life processes.

Carbohydrates and fats, which are composed primarily of carbon, hydrogen, and oxygen, are the major sources of the energy required to maintain the body and perform work. Fats are also needed to construct cell membranes, protect internal organs, and act as insulation.

Proteins form the substance of muscles, organs, antibodies, and enzymes. Proteins are essential for growth and development, particularly fetal development. Small units called **amino acids** combine in various ways to form larger protein molecules. There are twenty naturally occurring amino acids. Eleven can be synthesized by the body, but the remaining nine cannot always be synthesized at a sufficient rate and must be obtained through diet. These nine are known as essential amino acids. A complete protein, such as mother's milk, contains all the essential amino acids in approximately the correct proportions to meet human needs.

Complex carbohydrates, proteins, and fats, are known as macronutrients because they are needed by the body in large amounts. A healthy diet is high in complex carbohydrates with an adequate amount of protein and fats. A standard rule of nutrition is to obtain 60 percent of daily caloric intake from complex carbohydrates, 25 percent from fats, and 15 percent from proteins.

Micronutrients are substances the body needs in small, sometimes trace, amounts. Found in a wide variety of foods, micronutrients are necessary to regulate life processes such as transporting oxygen in the blood and maintaining the nervous and digestive systems. Iron and vitamins are important micronutrients.

Daily caloric requirements vary with gender, age, and factors such as activity level and temperature. According to the United Nations, the amount of food consumed daily by the average person globally should be equivalent to about 2,400 calories (Figure 10-1). A diet must also supply adequate fats and carbohydrates in order to spare protein for tissue building.

What Is the Current Status of Food Production?

Foods the Global Population Relies On

Although there are 80,000 potentially edible plants on earth, global agriculture is dependent upon a few plant species. We derive 95 percent of our nutrition from just 30 plant species, and a mere 8 crops supply 75 percent of the human diet. The top four are wheat, rice, maize (corn), and potato (Table 10-1). These crops are known as **staples** because of their importance in many people's diets. Other important food crops include barley, sweet potato, sorghum, oats, rye, and soybeans. Many of these are especially important for people in less-developed countries (Figure 10-2).

Humans rely on just a small number of animals. Nine domesticated animals—cattle, pigs, sheep,

FIGURE 10-1: Man and woman harvesting rice, Java, Indonesia. Men (ages 19 to 51+) require approximately 2,300 to 2,900 calories daily; women (ages 19 to 51+) require 1,900 to 2,200 calories. There is a normal variation in individual needs of plus or minus 20 percent.

horses, poultry, mules/asses, goats, camels, and buffalo—supply the bulk of the protein (meat, eggs, milk) from livestock. In order to produce foods and other products from animals, we use 7.4 billion acres (over 3 billion hectares) of grazing area, a much larger area than we use to plant crops. Although people in developing countries possess about half of the world's livestock, they consume only about 20 percent of the meat and milk they raise. The rest is exported to developed countries.

Table 10-1
World Food Production, Selected Crops
(thousand metric tons)

Crop	United States	Developed Countries	Developing Countries	World Total
Wheat	49,295	293,018	216,934	509,952
Rice	7,237	25,531	457,935	483,466
Corn	125,003	220,080	185,380	405,460
Potatoes	15,875	192,891	76,811	269,702
Barley	6,325	140,514	27,909	168,423
Sweet potatoes	537	2,095	128,259	130,354
Soybeans	41,876	46,598	45,734	92,332
Sorghum	14,670	17,462	44,325	61,787
Oats	3,175	36,584	2,264	38,848
Millet	—	2,790	28,745	31,535
Rye	382	28,182	1,435	29,617
Beans, dry	872	1,957	13,576.	15,533
Peas, dry	228	12,812	2,693	15,505
Lentils	41	240	2,267	2,507

Source: 1988 Food and Agriculture Organization Yearbook.

Wheat is the most important cereal worldwide, the staple for over one-third of the human population. Its protein content varies between 8 and 15 percent. Wheat is grown primarily in temperate climes and some subtropical zones.

Rice is the leading tropical crop in Asia. Because of wet-rice cultivation, which allows for continuous cropping, rice can support high densities of population. Its protein content ranges between 8 and 9 percent.

Maize, with a protein content around 10 percent, is a staple crop for the people in South America and Africa. The United States is the world's largest maize producer but the bulk of the crop is fed to livestock.

Potatoes, grown best in cool, moist, temperate climes, are a staple carbohydrate in many developed nations.

Barley is the fourth most important cereal crop. It is used primarily for animal feed and malting for beer and whiskey. For people in parts of Asia and Ethiopia, it is an important food crop.

Sweet potatoes, grown in moist tropical regions are generally used as a secondary food. Their chief value is as a source of starch.

Cassava is an important food crop in Africa. It has low protein content, but is drought-resistant.

Sorgum and millet are cereal grasses. They are staples in many drier parts of Africa and Asia. Because these grains lack gluten, they cannot be used to make bread.

Oats and rye are suited to cool, damp climes. Oats are grown chiefly for animal feed; rye is used principally for bread flour.

Pulses, especially **soybeans,** have high protein content, between 30 and 50 percent. In many poorer regions, they may be the people's chief source of protein.

FIGURE 10-2: Important food crops worldwide.

People in MDCs, in fact, consume twice as much meat as people in LDCs. Indeed, the largest tonnage of grains produced in MDCs goes to feed animals. Worldwide, approximately 40 percent of grain production is used to feed livestock, but in MDCs that figure can average 75 to 90 percent.

In and of itself, eating meat or meat products in conservative amounts is neither nutritionally nor environmentally unsound. Meat is an important source of amino acids, vitamins, and minerals. The consumption of grazing animals can be sustainable as long as they feed primarily on grasses, crop residues, and vegetation and are not allowed to overgraze an area. The efficiency of grazing animals is increased when their wastes are used to fertilize the soil. In this situation, grazing animals in effect "harvest" their own food and help to replenish the soil, restoring its fertility. Unfortunately, meat consumption *can* generate problems for human health and the environment when it is practiced to the extent and in the manner common in the United States and other MDCs. When cattle, pigs, chickens, and other livestock are fed grains, the efficiency of food production is greatly reduced, because a significant amount of energy must be invested to plant and harvest crops (Figure 10-3). Synthetic fertilizers, typically used to replenish the soil and keep yields high, increase the "energy investment" of meat production. The overgrazing of range and pasture by livestock (especially cattle and sheep) also incurs substantial environmental costs, particularly soil erosion and the pollution of nearby waterways. Figure 10-4 lists the potentially destructive environmental effects of meat-eating habits.

Food Security

Food security is the ability of a nation to feed itself on an ongoing basis. Food production and thus the potential for food security have grown tremendously in the last forty years. World grain output experienced unprecedented growth from 1950 to 1984, expanding nearly 2.6 times (3 percent per year), according to figures compiled by the Worldwatch

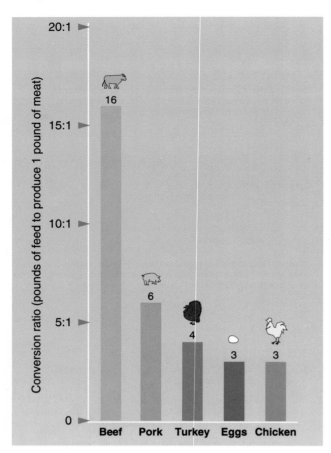

FIGURE 10-3: Efficiency of meat production. The production of animal commodities requires significant inputs of grain and soy. Beef requires substantially higher inputs than other meats (16 pounds of feed to produce 1 pound of meat). Given these constraints, it is little wonder that many people in LDCs cannot rely on meat or meat products as sources of protein.

Institute. During the 1950s and 1960s grain production exceeded population growth on every continent. But in 1985 and 1986 world grain output per capita increased only slightly, and then in the drought years of 1986 to 1988 it fell significantly (Table 10-2). The huge grain reserves of the early 1980s no longer exist. By 1990 the total supply of grain in reserves fell to one of the lowest levels in decades.

India, which tripled its wheat harvest between 1965 and 1983, has not increased its grain output since that time. China's wheat production, which enjoyed dramatic gains between 1976 and 1984, has leveled off. Similarly, Mexico quadrupled its wheat harvest between 1950 and 1984, but production has since fallen off. Both Latin America and Africa are experiencing grain deficits and are the most seriously affected by food shortages. In the 1970s grain production fell behind population growth in Africa. In the 1980s the same thing happened in Central America, Brazil, and Mexico. Gains have been made in producing more food per acre worldwide, but

these have been offset by numerous factors, including population growth, social inequities, and the increase in degraded cropland and cropland converted to nonfarm uses.

Physical Boundaries

Where Does Our Food Come From? Available Fertile Land

Of the earth's total ice-free land surface, 3.7 billion acres (1.5 billion hectares), about 11 percent, can be easily used to produce food. The rest, about 32 billion acres (13 billion hectares), is too wet, dry, cold, hot, poor in nutrients, or shallow in soil to cultivate easily. With intense management and a significant financial investment, another 3.7 billion acres (1.5 billion hectares) of land could be brought into food production. Even though the earth's fertility is not evenly distributed, properly managed, the earth's available fertile lands could and should meet the food needs of its human inhabitants.

Aquatic Harvest

We harvest most of our aquatic foods from shallow nearshore waters, estuaries, and the offshore conti-

• Length of time world's petroleum reserves would last if all human beings ate meat-centered diet	13 years
• Length of time world's petroleum reserves would last if all human beings ate vegetarian diet	260 years
• Pounds of beef that can be produced on 1 acre of land	165
• Pounds of potatoes that can be grown on 1 acre of land	20,000
• Percentage of protein wasted by cycling grain through livestock	90
• Number of people who could be adequately fed by the grain saved if Americans reduced their intake of meat by 10%	60 million
• Water needed to produce 1 pound of meat	2,500 gallons
• Water needed to produce 1 pound of wheat	25 gallons
• Amount of U.S. cropland lost each year to soil erosion	4,000,000 acres
• Percentage of U.S. topsoil loss directly associated with livestock raising	85

Source: Diet for a New America, by John Robbins, in *Carrying Capacity,* March 1989.

FIGURE 10-4: Potential adverse effects of meat-eating habits.

▶ Table 10-2
World Grain Production

Year	World Harvest			Per Capita		
	Total (million metric tons)	Change per Decade		Total (kilograms)	Change per Decade	
		Amount (million metric tons)	Percent		Amount (kilograms)	Percent
1950	631			246		
1960	847	+216	+34	278	+32	+13
1970	1,103	+256	+30	296	+18	+6
1980	1,441	+338	+31	322	+26	+9
1990	1,684[a]	+243	+17	316	-6	-2
2000	1,842[b]	+158	+9	295	-21	-7

[a]Assumes 1 percent increase over 1989 world harvest of 1,667 million metric tons as estimated by U.S. Department of Agriculture in October 1989.

[b]Assumes no appreciable gains or losses in world grain area and a rate of yield-per-hectare increase for world grain between 1990 and 2000 that will equal the 0.9 percent per year increase in Japan's rice yield between 1969–1971 and 1986–1988.

Source: State of the World, 1990. Worldwatch Institute.

nental shelf, areas that represent only a small part of the total water area of the earth. Fisheries are the source of roughly 10 percent of the food protein available to the world's peoples, but the aquatic harvest is not shared equally by people worldwide. About a third of the annual fish catch is not used to feed humans at all. Instead, it is reduced to meal to supplement the diets of livestock, pets, and zoo animals or it is processed to yield fish oil products.

Fish production has not appreciably increased in recent years (Table 10-3). Much of the global catch presently comes from depleted or fully exploited stocks, and most fisheries scientists believe we have reached the limit in exploiting traditional fish and shellfish species. In the past we have witnessed the collapse or near collapse of several fish or shellfish species due to a combination of overexploitation by humans and climatic or ecological changes. Examples include the collapse of the Peruvian anchovy fishery, the demise of California sardines, and severe reductions of Alaskan king crab, Atlantic striped bass, and Atlantic herring. Total fish catch could be dramatically increased if we start to exploit unconventional species (such as small fish, deep water fish, squid, and krill). Doing so could double or triple the global harvest. But an important question is: What would be the consequences to the marine food chain if we significantly increase our harvest of these species?

Social Boundaries

Suppose 300 jumbo jets crashed today, killing all on board. Then imagine that the same thing happened tomorrow and the day after and the day after that,

for an entire year. The scenario is so horrible that the mind rebels: Surely, the world's people would not allow such a thing to happen. And yet, each year an equal number of people—40 million, about half

▶ Table 10-3
Fisheries—Commercial Catch,[a] Selected Countries, 1984–1988 (billion pounds, live weight)

Country	1984	1985	1986	1987	1988
World, total[b]	183.2	190.6	203.7	206.7	216.9
Canada	2.8	3.1	3.3	3.4	3.5
Chile	9.9	10.6	12.3	10.6	11.5
China (mainland)	13.1	14.9	17.6	20.6	22.8
Denmark	4.1	3.9	4.1	3.8	4.3
Iceland	3.4	3.7	3.7	3.6	3.9
India	6.3	6.2	6.4	6.4	6.9
Indonesia	4.4	5.2	5.4	5.7	6.0
Japan	26.5	25.2	26.4	26.1	26.2
Mexico	2.4	2.7	2.9	3.1	3.0
North Korea (estimate)	3.6	3.7	3.7	3.7	3.7
Norway	5.4	4.7	4.2	4.3	4.0
Peru	7.4	9.2	12.4	10.1	14.6
Philippines	4.3	4.1	4.2	4.4	4.5
South Korea	5.5	5.8	6.8	6.3	6.0
Soviet Union	23.4	23.2	24.8	24.6	25.0
Spain	2.9	3.0	3.2	3.1	3.2
Thailand	4.7	4.9	5.6	4.9	5.2
United States	10.6	10.5	10.9	13.3	13.3

[a]Catch of fish, crustaceans, mollusks (including weight of shells), and other aquatic plants and animals, except whales and seals.

[b]Includes other countries not shown separately.

Source: Statistical Abstract of the United States, 1991.

FIGURE 10-5: A young Sudanese famine victim, 1991.

of them children—die of hunger or hunger-related diseases. Over one-third of the deaths, 15 million, are caused by starvation or problems directly related to malnutrition. Each day hunger and starvation kill 35,000 human beings; every two days the number of people who die from these causes is equal to the number who were killed instantly when the atomic bomb exploded on Hiroshima. Yet too few voices are raised in protest. With seeming regularity, the evening news is filled with images of starvation: The faces of listless young children, the blank stares of men and women without hope, the resignation of those dying far before their time. The haunting images of the dying awaken waves of sympathy and passion that briefly sweep the world. But, as always, the passion soon ebbs. The moment passes, the global media focus on more recent events, and the furor over the hungry fades (Figure 10-5).

What Is Hunger and How Does It Affect Human Health?

At first glance, hunger seems easy to define. Even a young child knows that being hungry means wanting something to eat. For most of us, being hungry simply means having an appetite. Few of us have known real hunger—an aching, gnawing desire for food that remains unabated day after day, week after week. Few of us have felt the effects of hunger—the sharp pain in the belly that gradually dulls but never leaves, the growing weakness, the increased susceptibility to disease and illness, the slow deterioration of mind and body. Indeed, few of us can even imagine chronic and persistent hunger. Perhaps that is part of the reason that hunger exists in our world,

for how can we fight an enemy that we do not know? Before we can eradicate hunger, then, we must learn to recognize it, and hunger is an enemy with many faces.

Starvation and Famine

Starvation, the most easily recognized manifestation of hunger, is suffering or death from the deprivation of nourishment; a person does not consume enough calories to sustain life. **Famine,** or widespread starvation, is typically the result of many interrelated factors. For example, the Ethiopian famine of the mid-1980s was a product of prolonged drought, a civil war that ravaged the country, and the production of cash crops. Famine can be precipitated by catastrophic events such as floods and earthquakes and by human activities such as war and the inequitable distribution of land (Figure 10-6).

Undernutrition and Malnutrition

Although famines are widely reported by the international community and media, they are only one aspect, and a relatively minor aspect at that, of hunger. Hunger is primarily a quiet killer. Of the 40 million people who will die this year because of hunger, most will die from diseases that afflict them because they are undernourished or malnourished.

Chronic undernutrition is the consumption of too few calories and protein over an extended period of time. People consuming fewer than 2,000 calories

FIGURE 10-6: Kurds fleeing hostilities in Iraq during the Persian Gulf War of 1991 approach a refugee camp in Isikveren, Turkey. Food and medical aid donated by nations around the world were provided to those who sought haven in refugee camps.

daily on a regular basis are considered to be chronically undernourished. From 2,000 to 2,400 calories, undernourishment may also occur, depending on a person's activity level. The undernourished gradually become weaker and less able to work or think productively, they may appear listless, and their weakened bodies are less able to ward off illness and disease. Chronic undernutrition is a slow but steady—and therefore often unnoticed—decline toward death.

Malnutrition is the consumption of too little (or more infrequently, too much) of specific nutrients essential for good health. Malnutrition may not kill, but it can maim and deform.

Malabsorptive hunger often accompanies undernutrition and malnutrition. The body loses the ability to absorb nutrients from the food consumed. Malabsorptive hunger can be caused by parasites in the intestinal tract or by a severe protein deficiency.

Nutritional Diseases

For children insufficient protein is a particularly serious and common health problem. A deficiency of protein inhibits mental and physical development. Severe protein insufficiency may cause mental retardation, physical wasting away, and death. Protein insufficiency contributes to kwashiorkor and marasmus, two of the most common childhood diseases related to a nutritionally deficient diet.

Kwashiorkor is associated with a low protein diet, a diet composed primarily of starches, often maize or sweet potato. Kwashiorkor, which means "displaced child," occurs when a child, typically from 1 to 3 years old, is weaned from mother's milk so that a new infant can be nursed. The child is deprived of a source of complete protein, and even though the child may consume a sufficient number of calories, protein insufficiency brings on the disease. Symptoms include skin rashes or discolored patches on the knees and elbows; swelling of the hands, feet, and face; liver degeneration; possible permanent stunting of growth; hair loss; diarrhea; mental apathy; and irritability.

Marasmus is caused by a diet deficient in both protein and overall number of calories. It is most common among infants or very young children who are not breast fed or who are fed an inadequate, high-starch diet after they are weaned. The child with marasmus typically has emaciated arms and legs, shriveled skin, wide eyes, and a gaunt, aged-looking face and suffers from diarrhea, muscle deterioration, and anemia. Kwashiorkor and marasmus can cause irreversible brain damage and mental retardation, especially in the very young, and both can be fatal. Fortunately, both can be reversed if the deficiency is not prolonged.

Annually, approximately 15 million infants and young children die due to hunger and hunger-related diseases. According to estimates by UNICEF, many of these deaths could be prevented through programs that promote breast feeding, oral rehydration therapy (ORT) to counteract diarrhea, and overall improved child care (Figure 10-7). The cost: $5 per child per year. Combating diarrhea, a leading cause of death in children, is particularly important; untreated episodes can lead to malnutrition and retarded growth and development.

Seasonal Hunger

For many people **seasonal hunger** is a way of life, a part of the annual cycle. Each year there is a time before the new harvest when the reserves from the previous year's harvest run out. The people go hungry, perhaps for days or weeks, but sometimes, if the previous harvest was small or if the new crop is late in maturing, for months. Seasonal hunger is so common in parts of Africa that the names of the months before harvest time reflect the expected lack of food. The Iteso people of eastern Uganda, for instance, refer to May as "the month when the children wait for food."

The United States has its own version of seasonal hunger. It is the period, from a few days to a few weeks, when the money runs out at the end of the month or when the food stamps are used up and there is little or no food. A particular kind of seasonal hunger can occur with a new political administration, which may eliminate or severely reduce assistance programs.

FIGURE 10-7: Malnourished children receiving food at a government feeding center in Moroto, Uganda, 1980.

Who Are the Hungry?

Where Hunger Occurs

For many people in developing nations, hunger and malnutrition are common. Most citizens of LDCs consume about 10 percent less than the minimum calories required for good health. In contrast, many individuals in MDCs consume about 30 to 40 percent more than necessary. On average, a person in the developing world consumes only two-thirds of the calories and one-fifth of the animal protein consumed by a person in the developed world (Tables 10-4 and 10-5). If we include all the grain fed to livestock, people in the developed world consume three times as much food, on average, as their LDC counterparts. Pets in Europe and North America receive more food (meat and calories) each day than most people in LDCs.

We often think of hunger and dietary diseases as problems only in the developing world. But that is not the case. As shown in Environmental Science in Action: Hunger in the United States (pages 186–189), many people in our nation go hungry. At the other end of the spectrum of dietary deficiency disease we have the "scandal of the waistline," the situation in which the wealthy are able to obtain so much excess food or so many rich foods with little nutrient value that they become fat while a large proportion of the world's people suffer from a lack of food. Illness caused by overeating or poor food choices is common and constitutes an often overlooked form of malnutrition. A high-fat diet low in nutrient value, common in the United States, is linked to heart disease, colon cancer, high blood pressure, and other illnesses that shorten lives. About 20 percent of all adults in the United States, some 33 million Americans, are significantly overweight. But the scandal of the waistline also occurs within other nations. In LDCs the wealthy class or large landowners often consume an inordinate amount and variety of foods while their fellow citizens starve or get by on meager diets. This widening gap often skews the statistics for average dietary consumption. In many LDCs the wealthiest 20 percent of the population are often ten to twenty times richer than the poorest 20 percent. Consequently, there is an unequal distribution of food, with the wealthy receiving an inordinate share.

In our world of stark and sobering contrasts, ironies abound. In 1989 India exported 32 million tons of wheat despite the fact that 300 million of its people (almost one-third of the world's hungry) are chronically hungry. Each year Americans spend billions of dollars on special diets that promise to help them "control" their appetite and "shed unwanted

▶ Table 10-4
Daily Caloric Consumption, 10 Highest and 10 Lowest Countries, 1988

Rank	Country	Average per Capita Daily Caloric Intake
10 HIGHEST		
1.	Libya	3,812
2.	Ireland	3,699
3.	East Germany	3,689
4.	Italy	3,688
5.	Greece	3,668
6.	United States	3,641
7.	Belgium	3,639
8.	Luxembourg	3,639
9.	Bulgaria	3,619
10.	Netherlands	3,617
10 LOWEST		
1.	Afghanistan	1,896
2.	Mali	1,893
3.	Mozambique	1,881
4.	Guinea	1,880
5.	Bangladesh	1,837
6.	Ethiopia	1,793
7.	Uganda	1,784
8.	Ghana	1,769
9.	Maldives	1,765
10.	Chad	1,762

Source: New Book of World Rankings, 1991.

pounds," while 500 million people worldwide are so undernourished that their bodies and minds are wasting away. In 1990 in the United States pets consumed more food per capita than the nearly 32 million Americans—including one-fifth of all children in the country—who live below the poverty line.

Worldwide, about one in every five people suffers from chronic hunger. Children account for about 40 percent of the world's hungry; the majority of the rest are women. Most of the world's hungry live within the "great hunger belt," a region encompassing various nations in Southeast Asia, the Indian subcontinent, the Middle East, Africa, and the equatorial region of Latin America. Just five nations—India, Bangladesh, Nigeria, Pakistan, and Indonesia—are home to at least half of the world's hungry.

Within the great hunger belt, hunger is a daily reality. The Food and Agricultural Organization (FAO) and World Bank estimate that 500 million people—one-tenth of the world population—consume less than the minimum critical diet needed to remain healthy and maintain body weight, even with little physical activity. Of these, more than two-thirds live in Asia; most of the rest live in Africa and

▶ Table 10-5
Meat Consumption, 10 Highest and 10
Lowest Countries, 1984

Rank	Country	Average per Capita Annual Consumption of Meat (kilograms)
10 HIGHEST		
1.	Argentina	77.0
2.	Uruguay	59.0
3.	United States	49.2
4.	Australia	41.6
5.	Canada	40.2
6.	New Zealand	39.8
7.	France	31.9
8.	Guam	30.8
9.	Soviet Union	28.0
10.	Swaziland	27.8
10 LOWEST		
1.	Ghana	1.3
2.	Zaire	1.1
3.	Vietnam	1.0
4.	South Yemen	1.0
5.	Indonesia	0.9
6.	Samoa	0.9
7.	Sri Lanka	0.8
8.	China	0.3
9.	Nepal	0.3
10.	India	0.1

Source: New Book of World Rankings, 1991.

Latin America. In other parts of the world, hunger is less visible and less widely recognized.

Statistics can tell us how widespread hunger is and where it occurs, but the hungry are not numbers. They are people who love and feel pain, people with diverse personalities and talents, people who laugh too seldom and weep too often, people who suffer and hope. Most of all, they are people who want not just food, but the means to make a living; people who want not just bread, but the land to grow wheat; people who want not just to survive the present, but to realize a better future for themselves and for their children.

Relationship Between Hunger and Poverty

No matter where they live, the hungry share a common characteristic: they are poor. In 1990, while more than 150 billionaires and 2 million millionaires inhabited the earth, over 100 million people were homeless and nearly 1 billion people were starving or suffering from hunger-related illnesses. Those that suffer the most from poverty are children. Most of the world's poor are under the age of 15; one-third of the world's children who live in poverty die before the age of 5; many who survive are physically or mentally impaired by the age of six months (Figure 10-8).

How poor is poor? By comparing income with the cost of buying food items, we can begin to answer that question (Table 10-6). If the price of a dozen eggs is more than one-half of a person's daily income (as it is in some parts of the world), daily existence becomes a struggle. In many countries people spend up to 70 percent of their daily income to find, purchase, and prepare food. In the United States, about 15 percent of a person's income is spent on food. When food prices increase because of worldwide increases in grain prices, the poor suffer disproportionately. For example, a doubling of the price of wheat worldwide might increase the cost of a loaf of bread in the United States by 5 cents, only a small percentage of the original cost. But this same

▶ Table 10-6
Cost of Food and Daily per Capita GNP (1985 U.S. dollars)

	Rice (2.2 pounds)	Wheat (2.2 pounds)	Milk (1 quart)	Eggs (1 dozen)	Chicken (2.2 pounds)	GNP per Capita per Day
Mali	$0.53	$0.58	$0.64	$1.91	$3.18	$0.41
Zimbabwe	0.79	0.66	0.40	1.10	2.61	1.86
Colombia	0.64	0.67	0.40	1.23	2.41	3.62
Honduras	0.91	0.55	0.19	0.92	2.04	1.97
India	0.28	0.17	0.61	0.70	—	0.74
Philippines	0.35	0.39	1.00	0.82	1.75	1.59
United States	0.47	1.03	0.60	0.86	1.65	45.73

Note: In Mali the GNP per capita per day is $0.41, making the cost of any one of the food items (rice, wheat, milk, eggs, or chicken) far more than the daily average income. In contrast, the price of all the food items for the average citizen in the United States is still only a fraction of daily income.

Source: Journey, publication of Plan International USA. Formerly Foster Parents Plan. Annual Report 1987.

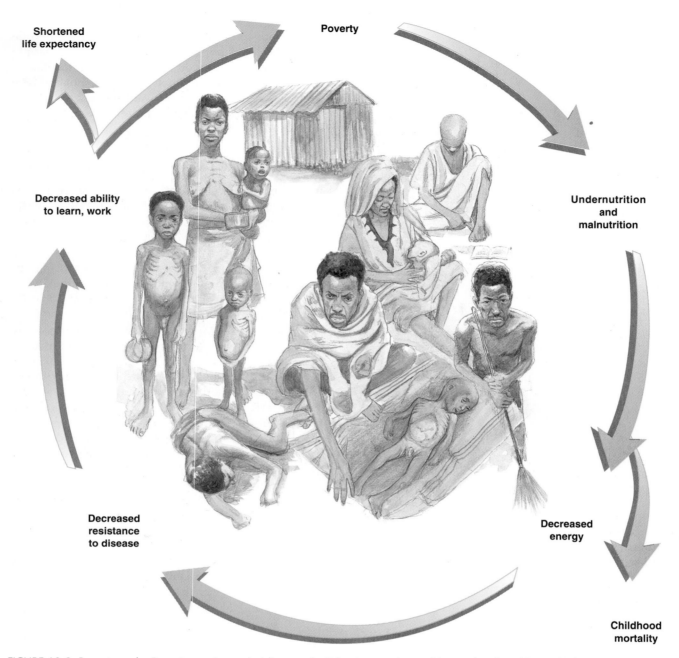

Poverty

Shortened life expectancy

Decreased ability to learn, work

Undernutrition and malnutrition

Decreased resistance to disease

Decreased energy

Childhood mortality

FIGURE 10-8: Poverty cycle. Poverty creates an insidious cycle. It leads to undernutrition and malnutrition, which in turn lead to decreased resistance to disease and decreased energy. Illness and lack of energy lead to decreased ability to learn and to work, which perpetuates poverty. Poverty also results in shortened life expectancy. Unless something is done to break the cycle of poverty, poor children are doomed to a life of hardship and want simply by virtue of having been born.

price doubling could double the cost of bread in LDCs, effectively putting bread beyond the reach of the poor.

Another measure of poverty is the Physical Quality of Life Index (PQLI), developed by Dr. Morris D. Morris, Director of the Center for the Comparative Study of Development at Brown University, for the Overseas Development Council. The PQLI is a composite index that is calculated by averaging three factors: infant mortality rate, life expectancy at age 1, and literacy rate. Countries are ranked on a scale of 0 to 100, with 100 the best. The PQLI depicts the reality of poverty—the powerlessness of the poor with respect to good health, normal life expectancy, and a decent education (Table 10-7).

Relief funds and health measures can provide immediate help to alleviate some symptoms of poverty, but such measures alone are not enough.

▶ Table 10-7
Physical Quality of Life Index (PQLI), 10
Highest and 10 Lowest Countries, 1985

Rank	Country	PQLI
10 HIGHEST		
1.	Australia	100
2.	France	100
3.	Iceland	100
4.	Finland	99
5.	Japan	99
6.	Netherlands	99
7.	Norway	99
8.	Sweden	99
9.	Switzerland	99
10.	Canada	98
	Cuba	98
	Denmark	98
	Italy	98
	Spain	98
	United States	98
10 LOWEST		
1.	Burkina	29
	Guinea-Bissau	29
	Somalia	29
2.	Gambia	28
3.	Guinea	28
4.	Mali	28
5.	Niger	28
6.	North Yemen	28
7.	Bhutan	26
8.	Sierra Leone	26
9.	Ethiopia	25
10.	Afghanistan	21

Source: *New Book of World Rankings*, 1991.

▶ Table 10-8
People Estimated to Be Living in Absolute
Poverty, 1989

Region	Number of People[a] (millions)	Share of Total Population (%)
Asia	675	25
Sub-Saharan Africa	325	62
Latin America	150	35
North Africa and Middle East	75	28
Total	1,225	23

[a]Estimates are best thought of as midpoints of ranges that extend 10 percent above and 10 percent below listed figures.
Source: Worldwatch Institute.

Unless poor, hungry, and sick people are eventually able to feed and care for themselves, unless they are given the tools to become self-sufficient, the poverty and hunger trap will never be eliminated. Economic development is imperative. In those parts of the developing world where economic development has taken hold, particularly eastern and southern Asia, the number of people living in poverty is expected to drop significantly (Table 10-8). Other areas are expected to show only modest decreases in the numbers of poor, and in sub-Saharan Africa, it is estimated that the number of poor will increase markedly. In these regions emergency relief efforts will again be urgently needed.

Relationship Between Poverty and Environmental Degradation

Although much of the world's environmental degradation is due to affluence in MDCs, poverty, especially in LDCs, causes environmental degradation

because people are forced to sacrifice their future to survive in the present. Lacking arable lands or decent paying jobs, the poor are forced to try to squeeze out a living on marginal lands where overworked soils cannot sustain cultivation; to overgraze grasslands and pastures; and to clear forests for fuel, food, or marketable resources. The degraded environment offers less and less to support its human inhabitants, resulting in still more poverty.

In LDCs poverty and environmental degradation are inseparable. Hungry people's concern for protecting the environment is far outweighed by their immediate need to survive. Consequently, they will search for food and fuel wherever they can find it. When the food or fuel is gone or the soil is depleted, they will move on to a new area if they have the strength or die in their homelands if they do not. Those living in extreme poverty, especially in rural areas and on the outskirts of many cities in Asia, Africa, and Latin America, increase the demand on agricultural systems and therefore on the environment for their incomes and survival. Forty percent of Guatemala's arable land has been lost to erosion. In Haiti practically no topsoil remains for cultivation. In Bangladesh, Nepal, and Java severe pressure on hillsides for fuel and growing crops has caused much erosion and flooding in lowland areas (Figure 10-9).

Why Does Hunger Exist in a World of Plenty?

Food production, especially of grains, is sufficient to feed the world's people. Moreover, if grain and root crops presently used to feed animals were instead used for human consumption, perhaps as many as twice the present global population could be adequately fed. Yet hunger continues to be a serious problem. In part, natural environmental conditions account for inequities in food production. Some

FIGURE 10-9: Poverty and pressure on the environment. Severe soil erosion in the Genangan area of Indonesia is a consequence of planting cassava—a basic food crop—every year without rotating crops.

areas are more fertile and have better growing conditions, while other regions are less productive and more vulnerable to injury.

Many people argue that hunger is primarily a result of social practices, politics, and economics. Hunger exists not because the earth cannot produce enough or because we lack the technical and scientific know-how to eradicate hunger; hunger exists because people lack the political and economic power to fight it. (See Focus On: Beyond Hunger—Extending Democracy.)

Various economic factors create or exacerbate poverty. In many LDCs the best lands are controlled by a small number of wealthy landowners. Poor peasants are forced to work for the wealthy landowners, rent good land at high prices, or try to make a living from marginal croplands.

The cash crop factor also contributes to hunger. Cash crops are those grown for export; typically, they are nonessential items such as coffee, tobacco, or cotton. The high prices these commodities often command compels many farmers to abandon food crops in favor of cash crops. As long as the market for cash crops remains strong, the farmers earn more money by growing crops for export than by growing food crops. With their earnings, they can purchase the foods they need as well as tools and technologies to improve food production. The country benefits, too, since the exports bring in critically important foreign exchange earnings which bolster the economy and aid development.

Cash crops tend to favor or support the establishment of large plantations. The lure of quick and handsome profits encourages wealthy landowners to buy up as much acreage as possible and convert it to cash crop production. Under these circumstances, small, independent farmers find it difficult to hold onto their land. In the Philippines, for example, sugarcane plantations are quite profitable, but the poor who work them are paid little and are often underfed and malnourished. Those who harvest the sugarcane do not reap the benefits of the sugarcane industry. Halfway round the world, in the Sahel region of northern Africa, high yields of cash crops, especially coffee, were recorded even as famine killed millions in the 1980s.

History of Management of Food Resources

How Have Humans Manipulated Food Resources in the Past?

For most of its history, the human animal was a hunter and gatherer. Gathering skills and a knowledge of plants and their uses eventually led to the domestication of plants and animals and the development of agriculture. **Domestication** is the result of the long-term selection of traits useful for survival in captivity over those traits necessary for survival in the wild. Early farmers domesticated certain species by selecting and planting the seeds or tubers of those plants that possessed desired characteristics, such as a particular taste or high yield. Consequently, the earliest domestications of plants occurred in the areas where the genetic diversity for a particular species was the greatest. These areas are known as centers of diversity or Vavilov centers, after the Russian botanist Nikolai I. Vavilov, who first recognized their existence.

Domestication gave way to **agriculture,** the purposeful tending of particular plant and animal species for human use. This system of deliberate food production began over ten thousand years ago, in a series of river basins around the globe: the Nile in Africa, the Euphrates-Tigris in the Middle East, the Ganges-Brahmaputra in India, the Indus in Pakistan, and the Yangtze in China. These river basins enjoyed year-round warmth and a constant and ample supply of water. Farmers' deliberate selections of certain strains resulted in **land races,** varieties of plants which are adapted to local conditions such as climate and soil type. In some parts of the world, a small number of traditional farmers continue to breed land races much as their ancestors did.

Changes in Food Production

Historically, farmers increased yields by simply cultivating more land. Agriculture was labor intensive; the power of humans and livestock was used to till

Focus On:
Beyond Hunger—Extending Democracy
Frances Moore Lappé

How we define any problem powerfully determines our search for solutions. So we must ask a seemingly grade-school question: "Just what *is* hunger?"

To that question, there's one very short answer—premature death. We read about it in numbers. Hunger kills as many as 20, or some even say 30, million people each year—two to three times as many people as died in hostilities each year during the height of World War II.

These statistics are staggering. They can shock and alarm. But numbers alone can numb. They can make remote something actually quite close to us. What would it mean instead to think of hunger in terms of universal human feelings, feelings that each of us has experienced at some time in our lives?

Consider three such emotions.

In his book *Witness to War*, Dr. Charlie Clements tells of a family he knew in El Salvador whose son and daughter had died from fever and diarrhea. "Both had been lost," he wrote, "in the years when Camila and her husband had chosen to pay their mortgage, a sum equal to half the value of their crop, rather than keep the money to feed their children. Each year, the choice was always the same," they told him. "If they paid, the children's lives were endangered. If they didn't, their land could be repossessed."

Being hungry thus means anguish. The anguish of impossible choices. But it is more.

In Nicaragua four years ago, I met Amanda Espinoza, a poor campesina who until recently had never had enough food to feed her family. She told me that she had endured six stillbirths and watched five of her children die.

To Amanda, being hungry meant watching people she loved die. It is grief.

Anguish and grief are thus part of what hunger means. But increasingly throughout the world, hunger has a third dimension.

In Guatemala, in 1978, I met two highland peasants who worked to teach their neighbors how to reduce erosion on the steep slopes to which they had been pushed by wealthy landowners in the valley. Two years later, I learned that one of these peasants had been forced into hiding. The other had been killed. To the big landowners, their crime was teaching better farming to the poor, making them less dependent on low-paying jobs on the plantations of the wealthy.

Increasingly, then, the third dimension of hunger is fear.

Anguish, grief, and fear. What if we were to refuse simply to count the hungry and instead tried to understand hunger as such universal emotions? We discover that how we understand hunger determines what we think are its solutions.

If we think of hunger as numbers—numbers of people with too few calories—the solution also appears in numbers—numbers of tons of food aid, or numbers of dollars in economic assistance. But once understanding hunger as real people coping with the most painful of human emotions, we can perceive its roots. We need only ask, when have we experienced any of these emotions ourselves? Hasn't it been when we have felt powerless to protect ourselves and those we love?

Hunger has become for me the ultimate *symbol of powerlessness*.

But if powerlessness lies at the very root of hunger, what are its causes? Certainly it is not powerlessness before nature's scarcity! Not when the world is awash with food—with enough in grain alone to make every person on earth fat!

Put most simply, the root cause of hunger isn't a scarcity of food or land; it's a scarcity of democracy. "Wait a minute!" you may ask, "What does democracy have to do with hunger?" In my view, everything. To me, democracy is not an institution; it is a set of principles.

First is the principle of *accountability*. Democratic structures are those in which people have a say in decisions that most affect their well-being. Precisely where decision makers are unaccountable—be they a government, corporate, or landed elite—do people go hungry. Unaccountable, such elites proceed to monopolize resources essential to everyone. Consider farmland. In most third world countries, less than 10 percent of the rural people now control more than half of the land. At the same time, one billion rural people throughout the third world are effectively landless. Similarly, governments unaccountable to the needs of the majority use the national budget to protect the privileges of a minority: spending on arms by third world governments quadrupled during the decade ending in 1980, and continues to increase.

A second, closely related, principle of democracy is the *sharing of power*. While in every social order some will always have more power than others, in a democracy no one is denied *all* power. But where some people lack even the minimal resources to sustain life, surely they are deprived of all power. For isn't survival the first priority of all living creatures? Thus, the very existence of hungry people—in India, in Africa, or in the United States—denies the existence of democracy.

Third, these principles apply equally in economic and political life. Indeed, as long as such democratic principles—accountability to those most affected by decisions and the sharing of power—are absent from economic life, people will continue to be made powerless. From the village (who owns the land?), through the national level (who allocates the national budget?), to the level of international finance (who incurs foreign debt and who is made to suffer to repay it?), we will witness the continued concentration of decision making over all aspects of economic life. And the world will know still *more* hunger.

If the roots of hunger lie deep in economic and political institutions which concentrate power so that some are left with none, then hunger cannot be ended by simply producing more food or providing more foreign aid. Citizens—here as in the developing world—can end hunger only through the democratization of the social order, incorporating the life-and-death question of economics into our concept of democracy.

and prepare the land, plant the seed, care for the growing plants, and harvest the crop. Machines were used, but they were relatively simple. Farmers grew a variety of crops, a strategy known as **polyculture,** and they allowed fields to lie fallow periodically in order to restore soil fertility.

Changes in the Developed World. Around 1950 agriculture in the United States underwent a profound change. (We will discuss the changes in U.S. agriculture as representative of changes in much of the developed world.) Agriculture became energy intensive, or more specifically, fossil fuel intensive. In 1950 an amount of energy equivalent to less than half a barrel of oil was used to produce a ton of grain. By 1985 the amount of energy needed to produce a ton of grain had more than doubled. Searching for ways to increase the yield of the lands already under cultivation, farmers began to rely heavily on inputs of water, on inputs of chemical fertilizers and pesticides (many of which are petroleum-derived products), and on high-yield strains of crops. In some areas, especially the drier regions of the Southwest, irrigation projects allowed marginal lands to be cultivated. Farmers began to concentrate on producing only one or two profitable crops; **monoculture** farming replaced traditional polyculture farming.

Since about 1960 the number of family farms in the United States has fallen by about half. Many of these farms have been purchased by large corporations, as traditional agriculture has given way to **agribusiness.** The primary difference between agriculture and agribusiness is that agriculture is a way of life, while agribusiness is an economic venture. The former is a rich tapestry of soils, crops, livestock, and humans; the latter is a balance sheet of debits and credits.

Corporations involved in agribusiness have virtually assumed control of the entire food production system—from growing the food to processing, distributing, and selling it—in the United States. Consider these facts. Fewer than 50 corporations account for more than two-thirds of all food processing in the United States. A few dozen firms dominate the food processing, manufacturing, and marketing industries. Since 1970 a handful of petroleum companies have taken over more than 400 small seed businesses, and only a few firms supply half of all hybrid seeds. In the past small businesses offered a variety of seeds with widely differing characteristics suited to diverse environmental conditions, tastes, and prices. The petroleum companies, however, concentrated their efforts on the production of crop seeds that require the intensive use of petroleum-based additives. Not coincidentally, these companies also control 75 percent of the chemicals and 65 percent of the petroleum-based products used in agriculture.

In the United States, the agribusiness corporations wield great power. Because they are involved in every phase of food production, they reap profits at each step. They can afford the costs of the chemical fertilizers and pesticides and energy characteristic of high-input farming. Lacking the capital available to corporations, family farmers are finding it increasingly difficult to compete in the marketplace as long as they try to compete at high-input farming. Family farmers and consumers alike can be hurt economically because these corporations exert such influence over price.

Changes in the Developing World. At about the same time that food production began to change dramatically in the developed world, significant changes took place in the developing world as well. The **green revolution** had its beginnings in Mexico in 1943. Special genetic strains of hybrid wheat were developed that eventually enabled Mexico to double production by 1958 and then double it again during the 1960s. By the end of the 1960s, 90 percent of Mexico's wheat crop was planted in the new high-yield varieties (HYVs). HYVs typically had short stalks, had an expanded planting season, and matured earlier than traditional varieties. Many countries were able to double food production in 30 to 40 years using green revolution varieties. By the 1970s HYV wheat had spread to Asia. Norman Borlaug won the Nobel Prize in 1970 for his work in bringing HYV wheat to Asia. From 1965 to 1972 India and Pakistan doubled wheat production by planting HYVs. India, in fact, eventually became a wheat exporting nation. Dramatic gains were also made in rice yields in Asia and the Philippines during the 1960s and 1970s.

Despite its successes, the green revolution has not ended hunger in LDCs. Productivity has biological limits, and recent world harvests seem to indicate that present HYVs have reached their peak of production. Moreover, the widespread use of HYVs has raised several significant problems for the farmers and nations who have come to rely on them. To achieve their full potential, HYVs require large amounts of fertilizer, carefully controlled irrigation, and pesticides. Thus, HYV farming is high input and expensive. Seeds are expensive to purchase, are expensive to grow, and generally require an extensive knowledge of appropriate farming techniques. Consequently, these varieties are often out of reach of poor farmers. In many countries the widespread use of HYVs has meant that the rich have become richer, and the poor have become poorer and hungrier. Because HYVs are genetically more similar than the native varieties they replace, they are vulnerable to disease caused by pests and insects. Lost harvests due to disease or pests spell hardships for farmers everywhere, but especially in LDCs, which do not provide a safety net, in the form of social programs, for their citizens.

Importance of Genetic Diversity

By the beginning of the twentieth century most areas in industrialized Europe and North America were planted to high-yield varieties. However, the areas in the world of greatest genetic diversity, most of them located in LDCs, were still relatively untouched by mid-century. Many of the world's most important crops originated in centers of diversity in the tropics and subtropics. With the onset of the green revolution, however, farmers stopped planting different types of crops, such as legumes, and local varieties of crops, which often had a higher protein, mineral, and vitamin content, were naturally more suited to local conditions, and required less input, in favor of the high-yield strains of cereals—wheat, rice, and corn. Thus, the rapid spread of HYVs diminished genetic diversity.

The need for increased productivity to feed ever-growing populations makes high-yield varieties important and necessary, but at the same time we cannot afford to lose the characteristics possessed by native varieties. Among these are adaptations to local climate and soil conditions, resistance to drought, resistance to diseases and pests, and nutritional qualities such as high protein content. The present and future value of traditional plant varieties lies primarily in their genetic code.

In recent years the international community has recognized the urgent need to protect the remaining genetic diversity of crop plants (Figure 10-10). An important component of that effort is the protection of wild lands within the centers of diversity. Preserving these lands can help to protect the wild relatives of crop plants as well as other potentially important, but currently uncultivated, crops. Because wild lands are disappearing so rapidly, however, protecting the remaining undisturbed lands cannot adequately ensure the preservation of genetic diversity. For that reason researchers and farmers in the developed and developing worlds alike are adapting an age-old tactic practiced by many traditional societies, including the Kayapo Indians of the Brazilian Amazon. For centuries the Kayapo have tended hillside gardens which contain representative samples of various food crops, particularly tuberous plants. The gardens' location protects the collections from flooding, thus preserving this important and

FIGURE 10-10: Many varieties of potatoes. In Peru, the center of diversity for potatoes, much of the genetic diversity is being lost because of the widespread use of high-yield varieties.

valued resource. Nations that establish gene and seed banks are doing much the same thing that the Kayapo do: preserving a commodity upon which society depends.

The Gene Bank. A **gene bank** is a place in which the germplasm (genetic material) of plant or animal species is preserved for future use. Samples of useful or economically valuable species are gathered and then maintained under controlled conditions. Plant genetic material is usually preserved outside of its natural environment in seed banks, although field gene banks, especially in the tropical regions, are becoming more common.

To maintain a seed sample in long-term storage, the seeds are dried to reduce their moisture content, preventing damage to the tissue. The sample is then stored at -4° F (-20° C). Under these conditions, samples can remain viable for up to 100 years. To test for viability, a subsample is periodically taken and germinated. If less than 85 percent of the seeds germinate, the entire sample is regerminated in order to maintain and protect its genetic variability. Because some genes are extremely rare in a population, even a minimal deterioration in a sample's viability could mean the loss of potentially valuable genes.

Not all plants can be preserved as seed in a gene bank. Root crops and plants that have recalcitrant seeds, which must remain moist and thus do not survive the drying process, can be preserved in one of three ways: field gene banks, *in vitro* preservation, and cryopreservation.

Field gene banks are plots in which plants to be preserved are grown. In a field gene bank, the plants are germinated annually. There are two major disadvantages in preserving plants in a field gene bank.

First, a large space is required to grow the plants and, second, the labor costs necessary to maintain the field can be high.

In vitro preservation, in which plant tissue is stored in a test tube, allows storage of many species for up to two years before the culture must be renewed. Samples saved in this manner are kept in windowless, air-conditioned rooms. Cool temperatures retard growth, and artificial lighting allows the operator to control day length. The major advantage of *in vitro* preservation is that many more samples can be stored in far less space than that required by field banks. For example, a 10-acre (about 4-hectare) field is required for 6,000 potato plants at the Huancayo substation of the International Potato Center (CIP, Centro Internacional de la Papa) in Peru. But at CIP headquarters in Lima, a duplicate collection is housed in facilities that require only 0.1 percent as much space.

In vitro preservation does have two major drawbacks. Certain materials appear to be genetically unstable in tissue culture and so cannot be stored reliably in this manner, and it is only a temporary means of storage. **Cryopreservation**—storing plant and animal materials in liquid nitrogen—may prove to be a longer term and safer method of preservation for certain species. The extremely low temperature of liquid nitrogen, -321° F (-196° C), should suspend genetic instability and change.

Issue of Access to Gene Banks. Of primary importance in the issues concerning gene banks is the LDC-MDC split over free access to naturally occurring species and patented hybrids. Naturally occurring species are considered part of the earth's common heritage; patented breeds, which are generally developed by corporations in industrialized countries, are not. Therefore, countries that want to use new breeds and hybrids are forced to buy seed at commercial prices, which they may not be able to afford. Many LDCs feel that this practice is exploitive, since their native plants are usually the source of the genetic material that yields the hybrids. On the other hand, breeders in the industrialized nations point to the years of research and development required to discover marketable hybrids. Anxious to protect their investment, these developers are against free access to patented breeds.

The principle of free exchange is occasionally violated, usually by seed banks under national control and typically for political reasons of "national security." Countries that restrict access to crop germplasm include Ethiopia (coffee), India (black pepper, turmeric), Taiwan (sugarcane), Jamaica (allspice), Ecuador (cacao), Iraq (date palm), Iran (pistachio), and Brazil (rubber).

Increasing Food Production

Adding Plants to Human Diet. With research, we will surely discover many plants to add to the human diet. For example, researchers are investigating the potential of plants in the genus amaranth, members of the cockscomb family. These wild plants vary in taste, appearance, and other traits, but all provide a protein-rich grain. Quinoa (or quinua, pronounced keen-wa), one member of the amaranth family, is a major source of protein for millions of people living in the highlands of Argentina, Bolivia, Chile, Colombia, Ecuador, and Peru (Figure 10-11). The protein in the quinoa grain contains a better amino acid balance than the protein in most true cereals. In fact, its protein is of such high quality that, in nutritional terms, it can take the place of meat in the diet. Quinoa is made into flour for baked goods, breakfast cereals, beer, soups, desserts, and livestock feed. It can also be prepared like rice or be used to thicken soups. The malted grains and flour can be used as a weaning food for infants. The grain holds particular promise for improving life and health for the poor in marginal upland areas, including highland regions in Ethiopia, Southeast Asia, and the Himalayas.

The preservation of many species and varieties of plants in gene banks may play an important role in the search for plants to add to the human diet and thus to increase production of food.

FIGURE 10-11: Quinoa in field plot. Quinoa (pronounced *keen-wa*), a member of the amaranth family, was such an integral part of the diet of the ancient Incas that they referred to it as the "mother grain" and considered it sacred.

Aquaculture. Aquaculture is the production of aquatic plants or animals in some type of controlled environment. It refers to a variety of activities such as managed harvesting, the use of artificial reefs to attract fish, the construction of fish pens, shellfish and marine fish hatcheries, pond and fish tank cultures, and integrated animal and fish farms. Aquaculture holds significant promise to give people access to excellent sources of nourishment (both protein and calories).

Until recently most research went into developing highly profitable operations in industrialized countries. In the United States, operations included cage-raised catfish in the South (especially Mississippi and Arkansas), trout and salmon in Washington, Oregon, and Idaho, and crayfish in Louisiana (Figure 10-12). According to the U.S. Department of Agriculture, aquaculture is the fastest growing sector of agriculture in the country. It is the fastest growing source of fish and shellfish in the United States, expanding from less than 1 percent of total harvest in 1970 to nearly 7 percent in 1987.

In the developing world aquaculture offers a good way to produce an excellent source of protein to supplement diets. India has a large number of village ponds, but only a small number are used for aquaculture. In contrast, the Chinese integrate aquaculture into their overall plan for providing food. They combine crop planting, animal husbandry, and fish raising into a single integrated system (Figure 10-13).

In Senegal, pond culture is being tried to raise tilapia, an African river and lake fish, to provide a source of much-needed protein. Farmers growing rice are being taught various aquaculture tech-

FIGURE 10-13: Aquaculture system between Kunming and Dali in Yunnan Province, southwest China. In China aquaculture is part of an integrated effort to produce food for local consumption. Ponds, cultivated fields and gardens, and animals are managed as a single food-production system.

niques—how to build ponds, maintain water supply, fertilize ponds, feed fish, and harvest fish. Unfortunately, because the Senegalese believe that plants are crops that need tending, but fish can grow on their own, yields of tilapia have not as yet met expectations.

Biotechnology. Biotechnology refers to a variety of processes in which organisms, or their parts or processes, are used to manufacture useful or commercial products. Promising research relating to food production falls into three main categories: improving plants and animals through hybridization (or selective breeding), genetic engineering, and tissue culture.

Until now, the green revolution was the most well-known result of agricultural biotechnology. Hybridization, or selective breeding, is the crossing of one or more varieties of a particular plant or animal species to produce a hybrid with special qualities. This hybridization process leads to a loss of reproductive capabilities. Parents must be continually crossed to get the desired hybrids (plants grown from the seeds of the hybrid are more likely to resemble one or the other parental type than the desired hybrid type, or in some cases the hybrids will be sterile). Through selective breeding, researchers attempt to produce plants with higher yields and other characteristics considered desirable. They will try to improve further on the cereal grains and eventually add many other plants.

Through **genetic engineering,** or the direct transfer of genes, characteristics from one organism may

FIGURE 10-12: Raising Atlantic salmon in an enclosure in Cutler, Maine. In the United States, aquaculture ventures tend to concentrate on specialty foods that bring high prices.

Environmental Science in Action:
Hunger in the United States
Pam Gates and Karla Armbruster

Many people in the United States have long heard of famines and starvation in poor developing countries, but they are just now becoming aware of the hungry people who live in our own country. An estimated 20 million Americans go hungry for at least some time during each month, and some are hungry every day. Unlike the hungry in the developing countries, Americans do not go hungry because of a scarcity of food or because of food distribution problems; instead, they simply cannot afford to buy what they need from the vast supply the country produces. Although many government and private programs try to help the hungry in the United States, political and social barriers prevent a significant number from receiving the assistance they need.

Physical and Biological Boundaries

Hunger hurts everyone who experiences it, but children and the elderly are at the greatest risk. Children need the proper nutrition and caloric intake to grow and develop, and they are likely to become ill if they don't receive it. A study conducted by Maine's Department of Human Services showed that poor children were three times as likely to die as other children, primarily from disease. Bread for the World, a nonprofit agency working against hunger, estimates that 27 U.S. children die every day from the health-related effects of poverty, often linked to inadequate diets (Figure 10-14).

Children of poor families are at risk even while they are still in the womb. Lack of adequate prenatal nutrition contributes significantly to infant mortality. The infant mortality rate in the United States is higher than that of most other industrialized countries, including Japan, Belgium, and the United Kingdom.

Many elderly people need medically prescribed diets to stay healthy. Eighty-five percent of people over 65 suffer from chronic conditions that influence their nutritional needs. Others are unable to absorb nutrients well and need special, highly nutritious foods. In addition, as people grow older, they usually become more susceptible to illness, and lack of nutrition exacerbates this vulnerability. Unfortunately, the millions of elderly who live below the poverty level are unlikely to be able to afford special diets when they need them.

Because hunger is a "silent epidemic," it is hard to pinpoint exactly who is hungry in America. However, we know that people who live in poverty are most likely to be hungry. The federal government defines poverty as living on less than the income level necessary to guarantee health and survival; in 1990 the poverty line was $13,359 for a family of four and $6,652 for one person. In 1990 roughly 33.6 million Americans were living below these income levels (Figure 10-15).

Some people wonder why the poor do not simply get jobs. In fact, many poor people do work; in 1989 more than 5.5 million people living below the poverty level belonged to households where someone worked full-time year-round. One adult working full-time for minimum wage cannot earn enough to bring his or her family out of poverty. In addition, over half of the poor are unable to work; approximately 40 percent are children and 10 percent are over 65. Some who can work must stay home with their children because 52 percent of poor families are headed by a single parent.

Social Boundaries

Even though the federal government has a variety of programs to help the poor avoid hunger, the prevailing government attitude toward hunger in the 1980s was that as the economy revived, poverty would disappear, and so would hunger. In the meantime, the hungry simply had to apply for government aid, learn to budget and eat nutritional food, and abstain from alcohol and drugs. In 1989 the wealthiest 20 percent of the U.S. population

FIGURE 10-14: Children living in poverty, McColl, South Carolina, 1990. In the United States, the hungry may not be a common sight, yet they exist. Children are overrepresented among the poor; one in every five children lives in poverty and is at risk of suffering chronic hunger.

received almost 48 percent of the national income, while the poorest 20 percent received less than 4 percent. Many citizens feel that federal programs that aid the poor and hungry are a drain on their tax money. But by ensuring that everyone has enough to eat, we avoid the economic and social costs associated with malnutrition, illness, and un- employment. For instance, when pregnant, low-income women participate in government food supplement programs, they have fewer infants with low birth weights and fewer premature births and infant deaths.

Looking Back

The federal government first became involved in fighting hunger in 1939, when it began a food stamp program. However, this program was aimed more at distributing farm surpluses than feeding the hungry, and it was curtailed in 1943 because of food shortages linked with World War II. By the

1960s approximately one-fifth of the U.S. population was living in poverty and at risk for hunger. At that time, the government established or revised many of the assistance programs used today.

How Federal Programs Help Combat Hunger

The federal food stamp program helps supplement the monthly food-purchasing power of the poor. This program is especially important because it can be used by the working poor as well as the unemployed. Stamps can be exchanged for food at grocery stores, but cannot be used to purchase alcohol, tobacco, imported food, or household products. Each household receives enough stamps so that, combined with a third of its income, it can afford the Thrifty Food Plan devised by the U.S. Department of Agriculture. For example, in 1987 the maximum allotment (given to households with 0 income) for a family of four was $268 per month. In 1989 18.8 million people received food stamps.

School breakfasts and lunches are major programs that help alleviate hunger among children. The government gives subsidies to public and private schools and day-care centers so they can provide meals for the students. Although some of these children pay for their meals,

those from poor families receive them free or at reduced prices. In 1989 24.2 million children participated in the lunch program, 3.8 million in the breakfast program, and 1 million in the child care food program. In 1990 slightly over $5 billion was budgeted by the Department of Agriculture for child nutrition programs.

Another important form of food assistance is the special supplemental feeding program for women, infants, and children, commonly known as WIC. WIC serves pregnant and postpartum women, infants, and children up to age 5 who live in low-income households. Participants receive vouchers for foods high in protein, iron, and calcium. In 1989 the program served over 4.4 million.

Other forms of assistance include the Temporary Emergency Food Assistance Program (TEFAP) and elderly nutrition programs, which help fund efforts like group meals and delivering meals to the housebound. Aid to Families with Dependent Children (AFDC), while not providing direct food assistance, gives poor families money which they may use to purchase food.

Hunger in the 1980s

Although assistance programs helped reduce hunger in the United States by the end of the 1970s, the

trend turned around in the 1980s. Concerned by the mounting federal budget, the Reagan administration cut $13 billion from food assistance programs between 1982 and 1985. These cuts eliminated nearly 1 million people from the food stamp program and reduced benefits to many others. By 1987 fewer people were participating than in 1980 even though over 6 million more had fallen below the poverty level in that time. Three million children were cut from school lunch programs and 500,000 from breakfast programs. Many summer food programs for school children lost funding. Although President Reagan claimed that these cuts in food assistance would not hurt the truly needy, two-thirds of spending reductions in food assistance programs came from cuts in aid to families below the poverty level.

The effects of the budget cuts began to show at private, nonprofit emergency food programs like soup kitchens and food banks. The nonprofit Center for Budget and Policy Priorities, based in Washington, D.C., reported that one-third of the country's emergency food organizations received 100 to 200 percent more requests for help in 1983 than they had in 1982. In 1984 the Washington, D.C.-based Food Research and Action Center (FRAC) surveyed 300 emergency food programs across the country and

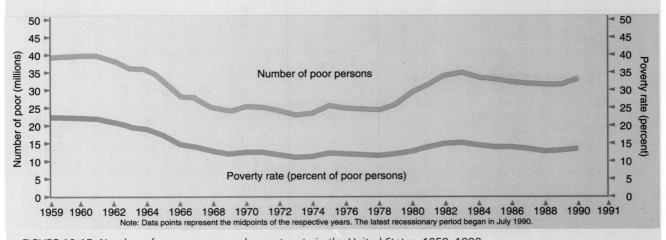

FIGURE 10-15: Number of poor persons and poverty rate in the United States, 1959–1990.

discovered that almost 75 percent felt that the private charities in their community could not meet local needs.

One of the most thorough studies of hunger in the United States during the 1980s was a report by the Physician Task Force on Hunger in America, a group of 22 physicians under the leadership of Dr. J. Larry Brown of the Harvard School of Public Health. Throughout 1984 the physicians visited cities and towns in the South, Northeast, Midwest, and Southwest to find out how widespread the hunger problem was and why it existed. Their conclusion was that "the problem of hunger in the U.S. is now more widespread and serious than at any time in the last ten to fifteen years."

Besides condemning the cutbacks in food assistance, the task force pointed out basic problems in the federal programs that caused unnecessary hunger and misery. One major problem in the food stamp program is the long waiting period for applicants to find out if they are eligible. The task force noted that it was not unusual for people to wait 45 to 60 days for a decision on their eligibility. Many food stamp offices prominently post signs warning applicants that they will be prosecuted for falsifying eligibility and thus create an intimidating atmosphere of suspicion that can discourage applicants. The complexity and length of food stamp applications also deter potential participants. In Mississippi an applicant must fill out 21 forms totaling 35 pages to apply for food stamps. In addition, the task force called the nutritional value of the Thrifty Food Plan into question. They claimed that the USDA's primary goal was to devise a food plan to match a predetermined expenditure level, not to find one to provide an adequate, nutritious diet.

How Private Organizations Help

Increasingly, private organizations are assisting the growing number of hungry falling through the holes in the federal government's "safety net" of assistance. Two types of private charities, food pantries and soup kitchens, distribute food directly to the poor. Food pantries are small-scale operations that distribute grocery items to needy individuals; soup kitchens provide prepared meals. While some of these organizations require applicants to meet eligibility standards, most try to give food to anyone who asks. Unfortunately, food pantries and soup kitchens frequently do not have enough food for everyone who comes to them.

Food banks collect food, often in the form of surpluses from corporations and retailers, and distribute it to food pantries. Many food banks use gleaning programs: volunteers go into fields, orchards, and packaging plants to collect food that is left unharvested or is not visually pleasing enough to be sold. Most food banks rely heavily on volunteers. For example, the Westside Food Bank in Surprise, Arizona, has a paid staff of only four or five, but its 5,000 volunteer gleaners and 600 other volunteers make its success possible. In 1987 Westside distributed food to 100 charities, which served 15,000 families.

One problem food banks face is not having enough vehicles or personnel to transport food items offered to them. For instance, the Potato Project, which distributes potatoes to food centers in 47 states, could not afford to ship 10 million pounds of the potatoes it had gathered in 1987. Food banks also frequently suffer from insufficient freezer, refrigerator, or storage space.

Second Harvest, the largest nongovernmental food distribution program in the United States, is an umbrella organization for food banks across the country. With headquarters in Chicago, Illinois, it distributes to 88 food banks and 112 of their affiliates in 46 states. These banks serve over 38,000 private charities, including day-care and senior centers, drug and alcohol treatment centers, and homeless shelters as well as food pantries and soup kitchens. Incredibly efficient, Second Harvest distributes $157 worth of food for every $1 it spends.

Other groups do not gather or distribute food, but concentrate on strengthening government programs that help the poor and hungry. One of the largest of these is the Food Research and Action Center (FRAC). FRAC's efforts include documenting hunger in the United States, monitoring public policy, keeping the public informed about hunger, working with concerned individuals and groups, and providing legal assistance on food aid issues.

How Recent Legislation Has Aided the Hungry

In the late 1980s the widespread hunger in the United States finally made an impression on the government. President Reagan signed the Hunger Prevention Act, providing $3 billion between 1988 and 1993 to strengthen domestic food assistance programs. As a result of these funds, the average food stamp recipient received a real (adjusted for inflation) increase of 1.5 cents per day in 1989, 5 cents in 1990, and 8 cents in 1991. In 1989 Congress also took a small step to help bring the working poor out of poverty. They voted to raise the minimum wage to $4.25 an hour by 1991. They also increased funding for WIC by $118 million, allowing the program to serve 200,000 additional people in 1990.

Looking Ahead

Despite recent legislation aiding the hungry, many believe that government action alone will not end hunger in the United States. Indeed, shortly after taking office in 1989 President Bush urged private citizens and charitable groups to assume a greater share of the burden in solving social problems like homelessness and hunger. Others maintain that, while private charities can help provide food, they cannot lift families out of poverty. Instead, they contend that we can end hunger only if we are willing to change the values and structures of our society.

The National Council of Churches, speaking out on hunger, calls for changes in our institutionalized and deeply ingrained values of greed, acquisitiveness, competition, uncontrolled growth, excessive individualism, and militarism. They not only recommend specific changes in legislation and programs, but also suggest ways that individuals can make a difference, such as organizing advocacy groups on specific issues relating to poverty and hunger.

Bread for the World, a Washington, D.C.-based citizens' group that concentrates on lobbying for actions to alleviate hunger worldwide, advocates that the United States cut mulitary spending by $20 billion (out of the $300 billion we spend currently) and use half of that for programs to combat hunger and its causes. The Institute for Food and Development Policy, also known as Food First, is a San Francisco-based nonprofit research and education organization dedicated to investigating and exposing the root causes of hunger worldwide. Food First encourages political involvement to develop policies that will combat hunger. According to Food First, there is too little direct citizen input on issues such as defense spending versus spending for programs to fight hunger.

Generally, learning about hunger raises more questions than it answers. Who is responsible for the poor and the hungry? Is it realistic to expect that people who are poor and hungry—sometimes homeless, often weak or ill—should "pull themselves up by their bootstraps"? Should children be consigned to a life of poverty merely for having the misfortune of being born to poor parents? Is it realistic to believe that local groups can shoulder the greatest share of the burden for caring for the poor and hungry? If the government—which is supposed to be, after all, a reflection of its citizens—will not help the poor and hungry, who will?

For more information on hunger in the United States and how you can help, contact the organizations listed in the Organizations and Publications section at the back of this book (page O-1).

---

be transferred to another organism. In this way, resistance to pests and disease; tolerance of drought, heat, and salt; ease of harvesting; and nitrogen-fixing capabilities could be transferred to food crop plants that do not possess them. Imagine pest-resistant wheat able to fix nitrogen while growing under drier, hotter climate conditions than wheat can now tolerate. Meat, egg, and milk production could be improved by inducing animals to produce more hormones or by producing the hormones through genetically engineered bacteria and then giving them to the animals.

Tissue culture allows whole plants to be produced from individual cells without planting seeds. This technique, called cloning, holds the promise of producing large numbers of tiny plants which could then be encapsulated with enough fertilizer, pesticides, and water to get a good start in growth.

Most of the money to finance this very costly biotechnological research is coming from multinational corporations that anticipate making huge profits by providing the results of research to the world's farmers. At present, more money is being spent on developing technologies to improve animals than on plants. Plant biotechnologists seem to be concentrating on improving flavors and sweetness in fruits and vegetables. Little money is being diverted toward helping farmers in LDCs improve production.

Future Management of Food Resources

As we approach the new millennium, food scarcity and inadequate and inequitable distribution are emerging as immediate consequences of environmental degradation and social and political conflicts. How will food be made available to feed the 900 million extra people who will populate the world by the year 2000? Nowhere is this predicament more evident than in Africa, which has the fastest-growing population of any continent ever. Overgrazing, deforestation, desertification, soil erosion, and drought have lowered per capita grain production. Social strife has interfered with distribution. This combination of unprecedented population increase, environmental degradation, and social strife could create a nightmarish scenario. To counteract this situation, strong measures must be taken to prevent resource abuse through conservation, protect human health and the environment, and preserve living systems.

What You Can Do:
Managing Food Resources

Individuals can take many actions to help better manage food resources. Here are some suggestions:

- Keep a "food diary" of everything you eat each day for one week. How many calories do you consume? What percentage of those calories come from fats, proteins, and carbohydrates? Have your diet analyzed by a dietician, nutritionist, or a friend studying nutrition or health. Ask them for suggestions. For example, if your diet is high in fat, you might try cutting down on fats and consuming a larger share of complex carbohydrates.

- Choose an average day from your food diary. For each item listed, try to identify when, where, and how it was produced. You'll probably find that you are unable to do so for many of the foods you eat. Try an experiment: for two weeks, modify your diet in order to eat more foods, particularly produce and meats, that are grown or produced locally.

- Learn how to combine grains, legumes, and other foods to assure adequate consumption of high quality proteins. Frances Moore Lappe's *Diet for a Small Planet* is an excellent reference.

- Consider cutting down on the amount of animal products in your diet. Jan Hartke, former President of the Global Tomorrow Coalition, notes that apart from limiting family size, eating fewer animals is probably the most significant step an individual can take to reduce environmental degradation and conserve natural resources.

- Identify and patronize local organic farmers who grow fruits and vegetables without the use of chemical fertilizers or pesticides. Ask local supermarkets to carry organic produce.

- Find out more about hunger in your area, the United States, and around the world. Who are the hungry? Why do they remain hungry in a world of plenty? Then translate your knowledge into action: volunteer at a local food bank, soup kitchen, or homeless shelter, or join an organization dedicated to alleviating hunger in the United States and worldwide.

- Learn as much as you can about issues related to food production and hunger so that you can make informed decisions at the grocery store, in your own backyard garden, and in the voting booth. Contact organizations concerned with improving agriculture, conserving natural resources (especially soil and water), and eliminating hunger. Ask them for publication listings; subscribe to newsletters so that you can keep up-to-date on developments such as proposed legislation or promising research.

- Write your representatives in Congress and inform them of your views on issues related to agriculture and hunger. Encourage them to support legislation that promotes sustainable agriculture and discourages environmentally harmful farming practices.

Prevent Resource Abuse Through Conservation

Food security can be attained only through a sustainable and sustaining agricultural system, and such a system depends on the safeguarding of arable lands. Conservation measures are needed worldwide to protect croplands from erosion and abuse. One such measure is the United States' Conservation Reserve Program, an integral component of the Food Security Act, which pays farmers to convert highly erodible cropland to woodland or grassland. Such measures have proven effective in reducing soil loss.

Agricultural lands worldwide must also be protected from conversion to nonfarm uses. Strategies to protect arable lands include giving tax benefits for continued agricultural use and adopting zoning plans that restrict urbanization, industrial, and other land uses to areas and soils not suited to agriculture. Other strategies to protect arable lands are discussed more fully in Chapter 16, Soil Resources. What

You Can Do: Managing Food Resources suggests measures individuals can take to safeguard food resources.

Protect Human Health and the Environment

Basic medical care, improved hygiene, education in nutrition and health, and family planning programs can help to protect the health of children and adults alike. When people are too poor, sick, or weak to care for themselves and their children, they lose their sense of human dignity and self-worth.

While emergency food aid can help people to survive physically in the short term, world hunger will not end unless reforms in food consumption and land use take place. Changing the patterns of food consumption, especially in MDCs, is an important step. Eating lower on the food chain—more grains, fruits, and vegetables—would help significantly to alleviate hunger and protect the environment.

Land reforms within LDCs are needed to ensure that all citizens have a chance to become fully productive members of society. Such reforms would recognize that human dignity is an integral part of good health. At the international level, debt reduction plans could be developed in exchange for land reforms and other policy changes that help the poor. If the poor are given the tools to become self-sufficient and raise their standard of living, they will not be forced to farm marginal lands, clear forests, and resort to other strategies that adversely affect the environment.

Preserve Living Systems

Historically, we have increased food production by increasing the amount of land used as cropland or pastureland. It is unlikely that any further, significant increase can occur without causing environmental degradation such as loss of rain forests and wetlands or without the expenditure of huge sums of money to convert marginal areas to productive lands. Consequently, to increase food production, we must raise the productivity of existing cultivated lands in environmentally sound ways and restore productivity to exhausted land.

While biotechnology, through the development of improved plant varieties, holds great promise for increasing food production, to realize the full potential of these technological advances in helping to solve the global food crisis, they must be made available to poor farmers. Research should be directed at developing strains that are labor intensive rather than high-input intensive. This goal will not be reached without some incentive for research beyond monetary gain. Perhaps some monies could be set aside by seed companies to support this research. The monies might be used for conservation efforts in LDCs, since traits that increase productivity are likely to be found in the germplasm of wild plants in the developing world.

Despite their widespread use, high-yield varieties are not universally applicable. Therefore, we must safeguard less costly, hardy native varieties and provide incentives for their use.

Farmers could also be taught better farming techniques which help to maximize yields of land races and other native strains and of standard and improved varieties. Chapter 16 provides a closer look at environmentally sound agricultural practices.

Summary

Carbohydrates and fats are the major sources of the energy required to maintain the body and perform work. Proteins are the "building blocks" of the body, forming the substance of muscles, organs, antibodies, and enzymes. Small units called amino acids combine in various ways to form larger protein molecules. Of the twenty naturally occurring amino acids, eleven can be synthesized by the body; the remaining nine, known as essential amino acids, cannot always be synthesized at a sufficient rate and must be obtained through diet.

Of the 80,000 potentially edible crops on earth, global agriculture is dependent upon just a few plant species. The top four, called staples, are wheat, rice, maize, and potatoes. Our agricultural system has also relied on just a small number of animals. Nine domesticated animals supply the bulk of the protein (meat, eggs, milk) from livestock.

Food security is the ability of a nation to feed itself on an ongoing basis. Building our reserve supply of grain is a task which is becoming increasingly more difficult because of population growth; reduced availability of cropland and irrigation water; diminishing returns in the use of fertilizers, herbicides, and pesticides; and the cumulative effects of environmental degradation.

Worldwide, more than 1 billion people suffer from chronic hunger. Children account for about 40 percent of the world's hungry; the majority of the rest are women. Most people in LDCs consume about 10 percent less than the minimum calories required for good health. In contrast, individuals in MDCs consume about 30 to 40 percent more than necessary.

The most extreme and dramatic effect of hunger is starvation, which occurs when a person does not consume enough calories to sustain life. Famine is widespread starvation. Each year, of the deaths attributed to hunger, most come not from starvation but from undernutrition, malnutrition, and hunger-related diseases. Chronic undernutrition results when an individual consumes too few calories and protein over an extended period of time. Malnutrition occurs when an individual consumes too little (or infrequently, too much) of specific nutrients essential for good health. Many children in LDCs suffer from kwashiorkor, caused by a diet deficient in protein, and marasmus, caused by a diet deficient in both protein and overall number of calories. Malabsorptive hunger—the loss of the ability of the body to absorb nutrients from the food consumed—often accompanies undernutrition and malnutrition. Seasonal hunger is the lack of adequate food for a period of time (days, weeks, or even months) after the reserves from the previous year's harvest have run out, but before the new crops have been harvested.

Meat production can cause problems when it is practiced to the extent and in the manner common in the developed world. When livestock are fed grains, the efficiency of food production is greatly reduced. Also, the overgrazing of range and pasture by livestock (especially cattle and sheep in the American West) causes soil erosion and water pollution.

Although much of the world's environmental degradation is due to affluence in MDCs, poverty, especially in LDCs, also causes environmental degradation. Lacking arable lands or decent paying jobs, the poor are forced to try to eke out a living on marginal lands where overworked soils cannot sustain cultivation, overgraze grasslands and pastures, or clear forests for fuel, food, or marketable resources.

For most of its history, humans were hunters and gatherers. Eventually, however, humans began to domesticate plants. Domestication is the result of the long-term selection of traits useful for survival in captivity over those traits necessary for survival in the wild. Domestication gave way to agriculture, the purposeful tending of particular plant species for human use. Farmers' deliberate selections of certain strains resulted in land races, varieties of plants adapted to local conditions such as climate and soil type.

In the past fifty years food production in the United States has come to rely heavily on inputs of energy, water, synthetic fertilizers, herbicides, and pesticides. Traditional agriculture has given way to agribusiness, and many family farms have disappeared. At about the same time that food production began to change dramatically in the developed world, the green revolution began in Mexico. Special high-yield strains of hybrid wheat were developed that enabled Mexico to more than double wheat production. However, despite its successes worldwide, the green revolution has not ended hunger in the developing nations.

In recent years the international community has recognized the urgent need to protect the remaining genetic diversity of crop plants and other species. An important component of that effort is the gene bank, a repository for the germplasm (genetic material) of plant species. Other efforts underway to increase world food production are aquaculture (the production of aquatic plants or animals in some type of controlled environment) and biotechnol-. ogy (a variety of processes in which organisms, or their parts or processes, are used to manufacture useful or commercial products).

Key Terms

agribusiness	aquaculture
agriculture	biotechnology
amino acids	chronic undernutrition
cryopreservation	malabsorptive hunger
domestication	malnutrition
famine	marasmus
food security	monoculture
gene bank	polyculture
genetic engineering	seasonal hunger
green revolution	staple
kwashiorkor	starvation
land race	tissue culture

Discussion Questions

1. Using the model for environmentally sound resource management (Chapter 2), develop a plan for the future sustainable management of food resources.

2. Imagine that you are one of the world's hungry people. What might your life be like? What factors might have contributed to your poverty?

3. How is democracy—or lack of it—related to hunger?

4. Why is germplasm management so controversial?

5. What are some benefits of growing high-yield crops? What are some disadvantages?

6. Of aquaculture and biotechnology, which do you feel has more potential to increase food production? Would one be more effective than the other under different circumstances? Explain.

7. Is there hunger in your community or city? What resources are available for the hungry in your area? What measures do you think should be taken to end hunger in the United States?

Energy Issues

All power pollutes.

Garrett DeBell

We are between a death and a difficult birth.

Samuel Beckett

Learning Objectives

When you finish reading this chapter, you should be able to:

1. Define energy and tell how it is measured.

2. Tell how energy resources are classified.

3. Define energy efficiency and summarize its environmental and economic benefits.

4. Describe how energy consumption patterns vary worldwide.

5. Identify five broad issues related to energy consumption.

Silent and unseen, energy fuels our bodies, our machines, our societies, and our planet. It is the common denominator among widely diverse activities. The beat of a hummingbird's wing, for example, like the fantastic journey of the Voyager II spaceship, is possible only through the conversion of energy from potential to useful forms. Similarly, a wildflower and a large urban shopping mall are both the product of harnessed energy. In this chapter we will look at the most important environmental issues concerning energy use.

Describing Energy Resources: Physical Boundaries

What Is Energy and How Is It Measured?

Energy is the capacity to do work. It can be measured in different units, including calories, British thermal units, and therms for heat energy and kilo-

watts for electrical energy (Table 11-1). Two forms of energy are kinetic and potential. **Kinetic energy** is energy in motion or action, such as light, heat, movement, or electrical current. **Potential energy** is energy in storage; it is energy that can be converted

Table 11-1
Units of Energy

Barrel is a measure of amount of crude oil or other liquid petroleum products; 1 barrel is equal to 42 gallons. Each barrel of petroleum motor gasoline produces 5.25 million Btus; each gallon produces 12,500 Btus.

Btu (British thermal unit) is the standardized measure of quantity of heat produced by various sources of energy. It is the quantity of heat required to raise the temperature of 1 pound of water 1° F. One Btu is the approximate energy equivalent of one burning wooden match tip.

Therm is a measure of quantity of heat derived from natural gas; 1 therm is equal to 100,000 Btus. It is roughly equivalent to the energy in 87 cubic feet of natural gas.

Cubic foot (cu ft) is a measure of amount of dry natural gas. One cubic foot of dry natural gas produces approximately 1,031 Btus of energy.

Kilowatt (kw) is a measure of electrical power; 1 kilowatt is equal to 1,000 watts; 1 megawatt (Mw) is equal to 1,000,000 (1 million) watts; 1 gigawatt (Gw) is equal to 1,000,000,000 (1 billion) watts. Watt is an electrical unit of power or rate of doing work or the rate of energy transfer equivalent to 1 ampere flowing under the pressure of 1 volt.

Kilowatt-hour (kwh) is a measure of electrical use, the energy equal to that expended by 1 kilowatt in 1 hour. For example, 1 kwh is the amount of electricity that an operating 100-watt light bulb would consume in 10 hours. It is equivalent to 3,411 Btus.

Quad is a measure of quantity of heat; 1 quad is equal to 1,000,000,000,000,000 (1 quadrillion) Btus. It is equivalent to the heat produced by 171.5 million barrels of oil.

Short ton is a measure of amount of coal; 1 short ton is equal to 2,000 pounds. (For comparison, 1 long ton is equal to 2240 pounds, and 1 short ton is equal to 0.907 metric tons.) One short ton of coal equals approximately 3.8 barrels of crude oil and 21,000 cubic feet of dry natural gas. There are 1.102 short tons in 1 metric ton; there are 1.120 short tons in 1 long ton.

to another form. For example, an automobile engine converts the potential energy stored in the chemical bonds of the gasoline into kinetic energy in the form of heat, light, and movement, eventually causing the car to move.

How Are Energy Resources Classified in Regard to Supply?

Energy resources are generally classified as nonrenewable, renewable, or perpetual. **Nonrenewable energy resources** exist in finite supply or are renewed at a slower rate than the rate of consumption. Fossil fuels—coal, oil, and natural gas—best exemplify nonrenewable energy resources. The fossil fuels are also known as **conventional resources** because they are so commonly used. Nuclear energy, as it is widely used, is also nonrenewable. In the United States 96 percent of the energy consumed is supplied by nonrenewable resources (Figure 11-1); worldwide, the figure is 90 percent. How long the world's supply of fossil fuels will last depends on many factors: energy demand, development of improved extraction techniques, discovery of new deposits, use of more energy-efficient technologies (especially automobiles), conservation practices, and development of other energy resources.

As the term implies, **renewable resources** are resupplied at rates faster than or consistent with use; consequently, supplies are not depleted. Renewable energy resources, such as biomass (wood, plant residues, and animal wastes), can become exhausted (and thus nonrenewable) when they are used faster than they can be regenerated. In contrast, **perpetual resources** originate from a source that is virtually inexhaustible, at least in time as measured by humans. Examples include the sun, tides, falling water, and winds.

Renewable and perpetual resources are often called **alternative resources** because they offer us a choice other than conventional fuels. They supply about 10 percent of the world's energy, but the proportion varies from 2 to 5 percent in many industrialized nations to 50 percent in some nonindustrialized countries. While most alternative resources are renewable or perpetual, our nation's reliance on conventional fuels is so heavy that almost any other energy resource (even one that is nonrenewable) can be considered an alternative.

What Is Energy Efficiency?

According to the first law of thermodynamics (see Chapter 4), also known as the conservation of energy law, energy can be neither created nor destroyed, but it can be changed or converted in form. Although the same amount of energy exists before and after the conversion, not all of it is in a useful form. The conversion of energy from one form to another always involves a change or degradation from a higher quality form to a lower quality form, a principle contained in the second law of thermodynamics. Ultimately, all forms of energy are degraded to heat.

For any system, device, or process that converts energy, the energy efficiency of the conversion can be determined. **Energy efficiency** is a measure of the percentage of the total energy input that does useful work and that is not converted into low temperature, low-quality, unwanted heat. Because heat is given off every time energy is converted from one form to another, the energy efficiency of any device or process is always less than 100 percent (Table 11-2).

The electricity produced by a coal-fired power plant illustrates a chain of three energy conversions and their associated efficiencies (Figure 11-2). When coal is burned, about 60 percent of its potential chemical energy is converted to high-quality

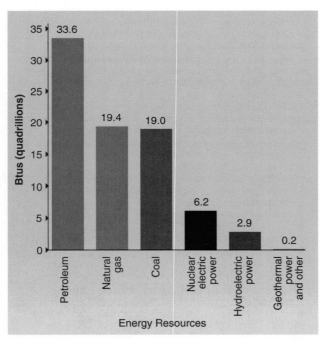

FIGURE 11-1: U.S. energy consumption, 1990. Nonrenewable resources meet almost all of the nation's energy demand. In 1990 petroleum, natural gas, coal, and nuclear electric power accounted for 78.2 quadrillion Btus of energy. (Note: Bars and figures do not represent a percentage of energy consumption and thus do not add up to 100.)

▶ Table 11-2
Efficiency of Selected Energy Conversion Devices

Device	Energy Conversion Function	Efficiency (%)
Incandescent bulb	Electrical to light	5
Standard fluorescent bulb	Electrical to light	20
Compact fluorescent bulb	Electrical to light	75–80
Automobile engine	Chemical to thermal to mechanical	25
Steam turbine	Thermal to mechanical	47
Home oil furnace	Chemical to thermal	65
Battery (charging)	Electrical to chemical	72
Battery (discharging)	Chemical to electrical	72
Electric generator	Mechanical to electrical	99

heat (thermal energy) used to produce steam; the rest is converted to low-quality heat and given off to the environment. As the steam is used to drive a turbine, its thermal energy is converted to mechanical energy. The efficiency of this conversion is 47 percent. The turbine, in turn, drives an electric generator; the conversion efficiency of mechanical energy to electrical energy is very high, 99 percent.

The **net efficiency** of a process or system that includes two or more energy conversions is found by determining the efficiency of each conversion. For every 1,000 units of potential energy in a specific amount of coal, only 600 units of thermal energy are produced. Only about half of the thermal energy in the steam actually drives the turbine, so a little less than 300 units of mechanical energy are produced. Nearly all of that, however, is converted to electrical energy by the generator. Thus, the 1,000 units of potential energy in the coal result in about 300 units of electrical energy. The net efficiency of the process is slightly less than 30 percent.

Our calculation of net efficiency does not include all the energy that is required to extract, transport, and pulverize the coal, nor does it include pollution control devices and other environmental costs. You can see that getting energy in a form that is convenient and useful—available at the flip of a light switch—is an inefficient undertaking! Moreover, when the electricity is used to provide light in a home, the energy efficiency of the entire process is much less. In conventional incandescent light bulbs,

for example, up to 95 percent of the electrical energy is converted to heat and only 5 percent to light. That's why the bulbs are so hot. New "cool lights" or compact fluorescent bulbs are much more efficient; using 13 to 15 watts of electricity, they deliver the equivalent light of a 60- or 75-watt incandescent bulb, an energy savings of about 80 percent.

Stop and think about all the devices found in the typical home that run on electricity (kitchen appliances, power tools, gardening equipment, grooming tools, toys, computers, entertainment equipment). A list of such devices might easily total a hundred or more. Using the most energy-efficient devices available can produce a significant savings in energy and conserve nonrenewable fuels. The purchase price of energy-efficient devices is usually higher than that of conventional models, but they are economical in the long run because they have a lower **life-cycle cost:** initial cost + lifetime operating costs. A 15-watt fluorescent bulb costs about $15. When used to replace a 75-watt incandescent bulb, it will save about $48 in electricity costs during its lifetime; because it lasts longer, it will also save about $7 in replacement bulbs. As electricity is used more efficiently, less of it must be generated (good news for the environment). Chapter 12 looks at some practical things individuals, communities, and industries can do to increase energy efficiency.

Biological Boundaries

What Is the Biological Significance of Energy Resources?

With the possible exception of humans, all species have evolved within a particular biological niche. Each species requires energy for growth, mainte

FIGURE 11-2: A coal-fired power plant in the Mojave desert. Worldwide, coal is used chiefly for electric production.

nance, and reproduction, as well as for the activities which it carries out within the ecosystem. If the energy requirements of a species are related to its biological niche, then the role of the contemporary human would seem to be one of using energy and materials with impunity. This is best illustrated by industrial societies, where the environment is extensively altered to meet wants and desires rather than needs, and progress is often measured by success in the pursuit of comfort and convenience. Sophisticated technologies convert energy resources in amounts far beyond that needed for ecological survival.

Social Boundaries

How Do Present Energy Consumption Patterns Vary Worldwide?

The United States, with roughly 5 percent of the world's people, accounts for over 33 percent of the world's energy consumption. In contrast, India, with roughly 15 percent of the world's people, uses less than 1.5 percent of the world's energy. Globally, just 20 percent of the human population uses 80 percent of the world's energy.

For most people in LDCs, energy is a daily, basic concern, similar to food and shelter, and they are quite aware of the origins of their energy (Figure 11-3). They may spend a major portion of their day collecting wood, plant residues, or animal dung or buying or bartering for the fuel that they need to power stoves and lanterns (Table 11-3). Because

▶ Table 11-3
Time Taken and Distance Traveled for Firewood Collection, Selected Areas

Place and Date	Firewood Collection	
	Time Taken	Distance Traveled (kilometers)
NEPAL		
Tinan (hills), 1978	3 hours/day	—
Pangua (hills), late 1970s	4–5 hours/ bundle	—
INDIA		
Chamoli (hills), 1982	4–5 hours/day	3–5
Gujerat (plains), 1980		
Forested	Every 4 days	—
Depleted	Every 2 days	4–5
Severely depleted	4–5 hours/day	—
Madhya Pradesh, 1980	1–2 times/week	5
Kumaon (hills), 1982	3 days/week	5–7
Karnataka	1 hour/day	3
Gashwal (hills)	5 hours/day	10
AFRICA		
Sahel, 1977	3 hours/day	10
1981	3–4 hours/day	—
Niger, 1977	4 hours/day	—

Source: Worldwatch Institute.

FIGURE 11-3: A Syrian Kurd gathers firewood, 1991. For women and children in many developing countries, the collection of firewood is an important daily chore.

they are so directly involved in the search for fuel, these people have an intimate knowledge of the sources of energy upon which they depend.

For the most part, we in the developed world are indifferent to the origins of the energy we use. Except for a brief period in the 1970s, when an embargo by the Organization of Petroleum Exporting Countries (OPEC) disrupted oil supplies, an inexpensive and plentiful supply of energy has always been available. We have largely taken this energy for granted, much as we have taken for granted clean water and fresh air. Think about your own energy supplies. Where does the electricity come from to illuminate your home, classroom, or office? What fuel heats your home? Where does it come from? Recall the last time you filled your car's gas tank. In what part of the world did that petroleum originate—the continental shelf, Saudi Arabia, Alaska?

These questions may seem irrelevant, but they are not. Ignorance of our energy sources has caused people in MDCs to make several dangerous assumptions. Chief among these is the assumption that energy will always be available at prices we can easily afford. Moreover, because we are generally unaware of the origins of our energy supplies and how they are located, extracted, and converted, we are typically unaware of the environmental effects of energy exploration, extraction, conversion, and use.

In the years ahead, we—as individuals and as a nation—will make many choices concerning energy use and conservation. Each choice will offer benefits, and each will exact environmental, social, economic, and political costs. Let's look at the major issues presented to us by these choices.

What Important Environmental Issues Are Related to the Use of Energy Resources?

Energy issues are complex, encompassing all aspects of cultural existence. For ease of study, we have grouped these issues into five categories: social changes, environmental and health effects, dependence on fossil fuels, nuclear power, and energy policy. However, in the real world, energy issues cannot be so neatly categorized; oftentimes, a single event, such as the Persian Gulf War of 1991, involves numerous issues that fall into most or all of these categories.

Carefully consider the specific questions that appear after each issue is presented. This is more than an academic exercise; very likely, you will have to address these and similar questions during the course of your lifetime. Bear in mind that there is no single correct answer; strong and convincing arguments can be made to support a variety of answers.

Issue 1: Social Changes

Our present industrial technological societies depend on the daily use of enormous amounts of energy. What social changes will be necessary if we begin to conserve our supplies of nonrenewable resources so that they can meet applications (food, medicines, plastics) other than the production of energy? How much are we willing to share energy resources with LDCs whose energy demands will escalate if their standards of living begin to approach our own? Will we forge important new alliances with energy-rich countries like Mexico as our own reserves become depleted or if Middle Eastern oil is no longer available to us? Finally, what price are we willing to pay to satisfy our energy demand if oil supplies are threatened? Many people believe that the Persian Gulf War was fought, at least in part, to protect the industrialized world's energy supply.

Per capita energy consumption in MDCs is four to seven times greater than that in LDCs. Even so, LDCs are outpacing MDCs in terms of *growth* of energy consumption. Between 1980 and 1986 population in MDCs grew 3 percent and energy consumption 8 percent, while population in LDCs rose 11 percent and energy consumption 22 percent. Developing nations are faced with two energy challenges. First, people must have enough energy to survive. Second, on the national level, there must be enough energy to fuel economic development. Economic development depends on having enough energy to support the transition from poverty to a better standard of living. When a consistent and reliable energy source is available to people in poor countries, it does help to alleviate poverty. The cost of conventional fuels, however, is usually too high for developing nations; they cannot depend on these fuels to help them break the cycle of poverty. In many developing nations renewable resources, especially biomass, are important energy sources. Unfortunately, supplies of fuelwood—a very important energy source in the developing world—are rapidly becoming depleted because of rising population, increasing demand, and widespread deforestation. Other local sources of renewable energy must be developed to ease the pressure on remaining fuelwood supplies.

Specific Questions
1. How might people living in industrialized countries modify their life-styles as nonrenewable energy resources become scarce and more expensive?
2. Should energy conservation be a high priority for government funding or should free market economics be used to determine price and availability of energy resources?
3. What measures (political, economic, and social) could be taken to help LDCs develop an environmentally sound and sustainable economy without exhausting their energy resources?

Issue 2: Environmental and Health Effects

All energy consumption affects the environment. These effects include the degradation of air, water, and soil; noise pollution; infringement on wilderness; and adverse effects on human health. The type and severity of effects, however, differ significantly according to the resource used. In general, renewable and perpetual resources are far more benign than fossil fuels. In Chapters 12 and 13 we will look more closely at the specific environmental effects associated with the use of both fossil fuels and alternative fuels. However, our continuing reliance on fossil fuels and the severity of the environmental and health effects associated with them justify a quick summary of those effects.

The by-products of fossil fuels combustion are a major concern. Power generation by internal combustion engines releases large quantities of carbon monoxide, hydrocarbons, and nitrogen oxides into the atmosphere. Electric power plants emit large quantities of carbon dioxide, sulfur oxides, and particulates. Many of these emissions are linked to global warming and acid precipitation. Over 5.4 billion tons of carbon are emitted into the atmosphere annually, causing the carbon dioxide concentration to increase 9 percent since 1960 and 30 percent since 1860. In Europe fossil fuel emissions have damaged an estimated 76.6 million acres (31 million hectares) of forestland. It has been speculated that, in the United States alone, emissions may cause as many as 50,000 premature deaths each year through the onset of such diseases as emphysema and lung cancer. Further, since about 90 percent of our nation's energy originates from fossil fuels, their low conversion efficiency—just 50 percent at best—is a serious health, environmental, and economic concern.

Specific Questions

1. How will the threat of global warming and acid precipitation change our present patterns of energy use?

2. Should threats to environmentally sensitive wilderness areas, habitat destruction, and disasters like the *Exxon Valdez* spill cause us to modify our methods of energy exploration, extraction, and transport, or should we remain committed to these methods and instead seek technological means to restore damaged ecosystems?

3. Should the pollution of food chains linked to the use of energy resources cause us to enact stricter antipollution standards on energy consumers, or should pollution standards be eased in order to stimulate the economy?

Issue 3: Dependence on Fossil Fuels

The use of fossil fuels pervades the typical American life-style. Many actions that we take for granted—switching on a light, turning up the setting on the thermostat on a cold winter morning, taking a hot shower, cooking dinner, driving to the store to pick up a newspaper—are made possible through the use of fossil fuels. Indeed, fossil fuels are the lifeblood of industrial societies. Many people argue convincingly, however, that reliance on any one energy resource, such as oil, compromises environmental safety and economic and national security. The largest oil spill ever occurred during the 1991 Persian Gulf War. Tens of millions of gallons of crude oil spilled into the waters of the Persian Gulf,

the result either of US bombing (according to Iraq) or of Iraqi attempts to thwart Allied forces (according to forces arrayed against Iraq) (Figure 11-4).

Our decades-long reliance on fossil fuels has greatly diminished available supplies. In 1859 the first successful oil strike occurred at a depth of only 69 feet. In 1974 the United States drilled to a record depth of 31,441 feet in search of oil and gas. Instead of enjoying the bounty of seemingly endless free-flowing rivers of oil as we did in the late 1800s, we are now devising methods to "squeeze" what remains from our known reservoirs. The harder we squeeze, the more expensive our energy becomes. What will happen when the fruits of our labor diminish to the point where it is no longer economically feasible to mine these energy resources?

Specific Questions

1. To compensate for dwindling oil supplies, should we increase our use of coal—a relatively abundant resource that has significant adverse environmental and health effects—or should we encourage the development of renewable and perpetual resources—energy supplies that are less polluting but may be fairly expensive (relative to coal) to develop?

2. Is it appropriate for governments to require increased efficiency in internal combustion engines, appliances, and electric power generation and transmission in order to conserve fossil fuel supplies?

3. If, as a society, we decide that the development of renewable energy resources is a worthy goal, what means should we take to accomplish this goal: the use of government funding and subsi-

FIGURE 11-4: The environment proved to be a casualty of the Persian Gulf War. The oil spill in the gulf was the worst ever, far exceeding the 1989 *Exxon Valdez* spill in Prince William Sound, Alaska.

dies or the free market? What are the advantages and disadvantages of each?

Issue 4: Nuclear Power

Although the wisdom of using nuclear energy to generate electricity has been debated for several decades, that debate has intensified in the past few years. Worldwide, nuclear power generation increased substantially in the period from 1980 to 1988 (Table 11-4). Proponents of nuclear energy argue that it is a clean, safe alternative to fossil fuels. They maintain that using nuclear energy to generate electricity significantly reduces emissions of carbon dioxide. Those opposed to the use of nuclear energy point to the possibility of accidents like the 1986 disaster at the Soviet Union's Chernobyl plant (and

▶ Table 11-4
World Nuclear Power Generation, by Region and Country, 1980–1988 (billion kilowatt hours)

Region and Country	Power Generation			Operable Reactors[b]
	1980	1985	1988[a]	
North America	287.0	440.8	605.1	126
Canada	35.9	57.1	78.2	18
United States	251.1	383.7	526.9	108
Central and South America	2.2	8.7	5.4	3
Argentina	2.2	5.5	5.0	2
Brazil	0.0	3.2	0.3	1
Western Europe	207.4	559.8	654.7	159
Belgium	11.9	32.8	40.6	7
Finland	6.6	17.8	18.2	4
France	63.4	211.2	260.2	55
Germany, West	43.7	125.9	137.3	21
Italy[c]	2.1	6.7	0.0	2
Netherlands	4.0	3.7	3.5	2
Spain	5.2	28.0	48.3	10
Sweden	25.3	55.8	65.6	12
Switzerland	12.9	20.1	21.5	5
United Kingdom	32.3	53.8	55.6	40
Yugoslavia	0.0	4.0	3.9	1
Eastern Europe and USSR	90.7	200.9	260.0	72
Bulgaria	5.9	12.5	11.7	5
Czechoslovakia	4.3	11.2	21.2	8
Germany, East	11.3	12.1	10.4	5
Hungary	0.0	6.1	12.7	4
USSR	69.3	159.0	204.0	50
Middle East	0.0	0.0	0.0	0
Africa	0.0	5.5	10.5	2
South Africa	0.0	5.5	10.5	2
Far East and Oceania	92.6	192.2	237.2	59
India	2.7	4.3	5.7	6
Japan	78.7	144.4	164.1	38
Korea, South	3.3	15.9	37.8	8
Pakistan	0.1	0.2	0.2	1
Taiwan	7.8	27.3	29.4	6
World total	679.9	1,407.8	1,772.8	421

Note: Net generation; does not include energy consumed by the generating unit. Sum of components may not equal total due to independent rounding.

[a]Preliminary.

[b]As of December 31, 1988.

[c]Under Italy's new National Energy Plan, existing nuclear plants have been mothballed or decommissioned and nuclear plants under construction are being converted to burn gas or coal.

Source: U.S. Department of Energy, *International Energy Annual 1988*, 1989, and *Commercial Nuclear Power 1989: Prospects for the United States and the World*, 1989.

Focus On:
Why Not Nuclear Power?
Christopher Flavin, Vice President for Research, Worldwatch Institute

When disaster struck the Three Mile Island (TMI) nuclear plant in March 1979, the global nuclear industry was running at full throttle with about 250 nuclear plants under construction worldwide. Now, TMI stands as a rusting, radioactive reminder of the world's first civilian nuclear disaster, and nuclear construction in most countries has been grinding to a halt. In early 1990 just 90 plants were being built. In the United States the total fell even faster, from 94 in 1979, to just two under active construction in 1990.

Three Mile Island graphically showed the world the hazards of this technology. But what made the accident pivotal is that its lessons have echoed again and again, notably in the Chernobyl disaster but also in the hundreds of mishaps that occur at the world's nuclear plants each year. Still unresolved is finding a safe means of disposing of nuclear wastes, preventing nations from using nuclear reactors to develop atomic weapons, and containing runaway costs that have made nuclear power uneconomical.

While nuclear power still provides nearly one-fifth of U.S. electricity demand, its future is cloudy at best. The last U.S. nuclear plant of the 20th century was slated to go into operation in 1991, the same year the clean-up at TMI was to be completed. Through the 1990s, the U.S. will see a slow drop in nuclear-generated electricity as older, accident-prone plants are retired.

Although the U.S. nuclear industry had begun to unravel before TMI, the accident heightened these problems. Public concern intensified over safety; knowledgeable scientists and local officials joined the chorus of opponents. California passed a law forbidding more nuclear plants in the state until a solution to the waste problem is found. Other states and localities put up roadblocks to individual plants.

The accident also had the effect of encouraging state commissions and federal regulators to take a hard look at nuclear power. The presidential commission on TMI put it starkly: "To prevent nuclear acci-

dents . . . fundamental changes will be necessary in the organization, procedures, and practices—and above all—in the attitudes of the Nuclear Regulatory Commission and . . . the nuclear industry."

The resulting crackdown shook the industry. Safety designs and regulations in the pre–Three Mile Island era were clearly inadequate. Everything from control room layouts and the training of operators to the standards for particular welds had to be altered—in many cases repeatedly—as new discoveries were made and standards tightened. Mountains of paperwork were required to verify each change.

Facing skyrocketing costs and slower growth in demand for power, utilities canceled billions of dollars of nuclear plants, many of them already well under construction. Between 1974 and 1987, 108 plants were scrapped, most of them after the TMI accident. Fully half of the nuclear power capacity planned in the nation in the mid-1970s has been canceled; no new orders have been placed since 1978.

Once touted as "too cheap to meter," nuclear power became too expensive for the balance sheets of most utility companies. Since 1986, 21 U.S. nuclear plants have been completed at an average cost of $3,700 per kilowatt of capacity, or more than $4 billion for a standard-sized plant. Meanwhile, operating costs have catapulted to an average of $140 million per year for a typical reactor. This is more than the average total fuel and operating costs of coal plants. When construction and operating costs are totaled, nuclear power is more than twice as expensive as competing energy sources such as gas-fired cogeneration or fluidized-bed coal plants.

At the same time that nuclear costs have gone up, technological improvements have allowed a reduction in the cost of other electricity sources. During the past five years the cost of installing solar photovoltaic cells has fallen from $10,000 to $5,000 per kilowatt, while that of wind power has fallen from $2,000 to $800. It is cheaper

still to "generate" electricity by saving energy through more efficient light bulbs, appliances, and other devices.

But nuclear proponents now are trying to light a new flame under a generation of "inherently safe" reactors intended to be impervious to meltdowns. And lawmakers, who are under pressure to come up with alternatives to fossil fuel plants that cause air pollution and global warming, are being lobbied to fund a rebirth of the nuclear industry.

Three major designs are being proposed, and while each has advantages over existing reactors, none is free of potential problems. Meltdowns are not the only kind of nuclear accident. Reactors can also be subject to hydrogen explosions (one of the fears at TMI) or a runaway reaction (as at Chernobyl). The concept of "inherent safety" may be an engineering mirage, implying a degree of certainty and flawlessness that is beyond the scope of foreseeable nuclear technologies. Indeed, the gas-cooled reactor is based on a plant in Colorado that, except for TMI, may be the most problem-plagued reactor in the U.S. industry. Since beginning operation, the reactor operated at just 14% of rated capacity before being shut down for good in 1989. The misfortunes of the Colorado plant and those of the entire nuclear industry show that engineers' best intentions and detailed calculations do not ensure reliability. None of the new designs would ameliorate the waste or proliferation problems, each of which would grow far more serious as nuclear plant numbers went from a few hundred to the thousands required if nuclear power were to help slow global warming.

The future requires we take a different path. There are safer and cheaper alternatives to provide us with energy services while containing global warming. Energy efficiency and some renewables are already cheaper than current nuclear plants and, most likely, any "new generation" plants. Nonnuclear energy sources are the only alternatives that can honestly be labeled "inherently safe."

Focus On:
The Arctic National Wildlife Refuge

The Arctic National Wildlife Refuge (ANWR) is one of the wildest places left on the planet, certainly the wildest in the United States (Figure 11-5). A vital component of ANWR is the coastal plain on the Beaufort Sea, summer calving grounds of the 200,000-member porcupine caribou herd and home to many other species. While other sections of the refuge are already designated wilderness, the coastal plain, specifically the "1002 section" (a thin slice of coastal plain 30 miles wide and 100 miles long), is not. Developing the area's oil resources would entail a work force of 6,000, 4 airfields (2 large and 2 small), 100 miles of pipeline, 2 desalinization plants, 7 large production facilities, 50 to 60 drilling pads, 10 to 15 gravel pits, 1 power plant, 1 seaport, and 300 miles of roads. The sights, smells, and sounds of airplanes and machinery would disturb the plain and other portions of the refuge, as well as Canada's Northern Yukon National Park to the east.

According to an Interior Department study, oil exploration and drilling would cause "widespread, long-term changes in the wilderness character of the region." A draft of the 1002 section report projected that the population of the porcupine herd would decline 20 to 40 percent, a loss of up to 80,000 caribou. That projection, along with all mention of the "unique and irreplaceable core calving grounds," was deleted from the final 1002 report. In fact, the Interior Department forbade U.S. Fish and Wildlife Service employees in Fairbanks to use the term "irreplaceable." Wildlife biologists unaffiliated with the Fish and Wildlife Service and conservationists in general consider the

report to be so mangled by political ploys and word games as to be meaningless.

Given the fuss over development, you might suppose that the coastal plain is rich in oil reserves. Alaska Senator Ted Stevens, a proponent of development, labels the area "the Saudi Arabia of North America." But the facts suggest that this label is a misnomer. Interior's own report estimates that there is only a 19 percent chance that any economically recoverable oil lies beneath the coastal plain. Even if oil is found, the report predicts a 95 percent chance of at least 600,000 million barrels and only a 5 percent chance of 9 billion. The mean estimate is 3.2 billion barrels—just a six-month supply at current U.S. rates of consumption! Moreover, many experts believe that Interior's estimates are high. According to an estimate by the Alaska Department of Natural Resources, there is a 95 percent chance that the plain contains no more than a 5-day supply of oil.

So, despite its own admission of "widespread, long-term changes in the wilderness character," the slim chances of finding economically valuable oil deposits, the oil glut on the world market, and recent administrations' policy of discouraging energy conservation measures, the Department of Interior touted oil development as "ultimately . . . in the best interest of preserving the environmental values" of the coastal plain. Interior's assessment ignores the fact that every other portion of the United States' 1,100-mile arctic coastline, onshore and offshore, is open to oil development.

Whether or not ANWR would be opened to the oil companies remained in doubt during George Bush's administration. Pressure to

FIGURE 11-5: Snowy owl, Alaska National Wildlife Refuge. Other residents of the refuge include the grizzly bear, Dall sheep, long-tailed jaguar, rock ptarmigan, snow goose, arctic fox, and musk-ox.

develop the refuge intensified with the 1990 Iraqi invasion of Kuwait, the subsequent Persian Gulf War, and fears of oil shortages. Even the war's end did not relieve development pressures. If the refuge is opened to exploration, it might well fall prey to the same problems oil development has wrought elsewhere in Alaska, with oil spills, stream siltation and contamination, and road construction disturbing pristine wildlife habitat. On the other hand, if the refuge is closed to development, it may be the most important achievement in U.S. conservation efforts and signal a change of spirit in the U.S. populace— a burgeoning willingness to forgo short-term economic incentives for preservation. According to the writer James R. Udall, "whatever Congress eventually decides—unless and until it acts, no development will be permitted—the 1002 vote will be one of the most important land-use decisions this country will make for decades."

the potentially devastating effects on human health and the environment) and the related problems of dismantling old nuclear plants and disposing of radioactive waste. Focus On: Why Not Nuclear Power? examines the current status of the U.S. nuclear industry and the problems associated with nuclear power.

Specific Questions
1. Should we accelerate our use of nuclear energy to offset the adverse environmental effects associated with the combustion of fossil fuels?

2. Is it appropriate to decrease or eliminate the substantial federal subsidies currently given to the commercial nuclear power industry in

(Text continues on page 205.) 201

Environmental Science in Action: The Alaska Pipeline

Steve Douglas

Alaska, the nation's largest state, contains more acreage than Texas, Montana, and California combined (Figure 11-6). Its population is relatively small. A significant proportion of the people outside its few major cities are native Americans who depend on the land for survival. However, millions of other Americans have come to rely on Alaska's North Slope because of the oil that lies below its surface. Retrieving that oil and transporting it through the Alaska pipeline illustrate many of the political, economic, and environmental issues that underlie U.S. patterns of energy consumption.

Physical Boundaries

Tundra is land covered with permafrost—a lower layer of permanently frozen muck and gravel covered by an insulating layer of soil, moss, and grass. The insulating layer helps to keep the permafrost frozen despite air temperatures that vary seasonally from 70° F (21° C) to -70° F (-56° C). If the overlying layer is stripped from the surface, the permafrost will melt when temperatures rise. Temperatures significantly higher than 70° F such as those emitted by hot oil or drilling rigs, can melt the permafrost layer even without stripping away the insulation. Once permafrost melts, it may never regain the delicate environmental balance of life-supporting soil above and frozen foundation below.

Approximately 0.9 to 1.8 miles (1.5 to 3 kilometers) beneath the permafrost lies the oil which has brought the North Slope into national prominence. The Atlantic Richfield Company first discovered oil in the North Slope in April 1968 at Prudhoe Bay. Original estimates predicted that the North Slope could produce 96 billion barrels of oil, enough to meet approximately 5 percent of the U.S. demand for 13 years. As of 1987 the North Slope oil fields contributed about 11 percent of total U.S. annual production.

Biological Boundaries

Many species of plants and wildlife thrive in the harsh Alaskan environment. In June, when the sun begins its period of continuous daylight, the surface above the permafrost thaws enough to allow a wide variety of herbaceous plants, flowers, and mosses to grow. Migrating moose and caribou return from more southern areas, and bears, squirrels, and other hibernating animals become active again.

Approximately 370,000 caribou summer on the North Slope. Its unspoiled nature allows the animals to follow the same migratory routes every year, and any disturbance or barriers in that environment, such as building crews or the pipeline itself, might affect migration. Caribou, for instance, are easily deterred by anything in their way; faced with a fence or high brush, they will turn and follow the impediment to its conclusion rather than attempt to jump over it or force their way through.

Fish also migrate to the North Slope area, although they reverse the patterns of the mammals and winter in Alaska's rivers. Salmon, for instance, winter in pools of the Sagavanirktok River and migrate to the Arctic Ocean with the onset of spring. From mid-July to mid-

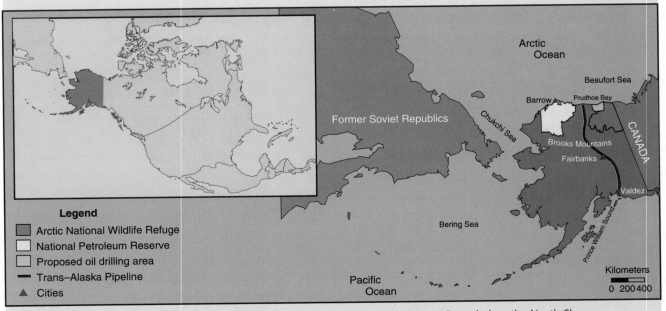

FIGURE 11-6: The North Slope extends from the Brooks range to the Arctic Ocean. Deep below the North Slope tundra and offshore beneath the waters of the Arctic Ocean lie petroleum reserves. The reserves may well determine the fate of Alaska's unspoiled wilderness areas.

August the salmon travel back up the river, where they spawn between September and October. Their migration cycles and spawning beds could easily be disturbed by construction activity such as laying sections of pipeline or roads across rivers and streams or diverting streams to lay pipe underneath.

Social Boundaries

Since the gold rush of 1899 brought the first major flux of outsiders to Alaska, the area has possessed a mythic aura for non-Alaskans. While some have valued Alaska for its vast tracts of unspoiled wilderness, most have seen it as a place to make a fortune through gold, trapping, hunting, fishing, or, more recently, oil drilling. But in their zeal to exploit Alaska's resources, newcomers consistently failed to consider the rights of the native Alaskans. At the time of the U.S. purchase of Alaska, most native Alaskans lived the traditional life-styles of their ancestors, hunting and fishing for a living and governing themselves through ancient tribal systems. Most white settlers and traders had little or no regard for the native Alaskans' traditions and territories because they considered themselves racially and culturally superior. They took whatever they wanted or needed from the native Alaskans, providing little or nothing in return. Some native Alaskans responded by adopting the unfamiliar life-styles and occupations of the outsiders, while others tried to withdraw from white civilization as much as possible in order to preserve their traditional ways.

In 1971 the United States took steps to protect the rights of native Alaskans. President Richard Nixon signed the Alaska Native Claims Settlement Act, which granted the natives 44 million acres, (17.8 million hectares) or approximately 10 percent of the state. This acreage was assigned to twelve native corporations, which were free to handle the land and associated resources as they saw fit. The act also reserved another 100 million acres (40.5 million hectares) for the native

Alaskans as federal protectorate land and granted them $962 million in cash.

At the same time that native Alaskans were gaining some control over their homeland, a group of oil companies were planning a pipeline to transport North Slope oil south to the Gulf of Alaska. Although their plans were stalled by a reluctant Congress, world politics soon altered the scene in the oil companies' favor. On October 6, 1973, Egypt and Syria invaded Israel. When the United States continued to support Israel, Saudi Arabia and the other countries in the Organization of Petroleum Exporting Countries (OPEC) protested by instituting a total embargo on oil exports to the United States. Because the United States relied heavily on OPEC countries for oil, the embargo precipitated steep rises in oil and gasoline prices in the United States. Under pressure to aid the development of domestic oil reserves, Congress quickly passed the Trans-Alaska Pipeline Authorizing (TAPA) Act, authorizing construction of the pipeline, on November 16, 1973.

Looking Back

Environmental considerations shaped the way the pipeline was constructed, and, once built, the pipeline and its related oil fields and tankers changed the Alaskan environment in ways that are still being discovered.

Planning and Constructing the Pipeline

The oil companies that formed the Trans-Alaska Pipeline System (TAPS), in February 1969, decided to extend the pipeline from the oil fields of Prudhoe Bay to the city of Valdez, a port on the Gulf of Alaska where oil could be loaded onto tankers and transported by sea. TAPS realized its pipeline would have to conform to the National Environmental Policy Act (NEPA) of 1969, which went into effect on January 1, 1970. Among other things NEPA required that an envi-

ronmental impact statement (EIS) be prepared for any project using federal funds or undertaken on federal grounds. The EIS is used to decide whether or not the project can be approved.

Besides having to comply with the new environmental regulations, TAPS faced a variety of lawsuits filed by groups who were opposed to the pipeline or objected to some aspect of its construction. For example, in early 1970 seven native Alaskan villages sued TAPS, withdrawing the waivers they had signed in order to allow the pipeline to be built on their territory. They felt that native Alaskans were no longer assured jobs associated with the pipeline as they had been promised when they signed the waivers. Three environmental groups, the Wilderness Society, Friends of the Earth, and the Environmental Defense Fund, also filed suit to reduce TAPS's right-of-way to 25 feet on either side of the pipeline, in accordance with a 50-year-old state law.

While TAPS struggled with political opposition and legal complications, its engineers faced perhaps the biggest challenge of all: how to build a pipeline that could transport oil at reasonably warm temperatures (since cold oil flows very slowly) without transferring heat into the surrounding environment and melting the permafrost. Originally, TAPS hoped to bury most of the pipeline. But adequately refrigerating underground pipe to prevent heat transfer proved to be costly and complicated except in areas where the permafrost was especially stable because of constant cold temperatures. Eventually, they decided to bury only about 372 miles of the almost 800-mile long pipeline. They refrigerated the pipe, insulated it with polystyrene, and surrounded it with cooling pipes which circulated refrigerated brine.

Over 400 miles of aboveground pipeline were laid on vertical support members (VSMs), which consisted of teflon-coated crossbeams held up by 18-inch diameter uprights drilled deep into the ground. The uprights were made of wood, a poor conductor of heat. Many uprights also contained sealed ammonia tubes to absorb any heat emitted from thawing permafrost and transmit it into the air. In order to allow for constriction and stretching with changing temperatures, each section of pipe was built at a slight angle to its adjacent sections (Figure 11-7).

In response to environmental regulations and concerns, the oil companies commissioned a variety of ecological studies on the areas to be affected by the pipeline. These studies resulted in various adaptations, such as constructing over 800 crossings for migratory animals. When temporarily diverting streams in order to lay pipes underneath, construction crews were also careful to work at places and times that would not disturb spawning grounds of salmon or other fish. Finally, after years of work, the pipeline was ready for use on June 20, 1977.

Assessing the Pipeline's Environmental Impact

According to most studies, the precautions taken by TAPS to preserve the environment surrounding the pipeline were successful. Almost all migratory animals seem to have adapted to crossing the pipeline, and the fish seem to be maintaining their natural cycles of behavior despite the disturbance to the rivers and streams where they spawn.

Even so, some environmentalists are wary of the long-range effects of the pipeline. Human activity in the Prudhoe Bay area has resulted in some obvious impacts. Most of the buildings, oil wells, and roads sit on gravel pads; these help prevent the permafrost from melting, but they also restrict the natural flow of water, causing excessive flooding in some arctic wetlands. Human activity in the arctic also causes and accelerates thermokarst, a localized thawing of ground ice which causes a depression in the ground. These depressions often fill with water, which increases the ground's heat absorption, leading to further thawing.

FIGURE 11-7: Trans-Alaska pipeline in winter, near the Alaska Range.

Although the degree to which the oil fields and finished pipeline have disturbed Alaska's environment is still a matter of debate, the 1989 oil spill in Prince William Sound dramatically illustrated the potential environmental costs of our quest for oil. When the *Exxon Valdez* ran aground on Bligh's Reef, spilling more than 11 million gallons of crude oil into the pristine Alaskan environment, the nation began to reevaluate its methods of energy use and development. An integral part of that ongoing reevaluation is closer scrutiny of the Alaska pipeline.

The highly publicized oil spill prompted Alyeska, the company which manages the pipeline, to fund a permanent citizens' watchdog group at an annual cost of $2 million. Established in 1990 to monitor Alyeska's shipping operations, the fifteen-member Regional Citizens Advisory Committee (RCAC) is made up of representatives from Alaskan towns, fishing organizations, and native groups. Alyeska has no power to cut off the group's funding or to choose members. If Alyeska disagrees with any of RCAC's recommendations, it must provide a written defense of its refusal. Disagreements are settled by arbitration; the results are binding, avoiding lengthy, expensive lawsuits. One of RCAC's first tasks was to review an oil spill response plan submitted by Alyeska to the State of Alaska.

Although RCAC was a positive step toward protecting Alaska's environment, another concern came into the public eye in 1990: the detection of external corrosion along the pipeline. The public became concerned that corrosion could eventually lead to a blowout. Alyeska maintains that a blowout precipitated by corrosion is impossible because detection equipment is so advanced that problems are quickly identified and corrected. Alyeska did replace 8.5 miles of underground pipe at Atigun Pass which was located in a floodplain and showed signs of external corrosion.

Looking Ahead

Although the pipeline itself seems to have left most of Alaska's environment relatively unchanged and the memory of the *Valdez* disaster is slowly fading, oil development still poses two types of threats. The first, of course, is more oil spills or oil blowouts. The second is the attempt by oil companies to obtain drilling rights in areas protected by the government as wilderness reserves.

To offset dwindling supplies, oil companies are clamoring to extend their drilling rights into the Arctic National Wildlife Refuge (ANWR). Environmentalists argue that any additional drilling in the North

Slope will simply contribute to overall oil consumption and thus add to related problems such as pollution and global warming. Others worry that, while the Prudhoe Bay activities have not had major consequences for the Alaskan environment, the effects of the existing and proposed oil fields could have a serious cumulative effect. Some scientists have suggested that the refuge seems to be more vulnerable to environmental damage than the Prudhoe Bay area. ANWR's summer population of caribou—fifteen times greater than that of the Prudhoe Bay area—along with great numbers of fish, waterfowl, and beaver, are important elements of the ecosystem as well as sources of food for neighboring tribes of native Alaskans. Exploring and constructing oil fields could seriously disrupt the life cycles of these populations. Drilling off the shores of the refuge could disturb populations of bowhead whales. On the other hand, prohibiting oil development in ANWR, and continuing to carefully monitor the oil pipeline and delivery systems, would serve to protect and preserve the Alaskan environment.

order to promote fairer competition with other alternative resources (such as solar energy) that receive little or no government support?

3. If, as a society, we decide to increase our reliance on nuclear energy, what steps should be taken to address concerns over reactor safety and the disposal of nuclear wastes?

Issue 5: Energy Policy

As we approach the new millennium, the United States must carefully craft an energy policy that deals with several problems: the potential for drastic changes in life-style caused by the scarcity of petroleum, the environmental cost of increasing reliance on coal or nuclear resources, the increased costs of extracting fossil fuels, the pros and cons of conservation measures to reduce fossil fuel consumption, and the costs necessary to develop and rely more heavily on alternative energies.

As you consider what energy policy to pursue, keep in mind several things. First, to reduce reliance on foreign oil supplies, the federal government's unofficial policy has been to "drain America first." Recent administrations have supported the petroleum industry's attempts to drill in environmentally sensitive areas like the North Slope of Alaska, the Arctic coastal plain, and the waters surrounding the Florida Keys. Opponents have managed to delay or prevent drilling in many of these areas, but the pressure to develop possible oil reserves continues. Without a coherent, long-range energy policy, the Arctic coastal plain and other areas will remain hostage to the fickle winds of politics. (See Focus On: The Arctic National Wildlife Refuge, page 201.)

Those who argue in favor of increased domestic oil production often claim that they are seeking to free the nation of its dependence on foreign supplies, and they cite national security as the primary reason for mining valuable and pristine wilderness areas. At the same time, recent administrations have eliminated incentives and support for fuel conservation and efficiency efforts and the development of alternative resources—two measures that could provide real security from disruptions in oil and gas supplies. Between 1980 and 1989 federal appropriations for renewable energy resources were slashed 80 percent. Similarly, funds for research and development of energy conservation measures fell from $1.2 billion in 1979 to $301 million in 1986. Corporate average fuel efficiency (CAFE) standards have been relaxed as well, from a high of 27.5 miles per gallon to 26 miles per gallon. By abandoning conservation programs and curtailing the development of alternative energies, the nation's leaders are following an *unofficial* and short-sighted policy aimed at increasing our reliance on the products of the fossil fuels industry at tremendous cost to our environment and our nation's future economic and social well-being.

Early planning is critical if we are to develop and implement an *official* national energy policy that is both effective and far-sighted. It takes many years to phase in a new energy resource. Supplies of petroleum and natural gas are likely to become virtually depleted about the middle of the next century. If we are to move smoothly from a dependence on oil and gas to a reliance on alternative fuels, we must begin the transition now. On the other hand, if we choose to rely on coal and nuclear power, we must be willing to risk environmental damage, develop means to mitigate or control that damage, and develop methods to enhance the safety of nuclear energy.

Specific Questions

1. Should the federal government develop an official national energy policy? Why or why not?

2. If, as a society, we decide that a comprehensive national energy policy is needed, which option or options should be pursued: (a) increased

development of domestic oil reserves and greater reliance on natural gas and readily available coal supplies, (b) increased development of nuclear energy resources, (c) a gradual phase-out of nonrenewable energy sources in favor of increased reliance on renewable energy sources and strict conservation of fossil fuels?

3. Is it appropriate for the federal government to increase mileage efficiency standards for automobiles, trucks, and buses and substantially increase taxes on automobile purchases and gasoline to help fund research and development for nonrenewable resources and more efficient technologies?

4. If the conservation measures described in question 3 are instituted, how might they affect the U.S. automobile industry? What steps could be taken, by the federal government or by the industry, to mitigate any adverse economic repercussions of fuel conservation measures?

Summary

Energy resources are generally classified as nonrenewable, renewable, or perpetual. Nonrenewable energy resources, such as fossil fuels, exist in finite supply or are renewed at a rate slower than the rate of consumption. Fossil fuels are also known as conventional resources or fuels because they have been so commonly used. Renewable resources are resupplied at rates greater than or consistent with use, so that supplies are not depleted. Perpetual resources originate from a source that is virtually inexhaustible, at least in time as measured by humans. Renewable and perpetual resources are called alternative resources because they offer us an alternative to conventional fuels.

In any energy conversion, although the same amount of energy exists before and after the conversion, not all of the energy remains useful. The conversion of energy from one form to another always involves a change or degradation from a higher quality form to a lower quality form. The energy efficiency of a system, process, or device that converts energy from one form to another is a measure of the percentage of the total energy input that does useful work and is not converted into low-temperature heat. Heat is given off with every energy conversion, and so the efficiency of any device or process is always less than 1. The net efficiency of a process or system that includes two or more energy conversions is found by determining the efficiency of each conversion. Because they use less energy, efficient devices have a lower life-cycle cost (initial cost + lifetime operating costs) than inefficient ones.

All energy consumption affects the environment. The type and severity of these effects differ significantly according to the resource used. In general, renewable and perpetual resources are far more benign than fossil fuels. The by-products of the conversion of fossil fuels—carbon monoxide, hydrocarbons, nitrogen oxides, carbon dioxide, sulfur oxides, and particulates—are a major concern.

Many of these emissions are linked to global warming, acid precipitation, and human health problems. The low conversion efficiency of fossil fuels is yet another serious environmental concern. Despite these problems, fossil fuels are the lifeblood of industrial societies. However, our decades-long reliance on fossil fuels has greatly diminished supplies.

The debate over using nuclear energy to generate electricity has intensified in the past few years. Proponents of nuclear energy argue that it is a clean, safe alternative to fossil fuels. Those opposed to its use point to the possibility of accidents, and the problems of decommissioning nuclear plants and disposing of radioactive waste.

To reduce U.S. reliance on foreign oil supplies, the federal government's unofficial policy has been to "drain America first." At the same time, recent administrations have eliminated incentives and support for fuel conservation and efficiency efforts and the development of alternative resources. Supplies of petroleum and natural gas are likely to be virtually depleted by the middle of next century.

Key Terms

alternative resource	net efficiency
conventional resource	nonrenewable resource
energy	perpetual resource
energy efficiency	renewable resource
life-cycle cost	

Discussion Questions

1. List the appliances in your home. What kind of energy does each one use, and where does the energy come from? If the energy supply were suddenly discontinued, how would you manage without each appliance?

2. Why should LDCs worry about saving energy when their most important concern is generating economic growth?

3. List some energy sources besides fossil fuels. What social and economic changes do you think need to be made for the United States to develop and exploit these alternatives?

4. Christopher Flavin argues that nuclear energy is neither inexpensive nor safe. Do you agree? Why or why not?

5. Imagine that you have been asked to help develop a national energy policy. What factors do you need to consider? What will you recommend?

6. What connections can you see between the history, development, and future of the Trans-Alaska Pipeline and the major energy issues presented in this chapter: social changes, environmental effects of energy use, dependence on fossil fuels, nuclear power, and energy policy?

Chapter

12

Energy: Fossil Fuels

Since the Industrial Age the world has increasingly depended on fossil fuels. Modern civilization is actually based on non-renewable resources. This puts a finite limit on the length of time our civilization can exist.

The Gaia Peace Atlas

If one has cut, split, hauled, and piled his own good oak, he will remember much about where the heat comes from, and with a wealth of detail denied to those who spend the weekend in town astride a radiator.

Aldo Leopold

Learning Objectives

When you finish reading this chapter, you should be able to:

1. Identify the three major types of fossil fuels and explain briefly how fossil fuels were formed.
2. Describe the current status (uses, availability, and environmental effects) of coal, petroleum, natural gas, oil shales, and tar sands.
3. Summarize how energy consumption has changed in the United States in the past 200 years.
4. Give examples to illustrate how industries, communities, and individuals can increase energy efficiency and conserve fossil fuels.

Many of the major issues surrounding the use of energy resources concern fossil fuels: their growing scarcity; their environmental impact, especially acid precipitation and global warming; and the role they play in world politics and economics. Historically, fossil fuels have been an important energy source, particularly in the industrialized nations (Figure 12-1). They have allowed us to enjoy a high quality of life by modifying our immediate environments for

personal comfort and by developing rapid, convenient transportation systems and sophisticated communications networks. In the United States, fossil fuels have supplied over 90 percent of the energy consumed. The amount of fossil fuels that humans use in just one year took about a million years to form. How long can we continue to rely upon our dwindling supplies?

FIGURE 12-1: First oil well at Titusville, Pennsylvania, in 1859. Edwin L. Drake (at right, in top hat) was responsible for the drilling of the world's first commercial oil well. The age of petroleum had begun.

In this chapter we describe the origin of fossil fuels and examine each of the major fuels, particularly their use, current and projected availability, and environmental effects. We also look at how these resources have been managed in the past and offer recommendations for their future management.

Describing Fossil Fuel Resources: Biological Boundaries

What Are Fossil Fuels?

The term **fossil fuels** suggests that these resources are ancient in origin, and indeed they are; they are the fossilized remains of organic matter. Coal, petroleum, and natural gas are the world's major fossil fuels; two other fossil fuels, oil shales and tar sands, are comparatively minor energy sources. The fossil fuels are also called conventional fuels or resources because they are the most commonly used.

Fossil fuels are composed chiefly of carbon and hydrogen. When a fossil fuel is burned, the chemical bonds that bind the carbon and hydrogen molecules are broken, and energy is released. This energy, whether it is contained in a tankful of gas or a lump of coal, was once sunlight captured by photoautotrophs. Each of the fossil fuels possesses unique characteristics and plays a different role in industrialized societies, but all originated from organic matter millions of years ago.

How Were Fossil Fuels Formed?

The greatest coal formation occurred some 300 million years ago during the Carboniferous period. Dead plant matter sank to the bottom of wetlands, was compressed by water and sediments, and was gradually transformed first into peat and later into coal (Figure 12-2). Similar processes were responsible for the production of petroleum and natural gas. But while the organic source of coal was largely plant material, petroleum originated from aquatic organisms (algae and plankton) that died and settled to the bottom of shallow, nutrient-rich seas. Over time, sediments accumulated over these deposits. Heat from both the earth's interior and the pressure of the sediments enhanced the oil's formation. Increasing amounts of heat, pressure, and anaerobic decomposition were required to transform organic matter into natural gas. Unlike coal and petroleum, natural gas is derived from all types of organic matter, even cellulose.

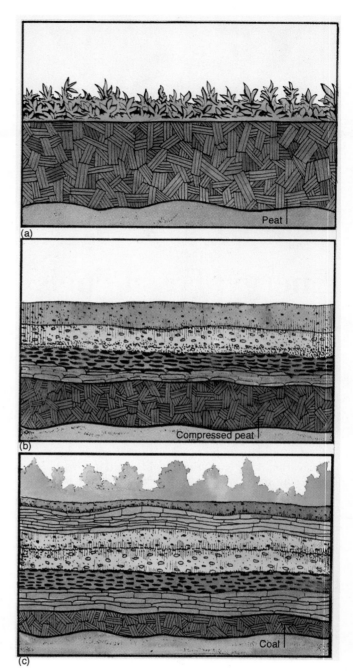

FIGURE 12-2: Development of coal. Coal formed in three main stages. (a) Giant ferns and other plants died and settled on the bottoms of swamps and other wetlands, forming thick layers of organic matter rich in carbon. Anaerobic decomposition, coupled with warm temperatures and the significant pressure exerted by the water, transformed the organic material into a low-quality fuel called peat. (b) The peat was buried and compressed beneath layers of sediments. (c) Additional pressure from settling sediments compacted the peat further, squeezing more water from the buried matter. Over millions of years the heat, along with the weight and pressure of the overlying sediments, turned the peat into coal.

The processes that formed fossil fuels continue today, as decaying plant materials accumulate in coal-forming environments like Florida's Everglades. But because those processes occur on a geologic time scale, the world's present deposits of coal, petroleum, and natural gas are all we can ever expect to use. Fossil fuels are thus classified as non-renewable and finite in supply.

Physical Boundaries

Where Are Fossil Fuel Deposits Located?

Fossil fuels are not distributed evenly beneath the earth's surface. Some areas are rich in deposits of coal; others are rich in deposits of petroleum or natural gas. This uneven distribution of fossil fuels exists because the conditions which gave rise to the preservation and fossilization of organic matter did not occur everywhere.

In the search for rich deposits, geologists use their knowledge of the origin and formation of fossil fuels. They explore areas that were exposed to the conditions favorable to the formation of coal or petroleum. Many of the areas most likely to yield deposits of fossil fuels have already been explored. Arid deserts, particularly those located on the edges of drying seas, have often been the site of major oil reserves. In geologic history such areas were likely covered by shallow, nutrient-rich seas, which fostered an abundance of algae and plankton. In recent years oil exploration has focused on the continental shelves, especially those located along partially land-locked seas such as the Baltic, the Gulf of Mexico, the Arctic Ocean, and the Bering Sea. Geologists believe that upwelling currents may have been present at these sites in the past 200 million years. Upwelling currents enrich seawater with nutrients, providing an ideal habitat for plankton and other microorganisms.

Exploration, location, and measurement of fossil fuels are ongoing processes. Estimates of the amount of each fuel are revised as new information about various deposits becomes known. **Proven** or **economic reserves** are deposits that have been located, measured, and inventoried. The fuel located in a proven reserve can be or is currently being extracted. In contrast, **subeconomic reserves** are deposits that have been discovered but cannot be extracted at a profit at current prices or with current technologies. **Indicated** or **inferred reserves** are deposits that are thought to exist and are likely to be discovered and to be available for use in the future. Estimates of fuel reserves are not static. For example, subeco-

nomic reserves may be reclassified as proven reserves if the price rises or if technologies are developed that make it possible to extract the resource at a profit.

Social Boundaries

What Is the Current Status of the World's Major Fossil Fuels?

Coal

Coal is a solid composed primarily of carbon (55 to 90 percent by weight) with small amounts of hydrogen, nitrogen, and sulfur compounds. Other elements or compounds may be present in trace amounts.

Coal is derived from deposits of organic plant matter. Depending on the amount of time that deposits were exposed to conditions of high pressure and temperature, different types of coal were formed. **Anthracite,** or hard coal, has the highest carbon content and lowest moisture content of all types of coal and, consequently, is the most efficient, releasing the largest quantity of heat per unit weight when burned. Anthracite's efficiency, and the fact that it is the cleanest burning coal, make it the preferred coal for heating homes and commercial buildings, but it comprises the smallest portion of U.S. reserves, less than 1 percent, making it very expensive.

Bituminous, or soft coal, is the most common coal, accounting for over half of U.S. reserves. It has a heating value slightly lower than that of anthracite. Bituminous coal has long been preferred for electric power generation and the production of coke, a hard mass of almost pure carbon used to make steel. When heated in air-tight ovens, coal is transformed into coke. Coke is then burned with iron ore and limestone to produce the pure iron required to make steel. Over 50 percent of bituminous coal reserves have a medium to high level of sulfur content. Many researchers have associated the burning of high sulfur coal to the production of acid precipitation.

Together, **subbituminous** and **lignite** coals account for 48 percent of our nation's coal reserves. Because both have low heating values, they must be burned in large amounts in order to heat effectively. One advantage of subbituminous and lignite coals is that they contain very little sulfur; all of the subbituminous and 90 percent of the lignite coals have a sulfur content of less than 1 percent. To meet state and federal pollution standards, many utility companies have switched to subbituminous or lignite coal.

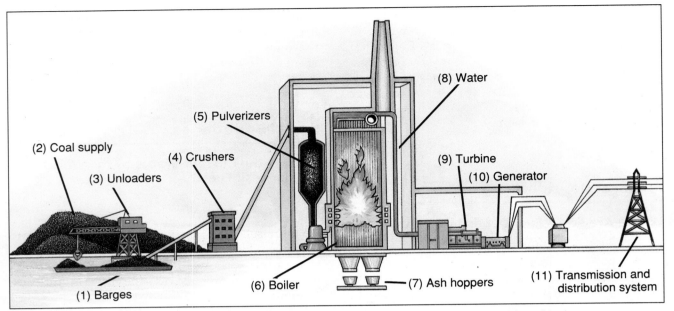

FIGURE 12-3: Coal-fired power plant. Electric utilities consume approximately 83 percent of the coal used in the United States. Coal supplies are brought in by barge or rail (1, 2), unloaded (3), then crushed and pulverized to form a fine dust (4, 5). The dust is blown into a furnace area inside a boiler (6), where it burns at very high temperatures. Waste ash collects in hoppers (7). The boiler is a huge structure made up of 30 miles of steel tubing. Purified water constantly circulates through the tubing (8) and is converted to steam by the intense heat of the furnace. The high-pressure steam turns the fanlike wheels of a turbine, spinning its shaft (9). The shaft turns the generator rotor (10), a large electromagnet, to produce electricity in coils of wire in the enclosure around the rotor. The electricity is then sent through a distribution system (11) to consumers.

Use. Coal consumption has declined over the last century, largely because it is a dirty fuel with serious environmental impacts (discussed below). Even so, it remains an important energy source, accounting for 22 percent of the United States' and 27 to 31 percent of the world's energy budget. In some nations the figure is much higher. China, for example, meets about 75 percent of its energy demand with coal. Worldwide, coal is used primarily to produce electric power (Figure 12-3). Coal is also heavily used in industry, principally to purify iron and produce steel. Increasingly, fuel oil and natural gas have been replacing coal as a heating source for residential and commercial structures. The rising cost of these fuels, however, has led some factories and commercial buildings to convert back to coal.

Availability. Coal is the most abundant fossil fuel on earth. Current world reserves are estimated to be between 650 to 1,500 billion short tons, representing about 93 percent of the world's fossil fuel resources. The United States, the former Soviet Union, and China have the majority of the world's coal reserves (Figure 12-4). U.S. coal reserves represent about 80 percent of the nation's energy resources. In terms of potential energy, they are com-

parable to Saudi Arabia's known oil reserves. The majority of these reserves lie in three broad regions: western or Rocky Mountain, Appalachian, and central (Figure 12-5). Wyoming leads all states in coal production, followed by Kentucky and West Virginia.

Estimates developed by the coal industry predict that, at present consumption rates, the United States' coal supply could last another 300 years. Most projections also call for world reserves to last between 300 and 400 years. These estimates are based on present consumption rates, but coal use is likely to increase as reserves of other fossil fuels, especially oil, dwindle. Naturally, increasing consumption would shorten the life expectancy of coal reserves. Some experts believe that if we were to become completely dependent on coal, the United States' proven reserves would last only 47 years and total estimated reserves only 75 years.

Environmental Effects. Of all energy resources in use today, coal has the most serious environmental and health impacts. The effects associated with mining are discussed more thoroughly in Chapter 17, but a brief synopsis is included here. Underground or subsurface mining produces **acid drainage.** When air and water come into contact with sulfur-bearing

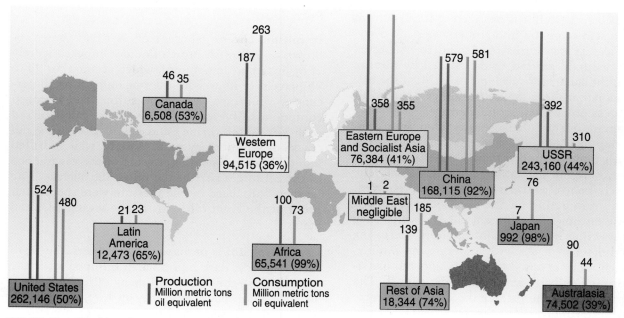

FIGURE 12-4: World coal reserves, 1988. The area of each country or region is drawn proportional to the amount of its coal reserves. Total amount (millions of metric tons) is shown next to country or region name. Anthracite and bituminous reserves are shown as a percentage of the total for each country or region.

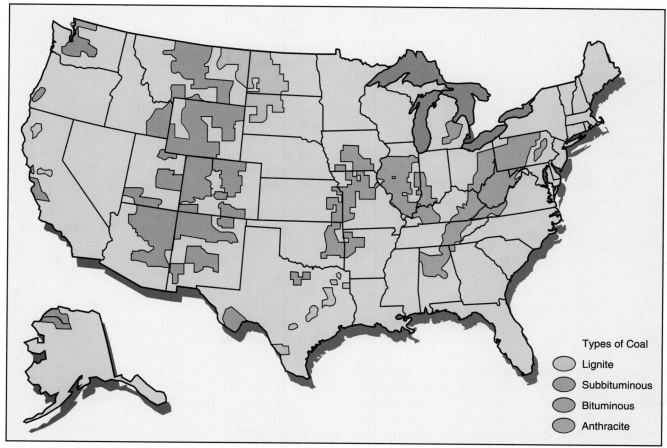

FIGURE 12-5: Distribution of coal reserves in the United States. Subbituminous and bituminous coal dominate in the western or Rocky Mountain region, which includes Montana, Wyoming, Colorado, New Mexico, Arizona, Utah, North Dakota, and South Dakota. The Powder River basin of northeastern Wyoming and southeastern Montana alone contains about 60 percent of the western minable coal reserves. Bituminous coal dominates in the Appalachian and central regions. Kentucky, West Virginia, Pennsylvania, and Illinois combine to have 80 percent of the coal reserves in these regions.

rock and coal, the sulfur is oxidized to form sulfuric acid. The acid can drain into groundwater and eventually make its way into streams and rivers. Additionally, as underground voids created by mining collapse, the land may subside or sink. **Land subsidence,** as it is called, makes buildings situated in mining areas prone to structural damage.

The environmental damage associated with strip surface mining is even more severe than that associated with underground mining. To strip mine an area, the overlying vegetation, soil, and rock layers, collectively known as the overburden, are removed. Wind and rain acting on the exposed overburden can cause erosion and sediment runoff, which may find its way into nearby streams or rivers, increasing their sediment load. As with underground mining, sulfur-rich rocks exposed to the air undergo oxidation, forming sulfuric acid and acidic runoff. The steeper the slope, the greater the danger of severe erosion and runoff. (The environmental consequences of strip mining are explored more fully in Environmental Science in Action: Strip Mining in Appalachia.) Environmental laws require that strip mined areas be reclaimed or restored after mining has ceased. Reclamation is difficult in some areas, particularly mountainous terrain.

Coal contains trace elements of heavy metals and other substances that pose serious environmental threats and health hazards. Arsenic, lead, mercury, sulfur, cadmium, selenium, and uranium, for instance, can be toxic to plants and animals alike. These substances can contaminate both air and water near coal mining sites.

Coal combustion also has serious environmental and health impacts. Coal contains varying amounts of sulfur, which has been implicated in acid precipitation. Eastern coal, in particular, has a high sulfur content. Technologies have been developed to remove sulfur from coal before or during combustion or from the flue gases that form after combustion. These technologies, however, are expensive. If we maintain or increase our use of high-sulfur coal, we can expect that it will continue to adversely affect our environment unless we take the necessary precautions. In other words, we must be willing to pay to finance the necessary pollution control practices.

Carbon dioxide, an end product of combustion and the primary greenhouse gas responsible for global warming, creates a more serious problem than sulfur because carbon dioxide emissions are not controllable with current technologies. The combustion of coal produces proportionately more carbon dioxide than the combustion of either oil or natural gas. If coal consumption rises, as seems likely given oil's relative scarcity and expense, carbon dioxide emissions will also increase.

Petroleum

Petroleum, also called crude oil, is a dark, greenish-brown or yellowish-brown liquid composed largely of hydrocarbon compounds, which account for 90 to 95 percent of its weight; the remaining 5 to 10 percent is a mixture of oxygen, sulfur, and nitrogen compounds. Typically, crude oil is found mixed with salt water (brine) and gas in the pores and cracks of sedimentary rock. When a well is drilled into oil-bearing rock, the pressure of the gas trapped in the oil forces part of the oil to the surface. The "gushers" produced by the pressure were spectacular proof of the great oil fields of the western United States.

Use. Petroleum is perhaps the most versatile fossil fuel. A gooey, dark liquid when pumped from the ground, it is refined to yield many different materials such as propane, gasoline, jet fuel, heating oil, motor oil, kerosene, and road tar. One-quarter of petroleum resources are used in the production of nonfuel substances and chemicals, which are used in the manufacture of plastics and medicines. Each day worldwide over $3 billion is spent on approximately 3,000 petroleum-derived products (Table 12-1).

Given petroleum's versatility, it is not surprising that annual global consumption of petroleum is approximately 23 billion barrels, an average of over 200 gallons of oil for every human being on earth. The United States consumes about 30 percent of this total, more than 1,100 gallons per person per year. Gasoline is by far the most consumed petroleum product in the United States. Annually, over 60 percent of the nation's oil is consumed by the transportation sector in the form of gasoline for automobiles, diesel fuel, and jet fuel.

Availability. In 1950 proven reserves of petroleum amounted to 76 billion barrels. This figure quickly grew to 664 billion barrels by 1973, an 874 percent increase in just 23 years. Since 1973, however, global proven reserves have increased by a paltry 5 percent. This drastic slowdown in proven reserves has generated concern about future supplies. Over 80 percent of all the oil discovered to date on the North American continent has already been burned. Proven U.S. reserves—estimated at 36 billion barrels—would meet domestic demand for a mere nine years at current consumption rates.

Estimates of proven global reserves range from 700 to 900 billion barrels. The largest portion of these reserves, two-thirds, is located in the Middle East (Figure 12-6). Other countries or regions that have considerable oil deposits are Latin America (12 percent of global reserves), the former Soviet Union (9 percent), Africa (8 percent), and North America (6 percent).

▶ Table 12-1
Petroleum Products, Selected List

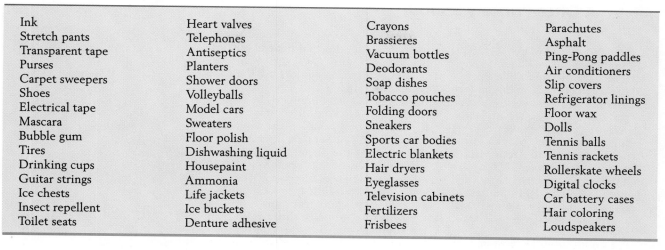

Ink	Heart valves	Crayons	Parachutes
Stretch pants	Telephones	Brassieres	Asphalt
Transparent tape	Antiseptics	Vacuum bottles	Ping-Pong paddles
Purses	Planters	Deodorants	Air conditioners
Carpet sweepers	Shower doors	Soap dishes	Slip covers
Shoes	Volleyballs	Tobacco pouches	Refrigerator linings
Electrical tape	Model cars	Folding doors	Floor wax
Mascara	Sweaters	Sneakers	Dolls
Bubble gum	Floor polish	Sports car bodies	Tennis balls
Tires	Dishwashing liquid	Electric blankets	Tennis rackets
Drinking cups	Housepaint	Hair dryers	Rollerskate wheels
Guitar strings	Ammonia	Eyeglasses	Digital clocks
Ice chests	Life jackets	Television cabinets	Car battery cases
Insect repellent	Ice buckets	Fertilizers	Hair coloring
Toilet seats	Denture adhesive	Frisbees	Loudspeakers

Even though global reserves seem substantial, they cannot last long given the world's appetite for oil. If the present annual global consumption rate remains constant, proven reserves would last approximately 32 years. Estimates that include subeconomic and indicated reserves are more optimistic. It has been estimated, for example, that current world resources equal approximately 1,800 billion barrels. But even if these estimates prove accurate, the annual global consumption rate places the life expectancy of the world's oil resources at 70 to 80 years.

No one can predict precisely when our oil supplies will become exhausted. What is certain, however, is that the world's oil supply is being drained from a pool that will eventually be emptied.

Environmental Effects. Oil production and use adversely affect the environment in many ways. Petroleum, like other fossil fuels, releases carbon dioxide when burned. In addition, unless pollution control devices are in place, the combustion of crude oil products releases carbon monoxide, sulfur oxides, nitrogen oxides, and hydrocarbons.

Perhaps the most disturbing aspect of oil development is the risk of oil spills like the 1989 *Exxon Valdez* disaster (see Focus On: A Once Princely Sound). Just three months after the disaster in Prince William Sound, smaller but still substantial spills occurred in Texas, Delaware, and Rhode Island all in the space of one weekend. As destructive as these spills were, all were dwarfed by the spill of crude oil into the Persian Gulf, an environmental casualty of the Persian Gulf War.

Natural Gas

Natural gas is composed of several different types of gases. Methane is by far the most abundant, approximately 85 percent of the typical deposit; other gases include ethane (10 percent), propane (3 percent), and small amounts of butane, pentane, hexane, heptane, and octane.

In most parts of the world most natural gas occurs in deposits along with petroleum; in this situation, it is known as **associated gas**. In contrast, in the United States 70 to 80 percent of the natural gas comes from deposits not associated with petroleum. Such **nonassociated gas** seeps through sedimentary rock until it becomes trapped against impervious rock layers.

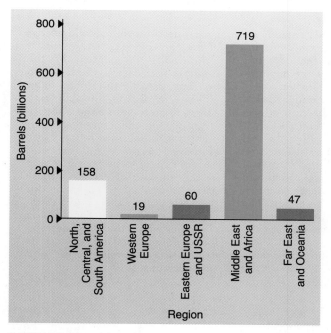

FIGURE 12-6: World crude oil reserves, January 1, 1990. The Middle East and Africa together comprise the region with the largest oil reserves, most of which are controlled by members of the Organization of Petroleum Exporting Countries (OPEC).

Focus On:
A Once Princely Sound

On the morning of March 24, 1989, the oil tanker *Exxon Valdez* rammed into the well-marked Bligh Reef in Alaska's Prince William Sound (see Figure 11-6). The result was the largest oil spill disaster in North America's history. Eleven million gallons of North Slope crude spewed into the cold Arctic waters. The spill affected over 1,300 square miles of highly sensitive habitat and damaged almost 1,200 miles of beaches, marshes, wetlands, tidal estuaries, salmon streams, islands, bays, and fjords.

The spill exacted a tremendous toll on the region's wildlife. Thousands of animals died in the days and weeks immediately following the disaster, many after ingesting the crude oil or after eating the oil-soaked carcasses of dead birds and fish or the oil-soaked kelp that washed onto beaches. By late summer the official death toll included over 33,000 seabirds, 146 bald eagles, and almost 980 sea otters (Figure 12-7). Those numbers probably represent only 10 to 30 percent of the actual numbers poisoned by the spill as the oil worked its way up the food chain. Not counted among the dead were mussels and starfish, residents of the intertidal zone. With little or no mobility, unable to escape the oil's onslaught, these organisms likely suffered severe mortality. Perhaps the most serious long-term blow to the ecosystem is the possibly devastating effect of the oil on the plankton which serve as the basis of the food chain of the sound and surrounding arctic waters. A change in the organisms at the base of the food chain may have a significant effect on the population dynamics of other organisms and the community structure of the ecosystem. Toxic compounds such as benzene, ingested by the plankton, may biomagnify in the food chain, ultimately affecting the area's famed salmon and many other species that depend on plankton.

The human inhabitants of the region also suffered as a result of the disaster. In May the state closed the sound's $12 million herring fishery. In mid-June, as the spill moved out of the sound and southward toward the Kenai Peninsula and Kodiak Island, the salmon fisheries of those areas directly in the spill's path were closed. In fact, oil kept salmon fishers from much of their fishing grounds throughout the summer, and they netted only half of what was expected to be a record year.

In the critical first days immediately following the spill, containment efforts were hampered by the oil industry's seeming lack of preparedness. The Alyeska Pipeline Company, the oil transportation consortium of which Exxon is a leading member, could not respond immediately because its only containment barge was stripped for repairs. Deep-water skimmers were buried under containment booms and other equipment. Fenders needed to allow a second tanker to pull alongside the *Exxon Valdez* and siphon off its remaining oil were buried under 14 feet of snow. As news of the hampered containment effort spread, citizens became increasingly alarmed at the industry's seeming disregard for the environmental effects of the disaster.

Early on, Exxon announced that it would conclude cleanup efforts by September 15, 1989. When it did so, 7 million gallons of oil remained in the sound—coated on and under rocks, welled into sand and pebble beaches, solidified into asphalt lumps, and congealed into a thick muck along shorelines. The once princely sound, a jewel of the Far North, had been dramatically altered.

In the weeks, months, and years following March 24, 1989, the

FIGURE 12-7: A rescue worker handles an oil-soaked otter in Prince William Sound, Alaska, April 1989. Following the *Exxon Valdez* spill, hundreds of sea otters died from exposure when the natural insulation of their fur was ruined by a coating of oil.

nation debated who should accept blame for the disaster: Joseph J. Hazelwood, the captain of the *Exxon Valdez*? Alyeska, for failing to be ready and able to respond to an emergency? Exxon, for failing to complete an adequate and effective cleanup? Or the federal government, for failing to mandate more effective emergency spill responses and for failing to intervene sooner in the cleanup? Certainly, these parties should all accept blame for their role in the *Exxon Valdez* disaster. And so must we. For it is the consumer whose thirst and demand for oil encourages and finances its exploration, development, and transportation.

Use. In the early days of petroleum exploration and exploitation, natural gas that was found in the same deposits as crude oil was burned off as a waste product at the wellhead. Later, technological advances in the handling of gas—welded pipelines to move gas effectively, improved storage systems, and methods for liquifying and shipping gas in tankers—made it an important resource on its own. The major reason for recent emphasis on natural gas has been the growing awareness of our finite oil supplies.

Since 1970 the most common application of natural gas in the United States has been in the industrial sector. Five key industries—chemicals, refineries, steel, paper, and cement—account for 40 percent of the total industrial gas use. Gas also provides heat for about half of all homes and industries in the United States. It is also gaining strength in the residential sector by way of home appliances like hot water heaters, stoves, and driers.

Natural gas systems are desirable in part because of their high conversion efficiency. A unit of natural gas can be produced and delivered to the end-user at an efficiency of nearly 89 percent. In contrast, the efficiency of generation and delivery of an equivalent unit of coal is only about 30 percent. In other words, when coal is used to generate electricity, 70 percent of the energy is wasted. Even with the use of a high-efficiency electric heat pump, the total efficiency of the coal-powered electric cycle rises to just 48 percent.

Increasingly, utilities must also consider the environmental aspects of power generation. The combustion of natural gas produces no ash or sludge by-products. In contrast, a 1,000-megawatt power plant, operating on high-sulfur coal and with technologies in place to remove sulfur from the flue gases, produces 700,000 tons of sludge per year and 250,000 tons of ash.

Natural gas holds significant potential in the transportation sector. According to a 1986 study by the Georgetown University Center for Strategic and International Studies, nearly 400,000 natural gas–powered vehicles (NGVs) are in use worldwide. Natural gas offers fuel cost savings, greater efficiency, and lower air emissions than conventional gasoline or other alternatives such as propane and methanol. An estimated 30,000 NGVs are cruising the roads and highways of the United States, mostly in utility service fleets and urban transit. Since the mid-1980s, the Texas Garland school district has operated buses on natural gas with such success (savings on fuel and maintenance) that the Texas legislature has ordered other larger school districts to convert their fleets. At the national level, President Bush, while introducing new clean air legislation in 1989, proposed to put 1 million vehicles capable of burning fuels other than gasoline on the streets of the most polluted cities by 1997.

Availability. Many countries are investing large amounts of capital and resources to determine the location and extent of their natural gas reserves. Much of this gas exists in unconventional sources such as sands, shales, coal seams, and geopressurized zones, and we currently lack the technology to extract it economically. Instead, we must rely on deposits of conventional associated or nonassociated gas.

Global reserves of conventional natural gas are relatively abundant, with the largest deposits in the former Soviet Union (38 percent) and the Middle East (30 percent) (Figure 12-8). Cumulative gas production in the United States is strongly concentrated in three regions: the Gulf Coast, the midcontinent, and west Texas–east New Mexico. The Gulf Coast region contains the largest deposits, about 43 percent of the U.S. resource base. According to most estimates, proven U.S. reserves will last only into the first decade of the next century. But since consumption of natural gas is expected to rise dramatically, it is difficult to estimate the life expectancy of proven reserves.

Environmental Effects. Environmental concerns provide a compelling reason for shifting from coal and oil to natural gas. When burned, coal and oil release 40 percent and 30 percent more carbon dioxide, respectively, than does gas. In fact, gas is the cleanest of all fossil fuels. Utilities using natural gas produce far less nitrogen oxides and sulfur oxides, the primary pollutants responsible for acid precipitation, than do utilities burning coal and oil. Even so, natural gas is not an entirely clean fuel. Its combustion does release some carbon monoxide, nitrogen oxides, and carbon dioxide.

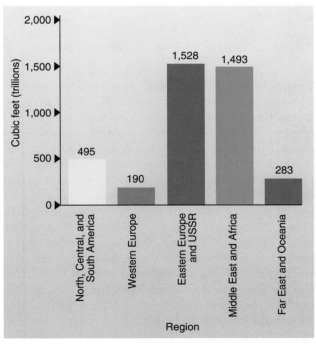

FIGURE 12-8: Natural gas reserves, January 1, 1990.

Minor Fossil Fuels: Oil Shales and Tar Sands

Oil shales and tar sands are deposits of fossil fuels that have played a minor role as energy sources. Despite their tremendous energy potential—just 1 percent of the total oil shale represents more oil than the world is expected to produce as petroleum—commercial production of these resources has generally been considered uneconomical.

Oil shales are the result of an interruption in the process that formed oil. In fact, the term is a misnomer; the shales do not contain oil. Rather, oil shales are fine-grained, compacted sedimentary rocks that contain varying amounts of a waxy, combustible organic matter called kerogen. Apparently, oil shales were not subjected to enough heat to complete the conversion process that produced petroleum.

The mined shale is first crushed and then heated to about 900°F (482°C). At these high temperatures, kerogen is vaporized, and the vapor is then condensed to yield oil. A heavy, slow-flowing liquid, shale oil must be refined to increase its flow rate and hydrogen content and to remove impurities. The resulting synthetic crude oil is then further refined to yield gasoline, heating oil, and other products. Most oil shale operations have been closed down for economic reasons.

Tar sands are sandstones that contain bitumen, a thick, high-sulfur, tarlike liquid, within their porous structure. High-pressure steam is used to force the bitumen from the sandstone. The bitumen is then purified and upgraded to synthetic crude oil. Tar sands are generally mined by surface mining techniques; at present underground mining is neither technically nor economically feasible. The world's largest deposits of tar sands are found along the Athabasca River in northern Alberta, Canada.

Environmental Effects. Many problems are associated with the development of oil shales and tar sands. Extracting and refining the shale oil and tar are energy-intensive processes. The energy equivalent of about 1 barrel of conventional crude oil is needed to produce 1 barrel of shale oil and 3 barrels of heavy oil from tar sands. Consequently, the net useful energy yield of these resources is negative, or at best low. Large amounts of water are also needed to extract the heavy oils and to process and refine them. Producing 1 barrel of shale oil, for example, requires 2 to 6 barrels of water, which is then contaminated. The world's richest deposit of oil-bearing shale is found in the Green River formation in Colorado, Utah, and Wyoming, a region with scarce water supplies (Figure 12-9).

The extraction of both shale oil and tar sands also disturbs large land areas and produces enormous

FIGURE 12-9: Oil shale exploration, western United States.

amounts of waste for each unit of oil produced. When shale rock is heated, it expands and breaks up; the broken rock, which may contain trace amounts of harmful substances, must be safely disposed of. Many areas under investigation as potential sites for oil shale extraction are located on federal public lands or in wilderness areas in the western United States. Development would very likely destroy the wilderness quality of these areas.

Projects operating in several countries are seeking to overcome the problems associated with the development of oil shales and tar sands. Since the early 1980s two plants in the Athabasca region have supplied about 10 percent of Canada's annual oil demand. The nation hopes to obtain a significant portion of its crude oil supplies from tar sands in the 1990s and beyond. Companies in Brazil and the United States have also experimented with oil shale production. Only the former Soviet Union and China, however, have commercial oil shale industries.

History of Management of Fossil Fuel Resources

How Have Fossil Fuels Been Used Historically?

Humans have used fossil fuels for thousands of years. The discovery that coal can be burned to produce heat was probably made independently by humans in various parts of the world during prehistoric times. By the A.D. 300s, the Chinese had developed a coal industry. They mined coal from surface deposits and used it to heat buildings and smelt metals. Commercial coal mining in Europe lagged behind development in China. Commercial mines were started in England and the area we now know as Belgium in the 1200s; the coal was used chiefly for smelting and forging metals. Europeans were slow to exploit coal because they considered it a dirty fuel; they preferred to use wood and charcoal made from wood. By the start of the seventeenth century, however, severe wood shortages in western Europe forced communities and factories to increase their production and use of coal. By the late 1600s, England accounted for 80 percent of total world coal production.

In North America, the Pueblo Indians were the first to use coal, digging the combustible material from hillsides to use in baking pottery. In the early eighteenth century, European colonists began using coal; the first commercial operation was established in Virginia in 1730. The first recorded commercial shipment of American coal, 32 tons, was transported from Virginia to New York in 1759. All large-scale early mining was done underground. The coal was extracted with picks and bars from solid beds and transported up from the mines in wheelbarrows, baskets, and buckets.

Like coal, petroleum has a long history. About 3000 B.C. the Sumerians of Mesopotamia used it to waterproof ships and wicker baskets and as a bedding compound for mosaic and inlay work. They mixed it with sand for use in architectural work and road construction. But perhaps the most interesting of their uses of petroleum were its medicinal applications. The Sumerians softened bitumen with olive oil and rubbed the mixture on sores, open wounds, and rheumatic joints. They drank bitumen mixed with beer as a cure for numerous ailments and burned the pitchlike substance to ward off evil spirits. Belief in the medicinal qualities of petroleum was also widespread in the days of the ancient Roman empire, and it endured over the intervening centuries. In the early part of the nineteenth century, enterprising persons collected and bottled oil which flowed from springs in western Pennsylvania into a tributary of the Allegheny River. They mar-keted the substance as "Seneca Oil," referring, most likely, to its medicinal use by the Seneca Indians. Despite petroleum's long history of use, little more than a century has passed since humans began to exploit its energy potential.

How Has Energy Consumption Changed in the United States?

For more than 250 years after the first English settlement was established on the North American continent, wood was the nation's primary energy source. By 1885, however, wood gave way to a more powerful fuel—coal. The Industrial Revolution, which reached the nation's shores in the early 1800s, increased coal's importance in both manufacturing and transportation. The steam engines that powered factories, ships, and railroads required large amounts of coal to fire their boilers. As industry and transportation grew, so did the production of coal, and by the late 1800s the United States had replaced England as the world's leading coal producer. In the early 1900s coal met 90 percent of energy needs in the United States.

The twentieth century brought rapid change to U.S. society, increasing the country's demand for more energy. Coal's years of dominance were numbered as the popularity of oil began to rise, in large measure because of the invention of the "horseless carriage" in 1892. Previously, the primary derivative of petroleum was kerosene, used for heating; indeed, much of the oil was typically discarded as useless. But with the development of the first American gasoline-powered engine, many found that "gasoline," a derivative of crude oil, was the most practical fuel for automotive transportation. Manufacturers, quick to appreciate the advantages of this new fuel, made gasoline-driven cars at an ever-increasing rate. This sudden increase in demand for gasoline caused the first severe petroleum shortage in the United States, from 1903 to 1911. As techniques for locating, recovering, and refining petroleum improved between the two world wars, supplies rose. The resulting glut of oil was so great that at times the selling price was driven to absurdly low levels, and many people became convinced that the supply of petroleum was limitless.

In addition to spurring on the development of the nation's modern transportation system, petroleum revolutionized agriculture. Farmers quickly found that horses and mules could not compete with tractors and mechanical harvesters. Because of their convenience, oil and gas also began to displace coal in the small industry and home heating markets.

In the decade from 1937 to 1947 more petroleum was taken from the earth than had been extracted in all previous history. With the end of gas rationing after World War II, Americans took to the roads with a vengeance, and motor vehicle registrations climbed by 50 percent. The use of farm tractors also increased by 50 percent, and the number of domestic oil burners doubled. As the consumption of all petroleum products rose to record levels, the rate of production began to outpace the rate of new oil discoveries. In 1947, for the first time ever, the United States became a net petroleum importer. President Eisenhower realized the importance of this transition and pledged in his 1952 inaugural address to make this nation energy self-sufficient, a pledge reiterated—but not achieved—by every succeeding president.

With increasing oil consumption, the use of natural gas also increased. Natural gas became a major fuel for home heating and for industry in the production of glass and other commodities. By 1946 oil and natural gas together displaced coal as the chief source of the nation's energy supplies, and by 1950 oil alone outdistanced coal to become the primary energy source.

In the years since 1960, oil, coal and natural gas have continued to play the major role in U.S. energy consumption. Together, these nonrenewable resources, along with nuclear energy, account for 96 percent of all energy consumed in the United States (Table 12-2).

What Was the 1973 OPEC Oil Embargo?

Continuing dependence on oil as its major energy source made the United States increasingly vulnerable to foreign suppliers, yet few people recognized the danger. However, a 1973 embargo by the Organization of Petroleum Exporting Countries (OPEC) soon shook the nation and the world from their complacency. The illusion of an abundant supply of cheap energy dissolved overnight. Global concern over the "energy crisis" brought national policy revisions, local community action, and individual efforts, all designed to conserve energy and discover reliable and workable ways to meet the threat to lifestyles in industrial societies. At the embargo's onset oil and gas accounted for three-quarters of the United States' energy supply. Although domestic production of the two fuels peaked during the embargo, severe fuel shortages, mile-long gas lines, and large price increases in consumer products were common. This situation was repeated during the Iranian revolution in 1978-1979.

The shortages of the 1970s forced the United States to think more realistically about its energy future. At the national level the energy crisis raised concerns over national security, trade deficits, and environmental degradation. Closer to home, oil shortages and rising energy costs resulted in worries about the family budget, transportation to and from work and school, and the heating and cooling of homes, schools, and offices. For the first time in peacetime, Americans were forced to curb their energy consumption drastically.

Events in the 1970s created, in effect, a **conservation revolution.** Conservation means reducing or eliminating wasteful use of a resource—in this case, energy—in order to protect, conserve, or extend supplies. Like the green revolution in agriculture, the success of the conservation revolution allowed the nation time to adjust to reduced energy consumption. Between 1972 and 1981 the industrial sector reduced consumption by 6 percent. The 1975 Energy Policy and Conservation Act established the corporate average fuel efficiency (CAFE) standards which mandated fuel efficiency levels for all new automobiles. These standards saved billions of gallons of gasoline. Energy-efficient technologies were developed and applied to home heating and lighting. People began to form car pools, turn out lights not in use, and lower thermostats. Conservation efforts saved approximately 6 million barrels of oil daily in the United States. Perhaps most importantly, the two decades since the onset of the conservation revolution have helped to dispel a widespread belief that economic growth and energy usage must increase hand-in-hand. Since 1973 the United States economy has grown 46 percent, but energy usage has risen just 8 percent, thanks to conservation and energy efficiency.

The oil shortages affected the national energy infrastructure in several ways. Conservation efforts reduced foreign imports. In 1977 imports reached 8.8 million barrels of oil each day; seven years later imports had dropped by almost half. Coal's share of the energy market increased, rebounding from a low of 16.9 percent in 1972 to 23.6 percent in 1983. This figure is expected to continue to rise, and many experts predict that coal will supply approximately 36 percent of our domestic energy needs by 1995.

Conservation efforts aimed at reducing dependence on foreign oil supplies also affected natural gas. From 1970 to 1985, industrial consumption of natural gas fell 27 percent. To reverse this decline, the federal government instituted partial deregulation of natural gas, resulting in an oversupply. Consumption slowly began to increase, and by 1988, it rose 11 percent. To encourage the industry's recovery, President Bush removed all remaining price controls on natural gas in 1989. Some economists predict that natural gas production and consumption could steadily increase over the next decade.

Since the last oil shortage in 1978–1979, supplies have been uninterrupted, even during the 1991 Persian Gulf War. While a steady supply of oil seems like good news, it has hurt conservation efforts.

▶ Table 12-2
Energy Overview for United States, 1960–1990

Activity and Energy Source	1960	1965	1970	1975	1980	1985	1990
Production	41.49	49.34	62.07	59.86	64.76	64.77	67.59
Coal	10.82	13.06	14.61	14.99	18.60	19.33	22.61
Natural gas[1]	12.66	15.78	21.67	19.64	19.91	16.91	18.05
Crude oil and lease condensate	14.93	16.52	20.40	17.73	18.25	18.99	15.46
Natural gas plant liquids	1.46	1.88	2.51	2.37	2.25	2.24	2.16
Nuclear electric power	0.01	0.04	0.24	1.90	2.74	4.15	6.19
Hydroelectric power	1.61	2.06	2.63	3.15	2.90	2.94	2.92
Other[2]	([3])	0.01	0.02	0.07	0.11	0.21	0.20
Imports	4.23	5.92	8.39	14.11	15.97	12.10	18.74
Natural gas	0.16	0.47	0.85	0.98	1.01	0.95	1.51
Crude oil[4]	2.20	2.65	2.81	8.72	11.19	6.81	12.67
Petroleum products	1.80	2.75	4.66	4.23	3.46	3.80	4.26
Other[5]	0.07	0.04	0.07	0.19	0.31	0.54	0.29
Exports	1.48	1.85	2.66	2.36	3.72	4.23	4.91
Coal	1.02	1.38	1.94	1.76	2.42	2.44	2.77
Crude oil and petroleum products	0.43	0.39	0.55	0.44	1.16	1.66	1.84
Other[6]	0.03	0.09	0.18	0.16	0.14	0.14	0.30
Adjustments[7]	-0.43	-0.72	-1.37	-1.07	-1.05	1.31	0.02
Consumption	43.80	52.68	66.43	70.55	75.96	73.95	81.44
Coal	9.84	11.58	12.26	12.66	15.42	17.48	19.05
Natural gas	12.39	15.77	21.79	19.95	20.39	17.83	19.41
Petroleum products	19.92	23.25	29.52	32.73	34.20	30.92	33.64
Nuclear power	0.01	0.04	0.24	1.90	2.74	4.15	6.19
Hydroelectric power	1.66	2.06	2.65	3.22	3.12	3.36	2.94
Other[8]	([4])	-0.01	-0.04	0.09	0.08	0.20	0.21

Note: Production, imports, exports, and consumption are shown for selected fuels. The table reveals some disturbing facts. In 1990, the last year for which data are given, coal, natural gas, petroleum, and nuclear resources—all nonrenewable—together accounted for 96 percent of all energy consumed in the United States. Moreover, imports meet about half of petroleum consumption.

[1]Dry natural gas.

[2]Includes electricity produced from geothermal, wood, waste, wind, photovoltaic, and solar thermal sources connected to electric utility distribution systems (see Note).

[3]Less than 0.005 quadrillion Btu.

[4]Includes imports of crude oil for the Strategic Petroleum Reserve, which began in 1977.

[5]Includes coal, coal coke, and hydroelectric power.

[6]Includes natural gas, coal coke, and hydroelectric power.

[7]A balancing item. Includes stock changes, losses, gains, miscellaneous blending components, and unaccounted for supply.

[8]Includes electricity produced from geothermal, wood, waste, wind, photovoltaic, and solar thermal sources connected to electric utility distribution systems (see Note) and net imports of coal coke.

Note: Data do not include the consumption of wood energy (other than that consumed by the electric utility industry) which amounted to an estimated 2.4 quadrillion Btu in 1987. This table also does not include small quantities of other energy forms for which consistent historical data are not available, such as geothermal, waste, wind, photovoltaic, or solar thermal energy sources except that consumed by electric utilities.

Note: Sum of components may not equal total due to independent rounding.

Sources: Energy Information Administration, *Annual Energy Review 1990.*

Since 1982 oil consumption has steadily inched upward. In July 1989 imports supplied more than half of the nation's petroleum for the first time since 1977 (Figure 12-10). The major reason for this turnaround is the price of oil. The price of oil rose from $2.10 per barrel in 1973 to $34 per barrel in 1982—a powerful incentive to conserve. By 1986, however, prices had declined to about $10 per barrel (Figure 12-11). An oil glut manufactured by the OPEC cartel was responsible for the lower prices.

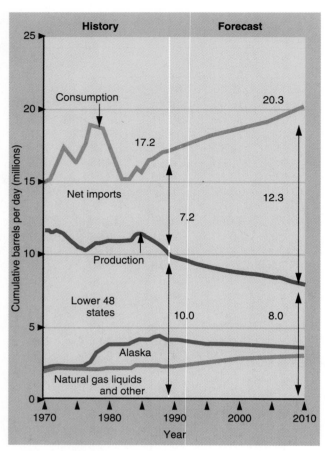

FIGURE 12-10: Petroleum supply, consumption, and import requirements to 2010. In the years between 1990 and 2010 consumption is expected to rise to over 20 million barrels per day. Imports will meet about 60 percent of demand, as domestic production falls from about 10 million barrels per day to 8 million barrels per day. Alaskan oil will account for ever smaller proportion of supplies.

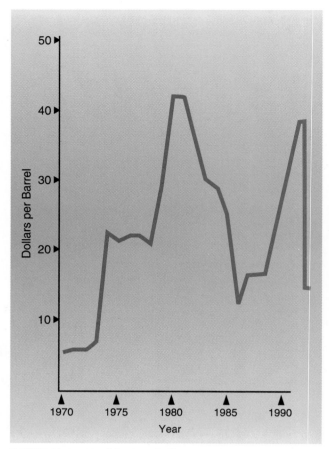

FIGURE 12-11: World price of oil, 1970 to 1991. Sharp increases in the price of oil in 1973 and 1979 corresponded to shortages caused by OPEC embargoes. The price of oil fell considerably in the mid-1980s as OPEC kept production high to discourage conservation and the development of alternative energy resources. It began to climb again toward the end of the decade and then rose sharply with the Iraqi invasion of Kuwait in 1990. By mid-1991, prices had again fallen.

Falling prices took the wind out of the sails of the conservation effort, lulling consumers into a false sense of security—and complacency—once again. Falling prices caused many people to abandon energy conservation measures. Gas-guzzling vehicles again became popular as energy consumption slowly crept back to the levels of the early 1970s.

Not surprisingly, the price of oil is again rising. By 1989 oil prices climbed to about $20 per barrel, then rose sharply after Iraq's invasion of Kuwait in August 1990. Although prices dropped somewhat after the conflict, there is a real danger that the United States will once again become vulnerable to foreign supplies and, correspondingly, that our economy will once again become vulnerable to maneuvers by OPEC. Many experts worry that if OPEC were to cut production sharply, enforce another embargo, or flood the market with oil, the result

could be the dismantling of the American oil industry and increased control over the U.S. economy by foreign sources. On the positive side, rising oil prices, coupled with increased concern about the environmental effects of fossil fuels, might spark a return of the conservation revolution of the 1970s and a renewed interest in energy efficiency.

How Can We Increase Energy Efficiency?

Researchers at the Rocky Mountain Institute, an energy think tank, estimate that the United States wastes about $300 billion a year because of drafty doors and windows, energy-guzzling cars and appliances, poorly insulated buildings, and other examples of energy inefficiency. Energy efficiency—obtaining greater productivity while using less energy—is our greatest untapped "source" of domestic oil (Figure 12-12). The energy efficiency campaigns

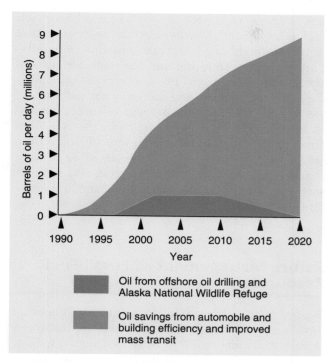

FIGURE 12-12: Estimated oil production from offshore drilling and the Alaska National Wildlife Refuge compared to estimated savings from efficiency.

ly. The following examples are just a few of the many possibilities that are being tried.

Cogeneration is the simultaneous production of two useful forms of energy, such as electricity and high-temperature heat, from the same process. For example, hot water or steam from an electric generator can be used to warm buildings, refine petroleum, or drive manufacturing processes.

The electricity-generating industry can realize significant energy gains by improving efficiency. Most coal-fired power plants send more than two-thirds of the energy straight up the smokestack along with significant amounts of pollutants. But with fluidized bed combustion, huge fans keep the powdered coal suspended in mid-air so that it burns cleaner and with less energy lost to the environment. Coal gasification systems turn coal into a gas (removing much of the sulfur in the process) and then use the gas to run two turbines, one powered by the hot combustion gases, the other fueled by steam. Add-on pollution control devices, or scrubbers, can remove up to 95 percent of the sulfur from flue gases. Scrubbers that remove nitrogen oxides have also been developed. A state-of-the-art 220-megawatt electric power plant, using the latest coal gasification technology, has been proposed for a site in central Florida. Funded in part through the Department of Energy's Clean Coal Program, it will be the nation's cleanest coal-powered plant when it comes on line in 1996. A variation developed in Sweden, known as the pressurized fluid bed, captures waste heat to achieve an energy efficiency of 85 percent with sulfur oxide and nitrogen oxide emissions that are just one-tenth and one-sixth, respectively, of those allowed in U.S. plants.

of the future will include reduced energy requirements of electric appliances, automobiles, and other technologies and more efficient and environmentally sound industrial processes. A simple axiom may well characterize the 1990s and beyond: *Conservation and efficiency mean building better things and doing things better, not doing without.* The following sections illustrate some of the creative, environmentally sound ways industries, communities, and individuals have found to make processes and products better and more efficient.

Industries. Japan, Italy, and Germany use only about half as much energy as the United States to manufacture an industrial product. They are able to do so because they have a newer industrial infrastructure (built after World War II), an economic climate more conducive to longer pay-back periods for efficiency investments, and a cultural attitude that values energy efficiency and its long-term benefits. Also, the governments of these nations are willing to subsidize industry's efforts to become more energy efficient. To remain competitive in the global market, U.S. industry is searching for ways to conserve energy resources and use them more efficient-

Communities. There are many diverse ways that communities can increase energy efficiency and enhance the quality of life for residents. Consider Osage, Iowa, a farm-based community of 3,800 people. Like many other agricultural communities, it has struggled to survive in recent years, hit hard by the recession of the 1970s, the farming crisis that began in the early 1980s, and the drought of 1988. But Osage is on sound economic footing primarily because of a unique economic development program based on energy efficiency. The program was spearheaded by Osage Municipal Utilities, a publicly owned company. Since 1974 the utilities company, under general manager Weston Birdsall, has led the state in establishing programs that help its electric and natural gas customers save energy, thereby reducing the amount of electricity and natural gas the town must buy from out-of-state suppliers.

One major program offers free infrared scans of citizens' houses. The scans spot energy leaks, allowing residents to insulate their homes more effectively. The improved insulation resulted in savings of up to 50 percent on many residents' heating bills. A second program requires that new homes meet minimum efficiency standards before they are hooked up to natural gas lines. This announcement spurred construction companies to improve insulation in attics and side walls and to install only energy-efficient windows. Another program distributes free water-heater jackets and compact fluorescent lightbulbs that can replace incandescent bulbs. Over the lifetime of these bulbs, they will save the energy equivalent of burning 120 tons of coal. Trees and a tree spade were donated to anyone who wished to plant shade trees around their homes. The trees are expected to reduce the need for air conditioning, lowering demand in the peak summer months. Many residents have also invested in new and more efficient home appliances, sometimes realizing savings of 40 to 50 percent on gas and electric bills.

Local businesses soon joined the utility company's approach to Osage economic development. Steele's Super Value, the local grocery store, put all its refrigerator compressors together in an insulated compartment and installed two fans and an insulated duct to carry the waste heat from the compressors into the main part of the store. The store saves $600 each month on heating, enough to keep prices low enough that Osage residents will not be tempted to take their business to a larger supermarket 15 miles away. The owners are also experimenting with new energy-efficient lighting that appears to use 22 percent less energy than the older system.

The energy efficiency programs have yielded significant benefits for the community. Residents are now far more knowledgeable about energy efficiency and conservation. The programs have enhanced community spirit, too. The local Jaycees, for instance, donated their time to help low-income residents install weather stripping and water-heater jackets. Efficiency measures have saved the community an estimated $1.2 million each year, money that would normally have left the area but is instead recirculating within the community. Osage has not had to issue any additional bonds, and the community has been able to retire all its old debts. The utility company has passed on the savings to consumers, cutting its rates five times over the last 10 years. Moreover, it has not had to add to existing generation capacity in 15 years.

Individuals. Can individual efforts really make a difference in the drive to conserve energy? Consider the following sobering facts. The amount of heat that leaks out of American windows and doors each year is equivalent to the amount of energy that flows through the Alaskan pipeline annually. Throwing away an aluminum beverage can instead of recycling it wastes as much energy as half-filling the can with gasoline and dumping the gasoline on the ground. And the same amount of energy is wasted by not recycling a daily edition of the *New York Times* or *Washington Post*. On an individual level the greatest energy savings can be realized by making changes in home use, transportation, and indirect energy consumption (see What You Can Do: Energy Efficiency).

Future Management of Fossil Fuel Resources

Energy resources are vital to a nation's economic development. Using resources wisely fosters a healthy economy *and* a healthy environment, both of which are critical for sustainable economic development. Fossil fuels are a valuable and integral part of the global economy, but they are finite, and their use has serious consequences for both human health and the environment. Managing these fuels in an environmentally sound manner requires a comprehensive management approach. The following suggestions should help to prevent overuse of fossil fuels through conservation, to protect environmental and human health from pollution and degradation, and to preserve living ecosystems.

The United States should abandon the unofficial "drain America first" policy, which has dominated government action toward domestic oil development. Instead, the nation needs a long-range comprehensive energy policy that is pursued at all levels of government from federal to local and that considers the environmental, economic, and social costs of various options. The cornerstones of a new energy policy should be conservation, improved energy efficiency, and the development of alternative and renewable energy resources. Government can lead by example by adopting energy conservation and efficiency practices.

Because the transportation sector accounts for over 60 percent of all oil consumed in the United States, conservation and efficiency measures in this area could result in significant savings. Raising the CAFE standards is the easiest and most significant measure this country can take in order to realize many substantial benefits: fuel conservation, decreased dependence on foreign oil supplies, improved air quality, and easing of global warming. Raising fuel efficiency by just 1 mile per gallon

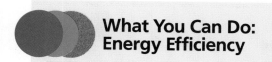

What You Can Do: Energy Efficiency

To Save Energy at Home

- The chief use of energy in the home is for space heating and cooling. Prevent or minimize heat loss by adding insulation, caulking, installing storm windows or covering windows with plastic. These simple measures can reduce annual energy consumption by 25 percent or more.

- If building a new home, superinsulate it by doubling the amount of insulation and using an airtight liner in walls. After superinsulating, some new homes in Minnesota required 68 percent less energy for heat.

- Use energy-efficient compact fluorescent lightbulbs in place of incandescent bulbs. A 15- to 18-watt compact fluorescent bulb (costing $15 to $28) replaces a 75-watt incandescent bulb. The compact fluorescent will save $40 in electrical costs and, because it lasts longer, $5 to $10 in replacement costs for incandescents. Because it requires less energy, over the course of its lifetime the compact fluorescent will prevent emissions of up to 2,000 pounds of carbon dioxide and 20 pounds of sulfur dioxide (from a coal-fired plant) or 25 milligrams of plutonium waste (from a nuclear-powered plant). It will also save the energy equivalent of about 500 pounds of coal or 1.3 barrels of oil.

- Use energy-efficient appliances rather than conventional models for tremendous energy savings. The following chart compares average conventional models, new energy-efficient models,

Appliance	Conventional	New Energy-Efficient	Best Available Technology	Potential Energy Savings[a]
Refrigerator[b]	1,500	1,100	750	87%
Central air conditioner[b]	3,600	2,900	1,800	75%
Electric water heater[b]	4,000	2,900	1,800	75%
*Electric range	800	750	700	50%
Gas furnace[c]	730	620	480	59%
Gas water heater[c]	270	250	200	63%
Gas range[c]	70	50	40	64%

[a]The potential savings listed is the percent reduction in energy consumption realized when the best available technology is used rather than today's conventional model.
[b]Kilowatt-hours per year.
[c]Therms per year.

and models that use the best available technology (BAT) for the greatest energy efficiency and the most significant energy savings.

To Save Energy in Transportation

- Walk or bicycle when traveling short distances. (Not only is it more energy-efficient to do so, it's also healthier!) For longer trips, especially those you make routinely (like trips to school or work), use public transit or organize car pools. Public transit and car pools also curb pollution and reduce congestion in crowded metropolitan areas. If those effects seem inconsequential, consider that Los Angeles freeways are so crowded that the average speed in 1988 was just 35 miles per hour. That figure is expected to drop to 19 miles per hour by 2010.

- Buy a fuel-efficient automobile. In 1991 the U.S. fleet averaged 27.5 miles per gallon. The Union of Concerned Scientists, in *Steering a New Course*, concluded that fuel economy could be raised to 40 miles per gallon without sacrificing safety or acceleration. Several automobile prototypes are obtaining close to 100 miles per gallon. Toyota's AUV, for example, averages 98 miles per gallon.

To Reduce Indirect Energy Consumption

- Reuse and recycle! Enormous amounts of energy are used to produce consumer goods. You can minimize energy waste by reusing or recycling goods such as paper products, aluminum and bimetal cans, glassware, and plastics. See Chapter 20 for ideas on what and how to recycle.

would cut carbon dioxide emissions by approximately 40 billion pounds, an amount roughly equivalent to closing six coal-fired power plants.

Other important measures that can be taken in the transportation sector include raising the gasoline tax to reflect the product's real worth and the environmental costs of its combustion. In many countries, including many European nations, Japan, Argentina, and the United Kingdom, the tax exceeds the price of the fuel itself. In the United States, however, the tax per gallon (about $0.34 to $0.35) is 26 to 30 percent of the price per gallon, depending on the state and the gasoline cost. Finally, a necessary long-range measure is to provide

Environmental Science in Action: Strip Mining in Appalachia

Marsha Shook

Appalachia is a region of great natural beauty, home to an interesting and unique culture. The area is rich in bituminous coal. It provides approximately 70 percent of the bituminous coal produced in the United States; 35 percent of that is obtained by strip mining. "King coal" and strip mining have significantly influenced both the region's natural heritage and its cultural fabric.

Describing the Resource

Physical Boundaries

North America's Appalachian Mountain system is 1,500 miles long, extending from southern Quebec, Canada, to northern Alabama. Its highest peak, at 6,711 feet, is Mount Mitchell, in North Carolina. Although the central and nonmountainous and plateau sections of the western coal-producing regions are generally flatter and easier to mine, Appalachia's coal remains competitive because much of it is high in Btu's, burns with low amounts of ash, and contains less than 2 percent sulfur. Some areas of Appalachia do produce bituminous coal which is relatively dirty and which has a high sulfur content. Ohio's coal is worst in this respect, often containing more than 3 percent sulfur. Mining companies may mix high-sulfur coal with a lower-sulfur variety or sell it to electrical plants equipped with scrubbers.

In Appalachia, where most areas have slopes of over 12 degrees, the chief method of surface mining is contour strip mining. It occurs primarily in southeastern Ohio, eastern Kentucky, West Virginia, and Alabama. Contour strip mining operations cut into the hillside, leaving a shelf or bench in the slope. These cuts follow the coal seam, producing a long, snakelike gouge that winds across the terrain. This type of mining creates a large pile of removed overburden, called a spoil bank, in front of the shelf and a high wall at the back.

Biological Boundaries

Appalachia lies within the temperate deciduous biome, characterized by hardwood trees such as oak, hickory, maple, and beech. The changing seasons enhance the area's natural beauty: forest-cloaked hills reveal the first blush of spring, the full bloom of summer, the rich tapestry of autumn, and the stark bareness of winter. The lush vegetation is made possible by abundant rainfall, evenly distributed throughout the year. The forests were once home to a variety of large mammals, including wolves, bears, and mountain lions, but today, most of these are gone, and the forest mammals are represented by deer, rabbits, squirrels, raccoons, opossums, and rodents. The largest predators (besides humans) are owls, hawks, a few black bears, foxes, occasional bobcats, and badgers.

Social Boundaries

Residents of Appalachia possess a unique culture, most well-known for its handcrafts, bluegrass and fiddle music, storytelling, strong family ties, and intimate knowledge of the environment. The region's social and economic structure and its environmental health have long been closely tied to the boom and bust cycle of the coal market.

Before World War II most coal in Appalachia was mined underground. At that time coal met over 50 percent of the country's energy needs. After the war the advanced earth-moving technology developed for military purposes was adapted to commercial enterprises, and strip mining became commonplace. Although it posed a much greater threat to the environment than underground mining, strip mining was relatively free from environmental regulations.

Strip mining allowed companies to extract coal more efficiently—80 to 90 percent of a coal seam rather than the 50 to 75 percent available in an underground mine. Seams that were too weak or thin to be mined underground became economically workable. Aboveground mining minimized threats to the health and safety of the miners, but the shift to machine-intensive strip mining left tens of thousands of them jobless. Between 1949 and 1985 domestic coal production doubled, but employment dropped from almost 400,000 miners to less than 200,000. An exodus from Appalachia took place, as many of the unemployed left the region and headed north to work in the factories of Detroit, Michigan, and Cincinnati and Hamilton, Ohio. Unemployment separated families and friends and left mountain people isolated in unfamiliar cities. The cultural and social fabric of Appalachian life began to tear.

Between World War II and the early 1970s coal's share of the domestic energy market dropped as the use of oil and natural gas increased. The 1973 OPEC oil embargo revived the coal industry, at least temporarily. By the end of the decade coal from some mines in Appalachia was selling for as much as $100 per ton. As a result of the increased demand, mining in the western region expanded considerably, and Wyoming soon overtook Kentucky and West Virginia as the nation's leading coal-producing state. Even so, by 1975, Appalachia provided about half of the coal that was obtained by strip mining in the United States.

The boom market of the 1970s collapsed in the early 1980s. The high prices of the 1970s had spurred foreign buyers such as Japan and Taiwan to turn to other coal-exporting nations, particularly Australia and Canada. U.S. coal exports, which had reached an all-time high in 1981, dropped 30 percent by 1984. Environmental concerns caused increasing numbers of utility companies to switch to cleaner-burning oil and natural gas as prices for these commodities dropped. Perhaps the most devastating blow for Appalachia was the decline of the nation's steel industry. Steel is produced using what is known as "met coal" (metallurgical coal), most of which is obtained from Appalachian coal fields. In 1974 consumption of met coal was 90 million tons (81.9 million metric tons) per year; by 1985 it fell to 40 million tons (36.4 million metric tons). By 1995 consumption is expected to be at just 32 million tons (29.1 million metric tons) per year.

Coal mining's legacy, which began generations ago but was forged during the period since World War II, can be seen in Appalachia today. Thousands of acres of land strip mined before the 1970s remain ruined, with once productive, balanced environments reduced to eroded, scarred wastelands. Among all states, West Virginia ranks 50th on the sulfur dioxide indicator, generating twice as much of the chief precursor of acid precipitation as does the next worst state, Indiana. Not surprisingly, rain pH levels in West Virginia are below 4.5. A fifth of the lakes and streams in Appalachia, as far south as Tennessee and North Carolina, have acidic pH levels. Kentucky and West Virginia also rank among the worst ten states for carbon emissions, occupational deaths, and high-risk jobs. Ohio and Pennsylvania also rank poorly in terms of sulfur and carbon emissions. West Virginia ranks last among all states in terms of surface water quality, 48th in overall water pollution, 43d in spending for water quality, and 46th in investment needed for adequate sewage facilities. Kentucky, West Virginia, and Alabama rank 46th, 47th, and 48th, respectively, in terms of community health.

What do all these numbers mean for the people of Appalachia? While strong unions have enabled working miners to secure relatively good wages and health benefits from employers, these benefits are little comfort in a region plagued by poor schools, severe pollution, weak enforcement of safety regulations, poverty, high unemployment, and a stagnant economy. Many families in West Virginia have no running water or septic systems. They still haul water from nearby creeks, which may be contaminated with mining, industrial, or household waste or with acid rain. A second exodus is taking place in Appalachia, but many of those now leaving the region are heading south for the manufacturing plants of North Carolina and Georgia.

Looking Back
Exploiting the Land

Contour mining made it easy for water to erode the earth in and around the pits. Once removed from a cut, overburden was placed in increasingly larger spoil banks which overflowed down the hillside. Spoil banks encouraged erosion and landslides that carried acid, silt, and metals into nearby surface waters. Contaminated water also eventually entered groundwater supplies. As sedimentation, acid drainage, and other pollutants reached streams and lakes, they caused widespread damage: eroding roads, clogging culverts, and preventing dams from controlling floods and storing water. Toxic substances contaminated wells, springs, and other sources of drinking water.

Contour mining also seriously damaged the ability of the land to support life. Even in rare cases where mining operations deposited spoil piles in previously mined cuts, the topsoil was usually completely destroyed. Because it was the first layer of overburden removed, the good topsoil was invariably ruined by the acid-contaminated earth piled on top of it. As a result, strip-mined land could almost never regenerate plant life.

The land disruption and water pollution resulting from contour mining adversely affected wildlife. Although most wildlife fled an area being mined, many burrowing animals were killed once mining operations began. Because the mining process destroyed their habitats, animals could almost never repopulate a strip-mined area. Strip-mined areas could also disorient migratory animals from their traditional travel patterns, and extensive mining operations substantially reduced feeding areas. Aquatic organisms were perhaps most affected by mining operations. High acid and silt levels turned their habitats into death traps.

Unregulated mining victimized not only the land, but the people as well. Mining brought noisy machinery and constant traffic. Sometimes whole communities had to move because of poisoned water sources and ruined farmland. Even in less extreme cases, mining companies left communities surrounded by a scarred wasteland. Abandoned mining areas often left behind dangerous piles of coal refuse which sometimes caught on fire, emitting toxic fumes into the atmosphere. Many states allowed "broad form" deeds, which gave the deed owner mineral rights to land owned by someone else. Most landowners agreed to such deeds before the advent of strip mining, thinking an underground mine would not disrupt their property. However, many of these deeds fell into the hands of strip-mining companies which simply unearthed the coal and left the owners with property unfit for any productive use. Some states, such as Kentucky in 1987, passed laws prohibiting the broad form deeds. But for residents who suffered such indignities as watching coal shovels unearth family graveyards to get at the coal beneath them, the legislation came too late.

Protecting the Land

The 1960s and early 1970s witnessed increasingly strident demands for federal protection against the exploitation of the land by mining companies. The Surface Mining Control and Reclamation Act (SMCRA), signed by President Carter in 1977, was a significant step forward for environmentally sound management.

Under the act each state established its own regulatory program. Regulations varied from state to state, but all programs protected certain areas, such as wilderness preserves, from strip mining and required companies to obtain permits to strip mine any area. Companies had to show what they would do to prevent damage to water sources and demonstrate that they had feasible plans for restoring the land to its premining level of productivity. The company had to post a bond for each site equal to the estimated cost of reclamation.

To comply with the new regulations, strip-mining companies quickly came up with simple adaptations that made contour mining much less devastating to the environment. Diversion ditches channeled water around the mining area. The slope reduction method spread the overburden over a larger area than before, reducing the slope and size of the spoil bank to prevent landslides and excessive erosion. The haul back technique deposited overburden into previously mined cuts, reducing spoil erosion by eliminating spoil banks and stabilizing the high wall.

Most states required that the land be restored to its premining productivity level, including a self-reproducing plant community. Many states, including Ohio, require back-to-contour reclamation. To be able to reclaim an area, companies must store the topsoil separately from the rest of the spoil; they can then deposit the uncontaminated topsoil over the land after it has been graded.

The final, and perhaps most difficult, step in reclamation is revegetation. Plants must be carefully chosen based on an area's soil type, chances of erosion, and proposed postmining use. Then vegetation must be carefully monitored to make sure a stable community has been established. Reclaimed land can be used for almost anything; currently, reclaimed strip mines have been used for agriculture, forestry, recreation, and wildlife habitats, as well as residential, commercial, and industrial buildings.

FIGURE 12-13: A reclaimed strip mine near Kanawah, West Virginia.

The Appalachian region is the national leader in number and diversity of postmining land uses (Figure 12-13).

The Surface Mining Control and Reclamation Act also established the Abandoned Mine Reclamation Fund, designed to encourage reclamation of strip mines abandoned before SMCRA went into effect. To obtain money for the fund, the government taxes both underground and surface coal production. Both state and federal governments award contracts to currently operating mining companies to reclaim old mine sites. For instance, Waterloo Coal, which operates in Jackson County, Ohio, has contracted to reclaim an abandoned area near its current site in Oak Hill, Ohio. A small, privately owned company, Waterloo is indicative of a growing trend toward socially and environmentally responsible mining. Waterloo maintains a good relationship with the surrounding communities. Many community members find work with Waterloo, from life-long employees who run mining equipment, to high school and college students who spend their summers

spreading topsoil, seeding, and helping with reclamation.

Looking Ahead

Despite progress made possible by the 1977 Surface Mining Control and Reclamation Act, current methods of pollution control and reclamation are still imperfect, and vast areas of Appalachia still remain unreclaimed. Reclamation is especially difficult in the more mountainous areas of Appalachia because the steep slopes encourage erosion and toxic runoff. The soil suffers from severe nutrient deficiencies and high toxic concentrations, and it is expensive to condition the soil to a point where it can sustain vegetation. The continuous erosion also prevents plant growth that would help retain soil.

Even though most strip mining operations store topsoil separately, simply digging it up breaks down the natural organic structure of the soil, destroying microorganisms. Once replaced, the topsoil is not as solidly packed as it used to be, and erosion occurs more easily. In addition, the soil below the topsoil is almost always very acidic; companies must frequently apply lime, fertilizer, and other soil amendments so plants can grow. Unfortunately, if they do not own the land, most companies abandon such efforts when their period of legal responsibility ends. Left on their own, some revegetated plant communities die out.

Some companies do not provide strip-mined areas with the biological diversity they had before mining, finding it easiest to revegetate with only one species of grass, legume, or tree. Most areas need more variety to become self-sustaining. Research will help find answers to problems with topsoil and vegetation. Some scientists are experimenting with different plant species to find types that are especially good at preventing erosion and growing in acidic soil, and others are testing compost and sewage sludge to determine if they can be used to replenish nutrients on reclaimed land.

But research is only part of the answer. The view of the land and its people as something to exploit for monetary gain is alive and well; many people still oppose SMCRA and other efforts to make strip mining safer for the environment. Powerful lobbyists for coal companies and other interest groups claim that some regulations are too harsh and unnecessary.

On the other hand, many major environmental groups, such as the National Wildlife Federation, insist that the government is not adequately enforcing SMCRA. In fact, as of 1987, approximately 4,000 strip-mining sites were operating without taking the required precautions, 2,400 operations had ignored warnings to halt illegal operations, and the Office of Surface Mining had failed to collect between $150 and $180 million in fines from violators. Critics claim that state enforcement is also ineffective because state governments are hesitant to offend the coal companies which provide them with so much tax revenue. In some cases, mining operations avoid reclaiming areas by forfeiting bonds which are substantially less than the cost of cleaning up.

One topic of current debate is mountain top removal, an increasingly popular alternative to strip mining in Appalachia. Mining companies find that the flat land that mountain top removal produces is more valuable for farmland and other developments than the original mountain terrain. Critics object to more than the esthetic presumption of turning a range of mountains into flat-topped mesas. Engineers worry that hollow fills—the valleys and hollows filled in with material from mountain tops—may not be stable enough to withstand the heavy rains common to Appalachia. If a hollow fill began to erode, it could result in a landslide of monumental proportions.

Environmental groups are trying to push the government to discover and punish violators of SMCRA as well as to raise current standards for environmentally sound strip mining. However, concerned citizens need to be on the alert for potential changes in regulations and enforcements that could damage the environment. For instance, in the late 1980s President Reagan proposed opening Appalachian national parks to strip mining. Only continual pressure on our federal and state representatives will reduce the environmental threat of strip mining to an acceptable level.

incentives to encourage mass transit in urban areas and the development of a national rail system similar to those in use in Europe.

In the commercial and residential sector, a number of conservation and efficiency measures could be instituted. Chief among these is revision of building codes to mandate efficiency standards for heating (insulation) and lighting in new buildings. Although superinsulated buildings cost 5 to 10 percent more to build than conventional buildings, they are less expensive in the long run because, with reduced energy consumption, utility bills for heating and cooling are lower. A nationwide public service campaign should be launched to alert consumers to the economic savings that can be realized through home insulation, the use of compact fluorescent bulbs, and the use of energy-efficient appliances.

The industrial sector provides the opportunity to realize substantial energy savings. Using a mix of tax incentives and disincentives (such as a carbon tax on the amount of carbon dioxide emitted annually), government could encourage industries and utilities to increase the efficiency of their processes and seek alternatives to coal and oil. Improving the efficiency of electric motors is one important option available in such diverse industries as paper, chemicals, and cement making. In the United States, electric motors consume a whopping 80 percent of all electricity used for industrial purposes. There is an added benefit to decreasing industrial energy consumption: because it reduces the cost to the producer, it can make U.S. products more competitive both at home and abroad.

Government could also revise its policy toward utility companies. Government has protected the utilities' right to earn the largest possible profit. Utilities (unless they are publicly owned) make money by selling electricity, and it is in their best interest to keep consumption high. A new regulatory policy should be adopted that requires utility companies to promote conservation and energy efficiency through a mix of programs.

It will be necessary to anticipate and seek ways to mitigate social and economic changes that will inevitably arise from efforts to reduce dependence on fossil fuels. For example, tax rebates and other economic incentives could encourage new businesses to locate in regions currently dependent on coal or oil production. Retraining and educational programs will be necessary for workers who lose their jobs as a result of changes in the nation's energy infrastructure.

Summary

Coal, petroleum, and natural gas, the world's major conventional fuels, are the fossilized remains of organic matter, as are oil shales and tar sands, two comparatively minor energy sources. All fossil fuels are composed chiefly of carbon and hydrogen. The processes that formed fossil fuels continue today, but because those processes are extremely slow, fossil fuels are classified as nonrenewable and finite.

Fossil fuels are not distributed evenly beneath the earth's surface. Proven or economic reserves are deposits that have been located, measured, and inventoried. Fuel in proven reserves can be or is currently being extracted. Subeconomic reserves have been discovered but cannot yet be extracted at a profit at current prices or with current technologies. Indicated or inferred reserves are thought to exist and likely to be discovered and be available for use in the future.

Coal is a solid composed primarily of carbon, with small amounts of hydrogen, nitrogen, sulfur compounds, and other trace elements or compounds. The four major types of coal are anthracite, bituminous, subbituminous, and lignite. Coal is the most abundant fossil fuel on earth. Consumption declined over the last century, but it remains an important energy source. In fact, coal is enjoying a resurgence in use, thanks to the relative scarcity of oil and natural gas.

Of all energy resources in use today, coal has the most serious environmental and health impacts. Underground or subsurface mining produces acid drainage, which can contaminate groundwater, streams, and rivers. Land subsidence makes buildings situated in mining areas prone to structural damage. Strip mining can cause erosion and sediment runoff. Coal combustion releases sulfur, which has been implicated in acid precipitation, and carbon dioxide, the primary greenhouse gas responsible for global warming.

Petroleum, also called crude oil, is a dark greenish-brown or yellowish-brown liquid composed largely of hydrocarbon compounds. When refined, petroleum yields propane, gasoline, jet fuel, heating oil, motor oil, kerosene, and road tar. Even though global reserves seem substantial, they cannot last long given the world's appetite for oil. However, no one can predict with absolute certainty when our oil supplies will become exhausted.

Oil production and use adversely affect the environment in numerous ways. Perhaps the most disturbing aspect is the risk of oil spills. Oil development and production can adversely affect environmentally sensitive areas. The use of petroleum products contributes to environmental degradation. Petroleum, like other fossil fuels, releases carbon dioxide, a greenhouse gas, and pollutants such as sulfur oxides, nitrogen oxides, and hydrocarbons.

Natural gas is composed mostly of methane. Worldwide, natural gas usually occurs with petroleum and is known as associated gas. The United States gets most of its natural gas from deposits of nonassociated gas. Natural gas has taken on new importance because of the growing awareness of finite oil supplies. Natural gas systems have high conversion efficiency and produce no ash or sludge by-products. Natural gas deposits are found on every continent, but many of them occur in unconventional resources such as sands, shales, coal seams, and geopressurized zones. Currently, we lack the technology to use the energy stored in these resources economically, so we must rely on collections of conventional associated or nonassociated forms of gas.

Oil shales are fine-grained, compacted sedimentary rocks that contain varying amounts of kerogen. Tar sands are sandstones that contain bitumen, a thick, high-sulfur, tarlike liquid, within their porous structures. Significant inputs of energy are necessary to extract the oil and tar from shales and sandstones and to refine it. The extraction, processing, and refinement of the shale oil and tar also disturbs large land areas, produces enormous amounts of waste, and requires large amounts of water.

Humans have used fossil fuels for thousands of years. In the United States wood was the nation's primary energy source historically. By 1885, wood gave way to coal. In the twentieth century the popularity of oil began to rise. With increasing oil consumption, the use of natural gas also increased. Continuing dependence on oil made the nation increasingly vulnerable to foreign suppliers. A 1973 embargo by the Organization of Petroleum Exporting Countries (OPEC) caused global concern about energy resources. National policy revisions, local community action, and individual efforts were implemented to conserve energy. Events in the 1970s created a conservation revolution. However, since 1982 oil consumption has steadily inched upward.

Key Terms

acid drainage	land subsidence
anthracite	lignite
associated gas	natural gas
bituminous	nonassociated gas
coal	oil shales
cogeneration	petroleum
conservation revolution	proven reserves
economic reserves	subbituminous
indicated reserves	subeconomic reserves
inferred reserves	tar sands

Discussion Questions

1. Imagine that you have been asked to help plan a power plant for your community. What fuel would you recommend? Why?

2. What are some of the immediate and potential long-term effects of the *Exxon Valdez* oil spill? Do you think the oil spill could affect you personally? Explain.

3. List the pros and cons of (a) increasing the nation's reliance on coal; (b) developing domestic oil supplies; and (c) switching to natural gas for as many applications as possible. In your opinion, should the United States expand its use of coal, oil, or natural gas (or any combination of the three)? Why or why not?

4. Develop an energy conservation program for your community. What major components would you include? What incentives would you provide to residents to encourage them to participate?

5. Imagine that a mining company has asked permission to strip mine coal near your community. Would you allow it? If so, what conditions, if any, would you place on the company to minimize environmental damage?

6. What economic and social effects do you think another oil embargo would have on the United States?

Chapter 13

Energy: Alternative Sources

The expanded use of renewables and a greater commitment to energy efficiency are the most environmentally sound and cost-effective approaches to mitigating many seemingly intractable problems.

Worldwatch Institute

Learning Objectives

When you finish reading this chapter, you should be able to:

1. Identify eight alternative energy resources, summarize their current use and future use, and explain the advantages and disadvantages of each.
2. Briefly summarize the role that alternative energy resources have played historically.
3. Describe how alternative energy resources might be developed to complement and eventually supplant fossil fuels.

In the ongoing effort to create a sustainable society, many renewable and perpetual energy resources and several nonrenewable resources represent alternatives to fossil fuels. They include nuclear energy, solar power, wind power, hydroelectric power, geothermal energy, ocean energy, biomass, and solid wastes.

Alternative fuels offer many advantages. They can supplement and eventually replace fuels that are dwindling or that will be too costly in the future. They can broaden our fuel base, allowing us to avoid dependence on any one fuel or on foreign suppliers. And, while alternatives do affect the environment to some degree, their impact is usually much less than that of fossil fuels.

There are limitations associated with alternative energy resources. Most are site-specific, that is,

unlike coal and oil, they cannot be transported from one area or nation to another. Solar, wind, hydro, and geothermal power are not constant, and we must be able to adapt or compensate for their variability if we are to make good use of them. Although perpetual and renewable resources are essentially free, the technologies needed to harness them are not. The costs associated with alternatives should drop, however, as technologies are improved and as fossil fuels become increasingly scarce and expensive.

Describing Alternative Energy Sources: Physical Boundaries

Alternative resources are a diverse lot. They range from nonrenewable geothermal energy to perpetual wind energy. The energy of the sun has been used by humans for millennia; the energy of the atom has only recently been harnessed. Humans have understood and exploited hydropower extensively; researchers have only begun to explore the many potential applications of energy generated by the world's oceans. Nuclear, solar, wind, hydro, and geothermal are physical forces, biomass is derived from once-living things, and solid waste is a by-product of human activity.

Nuclear Energy

Nuclear energy is the energy contained within the nucleus of the atom. The constituents of the nucleus—protons and neutrons—are held together by an attractive force called the **binding force.** The strength of the binding force varies from atom to atom. If the force is weak enough, the atom will split, or undergo **fission,** when bombarded by a free neutron. As the atom splits, it produces smaller atoms, more free neutrons, and heat (Figure 13-1). The release of particles initiates a chain reaction as

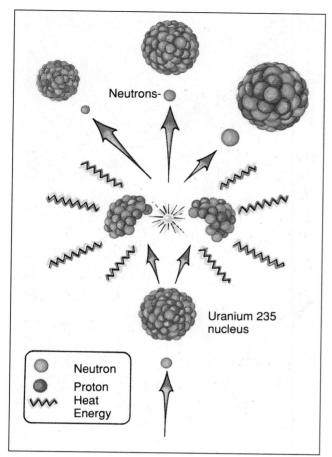

Neutrons-

Uranium 235
nucleus

Neutron

Proton

Heat
Energy

FIGURE 13-1: Fission reaction.

the free neutrons strike other nearby atoms, which also split.

Nuclear reactors are designed to sustain the fissioning process. Uranium 235 (U-235), a readily fissionable isotope of the element uranium, is the fuel that powers a nuclear reactor. An **isotope** is one of several forms of an element that have the same number of protons but different numbers of neutrons. U-235 is a relatively rare isotope of uranium, making up only less than 1 percent of natural uranium. U-238, an isotope that does not fission easily, is the common form (over 99 percent).

Before uranium can be used in a nuclear reactor, the impurities must be removed and the concentration of U-235 raised to increase the incidence of fissioning. The ore is milled to a granular form, and the uranium is dissolved out by use of an acid. The uranium is converted to a chemical form that exists as a gas at temperatures slightly above room temperature. The gas is purified. The concentration of U-235 is enriched to approximately 3 percent. The enriched gaseous uranium is then converted into powdered uranium dioxide. A highly stable ceramic material, uranium dioxide powder is subjected to high pressures to form fuel pellets, each about 5/8 inch long and 3/8 inch in diameter. About 5 pounds

of pellets are placed into a 12-foot long metal tube made of zirconium alloy, called a fuel rod, that is highly resistant to heat, radiation, and corrosion.

Fuel rods are grouped together to form a fuel assembly about 14 feet high and weighing about 2,000 pounds. Graphite or water is used around the fuel rods to slow the speed of the neutrons and increase the likelihood of fission. The fuel assembly also contains 16 vacant holes for the insertion of metal rods that absorb neutrons. By raising or lowering these control rods, operators control the rate of fission inside the reactor core. One hole within the assembly is left vacant for a probe that monitors temperature and neutron levels. Hundreds of fuel assemblies comprise the **reactor core** of a nuclear power plant, which is housed within a thick-walled **containment vessel**.

To get some idea of the tremendous energy of the atom, consider that the fissioning of one uranium atom releases about 50 million times more energy than the combustion of a single carbon atom. A small reactor pellet composed of trillions of atoms can release an enormous amount of heat energy, which in turn generates a substantial amount of electricity. Three reactor pellets generate approximately the same amount of electricity as 3.5 tons of coal or 12 barrels of oil, both carbon-based fuels.

Like generating plants that operate on fossil fuels, a nuclear reactor is designed to heat water to produce steam to run a turbine that generates electricity (Figure 13-2). The heat that boils the water is produced by the fission of the U-235 atoms within the fuel pellets. The heat energy released by the fission reaction is absorbed by a circulating coolant such as water, gas, or liquid metal (Figure 13-3). Water can be heated directly to steam, which can then be used to drive the generators that produce electricity. Coolants can also be kept under pressure and passed through heat exchangers to heat another body of water to its boiling point, creating steam to drive the turbines.

A number of different types of reactors are in use today. In a light-water reactor (LWR), the reactor core is immersed in water, which acts both as a moderator, to slow the rate of fission, and as a medium, to transfer the heat from the reaction to the turbines. Light-water reactors are the most common nuclear reactors in this country and worldwide, but several other types are in use. The former Soviet Union concentrated on reactors with graphite moderators, such as the one used at Chernobyl. The United Kingdom and France primarily have gas-cooled reactors, which use helium to transfer heat from the core to the turbine. Canada uses heavy-water reactors, which operate on nonenriched uranium. Using nonenriched uranium is much less expensive, but the incidence of fissioning is

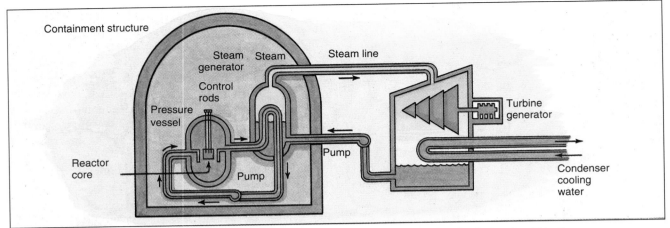

FIGURE 13-2: Essential components of a pressurized water reactor. Energy produced in the fuel rods within the reactor core heats water to steam, which then powers a turbine.

decreased. To compensate, deuterium oxide, or heavy water, which does not readily absorb neutrons, is used as a coolant.

Present and Future Use. In the early 1970s, before the OPEC oil embargo, nuclear energy produced about 5 percent of electricity in the United States and oil produced about 17 percent. Today, nuclear power accounts for about 20 percent of electric production, and oil for just 6 percent. In fact, nuclear power is our second largest source of electricity, after coal. The nuclear industry points out that, because of the nation's interconnected power system, virtually every American gets some electricity from nuclear energy. Most nuclear plants are in the east, with Vermont, South Carolina, and Maine relying most heavily on nuclear energy (Figure 13-4).

Just as the United States' dependence on nuclear power has grown, so has the world's. By 1990 over 400 nuclear power plants in 27 countries accounted for 20 percent of the world's electric production.

FIGURE 13-3: Cooling tower at Davis-Besse Nuclear Power Station, near Port Clinton, Ohio.

And while the United States has more plants on line than any other nation, over a dozen countries rely more heavily on nuclear energy for electric generation, led by France at 70 percent and Belgium at 66 percent (Figure 13-5).

The growth in the U.S. nuclear industry during the 1970s and early 1980s has tapered off in recent years. A number of factors are responsible for the industry's slowed growth: lower-than-predicted demand for electricity, increased costs due to longer lead times for licensing and construction of plants and higher financing expenses, rising interest rates and uncertain economic conditions, and increased concerns about safety and the disposal of nuclear wastes. Widely publicized accidents at Pennsylvania's Three Mile Island facility and the Soviet Union's Chernobyl plant have spurred opposition to nuclear power (Figure 13-6). Interestingly, in the debate over whether to increase or decrease reliance on nuclear energy, both opponents and proponents cite environmental concerns in their arguments.

Advantages and Disadvantages. Unlike fossil-fuel plants, nuclear plants do not emit carbon dioxide or sulfur and nitrogen oxides. Industry spokespersons are quick to point out that, each year, America's nuclear energy plants reduce emissions of greenhouse gases by 128 million tons, emissions of sulfur dioxide by 5 million tons, and emissions of nitrogen oxides by 2 million tons. During the 1980s, as France tripled its nuclear energy production in response to concerns over dependence on foreign oil, total pollution from the nation's electric power system fell by 80 to 90 percent. Proponents of nuclear energy maintain that the United States could realize the same benefits if it increased its reliance on this energy source. In answer to concerns over safety, proponents argue that modern nuclear

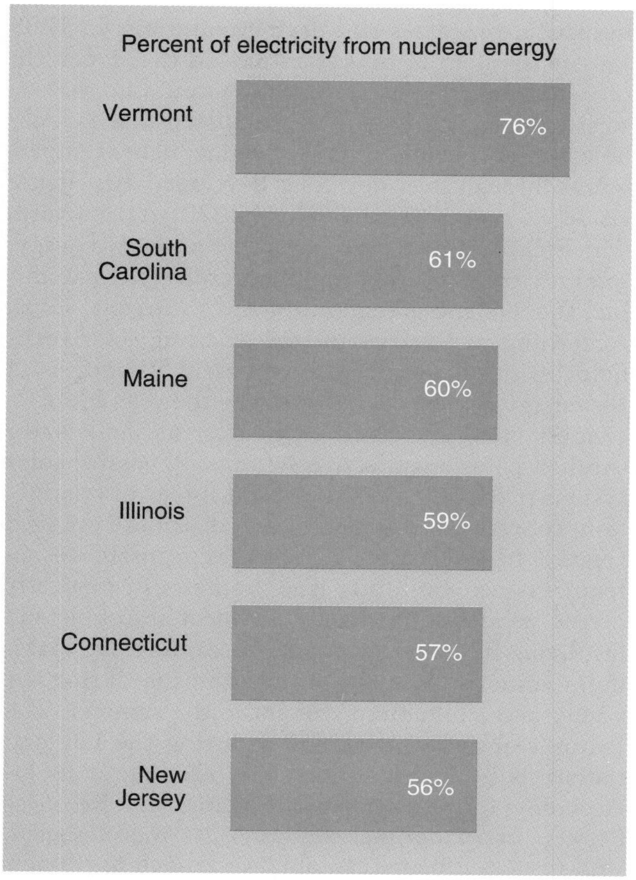

Percent of electricity from nuclear energy

State	Percent
Vermont	76%
South Carolina	61%
Maine	60%
Illinois	59%
Connecticut	57%
New Jersey	56%

FIGURE 13-4: States with the highest reliance on nuclear energy for electric production, late 1980s.

Percent of electricity from nuclear energy

Country	Percent
France	70%
Belgium	66%
Hungary	49%
Sweden	47%
Republic of Korea	47%
Taiwan	41%
Switzerland	37%
Finland	36%
Spain	36%
Bulgaria	36%
Federal Republic of Germany	34%
Japan	28%
Czechoslovakia	27%
United States	20%

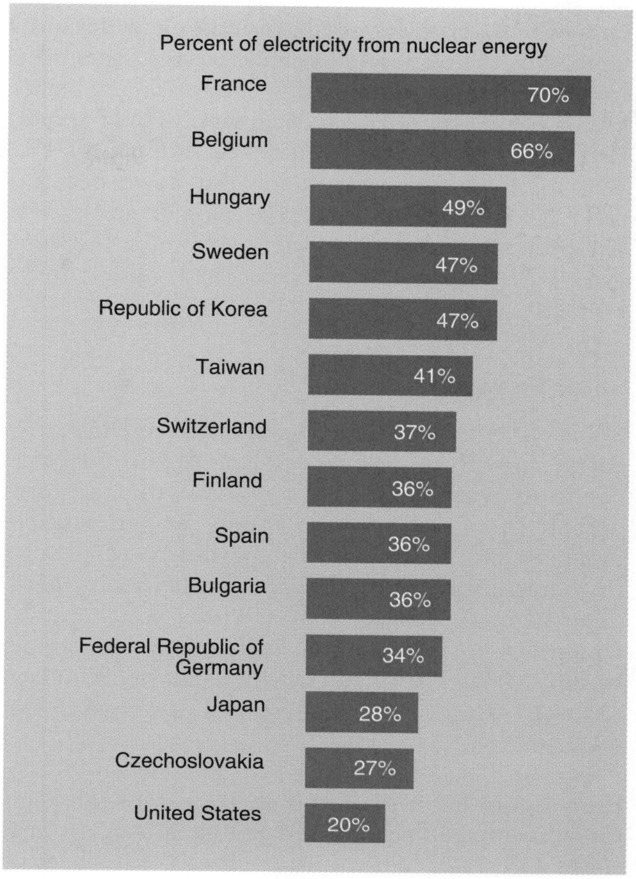

FIGURE 13-5: Countries with the highest reliance on nuclear energy for electric production, late 1980s.

power plants are inherently safe. The concentration of fissionable fuel (3 percent) is far less than that needed for a nuclear explosion (nuclear weapons have a 97 percent concentration of U-235). Improved reactor design and overlapping safety systems decrease the risk of accidents. Supporters also insist that increased reliance on nuclear energy means decreased dependence on foreign suppliers of oil and gas, and thus a stronger, healthier economy.

Those who oppose nuclear energy's further development argue that it is expensive, dangerous, and environmentally unsound. The fact that no new plants have been ordered since 1978 is largely a product of economics: in the United States, recently built nuclear plants generate electricity at a total cost of more than 13 cents per kilowatt-hour, twice the prevailing rate. Concern over safety is also a major issue. In 1987 the nation's power plants reported almost 3,000 mishaps, over 400 emergency shutdowns, and 104,000 incidents in which workers were exposed to measurable doses of radiation. No matter how safely a reactor is designed, opponents claim, human error cannot be eliminated. Because the consequences of a serious accident could be so grave, nuclear energy is a risk that we should not take.

A serious environmental concern is the large amount of cooling water nuclear plants require. Warmed water released back into its source can adversely affect the ecosystem. When cooling water is pumped from lakes, rivers, and oceans, aquatic organisms may be sucked in along with the water

FIGURE 13-6: Chernobyl nuclear reactor facility, reactor 4, shortly after it was sealed off.

and killed. An even more serious environmental concern is the disposal of nuclear wastes—spent fuel from functioning plants and decommissioned plants, which have a life of 30 to 40 years. (These important issues are discussed in detail in Chapter 18.) The high radiation levels in shut-down reactors make the retirement of plants a complex and expensive process. According to the Worldwatch Institute, estimates of its cost range from $50 million to $3 billion.

Solar Energy

Solar energy, the energy radiated from the sun to earth, provides the energy that all life on earth needs to survive, the energy base to transform organic material to fossil fuels and renewable biomass, and the energy that powers abiotic cycles in the atmosphere, hydrosphere, and lithosphere. Solar-generated heat creates winds which power windmills. Solar energy moves the hydrologic cycle, feeding streams and rivers, which can be harnessed to drive water mills and the turbines of hydroelectric plants.

Present and Future Use. Because much of the energy consumed in the United States is used to heat space and water and to dry materials, with heating of residential and commercial buildings alone accounting for one-third of all energy consumption, solar power holds enormous potential. According to studies by the U.S. Department of Energy (DOE), advanced solar technologies could meet up to 80 percent of a building's energy requirements.

Industry consumes 40 percent of the nation's energy for such applications as heating water for industrial processes and drying materials. In 1985 the Department of Energy estimated that 1 percent of industrial energy consumption could be met by solar energy. The DOE also speculated that as technology for achieving and maintaining higher temperatures through solar devices is developed, that figure could increase 20 percent by 2020. Agriculture, which consumes 2 percent of the nation's energy, requires energy to dry and cure crops, to heat animal shelters and greenhouses, and to irrigate crops. According to DOE estimates, existing solar technologies could supply approximately 50 percent of the energy needed for irrigation by the year 2000.

Solar energy is used to heat space and water through passive and active systems. A **passive solar system** relies only on the natural forces of conduction, convection, and radiation to distribute the heat (Figure 13-7). Passive systems incorporate design features into buildings and homes—strategically placed windows, overhangs, and heat-absorbing and insulating building materials—to capture the maximum amount of radiation from the sun during the winter and a minimum amount in the summer. The features absorb the radiation during the day and radiate it back throughout the building at night. According to the National Information and Resource Service, under optimal conditions, "a well-designed, well-insulated home can obtain as much as 90 percent of its heating and cooling needs through the use of passive solar design."

An **active solar system** uses fans or pumps driven by electricity to enhance the collection and distribution of the sun's heat. Air, water, or other types of fluids are pumped through solar collectors where they absorb energy from the sun. The heated substance can be used immediately or transferred to

FIGURE 13-7: Various passive solar designs. (a) Direct gain: The most widely used form of solar heating is sunlight entering a building directly through windows. Large amounts of solar heat can be absorbed and stored by heavy materials in the building for later use. (b) Thermal storage wall: Made of heavy materials such as concrete, stone, brick, adobe, earth, or containers of water, the thermal storage wall is placed behind windows. Radiant energy passing through the windows is stored in the wall. At night the moderately warm wall radiates warmth to the building. Vents at the top and bottom of the wall allow some of the heat entering the building to circulate directly by natural convection. (c) Thermal storage roof: Most thermal storage roofs consist of waterbed-like containers that cover much or all of the ceiling. The water absorbs the radiant energy of the sun and then conducts the heat energy throughout the ceiling, warming the building by radiation. Movable insulation serves a dual purpose: to cover the warm water in order to reduce heat loss when the sun is not shining and to prevent overheating. In mild, sunny climates like the American Southwest a thermal storage roof can provide 100 percent of a building's heating and cooling requirements. (d) Sunspace: A sunspace, such as a greenhouse, porch, or atrium, can provide considerable solar heat to a building and reduce heat loss. A south-facing wall permits maximum exposure to solar radiation. Heavy materials such as masonry and containers of water provide thermal mass in which heat is stored; when the ambient air temperature drops below the temperature of the thermal storage mass, the heat is radiated out into the space. To reduce heat loss, multiple layers of glass or plastic, or movable insulation (curtains or shutters) to cover the glass at night, must be used. (e) Convective loop: Hot air rises and as it does so, cooler air moves in to replace it. Known as convective air movement, this phenomenon can be exploited to heat buildings. In this system solar energy gained by a solar collector heats the air inside the collector. The warm air rises to enter the building while cool air from the building enters the collector, establishing a convective loop.

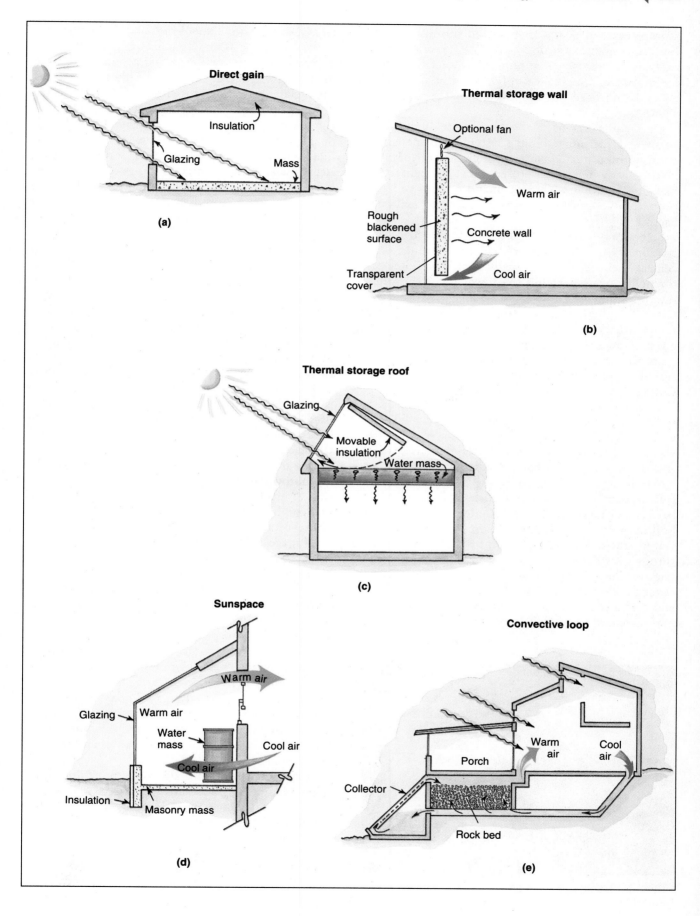

Direct gain

Insulation

Glazing

Mass

(a)

Thermal storage wall

Optional fan

Warm air

Rough blackened surface

Concrete wall

Transparent cover

Cool air

(b)

Thermal storage roof

Glazing

Movable insulation

Water mass

(c)

Sunspace

Warm air

Glazing

Warm air

Water mass

Cool air

Cool air

Insulation

Masonry mass

(d)

Convective loop

Warm air

Cool air

Porch

Collector

Rock bed

(e)

storage units and later distributed throughout the building. For example, hot water produced in rooftop collectors can be pumped through radiators for heating a home or it can be stored in an insulated tank and used for washing and bathing (Figure 13-8).

One major drawback of solar systems is that it is costly and difficult to build a system large enough to store enough heat to last through several days of cloudy weather. Therefore, most systems are installed together with on-demand sources of heat, such as woodburning stoves or conventional furnaces.

Solar energy also holds potential for electric generation. The "power tower" technology relies on a central receiving system, usually a tall tower. Reflector fields located around the tower focus the sun's rays on a gas- or fluid-filled receiver set atop the tower. The high concentration of sunlight produces temperatures reaching 2000° F (1093°C). The heated substance is then pumped through insulated pipes to a central power plant to drive conventional turbines. Solar One, located in the Mojave Desert in California, is currently producing electricity for Southern California Edison Company.

Another energy-producing method under investigation is the **solar pond,** a lined cavity filled with water and salt. Because salt water is denser than fresh water, it sinks to the bottom of the pond where it remains, trapped by the overlying layer of freshwater. The salt water absorbs and stores heat; because it cannot rise and evaporate, it can reach the boiling point even in the coldest winters. Heated water drawn from the bottom of the pond is sent through insulated pipes and used to generate electricity. The nation's largest solar pond, near Chattanooga, Tennessee, covers 1 acre, is 10 feet deep, and contains 2,000 tons of salt. In southern California, plans are being developed to build the world's

largest solar pond. The 48-megawatt capacity would supply some 40,000 households with electricity.

Perhaps the most exciting and promising development in solar technology is the **photovoltaic cell.** This technology generates clean, affordable electricity directly from sunlight, with no boilers, turbines, generators, pipes, or cooling towers. The cell relies on a process that occurs when light hits certain light-sensitive materials, called semiconductors. Atoms within the semiconductors absorb sunlight energy (photons) and liberate electrons. The movement of electrons produces a direct electrical current that is funneled into a wire leading from the cell. Most photovoltaic systems rely on wafer-thin slices of crystalline silicon.

Photovoltaic systems are based on technology discovered as long ago as 1839, but it was not until the mid-1950s that the first practical solar cells were manufactured. They were used to supply small amounts of power in such devices as remote weather equipment. As the technology advanced, its applications increased, and by the mid-1960s photovoltaic cells were used to power communication satellites. In 1973 they constituted the major power source for the orbiting Skylab. The technology is now widely used for other applications, including pocket calculators, photographic light meters, and portable electronic devices.

One of the most intriguing possible developments in photovoltaic technology is the large-scale generation of electricity. The U.S. Department of Energy has conducted several experiments in the use of photovoltaic cells. In 1978 Schuchuli, a village on the Papago Indian Reservation in southwestern Arizona, was electrified for the first time using photovoltaic cells. This community of 96 inhabitants now has fluorescent street lights (replacing kerosene lanterns) and a small electric water pump (replacing a diesel pump). The system produces 3.5 kilowatts—enough to power 47 fluorescent lights, a 2 horsepower water pump, 15 small refrigerators, a sewing machine, and a communal washing machine. Excess power is stored in batteries for use at night or on cloudy days. Other Energy Department experiments incorporating photovoltaic technology include a 25-kilowatt irrigation system, a 15-kilowatt system for a radio station, and a 60-kilowatt plant for an Air Force radar station.

Private companies are also actively exploring the many possibilities of solar technology. The world's largest photovoltaic power plant was built in 1985 in California by Arco Solar, Inc. The power generated is sold to Pacific Gas and Electric Company. The plant generates 16.5 megawatts, providing energy for 6,400 consumers. According to some estimates, photovoltaics has the potential to provide at least 30 percent of the world's electricity.

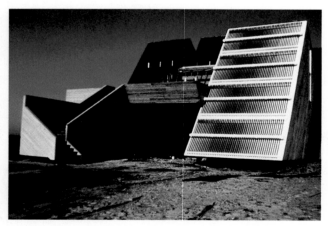

FIGURE 13-8: Solar collectors on a home on Long Island, New York.

Advantages and Disadvantages. Solar energy offers many important advantages. It exists in unlimited supply, is available everywhere (albeit intermittently in some areas), is nonpolluting, conserves natural resources, and is currently technologically available for widespread use. But solar power is not without disadvantages. Because solar energy is so diffuse, it has to be collected over large areas to make it practical to use, and because it is intermittent, it requires some means of storage. Its on-again, off-again nature is probably the greatest obstacle to its widespread use.

There are economic and environmental drawbacks to the use of solar power as well. The initial cost of building solar facilities can be high, although the life-cycle cost of a solar home, for instance, may be lower than that of a conventional, fossil-fuel–dependent home. In 1990 electricity generated by photovoltaic cells cost about 30 cents per kilowatt-hour compared to 3 cents per kilowatt-hour for conventional fuels (Table 13-1). Additionally, using solar power to generate sizable amounts of electricity requires a large land area. Generating 50 megawatts of electricity, enough to meet the demands of a town of 5,000 to 10,000 homes, requires 1 square mile (2.6 square km) of collectors. In order to supply enough electricity to the city of Pittsburgh, Pennsylvania, a 30-square-mile (78 - square - km) collector area would be needed. Considerable space is also needed for solar fields of photovoltaic cells, although the requisite area will drop as the efficiency of solar cells increases. The large amount of land required is probably solar power's most serious adverse effect on the environment and is likely to be the source of the strongest objections to its use, particularly in or near urban areas, where property values tend to be high, and particularly at a time when land use issues are so very important.

Wind Power

The unequal warming of the earth's surface and atmosphere by the sun causes the regional differences in pressure that are responsible for initiating winds. These flowing rivers of air can be harnessed to provide **wind power**—a safe, clean source of perpetual energy.

Present and Future Use. Like other renewable and perpetual energy resources, wind power has enjoyed a revival in popularity during the past several decades. During the 1970s dozens of small manufacturers sprang up, producing over 10,000 wind machines or **wind turbines** in 95 countries. These units were generally small and were used either to charge batteries or to produce minuscule amounts of electricity, usually less than 100 watts. In the 1980s

Table 13-1
Cost of Renewable Electricity (cents per kilowatt-hour)

Technology	1980	1988	2000	2030
Wind	32[a]	8	5	3
Geothermal	4	4	4	3
Photovoltaic	339	30	10	4
Biomass[b]	5	5	—	—

Note: All costs are averaged over the expected life of the technology and are rounded.
[a]1981.
[b]Future changes in biomass costs depend on feedstock cost.
Source: Worldwatch Institute, based on Idaho National Engineering Laboratory et al., *The Potential of Renewable Energy*, and various sources.

the market expanded into larger designs capable of producing significant amounts of electricity.

Worldwide, wind power is currently the second fastest growing source of electric power (after nuclear energy). Since 1974 some 50,000 wind power units have been installed worldwide, most of them in Denmark and California. The cost of installing wind power units for generating electricity is becoming increasingly competitive with conventional generating technologies. The cost of power from wind turbines, which was 25 to 32 cents per kilowatt-hour in the early 1980s, fell to about 7 to 9 cents by 1990. By 1995, the cost is expected to fall to 4 to 6 cents per kwh. As the cost drops, wind power should become more common. The Department of Energy estimates that wind power may generate 10 billion kilowatt-hours in the United States by 1995, and 70 billion by 2005.

Wind turbines typically require an average minimum wind speed of 13 miles per hour to produce electricity efficiently. Much of the research into wind power has focused on overcoming the limitations of wind speed through design modifications. Advances in aerodynamics and airfoil design have yielded a variety of experimental windmills of nontraditional design. Hundreds of research projects are concerned also with reducing costs and developing economical ways to store energy produced by wind turbines.

A wind farm is a cluster of wind turbines in a favorable geographic location, that is, where wind speeds average 14 to 20 miles per hour (Figure 13-9). The world's largest wind farm is in California's Altamont Pass, where 7,500 turbines are sited. Pacific Gas and Electric Company purchases power from a California wind farm composed of 36 large wind machines that generate 126,000 kilowatts of electricity, enough to power 63,000 homes. China has plans to establish wind farms capable of producing

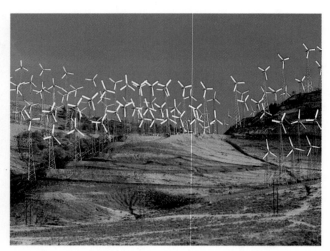

FIGURE 13-9: Wind turbines generating electricity at a wind-mill farm in southern California.

at least 100 megawatts by 1996. An ambitious project in the Netherlands plans to install 150 megawatts of capacity by 1992 and over 1,000 megawatts by the turn of the century.

Advantages and Disadvantages. Wind power is a clean and perpetual energy source, but it does have some disadvantages. It is unreliable and intermittent. It varies with climate, season, daily weather condition, geography, and topography. Average wind strength varies in different parts of the country, but the national average is just 10 miles per hour, under the average minimum for efficient production of large amounts of electricity. The equipment used to harness wind power is expensive, though its cost should decline with further development. Finally, the devices produce some noise and are very visible (covering entire hillsides in some parts of California, for example), so there is an esthetic cost to their construction.

Hydropower

Hydropower is the energy of falling water. Like wind power, it is an indirect form of solar energy. For centuries people have used the power of water cascading downstream and over natural falls or dams to turn paddles or turbines. Water at the top of the falls or dam is in a higher state of gravitational potential energy than water at the bottom of the dam. As the water drops from top to bottom, its potential energy is converted to kinetic energy; as the water strikes the blades of a turbine, spinning the turbine shaft, the kinetic energy is converted to mechanical energy. The mechanical energy of the turbine can be harnessed directly to run machinery or can be used to drive a generator to produce electric current.

Present and Future Use. Hydroelectric plants constructed in the industrialized nations in North America and Europe were initially large in scale and output and located in ideal sites, areas with a steep, narrow gorge through which water falls. By 1980 these nations had already developed 59 percent of the large-scale hydropower sites in North America and 36 percent of those in Europe. Most of the remaining sites were excluded from development as parkland or because of their natural beauty.

Some less-developed nations seem to be following the pattern set by the industrialized world, relying on large dams to supply the energy needed to fuel economic development. Presently, China and Brazil have the largest and most ambitious hydroelectric programs underway. China has under construction hydroprojects to generate 15,000 megawatts, with plans to double that amount by the end of this century. Brazil nearly tripled its resources in one decade, adding 21,535 megawatts between 1973 and 1983. The Itaipu Dam, currently under construction in Brazil, will be 5 miles long and half as high as the Empire State Building and will have a generating potential equal to 12 nuclear power plants.

Other developing nations, however, are taking a different route. Many are installing generators thousands of times smaller on remote rivers and streams. Such plants usually have an electrical capacity of 15 megawatts or less and provide power to sparsely populated communities and agricultural processing plants that are remote from electrical utility power lines. In China, Burma, Costa Rica, Guatemala, Guinea, Madagascar, Nepal, Papua New Guinea, and Peru, small hydropower potential exceeds the installed generating capacity of all other energy sources. China, with some 90,000 turbines supplying electricity to rural areas, leads all nations in this area.

Currently, hydropower generates almost one-fourth of the world's electricity, a contribution greater than that of nuclear power. Worldwide, it is responsible for the greatest proportion of electrical generating capacity of all renewable sources of energy. In the United States the proportion of hydroelectric production to total electric production has declined in the past several decades, despite the fact that hydroelectric production has grown. The proportional decline is a consequence of several factors, notably the lack of new sites to develop and growing resistance to large-scale projects from environmental groups and the public at large.

Advantages and Disadvantages. Proponents of large-scale projects argue that dams offer many advantages. Each year, dams displace over 530 million tons of carbon that would otherwise be emitted to the atmosphere. Dams are also multipurpose. They provide impounded water for recreational facilities, municipal water supplies, irrigation, and flood control, as well as generating electricity for surrounding communities and entire regions. But opponents favor free-flowing rivers and decry the destruction of lands flooded to create reservoirs. Many communities have banded together to fight proposed dams, which they contend will ruin local streams and waterfalls and destroy fishing in the area.

These concerns and others are well founded. Damming a river radically alters the surrounding ecosystems. For example, the reservoir of Egypt's Aswan High Dam on the Nile River provided a habitat for river snails that allowed their number to increase markedly compared to the population present before the dam was built. Snails are an intermediate host for the parasitic worm *Shistosoma*, or liver fluke, which causes a serious disease in humans called schistosomiasis. The flukes move from the snails to people swimming or bathing in the impounded water. Mosquitos, carriers of malaria and yellow fever, also breed in the still water.

Dams also cause siltation. Nutrient-rich sediment, which would normally provide food for downstream organisms or settle on agricultural floodplains, instead accumulates behind the turbines and dams. Large hydropower projects affect the temperature and oxygen content of downstream waters, altering the mix of aquatic and riparian species. Moreover, they displace people and wildlife. If China completes its proposed 13,000-megawatt station, known as the Three Gorge project, the station will displace several million people. One million individuals will be forced to move if India pursues its plan to build 3,000 dams in the Narmada Valley.

In regions where water is scarce, plans to construct dams and reservoirs can lead to international incidents. The Ataturk Dam on the Euphrates River, constructed by Turkey to expand its hydroelectric capacity, lowered water levels downstream, forcing the closure of five of Syria's eight 100-megawatt turbines at Tabaq Dam, built before the construction of the Ataturk Dam. Iran claims that it has been hurt in a similar manner by Syria.

Because of environmental and social concerns, the United States and other developed nations have begun to explore options other than large-scale construction, such as upgrading the efficiency of existing dams. At the Grand Coulee Dam on the Columbia River in Washington, operators plan to replace three older, less-efficient generators, each producing 1,150 megawatts, with newer superefficient generators capable of producing 3,900 megawatts each. The change will raise the dam's electrical output from approximately 3,500 megawatts to over 10,000 megawatts.

Hydroelectric projects offer many advantages, but they also require sound management that considers the entire watershed. Typically, various components of watershed management are parceled off to different agencies, each primarily interested in satisfying and fulfilling its own departmental responsibilities. Seldom are the agencies concerned with the complexity of the entire watershed. According to Professor Donald Worster of Brandeis University, "Everyone wants a piece of rivers, wants to siphon them off, dump waste into them, drink from them, or move barges along them, but no one has ever been given overall charge of protecting them and their renewability." Hydropower, as an environmentally sound source of alternative energy, will remain only a promise until and unless the varied concerns of flood control, irrigation, transportation, power production, forestries management, land use, fisheries management, and sanitation are coordinated within the primary goal of maintaining healthy and productive rivers.

Geothermal Power

Geothermal energy is heat generated by natural processes occurring beneath the earth's surface. Fifteen to 30 miles below the earth's crust lies the mantle, a semimolten rock layer. Beneath the mantle, intense pressure, caused by molten rock of iron and nickel and decaying radioactive elements, helps warm the planet's surface. Generally, the heat source lies too deep to be harnessed, but in certain areas, where the molten rock has risen closer to the earth's surface through massive fractures in the crust, underground deposits or reservoirs of dry steam, wet steam, and hot water have formed. These deposits can be drilled, much as oil deposits are, and their energy used to heat space and water, drive industrial processes, and generate electricity.

Dry steam is the rarest and most preferred geothermal resource. It is also the simplest and cheapest form for generating electricity. Steam released from a hole drilled into the reservoir is filtered to eliminate solid materials. The filtered steam is piped directly to a turbine, where it is used to produce electricity.

Wet steam, which consists of a mixture of steam, water droplets, and impurities such as salt, is more common than dry steam, but it is also more difficult to use. The water in a wet steam deposit is under such high pressure that it is superheated, that is, its temperature is far above the boiling point of water at normal atmospheric pressure. When wet steam is brought to the surface, a fraction of the water vaporizes instantly because of the dramatic decrease in pressure. The steam and water mixture is then spun at high speed in a centrifuge to separate the steam,

FIGURE 13-10: The largest wet steam geothermal electric power plant in the world, in Wairakei, New Zealand.

which is used to drive a turbine to generate electricity (Figure 13-10).

Hot water deposits are the most common source of geothermal energy. Iceland's tremendous geothermal resources provide 75 percent of the residential heat used by its population. Nearly all the homes and commercial buildings in its capital, Reykjavik (with a population of over 87,000), are heated by hot water geothermal deposits that lie deep beneath the city. Boise, Idaho, has used geothermal energy for space heating for over 80 years. The city's geothermal district heating system consists of over 400 wells which tap the 104° to 230°F (40° to 110°C) water. The system is currently being expanded to include city and state buildings as well as nearby residences.

Present and Future Use. People have always used hot springs for such purposes as heating, bathing, cooking, and therapeutic treatment. Today, over 20 countries are realizing a cumulative annual energy yield from this resource equivalent to 91 million barrels of oil. While geothermal energy does not presently play a major role globally, it does supply enough heat to warm over 2 million homes in cold climates and to supply 1.5 million homes with electricity. According to the Worldwatch Institute, reliance on geothermal energy is likely to increase by five to ten times by the end of the century. It will be used primarily for direct heating in industrial countries and for electric generation in LDCs.

Advantages and Disadvantages. Geothermal resources have numerous advantages. They are generally more environmentally benign than fossil fuels or nuclear energy. Costs are moderate, since most of the requisite technology has already been perfected by the oil industry. Advances in the design of drill bits should produce equipment better able to tolerate high temperatures and pressures. Research is currently underway to reduce the effects of corrosion on equipment caused by salts and silica picked up from the subterranean rocks.

There are several major drawbacks to geothermal energy. Like fossil fuels, geothermal deposits are nonrenewable on a human scale. There are relatively few easily accessible deposits. Finally, substances dissolved in the steam and water may affect air and water quality, and emptying underground reservoirs may affect land stability. A variety of gases may be released into the atmosphere when deposits are tapped; hydrogen sulfides, which smell like rotten eggs, cause the major complaints. Scrubbers, similar to those used on coal stacks, are rated as 90 percent efficient.

Ocean Power

The oceans are a relatively untapped energy resource. Continued research may prove many forms of "ocean power" to be significant, perpetual sources of clean energy. For instance, the flow and ebb of tides in and out of restricted areas can be harnessed to turn turbines for electrical energy and to drive water wheels for mechanical energy. Turbines might also be powered by the movement of waves, by the temperature differential between the sun-heated surface waters and the cold waters below, by floating and stationary "windmills" that use the natural airflow of onshore and offshore breezes, and by ocean currents, such as the Gulf Stream and Japanese Current. The means to use some of the near-surface heat of the earth's core in order to spin electrical generating turbines have already been developed; that near-surface heat under the oceans might be used to generate electricity.

Present and Future Use. Of the many forms of ocean power, only **tidal power** has been used commercially. Tidal electric power plants consist basically of barriers, or dams, built across inlets or estuaries in which a series of electrical generating turbines is housed. As the tide rises and the water moves inland, the water flows through gates in the barrier and is directed into the blades that turn the turbines. At the peak of the tide the gates are closed and the turbine blades are reversed. As the tide recedes, the gates are reopened and the water flows back over the turbine blades.

Other designs are based on capturing the water of the rising tide. The rising tide is allowed to fill a basin behind the dam. Once the basin is filled and the tide has reached its peak, the receding water is channeled through turbines to generate electricity.

In this type of design, electricity is generated only during the ebb tide.

Advantages and Disadvantages. Ocean power is nonpolluting, renewable, and, like other renewables, essentially "free." However, large facilities both inside and outside the tidal reservoir have such potentially adverse environmental effects as reduced tidal range (which would dry out the perimeter of the basin), reduced tidal current flow, altered sea levels, and the death of migratory fish species.

Biological Boundaries

Not all alternative resources are physical in nature. In the future the energy captured and stored by living things may also prove to be an important, renewable power source.

Biomass Energy

Biomass, derived directly or indirectly from plant photosynthesis, is one of the oldest and most versatile energy sources, capable of providing high-quality gaseous, liquid, and solid fuels. Forms of biomass that may serve as fuel are wood and wood processing residues, crop residues, animal waste products, and crops such as trees, seaweed, and kelp that are grown specifically to provide energy.

Present and Future Use. Worldwide, biomass provides over 15 percent of all energy consumed, with LDCs obtaining about 35 percent of their energy from biomass and MDCs just 3 percent (Table 13-2). In some LDCs biomass provides up to 90 percent of all energy consumed, with wood generally the principal fuel.

Crop residues, by-products of agriculture, are an increasingly important source of biomass. Examples include sugarcane and cotton stalks, rice husks, coconut shells, peanut and other nut hulls, fruit pits, and coffee and other seed hulls. In rice-growing nations of the developing world, rice husks are the most abundant crop residue. Every 5 tons of rice milled produces 1 ton of husk with the energy equivalent of 1 ton of wood. Rice-husk steam power plants are currently operating in India, Malaysia, the Philippines, Suriname, Thailand, and the United States. In Punjab, India, 20 tons of husks are burned each hour to fuel a 10.5-megawatt power plant.

Portions of the sugarcane industry are currently being subsidized by the sale of electricity generated from crop residues to local electric utilities. A case

▶ Table 13-2
Biomass Energy Consumption in the United States, 1987

Resource	Quads	Barrels of Oil Equivalent per Day
Wood and wood wastes		
Industrial	1.85	874,000
Residential	0.84	397,000
Commercial	0.022	10,400
Utilities	0.009	4,250
Subtotal	2.72	1,285,650
Municipal solid wastes	0.11	52,000
Agricultural and industrial wastes	0.04	18,900
Methane		
Landfill gas recovery	0.009	4,300
Digester gas recovery	0.003	1,400
Thermal gasification	0.001	500
Subtotal	0.013	6,200
Transportation Fuels		
Ethanol	0.07	33,100
Other biofuels	0.0	0
Total	2.95	1,395,850

Source: U.S. Department of Energy, *The Potential of Renewable Energy: An Interlaboratory White Paper,* March 1990.

in point is the sugarcane industry in Hawaii. In the wake of falling sugar prices in the late 1970s, sugar companies began installing plants with 150 megawatts of capacity which burn bagasse, the residue after the juice is extracted from the cane. Approximately 50 percent of the electricity produced is used to operate sugar-processing plants, and the remainder is sold to electric companies. The Hawaiian sugar industry supplies 10 percent of the state's electricity, saving about 2.7 million barrels of oil a year.

Using bagasse to generate electricity is fairly inefficient. For instance, a moderately efficient facility can produce only about 20 kilowatt-hours for every ton of cane. But with newly developed technology—a combined bagasse gasifier and steam-injected gas turbine—to cogenerate heat and electricity, electricity production could soar to 460 kilowatt-hours per ton of cane, 23 times greater than before.

The production of clean-burning alcohol fuels is another application of biomass energy technology. One of the most important biomass fuels is ethyl alcohol, or ethanol, which can also be used to make chemicals, solvents, detergents, and cosmetics. The use of alcohol as a fuel source dates back to the late nineteenth century. Ethanol fueled the engines of the first automobiles in the 1880s. To power tractors and other machinery, farmers in the 1920s mixed gasoline with ethanol, a mixture which became known as agrifuel. With the oil shortages of the 1970s agrifuel resurfaced under the name of gasohol, a mixture of 90 percent gasoline and 10 percent ethanol.

Alcohol fuels are considered a potentially significant supplement to our gasoline supplies. According to some estimates, ethanol and other alcohol-based fuels could replace as much as 7 percent of the U.S. gasoline consumption. These fuels are also gaining support because of concerns about worsening air pollution in urban areas. Over 60 cities in the United States failed to meet federal carbon monoxide and ozone standards in 1987. There are signs that states and municipalities may turn to alcohol-based fuels to combat the problem. Colorado now requires motorists in its major cities to use gasohol during the winter, when air pollution is worst. State officials expect carbon monoxide emissions to be reduced by 12 percent. A move is also underway to enact federal legislation requiring the use of gasohol nationwide.

The gases released by decaying plant matter and animal waste can also be captured and used as a boiler fuel. This fuel, called biogas, is a mixture of gases produced by the anaerobic microbial decomposition of organic materials. Methane is usually the chief component (typically 50 to 70 percent). Other gases include carbon dioxide (30 to 48 percent) and hydrogen sulfide. Several countries are successfully

FIGURE 13-11: Household biogas digester in Nepal. Biogas systems enable families to turn biomass into fuel. The manure of two to four pigs can supply a family of five with enough energy for lighting and cooking needs.

using biogas for diverse applications (Figure 13-11). In China, for example, more than 8 million biogas digesters convert manure and other organic waste into methane. These are especially useful in southern China, where higher temperatures stimulate decomposition.

Advantages and Disadvantages. Biomass is a readily accessible and fairly inexpensive resource. It is also a fairly clean fuel; the carbon dioxide released by burning plant matter can be offset by replanting. In some areas, however, biomass fuels have been exploited to such an extent that the supply of the resource and the land's fertility are threatened. The threat is particularly apparent in some LDCs, where wood is such a highly prized resource that deforestation has become a serious problem. Tree planting ventures by local communities have become more common as people have become aware of the need to reforest and protect their lands. But deforestation is not the only problem. Removing crop residues from agricultural lands deprives the soil of the nutrients needed to maintain fertility. According to a 1988 report by the Worldwatch Institute, the diversion of biomass from fields in Nepal has caused a 15 percent drop in grain yields.

Social Boundaries

Most alternative energy resources are physical or biological in origin. One alternative fuel, however, is a direct result of human activity: solid waste. This topic is discussed fully in Chapter 20, Unrealized Resources: Waste Minimization and Resource Recovery; but because it is presently being used in many areas as an energy source and holds significant potential for future exploitation, a summary of the major features of this resource is included here.

Solid Waste

Solid waste, or refuse, is material that is rejected or discarded as being spent, useless, worthless, or in excess. According to the U.S. Environmental Protection Agency, each person in the United States produces about 5 pounds of solid waste each day. Much of this refuse is derived from biomass, and like biomass, it can be tapped as an energy source. Using solid waste to produce energy is commonly called **trash conversion.**

Present and Future Use. One way to use refuse as a source of energy is to shred, screen, and pelletize it into a useable fuel. The pellets can be mixed with an equal amount of a standard fuel such as coal, and burned to supply heat and electricity to industries and surrounding communities. The Netherlands and Denmark are leaders in the use of municipal solid waste as a fuel. In the early 1980s these countries used an average of 40 percent of their solid waste to produce electricity. The United States, on the other hand, used less than 1 percent. Trash conversion has the potential to meet a much larger proportion of the nation's electricity demand. A $165-million, 47-megawatt plant on the Hudson River in Peekskill, New York, converts about half a million tons of refuse into electricity each year. Because it is able to generate twice the electricity the community needs, the plant sells the rest to Consolidated Edison Company, which channels the electricity into its power grid.

Burning refuse is not the only source of energy. Biogas digesters are now being installed in some landfills. These mountains of solid waste gradually decompose anaerobically, forming pockets and veins of methane gas. Pipes drilled into the landfills siphon the gas into storage tanks. As of 1991, there were 157 methane-recovery facilities at landfills in the United States (Figure 13-12).

Advantages and Disadvantages. Using solid waste as an energy source reduces the amount of material that must go into landfills. The nation is running out of room in existing fills, and new ones are difficult to site. Tapping the methane produced in landfills yields an energy supplement, potentially the equivalent of 20,000 to 85,000 barrels per day of oil. Another advantage of tapping methane is not so obvious. As the gas is siphoned off, the landfill settles and becomes smaller, allowing room for additional refuse and extending the useful life of the fill.

Major drawbacks to the use of refuse as a fuel are economic and environmental. To be economical a plant is limited in location to a site at which large amounts of refuse are available. If a municipality relies on trash conversion to generate electricity, local recycling and resource recovery efforts will be

FIGURE 13-12: Methane gas recovery system at Rumpke landfill, Cincinnati, Ohio. The recovery facility produces enough natural gas to meet the needs of 10,000 homes each year.

hampered. More energy can be saved through resource recovery than can be generated by burning recyclable materials. Emissions of mercury, lead, and other harmful substances are sometimes a result of trash incineration. (Environmental Science in Action: Columbus's Trash-Burning Power Plant, pages 248–250, illustrates both the benefits and limitations of trash conversion.)

History of Management of Alternative Energy Sources

For centuries people have used radiation from the sun, have tapped heat from the earth, and have harnessed the power of winds, water, and ocean tides. The Egyptians took advantage of wind power when they sailed their crafts down the Nile River some 5,000 years ago. The ancient Persians used water wheels for turning and grinding. In 85 B.C. the Greek poet Antipater celebrated the development of a water-powered gristmill, noting that it had liberated Greek maidens from the arduous task of grinding grain. Hot water from thermal springs heated pools used by Romans when they ruled England. The Japanese applied the basic mechanics of harnessing water in the seventh century. The ancient cliff dwellings of the Anasazi Indians, in Mesa Verde, Colorado, are excellent examples of the simplicity and effectiveness of passive solar use (Figure 13-13). In Hamburg, Germany, a tidal-powered sewage

FIGURE 13-13: Cliff Palace, one of the Anasazi *(left)*, dwellings in Mesa Verde National Park, Colorado. The Anasazi used the principles of passive solar heating to warm and cool their cliff dwellings. Massive rock overhangs provide shade from the high, hot summer sun, but permit the rays of the lower winter sun to penetrate the dwellings and bathe them in warm sunlight. The effect is seen in the photo *(right)* of the Tonto Ruins National Monument.

pump functioned until 1880. In the 1700s New Englanders built mills powered by tidal motion. Windmills, for centuries a source of mechanical power in parts of Europe, were also used by farmers and ranchers in the United States, particularly on the plains of the Midwest, to operate water pumps.

In the twentieth century people began to experiment with alternative energy sources to generate electricity. The world's first commercial hydroelectric plant started producing electricity at Niagara Falls on the United States–Canadian border in the early 1900s. In the 1930s Europeans began to investigate the use of wind power for large-scale electric generation. Soon after, the world's largest windmill was constructed near the town of Rutland, Vermont. A giant propeller-driven model, it had a 175-foot sweep and was capable of generating 1,250 kilowatts. Although it fed electricity into the local commercial power system for three years, Rutland's windmill was doomed by the same phenomenon that cut short interest in other perpetual and renewable energy sources—the age of fossil fuels.

The widespread availability of inexpensive supplies of oil, natural gas, and coal spurred utility companies, industry, and home owners to rely increasingly on fossil fuels to meet almost all energy demands. Only hydroelectric power, probably the least environmentally benign of all renewable sources, managed to grow during the age of fossil fuels. It was joined, in the latter half of the twentieth century, by a promising new resource: nuclear energy.

The world first witnessed the awesome power of the atom at the close of World War II, with the attacks on Hiroshima and Nagasaki. But with the 1954 Atomic Energy Act, the government stated its

commitment to developing civilian nuclear capabilities. This act essentially opened the nuclear industry to the private sector, although government research and activities still remained off limits. The first major developments within the civilian sector were in the areas of medicine, biological research, and agriculture. Then, as the 1950s came to a close, interest began to turn toward converting nuclear energy into electric power. This interest was spurred by lowered estimates of finite fossil fuel reserves, a growing demand for electrical power, and the 1956 war over the Suez Canal, a major route for transporting oil. By 1957 the world's first commercial nuclear reactor, the Shippingport nuclear plant, came on line at Pittsburgh, Pennsylvania.

The OPEC oil embargo of 1973 and the resulting oil shortages renewed interest in perpetual and renewable resources, especially for home heating and similar uses. Government funding for development of these alternative fuels rose, though not to the levels enjoyed by the nuclear industry, and peaked in 1980. That year saw the election of an administration that favored increased development of domestic oil reserves and nuclear energy over continued support for renewables. Funding dropped precipitously and has remained well below funding levels of the late 1970s (Figure 13-14).

As with other alternative fuels, interest in nuclear energy increased worldwide during the 1970s and 1980s as governments sought to reduce their dependence on foreign oil. Nuclear energy's share of electric production grew significantly during the period. Despite widespread support for nuclear energy among governments worldwide, a small but growing number of people continued to voice concerns about the safety and wisdom of increasing reliance on

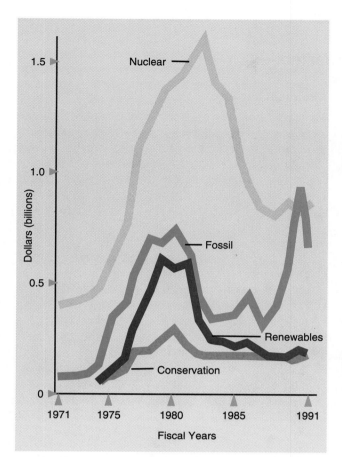

FIGURE 13-14: Research and development funding by the U.S. government for renewable energy resources, 1971–1991. The cumulative totals for the years from 1971 to 1991 were $19 billion to nuclear power, $8 billion to fossil fuels, $4 billion to renewable energies, and $2 billion to conservation.

nuclear energy. The issue of what to do with the by-products of nuclear power became a matter of great debate in the United States as the government searched for a geologically stable, and politically acceptable, storage site for the nation's mounting nuclear wastes. With the accidents at Three Mile Island and Chernobyl, many people became convinced that nuclear energy is not the solution to the world's energy problem. In the past decade public support for nuclear energy has fallen throughout North America and Western Europe.

Future Management of Alternative Energy Sources

Managing energy resources in the future calls for a mix of energy efficiency measures to extend fossil fuels supplies and alternative energy resources to replace conventional fuels and to eliminate their adverse environmental effects. The Rocky Mountain Institute (RMI), an energy think tank, blends energy-efficient technologies and alternative energy

resources to create a comfortable, attractive, and environmentally benign atmosphere for staff and visitors (see Focus On: Rocky Mountain Institute Headquarters—The Way to a New Energy Future).

As we look ahead to the twenty-first century, we will increasingly rely on perpetual and renewable energy resources to provide us with needed energy. Increased reliance on these alternatives can help us to prevent overuse of fossil fuels, protect human health and the environment, and preserve living systems.

Prevent Overuse of Fossil Fuels Through Conservation

A greater emphasis on energy efficiency (Chapter 12) and the development of alternative energy sources can help to reduce the demand for electrical power. Reductions in demand translate into a reduced need for power plants and, specifically, for nuclear power plants. Given that power generation is one of the costliest and potentially most dangerous applications of nuclear resources, it is prudent to avoid or minimize the use of nuclear energy for this purpose. Spending additional funds to develop alternative energies and mandating conservation measures (higher standards for fuel efficiencies of automobiles, increase in the gasoline tax, and increased use of recyclable materials) can help our society become energy self-sufficient for far less than the costs of nuclear energy. Moreover, achieving energy self-sufficiency through conservation and alternative energy resources would help safeguard national security while protecting the environment and human health. Other measures that might be taken to promote the development of perpetual and renewable resources include:

- Require that any construction project, partially or wholly funded with government monies, seek a bid from a contractor who uses designs that employ renewable energy resources like solar and wind power. Analysis of all project bids must be based on the life-cycle costs (including operating expenses for fuels) and not solely the initial construction costs.
- Support research to develop alternative fuels such as methane, gasohol, and solar power, which preliminary research has shown to be promising options to gasoline.
- Disseminate more efficient low-cost, prefabricated cooking stoves to people in developing nations in order to reduce fuelwood use by 25 to 50 percent. In areas hardest hit by deforestation, such as India and parts of Africa, make newly developed solar stoves available to the poor. The cost of these stoves could be paid for by a portion of a carbon tax collected from all nations.

Focus On:
Rocky Mountain Institute Headquarters— The Way to a New Energy Future

Hunter Lovins and Amory Lovins, Rocky Mountain Institute

A 4,000-square-foot castle is hard to hide, but many people drive right past the headquarters of Rocky Mountain Institute (RMI) in Old Snowmass, Colorado. They don't notice the building because it is made of native stone and most of its north wall is underground. Yet on the return trip down the winding country road, as the massive, curving walls and huge, slanted greenhouse windows come into view, even drivers who haven't seen any of the television shows about RMI can tell that this is no ordinary structure (Figure 13-15).

In fact, the RMI headquarters is one of the world's most energy-efficient buildings. It achieves this by being superinsulated, passive solar, and semiunderground and by using efficient lights and appliances. A home for 3, office for 34, and year-round greenhouse, it uses no heat and less than one-tenth the amount of electricity and less than one-half the amount of water of comparable buildings. The technologies that make this possible are commercially available and can be cost effective in conventional buildings and homes regardless of climate and outward appearance.

The first thing a visitor notices are the thick stone walls, plentiful and mainly south-facing windows, and the way the building nestles into the hill. But the superinsulation that keeps the building warm is harder to see. There is a cutaway in one wall showing the insulation system hidden inside the stone and concrete. A 4-inch thick sheet of foil-faced polyurethane foam is sandwiched between two 6-inch courses of masonry. The roof contains even thicker layers of the foam over a plastic-film vapor barrier. The insulation in the walls and roof is about twice normal levels. Superinsulation combined with airtight construction keeps the whole building warm with the heat generated by solar radiation through the

FIGURE 13-15: Exterior of the Rocky Mountain Institute in Old Snowmass, Colorado.

windows and by people, lights, and appliances, even in the Rocky Mountain winter that can reach -40°F. The two backup woodstoves are rarely fired up.

The building is air-tight but very well ventilated. Six air-to-air heat exchangers move stale, inside air out and bring fresh, outside air in. The stale air passes in channels alongside, but does not mix with, the incoming fresh air. Thus, when the fresh air enters the building, it has already been warmed by the heat of the exhaust air. This technology also works in hot climates to cool incoming air and is the answer to maintaining fresh, high-quality air in energy-efficient buildings at any time of year.

Usually, windows are the weakest link in a home's insulating armor. At RMI that is not the case, thanks to "superwindows." Most of the windows in the RMI building have an insulation rating twice as high as

standard triple-glazed windows. The extra insulation comes from clear Heat Mirror™ film suspended between two panes of glass and surrounded by argon gas. In effect, this makes 3 layers: glass, film, glass, with the gas filling the roughly half-inch spaces between. The windows lose only 19 percent as much heat as a single glass pane, but let in 75 percent of the visible light and 50 percent of the total solar energy. A few windows and a front storm-door add an extra layer of Heat Mirror™ to achieve insulation three times that of triple glazing. Recent innovations can reach even higher insulating levels.

Solar energy is very important to the RMI headquarters, and the entire building is designed to take advantage of solar light and heat. The curve of the walls allows sunlight and heat to penetrate the building from the east, southeast, and southwest, as well as the south.

The 900-square-foot greenhouse in the center of the building spreads heat and light through much of the building. The greenhouse provides a space for growing a garden that not only helps keep staff spirits high during the long Rocky Mountain winters, but also yields an abundance of fresh organic vegetables to eat. Fish, clams, and crayfish flourish in two ponds, recirculated through a waterfall whose pleasant sound masks noise in the otherwise silent building.

The sun heats water for the building through solar panels and preheating coils cast into the concrete arch at the back of the greenhouse. The 6-ton tank system for hot water, which is under the floor of a bedroom, in turn warms that part of the house. Water is used very efficiently, not only saving water but eliminating the need for a larger solar heater. Low-flow showerheads use a fourth to a third as much as conventional models, but give a vigorous tingly shower. Low-flow toilets use under 1 gallon per flush, compared to the 5 gallons used by conventional toilets. Faucet aerators cut use up to 70 percent. These devices reduce the building's total water use by 50 percent. Greater use of such efficient devices would solve many of the nation's water problems, and the devices would be more than paid for by the energy saved in heating water.

Other energy-using devices in the building are also superefficient. Dozens of new kinds of superefficient light bulbs are scattered throughout. Many of the light bulbs produce as much light as a standard 75-watt incandescent bulb, but use only 18 watts. Others replace 75 to 100 watts with 8 to 15 watts plus special reflectors to direct the light downward. Compact fluorescent lamps not only produce a better-quality light, tuned to the way the eye sees color, but also last 10 to 13 times longer than ordinary bulbs.

The efficient light bulbs are coupled with lighting systems that automatically turn off if no one is present, come back on when someone enters the room, and dim according to the available daylight. The result is lighting energy per square foot equivalent to 1 percent that of an efficiently lit normal office. Nationwide use of these superefficient lighting systems would save about $30 billion worth of utility fuel and lamp replacement each year and cut the nation's total electricity use by 20 to 25 percent at a cost lower than that of just running existing power plants. Exclusive use of all these lighting technologies in the United States would save the output of 120 Chernobyl-sized power plants. In comparison, the United States operates only 100 nuclear plants now.

RMI's kitchen has among the world's most efficient refrigerators and freezers. Each 16 cubic feet, the refrigerator and freezer use, respectively, about 8 and 15 percent as much electricity as conventional models, because of superinsulation and other design features. If this type of refrigerator were used throughout the United States, the energy savings would displace another 20 large nuclear plants.

Around the corner from the kitchen is a unique solar clothes dryer that saves 90 percent of the electricity of conventional dryers. The solar clothes dryer is a dark-painted, two-story closet with a Heat Mirror™ clerestory window at the top so the sun shines into it. An air-to-air heat exchanger brings in fresh dry air, while retaining most of the warmth, and a fan circulates the warm air. Clothes are hung on a ladder-like drying rack which is winched up and down. Clothes last longer with this system, do not shrink, and smell as sweet as clothes hung on an old-fashioned clothesline, but the dryer is right at hand to the washing machine and operates in any weather.

The technologies at RMI provide not only efficiency, but also convenience, comfort, beauty, and livability. They demonstrate that saving energy, water, and money does not mean changing your life-style or going without. In fact, efficiency can improve the quality of life. Weather stripping eliminates cold drafts. The efficient toilets are designed by sculptors to be beautiful and more comfortable to sit on. A greenhouse provides fresh food, greenery, warmth, nice smells, oxygen, and a bright place to sit on dark, winter days. Efficient showerheads allow everyone to get a warm shower in the morning without running out of hot water—and anyway, solar showers *feel* better.

New energy-efficient technologies and building techniques like those used in RMI's headquarters could cut the nation's $140 billion yearly electricity bill by three-fourths. The United States could produce as much as it does now, with greater comfort and higher quality goods, but use only one-quarter as much electricity. This is quite a lot to ask of the technologies demonstrated at the Rocky Mountain Institute, but it is true.

If you wish, you can come and see it for yourself, or write for more information. RMI gives tours of the building and publishes dozens of papers each year. For information and a free list of publications, write: Rocky Mountain Institute, 1739 Snowmass Creek Road, Snowmass, Colorado 81654-9199.

Environmental Science in Action:
Columbus's Trash-Burning Power Plant
Molly Grannan

The Columbus Refuse and Coal Fired Municipal Electric Plant, completed in 1983, is one example of how our society is slowly but surely exploring alternative energy sources (Figure 13-16). The idea for the plant was born in Columbus, Ohio, in 1973 in hopes of solving two of the city's problems at once: what to do with the growing amount of refuse and how to overcome the disadvantages of coal-derived electricity.

Describing the Resource
Physical Boundaries

Located on Jackson Pike, just south of downtown Columbus in Franklin County, Ohio, the plant occupies a 52-acre site. The site contains the main facility where refuse and coal are burned, a refuse shredder station, a warehouse, water and waste treatment facilities, and a circulating water pump house. The plant also leases a 180-acre lake adjacent to its property; water from the lake is used to collect ash and to dampen coal to reduce dust.

When burned in the main facility, the refuse and coal heat water in boilers above them, producing steam which powers turbines to generate electricity. Designers originally predicted that the plant would be capable of burning an average of 2,000 tons (1,820 metric tons) of refuse per day, generating 90 megawatts of electricity per day. In 1988 the plant burned an average of 1,500 tons per day (1,365 metric tons) and produced about 50 megawatts per day. By 1989 the plant served approximately 10,000 private and commercial customers in the greater metropolitan area and provided street and expressway lighting for areas surrounding the city.

Biological Boundaries

The greater Columbus area occupies about 187 square miles (485 square km), with a population of approximately 1,500,000. A ring of suburbs surrounds the metropolitan district, but woods and farmland begin to dominate within a thirty-minute drive from downtown.

Potentially, a trash-burning plant could adversely affect the environment and human health in several ways. The ash that is left after the refuse is burned is usually sent to a landfill. If the ash contains toxic substances, they could spread through the area and enter the groundwater. The Columbus plant mixes ash with Alkamet™, which neutralizes acidity and prevents any lead in the ash from leaching out, then sends the ash to a county sanitary landfill. The plant must also remove toxic waste from the water used to wash ash out of the boiler. The Columbus plant includes a water treatment facility to clean its water, and then the resulting waste goes to its waste treatment facility.

A trash-burning facility may produce toxic airborne emissions, usually because of metal and chlorine concentrations in the refuse and the sulfur content of the coal. Refuse incineration typically produces particulates, dioxins, furans, and acid-forming gases, such as sulfur dioxide, carbon monoxide, and nitrogen oxide. Dioxins and furans are suspected of causing birth defects and cancer, and the acid-forming gases can lead to acid precipitation.

Social Boundaries

Columbus is one of the fastest growing cities in the United States, and it has experienced many growing pains, including the problem of what to do with a seemingly endless supply of refuse. Garbage production in Franklin County increased by almost 400 percent between 1953 and 1983. In the early 1970s city officials realized that central Ohio landfills would not hold all the refuse that Columbus and its outlying areas would produce in the next decade. Studies that revealed the environmental drawbacks of landfills added to their concern, and the trash-burning power plant was conceived as a solution to this problem.

Another factor behind Columbus's decision to build a refuse-derived fuel (RDF) plant was the rising cost, dwindling reserves, and harmful effects of nonrenewable energy sources. As energy demands increase, communities like Colum-

FIGURE 13-16: Columbus Refuse and Coal Fired Municipal Electric Plant, 1990.

bus have come to realize that the only way they can ensure their continued growth and success is by long-range planning for energy that does not depend on nonrenewable resources.

Looking Back

Although many people now see refuse-derived fuel plants as important sources of alternative energy, people first thought of burning refuse simply as a way to reduce the amount of trash they had to dispose of. The first known facility in which refuse was burned and the resulting heat used to produce electricity was built in Oldham, England, in 1896.

Most of the current refuse-burning facilities can be divided into two types. About two-thirds of the facilities in the United States practice mass burning, the incineration of mass quantities of unsorted trash at very high temperatures. All mass burn facilities have as a primary purpose the reduction of the volume and weight of trash to be deposited in landfills; those which also generate steam or electricity do so as a second priority. The remaining one-third of the refuse-burning facilities incinerate refuse-derived fuel. RDF operations combine waste disposal with recycling and energy production. They allow the removal of recyclable and nonburnable items before incineration. Most RDF plants burn coal as well as refuse to ensure the steady production of electricity and to prevent corrosion in the pipes. Coal combustion produces sulfur oxide, which neutralizes the corrosive effect of chlorine gas, a potential by-product of burning refuse. The ratio of refuse to coal differs from plant to plant.

Designing an RDF Plant for Columbus

The 1960s and early 1970s witnessed the construction of RDF plants in many countries, including the United States, Austria, Switzerland, France, the Netherlands, and West Germany. The city of Columbus saw the relative success of most of these plants, and in 1973 proposed an RDF facility as a solution to the problem of solid waste. The ongoing energy crisis made the idea seem even more attractive. Although Franklin County voters voted down a bond issue to fund the building of the plant in 1976, the idea was not forgotten. The Columbus Council of South Side Organizations supported the project and conducted an intense public education program. The bond issue was placed on the ballot again in 1977 and passed by a margin of over 60 percent. Construction began in July 1979.

In designing the plant, engineers had to consider the environmental, health, and safety hazards of burning refuse as well as efficient operation. The plant included storage facilities for up to 140 tons of refuse, and all refuse had to pass through a shredder before being incinerated. Because volatile gases from the refuse can build up in storage areas and shredders, engineers had to guard against explosions. They surrounded the shredders with heavy concrete walls designed to withstand explosions and built the shredders without roofs so that pressure from volatile gases could not accumulate easily.

To minimize workers' exposure to airborne pollutants in coal dust, they designed the coal-burning operation so that water could be sprayed on the coal to control dust.

Because RDF plants burn fuel at much lower heats than mass burn facilities, they cannot burn objects containing metal, which account for approximately 3 percent by weight of incoming municipal solid waste. Designers of the Columbus plant included a magnetic separator to remove ferrous objects after they pass through the shredder. Flammable and explosive material is separated by hand. Aluminum and tin are recovered from the ash after the refuse has been incinerated.

The Columbus plant uses an electrostatic precipitator to remove acid-forming gases from the plant's emissions. The use of low-sulfur coal (less than 0.9 percent sulfur) also reduces the amount of sulfur and acid-forming gases produced.

Each of the plant's three emission stacks contains an EPA monitoring device to check emissions for unsafe levels of pollutants.

Making the Plant Work

Although the plant was ready for use in 1983, the city spent several years trying to overcome problems in efficiency and safety. The projected operating efficiency was drastically reduced by problems with the plant's fuel-feeding, fuel-processing, and ash-handling systems. Frequent breakdown of the ash-handling system allowed as much as a foot of waterlogged ash to pile up on the plant's 1-acre floor. Workers spent up to 75 percent of their time cleaning up and repairing the ash system. The shredders failed to remove such large items as stop signs, tires, and car hoods, which jammed the system and prevented efficient burning. Log jams of refuse also clogged up the system. The buildup of clinkers, the noncombustible material that remains after fuel is burned, blocked air flow into the cooling system and caused the boilers to overheat. All these problems kept the amount of refuse processed to an average of only 772 tons per day in 1984, far less than the 2,000 tons originally predicted.

The future of the "cash-burning plant," as it came to be called, became a local controversy. Some people wanted the city to cut its losses by closing the plant. In 1984 newly elected mayor Dana Rinehart commissioned experts from Ohio State University and Battelle Research Institute to see if they could find a way to make the plant work. The experts and a task force of city officials came up with a comprehensive plan, known as the "Big Fix," to rehabilitate the plant.

The Big Fix cost taxpayers about $12 million. It focused on repairing the problems that contributed to the plant's inefficiency. The entire ash-handling system was re-designed, stronger magnetic separators were installed to remove all ferrous objects, and the removal of large items for reshredding was improved. The fuel-feeding system was refitted with vibrating conveyor belts, which feed fuel in a smooth, even stream and eliminate jams. The burning of coal and refuse was separated to avoid the buildup of clinkers.

The plant had problems with environmental safety as well as efficiency. During the fall of 1985 a Franklin county landfill refused to accept ash from the plant because it contained unsafe amounts of lead and cadmium. While Battelle and two independent laboratories performed tests on ash samples, the plant was forced to burn only coal. The tests confirmed unsafe levels of lead, although cadmium levels were safe. The plant attempted to reduce the amount of lead in its ash by refusing to accept commercial garbage and asking residential customers not to throw away sources of lead, such as motor oil and batteries.

By all accounts the Big Fix was successful. By 1988 it was burning approximately 70 percent of the trash produced in Franklin County. In 1990, the plant burned 638,649 tons of RDF, an average of about 1,750 tons per day. Between 1986 and 1987, the plant reduced the ratio of coal to refuse that it burned, using 52 percent less coal and saving an average of $6,000 a day. Electric generation rose 5 percent during that time. Even so, the plant was unable to meet expectations concerning electric generation, and by 1990, its primary mission was restricted to waste reduction. The plant is now known as the Columbus Solid Waste Reduction Facility.

Looking Ahead

Improving Efficiency

An RDF plant almost always faces a certain lack of efficiency. Because the amount and type of refuse varies from day to day, it is difficult to maintain a steady fire and a steady output of electricity. As the Columbus experience has shown, if a plant cannot produce electricity at a steady rate, it cannot compete with conventional plants and must be heavily subsidized. In an attempt to combat the difficulty of unreliable output, in 1990 the Columbus plant began a pilot project to burn a mixture of natural gas and methane (obtained from old landfills) with its refuse. Plant officials believe natural gas will regulate flame and temperatures more effectively than coal; both refuse and coal take a long time to heat up and to cool down, but natural gas heat can be adjusted almost instantaneously. They hope this project will allow them to compete more effectively with coal-burning plants and eventually make their plant a profitable enterprise.

Protecting the Environment

Although the Columbus plant seemed to solve its toxic ash problem by adjusting the amount of lead in incoming refuse, many environmentalists worry that RDF plants threaten the environment because of unenforced regulations and unregulated emissions.

EPA tests on ash from RDF plants all over the country show that fly ash, the fine particulate matter trapped in air-control devices, almost always shows unacceptable levels of toxic metals such as lead and cadmium. However, bottom ash, which is washed from underneath the boilers, is unacceptable only 10 to 30 percent of the time. Plants make ash acceptable for disposal by mixing fly and bottom ash with Alkamet™. The composition of the refuse burned varies from day to day, and the ash mixture varies. To determine if ash is safe for regular landfills, where toxics could leach into groundwater, ash should be tested every day. However, many plants do not want to spend the time and money on daily testing, and they also want to avoid having to use toxic waste landfills, which can cost up to 15 times more than regular landfills. Because the EPA does not insist upon daily testing, some experts fear undetected toxic waste from ash is contaminating our environment.

Environmentalists point out that of the many toxic air emissions that refuse burning might produce, only lead, mercury, and beryllium emissions were controlled by government regulations as of 1987. Dioxin and furan, although suspected carcinogens, were not regulated by the federal government at all. Environmentalists claim that regulations about emissions overall need to be much more thorough and more strictly enforced if the ever-growing number of trash-burning plants in the country are not to damage air quality seriously.

Environmentalists are also concerned about the relationship between RDF plants and recycling. Although RDF plants can remove many recyclable objects before incineration, not all plants recycle as much as they can. By burning recyclables, they undercut recycling efforts in their areas even though studies have shown that recycling is often more economical than burning or landfilling. In addition, recyclable items add to the mass of ash which must be disposed of and can increase the plant's level of toxic emissions.

Some studies predict that by 2001 the United States will process half its trash in some way, including RDF plants, rather than sending it to landfills. Given this trend, RDF plants could play an important role. With improved technology plants like the Columbus Solid Waste Reduction Facility (Refuse and Coal Fired Municipal Electric Plant) could be environmentally sound as well as economically efficient. To provide the maximum benefit to the environment and to society, RDF plants must monitor and control ash and airborne emissions and must strive to recycle as much as possible.

Protect Human Health and the Environment

Based on expected declining trends in the demand for nuclear power, experts predict that the United States and many other countries will become less and less dependent on nuclear power as aging plants are decommissioned and no new plants are built. Nuclear power will probably become popular again only in the event that demand for electricity rapidly rises or scientists develop technology to make nuclear power plants safer and to dispose of radioactive waste.

With regard to the management of alternative fuels, two important steps can be taken to protect human health and the environment. The first is to encourage industrialized nations to switch from coal to gas and alternative sources in order to generate electricity. A carbon tax should make nations more amenable to such a measure. Based on each nation's emission of carbon dioxide, a carbon tax would be paid to an international governing body (such as the United Nations, the International Union for the Conservation of Nature, the World Wide Fund for Nature, or a joint committee of these and other international organizations) and used to fund research aimed at slowing global warming.

Protecting human and environmental health also requires a change in the nature of development projects in LDCs. Truly sustainable development projects should incorporate energy-efficiency programs and programs designed to take advantage of renewable energy sources. For example, North Africa and the Caribbean islands offer promising sites for wind power projects. Monies should be diverted from or denied to energy development proj- ects that are based on fossil fuels or are environmentally questionable (such as the huge hydro projects currently underway in Brazil and China). Stopping such projects would yield two benefits. First, it would reduce global emissions of carbon dioxide and other greenhouse gases. Second, it would reduce or eliminate developing countries' dependence on costly oil and natural gas, which most of them must import (and can scarcely afford to do). Energy self-sufficiency is a challenging but necessary step in sustainable development for LDCs.

Preserve Living Systems

The greatest benefit to be gained by increased reliance on perpetual and renewable energy resources is that, when managed properly, they can provide the power we need for development while preserving the living systems which are the basis for sustainable development and all sound economic activity. For this reason, if for no other, these alter-native resources are preferable to nuclear energy. Nevertheless, because nuclear energy is an important source of electric power generation worldwide, and because its continued use seems likely (at least for the next several decades), nations must work together to solve its problems within a global context.

- Mandatory standards should be set to govern construction of nuclear reactors worldwide.
- International teams must be created to respond to nuclear crises.
- Quick and concise reporting procedures must be established for all nuclear accidents.
- International licensing requirements for all operators of nuclear facilities should be defined.
- International monitoring and inspections programs must be initiated.

Nuclear energy is one of the most controversial issues of our era, and with good reason. We can all benefit from the many uses of fission energy just as we can all be harmed by radiation from fallout, nuclear waste, and nuclear accidents. While each of us must weigh the benefits and risks of this energy source, we need to work together to ensure that governments take the necessary measures both to protect human health and to preserve the living systems upon which, ultimately, we all depend.

Summary

Numerous energy sources may prove to be important alternatives to fossil fuels. These include nuclear energy, solar energy, wind power, hydropower, geothermal energy, ocean energy, biomass, and solid waste. Nuclear energy is the energy contained within the nucleus of the atom. When an atom splits into smaller particles, it releases neutrons and emits heat. The released neutrons may bombard other fissionable atoms, causing a chain reaction. To harness the energy of the atom, nuclear reactors are designed to sustain the fissioning process. Uranium 235 (U-235), a relatively rare uranium isotope, is the fuel that powers a nuclear reactor. The energy from the nuclear reactor heats water to produce steam to run a turbine that produces electricity. Nuclear energy accounts for a modest proportion of the United States' and the world's energy supply. In terms of electric generation, however, nuclear power's share is about 20 percent in the United States and worldwide. France and Belgium rely on nuclear energy to produce as much as two-thirds of their electricity. The significant growth in the nuclear industry in the United States during the 1970s and early 1980s has tapered off in recent years, primarily because of safety concerns and less-than-anticipated demand for electricity.

The sun provides the energy that all life forms need to survive. Currently, solar energy is used primarily for space and water heating. There are two major types of solar systems. Passive solar systems rely on natural forces to distribute the heat; active systems use fans or pumps driven by electricity to enhance the collection and distribution of the sun's heat. All solar systems incorporate design features of buildings to capture the maximum amount of radiation from the sun during winter months and a minimum amount of radiation in the summer months. In addition to space heating, solar energy also holds potential for electric generation. Photovoltaic cells generate electricity directly from sunlight. Atoms within these semiconductors absorb sunlight energy and liberate electrons, which produce a direct electrical current.

The unequal warming of the earth's surface and atmosphere by the sun causes regional pressure differences that produce winds. Winds can be harnessed to provide a safe, clean, perpetual source of energy. Wind turbines clustered in favorable geographic locations make up wind farms.

Hydropower is the energy produced as flowing water strikes the blades of a paddle or turbine. The energy can drive an electric generator or run machinery mechanically. Currently, hydropower generates almost one-fourth of the world's electricity, a contribution greater than that of nuclear power. Worldwide, it is responsible for the greatest proportion of electrical generating capacity of all renewable sources of energy.

Geothermal energy is heat generated by natural processes beneath the surface of the earth. Dry steam deposits, the rarest geothermal resource, are also the simplest and cheapest for generating electricity. Wet steam deposits are more common than dry steam deposits, but are also more difficult to harness. Hot water deposits are the most common type of geothermal energy. For all three forms, a hole is drilled into the reservoir, the released deposits are filtered as necessary, and the deposits drive a turbine.

Ocean power is a relatively untapped energy resource. Researchers believe that the oceans contain numerous different energy sources, including the tides, waves, deep ocean currents, and on- and off-shore winds. Of these, only tidal power has been used commercially.

Not all alternative energy sources are physical in nature. The energy captured and stored by living things may also prove to be an important, renewable source of energy. Biomass, derived directly or indirectly from plant photosynthesis, is one of the oldest and most versatile energy sources. Among its primary sources are wood and wood processing residues; crop residues; animal waste products; garbage; and energy crops like trees, seaweed, and kelp. Another application of biomass energy technology is the production of clean-burning alcohol fuels, which are considered a potentially significant supplement to gasoline supplies. The gases released by decaying plant matter and animal waste, called biogas, can also be captured and used as a boiler fuel. Biomass holds significant potential as a renewable energy resource, especially in the LDCs. It is a clean, readily accessible, and fairly inexpensive resource. In some areas, however, biomass fuels have been exploited to such an extent that the land's fertility is threatened.

Most alternative energy sources are physical or biological in origin. One unique alternative fuel, however, is a direct result of human activity: solid waste. Some communities have experimented with burning solid waste to produce energy, commonly called trash conversion. Biogas digesters are also being installed in urban landfills.

For centuries people have used radiation from the sun, heat from the earth, and the power of winds, water, and ocean tides. However, the widespread availability of inexpensive supplies of oil, natural gas, and coal spurred utility companies, industry, and residential users to rely on fossil fuels to meet almost all energy demands until the latter half of the twentieth century. The emergence of nuclear power, the OPEC oil embargo of 1973, and awareness of the dwindling supplies of fossil fuel stimulated renewed interest in perpetual and renewable resources like solar power, wind power, and geothermal energy, as well as nuclear energy.

Key Terms

active solar system	passive solar system
binding force	photovoltaic cell
biomass	reactor core
containment vessel	solar energy
fission	solar pond
geothermal energy	tidal power
hydropower	trash conversion
isotope	wind power
nuclear energy	wind turbine

Discussion Questions

1. Explain, in general terms, how a nuclear power plant generates electricity.

2. What are some of the major advantages and disadvantages of each of the perpetual or renewable energy resources described in this chapter? Which seem to be most desirable from an environmental standpoint? Why?

3. What are some of the energy-saving features of the Rocky Mountain Institute headquarters? Which of these features might you incorporate into your own home?

4. Would you vote for or against the building of a trash-burning power plant in your area? Why?

5. Imagine you are planning a new home for yourself. What kind(s) of energy would you want to use? What energy-saving features would you incorporate into your design? Explain how they would function.

An Environmental Triad

Protecting Biospheric Components

Air Resources

The Atmosphere

The bottom line of global warming is that we insult the environment at a faster rate than we understand the circumstances.

Stephen H. Schneider

The atmosphere is the last symbol of global interdependence we have. If we can't solve some of our problems in the face of threats to this global commons, then I can't be very optimistic about the future of the world.

Margaret Mead

Learning Objectives

When you have finished reading this chapter, you should be able to:

1. Describe the composition of the atmosphere and how it is maintained by living systems.

2. Distinguish between primary and secondary air pollutants.

3. Identify and briefly discuss six primary air pollutants and two major secondary pollutants.

4. Identify and describe the characteristics that affect air pollution levels.

5. Describe the effects of air pollutants on environmental and human health.

6. Summarize the methods used to control air pollution.

Air is one of our essential resources, sustaining life even as it arouses and delights the senses. Although invisible to the human eye, it makes possible sights such as beautiful sunsets and dazzling rainbows. It carries thousands of scents, both pungent and subtle: approaching rain, salty ocean breezes, perfumes from blooming flowers, and sulfurous gases from underground vents. Air enables us to taste dust or strong chemicals carried along by wind currents and to feel cooling breezes on our bare skin or strong winds that blow rain and snow into our faces. Most importantly, air fulfills our body's need to take in oxygen. Although air surrounds us, we are seldom aware of it until pollution, natural or human-induced, degrades its quality. The atmosphere is critically important, for in concert with living organisms it maintains the planet's climate and allows life on earth to flourish.

In this chapter we look closely at the air resource and how it interacts with living organisms. We examine the major types of air pollution, their effects on environmental and human health, and factors such as weather and topography that influence pollution levels. In looking at how the air resource has been managed historically, we review air pollution control measures in the United States and explore current efforts to improve air quality in both the United States and throughout the world. Finally, we recommend steps to clean up and safeguard air resources in the future.

Describing Air Resources: Physical and Biological Boundaries

What Is the Atmosphere?

The **atmosphere** is the layers of gases that surround the earth (Figure 14-1). The innermost layer, called the **troposphere,** extends to about 7 miles (11.2 kilometers) above the earth's surface and contains the gases that support life. The **stratosphere** extends from roughly 7 miles to 31 miles (50 kilometers) above the earth's surface. **Ozone** gas (O_3) present in the stratosphere acts as a shield, preventing harmful amounts of ultraviolet radiation from penetrating the troposphere and reaching the earth's surface. The **mesosphere** extends from the top of the stratosphere to about 56 miles (90 kilometers) above the earth, the **thermosphere** from 56 miles to outer space.

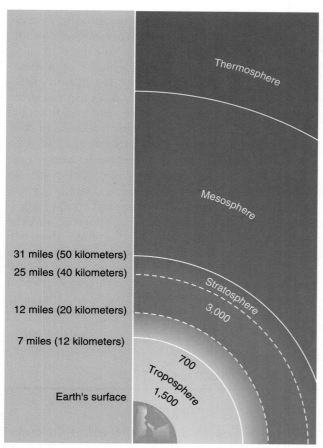

FIGURE 14-1: The atmosphere. The earth's atmosphere consists of the troposphere, stratosphere, mesosphere, and thermosphere. The numbers given indicate the concentration of ozone (in billion molecules per cubic centimeter of air). Atmospheric ozone is found at its densest concentration, about 3,000 billion molecules per cubic centimeter, in the stratosphere at 12 to 25 miles above the earth. There, ozone acts as a shield to screen out about 90 percent of harmful ultraviolet radiation from the sun.

Even though the wind patterns of the upper atmosphere influence weather conditions, weather occurs in the troposphere. Weather is the day-to-day pattern of precipitation, temperature, wind (direction and speed), barometric pressure, and humidity. In contrast, climate is the long-term weather pattern of a particular region. Climatic factors such as temperature and precipitation are expressed as averages observed over time. Although short-term predictions of weather trends (up to 5-day forecasts) have become increasingly accurate, long-term patterns and subtle shifts in climate—which can significantly affect such things as growing seasons, drought, sea level, and average temperature—remain difficult to predict.

Clean, dry air is a mixture of gases containing 78 percent nitrogen and 21 percent oxygen. The remaining 1 percent is composed of 0.03 percent carbon dioxide and rare gases such as helium, argon, and krypton. But air is rarely clean or dry. Air over

the tropics may contain as much as 5 percent moisture, and air over the earth's poles may contain less than 1 percent. Typically, air contains varying amounts of water (in the form of vapor, droplets, hail, or snowflakes), tiny particles of solids such as dust and soot, other gases and elements such as methane and sulfur, living organisms such as bacteria and molds, and reproductive cells such as pollen and spores.

How Does the Atmosphere Help to Maintain the Earth's Climate?

The atmosphere plays a critical role in maintaining the earth's temperature, which in turn has a profound effect on the earth's climate. Much of the incoming solar radiation is reflected away from earth by particles, water vapor, and ozone in the atmosphere and by the reflectivity, or **albedo,** of the earth's surface. Lighter colored areas of the earth—the polar ice caps, deserts and rangelands, and areas with heavy cloud cover—have a high albedo. A low albedo is associated with dark surfaces, such as forests and ocean areas rich in phytoplankton and other plant life. Because areas with a high albedo reflect more solar energy into space than do areas with a low albedo, they retain less heat. Dark areas absorb more solar radiation than they reflect; the absorbed radiation is degraded to long-wave infrared radiation or heat, thus raising the temperature of the areas. Eventually, this long-wave heat is radiated back into space. The capacity of a surface to radiate heat is its **emissivity.** It is equal to the amount of solar radiation that is absorbed, degraded to heat energy, and radiated back into space. The combined effects of the earth's albedo and emissivity help to maintain the average global temperature at about 59°F (15°C).

Gases present in the earth's atmosphere also help to maintain the earth's temperature and climate. Biogeochemical processes cycle materials such as carbon, nitrogen, and sulfur through the lithosphere, hydrosphere, and atmosphere (see Chapter 4 for a description of these processes). Carbon dioxide (CO_2), chlorofluorocarbons (CFCs), methane (CH_4), tropospheric ozone, and nitrous oxide (N_2O) allow incoming solar radiation to pass through the atmosphere, but absorb long-wave infrared rays (heat) which would otherwise be radiated back into outer space from the earth's surface (Figure 14-2). The trapped infrared rays raise the temperature near the earth's surface, in much the same way that heat trapped by the glass of a greenhouse raises the temperature inside. For this reason, carbon dioxide and the others are referred to as greenhouse (infrared-absorbing) gases and their effect on temperature is called the **greenhouse effect.**

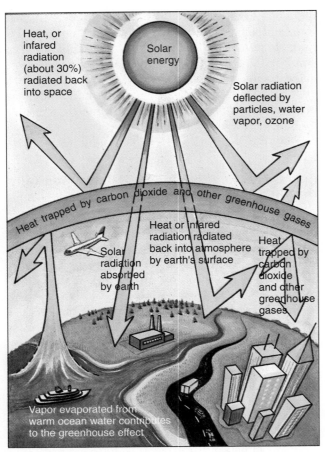

FIGURE 14-2: Greenhouse effect. The greenhouse effect makes possible all life on earth, but increased levels of greenhouse gases in the atmosphere have potentially disastrous consequences, such as changes in rainfall patterns, growing seasons, suitability of agricultural lands and a rise in sea levels causing coastal inundation.

To understand the effect of greenhouse gases, we need only look at our planetary neighbors. Venus, nearer the sun and with a much higher proportion of greenhouse gases, is far too hot to sustain life, with an average temperature of 837°F (447°C). The thin atmosphere of Mars, although predominantly composed of carbon dioxide, does not retain enough of the sun's heat, and Mars is far too cold to sustain life; its average temperature is -63°F (-53°C). Earth benefits from the Goldilocks effect: its concentration of greenhouse gases is "just right" to enable life to flourish, and living organisms—specifically green plants, which require carbon dioxide for photosynthesis and thus remove the gas from the atmosphere—help to maintain the proper atmospheric concentration of greenhouse gases (Figure 14-3). According to the Gaia hypothesis, developed by British atmospheric scientist James Lovelock, the abiotic component of the earth and its living organisms work together to create and maintain the conditions (including climate) necessary for life.

Atmospheric carbon dioxide can also dissolve in the oceans, combining with calcium and magnesium to form limestone or dolomite precipitates. Because oceans and forests absorb and store carbon dioxide, they are known as carbon dioxide sinks.

While greenhouse gases play a vital role in maintaining the necessary conditions for life on earth, too high a concentration of these gases can cause the atmosphere to become a heat trap and lead to **global warming.** Greenhouse gases are increased in four major ways. (1) Increased combustion of fossil fuels releases carbon dioxide, nitrogen oxides and ozone. (2) Increased clearing and burning of forests release carbon dioxide. When vegetation is cleared, burned, or left to decay, carbon is released to the atmosphere. Where vegetation has been cleared and the environment degraded (e.g., deforestation, desertification), carbon dioxide is no longer removed from the atmosphere at the same rate as when the vegetation sink was intact. (3) Increased decomposition of organic matter releases carbon dioxide and methane. (4) Increased activity of ruminants (cows, sheep) and termites releases methane.

Social Boundaries

What Is Air Pollution?

Any substance present in or released to the atmosphere that adversely affects environmental or human health is considered an **air pollutant.** Pollutants emitted by natural sources, such as volcanoes, forest fires, and living organisms, are part of the complex biogeochemical cycles that regulate the cycling of materials between the land, sea, and air.

FIGURE 14-3: Woody tropical vines, known as lianas, in a forest canopy in Panama. Trees play an important role in the carbon cycle. They absorb carbon dioxide from the atmosphere and release oxygen. Because carbon is incorporated into the tissue of trees as they grow, forests act as a carbon sink.

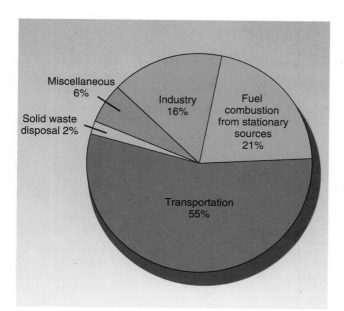

FIGURE 14-4: Cultural sources of air pollutants in the United States. Transportation is the largest single source of cultural pollutants, with fuel combustion from stationary sources, such as coal-fired electric generating plants, the second-largest source. Clearly, the combustion of fossil fuels is by far the most significant source of anthropogenic, or human-caused, air pollution.

However, cultural sources, such as automobiles, power plants, and manufacturing industries, contribute a significant amount of materials to the atmosphere (Figure 14-4). Historically, natural sources emitted more substances to the atmosphere than did cultural sources. This situation still exists in most of the Southern Hemisphere. In the Northern Hemisphere, however, human activities are adding more pollutants to the atmosphere than natural sources.

There are two categories of air pollutants, primary and secondary. **Primary pollutants** are emitted directly into the atmosphere where they exert an adverse influence on human health or the environment. Of particular concern are six primary pollutants emitted in large quantities: carbon dioxide, carbon monoxide, sulfur oxides, nitrogen oxides, hydrocarbons, and particulates. Table 14-1 summarizes the sources and environmental and human health effects of the "Big Six." Once in the atmosphere, primary pollutants may react with other primary pollutants or atmospheric compounds such as water vapor to form **secondary pollutants.** Acid precipitation, for example, is a secondary pollutant formed when sulfur or nitrogen oxides react with water vapor in the atmosphere.

In the 1970s pollution control measures focused on primary pollutants emitted from point sources, such as particulates from smokestacks. In some respects the nation's air is cleaner now than it was

two decades ago, but in other respects control efforts have fallen far short of goals. For example, smog still cloaks many urban centers. Additionally, in the 1980s the United States and the global community became aware of new, and perhaps more deadly, threats to air quality such as increased levels of greenhouse gases and stratospheric ozone depletion. In the following sections we examine six air quality problems that warrant special attention: global warming, acid precipitation, photochemical smog, stratospheric ozone depletion, indoor air pollution, and airborne toxins.

Global Warming. Carbon dioxide is responsible for about 50 percent of global warming; chlorofluorocarbons (CFCs), chemicals found in aerosol propellants, coolants, and solvents, about 20 percent; methane about 16 percent; ozone about 8 percent; and nitrous oxide about 6 percent.

Since the dawn of the Industrial Revolution cultural activities have slowly but steadily increased the transfer of carbon dioxide from the land to the atmosphere. Carbon dioxide bound up in coal, oil, and natural gas has been released to the atmosphere as these fuels have been burned to power industry, heat homes, and drive automotive engines. So much carbon dioxide has been released, in fact, that the oceans and forests can no longer absorb and use the excess amount. The continuing loss of both temperate and tropical forests—important carbon sinks— exacerbates the problem. In little over a century atmospheric carbon dioxide increased about 25 percent, from approximately 260 parts per million (ppm) in 1860 to 350 parts per million in 1990. Just in the past thirty years the concentration has risen almost 10 percent (Figure 14-6, p. 260).

Many scientists have linked the increase in atmospheric carbon dioxide with the rise in the earth's average temperature over the last century. A NASA study concluded that global temperature increased by about 1°F (0.56°C) in the past century and 0.27°F (0.15°C) since the 1960s. Eight of the nine warmest years on record took place since 1980 (1991, 1988, 1987, 1983, 1981, 1980, 1986), with 1990 the warmest year since reliable records have been kept (about 100 years). Some experts predict that if present trends continue, atmospheric CO_2 concentrations will double from preindustrial levels by around 2075. According to the National Academy of Sciences, a doubling of atmospheric carbon dioxide would lead to a 9°F (5°C) increase in average world temperature within 20 to 30 years (Figure 14-7, p. 261). Such an increase would make the earth warmer than at any time since the dawn of civilization. Scientists warn that the earth's temperature will not increase evenly, rising most at the poles and least at the equator.

Table 14-1
The "Big Six": Primary Pollutants

Pollutant	Environmental Effects	Health Effects	Sources
Carbon dioxide (CO_2)	Major greenhouse gas responsible for global warming Known to stimulate plant growth	No known direct adverse health effects	Fossil fuel combustion Burning of forests
Carbon monoxide (CO)	Not thought to have general effects at average concentrations found in urban atmospheres Contributes to photochemical smog	Has greatest effect on people who suffer from cardiovascular disease, especially those with angina or peripheral vascular disease Interferes with the oxygen-carrying capacity of the blood because it has greater affinity than oxygen for binding sites on the hemoglobin molecule Can aggravate cardiovascular disease Slows reflexes, causes unconsciousness, impairs perception and thinking Causes headaches Thought to affect fetal growth and mental development	Internal combustion engines, blast furnaces of steel industry, space heaters, improperly adjusted oil and gas burners, and charcoal grills 70% from mobile sources (cars, trucks, planes, etc.) Cigarette smoking
Sulfur oxides (SO_x)	Injure foliage of plants (continued chronic injury to perennial foliage results in early death) Make plants more susceptible to disease Major precursors of acid precipitation	Children, elderly, asthmatics and allergy sufferers at greatest risk Irritates nose, nasopharynx, and bronchi and increases likelihood of lung cancer Aggravates heart and lung disease, obstructs breathing Headaches, coughing, throat irritation, emphysema and chronic bronchitis Associated with premature death (as many as 50,000 deaths a year in the United States, especially when SO_2 appears with particulates)	Burning of fossil fuels, especially coal
Nitrogen oxides (NO_x)	Precursors of acid precipitation and photochemical smog Nitrous oxide (N_2O) contributes to greenhouse effect	Has greatest effects on children and those with chronic bronchitis, asthma, and emphysema Headaches, dizziness and labored breathing, impaired lung function, death	Transportation vehicles (cars, trucks, trains, buses, ships) Combustion of fuels used in industry for generating electricity and the heating of homes
Hydrocarbons (HC), such as benzene, terpene (Figure 14-5)	Photochemically reactive Contribute to photochemical smog Contribute to a number of intermediate products that may be additionally toxic to plants and animals	Greatest health risk associated with benzene, which is a suspected carcinogen Adhere to particulates and enter lungs	Benzene from cigarettes Mobile sources Petroleum refineries, petrochemical plants, paint and printing solvents Combustion of fossil fuels (coal, oil, gas, refuse burning, motor vehicles, catalytic cracking of petroleum)

Pollutant	Environmental Effects	Health Effects	Sources
Particulates	Responsible for atmospheric haze Causes damage to materials, soiling	Has greatest effect on children, the elderly, and those with chronic obstructive pulmonary disease, cardiovascular disease, influenza, or asthma Likelihood of increased respiratory disease, reduced lung function in children during exercise Aggravation of existing respiratory and cardiovascular disease Decreased gas exchange capacity of lungs Can absorb gaseous pollutants and deliver these directly to the lungs (e.g., SO_2 condensed on fly ash particles converts to sulfuric acid when inhaled) Fine particulates (under 0.2 microns) most dangerous; can contain up to 14 carcinogens	Major sources of fine particulates: auto exhaust, road dust, oil refining, coal-fired power plants, wood-burning stoves, and restaurant grills

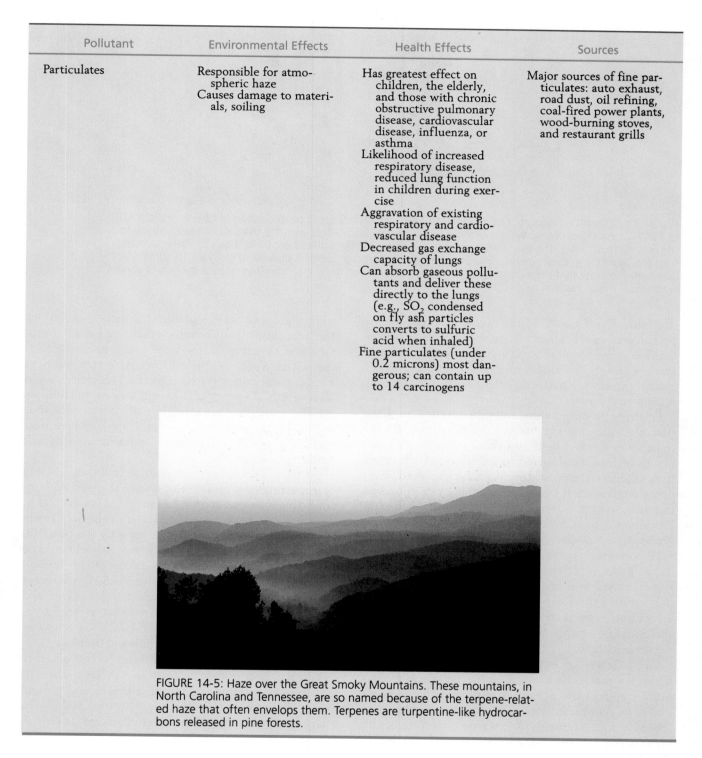

FIGURE 14-5: Haze over the Great Smoky Mountains. These mountains, in North Carolina and Tennessee, are so named because of the terpene-related haze that often envelops them. Terpenes are turpentine-like hydrocarbons released in pine forests.

It does not take a dramatic shift in temperature to have a substantial effect on the earth's climate. The average worldwide temperature of the last ice age was only about 16°F (9°C) lower than the average temperature today. The most likely and most worrisome effects are changes in rainfall patterns (some places will become much dryer and others much wetter), changes in growing seasons, changes in arable land, and rising sea levels. The severity of each, however, is a matter of debate. Southeast Asia, for instance, is projected to be wetter, while parts of North America and Asia could be much drier. If rainfall patterns do change, parts of North America's grain-producing areas might suffer permanent

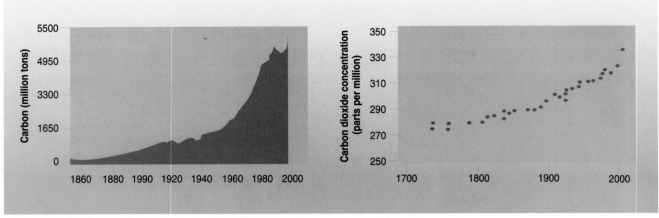

FIGURE 14-6: Global emission of carbon, 1860 to 1990, and atmospheric concentration of carbon dioxide, 1760 to 1990. Global emissions of carbon (left) have risen steadily since 1860, with the sharpest increase coming in the past 40 to 50 years. The concentration of carbon dioxide in the atmosphere (right) has also risen. Records from the past 100 years or so also show an increase in the average global temperature of about 1°F (0.56°C), which may indicate that the increasing concentration of carbon dioxide and other greenhouse gases is causing a warming of the planet's climate.

drought. The United States might not be able to produce as much grain as it currently does, while Canada, the former Soviet states, and northern Europe might produce more. In general, then, changing rainfall patterns might shift agricultural regions northward.

Changes in temperature and precipitation are likely to alter growing seasons and the suitability of land to produce crops. Plants and crops in currently temperate regions would need to be heat resistant, even if enough irrigation water could be supplied. Evidence suggests that trees are beginning to respond to the increase in atmospheric carbon dioxide. Ian Woodward of Cambridge University compared leaves of seven species of trees growing in England with preserved specimens collected 200 years earlier in the same locations. He discovered that modern leaves have fewer stomata, the microscopic pores that control the flow of water and gas from leaves. Conceivably, they are better able to conserve water and thus better able to survive drought as the temperature increases.

Ocean levels are expected to rise as warmer temperatures accelerate melting of the polar ice, adding more water to the oceans. Coupled with the thermal expansion of the upper layers of the seas (because water expands as it warms), melting of the polar ice caps will hasten the rise of sea levels throughout the world. Over the last 100 years sea levels have risen by 4 to 6 inches (10.16 to 15.24 cm). Because the oceans are so vast, sea level changes will be gradual (the oceans are presently rising at 0.4 inch (1.02 cm) per decade, lagging behind the carbon dioxide increase by about 20 to 25 years.

Some researchers are predicting a 16.4 to 23 feet (5- to 7-meter) rise in ocean levels by 2050. But even a gradual change of about 3 feet (1 meter) by the year 2100 would cause widespread coastal erosion and flooding of low-lying areas, greatly affecting cities such as New Orleans, states such as Florida, and countries such as the Netherlands and Bangladesh. As many as 200 million people could be made homeless. Vast stretches of coastal wetlands would be inundated and groundwater reserves near coastal zones would become contaminated by salt water encroachment.

At this point it is impossible to link conclusively the rise in global temperatures over the past century and the rise in atmospheric greenhouse gases. Because we have records of atmospheric carbon dioxide and temperature levels for only the past 100 years or so, the planet's rising temperatures could be the result of natural cycles rather than human actions. A study conducted by the U.S. Department of Agriculture's Agricultural Research Service and the University of Arizona showed that average temperatures decreased by 0.45°F (0.25°C) between 1920 and 1984 at 960 official weather stations. Researchers blamed perceived rises in temperature on "urban heat islands," a term that refers to the way urbanization raises the temperature in and around large cities because of the concentrated body heat, large energy use, and large areas of pavement (which decrease albedo).

Scientists agree that there are three "greenhouse knowns": the trapping of heat by atmospheric gases is a reality; the atmospheric concentration of greenhouse gases is rising rapidly; and changes in the concentrations of these gases is closely correlated with changes in the Earth's surface temperature. Moreover, there is consensus within the scientific community that the greenhouse effect will act to warm

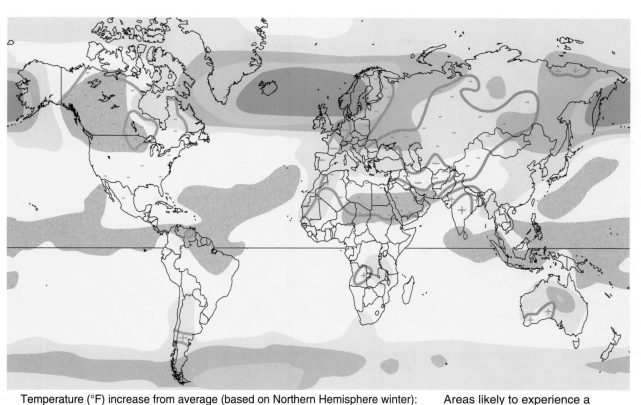

Temperature (°F) increase from average (based on Northern Hemisphere winter):

18° and over	10.7°–14.3°	3.6°–7°
14.4°–17.9°	7.1°—10.6°	3.5° or less

Areas likely to experience a change in soil moisture:

+ Increase of over 40%   – Decrease of over 40%
+ Increase of 20–40%   – Decrease of 20–40%

FIGURE 14.7: Estimated extent of global warming with a doubling of atmospheric carbon dioxide. Temperatures would increase from 3.6° to 7°F (2° to 4°C) across much of North America, Africa, and South America. Temperature increases would be even higher across much of Asia, Europe, and Australia. The far Northern Hemisphere would experience the most significant increase and the equatorial regions the least. Soil moisture would drop throughout much of North America, Africa, Asia, and Europe.

the global climate, if indeed it has not already begun to do so. What is unknown, however, is the role that ocean currents, cloud cover, and vegetation may play in either exacerbating or modifying the warming effect. High cirrus clouds, for example, tend to trap heat, while low- to mid-level stratus and cumulus clouds tend to reflect heat. The amount and kind of clouds produced is a result of atmospheric and ocean circulations, factors which may well change with global warming. Further, pollutants in the atmosphere might also reflect solar radiation, thereby modifying global warming.

Despite the uncertainties associated with global warming, many scientists remain concerned. In a May, 1990 report to the United Nations Intergovernmental Panel on Climatic Change, 300 of the world's top experts asserted that they were certain that cultural emissions are substantially increasing atmospheric concentrations of the main greenhouse gases, and that these increases will lead to a warming of the earth's surface. According to their best estimate, the world on average will be 2.34°F (1.3°C) warmer in 2020, rising to 5.4°F (3°C) warmer by 2070.

Acid Precipitation. For the past two decades few environmental issues have spurred more study, debate, or controversy than acid precipitation. And though the debate continues over what remedial actions, if any, should be taken to protect human health and the environment, there is growing agreement on what acid precipitation is and how it is formed.

Acid precipitation is rain, snow, fog, mist, or dust that contains enough sulfuric acid, nitric acid, or their precursors to raise the acidity of the precipitation above normal. Sulfuric, nitric, and carbonic acids are formed in the atmosphere when sulfur dioxide, nitrogen oxides, and carbon dioxide combine with water. Carbonic acid is weak enough that most of the damage attributed to acid precipitation has been traced to sulfuric and nitric acids. Normal precipitation is slightly acidic and is not harmful to plant or animal life. Acid precipitation poses a threat to living organisms and to buildings, monuments, and other objects (Figure 14-8).

To measure level of acidity, we use a pH scale of 1 to 14. Substances with a pH from 1 to 6 are called acid, from 8 to 14 alkaline, and 7 neutral. Sulfuric

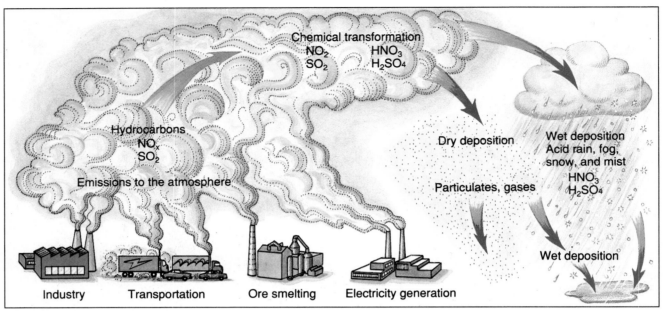

FIGURE 14-8: Formation of acid precipitation. Major sources of sulfur oxides and nitrogen oxides include electric generating plants, vehicles (cars, buses, trucks), factories, and ore-smelting facilities. Sulfur dioxide is transformed into sulfate particles and nitrogen oxides into nitrate particles; when the sulfate or nitrate particles react with water vapor, they form sulfuric acid and nitric acids, which then fall to the earth as wet deposition—rain, snow, mist, or fog. Sometimes the sulfates or nitrates fall to the earth as dry deposits, then combine with water in a stream, lake, or pond to form acids.

(battery) acid has a pH of 1 (a strong acid), tomato juice 4 (a moderate acid), cow's milk 6 (a weak acid), baking soda 8 (a weak alkali), and lye (sodium hydroxide) 14 (a strong alkali). Because the pH scale is logarithmic, each change in whole numbers represents a tenfold increase or decrease in acidity. For example, a solution with a pH of 4 contains 10 times more acid than one with a pH of 5, and 100 times more acid than a solution with a pH of 6. Precipitation with a pH as low as 5.6 to 5.9 is considered normal, but precipitation with a pH of 5.5 or lower is labeled acid and is a cause for concern. Readings taken in pristine, remote areas of the world and in core samples of ice from glaciers in Greenland and Antarctica suggest that prior to the Industrial Revolution, the pH of precipitation was rarely lower than 5.0. Industrial processes are commonly believed to be responsible for the formation of highly acidic precipitation.

An estimated 60 million tons (54.6 million metric tons) of sulfur oxides and nitrogen oxides are released by cultural sources over North America each year, with over 80 percent originating in the United States. In northeastern United States between 75 and 80 percent of acid precipitation can be traced to sulfur oxides. The leading emitters of sulfur oxides are electric utility plants in the Midwest, coal-fired industrial boilers, and metal-smelting operations in Canada, Mexico, and the Southwest. Ohio, Indiana, Illinois, and Kentucky release about 25 percent of all the sulfur dioxide

produced in the United States each year. Ohio emits more sulfur dioxide than any other single state—more than all of the New England states combined.

Motor vehicle exhaust accounts for nearly 50 percent of the nitrogen oxides released into the atmosphere. Industrial processes that involve high-temperature combustion and electric utility processes account for most of the remaining nitrogen oxide emissions. In the western states, nitrogen oxides from automobile exhaust are the major precursor of acid precipitation. Acid fogs at pH 1.7, 10,000 times more acidic than normal precipitation, have been measured over Los Angeles. The pH of this fog is comparable to the pH of commercial toilet bowl cleaner.

Acid precipitation originates, and to some extent, falls, anywhere sulfur or nitrogen oxides are produced in abundance (Figure 14-9). If the emissions are deposited near their source, a smaller amount of acid has a chance to form than if they remain aloft in the atmosphere for several days. Over several days sulfur and nitrate particles can be transported great distances, allowing ample time for the formation of sulfuric and nitric acid. After three days aloft as much as 50 percent of available sulfur dioxide is converted to acids. The long-distance transport of pollutants is made easier by the construction of tall smokestacks. Pollutants that enter the atmosphere at greater heights are likely to travel farther before they are eventually deposited. Beginning in the 1970s many utilities and industries, in an attempt to allevi-

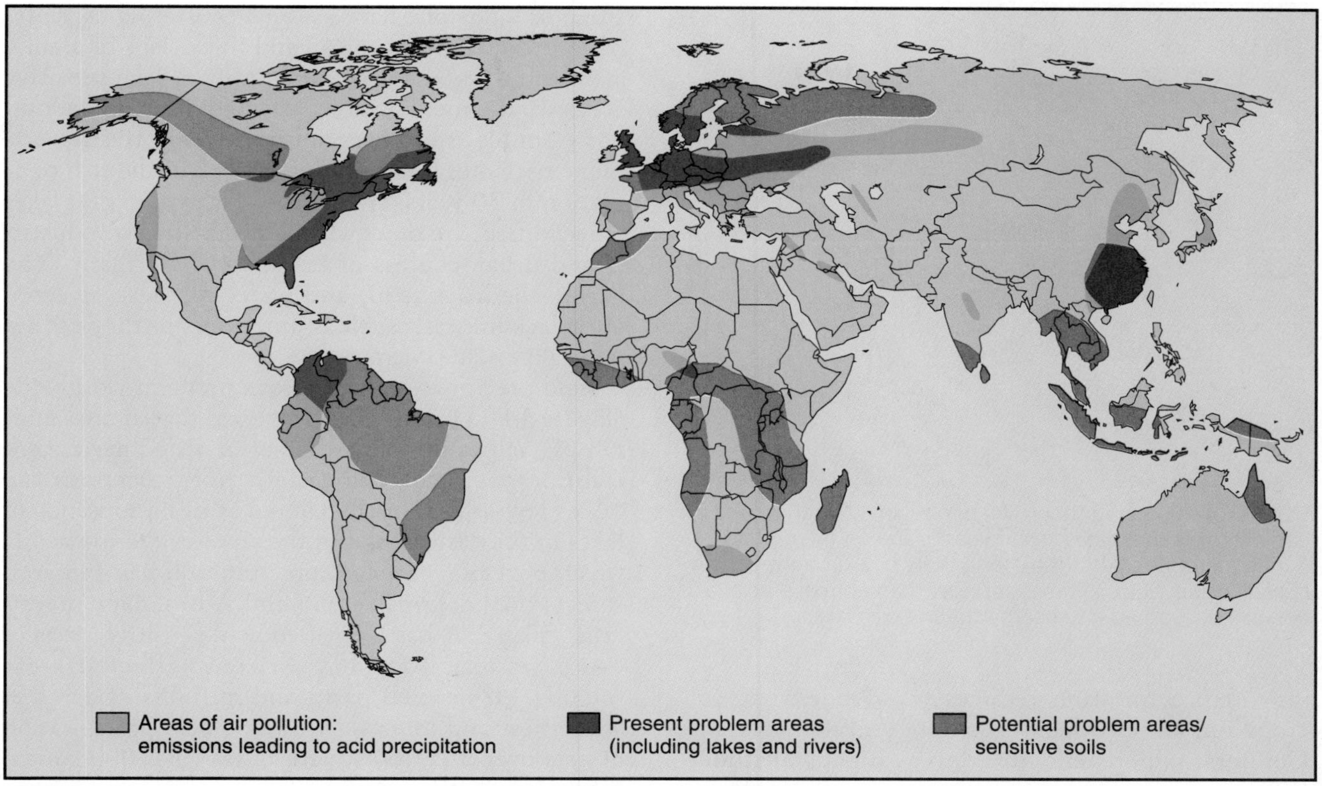

Areas of air pollution:
emissions leading to acid precipitation

Present problem areas
(including lakes and rivers)

Potential problem areas/
sensitive soils

FIGURE 14-9: Acid precipitation and point of origin of its precursors worldwide. Over 60 percent of the world's cultural emissions of the precursors of acid precipitation comes from North America and Europe. Problems caused by acid precipitation are especially severe in the eastern United States, eastern Canada, Scandinavia, northern and central Europe, the United Kingdom, and China.

ate local pollution problems, constructed extra tall smokestacks, which emit hot, gaseous pollutants 500 feet (150 meters) or higher into the atmosphere. Canada's single largest source of sulfur dioxides, International Nickel's smelter at Sudbury, Ontario, uses a 1,250-foot (375 meters) smokestack to spew its emissions as high as possible.

Research has shown that prevailing winds out of the Mississippi and Ohio valleys pick up sulfur dioxide and nitrogen oxides in Missouri, Ohio, Kentucky, Indiana, Illinois, West Virginia, and Pennsylvania and carry them toward the north and the northeast, where they are deposited over New York, New Jersey, New England, and Canada. Tracer experiments have shown that half of the sulfur dioxides in the Northeast are produced locally and about half are produced in the Midwest. The same is true of sulfur dioxides in Canada; about half originate in Canada and half in the United States. Because the sulfur dioxides not produced locally have been aloft for several days, they tend to produce about 75 percent of the actual acid precipitation that falls in the Northeast and Canada.

Acid precipitation damages, and in some cases destroys, the aquatic ecosystems of lakes and streams. An acidified lake or stream undergoes well-documented changes in vegetation, food chains, and fish populations. Small invertebrates are often the first affected; when they begin to disappear, the organisms (fish, frogs, and other vertebrates) that feed upon them suffer. Populations dwindle as the food supply decreases. Moreover, different species of fish stop breeding at different levels of acidity. Sensitive fish species such as salmon, trout, minnows, and arctic char are directly affected by heavy metals (mercury, aluminum, manganese, lead, and zinc) which are leached out of soils by acid precipitation and then washed into streams and lakes. Aluminum, which is most toxic at a pH of about 6, clogs fish gills, causing suffocation.

Acid surges are periods of short, intense acid deposition in lakes and streams. A spring snow melt or a rainfall after a prolonged drought can transfer acids previously locked in snow or soil into nearby waters, resulting in a sudden surge of acidity. This temporary rise in acidity can have a devastating effect on animal species that reproduce in the spring. Often the acid surge is enough to deform or kill young fish and newly hatched salamanders or frogs. Acid surges can also kill adult fish and decimate populations of aquatic insects, crayfish, and snails that serve as food for fish (Figure 14-10).

FIGURE 14-10: Spring thaw. Dry deposition of acid particulate matter accumulates in snowpacks and on the ground during a drought. After the spring thaw or a drought-ending rainfall, acidity 3 to 20 times stronger than normal precipitation can be washed into nearby streams and lakes.

Acid precipitation reduces the productivity of crops and harms other vegetation, especially forests. Conifers, which tend to grow at higher altitudes than hardwoods, are particularly affected because precipitation is most likely to form when air is lifted and cooled as it travels over mountain ranges. Many mountain peaks are shrouded in acid mists, covered by acid snows, and pelted by acid rains throughout the year. Some scientists believe that this phenomenon contributes to mountain forest decline and fish mortality in mountainous lakes. Moreover, needles of conifers are also more likely to collect particles and gases from the atmosphere than are the leaves of deciduous trees, and so they absorb more sulfuric and nitric acid from the acid deposition they experience. (Environmental Science in Action: Acid Precipitation in the Adirondack Mountains, on pages 282–285, presents a detailed discussion of the way acid precipitation affects forests.)

Acid precipitation also corrodes marble, limestone, sandstone, and bronze, defacing or destroying statues, monuments, gravestones, and buildings. In the late 1980s, the Organisation for Economic Cooperation and Development (OECD), whose members include many European countries, the United States, Canada, and Japan, reported an estimated $20 billion in damage to metals, buildings, and paint annually.

Soil type is a major factor in determining how severely acid precipitation affects a region. Most of midwestern United States and much of southeastern England, for example, are able to resist acidic precipitation because they have alkaline soils; the calcium in the soil buffers, or neutralizes, the acids, rendering them harmless. Lakes having beds of lime-

stone or sandstone have a similar buffering capacity. In contrast, thin glacial soil and thick slabs of granite are unable to buffer acidic fallout. These sensitive areas are the most severely affected by acidic fallout. For example, southern Norway, where the government estimates a loss of more than half the fish population in 30 years, and Sweden, where 18,000 lakes are acidified, are downwind from the major industrial and urban centers of Britain and Germany. The granite bedrock, soil, and water of these affected areas lack minerals such as limestone, so they cannot neutralize acid precipitation.

Acid precipitation is a serious problem worldwide (Figure 14-11). The OECD reports that annual rainfall pH values in polluted areas in Scandinavia, central Europe, Japan, and eastern North America can fall as low as 3.5. In the United States, thousands of lakes in the eastern half of the country are too acidic to support life, and acid precipitation has damaged forests in the Appalachian and Adirondack mountains. The EPA has identified acid-sensitive areas in every western state but Arizona; affected areas include 11 national parks and millions of acres of wilderness and forests. In Europe acid precipitation affects over 345,960 square miles (900,000 square kilometers) of European Russia. With the dismantling of the Communist Bloc, researchers are finding that acid precipitation has taken a significant toll on historic buildings, forests, lakes, and streams in the former German Democratic Republic (East Germany), Poland, Czechoslovakia, and Hungary. In much of China, which relies heavily on high-sulfur coal, acid precipitation poses environmental problems. In Africa and South America the burning of grasslands and forests to prepare the land for agriculture and the lack of controls on industrial emissions contribute to acid precipitation.

Acid-forming emissions increased in the 1980s over North America, despite the use of pollution control technologies and a reduction in the use of high-sulfur coal. Several factors contributed to the increase. Perhaps most importantly, nitrogen oxide emissions have risen over the past decade because of increased automobile use. Americans' love affair with the automobile, which began to wane during the oil embargoes of the 1970s with rising gasoline prices and long lines at the gas pump, is now stronger than ever. Relaxing automobile fuel-efficiency standards has also increased pollutant emissions. Another factor in the increased production of acid precipitation is a shift in the way coal is used by utility plants. Even though coal use has not increased appreciably on a yearly basis, utilities are using more coal during the summer months, when there is a greater demand for electricity to be used for cooling purposes. The heat and humidity of summer hasten the transformation of sulfates and

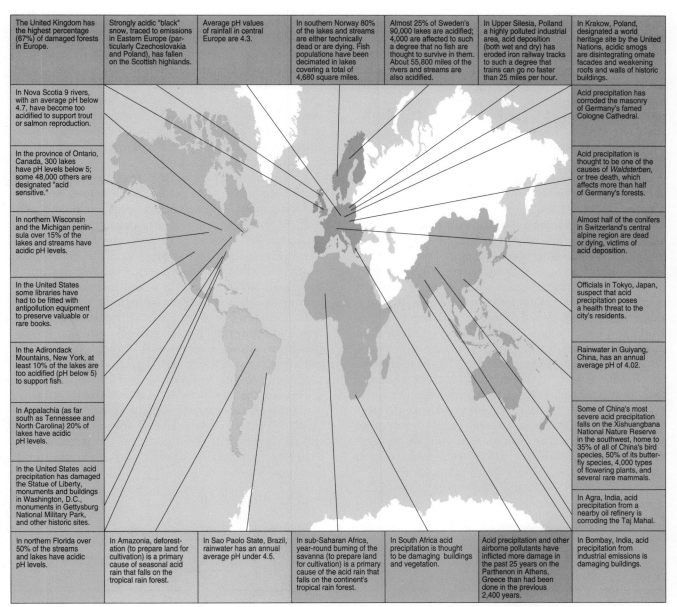

The United Kingdom has the highest percentage (67%) of damaged forests in Europe.

Strongly acidic "black" snow, traced to emissions in Eastern Europe (particularly Czechoslovakia and Poland), has fallen on the Scottish highlands.

Average pH values of rainfall in central Europe are 4.3.

In southern Norway 80% of the lakes and streams are either technically dead or are dying. Fish populations have been decimated in lakes covering a total of 4,680 square miles.

Almost 25% of Sweden's 90,000 lakes are acidified; 4,000 are affected to such a degree that no fish are thought to survive in them. About 55,800 miles of the rivers and streams are also acidified.

In Upper Silesia, Polland a highly polluted industrial area, acid deposition (both wet and dry) has eroded iron railway tracks to such a degree that trains can go no faster than 25 miles per hour.

In Krakow, Poland, designated a world heritage site by the United Nations, acidic smogs are disintegrating ornate facades and weakening roofs and walls of historic buildings.

In Nova Scotia 9 rivers, with an average pH below 4.7, have become too acidified to support trout or salmon reproduction.

In the province of Ontario, Canada, 300 lakes have pH levels below 5; some 48,000 others are designated "acid sensitive."

In northern Wisconsin and the Michigan peninsula over 15% of the lakes and streams have acidic pH levels.

In the United States some libraries have had to be fitted with antipollution equipment to preserve valuable or rare books.

In the Adirondack Mountains, New York, at least 10% of the lakes are too acidified (pH below 5) to support fish.

In Appalachia (as far south as Tennessee and North Carolina) 20% of lakes have acidic pH levels.

In the United States acid precipitation has damaged the Statue of Liberty, monuments and buildings in Washington, D.C., monuments in Gettysburg National Military Park, and other historic sites.

Acid precipitation has corroded the masonry of Germany's famed Cologne Cathedral.

Acid precipitation is thought to be one of the causes of Waldsterben, or tree death, which affects more than half of Germany's forests.

Almost half of the conifers in Switzerland's central alpine region are dead or dying, victims of acid deposition.

Officials in Tokyo, Japan, suspect that acid precipitation poses a health threat to the city's residents.

Rainwater in Guiyang, China, has an annual average pH of 4.02.

Some of China's most severe acid precipitation falls on the Xishuangbana National Nature Reserve in the southwest, home to 35% of all of China's bird species, 50% of its butterfly species, 4,000 types of flowering plants, and several rare mammals.

In Agra, India, acid precipitation from a nearby oil refinery is corroding the Taj Mahal.

In northern Florida over 50% of the streams and lakes have acidic pH levels.

In Amazonia, deforestation (to prepare land for cultivation) is a primary cause of seasonal acid rain that falls on the tropical rain forest.

In Sao Paolo State, Brazil, rainwater has an annual average pH under 4.5.

In sub-Saharan Africa, year-round burning of the savanna (to prepare land for cultivation) is a primary cause of the acid rain that falls on the continent's tropical rain forest.

In South Africa acid precipitation is thought to be damaging buildings and vegetation.

Acid precipitation and other airborne pollutants have inflicted more damage in the past 25 years on the Parthenon in Athens, Greece than had been done in the previous 2,400 years.

In Bombay, India, acid precipitation from industrial emissions is damaging buildings.

FIGURE 14-11: "Acid Earth." Forests, lakes, streams, croplands, buildings, and monuments around the world are being damaged or destroyed by acid precipitation. The map indicates just some of the affected areas.

nitrates to acids, increasing the amount of acid produced per unit of coal.

Photochemical Smog. In the presence of sunlight volatile organic compounds (VOCs), or hydrocarbons, react with nitrogen oxides and oxygen to form chemicals such as ozone and peroxyacetyl nitrate (PAN). Because this reaction takes place in the presence of sunlight, it is called a **photochemical reaction.** The result is known as **photochemical smog.** Photochemical smog is also known as brown smog because of its characteristic color, given to it by nitrogen dioxide (NO_2), a brown gas. The incomplete combustion of fossil fuels, particularly from automobiles, is the primary cultural source of

volatile organic compounds, so it is not surprising that brown smog cloaks many urban centers, including Los Angeles; Ankara, Turkey; New Delhi, India; Melbourne, Australia; Mexico City, Mexico; and Sao Paulo, Brazil.

The primary component of photochemical smog is ozone. Ozone is toxic; less than 1 part per million in air is poisonous to humans. There is mounting evidence that ozone adversely affects health even at concentrations considered harmless—0.12 parts per million, the standard set by the EPA. Ozone occurring at ground level (anywhere between several feet and several miles high) irritates and possibly damages eyes, skin, and lungs; dries out the protective, moist mucous membranes of the nose and throat;

interferes with the body's ability to fight infections; and causes short-term breathing difficulty and long-term lung damage. Joggers and runners can damage lung tissue by exercising even when ozone levels are acceptable. Moreover, a study published in 1991 found that ozone can double a person's sensitivity to the allergens that cause asthma attacks. Ozone also adversely affects the environment. It damages crops and forests and destroys rubber and some other synthetic fibers and dyes. According to the EPA, crop yield losses in the United States attributed to ozone damage amount to several billion dollars annually.

Ozone levels are generally higher in the afternoons, after radiation from the sun begins to take effect. Smog buildup is also greater when air is trapped close to the ground for long periods of time and in valleys or basins, areas where large numbers of people tend to live and work. The 12 million inhabitants of Los Angeles, for example, live in a valley nearly surrounded by mountain ranges (Figure 14-12). The city has 8 million cars, 10 million buses, trucks, and trains, and thousands of stationary sources of pollution, including 12 oil refineries. Its climate is hot, with little air-cleansing rainfall each year. Consequently, Los Angeles has the highest ozone levels of any city in the United States (Table 14-2). Ozone is a year-round problem in warmer climates. It occurs more seasonally in temperate zones, reaching its highest levels in the Northern Hemisphere during the warmer months from April through September.

Stratospheric Ozone Depletion. Ozone has a split personality as far as air quality is concerned. Near ground level it is a health threat, but in the upper atmosphere it acts as a shield, protecting us from the sun's lethal ultraviolet (UV) radiation. This blue-tinted gas is the only atmospheric gas capable of screening out ultraviolet rays.

The ultraviolet radiation that does reach the earth's surface is responsible for cataracts, sunburn, snow blindness, and aging and wrinkling of the skin. It is the primary cause of skin cancer, which claims some 12,000 lives each year in the United States alone (Figure 14-13). Exposure to UV radiation also suppresses the immune system, enabling cancers to become established and grow and increasing susceptibility to diseases such as herpes. In addition, UV radiation slows plant growth, delays seed germination, and interferes with photosynthesis. Phytoplankton show a particular sensitivity to UV radiation, raising the possibility that increased UV radiation might interrupt marine ecosystems, decreasing the productivity of marine food chains.

Ozone occurs naturally in the upper atmosphere up to 37 miles (60 kilometers) above the surface of the earth, but it is most dense in the stratosphere, between 7 and 12 miles (12 and 20 kilometers).

▶ Table 14-2
Ozone Levels in Metropolitan Areas, 1986–1988

Metropolitan Area	Number of Days Exceeding Standards	
	1986–1988 Average	1988
Los Angeles-Anaheim-Riverside (California)	145.4	148.0
Bakersfield (California)	37.8	54.0
Fresno (California)	31.6	30.3
New York-Northern New Jersey-Long Island (New York, New Jersey, Connecticut)	18.0	19.4
Baltimore (Maryland)	13.3	19.3
Philadelphia-Wilmington-Trenton (Pennsylvania, Delaware, New Jersey, Maryland)	8.9	18.2
Worcester (Massachusetts)	7.3	17.1
Parkersburg-Marietta (West Virginia, Ohio)	7.2	17.0
Chicago-Gary-Lake County (Illinois, Indiana, Wisconsin)	21.2	16.2
Knox County[a] (Maine)	8.1	15.6
Sacramento (California)	8.9	15.5
Sheboygan (Wisconsin)	9.5	15.5
Pittsburgh-Beaver Valley (Pennsylvania)	6.6	14.8
Milwaukee-Racine (Wisconsin)	9.1	14.2
Cincinnati-Hamilton (Ohio, Kentucky, Indiana)	5.4	14.1
Washington, D.C. (Maryland, Virginia)	8.3	13.8

[a]Not a metropolitan area.
Source: U.S. Environmental Protection Agency, 1989.

(a)

(b)

FIGURE 14-12: The photochemical cloak in Los Angeles. (a) Los Angeles has the worst air quality in the United States. (b) On a clear day, however, it is easy to imagine that the area was once beautiful.

Even at its most dense concentrations, only 1 molecule in 100,000 is ozone. If we could collect all of the stratospheric ozone and spread it evenly over the earth's surface, it would form a layer just 0.117 inch (3 millimeters) thick. Thus, even small changes in the concentration of ozone in the stratosphere can profoundly alter its ability to screen UV radiation.

Although stratospheric ozone is constantly being produced and destroyed by natural processes, pollutants emitted by cultural sources increase the rate at which it is destroyed. The most important of these is chlorine; others include nitrogen oxides, fluorine, and bromine. Atmospheric chlorine and fluorine have been traced to the chemical breakdown of chlorofluorocarbons (CFCs), and bromine has been identified with the breakdown of halons, chemicals most frequently used in fire extinguishers.

Chlorofluorocarbons are useful and versatile chemicals. First developed as coolants, they were used as aerosol propellants during World War II to spray pesticides in the fight against malaria. After the war they were widely used in a variety of products and applications: as propellants in aerosol sprays, as coolants in home and automobile air conditioners, as solvents in the electronics industry, and as agents in molding polystyrene plastic foams such as Styrofoam. In some industrial nations, including the United States, manufacturers no longer use CFCs as propellants in aerosols, and some plastic foam makers have phased out CFCs.

The chemical stability that makes CFCs so useful in industry enables them to destroy stratospheric ozone. It takes about eight years before CFCs reach the stratosphere; once there, they may persist for decades before they break down. CFC-12, for example, which accounts for 45 percent of ozone deple-

FIGURE 14-13: College students on spring break, South Padre Island, Texas. Depletion of stratospheric ozone is expected to precipitate an increase in cases of skin cancer worldwide, particularly among lighter-skinned peoples, who tend to be more susceptible to the effects of damaging ultraviolet radiation. In the United States alone it may cause 3 million to 15 million new cases of skin cancer by 2075. During that same period ozone depletion may be responsible for roughly an additional 800,000 cancer deaths, according to an estimate by the U.S. Environmental Protection Agency.

tion, has a lifetime in the atmosphere of 111 years. Eventually, however, the sun's ultraviolet energy destroys the CFC molecule, releasing chlorine and fluorine atoms. Chlorine reacts with ozone, converting it to ordinary oxygen. Because chlorine acts only as a catalyst and does not undergo permanent change in the reaction with ozone, a single chlorine atom can catalyze the destruction of as many as 100,000 ozone molecules.

The destructive potential of CFCs was first discovered in 1974 by F. Sherwood Rowland and Mario Molina of the University of California at Irvine. Other researchers and the CFC industry initially resisted their findings, but Rowland and Molina publicized their results and by 1978 CFCs had been banned from use as aerosol propellants in the United States, Canada, Norway, and Sweden. In spite of this measure the use of CFCs significantly increased, especially in the production of computer circuit boards, as a coolant, and as an important ingredient in plastic foams.

In 1985 British scientists reported that a dramatic change in the ozone levels had occurred over Antarctica. A hole, devoid of ozone, formed each September and October. Each year the size of the hole increased, until in 1987 it was as large as the continental United States. Though summer winds eventually caused it to disappear, the hole lasted longer than usual and spread toward Australia and South America. In 1986 researchers discovered high levels of chlorine in the atmosphere over Antarctica, which, they concluded, came from the breakdown of CFCs. In 1988, ozone depletion over the Antarctic was comparatively moderate, but in 1989, the seasonal depletion was once again severe, with a decline nearly equal to that which occurred in 1987. A decline in stratospheric ozone has also been observed over other parts of the globe (Figure 14-14). Because chlorofluorocarbons persist so long in the atmosphere and because such a large amount already exists in discarded items that continue to release CFCs into the air, it will be hundreds of years after the last CFC is manufactured before the atmosphere is free of them.

Indoor Air Pollution. Indoor air pollution became a topic of serious concern in the 1980s as Americans learned that the air inside homes, schools, and offices might pose a greater health threat than that found outside. (Box 14-1: Air Pollution and Human

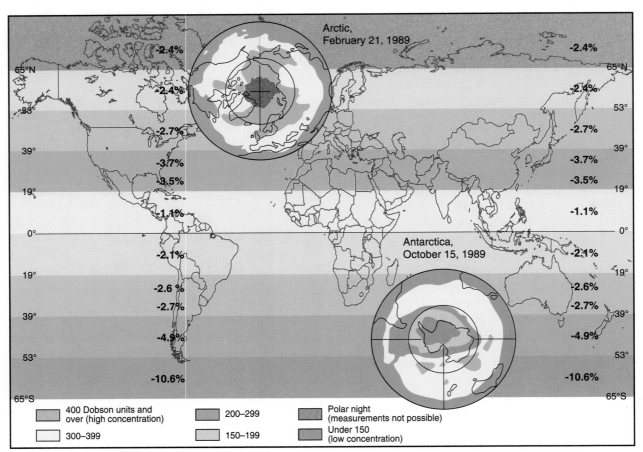

FIGURE 14-14: Thinning of stratospheric ozone. Between 1978 and 1987 the concentration of stratospheric ozone declined 1.1 percent at the equator, from 2.7 to 3.7 percent over the United States, and 10.6 percent over Antarctica. The polar insets show the concentration of stratospheric ozone at the dates shown (refer to color key).

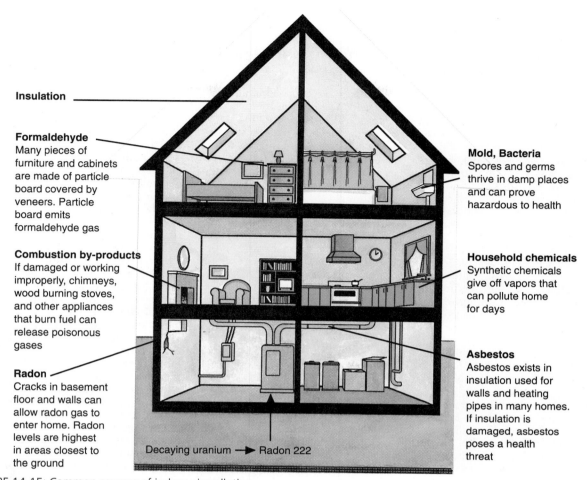

Insulation

Formaldehyde
Many pieces of furniture and cabinets are made of particle board covered by veneers. Particle board emits formaldehyde gas

Combustion by-products
If damaged or working improperly, chimneys, wood burning stoves, and other appliances that burn fuel can release poisonous gases

Radon
Cracks in basement floor and walls can allow radon gas to enter home. Radon levels are highest in areas closest to the ground

Mold, Bacteria
Spores and germs thrive in damp places and can prove hazardous to health

Household chemicals
Synthetic chemicals give off vapors that can pollute home for days

Asbestos
Asbestos exists in insulation used for walls and heating pipes in many homes. If insulation is damaged, asbestos poses a health threat

Decaying uranium ➝ Radon 222

FIGURE 14-15: Common sources of indoor air pollution.

Health, page 270, looks at the health effects of indoor air pollution.) The chief indoor contaminants are formaldehyde, radon 222, tobacco smoke, asbestos, combustion products from gas stoves and poorly vented furnaces (carbon monoxide, nitrogen oxides, sulfur dioxide, and particulates), chemicals used in building and consumer products (pesticides and household chemicals), and disease-causing organisms (bacteria, viruses) or spores (Figure 14-15). Inadequate ventilation can lead to carbon monoxide and nitrogen dioxide levels that exceed standards set for outdoor levels of those gases. In the homes of cigarette smokers, levels of carbon monoxide, particulates, and benzene may be considerably higher than standards set for outdoor air—up to 50 to 100 times higher! Mobile homes are particularly susceptible to indoor air pollution because they often have lower air exchange rates than conventional homes, are more likely to use propane for heating and cooking, and feature more plywood and other synthetic construction materials, many of which are made with formaldehyde.

In 1988 the EPA published the findings of its first in-depth study of indoor air in the workplace and public buildings. Among the chemicals found were benzene, a known carcinogenic substance emitted by synthetic fibers, plastics, and some cleaning products; trichlorethylene, a carcinogenic organic solvent used for cleaning and degreasing; carbon tetrachloride, a carcinogenic substance found in cleaning solutions; and paradichlorobenzene, an animal carcinogen found in mothballs and air fresheners. Formaldehyde and chloroform were also found. Chloroform, known to cause cancer in laboratory animals, is vaporized from the water in hot water sources such as hot showers. It is formed when chlorine, which is used to kill bacteria in water, interacts chemically with organic particles in the water. Even the EPA's headquarters in Washington, D.C., came under scrutiny. Two hundred of 6,000 employees reported having headaches, dizziness, sore throats, and burning eyes from working in the building. The most significant findings were harmful fumes given off from latex carpet backings.

Generally, federal and state health and safety standards regulate pollutants in workplace, school, and public buildings, but there are no laws that regulate pollutants in homes. Ironically, buildings constructed to be energy-efficient often have restricted air flow or the continual recirculation of the same

Box 14-1: Air Pollution and Human Health

Air pollution usually makes headlines only when something dramatic occurs, such as on December 2, 1984, when the accidental release of methyl isocyanate from Union Carbide's plant in Bhopal, India, killed as many as 3,000 people and harmed up to 50,000. And yet every day, many of us breathe polluted air that can adversely affect our health and may ultimately prove lethal.

Each time we breathe we take about 0.26 gallons (1 liter) of air into our lungs. That translates into some 7,800 gallons (30,000 liters) of air that pass through the lungs daily. When the lungs are functioning properly, they take in oxygen and expel carbon dioxide. Most of the time the body's defense system can cope with pollutants that enter the lungs. Mucus traps foreign particles, protecting delicate lung tissue. Cilia, minute, hairlike structures that line air passageways, help to move the mucus away from lung tissue. Coughing expels the mucus and impurities. Microorganisms and chemicals that escape or overwhelm initial defense mechanisms may cause infections or cell damage. In these cases, the body's immune system helps to fight the invaders and restore health. Pollutants too powerful for the body's defenses can affect health in a number of ways, including choking, labored breathing, burning and watery eyes, and, in the most severe cases, death.

Although the link between air pollution and human health is difficult to establish definitively, many experts believe that air pollution is responsible for as many as 50,000 premature deaths in the United States each year. Most of these deaths are believed to be caused by pollutants released when fossil fuels are burned. The effects on human health depend on the type of pollutant, its concentration, and length of exposure. An EPA study suggests that up to 1,700 cancer deaths each year are attributable to toxic air pollution. This study covered only a small fraction of the air toxins released to the air; it did not consider the synergistic effects of combined pollutants, accidental releases of chemicals, or secondary pollutants. Over long periods of time even mildly polluted air can have damaging effects: lung cancer, chronic bronchitis, emphysema, asthma, and lead or other heavy metal poisonings of the blood, nervous system, and kidneys (Figure 14-16).

Air pollution takes the greatest toll on the very old, very young, and persons already suffering from respiratory or circulatory disease. Researchers at the University of California at Irvine have shown that the risk of health problems due to air pollution is six times greater for children than for adults. Because they have smaller air passages and must breathe more air per unit of body weight to maintain their metabolism, children tend to collect more pollutants than do adults. Further, children often breathe through their mouths and so do not benefit from the filtering effect of the nasal passages.

Air pollution also has a greater effect in urban and suburban areas than in rural areas and a greater effect on cigarette smokers than on nonsmokers. A smoker living in a large city is at greatest risk, especially if that person works in a dusty environment or has impaired respiratory functions. Indoor air in

air, conditions that allow pollutants to build up to dangerous levels. Recent studies show that most U.S. citizens spend 90 percent of their time indoors, thus dramatically increasing their exposure to indoor pollutants.

Formaldehyde. Commonly recognized as a preservative for biological specimens, formaldehyde is also used in numerous other products, largely because it prevents bacteria and mold growth: foam insulation, permanent-press clothes, carpets, toothpaste, shampoos, some paper products, some medicines, and pressed wood products (plywood, particle board, and wood paneling). Formaldehyde emits fumes that can irritate the eyes, nose, and throat and cause shortness of breath, nausea, and headaches. This steady release of noxious fumes, called outgassing, is greatest when products are new, but materials can continue to release fumes for years. Because use of formaldehyde is so widespread, we may be exposed to it more frequently than to any other potentially dangerous chemical. A known carcinogen in rats, it causes mutations in bacteria cells and cell changes in primates, and some evidence links formaldehyde to skin cancer in humans. The EPA has not regulated the use of formaldehyde in consumer products, despite recommendations to do so from health scientists.

Radon 222. **Radon 222** is a naturally occurring colorless, odorless, and inert radioactive gas. It is formed by the disintegration of radium, itself a byproduct of the decay of uranium 238. Uranium 238 is an element found naturally in the earth's crust. Radon can be found in high concentrations in soils and rock containing uranium, granite, slate, phosphate, and pitchblende. It can also be found in soils contaminated from wastes of uranium or phosphate mining. Some of it filters through cracks in rocks and pores in the soil to the surface and passes into the atmosphere. Radon 222 can also seep through cracks in basement walls or around improperly sealed pipes and enter homes.

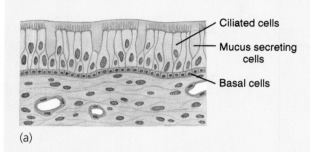

- Ciliated cells
- Mucus secreting cells
- Basal cells

(a)

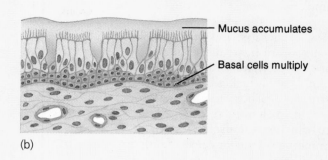

- Mucus accumulates
- Basal cells multiply

(b)

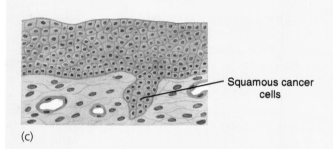

- Squamous cancer cells

(c)

FIGURE 14-16: A brief history of lung cancer. (a) In a healthy lung the lining of the respiratory passages includes both ciliated and mucus-secreting cells. In a smoker's lung, the cilia become partially paralyzed and mucus accumulates on the irritated lining. (b) In an early cancerous state the underlying basal cells divide rapidly and displace the columnar ciliated and mucus-secreting cells. (c) In later stages most of the normal, columnar cells are replaced by the cancerous cells. If the disease reaches an advanced stage, clusters of cancerous cells may be carried away by the lymphatic system and spread to other parts of the body.

improperly vented buildings, as well as air polluted by improperly operated wood-burning stoves, can also play an important role in the development of the chronic or acute effects of air pollution. Breathing difficulties, increased susceptibility to respiratory infection, headaches, irritation of the nose, throat, eyes and sinuses, nausea, drowsiness, colds, coughs, and impaired judgment are often good indications of dangerous levels of pollutants.

Radon 222 is stable only for a few days, after which it undergoes a series of four changes to form chemically active solids. These solids can attach to dust particles or clothing. When inhaled, they can penetrate lung tissues, emitting dangerous radiation that may cause lung cancers. Next to smoking, radon exposure is the most common cause of lung cancer. EPA studies indicate that exposure to radon may contribute to or cause as many as 5,000 to 20,000 cases of lung cancer each year and that radon might be a threat in 1 of every 8 homes. In homes with increased radon levels, those at greatest risk are smokers, children, and people who spend a lot of time indoors.

Geographic locations having greater amounts of uranium typically emit increased amounts of radon 222. In Colorado radon exposure has been traced to working in uranium mines or other hard rock mines and to proximity to uranium mill tailings, the waste rock left over after the ore is processed. Uranium tailings were used as fill under the concrete slabs on which homes are built, as backfill around foundations, as heat sinks in fireplaces, and as an accidental component of concrete or mortar. In east central Pennsylvania radon gas has been discovered in high levels in some homes built over naturally occurring "hot spots" in an area called the Reading Prong. Abnormally high levels were first noted in 1984, when the radiation detection badge of an atomic energy employee registered high radiation levels from his home and not from his workplace. This finding led to the realization that radon was a threat in homes and not just near uranium mines.

Tobacco Smoke. Smoking poses numerous health risks. Cigarette smoke contains such toxics as carbon monoxide, nitrogen oxides, hydrogen cyanide, and cadmium, as well as carcinogenic substances such as formaldehyde, vinyl chloride, nitrosamines, nickel carbonyl, and benzopyrene. Compared to nonsmokers, smokers suffer more ordinary respiratory illnesses and have been shown to have more complications, such as pneumonia, after surgery. They also suffer death rates that are twice as high as those of non-smokers at any age.

A plethora of health effects are associated with smoking, among them cough, loss of appetite, short-ness of breath, abdominal pains, headaches, weight loss, bronchitis, and insomnia, but the two greatest dangers are heart disease and lung disease. Smokers are about twice as likely to suffer an immediately fatal heart attack as nonsmokers. Researchers specu-late that carbon monoxide, which binds to hemo-globin in the blood, reduces the amount of oxygen to the heart muscle, forcing it to pump harder. Smoking is a major cause of emphysema, a disease in which the alveoli, tiny sacs found at the end of each bronchiole in the lungs, do not expel air as they should. Emphysema sufferers have difficulty expelling used air from their lungs in order to take a fresh breath. Cigarette smoking is also a major cause of lung cancer; an estimated 85 percent of all lung cancer cases are attributable to smoking.

Passive, or second-hand, smoke, can also be dan-gerous. Nonsmokers, including children, who are exposed to cigarette smoke may be at higher risk for developing respiratory illnesses and lung cancer. A 1990 University of California at San Francisco report summarizing research on second-hand smoke for the EPA, found that the health effects on non-smokers include as many as 32,000 heart-disease deaths annually.

Asbestos. Found in insulation, vinyl floor tiles, cement, and other building products, asbestos con-tributes to lung cancer and asbestosis, an ailment in which breathing is made difficult because of the presence of asbestos fibers in the lungs. In general, it poses a problem only in older buildings; newer building materials are typically made without asbestos. Because it becomes a health threat only if the asbestos fibers are released to the air and inhaled, it is thought to be relatively safe as long as the materials containing asbestos are not disturbed.

Airborne Toxins. **Airborne toxins** are a large and varied group of chemical pollutants that pose a sig-nificant hazard to human health and the environ-ment (Table 14-3). By volume, they are the largest source of human exposure to toxic substances.

Airborne toxins are a problem nationwide. Atmo-spheric deposition may account for 20 percent of toxic water pollution in the United States. Atmo-spheric mercury is a contaminant in inland lakes in northern Minnesota, Wisconsin, and Michigan. Almost one-third of the heavy metals entering the Chesapeake Bay come from the air. Airborne toxins are one of the largest sources of contaminants, including lead, zinc, and arsenic, for all of the Great Lakes. For example, toxaphene is an insecticide used primarily on cotton and secondarily on soybeans, peanuts, and coffee. Although 86 percent of the croplands treated with toxaphene are in the South, the chemical has been detected in Great Lakes fish since the 1970s.

Just as airborne toxins can pollute water systems, contaminated water supplies can give rise to toxic air pollution. Each day sewer systems receive thou-sands of gallons of cleaning fluids dumped by met-al finishing operations, sludges collected from smokestack scrubbers and dumped by chemical plants, and household chemicals like paint thinner. These and other toxins can revaporize at sewage treatment plants and return to the air. An EPA study concluded that 150 tons (136.5 metric tons) of six known carcinogens, including benzene, chlo-roform and 1,2 dichloroethane, volatilize annually from the secondary aeration tanks of Philadelphia's sewage treatment plant.

The Emergency Planning and Community Right to Know Act of the 1986 Superfund Amendments and Reauthorization Act requires major industrial producers to determine how much of each toxic chemical they release to air, water, land, public sew-ers, or injection wells. They report these emissions to the EPA, which releases the information to the public in a report called the Toxic Releases Invento-ry (TRI). In 1988 U.S. industry spewed 2.5 billion pounds (1.125 billion kilograms) of toxic substances into the air—10 pounds (4.5 kilograms) per person. The ten states reporting the highest totals of toxic emissions were Texas, Ohio, Tennessee, Louisiana, Virginia, Utah, Indiana, Illinois, Alabama, and Michigan. Chief polluters are chemical producers, steel and other metal manufacturers, paper makers, and the auto industry. Because the TRI covers only manufacturing processes and excludes utilities, gov-ernment-owned plants, mining operations, waste management firms, and minor industrial producers, the total amount of toxic emissions is probably much higher.

It is important to realize that many of the toxins emitted to the nation's air are done so legally. Although the EPA is required by law to regulate emissions of toxic substances, by 1990 the agency had set emission standards for only a handful of more than 300 toxic chemicals. Specific examples provide some idea of the scope of the problem. The AMAX Magnesium complex in Tooele, Utah, for instance, emits 110 million pounds (49.5 million kilograms) of toxins annually, making it the nation's single largest industrial source of toxins. In south-east Texas, residents of Port Neches run a 1-in-10 risk of getting cancer because of exposure to the 1 million pounds (450,000 kilograms) of butadiene emitted by Texaco's Neches West Chemical Plant. In 1986 chemical plants spewed 200 million pounds (90 million kilograms) of toxic chemicals into the

▶ Table 14-3
Airborne Toxins

Airborne Toxin	Environmental Effects	Health Effects	Sources
Lead	Bioaccumulates in food chains	Destroys blood-cell forming tissues in bones Damages liver and kidney Damages brain and central nervous system, which can cause intelligence and behavioral problems Children and fetuses especially vulnerable	Leaded fuels Paint in old homes Lead solders in old pipe Industrial sources such as lead smelting, lead battery manufacturers, metal casting foundries
Copper	Adversely affects vegetation and ecosystems. For example, fumes from the now closed Anaconda Copper Company's smelting operations in Millcreek, Montana, have destroyed vegetation and disturbed ecosystems around the plant	Arsenic, a residue of copper smelting, believed to cause skin and lung cancer as well as nervous system and gastrointestinal problems	Copper smelting and refining plants
Cadmium	Can be toxic to wildlife Accumulates in plant tissues and is biomagnified at each step in the food chain	Interferes with body's use of calcium; prolonged exposure results in loss of calcium from bones, which can be painful; bones become brittle and break easily. After industrial effluents contaminated the rice supplies of a Japanese village, many residents were stricken with cadmium poisoning, known as "itai-itai disease," which means "it hurts, it hurts." Inhaling dust or vapors can cause fluid to accumulate in the lungs, a condition known as pulmonary edema, and death Chronic cadmium poisoning damages the kidneys and heart. May contribute to high blood pressure and heart disease	Can be found as a trace element in the ash from burned coal and cigarettes Industrial sources such as zinc and lead smelting, since it is usually present in zinc and lead ores; metal plating, plastics manufacture
Dioxins (family of 75 closely related compounds; most well known: tetra-chlorodibenzo-paradioxin)		Causes a skin disease known as chloracne Suspected of causing cancer and birth defects in humans Carcinogenic in animal tests	Contaminants produced during manufacturing processes involving chlorinated phenols Produced by incomplete incineration of chlorinated wastes and domestic rubbish, especially plastics such as PVC
Vinyl chloride Inflammable, gas		Painful vasospastic disorders of the hands Increased risk of malformations, particularly of the central nervous system Narcotic effect when inhaled in quantity Causes a rare form of liver cancer, angiosarcoma, in workers exposed to high levels of vinyl chloride	Plastics production
Polychlorinated biphenyls		Skin damage Possible gastro-intestinal damage Possible carcinogen	Electric transformers and capacitors, fluorescent lights, hydraulic fluid

FIGURE 14-17: Exxon refinery near Baton Rouge, Louisiana, one of the world's largest. It is located in what is often referred to as Cancer Alley, a stretch of land along the Mississippi River from Baton Rouge to New Orleans that is home to many petrochemical plants and industrial sites. Residents of Geismer, St. Gabriel, and Carville, towns along Cancer Alley, suffer an unusually high cancer rate.

skies over the residents of Geismer, St. Gabriel, and Carville, Louisiana. Most were released under permits granted by the Louisiana Department of Environmental Quality (Figure 14-17).

What Factors Affect Air Pollution Levels?

Air pollutants are dissipated, concentrated, and transported as a result of the interactions of weather and topography. These factors may act singly or in combination to affect the air quality in a particular locale.

Weather. Air in the atmosphere is rarely still. Air masses warmed by the sun rise at the equator and spread toward the colder poles where they sink, eventually making their way back to the equator. The earth's rotation causes major wind patterns to develop near the earth's surface. During the day the land warms more quickly than the sea; at night it cools more quickly. Differential warming and cooling rates between the land and adjacent bodies of water direct the movement of local wind patterns. Typically, onshore breezes bring cooler, denser air from over the waters onto land during the day; offshore breezes move cooler, denser air from over the land masses out over the waters during the night. Wind patterns also affect precipitation. Warm, moisture-laden air rising from the oceans is carried inland, where the air masses eventually cool, causing the moisture to fall as rain, snow, sleet, or hail.

Hot, sunny, calm weather with stagnating high pressure cells and warm fronts usually favors the buildup of pollutants close to the ground, contributing to worsening air quality. Cool, windy, stormy weather with turbulent low pressure cells and cold fronts favors the upward mixing and dissipation of pollution.

Weather affects pollution levels in several ways. Precipitation helps cleanse the air of pollutants, although those pollutants are then transferred to the soil, lakes, or streams. Winds transport pollutants from one place to another, usually moving west to east, and sometimes traveling as far as 500 miles (805 kilometers) in a 24-hour period. For example, in 1982 sandy dust registered as particulate fallout in Indianapolis, Indiana, was traced to a severe windstorm with southwest winds out of Texas. Winds and storms may dilute pollutants with cleaner air, making pollution levels less troublesome in the area of their release. Air heated by the sun or air in a low pressure cell rises, carrying pollution with it. When wind accompanies the rising air mass, the pollution is carried aloft and diluted with clean, fresh air. Sometimes air sinks toward the ground, as in a high pressure cell. In the absence of winds, pollutants are then trapped and concentrated near the ground, often leading to serious air pollution episodes.

Weather can also affect pollution levels through chemical reactions. Winds and turbulence mix pollutants together in sort of a giant chemical soup in the sky. Energy from the sun, moisture in clouds, and the proximity of highly reactive chemicals may cause the formation of secondary pollutants, many of which may be more dangerous than the original pollutants.

Topography. Topography, the shape of land formations, helps to direct pollution movements. For example, valleys can act as sinks for pollutants, and mountains can act as barriers to air flow. Cities like Pittsburgh, Pennsylvania, and Cincinnati, Ohio, suffer poor air quality because air masses often stagnate in the deep river valleys which these cities occupy. Large urban areas like Los Angeles and Mexico City, which are located in valleys ringed by mountains, experience even more serious pollution episodes. On average, Los Angeles has the worst air pollution in the United States, and Mexico City has the worst in the world. Why? During the night and early morning downslope winds push cooler, denser air from the mountains into the valley. Throughout the day the sun warms the air in the valley and the lighter, heated air rises, accompanied by upslope winds. But if the upslope winds are not strong enough to carry the air mass over the mountains during the day, the pollution that the air mass holds does not leave the valley. Instead, the pollution keeps moving up and down with the rising and falling air masses, and each day the total pollutant load in the air increases.

Mountains affect the movement of air masses and the deposition of pollution in other ways as well. When an air mass encounters a mountain, it begins to rise. As it rises, the air cools and precipitation forms. Often, the air mass will drop its moisture—and whatever pollutant load it carries—on the windward side of the mountain. Once it clears the mountain, the air mass has lost its moisture. Consequently, the lee side of the mountain receives little or no precipitation. This phenomenon is known as the **rain shadow effect,** and it explains why one side of a mountain may experience severe damage from air pollution while the other side of the same mountain remains essentially unharmed (Figure 14-18). Prevailing winds along the Pacific coast transport nitrogen oxides hundreds of miles inland. When air masses reach the Sierras and the Rocky Mountains, they drop their moisture load on the windward (west-facing) side of the mountains, primarily as acid snow. Many of Colorado's high mountain lakes are now threatened by acid precipitation, and damage to Colorado blue spruce is similar to that found in acidified forests in southern Germany.

Temperature Inversions. Usually, air temperature decreases with distance from the ground. Heated gases emitted from the surface and air heated by normal reradiation from the surface rise until their temperature cools to the same temperature as the surrounding air. Occasionally, however, because of specific weather, topographic, or climatic features,

cooler air becomes temporarily trapped beneath a layer of warmer air (Figure 14-19). Any hot gases emitted at ground level pass through the cold layer next to the ground and into the warmer inversion layer above. When their temperature equals the temperature of the air around them, they stop rising. In this way, the inversion layer acts as a cap to prevent the further upward dispersal of the pollutants. Typically, inversions last less than a day, and the pollution quickly disperses. Under special conditions, however, pollution can remain trapped near the ground for up to several weeks.

Temperature inversions may be caused by radiational cooling, downslope movements in valleys, and high pressure cells.

Radiational cooling occurs after the sun sets. The surface of the ground reradiates heat faster than the air above it. Therefore, the ground tends to cool faster than overlying air, and lower air masses tend to cool faster than the air above them. This denser, colder air usually remains close to ground level. By morning, before the sun has had a chance to warm the earth's surface, a cool air mass remains under a warmer air mass above it. This type of radiational cooling is common in the temperate zone during the cool, clear nights of early spring and fall.

Cold air descends from higher valley walls or mountain slopes during the night and tends to rise again during the day. Inversion layers caused by downslope air movements tend to be deeper and longer lasting than radiational layers. Consequently, larger amounts of pollution can accumulate in them and be trapped for longer periods of time. In the fall

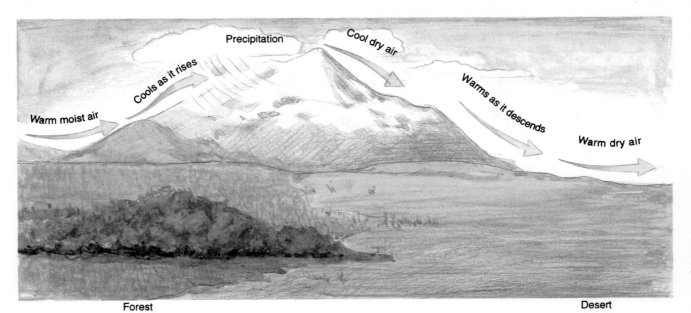

FIGURE 14-18: Rain shadow effect. As a warm, moist air mass approaches a mountain, it rises. As it does so, the air cools, causing the moisture to condense and precipitate out as rain, snow, fog, or mist. After the now cool and dry air mass clears the top of the mountain, it begins to descend and warm the other side. Because the air mass contains little or no moisture, the land in the lee of the mountain receives little precipitation.

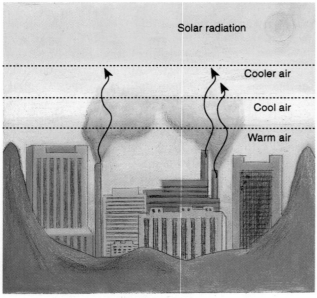

Normal pattern

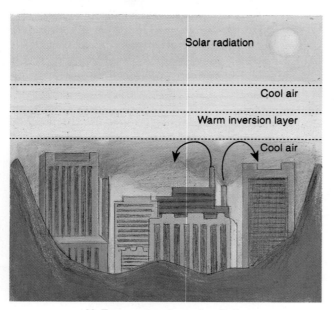

Air Temperature Inversion Pattern

FIGURE 14-19: Temperature inversion. Normally, hot gases released from stationary and mobile sources rise and disperse. Unusual weather patterns, however, can create a temperature inversion, in which a layer of lighter, warm air acts as a cap, holding cooler, denser air closer to the ground. Hot gases begin to rise through the cooler air but as they do, they cool and do not rise further. Consequently, the total load of pollutants near ground level increases.

and early spring a combination of radiational cooling and downslope air movements can intensify air pollution.

High pressure cells are generally characterized by clear, calm weather. Air masses in high pressure cells sink toward the ground, and air pressure at

FIGURE 14-20: Pittsburgh, Pennsylvania, fall 1945. Before the introduction of air pollution controls, stagnating air masses translated into poor air quality for the residents of industrial towns such as Pittsburgh, home to many steel manufacturers. This photo was taken at 9 a.m., looking east on Liberty Avenue.

ground level increases. As air becomes compressed, its temperature rises, often trapping a layer of cold air between the compressing air and the ground surface. This gives rise to an inversion layer that is generally higher in the air column than other inversion layers. When high pressure cells stagnate over an area for several days to a week, pollution can build up to dangerous levels (Figure 14-20). This type of weather pattern can occur in the fall, compounding the effects of radiational cooling and downslope air movements. More often, it occurs during summer when stationary weather systems trap pollutants close to the ground for extended periods of time. Because the effects of sunlight are greater during summer and weeks may pass without cleansing rain or winds, photochemical smogs can become even more dangerous.

History of Management of Air Resources

Through the middle of the twentieth century, the United States was characterized by growing industrialization and a rising standard of living. By the 1950s, however, Americans began to realize that life-styles which included plentiful electricity and a car (or two) in every garage meant paying the price of worsening air quality (Figure 14-21). In 1955 the first federal air legislation in the United States was

FIGURE 14-21: A change in life-style can help combat life-threatening air pollution. (a) A freeway in Los Angeles. (b) A cyclist in Chicago.

passed, enabling the U.S. Public Health Service to gather data, perform research, and assist state and local governments in their effort to understand and mitigate the effect of air pollutants. While this legislation did little to improve air quality, it was the first step in the process of identifying and regulating pollutants and safeguarding the nation's air resources. It set the stage for the Clean Air Act of 1963, the first major piece of legislation aimed at improving air quality.

What Is the Clean Air Act?

The Clean Air Act called for the states to develop air quality standards and to implement plans to curb pollutants. State and local governments were slow to respond, however, and in 1970 and again in 1977, growing public concern and increased environmental activism led Congress to pass tough amendments to the act.

Setting Emissions Standards. In 1970, Congress directed the EPA to develop standards limiting the quantities of air pollutants that could be emitted by specific sources. **Emission standards,** levels of pollutants that could be legally emitted, were established for stationary sources such as new power plants and factories. Standards were developed for six "criteria" pollutants: lead, particulates, sulfur oxides, carbon monoxide, nitrogen oxides, and volatile organic compounds. These pollutants either directly affect human health or give rise to other pollutants, such as ozone, that affect human health. New plants or factories could not be approved for use unless they

were shown to remove 55 percent of sulfur dioxide emissions, 20 percent of nitrogen oxide emissions, and 70 percent of particulates. Of particular concern was the sulfur found in coal. The standards called for a 70 percent removal of sulfur from all coal and a 90 percent removal from high sulfur coal. The idea was that as new plants replaced older plants, air quality would improve. But many older plants remained in operation, and because they were exempt from emission standards, they continued to pollute the atmosphere.

To determine the effectiveness of emission standards, the EPA established standards for the maximum allowable level of lead, total particulate matter, sulfur dioxide, carbon monoxide, nitrogen dioxide, and ground-level ozone, averaged over a specific time period, for ambient air. **Ambient concentrations** are those in the outside air that people breathe. The recommended limits are called national ambient air quality standards (NAAQS). Primary air quality standards are designed to protect human health and welfare; secondary standards protect the environment (by maintaining visibility and protecting structures, plants, animals, ecosystems, water supplies, and so forth). Deadlines were established for the attainment of primary standards but not for the attainment of secondary standards.

Continuous air sampling provides data to determine how often primary air quality standards are violated. Generally, the pollutant with the highest measured level is used to determine the air quality index, a measure of the frequency and degree of air pollution in a particular city or region. According to the ambient air quality standards, an index of 100 should not be exceeded more than once a year. An air pollution alert can be called when the index reaches 200 or higher.

During an alert persons with existing heart or respiratory illness are advised to reduce physical exercise and outdoor activities. In 1988 New Jersey violated federal standards for ground-level ozone on 45 days, one of the worst records in the country. During a particularly hot week in August, New Jersey residents with respiratory problems were advised to remain indoors; hospitals throughout the state reported a dramatic increase in admissions of patients with lung problems. As air quality worsens, people with health problems, the elderly, and the very young are also advised to curtail physical activity and stay indoors. At the emergency level even healthy persons may be advised to stay indoors and curtail physical activity. If emergency conditions persist, stationary sources such as factories and foundries can be ordered by the EPA to curtail any or all activities that would add pollutants to the atmosphere.

Each of the air quality control regions established across the United States by the EPA was ordered to reduce pollution and meet primary standards by 1982. To do so, each state had to establish specific emission regulations restricting the amount of pollution that stationary sources within its jurisdiction were allowed to release and deadlines to force rapid attainment of the standards in highly polluted areas. Extensions were granted through 1989 for complete attainment in some regions. Regions that did not meet deadlines for one or more primary standards were designated nonattainment regions; no plants could be built or expanded unless the region could demonstrate progress toward attainment of primary standards. If a state failed to develop EPA-approved plans, the federal government could withhold funds for major new construction or highway funds.

Meeting Emissions Standards. To meet the requirements of the Clean Air Act, power plants, manufacturing industries, and the automotive industry developed technologies to remove pollutants from emissions or to convert pollutants in emissions to harmless substances.

Devices to remove pollutants from emissions of stationary sources include electrostatic precipitators, which are commonly used to remove particulates in coal-fired electric generating facilities, and scrubbers, which remove sulfur dioxide but are not yet widely used (Figure 14-22). These devices collect toxic liquids or solids which must be properly disposed of; particulates can contain harmful trace elements, and sludge from scrubbers is often high in sulfur compounds.

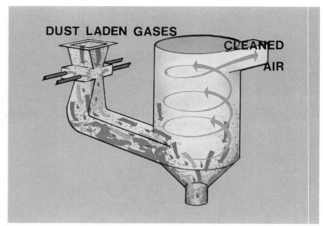

FIGURE 14-22: Venturi scrubber. This device allows electric generating facilities to burn sulfur-containing coal without emitting sulfur dioxide to the atmosphere.

Another way to reduce sulfur dioxide pollution is to use "clean" coal technologies. Sulfur can be "washed" or removed from coal before it is used. New techniques for burning coal can make the combustion more efficient so that less coal is used and can reduce the amount of sulfur produced. In the late 1980s the federal government and private industry spent billions of dollars on research and demonstration projects designed to reduce the main precursors of acid precipitation. Many of the promising demonstration projects employ new techniques that can be fitted to older power plants. One actually removes sulfur dioxide while leaving behind a commercial grade of sulfuric acid that can be sold, thus helping to offset the cost of installation.

In Japan and the former West Germany clean coal technologies developed in the United States have been used since 1974 to reduce nitrogen oxide emissions. Both Japan and Germany have strict nitrogen oxide control regulations for stationary sources. Costs for clean coal technologies have dropped and are now lower than scrubbers. Many utilities are arguing for the right to install the most cost-effective technologies, while environmentalists want to set strict emission standards and then force compliance no matter what the cost. Clean coal technologies also allow older electric utility plants to retrofit their facilities, resulting in as much as double the output of electricity with lower emissions. This is important because as many as one out of four power plants were 30 years old or older by 1990. Power plants become less efficient after 30 years and must be replaced at great cost or retrofitted to extend their productive years.

Technological advances such as the catalytic converter help to reduce pollution from mobile sources. Catalytic converters are attached to the exhaust systems of cars; as exhaust gases pass through the converter, carbon monoxide and hydrocarbons are converted to water and carbon dioxide. Catalytic converters made possible the drastic cuts in automobile emissions required by the Clean Air Act. Future gains in emission reductions will probably be made through the use of refined converters, technologies that allow motor vehicles to burn cleaner fuels such as methanol, and particulate traps installed on new diesel-powered vehicles and retrofitted on older ones. Eventually, new technologies might lead to replacing diesel-powered buses in cities and urban areas. New EPA regulations to reduce pollutants went into effect for new urban transit buses in 1991 and will apply to other new vehicles, such as large trucks, by 1994.

How Did the Clean Air Act Affect Air Quality in the United States?

By 1990, all six of the criteria pollutants showed significant declines in emissions, with the largest reductions in lead and total particulate levels, which decreased 97 percent and 59 percent, respectively (Figure 14-23). Nevertheless, many urban areas had not met some of the primary ambient air quality standards. In 1990, approximately 74 million people in the United States lived in counties which failed to meet at least one air quality standard. Levels of ozone and carbon monoxide around major urban areas remain the biggest problems. Ground-level ozone became particularly serious during the unusually hot, drought-prone years of the late 1980s. In 1988 over half of the nation's population lived in areas where ozone levels exceeded the national standard at least part of the time. Ozone problems persist in large measure because the number of vehicles on the nation's roads continues to grow (Figure 14-24). In fact, an ever-growing number of vehicles remains the *single* largest source of the precursors of smog—volatile organic compounds, carbon monoxide, and nitrogen oxides.

Toxins also continue to contaminate the nation's air. Although the 1970 amendment to the Clean Air Act required the EPA to establish standards for hazardous air pollutants emitted from stationary sources, a 1989 EPA report indicated that of the more than 300 such substances identified, standards had been developed for only 9—asbestos, arsenic, beryllium, mercury, vinyl chloride, lead, benzene, sulfuric acid, and radioactive isotopes. After 20 years of official concern, hundreds of dangerous chemicals remained unregulated, including chlorine, dioxin, and formaldehyde.

In addition to air toxins from stationary sources, the EPA was also directed to establish national emission standards for pollutants from motor vehicles.

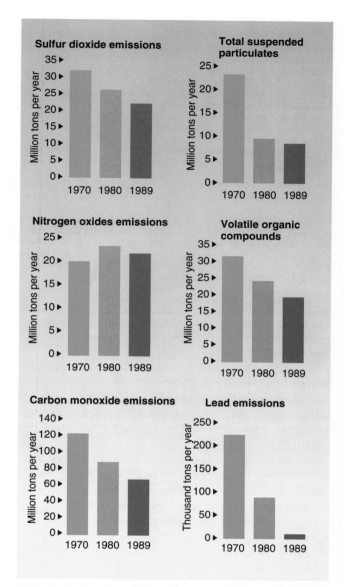

FIGURE 14-23: Air quality trends, 1970–1989. The Clean Air Act of 1970 significantly reduced emissions of the Big Six primary pollutants with the exception of nitrogen oxides.

Congress mandated that vehicle manufacturers meet these standards by 1982. After lobbying by the automotive industry, which claimed that the standards were too strict and imposed economic hardships on manufacturers, a series of extensions was granted to car and light truck manufacturers, pushing attainment back to 1988–1989. Ironically, manufacturers were able to meet the 1975 emission standards for automobiles sold in California, a state with stricter controls than the federal government.

While industrial sources and automobiles account for our most serious air pollution problems, other sources also contribute to worsening air quality. In the United States the largest unregulated source of air pollution is wood-burning stoves. Because most wood-burning stoves are airtight, they can burn

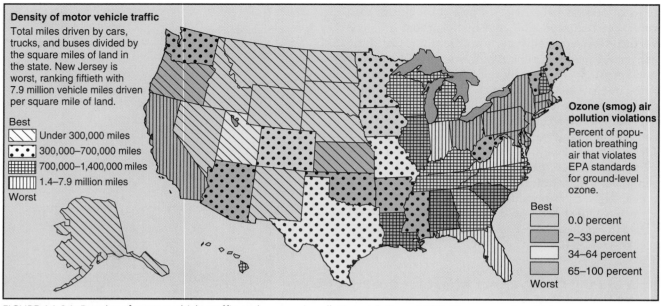

Density of motor vehicle traffic

Total miles driven by cars, trucks, and buses divided by the square miles of land in the state. New Jersey is worst, ranking fiftieth with 7.9 million vehicle miles driven per square mile of land.

Best

- Under 300,000 miles
- 300,000–700,000 miles
- 700,000–1,400,000 miles
- 1.4–7.9 million miles

Worst

Ozone (smog) air pollution violations

Percent of population breathing air that violates EPA standards for ground-level ozone.

Best

- 0.0 percent
- 2–33 percent
- 34–64 percent
- 65–100 percent

Worst

FIGURE 14-24: Density of motor vehicle traffic and ozone air pollution violations, 1990. Those areas in which motor vehicle traffic is greatest are also the areas in which the highest proportion of the population breathes air that violates EPA standards for ground-level ozone.

wood slowly with minimum air flow. While burning this way tends to conserve wood, it produces more carbon monoxide, volatile organic compounds, particulates, creosote, and carcinogens. In the mid-1980s, the country's more than 24 million wood-burning stoves and fireplaces produced millions of tons of unregulated pollutants. On windless winter days or days when a temperature inversion occurs, wood smoke envelops many communities in a choking mass of gas and particulates. Many cities, such as Denver, Colorado, regulate wood burning in an effort to improve the quality of their winter air. The states of Oregon and Colorado have passed regulations to control emissions and prevent the sale of wood stoves that do not meet their emission standards. Cities such as Medford, Oregon, and Albuquerque, New Mexico, issue citations to wood-stove users during air alerts and under certain weather conditions.

Indoor air pollution is receiving growing attention from regulatory agencies. Testing by both state and federal agencies has revealed that 17 states and the District of Columbia contain areas where radon levels exceed the EPA's safe exposure standard. According to the EPA testing program, more than 45 percent of the homes tested in Minnesota and more than 60 percent of the homes tested in North Dakota have levels over the standard. In 1988 the EPA and the Surgeon General's Office issued a statement calling for radon testing in virtually all homes in the United States. Radon test kits are commercially available and easy to use. Corrective measures, such as cementing cracks and repairing seals around pipes and ventilators, are usually inexpensive to moderately priced.

What Are the Provisions of the 1990 Amendment to the Clean Air Act?

After thirteen years of political wrangling, Congress finally amended and reauthorized the Clean Air Act in 1990. Its major provisions concern acid precipitation, airborne toxins, and smog. The country as a whole must cut sulfur oxide emissions by 50 percent by 2000 and reduce nitrogen oxide emissions by 33 percent starting in 1992. The most significant cuts are to be made by 1995 by over 100 of the dirtiest utility plants in 21 states (most of them in the Midwest). Utilities can choose between installing scrubbers or switching to low-sulfur coal; plants reducing emissions below required levels may sell their emission permits to other plants. A proposal that would have required utilities to reduce emissions of the heavy metals mercury and cadmium, which are released from coal-fired plants, was deleted from the final draft of the legislation. In addition, the steel industry was given until 2020 to eliminate cancer-causing emissions from coke ovens.

The 1990 amendment requires industrial emitters of 189 airborne toxins to install "maximum achievable control technology" by 2003. It also covers, to varying degrees, coke ovens, utilities, and municipal incinerators, sources previously exempt from toxic emission regulations.

Almost 100 cities have not complied with existing regulations mandating reductions in the emissions of the precursors of smog. According to the 1990 amendment, all but nine of those cities must comply with regulations by 1999, and another six must comply by 2005. The nation's most polluted cities have been granted longer compliance times: Baltimore and New York have until 2007 and Los Angeles until 2010.

Other regulations were also adopted to combat urban smog. Smaller industrial polluters, such as dry cleaners, printers, and automobile repair shops, must limit emissions of smog-forming chemicals. Moderately or severely polluted urban areas must achieve annual reductions in ozone—15 percent within six years and 3 percent each year after that—until the law's standards are met. The amendment establishes strict standards for tail pipe emissions beginning with 1994 models and requires that the standards be maintained over a vehicle life of 10 years or 100,000 miles (161,000 km).

Industries most affected by the Clean Air Act, such as utilities and automakers, have balked at the $25 to $35 billion a year reforms will probably cost, but the costs will eventually be passed on to consumers. Moreover, environmentalists maintain that it is a small price to pay; not only is clean air priceless, but the damage to human health and the environment caused by air pollution could have cost twice that much if the act had not been passed.

What Initiatives Are Being Undertaken to Improve Air Quality in the United States?

Promising initiatives are underway throughout the United States to protect air quality. The situation remains so bad in Los Angeles that it must now consider drastic, nontraditional steps to reduce its air pollution and meet tough new federal and state standards. California's south coast air basin, despite enforcement of the nation's strictest air quality control programs, has the worst air quality problems in the United States. Its maximum ground-level ozone concentration is three times the federal standard. It is the only air in the country that consistently fails to meet the nitrogen dioxide standard, it has the highest carbon monoxide levels in the country, and it has one of the highest levels of fine particulates. Mobile sources still produce 60 percent of the total pollution.

Some of the measures under consideration include regulating the use of household cooling appliances, lawn mower engines, consumer aerosol products, and outdoor grills. More stringent standards will be imposed on auto emissions to reduce nitrogen oxides. Federal employees will use alternative fuel vehicles. Other measures include improving vapor recovery at service station pumps, the development of an electric van with a range of 120 miles (193 km) and top speeds of 60 miles (97 km) per hour, and a comprehensive ride sharing program. These nontraditional approaches can be accomplished only with changes in life-style and political will.

A promising initiative is being undertaken by some members of the chemical industry. Monsanto is going to try to eliminate 90 percent of the emissions of air toxins from its factories by 1992. It will need to develop new technologies and redesign some of its processes. For example, a switch from a wet to dry process in a moth crystals plant will reduce toxic emissions by 750,000 pounds (337,500 kg) per year from that one plant. Dow Chemicals U.S.A. is working on new technologies to control emissions of highly volatile chemicals through source reduction and recycling.

What Initiatives Are Being Undertaken to Improve Global Air Quality?

In recent years progress on global air quality issues has been mixed. The chief victory came in 1987, when the Montreal Protocol on Substances that Deplete the Ozone Layer was signed under the guidance of the United Nations Environment Programme (UNEP). Its aim was to reduce the consumption of CFCs 50 percent by 1998. By 1990 mounting evidence about the dangers of stratospheric ozone depletion compelled the participating countries, which include the United States, to strengthen the protocol. They agreed to eliminate consumption of CFCs by the year 2000 and to phase out the use of halons by the year 2000, beginning with a 50 percent reduction in 1995. Exceptions are granted for essential uses of halons in high-hazard situations. Controls were also introduced for carbon tetrachloride and methyl chloroform, two other ozone-depleting chemicals. Excluding their use as a feedstock in the production of other substances, the consumption of carbon tetrachloride and methyl chloroform will be eliminated by 2000 and 2005, respectively. Perhaps most significantly, the parties to the protocol established a special fund to enable less-developed countries to afford less polluting, but more expensive technologies. The LDCs have a 10-year grace period to install the newer technologies. Largely because of this provision, India and China, the world's two most populous countries and the most significant countries still outside the protocol, indicated that they would soon sign the agreement.

While no agreements have been reached concerning global warming, intensified research and increased public awareness may soon result in action at the international level. One of the major goals of the 1992 United Nations Conference on Environment and Development, held in Brazil, was to negotiate an international climate control treaty to limit countries' emissions of greenhouse gases. Proposed strategies include reducing the consumption of fossil fuels through energy efficiency and conservation measures for buildings and vehicles, developing alternative fuels, placing a tax on the carbon content of fuels, eliminating natural gas leaks from pipelines, and replanting forests. It has been estimated that 1

Environmental Science in Action:
Acid Precipitation in the Adirondack Mountains
Debbie Lokai

The Adirondack region of upstate New York, once renowned for its unspoiled wilderness areas and hundreds of lakes teeming with fish, has achieved notice of a more dubious nature. It is one of the first regions in the United States to suffer significant and widespread damage from acid deposition, which includes acidic rainfall, snow, mist, and dry forms of precipitation.

Describing the Resource
Physical Boundaries

The Adirondack Mountains stretch from the Mohawk River valley in the south to the St. Lawrence River in the north and from Lake Champlain in the east to the Black River in the west (Figure 14-25). Although near the heavily populated eastern seaboard, the Adirondacks encompass twelve wilderness areas so remote that they can be reached only by canoe or foot and a northwest stretch of 50,000 acres (20,235 hectares) that has never been touched by loggers. The region has remained wild because 6 million acres (2.43 million hectares) of the mountains are in the Adirondack State Park. The largest park in the lower 48 states, it contains more land than the state of Massachusetts. Unfortunately, this vast preserve is vulnerable to the effects of acid precipitation because of its location, altitude, and geologic formation.

The northeastern region of the United States is subject to an unusually large amount of acid-forming emissions. About 90 percent of the sulfur, nitrogen, and other acid precursors found in the northeast United States results from human activities. Approximately 75 percent of the acid deposition in the Adirondacks originates in Pennsylvania and the Midwest, particularly Ohio, Indiana, and Illinois. Prevailing air currents transport sulfur dioxide from these areas hundreds of miles eastward. When air masses reach the Adirondack Mountains, the first high-altitude area in the eastward path, they

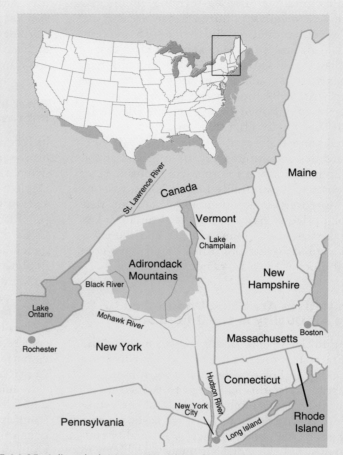

FIGURE 14-25: Adirondack Mountains. The Adirondack region of upstate New York is one of the regions most affected by acid precipitation worldwide.

ascend and cool, often resulting in rain or snow. Thus, air currents regularly drop the bulk of their acid load on the Adirondacks as they move to the east. Because the higher peaks receive more acid deposition, lakes and vegetation in the more remote, high-altitude regions of the Adirondacks were the first to show adverse effects.

The Adirondacks are also sensitive to acid deposition because of their geologic structure. As an extension of the Canadian shield, a large sheet of Precambrian granite, they lack the thick soil cover that could neutralize at least some of the acids from the precipitation. Consequently, much of the acid that is deposited can reach vegetation and lakes.

Biological Boundaries

The Adirondack State Park has provided a safe, natural haven for a wide variety of plants and animals. The forests contain large numbers of northern hardwoods, hemlock, and white pine. Conifers, especially spruce and balsam, flourish at higher elevations. The 2,200 ponds and lakes are home to many types of sporting fish, including trout and salmon, as well as other aquatic life. Deer, black bears, small game animals, and coyotes range through the forests, and a healthy beaver community has reestablished itself since it was almost trapped into extinction in the mid-1800s. However, all these populations have become potential victims of acid deposition.

How Acid Deposition Affects Soil and Vegetation. As acid rainfall filters through soil, it affects it in ways that then affect vegetation. By bonding with calcium, magnesium, potassium, and other soil nutrients essential to plant growth, acids make these minerals unavailable to vegetation. In the Adirondacks, where soil contains small amounts of these nutrients, this process severely retards the growth of trees and other plants. Acid precipitation can also leach metals from soil and bedrock, thus freeing them and making them available for absorption by plants. Metals such as aluminum and manganese can kill plant roots or damage their ability to absorb nutrients.

Acid deposition attacks trees and plants from above as well as from below. Sulfur dioxide gases entering trees through stomata on leaves and needles can erode cuticle wax and leave the trees susceptible to insect infestation and loss of moisture. The acid also affects buds and bark and can remove nutrients directly from needles and leaves. Trees and plants not killed outright may be so weakened that they are more likely to succumb to natural stresses such as insects, disease, and harsh weather. Studies show that forests at high altitudes in the Adirondacks are growing at slower rates and losing leaves and needles. Red spruce, especially at risk, have experienced a substantial decrease in growth rate in the past 25 years (Figure 14-26).

How Acid Deposition Affects Water and Aquatic Life. Most lakes receive water from direct rainfall and from runoff that filters through the surrounding soil. As acid runoff leaches metals from the thin soil, it washes them into streams and lakes where they can damage aquatic life. Aluminum is especially deadly for fish because it damages their gills, ruining their ability to absorb oxygen from the water. Other metals, such as mercury, accumulate in fish tissues and can be harmful for humans or other animals that eat the fish.

Although runoff picks up toxic metals as it filters through soil, the filtering process can also help to buffer the acids before they reach the lake. Soil rich in minerals such as calcium tends to neutralize acid because the minerals bond with the sulfate or nitrate component of the acid. A 1977 study showed that three lakes in the Adirondacks that received the same amounts of acid deposition varied widely in pH. The lake that retained a neutral pH was surrounded by soil eight times thicker than the soil around the highly acidic lake. Scientists concluded that the thicker soil gave the acid more opportunity to bond with minerals. However, many plants and animals need these same minerals as nutrients, and if the minerals are being tied up by molecules from acid rain, the plants and animals can be deprived. For example, studies of fish in acidified lakes show they lack calcium, a deficiency that sometimes can result in such weak bone structure that their muscles pull their skeletons out of shape.

The acid that remains in groundwater and the acid that falls directly into surface water harms aquatic life forms in ways that are still being explored. One study showed that comparatively low levels of acid (pHs between 5.0 and 5.5) seemed to prevent fish from smelling odors that help them migrate and find mates. Thus, even low levels of acid could virtually end reproduction in entire populations of fishes.

If acid deposition continues to assault a lake, eventually everything in it is destroyed. Almost no forms of aquatic life can live in high concentrations of acid, and those few that can survive may die because of lack of food. Although estimates vary from study to study, acid deposition has rendered approximately 200 lakes in the Adirondack region incapable of supporting fish populations. A study comparing acid levels in 274 lakes in 1989 to measurements taken in a rare study conducted 50 years earlier showed that 80 percent of the lakes had acidified in that time period.

How Acid Deposition Affects the Human Population. The heavy acid deposition in the Adirondacks may pose a danger to the 120,000 people who live in the region year-round as well as the additional 90,000 summer residents. Crops, fish, and meat may be contaminated with toxic metals or lacking in nutritional value. High concentrations of sulfur dioxide in the air can cause sore throats, coughing, and lung irritation and tissue damage. The Adirondack region experienced a mysterious outbreak of cases of gastroenteritis in the mid-1980s.

FIGURE 14-26: Damaged red spruce, Whiteface Mountain, New York. The technician is climbing the spruce tree to obtain samples for laboratory analysis. Conifers such as the red spruce are the first trees to show damage from acid deposition (loss of branches, discolored needles) because they grow at higher altitudes than hardwoods and are thus on the front line of attack. Needles are also more likely to collect particles and gases from the atmosphere than are the leaves of deciduous trees, making them more vulnerable to sulfuric and nitric acids.

Investigators later traced the outbreak to a normally rare form of bacteria. This bacterium was especially resistant to acid, and temporarily flourished when acid deposition killed off competing life forms.

All water sources in the Adirondacks are in danger of acidification as well, and consuming water with high acid levels is known to increase the chance of cardiovascular disease. An even more immediate problem with acidified water is created when it runs through lead pipes; as sulfuric acid corrodes the pipes, the water can become contaminated with lethal concentrations of lead.

Social Boundaries

The Adirondacks were explored early in the history of the United States. Most of the forests had been burned or cut down at least once by 1885. At this time people became concerned about the effect of the activities of increasing numbers of tourists, settlers, loggers, and others on the mountains. The New York State surveyor at that time, Verplanck Golvin, had developed a special interest in the area, which he had explored extensively. His interest and the public concern led to the creation of the Adirondack Forest Preserve in 1885 and the Adirondack Park in 1892. Two years later a constitutional amendment declared that forest preserve land would be protected as wild forest forever. Today, more than a million people enjoy hiking and camping in the Adirondacks every year.

Looking Back

Few records of acid levels in the Adirondacks in the past exist, but researchers believe that acid deposition may have begun to affect the area as early as the beginning of the twentieth century. For instance, Lake Awosting had a pH as low as 4.5 in 1930 and was empty of fish life as early as 1915; scientists theorize that increased activity in railroads, tanneries, iron forges, and other industries in the area during the late 1800s could have been responsible for killing the fish population and acidifying the lake.

Damage to lakes and trees in the Adirondacks could easily have occurred long before anyone noticed, since the first damage occurs at the highest, most inaccessible altitudes. Acids also tend to kill smaller, weaker fish first, so fishermen who look only for the bigger fish would not notice any change until the whole population had been affected. However, by the 1970s people were concerned about obvious changes, and scientists began to study why and how these changes were taking place.

The Clean Air Act Controversy

By the time the Clean Air Act came up for congressional reauthorization in 1977, scientists believed that acid precipitation in the Adirondacks would not end until midwestern electric plants curtailed acid-forming emissions. However, opponents of emissions controls argued that more study was needed, beginning a debate that would last fourteen years.

In 1984 Representatives Waxman and Sikorski proposed legislation to require the 50 main plants responsible for acid-forming emissions to install scrubbers. Although pollution control in the United States has traditionally been based on the idea that "the polluter pays," the proposed bill mandated that the new scrubbers be funded by a tax on fossil fuel energy in all 48 contiguous states. Representatives of the nonpolluting states, especially those from states where utility customers had been footing the bill for "clean" energy for years already, objected. They were opposed by the representatives of the 19 states where the 50 sources are found, especially Pennsylvania, West Virginia, Georgia, Ohio, Indiana, Kentucky, Tennessee, and Missouri, which would be most affected. Elected officials, industry representatives, and the general public in those states maintained that they should not be forced to bear the full costs of cleaning up emissions. They warned that such a move would unfairly target industry in their states and wreak havoc on state economies. Not surprisingly, Congress never passed the bill. Other legislation that would allow plants to choose their own methods of emissions control also failed. States that produce high-sulfur coal, such as Ohio, were afraid that plants would opt for burning low-sulfur coals rather than installing expensive scrubbers, thus eliminating jobs in their coal industries.

Opponents of emissions regulations stalled decisive action by insisting on even more research. They claimed that no one knew if expensive measures to control emissions really would reduce acid precipitation. President Reagan assisted this coalition in 1980 by sponsoring the National Acid Precipitation Act, which called for a 10-year program of research into the causes and consequences of acid deposition. By this time study and attempts at reducing emissions in other countries, such as Great Britain, had shown a definite correlation between limiting pollution and reducing acid deposition. Throughout the 1980s most experts felt that, while more investigation might reveal ways to mitigate the damage of acid rain, waiting to reduce emissions could only lead to increased and possibly irreversible damage in hard-hit areas.

Finally, Congress reauthorized the Clean Air Act in 1990. The act stipulates that sulfur dioxide production must be reduced to 10 million tons (9.1 million metric tons) per year, half of current output, by the year 2000. Utility plants affected by this bill (in Pennsylvania, West Virginia, Georgia, Ohio, Illinois, Indiana, Kentucky, Tennessee, and Missouri) must either install scrubbers or switch to low-sulfur coal. The legislation also mandated that nitrogen oxide production be reduced by 33 percent (to 4 million tons or 3.6 million metric tons per year) beginning in 1992.

How Scientists Are Attempting to Help

Although the only way to prevent acid precipitation is to eliminate acid-forming emissions, scientists are looking for ways to make the effects of acid deposition less harmful to life in threatened areas. Some researchers are concentrating on stocking lakes with fish that are especially good at surviving in acidic waters; geneticists are trying to breed strains of acid-resistant fish. However, some environmentalists point out that even if the fish can tolerate acidity, they must also be able to survive high metal concentrations. In addition, they may die from lack of food if the rest of the aquatic population has disappeared. Other environmentalists point out that finding acid-resistant fish would simply give the polluter states and power plants an excuse not to clean up emissions.

A more widely accepted area of experimentation is liming lakes that have high acid levels on the assumption that the highly alkaline lime will neutralize the acid. The New York State Department of Environmental Conservation and some private organizations have been conducting studies on liming lakes in the Adirondacks since 1959. They have found that liming usually restores a normal pH and so can be used in conjunction with restocking programs to restore lakes that have become acidified. Skeptics point out that acid deposition damages the entire watershed, not just the lakes. They emphasize that liming can neither repair watershed damage nor bring back the rest of the biotic life which has been lost. Moreover, liming is expensive. Estimates predict that it would cost $10 to $20 billion to lime several hundred lakes in the Adirondacks over five years; restocking and monitoring efforts would cost even more. Of course, if acid deposition continues, a limed lake soon loses its normal pH; some lakes have re-

acidified in as few as six months after they were limed.

Looking Ahead

The reauthorized Clean Air Act goes a long way toward starting the fight against acid precipitation. In addition to its restrictions on utilities and other stationary sources of acid precipitation, the act places limits on emissions of passenger vehicles. However, other measures that could make an important difference remain to be taken. For example, raising the corporate average fuel efficiency standards on new automobiles could drastically cut emissions of nitrogen oxides as well as other pollutants such as hydrocarbons. In other words, the United States has taken an important step, but it is not enough to solve the problem. Also, because acid precipitation is a global problem, we must work vigilantly with other countries to reduce the production of acid-forming emissions worldwide.

acre (0.41 hectare) of tropical trees can absorb an average of 4.4 tons (4 metric tons) of atmospheric carbon dioxide per year. To remove 1.1 billion tons (1 billion metric tons) would require the reforestation of 400,000 square miles (1,036,840 square kilometers) of the tropics. The cost to do this, approximately $160 per acre ($395 per hectare) is not so imposing when we look at costs associated with results of the greenhouse effect, such as sea level rise and agricultural dislocation.

Future Management of Air Resources

There is little doubt that international concern over air quality will help to reduce global air pollution in the future. As a society, we can work toward cleaner air and a safer atmosphere by preventing the abuse of air resources, protecting human health and the environment from the detrimental effects of air pollution, and preserving living systems threatened by air pollution.

Prevent the Abuse of Air Resources Through Conservation

Within the United States the federal government needs to make sure industries affected by the reauthorization of the Clean Air Act meet deadlines for compliance. We need to go even further in mandating measures that will allow us to achieve an ultimate objective of zero emissions of toxic air pollutants. Such measures would strengthen restrictions on automobile emissions to counter the effects of future increases in the number of automobiles and increases in the number of miles driven. The measures should set tougher standards for secondary emissions. In business and industry, standards need to be established for hazardous indoor pollutants. Indoor air pollution can be further reduced by banning the use of formaldehyde in building materials and textiles.

In addition to passing legislation to protect air quality, the government should provide incentives for cities to meet clean air deadlines and support

What You Can Do:
To Safeguard Air Resources and Protect Yourself Against Air Pollution

- Install air exchange systems to reduce indoor air pollution buildup in tight, energy-efficient homes.
- Put catalytic converters on wood-burning stoves.
- Test your home for radon.
- Avoid the use of toxic chemicals at home, in school, and in the office.
- Become aware of the danger of skin cancer from ultraviolet rays,

a danger that increases as stratospheric ozone thins.
- Help improve global air quality by reforesting the earth: plant trees!
- Let your legislative representatives know you think it is important to pass laws to reduce reliance on fossil fuels and to pass stricter ground-level ozone laws.
- Find out about organizations

that are active in attaining and maintaining air quality. Write to:
National Clean Air Coalition
530 Seventh Street, S.E.
Washington D.C. 20003
(202) 523-8200

research in specific projects, such as packaging materials to replace polluting polystyrene plastic. Utilities and industries should be urged to reduce emissions at the source. We should also establish a national program encouraging individuals and businesses to reduce their use of energy derived from fossil fuels as much as possible. (See What You Can Do: To Safeguard Air Resources and Protect Yourself Against Air Pollution for suggested actions individuals can take to protect air resources and safeguard themselves against the effects of air pollution.)

Protect Human Health and the Environment

In order to protect human health and the environment from the effects of air pollution, we must encourage research that determines the total effect of exposure to the sum of all air pollutants. We should also provide more money for research into improved air pollution technology, energy conservation practices, and alternative energy sources. Emission standards for both stationary and mobile fuel combustion sources should be strict enough to eliminate dangers to human health. Manufacturers should be required to reduce or eliminate toxins in household products and to warn consumers about the toxins they include.

Preserve Living Systems

Because air is truly an international resource, moving freely and quickly around the globe, international efforts to protect air quality are especially important. The United States can participate in global initiatives by joining other countries in reducing sulfur dioxide and nitrogen oxide emissions by

50 percent by 1995 and by phasing out CFCs and other ozone-depleting chemicals as soon as possible. An agreement by the United States to limit carbon dioxide emissions is essential to international efforts to minimize global warming. Finally, we need to stop excessive cutting and burning of forests in the United States and support efforts to prevent tropical deforestation.

Summary

The atmosphere is all the gaseous matter that surrounds the earth. The innermost layer, called the troposphere, contains the gases that support life. The stratosphere contains a protective layer of ozone. Weather, the day-to-day patterns of precipitation, temperature, wind (direction and speed), barometric pressure, and humidity, occurs in the troposphere. The atmosphere plays an important role in influencing climate, the long-term weather pattern of a particular region.

Atmospheric gases allow some of the incoming solar radiation to pass through the atmosphere, but trap heat that would otherwise be reradiated back toward space. This phenomenon is known as the greenhouse effect. The greenhouse gases are essential to maintaining the conditions necessary for life on earth, but increased levels of carbon dioxide, methane, chlorofluorocarbons, and nitrous oxide can lead to global warming.

Any substance present in or released to the atmosphere that adversely affects human health or the environment is considered an air pollutant. Primary pollutants are emitted directly into the atmosphere. They may react with other primary pollutants or atmospheric compounds, such as water vapor, to form secondary pollutants. Of the thousands of primary pollutants, six are of special concern because they are emitted in particularly large quantities: carbon dioxide, carbon monoxide, sulfur oxides, nitrogen oxides, hydrocarbons, and particulates. In the 1970s efforts at cleaning up air pollution in the United States focused on primary pollutants. Currently, increased attention is being given to six air pollution issues that pose a particular threat: global warming, acid precipitation, photochemical smog, stratospheric ozone depletion, indoor air pollution, and airborne toxins.

Global warming is expected to cause changes in temperature, precipitation, world climate, and sea levels if not stopped in time. However, not everyone agrees that global warming is actually occurring or that it is being caused by human activity.

Acid precipitation damages or destroys aquatic ecosystems of lakes and streams, harms vegetation, reduces the productivity of crops, damages human health, and destroys statues and buildings. Acid surges are periods of short, intense acid deposition in lakes and streams. Damage caused by acid precipitation is widespread, both in the United States and globally.

Photochemical, or brown, smog is formed when hydrocarbons react with nitrogen oxides and oxygen to form chemicals such as ozone and peroxyacetyl nitrate (PAN). Photochemical smog is a serious problem in urban areas worldwide. Its primary component, ground-level ozone, is toxic; it can adversely affect human health in many ways and can also damage crops and forests.

Ozone in the stratosphere plays a critical role in protecting the earth from harmful ultraviolet radiation. The stratospheric ozone layer is extremely thin, so even small decreases are expected to cause significant increases in cancer rates among humans and decreases in marine productivity. Nitrogen oxides, fluorine, and bromine influence the speed at which ozone is destroyed. Atmospheric chlorine and fluorine have been traced to the chemical breakdown of chlorofluorocarbons (CFCs). In 1987 parties to the Montreal Protocol, amended in 1990, agreed to phase out consumption of CFCs by the year 2000.

Major indoor air pollutants are formaldehyde, radon 222, tobacco smoke, asbestos, combustion products from gas stoves and poorly vented furnaces, chemicals used in building and consumer products, and disease-causing organisms or spores.

Airborne toxins include over 300 chemicals—among them lead, cadmium, copper, dioxin, and vinyl chloride—that pose a significant hazard to human health and the environment.

Air pollution takes the greatest toll on the very old, the very young, and persons already suffering from respiratory or circulatory disease. In general it is worse in urban and suburban areas and can be concentrated in improperly ventilated buildings. Weather and topography affect pollution levels. Energy from the sun, moisture in clouds, and the presence of highly reactive chemicals in the atmosphere may cause the formation of new pollutants. Valleys act as sinks for pollutants, and mountains act as barriers to air flow that could carry pollutants away from certain locations.

In the United States air pollution is regulated chiefly by the Clean Air Act, first passed in 1963 and amended in 1970, 1977, and 1990. The Clean Air Act directed the EPA to develop national ambient air quality standards (NAAQS), emission standards that specify the quantities of air pollutants that can be emitted by specific sources. Major provisions of the 1990 amendment concern acid precipitation, airborne toxins, and smog.

Key Terms

acid precipitation	mesosphere
acid surge	ozone
airborne toxins	photochemical reaction
air pollutant	photochemical smog
albedo	primary pollutant
ambient concentration	radon 222
atmosphere	rain shadow effect
chlorofluorocarbons	secondary pollutant
emission standards	stratosphere
emissivity	thermosphere
global warming	troposphere
greenhouse effect	

Discussion Questions

1. Explain the greenhouse effect. How is it related to global warming?

2. Distinguish between primary and secondary pollutants, and give three examples of each.

3. What factors are contributing to the destruction of the stratospheric ozone layer? What effects could this destruction have?

4. Why is acid precipitation such a widespread problem? What political problems are involved with regulating it?

5. Explain the role of the EPA in reducing air pollution. Are its measures always effective? Why or why not?

6. Imagine that your community faces an air pollution problem. What steps would you recommend to reduce the pollution? Explain your choices in terms of the size, location, and demographic composition of your community.

Water Resources

The Hydrosphere

*All the rivers run into the sea, yet the sea is not full;
to the place from which the rivers come, there they
return again.*

Ecclesiastes

*The sober citizen who would never s ubmit his watch
or his motor to amateur tamperings freely submits his
lakes to drainings, fillings, dredgings, pollutions, stabi-
lizations, mosquito control, algae control, swimmer's
itch control, and the planting of any fish able to swim.
So also with rivers. We constrict them with levees and
dams, and then flush them with dredgings, channel-
izations, and the floods and silt of bad farming.*

Aldo Leopold

Learning Objectives

When you finish reading this chapter, you should be
able to:

1. List properties of water that enable it to support
    life.

2. Identify how water resources are classified and
    briefly describe each classification.

3. Describe worldwide water consumption pat-
    terns.

4. Identify eight broad categories of water pollu-
    tion.

5. Describe how people have managed water
    resources in the past.

6. Describe how both drinking water supplies and
    wastewater are treated.

7. List some of the current threats to groundwater,
    lakes, rivers, and marine waters.

Earth, the water planet. How incongruous it seems
that the word *earth* is both another word for soil and
the name of our planet when the seas cover 71 percent of the planet's surface! In addition to the water
in the seas are the water held in lakes, ponds, rivers,
streams, marshes, bays, and estuaries and the water
locked in polar ice caps and glaciers. Water so dominates the surface that it is responsible for the planet's familiar image as seen in satellite photos: a
beautiful blue ball existing in stark contrast to the
blackness of space.

Vast amounts of water also exist in liquid form
underground and as water vapor in the atmosphere.
All of this water is linked in a continuous cycle of
evaporation (from the surface to the atmosphere),
precipitation (from the atmosphere to the surface),
and runoff (from the land or beneath the land's surface). Less than 0.1 percent of the evaporated water
returns to lakes, streams, and rivers or seeps underground; the rest returns to the oceans or is frozen in
the polar ice caps. The hydrological cycle (see Chapter 4) is driven by the sun's energy and has been
recycling the same water through the millennia of
the earth's existence. The water we bathe in today
might have quenched the thirsts of dinosaurs millions of years ago.

In this chapter, we look at how water supports
life, how much clean, fresh water there is, how we
classify water, how we use water, and how those
uses contribute to water pollution. We look at the
history of water management, with special attention
to the legislation and practices designed to manage
drinking water supplies and wastewater in the United States. We discuss the problems of managing and
protecting drinking water, wastewater, groundwater,
fresh surface water, and marine waters. Finally, we
look at ways we can safeguard water resources in the
future.

Describing Water Resources: Biological Boundaries

Water is synonymous with life. It is the largest constituent of living organisms, and it is also a habitat in which life evolved on earth and exists today (Figure 15-1). Water is able to accomplish these dual roles because of unique physical and chemical characteristics (Table 15-1).

For most of the earth's existence, living organisms have been able to capture some of the water in the hydrological cycle to maintain their internal environments. Human bodies are composed primarily of water (about 65 percent); there are about 43 quarts (41 liters) of water in a 150-pound (67.5 kilogram) person. Water makes up 83 percent of the blood, helps digest food, transports body wastes, and lubricates joints. The body's internal water supply must remain constant and free from impurities to maintain health. Like all living organisms, we are intricately bound to the hydrological cycle.

Physical Boundaries

How Much Clean Fresh Water Is There?

Most of the earth's water (97 percent) is salty; only a small portion (less than 3 percent) is fresh. Three-quarters of all fresh water is found in polar ice caps and glaciers, and nearly one-quarter, known as

(a)

(b)

(c)

(d)

FIGURE 15-1: Diversity of water habitats. As a life-giving environment, water shows remarkable diversity: (a) The stark beauty of St. Mary's Lake in Glacier National Park, Montana; (b) the life-filled waters of northern Ohio's Old Woman Creek National Freshwater Estuary; (c) warm tropical waters that nourish coral reefs; and (d) a running stream in the Great Smoky Mountains, Tennessee. Each illustrates that aquatic ecosystems are far more than a collection of the physical and chemical properties necessary to sustain individual species; these habitats are alive.

▶ Table 15-1
Physical and Chemical Properties of Water

Superior solvent. More materials can be dissolved in water than in any other solvent. Water can permeate living cell membranes; dissolved materials also diffuse through the membranes or are "pumped" through using respiration energy.

Strong attractive force. The strong attractive force between water molecules allows them to be transported through the spaces in soil, into roots, and through the conducting tissue in plants and then to be transpired through the leaves and needles back into the atmosphere.

State changes. In the hydrological cycle the sun supplies the energy that vaporizes water to a gaseous state. When water vapor condenses to a liquid, energy is given off, helping to distribute the sun's energy across the earth.

High specific heat. It takes substantial amounts of energy to raise the temperature of a body of water, and, conversely, water is slow to cool. This property protects aquatic environments against rapid temperature changes. The different heating and cooling rates of bodies of water and adjacent land masses help to circulate air, create winds, and establish weather patterns.

Expansion at freezing. Water is the only common substance to expand when it freezes. Other substances contract when they freeze.

Density at freezing. Water reaches its maximum density at 46° F (4° C), 14 degrees above the freezing point. As water approaches freezing, it becomes lighter and moves toward the surface. When frozen, it floats on the surface as ice. Thus, streams, lakes, and ponds freeze from the top down, protecting aquatic environments from freezing solid in the winter.

Oxygen- and nutrient-holding capacities. More oxygen dissolves in cold water than in warm water, but warm water generally holds more nutrients. The largest populations of aquatic organisms are found where nutrient-rich warm waters and oxygen-rich cold waters mix.

Flow. Water tends to move easily downhill over land toward streams, rivers, lakes, ponds, and oceans. Bodies of water tend to flow downhill and toward larger bodies of water. Water also percolates through the ground, sometimes collecting in huge, slowly moving underground aquifers.

Esthetic quality. Flowing water, shimmering water, cascading water, waves, small drops of dew collected on a spider's web, rainwater pattering on a roof or running down a windowpane, puddles, a single drop of water from a pond teeming with life, a tear: each affects the human psyche in ways science is unable to explain.

groundwater, is found underground in water-bearing porous rock or sand or gravel formations. Only a small proportion (0.5 percent) of all water in the world is found in lakes, rivers, streams, and the atmosphere. Even so, if this small amount were free from pollutants and were distributed evenly, it could provide for the drinking, food preparation, and agricultural needs of all of the world's people.

How Is Water Classified?

Water is classified as either salt (marine) or fresh depending upon its salt content, or saline concentration. The saline concentration of marine waters is generally fairly consistent, about 35 parts per thousand (ppt). On average the saline concentration of fresh waters is 0.5 parts per thousand. The saline concentration of fresh waters tends to vary more than that of marine waters, because lakes, rivers, and streams are much more dominated by local environmental conditions such as the soils they drain and the rate of evaporation.

Fresh Water. Fresh water is found on land in two basic forms: surface water and groundwater. The two are not entirely distinct. Some water from lakes, streams, and rivers may percolate downward to groundwater supplies. Similarly, during dry spells, when surface runoff may be unavailable, groundwater may help to maintain the flow or level of rivers, streams, or lakes.

Surface Water. Surface waters include all bodies of water that are usually recharged by precipitation that flows along land contours from high elevations to low elevations as **runoff.** Runoff makes its way into streams, rivers, ponds, and lakes, eventually reaching the oceans and seas. The entire runoff area of a particular body of water is known as its **watershed.** A watershed contributes water and dissolved nutrients and sediments to the body of surface water toward which it drains. The fertility of streams, rivers, ponds, lakes, and, ultimately, the oceans depends on the fertility of the land of the watershed.

Surface water ecosystems include both standing water habitats (ponds, lakes, reservoirs, and in some cases wetlands) and running water habitats (springs, streams, and rivers). **Standing water habitats** are relatively closed ecosystems with well-defined boundaries; generally, they contain both inlet and outlet streams. Standing water generally has a lower oxygen level than running water because there is less mixing of water. Water does move, or flush, through ponds and lakes, but at a slow rate—from 1 year or less in small lakes to 200 years in large lakes such as Lake Superior. Pollutants have more time to build up to dangerous levels in standing water and to settle in sediments.

Lakes may be categorized according to the amount of dissolved nutrients they contain (Figure 15-2). **Oligotrophic** lakes contain a relatively low amount of dissolved solids, nutrients, and phytoplankton and are thus clear and deep blue in color. They lie in infertile watersheds, are cold and deep,

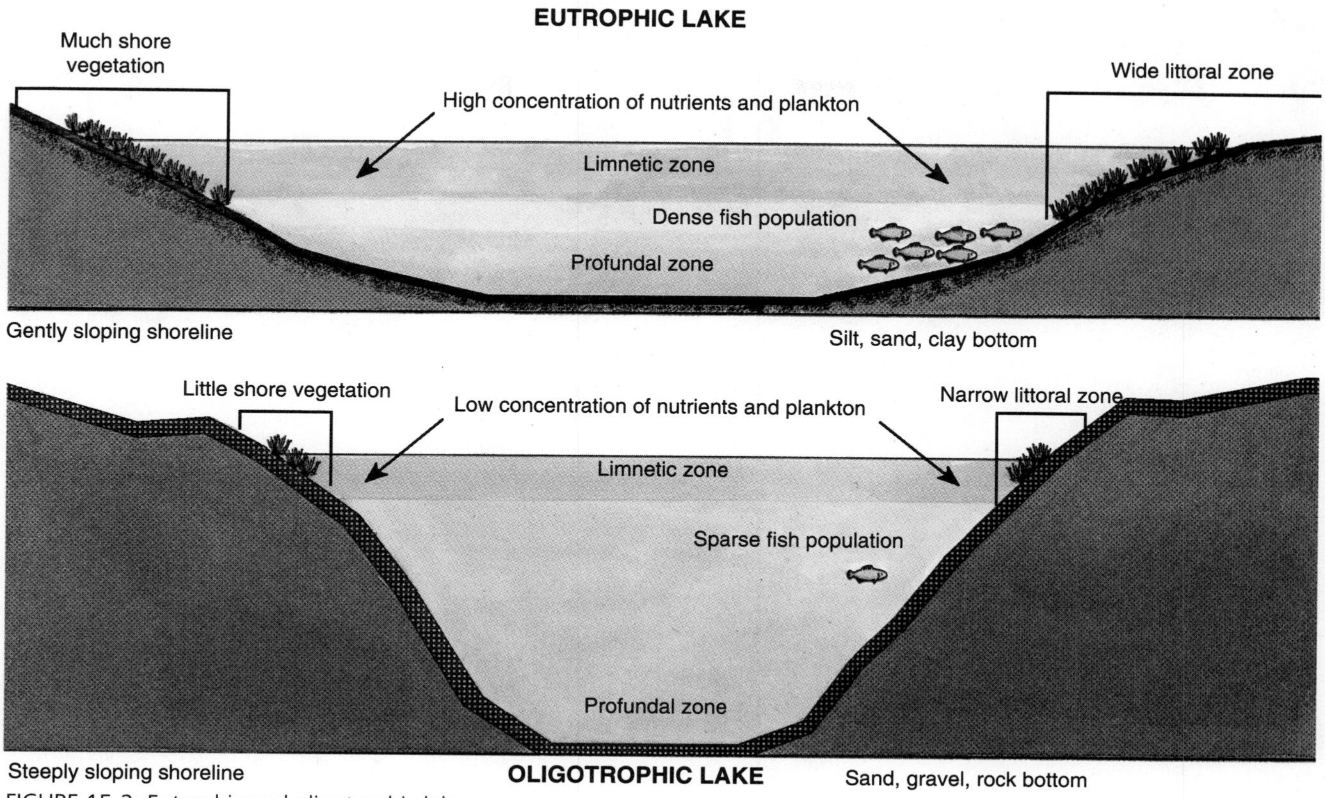

EUTROPHIC LAKE

Much shore
vegetation

Wide littoral zone

High concentration of nutrients and plankton

Limnetic zone

Dense fish population

Profundal zone

Gently sloping shoreline

Silt, sand, clay bottom

Little shore vegetation

Narrow littoral zone

Low concentration of nutrients and plankton

Limnetic zone

Sparse fish population

Profundal zone

Steeply sloping shoreline

OLIGOTROPHIC LAKE

Sand, gravel, rock bottom

FIGURE 15-2: Eutrophic and oligotrophic lakes.

have a high oxygen content, and usually have rocky bottoms. The low nutrient level of oligotrophic lakes results in a low production of organic matter, particularly phytoplankton, relative to total volume. Some oligotrophic lakes, such as Lake Tahoe (on the California-Nevada border), Lake Superior, and Lake Baikal (in Siberia), are moderately productive.

Compared to oligotrophic lakes, **eutrophic** lakes are warmer, are more turbid, have a lower oxygen content, often have muddy or sandy bottoms, and are far more productive. Fertile watersheds supply eutrophic lakes with abundant nutrients, enabling them to support a large phytoplankton population and a rich diversity of organisms. As we learned in Chapter 5, lakes undergo succession, becoming more eutrophic over time. Eventually, as eutrophic lakes and ponds grow warmer and shallower, they become weed-choked and more like marshes and dry land.

Lakes are made up of internal zones based on depth and light penetration. The **littoral zone** is a shallow, near-shore area where rooted plants grow because light can penetrate to the bottom. The **limnetic zone** is deeper water where light can penetrate and support populations of plankton. Below the limnetic zone is the **profundal zone,** into which light does not penetrate enough to allow photosynthesis

to occur. Organisms of the profundal zone depend on food and nutrients filtering down from above.

Large lakes in temperate zones undergo **thermal stratification** (Figure 15-3). In the summer surface waters heated by the sun become lighter and less dense and form a top layer or **epilimnion.** Colder, more dense water sinks to the bottom to form the **hypolimnion.** A sharp temperature gradient, or **thermocline,** exists between the upper and lower layers. Sometimes, especially in eutrophic lakes, all photosynthesis and oxygen production occur in the warm, upper waters. The thermocline prevents the two layers from mixing, and the hypolimnion can become depleted of oxygen. Occasionally, late in the summer the shallower areas of eutrophic lakes and ponds can also become completely devoid of oxygen. The lack of oxygen can stress fish populations and result in fish kills. Generally, this phenomenon can be traced to very warm weather, with little wind to mix the surface water, and a high decomposition rate of algae or other organic matter that has been produced throughout the summer. In the autumn, when seasonal changes help to cool the surface water, the thermocline disappears and the top and bottom layers mix, returning oxygen to the bottom layers. This mixing is known as fall turnover. If the

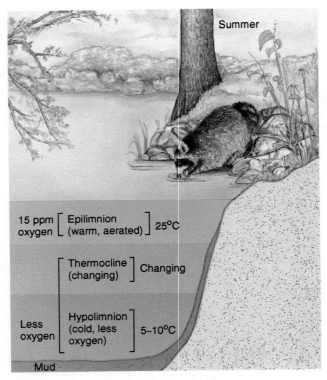
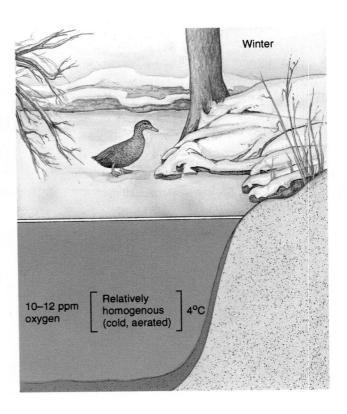

FIGURE 15-3: Thermal stratification in lakes.

winter is cold and long, and snow covers the ice for long periods, oxygen can be depleted in the deeper areas. Spring warming causes the water to turn over once again, returning oxygen to the deepest levels and circulating both nutrients and the overwintering stages of organisms to the warmer, light-filled waters.

Running water habitats include streams and rivers, continuously moving currents of water that cut channels or beds through the land's surface. The speed of the current influences both the composition of the channel (rock, sand, gravel, mud) and the oxygen content, and, consequently, the composition of the organisms found there. Fine-grained particles tend to collect in pools or shallows, while large rocks and boulders are found on the stream bed of fast-flowing waters. The oxygen content of fast-flowing streams is higher than that of slow-flowing ones.

Temperature also affects the composition of the organisms in streams and rivers. In general, the temperature of small streams tends to rise and fall with the air temperature. Rivers with large surface areas exposed to the sun are generally warmer than those with trees and shrubs on their banks. Often, when streams are channelized or straightened or the land surrounding them is cleared for farming, they go through drastic changes in temperature with subsequent drastic changes in species composition.

A look below the surface of a river or stream can reveal many interesting habitats and associations of living organisms. Fast-moving rivers and streams usually contain two kinds of habitats: riffles and pools. Riffles, with a high oxygen content, tend to house the producers of biomass, and pools tend to contain consumers and decomposers.

All streams and rivers are a product of their watershed. Consequently, both the quality and quantity of their waters reflect the land management practices used there. (See Focus On: The River Restoration Method of George Palmiter, page 294.)

Groundwater. Imagine that you could contain all of the fresh surface water in the United States in one lake. Holding the water from all rivers, streams, ponds, and lakes, our imaginary lake still would not contain the amount of water that exists underground. Our lake would have to be 20 times larger to hold all of the water stored underground! This huge amount, some 30 to 60 quadrillion gallons (114 to 227 quadrillion liters), represents 96 percent of the United States' fresh water. We depend on groundwater for drinking, to help produce food, to keep factories and businesses going, and to help heat and cool buildings (Figure 15-4).

Groundwater percolates downward through the soil after a rain or snow or seeps downward from surface water and is stored in an aquifer. An **aquifer**

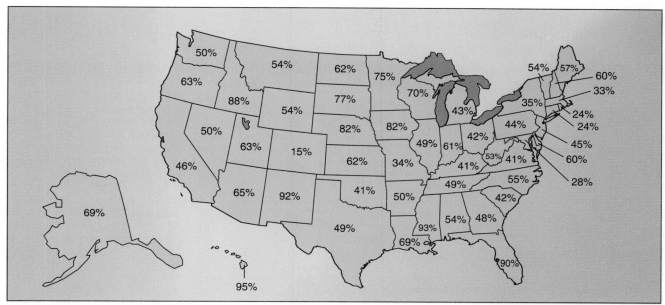

FIGURE 15-4: Percentage of state populations served by groundwater. A proportion of each state's population is served by groundwater. Colorado has the lowest at 15 percent and Hawaii the highest at 95 percent.

is a water-bearing geologic formation composed of layers of sedimentary material such as sand, gravel, or porous rock. Water fills the cracks and crevices of the rock and the pores between the particles of sand and gravel. The depth at which the aquifer begins is known as the **water table.** Before it reaches the aquifer, water passes through an unsaturated zone, where pores contain both water and air. Plants remove some of this water; the rest continues to move downward to the saturated zone.

Aquifers can cover a few miles or thousands of square miles. The Ogallala, the largest aquifer in the United States, stretches from South Dakota to central Texas (Figure 15-5). Because it was formed long ago, when the earth's rainfall pattern was much greater, and it is not significantly recharged by present precipitation patterns, the water in the aquifer is

FIGURE 15-5: The Ogallala aquifer.

known as fossil groundwater. Current rates of withdrawal are so high that the Ogallala is being depleted.

Aquifers can be several feet to several hundred feet thick; they may occur several hundred feet underground or quite close to the surface, perhaps emerging as a free-flowing spring or contributing water to a stream, river, lake, or wetland. If the material above the aquifer is permeable, allowing water to freely move downward to the water table, the aquifer is said to be confirmed. If an impermeable layer exists above the water table, restricting the flow of water from above, the aquifer is said to be confined. Confirmed aquifers are more susceptible to contamination than confined ones because they are not protected by an impermeable rock or clay layer.

Groundwater does not flow like a river. Instead it percolates downward, moving from high elevations to lower elevations, at a variable rate (Figure 15-6, p. 296). It sometimes moves very slowly and sometimes surprisingly fast, from a fraction of an inch to a couple of feet a day. Groundwater flow is determined by the slope of the water table as well as by the permeability of the material through which the water is moving. The steeper the slope and the more permeable the substrate, the more rapidly the water flows. Usually, but not always, groundwater flow parallels the aboveground flow. This may not be true, however, in the case of a confined aquifer, where the surface water and groundwater are separated by an impermeable layer.

Focus On:
The River Restoration Method of George Palmiter
Gene Willeke, Director, Institute of Environmental Sciences, Miami University

During the 1950s and 1960s Dutch elm disease took a dramatic toll on elms in the United States. The resulting masses of dead trees clogged many rivers. The especially large logjams in the St. Joseph and Tiffin rivers in northwestern Ohio posed obstacles for recreational users of the rivers, caused bank erosion that cut into agricultural fields, and caused upstream flooding, resulting in property damage. The logjams also altered the placement of sand and gravel bars, so that the bars became additional obstructions.

Channel obstructions, such as logjams, trees embedded in the stream, and bars were normally cleared by snagging and dredging. Bars and sediment deposits were removed by draglines extending from the shore or by boat-mounted dredges. Both types of heavy equipment removed bank vegetation along with the obstructions. Once cleared, many rivers were channelized, that is, the channel was straightened and deepened to increase flow capacity and navigable depths.

Though widely practiced, clearing and channelization were controversial because they destroyed aquatic habitats, left barren river banks in place of a canopy of trees, and provided little if any flood protection, sometimes even increasing downstream flooding. Advocates of the methods argued that they helped control flooding and that channelization, in some cases, provided new aquatic habitats better than the original ones.

George Palmiter, a resident of northwestern Ohio, took on the problem of clearing obstructions. An avid canoeist and duck hunter, Palmiter was disturbed that the St. Joseph and Tiffin rivers were so hard to traverse. Lacking access to heavy equipment, he set about developing a new way to clear the jams. For several years Palmiter observed the rivers and experimented with different ways to remove obstructions. Because the simple techniques he eventually developed are based on principles of river behavior, they require only manual labor and hand tools. Palmiter harnessed the river to do much of the

work that had required high levels of energy and expensive equipment in conventional dredging and snagging. Palmiter's techniques came to be known as the Palmiter river restoration method.

In the St. Joseph and Tiffin rivers Palmiter removed logjams by making well-placed cuts in the logs on the downstream side of the jam during low-water periods. When the water level rose, the stream flow carried the debris away and deposited it on the floodplains below the jam.

Palmiter selectively removed bars by pulling out any trees and shrubs that had grown up and cutting pilot channels through the bars at a time of low water. High water simply carried the bars away. He accelerated bar removal by placing brush piles and tree trunks in the channel to deflect the flow of the stream toward the bars.

The Palmiter method also employs brush piles and tree trunks to divert flow away from eroding banks. Although all rivers normally erode their banks to some degree, obstructions can cause excessive

Aquifers are slow to recharge after water is removed and slow to cleanse after they are contaminated. Contamination is becoming more significant as aquifers are increasingly threatened by pollution from municipal and industrial landfills, oil drilling, mining, highway de-icing salts, urban and agricultural runoff, leaking underground storage tanks for petroleum and gasoline, improper storage and disposal of hazardous waste, and private septic tanks and municipal sewage systems. These problems loom larger when we see just how important groundwater supplies are.

Aquifers supply drinking water for half of the U.S. population, including almost all of the rural population. While groundwater is used to some extent in every state, it is the major source of drinking water in about two-thirds of them. In addition to its use as a source of drinking water, groundwater supplies over 95 percent of rural household needs, 26 percent of industrial demand, and 40 percent of agricultural demand. It helps to maintain water levels and the productivity of streams, lakes, rivers,

wetlands, bays, and estuaries. Mineral-rich groundwater helps to supply nutrients that nourish such aquatic ecosystems as Tarpon Springs and the Apalachicola River in Florida. These systems often support unique plants and animals and provide excellent fishing and recreation opportunities.

Marine and Coastal Resources. Cradles of life, crucibles of diversity, and nurturers of civilizations, the oceans influence nearly everything we do. The oceans are one huge living system. Constantly moving water currents operate under the influence of the sun's energy, the earth's rotation, and the moon's gravitational pull. Currents move huge masses of water around the continents and across the vast open reaches, continually circulating nutrients that have washed in from the land.

The interaction of the oceans and the atmosphere affects heat distribution, weather patterns, and concentrations of atmospheric gases throughout the world. Because they are so expansive and deep, the oceans moderate climate and provide a sink for dis-

erosion or cause it to occur in undesirable locations, such as the edges of agricultural fields. Carefully placed deflectors can limit such erosion.

The Palmiter method includes planting the river banks with trees, often water-tolerant, fast-growing willows. These trees, the first stage of plant succession on the banks, provide roots to hold the soil, detritus for the animals that live in the stream, and shade for the waters near the shore. Later stages of succession can be accelerated by planting late stage trees after the willows take hold; these larger trees develop soil-holding root systems and provide shade for even more of the river.

The team or person who wants to use the Palmiter method must carefully observe the behavior of the river to be cleared. No scientific exposition of the physical and biological principles behind the method exists; use of the method must be based on an intuitive knowledge of river properties. Like the pool player who cannot articulate the principles of physics behind

the game, but whose skill is based on successful adherence to those principles, the person who employs the Palmiter method must be able to sense how the river will respond to different conditions.

This intuitive knowledge of the river grows out of observation. The person responsible for planning the project usually examines the river from its banks, from a boat or canoe, and in aerial photos. For each section of the river where work must be done, the planner provides marked photos or sketches showing field crews how to remove trees, cut pilot channels, remove obstructions, and construct deflectors.

Overall, the Palmiter method provides more benefits at less cost than do conventional methods. By returning the channel's flow capacity to its natural state, the Palmiter method reduces the risk of floods for agricultural lands. It maintains or improves aquatic habitats by shading (which prevents excessive warming of the water), by adding detritus to the food supply of the stream, and by providing refuge for

fish in brush piles and other nearshore vegetation. The cost of the Palmiter method is low because it uses unskilled or semiskilled labor and inexpensive hand tools and relies on the river for much of the energy needed to move debris and sediment.

Although full evaluations of the Palmiter method would require extensive, long-term observation and have not been undertaken, less formal evaluations are highly favorable. Palmiter worked on the St. Joseph and Tiffin rivers in the mid-1970s, and today the bank stabilization seems to be successful, no new major logjams have developed, the tree canopy has returned, and neither river has required much maintenance. Landowners and county and state officials who have undertaken Palmiter method projects in other parts of the country have been well satisfied. They especially like the short planning time required (usually a matter of weeks for the river survey and preparation of work plans) and the low cost.

solved solids and gases. Beneath the ocean floors lies a rich storehouse of minerals, petroleum, and natural gas. Last but not least, hundreds of millions of people live near the ocean shores and on coastal plains. Even greater numbers depend upon ocean fisheries, energy, and minerals for their sustenance.

The ocean is divided into zones (Figure 15-7). The **littoral zone** is the area of shallow waters near the shore. The **euphotic zone** is the area of light penetration; here, phytoplankton exist in great abundance. Where light can penetrate to the ocean floor, attached plants like the giant kelps are found. The **neritic zone,** comprising 7 to 8 percent of the oceans, is the part of the euphotic zone over the continental shelf and near-shore islands. The greatest variety of life is found in the neritic zone, where ocean waters mix with waters from the land, and it supports most of the ocean fisheries. The **pelagic zone** is the deep water zone of the open oceans. Below it lies the **abyssal zone,** the deepest part of the ocean.

Four biologically productive ecosystems are

responsible for most of the oceans' primary production. Three are coastal ecosystems: estuaries, coastal wetlands, and coral reefs. The fourth is the offshore continental shelves. Together, these four ecosystems are 16 to 26 times more productive than the open oceans.

Estuaries. Estuaries are shallow, nutrient-rich, semienclosed bodies of water. Although freshwater estuaries, such as Old Woman Creek Estuary on Lake Erie, do exist, most estuaries have a direct connection to the ocean. Tides constantly mix salt water with fresh water from the surrounding land. Concentration allows the estuaries to serve as habitats for larval stages of many ocean fishes and shellfishes. These larvae are washed out to sea with outgoing tides, where they feed and grow to adults. Estuaries produce 80 million tons (73 million metric tons) of fish and shellfish per year, more than any other ocean ecosystem. Representing 2 to 3 percent of total ocean areas, estuaries alone are responsible, either directly (adult fish and shellfish) or

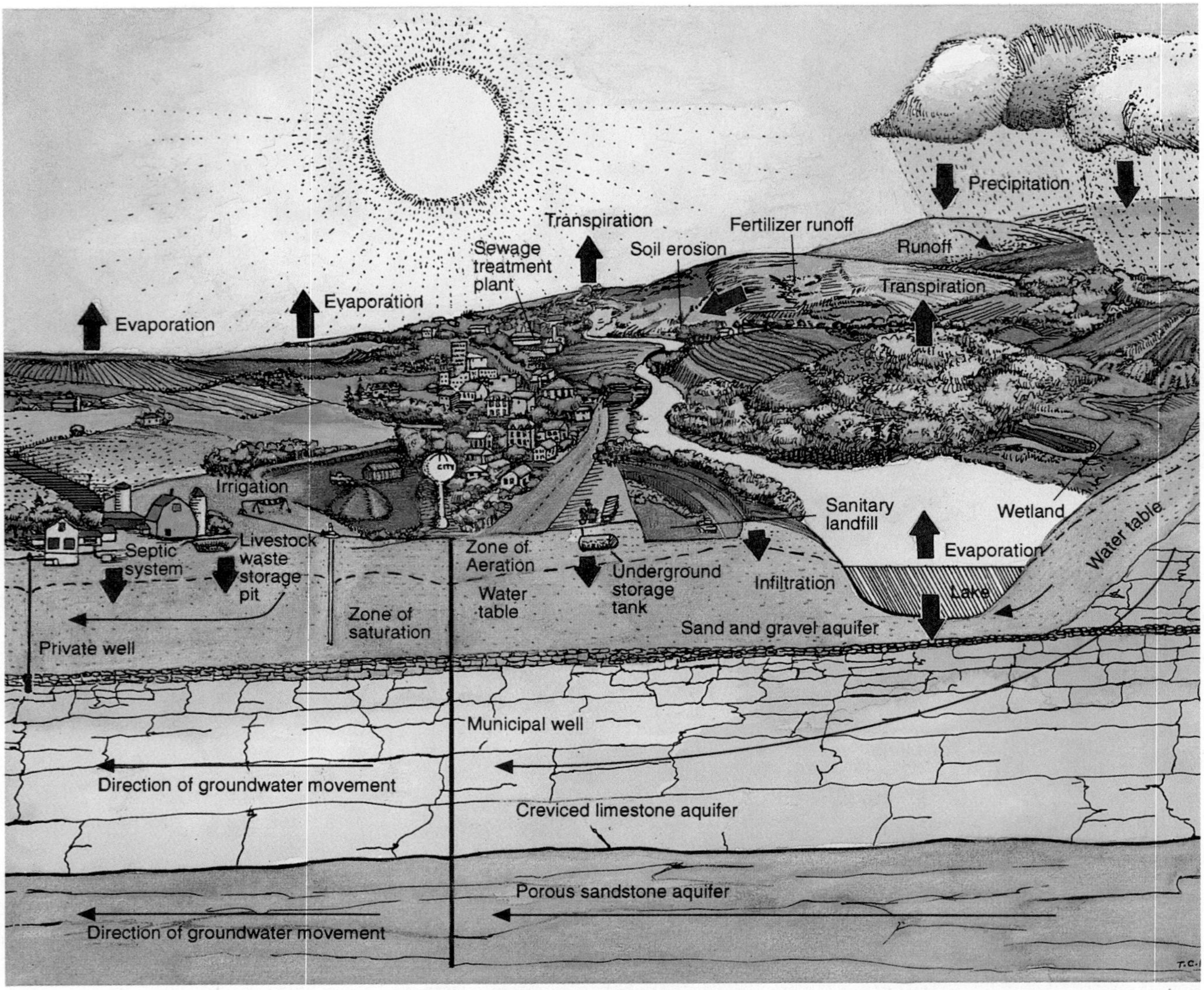

FIGURE 15-6: Generalized diagram of groundwater. Aquifers are recharged by precipitation that percolates through the ground or water that seeps down from lakes, rivers, streams and wetlands. Potential sources of pollution include underground storage tanks, septic systems, livestock waste, landfills, and agricultural and urban runoff.

indirectly (nurseries for fish and shellfish), for 75 to 90 percent of all fish and shellfish caught. (Chapter 4 presents an in-depth look at the Chesapeake Bay, a highly productive estuary located on the Atlantic coast of the United States.)

Coastal Wetlands. Wetlands exist on a continuum between aquatic and terrestrial habitats. Those dominated by grasses are called marshes, and those dominated by woody plants are called swamps. Coastal wetlands serve as nursery grounds for many species of fish, provide habitat for many other animals and plants, and buffer coastal areas from storms and flooding.

Like wetlands throughout the nation, coastal wetlands are under siege. Over 3.7 million acres (1.5 million hectares) of coastal wetlands, about 41 per-

cent of all wetlands in the United States, are located in Louisiana alone (Figure 15-8). They evolved on sediments deposited by the Mississippi River. The river's primary channel has moved at least six times in the past 10,000 years, and as it has, it deposited sediments first at one site and then at another. A growing river-mouth delta formed at each site, but eventually, as the Mississippi waters found a more direct route to the Gulf of Mexico, the waters retreated and a wetland was formed. Presently, however, these irreplaceable wetlands are rapidly deteriorating and being converted to open water. Levees along the Mississippi, which eliminate overbank flooding, prevent additional sediments from being deposited on existing wetlands, effectively starving them of needed nutrients and materials. Along with other control structures, the levees also prevent the

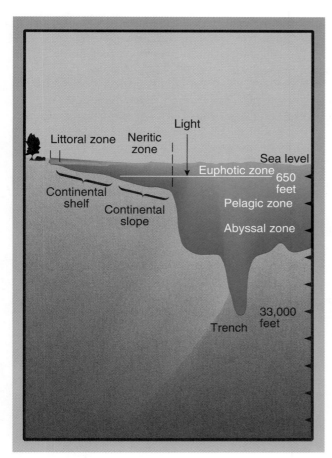

FIGURE 15-7: Zones of the ocean.

river waters from seeking other routes to the gulf, thus halting the formation of river deltas and ultimately new wetlands. Additionally, thousands of miles of dredged canals crisscross the wetlands; these encourage the intrusion of salt water from the gulf. Wetland plants, which are adapted to a freshwater environment, are devastated by saltwater intrusion.

Coral Reefs. The greatest diversity of life in the coastal zone, and in aquatic environments as a whole, is found in the coral reefs of shallow tropical seas. Composed of hard, limy (calcium carbonate) skeletons produced by coral animals, as well as the living coral animals themselves, coral reefs support about one-third of all marine fish species. As Chapters 6 and 23 indicate, coral reefs are subject to destruction by numerous human activities, including dynamite blasting for development, collection of corals for sale, and siltation due to erosion from nearby land development sites.

Continental Shelves. Like coastal ecosystems, near-shore areas over continental shelves are also productive. Where cold currents from the polar regions meet warm tropical currents, upwellings of turbu-

lent nutrient-rich waters nourish billions of phytoplankton. These plankton are food for zooplankton, small fish, larger fish, and marine mammals in a complex food web so vast that it supports over half of the world's biomass and most of the world's important fisheries. For example, the southern oceans of Antarctica, near the great convergence of cold and warm currents, support huge schools of krill, a shrimplike crustacean, which is a major food source for fish and marine mammals, especially whales, and is harvested for human consumption.

Social Boundaries

How Do We Use Water?

We use water in countless ways: to drink, to bathe, to dispose of wastes, to irrigate crops, to support industry, and to generate power. Some of the world's most important transportation corridors are waterways; many play a vital role in global economics and politics. Water provides many of our favorite recreational opportunities. Finally, water is a source of inspiration for poets, writers, artists, and others whose souls are stirred by the sight and sound of this most basic of all compounds.

Globally, agriculture is the single greatest drain on water supplies; about 70 percent of the water withdrawn from the earth is used to produce food. By 1990 the total amount of land under irrigation worldwide was approximately equal to an area the

FIGURE 15-8: Wetland in Louisiana's Barataria Bay near Grand Isle, southeast coastal Louisiana. Between 1940 and 1985 over 40 percent of the wetlands in Barataria Bay were lost, a rate common in the region's delta plain.

size of India. Industry accounts for about 23 percent of water use worldwide, and municipalities and households account for 7 percent.

Water uses may be nonconsumptive or consumptive. Nonconsumptive uses remove water, use it, and return it, usually altered in some way, to its original source. Water is removed from a stream or lake, used for cooling, power generation, or drinking, then returned to that same stream or lake. Municipal and industrial uses fall into this category. About 90 percent of the water used by industry and homes is available for reuse. In contrast, consumptive uses remove water from one place in the hydrological cycle and return it to another. For example, irrigation removes water from a stream, lake, or aquifer to a place where it doesn't return to its original source. In fact, on average worldwide, less than 40 percent of irrigation water is actually taken up by crops; the rest is lost (as runoff or evaporate, for example). Consequently, agricultural removal over the long term tends to deplete local water supplies.

How Do Water Consumption Patterns Vary Worldwide?

Water resources are not evenly distributed; some areas have abundant water and others have little (Figure 15-9). In freshwater-poor areas, such as the Middle East and North Africa, **desalination**—the process of removing dissolved salts from sea or ocean water or brackish groundwater—is used to provide citizens with water for drinking, cooking, and other needs. Even highly developed countries with seemingly adequate water resources can suffer from lack of rainfall and groundwater resources or pollution that renders water supplies unfit for use.

Water is fast becoming a resource Americans can no longer take for granted. Each American uses an average of 120 gallons (455 liters) of water per day. Approximately 85 gallons per day (322 liters) are used in the home, most of these to flush toilets. Figure 15-10 shows the daily water use of a typical U.S. family of four. If we calculate industrial and home use combined, each American consumes over 500 gallons (1,900 liters) of water per day. Providing a single fast food order of hamburger, fries, and a soft drink in the United States requires over 1,500 gallons (5,700 liters) of water; a typical Thanksgiving meal, according to the Freshwater Foundation, requires 42,674 gallons (162,000 liters). These figures include the "hidden" costs (at least to the consumer) of water needed to produce and process food as well as the water needed to prepare the meal. For example, about 16,300 gallons (62,000 liters) of water are needed to raise, process, and cook a 20-pound turkey; 6,004 gallons (22,700 liters) are needed to put stuffing on the holiday dinner table; green beans, carrots, and milk for four people require 1,000 gallons each (3,790 liters); pumpkin pie requires 1,240 gallons (4,700 liters).

The World Health Organization (WHO) considers 5 gallons (19 liters) per day as minimal for basic hygiene. A more realistic figure is 15 gallons (57 liters). Many of the world's people do not have

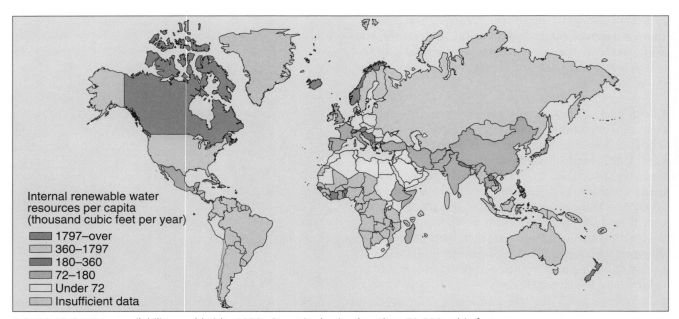

FIGURE 15-9: Water availability worldwide, 1989. Countries having less than 72,000 cubic feet per person per year are considered to suffer from a chronic water shortage. The figures given do not include river flows into countries from other nations. These flows can make a significant difference, as in the cases of the Nile in Sudan and Egypt, the Danube in Bulgaria and Hungary, and the Tigris and Euphrates in Iraq.

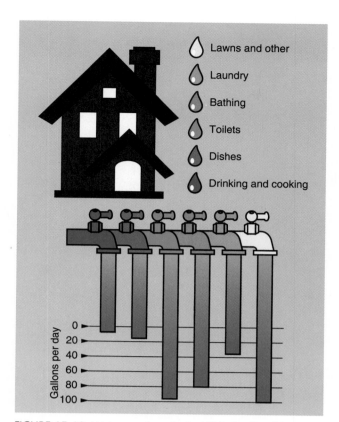

FIGURE 15-10: Water use by a typical U.S. family of four.

Lawns and other

Laundry

Bathing

Toilets

Dishes

Drinking and cooking

Gallons per day
0
20
40
60
80
100

access to even 5. One flush of a standard toilet in the United States uses more water than most of the world's people use in a day. In many less-developed countries clean drinking water and adequate sanitation facilities are unknown luxuries (Figure 15-11). Many areas do not have toilets, latrines, or proper drains; wastes are disposed of near or in the same river, lakes, or wells used for drinking and food preparation. WHO estimates that 2 billion people in LDCs (other than China) are at risk of drinking contaminated water. About three in five do not have access to safe drinking water, and only one in four has any sanitary treatment of waste. At least 25 million people per year in LDCs die from contaminated water; three-fifths are children. Worldwide, every hour 1,000 children die from diarrhea-related diseases.

On November 10, 1980, the United Nations declared the International Drinking Water Supply and Sanitation Decade (1981–1990) to work for the improvement of the human condition through the development of safe drinking water supplies and adequate sanitation. During the decade some progress was made in several countries, including Mexico, Indonesia, India, and Ghana, but by 1990 the proportion of the global population without

access to safe sanitation and adequate supplies of clean drinking water was about the same as it had been in 1981.

WHO maintains that the total number of water taps per 1,000 people is a better measure of health than the total number of hospital beds per 1,000 people. To increase water availability and to provide basic waste treatment require large sums of money and cooperation among nations. But the results—better health, reduced infant mortality, and population stability—yield a much better return than dollars spent on medical treatment. Disease prevention through an adequate supply of clean water is the key to improving health standards. The greatest success has come in countries where grass roots movements have led to better education about sanitation, the development of self-help programs, and low-energy, appropriate technology.

What Kinds of Water Pollution Are There?

Water pollutants can be divided into eight general categories: organic wastes, disease-causing wastes, plant nutrients, toxic and hazardous substances, persistent substances, sediments, radioactive substances, and heat. These categories are somewhat subjective and overlap, but they indicate the diversity and complex nature of water pollution. One source may be responsible for more than one type of pollutant. For example, improperly treated sewage may contribute organic wastes, disease-causing wastes, plant nutrients, toxic substances, and persistent substances to a waterway. One pollutant may fit in more than one category. For example, mercury and polychlorinated biphenyls (PCBs) are both hazardous and persistent. And, one type of pollutant may enter water attached to another type. For example, organic chemical pollutants often adhere to sediments. Many water systems are assaulted by pollutants from all eight categories, and the difficulty of cleanup is compounded by the magnitude of the total pollutants and the need to use different techniques for different categories. A single pollutant may not be found in amounts necessary to kill organisms, but many pollutants acting together or synergistically may be deadly. Unless steps are taken to alleviate all categories of pollution, aquatic systems will continue to be degraded, with the result that human and environmental health suffer.

Organic Wastes. Organic or oxygen-demanding wastes are small pieces of once-living plant or animal matter. They are usually suspended in the water column, but can sometimes accumulate in thick layers on the bottom of lakes or streams, thus contributing to the sediments. Most suspended organic matter comes from human and animal wastes or

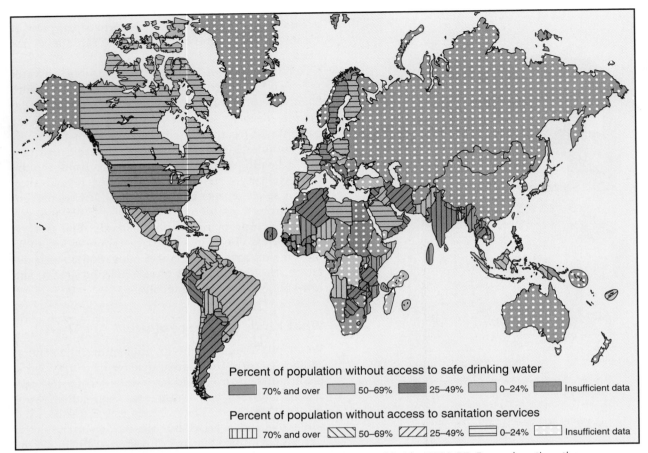

FIGURE 15-11: Access to safe drinking water and sanitation services worldwide, 1980-85. For each nation, the map indicates the percent of the population that does not have access to safe drinking water and basic sanitation services. The nations with the greatest need are primarily in the developing world.

plant residues. Aerobic bacteria in the water use the organic matter as an energy source, a process called decomposition. The bacteria use dissolved oxygen from the water as they decompose or consume the organic matter. Anaerobic bacteria, which often operate in the sediments, do not require oxygen to decompose organics; they may emit noxious gases, such as hydrogen sulfide or methane, as a by-product of decomposition.

Decomposition by aerobic bacteria removes dissolved oxygen from the water to the detriment of other aquatic organisms, such as fish and shellfish. As the oxygen is depleted, the species composition of the area may change dramatically, as higher aquatic organisms (such as fish, oysters, and clams) die off or leave the area, and organisms that can tolerate low oxygen levels, such as sludge worms and rattailed maggots, proliferate. If the oxygen depletion is drastic enough, it may cause many fish to die off, an event known as a fish kill.

Biological oxygen demand (BOD) is a measure of the amount of oxygen needed to decompose organic matter in water. A stream or lake with a high BOD will have a low concentration of dissolved oxygen because oxygen is being used up by bacteria to decompose organic matter. Accordingly, a body of water with a high BOD is also high in organic matter. Dissolved oxygen is a major limiting factor in the aquatic habitat. Thus, oxygen depletion is a major contributor to the degradation of a stream or lake through loss of species diversity.

Resources for the Future reports that about half of the oxygen-demanding pollutants come from non-point sources. Animal waste used as fertilizer is the most conspicuous nonpoint source. Sewage treatment plants, the food-processing industry, and the paper industry are the most conspicuous point sources. We have concentrated our efforts on these point sources, and they are making strides in preventing organics from entering waterways.

Disease-Causing Wastes. The entrance into an environment of untreated human or other animal wastes increases the chance that infectious organisms (bacteria, viruses, protists) will spread disease to humans. This occurs commonly in LDCs, where human sewage is treated improperly, if at all, before it enters an aquatic environment (Figure 15-12).

FIGURE 15-12: Kano, Nigeria. A stream running through the village has been converted to a sewage and garbage conduit.

Since millions of people in LDCs have no basic sanitation, providing adequate supplies of disease-free water presents a major challenge. Cholera, typhoid fever, hepatitis, dysentery, and giardiasis are the major diseases transmitted through drinking water supplies.

Even in the United States, where the chlorination of drinking water helps to prevent major outbreaks of these infectious diseases, contaminated water leads to some outbreaks each year. Giardiasis, a disease caused by the protozoan *Giardia lamblia*, is the most common waterborne disease in the United States today. It is known as a camper's, hiker's, or backpacker's disease because of its high incidence in those groups. Even the most pristine lakes and streams can harbor these infectious protozoans, which enter the water via animal wastes. In the late 1980s an alarming rise in the incidence of giardiasis was traced to antiquated reservoirs that serve as sources of municipal water. For example, residents of Fairchance, Pennsylvania, were instructed to drink bottled or boiled water after *Giardia lamblia* was discovered in their reservoir.

Many other diseases are also transmitted by organisms in water. For example, mosquitos (whose larvae live in water) transmit the protozoan that causes malaria, and snails transmit the fluke that causes schistosomiasis. Both of these diseases are serious threats to human health in LDCs; about 200 million cases of schistosomiasis occur annually.

The test for disease-causing organisms in water is the presence of intestinal fecal coliform bacteria. The presence of coliforms, bacteria normally present in high numbers in the large intestines of humans and other animals, indicates the likelihood of contamination by disease-causing bacteria. Coliform standards for drinking water are stricter than those for water used for recreation. WHO recommends that drinking water have a coliform count of 0 per 100 milliliters. Many rivers in Latin America have coliform counts of more than 100,000 per 100 milliliters. In the United States, despite our attention to proper sewage treatment, many beaches are closed each summer because of high coliform counts.

Plant Nutrients. Algal and aquatic plant growth is normally limited by the amount of nitrogen and phosphorus in the water. The large amounts of these nutrients contained in sewage, including the phosphates in some soap detergents and agricultural and urban runoff, stimulate massive, rapid reproduction and growth in algae, known as blooms. Algae impart a green color to the water and form a green scum on the surface and on rocks near shore. When algae die and decompose, more nutrients are added to the water, increasing the biological oxygen demand. Fast-moving water is generally not as affected by algal growth as slow-moving water. Ponds, lakes, and bays are most affected, through the process of eutrophication (see Chapter 3, Environmental Science in Action: Lake Erie).

Excess phosphorus in the water is generally not a human health hazard, but excess nitrogen, in the form of nitrates, is. Found in fertilizers and organic waste from livestock feedlots, nitrates are soluble in water and do not bind to soil particles, making them highly mobile; they wash into surface water supplies and percolate into groundwater. Nitrates in drinking water pose a significant health threat. In the intestinal tract of infants, they are reduced to nitrites, which oxidize the hemoglobin in the blood, rendering it unable to carry oxygen. This condition, called methemoglobinemia, can result in brain damage or death. Nitrates and nitrites can also form toxic substances called nitrosamines, which have been found to cause birth defects and cancer in animals. Studies have established a link between high nitrate levels and stomach cancer in humans. In Europe and the United States between 5 and 10 percent of all wells that have been tested show nitrate levels that are higher than the WHO-recommended value of 45 milligrams per liter.

Sediments. By weight, sediments are the most abundant water pollutant. Every time it rains we need only to look at our muddied streams or rivers for a visible reminder of the millions of tons of sediments that wash from the land into aquatic systems (Figure 15-13). Particles of soil, rock, sand, and minerals run off the land and enter waterways, filling in lake bottoms, river channels, and reservoirs.

FIGURE 15-13: Sediment load in an aquatic system in Peru. In this 1988 photo, a clear tributary flows into Peru's sediment-laden Palcazo River.

Sediments contain nutrients; natural erosion thus helps to maintain the fertility of an aquatic ecosystem. However, a dramatic increase in the sediment load—from poorly managed agricultural lands and urban construction sites, for example—can cause problems. While the threat to human health is minimal because filtration easily removes sediments from drinking water, aquatic life is often adversely affected. Excess sediments smother fish eggs and prevent light from penetrating to rooted aquatic plants. Chemicals may adhere to sediments and accumulate on the bottom of bodies of water, where they can adversely affect aquatic habitats. Some dredged sediments are so laced with dangerous chemicals they must be treated like hazardous wastes. Polyaromatic hydrocarbons (PAHs), first identified in cigarette smoke, are carcinogens that bind to sediment particles. Emissions from coke ovens, creosote plants, and coal gasification plants; sewage containing petroleum products; and oil leaked from boats and automobile exhaust are major sources of PAHs. Researchers at Puget Sound in Washington have identified sediments near urban centers that contain 150 times more PAHs than sediments near rural areas do.

Toxic and Hazardous Substances. Substances that are injurious to the health of individual organisms or their cells are called hazardous or toxic. Toxic organic substances include oils, gasoline, greases, solvents, cleaning agents, biocides, and synthetics. Thousands of organic chemicals enter aquatic ecosystems every day. Most are by-products of industrial processes or are present in hundreds of thousands of commonly used products. Some of them are carcinogenic, for example, dioxin, polychlorinated biphenyls (PCBs), and trichlorethylene (TCE). TCE has been discov-

ered in contaminated groundwater and surface water.

Inorganic hazardous substances include acids from mine drainage and manufacturing processes, salt from roads and irrigation ditches, brine from oil and natural gas wells, and heavy metals such as lead and mercury from manufacturing and industrial processes. Old pipes and lead solder joints also contribute to the incidence of lead found in public drinking water supplies. Enough lead contamination has been found to cause the EPA to release a lead advisory explaining how consumers can reduce the lead content in their drinking water until new pipes and joints can be installed.

Many rivers, lakes, and bays have thousands of toxic and hazardous chemicals in their sediments. The increased incidence of cancers discovered in bottom-dwelling fish has been traced to contaminated sediments. Particularly affected are New York Harbor, Boston Harbor, Puget Sound, New York's Buffalo River, Ohio's Black River, and Michigan's Torch Lake.

The increasingly frequent discovery of hazardous chemicals in dumps and landfills is one of the most serious environmental problems in the United States and other industrialized countries. As hazardous contaminants leach from dumps and landfills, they make their way into streams, rivers, and groundwater supplies, and hazardous leachate is becoming a more frequent contaminant in drinking water.

Persistent Substances. Some pollutants are not normally changed or degraded to harmless substances and persist in their original form in the environment. Pesticides such as DDT and chlordane, metals such as mercury, and organic chemicals such as PAHs and PCBs do not break down easily and tend to magnify through the food chain. Consequently, organisms at the higher trophic levels, such as the bald eagle and peregrine falcon, suffer the most serious effects of these persistent, or nonbiodegradable, substances.

Many plastic products, such as bags, monofilament line, and beverage six-pack rings, have a life expectancy of hundreds of years. Plastic pollution is particularly troublesome in marine systems (Figure 15-14). Plastic bags and balloons become entwined in the stomachs and intestines of sea turtles and mammals. Polystyrene plastic foam breaks up into pellets, resembling food. When consumed by sea turtles, its buoyancy keeps them from diving; it can also clog their systems, causing them to starve to death.

Radioactive Substances. Radioactive isotopes, such as strontium 90, cesium 137, and iodine 131, enter water from several sources: the mining and process-

FIGURE 15-14: Monofilament gill net around the neck of a California sea lion, *Zalophus californianus*. Because they do not biodegrade, plastics pose a significant threat to wildlife. For example, an estimated 6 million seabirds and 100,000 marine mammals die each year after becoming entangled in beverage six-pack holders.

ing of radioactive ores; the use of refined radioactive materials for industrial, scientific, and medical purposes; nuclear accidents; the production and testing of nuclear weapons; and the use of cooling water in nuclear power plants. Many of these isotopes magnify in food chains and can cause cell mutations and cancers. Soviet authorities reported extensive groundwater contamination in the area around the Chernobyl nuclear accident. They sealed off over 7,000 wells within a 50-mile (81-kilometer) radius of the plant, drilled emergency wells into bedrock to provide water to bakeries and milk-bottling plants, and constructed an emergency pipeline from the uncontaminated Desno River.

Heat. Changes in water temperature can cause major shifts in the structure of biotic communities. Water is used for cooling purposes in manufacturing, industrial, and electric power generation plants. The heated water is then returned to its source. Hot water holds less oxygen, speeds up respiration in aquatic organisms, and tends to accelerate ecosystem degradation. The relatively still waters of lakes and bays are particularly vulnerable to thermal pollution.

History of Management of Water Resources

How Have Humans Managed Water Resources in the Past?

People have always settled close to waterways. Waterways could supply water to drink and to irrigate crops, transport goods, supply food, and carry off wastes. So long as settlements were small, they had enough clean water to meet their needs. As civilizations became more complex and as cities began to grow, the demand for water increased and water delivery systems became more complicated. Some cultures flourished or disappeared because of the ways in which people managed their water supply. For example, historians believe that the ancient Sumerian civilization, located in the rich Tigris and Euphrates Valley, disappeared because its irrigation system deposited salt on fertile farm lands, eventually rendering them unfit for agriculture.

For centuries humans have followed two axioms concerning water management. First, draw drinking water from upstream sources, and dispose of wastes into downstream sources. Second, if the habitat is dry, make it wet, and if the habitat is wet, make it dry. The consequence of the first axiom is that as populations increase, the volume of wastes deposited by communities upstream makes it impossible for communities downstream to receive an adequate supply of clean, fresh water without costly treatment. In following the second axiom, societies construct water delivery systems—dams, reservoirs, canals, and pipelines—to supply water wherever and in whatever quantity desired, and they drain wetlands to provide land for development. Water management and the development of civilization have gone hand in hand, as governments have often undertaken major water projects to encourage agricultural and economic expansion.

How Have Water Resources Been Managed in the United States?

Europeans who sailed for the New World left behind fetid cities, where human excrement was dumped in the streets to be washed away by rainwater and where disease epidemics were common. In North America they found a seemingly endless supply of clean water. They could not imagine that water would ever be a problem, and as long as the volume of waste produced could be diluted and recycled by natural processes, the waterways did remain clean. But with increasing populations, especially in cities, water pollution became a threat to the environment and to human health.

Let's look at the cities of New York, Los Angeles, and New Orleans to see how they attempted to increase the quantity and to protect the quality of water supplies.

New York. Prior to the 1830s New York City relied on wells, streams, and ponds for its drinking water. But with the city's phenomenal growth in the early nineteenth century, its water sources became inadequate and so polluted that they could no longer be used safely. An 1832 cholera epidemic claimed many lives, especially among the poor, who could not afford to escape to the country or purchase clean water brought to the city by water wagons. Then, in 1835 a devastating fire burned 100 square blocks of industrial and business property. These two events underscored the inefficiency of New York's water supply and led to the construction of a massive aqueduct and reservoir system to transport water from nearby Westchester County to the city.

Though costly and disruptive to many farmers in the surrounding countryside, the system alleviated the city's water problems for a time. But by the late nineteenth century, New York's continued growth necessitated the construction of protected reservoirs in the Catskill Mountains and the Delaware River valley (Figure 15-15). Two large tunnels, 24 feet (7.2 meters) in diameter and located up to 800 feet (240 meters) underground, brought more than 1 billion gallons (3.79 billion liters) of water to New York City. Many small towns were inundated to create these reservoirs, and the construction and maintenance costs were high. Still, the economic and health benefits to the city were enormous, and the system has served the city well; New York is known for the quality and good taste of its water. Currently, the system and many of the pipes that deliver water within the city are in need of costly repairs. New York is investing heavily in a third tunnel so that one of the older ones can be shut down for repairs without interrupting water delivery.

Like all large cities, New York's water problems did not end with ensuring an adequate supply of clean water. Waste disposal is the other part of the water equation. By the start of the twentieth century the implementation of waste disposal systems began to eliminate the spread of waterborne contagious diseases. Sewer pipes leading to sewage treat-

FIGURE 15-15: Asholcan Reservoir, Eastern Catskill Forest Preserve. The reservoir is owned and used by New York City.

ment facilities removed wastes from homes and factories. However, the sewer pipes were coupled to the same pipes that removed rainwater from the streets after storms, and when the combined sewers overflowed, raw sewage poured directly into rivers and oceans. By the 1920s sewage from upstream cities began to appear in the drinking water supplies of many downstream cities in the United States, resulting in a resurgence of waterborne diseases. Drinking water filtration, treatment, and chlorination systems were built to remove pathogens and contaminants from polluted drinking water. For a time the treatment facilities prevented the spread of waterborne diseases caused by poor sanitation, but unprecedented population growth again brought problems. While New York's drinking water supply was drawn from a protected reservoir, its sanitation system was eventually overtaxed. As it became overloaded, billions of gallons of partially treated or untreated sewage were allowed to flow through the system to enter and contaminate waterways (see the description of the New York Bight, pages 000–000).

Los Angeles. Located in a valley framed by mountain ranges, Los Angeles is renowned for its fine beaches and a Mediterranean-like climate that promotes year-round outdoor activities. The city itself has a population of over 7 million, and the entire southern California area, which includes Los Angeles and San Diego, is home to over 15 million people. Southern California business enterprises play a significant role in the entire state's economy. Given its population and economic importance, the area exerts a great amount of influence on California politics.

The one drawback of southern California was—and is—inadequate rainfall. Early in its history Los Angeles did not have enough fresh water to meet the needs of a population of 100,000. A series of water development projects allowed it to become the nation's second largest city and the influential metropolis that it is today.

Los Angeles built its first aqueduct in the early 1900s. Since that time an expanding network of reservoirs, canals, and tunnels has been built to transport billions of gallons of fresh water over hundreds of miles from the Sierra Nevada in northern and eastern California. Additional water is diverted from the Colorado River in the western Rockies. For southern California's agribusiness, industry, and real estate developers, the water has meant prosperity and power.

However, Los Angeles is built and prospers on borrowed water. To discover the true cost of southern California's prosperity, we must look not at southern California, but at those areas where the water originates. For the small farms and towns of the Owens and San Fernando valleys in northern California, the lost water—and the tremendous growth of the south—has meant fewer economic opportunities and diminished political power. For Mono Lake, site of the second largest expansion of southern California's aqueduct in the 1930s and 1940s, it has meant a dramatic drawdown, which has increased salinity and made the lake much shallower. The unique biota of Mono Lake is severely threatened, prompting ecologists and conservationists to urge the imposition of a limit on the amount of water drawn from the lake. For Arizona, attempts by Los Angeles to support continued growth through water development schemes have meant a court case, heard by the U.S. Supreme Court, which forced California, in 1985, to share its Colorado River water with Arizona.

There are signs that southern Californians may be recognizing the folly of fueling growth with borrowed water. Recent attempts to construct yet another series of canals, designed to transport even more water from the north, were defeated by a statewide referendum. Even most southern Californians opposed the measure on the basis of cost and environmental concerns.

New Orleans. New Orleans, a river town, has plenty of water. Sometimes, because of violent storms and river floods, it has far too much. Built on a low delta and surrounded by wetlands, the city is subject to the vagaries of one of the world's foremost rivers, the Mississippi. Unlike the protected watershed reservoirs of New York City and Los Angeles, its water source is downstream from millions of people and several large cities, including Chicago, Pittsburgh, and St. Louis. Because of its location, flood control and the decontamination of river water are particularly formidable tasks. Over the years New Orleans controlled flooding through the construction of a complicated network of canals, levees, and pumps. To provide safe drinking water, the city built water treatment plants to clean and disinfect the muddy Mississippi waters of disease-causing organisms.

Then, in 1974 the EPA announced that high concentrations of 66 nontraditional human-made chemical contaminants, including six suspected carcinogens, were discovered in the drinking water. The contaminants included heavy metals such as chromium, zinc, and lead and organic chemicals such as benzene and carbon tetrachloride. Hundreds of other chemicals were found in trace amounts. Many of the contaminants were traced to the petrochemical industry operating upstream from New Orleans, agricultural operations, and numerous other industries on the river and its tributaries. At the same time researchers discovered that New Orleans had one of the highest rates of cancer in the nation.

These discoveries focused attention on an entirely new water purification problem—removing substances toxic to human health. Water treatment plants designed to remove silt and disease-causing organisms were not adequate to treat river water laced with chemical contaminants. Problems with toxic contamination have continued. In 1991 the 150-mile (242-kilometer) corridor of the Mississippi between Baton Rouge and New Orleans earned the dubious distinction of receiving more toxins than any other stretch of water in the country, sobering news for the many residents who rely on the river for drinking water.

What Legislation Protects the Nation's Water Resources?

Legislation, or the lack of it, determines how water is managed in the United States. Let us briefly look at the provisions of the Safe Drinking Water Act and the Federal Water Resources Protection Act, the two major pieces of legislation specifically designed to protect the nation's water resources.

Safe Drinking Water Act. In part, the condition of New Orleans' drinking water and the reluctance of the state to force Louisiana's important petrochemical industry to remove pollutants from its wastewater helped spur passage of the Safe Drinking Water Act of 1974. The act set national drinking water standards, called maximum contaminant levels (MCLs), for pollutants that might adversely affect human health; three years later, the first standards went into effect. It also established standards to protect groundwater from hazardous wastes injected into the soil, once a common waste disposal practice. Over 200 contaminants have been identified in groundwater alone.

In 1982 and 1983 the EPA established a priority list for setting regulations for over 70 substances. These substances were listed because they are toxic and likely to be found in drinking water. In 1986, when Congress reauthorized the Safe Drinking Water Act, it directed the EPA to set standards within 30 years for all 70 substances on its priority list. As of 1988, the EPA had set standards for 22 of them. The 1986 reauthorization also instructed the EPA to monitor drinking water for unregulated contaminants and to inform public water suppliers which substances to look for.

Federal Water Pollution Control Act. In 1972 Congress enacted the Federal Water Pollution Control Act, commonly known as the Clean Water Act. Amended in 1977, the act divided pollutants into three classes: toxic, conventional, and unconventional. It stipulated that industries must use the best available technology (BAT) to treat toxic wastes before releasing them into waters. Conventional pollutants, such as municipal wastes, must be treated using the best conventional technology (BCT). All other pollutants, classified as unconventional, must meet BAT standards, though waivers can be granted for pollutants in this class. If the BAT will not protect certain waters, stricter standards (including "no discharge") must be enforced.

The Clean Water Act contained several stipulations pertaining to industry. It established pretreatment standards for industrial wastes that pass through sewage treatment plants. These standards were set in order to force industries to handle toxic wastes rather than simply dump them into municipal facilities. Such dumping in many cases had resulted in the production of toxic sewage sludge. The act also mandated that the EPA or state must grant a special permit to allow the discharge of any pollutant into navigable waters. These permits define and limit the amount of pollutants that can be included in the wastewater and form the primary means for implementing the Clean Water Act. Discharges must meet the appropriate toxic, conventional, or unconventional standards. Unfortunately, the goal—to phase out toxic discharges by 1985—was not met. In 1987 the EPA reported that 250 city sewage facilities and 627 industrial operations were still dumping toxic wastes into U.S. waters.

The Water Quality Act of 1987, a reauthorization of the original Clean Water Act, provided $20 billion to curb water pollution, primarily through the construction of wastewater treatment plants. The federal government provides 20 percent of the money needed; local sources provide the remaining funds. For many small communities the cost of building or remodeling existing sewage treatment plants (80 percent of the total cost) is prohibitive. The 1987 act also provides money for nonpoint pollution programs. It provides appropriations to help state and local governments improve the water quality of bays and estuaries. Through a National Estuary Program, money will help to clean up and protect San Francisco Bay, Puget Sound, and Boston Harbor. In addition, special appropriations have been made to monitor and control pollutants entering the Great Lakes and Chesapeake Bay.

How Are Drinking Water Supplies Treated?

After water is withdrawn from a lake, river, or aquifer, it is treated before being distributed to local destinations such as homes, businesses, schools, and hospitals. Most water systems, large or small, include certain basic steps. A chemical such as aluminum sulfate (alum) is added to water supplies to create small gelatinous particles, called floc, which gather dirt and other solids. Gentle mixing of the water causes floc particles to join and form larger particles; floc and sediment fall to the bottom and are eventually removed as sludge. The water is then filtered through a granular material such as sand or crushed anthracite coal (carbon). Chlorine is added to kill bacteria and other microbes, and a chemical such as quicklime is added to reduce acidity in water and prevent corrosion in city and household pipes. Many municipal treatment plants also add fluoride to water supplies to prevent tooth decay.

Treated water is sent through a network of pipes to consumers. To ensure its quality, water must be monitored and tested throughout the treatment and delivery process. Generally, surface water is more complicated to treat than groundwater because it is more likely to be contaminated.

How Is Wastewater Treated?

In rural areas (and some suburban areas) that have suitable soils, sewage and wastewater from each home are usually discharged into a **septic tank**, an underground tank made of concrete. Solids settle to the bottom of the tank. Grease and oil rise to the top where they are trapped and periodically removed to prevent them from clogging the tank. Bacteria in the wastewater feed on the sludge and scum and liquify the waste products. The wastewater filters out into a drain field. This process requires time, and septic systems must be large enough to accommodate the expected flow from each home (Figure 15-16).

Wastewater in urban areas must be managed much differently. Underground sewers collect wastewater from residences, businesses, schools, hospitals, industrial sites, and other buildings and transport it to sewage treatment plants. These plants are designed to make wastewater safe for discharge into streams or rivers or to make it acceptable for reuse.

There are two kinds of sewer systems: combined and separate. A **combined sewer system** carries both wastewater and rainwater. Pipes that collect each type of water connect with an interceptor pipe that leads to a treatment plant. The interceptor pipe is large enough to hold several times the normal combined flow. But during storms the rainwater flow might increase by a factor of 100. Since the combined sewer system is designed to protect the treatment plant, which can handle only a certain amount of flow, and to prevent combined flow from backing up into buildings, some of the combined flow (raw sewage and rainwater) may bypass the treatment plant and go directly into a receiving stream. The combined system does not protect the environment,

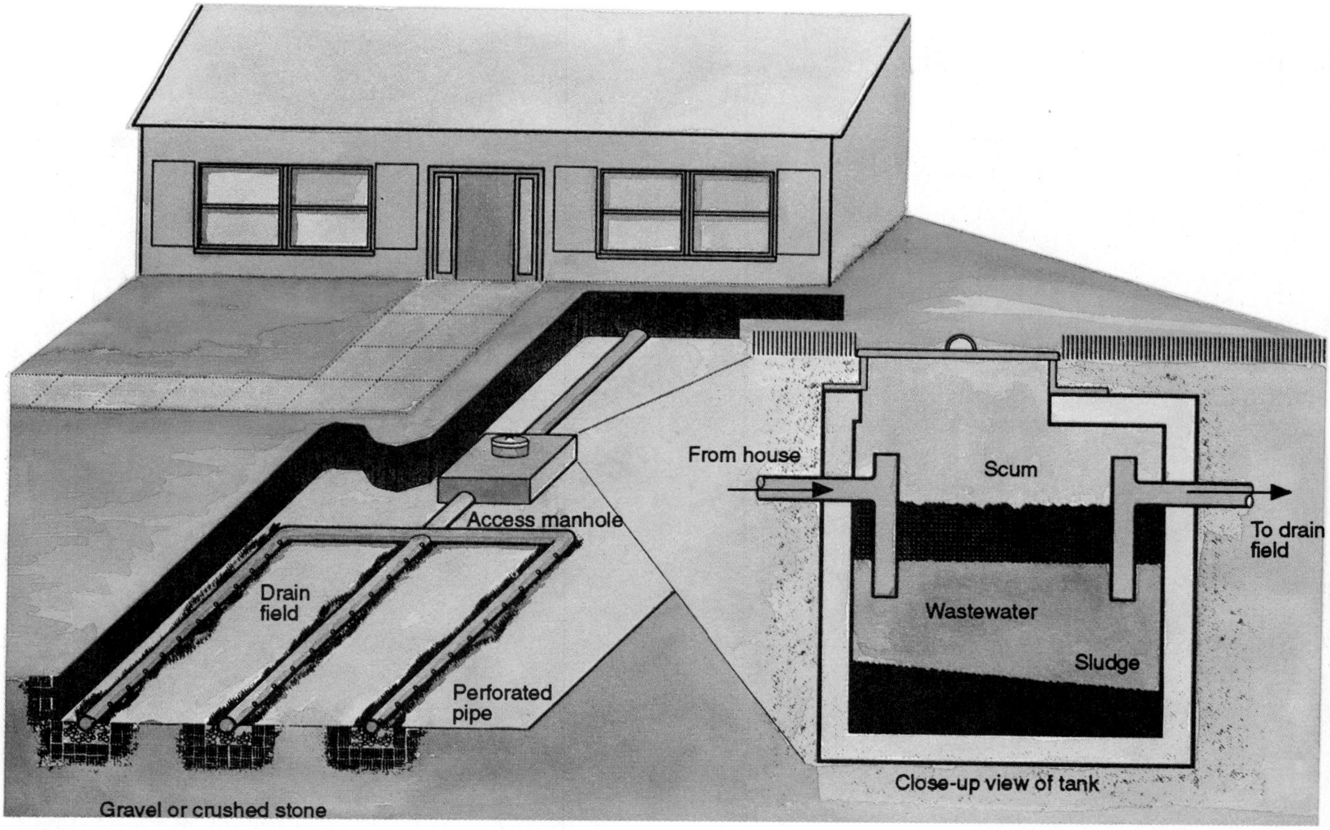

FIGURE 15-16: Septic tank system. In a properly functioning septic tank, only treated wastewater filters into the drain field. The drain field should be constructed so as to allow the water to drain properly and the soil to absorb the water and nutrients. For this reason, drain fields should not be placed in clay or sandy soils. When septic systems are not functioning properly, effluent can be seen rising to ground level and drains or toilets operate slowly or not at all. Odors can often be detected near or below drain fields.

which receives the full assault of the combined flow, with its load of sewage and other contaminants.

As the name implies, a **separate sewer system** consists of sanitary sewers, which carry only wastewater, and storm sewers, which carry only rainwater and melting snow. Storm sewer water follows a different route from that of wastewater, bypassing the treatment plant and going directly into a receiving stream. Separating sanitary and storm water flow is costly, but it does significantly lower the volume of pollutants that enter a receiving stream after a storm. Unfortunately, separate systems have one significant flaw. Storm water often carries sediments, oils, greases and other nonpoint source pollutants, and because storm water is not treated, these pollutants are transported directly into the aquatic environment.

Once the flow enters a sewage plant, it usually goes through a multistage process to reduce it to an acceptable effluent, the water that leaves the plant (Figure 15-17). Properly treating sewage maintains the integrity of the stream or river receiving the effluent.

Primary treatment is a physical process to remove undissolved solids. Screens remove sticks, rags, and other large objects. In some plants the remaining sewage is chopped or ground into smaller particles, then passed to a grit chamber where dense material such as cinders, sand, and small stones settle out and are removed. Undissolved suspended materials, including greases and oils, are removed in a settling tank or primary clarifier. Greases and oils are skimmed off of the top and the undissolved organic material sinks to the bottom, where it forms a mass called raw sludge. This sludge is drawn off to a sludge digester where it is further reduced by bacteria.

In plants that provide only primary treatment, the remaining liquid, called primary effluent, may be chlorinated and released to the receiver stream. Primary treatment is only about 50 percent effective at purifying wastewater. It cannot remove excess nutrients, dissolved organic material, or bacteria. In large cities the sewage volume is so great that most of the sewage receives only primary treatment. Consequently, treatment plants release billions of gallons of partially treated sewage into the environment, in addition to billions of gallons of untreated sludge.

Secondary treatment is a biological process. Primary effluent is treated in one of two ways: trickling

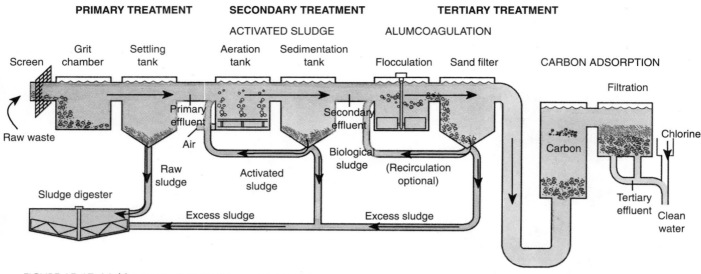

FIGURE 15-17: Multistage wastewater treatment process.

filters or activated sludge. Plants with trickling filters have tanks with beds of stones several feet thick. As wastewater passes, or trickles, through the beds, bacteria living on the stones consume most of the dissolved organic matter. After leaving the filter, wastewater is allowed to settle in a secondary clarifier and then may be disinfected with chlorine and released to the environment.

In plants using activated sludge, primary effluent is pumped into an aeration tank (in which oxygen is added) and combined with biological sludge, creating a suitable environment for bacteria to digest the dissolved organic material still present in the wastewater. After aeration, the secondary effluent passes to a sedimentation tank or secondary clarifier where biological sludge is removed. Part of the sludge is recycled back to the aeration tanks, and the remainder is removed to the digester. The secondary effluent can then be chlorinated and released at this point. Secondary treatment is usually 85 to 90 percent effective.

Tertiary treatment may be a physical or chemical process. It is designed to remove ammonia, nutrients, and organic compounds still remaining in the wastewater. Sand filters, activated carbon absorption, or alum coagulation can remove nearly all of the remaining solids or chemicals. Tertiary effluent can then be disinfected and released to the environment. Tertiary effluent is considered to be 90 to 95 percent treated. In many instances, it is of a higher quality than the water in the receiver stream and is near drinking water quality. If phosphates are removed during this stage, such advanced treatment can render the water as much as 97 to 99 percent treated.

Management Issues Associated with Wastewater Treatment. Proper sewage plant construction and operation are expensive. Upgrading the waste disposal facilities in the 43 municipalities that surround Boston Harbor will cost more than $3 billion, an expenditure that will quadruple the water and sewage bills for residents. Nationwide, an estimated $75 to $100 billion is needed to upgrade sewage plants that discharge effluent into coastal areas.

In the 1970s the federal government bore 80 percent of the cost of sewage system upgrades; 20 percent was contributed by local authorities. By the late 1980s those figures had been reversed. Consequently, for many towns and municipalities the costs of upgrading are prohibitive. Large metropolitan areas are also struggling to pay for new plant construction to accommodate increased populations.

Perhaps it is time to reevaluate federal funding priorities. Upgrading treatment plants is essential in order to prevent sewage and contaminants from entering our waterways. Moreover, effluent and properly treated sewage sludge should be considered resources to be used for constructive purposes. (Chapter 20 explores ways in which effluent and sewage sludge can be recycled.)

While government's role in improving the management of wastewater is a vital one, individual and industrial efforts are no less important. Individuals must begin to realize that, once flushed, sewage should not be "out of sight, out of mind." Toilets should not be used to dispose of garbage. Moreover, we can all make an effort to reduce the amount of water we flush. On average, each U.S. citizen uses 12,000 gallons (45,500 liters) of water per year to flush away 150 gallons (570 liters) of sewage.

Industry can do its part through a willing compliance with the pretreatment of industrial wastes. According to a 1988 EPA report, three-quarters of all wastewater treatment plants surveyed cannot remove industrial toxins. The Clean Water Act requires manufacturers to treat their waste to remove chemicals, but the EPA allows local municipalities to enforce compliance. The EPA estimates that between 100,000 and 200,000 industrial concerns are releasing their wastes untreated directly into the municipal sewer systems. In effect, we are engaged in a totally irrational practice: allowing wastes to enter our waterways and then spending billions of dollars to remove them.

What Problems Surround the Use of Water Worldwide?

Worldwide, water resources are the source of many environmental and economic problems (Table 15-2). Water scarcity is at the heart of most problems. Groundwater depletion is a problem in many areas; parts of India and China, for example, are experiencing severe shortages as demand exceeds supply. Competition for the use of surface waters has also become increasingly intense. In 1960 the Aral Sea was the fourth largest inland body of water (in surface area) in the world, behind the Caspian Sea, Lake Superior, and Lake Victoria. By 1990 it had lost over 40 percent of its area and 67 percent of its volume, while its mineral concentration (including its saline concentration) nearly tripled. The principal cause of the Aral Sea's decline is irrigation, which withdraws waters from the sea's two major freshwater tributaries, the Amu Dar'ya and the Syr Dar'ya (Figure 15-18).

Many areas around the world have been subjected to prolonged droughts. Streams and waterholes dry up, as do sources of groundwater, as the water table in the area is lowered. Prolonged drought chronically affects such areas as the Sahel region of Africa, southwest Texas, and parts of California. During a devastating drought in the summer of 1988 at least 30 states from California to Georgia and from the Canadian border to Texas declared federal emergencies. The drought cost billions in lost crops and livestock, whose market value plummeted because inadequate pasturelands and a lack of water forced ranchers to sell them early. The drought triggered forest fires in Montana, Idaho, and California, and low water caused massive barge traffic jams on the Mississippi River. Nationwide, cities and towns that rely on rain-recharged groundwater for their drinking water supplies suffered serious water shortages.

While drought poses a serious problem, many of the difficulties associated with too little water are of our own making. Agricultural ventures and urban growth in arid climates such as the American West and Southwest place a great strain on available water supplies. Communities in these locales have always had too little water, and continued growth only exacerbates the problem. The struggle over water

Table 15-2
Water Scarcity, Selected Countries and Regions

Country or Region	Observation
Africa, north and east	Ten countries are likely to experience severe water stress by 2000. Egypt, already near its limits, could lose vital supplies from the Nile as upper-basin countries develop the river's headwaters.
China	Fifty cities face acute shortages. Water tables beneath Beijing are dropping 3 to 6 feet (1 to 2 meters) per year. Farmers in the Beijing region could lose 30 to 40 percent of their supplies to domestic and industrial uses.
India	Thousands of villages throughout India face shortages. Large portions of New Delhi have water only a few hours a day. Plans to divert water from the Brahmaputra River have heightened Bangladesh's fear of shortages.
Mexico	Groundwater pumping in parts of the valley containing Mexico City exceeds recharge by 40 percent, causing land to subside. Few options exist to import more fresh water.
Middle East	With Israel, Jordan, and the West Bank expected to be using all renewable sources by 1995, shortages are imminent. Syria could lose vital supplies when Turkey's massive Ataturk Dam is completed in 1992.
Former Soviet Union	Depletion of river flows has caused volume of Aral Sea to drop by two-thirds since 1960. Irrigation plans have been scaled back. High unemployment and deteriorating conditions have caused tens of thousands to leave the area.
United States	One-fifth of the total irrigated area is watered by excessive pumping of groundwater. Roughly half of western rivers are over-appropriated. To augment supplies, cities are buying farmers' water rights.

Source: Worldwatch Institute.

often pits community against community, farmers against urban dwellers, and ranchers against wilderness preservationists. Nowhere is that struggle more apparent than in California. By 1991, as residents endured the fifth year of a severe drought, water storage in the state's reservoirs fell to about 30 percent of normal, and water rationing plans went into

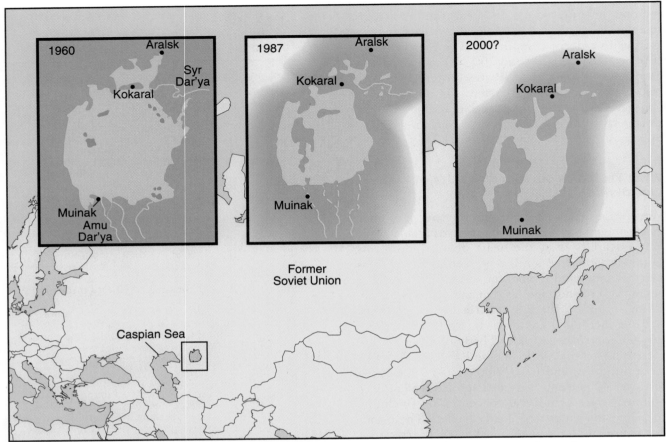

FIGURE 15-18: The vanishing Aral Sea. In the late 1980s, the Institute of Geography of the U.S.S.R. Academy of Sciences estimated that 80 percent of the depletion of the Aral Sea over the previous three decades was due to human activity; just 20 percent was due to natural climate variations. Between 1950 and 1990 irrigated land in the region rose from 7.2 million to 18.5 million acres (2.9 million to 7.5 million hectares). Towns once located on the shores of the Aral Sea now lie in a desert. Drinking water is a critical problem, as are unemployment and other social changes that are brought about by the collapse of traditional marine trades. Animal life in the basin has been profoundly disturbed; as lakes and swamps in the river deltas have disappeared, 75 to 80 percent of the animal species have become extinct.

effect in northern California and Los Angeles. Allotments of irrigation water to farmers, which had been reduced over the previous years of the drought, were completely cut off to save dwindling supplies for the cities.

Water is power. The development of water resources has been and is a significant force in shaping society, particularly in the western United States, where, as Mark Twain once noted, "whiskey's for drinken [sic], water's for fightin." Today's water wars are fought with legal suits, research papers, and blueprints rather than with six shooters and dynamite. For example, El Paso, Texas, sued for the right to sink wells into aquifers that have traditionally served other interests in New Mexico. The Central Arizona pipeline project, slated for completion in the 1990s, will divert billions of gallons of Colorado River water to Phoenix and Tucson to relieve the state's serious groundwater shortage. The Colorado, a once mighty river, now trickles to the sea, laden

with silt, pollution, and salts (Figure 15-19). From Utah to California, cities and irrigation projects are drinking that river dry. Other controversial diversion projects in the United States include the Garrison Diversion Project (North Dakota), the Colorado River/San Diego Diversion and River Aqueduct Expansion Project, the Pinelands (New Jersey) Groundwater Diversion project, and the Great Lakes Diversion Project, which would draw water from our inland seas to fuel continued growth in the Southwest.

From the one-hundredth meridian to the Pacific Ocean, rainfall averages less than 20 inches (50.8 centimeters) per year (with the exception of the Pacific Northwest and high mountains). Westerners live in a near perpetual drought. While Chicago, Illinois, receives about 40 inches (101.6 centimeters) of rain per year, Los Angeles averages just 9 inches (23 centimeters) and Phoenix, Arizona, about 8 inches (20.3 centimeters). Many western areas do

FIGURE 15-19: The Colorado River as it empties into the ocean. Diversion projects have reduced water flow in the Colorado River to such an extent that the once-mighty river is scarcely recognizable by the time it reaches the ocean.

not receive enough rain to sustain agribusiness or rapid increases in urban growth. Only because the federal government subsidizes water provision projects can cities like Palm Springs, California, continue watering 74 golf courses, Arizona farms grow alfalfa in the desert, and dry towns like Phoenix experience phenomenal growth. The government's support for large water development projects makes water available at a low consumer cost. As long as the price remains low, users have little incentive to evaluate future projects in terms of the actual availability of water (without major engineering intervention) and to practice conservation of water from already established projects. For example, drip irrigation systems are more efficient than conventional

systems, but because they are costly to install they are used less frequently (Figure 15-20).

Where the climate is hot and dry, but water comes cheaply, an oasis mentality prevails. We raise extensive gardens and build great cities, and congratulate ourselves for our cleverness. But our gardens and cities exist at the expense of free-flowing rivers, underground aquifers, and fertile soils. Problems have only begun to surface. Perhaps most troublesome is the continued decline in the water table of many aquifers. Water levels in the Ogallala aquifer, for example, are dropping by 3 to 5 feet (1 to 1.5 meters) per year in some areas.

As demand continues to escalate, the cost of pumping and transporting water will continue to rise. Moreover, future generations will not have the easy options that have been available to us: aquifers to tap, rivers to dam, and lakes to drain. They will be faced with harder choices concerning water allocations. Will they think our great cities and extensive desert gardens so clever?

How Is Groundwater Managed?

Groundwater withdrawal in the United States has increased 300 percent since the 1950s. As demand increases, both the quantity and quality of this vast resource are threatened. Our past use of groundwater supplies was based on two popular misconceptions: (1) there is an inexhaustible supply of groundwater; and (2) groundwater is purified as it percolates through the ground. Let us examine both more closely.

It is hard to imagine that the total groundwater supply in the United States, some 30 to 60 quintrillion gallons, could be exhausted, but because it is

(a)

(b)

FIGURE 15-20: (a) Trickle drip irrigation system in a California vineyard and (b) conventional system in a wheat field in eastern Washington.

not evenly distributed and because all aquifers do not recharge in the same way, many aquifers experience overdraft. An **overdraft** is the withdrawal of water from a source faster than it can recharge; it is akin to writing a check for more money than you have in your bank account. Recharge is also adversely affected by draining wetlands, clearing forests, and diverting streams, all of which can reduce the amount of water absorbed by the soil. Paving for buildings, parking lots, and roads prevents water from entering the soil, as do prolonged drought and flooding.

More than two-thirds of all groundwater extracted in the United States is used for agricultural purposes, particularly irrigation. Irrigation uses seven times more than all of our city water systems combined, some 150 billion gallons (569 billion liters) per day. Severe overdrafts, called **water mining,** can lead to depletion, often lowering the water table so drastically that further extraction is no longer economically feasible. Thirty-five states have reported declining groundwater levels as a direct result of irrigation, with the most dramatic declines in Arizona, Kansas, New Mexico, Texas, and California. One acre of corn in western Kansas uses 400,000 gallons (1.5 million liters) of water in one season during a dry year. Projected population increases in the South and Southwest will put even more pressure on groundwater reserves in those areas.

Overdrafts can lead to **land subsidence.** Water pressure in the pores of aquifers helps to support the overlying material. When large volumes of water are removed without adequate recharge, the pores collapse, allowing the rock, sand, or clay particles to settle. If the affected area is large, the ground can collapse or subside, forming sink holes, cracks, and fissures of varying sizes. Once the ground subsides, there is little chance for it to hold water again, as it generally becomes too tightly compressed to expand with new water. Dramatic effects of subsidence have occurred nationwide. In Winter Park, Florida, sink holes have swallowed trees, cars, and buildings. In New Orleans many places have sunk as much as 2½ feet (0.75 meter), and in the San Joaquin Valley of California, where water for irrigation is pumped by the millions of gallons, many places have sunk as much as 30 feet (10 meters) (Figure 15-21)!

In addition to problems of quantity, problems of groundwater quality are also becoming critical. For a long time we erroneously believed that groundwater was safe from surface pollutants. We thought that aquifers were deep enough so that contaminants would not reach them. For many years landfills for both hazardous and municipal waste were routinely located in abandoned gravel pits along rivers, where chemicals could leach into both surface and groundwater. Moreover, we thought that pollutants

FIGURE 15-21: Land subsidence in Florida. Pauline Bennett looks at what is left of her home in Frostproof, Florida, after a sinkhole opened up under the house Friday, July 21, 1991.

dumped on or near the ground's surface would be filtered out of the water as it percolated through soil, sand, and rock layers, so that the water would be purified by the time it reached the water table. Because groundwater usually moves slowly, the results of our carelessness took a long time to show up. But they have shown up. Tragically, most water treatment plants and nearly all private wells sunk into groundwater sources are not equipped to remove pollutants. When groundwater becomes contaminated, it is a costly, slow, and sometimes impossible task to remove pollutants.

Groundwater contamination has reached such magnitude that surveys indicate it is a concern to people in all 50 states. Pollution by bacteria from human and nonhuman waste sources is the number one contaminant in rural water supplies, and a dramatic increase in toxic chemicals, especially synthetic organic chemicals, is being noted in both rural and urban groundwater supplies. A leak of a single gallon of a hazardous substance can contaminate millions of gallons of groundwater. For instance, in Belview, Florida, 10,000 gallons (37,900 liters) of gasoline leaked from an underground service station tank from 1979 to 1980. The gasoline was detected in the city's water supply in 1982. The gasoline contaminated the entire drinking water supply for a population of 2,500 people. Eventually, Belview constructed a new water system from a supply located away from the town.

Thirty-four states have reported serious groundwater contamination from toxic chemicals. Pesticides, which have been found in groundwater in 25 states, are the major contaminants in big agricultural states such as Iowa, Florida, Nebraska, Mississippi, Minnesota, Wisconsin, and Idaho, all of which are among the ten states most dependent on groundwater for drinking water. The EPA estimates that 1.5 trillion gallons (5.7 trillion liters) of toxic substances leak into groundwater each year. According to the California Board of Health, one in four wells in that state is polluted by chemicals, chiefly industrial chemicals and pesticides. The Michigan Department of Natural Resources estimates that as many as 50,000 sites statewide could lead to groundwater contamination and cost $3 billion to clean up. According to a 1986 EPA report, as many as 300,000 underground gasoline storage tanks are leaking. Further, 30 to 40 percent of the estimated 1.4 to 2.5 million underground storage tanks (USTs), containing billions of gallons of many different potentially dangerous substances, are leaking. For example, trichlorethylene (TCE) has contaminated drinking water supplies in such widely disparate areas of the country as Bedford, Massachusetts; Fort Edward, New York; Manassas, Virginia; Denver, Colorado; and the San Gabriel Valley, California.

How Are Fresh Surface Waters Managed?

Pollution and development pose serious challenges to those responsible for managing the nation's tens of thousands of lakes. The Clean Water Act brought dramatic improvements in the treatment of wastewater (principally organic wastes), but toxic pollution remains a serious threat. Airborne toxics are thought to contribute significantly to the total toxic contamination in the Great Lakes and are responsible for the acidification of lakes and streams in the Adirondack Mountains and parts of eastern Canada.

Accelerated eutrophication is also a problem in many lakes. Wastewater and agricultural runoff carry nutrients into streams and lakes, encouraging algal and plant growth. In some lakes, such as Lake Erie, eutrophication has been slowed by banning phosphates in detergents, encouraging farming practices that reduce runoff, and improving wastewater treatment.

Development of coastal areas, especially along the Great Lakes and the Gulf of Mexico, is also of concern to resource managers. Development can cause soil erosion, contributing to water quality problems.

Rivers throughout the country are besieged by numerous threats. Eastern United States has thousands of rivers and streams that flow year-round. The Midwest also has many rivers, though some streams are dry for part of the year. But in the West and Southwest many waterways are dry beds for significant periods of time each year.

In the West and Southwest the Bureau of Reclamation began a desert reclamation project in 1902. Under the Department of the Interior, the bureau was charged with developing water projects in 17 western states to provide storage and water conveyance for irrigation projects. Proponents of the desert reclamation project maintained that it would pay for itself via increased use of public lands and a growing population. Unfortunately, low interest rates, water subsidies, and procedural loopholes delayed the payback. Further, the cheap water discouraged conservation.

Dams and water diversion became the order of the day. On the positive side, dams do control flooding, generate hydropower, and store water for a variety of uses. But they were often built for political purposes, and they usually did more harm than good. Generally, larger water diversion projects were expensive to construct. Tax monies subsidized construction, water use, and hydropower generation with little return to the federal coffers or to the public.

Dams, diversions, channelization, and pollution have taken their toll on our rivers. A National Park Service study found that only 312 significant streams in the lower 48 states (12,700 miles out of 3.25 million miles, or 20,447 kilometers out of 5.2 million kilometers) are free flowing and undeveloped in their entirety. In 1982 only 2 percent of river mileage was undammed, undeveloped, and clean enough to be eligible for the Federal Wild and Scenic Rivers program of 1968 (Figure 15-22). The program protects free-flowing rivers from federal water projects and establishes a national policy for river conservation. It was enacted to be a check on the building of too many and perhaps unnecessary dams by the federal government. However, of some 62,000 miles of river (99,820 kilometers) that were originally proposed for inclusion in the program, only about 7,000 miles (11,270 kilometers) are designated wild and scenic and thus are afforded federal protection.

Much of the blame for the rarity of free-flowing rivers must rest with federal policy, which puts power generation first and environmental consequences of altered habitat last. The Federal Energy Regulatory Industry grants licenses for hydroelectric plants on free-flowing rivers. It can also override states' designations of scenic rivers, granting permits to build dams to generate hydroelectric power.

Currently, water rights are guaranteed by law in western states for diversions for hydroelectric, irrigation, municipal, and industrial use. Streams can be diverted to the extent that no water remains and the stream bed dries up. In contrast, no water rights are granted for environmental purposes. Water for wildlife, wilderness, human recreation, and replenishing of groundwater is defined as "wasted." The situation in the western United States mirrors what is occurring in many regions of the world. There is

7. Columbia and Snake Rivers (Oregon and Washington): Once the biggest producer of salmon in the world, the Columbia River system is now threatened by hydropower dams, logging, water withdrawals, overfishing, and other problems.

3. American River (California):The primary threat is a 498-foot high dam recommended by the Army Corps of Engineers, ostensibly to protect Sacramento from flooding. It would provide the city with three times the necessary flood protection and could easily be converted into a full-time hydropower and water supply dam —uses that would drown up to 10,000 acres of outstanding river and canyons and destroy up to 48 miles of free-flowing river on the north and middle forks, which lead into the Sierra Nevada.

8.Gunnison River (Colorado): Descending from the western slope of the Rocky Mountains some 250 miles to the Colorado River, the Gunnison flows through the Black Canyon of Gunnison National Monument. The proposed AB Lateral Hydropower Project would divert immense amounts of water to the Uncompaghre River. Reduced flows in the Gunnison would result in higher water temperature, damaging the trout fishery, and would reduce the river's value to white-water rafters. The Gunnison is also threatened by the proposed Union Park Transmountain Diversion Project, which would divert at least 100,000 acre-feet (an acre-foot is the amount of water needed to cover one acre to a depth of one foot) of water to quench the thirst of the population of Colorado's eastern slope.

6. Upper Mississippi River (Midwest): The upper Mississippi—the 800-mile stretch between St. Louis and Minneapolis—flows through more than 500 miles of national wildlife refuges and harbors at least six federally endangered species. It is threatened by runoff from farms and cities, new environmentally destructive hydropower projects, and development pressure. One billion gallons of oil and other hazardous materials are transported on the river each year, and thus the potential exists for a catastrophic spill.

4. Penobscot River (Maine): Although the State of Maine and the U.S. Fish and Wildlife Service have restored the river as an Atlantic salmon spawning ground, the Bangor Electric Company's proposed dam would eliminate 3.6 miles of free-flowing river and pose a new impediment to fish migration.

9. Passaic River (New Jersey): Although the Passaic River basin has been identified as the most flood-prone river valley in the eastern United States, thousands of structures have been built on the flood-plain. The Army Corps of Engineers proposes to build a massive underground tunnel to draw down storm water and deposit it further downstream. This plan would destroy 600 acres of wetlands, damage the intertidal zone in the lower Passaic, draw down highland bogs (threatening clams and salamanders), change the hydrogeology of the area, and require the construction of 15 to 18-foot high walls along the banks of the river.

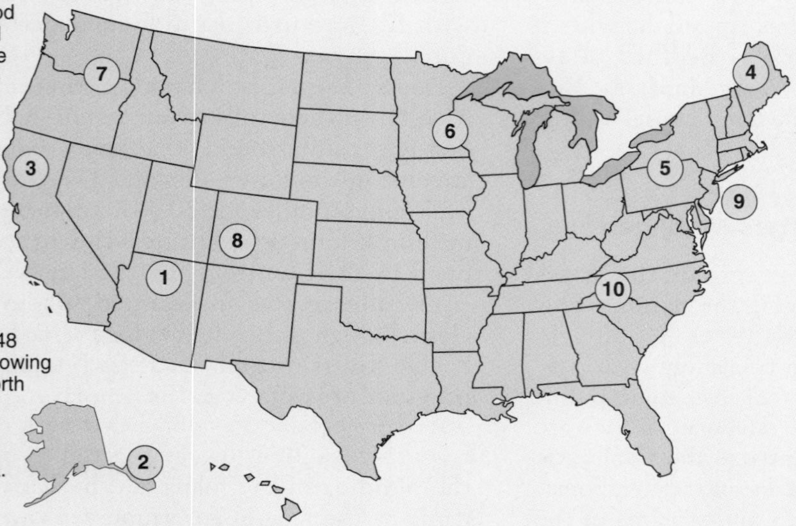

5. Susquehanna River (Pennsylvania): The proposed 17-foot high Dock Street Dam, designed to provide electrical power and revenue for the city of Harrisburg, would create a 3,800-acre reservoir. It would set back efforts to restore American shad to the river, inhibiting upstream passage and destroying over 8 miles of spawning habitat. It also would inundate about 208 acres of island and other river wetland habitats, and eliminate approximately 2,310 acres of diversified aquatic habitat.

10. New River (North Carolina): The New River, the second oldest river in the world according to geologists, is a favorite of white-water rafters. Land development, spurred by the area's increasing popularity as a vacation home site, could cause the river to become so degraded it would lose its designation as a National Wild and Scenic River segment.

2. Alsek and Tatshenshini rivers (Canada and Alaska): Known as North America's wildest river system, the Alsek and Tatshenshini rivers form the northern boundary of Glacier Bay National Park. In the Tatshenshini Valley, the only unprotected area in this zone of wilderness, a mining company has proposed the largest open-pit copper mine in North America. With the mine would come acid mine drainage decimating salmon fisheries and damaging water quality in general, roads ruining the wilderness area, and loss of white-water recreational opportunity.

1. Colorado River (Arizona): The Glen Canyon Dam, 16 miles upstream from Grand Canyon National Park, seriously threatens the Colorado. The dam releases water in wildly fluctuating spurts, making river flow unstable. The level of the river can rise and fall 13 feet in a single day! Such fluctuations damage fragile beaches and vegetation, habitat for a number of species (including the federally endangered humpback chub),a prized trout fishery, and native American archaeological sites. They also impair river recreation, especially white water rafting. Diversion projects are a significant drain on the Colorado.

Note: Unclassified and transitional areas are not shown.

FIGURE 15-22: Ten most endangered U.S. rivers, 1991.

intense competition for the waters of such international rivers as the Nile, the Jordan, and the Ganges.

Human activities not directly related to water supply have significantly altered most U.S. rivers and a large number of our streams. Most stream pollution stems from nonpoint sources such as acid drainage and agricultural runoff. For example, acid drainage from mining operations of all kinds pollutes over 12,000 miles (19,320 kilometers) of streams. Coal mines are the biggest culprits, but copper, lead, zinc, iron, uranium, and gold mines also contribute (Figure 15-23).

FIGURE 15-23: Mine drainage from abandoned coal mine near Pittsburgh, Pennsylvania. In the past, iron-laden water from the abandoned Harmar coal mine posed a threat to Pittsburgh's drinking water supply, which is drawn from the Allegheny River only 4 miles from the mine drainage site.

Measures to Protect Fresh Surface Water. Across the nation, grass roots activists and responsible corporate citizens are rising to the defense of the nation's rivers, streams, and lakes. For example, Milwaukee's two sewage treatment plants flush approximately 200 million gallons (758 million liters) of wastewater into Lake Michigan each day. Hundreds of pounds of toxic chemicals, generated by the more than 500 industries that dump waste into the Milwaukee District Sewage System, are contained in the wastewater. To begin addressing this problem, Milwaukee has developed a voluntary program that calls for companies to halt or minimize their toxic waste discharges. Participating companies are realizing economic savings while lessening their environmental impact. For example, Powder Finishers, Inc., has installed equipment that recovers chromium waste for recycling so that none of it leaves the plant. Another company, Electrotek Corp., a manufacturer of printed circuit boards, captures and recycles about 100 pounds (45 kilograms) of copper each month.

Cities, states, and regions are fighting to save their rivers from dams, diversions, channelization, and pollution. (See Environmental Science in Action: The Willamette River Basin, pages 318–321.) Rivers and streams tend to recover more quickly from pol-

lution, especially organic and thermal pollution, than do lakes. In a first-of-its-kind action, the Pillsbury and Midbury Coal Mining Company donated its water rights (worth $7.2 million) for conservation purposes to help maintain flow through Black Canyon on the Gunnison River in Colorado. Black Canyon is considered to be one of the best natural trout streams in the West, is noted for its scenic beauty, and is home to river otters and eagles. Cities and towns have used varied river corridor management programs to create greenways, parkways, and bikeways; to regulate flood plain development; to hold storm water runoff; and to rejuvenate manufacturing areas no longer in use.

Environmentalists, conservationists, and fishers are fighting to maintain flows in river systems such as the Platte River in Nebraska, Wyoming, and Colorado, where nearly 70 percent of the river is already diverted by various dams and diversion projects; the West Branch of the Penobscot in Maine, which is threatened by the proposed Great Northern Paper Company's hydropower dam; and the Meramec River in the Ozarks, currently free flowing from the Ozarks to the Mississippi but threatened by flood control dams proposed by the federal government. If rivers reflect the health of the land, and if they can be used as environmental monitors, then endangered rivers brought back to ecological health can act as a measure of our commitment to keep our environment clean and diverse.

How Are Marine Waters Managed?

Abuse of the ocean ecosystem can come from both pollution and overuse. Just as the ocean is a vast sink for minerals, nutrients, and gases, it is also a sink for pollutants. Eighty-five percent of ocean pollutants come from the land; 90 percent directly affect vital coastal ecosystems (estuaries, coastal wetlands, and coral reefs) and continental shelf ecosystems. Without thinking, we dump a cauldron of thousands of chemicals into the most biologically productive of the oceans' zones, where they slowly accumulate in aquatic food chains.

About 7 million tons (6.4 million metric tons) of sewage sludge per year is dumped into coastal waters. All of our coastal ecosystems are under severe stress. Particularly hard hit are the New York Bight, Boston Harbor, and the New Jersey coastline. New rules and procedures, which require that sludge and garbage be dumped farther out to sea, may not be ecologically safe and may not prevent the degradation of the ocean ecosystem. Other hard-hit areas include Puget Sound, Washington; San Francisco and Santa Monica bays, California; Albemarle and Pamlico sounds, North Carolina; Long Island Sound, New York; Narragansett Bay, Rhode Island; and Buzzards Bay, Massachusetts.

Cheap disposal of wastes, indiscriminate use of pesticides, unwillingness to prevent erosion, and a careless attitude toward the ocean have begun to exact a tremendous ecological toll on its vital life processes. In the United States, we have dredged and filled 769 square miles (2,000 square kilometers) of wetlands, and we have eliminated entire communities from productive estuaries. We have deposited billions of tons of sewage sludge, garbage, and dredge spoils into coastal and shelf waters—so much that it has begun to return as "gifts from the sea," closing beaches and destroying shellfish beds all along our coasts. Who can forget the infamous barge loaded with trash from New York's Long Island wandering along the Atlantic Coast and around the Gulf of Mexico as its captain searched for a disposal site (Figure 15-24)? It is time to realize that we all have a stake in the ocean's future, that what we do in Ohio or Utah or Ontario affects the quality of the world's largest and most influential ecosystem, which, in turn, ultimately affects us.

FIGURE 15-24: The infamous garbage barge was at sea for months, trying to find a country that would accept its cargo. Eventually, the barge returned to its port of origin on New York's Long Island.

What You Can Do: To Protect Water Resources

- Locate and correct leaks around your home. An estimated 50 percent of all households have some kind of plumbing leak. Worn out washers and faulty tank valves are the prime culprits.

  Look for a silent leak in the toilet, which can waste hundreds of gallons of water each day. To test for leaks, add food coloring to the toilet tank. If there is a leak, color will appear in the bowl within 30 minutes.

  Look for dripping faucets. Consider that it takes about 11,600 average-size droplets to fill a 1-gallon container. If your faucet is leaking at the rate of 1 droplet per second, and those droplets are of average size, about 7.5 gallons (28 liters) of water will be wasted each day. Over a year the leak will send 2,700 gallons (10,200 liters) of unused water down the drain. Leaks are costly: a hot water leak increases your utility bill, and since sewer charges are based on water consumption, leaks increase your sewer charge.

  Read your water meter before and after a period of hours when no water is used in the house. If the meter shows a change, there is a leak in the house. Reading meters is easy: most record gallons much as a car's odometer records mileage. For meters that show cubic feet of water used, you can convert it into gallons by multiplying the figure shown on your meter by 7.5, the approximate number of gallons in one cubic foot.

- Install water-saving devices in your home or business:

  Toilet dams block off a portion of the toilet tank, preventing water from behind the dam from leaving the water closet. Properly installed, they reduce water use by about 2 gallons (7.6 liters) per flush.

  Low volume toilets use about 3 gallons (11.4 liters) of water per flush, compared to 5 or 6 gallons (19 to 23 liters) for conventional toilets.

  Water-saving showerheads have a flow capacity of about 3 gallons per minute, compared to 6 to 9 gallons (23 to 34 liters) for conventional showerheads. Water-saving showerheads can be installed easily with a pair of pliers.

  Flow restrictors reduce the size of faucet openings, conserving water while maintaining the same pressure. They are inexpensive and easy to install on most faucets and showerheads.

- Do not use detergents containing phosphates.
- When landscaping, use indigenous plants. For example, if you live in a dry climate, forgo green lawns in favor of plants that do not require constant watering.
- Never throw garbage into a body of water.
- Get to know a stream, river, pond, lake, estuary, or bay through observation and study; help to protect watershed or drainage basins.
- Become involved with a group or organization involved in restoring a body of water near your home.
- Make local and national politicians aware of your views.

Source: Adapted from a variety of sources, including authors' own notes and Special Issue: Cleaning Our Water, *Chemecology*, October 1991.

Measures to Protect Marine Waters. Until recently little thought was given to managing and preserving marine environments. In fact, the most promising current trend in water resources management is that we have begun to recognize that coastal zones and the ocean are living ecosystems. We have begun to realize the important role that each plays in maintaining the integrity of the biosphere, and we are beginning to propose management strategies for these systems *because nature matters*. For example, ongoing research using living organisms to biomonitor ocean and estuarine ecosystems appears promising. One program uses living mussels and oysters to record levels of pollution from such contaminants as heavy metals and petroleum hydrocarbons. Because these organisms concentrate and retain the pollutants in their tissues, they can make detection easier. We then can determine the effects of these pollutants on other constituents of the ecosystem.

Most scientists agree that we cannot afford to lose any more wetlands, estuaries, bays, or coastal zones. In Florida, where the population increases by some 6,200 new residents each day, most of whom want to live near the coast, almost unrestricted development has destroyed a third of the state's sea grass beds and more than half of its mangrove swamps. Since 1981 Florida has taken several actions to reverse these trends, including purchasing 70,000 acres (28,328 hectares) of shoreline to ensure their preservation and replanting mangroves in northern Biscayne Bay. State authorities are also considering purchasing half-finished, bankrupt development sites and allowing them to return to their natural state.

In recent years the nation, and the global community witnessed a series of horrors on our oceans: hypodermic needles, catheter bags, and blood vials (some of which tested positive for infectious hepatitis and the AIDS virus) on beaches along the East Coast and the Great Lakes; the feces-filled waters of Boston Harbor; fish with tumors and rotted fins and crab and lobsters with "burn holes" in their shells; hundreds of dead dolphins; and marine mammals and sea birds maimed and killed by plastic debris. Perhaps these and other horrors will finally spur the individual, national, and international action necessary to protect the earth's life-giving waters.

Future Management of Water Resources

As supplies of fresh water become increasingly scarce, nations around the world need to take action. The following suggestions incorporate and exemplify strategies necessary to manage water resources in environmentally sound ways. Some suggestions

FIGURE 15-25: Government poster in China promoting water conservation. In the early 1970s the Chinese used the availability of fresh water, among other factors, to determine that their country's resource base could support a population of about 650 to 700 million people. The country's 1991 population was 1.15 billion. Not surprisingly, water is recognized as a precious commodity, and the government encourages wise water use.

apply broadly to water resources and others to a particular category or living system (Figure 15-25).

Prevent Overuse Through Conservation

The federal government should make water conservation a true priority by eliminating subsidies that encourage wasteful water use (especially from federally funded water diversion projects). Instead, it should institute subsidies for practices that conserve water, such as efficient irrigation methods, reuse and recycling of water in industry, and water-saving devices in homes (see What You Can Do: To Protect Water Resources).

Federal policy should discourage large new water projects, especially those that may damage the environment. Our government should develop a national groundwater policy to prevent the pollution and depletion of aquifers.

Environmental Science in Action:
The Willamette River Basin
Cynthia Tufts

The Willamette River basin, covering just 12 percent of Oregon, is home to nearly 70 percent of the state's residents. Settlers were attracted by the basin's fertile soil, by its productive forests, and, most importantly, by the river itself. By the early 1900s overuse and pollution had taken their toll, and many people felt that the Willamette River would never recover. However, as early as the 1920s local residents made improving river management a priority. Because of their efforts, the Willamette basin is once again a popular and productive region for agriculture, industry, and residence.

Describing the Resource

Physical Boundaries

Roughly 150 miles (242 kilometers) long and 75 miles (121 kilometers) wide, the Willamette Basin is defined by the Coast Range on the west, the Cascade Range on the east, the Columbia River on the north, and the Calapooya Mountains on the south (Figure 15-26). The river begins in the Calapooyas, at the confluence of the Coast Fork and the Middle Fork. From there it runs 187 miles (301 kilometers), emptying into the Columbia River near Portland. The Willamette is the twelfth largest river in the United States and one of the few major North American rivers to run northward.

Temperatures in the basin are moderate, ranging from an average 38° F (3° C) in the winter to 67° F (20° C) in summer. The basin maintains this pleasant range because the Coast Range blocks many of the more violent ocean storms coming from the west, and the Cascade Range blocks air of more extreme temperatures from the east. While temperature extremes are rare, precipitation extremes are not. Forty-eight percent of the precipitation in the basin falls during the winter months of November, December, and January, while only

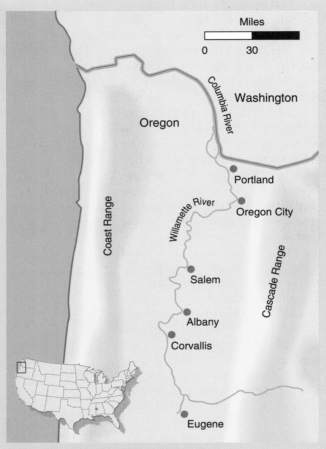

FIGURE 15-26: Map of Willamette River basin.

2 percent falls in July and August. This variation occurs because of the interaction between ocean air and the land. During the winter as the ocean air rises to cross the Coast Range, it cools and forms precipitation as rain or snow. During the summer, when the temperature of ocean air is cooler than land, the air warms as it moves inland; this warming offsets the cooling effect of rising, and the air retains most of its moisture.

Variable rainfall in the basin affects the natural flow rate of the Willamette River. Before the state began to regulate flow, the Willamette's rate varied from summer flows as low as 2,500 cubic feet per second (67.5 cubic meters per second) to winter flood conditions of 500,000 cubic feet per second

(13,500 cubic meters per second). The high-velocity flows of winter and spring made flooding a danger, and the low summer flows reduced water available for irrigation. In addition, the slower movement of the water allowed sludge and bacteria to build up, decreasing dissolved oxygen and degrading water quality. Today, controlled releases of water from reservoirs minimize seasonal variation; the yearly average rate of flow at Salem is 23,000 cubic feet per second (621 cubic meters per second).

The Willamette River valley, the 3,500 square miles (9,072 square kilometers) of the basin under 500 feet (150 meters) above sea level, is ideal for agriculture. Besides a ready supply of water, the valley has rich soil, a legacy of the last ice age (Figure 15-27). The valley floor is flat and relatively well-covered with grasses and trees, so erosion is not a major problem. In the mountains forest cover prevents erosion.

Biological Boundaries

Although forests cover only about 30 percent of the land in the valley (most land is used for agriculture or urban development), they cloak nearly 75 percent of the total land in the basin. Conifers dominate; most occur in the mountains and foothills above 1,000 feet (300 meters). Fish are an important resource of the Willamette basin; many fish, such as salmon and trout, migrate from the Pacific Ocean and Columbia River to the Willamette and its tributaries to spawn. The basin is also home to a variety of wildlife, including big-game species such as black bear, Roosevelt elk, and black-tailed deer. Wintering and migratory birds, such as Canada geese, depend upon the streams, lakes, and reservoirs of the Willamette basin for resting and feeding grounds.

Social Boundaries

European explorers and trappers ventured into the Willamette basin as early as the late 1700s, but they did not begin to settle the region in earnest until the 1840s. The rich natural resources of the area quickly gave rise to a variety of industries.

Agriculture became a major source of food and income in the valley. Currently, approximately one-third of the basin's 8 million acres (3.2 million hectares) are used for crops and grazing. Not surprisingly, a food-processing industry also developed in the area, and today there are nearly 400 of these companies in the basin.

Logging also became a big industry in the region. The Willamette and its tributaries were used to transport cut lumber. A ready supply of wood in the vast forests and inexpensive hydropower gave birth to a thriving pulp and paper industry along the river banks.

Because rivers were such important modes of transportation, several major cities developed along the Willamette. Portland, at the junction of the Willamette and Columbia rivers, is now the largest freshwater port on the Pacific Coast and has become a major railroad center. Salem became the state capital of Oregon, and Eugene grew into the dominant center of trade for all of southwest Oregon.

Agriculture, industry, transportation: each contributed to the economic well-being of the basin, yet each contributed to the degradation of the Willamette River. However, two social characteristics of the basin combined to help save the river over the course of the 1900s. First, almost all of the basin lies within the state of Oregon, so the state can control what happens to the river. Second, intense settlement did not begin along the river until the mid-1800s, much later than in many other areas. When citizens of the basin became concerned about pollution along with the rest of the United States, their river was not as irretrievably damaged as many others.

Looking Back

Abusing the Resources of the Basin

Uncontrolled use of a resource often results in abuse, and such was the case with the Willamette. Logging and agricultural development resulted in massive timber cuts, which increased soil erosion and sedimentation. Agricultural irrigation reduced low summer flows even further, degrading water quality, and dams drastically reduced salmon runs. However, by far the most serious problem for the Willamette River was domestic and industrial waste.

Until 1939 domestic and industrial waste was commonly dumped untreated into the river and its tributaries. Only 52 percent of the people living in the basin were connected to a sewer system, and only 17 percent of these systems had any sort of water treatment facilities. Most of this waste, including wood fiber from the pulp and paper factories, substantially raised the biological oxygen demand (BOD) in the river. The high concentration of fecal wastes made the polluted water a potential source of disease. Poor water quality also reduced fish and other aquatic populations.

FIGURE 15-27: The Willamette River, as seen from the Buena Vista ferry crossing between Salem and Albany. The island in the photo is a park.

Increasing Awareness of the Problem

By the late 1920s widespread recognition of the river's degradation led to the formation of civic groups such as the Anti-Pollution League to explore and correct the problem. At the request of another civic group, the League of Municipalities, the Oregon Agricultural College (now Oregon State University) conducted a preliminary study of Oregon streams in 1929. Their report catalyzed an in-depth study of water quality during the summer of that year. The resulting report confirmed that the Willamette was in trouble. Although dissolved oxygen content was high in the upper (upstream or southern) river, two significant decreases occurred just below (north of) Salem and below Newberg, where waste from households and industries increased the biological oxygen demand. By the time the Willamette reached the Columbia River, dissolved oxygen in the water was less than 0.5 parts per million, significantly less than the minimum 5.0 parts per million needed to maintain healthy populations of aquatic life. The same study found that concentrations of coliform bacteria shot up from 0 or 1 per milliliter to 100 per milliliter below small towns and 1,000 per milliliter below the city of Salem.

Despite these sobering findings, little was done to alleviate the problem until 1938, when the people of Oregon initiated the Water Purification and Prevention of Pollution Bill. This legislation set standards for pollution control and created the State Sanitary Authority to help take remedial action against pollution, but a lack of funds, unenforced regulations, and the advent of World War II combined to prevent any real improvement in the state of the river. In 1944 a new study by Oregon State University showed that the river's condition had worsened since 1929. By midcentury, dissolved oxygen in Portland Harbor was 0, and the upstream concentration at Salem was only 3.6 parts per million.

Reviving the River

Eventually, the State of Oregon took measures that finally reversed the steady decline of water quality. Although the State Sanitary Authority had required municipalities to begin plans for sewage treatment facilities by the early 1940s, they were not all completed until 1957. Meanwhile, the authority discovered that the pulp and paper industry posed an even greater threat to water quality. Sulphite waste liquors dumped into the Willamette were responsible for about 84 percent of the biological oxygen demand on the river. In 1950 the authority ordered these factories to develop treatment facilities for those wastes and to limit discharges during the low-flow months of June through October.

When further studies in 1959 showed no improvement in DO, BOD, or coliform count, the authority ordered the cities of Eugene, Salem, and Newberg to install secondary sewage treatment facilities and required Portland to accelerate completion of facilities under construction. The authority also further reduced the amount of waste that pulp and paper operations could dump into the river. By 1967 Oregon required secondary treatment for any wastes dumped into the river or other public waters and placed strict limits on the BOD that pulp and paper wastes could impose on the river. By 1972 all mills had installed secondary treatment equipment, which reduced BOD 80 to 94 percent from original levels. By 1980 all types of wastes were being reduced by 91 percent before they were discharged into the river.

Another strategy the State Sanitary Authority used to reverse degradation was to regulate the rate of flow in the river. In the early 1960s the state began regulating flow by releasing controlled amounts of water from reservoirs during the summer. In winter the reservoirs helped to contain flood waters. By augmenting summer flows and substantially reducing municipal and industrial wastes,

Oregon raised water quality to acceptable standards at least 75 percent of the time between 1976 and 1980. This was achieved through well-enforced regulations and the cooperation and concern of the industries in the Willamette basin. For example, during a 1977 drought major industries voluntarily reduced their waste output by 15 percent.

Graduating from Cure to Prevention

By the 1970s the citizens of the basin were able to turn their attention from cleaning up old problems to preventing new ones. Not only had water quality improved, but increased awareness helped solve other environmental problems as well. The installation of fish ladders at dams helped to increase the fall chinook salmon run from a count of 79 in 1965 to more than 22,000 in 1973. Logging operations had begun to use sustained yield practices, reducing erosion and sedimentation problems.

One program that helps maintain the quality of the Willamette River is the Willamette Greenway Program, a river park system established by the 1967 Greenway Statute. By ensuring the existence of vegetated areas along the river, the statute protects natural resources while allowing the public to enjoy and learn about them. The vegetation helps prevent erosion, provides habitats for wildlife, and enhances the beauty of the river.

Because the riverside was already heavily populated, the greenway is a conservation corridor that is primarily privately owned with public park areas and river access points. The program involves 510 riverbank miles (821 kilometers), stretching from the Columbia River to the Dexter Dam on the Middle Fork and the Cottage Grove Dam on the Coast Fork. The program safeguards all existing uses of the land, especially farmland.

During 1971 the Oregon Department of Transportation began acquiring land in the greenway area for five major state parks. The state parks, three of which are located near heavily populated areas, encourage short trips for activities such as canoeing and wildlife observation. In addition to the parks, each of the 19 cities and 19 counties involved has placed land available for public use into five categories, ranging from preservation (sensitive areas protected as wildlife and vegetation preserves) to recreation (areas that can be used intensively by the public). Between 1973 and 1979, 43 greenway sites were constructed or improved.

Looking Ahead

The efforts of the people of Oregon to clean up the Willamette River are a true success story. The Willamette is widely hailed as an example of what can be accomplished when the public, government, and industry work together to solve environmental problems. But continued diligence is essential. Authorities must continue to monitor water quality. Although municipal and industrial wastes are under control, pollution can come from less obvious sources, such as fecal matter from grazing livestock and agricultural runoff containing fertilizers. Another danger is toxic materials, which may accumulate in the food chain.

The biggest challenge that Oregon will have to meet is the basin's growing population. A larger population will produce more and more waste, increasing the potential for pollution. New ways to use or recycle waste, such as using it for livestock feed, fertilizer, or energy, should be explored. An expanding population also threatens the existence of vegetated areas near the river. The public must remain interested in and educated about the greenway so they will continue to support it. The public must remain informed and interested in water quality. Citizens were the driving force behind cleaning up the Willamette, and their support will be necessary for continued monitoring and protection of the river. Government agencies and civic groups can help secure that support by continuing their efforts to educate and motivate the public about water resources.

Although government policy must play an important role in water conservation, we can make our economic system work to prevent overuse as well. Water utilities should increase the cost of water to reflect the true cost of this valuable resource, thus encouraging conservation. In addition, they can make it economically attractive for consumers to use water efficiently by charging them on a sliding scale: the more they use, the more they pay per gallon.

Water conservation will require the development and adoption of new, more efficient technology in some areas. Such areas include the reuse of wastewater and the development of an alternative to the flush toilet. In addition, some existing systems, such as old and leaking water supply systems in many urban areas, could be made more efficient by repairing them.

One crucial step in encouraging water conservation is education. We must make an effort to teach children and adults alike the importance of wisely using water resources.

Protect Human Health and the Environment

As in conservation, the federal government could substantially help to mitigate the negative effects of water pollution on the environment and human health. It should eliminate measures that encourage degradation of coastal areas and flood plains (for example, federally funded subdivisions and federal insurance for building on flood plains). It should discourage building and development on tidal lands and require that any one who does develop such land post bonds that can be used to offset any unexpected environmental problems caused by the project.

Because runoff from agricultural fields adds so much silt and other pollutants, the federal government should strengthen programs designed to encourage farmers to use methods to reduce erosion and minimize agricultural runoff. To reduce water pollution in general, the government should stiffen enforcement of environmental laws and regulations that relate to pollution discharges, dumping of garbage and sludge, dredging of rivers, and filling of wetlands. Also important are the reauthorization and strengthening of the Clean Water Act, a process that began in 1992.

To protect human health, the federal government should provide funds to upgrade and construct adequate sewage treatment plants to meet federal clean water standards. In addition, we need to encourage the development of improved technologies for septic tank waste treatment and eliminate the need, when possible, for households to rely on septic systems.

Because ocean pollution has become such a large and global problem, all sectors of society must eliminate ocean dumping of sewage sludge and garbage, especially plastics. To help clean up the severe pollution that already exists in many coastal zones and bays, we should create a marine cleanup fund.

Preserve Living Systems

Perhaps the most important step our society could take in preserving living aquatic systems is to develop a true realization that they *are* alive and cannot be treated as never-ending wells or all-absorbing sewers. Again, education is vital to make people more water aware. In addition to preserving aquifers, lakes, rivers, and oceans, we need to pay attention to watersheds and wetlands. In general, we should work to protect watershed and drainage basins and restore degraded wetlands, watersheds, rivers, and lakes through federal, state, local, and private initiatives.

Summary

Most of the earth's water is salty. Only 0.5 percent of all water in the world is found in lakes, rivers, streams, and the atmosphere. Water is classified as either salt (marine) water or fresh water depending upon its salt content. Fresh water is found on land in two basic forms: surface water and groundwater.

Lakes are categorized according to the amount of dissolved nutrients they contain. Oligotrophic lakes are cold, blue, and deep, often have rocky bottoms, and have a high oxygen content. They contain low amounts of dissolved solids, nutrients, and phytoplankton. Eutrophic lakes are warmer and more turbid, often have muddy or sandy bottoms, and have a lower oxygen content. They are far more productive.

Lakes are made up of internal zones based on depth and light penetration. These include the littoral zone, the area where rooted plants grow; the limnetic zone, the zone of deeper water where light can penetrate and support populations of plankton; and the profundal zone, where light does not penetrate enough to allow photosynthesis to occur. Large lakes in temperate climates undergo thermal stratification. In the summer surface waters heated by the sun become lighter and less dense, rise, and form the epilimnion. Colder, denser water sinks to the bottom to form the hypolimnion. A sharp temperature gradient, or thermocline, exists between the upper and lower layers. In the autumn the thermocline disappears and the epilimnion and hypolimnion mix, a process known as fall turnover.

Running water habitats include streams and rivers. The speed and temperature of the current affects the kinds of organisms that live in a habitat. Fast-moving rivers and streams usually contain two kinds of habitats: riffles, with high oxygen content, tend to house the producers of biomass, and pools to contain consumers and decomposers.

The United States contains about 20 times more groundwater than surface water. Groundwater percolates downward through the soil after precipitation or from surface water and is stored in an aquifer, a water-bearing geologic formation composed of layers of sedimentary material. The depth at which the aquifer begins is known as the water table. Groundwater is the major source of drinking water in about two-thirds of the states. In addition, it helps to maintain water levels and the productivity of streams, lakes, rivers, wetlands, bays, and estuaries.

The world's oceans support over half of the world's biomass. The ocean is divided into zones. The littoral zone is the near shore zone of shallow water. The euphotic zone is the area of light penetration. The neritic zone is the part of the euphotic zone over the continental shelf and near-shore islands; it supports most of the ocean fisheries. The pelagic zone is the deep-water zone of the open oceans. The abyssal zone is the deepest part of the oceans. Four biologically productive ecosystems are responsible for most of the oceans' primary production: estuaries, coastal wetlands, coral reefs, and offshore continental shelves.

Water uses may be consumptive or nonconsumptive. Nonconsumptive uses remove water, use it, and return it to its original source. Consumptive uses remove water from one place in the hydrological cycle and return it to another. Worldwide, agriculture is the greatest use of water.

Water pollution can be divided into eight general categories: organic or oxygen-demanding wastes, disease-causing wastes, plant nutrients, toxic and hazardous substances, persistent substances, sediments, radioactive substances, and heat. Biological oxygen demand (BOD) is a measure of the amount of oxygen needed to decompose the organic matter in water.

Legislation, or the lack of it, determines how water is managed in the United States. The Safe Drinking Water Act of 1974 set national drinking water standards, called maximum contaminant levels (MCLs), for pollutants that might adversely affect human health. It also established standards to protect groundwater from hazardous wastes injected into the soil. The 1986 reauthorization of the act also instructed the EPA to monitor drinking water for unregulated contaminants and to inform public water suppliers concerning which substances to look for. The 1972 Federal Water Pollution Control Act, commonly known as the Clean Water Act, divided pollutants into three classes: toxic, conventional, and unconventional. The act stipulated that industries must use the best available technology (BAT) to treat toxic wastes before releasing them into natural waters and the best conventional technology (BCT) to treat conventional pollutants, such as municipal wastes. Unconventional pollutants must meet BAT standards.

After water is withdrawn from a lake, river, or aquifer, it is treated before being distributed. In areas that have suitable soils, sewage and wastewater from each home is usually discharged into a septic system consisting of an underground tank made of concrete and a drain field. Wastewater in urban areas must be collected in underground sewers and treated in sewage treatment plants. These plants are designed to make wastewater safe for discharge into streams or rivers or to make it acceptable for reuse. Most sewage plants employ a multistage process to reduce wastewater to an acceptable effluent.

Fresh and marine waters worldwide are besieged by numerous threats. Many groundwater aquifers experience overdraft, the withdrawal of water faster than the aquifer can be recharged. Overdrafts can lead to land subsidence. When groundwater becomes contaminated, it is a costly, slow, and sometimes impossible task to remove pollutants. Toxic pollution and accelerated eutrophication are serious threats to the nation's lakes. Threats to rivers include dams, diversions, channelization, and pollution. Abuse of the ocean ecosystem can come from pollution and overuse. All of our coastal ecosystems are under severe stress. Until recently, little thought was given to effectively managing and preserving marine environments.

Key Terms

abyssal zone

aquifer

biological oxygen demand
(BOD)

combined sewer system

desalination

epilimnion

euphotic zone

eutrophic

groundwater

hypolimnion

land subsidence

limnetic zone

littoral zone

neritic zone

oligotrophic

overdraft

pelagic zone

profundal zone

running water habitat

runoff

separate sewer system

septic tank

standing water habitat

thermal stratification

thermocline

water mining

water table

watershed

Discussion Questions

1. Why is water so important?

2. More than 71 percent of the earth's surface is covered with water, yet many ecosystems and people suffer from lack of water. Explain how this is possible.

3. Describe the important features of the Palmiter river restoration system. Why is this system attractive to both conservationists and public officials?

4. Briefly describe the basic stages of municipal water treatment, telling what substances each stage removes and how.

5. Name five things individuals can do to help protect water resources.

6. What factors contributed to the degradation of the Willamette River basin? What factors contributed to the success of the cleanup effort?

Soil Resources

The Lithosphere

Poundage or tonnage is no measure of the food-value of farm crops; the products of fertile soil may be qualitatively as well as quantitatively superior. We can bolster poundage from depleted soils by pouring on imported fertility, but we are not necessarily bolstering food-value. . . . The marvelous advances in technique made during recent decades are improvements in the pump, rather than the well. Acre for acre, they have barely sufficed to offset the sinking of the well.

Aldo Leopold

Moving together in a perfect rhythm, without a word, hour after hour, he fell into a union with her which took the pain from his labor. He had no articulate thought of anything; there was only this perfect sympathy of movement, of turning this earth of theirs over and over to the sun, this earth which formed their home and fed their bodies and made their gods . . . . They worked on, moving together—together—producing the fruit of this earth—speechless in their movement together.

Pearl S. Buck

Learning Objectives

When you finish reading this chapter, you should be able to:

1. Describe the major components of soil and explain how soil is formed.

2. List the major uses for land and describe how land use affects agriculture.

3. Compare conventional and sustainable agricultural practices.

4. Discuss methods to prevent soil erosion and maintain soil fertility.

Like air and water, soil is essential for life, and when used wisely, soil is a renewable resource. Unfortunately, most of us take this precious resource for granted. We are aware of it only when we track it into our homes, wash it from our clothes, or clean it off our automobiles. In part, our ignorance about soil resources stems from the fact that few Americans (about 4 percent) live on farms, where the connection between the life-sustaining soil, plants, and animals, including humans, is most obvious. Nevertheless, most of the food we eat, the clothes we wear, the medicines we use, and the materials in our homes and automobiles originate, directly or indirectly, in the soil (Figure 16-1).

In this chapter we begin by discussing what soil is, how living organisms help to maintain soil fertility, and the physical characteristics and types of soil. We look at agricultural land use in the United States and discuss some of the problems associated with conventional agriculture. We also explore forms of alternative agriculture and recommend future action to protect soil resources.

Describing Soil Resources: Biological Boundaries

What Is Soil?

Soil is the topmost layer of the earth's surface. It is the layer in which plants grow. It is an ecosystem composed of both abiotic and biotic components—inorganic chemicals, air, water, decaying organic material, and living organisms. As such, it is subject to the dynamics that operate in all ecosystems.

The abiotic composition of soil varies from one location to another, but on average it is 45 percent minerals (particles of stone, gravel, sand, silt, and clay), 25 percent water (the amount varying with rainfall and the soil's capacity for holding water), 25

FIGURE 16-1: Pine barrens, New Jersey. Several promising antibiotics have recently been discovered in the pine barren soils.

percent air, and 5 percent **humus,** or partially decomposed organic matter (dead producers, consumers, and decomposers). Humus is an essential component of fertile soil. The humus content of soil accumulates over many hundreds, or even thousands, of years and remains at a fairly constant level unless it is carried away by erosion. Humus helps to retain water and to maintain a high nutrient content, thus enabling soil to remain fertile.

Soil fertility refers to mineral and organic content, while **soil productivity** refers to its ability to sustain life, especially vegetation. A primary factor in productivity is availability of water. Desert soils, for example, have high fertility but low productivity because of a lack of water. Rain forest soils have low fertility, since the nutrients are bound up in the vegetation, but high productivity because of ample rainfall.

Maintaining soil productivity depends on maintaining soil as a living, renewable resource. Like air and water, soil can be degraded and depleted to such an extent that it becomes, for all practical purposes, nonrenewable. We degrade our soil when we cover it with concrete and asphalt, poison it with pesticides, herbicides, and toxic wastes, allow it to wash or blow away, and allow it to become salty or waterlogged. We can see the effects of all these forms of abuse in our everyday lives. For example, excessive use of pesticides meant to bolster the productivity of worn-out soil may threaten our health when we eat the vegetables and grains they were used on.

How Do Organisms Maintain Soil Fertility?

Soil teems with life (Figure 16-2). Every teaspoonful of the upper layer of fertile soil contains billions of

beneficial organisms. Most of these are soil microbes (bacteria, fungi, nematodes, and viruses). Nitrogen-fixing bacteria are microorganisms that convert the nitrogen in the air spaces of the soil into nitrates, a form that can be used by plants. Soil microorganisms maintain the fertility and structure of healthy soil as they decompose organic material (dead plants and animals), recycle the constituent nutrients, and produce humus. Some microbes provide us with antibiotics, such as penicillin and streptomycin.

Larger soil organisms also play an important role in soil fertility. Earthworms turn over an estimated 11 tons of soil per acre per year, mixing fertile topsoil with deeper, less fertile soils and helping to aerate the soil. The digging and burrowing of small mammals such as moles, insects and their larvae, and other arthropods also aerate and drain the soil. Larger animals play a role in the nitrogen cycle, since their feces contain many forms of nitrogen that are readily converted into substances plants can use. Additionally, the matter bound in animal tissues is a reservoir of nutrients that is released gradually as the organisms decompose.

Physical Boundaries

What Is Soil Texture?

Since soil is composed mostly of minerals, it is the mineral content that determines the texture, or feel, of the soil. The mineral content consists largely of

FIGURE 16-2: Soil decomposers at work in a compost pile.

sand, silt, or clay particles. Sand particles range from 0.002 inches to 0.08 inches (0.05 to 2.0 millimeters), about the thickness of a paper clip, and feel gritty to the touch. Silt particles range from 0.00008 to 0.002 inches (0.002 to 0.05 millimeters); rubbed between the fingers, silt particles feel like flour. Clay particles are even smaller in diameter; they feel like corn starch when dry, but are sticky when wet. Soil texture is determined by the proportions of sand, silt, and clay present (Figure 16-3).

Texture determines the oxygen-holding and water-holding capacity of a soil. There are three broad categories of soil based on texture: loams, clays, and sands. Loams, comprised of about 40 percent silt, 40 percent sand, and 20 percent clay, have the best texture for growing most crops. These soils provide adequate air spaces, and they allow good drainage while retaining enough moisture for plant growth. Clayey soils may retain too much water and become waterlogged. They may also become deficient in oxygen, either because the spaces between particles become filled with water or because the soil becomes compacted. Sandy soils are typically too porous to retain sufficient moisture for plant growth.

What Is Soil Structure?

The texture of a soil helps determine the soil's structure. Soil structure, or **tilth,** is the arrangement of soil particles, or how they cling together to form larger aggregates such as crumbs, chunks, and lumps. The structure of a particular soil is strongly influenced by the amount of clay and the amount of organic material it contains. Because of their physical and chemical properties, clay particles and organic matter are able to form links with other particles, thereby forming larger soil aggregates. For instance, the presence of clay particles, being small and numerous, creates a large amount of surface area in a given amount of soil; water molecules and nutrients adhere to the surfaces. Soils that lack clay or organic matter have an unstable structure and are likely to form dusts or loose sands which can easily blow or drift away.

How Is Soil Formed?

All soil types result from the physical, chemical, and biological interactions in specific locations. Soils of the tundra and rain forest differ vastly from each other and from soils of the prairie and deciduous forest. Just as vegetation varies among biomes, so do the soil types that support that vegetation. Scientists typically recognize ten major soil types, or orders, but there are many variations of these (Table 16-1). An estimated 100,000 different soil types have been formed through a combination of five interacting factors: parent material, climate, topography, living organisms, and time.

Parent material, the raw mineral material from which soil is eventually formed, has the greatest effect on the texture and structure of soil. Soil begins to form when parent materials (rocks or other mineral materials) are broken down into smaller particles by weathering, abrasion, and dissolution. Weathering occurs when water from rain or snow seeps into the cracks of rocks, freezes and expands, and cracks the rocks into smaller pieces. The abrasive action of moving glaciers, ocean waves, rivers and streams, and wind also breaks down parent material. The minerals in parent material can be dissolved by the action of acids deposited on land from the atmosphere. For example, water and carbon dioxide form carbonic acid, which reacts with calcium to form carbonates, which are dissolved in water passing over the rocks and are carried away.

Climate is an important factor in the formation of soil. The macroclimate of a region, or the average

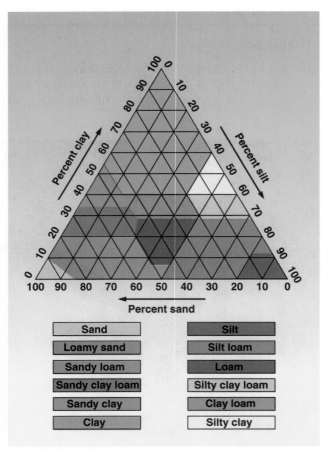

FIGURE 16-3: Soil texture classes. The U.S. Department of Agriculture classifies soils according to the percentages of sand, silt, and clay they contain. Loams are the best soils for cultivation.

▶ Table 16-1
Ten Major Soil Orders

Soil Order	Description	Example Location
Alfisol	Moderately acidic soil formed under forests and mixed trees and grasses in humid, mild climates. Moderate fertility and good moisture supply make it favorable for agriculture.	Ohio
Aridisol	Sandy soil formed under desert shrubs in arid or semiarid regions. Low humus content (due to sparse vegetation), high mineral salts, and lack of water usually make it unsuitable for agriculture. If mineral accumulation is low, irrigation may allow cultivation.	Sonoran desert in southwestern United States
Entisol	Newly formed soil (for example, those that arise from deposits left by rivers) that contains mostly minerals and sand and has no noticeable horizons. Productivity varies greatly.	Valley of Tigris and Euphrates rivers in Iraq
Histosol	Highly acidic, poorly drained soil with a high content of organic matter formed in climates ranging from tropic to polar. This bog soil can be productive if drained (climate permitting).	Western Ireland
Inceptisol	Young soil just starting to layer, usually on geologically young materials or recently eroded surfaces, formed in climates ranging from humid to sub-humid, tropical to subpolar. Productivity varies from excellent to poor.	
Mollisol	Soft, dark soil usually formed under temperate grasslands in semiarid climates. High humus content and workable texture make it ideal for agriculture.	Corn and wheat belts in western United States; pampas of Argentina
Oxisol	Highly acidic soil with high concentrations of iron and aluminum and a limited A horizon formed in tropical and subtropical climates. Although easily worked because of absence of expanding clays, low humus content and poor cation-exchange capacity limit fertility.	
Spodosol	Acidic soil formed from sandy parent materials, usually under coniferous forests, in humid climates ranging from tropical to arctic.	Great Lakes region
Ultisol	Formed under forests in humid tropical and subtropical climates. Heavy leaching of nutrients results in low fertility.	South of the limits of glaciation in southeastern United States
Vertisol	Mature soil with a high clay content. It is hard to farm even though fertility may be high because cracking clays shrink and swell with changing moisture content.	Northern Ethiopia

weather pattern, including temperature and precipitation, affects the soil because it regulates both the amount of water entering the soil and the amount that can be evaporated from its surface. And within a region, soils may vary because of variations in microclimate. The microclimate, or the weather conditions just above the surface of the ground, influences soil temperature, rates of chemical and microbial activity, and plant growth.

The topography of the land influences the natural drainage and amount of erosion that occur in a particular region. Steep slopes, which are usually well drained, often have little topsoil because much is carried away by wind and water. Valleys, usually poorly drained, often contain thick layers of soil that have been washed or blown there from other areas.

Living organisms participate in soil formation in several ways. The pioneer lichens and mosses secrete mild acidic solutions that dissolve rock. The roots of herbs and shrubs penetrate cracks in rock, eventually expanding the cracks and breaking the rock into smaller particles. As organisms die, they add organic matter to parent material, improving moisture retention and nutrient availability. Eventually, complex food webs form that maintain soil formation and fertility.

The final factor affecting soil formation is time. Because parent material changes into soil only very slowly, it may take hundreds or even thousands of years for a mature soil to develop; 10,000 years may be needed to form a layer of soil 1 foot thick. There are places where soil has accumulated to a thickness of several feet and places where it is only a thin band less than an inch thick.

What Is a Soil Profile?

As soils develop, they form distinct horizontal layers called **soil horizons,** each of which has a characteristic color, texture, structure, acidity, and composition. A vertical series of soil horizons makes up a **soil profile.** Although soil profiles differ greatly from region to region, in general soils have three to five major horizons. Their thickness depends on the interaction of the physical, chemical, and biological components of the soil as well as the climate, major vegetation type, and length of time the profiles have had to develop (Figure 16-4).

The top layer, or O horizon, consists mostly of surface litter and partially decaying organic matter. The O horizon protects the lower horizons from the compacting effect of rain and the drying effect of

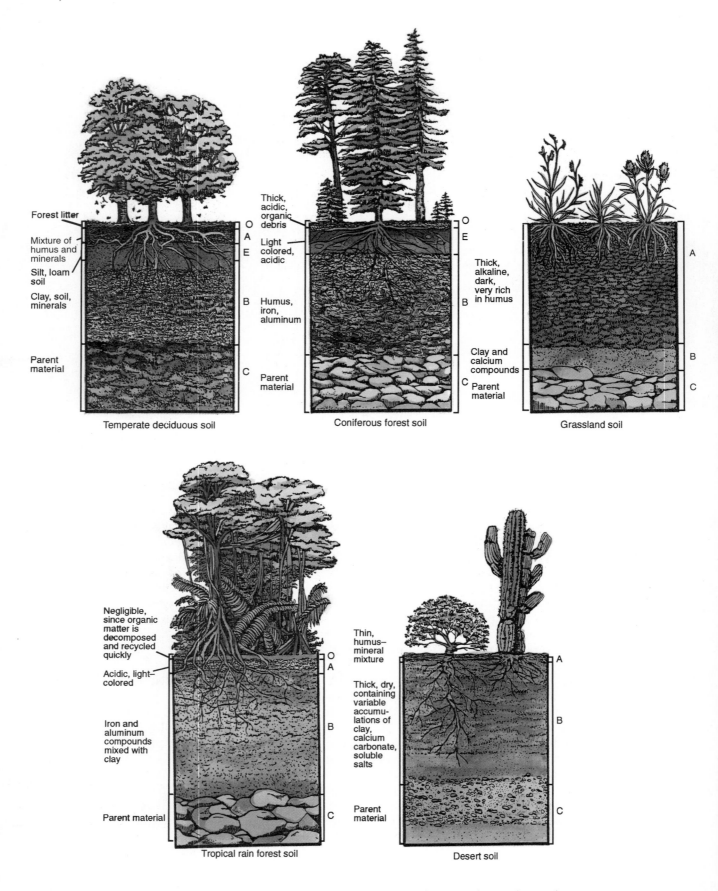

FIGURE 16-4: Generalized soil profiles found in five major ecosystems. Depending on soil type, the number, composition, and thickness of the soil horizons vary.

wind. When there is no surface litter, the exposed soil is extremely susceptible to erosion.

Directly below the surface litter is the *A* horizon. Commonly called the **topsoil,** the *A* horizon contains the essential organic component humus, living organisms, and some minerals. It is the most fertile part of the soil. Plants with shallow root systems obtain the nutrients they need directly from the topsoil, which usually also retains adequate water and oxygen for their needs.

The *E* horizon, or zone of leaching, lies between the fertile topsoil and the less fertile layers below. Water and soluble minerals pass through this layer. Not all soil types possess an *E* horizon.

The layer that lies below either the *A* or *E* horizon is known as the *B* horizon. Soluble minerals, such as aluminum and iron, may accumulate here. The *B* horizon is not as fertile as the topsoil, but deep-rooted plants can withdraw minerals and oxygen that have leached into it. By transporting nutrients out of the *B* horizon, these plants keep the minerals in circulation in the ecosystem, since the minerals are returned to the topsoil once again after the plant has died and decomposed (assuming that the plant with its constituent nutrients is not harvested and removed from the ecosystem altogether).

The lowest soil layer is the C horizon; it consists mostly of parent material. Too far underground to contain any organic material, the C horizon lies above the impenetrable layer of bedrock, sometimes called the *R* horizon.

Social Boundaries

How Do We Use Land?

We use land for many different purposes—to grow plants and livestock for food and fiber, to build houses and communities, to pursue recreational activities, to dispose of wastes, and to build roads, highways, and airports (Figure 16-5).

Very simply, the way in which a particular piece of land is used is known as its **land use.** Depending on the soil type and the terrain, there may be many potential uses for a specific parcel of land. The Soil Conservation Service has developed a system to classify land for different uses based on the limitations of the soil, the risk of damage, and the soil's predicted response to management techniques. As Table 16-2 shows, classes I to IV are capable of sustaining crops or pastureland, grassy areas suitable for grazing. Classes V to VIII are better suited to forestry, conservation, and rangeland. Rangeland differs from pastureland in that its vegetation tends to be sparser and the land tends to be drier. In the United States approximately 54 percent of the land is used for crops and livestock, 43 percent is used for

FIGURE 16-5: A cloverleaf interchange near Arlington Heights, Illinois. A Transportation corridor is one type of land use. Because we all use land in many ways, both directly and indirectly, we are all responsible for the ways those uses affect soil resources and agriculture.

forests and natural areas, and 3 percent is used for urban centers and transportation corridors.

How Do Land Uses Affect the Soil?

The purposes for which we use land have an immediate and often long-lasting effect on soil resources. Sometimes the effect is obvious: land paved for a parking lot, for instance, cannot be farmed. The spread of urban areas, known as urban sprawl, consumes over a million acres a year in the United States, much of it prime farmland (Figure 16-6). By the mid-1980s, 493 of the 619 counties which produced the majority of the United States' crops were either part of or adjacent to urban areas. According

FIGURE 16-6: Farmland in southwestern Ohio being converted to a subdivision. Conversion of farmland to nonfarm uses poses a serious threat to agricultural production and places increasing pressure on marginal croplands.

Table 16-2
Land Use Classification System

Class	Recommended Use	Actual Use
CLASSES SUITABLE FOR CULTIVATION		
I	Few limitations	Predominantly agriculture; also recreation, wildlife, and pasture
II	Moderate limitations due to soil, slope, or drainage that reduce choice of plants and require precautions such as contour plowing and drainage	Predominantly agriculture and pasture; also recreation and wildlife
III	Severe limitations due to soil, slope, or drainage that reduce choice of plants and require more extreme precautions such as contour terracing	Predominantly agriculture, pasture, and watershed; also recreation, wildlife, and urban/industrial use
IV	Very severe limitations due to soil, slope, or drainage that reduce choice of plants and make occasional or limited cultivation preferable	Predominantly pasture, tree crops, agriculture, and urban/industrial use; also recreation, wildlife, and watershed
CLASSES NOT SUITABLE FOR CULTIVATION		
V	Not likely to erode, but other limitations that are impractical to remove	Predominantly forestry, range, and watershed; also recreation and wildlife
VI	Severe limitations that make land unsuitable for cultivation and require precautions for forestry or grazing	Predominantly forestry, range, watershed, and urban/industrial uses; also recreation and wildlife
VII	Very severe limitations that make it unsuitable for cultivation and suitable for forestry or grazing only on a limited basis with precautions	Watershed, recreation, wildlife, forestry, range, and urban/industrial uses
VIII	Limitations that nearly preclude its use for commercial crop production, forestry, or grazing	Recreation, wildlife, watershed, and urban/industrial uses

Source: Adapted from U.S. Department of Agriculture.

to the Census Bureau, urban areas almost doubled in the twenty years from 1960 to 1980, increasing from 25 million to 47 million acres (10 million to 19 million hectares).

Worldwide, expanding urbanization poses a long-term threat to soil resources. As we learned in Chapter 8, the majority of future world population growth is expected to occur in and around cities. Consequently, more and more land will be needed to accommodate housing, schools, hospitals, roads, commercial development, and dumps and landfills. Without careful management it is likely that urban growth will continue to result in the loss of productive soils—soils that will be increasingly needed to keep pace with the world's rapidly growing population and the ever-rising demand for food.

While rampant urbanization is a poor and unwise use of productive soils, a greater threat to soil fertility and conservation is erosion. Erosion by wind and water is a natural process, and when undisturbed, soil is usually replaced faster than it erodes. The amount of soil that can be lost through erosion without a subsequent decline in fertility is known as the **soil loss tolerance level (T-value)** or replacement level. Depending on the type of soil, land may have a T-value between 2 and 5 tons per acre per year,

with natural processes compensating for that loss with the production of new topsoil. But natural processes cannot offset soil loss as it accelerates above the T-value because of human activity, nor can natural processes make up for the declining productivity that results.

Of all the soil that is carried off by erosion, some is removed by runoff from construction sites and urban areas. Deforested areas also often suffer heavy losses to erosion. (For a discussion of issues of soil conservation in forested areas, see Chapter 21.) However, as much as 85 percent of eroded soil originates from croplands. Agricultural soil erosion is a serious problem worldwide. Many scientists believe that as much as one-third of the croplands of the earth are eroding faster than natural processes can replace them. The two major causes of agricultural soil erosion are the cultivation of marginal or poor cropland and the use of poor farming techniques on good cropland. Farmers worldwide have attempted to maintain or increase production by cultivating steep slopes, rain forests, wetlands, and other areas that are not suited to agriculture. Inappropriate farming techniques, such as allowing fields to lie bare and exposed to the wind, rain, and snow, also take their toll on soil resources.

Because agricultural erosion is such a serious problem, we examine it in more detail in the following section. As we review the way land has been used in the United States, we focus on the effect those uses, especially agriculture, have had on soil resources.

History of Management of Soil Resources

What Has Been the History of Agricultural Land Use in the United States?

Even though most people in the United States are now urban dwellers, the image of the family farm and the rural community and the way of life those images represent are a cherished part of the American culture. The qualities we associate with that way of life—hard work, independence, and resilience—are qualities we like to believe define our national character. The farm crisis in America, which had been building for years but came to its apex in the 1980s, saddens and frustrates people precisely because it shakes their beliefs in such a national identity. If the family farm and the small independent farmer are doomed to failure, then perhaps everything they represent may be lost as well.

Many historians, philosophers, and social scientists have attempted to trace the roots of our nation's farm crisis by studying how agriculture has changed during the last century. Although agricultural practices differ widely from one area of the country to another, we can detect some overall trends. The first European settlers to the North American continent used standard European farming practices such as rotating crops, fertilizing fields with livestock manure, and liming to neutralize acidic soils. For a time the soil remained fertile despite cultivation. But as land became increasingly scarce, and the soils in some areas became less productive from year to year, settlers began to move westward in search of more land. By the late 1700s agriculture moved through New York to Ohio, and by the mid-1800s the wave of settlers had reached the Midwest.

To those settlers the vast, fertile North American continent must have seemed an endless resource. Most farmers had access to plentiful land, but little manpower or horsepower. It was the nation's small population, in fact, that helped spur the development of labor-saving devices. The first horse-drawn reaper came into use in the 1830s; horses gave way to tractors between the two world wars. The 1950s saw the rise of the combine, which could harvest and thresh crops at once. The average farmer, who could produce food for 5 people in 1870, could now feed 40 people.

When settlers reached the Great Plains, they encountered land far different from eastern lands. The rich grassland soils were resistant to soil erosion and slow to lose their fertility. They continued to produce large yields year after year. But even these soils were not inexhaustible. As nutrients absorbed by the crops were not replenished, soil fertility gradually declined. A cycle of dry spells in 1890 and 1910 further decreased crop productivity and exposed more bare soil.

In an attempt to capitalize on the high grain prices after World War I, farmers planted millions of acres with wheat in the short-grass environment of a 150,000-square-mile area (388,815 square kilometers) in Kansas, Colorado, New Mexico, Oklahoma, and Texas (Figure 16-7). To avoid losing out on a season's profits, farmers did not allow the land to lie fallow periodically. This practice, coupled with the fact that wheat is a medium grass not suited for the short-grass prairie, began to take its toll on the land. Then, in 1931 a drought surpassing all previous droughts in its severity returned to the Great Plains and turned that 150,000 square miles into the Dust Bowl. The bare, dry topsoil was easily swept up by the wind and carried thousands of miles, darkening the skies of the eastern United States and rendering life impossible for livestock and people in many parts of the Great Plains (Figure 16-8). The drought severely damaged over 10 million acres (4 million hectares) of farmland. In some places as much as 12 inches (30.5 centimeters) of topsoil blew away, exposing the infertile subsoil. Thousands were forced to leave their homes in Kansas, Oklahoma, and Texas. In a land suffering from an extensive economic depression, this exodus of homeless farmers increased the hardships of all.

It is important to keep in mind that the drought alone did not cause the Dust Bowl. Intense cultiva-

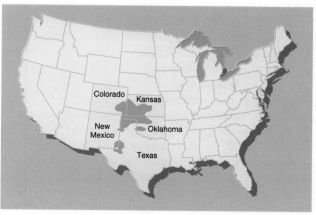

FIGURE 16-7: The Dust Bowl of the 1930s.

FIGURE 16-8: Dust storm, Clayton, New Mexico, 1937.

tion (especially continual cropping), allowing fields to lie bare of vegetation during the winter, and an inappropriate choice of crops set the stage for this disaster.

The devastation of the Dust Bowl brought the issue of soil conservation into the open and forced the government to take action. On August 25, 1933, Congress formed the Soil Erosion Service, later renamed the Soil Conservation Service (SCS). One of its first projects was to conduct a nationwide soil survey in order to produce maps and data for erosion control. Armed with this information, it recommended the best conservation practices for particular areas and provided technical assistance when needed. In the following decades the government implemented programs of economic incentives to keep erodible land out of production and to protect land under cultivation.

In the 1950s and 1960s agricultural research efforts were increased in order to raise food production, largely in response to rapid population growth worldwide. Agriculture became energy-intensive rather than labor-intensive, requiring high inputs of chemical fertilizers and pesticides (which require the use of fossil fuels in their manufacture), large machinery (which also requires substantial amounts of fossil fuels), and hybrid strains of crops, especially grains like wheat and corn. These changes raised production dramatically. From 1790 to 1930, for instance, American farmers produced an average 22 to 26 bushels of corn per acre per year; that figure jumped to 80 bushels per acre by 1968. Between 1949 and 1969 American agriculture increased output by 50 percent. But the technological advances that made this "agricultural miracle" possible rendered croplands less and less like natural ecosystems. Unlike natural systems, which are self-supporting

(given energy from the sun), modern croplands can be maintained only through external inputs, especially chemical fertilizers and pesticides. And as the experience of the past 10 to 15 years has shown, conventional or **high-input agriculture,** besides being increasingly expensive, can cause serious environmental problems.

What Problems Are Associated with Conventional Agriculture?

The problems cited most often in connection with conventional agriculture are soil degradation, soil erosion, reliance on fossil fuels, reliance on synthetic agricultural chemicals, groundwater depletion, overirrigation, loss of genetic diversity, and socioeconomic concerns.

Soil Degradation. Soil degradation is one of the most serious problems associated with high-input farming. Degradation, a deterioration of the quality and capacity of the soil's life-supporting processes, usually stems from the hallmarks of conventional intensive farming: erosion, soil compaction, loss of nutrients and biotic activity, and poisoning by chemicals, salts, acids, or bases.

Although any plant community withdraws nutrients from the soil, natural ecosystems replenish the soil. The organic matter from dead plants and animals stays near the top of the soil, where microorganisms can break it down to release nutrients for growing plants to absorb. Modern agriculture generally removes most of the plant material from the soil; because the nutrients cannot return to the topsoil, the farmer compensates for the loss with fertilizers. Natural systems also include nitrogen-fixing bacteria associated with the roots of certain plants; domestic varieties of such plants include the legumes alfalfa, soybeans, and peas. As modern agriculture has become more specialized, fewer and fewer farmers rotate nitrogen-depleting crops with nitrogen-fixing ones. Rather, they make up the deficit by applying more fertilizer. Because most modern crop farms have no livestock, manure is not readily available and chemical fertilizers are the rule. The heavy machinery used to plow and harvest also compacts the soil, lessening its ability to retain water, which can then simply run off the hard-packed surface.

Soil degradation is part of a vicious circle, because as land loses its fertility, farmers must find new land to cultivate. Much of the new land is less than optimally suited for farming, so cultivation leads to rapid degradation. In many dry regions marginal lands have been cultivated by practicing extensive irrigation. Over time irrigated land can suffer from **salinization,** the buildup of mineral salts, which eventually ruins the soil's ability to produce. Farmers also expand arable land by clearing forests, and deforestation accelerates erosion.

Soil Erosion. Many of the practices that degrade soil also lead to erosion, which literally removes the fertile topsoil from the land. Soil experts estimate that the arable land in the United States lost one-third of its topsoil layer between 1776 and 1976. By 1980, three decades of high-input, conventional agriculture had already taken a serious toll on the nation's farmland. For example, in 1974 the SCS found that in Iowa, the heart of the Midwest cornbelt, some areas were losing 40 to 50 tons of soil per acre, with erosion reaching 100 to 200 tons per acre on unprotected slopes. Some topsoils, once 12 to 16 inches (30 to 41 centimeters) thick are now just 6 to 8 inches (15 to 20 centimeters) thick. In western Tennessee some farms have lost as much as 3 feet (1 meter) of farmable soil; only 6 inches (15 centimeters) remains. If farms lose even 1 inch (2.5 centimeters) of soil per decade—an amount scarcely detectable on a daily basis—those with 6 inches left will deplete their productive soil within one human lifetime.

A 6-inch loss of soil can reduce crop productivity by as much as 40 percent. According to the Worldwatch Institute, for every inch of topsoil lost, yields of wheat and corn drop by about 6 percent. Researchers at Resources for the Future, a nonprofit organization based in Washington, D.C., reported that areas with the highest rates of erosion between 1950 and 1980 were the areas with the lowest rate of crop yield increase.

Conservation programs instituted in the 1980s have significantly reduced cropland erosion. Even so, of the approximately 400 million acres (162 million hectares) of productive farmland in the United States, about 90 million acres (36 million hectares) are eroding at one to two times the replacement level and 100 million acres (40.5 million hectares) are eroding at two times the replacement level. Some croplands on highly erodible land are being lost at four times the replacement level. The *1991–1992 Green Index*, a state-by-state guide to environmental conditions prepared at the Institute for Southern Studies, reports that up to one-third of the nation's cropland is eroding at an annual rate of 4.4 tons per acre and a cost of $2 billion. States with the highest erosion rates, about 10 to 15 tons per acre annually, include many in the Southwest and Rocky Mountain regions—Nevada, Texas, Colorado, Arizona, New Mexico, Montana, Washington, and California—as well as those in the country's farmbelt and tobacco-producing regions—Iowa, Missouri, Mississippi, Minnesota, Oklahoma, Nebraska, Tennessee, and Kentucky.

Although experts say that soil is disappearing 25 percent faster than in the days of the Dust Bowl, the loss is not attracting as much attention because more is being removed by water erosion and less by wind. Water causes two-thirds of the erosion on U.S. farmland; wind accounts for one-third. Water erosion occurs on cropland almost exclusively because of a lack of plant cover and root systems; after the crop is harvested, many farmers plow the fields, leaving bare soil exposed. Root systems not only hold soil in place, but also pull down water so it washes away less soil. A heavy rain can wash away precious topsoil and even newly planted seeds or seedlings which haven't put down firm roots (Figure 16-9).

Soil erosion occurs in several ways. Sheet erosion occurs when rain falls faster than it can be absorbed by soil, and sheets of water sweep away soil particles dislodged by the raindrops. Heavy sheets, or runoff, can cut grooves, or rills, into the soil. As the runoff carries away more and more soil, the rills may form large gullies.

Wind erosion is always a problem in arid areas like the American western and central plains. Severe wind erosion can lead to desertification. In the United States about 5 billion acres (2 billion hectares) have been identified as at high risk of desertification. Not only does desertification ruin land, but the resulting dust damages buildings and chokes water sources as well.

The practices that lead to soil erosion can also adversely affect wildlife. When soil fertility declines, more acreage is often brought under the plow to maintain output, resulting in diminished habitat for wildlife. Streams filled with silt from water erosion

FIGURE 16-9: Water erosion on a farm field in northern Ohio. When farmers plow their fields in the fall and leave the fields bare, the soil is unprotected from erosion by wind, rain, and melting snow throughout the winter.

can reduce fish populations, and fall plowing eliminates ground cover and possible grazing material. Research in Manitoba's pothole country, a region dotted with hundreds of small ponds and water holes produced by the receding glaciers, showed that duck production was four times greater on fields that had not been tilled than on fields that had been tilled.

Reliance on Fossil Fuels. Even if conventional agriculture did not cause serious problems with soil degradation and erosion, its sustainability would still be in doubt because it is dependent on fossil fuels. The huge labor-saving machines used on conventional farms are powered by petroleum, natural gas, or electricity (coal). In addition, many chemical pesticides, herbicides, and fertilizers are petroleum-based. These expensive petroleum substances are a major drain on the accounts of small farmers, and since the fossil fuels are essentially nonrenewable, they are likely to become more expensive as supplies dwindle, placing farmers under even greater financial constraints. Fossil fuels carry an environmental as well as an economic cost. The extraction, processing, delivery, and burning of these fuels all pose threats to the environment.

Reliance on Synthetic Agricultural Chemicals. The intensive use of agricultural chemicals, or agrichemicals—fertilizers and pesticides (including herbicides, insecticides, and fungicides)—causes serious environmental problems. Research can scarcely keep pace with the development of resistant strains of pests. In the 1950s, as the use of pesticides became widespread, organisms began to show resistance to the chemicals (Table 16-3). Pesticides can also kill off beneficial organisms that prey on pests, increasing populations of old pests or making hazards out of organisms that had not threatened crops before.

Pesticides also pose a threat to animal populations. Through inefficient application methods a large amount of pesticides enters the environment. Pesticides enter waterways through agricultural runoff and erosion. In the United States 45,000 people are treated for pesticide poisoning each year and approximately 50 die. People exposed to pesticides, especially farmers and their families and migrant workers, often complain of nausea, vomiting, and headaches. Certain pesticides are suspected of causing birth defects, damage to the nervous system, and cancer. Traces of pesticides are often found in dead and deformed animals.

Fertilizers also run off cropland with water erosion. While less hazardous than pesticides, fertilizers can encourage the growth of algae and other aquatic plants, hastening eutrophication of lakes or ponds.

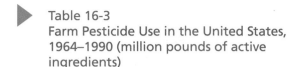

Table 16-3
Farm Pesticide Use in the United States, 1964–1990 (million pounds of active ingredients)

Year	Herbicides	Insecticides	Other	Total
1964	76	143	72	291
1966	112	138	79	329
1971	207	127	130	464
1976	374	130	146	650
1982	451	71	30	552
1986	410	59	6	475
1987	365	57	7	429
1988	372	60	8	440
1989	394	61	8	463
1990	393	64	8	465

Note: For the years 1964, 1966, 1971, and 1976 estimates of pesticide use are for total use on all crops in the United States. The 1982 estimates are for major field and forage crops only and represent 33 major producing states, excluding California. The 1986 to 1990 estimates are for major U.S. field crops. Data for 1990 are projections.
Source: U.S. Department of Agriculture.

The washing of agrichemicals and silt into waterways incurs an economic as well as environmental cost. According to the World Resources Institute, soil erosion causes over $10 billion in damage to waterways each year, an estimated 36 percent of it due to erosion from croplands (Table 16-4).

Groundwater Depletion. Besides polluting water with chemicals and silt, agriculture consumes a significant amount of our water supply. The Worldwatch Institute estimates that farming accounts for roughly 70 percent of water use worldwide. In the United States the use of groundwater for agricultural irrigation increased 300 percent between about 1955 and 1985. Currently, irrigation is the major use for the nation's groundwater. Most irrigation systems are grossly inefficient; in many cases, 75 percent of the water is not used by the crop.

About 10 million acres (roughly 4 million hectares), which make up about one-fifth of the nation's irrigated area, are watered by overpumping aquifers, that is, removing the groundwater at a rate faster than the aquifers can be recharged. By the early 1980s the depletion of aquifers had become particularly severe in four important food producing states: Texas, California, Kansas, and Nebraska. (Groundwater depletion is discussed in detail in Chapter 15.)

Overirrigation. Overirrigating farmland can cause air spaces in the soil to fill with water for too long a period of time. The excess water leaches away nutrients needed by crops, prevents roots from getting

Table 16-4
Annual Off-Site Damage from Soil Erosion in the United States (millions of dollars)

Damage Category	Cost of Damage
Freshwater recreation	2,404
Marine recreation	692
Water storage	1,260
Navigation	866
Flooding	1,130
Roadside ditches	618
Irrigation ditches	136
Freshwater commercial fishing	69
Marine commercial fishing	451
Municipal water treatment	1,114
Municipal and industrial use	1,382
Steam power cooling	28
Total	**10,150**

Source: World Resources Institute, *Paying the Farm Bill,* March 1991.

oxygen, kills soil organisms, and promotes mold growth.

Prolonged irrigation may result in salinization, decreasing the soil's ability to support crops and eventually leading to impoverished farmlands. About one-fifth of irrigated land in the United States is affected by salinization.

Loss of Genetic Diversity. As farmers have been relying on fewer and fewer hybrid strains of crop species, the risk that those crops will be destroyed by pests or disease has increased. In natural systems a wide variety of strains helps ensure that some will be resistant to specific pests or diseases. As natural habitats are disrupted, wild relatives of food crops are often wiped out, thus depleting the genetic diversity which could be used to strengthen agricultural strains. The value of wild strains became clear in 1970, when severe southern corn leaf blight struck the U.S. corn crop. At that time 70 percent of the corn seed came from only six types. Because there was so little genetic diversity, a huge percentage of the corn crop was vulnerable to the blight. One-seventh of the entire crop was lost, raising prices by 20 percent and resulting in a loss of approximately $2 billion. The damage was finally brought under control by crossbreeding with blight-resistant germ plasm that originated in Mexico. As awareness of the importance of genetic diversity has become more widespread, researchers have become more interested in preserving rare and endangered plant species in gene banks (see Chapter 10). In addition, increased effort is being made to preserve older strains of crop species, known as heirloom varieties. These plants, which have been handed down from generation to generation, have often been supplanted by hybrid strains (see Chapter 23).

Socioeconomic Concerns. There are significant social and economic problems associated with conventional agriculture. Traditionally, farming was a way of life rather than a business. Gradually, U.S. agriculture was transformed from subsistence farming—farming to support one's family—into farming for profit. This change began in the 1800s, when the building of roads, canals, and railroads, led to the rise of commercial agriculture. Farming as a business venture, in turn, gradually developed into high-input agriculture and agribusiness by the latter half of the twentieth century.

By the 1970s **agribusiness**—the ownership and commercial operation of large farms, often by corporations—became the model for U.S. agriculture, and "bigger is better" became the rule. Farms became increasingly larger and more specialized. Farms that featured animal husbandry and the cultivation of multiple, diverse crops gave way to farms characterized by the cultivation of a single cash crop. The trend toward bigness was stimulated by the desire to take advantage of economies of scale: the expensive and often specialized equipment characteristic of conventional farming must be used on extensive fields if it is to be a profitable investment. In 1985, 30 percent of farmland was controlled by a mere 1 percent of farm owners. In addition, more and more farmers began to rent land from absentee landlords; having no stake in the long-term fertility of the soil, such farmers have little incentive to prevent soil degradation, especially if such prevention would interfere with short-term profits (Table 16-5).

Table 16-5
Number of Farms in the United States and Total Acreage, 1960–1989

Year	Number (thousands)	Acreage (million acres)
1960	3,963	1,176
1965	3,356	1,140
1970	2,949	1,102
1975	2,521	1,059
1980	2,440	1,039
1985	2,293	1,018
1989	2,173	991

Note: 1975 and later based on 1974 definition; 1989 preliminary estimate.
Source: 1990 U.S. Statistical Abstracts.

How Has the U.S. Government Responded to the Farm Crisis?

The U.S. Department of Agriculture and government programs have supported the trend toward highly specialized, energy-intensive, and environmentally questionable farming practices (Table 16-6). Perhaps the best example of this are federal commodity programs, which about 80 to 95 percent of farmers take advantage of. These programs base eligibility for benefits on the number of acres planted in one crop for the previous five years. Planting acreage in a different crop disqualifies a farmer from receiving government payments for that acreage. The government's commodity programs also encourage farmers to reap the highest per acre yield each year, a short-term goal that necessitates the heavy use of fertilizers and pesticides. Using large amounts of these substances is a significant drain on the financial resources of family farmers. Unfortunately, if they do not use fertilizers and pesticides, their harvests—and the subsidies they earn—may drop.

Although government policies such as subsidies and price supports can adversely affect soil quality, the government has also taken steps to try to protect soil resources. U.S. Soil Conservation Service efforts led to the passage in 1980 of the Soil and Water Resources Conservation Act. It included strategies to make soil conservation mandatory for any farm receiving federal farm aid, including commodity price supports for corn, wheat, or milk. This controversial plan, called cross-compliance, was generally

Table 16-6
Economic and Environmental Effects of Government Programs

Type of Program	Mechanics of Program	Economic Effects	Environmental Effects
PRICE SUPPORTS			
Government purchases	Government sets prices above those an unrestrained market would produce. Higher prices often result in consumers buying less than is produced, so government buys surplus to maintain support price.	Surplus cannot be sold on market without driving price down, so government must dispose of it in other ways (give to citizens or other nations in need, sell at low cost, etc.). Consumers pay higher prices and pay for surplus purchases and storage through taxes.	Artificially high demand encourages overproduction. Farmers cultivate intensely, contributing to soil erosion, wasteful irrigation, and chemical contamination. They may also bring marginal land under cultivation
Nonrecourse loans	Farmers use crops as collateral to get loans at harvest time from Commodity Credit Corp. Amount loaned per unit of crops (e.g., $3 per bushel) becomes support price. Farmers place crops in storage for up to 9 months. If market price rises above loan per unit plus interest, farmers can sell crops; if market price falls below, farmers can default and government effectively buys crops.	Loans have same effect as government purchase, even for farmers not actually participating. By effectively buying crops from defaulters, government sets market price at level of amount loaned per unit of crop.	Loans have same effect as government purchase.
ACREAGE RESTRICTIONS	Farmers are encouraged to take acreage out of cultivation to curb wasteful surplus and cultivation of marginal land. Farmers may receive subsidy in return or may be required to cut back in order to receive benefits of other farm programs.	Farmers cultivate acreage they can use by more intensive methods using more inputs, thus producing at higher cost than if they had used more land. There is no guarantee that surplus will be eliminated; government may still have to purchase and store some crops. Consumers pay higher prices and pay for any subsidies, surplus purchases, and storage costs.	Farmers take least fertile land out of production first. They cultivate remainder more intensively, applying more chemicals and water, and increase chance of exhausting soil through overuse.

not favored by independent-minded farmers, and it was not considered to be an important part of the Reagan administration's farm policy. Ironically, while that administration did little to support small farmers and promote environmentally sound agriculture, the 1985 farm bill, passed during Reagan's tenure, has been hailed as the cornerstone of the most progressive agricultural policy worldwide.

The 1985 Farm Bill. The 1985 Farm Bill, or the Food Security Act of 1985 as it is officially known, included four major new provisions popularly called swampbuster, sodbuster, conservation compliance, and CRP (conservation reserve program). The first two deny valuable federal price or income supports, crop insurance, and federal loans to farmers who

cultivate environmentally fragile lands. Conservation compliance, which was required after 1990, mandates that farmers who crop highly erodible land implement conservation practices to minimize erosion. Farmers who fail to comply become ineligible for all farm programs on all of their land, not just their highly erodible land.

The CRP is a voluntary program in which farmers receive annual "rent" payments of about $50 per acre to take highly erodible land out of production. Farmers agree to plant the land in trees or a grass or legume cover for 10 years, with the costs of the planting to be shared by the federal government. The Agricultural Stabilization and Conservation Service (ASCS) and the Soil Conservation Service, in conjunction with local county agencies, helps

Type of Program	Mechanics of Program	Economic Effects	Environmental Effects
TARGET PRICE SUBSIDIES	Government guarantees farmers a certain price per unit for their crops. If market rate is less than guaranteed rate, government pays farmers the difference; if market rate is more, there is no subsidy.	All output is sold, eliminating the surplus problem. Consumers pay market price and any subsidies that are necessary.	Farmers have no incentive to cut back production since they are guaranteed a certain price. They may cultivate too intensely.
TARIFFS AND QUOTAS Tariffs	Government taxes imported crops to reduce competition with domestic production.	Tariffs keep domestic prices high and set up a barrier to cheaper foreign produce. Consumers pay higher prices.	Tariffs encourage more production than a free international market would. When other countries impose tariffs, demand decreases. However, domestic farmers still do not decrease production because of government support programs, with corresponding environmental damage.
Import quotas	Government limits quantity or value of imports to keep prices high and domestic production profitable.	Quotas have same effect as tariffs.	Quotas have same effect as tariffs.
Production quotas	Government limits quantity domestic farmers may produce and guarantees a price for it.	Production quotas keep domestic prices high; tariffs or import quotas are usually required to maintain production quotas. Consumers pay higher prices.	Quotas reduce incentive to overcultivate or to cultivate marginal lands.

farmers develop and secure funds for the 10-year plan.

The CRP has yielded mixed results. From 1985 to 1990 erosion was reduced by one-third, from 1.6 billion tons per year to 1 billion tons. On the most highly erodible land, erosion was cut from roughly 21 tons per acre annually to just 2 tons per acre. Eight southern states have taken greatest advantage of the CRP, enrolling over half of their eligible land. Unfortunately, few farmers in the Rocky Mountain and southwestern states participate in the CRP. Not surprisingly, these states lead the nation in soil erosion rates.

The 1985 farm bill, reauthorized in 1990, contains one of the most potentially significant conservation programs in U.S. history. It calls for a reduction of sedimentation in streams, an increase in wildlife habitat, and a reduction of highly erodible lands planted to crops. It provides for cost-share funds to accomplish these goals. Conservation farming, especially removing marginal land from production and alternative methods of tillage, has higher immediate costs, but these are outweighed by its long-term benefits. By 1995, the CRP is expected to reduce erosion from croplands by an additional 450 million tons per year, so that total remaining excessive soil erosion will equal about 550 million tons annually.

LISA. In the mid-1980s the U.S. Department of Agriculture launched a research and extension program known as LISA, or low-input sustainable agriculture. LISA was designed to assist farmers in converting to low-input farming, which is characterized by a limited use of synthetic inputs. Practitioners of LISA use some synthetic fertilizers and pesticides, but only sparingly. Chemical pesticides, for instance, are typically used as a last resort. As LISA helps farmers reduce their dependency on these inputs, it helps them protect their soil at the same time. By 1990 the program had a $4.5 million budget. Though that is not a significant amount in terms of government spending, the establishment and continued funding of the program are encouraging, since the program represents a departure from the government's decades-long practice of promoting the use of chemical and energy inputs and monocultures.

An estimated 30 to 40 percent of the nation's farmers have taken simple steps to reduce the use of chemicals and other inputs on their farms. For most, economics—the need to reduce costs—is the driving force behind the switch to low-input farming. For other farmers, however, LISA does not go far enough. They feel that it is not the answer to the plethora of interrelated economic and environmental problems associated with conventional farming.

Backed by the support of a growing cadre of researchers and soil scientists, these farmers are searching for alternative forms of agriculture.

What Is Sustainable Agriculture?

The search for effective methods to protect the soil and restore its fertility is resulting in the development of various systems of alternative or **sustainable agriculture**—ways of farming that both sustain and protect the soil and productivity. Because sustainable agriculture safeguards soil resources, it can in turn be sustained by the living soil over time.

Sustainable agriculture refers to a broad group of techniques and practices, such as growing a variety of crops rather than one or two, rotating crops, using organic fertilizers such as animal manure and crop residues, allowing croplands to lie fallow periodically, planting cover crops to protect the soil between crops or during the winter, and encouraging the natural enemies of crop pests. At one end of the spectrum are practitioners of LISA, who use some synthetic inputs; at the other end are organic growers, who disavow the use of any synthetic fertilizers or pesticides. All forms of sustainable agriculture, however, share a common dual goal: to prevent erosion and to maintain soil fertility and structure. According to Department of Agriculture estimates, by 1980 about 20,000 to 30,000 of the nation's farmers were using alternative methods, a figure that many experts feel has probably risen to about 50,000 to 100,000 since then.

Most of the practices that define sustainable agriculture are not new. Until the middle of this century, for instance, crop rotation and the use of manures and cover crops were common. With the development of pesticides, hybrid strains, and other agricultural advances, some farmers abandoned these practices. They were retained by some cultural groups, for example, the Amish people, whose farms prospered throughout the crisis affecting the majority of American farmers. The prosperity of Amish farms is due in part to their farming methods (small-scale farms with little use of synthetic fertilizers or pesticides) and in part to their accounting system. The Amish believe that land ownership is a privilege and a reward, unlike agribusinesses, which count the "cost of ownership" of the land as a fixed cost. The Amish hire no additional labor and consider their own labor as part of their profit, not their cost. (Because of their economic implications, the Amish accounting and farming techniques are discussed more thoroughly in Chapter 26, Politics and Economics.)

Measures to Prevent Excessive Soil Erosion. The most effective way to prevent excessive erosion is simply not to farm marginal croplands, especially slopes and soils with structures not suited to cultivation. To prevent excessive soil erosion on croplands, one of the most effective measures is to plant a **cover crop**, a crop that is grown when the land is not planted with a main crop or that is grown to provide protection for the soil. The roots of the cover crop anchor the soil, protecting it from the effects of rain, snow, and wind. When it is time to plant the main crop, the cover crop litter and residues can be left on the ground or tilled into the soil to act as a **green manure**. Planting trees and shrubs along the windward side of fields is an effective defense against wind erosion. The most effective windbreaks, or shelterbelts as they are sometimes called, are wide enough and tall enough to deflect the wind upward and over the cropland while slowing its velocity.

Several alternative methods of tilling can help prevent soil erosion. **Ridge tilling** is planting the crop on top of raised ridges (Figure 16-10). **Conservation tillage** includes both low-till and no-till methods of planting. Low-till planting consists of tilling the soil just once in the fall or spring, leaving 50 percent or more of previous crop residue on the ground's surface. No-till planting, or stubble mulch farming, is done with a no-till planter or drill which can sow seeds without turning over the soil. Crops are planted amid the stubble of the previous year's crop, which acts as a mulch to fertilize the soil and prevent it from drying out. One disadvantage of conservation tillage is that the previous year's mulch provides food and cover for insect pests and allows weeds to become established. Consequently, conser-

vation tillage may require the use of pesticides in order to combat pests and undesirable plants.

The risk of erosion is greatest on sloped land, but there are a number of techniques that can reduce that risk. **Strip cropping** is the alternation of rows of grain with low-growing leaf crops or sod. It offers greater protection to the soil than conventional planting, and the strips of legumes or grasses also help to enhance the structure and organic content of the soil. Strip cropping can be used in conjunction with contour plowing or terracing to give increased protection to land at risk of erosion. **Contour plowing** is tilling the soil parallel to the natural contours of the land rather than in the straight rows and square fields characteristic of conventional fields. Contour plowing helps to keep the soil from washing down the hillside. On extreme slopes **contour terracing** can be used. Contour terracing is building broad "steps" or level plateaus into the hillside (Figure 16-11). Swales, or trenches, located at the edges of the terraces act as catch basins for rain, channeling it along the hillside. Swales planted in a grass, called grassed waterways, help to both slow the course of the water and absorb some of the excess water and any eroding soil.

Whether the land is flat or sloped, improving the structure of the soil is imperative to prevent the long-term loss of soil. Soil with a healthy and stable structure is less prone to washing or blowing away because the soil particles tend to stick together. Mulching and fertilizing with crop residues and animal manure improve the structure of the soil and enhance its fertility.

Many farm communities across the nation are realizing the benefits of these erosion control methods and are establishing programs to help farmers apply them. One such community is the focus of this chapter's Environmental Science in Action: Preserving Agricultural Land in Marathon County, Wisconsin (pages 344–346).

Measures to Restore and Maintain Soil Fertility. Four of the most effective measures to preserve soil quality by enhancing its fertility are allowing fields to lie fallow periodically, rotating crops, using organic fertilizers, and using good irrigation techniques.

A field which is allowed to lie fallow is planted in a cover crop such as hay or clover. The cover crop protects the soil from the erosive effects of wind and water. It can be plowed into the soil at the end of the growing season, thereby acting as a soil amendment, or green manure.

Crop rotation is changing the type of crop grown on a field from year to year. Rotating a grain crop, which uses nitrogen, with a legume crop, which fixes nitrogen, is especially effective.

FIGURE 16-10: Ridge tilling on a potato field in Idaho. In addition to minimizing erosion by creating catch basins for water and soil, ridge tilling minimizes the use of pesticides and fertilizers since these substances can be applied directly to the plants atop the ridge rather than being applied to the entire field.

FIGURE 16-11: Contour terraces above Uttarkashi on the Bhagirathi River, Himalayas, India.

Organic fertilizers are an essential ingredient in alternative agriculture, and they offer some important benefits over synthetic fertilizers. The most obvious advantage is that many can be obtained free or at low cost, while synthetics are increasingly expensive. Organic soil amendments release nutrients slowly over a longer period of time. In contrast, many synthetic fertilizers release a quick burst of nutrients just after they are applied, but their benefits are not long lasting. Organic fertilizers tend to become an integral component of the topsoil, while synthetics often wash away in rains. And, unlike synthetic fertilizers, organics stimulate the growth and proliferation of soil microorganisms, which are essential to maintaining the fertility of the soil over the long term. Although gathering and spreading manure and planting cover crops for residues require more labor than synthetics, they can make up for this drawback with lower cost, better soil fertility, and less risk of environmental damage.

Trickle drip irrigation can protect soil fertility from the salinization that accompanies the long-term use of large quantities of irrigation water. Trickle drip irrigation is the delivery of water through permeable or perforated pipes directly onto the soil surrounding the base of the plants. It uses far less water than traditional irrigation methods, resulting in less evaporation and thus leaving fewer salts behind to salinize the soil.

Organic Farming. Although techniques for preventing erosion and enhancing soil fertility can be used on all types of farms, including conventional ones, they work best when integrated into a program designed for overall sustainability, such as organic farming.

Of all forms of agriculture, organic farms most closely resemble natural systems. They typically feature polyculture, the planting of a variety of crops suited to the particular climate and soil of the area. Synthetic chemicals are not used to give a nutrient boost to crops, but rather natural fertilizers, such as animal manure, green manure, and fish emulsions, are used to enhance soil fertility. Carefully timed planting and cultivating regimens minimize problems caused by pests and weeds. Killing weed seedlings in the spring with a rotary hoe, for instance, before the undesirable plants germinate and go to seed, is far more effective than trying to weed a field after the plants have taken hold. Weed growth is discouraged by mulch between plant rows and by certain biological controls. For example, some organic farmers keep geese and ducks, which eat weeds.

Insect pests are controlled through a variety of methods collectively known as **integrated pest management (IPM)**. If the plot is small enough, pests can be picked off by hand. Traps (typically baited with pheromones or other substances) attract and capture pests. Biological control can be achieved in one of several ways. Microbes, insects, birds, and animals that feed on weeds or pests can be introduced to a field or encouraged to take up residence nearby. Some species of predatory insects can act as a check on pest populations without harming crops. The ladybug, for instance, preys upon crop-damaging aphids. A simple, preventative measure is to provide bird houses and water (in the form of a small pond, for example) in order to encourage the presence of martins and other insect-eating birds. Rows of alfalfa, interspersed with rows of cotton, for example, will attract lygus bugs, which would otherwise attack a cotton crop. Sterile Mediterranean fruit flies have been introduced to fields in order to slow down and reverse the population growth of this potentially devastating pest.

Crop rotation prevents the buildup of specific pest populations. Pest species that prey on a crop and can overwinter from growing season to growing season will accumulate if that crop is planted year after year. Crop rotation also helps to avoid depleting the soil of specific nutrients that crops need and therefore can result in higher yields. For example, potatoes yield best when planted on a plot in which corn was grown the year before. Figure 16-12 illustrates how crops might be rotated on a plot in succeeding years in order to maximize benefits to the soil and enhance the yields of various crops.

To compare conventional and organic agriculture, we must consider both the short- and long-term economic performance of the farm as well as the environmental effects of the farming techniques. The yields of crops with high nitrogen requirements, for example, corn, wheat, and potatoes, are usually somewhat lower when organic techniques are used.

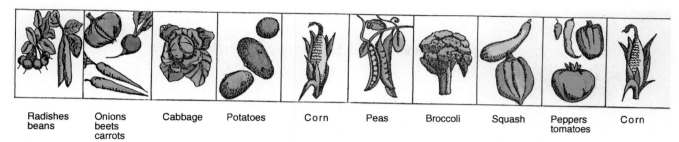

| Radishes beans | Onions beets carrots | Cabbage | Potatoes | Corn | Peas | Broccoli | Squash | Peppers tomatoes | Corn |

FIGURE 16-12: Ten-year crop rotation plan devised by organic farming expert Eliot Coleman. The diagram represents the first year of a 10-year rotation in 10 plots. Each succeeding year, the crops are moved one plot to the right, with the crop from the tenth plot moving to the first. Rotation is based mainly on the nutrient requirements of the plants. For example, corn, a heavy feeder with high nitrogen requirements, follows legumes (peas and beans), lighter feeders that fix nitrogen.

But for some crops, notably alfalfa, soybeans, and oats, yields may be higher. Conventional farms may produce more in years with adequate rainfall, but organic farms seem to score better in dry years. More importantly, because organic farmers have lower capital costs for equipment and machinery and do not have to pay for synthetic chemicals, their profits may exceed those of conventional farmers.

Organic farming offers several important environmental advantages over conventional farming. It improves the structure of the soil, making it easy to work and enhancing plant growth. Maintaining good soil structure also minimizes runoff, reducing the loss of topsoil after a heavy rain and the leaching of nutrients. Organic agriculture enhances soil fertility, eliminating the need for inputs of synthetic fertilizers. Finally, organically grown produce eliminates the health and environmental risks associated with pesticides.

Despite its many advantages, organic agriculture still has far fewer practitioners than conventional, high-input farming. About 1 to 2 percent of the nation's 2.2 million farmers are organic growers. One reason is economic: limited consumer demand. Many consumers have grown accustomed to picture-perfect produce and have a prejudice against buying produce with any blemishes or color that is less than ideal. This attitude seems to be changing, however, as people begin to realize that slight imperfections are perfectly normal, are not harmful, and do not signal inferior quality. In 1989 a public outcry about use of the suspected carcinogen Alar™ on apples caused a number of large supermarket chains to refuse to carry apples sprayed with Alar™ and apple juice and other products made from sprayed apples. Other recent developments also indicate a change in public attitudes toward pesticides. By 1990 several supermarket chains announced that they would begin to test certain fruits and vegetable for pesticide residues. In a move of particular import to parents and child care advocates, the H. J. Heinz Company announced that it would no longer make baby food from produce that contained traces of certain pesticides. The increasing objections to pesticides should help people to put the cosmetic appearance of their produce in proper perspective. Given the growing consumer demand for organic produce, profits are likely to increase in the coming years.

Organic farming also has fewer practitioners because it requires training, willingness to deal with a new market, and time—an average three to five years—to convert a farm from conventional techniques to organic practices. It generally takes a number of years for the soil to cleanse itself of the residues of chemical fertilizers. Some states mandate that during that time produce cannot be marketed as "organic." Legal definitions of "organic" farming vary from state to state, and it is not always clear just what the "organically grown" label means. To clear up the confusion, some states, such as California and Oregon, where organic farming is more popular, have established watchdog agencies to define organic standards. These agencies certify growers who meet their requirements and offer assistance to farmers attempting to make the transition from conventional farming. That assistance is important, for there are few information resources for farmers making the transition. Organic farming requires an intimate knowledge of the soil and a sophisticated understanding of ecological processes. That sort of knowledge does not accumulate overnight.

Polyculture. An innovative approach to making agriculture less environmentally harmful is polyculture. While most conventional farms grow one or perhaps two crops, an alternative approach is to grow a number of different crops. The diversity of a polyculture helps provide stability to a farm; if one crop fails in a year, the other crops may produce well enough to compensate for it. A polyculture is the agricultural equivalent of the saying "Don't put all your eggs in one basket."

Perennial polyculture—growing a mixture of self-sustaining, or perennial, crops—is a new area of

Focus On:
A New Relationship Between Agriculture and the Earth
Wes Jackson, Land Institute

Wes Jackson and Dana Jackson direct the Land Institute, an education and research institute in Salina, Kansas. Located on 277 acres, the Land Institute is devoted to working toward a sustainable agriculture, chiefly through developing grains that can grow in perennial polycultures, much like the native prairie. Each year the institute accepts 8 to 10 interns who take classes, assist in research, and write articles.

In the late 1800s a Scandinavian-born farmer of the northern plains looked up to discover an old Sioux watching him as he plowed his field. When the farmer stopped to smoke a cigarette, the Sioux knelt and examined the sod and the buried grass. Eventually he stood up and told the farmer, "Wrong side up."

Farmers have traditionally told this story to laugh at the Sioux's ignorance of proper farming methods, but many people are beginning to realize that the Sioux was right. To solve the agricultural problems threatening our soil, we don't need a technological fix. We need a different way of thinking about our relationship with the earth.

We live in a "fallen" world in the sense that we have alienated ourselves from the natural world that produced us and fed and clothed us when we were gatherers and hunters. During the 10,000 years that we have practiced agriculture, we have shaped nature to our own ends, using human cleverness—and ignoring the wisdom of nature. Natural systems, including the crop plants we have bred from wild species, have been shaped by climatic and evolutionary histories that are beyond complete human comprehension.

This split between humans and nature has encouraged us to regard livestock and crops more as human property than as the relatives of wild things that have evolved in a context not of our making. We have ignored the way an individual species interacts with other plants and its own ecosystem in favor of altering that species to meet our needs by making it dependent on massive inputs of energy, fertilizer, and other chemicals. This mindset has allowed the industrialization of farming, which treats the potentially cyclic and renewing economies of nature as resources to be used up for the maximum immediate profit.

If we are to save our soil and ensure that we can continue to grow food in that soil a hundred years from now, we must return to nature and use it as a standard to create a sustainable agriculture. Compare the traditional wheat field with the prairie it has largely supplanted. With the wheat field, an annual monoculture, come industry-produced pesticides and fertilizer, dependence on fossil energy, soil degradation, and soil erosion. The prairie, a polyculture which features perennials, counts on species diversity and genetic diversity within species to avoid epidemics of insects and pathogens. The prairie sponsors its own fertility, runs on sunlight, and actually accumulates capital, accumulates soil.

The key to the prairie's success at accumulating soil is its roots. Prairie species feature a diversity of roots—some tap, some fibrous, some hard to describe, but overall a dense and deep network quite unlike our

research. Fields would have a greater number of species than a traditional or even organic crop field. Because perennial plantings do not require that the ground be turned over every year and maintain their root systems in the ground, they reduce soil erosion and allow soil to accumulate. Research on perennial polycultures is being conducted by Wes and Dana Jackson and their colleagues at the Land Institute, Salina, Kansas (see Focus On: A New Relationship Between Agriculture and the Earth). They are working on developing new strains of wild grain-producing perennials (Figure 16-13).

Future Management

Managing soil resources in the 1990s and beyond presents society with one of its most important challenges. The future of the United States' land base and its valuable soil should be a matter of concern to all citizens since land use and the welfare of agriculture are issues that affect everyone—urbanites and

FIGURE 16-13: Interns at the Land Institute in Salina, Kansas, sample prairie vegetation for their research on perennial polyculture.

annual crop plants which put down tenuous roots only to have them destroyed with the next plowing. Prairie soil still erodes, but the forces of erosion cannot compete with the power of this living net which holds the soil in place. The small amount of soil lost each year is more than replaced because the roots catch and hold the hard-won nutrients released when the inevitable death comes to all living things—when production becomes the future fertility we call soil.

Based on this comparison, we at the Land Institute believe that the best way to develop a sustainable soil policy is to grow perennial polycultures of grain-producing plants, to create domestic prairies that closely reflect natural ecosystems. Based on the advances made in biology over the past half-century, there is every reason to believe that the scientific community has the knowledge to develop these crops.

Once we begin growing grain-producing perennials in polycultures, we can transform the face of agriculture. The diversity will make plants less vulnerable to insects and

disease. The more efficient retention and use of soil moisture by perennial roots will make the ecosystem more resilient. With soil loss cut significantly, soil could begin to accumulate.

A few years ago 316 million acres (128 million hectares) in the United States were devoted to the top 10 crops. Were this acreage planted to mixed perennials and one-fifth replanted every five years, 253 million (102 million hectares) of the 316 million acres in any given year would remain unplowed. Assuming a soil loss of 5 tons per acre on conventionally cultivated cropland, that would save 1.27 billion tons of soil. Given a 12-ton-per-acre loss, the savings would exceed 3 billion tons a year, an amount equal to the annual loss for all lands in the United States when the Soil Conservation Service was established.

These cultures could be successfully planted on our marginal lands, which are degraded in only a few years when cultivated with traditional farming methods. Because all inputs would be greatly reduced, including fertilizers, pesticides, labor, and energy, small farmers

with little capital would be able to earn a living again. Take away the fossil fuels for traction and fertility, and these future perennial polycultures may compete favorably with conventional crops. In addition, they could cause soil erosion beyond replacement level to go to zero.

The seed, chemical, and fertilizer industries would have us believe that anything but the maximum crop yield is unacceptable. And yet our progress as a species does not have to be defined in terms of wealth or material and physical growth any more than our progress as individuals has to be defined in terms of physical growth. We need to look at the long-term economic benefits of soil conservation, energy savings, and sustainability. We need to tap the vast knowledge nature has accumulated over the centuries and use it to devise an alternative to the catastrophe our present agricultural practices are causing rather than sitting back and passively watching the tragedy unfold.

rural dwellers, consumers and producers alike. (See What You Can Do: To Protect Soil Resources.)

The goal of environmentally sound management—to sustain the soil ecosystem as a living resource—requires a comprehensive management approach that both guides land use and safeguards invaluable arable land. Accordingly, environmentally sound management strategies should prevent overuse of soil resources through conservation, protect environ-

mental and human health from pollution and degradation of soil resources, and preserve living soil ecosystems.

Prevent Overuse Through Conservation

Agriculture is arguably the most essential of all land uses, and since prime arable land accounts for only a portion of the U.S. land mass, the conservation of

What You Can Do: To Protect Soil Resources

- If you have a vegetable garden, try biological pest controls, crop rotation, and other alternatives to conventional pesticides and fertilizers.

- Support farmers who use organic and other alternative methods

of agriculture by buying their fruits and vegetables or by encouraging your supermarket to buy from them.

- Get involved in your area's land use planning meetings.

- Write to your legislative repre-

sentatives to express your support for measures such as the 1995 reauthorization of the Food Security Act which will help protect soil resources as well as human health and the environment.

Environmental Science in Action:
Preserving Agricultural Land in Marathon County, Wisconsin
Michael Fath

Marathon County, located in central Wisconsin, is the largest of the Wisconsin counties and one of the state's biggest agricultural producers (Figure 16-14). Marathon farmers grow a wide variety of grains, but they are best known for their output of dairy products. Of all dairy-producing counties in the United States, Marathon ranks fifth in total number of dairy cows and seventh in total milk production.

Because of this heavy reliance on agriculture, particularly dairy farming, Marathon County residents realize that maintaining the fertility of their soil is vitally important. When studies in the early 1970s revealed that at least some Marathon County farms were losing soil much faster than it could be replaced, in accordance with Wisconsin's long history of soil conservation and management, Marathon County established a number of programs aimed at preserving farmland, conserving soil, and reducing water pollution caused by erosion.

Describing the Resource

Physical Boundaries

Marathon County occupies approximately 1 million acres (405,000 hectares) in the heart of Wisconsin. Its agricultural success is due to its climate of long, relatively cold winters, warm summers, and steady, ample rainfall and to its rich, fertile soil—the most precious resource for Marathon farmers.

The activity of a series of glaciers is largely responsible for the character of Marathon County's soil. Early glaciers affected the entire county, flattening hills and depositing drift that became the basis for today's soil. The western part of the county, untouched after this process, is a flat to gently rolling area with rich soils that make it the major agricultural area of the county (Figure 16-15). The eastern part of the county was covered by a later glacier which left a pattern of swamps, hills, and scattered pothole lakes.

Soil in Marathon County has also been created by the weathering of the Canadian shield, a huge sheet of igneous and metamorphic bedrock

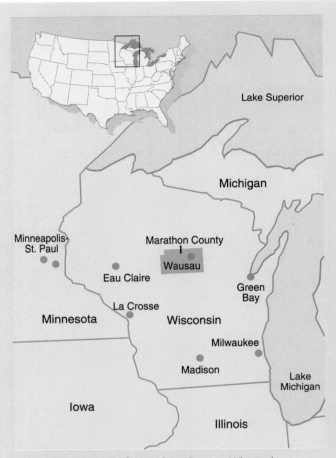

FIGURE 16-14: Map of Marathon County, Wisconsin.

that underlies most of Canada and the northern United States. Wind and rain wear away exposed areas of bedrock, eventually producing soil. Additional soil has been carried into the area by water and rain.

The major soil order found in Marathon County is alfisol. The soil can be generally classified as loam, with silty loams in the western part and sandy loams in the eastern. Humus content ranges from 3 to 4 percent in the west to 2 percent or less in the east. The limiting factors in the soils are nitrogen (macronutrient) and boron (micronutrient). The soils are acidic, with a pH range of 4.5 to 6.5. To neutralize the soil's acidity, farmers must lime their fields heavily. Although much of the soil is suitable for growing crops, approximately 25 percent is better suited to less intense uses, such as grazing herds of cattle.

Erosion presents a substantial threat to the quality of Marathon County soil. An estimated 64 million tons (58.2 million metric tons) of soil are lost from Wisconsin cropland annually. Wind erosion carries away about one-fourth of this soil, and the remaining 48 million tons (43.7 million metric tons) are lost to water erosion. Runoff from construction sites contributes to eroded soil, but croplands are the major source, responsible for as much as 85 percent of the eroded soil. Studies link an increase in erosion with an increase in the amount of land planted to row crops, especially corn and soybeans. Losses in Wisconsin may be higher than the national average because much of the cropland has relatively thin soil or is on sloping land. In Marathon County actual erosion rates range from 1 to 16 tons per acre per year, depending on the topography and crops grown, with an average soil loss of 2.7 tons per acre. Replacement level (T-value) averages 3 to 4 tons per acre. About 5 percent of the county's total cropland (approximately 16,900 acres or 6,800 hectares) is eroding at rates greater than the T-value.

Biological Boundaries

Farmland and forests account for 9 out of 10 acres in Marathon County, allowing a wide variety of plants and animals to flourish. The forested areas contain mainly second-growth hardwoods and evergreens. The county's 197 lakes and 256 rivers and streams are populated by hundreds of fish species, including popular sportfish such as trout, bass, pike, and panfish. These waters are also home to many types of waterfowl and attract flocks of migratory geese. Mammals, ranging from chipmunks and rabbits to white-tailed deer and black bears, can also be found in the county.

Farms occupy over 50 percent of the land in Marathon. While many of these farms yield large crops of oats, corn, hay, barley, potatoes, and wheat, the grain is used primarily to feed dairy cows. Some farm acreage is also used to graze herds

FIGURE 16-15: Contour plowing on the gentle slopes of western Marathon County.

of cattle. The county's dairy industry boasts an annual output of over 1 billion pounds (450 million kilograms).

Social Boundaries

Although Marathon County is primarily an agricultural district, somewhere between 125,000 and 150,000 acres (51,000 and 61,000 hectares) are reserved as open space and recreation areas. Only about 32,000 acres (13,000 hectares), 3.2 percent of the county's land, is used for nonfarm development. Roughly half of the 112,000 people who live in the county live in the Wausau urban area, which includes the cities of Wausau, Schofield, Rothschild and the townships of Weston, Rib Mountain, and Stettin.

Despite the rural nature of the county, a trend toward nonfarm uses that began in the 1960s poses a threat to its prime farmland. Land that once produced 150 bushels of corn per acre has been converted to urban development, roads, and housing. In 1982 Marathon County had 3,300 working farms. By 1987 a combination of economic factors, urban development, and consolidation had brought that number down to 3,000. Although the size of the average farm has been increasing as the number of farms drops, there has been an overall decrease in land used for agriculture. Between 1961 and 1981, farm acreage in Wisconsin declined by about 3.5 million acres (1.4 million hectares), 15 percent. While corporate farms and farm partnerships have increased, over 90 percent of Marathon's farms are still family owned, with less than 10 percent tenancy or absentee ownership.

Looking Back

Marathon County is a microcosm of agricultural United States. It has all the problems facing farm areas throughout the country: soil erosion, runoff laden with fertilizer and animal waste, encroachment by urban development, and the trend toward specialization and conglomeration. However, Marathon's history of extensive farmland preservation efforts contains a model for coping with these problems.

Tradition of Soil Conservation

Agriculture has been the dominant influence on Marathon County since the late 1800s. In the 1850s the first wave of German and Polish farmers arrived in Marathon County anxious to create pastoral landscapes similar to those they had left behind. And the fact that cleared land was considered of less value for tax purposes also hastened the transition of the county from forestland to farmland.

Continuing a state tradition of interest in soil conservation, in 1931 the Wisconsin Agricultural Experiment Station joined with the U.S. Department of Agriculture to study erosion and ways to control it, establishing a Soil Conservation Experiment Station on Grand Dad's Bluff at La Crosse. In 1933 nine Civilian Conservation Corps (CCC) camps were set up in Wisconsin to fight erosion, and the Soil Conservation Service (SCS) began the nation's first watershed demonstration project on 90,000 acres (36,000 hectares) in Coon Valley. Coon Valley had lost approximately 3 inches (8 centimeters) of soil from fields that had been farmed for 80 years. Persuaded by low yields and incomes, area farmers agreed to work with the SCS to develop and apply whole-farm conservation plans.

In 1937 the Wisconsin legislature enacted a law providing for the organization of each county into a soil conservation district to help county residents work against erosion. All districts were to have working agreements with the Soil Conservation Service and other groups that could give technical, financial, and educational assistance. The Marathon County Soil and Water Conservation District (SWCD) was established in 1941.

Protecting the Soil Today

The amount of land used for agriculture in Marathon County expanded steadily until the early 1950s; even land only marginally suited for agriculture was cultivated. By the end of the 1950s advances in machinery, seed quality, fertilizers, and pesticides made farming more profitable and less physically demanding. Marginal lands were taken out of production.

The State of Wisconsin passed the Farmland Preservation Act in 1977. This act created a program of tax incentives for landowners to preserve agricultural land. Farmers could become eligible for tax credits in two ways: if their land is zoned for exclusive agricultural use or if they sign a contract agreeing not to develop their land for a specific amount of time. Qualification through zoning was later eliminated, and currently, only the contract option is available. To qualify for a contract, farmers must have at least 35 acres (14 hectares) which produced at least $6000 worth of farm products in the last year or $18,000 worth in last three years. All participants must conform to soil and water conservations standards adopted by the county.

Marathon County completed its own Farmland Preservation Plan in 1982. Its goals are the preservation of prime agricultural land with soil productivity as the basis of preservation, support of family farm ownership, reduction of the erosion of topsoil, and the wise use of farmland in urban fringe areas. A comprehensive map of the county was prepared to delineate farmland preservation areas, urban growth areas, and special environmental areas (wildlife habitat, wetlands, scenic and historic sites).

Participation in the state program is optional, but as of 1987, if farmers do participate, they must adopt a comprehensive soil management plan that reduces soil loss to T-val-

ues before tax incentives are granted. From 1982 to 1987 over 600 Marathon farms took advantage of the state's program, which is recognized as one of the most effective and innovative in the United States.

Marathon County is also participating in the "T by 2000" erosion control program passed by the state legislature in 1982. By the year 2000 soil erosion rates on Wisconsin cropland are to be reduced to the soil loss tolerance level. Erosion control practices include conservation tillage, strip cropping, water runoff diversions, terraces, windbreaks, and permanent vegetative cover. Use of these practices can reduce erosion to 10 percent or less of the erosion experienced before the practices are introduced.

Approximately 1,000 farms in Marathon County have met conservation compliance requirements because of stipulations in the federal Food Security Act. This act requires those farmers who participate in federal farm commodity programs and have land identified as highly susceptible to erosion to prepare and implement a conservation program which includes a variety of conservation measures, including conservation tillage, strip cropping, terracing, and grassed waterways. The roughly 2,000 farms not included in federal programs also contain some highly erodible land, but these farmers can choose whether or not they will make an effort to preserve soil.

In the late 1980s Marathon County residents became concerned about the pollution of streams and lakes by runoff containing animal wastes, principally from dairy cows. An innovative program was designed to address the problem. In some cases, barnyards were redesigned to contain the 4,000 tons (3,640 metric tons) per day of ma-

nure produced by the county's 88,000 cows.

Faced with such a variety of programs designed to conserve soil and prevent runoff from polluting water, Marathon County established a Soil Erosion Control plan in 1988 to coordinate activities. The plan's purpose is to coordinate and implement the federal, state, and county programs and also to continue to identify land on which erosion is greater than replacement levels.

Looking Ahead

Marathon County certainly does not have the worst erosion problems in the country. Counties in California, Iowa, Washington, and Tennessee have experienced greater soil loss, both in total tons of topsoil and in acres of prime farmland. However, Marathon land has the potential for severe soil fertility loss because its topsoil is not naturally deep, averaging just 8 inches (20 centimeters). Any sustained loss of topsoil will have serious effects on soil fertility and productivity and thus have a devastating impact on the county's economy.

Consistent and timely concern by both the state and the county have prevented severe erosion from taking place so far. Through its various programs, Marathon County provides an example of the kind of soil management and farmland preservation efforts needed to combat erosion, maintain soil productivity, maintain family farms, and protect waterways, wetlands, and historic places. Of course, erosion will always be a threat, so continued vigilance is essential even after soil loss has been reduced to replacement levels throughout the county.

these lands is of the greatest importance. The overuse of arable lands can be avoided by encouraging farmers to adopt measures that protect the soil from erosion and enhance its fertility. Cover crops, conservation tillage, contour plowing, contour terracing, and other techniques can help to prevent

excessive erosion. Allowing fields to lie fallow periodically and applying organic soil amendments such as animal and green manure can help to maintain soil fertility. Measures that enhance soil fertility and structure also contribute to its resistance to erosion. Since many farmers lack knowledge of these and other techniques, it is imperative that the Department of Agriculture increase the budget allocated to its low-input sustainable agriculture (LISA) program. The increase would be more than offset by the savings realized in terms of soil conserved and reduced costs for synthetic fertilizers.

Conservation also means adopting irrigation practices that are more beneficial to the soil. For example, trickle drip irrigation uses less water by delivering it directly to the ground surrounding plant roots. The use of less water results in less salinization of the soil and also conserves our water resources.

Protect Human Health and the Environment

Preventing excessive soil erosion and limiting the use of fertilizers and pesticides help to protect human and environmental health. Erosion pollutes streams and waterways with sediments and chemicals. Limiting the use of synthetic inputs can only benefit the soil, aquatic ecosystems, wildlife, and humans.

Encouraging farmers to use integrated pest management (IPM) and organic techniques is critical; again, this encouragement must come from Department of Agriculture extension workers. Since some farmers will continue to use synthetic pesticides, attention must also be given to the safety and reliability of these substances. In its 1989 report called *Alternative Agriculture*, the National Academy of Sciences found that current regulations are stricter for new alternative pesticides than for those substances approved for use before 1972, even though many of the older pesticides are more toxic. Because these regulations can discourage farmers from switching to the newer and more environmentally benign alternative pesticides, they should be reexamined and revised where appropriate. The 1989 report also recommended revising federal grading standards that impose cosmetic criteria (for fruits and vegetables) that have little or no relevance to nutritional quality. The report concluded that alternative farming is both productive and profitable and that the nation's economy and environment would benefit if more farmers turned to organic agriculture.

Preserve Living Systems

An important step in preserving arable lands is to encourage polyculture and greater diversity on farms and to encourage removing erodible land from production. To do so will require removing government incentives that encourage monoculture and emphasize high per-acre yields. Currently, subsidies are offered only to farmers who grow the same crop on the same field for five consecutive years. Remember that subsidies are paid for with tax monies, and although some subsidy programs do lower the price of some agricultural commodities at the grocery store, consumers still pay for those lower prices when they pay their taxes. Existing commodity programs could be revised to allow farmers flexibility in diversifying their farms and rotating their crops. The programs could be revised to emphasize low per-unit-cost production rather than high per-acre yields. By encouraging farmers to keep down their costs, the government would enable those who choose to do so to convert to organic or LISA methods. At the same time the decreased use of fertilizers and pesticides would reduce pollution of water supplies. Subsidies could also be limited to farmers with annual incomes less than $100,000; this restriction would disqualify farms owned by large corporations, while ensuring that family farmers are provided with the help they need.

Making participation in the Conservation Reserve Program mandatory for all farmers is another means to help preserve the soil. While this proposal is likely to meet with resistance by many farmers, it is not an unreasonable stipulation. After all, agriculture is a business and, like any other industry, can and should be regulated when its activities pollute or otherwise degrade common resources. Hence, it is appropriate that the control of soil erosion, which pollutes rivers, lakes, streams, and air, be mandatory.

Summary

Soil, the topmost layer of the earth's surface in which plants grow, is an ecosystem composed of abiotic and biotic components. Humus, which consists of partially decomposed organic matter, helps to retain water and to maintain a high nutrient content, thus enabling soil to remain fertile. Soil contains billions of organisms that help to maintain soil fertility by aerating it and adding nutrients. The mineral content of soil determines its texture, which determines the air spaces within it and its ability to retain water. Soil structure, or tilth, is how soil particles are arranged, that is, how they cling together.

Scientists typically recognize ten major soil orders and an estimated 100,000 soil types. The soil types have been formed through the interaction of five factors: parent material, climate, topography, living organisms, and time. As soils develop, they form distinct horizontal layers called soil horizons. These layers include the O horizon, or litter layer; the *A* horizon, or topsoil; the *E* horizon, or zone of leaching; the *B* horizon, or subsoil; and the C-horizon, or parent material. The C horizon lies above the impenetrable layer of bedrock, sometimes called the *R* horizon. A vertical series of soil horizons in a particular location is a soil profile.

The way in which a particular piece of land is used is known as its land use. Worldwide, expanding urbanization poses a long-term threat to soil resources. An even greater threat to soil fertility and conservation is erosion. Erosion by wind and water is a natural process, but when undisturbed, soil is usually replaced faster than it erodes. The amount of soil that can be lost through erosion without a subsequent decline in fertility is known as the soil loss tolerance level (T-value), or replacement level.

In the United States no land use has been more important historically than agriculture. When the Europeans first arrived on the North American continent, they cleared seemingly inexhaustible forests for farmlands. As land became increasingly scarce and the soils in some areas became less productive, settlers began to move westward. The rich grassland soils of the Great Plains produced large yields for years, but soil fertility gradually declined. High grain prices after World War I and years of sufficient rainfall encouraged the plains farmers to plant millions of acres with wheat and to practice continuous cropping, or growing the same crop year after year without allowing the land to lie fallow periodically. In 1931 drought returned to the Great Plains, severely damaging over 10 million acres (4.1 million hectares) of farmland and creating the Dust Bowl.

On August 25, 1933, Congress formed the Soil Erosion Service, later renamed the Soil Conservation Service (SCS). One of its first projects was to conduct a nationwide soil survey to produce maps and data for erosion control. In the following decades, the federal government implemented programs that paid farmers to keep erodible land out of production.

In the 1950s and 1960s agricultural production was raised by the use of high inputs of chemicals, large machinery, and hybrid strains of crops. But high-input or conventional agriculture can cause serious problems, including soil degradation, soil erosion, reliance on fossil fuels, reliance on agrichemicals, groundwater depletion, overirrigation, loss of genetic diversity, and socioeconomic concerns. U.S. agriculture was gradually transformed from subsistence farming into agribusiness, large-scale farming, often with corporate land ownership, for as much profit as possible. For the most part, the U.S. Department of Agriculture and government programs have supported the trend toward highly specialized, energy-intensive, and environmentally questionable farming practices.

The 1980 Soil and Water Resources Conservation Act includes conservation strategies that would make soil conservation mandatory for any farm receiving federal farm aid. The Food Security Act of 1985 has been hailed as the cornerstone of the most progressive agricultural policy worldwide. In the mid-1980s the Department of Agriculture launched a research and extension program known as LISA, or low-input sustainable agriculture, to assist farmers in converting to low-input farming.

The search for effective methods to protect the soil and restore its fertility is resulting in the development of various systems of alternative or sustainable agriculture. Sustainable agriculture involves techniques such as growing a variety of crops, rotating crops, using organic fertilizers, allowing croplands to lie fallow periodically, planting cover crops, and encouraging the natural enemies of crop pests. Alternative methods of tilling help prevent soil erosion. Ridge tilling, planting crops on top of raised ridges, minimizes the use of pesticides and fertilizers. Conservation tillage includes both low-till and no-till methods of planting. Low-till planting is tilling the soil just once in the fall or spring, leaving 50 percent or more of previous crop residue on the ground. No-till planting, or stubble mulch farming, is sowing seeds without turning over the soil.

A number of techniques for sloped land can be used to reduce erosion, slow the flow of water, and add to organic content. Strip cropping is planting alternating rows of grain and low-growing leaf crops or sod. Contour plowing is tilling the soil parallel to the natural contours of the land rather than in straight rows. Contour terracing is building broad steps or terraces into a hillside. Swales are trenches along the edges of terraces; when planted with grass, they are called grassed waterways.

Of all forms of agriculture, organic farms most closely resemble natural systems. They are characterized by diversity. Soil texture and fertility are enhanced by organic soil amendments. Insect pests are controlled through a variety of methods collectively known as integrated pest management, or IPM. Biological pest control involves the introduction of insects, birds, animals, or microbes that attack pests. Weeds are controlled by mulching and cultivating techniques. Crop rotation prevents the buildup of pest populations and the depletion of soil nutrients. A new agricultural approach called perennial polyculture, in which a number of perennial crops are established in an area, also renders agriculture less environmentally harmful.

Key Terms

agribusiness	perennial polyculture
agrichemical	ridge tilling
conservation tillage	salinization
contour plowing	soil fertility
contour terracing	soil horizon
cover crop	soil loss tolerance level (T-value)
crop rotation	soil productivity
green manure	soil profile
high-input agriculture	strip cropping
humus	sustainable agriculture
integrated pest management (IPM)	tilth
land use	topsoil
parent material	trickle drip irrigation

Discussion Questions

1. Describe the texture, structure, and profile of a fertile soil.

2. What factors affect the formation of soil? Briefly describe how these factors interact to form soil.

3. What effects can land use have on soil resources? On water resources? On air resources? What conclusions can you draw about these three resources?

4. Briefly describe some of the problems associated with high-input agriculture. How can sustainable farming techniques reduce or solve these problems?

5. Explain the concept of perennial polyculture. What are the benefits of this form of agriculture?

6. Imagine you have been asked to design a farm preservation program for a farming community in your state. Would you model your program after the one in Marathon County? Why or why not?

An Environmental Pandora's Box

Managing the Products of Industrial Societies

Mineral Resources

The entire lithosphere is an intricately interwoven fabric of many crystals: rocks and sand, gold and tin, diamonds and ice. Crystallization is perhaps the most important single process that creates the world we know. It is an expression of the supremely logical structure underlying all things—an orderliness usually hidden from our sight beneath the ever-changing masses of clouds and soil and sea.

Louise B. Young

At the end of the minerals- and energy-intensive development path taken by today's industrial nations lies ecological ruin. Mining enough to supply a world that has twice as many people, all using minerals at rates that now prevail only in rich countries, would have staggering environmental consequences.

John E. Young

Learning Objectives

When you finish reading this chapter, you should be able to:

1. Explain how minerals formed in the earth's crust.
2. Distinguish between proven and recoverable mineral resources.
3. Describe the steps and environmental consequences of the mining process.
4. Compare mineral availability and use in more-developed and less-developed countries.

No new matter is ever created or destroyed; the material of the earth is merely transformed. Geologic processes transform minerals over thousands and millions of years. Humans, too, have learned to transform earth materials. We extract minerals from the earth and fashion many goods from them. Consider the automobile you drive, the brick walls of your school, the concrete roadway you drive on, and the silicon chips that empower your computer. The matter of which these objects are made was present at the birth of the planet, and it will be present one hundred years, five hundred years, a million years, from now.

In this chapter we look at the formation of the earth, the nature and formation of minerals, the use and classifications of minerals, and patterns of mineral production and consumption worldwide. We then examine the mining process and how it affects the environment and human health. We look at the historical significance of minerals, economic factors affecting their production, the international minerals trade, and U.S. management of its domestic and foreign mineral supplies. We close with a discussion of ways in which we can conserve mineral resources.

Describing Mineral Resources: Physical and Biological Boundaries

How Did the Earth Form and How Does It Change?

We tend to think of inorganic matter as immutable, unchanging. A rock is a rock is a rock, so to speak. But nothing could be further from the truth. A witness to the birth and development of the earth would tell a story of change and transformation, a tale of process and becoming (Figure 17-1).

Scientists theorize that when our solar system developed from a cosmic cloud about 4.6 billion years ago, small concentrations of matter began to condense at different distances from the newly formed sun, giving rise to the planets. The composition of the planets depended upon the matter that could condense at the different distances. The lighter, more volatile elements, unable to condense

FIGURE 17-1: Mount Pinatubo in the Philippines. Events such as the 1991 eruption of Mount Pinatubo remind us that the transformation of the earth is an ongoing process.

within the earth's crust or within the upper reaches of the mantle, the thick band that lies between the planet's core and crust.

Igneous rock is one of two predominant rock types on earth; the other is sedimentary rock. **Sedimentary rock** is formed by the deposit of small bits and pieces of matter, or sediments, that are carried by wind or rain and then compacted and cemented to form rock. Sediments are continually eroded from the continents by atmospheric forces (water, wind, and ice) and chemical action.

Most sediments are deposited in the seas along the continental margins. As the piles of sediments grow larger, increasing pressure and rising temperatures produce physical and chemical changes in the underlying sedimentary rock. If the sedimentary pile is thick enough, material near the bottom may melt and form magma. Since the magma is less dense than the parent (sedimentary) rock, it tends to rise, and is forced through the rock from which it origi-

at high temperatures, were driven outward by the sun's heat. Eventually these elements and compounds cooled and became part of the planets in the far reaches of the solar system. The inner planets—Mercury, Venus, Earth, and Mars—were formed of denser, less volatile particles of stardust. Mercury and Venus formed in particularly hot portions of the cosmic cloud where water was unable to condense. Earth formed in a portion of the cloud where water was able to condense into a liquid form and remain on the planet's surface in that state.

Throughout its first few million years, the earth underwent rapid changes. Gravity drew the planet's matter toward the center. The tremendous heat released by decaying radioactive elements within the earth's mass caused the solid materials to melt, and the most dense minerals, such as iron, flowed inward to form the planet's metallic core. **Magma,** or melted rock, which contained less dense minerals, floated to the top, where it cooled and solidified, or crystallized, into **igneous rock.** These rocks formed the earth's solid crust (Figure 17-2). Materials that cooled below the earth's surface solidified more slowly.

In some places such intense pressure builds up within the earth that magma breaks through the surface. We call these sites volcanoes, and the magma that erupts from them is known as lava. Igneous rock is continually being formed by the cooling and crystallization of magma from deep

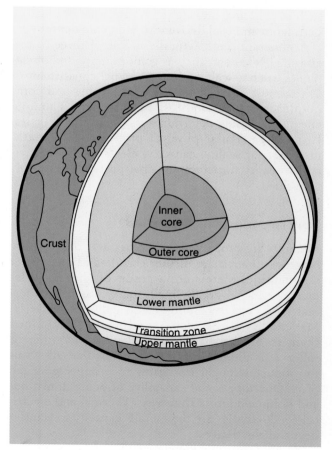

FIGURE 17-2: Internal structure of the earth. The planet's mantle and crust surround its dense, metallic core. The mineral deposits exploited by humans lie within the earth's crust.

nated. As the magma rises, it cools, forming new igneous rocks.

A third, less common type of rock is metamorphic rock. **Metamorphic rock** forms when rocks lying deep below the earth's surface are heated to such a degree that their original crystal structure is lost. Different crystals form as the minerals that compose the rock cool. For example, limestone that has been heated and recrystallized forms marble.

Rock-forming processes demonstrate that over hundreds of millions of years the earth refigures itself, transforms itself anew, so that the composition of rock changes over the ages. Even now, the earth is changing—mountains are forming and the pressure that gives rise to earthquakes and volcanoes is building.

What Are Minerals?

Minerals are nonliving, naturally occurring substances with a limited range in chemical composition and with an orderly atomic arrangement. This definition encompasses a wide range of substances— basic materials such as granite and sand, nutrients such as phosphorus and nitrogen, precious metals such as gold and silver, and energy resources such as petroleum and coal derived from organic matter.

A mineral occurs naturally in one of three forms: as a single element, such as gold; a compound of elements, such as calcite (the principal constituent of limestone); or an aggregate of elements and compounds, such as asbestos. Each mineral's unique chemical formula determines its physical properties. It is for particular physical properties—strength, insulating or sealing capacity, electrical conductivity, or beauty for example—that particular minerals are valued.

How Are Minerals Classified?

Minerals are broadly classified as fuels or nonfuels. Nonfuel minerals are further classified as metallic or nonmetallic. Metals are a group of chemical elements characterized by malleability, ductility, thermal conductivity, and electrical conductivity. Malleability is the property of metals that allows them to be shaped or worked with a hammer or roller. It is the malleability of metals that allows us to produce wrought iron gates and copper pots and pans. Ductility is the capability of being drawn out and fashioned into a thin wire. Thermal and electrical conductivities are the metals' capacity to transfer heat and electricity.

Metallic minerals are classified as ferrous or nonferrous. **Ferrous metals** contain iron or elements alloyed with iron to make steel. **Nonferrous metals** contain metallic minerals not commonly alloyed with iron.

Metals are also sometimes classified by their abundance or scarcity in the earth's crust. Just eight elements make up more than 99 percent, by weight, of the crust. Oxygen, found in combination with other elements to form minerals, accounts for about 46 percent of the earth's crust by weight. The next seven most common elements are silicon, 28 percent; aluminum, 8.3 percent; iron, 5.6 percent; calcium, 4.2 percent; sodium, 2.4 percent; magnesium, 2.3 percent; and potassium, 2.1 percent. **Abundant metals** are those which account for more than 0.1 percent of the earth's crust, by weight. They include aluminum, iron, magnesium, and manganese. Copper, lead, zinc, tin, tungsten, chromium, gold, silver, platinum, uranium, and mercury comprise less than 0.1 percent of the earth's crust and are known as **scarce metals.**

Nonmetals are simply those elements that do not possess the characteristics of metals. They encompass a wide variety of resources. Industrial materials include sulfur, salts, fertilizers (composed of nitrogen, phosphorus, and potassium), abrasives, asbestos, and industrial diamonds. Structural materials include stone, cement, sand, and gravel.

How Are Mineral Deposits Formed?

Geologic processes are responsible for the formation of minerals. As magma rises from the mantle or from deep within the earth's crust, it begins to cool. The various minerals constituting the magma crystallize at different temperatures; materials with a high melting point crystallize first. The heavier crystals sink and gradually form solid mineral deposits. As the lighter liquid magma continues to rise, it continues to cool, and other minerals begin to crystallize. In this way varied deposits of minerals form throughout the mantle and crust.

Mineral deposits are not found in the crust in as orderly an arrangement as the preceding description might indicate. Deposits of one mineral are often mixed with deposits of others. There are several reasons for the dispersion of mineral deposits. Fractures in the earth's crust divide it into six huge tectonic plates and a number of smaller ones. As newly formed igneous rock rises to the crust's surface at major fracture zones beneath the oceans, it pushes the plates apart. The side of the plate farthest away from the spreading zone eventually bends and slides beneath the plate it is being pushed against by the movement. Although scientists do not completely understand the details of this process, many believe that the distribution of mineral deposits is related to the past and present fractures that are the boundaries of the moving plates.

Weathering and erosion disperse mineral deposits, transporting some particles far from their parent rock. Chemical separation is responsible for forming many mineral deposits as substances are chemically dissolved and removed in solution. Gravity also plays a part. Heavier minerals do not remain suspended in solution as long as lighter minerals and particles. Instead, they tend to sink to the bottom of the fluid in which they are trapped and form deposits.

Where Are Minerals Found?

Minerals can be found everywhere—the oceans, the highest mountain peaks, the air—but they are typically bound up in rock within the earth's crust. Minerals are not evenly distributed; they differ in kind and amount from place to place.

To be useful, a mineral must be profitable to extract. Most are present in most places in concentrations far too low to make their extraction profitable. Where a concentration is high enough to make mining economically feasible, the mineral deposit is known as an ore. Bauxite, for instance, is an aluminum-containing ore. If the amount of mineral per given volume is high, it is a high-grade ore; if the amount of mineral per given volume is low, it is a low-grade ore. Naturally, high-grade ores are more profitable to mine than low-grade ores.

Some nations are mineral-rich and others are mineral-poor. The United States, the former Soviet Union, Canada, Australia, and South Africa are the world's major producers of many nonfuel minerals (Table 17-1). Less-developed countries in the Andean region and west coast of South America, central Africa (especially a belt running through eastern Zaire and central Zambia), and Southeast Asia are the leading producers of tin, manganese, cobalt, and chromium. Mineral production is not limited to these regions (Jamaica and Morocco, for example, are major producers of bauxite and phosphates, respectively), but these areas do have large concentrations of important minerals.

Table 17-1
Major Mineral-Producing Countries, 1990

Mineral	Country	Share in World Production (percent)
Bauxite	Australia	37
	Guinea	16
Chromium	South Africa	32
	Soviet Union	32
Cobalt	Zaire	58
	Zambia	16
Copper	Chile	17
	United States	17
Gold	South Africa	30
	United States	15
Iron Ore	Soviet Union	26
	Brazil	17
Lead	Australia	16
	United States	15
Manganese	Soviet Union	36
	South Africa	16
Molybdenum	United States	53
	Chile	15
Nickel	Soviet Union	23
	Canada	22
Phosphate Rock	United States	28
	Soviet Union	24
Platinum Group	South Africa	48
	Soviet Union	45
Silver	Mexico	17
	United States	14
Tin	Brazil	24
	Malaysia	14
Titanium	Soviet Union	46
	Spain	25
Tungsten	China	52
	Soviet Union	21
Zinc	Canada	17
	Australia	13

Source: U.S. Bureau of Mines, *Mineral Commodity Summaries 1991* (Washington, D.C.: 1991).

Social Boundaries

How Are Minerals Used?

Minerals play vital roles in technological processes: iron and coal in steel production; aluminum in the manufacture of such diverse products as cans and automotive parts; nitrogen and phosphorus as ingredients in fertilizers; titanium, manganese, cobalt, magnesium, platinum, and chromium in industrial processes and as components in aircraft, automobile engines, and other high-tech applications; limestone, gravel, sand, and crushed rock in construction and transportation (Table 17-2).

The more-developed countries consume the greatest share of mineral resources. Together, the MDCs, with just about a quarter of the world's population, use about three-quarters of the annual global production of nonfuel minerals. On average, a person in the developed world uses 50 times as much minerals per year as a person in the developing world. The United States alone, having just 5 percent of the world's population, produces about 11 percent, and consumes more than 13 percent, of the global nonfuel mineral production (Figure 17-3). If fuel minerals are figured into the calculation, U.S. consumption of global mineral production rises to 30 percent. Iron is one exception; iron consumption is more evenly spread among nations, in part because several developing countries, including Brazil, India, and Korea, are major iron and steel producers. By

▶ Table 17-2
Principal Industrial Uses, Selected Minerals

Mineral	Industrial Uses
Aluminum	Packaging (39%), transportation (20%), building (14%), electrical (8%), consumer durables (8%), other (11%)
Beryllium	Nuclear power, aerospace (40%), electrical (36%), electronic components (17%), other (7%)
Chromium	Metallurgical (52%), chemical (33%), refractory (15%)
Cobalt	Superalloys (37%), magnetic materials (16%), driers (11%), catalysts (10%), cutting and mining bits (7%), other (19%)
Copper	Refined metal fabrication (80%), other (20%)
Industrial diamonds	Machinery (27%), stone and ceramic products (22%), abrasives (16%), construction (13%), mineral service (8%), transportation (6%), other (8%)
Gold	Jewelry and art (61%), industrial (29%), dental (9%), small bars (1%)
Lead	Batteries and gasoline additives (75%), construction, paint, ammunition (2%), other (5%)
Molybdenum	Iron and steel production (75%); machinery, oil and gas industry, transportation, chemical, electrical (25%)
Nickel	Stainless and alloy steel production (45%), nonferrous alloys (30%), electroplating (15%), other (10%)
Platinum group metals	Automotive (33%), electrical (28%), chemical (15%), dental (9%), other (15%)
Silver	Photographic (39%), electrical (29%), silverware and jewelry (14%), alloys and solders (7%), other (11%)
Tin	Containers (25%), electrical (17%), construction (13%), transportation (14%), other (32%)
Tungsten	Metalworking and construction machinery (72%), transportation (11%), lighting (8%), electrical (5%), other (4%)
Zinc	Construction (40%), transportation (20%), machinery (12%), electrical and chemical (15%), other (13%)

Source: U.S. Bureau of Mines, Mineral Commodity Summaries, 1983.

the year 2000 global demand for most major minerals is expected to double. Two important factors will influence the rate of increase of global demand: the growing human population and its reliance on mineral resources to support a rising standard of living worldwide.

The more developed countries rely on about 80 minerals. Three-quarters of these either exist in abundant supply to meet our anticipated needs or can be replaced by existing substitutes. **Critical minerals** are those considered essential to a nation's economic activity; **strategic minerals** are those considered essential to a nation's defense. At present, there are no suitable alternatives for critical or strategic minerals. Cobalt, for example, is needed to produce high-strength, high-temperature alloys used in the aerospace industry, and platinum is unrivaled as a catalyst.

Scarcity of minerals is largely due to technical and economic constraints rather than to the absolute constraint of finite resources. For example, the United States relies on imports of four critical and strategic minerals (chromium, cobalt, manganese, and the platinum group metals) from countries in politically volatile regions such as central and southern Africa. Because the United States has no developed domestic reserves of these minerals, the possibility of a sudden disruption in supply always exists. Environmental Science in Action: Critical and Strategic Minerals (pp. 363–367) examines these four vital mineral resources.

How Is the Size of Mineral Deposits Estimated?

An estimate of the total sum of a mineral found in the earth is called the **resource base**. Because of the earth's tremendous mass, the total tonnage of any one mineral, even those defined as scarce, is great. But most of these minerals are inaccessible, for they lie within the core and mantle of the planet. Hence, the resource base of a given mineral is a highly theoretical figure, and for most minerals it does not indicate how much is ever likely to be available for use.

A mineral deposit that can be extracted profitably with current technology is called a **proven reserve** or economic resource. Proven reserves have been explored, measured, and inventoried. We know where to find proven reserves, we know how to extract and exploit them, we know how much it will cost to recover them, and we know we can make a profit on the recovery. **Subeconomic resources** are reserves that have been discovered, but cannot yet be extracted at a profit at current prices or with current technologies.

Proven reserves are an estimate of what is available and profitable now. An estimate of the total amount of a given mineral that is likely to be available for future use is called the **ultimately recoverable resource**. However, because it is based on assumptions about discovery rates, future costs, market factors, and advances in extraction and pro-

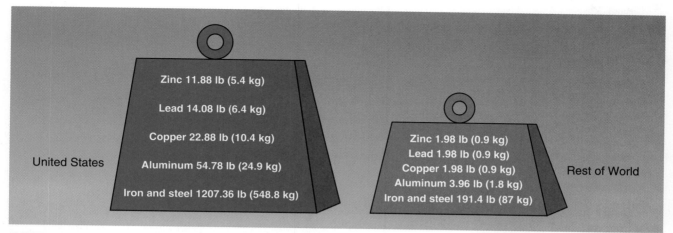

FIGURE 17-3: U.S. annual per capita consumption of selected major minerals compared with consumption by the rest of the world.

cessing technologies, such an estimate is difficult to determine.

How Long Will Mineral Reserves Last?

It is also difficult to estimate how long mineral supplies will last because of the variables involved. For salt, magnesium metal, lime, and silicon, for instance, the estimated life expectancy is at least in the thousands of years. But as Table 17-3 indicates, some mineral resources may shrink significantly by the year 2030 if the estimated world population of 10 billion consumes resources at current U.S. rates.

It is unlikely that we will exhaust mineral resources. First, as proven reserves become depleted, the cost is likely to rise. High prices and high demand encourage more exploration, leading to the discovery of more ores and increasing the size of proven reserves. They also encourage conservation measures, the search for technological advances to make it possible to mine previously unprofitable deposits, and the search for suitable substitutes. These activities ease pressure on dwindling mineral supplies.

What Are the Steps in the Mining Process?

Optimists claim that newly discovered deposits, substitutes, and technological advances will ensure the continuation of our industrial way of life. But the exploitation of lower grade ores and less accessible deposits requires more energy and water and causes more environmental damage. For example, four hundred years ago, copper ores commonly contained about 8 percent metal. Today, the typical copper ore contains less than 1 percent.

It is a long way from a raw mineral to a finished product. The mineral ore must be located, extracted, and processed before it is ready for use. Let us look at the process of mining in order to understand the associated economic, energy, and environmental costs (Table 17-4).

Location. The image of the old time prospector, armed with a pick and trusty mule, is part of our frontier folklore, but the prospectors of yesteryear would certainly not recognize contemporary exploration techniques. Modern exploration relies on a

▶ Table 17-3
Estimated Years Until Depletion, Selected Minerals

	Based on Current Consumption Rates		Based on 2030 Projections[a]	
	Reserves	Resources	Reserves	Resources
Aluminum	256	805	124	407
Copper	41	277	4	26
Cobalt	109	429	10	40
Molybdenum	67	256	8	33
Nickel	66	163	7	16
Platinum group metals	225	413	21	39

[a]Estimates based on the assumption that in 2030 a population of 10 billion will consume resources at current U.S. rates. If LDCs increase their consumption to match that of MDCs, which they will do if they follow the development path of the industrialized nations, global stocks of minerals will dwindle rapidly.

Source: Adapted from illustration (modified) by Ed Bell from page 146 of "Strategies for Manufacturing" by Robert A. Frosch and Nicholas E. Gallopoulos. Copyright © 1989 by Scientific American, Inc. All rights reserved

▶ Table 17-4
Costs Associated with Locating, Extracting, and Processing Minerals

Exploration

Has no direct effects on human health

Has minimal direct effects on the environment

Requires energy to locate mineral deposits

Includes drilling for test samples, which can have adverse effects on wildlife and water tables in sensitive ecosystems

Extraction

Puts miners at risk for diseases caused by the substances they work with (for example, black lung from coal dust)

Puts miners at risk from mine collapse, underground explosions and fires, and other safety hazards

Disturbs soil and overlying vegetation, which disrupts ecosystems, reduces productivity, and leads to soil erosion

Causes siltation of streams, lakes, and rivers by eroding soil

Can diminish or reduce productivity of land

Degrades water used in mining processes

Produces tailings and mine drainage that can contaminate soil and water

Processing

Puts workers at risk for certain kinds of cancer in some smelting industries

Produces tailings that can pollute air, soil, and water

Produces emission of pollutants into air

Contaminates large quantities of water

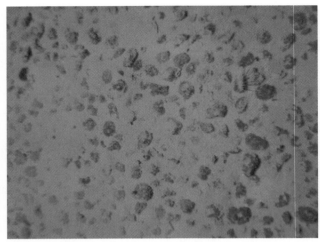

FIGURE 17-4: Manganese nodules on the floor of the Pacific Ocean.

knowledge of geology, particularly of crustal movements and the formation of mineral deposits, and sophisticated instruments. Some instruments, such as remote sensing devices carried aboard satellites, can identify rock formations or mounds indicative of the presence of mineral deposits. Others can detect changes in the earth's magnetic field or the field of gravity that may be caused by concentrated mineral deposits. After a potential site is identified, rock samples are taken and analyzed for their mineral content.

Most of the high-grade ores, particularly those located in industrial nations, have already been identified and exploited. The ocean floor and open waters are two potential sources of minerals, but the latter is not a likely candidate for profitable mining soon. With few exceptions, such as sea salt, minerals are not present in high enough concentrations in seawater to make their extraction profitable or practical. In contrast, the ocean floor holds real potential for profitable mining. Manganese nodules approximately the size of potatoes have been found in abundance on the ocean floor in the deep waters of the Pacific (Figure 17-4). The nodules also contain lesser quantities of iron, cobalt, and copper. Concentra-

tions of nickel have also been identified in the seabed.

Antarctica is another potentially significant source of minerals. Substantial deposits of petroleum and iron ore have been located; methane, copper, silver, and nickel deposits may also be locked below the continent's surface, which lies beneath ice sheets up to a mile thick. Exploiting the continent's resources would be difficult, however. Antarctica has the harshest climate in the world; the average temperature in the interior during the coldest months of the year is -96°F (-71°C). It may be too difficult or too expensive to conduct mining operations in such an inhospitable environment. Even if the economics proved favorable, the environmental consequences of mining in Antarctica may be devastating. Extracting and transporting oil in the antarctic seas, the world's roughest, pose the danger of oil spills, which would threaten most of the marine environment's animal life (krill, seals, and penguins, for example). At such cold temperatures oil takes far longer to decompose. Also, a spill on ice would increase the ice's heat absorption, possibly causing it to melt. In 1991 environmental concerns prompted the countries that jointly govern the continent to continue a moratorium on mineral development (a topic covered more fully in Chapter 26), but many people fear that rising prices and minerals scarcity could lead to pressure to develop Antarctica's mineral wealth in the future (Figure 17-5).

Exploration is the most benign step in the mining process; the effect on the environment is usually minimal. However, energy is needed to locate mineral deposits. The more remote or inaccessible an area, the more energy required for exploration and study. Further, excavating and drilling for test samples, especially if done to a large extent in a limited area, can have adverse effects on sensitive ecosystems and on the local water table.

FIGURE 17-5: Antarctic landscape. Scientific research and tourism are the only activities currently permitted on the Antarctic continent. Many people are hoping to make Antarctica a world park off limits to war and commerce.

Extraction. Extracting mineral ores is the step we commonly think of when we hear the word *mining.* **Extraction** is the process of separating the mineral ore from the surrounding rock in which it is embedded. An ore may be extracted by surface mining or subsurface mining techniques.

Surface mining techniques are used to extract deposits located relatively near the earth's surface. About 90 percent of the ores mined in the United States are extracted by surface mining. The overlying vegetation, soil, and rock layers, collectively known as the **overburden,** are removed to expose the ore deposit. Surface mining results in more waste than does subsurface mining. The U.S. Bureau of Mines reported that, in 1988, surface mines produced 11 times as much waste per ton of ore as did subsurface mines. Further, of the 3.3 billion tons (3 billion metric tons) of material handled at nonfuel mines in 1988, over one-third of it was overburden.

There are three types of surface mining: open pit surface mining, area strip mining, and contour strip mining.

As its name implies, **open pit surface mining** consists of digging a large pit and removing the exposed ore. This technique is commonly used to extract sand, stone, gravel, copper, and iron. At 2,322 feet (696 meters) deep, Utah's Bingham Canyon copper mine is the largest human excavation in the world; about 3.3 billion tons (3 billion metric tons) of material have been removed from the mine. The extensive pits created in the process of extracting limestone, granite, and marble are known as quarries.

Area strip mining is typically used in flat or rolling terrain. It is most often used to mine coal in the West and Midwest and phosphate rock in Florida, North Carolina, and Idaho. Bulldozers dig a trench and power shovels remove the ore. A second trench is then dug parallel to the first, and the overburden is removed and deposited in the first trench. Once the ore has been removed, a third trench, parallel to the second, is dug, and the process repeated. If the land is not reclaimed, the result is a series of rolling hills, known as spoil banks, formed of rubble. The spoil banks, which resemble ocean waves or swells, are highly erodible.

Contour strip mining is used to extract deposits in hilly or mountainous terrain. In the United States, contour strip mining is used chiefly by coal miners in the Appalachian region. A power shovel cuts a series of terraces into a hillside, and the ore is removed at each terrace. The overburden from each new terrace is dumped onto the terrace below. If the land is not reclaimed, the result is a steep and highly erodible bank of rock and soil fronted by a wall of dirt. If the overburden is too thick to remove economically, huge drills called augurs are used to burrow horizontally into the mountainside.

When mineral deposits lie deep underground, **subsurface mining** techniques must be used. For coal and metallic minerals a vertical shaft is dug to reach the level of the deposit, tunnels and rooms are blasted in the rock, and the ore is extracted and hauled to the surface. For oil and natural gas a well is drilled into a reservoir deep underground, and the gas or oil, which is under pressure, rises to the surface. As an oil reservoir becomes depleted, heated water or a gaseous substance may be injected into the well to force the remaining oil to the surface.

Extraction degrades the environment in many ways. Mining disturbs the overlying soil and vegetation and can cause erosion. Eroded soil carried into nearby ponds or streams causes siltation, which poses a threat to fish populations. Because it rearranges the top layers of soil, strip mining can lower the productivity of the land. Many types of mining require large amounts of water, which is unfit for consumption or use in agriculture or recreation after it is used in the mining operation.

As miners remove an ore deposit from the ground, they discard the rock in which it is embedded. These materials, called **tailings,** are dumped in large unsightly piles or in nearby ponds. Tailings may contain materials, such as asbestos, arsenic, lead, and radioactive substances, that are a health threat to humans and other living organisms. Mine drainage may contain hazardous substances, such as sulfuric acid or cyanide, that can contaminate nearby surface waters and move into groundwater (Figure 17-6).

Extraction can also adversely affect human health. Miners are at risk for developing diseases caused by the substances with which they come into contact,

FIGURE 17-6: Castle Mountains in the Mojave Desert, California. Mining companies have proposed mining low-grade gold deposits in the Castle Mountains by a process known as heap leaching. Huge amounts of ore are drenched in cyanide, which draws, or leaches, gold flecks out of the ore. Toxic contamination is a serious concern, since cyanide is a deadly poison.

for example, the black lung disease of coal miners and the lung disease and bronchial cancer of asbestos workers. The excavating operation is hazardous, with risks of underground explosions, fires, mine collapse, and accidents associated with heavy equipment. Modern techniques and safety precautions have reduced, but not eliminated, these hazards.

Processing. Processing consists of separating the mineral from the ore in which it is held and concentrating and refining the separated mineral. Processing plants, called smelters, are usually located near mining sites to reduce the cost of transporting raw ores.

The ore is physically or chemically treated to yield the desired mineral. For example, after a metal ore is mined, it is crushed, run through a concentrator to physically remove impurities, and reduced to a crude metal (that is, melted) at high temperatures. Refining removes any remaining impurities. The wastes from refining, like the wastes from extraction, are called tailings and are disposed of in the same way. Tailings removed by refining are typically fine-grained particles. When piled on the ground in the open air, they can become airborne, contributing to air pollution. Because the material is so finely ground, contaminants that were bound up in solid rock—arsenic, cadmium, copper, lead, and zinc, for example—are released to air and water. Tailings may also contain traces of toxic organic chemicals used in ore concentrators. Smelters also emit substantial amounts of air pollutants, particularly soot, sulfur

oxides, and toxic particulates, such as arsenic, lead, and cadmium. In the western United States, pollutants released from copper smelters have been shown to contribute to acid rain and to reduce visibility in many national parks and recreational areas (Figure 17-7).

Processing ores causes soil and water pollution. Contaminants from tailings leach into the ground and water supplies. The huge amount of water used to process ores is contaminated with impurities. The liquid and solid hazardous wastes that are by-products of ore processing must be safely disposed of or converted into less harmful materials.

Processing also affects human health. Workers in some smelting industries are at increased risk for certain kinds of cancer. Arsenic smelter workers, for instance, have a lung cancer rate three times greater than average. Workers in cadmium smelters have twice the average lung cancer rate, and workers in lead smelters have higher than average rates for both lung and stomach cancer.

Summary: Environmental Impact of Mining. Mining significantly disturbs large areas of land, degrades water and air resources, and poses human health risks. Countries with a long history of mining bear inevitable scars. In the United States, for example, 48 of the 1,189 sites on the Superfund cleanup list are former mining sites. The largest Superfund site, the Clark Fork Basin site in Montana, stretches 136 miles (220 kilometers) along Silver Bow Creek and the Clark Fork River. For more than 100 years, copper mining and smelting were conducted at the site.

FIGURE 17-7: Copper mine and smelter, Morenci, Arizona. Atmospheric deposition from copper smelters contributes to air pollution in the Golden Circle of the Southwest, a region encompassing five national parks (Grand Canyon, Bryce Canyon, Canyonlands, Capitol Reef, and Arches) and two national recreational areas (Lake Mead and Glen Canyon).

Table 17-5
Selected Examples of Environmental Impacts of Minerals Extraction and Processing

Location/Mineral	Observation
Ilo-Locumbo Area, Peru copper mining and smelting	The Ilo smelter emits 660,000 tons (600,000 metric tons) of sulfur compounds each year; nearly 1.41 billion cubic feet (40 million cubic meters) per year of tailings containing copper, zinc, lead, aluminum, and traces of cyanides are dumped into the sea each year, affecting marine life in a 49,420-acre (20,000-hectare) area; nearly 880,000 tons (800,000 metric tons) of slag are also dumped each year.
Nauru, South Pacific phosphate mining	When mining is completed—in 5-15 years—four-fifths of the 5,190-acre (2,100-hectare) South Pacific island will be uninhabitable.
Pará state, Brazil Carajás iron ore project	The project's wood requirements (for smelting of iron ore) will require the cutting of enough native wood to deforest 123,550 acres (50,000 hectares) of tropical forest each year during the mine's expected 250-year life.
Russia, Former Soviet Union Severonikel smelters	Two nickel smelters in the extreme northwest corner of the republic, near the Norwegian and Finnish borders, pump 330,000 tons (300,000 metric tons) of sulfur dioxide into the atmosphere each year, along with lesser amounts of heavy metals. Over 494,200 acres (200,000 hectares) of local forests are dying, and the emissions appear to be affecting the health of local residents.
Sabah Province, Malaysia Mamut Copper Mine	Local rivers are contaminated with high levels of chromium, copper, iron, lead, manganese, and nickel. Samples of local fish have been found unfit for human consumption, and rice grown in the area is contaminated.
Amazon Basin, Brazil gold mining	Hundreds of thousands of miners have flooded the area in search of gold, clogging rivers with sediment and releasing an estimated 110 tons (100 metric tons) of mercury into the ecosystem each year. Fish in some rivers contain high levels of mercury.

Source: Worldwatch Institute

Even with regulations in place to control its environmental impact, mining continues to adversely affect human and environmental health. In countries that do little to regulate mining operations, particularly less-developed countries, mining creates environmental disaster areas. Table 17-5 presents selected examples of the environmental impact of minerals extraction and processing worldwide.

History of Management of Mineral Resources

What Is the Historical Significance of Minerals?

The discovery and use of minerals parallel the development of civilizations. Humans have used minerals since the day they discovered that they could fashion tools out of stone. In fact, many eras in human prehistory are named after the minerals, especially metals, that were first developed and widely used during that period—the Stone Age, the Bronze Age, the Chalcolithic (copper) Age, and the Iron Age. Many famed achievements of the world's civilizations are directly related to the use of minerals: the pyramids of Egypt, the aqueducts of Rome, and the artifacts of the ancient Mayan and Incan civilizations.

The Industrial Revolution ushered in an age of intense use of both fuel and nonfuel minerals. Between 1750 and 1900, global minerals use rose tenfold; since 1900, it has increased over thirteenfold. The world's yearly production of pig iron, the crude metal typically converted into steel, is now 22,000 times greater than it was in 1700. The current production of copper and zinc are 560 and 7,300 times greater, respectively, than output in 1800.

We are currently in the "Energy Age," an era supported by our exploitation of fuels. Our life-styles depend on the use of fuels and other minerals, and our consumption of minerals depends on our attitude toward the earth's resources. The "throwaway" mentality that dominates Western societies encourages wasteful use of these nonrenewable resources. (Changing that mentality in order to make the transition to a sustainable society is the focus of Chapter 20, "Unrealized Resources: Waste Minimization and Resource Recovery".)

How Do Economic Factors Affect Mineral Production and Consumption?

Market supply and demand determine the price of minerals. Supply and demand, in turn, are influenced by the needs and uses of society, technological constraints, and economic forces. For example, after 1974 the demand for most metals slowed. Producers of metals did not anticipate this slowed demand, and as a result, supply exceeded demand and the price of metals fell.

While metal consumption is still growing, it is growing at a slower rate than before 1974. Two factors are generally given to explain the downturn in global demand for metals. First, slow world economic growth has reduced the demand for metal products such as heavy machinery and equipment. Second, technological changes have reduced the amount of metal used in goods; the composition of many products is changing as the use of plastics and fiberglass increases. Additionally, improved technologies use less metal to develop a product of equal or superior quality; metals use is thus becoming more efficient. While efficiency will help to conserve metal supplies, producers must readjust their production strategies to remain competitive.

Supply and demand are also influenced by resource scarcity. Resource scarcity is related to political, economic, and technical factors, not actual reserves. The oil embargo of 1973 by the Organization of Petroleum Exporting Countries raised fears of similar crises involving other "minerals cartels," but they did not materialize. Most experts agree that producer nations are unlikely to form cartels since they often depend on their mineral exports to secure foreign exchange. Botswana, Zambia, Liberia, Jamaica, Togo, Mauritania, Bolivia, Zaire, Chile, Peru, and Papua New Guinea derive a substantial portion of their foreign exchange from mineral exports. Such nations are unlikely to take any action that could disrupt exports, since a decline in export trades, caused by decreased demand due to rising prices, would almost certainly damage their economies.

Mineral prices tend to fluctuate wildly. For instance, after the collapse of the International Tin Council in late 1985, the price of tin fell from $12,000 per metric ton to just over $5,000 per metric ton in just six months. However, there does seem to be a long-term trend in overall prices. The economic value attached to commodity exports like minerals and agricultural products has fallen relative to the value of manufactured goods; in other words, the price of manufactured goods has risen more than the price of minerals. Consequently, mineral-exporting nations, particularly LDCs, must sell increasing amounts of minerals to "pay for" items such as tractors and fertilizers, usually exported by industrial nations. Strategic and critical minerals are exceptions to this long-term trend; because there are no substitutes for them at present, their price is likely to rise.

What Is the International Minerals Industry?

The geologic dispersal of minerals ties together the economies of producer nations and consumer nations. The international minerals industry refers to the global trade in these resources.

Many industrialized nations import most or all of many important mineral resources. For instance, the United States, Japan, and the nations of the European Economic Community (EEC) rely on imports to meet more than 50 percent of their needs for copper, nickel, lead, tin, zinc, cobalt, iron ore, manganese, and chromium. In part because the former Soviet Union was nearly self-sufficient in minerals, the Soviet Union and eastern Europe were not major participants in the international minerals trade, but that situation may change with the dismantling of the eastern bloc and the Soviet reorganization.

There are two distinct groups of mineral-producing nations. The United States, Canada, Australia, and South Africa are all part of the industrialized world; they are major producers of a wide range of nonfuel minerals. The other mineral producers are LDCs, generally with one or two minerals as their primary source of income, accounting for the bulk of their export earnings. For example, copper accounts for over 90 percent of Zambia's export earnings. Chile, Peru, and Zaire also rely heavily on copper exports. Indonesia, Bolivia, and Malaysia depend heavily on tin exports. Because their economies are so closely tied to the world minerals industry, falling demand and declining prices pose a threat to the economic, political, and social stability of these nations. In the early 1980s, when the price of copper dropped precipitously, Zambia—which obtained about 86 percent of its export revenue from copper sales—was seriously hurt. The nation's economy went into a tailspin; suffering was widespread. Twice as many of the nation's children died from malnutrition in 1984 as in 1980.

Some of the mineral-producing LDCs do have diversified economies. They include India, Zimbabwe, and Thailand. Brazil is among the world's largest producers of iron ore and produces substantial quantities of petroleum, coal, bauxite, and other minerals, but much of this production is targeted for domestic trade. Minerals account for less than 1 percent of Brazil's GNP and export earnings.

How Is Seabed Mining Overseen?

As mineral exploration shifts from sovereign nations to global commons such as Antarctica and the oceans, the question of who will gain control of potential deposits becomes more important. In 1980 the United Nations set up a convention to draft a treaty specifying who controls the ocean and its resources. The proposed Law of the Sea establishes a 200-mile (322-kilometer) exclusive economic zone (EEZ) within which coastal nations control all fishing, marine life, and mineral rights. To oversee the mining of seabed minerals in the open ocean, the law establishes an International Seabed Authority (ISA), which would have the authority to license companies to mine seabed deposits and to collect taxes on the minerals produced. The taxes would be used to finance projects in LDCs.

By 1990 the Law of the Sea had been ratified by 40 or so of the 150 countries participating in the convention; 60 must ratify it before it takes effect. Although most of the treaty is acceptable to the United States, the country has not ratified the agreement. It argues that countries with heavy mining interests should have greater control over decisions affecting deep-sea mining.

How Does the United States Manage Mineral Supplies?

The laws, regulations, and agreements a nation establishes in order to govern the production, use, and commerce in minerals are known as a minerals policy. Many people argue that the United States has no unified minerals policy, and there is no established policy to direct the consumption, conservation, or recycling of minerals and mineral-based products. Let us look more closely at how both domestic and imported mineral supplies are managed.

Domestic Mineral Supplies. In the last century mining was seen as a primary use for most public lands in the West, a belief that persists in some quarters today. The Mining Law of 1872 granted title to certain public lands as long as the claimant proved that the minerals were "valuable," conducted minimal mining activities, and paid the government $2.50 per acre. This law is still in effect for gold, iron, copper, and others, and valid claims remain on some 28 million acres (11 million hectares) of public land (see Chapter 21). Those who develop claims under the 1872 law pay no royalties on the minerals they extract. Coal, oil, gas, potash, and other sedimentary deposits, originally covered by the Mining Law of 1872, are currently managed under the Mineral Leasing Act of 1920, which allowed private individuals and companies to lease the rights to develop these resources, but discontinued the practice of selling the land outright. This act gave the Department of the Interior the discretion to approve or deny leasing requests, and it required current leasers to pay royalties to the government.

In an effort to protect public lands, conservationists have pushed Congress, unsuccessfully, to repeal the Mining Law of 1872 and replace it with a leasing system. Public lands came under increased assault in late 1988, when the Department of the Interior announced plans to permit strip mining for coal on 52 million acres (21 million hectares) of protected lands (Figure 17-8). Under the 1977 Surface Mining Act, coal companies can mine in protected areas as long as they demonstrate that they held "valid existing rights" prior to the law's enactment. Interior's proposed rule change would make it easier for coal companies to claim rights on public lands that Congress and the public already consider protected. The federal government would have to spend more than $800 million to buy back the rights to prevent mining.

In the 1990s the debate between mineral development and preservation on the public lands has intensified. Wilderness designation and permanent protection for national parks are just two of the many issues involved. The focus of the debate will probably shift to Alaska, where sizable reserves of many important nonfuel minerals, such as copper and nickel, have been identified and developed. The state also has several large deposits of critical and strategic minerals, including cobalt, the platinum group metals, chromite, tantalum, and columbium. In the past high energy costs, the difficulty of mining in the Alaskan environment, and the lack of suitable infrastructures discouraged mineral development. But numerous factors—a disruption in foreign supplies, a substantial price increase on the

FIGURE 17-8: Great Smoky Mountains National Park. The decision by the Department of the Interior to allow strip mining in protected lands—from areas near homes, schools, churches, and cemeteries to national wildlife refuges and parks, including the Great Smoky Mountains—was widely considered a setback for public lands.

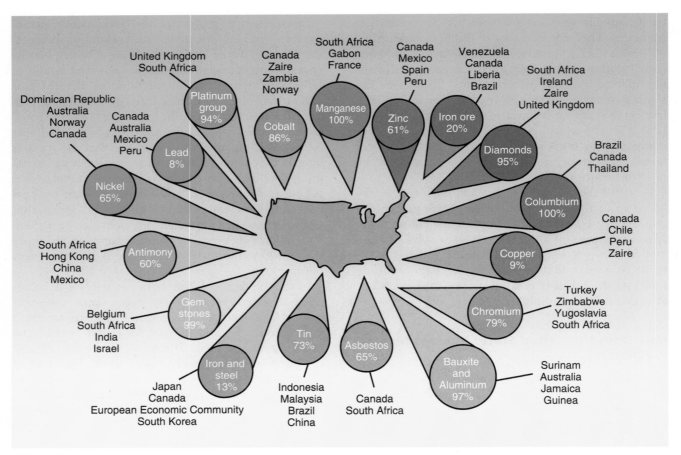

FIGURE 17-9: The United States' net import reliance, as percent of apparent consumption, selected nonfuel minerals, 1989. Domestic reserves meet U.S. demand for minerals such as lead, iron ore, and copper, but for many critical and strategic minerals, such as manganese, columbium, and the platinum group metals, the nation relies heavily on foreign supplies.

metals market, or a push to diversify the Alaskan economy, which is highly dependent on oil—could increase pressure to develop these mineral deposits.

Imported Mineral Supplies. The United States relies on imports for supplies of many critical and strategic minerals (Figure 17-9). For some, production from domestic reserves cannot meet demand; for others, it is cheaper for the United States to import supplies from nations with high-grade ores than to develop its own lower-grade domestic reserves. In an effort to guard against shortages, the United States—the world's largest consumer and importer of critical and strategic minerals—began to stockpile important minerals soon after World War II. For most minerals, stockpile targets are set at a three- to four-year supply. In addition to providing a reserve of supplies in case of a sudden cutoff or embargo, stockpiles help to even out sharp fluctuations in mineral prices. In 1980 the stockpiles of many strategic minerals were short of target goals. Because

of the Reagan administration's emphasis on defense, an effort was made to increase stockpiles to the target level, and in most cases this objective was accomplished. Currently, the United States has stockpiled enough reserves of critical and strategic minerals to last for several years.

What Can We Do to Conserve Mineral Resources?

There are three basic ways to conserve mineral resources: find substitutes for minerals that are in short supply, reuse products and materials, and recycle.

In the area of substitutes, the emphasis is on **advanced materials,** materials that can replace traditional materials and have more of a desired property (such as strength, hardness, or thermal, electrical, or chemical properties) than traditional materials. Though advanced materials do use metals and other minerals, they generally use these substances in less-

(Text continues on page 368.)

Environmental Science in Action: Critical and Strategic Minerals

Doug Davidson

Of the numerous minerals essential to U.S. defense and civilian industries, four are considered particularly important: chromium, cobalt, manganese, and platinum group metals. Interest in these four is high because the United States lacks reserves of these minerals and must rely on imports from politically volatile regions. Currently, there is no supply crisis for these minerals, but some experts are concerned that we are too vulnerable to political situations that might disrupt supplies in the future.

Describing the Resource

Physical Boundaries

Although the United States recycles varying proportions of chromium, cobalt, manganese, and platinum group metals, it is almost completely dependent on imports for new supplies. World production of these metals is dominated by a few countries, including South Africa and Zaire.

Chromium, a bluish-white metal, is not found naturally in its pure form, but as chromite ore. Chromite must be processed into ferrochromium before it can be used by industry. In 1988 the United States imported 75 percent of its chromium (recycling supplied the rest). Cobalt, a silver-white metal with exceptional properties of strength and hardness, is rarely found by itself, but is mined as a by-product of copper or nickel sulfides. In 1988 the United States imported 84 percent of its supply. Manganese, a silver-white metal, is commonly found in the form of oxides, such as manganese dioxide. The United States imports 100 percent of its manganese supply. Statistics show that U.S. reliance on South Africa, the major world supplier of manganese, is growing. Platinum group metals, known as the "noble" metals, have similar properties and often occur together in nature. They are platinum, iridium, osmium, palladium, rhodium, and ruthenium. They are usually by-products of nickel mining. In 1988 the United States imported 93 percent of its platinum group metals.

Biological Boundaries

Low-grade deposits of chromium, cobalt, and platinum group metals exist domestically, but they are generally not mined because it is more economical to buy them abroad. Many deposits lie on public lands, where mining is controversial. It damages wildlife habitat and can pose a significant threat to populations of endangered species, such as the grizzly bear. Roads, traffic, and the mining operations disrupt animals' life-styles, hamper reproduction, and even endanger their lives. Wild animals that become overexposed to humans and human activities lose their natural fear, becoming potentially dangerous to humans and more likely to fall prey to hunters.

Social Boundaries

Chromium, when mixed with other metals, gives them the sought-after qualities of hardness and resistance to high temperatures, wear, and oxidation. It is an essential ingredient in stainless steel, forming a chromium oxide film on the surface that affords protection against corrosion and oxidation. Combined with nickel, aluminum, cobalt, or titanium, chromium forms superalloys able to withstand very high temperatures. These alloys are used for products such as jet engine casings and turbine blades, ball bearings, and high-speed drills. Chromium is also a component of exhaust systems and some dyes and pigments and acts as an agent in chemical processing, gas and oil production, and power generation.

Cobalt, like chromium, is known for its strength and resistance to corrosion. It is an essential component of jet engines; each U.S. fighter plane uses about 1 ton (0.91 metric ton) of cobalt (Figure 17-10). It is also used in nuclear control rods, computers, magnets, and many types of electrical equipment and acts as a catalyst in petroleum refining and other chemical processes. Cobalt alloys are used to make high-speed tools and are popular for medical and dental work because the body rarely rejects implants made of cobalt. Cobalt is also added to glass, ceramics, livestock food, and fertilizers.

Manganese is essential for producing steel; 90 percent of the manganese consumed in the United States is used for the alloying and processing of steel. Manganese helps prevent the formation of iron sulfides, which weaken steel, by combining with the sulfides. It also removes oxygen and helps harden the final product. Because it becomes harder with pounding, manganese is also an important part of railroad equipment, rock crusher parts, and the teeth of power shovels. As manganese dioxide, it is used in dry cell batteries, photo development, and the production of rubber and plastic. Forms of manganese are also added to livestock feed and fertilizers.

Platinum is probably best known for its use in jewelry, but platinum group metals are also an important part of catalytic converters, which control auto emissions. Platinum group metals also serve as catalysts in chemical and petroleum refining processes. Platinum is frequently a component in equipment for the electronics and communications industries. Because platinum group metals are so expensive, they are generally used only when there are no satisfactory substitutes.

Looking Back

One reason the United States has been so successful in industrialization is its abundant supply of natural resources. However, World War I made the United States realize that it depended on some minerals that could not be produced domestically in significant amounts. Since that time, opinions on the strategic mineral situation have fluctuated between the extremes of panic and complacency.

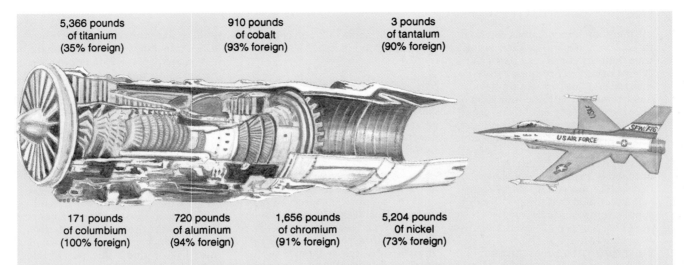

FIGURE 17-10: Amount of strategic minerals used in one jet-fighter engine and percentage supplied by imports. The diagram shows the materials used for engines on F-15 and F-16 warplanes.

5,366 pounds
of titanium
(35% foreign)

910 pounds
of cobalt
(93% foreign)

3 pounds
of tantalum
(90% foreign)

171 pounds
of columbium
(100% foreign)

720 pounds
of aluminum
(94% foreign)

1,656 pounds
of chromium
(91% foreign)

5,204 pounds
0f nickel
(73% foreign)

How the United States Developed a Stockpile

In 1921 the U.S. War Department drew up the Harbord list, a list of 28 minerals that had been in short supply during the war. In 1939 the government gave the navy $3.8 million to purchase reserves of important materials like tin, ferromanganese, tungsten, chromite, optical glass, and manila fiber. Then, in 1939 Congress passed the first stockpiling act, granting $70 million to establish a backup supply of strategic materials including chromium and manganese. To further strengthen stockpiling efforts, President Franklin Roosevelt ordered the Reconstruction Finance Corporation in 1940 to make large-scale purchases of materials deemed necessary for war. One branch, the Metals Reserve Corporation, was in charge of building up supplies of strategic metals.

After World War II, in 1946, Congress passed the Strategic and Critical Materials Stockpiling Act. Because of all the new technology developed during the war, the list of raw materials needed for defense and industry changed and grew, increasing the amount of materials the United States would need to import. In 1950 Congress allocated $8 million to restock the reserve, continuing the buildup through the decade. By the end of the 1950s the United States became even more dependent on foreign sources for strategic minerals. During that period all chromium production in the country ceased and the large Blackbird cobalt mines in Idaho, the major domestic source of cobalt, closed because the low-grade deposits were uneconomical to mine.

The Cold War Era and Emerging Nations

Before the Berlin Crisis of 1949, when the United States and the USSR confronted each other over control of West Berlin, the United States imported 31 percent of its manganese, 47 percent of its chromium, and 51 percent of its platinum from the USSR. Following the confrontation, the Soviets cut off this supply, but the United States simply turned to small producers such as India and Turkey, which expanded production to meet their new market.

After World War II, many colonial countries, particularly in Africa, gained their independence. By the 1970s over 100 new nations had emerged. Many of these new nations exported raw materials, including strategic ones, to the United States. Supplies were not always steady. Many of the countries were plagued by civil unrest, disorganization, and poverty. For instance, in 1978 antigovernment guerillas invaded Zaire's Shaba province, the principal location of cobalt production, preventing any cobalt from leaving the country. As a result, the price of cobalt rose from $6 to over $45 per pound.

While the temporary disruption in cobalt from Zaire was due to unrest, a different sort of political problem occurred when Rhodesia (Zimbabwe) announced its independence from Great Britain in 1965. Because Rhodesia vowed to continue white minority rule, the United Nations passed a resolution requiring all members to refrain from trade with Rhodesia. As a result, the United States could not buy Rhodesian chromium until the ban was lifted in 1971. The government did not need large amounts of chromium at that time and managed to maintain its supply by buying from Turkey and the Philippines. Also, the steel industry developed a new method of producing stainless steel which used high-carbon ferrochromium rather than the more expensive low-carbon variety normally purchased from Rhodesia.

South Africa, the United States's single greatest source of chromium, manganese, and platinum group metals, has always been a steady supplier. However, like Rhodesia in 1966, South Africa maintains a racist government through apartheid. In 1985 South Africa threatened to stop mineral sales to any country imposing economic sanctions because of apartheid. Because the United States as a nation has never imposed economic sanctions on South Africa, its supply of strategic minerals continues to flow from that source. However, increased internal resistance to the apartheid system and a growing global disapproval of any trade with South Africa may well threaten our reliance on that country for strategic minerals in the future.

The 1980s: Rebuilding the Stockpiles

Despite the cold war and conflicts in minerals-producing nations, the United States sold parts of the stockpile deemed unimportant during the 1960s, including aluminum, nickel, copper, and cobalt. Sixty million pounds (27 million kilograms)—60 percent of the amount in the stockpile—of cobalt were sold between 1964 and 1976. Eventually, Presidents Ford and Carter wanted to replenish the stockpile, but Congress would not authorize it. In 1980 President Reagan created a strategic materials task force, the National Materials and Minerals Program Board, to evaluate the situation. On the recommendations of the board, Reagan ordered stockpile administrators to supplement supplies of 13 materials in 1981. The highest priority was given to cobalt, of which 5.2 million pounds (2.3 million kilograms) was purchased. Throughout the 1980s the government continued to add to the stockpile.

In the early 1980s an increased awareness of the importance of

strategic minerals and the instability of their supply resulted in public debate over how precarious the situation really was. Those who believed a crisis was at hand pointed out that supplies were limited and could become inaccessible at any moment, while demand was growing. They even suggested that foreign sources might form a mineral cartel, similar to OPEC, and raise prices or cut off supplies to the United States. Others pointed out that comparing oil to other minerals was unfounded. Minerals cost much less than oil, they require far smaller quantities per unit of output, and most countries supplying minerals to the United States could not afford to halt sales for long. In addition, they claimed that the stockpile was adequate defense against any supply cutoff; even though private industry could not use the government stockpile, most companies that relied on large amounts of strategic minerals had their own 6- to 12-month backup supplies. In addition to stockpiles, conservation and recycling could stretch supplies, and new technology could augment them by producing replacement materials and making it possible to extract previously uneconomical ores. Even without a supply crisis, the market possesses an amazing power of self-adjustment. U.S. imports of chromium fell between 1950 and 1970 due to recycling and exporting scrap; at the same time, total world reserves of chromite "rose" 675 percent because of growing technical advances and geologic knowledge spurred by rising demand and price.

Many people began to look for previously unexplored deposits of strategic minerals, and showed great interest in the potential of the sea, particularly for cobalt and manganese. For example, metallic nodules found in abundance on the ocean floor contain 30 to 40 percent manganese, as well as small amounts of cobalt and other metals. While the expense of ocean mining makes the exploitation of these resources impractical, a rise in price and demand or a technological innovation that reduced the expense

could make manganese nodules and other undersea deposits an economically feasible source of some strategic minerals.

The debate over strategic minerals also brought up the issue of mining on public lands. For instance, the Absaroka-Beartooth Wilderness in Montana contains reserves of chromium as well as a platinum-palladium belt that represents 70 percent of domestic reserves of platinum group metals. Deposits of cobalt near the Blackbird Mine in Idaho extend into the River of No Return Wilderness. Although Congress tried to leave areas rich in minerals out of the wilderness area and declared 39,000 acres (15,783 hectares)a special mining management zone, 1,900 claims were staked in the wilderness area before it was closed to exploration. Alaska, home to more wilderness areas than any other state, also contains deposits of cobalt, copper, nickel, platinum group metals, and chromite.

Those who predicted a strategic minerals crisis claimed that opening up more federally protected wilderness areas to mining was the country's only option. They cited the Mining Law of 1872: "all valuable mineral deposits in land belonging to the United States. . . shall be free and open to exploration and purchase under regulations prescribed by law." However, the 1872 law is contradicted and limited by more recent laws, including the Endangered Species Act of 1973, which prohibits mining operations that would threaten endangered species, and the Wilderness Act of 1964, which prohibits mineral ex- ploration and filing of claims on wilderness lands after December 31, 1983. While many companies established mineral rights on federal lands before that deadline, the Wilderness Act specifies that mining in these wilderness areas must be "substantially unnoticeable" and prohibits roads, power lines, mechanical equipment, and air transportation to and from the wilderness. Environmentalists contend that the Bureau of Land Management, which manages many wilderness areas, has tended to interpret the act in favor of mining interests.

Looking Ahead

The debate over mining development versus wilderness preservation is only tangential to the effort to secure a long-term, stable supply of critical minerals. In the end, we will have to solve the strategic minerals problem through conservation and the development of substitute materials.

In 1982 Congress allocated money for conservation and substitution efforts in addition to strategic materials production. Although we derive 10 to 15 percent of our chromium supply from recycling, the Office of Technology Assessment (OTA) estimates that we could add another 3.5 percent annually with greater efforts. Significant amounts of cobalt are currently lost or downgraded as scrap and waste, and recent research has investigated new ways to extract cobalt from lead smelter waste. Although manganese is difficult to recycle, improvements in steel processing could reduce the U.S. demand by 45 percent by the year 2000. In addition, the OTA predicts that consistent recycling of catalytic converters could recover significant amounts of platinum group metals annually by the mid-1990s.

Advanced production technology could also "expand" supplies of

> Table 17-6
> Promising Technological Approaches for Reducing U.S. Reliance on Imports of Strategic Minerals

Approach	Potential Benefits[a]	Barriers to Implementation
CHROMIUM		
Conservation	Expanded recycling of scrap and waste could provide at least 20,000 tons of chromium beyond current recycling levels.	Barriers to chromium recycling are economic, not technical.
Substitution	Direct substitution could now reduce U.S. chromium needs by one-third. Another one-third reduction may possibly be achieved through a 10-year R&D program.	Low cost of chromium alloys deters use of substitutes. Lack of information on substitutes slows their use in times of shortage. Need for further tests and experience limits near-term potential for substitution to one-third of consumption.
	Advanced materials may displace chromium alloys in certain aerospace and industrial applications.	Basic and applied research is needed to improve properties and reliability of advanced materials. Designers and engineers need better understanding of properties and limitations of advanced materials. Tests and standards need to be developed for these materials.
Production	Development of alternative foreign sources could provide about 30,000 to 60,000 tons, about 10 to 20 percent of current U.S. demand.	At current prices for chromium, there is no economic incentive to diversify suppliers. Government assistance would be required to make development of alternative suppliers more attractive.
COBALT		
Conservation	Recycling could recover much of the cobalt in scrap and waste that is currently lost or downgraded.	Principal barrier is economic. In addition, extensive recovery of superalloy scrap may require use of technology that is now limited to laboratory testing.
	Process improvements now being adopted may make significant reductions in the amount of cobalt used to make jet engine components.	There are no economic barriers; economic factors favor the adoption of process improvements.
Substitution	Direct substitutes under development could reduce the need for cobalt by 50 percent or more in some critical superalloy applications.	Industry has little or no incentive to expend the time and money needed to qualify alternative alloys except when there are significant performance advantages.
		Barriers to adoption of advanced materials to reduce cobalt consumption are the same as for chromium, as described above.

strategic minerals (Table 17-6). Pratt & Whitney, a manufacturer of aircraft engines, has developed a method that uses less metal in jet engines, minimizing the amounts of strategic metals needed. General Electric uses less heat-resistant cobalt in engine blades by drilling tiny holes in the blades to encourage air cooling.

Ceramic materials and plastics play significant roles as mineral substitutes. Ceramics can withstand higher temperatures and more hostile environments and are gaining use in cutting tools, seals, bearings, and sandblasting nozzles. Pratt & Whitney is experimenting with jet engine combusters (the fuel-burning chamber) composed of nickel with a ceramic coating rather than cobalt. Plastics can also withstand high temperatures (up to 932°F or 500°C) and have been used in rotor blades of gas turbine engines. Scientists are currently working on plastic auto engines. Carbon-carbon composites are being developed in hopes of replacing superalloys in jet engines. They consist of carbon fibers held together by a resinous matrix; when the matrix dries, the material becomes a solid compound.

Approach	Potential Benefits[a]	Barriers to Implementation
Production	Domestic production from three sites could produce up to 8 million pounds of cobalt per year. New foreign production could provide almost 15 million pounds of cobalt per year.	Current prices for cobalt and co-product metals are too low to justify investment without federal subsidies. Investments are being postponed until cobalt prices rise. Lead times of 2 to 5 years are needed to bring deposits into production.
MANGANESE		
Conservation	Process improvements could reduce need for imported manganese in steel by 45 percent by year 2000.	Adoption of improvements will depend on incremental upgrading of domestic steel-making facilities.
Production	Alternative suppliers to South Africa and the former Soviet Union could increase production after 2 to 3 years to expand facilities.	Assured market for increased production is needed to justify investment in production and transportation facilities. United States would be in competition with other consumers for new production. Facilities for processing ore into ferromanganese must also be available.
PLATINUM GROUP METALS		
Conservation	Recycling of catalytic converters could recover 500,000 troy ounces of platinum group metals annually by 1995.	There are no significant barriers. Several years will be needed to develop collection and processing infrastructure.
Production	Development of the Stillwater deposit could produce 175,000 troy ounces of platinum group metals in the near term. Additional development is possible.	Domestic production will require slightly higher prices for platinum and palladium and evidence of increased demand.

[a]The benefits accruing from the various approaches are not cumulative. For example, as scrap generation in manufacturing is reduced through improved processing techniques, the potential benefits of recycling are also reduced.
Source: U.S. Office of Technology Assessment.

What You Can Do:
To Conserve Mineral Resources

- Purchase durable items that last, not disposable ones, and items that will not become obsolete. Repair items or replace worn parts in order to extend the useful life of objects.

- Reuse objects whenever possible.

- Recycle cans, bottles, paper, and other materials.

- To dispose of a car battery, take it to a dealer who will recycle it for its lead content.

- Write your Congressperson and ask him or her to work for the repeal of the Mining Law of 1872 in favor of a leasing system.

- Support corporate efforts to conserve minerals and develop

safe substitutes. One way to do this is to patronize companies that are implementing conservation programs or that are developing alternatives to minerals.

---

er amounts than traditional materials or they use minerals that have not been widely used. Advanced materials offer the promise of energy conservation, better performance at lower prices, and reduced dependence on imports of critical and strategic minerals.

Composites are the fastest growing of the new advanced materials. A composite is a matrix of one material reinforced with fibers or dispersions of another. Other advanced materials include superconducting substances; advanced ceramics; and medical, dental, electronic, magnetic, and optical materials. Advanced materials have numerous important applications, particularly in the aerospace, automotive, packaging, and communications industries. They have been used successfully in many cases to replace traditional materials (Figure 17-11).

While advanced materials hold potential for reducing consumption of traditional materials, reusing and recycling materials are the two factors over which individuals have the greatest control. Reusing materials whenever possible—glass containers, for example—saves energy as well as mineral resources. Many materials that cannot be reused, such as an aluminum beverage can, can be recycled. Recycling conserves material resources and energy and reduces solid waste.

Only about 10 percent of recyclable mineral resources are recycled in the United States, compared to rates as high as 40 percent in the Netherlands, Japan, and Germany. There are many reasons we in the United States do not take advantage of the potential of recycling. We have a "throwaway mentality" that leads us to think nothing of using things once and then discarding them. To satisfy our defense needs, laws were written during World War II to encourage domestic exploration and expansion of mineral supplies. Although the need for the laws has clearly passed, they remain in place, discourag-

ing materials reuse and recycling at the expense of the environment. Finally, we have not yet adjusted our thinking to a world in which there are limits— limits to development, limits to energy sources, limits to natural resources.

Future Management of Mineral Resources

Minerals are essential to industrial societies and play a critical role in the economies of many nations. The sustained and sustainable use of minerals is an important goal for peoples and governments worldwide. To ensure the sustainability of mineral resources, we should take action in three broad

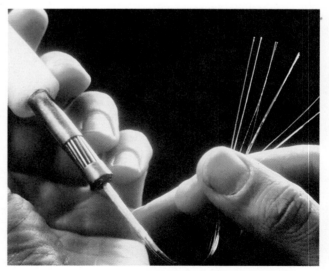

FIGURE 17-11: Cable using optical transmission fibers as a replacement for copper telephone wires. Advanced materials can replace traditional materials that may become scarce or expensive. Many have qualities that make them superior to the materials they replace.

areas: prevent overuse through conservation, protect human and environmental health from pollution and degradation, and preserve living systems.

Prevent Overuse Through Conservation

Conservation—reusing and recycling materials—can prevent the overuse and depletion of minerals. By emphasizing conservation, we can also preserve and protect natural systems, eliminate the adverse effects associated with mineral exploration, extraction, and processing, and reduce the amount of materials in the solid waste stream.

Congress can lead the way in conservation measures by eliminating subsidies and laws that encourage mineral exploration over recycling and other conservation methods. Repealing the Mining Law of 1872 and instituting a leasing system for claims are important first steps. Communities can also take action to promote conservation. Suggestions for individuals are offered in What You Can Do: To Conserve Mineral Resources.

Protect Human Health and the Environment

Strictly enforcing and periodically updating the Surface Mining Act and provisions of the Clean Air and Clean Water acts that affect mining operations can help protect human health and the environment.

Preserve Living Systems

As minerals become increasingly scarce, and exploration shifts to previously inaccessible areas, the energy and environmental costs of mining are likely to rise. To prevent or minimize damage in these areas, the United States and other nations, either jointly or alone, should identify those materials that are in shortest supply and fund research to develop substitutes. Congress should expressly set national parks and wilderness areas off limits to mineral exploration and development. The nations of the world must agree to preserve the living systems of Antarctica and the seabed, as they become increasingly attractive targets for exploitation. The United States should ratify the Law of the Sea and urge other nations to do so. This legislation will allow us to benefit from the mineral resources of the oceans while minimizing environmental problems.

Summary

Minerals are nonliving, naturally occurring substances with a limited range in chemical composition and with an orderly atomic arrangement. A mineral may occur as a single element, a compound of elements, or an aggregate of elements and compounds. A mineral's unique chemical formula determines its physical properties. Minerals are broadly classified as fuel or nonfuel minerals; nonfuel minerals are further classified as metallic or nonmetallic. A metal is any of a group of chemical elements characterized by malleability, ductility, and conductivity of heat or electricity. Ferrous metals contain iron or elements alloyed with iron to make steel; nonferrous metals contain metallic minerals not commonly alloyed with iron. Nonmetals are those mineral substances that do not possess the characteristics of metals. Nonmetallic minerals encompass a wide variety of resources including salts, fertilizers, and industrial diamonds.

Minerals are not evenly distributed throughout the earth's crust. Additionally, deposits of one mineral are often mixed with deposits of others. The dispersion of mineral deposits is caused by fracturing of the earth's crust, weathering and erosion, chemical separation, and gravity. A useful mineral is one that occurs in a concentration high enough to make it profitable to extract; a deposit of such concentration is known as an ore.

The greatest share of mineral resources is consumed by the more-developed countries. Global demand for most major minerals is expected to rise significantly as the human population continues to grow rapidly and societies continue to rely on mineral resources to support rising standards of living. Industrial nations rely on about 80 minerals, 75 percent of which exist in abundant supply to meet anticipated needs or can be replaced by existing substitutes. Strategic minerals are those vital to national defense; critical minerals are those considered essential to industry.

The resource base is an estimate of the total sum of a mineral in the earth's crust; it includes deposits that we will probably never be able to extract and is thus not a very useful figure. A deposit that can be extracted profitably with current technology is called a proven reserve or economic resource. Subeconomic resources are known reserves that cannot be extracted at a profit at current prices or with current technologies. An estimate of the total amount of a given mineral that is likely to be available for future use is called the ultimately recoverable resource. It is based on assumptions about discovery rates, future costs, market factors, and future advances in extraction and processing technologies.

It is unlikely that mineral resources will become exhausted. As reserves become depleted, the cost is likely to rise. High prices and high demand encourage more exploration and lead to the discovery of more ores, increasing the size of proven reserves. Other factors likely to ease the pressure on dwindling supplies include conservation measures, decreasing demand, technological advances that make it feasible to mine previously unprofitable deposits, and the development of substitutes.

Mining includes three major steps: location (or exploration), extraction, and processing. All can adversely affect human and environmental health. Location is the most benign step, extraction the most harmful. Acid mine drainage, water and air contamination from tailings and emissions, disturbed vegetation and soil, and erosion are several of the most serious environmental consequences of surface mining. Exposure to harmful substances and dangerous physical conditions associated with subsurface mining and the heavy equipment of surface mining are threats to human health.

Mineral prices tend to fluctuate wildly in the short term, but over time the economic value attached to commodity exports like minerals has fallen relative to the value of manufactured goods. Mineral-exporting nations, particularly LDCs, must sell increasing amounts of minerals to pay for imports such as tractors and fertilizers. Critical and strategic minerals are exceptions to this long-term trend of falling value; because there are no substitutes for them at present, the price of these minerals is likely to rise.

In the United States the development of domestic mineral deposits on public land is governed by the Mining Law of 1872, which granted title to certain public lands as long as the claimant proved that the minerals were "valuable," conducted minimal mining activities, and paid the government $2.50 per acre. Although many people favor the repeal of the Mining Law of 1872, it is still in effect. Some mineral deposits, including coal, oil, gas, potash, and other sedimentary deposits, are currently governed under the 1920 Mineral Leasing Act, which allows private individuals and companies to lease the rights to develop these resources, but not buy the land outright. Leasers must pay royalties to the government, and the Department of the Interior has the authority to approve or deny leasing requests.

The United States imports most critical and strategic minerals. To guard against shortages, it stockpiles important minerals. Besides ensuring a supply of important minerals, stockpiles even out sharp fluctuations in mineral prices.

Minerals can be conserved by finding substitutes for those that are in short supply, reusing products and materials, and recycling materials. Advanced materials, which exhibit greater strength, greater hardness, or superior thermal, electrical, or chemical properties compared to traditional materials, are important substitutes. Composites, a matrix of one material reinforced with fibers or dispersions of another, are the fastest growing of the new advanced materials. Others include superconducting substances, advanced ceramics, medical and dental materials, and electronic, magnetic, and optical materials.

Key Terms

abundant metal	ore
advanced material	overburden
area strip mining	processing
contour strip mining	proven reserve
critical mineral	resource base
extraction	scarce metal
ferrous metal	sedimentary rock
igneous rock	strategic mineral
magma	subeconomic resource
metamorphic rock	subsurface mining
mineral	surface mining
nonferrous metal	tailings
open pit surface mining	ultimately recoverable resource

Discussion Questions

1. What is the significance of understanding how minerals form in the earth's crust?

2. Should mineral exploration and development in less accessible areas, such as Alaska and Antarctica, be encouraged? Why or why not?

3. Suggest several ways that the availability of minerals might be extended.

4. What would you include in a strategic minerals policy developed by the United States to ensure both an adequate supply of strategic minerals and the environmental integrity of public lands?

5. How might the more-developed countries help less-developed countries become less dependent on minerals exports? What programs or actions could MDCs take in order to share, more equitably, the profits gained by exploiting the mineral wealth of LDCs?

Chapter 18

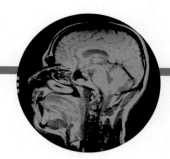

Nuclear Resources

You're always balancing risks against possible benefits.

Dr. Joseph R. Castro

Even if no more nuclear waste were created, addressing that which already exists will require attention and investments for a period that defies our usual notion of time. The challenge before human societies is to keep nuclear wastes in isolation for the millennia that make up the hazardous life of these materials. In this light, no matter what becomes of nuclear power, the nuclear age will continue for a long, long time.

Nicholas Lenssen

Learning Objectives

When you finish reading this chapter, you should be able to:

1. Describe the different applications of nuclear technology.

2. Explain two types of exposure to radiation and the possible health effects.

3. Identify four types of waste produced by a nuclear power plant.

4. Identify the issues involved in long-term storage of nuclear waste.

A cusp is a point at which two curves meet. For the poetic among us, to approach a cusp is to come to a place where the path suddenly, drastically changes, where life is forever after altered. Individuals can reach and cross a cusp; so, too, can the human race. The use of fire, the development of the wheel, the first use of tools, all signaled dramatic changes in human history. Another such cusp occurred in 1945.

During World War II the United States, in a race with Nazi Germany to develop atomic weapons, sponsored the Manhattan Project, a top-secret effort to create the world's first nuclear bomb. In 1945 Manhattan Project scientists succeeded in developing the bomb before the Germans, but the United States never used its new weapon against Hitler's regime. Instead, in the early morning hours of August 6, 1945, a U.S. warplane named the Enola Gay released the first atomic bomb on the city of Hiroshima, Japan. The ensuing destruction clearly demonstrated that humankind was capable of destroying life on a scale previously unimagined.

The attack on Hiroshima and a similar attack on the Japanese city of Nagasaki several days later bore witness to the awesome power of the atom. In the years that followed, researchers began developing other applications for nuclear power. Nuclear resources yielded significant benefits for medicine, agriculture, industry, and science (Figure 18-1). Nuclear power was used to generate electricity, avoiding the release of greenhouse gases and sulfur oxides that accompany the combustion of coal and oil. Ironically, the same resource that killed and maimed thousands of Japanese in an instant on that day in 1945 also saved the lives of untold numbers of cancer patients, improved the overall quality of life for many in the industrialized world, allowed researchers to push forward the frontiers of science, and provided electricity for millions.

The gains made through the use of nuclear resources have not come without cost: nuclear wastes continue to accumulate; water, land, and air have been contaminated by radioactivity escaping from weapons facilities and power plants; and the threat of accidents or nuclear war darken the collective heart of humanity.

In this chapter we discover what nuclear resources are, what radiation is, and how radiation can

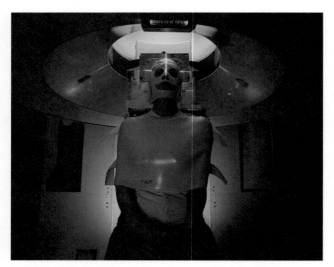

FIGURE 18-1: A cancer patient is treated using radiation therapy. Among the most important beneficial uses of nuclear resources are their medical applications.

adversely affect human health and the environment. We describe both the beneficial and destructive applications of nuclear resources. We look at the historical uses of nuclear resources and the management problems they present. Finally, we see how these resources can be managed in environmentally sound ways in order to allow us to realize their benefits while minimizing associated risks.

Describing Nuclear Resources: Physical Boundaries

What Are Nuclear Resources?

Nuclear resources are derived from atoms, their energy, and the particles they emit. **Nuclear energy** is the energy released, or radiated, from an atom; energy released from an atom is called **radiation**. Radiation takes two basic forms: ionizing and nonionizing. **Ionizing radiation** travels in waves (x-rays, gamma rays) or as particles (alpha particles, beta particles). The energy level of ionizing radiation is high enough to remove electrons from atoms, creating charged particles called **ions.** These ions can then react with and damage living tissue. **Nonionizing radiation** (heat, light, radio waves) can also affect atoms, but its energy level is not high enough to create ions.

Ionizing radiation is released into the environment from both cultural and natural sources. Cultural sources such as nuclear testing, nuclear power plants, medical x-rays, luminous watch dials, color television sets, smoke detectors, and metals production account for only an estimated 18 to 32 percent of the radiation in the environment. The remaining radiation, about 68 to 82 percent, originates from natural sources, including the sun, outer space, soil, air, water, food, and rocks (Table 18-1). Radon, as we saw in Chapter 14, is a gas produced during the decay of uranium. Found in small quantities in water, soil, rocks, and various building materials, it can enter buildings through cracks in the foundation or basement. Once inhaled, it releases particles that can cause lung cancer. Many scientists believe radon gas is the single largest source of natural radiation exposure. Because the gas is odorless, invisible, and tasteless, only special tests can detect its presence. The discovery of radon in many homes in the late 1980s precipitated nationwide concern about the dangers of this natural source of radiation.

How Is Nuclear Energy Released?

Nuclear energy is released through three types of reactions: spontaneous radioactivity, fission, and fusion.

Spontaneous Radioactivity. Spontaneous radioactivity occurs when unstable atoms release mass in the form of particles (particulate radiation), energy in the form of waves (electromagnetic radiation), or both. The unstable nucleus emits particles and energy during a process called **radioactive decay.** For example, uranium 238 decays into thorium 234, then protactinium 234, and eventually into lead 206 (Figure 18-2). Each radioactive element goes through its own unique sequence to reach stability.

Table 18-1
Sources of Radiation

Source	Dose (rem)[a]
Natural sources (67.6% to 82% of our radiation exposure per year)	
Concrete homes	.07–.10
Brick homes	.05–.10
Wooden homes	.03–.05
Cosmic rays	.045
Air, food, water	.025
Soil	.015
Human-made sources (18% to 32.4% of our radiation exposure per year)	
Diagnostic x-rays (per x-ray)	.02
Watches, color television sets, smoke detectors	.004
Air travel (round trip to London)	.004
Nuclear plant vicinity	.001

[a]Röentgen equivalent, man; the dose of an ionizing radiation that will have the same effect on living tissue as 1 roentgen of gamma ray or x-ray.
Source: Factsheet: Radiation

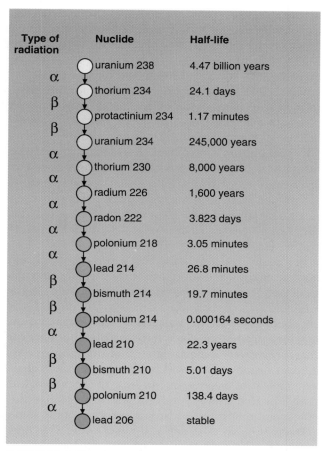

Type of radiation	Nuclide	Half-life
α	uranium 238	4.47 billion years
β	thorium 234	24.1 days
β	protactinium 234	1.17 minutes
α	uranium 234	245,000 years
α	thorium 230	8,000 years
α	radium 226	1,600 years
α	radon 222	3.823 days
α	polonium 218	3.05 minutes
β	lead 214	26.8 minutes
β	bismuth 214	19.7 minutes
α	polonium 214	0.000164 seconds
β	lead 210	22.3 years
β	bismuth 210	5.01 days
α	polonium 210	138.4 days
	lead 206	stable

FIGURE 18-2: The decay of uranium 238.

Atoms of the same element which have different numbers of neutrons are known as **isotopes** of the element. An element may have many isotopes. Hydrogen has two isotopes: deuterium, which has two neutrons, and tritium, which has three neutrons. The usual isotope of carbon is carbon 12 (6 neutrons, 6 protons); a carbon atom with 8 neutrons is the isotope carbon 14 and is radioactive.

Isotopes that release particles or high-level energy are called **radioisotopes.** Radiation emitted by radioisotopes is an example of ionizing radiation. The most common types of ionizing radiation are alpha, beta, and gamma radiation. **Alpha particles** consist of two protons and two neutrons and carry a positive charge. They are emitted at high speeds, but travel only short distances before losing energy. Alpha particles can be stopped by a sheet of paper or by the skin, but they are dangerous when inhaled or when ingested through food or water. **Beta particles** are negatively charged particles that are emitted from nuclei. They are equivalent to electrons, but contain more energy. Beta particles arise when neutrons in the nucleus are converted into protons. The small amount of energy that is lost in the process is the energetic beta particle. Beta particles can be stopped by a piece of wood an inch thick or a thin sheet of aluminum, but they can penetrate the skin. They are harmful when emitted inside the tissues of a living organism. **Gamma radiation** is a powerful, high-energy wave; gamma rays are the most common form of ionizing electromagnetic radiation released from radioisotopes. Even a thick piece of lead or concrete will not stop all gamma rays. They can pass through the human body, causing damage as they do so. **X-rays,** a form of cosmic radiation, can also be produced by firing electrons at a target made of tungsten metal. When the electrons hit the metal, they give up their energy in the form of x-rays. Like gamma rays, x-rays can ionize atoms in living tissue, but they have considerably less energy and so are less penetrating (Figure 18-3).

Each stage in a decay sequence exists for a specific amount of time, measured in half-life. **Half-life** is the length of time it takes for any radioactive substance to lose one-half of its radioactivity. The half-life of a substance may vary from a split second to billions of years (Table 18-2). The half-life of a

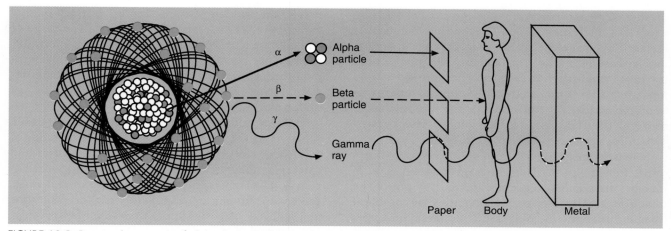

FIGURE 18-3: Penetrating power of alpha, beta, and gamma radiation. The three types of radiation possess different amounts of energy that determine their ability to penetrate different substances.

Table 18-2
Half-Life of Selected Radioactive Materials

Radioactive Material	Half-Life (years)
Cesium	13,730
Neptunium	2,372,000,000
Plutonium 239	24,110
Strontium	9,029
Uranium 233	160,000
Uranium 235	710,000,000

Numbers following elements represent isotopes of those elements.

radioactive substance must not be confused with the amount of time before that substance can safely be released into the environment. Even after one half-life has gone by, 50 percent of a radioactive substance's atoms remain unchanged and hazardous.

Fission. A **fission reaction** occurs when an atom is split into two or more new atoms. When a large atom, such as uranium 235, is struck by neutrons, energy and one or more neutrons from the atom's nucleus are released. The atoms formed when uranium in a nuclear reactor is split are called **fission products**. Fission products are almost always radioactive. The neutron released by each fissioned atom can then split other uranium 235 nuclei. **Critical mass** is the minimum amount of material that can sustain a chain reaction. A **chain reaction** is a self-perpetuating series of events that occur when a neutron splits a heavy atom like uranium 235 or plutonium 239, releasing additional neutrons to cause other atoms to split in a like manner, thus releasing tremendous amounts of energy.

Fusion. A **fusion reaction** is the opposite of a fission reaction; nuclei are forced to combine. The sun is essentially a nuclear fusion reactor: two hydrogen nuclei fuse into a helium nucleus. The nuclei tend to repel one another because they are positively charged; therefore, a great amount of energy is required to fuse the nuclei. Once the repulsive force is overcome and the nuclei are joined, a huge quantity of high-energy radiation is released.

Scientists have tried to fuse various combinations of deuterium and tritium in the laboratory, but so far they have failed to get as much energy out of the reaction as they have put into it. One problem plaguing fusion research is difficulty in finding a material for the containment vessel. The material must be highly resistant to radiation and the extremely high temperatures (100 million °C or

higher) that are required to form the plasma (an electrically neutral mixture of nuclei and electrons) in which the reaction takes place. Another difficulty is how to keep the plasma from striking the containment vessel's walls and losing the plasma particles' energy. One of the most promising containment technologies is magnetic containment, in which the plasma is suspended in a magnetic field. In November 1991 scientists at the Joint European Torus (JET) laboratory in Culham, England, achieved the fusion of deuterium and tritium nuclei in a magnetically confined plasma. The reaction generated about 1.7 million watts of power in a burst lasting nearly 1 second. Although the experiment consumed ten times more energy than it produced, it demonstrated that fusion is scientifically feasible.

For many, fusion holds the promise of "clean" nuclear energy, and some scientists predict that construction of a demonstrator reactor could begin around 2010, costing approximately $6 billion. Others speculate that we will have commercial fusion reactors producing electricity within the next 50 years. However, experts agree that building a commercial fusion reactor will be the world's most challenging feat of engineering. In fact, the technological problems that must be solved before we realize the practical applications of fusion energy prompt some detractors to say that day will never come. Furthermore, they caution that fusion energy, like fission, could easily be accompanied by problems we cannot even imagine today.

Biological Boundaries

How Does Radiation Affect Human Health?

As ionizing radiation penetrates living tissue, it can destroy cells or alter their genetic structure. For example, when high-energy alpha particles come in contact with cells, which are largely composed of water, the sudden introduction of so much energy excites the water molecules. The positive charge of the alpha particles strips electrons from the water molecules, destabilizing their relationships with neighboring water molecules. Eventually, the destabilization may become so great that the cells can no longer function.

Ionizing radiation can also break or scramble DNA molecules in the nucleus. Damaged strands of DNA can cause chromosomes to break apart and then recombine in an abnormal fashion. If the radiation dosage is high enough and the resulting damage too severe for the organism's system to repair, the cell will die, often within hours. Those cells receiving nonlethal doses can continue to exist with altered DNA, reproducing abnormally for years, spawning cancerous cells and eventually tumors.

The cumulative effect of cellular change caused by ionizing radiation can lead to a wide range of health problems, including cancer, leukemia, degenerative diseases such as cataracts, mental retardation, chromosome aberrations, genetic disorders such as neural tube defects, and a weakened immune system.

Effects of Different Amounts of Ionizing Radiation. One factor that determines how radiation affects living organisms is the amount of exposure they receive. Amounts are typically measured in two dosage units. The amount of radiation absorbed per gram of tissue is expressed by the *rad* (*radiation absorbed dose*). The international unit, which is replacing the rad, is called the *gray*; it is equivalent to 100 rads. The damage potential caused by this amount (or dose) of radiation is expressed by the *rem* (*Röentgen Equivalent, Man*) and *millirem* (one-thousandth of a rem). The sievert is the international unit used to express damage potential. One sievert is equivalent to 100 rems.

The government has established guidelines for the amount of radiation a person can safely withstand. The Nuclear Regulatory Commission (NRC) has set 0.17 rem per year (not including the 0.2 rem of background radiation most people receive from natural sources in a year) as the maximum safe exposure for the general public. However, NRC limits for those working in the nuclear industry are almost thirty times higher—3 rems within a 13-week period or a total of 5 rems annually (again excluding background radiation).

Doses of radiation can be classified as low or high level. Low levels, which range from 1 to 5 rems per year, can still be health hazards because even small amounts of radiation can cause cell damage which can accumulate over time. These damaged cells continue to grow, creating the possibility of cancer years later. Some scientists believe that any amount of radiation, no matter how minute, affects biological systems in some way, even though it could be so slight as to be undetectable (Table 18-3).

Doses of 5 to 10 rems per year and over are considered high-level radiation. This amount of radiation kills massive numbers of cells, causes radiation burns, destroys bone marrow, and damages internal organs. Radiation sickness is caused by doses ranging from 50 to 250 rems. Symptoms include nausea, vomiting, and a decrease in white blood cells. Doses of radiation over 500 rems are almost always fatal.

When a reactor at the Soviet Union's Chernobyl nuclear power plant exploded in 1986, it emitted tons of radioactive material into the environment. Victims living within a 3- to 4-mile radius of the plant stood only a 50 percent chance of surviving; those who survived suffered bone marrow and gastrointestinal damage. Even those living 60 miles

▶ Table 18-3
Health Effects, Exposure to Radiation

Sudden, whole body exposure to radiation can cause the following general effects:

1–100 rem	Nausea, vomiting
100–200 rem	Moderately depressed white blood cell count, not immediately fatal but long-term cancer risk
200–600 rem	Heavily depressed white blood count, blotched skin in 4 to 6 weeks, 80 to 100 percent possibility of death
600–1,000 rem	Diarrhea, fever, blood-chemical imbalance in 1 to 14 days, almost 100 percent probability of death

Radiation exposure can also affect specific organs and systems:

1–10 rem	Organs and tissues of high sensitivity: bone marrow, colon, stomach, breast, lung, thyroid
	Organs and tissues of moderate sensitivity: ovary, intestines, pancreas, liver, esophagus, lymph, brain
	Organs and tissues of low sensitivity: bone, kidney, spleen, gallbladder, skin
50–500 rem	Brain and central nervous system: delirium, convulsions, death within hours or days
	Eye: death of cells in lens leading to opaqueness and cataracts that impair sight
	Gastrointestinal tract: nausea and vomiting within a few hours, bleeding of the gums, mouth ulcers, intestinal wall infection leading to death
	Ovaries and testes: acute damage can affect the victim's fertility or offspring
	Bone marrow: retards the body's ability to fight infections and hemorrhaging; leukemia.

Source: Adapted from *Time,* May 12, 1986; *National Geographic,* April 1989.

away were put at increased risk of leukemia and other forms of cancer for the next 30 years. International groups concerned with nuclear issues estimate that 12,000 people will die over the next 70 years as a result of the accident.

How Age and Gender Affect Susceptibility to Radiation. Many studies have shown that people of different ages and genders respond differently to similar amounts of radiation. For example, women are more likely to develop thyroid cancers than men. Infants and fetuses are more vulnerable to all types of cancers and disabilities than adults because of their faster rate of cell division.

▶ Table 18-4
Effects of Radiation Exposure on the
Unborn

Central nervous system defects, most of which lead to moderate to extremely severe mental retardation in the live-born child. These defects can include failure of development of some segments of the brain itself and associated abnormalities of skull development, including small head circumference and elongated shape of the skull.

Abnormalities of the sensory organs, in many different forms and affecting all sensory organs, such as defects in the structure of the outer and inner ears, eyelids, eye muscles, iris, eyeball

Heart and great vessel defects

Cleft palate, a common deformity

Skeletal abnormalities, such as congenital dislocation of the hip, the malformation and even absence of such structures as the kneecap

Polydactyly, an abnormal number of fingers and toes

Defects in the gastrointestinal canal, such as failure of perforation of the anal opening

Failure of urinary system organs to function properly or to develop at all, including kidneys, ureters, bladder

Cryptorchidism, the failure of the testes to descend, a common abnormality

Deformities of the vertebral column and ribs

Hernias

Failure to thrive. This is a functional description of many children born with serious malformations. When they survive the embryonic and fetal stages and are live-born, they simply do not develop normally in early infancy.

Source: Adapted from Dr. John W. Gofman, *Radiation on Human Health,* Sierra Club Books, San Francisco, 1981.

Fetuses are especially likely to suffer severe effects from radiation exposure. If doses are high enough, a fertilized egg can develop lesions that will prevent its implantation in the uterus, thus causing a spontaneous abortion. Lower doses of radiation allow the fertilized egg to implant, but can damage the fetus's developing organ systems. Such fertilized eggs can develop into live-born children with mild to severe defects (Table 18-4).

How Does Radiation Enter the Environment?

Direct exposure is exposure to the original radioactive source (a power plant, a nuclear explosion). **Radioactive fallout** refers to dirt and debris contaminated with radiation; it can be produced by nuclear testing, explosion of a bomb, or an accident at a nuclear power plant. Winds and rain can spread radioactive fallout far from its original source.

Indirect exposure can also be dangerous, especially since it is often less obvious (Figure 18-4). Radiation contamination may travel from its original source through food chains. Grains grown in con-

taminated areas can pass on radioactivity. For example, milk can become radioactive if cows eat grain growing in areas exposed to radioactive rain. One major study documented indirect exposure through food chains resulting from atmospheric nuclear testing between 1945 and 1980, when the United States, the Soviet Union, France, the United Kingdom, and China detonated a total of 423 above-ground nuclear devices. All these explosions released massive amounts of radioactive debris into the upper atmosphere, where air currents distributed it all over the world. As the debris returned to earth, increasing levels of radiation began to enter various food chains (Figure 18-5).

What Is Nuclear Winter?

In the mid-1980s scientists investigating the potential environmental and human health effects of a nuclear war raised the probability of what some call the ultimate environmental disaster: **nuclear winter.** The explosions and fires following a nuclear war would pump vast quantities of smoke, soot, and debris into the atmosphere, effectively preventing sunlight from reaching huge areas of the globe. This sun block could result in temperature drops of 36° to 72°F (20° to 40°C). Given that drops of only a few degrees can drastically reduce photosynthetic rates, such a huge change would reduce agricultural yields and result in the mass extinction of countless plant and animal species, possibly including humans. Although scientists disagree over how severe and long lasting the effects would be from any one nuclear war scenario, they generally agree that some sort of nuclear winter would take place.

Social Boundaries

How Are Nuclear Resources Used?

Nuclear resources are used in a variety of ways, including medical applications, food preservation, power sources for satellites, production of unique metal alloys, generation of electricity, and military applications.

Medical Applications. Nuclear resources play a prominent role in the diagnosis and treatment of illness and the sterilization of supplies. They are so widely used that medical uses represent the most significant human-made source of radiation exposure.

Nuclear technology has allowed diagnostic medicine to progress further in the last few decades than in the entire previous history of medicine. Radioisotopes have played an important role in

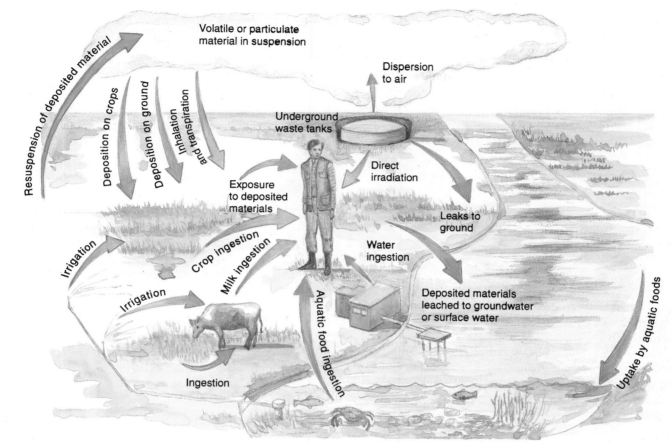

FIGURE 18-4: Major pathways of the environmental dispersion of radionuclides. Humans may be exposed to radiation directly, through contact with radioactive materials in the air or on the land, or indirectly, through food chains.

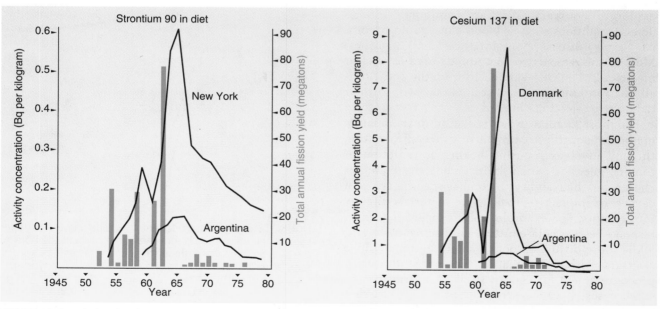

FIGURE 18-5: Relationship between nuclear testing and food-chain contamination by strontium 90 and cesium 137 produced during nuclear blasts. Concentrations of strontium 90 and cesium 137 (measured in bequerels per kilogram) in the total diet (graphed line, in black) are compared with the annual yield from atmospheric nuclear tests (red bars) measured in megatons. Exposure to the radioactive substances was much higher in the Northern Hemisphere (New York and Denmark) than in the Southern Hemisphere (Argentina).

Table 18-5
Widely Used Diagnostic Procedures
Incorporating Nuclear Imaging Technology

Brain scan. Investigates blood circulation and diseases of the brain such as infections, strokes, and tumors

Lung scan. Most commonly used to detect blood clots

Bone scan. Detects areas of bone growth, fractures, tumors, and infections

Liver and gallbladder imaging. Detects liver disorders, such as cirrhosis and tumors, and gallbladder diseases

Cardiac imaging. Shows blood flow to the heart (coronary artery diseases), heart functions, and evidence of recent heart attacks

Source: About Nuclear Medicine: Its Role in Diagnosis and Treatment of Disease (South Deerfield, Mass.: Channing L. Bete, 1989.)

developing safe, accurate methods of diagnosing medical conditions.

The most common form of diagnostic nuclear medicine is scanning or imaging (Table 18-5). During these procedures, the patient swallows, inhales, or is injected with radioactive compounds that travel through the body, continuously giving off gamma rays. Specially designed equipment detects and records the gamma rays, creating pictures or images of the body. Computers then interpret the test results.

These scans enable physicians to see inside the body without having to conduct traumatic exploratory surgery. They are commonly used to detect conditions such as blood clots, tumors, coronary artery disease, bone fractures, and infections. Most of the radioactive compounds used in nuclear medicine are quickly eliminated from the body, usually within hours or at most a day or two, so their radioactive effects are limited.

The most familiar use of radiation in therapeutic medicine is in the treatment of cancers. Radiation therapy involves focusing beams of radiation at tumors, injecting patients with radioactive materials chosen for their ability to combat specific types of cancer, or implanting radioactive materials directly into tumors. For example, solid cancerous tumors respond best to beta-emitting radioactive materials, while more diffuse tumors respond better to alpha emitters. Although radiation therapy is not a foolproof treatment for cancer, studies confirm that, in some cases, it can destroy or inhibit the further growth of a cancerous tumor. Radioactive materials can also be used as tracers to monitor the success of treatments for some disorders, usually to tell if specific dosages of medicines have been effective. Radiopharmaceuticals such as iodine 131 correct

some types of thyroid conditions, such as overactivity and tumors. Continued refinement of radiation therapies promises to improve cancer treatment in the future (Figure 18-6).

Using radiation to disinfect and sterilize medical instruments and supplies, such as surgical dressings, sutures, catheters, donor organs and tissues, and syringes, is a rapidly growing practice. Instead of the

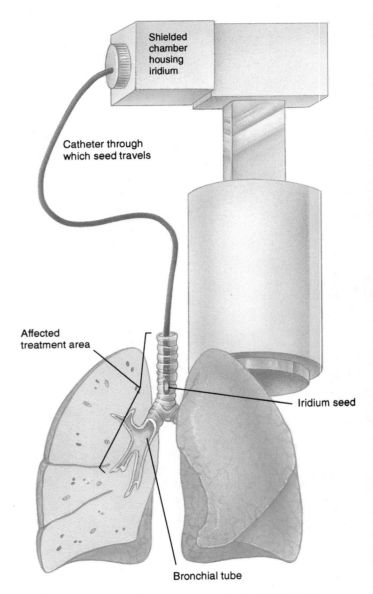

FIGURE 18-6: High dose rate remote afterloader (HDR) therapy. A catheterlike tube is sent through a cavity in the body to the affected area (here, the trachea of the respiratory system). After the passage is tested and cleared, the physician programs the HDR machine to send the iridium seed through the tube to the site. A computer precisely controls the movement of the seed, thereby more closely confining the radiation to the cancerous cells in its path than can be done in other methods of radiation therapy. Because the seed is so powerful, the prescribed dose is delivered quickly, thus reducing the time of exposure compared to other methods of radiation therapy.

traditional method of heating objects to kill contaminating bacteria and spores, medical supply companies use gamma rays to penetrate hermetically sealed packages, thereby sterilizing their contents without risking contamination during the packaging process. Sterilization by radiation, often referred to as the cold process, offers the only method of sterilizing such heat-sensitive materials as plastic heart valves and medicines.

Food Preservation. The use of low-level gamma radiation to kill pathogenic bacteria in foods is gaining acceptance with the Food and Drug Administration. In the irradiation process gamma rays emitted by radioactive cobalt 60 kill bacteria, insects, and fungi on produce and extend shelf life by inhibiting enzyme activity. The food itself does not become radioactive. The FDA approved irradiation for wheat flour in 1963, white potatoes in 1964, and fruits and vegetables in 1986. In 1985 the FDA approved irradiation in pork processing to destroy *Trichinella spiralis*, a microscopic roundworm found in the muscle tissue of hogs.

Those who support irradiation argue that it ensures a safe, wholesome food supply. Critics contend that more study and research are needed to determine if treating food with low-level radiation is safe for humans who consume the food. Gamma rays destroy some nutrients, especially vitamins A, C, E, and certain B's, and they also produce new chemicals in irradiated food. Many of these have not been identified or tested for toxicity, but some potentially dangerous chemicals have been identified, such as carcinogenic formaldehyde, benzene, mutagenic peroxides, and formic acid. However, these chemicals are formed in extremely small quantities; some nonirradiated foods contain larger quantities of these chemicals than irradiated foods. Other critics are worried that irradiation will be used to cover up serious bacterial contamination resulting from inadequate plant sanitation or product mishandling.

Power Sources for Satellites. Launched in 1977, the Voyager I and II satellites began their journeys to observe and photograph the planets and moons in our solar system. Because the Voyager satellites would travel too far from the sun to use the solar cells used to power previous satellites, NASA had to develop a power source capable of functioning effectively for many years. It chose small plutonium reactors. In 1992 Voyager II's reactor was still functioning as the satellite continued its journey outside our solar system.

Some people protest that nuclear-powered satellites could endanger the planet if any accidents occurred during launching or orbit. For these rea-

sons, antinuclear groups protested the 1989 launching of Galileo, a nuclear-powered satellite designed to orbit Jupiter and its moons. Galileo took off safely, but fear of accidents is not unwarranted. In 1983 the Soviet's Cosmos 1402 satellite fell from orbit and crashed to earth. Fortunately, the nuclear reactor disintegrated harmlessly upon reentry into our atmosphere.

Radioactive materials are also used as power sources for robots that are designed to operate in environments hostile to humans, such as nuclear power plants, burning buildings, outer space, and the ocean floor.

Production of Metal Alloys. Small amounts of the radioactive element thorium added to other metals can produce stronger, lighter weight, and more heat-resistant metal alloys. These alloys are most commonly used in aircraft engines and airframe construction. Some forms of uranium have been used in armor plating for tanks.

Generation of Electricity

How Conventional Nuclear Fission Reactors Work. Conventional fission reactors work by sustaining the fissioning process. Typically, the fuel used is uranium 235 (U-235), a readily fissionable isotope of the element uranium. The concentration of U-235 in uranium ore is only about 0.07 percent, too low to sustain a chain reaction; uranium 238, a nonfissionable isotope, is the principal constituent of uranium ore. Before uranium can be economically used in a reactor therefore, impurities must be removed and the concentration of U-235 raised to increase the incidence of fissioning. These tasks are accomplished by milling the ore to a granular form and using an acid to dissolve out the uranium. After it is converted to a gas, the fuel is purified. Its concentration of U-235 is enriched to approximately 3 percent, and the enriched gaseous form of uranium is converted into powdered uranium dioxide. The uranium dioxide powder is subjected to high pressures to form fuel pellets about ⅝ inch long and ⅜ inch in diameter. The U-235 in just three reactor pellets generates about the same amount of electricity as 3.5 tons of coal or 12 barrels of oil. The pellets are then placed into a 12-foot-long metal tube (about 5 pounds of pellets per tube), or fuel rod, which is made of a zirconium alloy that is highly resistant to heat, radiation, and corrosion.

The actual nuclear reaction occurs within the reactor core of a power plant, which is housed in a thick-walled structure called a **containment vessel** (Figure 18-7). Within the core are hundreds of fuel assemblies, each composed of groups of fuel rods. A material called a moderator is placed or circulated

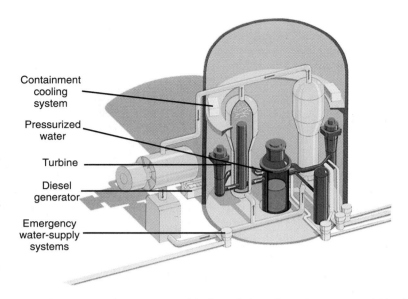

FIGURE 18-7: How a nuclear power plant operates. (1) The splitting of uranium atoms within the reactor core releases heat. (2) Pressurized water in the primary (coolant) loop carries heat to the steam generator. (3) Steam generator vaporizes water in a secondary (power-generating) loop to drive the turbine.

between the rods and assemblies to slow down the neutrons emitted by the fission reaction, increasing the likelihood of fission and sustaining the chain reaction. Most commercial reactors worldwide use ordinary water, called light water, as a moderator. Others, including about half of the reactors in the former Soviet republics, use solid graphite, a form of carbon. Some reactors use heavy water as a moderator. Heavy water is a compound in which the hydrogen atoms are replaced by deuterium, the hydrogen isotope containing two neutrons.

A coolant circulated throughout the reactor core absorbs and removes heat to prevent the fuel rods and other materials from melting. Water is used as a coolant in most water-moderated and graphite-moderated reactors. In some reactors a gas or liquid metal is used as a coolant. If the coolant is water, it may be heated directly to steam, which can then be used to drive the generator turbines that produce electricity. In other nuclear reactors coolants are kept under pressure and passed through heat exchangers to heat another body of water to its boiling point, creating steam to drive the turbines. Operators can control the rate of fission inside the reactor core by raising or lowering metal control rods that absorb neutrons into vacant holes within each fuel assembly (16 vacant holes per assembly).

Reserves of Radioactive Materials. Currently, the nuclear power industry is dependent on the element uranium, a finite resource. The United States contains the largest uranium reserves in the world, with the richest deposits located in western states such as New Mexico, Wyoming, Colorado, Utah, Arizona,

and Texas. Existing uranium reserves worldwid estimated to last about 100 years.

Thorium is another natural radioactive mat being explored for potential uses in the comm production of electricity. Thorium is three t more abundant in the earth's crust than uraniu has been found in soils worldwide, but is m concentrated in the sands of India, Brazil, and lon. Major deposits within the United State found in Florida, Idaho, and the Carolinas.

Breeder Nuclear Fission Reactors. To compensat limited supplies of U-235, researchers have d oped the breeder nuclear fission reactor. Bre reactors are similar to conventional fission rea with one important distinction: they produc extremely toxic, fissionable material plutonium (Pu-239) from uranium 238, the common, nc sionable isotope of uranium. Some of the neu released by the fissioning of Pu-239 cause othe 239 atoms to split, while others "breed" more 239 from U-238. Theoretically, for every 100 a of Pu-239 consumed in the fission reactions, atoms of Pu-239 are produced. However, few b er reactors are in use today because they cost tv three times as much as conventional rea because of safety concerns associated with the l ly toxic plutonium.

Military Applications. Nuclear bombs and warl are probably the most widely recognized mil use of nuclear resources. An atomic or nuclea sion bomb releases a tremendous amount of er in a split second through an uncontrolled chain tion. These weapons use plutonium, which the tary produces at several reactors in the United States. Nuclear reactors (a controlled chain reaction) are also used to power submarines and ships, eliminating the need to refuel frequently, but posing a danger of contamination if the vessel is destroyed. Because military applications are intricately tied to the historical development of nuclear resources, they are discussed in further detail below.

History of Management of Nuclear Resources

How Have Nuclear Resources Been Managed Historically?

It has been only fifty years or so since humankind first learned to tap the tremendous potential of nuclear resources. Since World War II our understanding of the potential benefits, dangers, and complexity of nuclear resources has changed dramatically. Many people argue that we have not yet learned to manage these resources in an environmentally safe manner.

Early Uses of Nuclear Resources. Before the United States dropped the first atomic bomb, scientists could only speculate on how such a huge concentration of radiation would affect living organisms, including humans. Of course, those near the bombs were killed instantly; the number of people killed at Hiroshima and Nagasaki soared to over 200,000 almost immediately. Only after years of study and experimentation, however, have we begun to realize the true range of radiation's devastating effects (Figure 18-8).

Our ignorance about the effects of radiation led the military and government to conduct studies which today would be considered criminally careless. On several occasions troops were stationed within blast zones of aboveground nuclear explosions to test their ability to negotiate radiation-contaminated terrain. The Atomic Energy Commission, predecessor of the Nuclear Regulatory Commission, conducted tests such as injecting subjects with uranium, feeding subjects fallout from atomic bomb tests, and having subjects breathe radioactive air. The Soviets conducted similar tests.

The Cold War and the Arms Race. The attacks on Hiroshima and Nagasaki ended the war in the Pacific, but they signaled the dawn of an era marked by fears of a nuclear confrontation between the United States and the Soviet Union. In 1947 the United States established the Atomic Energy Commission (AEC) to control the use and disclosure of information on atomic power. When the Soviet Union deto-

FIGURE 18-8: Kunizo Hanataka and his mentally retarded daughter, Yuriko. Yuriko was conceived 3 months before the bombing of Hiroshima. Her mother was about 2,400 feet from the hypocenter of the bomb. The radiation caused severe mental retardation in Yuriko. By studying Yuriko and others like her in Hiroshima and Nagasaki, researchers have been able to establish the window in pregnancy of 8 to 15 weeks during which the developing fetus is extremely susceptible to radiation exposure. Kunizo operates a small barber shop from his home so that he can care for Yuriko, who requires full-time attention.

nated its first atomic bomb in 1949, the nuclear race between the two superpowers began, and strict control of the secrets of nuclear technology was deemed even more important to national security.

In 1963 concern over radioactive fallout spurred 105 nations to sign the Nuclear Test Ban Treaty, agreeing to halt nuclear testing in the atmosphere, in outer space, under water, and anywhere else where fallout might spread beyond the borders of the country testing the weapon. This treaty eliminated and restricted some testing, but the United States and other countries continued underground testing.

One of the first attempts to halt the spread of nuclear weapons technology was the Nuclear Non-Proliferation Treaty, signed by 17 nations in 1968. This treaty stipulated that the United States and the Soviet Union would not provide nuclear weapons to other countries, nor would they assist other nations in developing these weapons. Further, the signatories agreed to facilitate the development of peaceful uses of nuclear energy. Despite global concern arsenals of nuclear weapons continued to grow, since both the United States and the Soviet Union argued that the only way to prevent a nuclear war was to be well prepared against one. As of 1987, the United Kingdom, France, China, and India, in addition to the United States and the Soviet Union, had all built and tested nuclear weapons, and at least 21 other countries were known to have the capability to do

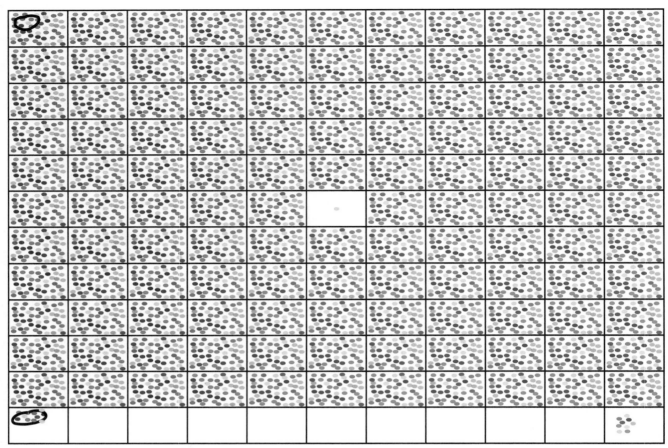

FIGURE 18-9: Graphic depiction of the extent of the world's nuclear firepower during the Cold War era. The dot in the center square represents all the firepower of World War II, equal to 3 megatons. The other dots represent the firepower in nuclear weapons in 1987, 18,000 megatons (equal to 6,000 World War IIs). About half belonged to the former Soviet Union, the other half to the United States. The circle at the top left represents the weapons on just one of the 31 U.S. Poseidon submarines operational in 1987, 9 megatons—enough to destroy more than 200 of the largest Soviet cities. The circle at the bottom left represents one of the newer Trident submarines, 24 megatons—enough to destroy every major city in the Northern Hemisphere. Place a dime on the chart; the covered dots represent enough firepower to destroy all the large and medium-size cities in the entire world. Since 1987, efforts have been made to reduce the world's nuclear arsenal. Poseidon submarines are being phased out, for example, with the United States Planning to rely on a fleet of 18 Trident submarines.

so. Some experts say any country with the capability to build a nuclear reactor can build an atomic bomb. For these reasons, people still fear the consequences should such weapons, or the capability to produce them, fall into the hands of unstable governments, terrorists, or guerilla groups (Figure 18-9).

The 1991 breakup of the Soviet Union offered real hope for the dismantling of the superpowers' nuclear arsenals. Within months of the breakup both Soviet and U.S. officials revealed plans to cut back on certain types of nuclear weapons, easing fears of a nuclear confrontation. That fear has not been eliminated, however, and the buildup of nuclear weapons remains a legitimate concern. This fear was brought home in 1991, when a team of United Nations observers investigated the possibility that the Iraqi government, under Saddam Hussein, had developed nuclear weapons. After months of investigation, the extent of Iraq's nuclear capability remained unclear.

Development of Peaceful Uses of Nuclear Power. Worried that the air of secrecy and the public concern over the bomb's destructive powers would give atomic energy a negative image, the federal government decided to encourage the creation of more constructive uses for nuclear resources. In March 1953 the AEC made the development of economically competitive nuclear power a goal of national importance. That same year President Eisenhower, in a speech entitled "Atoms for Peace," emphasized the need for productive uses of nuclear power.

The government's commitment to developing civilian nuclear capabilities led to the 1954 Atomic Energy Act, which opened the nuclear industry to the private sector. The first major developments were in the areas of medicine, biological research, and agriculture, but at the end of the 1950s interest began to turn toward using nuclear energy to produce electricity. This interest was spurred by lowered estimates of fossil fuel reserves, a growing demand for electrical power, and the 1956 war over the Suez Canal, a major route for transporting oil. By 1957 the world's first commercial nuclear reactor, the Shippingport nuclear plant, came on line at Pittsburgh, Pennsylvania.

Nuclear-powered production of electricity increased slowly but steadily until the early 1970s, when an oil embargo by the Organization of Petroleum Exporting Countries (OPEC) compelled the United States and other industrialized nations to turn to alternative fuel sources. Before the OPEC embargo nuclear energy produced only about 5 percent of electricity in the United States (compared to 17 percent for oil); by 1990 it was our second largest source of electricity after coal, accounting for 20 percent of electric production (oil's contribution had fallen to just 6 percent). Global dependence on nuclear power has also grown, accounting for 20 percent of electric production worldwide. Today, 11 countries rely more heavily on nuclear power than does the United States, led by France, at 70 percent of electric production, and Belgium, at 66 percent. (Reliance on nuclear power in the United States and worldwide is discussed more thoroughly in Chapter 13, Energy: Alternative Sources.)

Growing Opposition to the Use of Nuclear Resources. While the U.S. nuclear industry grew significantly during the 1970s and early 1980s, it has tapered off in recent years for a number of reasons. The demand for electricity is lower than predicted. Costs have risen because of longer lead times for licensing and construction of plants and higher financing expenses. Accidents at the Three Mile Island nuclear facility near Harrisburg, Pennsylvania, and the Soviet Union's Chernobyl nuclear power plant have raised concerns about the safety of nuclear power plants. Moreover, the environmental impact of nuclear waste disposal, decommissioning of old nuclear power plants, and the secrecy surrounding nuclear resource issues has fueled public concern about the long-term consequences of using nuclear resources.

What Environmental Problems Are Associated with the Use of Nuclear Resources?

The use of nuclear resources involves serious problems, among them radioactive leaks from military and commercial facilities, accidents at nuclear power plants, disposal of nuclear wastes, decommissioning of old plants, and the secrecy surrounding nuclear activities.

Radioactive Leaks. Although military and commercial nuclear facilities have accidentally leaked radioactive substances over the years, many of these incidents have gone unreported to the general public. Only recently have we begun to realize the scope of environmental contamination that has taken place from military and commercial nuclear facilities (Figure 18-10).

For example, recent reports have disclosed that between 1952 and 1954 the Hanford Federal Nuclear Facility in the state of Washington released large quantities of radioactive ruthenium into the atmosphere on nine different occasions. (See Environmental Science in Action: The Hanford Federal Nuclear Facility, pages 390–393.) Other government reports have discovered releases of over 395,000 pounds (177,750 kilograms) of uranium into the atmosphere from the Feed Materials Production Center at Fernald, Ohio, just northwest of Cincinnati. One study commissioned by residents in the Fernald area estimated that those living near the plant have at times inhaled air containing almost 70 times the levels of uranium prescribed by the Department of Energy's safety guidelines.

Another frightening discovery occurred at the Rocky Flats munition plant, only 50 miles upwind from the 1.4 million population center of Denver. Smokestack monitors have registered plutonium emissions 16,000 times greater than the standard levels. Some environmentalists are referring to this situation as the "creeping Chernobyl."

Accidents at Nuclear Power Plants. Even more frightening than radioactive leaks is the specter of a major nuclear accident—a **meltdown** in which the reactor core becomes so hot that the fuel rods melt,

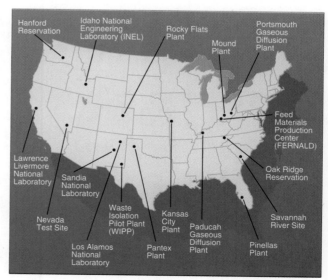

FIGURE 18-10: Department of Energy Weapons Complex, 1992. The map shows those sites where research and development, testing, and production of nuclear weapons occur. Although weapons are no longer produced at the Hanford site, the reservation has not been dismantled and cleaned up. That process is now ongoing.

possibly burning through the containment vessel and the underlying concrete slab and boring into the earth.

Researchers have investigated the probability of major nuclear catastrophes. A 1982 study conducted by the Sandia National Laboratory estimated that the chance that a catastrophic accident will occur in a 1-year-old reactor is 20 million to 1. However, other studies, which consider factors such as human error that make accidents much more likely, estimate that a catastrophic accident will occur every 8 to 20 years.

As nuclear power plants age, the probability of accidents increases. The average life span of nuclear plants is estimated to be 30 to 40 years. A large number of the world's nuclear plants are approaching this limit. By the year 2000, 150 plants will have exhausted their usefulness. Aging plants are often plagued by corrosion of steam generators and the embrittlement of steel pressure vessels through neutron bombardment. These conditions are difficult to remedy and greatly increase the possibility of disaster in the case of overheated reactors. The Yankee Rowe power plant in western Massachusetts, one of the nation's oldest nuclear plants, was shut down in 1991 through the actions of the Union of Concerned Scientists, who charged that the reactor vessel had become so embrittled that any sudden temperature change could cause it to rupture, possibly leading to a meltdown.

History seems to confirm that catastrophic accidents are more likely than we would wish. In 1979 the Three Mile Island nuclear facility suffered a partial meltdown when a loss of coolant in the nuclear core caused reactor fuel to overheat. Luckily, no immediate casualties were reported and the amount of radioactive material escaping into the environment was minimized by the containment vessel surrounding the nuclear core.

Just seven years later, one of the most devastating nuclear accidents ever rocked the Soviet Union. On April 26, 1986, operators at the Chernobyl nuclear plant turned off the plant's safety systems in order to conduct some unauthorized tests; while the safety systems were down, the plant's reactor number 4 overheated and exploded, spewing tons of cesium, iodine, uranium fuel, and other radioactive contaminants about 3 miles (5 kilometers) into the atmosphere. The graphite moderator caught fire, further spreading contamination through smoke and flames. Thirty-one people died as a direct result of the accident, and 237 received severe radiation injuries. Whole communities within an 18-mile radius had to be evacuated, with no hope of returning for several generations. In all, 135,000 people from 179 villages were evacuated, and topsoil and trees were stripped away and moved to an unspecified site. Radiation was detected by countries around the world (Figure 18-11).

In 1975 the U.S. Nuclear Regulatory Commission published its estimates of the worst damage a reactor accident could possibly cause: 3,300 immediate human deaths, 45,000 human deaths from cancer, 45,000 radiation sickness victims requiring hospitalization, 240,000 people suffering from thyroid tumors, and 5,000 children born with genetic defects in the first generation following the accident. Accidents at Chernobyl, Three Mile Island, and

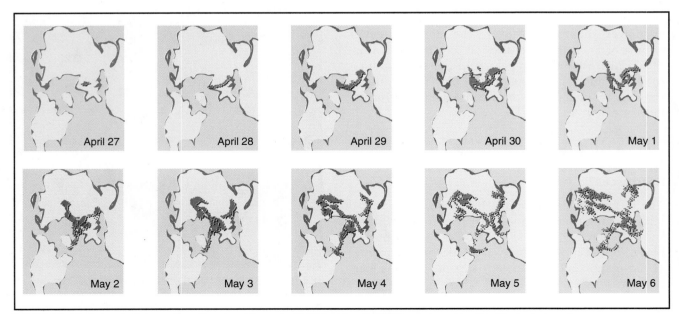

FIGURE 18-11: Spread of the radioactive cloud from April 27 to May 6, 1986, produced by the Chernobyl explosion. The maps provide a good qualitative picture of the cloud's dispersion, but contain no information about terrain, particle size, and vertical distribution.

▶ Table 18-6
Selected Nuclear Power Plant Accidents

December 12, 1952. A partial meltdown occurred at Chalk River, Canada. This accident represents the first known malfunction of a nuclear power plant. No injuries were reported.

October 7, 1957. A plant similar in design to Chernobyl caught fire north of Liverpool, England. Two hundred square miles (518 square kilometers) of countryside were contaminated, and at least 33 cancer-related deaths were reported.

January 3, 1961. A steam explosion occurred at a military experimental reactor near Idaho Falls, Idaho. Three servicemen died. Their deaths were the first casualties in the history of U.S. nuclear reactor production.

March 28, 1979. A loss of coolant at the Three Mile Island plant near Harrisburg, Pennsylvania, caused the radioactive fuel to overheat. The buildup of heat led to a partial meltdown and the release of radioactive material into the atmosphere. The hazards posed to local inhabitants are still being debated.

March 8, 1981. Radioactive waste leaked from the storage tanks of a plant in Tsuruga, Japan. The leak was not disclosed until radiation was detected in the city's bay six weeks later.

January 6, 1986. One worker died and 100 were hospitalized when an improperly heated, over-filled container of nuclear material burst at a uranium-processing plant in Gore, Oklahoma.

April 26, 1986. The Chernobyl number 4 reactor exploded, spewing tons of radioactive materials 3 miles (5 kilometers) into the atmosphere. Thirty-one workers died as a direct result of the explosion, and 237 received severe radiation burns. Surrounding communities were completely contaminated with radiation, which was also detected around the world.

other plants prove that such a disaster could occur; some experts say the worst case scenario has been avoided so far only by luck (Table 18-6).

Disposal of Nuclear Wastes. Almost all of the waste produced by the nuclear industry is radioactive to some degree and therefore cannot be disposed of in the same manner as solid or even other hazardous wastes. Four classes of waste are produced—uranium mill tailings and low-level, transuranic, and high-level wastes—each of which poses a unique threat to the environment.

Uranium mill tailings are created when uranium ore is milled or processed into an enriched form which can be used in a reactor. Extracting just 1 pound of finished reactor fuel generates approximately 500 to 1,000 pounds (225 to 450 kilograms) of tailings. Because the amount of tailings is so huge, the tailings are frequently piled in the open. There, wind and rain can disperse them in the environment. Over 200 million tons (182 million metric tons) have accumulated throughout the western United States alone. Tailings have also found their

way into such everyday objects as road pavement, fill material for construction sites, and sand for golf course sand traps and children's sandboxes.

Low-level radioactive wastes are radioactive liquids and solids that will remain dangerous for a few hundred years or less. They include air filters, rags, protective clothing, and tools used in routine plant maintenance and fuel fabrication, as well as liquids used to cool reactors.

For several decades much of the low-level waste produced in the United States was mixed with concrete, encased in steel drums, and dumped into the ocean. Although the EPA originally condoned this practice as safe, they reversed their opinion in the mid-1970s. Scientists found that many of the barrels had begun leaking, possibly eaten away by the corrosive salt water, and that radioactive leakage did not safely dilute and disperse, but rather left long-lived hot spots of radioactivity. Although this discovery led to the halt of ocean dumping by the United States in 1983, over 90,000 barrels of radioactive materials had already been tossed into the Atlantic and Pacific oceans, some near commercial fishing zones.

In the past U.S. nuclear facilities also disposed of some solid low-level waste by burying it in long trenches, usually without any type of lining to protect against leaching. Materials were often packed in cardboard, wooden, and metal containers which were sometimes damaged in the dumping process, allowing the radioactive material to spill directly into the trench. As groundwater moved through the trenches, radioactive leachate would slowly migrate from the site.

The Low-Level Radioactive Waste Act was passed in 1980 to control the growing number of cases of groundwater contaminated by low-level waste. This act requires each state to either establish a low-level waste site or gain access to another state's facility by 1992. Although researchers have investigated all sorts of disposal methods, including shallow-land burial, aboveground vaults, modular concrete canisters, earth mound bunkers, mined cavities, and augered holes, they have yet to determine that any one method is most effective.

Transuranic radioactive waste contains human-made radioactive elements or any element with an atomic number higher than that of uranium. These elements, which include plutonium (half-life 24,000 years), americium (half-life 430 years), and neptunium (half-life 2 million years), must be handled and stored with extreme care. Just 1 ounce of plutonium could kill every human on earth if it were dispersed widely enough. Inhaling or absorbing even a tiny speck of plutonium can cause cancer or death. Unfortunately, the nearly 12 million cubic feet (339,000 cubic meters) of transuranic waste that have been generated over the last 30 years have often not been properly isolated from the environment.

High-level radioactive wastes remain radioactive for tens of thousands of years or more. They consist primarily of spent or used reactor fuel and fission products created when spent fuel rods are reprocessed. The uranium 235 in each fuel rod lasts about three to four years, after which it can no longer sustain the fissioning process and must be removed and replaced. Most commercial plants store used rods on site. Studies predict that the United States will accumulate roughly 43,000 tons (39,130 metric tons) of spent fuel by the year 2000. Many military reactors reprocess the rods by separating the remaining uranium 235 and plutonium 239 (which is produced during the fission reaction and can be used as fuel for atomic weapons) from the other fission products.

Spent fuel rods and the fission products produced during reprocessing pose major challenges for permanent, safe disposal. Presently, there is no satisfactory procedure for isolating these highly radioactive wastes from the environment for the tens of thousands of years necessary to guarantee the safety of future generations. Much of this waste is stored in concrete and steel silos and tanks or in metal drums which are often exposed to various climatic conditions. The combination of outside weathering and internal corrosion has already led to alarming amounts of leakage from these containers.

Long-Term Nuclear Waste Disposal. As the problems involved in storing radioactive wastes mount, scientists search for a permanent method of disposal. The ideal solution would be to alter the wastes chemically so that they are no longer harmful, but experts doubt if we will ever find a way to accomplish such a transformation. Another method proposed in the past was to shoot containers of waste deep into space. However, this path was never pursued because of concerns about the possibility of accidents during launching and a distaste for polluting the universe with such hazardous substances. Today, the two most often proposed means of permanent disposal are ocean-bed repositories and deep underground sites.

Advocates of burying waste deep below the ocean's floor point out that this method offers the double advantage of numerous appropriate sites and the removal of radioactive waste from our immediate surroundings. Ocean-bed repositories would also make use of areas without "resource potential," traditionally defined as having the ability to produce food or other resources useful to humans. Although the Department of Energy rejected a proposal to develop ocean-bed repositories in 1982, they are currently reexamining the possibility (Figure 18-12). However, political opposition and high cost esti-

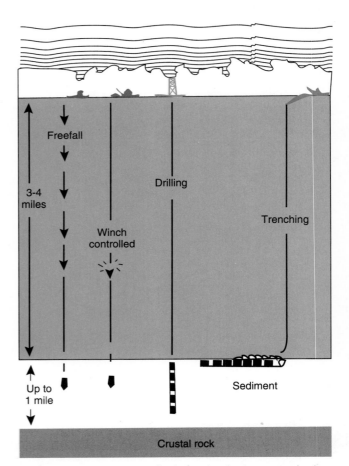

FIGURE 18-12: Various methods for developing ocean-bed repositories. Material may simply be dropped to the ocean floor in a freefall, or its fall may be controlled by a winch attached to a transport ship. Drilling and trenching are two other proposed methods.

mates may stand in the way of implementing this solution.

Land-based repositories seem to present our best option for permanent disposal so far. Although past plans have been abandoned due to uncertainty as to how well the waste could be contained under specific geologic conditions, Department of Energy scientists feel that they are capable of constructing safe, permanent waste disposal sites underground. Military wastes are already being stored in an underground site near Carlsbad, New Mexico. Some experts worry that water can enter the repository and carry highly radioactive plutonium 239 to the Pecos River, but others maintain that the site is safe.

One of the more difficult questions facing the planners of such a site—for reasons of public concern as well as geologic criteria—is where it should be located. Fear of radioactive waste leads to a "Not in my backyard" (NIMBY) mentality. Most individuals see a need for safe, permanent storage facilities for radioactive waste, but no one wants a facility in his or her neighborhood. Others go even further to a NIABY position, "Not in anybody's backyard."

In 1982 Congress attempted to address the problem of location by passing the Nuclear Waste Policy Act (NWPA). The NWPA directed the Department of Energy to select two land-based repositories (one in the eastern half of the country and one in the western) by investigating and comparing numerous potential sites. The first site was meant to be operational by the mid-1990s. Once a site was established, machinery would lower canisters of waste packed in lined, filled cells 2,000 to 4,000 feet (600 to 1,200 meters) below ground. Then special vehicles would transport the cells through tunnels to a permanent storage spot. Once filled, the entire maze of holes, tunnels and shafts would be backfilled, sealed, and protected from drilling and mining by law (Figure 18-13).

From three potential sites in the west, the Department of Energy chose an area at Yucca Mountain, Nevada, characterized by volcanic rock. Critics point out that the department never went through the comprehensive selection process spelled out by the NWPA, focusing instead on sites it had had in mind all along. Despite the prospect of more jobs and a government incentive of $1 million per year, Nevada residents were outraged at being designated the country's nuclear waste dump with little choice in the matter. The indefinite postponement of a search for the eastern site only made matters worse.

Although the Nevada state government fought to prevent the creation of the repository, the site will eventually be installed as planned. However, the delays and complications have multiplied original cost estimates by 1,500 percent, and burial will not begin until 2010 at the earliest. While land-based repositories may be a satisfactory solution to our radioactive waste problem from a technical viewpoint, the political ramifications may very well impede their widespread use in the future.

Decommissioning Old Plants. Perhaps the most overlooked cost associated with nuclear power is that of disposing of these plants once they are too old to function. To date, no major reactors have been retired, so the methods and their costs are still a matter of speculation. In 1989 the Shippingport plant, a commercial prototype capable of producing only 7 percent of the power of an average reactor, was dismantled and the reactor buried at the Hanford facility in Washington State. This small operation cost $98 million.

TUFF FILE
Formed of compacted volcanic ash and dust, tuff from excavations would be piled at the surface - then used later to reseal the tunnels.

SURFACE FACILITY
Delivered by rail or truck, nuclear waste would be unloaded at two buildings, resealed in steel or copper canisters, then transported via ramps to the storage area below.

FINAL BURIAL
Lowered into boreholes, canisters would be monitored for leaks for 50 years. Then tunnels, shafts and ramps would be filled and sealed.

TUNNEL NETWORK
Covering about 1,400 acres, the underground mines would have room to store a total of 77,000 tons of radioactive waste.

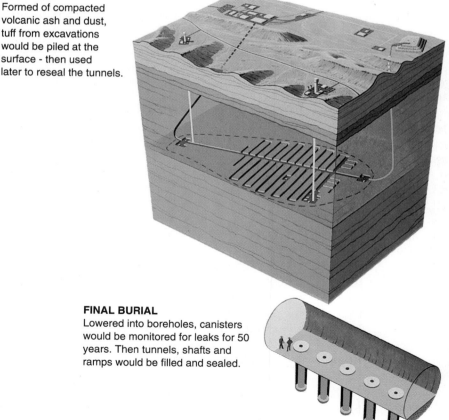

FIGURE 18-13: Land-based repository, Yucca Mountain, Nevada. The federal government proposes to place sealed canisters of radioactive waste in bore holes 1,000 feet below the mountain.

What is the best way to deal with the host of much larger commercial reactors that will have to be retired in the next few decades? Plant owners can simply entomb the entire reactor building in concrete and bury it, although they cannot be certain that the concrete will be able to contain the contamination over the centuries. The preferred method is to decommission the plant by decontaminating and dismantling the reactor and then disposing of the resulting radioactive parts and waste at an appropriate facility. Estimates for the cost of decommissioning an average-size plant range from $50 million to $3 billion—sometimes more than building a new plant. Experts predict that a dismantled plant will produce approximately 653,176 cubic feet (18,000 cubic meters) of low-level waste, enough to cover a football field 13 feet (4 meters) deep.

Secrecy Surrounding Nuclear Activities. Secrecy has played a prominent role in the development of nuclear resources. When the U.S. government began developing nuclear power in the early 1940s, it stressed the importance of withholding information for reasons of national security. Scientists working on the Manhattan Project were not permitted to share information among their peers in the scientific community. All discoveries and developments became the secret property of the federal government. In 1946 Congress even passed legislation establishing the death penalty as the punishment for anyone revealing atomic secrets.

Perhaps the greatest safeguard of government secrecy has grown out of the nature of the Atomic Energy Commission (AEC). This government agency was set up to promote, regulate, and license the nuclear industry and also to control the flow of information to those outside the agency. Because its military operations were deemed vital to national security, the AEC was also granted the unprecedented power to virtually regulate itself. Self-regulation exempted the AEC from outside reviews of its spending and from external regulation of defense-related impacts on the environment. By 1974 the AEC had evolved into the Nuclear Regulatory Commission, responsible for licensing and regulating commercial nuclear facilities, and the Energy Research and Development Administration (ERDA), responsible for energy research and the construction of nuclear weapons. In 1977 ERDA became the Department of Energy (DOE), which inherited the privilege of self-regulation originally granted to the AEC.

Despite protests by scientists and the private nuclear industry, the government has consistently hidden the nature of its nuclear activities from the public—even in cases where leaks and accidents threatened the lives of workers and people living in the area. A recent congressional report shows that the federal government failed to inform workers at federal weapons plants of studies conducted in the 1940s which involved substantial releases of radiation. Although most releases were accidental, at least one that occurred at the Hanford plant in 1949 was a planned military experiment, the details of which are still secret. Over the last 45 years, more than 600,000 people have worked at U.S. weapons facilities without knowing whether or not they were being exposed to harmful releases of radiation.

Countless similar situations prompt critics to point out that the government has used its privilege of secrecy to cover up incompetence and carelessness which have put the lives of countless citizens at risk. Most of these cover-ups have come to light recently only because of public pressure on the federal government to release information pertaining to government nuclear facilities.

The problem of secrecy is not confined to the United States. The Soviet government attempted to cover up the 1986 Chernobyl disaster, but was unsuccessful because of the extent of radioactive contamination. Thirty-six hours after the accident, technicians at a Swedish nuclear power plant noted disturbing signs of high radiation, the first indication to the rest of the world that a nuclear accident had taken place. The USSR's reluctance to reveal information about the disaster prevented other countries from sending immediate aid and put the lives of people in nearby countries at risk without their knowledge.

Future Management of Nuclear Resources

The 1990s will be a crucial decade for nuclear resources. By all indications, we are standing at a crossroads in our development of nuclear energy technology: we can pursue greater reliance on it, maintain it at current levels, or phase it out. Currently, the growth of the nuclear power industry is tapering off because of decreased demand, rising costs, and public concern over accidents and waste disposal. Between 1991 and 1995 new sources of nuclear power are expected to add only 18,400 megawatts to the world output, only 20 percent of the amount added between 1981 and 1985. The use of nuclear resources for other applications, particularly medical, may well expand. But in these areas, too, it is essential to proceed carefully. In order to ensure the environmentally sound management of nuclear resources for both energy and nonenergy applications, it is important to prevent overuse through conservation, protect human health and the environment, and preserve living systems. What You Can Do: To Encourage Safe Use of Nuclear Resources suggests actions individuals can take.

What You Can Do:
To Encourage Safe Use of Nuclear Resources

- Conserve energy. The higher the demand for electricity, the greater the need for additional power sources such as nuclear energy. You can conserve energy by installing energy-saving devices on appliances in your home or by simply using less energy.
- Make an effort to learn about alternative, "clean" energy sources. If possible, switch over to these energy sources. Chapter 13, Energy: Alternative Sources, contains numerous suggestions for conserving fossil fuels and developing alternative energy sources.
- Make an effort to learn about the applications of nuclear resources and the advantages and disadvantages associated with each application.
- Join a community or national organization committed to protecting the environment. Such organizations can help raise public consciousness and compel government accountability.
- Make local and national politicians aware of your views.

Prevent Overuse Through Conservation

Conservation measures can help to reduce the demand for electrical power. Reductions in demand translate into a reduced need for power plants, specifically for nuclear power plants. Given that nuclear power generation is costly and potentially dangerous, it is prudent to minimize the use of nuclear resources for this purpose. Spending additional funds to develop alternative energies and mandating conservation measures (such as increased fuel efficiencies for automobiles, a substantial increase in the gasoline tax, and increased recycling of materials like aluminum) can help our society to become energy self-sufficient for far less than the costs of nuclear energy. Achieving energy self-sufficiency through conservation and alternative energy sources would help safeguard national security (a goal often espoused by proponents of nuclear energy), while protecting the environment and human health.

Protect Human Health and the Environment

Based on an expected decline in the demand for nuclear power, experts predict that the United States and many other countries will become less dependent on nuclear power as aging plants are retired and no new plants are built. Nuclear power will probably become popular again only in the event that demand for electricity rises rapidly or scientists develop technology to make nuclear power plants safer and to dispose of radioactive waste safely. However, regardless of the course we take, we should first confront several serious problems which increase the dangers posed by nuclear technology.

First, we must work toward a truly safe, permanent way to dispose of nuclear waste. Radioactive waste is a long-term problem, living on for centuries after its creators, and so requires a careful, well-reasoned, long-term solution. To achieve this solution, we could divert money and effort from nuclear weapons to waste management. Only when we have developed adequate methods to dispose of our nuclear waste will it make sense to create more from any source. After all, no government or human convention has lasted anywhere near as long as the half-life of many nuclear waste products.

Second, we must address the issue of secrecy by our government and private industry. In order to ensure public access to information on how nuclear technology is being used and to prevent the government from covering up activities that endanger citizens' lives in the name of national security, we must create a regulatory agency that will have the public's interest at heart. One option would be to give one or more already existing regulatory agencies increased jurisdiction over the Department of Energy. Even today, state governments and the federal EPA are making strides in forcing the Department of Energy to clean up radioactive waste. Another possibility would be to create an agency completely free from any federal association, yet empowered by the people to oversee our country's nuclear activities.

Preserve Living Systems

Because the problems of nuclear technology are rarely contained within one country, we must also work to solve our problems in a global context. Mandatory standards should be set to govern construction of nuclear reactors worldwide; international teams must be created to respond to nuclear crises; quick and concise reporting procedures must be established for all nuclear accidents; international licensing requirements for all operators of nuclear facilities should be defined; and international monitoring and inspections programs must be initiated.

Nuclear energy is one of the most controversial issues of our era, and with good reason. We can benefit from the many uses of fission energy just as we can be harmed by radiation from fallout, nuclear

(Text continues on page 394.)

Environmental Science in Action: Hanford Federal Nuclear Facility

Tom Pedroni

During World War II the United States established the 560-square-mile (1,452-square-kilometer) Hanford Reservation in southeastern Washington, a facility for producing plutonium as part of the Manhattan Project. Hanford plutonium initiated the world's first atomic explosion—the Trinity test in Alamogordo, New Mexico—and the bomb dropped on Nagasaki, Japan.

Activities at Hanford eventually expanded beyond plutonium production into areas such as electrical power generation, research, and development. Decreased demand for plutonium, a growing concern about plant safety, and decades of inadequate waste disposal practices brought plutonium processing at Hanford to a halt. Hanford's sole mission is now waste management and cleanup.

Describing the Resource

Physical Boundaries

The Hanford site lies on a semiarid desert plain in Washington State. The Columbia River flows through the northern portion of the reservation. Downriver from Hanford lie the "Tri-Cities" of Richland, Kennewick, and Pasco, with a combined population of about 144,000 (Figure 18-14).

The geology of the area has unique characteristics which affect the movement of groundwater. The ground beneath Hanford is made up of three major layers. Directly below the soil layer lies a gravel layer, through which water and other liquids percolate readily. The third layer of basalt, or hardened lava, is relatively impermeable, slowing groundwater movement toward the Columbia River. For millions of years before the glaciers deposited the gravel layer, wind and retreating floodwaters carved channels in the exposed sedimentary rock, channels that were later filled by the gravel. Some of these channels lie below the water table, funneling groundwater to the Columbia much more quickly than it could travel through the sedimen-

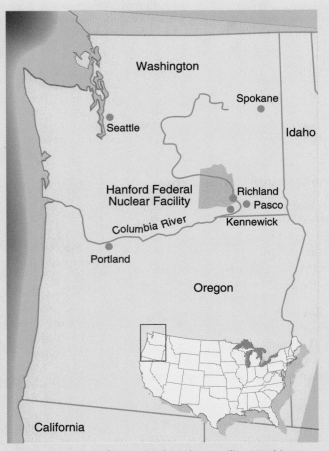

FIGURE 18-14: Hanford Federal Nuclear Facility, Washington.

tary rock. Over 200 square miles (518 square kilometers) of groundwater have been contaminated by radioactive and toxic chemical wastes from the Hanford plant.

Biological Boundaries

Besides threatening to spread radioactive waste and other hazardous materials through the groundwater, the Hanford Reservation has affected the resident biotic community. The reservation is home to eight major shrub communities, with the sagebrush-cheatgrass-bluegrass association the most prominent. Hanford has had many problems with radioactive tumbleweeds (sagebrush) spreading contamination across the reservation and probably

off site. The long tap root of sagebrush (15 feet) provides a ready pathway for the uptake of radioactive liquids in the arid environment. Hanford manages a large program of spraying defoliants to prevent plant growth on contaminated soil sites.

Social Boundaries

The native Americans (Yakimas) who originally inhabited the Hanford area ceded most of their territory to the federal government in 1855, retaining the right to fish, hunt, gather plants, and graze horses and cattle on the land. Later, European farmers and ranchers moved in. World War II radically altered the character of the Hanford area. In 1943 leaders of the Manhattan Project found Hanford ideal for the production of plutonium for nuclear weapons because of the sparse population, the ample supply of water for cooling the reactors (provided by the Columbia River), and the easy access to electrical power from the Grand Coulee Dam. Approximately 1,500 people who had been living in the town of Hanford on the proposed site were relocated.

Disregard for or lack of understanding of environmental and human safety was compounded by the assumption that operations at Hanford would cease after the war. After World War II ended, however, plutonium production actually increased during the arms buildup of the cold war. By 1964 six additional reactors had joined the three built in 1944.

Looking Back

Hanford was a major link in the military's chain of plutonium production facilities (Figure 18-15). Hanford received enriched uranium from other government sites and refined it into plutonium, which was then fashioned into hydrogen bomb triggers at other facilities located elsewhere in the United States. Eight of the nine plutonium production reactors were deactivated between 1964 and 1971 because of decreasing demand for plutonium. The ninth, the N reactor, operated alone after 1971 until it was placed on standby in 1988 because of potential safety concerns resulting from the Chernobyl accident.

The plutonium-uranium extraction (PUREX) facility, a reprocessing plant where plutonium and uranium were separated from other fission products, stayed in operation after the N reactor ceased production. In fact, PUREX had a big enough backlog to continue extracting plutonium and uranium until around 1995. However, after PUREX experienced steam pressure problems in late 1988, it was shut down. It has remained inoperative because of public concern over high-level waste, low-level liquid

waste disposed of in the ground, and plant safety.

How Plutonium Production Affected the Environment

During Hanford's early years radioactive wastes were disposed of with little accurate knowledge as to how they would affect the environment. Even as knowledge improved, environmental threats persisted due to the sheer volume of waste produced. Over 200 billion gallons (758 billion liters) of waste requiring storage and treatment were discharged to the environment from Hanford reprocessing plants between 1945 and 1985. Experts agree that Hanford is the most heavily contaminated of all facilities operated by the Department of Energy (DOE).

Inadequate disposal of high-level waste has been a major source of contamination at Hanford. Most of Hanford's high-level waste has been produced by PUREX and earlier reprocessing facilities. These facilities separate usable uranium and plutonium from fission products such as radioactive isotopes of iodine and tritium, a radioactive isotope of hydrogen. During its years of operation PUREX produced approximately 1 million gallons (3.79 million liters) of this waste annually.

Plant operators believed that high-level wastes could be effectively stored in single-shell carbon steel tanks buried a few meters below ground. Between 1943 and 1980 Hanford officials pumped 46 million gallons (174 million liters) of high-level waste into 149 single-shell carbon steel tanks. Although these tanks were intended for temporary storage, it was estimated that they could last up to 300 years if necessary. But in 1959, after only 16 years, Hanford officials found that some tanks were already leaking. An estimated 1 million gallons of high-level waste have leaked from the tanks into the soil. Liquids have been removed from most of the leaking tanks. However, even after removing as much liquid as possi-

ble, DOE estimates that 50,000 gallons (189,500 liters) of liquid could remain in a tank, not including wastes in the form of solids or sludges. Undamaged single-shell tanks still contain 46 million gallons (174 million liters) of high-level waste. Twenty-six double-shelled tanks constructed since 1968 contain about 14 million gallons (53 million liters). Unfortunately, the new double-shelled tanks, designed to last 50 years, have not performed as expected; pits of rust have been found in their stainless steel liners.

The nearby Columbia River has been a frequent victim of inadequate waste disposal. Between 1943 and 1964 cooling water containing low-level radioactive waste from the first eight plutonium production reactors was dumped directly into the river after an initial period of decay. This cooling water frequently contained pieces of uranium fuel from within the reactor. Often, the length of time allowed for decay was inadequate, especially when the plant was in full operation.

Officials did not worry about dumping liquid waste on the soil or in cribs, trenches, ditches, and ponds because they believed it would take the liquid at least fifty years to reach the Columbia River during which time radioactive compounds would decay or be filtered out of the liquid by the soil. However, they were unaware of the gravel-filled channels in the sedimentary rock below the water table, which allowed for faster movement. In 1963 Hanford scientists found that radioactive tritium from cooling water from the PUREX plant had already traveled to the Columbia River, just seven years after the cooling water was released.

In July 1984 liquid waste from uranium extraction operations was routed to a newly constructed crib. Within a few months uranium concentrations in nearby groundwater monitoring wells increased to 170 times their previous levels and more than 5,000 times the current EPA drinking water standard. Scientists found that the liquid deposited in the new crib had percolated down over 100 feet through a forgotten and unmarked reverse well. This old reverse well had previously pumped liquid waste deep into the ground. Although a $1 million remediation effort was initiated after this incident, uranium levels remained more than 1,300 times greater than the EPA drinking water standard.

The approximately 30 billion gallons (113.7 billion liters) of low-level liquid waste dumped into the ground and spills from waste transfers and other accidents have left Hanford's soil widely and heavily contaminated. The exact remediation method for the soil has not been decided upon. Some soil might be removed, some might be stored on site, some might be incinerated, and some might be encased in glass.

Some area residents and environmental groups are worried that the DOE is compounding contamination problems at Hanford by using it to store waste generated off site as well as the waste produced on the reservation itself. They fear that the area will become a "nuclear junkyard." Eight 1,000-ton reactor sections from retired Polaris nuclear-powered submarines are already buried on the premises. The navy plans to retire 92 more of these subs over the next 20 to 30 years, and Hanford, which has the space to accommodate them, could easily become the disposal site. Also interred there is the country's first commercial nuclear reactor, the Shippingport Atomic Power Station.

Although the DOE records maintained on the site are classified as secret, the public is slowly learning how activities and accidents at the facility have endangered the health of workers and area residents. For example, public pressure in 1986 moved the DOE to reveal that emissions from the stacks of reprocessing plants carried radioactive iodine into the environment during the first 12 years of Hanford's operation. The Technical Steering Panel (TSP) of the Hanford Environmental Dose Reconstruction project released preliminary estimates from a study conducted by Battelle Pacific Northwest Laboratory for the Centers for Disease Control (CDC) in July 1990. The study estimated that in 1945 infants in nearby Pasco consuming local milk might have received radiation doses to the thyroid at levels as high as 2,300 rems per year. Infants as far away as Spokane would have received doses of up to 256 rems that year. By comparison, current health studies suggest that radiation doses of 20 rems or less can cause damage to the thyroid, and Washington state law today restricts doses of radiation to the thyroid gland and other critical organs to .075 rem per year.

Although public dissatisfaction with Hanford's history of waste contamination and accidental leaks is steadily growing, a new area of concern surfaced when the world learned of the disaster at Chernobyl. In many respects the 23-year-old N reactor at Hanford bore an uncanny resemblance to the ill-fated Soviet plant. The designs of both reactors allowed them to produce plutonium for military use and steam for generating electricity simultaneously. More importantly, both used graphite to moderate the rate of fission. The N reactor is the only U.S. reactor to use the dangerously flammable substance which caught fire and spread radioactive contamination during the Chernobyl accident. Both reactors also lacked the customary high-density concrete containment structures that guard against the release of radioactivity to the environment in case of an accident. Again, the N reactor is the only major reactor facility in the United States without such a containment structure.

Despite efforts by DOE officials to assure the public that a major accident was unlikely to occur at Hanford, the public and environmentalists called for an immediate investigation of N reactor safety. In late 1986 six consultants to the DOE recommended a six-month, $50 million safety overhaul for the plant, and the plant was closed pending repairs. The lack of important safety features such as a containment shield led environmental organizations to announce that they would file suit for a full environmental impact statement if the DOE tried to restart the facility after the repairs. The N reactor remained inactive into early 1988, when the Reagan administration

FIGURE 18-15: The Hanford Reservation in 1988. The Columbia River can be seen at the top of the photo. Weapons production has been halted at Hanford; cleaning up—the only activity currently taking place—is likely to take years.

announced that the facility would be "mothballed"—maintained in working condition until national security called for the production of more plutonium. However, in August 1991 the DOE announced that it would permanently close the N reactor.

How the Government Plans to Clean Up Hanford

In response to legislation passed by Congress in 1983, the Department of Energy proposed the Defense Waste Management Plan, a blueprint for the cleanup and long-term waste management of its major nuclear sites. The plan included burying transuranic wastes at a New Mexico nuclear waste repository; solidifying radioactive sludge from high-level tanks into glass (vitrification) and burying it in a deep repository planned for Nevada; and grouting low-level waste (mixing it with cement and pouring it into concrete-lined pits).

However, the DOE was slow to implement this plan. When Congress reauthorized the Resource Conservation and Recovery Act (RCRA) in 1984, states received new authority over solid and hazardous liquid waste problems. Responding to the lack of cleanup effort by the DOE, the Washington Department of Ecology decided to confront the Hanford facility on its failure to comply with state standards. When the department identified 20 violations of RCRA requirements, the DOE pointed out that almost all the wastes at Hanford were "mixed," that is, they contained radioactive as well as hazardous substances. The DOE claimed that RCRA did not apply to mixed wastes because the Atomic Energy Act gave the DOE complete control over any radioactive waste produced by federal facilities. Wastes containing radioactive substances were completely under jurisdiction of the DOE. Although the EPA insisted that RCRA covered mixed wastes as well as simply hazardous ones, the law was inconclusive.

Several members of Congress proposed bills that would have given the EPA definite jurisdiction over mixed wastes, but the bills did not pass. Then, in 1986 the state of Tennessee won a lawsuit against a DOE plant in Oak Ridge, setting a precedent for applying RCRA to mixed wastes. Soon after, Washington's Department of Ecology fined the DOE for its violations and, along with the EPA, demanded compliance on the five worst violations. The DOE agreed to comply but not to pay the fines.

On May 15, 1989, the Washington Department of Ecology, the EPA, and the DOE negotiated the Tri-Party agreement, an unprecedented cleanup program that calls for spending $50 billion over the next 30 years to eliminate Hanford's worst problems. The EPA is responsible for enforcing cleanup of old waste, and the state is involved in controlling wastes currently being produced. At Hanford both wastes contaminate the same soil and groundwater, and the Tri-Party agreement was intended to determine which agency would have authority at which waste site, eliminating duplication of efforts and conflicting decisions. The agencies estimate that they will need $2.8 billion in the next 5 years to remove liquids from single-shell tanks, study how to remove radioactive sludge from the bottom of the tanks, install new groundwater monitoring wells, investigate old waste sites, and begin to grout and vitrify waste from double-shell tanks.

Unfortunately, the removal of liquids from the single-shell tanks will not happen within 5 years, if ever (some of the chemicals in the tanks become explosive when dry or hot). Also, the DOE is dragging its feet in researching methods of removing solids from the tanks. Grouting has been delayed two years and the DOE is expected to announce shortly that there will be further delays. In December 1991 the DOE and Washington's Department of Ecology agreed to build a waste vitrification plant at Hanford. Con-

struction was scheduled to begin in April 1992, and the plant should be operational in December 1999.

While the Tri-Party agreement is a positive step, critics note that there is no way of ensuring that Congress will continue to appropriate funds for the cleanup. Also, many people at public hearings strongly objected to a provision which would allow PUREX to continue operating until 1995. As a result, the three agencies agreed to a separate, 14-month investigation of liquid waste from PUREX and other facilities to determine if some or all of the discharges should be halted.

Looking Ahead

In addition to the massive cleanup initiated by the Tri-Party agreement, all nine nuclear reactors and PUREX will eventually need to be decommissioned. These facilities contain vast amounts of radioactive materials which must be prevented from contact with humans or the environment. Disposing of them safely will cost billions of dollars.

While Hanford is a costly reminder of our carelessness and ignorance in handling nuclear resources, the mistakes that occurred there can serve a positive function: preventing a similar scenario in the future. By participating in the Tri-Party agreement, the state of Washington has joined many other states that are attempting to force the DOE to reform its practices. These states want Congress to pass legislation that will leave no doubt that all federal agencies, not just the DOE, must comply with the same laws as other businesses and institutions, especially when it comes to human and environmental safety.

In 1991 Congress passed legislation that would force the DOE to pay environmental fines. The Federal Nuclear Facilities Environmental Response Act established a trust to finance clean-up efforts, decommissioning processes, environmental compliance, and long-term monitoring of federal facilities.

waste, and nuclear accidents. While each of us must weigh the benefits and risks of this energy source, we need to work together to ensure that governments take the necessary measures to both protect and preserve the living systems upon which we all depend.

Summary

Nuclear resources are derived from atoms, their energy, and the particles they emit. Nuclear energy is the energy released, or radiated, from an atom. Energy released from an atom is called radiation. Radiation takes two basic forms: ionizing and nonionizing.

Nuclear energy is released through three types of reactions: spontaneous radioactivity, fission, and fusion. Spontaneous radioactivity occurs when unstable atoms release mass in the form of particles, energy in the form of waves, or both during radioactive decay. Isotopes that release particles or high-level energy are called radioisotopes. The most common types of ionizing radiation are alpha, beta, and gamma radiation. Alpha particles consist of two protons and two neutrons and carry a positive charge. Beta particles are negatively charged particles emitted from nuclei. Gamma radiation is a powerful electromagnetic wave. X-rays, a form of cosmic radiation, can also be produced by firing electrons at tungsten metal. Half-life is the length of time it takes for any radioactive substance to lose one-half of its radioactivity.

A fission reaction occurs when an atom is split apart into two or more new atoms. The atoms formed when uranium in a nuclear reactor is split are called fission products. Critical mass is the smallest amount of fuel necessary to sustain a chain reaction. A chain reaction is a self-perpetuating series of events that occur when a neutron splits a heavy atom, releasing additional neutrons to cause other atoms to split. In a fusion reaction nuclei are forced to combine, or fuse. For many people, fusion holds the promise of "clean" nuclear energy. Fusion is not yet feasible for commercial electric generation.

As ionizing radiation penetrates living tissue, it can destroy cells or alter their genetic structure. The effects of radiation on living organisms depend on several factors: amount of exposure, age and gender of the organisms, and type of exposure. Direct exposure is exposure to the original radioactive source. Indirect exposure is exposure to radioactive substances through food chains. Radioactive fallout is dirt and debris contaminated with radiation.

Nuclear resources are used in a variety of ways, including medical applications, food preservation, power sources for satellites, production of metal alloys, generation of electricity, and military applications.

In 1947 the government established the Atomic Energy Commission (AEC) to control the use and disclosure of information on atomic power. In March 1953 the AEC made the development of economically competitive nuclear power a goal of national importance. The 1954 Atomic Energy Act opened the nuclear industry to the private sector.

When the Soviet Union detonated its first atomic bomb in 1949, the nuclear arms race between that country and the United States began. In 1963, 105 nations signed the Nuclear Test Ban Treaty, agreeing to halt nuclear testing anywhere fallout might spread beyond the borders of the country testing the weapon. One of the first attempts to halt the spread of nuclear weapons technology was the Nuclear Non-Proliferation Treaty, signed by 17 nations in 1968.

The use of nuclear resources has been accompanied by many environmental and social problems. Among the most serious are radioactive leaks, accidents at nuclear power plants, disposal of radioactive wastes, decommissioning of old plants, and the secrecy surrounding nuclear activities.

Key Terms

alpha particle	isotope
beta particle	low-level radioactive waste
chain reaction	meltdown
containment vessel	nonionizing radiation
critical mass	nuclear energy
direct exposure	nuclear resources
fission product	nuclear winter
fission reaction	radiation
fusion reaction	radioactive decay
gamma radiation	radioactive fallout
half-life	radioisotope
high-level radioactive waste	spontaneous radioactivity
indirect exposure	transuranic radioactive waste
ionizing radiation	uranium mill tailings
ions	x-ray

Discussion Questions

1. What is nuclear energy, and how is it produced?
2. What are some of the immediate and long-term health effects of exposure to radiation?
3. Explain the difference between direct and indirect exposure to radiation, using the accident at Chernobyl as an illustration.
4. Describe four applications of nuclear technology other than electric power generation.
5. What are some of the social and environmental problems involved with the disposal of nuclear waste?

Toxic and Hazardous Substances

I do wonder whether there will come a time when we can no longer afford our wastefulness—chemical wastes in the rivers, metal wastes everywhere, and atomic wastes buried deep in the earth or sunk in the sea.

John Steinbeck

In an era when a chemical spill can kill thousands, when wastes buried 20 years earlier can contaminate a community's water supplies, when our environmental ignorance can plant the seeds for later diseases, environmentalists and industrialists cannot continue to draw false battlelines.

Jay Hair

Learning Objectives

When you finish reading this chapter, you should be able to:

1. Explain the difference between toxic substances and hazardous substances.

2. Describe at least three different methods of disposing of hazardous waste.

3. Define the concept of waste minimization and explain why it has become so important.

4. Identify three major pieces of legislation on regulating hazardous waste.

By the mid-1970s the fledgling environmental movement in the United States had achieved noticeable improvement in cleaning up the nation's air and water. Controls on the release of smoke and particulates resulted in cleaner air, and a reduction in the discharge of municipal sewage led to cleaner waterways and the reopening of beaches previously closed because of high bacteria counts. It seemed as if American ingenuity and resolve were more than up to the task of restoring the nation's environment. But the flush of those early successes was short-lived, for by the end of the decade a new and greater danger had come to light. Almost overnight, the country had awakened to the threat of hazardous and toxic substances, a threat that was embodied in the name Love Canal.

From 1942 to 1953 Hooker Chemicals and Plastics Corporation, a producer of pesticides and plastics, had dumped its industrial wastes into the Love Canal, an excavated but never-used canal near Niagara Falls, New York. A total of more than 20,900 tons (19,000 metric tons) of highly toxic wastes, many of which contained suspected or known cancer-causing agents, were placed in the dump. The company then covered the dump with a clay cap and topsoil. In 1953 Hooker sold the entire canal area to the Niagara Falls school board for a token fee of $1 on the stipulation that the company would not be held responsible for any future injury or property damage caused by the dump's contents. According to Hooker officials, the school board was warned against developing the site for residential use. Nevertheless, it is unclear whether the contents of the dump and the potential health hazards they posed were clearly revealed. In time, a housing project and an elementary school were built in the area.

By the late 1970s signs of trouble had appeared in Love Canal. Residents began to notice a pervasive smell of chemicals, and children playing in the canal proper suffered chemical burns. Soon, corroded barrels containing hazardous wastes began to leak. Toxic chemicals migrated into groundwater supplies, basements, storm sewers, backyards, gardens, and the local school playground. Complaints about the chemicals were ignored by city officials. Concerned residents, many of them full-time mothers in this traditional, low- to middle-income community, took informal surveys and discovered what appeared to be an unusually high number of health complaints—from symptoms such as a burning sensation in the eyes and nose, rashes, and migraine headaches to more serious ailments including various forms of cancer; nerve, kidney, and respiratory disorders; miscarriages; and birth defects. The more serious ailments were reported largely among those residents living in the area immediately surrounding the dump.

Though the residents continued to complain to city and health officials, little action was taken until unfavorable press coverage—and the persistence of area residents—convinced state officials to conduct formal health surveys and tests. Test results indicated toxic contamination of the soil, air, and water of the canal area and the basements and backyards of local homes. Preliminary official health surveys seemed to support residents' suspicions about the adverse health effects on people living in the area. Before long, Love Canal became the topic of national debate as the public and local, state, and federal governments confronted the adverse health effects of hazardous wastes. Eventually, the state closed the school and permanently relocated the 239 families living in the area adjacent to the dump (Figure 19-1). The area was bulldozed and fenced off. Several years later, in 1980, President Carter declared Love Canal a federal disaster area, and more than 700 families living in the surrounding area were temporarily relocated. Federal and state funds were used to purchase the homes of those who wished to move permanently.

After a 12-year, $250 million effort to clean up Love Canal, the EPA concluded in the spring of 1991 that four of its seven sites are "habitable." The other three areas are slated to become industrial areas and parkland. A total of 256 houses are scheduled to go on the market, and eager buyers await. A state-of-the-art containment system has sealed off the 16-acre dump with dense clay walls and two 3-foot-thick caps, one spanning 22 acres and the other 40 acres. The 239 houses immediately surrounding the dump have been demolished; the entire area is sealed off by a chain link fence. Houses are offered at discount prices, 20 percent below market value. Environmentalists fear that the inexpensive housing will attract young families looking for starter homes; pregnant women and children are at greatest risk from toxins.

Many things have changed since the Love Canal story first broke. Congress has enacted major legislation to fund the cleanup of abandoned hazardous waste sites and has significantly strengthened legislation regulating the storage and disposal of these

FIGURE 19-1: The Love Canal community in the immediate area of the dump, shortly after residents were evacuated. Toxic wastes turned Love Canal, once an average working-class neighborhood, into a ghost town.

materials. Unfortunately, toxic contamination continues. For example, traces of dioxins, the most toxic family of synthetic chemicals, can be found in the breast milk of many lactating mothers in the Northern Hemisphere; residues of DDT, a deadly pesticide banned in the United States over 20 years ago, can be found in the muds of Lake Siskiwit, a remote wilderness area on Lake Superior's Isle Royale; certain species of fish in the Great Lakes are deemed too dangerous for unlimited consumption because of toxic contamination, and children and nursing mothers are advised to avoid eating those species altogether; and hazardous wastes from industrialized nations are sometimes shipped to less-developed countries where there are few or no regulations to ensure proper disposal.

Just as the threat of toxic contamination continues, the shroud of mystery and fear surrounding hazardous and toxic substances persists as well. In this chapter we look at what these substances are, how they enter the environment, how they affect both environmental and human health, who uses and regulates them, and how they affect communities. After discussing environmental problems caused by past management, we examine promising strategies to manage hazardous resources and the legislation that governs their production, use, and disposal. Finally, we discuss how toxic and hazardous substances can be managed in environmentally sound ways in the future.

Describing Toxic and Hazardous Substances: Physical Boundaries

Worldwide, over 70,000 different chemicals are used daily, and each year between 500 and 1,000 new synthetic compounds are introduced. More than 6 billion tons (5.5 billion metric tons) of waste are disposed of annually in the United States. Of that, 270 million tons (245.7 million metric tons)—enough to fill the New Orleans Superdome 1,500 times—are hazardous. These figures underscore the difficulty of managing these substances adequately, from testing potential toxins for their health and environmental effects to disposing of wastes safely.

What Are Toxic and Hazardous Substances?

Toxic and hazardous substances are chemicals that can adversely affect human health and the environment. This broad definition includes elements such as lead, compounds such as polychlorinated biphenyls (PCBs), and the products of infectious agents like bacteria and protozoans. The term *hazardous* may conjure images of leaking, corroded barrels and abandoned, desolate waste dumps, but these images represent only one part of the range of hazardous and toxic substances (Figure 19-2). Indeed, hazardous and toxic chemicals can be found virtually everywhere. Many toxic substances occur naturally. Some are as old as the earth itself; lead and radium 222 (contained in uranium rock), for instance, are found in the earth's crust. Others are bound up in the planet's biota; many plants and animals, for example, manufacture substances that are poisonous in varying degrees to humans and other animals. Toxic substances may also be synthetic, manufactured by humans through physical or chemical processes. Synthetic chemicals, many of which are highly toxic, are used in industrial processes to manufacture products, such as plastics, inks, and pharmaceuticals, for industrial and home use. Pesticides, often used to ensure the cosmetically attractive fruits and vegetables consumers demand, contain many potentially dangerous chemicals (Table 19-1). Common household products—cleaners, shoe polish, used motor oil, antifreeze, insecticides—fill the American home with an amazing array of harmful chemicals.

A toxic substance that is disposed of or managed in such a way that it poses a threat to human health or the environment is known as a **hazardous waste.** Industrial and agricultural chemicals account for most of the hazardous wastes. The safe disposal of these wastes and the cleanup of hazardous waste sites are among the most serious of all environmental problems.

What Is the Difference Between Toxic and Hazardous Substances?

Although the terms toxic and hazardous are often used interchangeably, they are not synonymous. The term **toxic** implies the potential to cause injury to living organisms. Almost any substance can be toxic under the right conditions, if the concentration is high enough or if an organism is exposed to the substance for long enough. The term **hazardous** implies that there is some chance that an organism will be exposed to a substance and that that exposure will result in harm. Substances can be considered hazardous only when a possibility exists that plants or animals will be exposed to them. Imagine a dangerous chemical stored in a sealed jar on a laboratory shelf. The chemical is toxic because it would harm anyone who inhaled its fumes, but it is not hazardous because there is no chance that anyone will. If the jar falls from the shelf and breaks, however, the chemical becomes hazardous; there is now some risk—high or low—that injury will occur.

Determining whether a waste is *legally* hazardous or toxic is difficult because legal definitions of hazardous and toxic substances vary. In the United States there are no less than 10 different federal laws to regulate materials that are hazardous or toxic to human and environmental health. But with the Resource Conservation and Recovery Act (RCRA) of 1976, Congress provided the EPA with a statutory framework for defining hazardous wastes. A hazardous waste is defined as any solid, liquid, or gaseous waste which, due to its quantity, concentration, or physical, chemical, or infectious characteristics, may cause or significantly contribute to an increase in mortality or serious illness; or pose a substantial present or potential hazard to human health or the environment when improperly stored, transported, disposed of, or recycled.

Congress allowed for exceptions to the legal definition of hazardous waste, automatically exempting certain wastes from being regulated as hazardous.

FIGURE 19-2: The former Chem-Dyne hazardous waste site in Hamilton, Ohio. The site was cleaned up under the Superfund program of the Environmental Protection Agency.

▶ Table 19-1
Potential Hazards and Common Uses, Selected Pesticides

Pesticide	Potential Hazards	Common Uses
Acephate	Cancer, mutagenicity, reproductive toxicity (interferes with an organism's ability to reproduce)	Bell peppers, celery, green beans
Aldicarb	Severely toxic at small doses	Potatoes
BHC	Cancer, reproductive toxicity	Cabbage, sweet potatoes
Captan	Cancer, mutagenicity	Apples, cherries, grapes, peaches, strawberries, watermelons
Chlordane	Cancer, birth defects, reproductive toxicity, mutagenicity	Potatoes
Chlorothalonil	Cancer, chronic effects, mutagenicity	Cantaloupes, cauliflower, celery, green beans, tomatoes, watermelon
Cyhexatin	Birth defects, chronic effects	Pears
DDT	Cancer, liver effects, reproductive toxicity	Carrots, onions, potatoes, spinach, sweet potatoes
Dieldrin	Cancer, birth defects, reproductive toxicity	Carrots, cucumbers, corn, potatoes, sweet potatoes
Lindane	Cancer, chronic effects	Corn
Methyl parathion	Cancer, chronic effects, mutagenicity	Cantaloupes, strawberries
Parathion	Cancer, mutagenicity	Broccoli, carrots, cherries, oranges, peaches
Permethrin	Cancer, reproductive toxicity	Cabbage, lettuce, tomatoes
Phosmet	Cancer, mutagenicity	Apples, pears, sweet potatoes
Sulfallate	Cancer, mutagenicity	Corn
Trifluralin	Cancer, chronic effects	Carrots

Source: Lawrie Mott and Karen Snyder, *Pesticide Alert* (San Francisco: Natural Resources Defense Council, Sierra Club, 1987).

For example, all household waste, even if it contains hazardous materials, is exempt from hazardous waste regulations.

How Do Toxic and Hazardous Substances Enter the Environment?

Toxic and hazardous substances are released into the environment either as by-products or as end products. The chemical and petroleum industries refine crude oil (their basic raw material) into fuels and solvents and manufacture it into products and raw materials for innumerable other products. Chemical and petroleum-refining companies are the major producers and users of toxic and hazardous chemicals. These industries, which generate two-thirds of the total regulated hazardous waste stream, are the prime suppliers of raw materials to the nation's manufacturing and service sectors.

Modern manufacturing industries use and discard enormous amounts of chemicals in the production of fabricated metals, plastics, paper and other wood products, and many consumer goods. Manufacturing operations generate large quantities of chemical by-products which, in the eyes of the manufacturer, possess little or no economic value. After a chemical or chemically based substance is produced, it enters the marketplace to be sold for profit, and the by-products of its production are disposed of as wastes or released into the environment. The release of

toxic by-products may occur in a number of ways (Figure 19-3).

Toxic and hazardous substances also enter the environment through the use and disposal of manufactured goods. For example, the worldwide use of

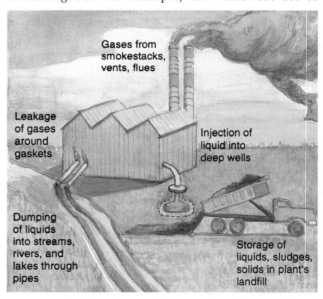

FIGURE 19-3: How toxic materials enter the environment from industrial sources. Industries release toxic materials to the environment in liquids dumped into nearby waters and in gases from smokestacks, vents, and flues. Unless carefully planned and properly executed, disposal methods such as burying hazardous wastes in a landfill or injecting them deep underground can also allow toxins to enter the environment.

pesticides results in the release, each year, of millions of tons of these toxic chemicals into the soil, air, and water. Many other products are simply used and disposed of. An auto parts manufacturer, for instance, might use solvents to degrease metal parts. After the solvents are used, they are no longer useful to the manufacturer, so they are disposed of. Residential users also release hazardous chemicals into the environment. Many common household products contain harmful substances; when these products are thrown into the garbage, they eventually end up in municipal landfills, which are not designed to contain hazardous substances securely. And people often unwittingly dump hazardous chemicals down the drain or on the ground, unaware that the material may contaminate local waters and soil.

Biological Boundaries

Toxic and hazardous substances are of serious concern because of their potential and suspected adverse effects on ecosystems and living organisms, including but not limited to humans. Of the almost 70,000 chemicals in use in the United States today, less than one-third have been adequately studied for their effects on human health. There is even less knowledge about how toxins affect ecosystems and biotic communities. Managing these chemicals in environmentally sound ways requires far more research at both the species and ecosystem level.

How Do Toxic and Hazardous Substances Affect Environmental Health?

Concern over toxic and hazardous substances has focused on how these compounds affect human health, but they also pose a substantial threat to the environment, the living systems upon which we all depend. Substances that adversely affect the environment will also affect the inhabitants of that system.

Toxins can contaminate surface water and groundwaters, causing fish kills and destroying other aquatic life. Sensitive species are the first to be affected, and species diversity in the contaminated area typically decreases. Toxins can also pollute the soil and air, sometimes dramatically upsetting the dynamic balance within the ecosystem. One of the major problems with pesticides, for example, is that they kill beneficial insects as well as pests. After the target species builds up resistance to the pesticide, its numbers can increase dramatically because the populations of its natural predators have been greatly reduced. Greater attention must be paid to identifying the ways in which toxic and hazardous substances adversely affect the environment and how these adverse effects can be mitigated.

How Do Toxic and Hazardous Substances Affect Human Health?

The health effects of toxic and hazardous substances are almost as numerous and diverse as the substances themselves. Some chemicals attack a single organ; others have systemic effects. Much is known about the effects of acute exposure to high concentrations of some chemicals, but little is known about the effects of chronic exposure to low concentrations of most chemicals. Moreover, little is known about how the effects actually occur.

Acute toxicity is the occurrence of serious symptoms immediately after a single exposure to a substance. Methyl isocyanate, the substance that precipitated the Bhopal, India, disaster (see page 403), is acutely toxic. A single exposure can maim or kill. **Chronic toxicity** is the delayed appearance of symptoms until a substance accumulates to a threshold level in the body after repeated exposures to the substance. This may occur weeks, months, or years after the initial exposure. Lead is chronically toxic; the longer the exposure, the greater the health risks (especially to children, who are particularly susceptible to lead poisoning). These include behavioral disorders, hearing problems, brain damage, and death.

There are a number of general ways in which hazardous and toxic substances affect long-term human health. **Carcinogenic** substances cause cancer in humans and animals. **Infectious** substances contain disease-causing organisms. **Teratogenic** substances affect the unborn fetus; they may cause birth defects or spontaneous abortions or otherwise damage the fetus. **Mutagenic** substances cause genetic changes or mutations, which then appear in future generations. Most, but not all, mutagens turn out to be carcinogenic; so far, all known carcinogens have proven to be mutagens as well.

To express the risk that a substance may cause cancer in humans in a way that is understandable but not too oversimplified, the U.S. government uses a statement containing three basic elements. The three elements are the quality of the evidence that a substance is a human carcinogen, the numerical probability that a person who has ingested the substance will get cancer, and the degree of the risk. For example, in order to describe the risk of pesticide residues on food, the EPA might release a statement like "The EPA considers chemical X to be a possible human carcinogen and estimates the highest lifetime dietary risk of cancer to be one in a million." This statement says chemical X is at present a suspected but not a proven carcinogen; the numerical risk of getting cancer is at greatest 1 in 1,000,000, but the risk could be much lower.

Toxic and hazardous substances can cause the onset of **multiple chemical sensitivity** (MCS), an illness characterized by an intolerance of one or more

FIGURE 19-4: Janet Bennett, of Wimberley, Texas, suffers from multiple chemical sensitivity (MCS) and must remain isolated in a porcelain box. Here she "visits" with a friend. Although the MCS diagnosis is controversial, almost all MCS sufferers experience one or more symptoms associated with the central nervous system: tension, memory loss, fatigue, sleepiness, headaches, confusion, and depression. Problems associated with the gastrointestinal tract such as nausea, bloating, cramps, indigestion, and constipation are also common, as are respiratory ailments such as frequent colds, bronchitis, congestion, asthma, and shortness of breath.

classes of chemicals. A **chemical sensitizer** is a substance that can cause an individual to become extremely sensitive to low levels of that particular substance or to experience cross-reactions to other substances. A number of substances are suspected of being chemical sensitizers: acrylic resins, formaldehyde, isocyanates, mercury compounds, pesticides, and solvents. People who suffer from MCS cannot tolerate any exposure to the offending chemicals; in some cases, they must live in near isolation from synthetic compounds (Figure 19-4).

Social Boundaries

While it is true that many toxic substances occur naturally, it is equally true that widespread toxic contamination and hazardous waste sites are a product of industrialized societies. Because toxic and hazardous substances are now widely distributed throughout the world, they are a pressing global concern. The issues associated with toxins—from adverse health effects to hazards posed by transporting these materials to siting of hazardous waste landfills and incinerators—are social in nature.

Who Produces and Uses Toxic and Hazardous Substances?

Worldwide, production and use of toxic substances are concentrated in the industrialized nations, although the less-developed countries use significant quantities of pesticides (many of them imported from the United States and Europe, where their use may be banned). In the United States toxic substances are produced and used in every state in the country by businesses, industries, and individuals. Almost two-thirds of those substances, however, are produced in just ten states (listed in decreasing order of production): Texas, Ohio, Pennsylvania, Louisiana, Michigan, Indiana, Illinois, Tennessee, West Virginia, and California.

Worldwide, most hazardous wastes are generated and disposed of by the chemical, petroleum-refining, and metals-processing industries. In fact, in the United States 100 industrial waste generators are responsible for creating 87 percent of our hazardous waste. Table 19-2 illustrates the wide variety of businesses and other organizations that use and dispose of toxic and hazardous substances.

Toxic and hazardous substances often begin as pure raw materials produced by the chemical and petroleum industries. They are then used in manufacturing or servicing processes. For example, acids and bases are used to create plastics, pesticides, and other chemical substances. Cyanide solutions are used in electroplating and metal heat-treating operations. Solvents such as xylene, acetone, and methanol are used to clean and degrease metal and glass. Chlorinated organics are used to manufacture pesticides.

What Are Household Hazardous Wastes?

Toxic chemicals can be found in garages, under kitchen sinks, and in basements all across the United States (Figure 19-5). It is important to note that hazardous household items include those which are hazardous when used or disposed of. When we use and improperly dispose of these products, we release their dangerous chemicals into the environment. Many would legally be considered "hazardous waste" if disposed of in large quantities by industry.

According to the EPA, the average American family of four produces about 4 tons (3.64 metric tons) of solid waste each year. An estimated 400 pounds (180 kilograms) of that trash is hazardous. But because Congress exempted all household wastes from regulation under RCRA, these toxins can be legally hauled off to the local municipal landfill or solid waste incinerator for disposal. Every day toxic substances such as cyanide, chlorine, mercury, and lead are thrown out like yesterday's newspaper and

▶ Table 19-2
Users of Toxic and Hazardous Substances

Type of User	Types of Toxic and Hazardous Substances Used and Disposed Of
Automobile maintenance and repair	Acids, bases, heavy metals, inorganics, flammables, lead-acid batteries, solvents
Chemical manufacturers	Acids, bases, cyanide wastes, heavy metals, inorganics, flammables, reactives, solvents
Cosmetics	Acids, bases, heavy metals, inorganics, flammables, pesticides, solvents
Schools (including vocational schools)	Acids, bases, flammables, pesticides, reactives, solvents
Funeral services	Solvents, formaldehyde
Furniture manufacturing, refinishing, and preserving	Flammables, solvents, preserving agents
Hospitals	Flammables, solvents, infectives
Laboratories	Acids, bases, heavy metals, inorganics, flammables, reactives, solvents
Laundries and dry cleaners	Dry cleaning fluids, filtration residues, solvents
Metal manufacturing	Acids, bases, cyanide wastes, heavy metals, inorganics, flammables, reactives, solvents, spent plating wastes
Pesticide formulators and users	Acids, bases, cyanide wastes, heavy metals, inorganics, flammables, reactives, pesticides, solvents
Printing and allied industries	Acids, bases, heavy metals, inorganics, ink sludges, spent plating wastes, solvents

potato peels. Since municipal landfills were not designed to contain them, these household hazardous wastes are slowly starting to leak out, poisoning groundwater and surface waters in many locations.

A gallon of gasoline leaking out of a municipal landfill can contaminate a million gallons (3.79 million liters) of groundwater. Since half of all Americans rely on groundwater for their drinking water supplies, the problem is painfully clear. As consumers of chemical-based products, products that require chemicals in their manufacture, and products packaged in chemical-based materials, we are all contributing to the problem of hazardous and toxic pollution, and we must all take responsibility for resolving this problem.

How Do Toxic and Hazardous Substances Affect Communities?

Earlier, we learned that toxic and hazardous substances can affect human health and the environment in serious ways. The production, use, and disposal of these materials have important social consequences as well. Often, those at greatest risk from toxic contamination are the poor or those with little political clout: migrant laborers who work in fields sprayed with dangerous pesticides; factory workers who are exposed to low levels of toxic heavy metals over a prolonged period of time; residents of inner cities or impoverished neighborhoods who live under conditions that pose a constant

threat to health—from closeness to manufacturing and processing plants that pollute the air to rundown apartments covered with peeling lead-based paint (Figure 19-6).

It is no coincidence that hazardous waste sites and production plants that use or manufacture toxic substances are usually located in poorer areas or rural communities, where resistance is likely to be poorly organized or minimal. For instance, in an economically depressed community desperate for jobs, the construction of a pesticide production factory promises employment and economic development. Similarly, a hazardous waste incinerator or a new "super landfill" translates into additional jobs and, perhaps more importantly, an industrial "resident" of the community which will augment the area's tax base—a boon to rural areas where the tax base is generally low.

Another important social issue is the transport of toxic materials. Every day many toxins—like anhydrous ammonia, caustic soda, sulfuric acid, chlorine, gasoline, and radioactive nuclear fuels, among others—are transported across the nation's roads, rail lines, and waterways. Almost 60 percent of hazardous materials are shipped by truck, over 33 percent by water, about 5 percent by rail, and less than 1 percent by air. Many communities are concerned about the safety hazards associated with shipping these materials by truck or rail through or near residential areas (Figure 19-7).

There are international social consequences associated with toxic substances as well. As regulations

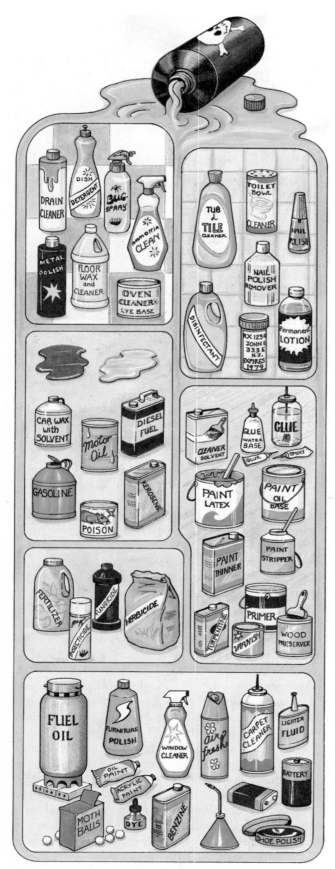

FIGURE 19-5: The plethora of toxic and hazardous substances commonly found in a typical American home.

governing the production, use, and disposal of hazardous wastes have become more stringent in the industrialized world, many chemical corporations have moved their operations to LDCs, where little or no regulations exist. Until recently, LDCs were also often the recipients of hazardous wastes that could not be easily disposed of in the countries of origin (Figure 19-8). One such country, the Congo Republic, recently canceled a contract to accept 20,000 to 50,000 tons (18,200 to 45,500 metric tons) of pesticide residue and sludge waste a month from a New Jersey firm. Djibouti has also rejected shipments of hazardous wastes. LDCs that currently accept wastes, including Morocco, Senegal, Benin, and Gabon, may reconsider their contracts as well. In densely populated Europe landfills and other disposal methods are impractical or economically undesirable, largely because land is scarce. The continent's leading waste exporters are Italy, France, Germany, Switzerland, Belgium, Luxembourg, and the Netherlands. Prior to unification, West Germany shipped much of its hazardous wastes to East Germany.

Many incidents in recent years illustrate the social dimensions of the production and use of toxic

FIGURE 19-6: Migrant workers harvest carrots in a field in California. Chronic exposure to pesticides puts these workers at increased risk for many diseases and conditions.

FIGURE 19-7: Southern Pacific train spill, Dunsmuir, California, July 14, 1991. A tank car carrying metam-sodium, which can be deadly when it reacts with water and may cause birth defects in the children of pregnant women exposed to its vapors, jumped the track and fell from the bridge, spilling part of its load into the Sacramento River. Wildlife and vegetation were destroyed along a 45-mile stretch of the river. No human deaths were attributed to the spill, but 300 local residents and emergency workers sought medical care.

FIGURE 19-8: The mystery of the *Khian Sea*. In 1988, the crew of the barge *Khian Sea* began unloading incinerator ash from a municipal incinerator in Philadelphia, Pennsylvania, on a beach in rural Haiti. The ash contained dioxins, furans, and heavy metals. The crew claimed that the material was fertilizer, which the barge had a permit to unload. Before the authorities intervened and ordered that the ash be put back on the barge, the crew had already unloaded 3,000 tons of the more than 13,000 tons the barge carried, piling it in a heap on the beach, only yards from incoming waves. Instead of reloading the ash, the *Khian Sea* left port under cover of night. It reappeared in Philadelphia, crossed the Atlantic Ocean to the African coast, crossed the Mediterranean Sea, and passed through the Suez Canal into the Indian Ocean, unable to find a port that would accept its cargo. Eventually, the barge entered Singapore under a new name, the *Pelicano*. Its hull was empty. Its hazardous cargo is believed to have been dumped at sea. The City of Philadelphia has denied responsibility for the matter, as have the middleman and the barge owners. It will be up to a judge to untangle the tale of the *Khian Sea* and determine the responsible parties. Meanwhile, the heap of ash in Haiti—that part of it that hasn't blown away or been washed out to sea—remains on the beach, partially enclosed in barrels.

chemicals. Consider these human-made disasters: Love Canal, New York; Times Beach, Missouri; Seveso, Italy; Minamata Bay, Japan; Schonnbrun, East Germany. The twentieth-century nightmare of severe contamination by toxic substances is perhaps best illustrated by the case of Bhopal, India.

Bhopal, India. On the night of December 2, 1984, a contaminant, probably water, leaked into storage tank 610 at Union Carbide of India's pesticide production facility in Bhopal, India. Forty-five tons (40.95 metric tons) of methyl isocyanate (MIC) were stored in the tank. Used to make the insecticide Sevin, MIC is one of the most toxic of the chemicals used in the industry. The MIC reacted violently with the contaminant and quickly began to convert from a liquid to a gas. The tank's pressure raced from 2 pounds per square inch at 10:20 P.M. to 30 pounds per square inch at 12:15 A.M. By 1:00 A.M. the rapidly expanding MIC vapor ruptured the tank's safety valve and escaped into the night air. A temperature inversion trapped the lethal gas near the ground, preventing it from dissipating.

The dense MIC cloud first drifted through the shantytowns just outside the plant's gates, instantly killing hundreds in their sleep. The path of the gas took it through homes, barns, a railway station, and a hospital, spreading death and sickness across a 25-square-mile (65-square-kilometer) area of Bhopal. As people awoke gasping for breath, they flooded into the streets. Thousands attempted to flee the invisible, odorless vapor. In the chaos some ran toward the plant. Others, blinded by the gas, stum-

bled into one another and were trampled by the crush. By morning, over 1,000 people were dead. An estimated 200,000 were exposed to the gas (Figure 19-9).

We will never have an accurate count of the human fatalities. The Indian government's official count is 3,000; early press reports said 10,000 dead. The government lists 30,000 victims as permanently disabled with another 25,000 severely injured. Many of the dead were children and elderly people. Their lungs, wracked by the MIC, filled with fluid, causing a death analogous to drowning. Many survivors suffered permanent blindness, paralysis, or severely diminished lung capacity. Pregnant women miscarried or delivered premature, deformed babies.

The protracted negotiation between Union Carbide and India over compensation for the victims of the world's most deadly chemical accident lasted over four years. The settlement, announced unex-

FIGURE 19-9: A Bhopal survivor, 1987. An aging woman, whose eyes were damaged by the Bhopal gas disaster, protests against a proposed out-of-court settlement between the Indian government and Union Carbide.

pectedly on February 14, 1989, called for Union Carbide to pay $470 million in damages (India had been seeking $3.3 billion). Based on the Indian government's official count, this works out to just over $8,000 for each man, woman, and child killed, disabled, or severely injured in Bhopal. The Indian Supreme Court unanimously upheld this settlement in 1991, but rejected the part of the settlement granting Union Carbide immunity from criminal prosecution. If the $470 million does not cover damages for all victims, the Indian government will pay the rest.

History of Management of Toxic and Hazardous Substances

During World War II the armed forces conducted a great deal of research into synthetic chemicals and materials needed for the war effort. At the end of the war, researchers turned their attention to the application of synthetic chemicals to peaceful uses, signaling the beginning of the modern industrial chemical age. The amount and toxicity of industrial wastes produced in the United States and other developed countries increased dramatically; between 1945 and 1985 the total production of synthetic organic substances increased from 7.37 tons to 112.2 million tons (6.7 to 102 million metric tons) in the United States. In 1990 the production of the 50 largest-volume chemical products rose 2 percent from 1989 to a new total of 216.82 billion pounds (97.57 billion kilograms).

The way in which society managed toxic and hazardous substances did not keep pace with the chemical revolution. At best, wastes were stored in metal barrels, placed in a dump, and covered with a layer of topsoil. At worst, toxic wastes were deposited in unlined evaporation pits, dumps, oceans, streams, rivers, and lakes. In fact, for the past fifty years or so the "management" of toxins has consisted solely of various disposal methods, both legal and illegal, and, unfortunately, disposing of toxic substances is still the management option most commonly used by business, industry, and households. Understanding the advantages and disadvantages of the major disposal methods—incineration, landfilling, and deep-well injection—can help us to develop more effective management strategies for the misplaced resources known as hazardous wastes.

What Is Incineration?

The burning or incineration of wastes—purification by fire—dates back to biblical Jerusalem. The Hebrew word for hell, *Gehenna*, was derived from the location of Jerusalem's smoldering town dump, *geben Hinnom*, or the valley of the son of Hinnom. The world's first large municipal waste incinerator was constructed in Nottingham, England, in the late 1800s. Crude incinerators can still be found in the backyards of many American homes, in the form of rusty 55-gallon drums. Used to burn household paper and plastic wastes, these burn barrels, like the larger but no more sophisticated early municipal units, are a cheap method of waste disposal, but they emit a substantial amount of particulates and greenhouse gases.

Modern large-scale incineration, which was first used to dispose of industrial hazardous wastes in the 1960s, is a complex and highly refined process (Figure 19-10). During incineration organic (carbon-based) materials are burned at high temperatures (typically 1,652 °F or 900 °C or greater) to break them down into their constituent elements, chiefly hydrogen, carbon, sulfur, nitrogen, and chlorine. These elements then combine with oxygen and are released into the atmosphere.

Combustion forms gases completely different from the original organic materials. The incinerator channels these gases from the primary combustion chamber into an even hotter afterburner, where they are further combusted before moving on to the pollution control system. Incinerators can use a variety of air pollution control systems, ranging from

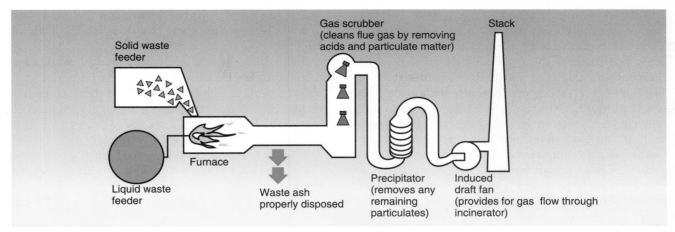

FIGURE 19-10: Standard incineration system. Solid and liquid wastes are fed into a furnace. High-temperature combustion destroys many of the toxic and hazardous substances. A gas scrubber removes acids and particulate matter from the resulting gas, and a precipitator removes any remaining particulates. Incinerators have several major drawbacks: Hazardous products formed during combustion may not be removed by the scrubber or precipitator, and the bottom ash may contain metals (which are not destroyed by incineration) and other toxic substances.

virtually nothing to those which effectively eliminate certain hazardous substances.

▶ Table 19-3
U.S. Municipal Waste Incinerators, Estimated Stack Emissions for Selected Pollutants

Pollutant	July 1987 (pounds per year)	Projected 1992–1995 (pounds per year)
Arsenic[a]	5,940	28,600
Beryllium	220	1,276
Cadmium[a]	22,800	43,780
Carbon monoxide	44,000,000	185,900,000
Chlorobenzenes	8,316	5,522
Chlorophenols	11,770	7,700
Chromium +6[a]	1,320	5,610
Chlorinated dioxins and dibenzofurans[a]	53	46
Formaldehyde[a]	128,920	261,800
Hydrogen chloride	104,940,000	427,900,000
Lead	750,200	2,772,000
Mercury	149,600	325,600
Nitrogen oxides	67,100,000	294,800,000
Particulate matter	22,880,000	58,740,000
Polychlorinated biphenyls (PCBs)[a]	10,934	46,640
Polycyclic aromatic hydrocarbons (PAHs)[a]	2,101	8,954
Sulfur oxides	28,600,000	138,820,000

[a]Known or suspected carcinogens.
Source: U.S. Environmental Protection Agency.

Incineration has one major advantage as a method of managing organic hazardous substances: it permanently reduces or eliminates the hazardous character of the waste. And while many of the gases formed in the incinerator are hazardous as well, pollution control devices can destroy and remove some of them. In order to receive an operating permit, an incineration facility must demonstrate a 99.99 percent destruction and removal efficiency for each principal organic hazardous constituent in the feed material. Additionally, the incinerator must attain a 99.9999 percent efficiency to burn materials containing polychlorinated biphenyls (PCBs).

Ideally, air pollution control systems would allow only carbon dioxide and water vapor to escape into the atmosphere. Unfortunately, the combustion process is complex, involving thousands of physical and chemical reactions. Depending on the composition of the waste and the particular combustion conditions, an infinite number of combustion products may form—some of them hazardous. Over 200 toxic or potentially toxic chemicals have been identified in municipal waste incinerator emissions and ash residues, including aldrin, benzene, benzopyrene, dieldrin, formaldehyde, PCBs, and vinyl chloride (Table 19-3). If some of the waste does not reach a high enough temperature to be completely destroyed, it can form organic compounds known as dioxins and furans, which have been linked to high cancer rates, miscarriages, birth defects, liver disease, neurological damage, and immune system disorders. If they escape the combustion chamber, these organic compounds may not be removed by pollution control systems.

Another serious drawback of incineration is that it cannot destroy metals present in the waste stream. They are typically emitted in the form of particulates, but some of the more volatile elements, like mercury and selenium and their reaction products, may be released as vapors. Arsenic, barium, beryllium, cadmium, chromium, lead, mercury, nickel, and zinc are frequently found in hazardous waste and, if released in emissions, may have adverse health effects on humans and the environment.

Besides releasing airborne substances, incinerators also generate large amounts of unburned residues, or ash, which often contains high concentrations of heavy metals and other toxins and must be disposed of on land in an approved hazardous waste landfill. When considering incineration as a disposal option, the disadvantages—uncontrollable emissions and bottom ash—must be balanced against the advantages—total waste volume reduction and the destruction of most organic contaminants.

What Is Landfilling?

Today's technology has made considerable progress since the time of the open dump. Municipal refuse and industrial waste that is not deemed hazardous or toxic are placed in **sanitary landfills.** Sanitary fills are legally prohibited from accepting hazardous substances, except that which comes from municipal solid waste, because they are not capable of isolating these wastes for long periods of time. Instead, hazardous wastes must be disposed of in **secure landfills.** Secure fills are specially engineered to prevent the escape of **leachate**—the potentially dangerous liquid formed when rainwater percolates downward through soil and wastes—into the surrounding environment.

Secure sites are chosen on the basis of their geology and relation to surface waters and aquifers. The best sites have a thick layer of dense clay or chalk between the landfill and the aquifer. Such a layer resists the flow of water and helps protect the groundwater. The bottom of the landfill is compacted and slightly graded toward a lower, central area. A thick sheet of waterproof plastic lines the sides and bottom of the landfill. If it is not broken, the liner will prevent leachate from flowing out of the landfill. A layer of sand or gravel covers the liner, along with a network of perforated pipes which are part of the leachate collection system. Gravity draws leachate through the sloping pipes and down to a sump pump situated in the low-lying center. Once the leachate is pumped from the landfill, it is treated before being disposed of. A leachate detection and collection system located beneath the plastic liner is separated from the clay bottom by a second plastic liner, which further minimizes the possibility of toxic leachate escaping from the landfill (Figure 19-11).

Before being placed in a secure landfill, wastes are sorted into groups that would not chemically react if allowed to mix. The landfill is divided into compartments; each compartment accepts only one group of materials. Landfill operators are required to maintain records of exactly what materials are in what part of the landfill. These records are important in the event that problems with the landfill develop in the future.

After the landfill is completely filled, it is covered with a thick clay cap and possibly a layer of waterproof plastic (much like the bottom liners). The entire landfill is then covered with earth and planted with vegetation to prevent erosion. The cap is sloped to cause rainwater to drain away from the landfill, thereby reducing the amount of leachate. Clay caps are subject to many stresses which may lead to their failure: uneven settling of the landfill contents, animal and human activity, and natural processes such as freezing and thawing and geologic movement. Consequently, the landfill operator must monitor the leachate collection system indefinitely for any increase in leachate production that would signal that the landfill cap had failed.

A state-of-the-art 2,400-acre (970-hectare) secure landfill exists in Emelle, Alabama. Owned by Chemical Waste Management of Oak Brook, Illinois, it is the United States' largest secure landfill. Located in economically depressed, rural Sumter County (population 17,000), the Emelle site appears to be geologically ideal for such disposal. A 700-foot (210-meter) layer of dense limestone known as Selma chalk separates the landfill from the aquifer below, which happens to be the primary source of drinking water for western Alabama. Chemical Waste Management calculates that it would take 10,000 years for liquids that may escape from the landfill to migrate into the aquifer.

The Emelle site is a major part of the EPA's Superfund program, which guides the cleanup of this country's worst hazardous waste sites. Wastes from many Superfund sites have been buried in Emelle, and the landfill will probably play an even greater role as cleanup activities continue. Yet Emelle, and every other secure landfill in the country (like all landfills), may eventually leak. Since the beginning of Superfund in 1980, the EPA has relied on landfills for the disposal of contaminated materials removed from hazardous waste sites. In effect, they have simply moved these hazardous materials from one site to another. Many people argue that landfills are not a permanent solution for hazardous and toxic materials carelessly disposed of twenty years ago *or* for those produced today.

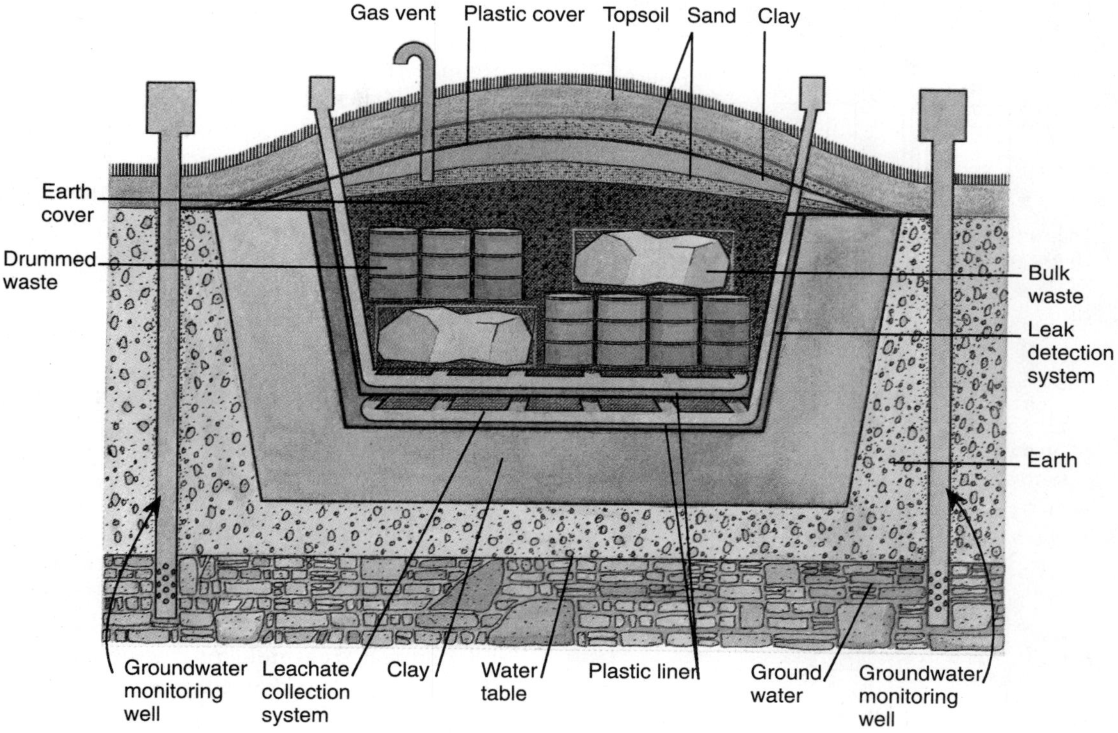

FIGURE 19-11: Secure landfill. A secure landfill is designed to contain hazardous wastes, preventing hazardous leachate from contaminating groundwater, nearby surface waters, and soils.

What Is Deep-Well Injection?

Since the 1880s the petroleum industry has used deep-well injection to dispose of brine (salt water produced when drilling for oil). Disposing of hazardous liquids by deep-well injection, however, is a relatively recent development. The EPA estimates that about 11 percent of all liquid hazardous wastes produced in the United States (11.5 billion gallons or 43.59 billion liters) is injected deep into the ground. There are approximately 250 hazardous liquid injection wells in the country, most of them located in the Great Lakes region and along the Gulf Coast.

Deep-well injection involves pumping liquid wastes deep underground into rock formations that geologists believe will contain them permanently (Figure 19-12). A high-pressure pump forces the hazardous liquids into small spaces or pores in the underground rock, where they displace the liquids (water and oil) and gases originally present. Sedimentary rock formations like sandstone are used because they are porous and allow the movement of liquids. However, this sandstone disposal layer must

be well below the lowest level of the freshwater aquifers and must be separated from these aquifers by an adequately thick layer of rock impervious to water, such as shale.

Theoretically, deep-well injection systems, when properly constructed, operated, and monitored, may be the most environmentally sound disposal method for hazardous and toxic materials that cannot be recycled. If the substances are injected into a stable and receptive layer that is below the lowest drinking water aquifer, that aquifer is less likely to become contaminated than if the substances are placed in a landfill above it. However, wells can fail for a variety of reasons. A leak at the entrance to the well or in its casing would pose a direct threat to groundwater. Unknown fractures in local geology pose a more indirect, less easily detected threat. Liquid injected under extreme pressures seeks the pathway of least resistance, and a fracture that leads toward the surface could provide an unseen pathway for wastes to travel. If the fracture extends into a freshwater aquifer or a porous subsurface stratum that lies below an aquifer, contamination could occur unchecked for an indefinite amount of time.

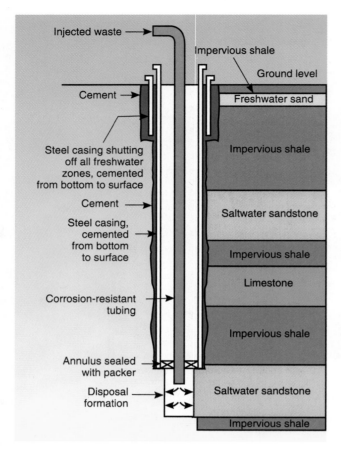

FIGURE 19-12: Deep-well injection. Liquid hazardous waste is injected into corrosion-resistant tubing within a steel casing. Cement around the casing further isolates the injection system from the environment. The waste is pumped under high pressure deep underground into permeable rock formations, such as sandstone, which permit the movement of liquids. As long as the sandstone disposal layer lies far below the lowest level of nearby aquifers and is separated from them by impervious layers of rock, such as shale, the wastes should be properly contained.

The history of hazardous waste injection wells illustrates the potential dangers of this disposal method. In 1968 an overpressurized well in Erie, Pennsylvania, erupted. Roughly 4 million gallons (15 million liters) of injected hazardous material escaped over a period of three weeks. In the late 1970s an injection well constructed by Vesicol Chemical Corporation contaminated a drinking water aquifer under Beaumont, Texas. The well lacked a separate injection tube, allowing the hazardous materials to corrode both the inner and outer casings. Probably the worst known case of environmental damage due to the failure of a hazardous waste injection well took place in Vickery, Ohio, in 1983. Inadequate design, operation, and maintenance procedures on the part of the operator, Ohio Liquid Disposal, led to the massive migration of about 20 million gallons (75.8 million liters) of hazardous substances from six injection wells. The

Ohio EPA ordered the owner, Chemical Waste Management, to pay $12 million in federal and state fines and to spend an additional $10 million to upgrade the site.

Even with improved management of deep wells and stricter regulatory oversight, the lack of precise knowledge about the exact fate of substances after injection prevents this management option from being a final solution to the disposal of nonrecyclable hazardous liquids. It may take up to 50 years before the success or failure of a specific well can be determined.

How Are Toxic and Hazardous Substances Treated to Reduce the Risk of Environmental Contamination?

Prior to disposal, most hazardous and toxic materials can be treated to reduce their hazardous nature and the threat that they will contaminate the environment if released. Methods of treatment can be classified as chemical, physical, or biological.

Chemical treatment processes alter the chemical structure of the constituents to produce a waste residue that is less hazardous than the original waste (Table 19-4). The altered constituents may also be easier to remove from the waste stream. Treatment is often accomplished by the addition of other chemicals to the waste.

Physical treatment, which does not affect the chemical structure of the wastes, can take many forms. **Phase separation processes,** such as filtration, centrifugation, sedimentation, and flotation, separate a waste into a liquid phase and a solid phase, both of which can then be more easily treated. **Component separation processes** remove particular hazardous materials from a waste stream without the use of chemicals. Some have been developed to remove inorganic components; others remove volatile organics. A third type of physical treatment, solidification, is becoming increasingly popular because it minimizes the chance that leachates from hazardous materials will contaminate the environment. **Solidification** transforms a liquid, semisolid, or solid hazardous waste into a more benign solid product. Solidification does not destroy the hazardous components in the waste nor does it make them available for recovery and recycling, but the technique does entomb these components in a concrete block, thus isolating them from the environment for an extended period of time.

A new way to remove dioxins from industrial waste water and detoxify it has been developed by researchers at the University of Michigan. In this process humic acid, an organic soil compound, is bound to a clay base with an aluminum compound. When the humic acid and clay compound is intro-

▶ Table 19-4
Chemical Treatment Methods

Electrolysis is the reaction of either oxidation (loss of electrons) or reduction (gain of electrons) taking place at the surface of electrodes. Electrolytic processes can be used for reclaiming heavy metals from contaminated aqueous wastes.

Neutralization is a reaction that adjusts the pH of either an acid (low pH) or a base (high pH) by the addition of the other to achieve a neutral substance (pH 7). This commonly used treatment method may not always make a substance less hazardous, but it can aid in the removal of heavy metals, prevent metal corrosion, or provide a more manageable material to be recycled.

Oxidation is used to detoxify wastes by causing the transfer of electrons from the waste chemical being oxidized to an oxidizing agent. Some common oxidizing agents and their uses are:

Ozone is used in the oxidation of cyanide to carbon dioxide and nitrogen and the oxidation of phenols to less toxic compounds. Ozone is an unstable molecule, which makes it a highly reactive oxidizing agent. In the presence of ultraviolet light ozone can oxidize halogenated organic compounds that are otherwise resistant to ozone oxidation.

Hydrogen peroxide, also a powerful oxidizing agent, can be used to treat wastes that contain phenols, cyanides, sulfur compounds, and metals.

Potassium permanganate can oxidize aldehydes, mercaptans, phenols, and unsaturated acids. It has been used to destroy organics in wastewater and drinking water.

Precipitation is a process that converts soluble metal ions into insoluble precipitates, which can be easily removed from the waste stream by filtration. It involves the addition of substances that change the chemical equilibrium of the waste stream and induce precipitation. Some chemical precipitators are:

Alkaline agents such as lime or caustic soda which, when added to the waste stream, raise its pH, causing a decrease in the solubility of certain metal ions.

Sulfides are used to precipitate heavy metals ions as insoluble metal sulfides. Hydrogen sulfide, sodium sulfide, and ferrous sulfide are most commonly used for this purpose. Sulfide precipitation is very pH sensitive.

Sulfates of zinc and iron (II) can be used to remove cyanide from an aqueous waste stream. However, the resulting sludge is an extremely hazardous mixture of cyanide complexes that must be handled appropriately.

Carbonate precipitation of metal ions is common because carbonates can easily be filtered out of aqueous waste.

Although precipitation with chemicals is an effective and reliable treatment process, the resulting sludge contains high levels of metals and other toxic materials. After the recoverable metals are removed, the sludge requires ultimate disposal. Industries that use precipitation include inorganic chemicals, metal finishing, copper forming, foundries, pharmaceutical manufacturing, and textile mills.

Reduction is the chemical opposite of oxidation; electrons are transformed from a reducing agent to the waste chemical being reduced. This process may reduce toxicity or encourage precipitation. There are many common reducing agents. A major application of chemical reduction uses sulfur dioxide, sodium bisulfites, or ferrous salts to reduce the extremely toxic chromium (VI) to the much less toxic chromium (III), which can then be removed by precipitation as hydroxide.

duced into a stream of industrial waste water, it absorbs dioxins, lowering the concentration of these chemicals below current detection levels.

Biological treatment, or bioremediation, is fast becoming a preferred option for hazardous waste treatment. Biological treatments often cost far less than disposal options, and in some cases they can be so successful that no disposal is needed. However, unlike incineration, which takes seconds to destroy a hazardous material, biological treatment can take months or years to complete.

Bioremediation uses both naturally occurring and genetically engineered microorganisms to decompose organic chemical wastes into harmless by-products. For example, bacteria can be used to convert hazardous organic chemicals such as PCBs into water, carbon dioxide, and simple salts. Bacteria have been used for decades to treat municipal sewage, but relying on them to break down highly toxic substances is a relatively recent innovation. Biological treatment methods include activated sludge, aerated lagoons, anaerobic digestion, composting, trickling filters, waste stabilization ponds,

and landfarming. Petroleum refineries often use landfarms successfully to treat their oily wastes. In this process, operators spread wastes on a parcel of land and till them into the soil. Microorganisms in the upper 12 inches of the soil decompose the organic chemicals in the waste. Nutrients like nitrogen and phosphates may be added to bolster the microorganism population.

Bioremediation has been used recently on three major oil spills in the United States, the *Exxon Valdez* spill in Prince William Sound in March 1989, the *Mega Borg* accident in the Gulf of Mexico in June 1990, and the spill in Galveston Bay in June, 1990. The Alaskan beaches were sprayed with a garden variety of bacteria and then sprayed with a water-soluble fertilizer to stimulate the bacteria's growth (Figure 19-13). These organisms fed on the oil that had seeped nearly 2 feet (0.6 meter) into the gravel and sand. Within three weeks the fertilized areas were free of oil for 12 inches (30.5 centimeters), while untreated areas were still coated. In the Gulf of Mexico laboratory-grown bacteria and fertilizer were applied directly on the *Mega Borg*

FIGURE 19-13: Bioremediation in Alaska's Prince William Sound after the *Exxon Valdez* spill in March 1989.

spill. The EPA is still examining the results. Success in the bioremediation process is difficult to analyze because of uncontrollable outside variables such as shifting currents and winds.

Bioremediation appears promising because it can be used on both soil and water at lower costs, with less disruption to the site and without generating by-products that require disposal. However, its effectiveness is limited by the amount of oxygen and other nutrients in the contaminated zone.

What Is Waste Minimization?

Soaring disposal costs, shrinking disposal capacity, the increasing threat of financial liability for unsound disposal methods, and the rising cost of raw materials have led industries to turn to waste minimization. **Waste minimization** is an umbrella term that refers to a variety of strategies to reduce the amount of toxic and hazardous substances used and the volume of hazardous wastes that must be disposed of. Waste minimization strategies are based on the premise that toxic and hazardous substances are resources that can and should be managed in environmentally sound ways. While they cannot eliminate all hazardous wastes, these strategies can enable business, industry, and households to mimic natural systems, thereby becoming less wasteful and more efficient. Waste minimization strategies can be grouped into two broad categories. "Front door" strategies reduce the amount of hazardous substances used and the volume of wastes produced; "back door" strategies reduce the volume of hazardous wastes that must be disposed of.

Reducing the Amount of Hazardous Substances Used and the Volume of Wastes Produced. Keeping

machinery and pipe systems in good order and working properly minimizes the volume of hazardous material that must be used in a given process. So do distillation systems, which allow solvent vapors to be captured (rather than being emitted to the atmosphere) and reused, thus extending the useful life of the solvent. Over time they can significantly reduce the total amount of solvent that is used. Modifying manufacturing processes can also reduce the volume of hazardous resources used. Modifications include substituting water for organic solvents, altering rinsing procedures, and using better housekeeping measures in general. American Telephone and Telegraph, for instance, now uses a cleaning solution based on citrus oils rather than chlorofluorocarbons (CFCs), which deplete the ozone layer when emitted into the atmosphere. General Motors switched from solvent-based to water-based paints and installed computer-controlled robotic painting systems, changes that have reduced emissions of volatile hydrocarbons by 70 percent since 1977 and sharply reduced the amount of hazardous paint sludge generated, saving disposal fees.

Another front door minimization strategy is **source segregation,** the practice of keeping used hazardous resources apart from nonhazardous resources. Source segregation prevents hazardous material from contaminating nonharmful substances. Training employees to keep contaminated safety clothes, rags, or other materials separate from noncontaminated ones, for example, helps to reduce the volume of contaminated materials produced.

Reducing the Volume of Hazardous Wastes That Must Be Disposed of. Various strategies have been developed to minimize the volume of hazardous wastes that must be "sent out the back door," that is, disposed of. These strategies recover resources, either materials or energy, from the waste stream. Examples include the reclamation of titanium from the residues of utility company pollution control systems and the recovery of precious metals from plating operation liquids. While these are positive developments, it is important that proposed recovery operations be carefully scrutinized and regulated. For example, used organic solvents with high energy (Btu) contents and low concentrations of toxic inorganic contaminants may be used as fuel for heating and steam production. Since this resource recovery strategy is a form of incineration, it should be subjected to the same stringent pollution control regulations that apply to hazardous substance incinerators.

Most resource recovery efforts involve the reclamation of organic solvents by highly efficient distillation techniques. Solvents contaminated with metals and organics are heated to produce a liquid phase, or component, and a vapor phase. Lighter components with high volatilities rise to the top of the liquid phase and begin to vaporize. By carefully controlling the waste mixture's temperature, the desired substance can be vaporized and recovered by condensation, leaving the heavier contaminants behind. What remains is a concentrated, highly toxic mixture, far reduced in volume, referred to as **still bottoms.** Bottoms may contain usable metals and other solvents. The development of improved distillation techniques will allow increased recovery of these materials.

One of the corporate leaders in reducing the amount of hazardous waste to be disposed of is Minnesota Mining and Manufacturing (3M), best known as the producer of Scotch tape and video cassettes. In 1991 the company installed $26 million worth of equipment to recycle and burn off solvents from two plants in an area that already met EPA standards (Figure 19-14). After pumping 25 million pounds (11.25 million kilograms) of volatile organic compounds into the air in 1989, the company claims to have achieved 95 to 98 percent reduction in emissions. The Franklin Research and Development Corporation, a Boston-based investment firm that rates companies on social responsibility, has given 3M the highest ranking on environmental issues on a scale of 1 (best) to 5 (worst). Most companies rate around a 3. At factories around the country 3M is forfeiting rather than selling federal pollution-reduction credits, a move it says will reduce pollution overall rather than merely allowing other companies to increase their emissions of pollutants. Credits alone could have been worth $1 million, experts say. While some companies have occasionally turned in credits, 3M is the only one to have a stated policy of doing so. For years it has been cleaning up operations before state or federal regulations force it to do so, often adopting a "front-door" instead of a "back-door" approach to waste minimization.

What Are Waste Exchanges?

Hazardous resources can be recovered on site by the generator or an outside contractor or sent to a commercial resource recovery operation. Another option becoming increasingly popular is a waste exchange system, which locates and brings together companies that have waste and companies that want to recover and reuse these resources.

The first waste exchange began in 1972 in Norway. Currently, almost 30 regional waste exchanges operate in the United States alone. One such exchange is the Waste Systems Institute of Michigan, Inc., a nonprofit organization that publishes a listing of offers of and requests for used resources in the Great Lakes region. This information clearinghouse promotes the reuse of substances that would other-

FIGURE 19-14: 3M plant. Through an aggressive waste-minimization program called Pollution Prevention Pays, the Minnesota Mining and Manufacturing (3M) Company claims that it has achieved a 50 percent reduction in potential hazardous waste generation in 10 years. In addition to the environmental benefits, the program saved the company $300 million.

wise be disposed of at great cost to the generator and significant danger to the environment. Profit-making waste exchanges have proven successful in the United Kingdom, where they serve as resource brokers, aggressively seeking out both exchangeable resources and markets for them. Also called material exchanges because they actually handle the resource materials, these for-profit enterprises are becoming more popular in the United States.

What Legislation Affects the Management of Toxic and Hazardous Substances?

The government has been slow to provide funds in support of waste minimization strategies despite the facts that Congress declared the reduction or elimination of hazardous waste generation a national policy and the EPA lists waste reduction as its preferred hazardous resource management strategy. Federal and state spending has concentrated on controlling environmental harm from waste already generated, rather than on reducing the generation and disposal of these materials to begin with.

Of the federal laws drafted to address the management of hazardous and toxic materials, a few are of primary importance. The Comprehensive Environmental Response, Compensation, and Liability Act (CERCLA), more commonly known as Superfund, and the Resource Conservation and Recovery Act (RCRA) address issues concerning hazardous wastes. The Toxic Substances Control Act (TSCA) differs from Superfund and RCRA because it concerns the regulation of toxic and hazardous substances rather than wastes.

Superfund: The Comprehensive Environmental Response, Compensation, and Liability Act. Superfund was developed in response to several environmental crises in the late 1970s, particularly Love Canal. Enacted in 1980, it empowered the federal government to clean up hazardous waste sites and to respond to uncontrolled releases of hazardous substances into the environment.

Superfund authorizes the federal government to respond directly to releases or threatened releases of hazardous substances into the environment. The act has several key objectives: to develop a comprehensive program for cleaning up the worst hazardous waste dump sites; to force those parties responsible for creating a site to pay for its cleanup; to create a $1.6 billion Hazardous Substance Response Trust Fund (called Superfund) to pay for the remediation of dump sites where no responsible parties can be held accountable and to respond to emergency situations involving hazardous substance releases from these sites (this program was funded through 1985); and to advance scientific and technological capabilities for the environmentally sound management, treatment, and disposal of hazardous resources.

The EPA manages Superfund, which is financed by taxes on the manufacture or import of crude oil and commercial chemical feedstocks (the raw chemical materials used in the manufacture of chemical and chemical-based products). If the EPA decides that a site poses significant potential harm to human or environmental health, the agency can use Superfund money to respond to the crisis in a number of ways. It may immediately remove hazardous substances from the site, conduct remedial actions such as on-site incineration of contaminated soil, water, and debris, and take measures to protect public health (including providing alternate water supplies and temporarily or permanently relocating affected residents). Whenever possible, those responsible for the release of hazardous wastes are forced to reimburse the fund for such expenses (Figure 19-15).

Superfund was envisioned by many to be a one-time, short-term effort at cleaning up the country's most polluted hazardous waste disposal sites. Unfortunately, the first five years of the Superfund program were marked by few victories and many setbacks. Of the thousands of potentially dangerous waste sites in existence, the EPA was successful in cleaning up only six. Many argued that even those cleanups were inadequate. Henry Cole, science director for the Clean Water Fund, a nonprofit environmental advocacy group based in Washington, D.C., contends that the Industrial Excess Landfill (IEL) in Uniontown, Ohio, illustrates what can happen to any community with a Superfund site.

Industrial Excess Landfill resulted from the disposal of industrial chemical wastes during the

FIGURE 19-15: Workers in protective clothing help clean up a hazardous waste site at Hamilton, Ohio.

1950s, 1960s, and 1970s. Toxic chemicals, including carcinogens, seeped from the landfill into the groundwater, which is the sole source of public water in the area. The EPA ordered a replacement municipal water supply for 100 families living to the west of the site; however, these families waited years for the alternate water. People living nearest the site have had to contend with an even greater hazard—a deadly mixture of methane gas and volatile organic chemicals generated in the landfill. The methane forms from the decay of municipal wastes that were also placed in the landfill. In 1986 testing revealed explosive gas levels in the crawl spaces below homes adjacent to the landfill. The EPA installed a gas pumping system to prevent the outward migration of landfill gases. However, this system has malfunctioned several times and in the words of one resident, "We live a mechanical heart beat away from disaster." Moreover, these residents now know that they were exposed to cancer-causing compounds present in the gases before the gas extraction system was installed. The families are faced with a terrible dilemma; they may want desperately to move but cannot afford to buy a second house without selling and are unwilling or unable to sell because of severely reduced property values. Only after enormous public outcry and pressure from elected officials did the EPA offer to move the dozen or so "border families" in Uniontown. Superfund, designed mainly as a cleanup program, did little to address community disruption.

The EPA has begun a second and final phase of remedial action at this site (the first being the provision of an alternate water supply to homes west of the site). The second action consists of installing a cap over the site to prevent surface water infiltration, expanding the existing methane venting system (to accommodate the potentially greater volume of gas that may collect as a result of the cap), extracting and treating 256 million gallons (970 million liters) of contaminated groundwater, treating surface water from ponds if necessary, and dredging contaminated sediment from nearby ponds and ditches and placing them under the cap. Future use of this site will be restricted. Initial construction for this phase is underway. The project is estimated to cost $18.5 million.

The Superfund Amendments and Reauthorization Act (SARA) was approved in October 1986. SARA authorized a new $8.5 billion fund through 1991, providing for a nearly sixfold increase in funding over the original law. More than 1,200 uncontrolled hazardous waste sites, most of them in the Northeast and Great Lakes regions, were included on the EPA's National Priorities List (NPL) of sites requiring major cleanup actions. By 1991, however, only 34 sites had been cleaned up and removed from the list. Another 65 had been cleaned and were in the process of being taken off the list. With an average price tag of between $21 and $30 million per site, the total cleanup cost for existing NPL sites is almost $30 billion dollars. Over 31,000 sites were still under investigation for possible inclusion on the list.

In October 1990 Superfund was again reauthorized and allotted a maximum of $212.5 million per year from general revenues. This reauthorization required the EPA to begin 200 new cleanups during 1990 and 1991, but soon after the reauthorization the EPA announced it would not be able to meet this goal. Beginning in 1992, the EPA was no longer required to initiate a certain number of new Superfund cleanups each year.

The shortcomings of the Superfund process spurred a promising alliance of strange bedfellows. Since 1984 Clean Sites, Inc., a nonprofit organization composed of representatives of industry and environmental groups, has been working to improve the process of cleaning up hazardous waste sites. (Clean Sites, Inc., is the subject of this chapter's Environmental Science in Action, pages 416–418.)

Resource Conservation and Recovery Act. Whereas Superfund cleans up sites polluted by the unwise disposal practices of the past, the Resource Conservation and Recovery Act (RCRA), enacted in 1976 and amended in 1984, establishes a legal definition for hazardous wastes and sets guidelines for managing, storing, and disposing of hazardous resources in an environmentally sound manner. Congress granted the EPA leeway in determining which substances are to be regulated as legally hazardous and which are not. RCRA is intended to prevent the creation of new Superfund sites by severely limiting the land disposal of hazardous wastes and by regulating the generators of these materials.

Regulating Hazardous Wastes. The EPA was given authority to label substances as hazardous and therefore subject to hazardous waste regulations. According to RCRA, if a waste has been found to be fatal to humans at low doses or has been shown through animal studies to be dangerous to humans, the EPA can place it on its official list of hazardous wastes.

RCRA also specified four hazardous waste characteristics: ignitability, corrosivity, reactivity, and toxicity characteristic leaching procedure (TCLP). Ignitable wastes are easily combustible or flammable during routine handling. Examples of ignitable wastes are gasoline, paint, degreasers, and solvents. Corrosive wastes dissolve steel or burn skin and can corrode containers and escape into the environment. They are strong acids or bases such as are found in used rust remover, acid or alkaline cleaning solutions, and battery acid. Reactive wastes are unstable under normal conditions and undergo rapid or violent chemical reactions with water or other materials. Examples are cyanide plating wastes, bleaches, and other oxidizers. TCLP wastes are identified by a standard test designed to simulate the release of toxic contaminants by leaching. If the waste contains any of 39 heavy metals or organic compounds above their maximum concentration levels, the waste is legally hazardous.

If a waste exhibits one or more of these four characteristics, the EPA can place it on the list of hazardous wastes. Even if the waste is not included on the EPA's list, simply displaying one of the characteristics is enough to make it legally hazardous.

It is relatively easy to determine if industrial wastes exhibit any of the four characteristics specified in RCRA. However, the characteristics are not adequate to identify wastes that are hazardous due to biological properties—carcinogenicity, mutagenicity, teratogenicity, or infectivity—which are much more difficult to detect. To be considered hazardous and therefore subject to regulation, substances with these properties must be specifically identified as hazardous by the EPA. Testing a substance for such biological characteristics is time consuming and expensive. Consequently, the disposal of many potentially dangerous substances is not regulated because the EPA has not yet performed sufficient testing to determine whether to add them to its hazardous waste lists.

Cradle to Grave Management of Hazardous Wastes. Prior to RCRA, there was no regulatory system that established accountability for hazardous waste management practices. Consequently, those who disposed of hazardous waste often used unsound methods and kept no records of what the wastes

- Read labels and choose products that are biodegradable and environmentally friendly. Substitute nonhazardous alternatives for hazardous substances whenever possible. Alternatives to household hazardous substances are listed in the adjacent table.

- If you buy a hazardous product, buy only what you need, use it according to the label directions, and use all of it. This prevents having to store or dispose of unwanted excess chemicals. If you can't use all of the product, give it to someone who can.

- Use recycled and unbleached paper products and water-based latex paint instead of oil-based paint.

- Ask your grocer to stock pesticide-free, organically grown foods. Educate others about the dangers of pesticide residues. Realize that fruits and vegetables do not have to look perfect to taste good. If we change our buying habits in this way, the nature of industrial "waste" generation will change, too.

- Practice safe hazardous substance disposal in the home. Hazardous or toxic materials should *never* be placed in the trash. Taking advantage of household hazardous waste collection programs is one obvious way to minimize the dangers

Common Toxins	Alternative Products
All-purpose cleaner	Mild: 1 gallon hot water 1/4 cup sudsy ammonia 1/4 cup vinegar 1 tablespoon baking soda Strong: Double all ingredients except water
Laundry detergent	Soap
Carpet cleaner	Mixture of cornmeal and borax
Copper polish	Lemon juice and salt or hot vinegar and salt
Oven cleaner	Ammonia or mixture of strong all-purpose cleaner and baking soda
Drain cleaner	Weekly applications of mixture of baking soda, salt, and cream of tartar. Pour 1/4 cup of mixture into drain, follow with a pot of boiling water, and then rinse with cold water.
Tub and tile cleaner	Firm-bristled brush with either baking soda and hot water or the mild all-purpose cleaner
Dish detergent	Soap, with vinegar added to remove grease
Furniture polish	Mixture of lemon oil and mineral oil
Window cleaner	Soap and water, with dilute vinegar rinse Mixture of ammonia, vinegar, and very warm water (applied with spray bottle)

associated with toxic products. These programs often recycle collected materials like used batteries, paint, and used motor oil. The public can bring unwanted hazardous materials to a collection site staffed by personnel trained in the safe handling of dangerous chemicals. Workers segregate the wastes by chemical type and remove materials that can be recycled. The remaining wastes are drummed, labeled, sealed, and transported to a licensed hazardous waste management facility for treatment and ultimate disposal.

were, where they came from, or who had generated them. RCRA established a system of "cradle-to-grave" regulation of hazardous wastes. From the time a hazardous waste is generated, a record must be kept of what the waste is, who generated it, who transported it, who treated it, and who disposed of it. This record, called a manifest, provides the basis for assuring that wastes are handled, treated, and disposed of in compliance with RCRA regulations. Under RCRA, facilities that treat, store, or dispose of hazardous waste must obtain an operating permit from the EPA or the state.

Toxic Substances Control Act. The Toxic Substances Control Act (TSCA) concerns toxic *commercial products*, rather than hazardous *wastes*. TSCA regulates the production, distribution, and use of chemical substances that may present an unreasonable risk of injury to human health or the environment. The EPA determines whether new chemical products present unreasonable risks by weighing their expected environmental, economic, and social benefits.

Manufacturers of chemicals are required to notify the EPA before producing new chemical substances or processing chemicals for significant new uses. This allows the EPA to review and evaluate information about the potential health and environmental effects of new products before they can cause environmental damage. The EPA has the authority to take regulatory action to control new, potentially risky chemicals.

Common Toxics	Alternative Products
Air freshener	House plants; baking soda in refrigerator or garbage can
Garden and houseplant insecticides	Soap solution Hot pepper solution Tobacco water Natural insect predators such as ladybugs, praying mantis, dragon flies, spiders, toads, birds
Herbicides	Hand weeding
Household insecticides	Ants: Lemon juice and lemon peel at entry points Cockroaches: Light dust of borax around the refrigerator, stove, and ductwork; plug all small holes Flies: Flypaper; close windows before sun hits them Moths: Camphor; keep vulnerable clothes dry and well aired

Greenpeace compiled the information in this table from a variety of sources and assumes no responsibility for the effectiveness of the suggestions. Caution is urged in the use of cleaning solutions and pest-control substances. **Keep them out of the reach of children.** For further and more detailed information, visit your library.

- If your community does not have a household hazardous waste collection program, encourage your local officials to start one or start it yourself. For example, concerned citizens were responsible for starting household hazardous waste programs in Madison, Wisconsin; Sacramento, California; and Seattle, Washington. Here are some suggestions on how to set up a program in your community:

1. Let your local officials know that you are concerned.
2. Form a planning committee. Approach community groups, local industries, and service organizations to enlist volunteer help and garner financial, technical, and community support for your project.
3. Raise funds. Monies will be needed for public education and to pay the licensed hazardous waste firm that will be removing the materials. (If the firm is local, try to work out a community service arrangement with them.)
4. Contact your state hazardous waste agency. It will be a valuable source of information and may aid in finding funding.
5. Contact local organizations that will accept particular hazardous resources for free. These might include gas stations that recycle used motor oil or car batteries, theatre groups that can use leftover paint and thinner, dentists who recycle waste mercury, and so on.
6. Write or phone organizations that have already started programs:

   Citizens for a Better Environment
   111 King Street
   Madison, WI 53703
   (608) 251-2804
7. Design your program to be more than a one-time effort. Look for ways to make it a permanent part of your community.

Future Management of Toxic and Hazardous Substances

In large measure toxic substances underpin much of industrialized society. Research after World War II promised "better living through chemistry," and for a time the chemical age did indeed provide a higher standard of living than any previous generation had enjoyed. But as the dangerous environmental and health effects of many synthetic materials have become apparent, people have begun to think about the consequences of indiscriminate use and careless disposal of chemicals.

Many people argue that the dangers of toxic substances outweigh the benefits they provide. Others contend that it is impossible to turn back the clock, that most people will not forgo the conveniences and comforts of the modern industrial world. But this need not be an "all or nothing" issue. Perhaps it is time to refocus the debate to define a new and realistic challenge for our society: to produce, use, and manage these substances in environmentally sound ways in order to ensure a high standard of living while safeguarding human and environmental health. To meet this challenge, action can be taken to prevent overuse of resources, protect human and environmental health from pollution and degradation, and preserve living systems.

Prevent Overuse of Resources

As a society, we should strive to reduce the demand for toxic substances and find alternatives whenever

Environmental Science in Action:
Clean Sites, Inc.
Lindsay Koehler

Clean Sites, Inc., a private, nonprofit organization dedicated to accelerating the cleanup of unsafe hazardous waste sites in the United States, was founded in 1984 through the efforts of the Conservation Foundation and the National Wildlife Federation. Among the environmentalists and business leaders who helped found Clean Sites, Inc., are William K. Reilly, former president of the Conservation Foundation and current executive director of the Environmental Protection Agency; Jay Hair, executive director of the National Wildlife Foundation; and Louis Fernandez, chief executive officer of Monsanto, Inc. Recognizing that adversarial relationships between business and environmental organizations can hinder necessary action, the founders of Clean Sites decided to create an organization in which these two groups can collaborate to solve problems. Since its founding Clean Sites has made significant progress in initiating and assisting in Superfund site remediations and in developing a core of information for use by others involved in hazardous waste cleanup.

Describing the Resource
Physical and Biological Boundaries
Clean Sites targets hazardous waste sites included on the Superfund National Priority List (NPL). Most sites are industrial sites where hazardous wastes were carelessly or unknowingly deposited; others are hazardous waste storage facilities such as Pollution Abatement Services (PAS) of Oswego, New York. PAS collected hazardous waste from a variety of customers and stored it in drums at a main facility and eight satellite operations in New York; the main facility was ranked the seventh most dangerous site on the NPL and the most dangerous site on the New York State list.

Social Boundaries
Historically, industrial and economic development has taken priority over the possible environmental hazards associated with industrial activity. However, legislation like Superfund and organizations like Clean Sites are part of a growing attempt to bring the two considerations into balance by making those who produce toxic waste responsible for the environmental consequences.

One major reason responsible parties failed to clean up NPL sites after the passage of Superfund was economic confusion: what parties were responsible for paying for a cleanup, and how much did they owe? Before Superfund, companies would send hazardous waste to a recycler or a landfill and never expect to worry about it again. However, Superfund made the producers of hazardous waste responsible for all of its environmental impacts, even after they had passed it on to someone else. As a result, many companies were surprised to discover that they had to pay for expensive cleanups of waste they thought was out of their hands forever. Many attempted to shift the blame to other parties or simply avoided paying. Because of these sorts of problems, in 1984, four years after th passage of Superfund, only six NPL sites had been cleaned up.

Looking Back

On August 8, 1982, an article by Christopher Palmer of the National Audubon Society appeared in the *Washington Post*. He called for business leaders and environmentalists to work together to preserve land, air, and water resources. In response, Louis Fernandez, Chief Executive Officer, of Monsanto Company, formed a steering committee to explore the problem of hazardous waste cleanup, an issue of special interest to chemical companies; the committee included representatives from the Conservation Foundation, the National Wildlife Federation, Exxon Chemical Co., the Chemical Manufacturers Association, and E. I. duPont de Nemours & Co. Between 1983 and 1984 this diverse group worked on the problems of hazardous waste site cleanup. Given the failure of responsible parties to respond to Superfund, the committee concluded that the situation required more attention than the EPA could give it. They proposed forming a third-party organization which could act as a catalyst to encourage cleanup at inactive dump sites, especially those where potentially responsible parties (PRPs) could be identified.

How Clean Sites Began
On May 31, 1984, the resulting organization was introduced to the public as Clean Sites, Inc. Clean Sites officials stressed that the group would supplement, not replace, the efforts of the EPA and Superfund. With the support of EPA administrator William Ruckelshaus, Clean Sites hoped to facilitate cleanup of NPL sites by mediating negotiations among PRPs and between PRPs and government agencies. Clean Sites's goal was to begin working toward cleanups at 20 sites in the first year. Because cleanup of the thousands of dangerous hazardous waste sites in the United States would require the efforts of many groups besides Clean Sites, it also resolved to provide those groups (including environmental consulting firms and PRPs which undertake cleanups on their own) with tested and proven models of how to allocate responsibility among PRPs, how to negotiate settlements, and how to conduct remedial cleanup activities.

Clean Sites almost lost the chance to realize its goals because of a problem few had anticipated: in the summer of 1984 the private environmental liability insurance market dried up, leaving Clean Sites unable to find insurance coverage despite a worldwide search. If Clean Sites became active, its directors and officers would face personal liability in case of a lawsuit stemming from the group's actions as a facilitator or advisor.

At that point Charles Powers, then president of Clean Sites, told the EPA that he gave the organization a life expectancy of one month. However, in February of 1985 the EPA agreed to cover Clean Sites's liability for up to $5 million per site and a total of $10 million per year with money from the Superfund trust fund. This indemnification covered Clean Sites's negotiating and consulting services. In cases where it actually worked on a site, such as project management, the PRPs had to provide insurance for those activities. The agreement required that Clean Sites seek EPA approval for each site before beginning any sort of negotiations or work and that Clean Sites continue to search for another source of insurance. Also, Clean Sites could not accept payment from responsible parties for settlement, technical, or management services that it provided at sites covered by the EPA.

FIGURE 19-16: Clean Sites participated in the clean-up of this hazardous waste site. Twelve hundred drums were excavated and incinerated. The men in the photo are preparing to clean out and decontaminate a 15,000-gallon storage tank.

How Clean Sites Obtains Funds for Activities and Cleanups

Prevented from charging for its services and with no government money for operating costs, Clean Sites faced an estimated $5 million in operating costs in 1984 and between $12 and $15 million for the next few years after that. Members of the Chemical Manufacturers Association agreed to provide up to half of Clean Sites's operating funds for its first three years. The organization hoped to obtain the other half from foundations and other industries involved with hazardous waste, including petroleum, paper products, steel, nonferrous metals, electronics, rubber, and health care companies. Some environmentalists worried that heavy subsidization of Clean Sites by the chemical industry would affect the group's objectivity, but careful selection of the board of directors, which included officials from groups like the National Wildlife Federation and Audubon Society, helped guard against conflict of interest.

In 1986 the EPA allowed Clean Sites to begin charging responsible parties for its services. By 1989 Clean Sites earned over half of its income from PRP payments for ser-

vices and received less than a third of its funds from chemical companies. Despite the organization's efforts to remain neutral, however, some environmentalists have begun to question its neutrality, arguing that recent meetings and reports reflect a proindustry bias. Jay Hair disagrees, claiming that detractors have misunderstood its willingness to work with industry on cleanups and have not considered the organization fairly.

How Clean Sites Operates

Clean Sites does not actually clean up sites; rather, it provides a variety of services including settlement, technical review, and project management (Figure 19-16). In most cases a responsible party at an NPL site will contact Clean Sites, often because it has been notified of its potential responsibility by the EPA.

One of the organization's most sought-after services is settlement, the negotiation of cleanup agreements among PRPs. Clean Sites identifies as many parties responsible for contamination at a site as possible, then assists in assigning responsibility for the hazardous waste among them. For example, Clean Sites helps the owners and users of hazardous waste sites

reconstruct how much hazardous material each had dumped over the years. It also helps all responsible parties agree on a payment plan. Because sites can involve hundreds of parties (the Rose Chemical site in Missouri involved 750), this process might be almost impossible without a nonpartisan mediator like Clean Sites.

In some cases responsible parties will request the technical review service, which helps develop a site-specific program. Usually, responsible parties hire contractors to conduct two major studies of a site: a remedial investigation to determine contamination levels and pathways and a feasibility study to evaluate the different ways to clean up the site. Clean Sites examines the studies to make sure they meet government requirements because each must be approved by the EPA.

At some sites, especially those involving many PRPs, Clean Sites is hired to manage the actual cleanup. The project management service coordinates the use of funds, organizes the operation of the cleanup, and selects and monitors contractors who do the clean-up work.

Clean Sites can fulfill other functions as well: keeping communities near a site informed about progress in cleanup activities; assisting state and federal agencies in preparing methods for hazardous waste assessment and cleanup; and participating in educational programs, such as teaching citizens how to apply for EPA grants.

Since its founding Clean Sites has progressed both in the number of sites where it is active and in the variety of services it provides. At the end of its first year it was involved at 19 sites; by 1989 that number reached 58. Thousands of NPL sites still remain polluted, however, and tens of thousands of nonlisted sites may need attention in the future. Clean Sites has played an important role in dealing with site remediation, but it cannot clean up all of the thousands of potentially harmful sites by itself. To help the EPA, other groups, and responsible industries carry on their own remediation efforts, Clean Sites also provides models, such as the 1987 booklet *Allocation of Superfund Site*

Costs Through Mediation. President Thomas Grumbly feels that providing information on how to run effective cleanups is one of the best ways Clean Sites can combat the overwhelming number of untouched hazardous waste sites, and he anticipates that Clean Sites will expand this role in the future. He compares the role to that of a research hospital which not only helps specific patients, but also learns more about the processes involved so it can pass the information on to others in the health care field, policymakers, and other involved groups.

In 1990 the public policy efforts of Clean Sites centered on improving the remedy selection process. A comprehensive study brought together over 100 experts from government, industry, and public interest groups in the fields of technology, policy making, law, and academics. Funded by a $135,000

grant from the EPA and a matching amount from the Andrew W. Mellon Foundation, the study concentrated on devising alternatives to the current process of selecting remedies for site cleanups.

Looking Ahead

Clean Sites must maintain its objectivity and nonpartisan status, the invaluable credentials that convince both responsible parties and the government to accept it as a mediator. To preserve these credentials, Clean Sites must be careful to maintain a balance of environmental organizations and private industry on its board and among its supporters. Also, Clean Sites should become more involved in shaping public policy on hazardous waste cleanup and should help the EPA manage NPL cleanup sites more efficiently.

possible. When toxic substances are produced and used, they should be reused and recycled. Increased government support for waste minimization strategies would help to increase the profitability of alternatives to waste disposal. Ultimately, however, finding solutions to the problem of hazardous wastes rests with ordinary individuals. The majority values and decisions of the public are the forces that spur business, industry, and the government to action. Until enough people become concerned about this issue, and are willing to show that concern in the marketplace and the voting booth, toxic and hazardous substances will continue to pose a serious threat to human health and the environment. (See page 414, What You Can Do: To Minimize Dangers of Toxic and Hazardous Substances for suggestions for individual action.)

Protect Human Health and the Environment

Protecting human health and the environment from pollution and degradation depends on strong and rigorously enforced legislation. While some businesses and industries are making concerted and sincere efforts to reduce their use and generation of hazardous substances, many others are attempting to maintain the status quo. Industry lobbying groups

traditionally have opposed strong hazardous waste legislation, and they are influential in Washington, D.C., and many state capitals. To counteract the effects of these lobbies, individuals can write and call their elected officials.

Although RCRA addresses deep-well injection of hazardous wastes, legislation is needed to initiate a comprehensive regulation program. The EPA should begin a major study of the safety of deep-well injection. Too little is known about the fate of materials once they have been injected underground. Once an aquifer is polluted, it may never become clean again. The fate of many drinking water sources across the country may depend on how we proceed with deep-well injection.

Congress should also continue to reauthorize Superfund and work to improve its operation. It is essential that cleanups not be interrupted every time Superfund comes up for reauthorization. A strong, consistent, and serious approach to hazardous waste site cleanup must be supported by government, industry, and the public.

Preserve Living Systems

Too often, the debate over the production, use, and disposal of toxic and hazardous substances focuses narrowly on their effects on human health. We must remember that all life depends upon healthy air, water, and soil. Accordingly, standards for releases of toxic emissions into air and water should be revised to give more adequate protection to the environment.

Perhaps the most important step we can take to preserve living systems is to use nonhazardous alternatives whenever possible. Doing so depends on government and industry support for the research and development of safe alternatives. For example, Department of Agriculture and Soil Conservation Service programs should promote the use of biological pest controls over hazardous synthetic pesticides. Conserving toxic and hazardous substances will minimize the production and disposal of these materials, thus preserving the integrity of living systems. Substances that cannot be reused or disposed of safely should not be produced.

mediation, uses microorganisms to decompose organic chemical wastes into harmless by-products.

Waste minimization refers to "front door" strategies to reduce the amount of hazardous substances used and "back door" strategies to reduce the volume of hazardous wastes that must be disposed of.

Hazardous and toxic substances are managed according to the provisions of three major laws. The Comprehensive Environmental Response, Compensation, and Liability Act (CERCLA), called Superfund, was enacted in 1980 and is administered by the EPA. It authorizes the federal government to respond directly to releases or threatened releases of hazardous substances into the environment. The Resource Conservation and Recovery Act (RCRA), enacted in 1976, defined hazardous wastes; set guidelines for managing, storing, and disposing of hazardous resources; and established a cradle-to-grave regulation system, in which manifests must be kept of every stage of waste from its generation to its disposal. The Toxic Substances Control Act (TSCA) regulates the production, distribution, and use of toxic commercial products that may present an unreasonable risk of injury to human health or the environment.

Summary

Many toxic substances occur naturally in the earth's crust and biota; others are manufactured by industrial processes. Toxic chemicals have the potential to cause injury to organisms. Substances are called hazardous when a possibility exists that plants, animals, or humans will be exposed to them. Toxic and hazardous substances are released into the environment as by-products, as end products, or through the use and disposal of manufactured products.

Acute toxicity occurs when a substance triggers a reaction after a single exposure. Chronic toxicity occurs when a compound does not produce symptoms until it accumulates to a threshold level in the organism. Carcinogenic substances cause cancer in humans and animals. Infectious substances contain disease-causing organisms. Teratogenic substances affect the fetus. Mutagenic substances cause genetic changes, or mutations. Multiple chemical sensitivity, or MCS, is an illness characterized by intolerance of one or more classes of chemicals. A chemical sensitizer causes an individual to become extremely sensitive to low levels of that particular substance or to experience cross-reactions to other substances.

Wastes are disposed of by burning them, burying them, or pumping them underground. Incineration consists of burning organic materials at high temperatures to break them down into their constituent elements, which combine with oxygen and are released into the atmosphere. Secure landfills prevent the escape of hazardous and toxic substances into the surrounding environment. Deep-well injection involves pumping liquid wastes deep underground into rock formations which geologists believe will contain them permanently.

Prior to disposal, most hazardous and toxic materials can be treated to reduce their hazardous nature. Treatment can be chemical, physical, or biological. Chemical treatment processes alter the chemical structure of the constituents to produce a residue that is less hazardous than the original waste. Physical treatment, which does not affect the chemical makeup of the wastes, can take many forms, including phase separation, component separation, and solidification. Biological treatment, or biore-

Key Terms

acute toxicity	mutagenic
bioremediation	phase separation process
carcinogenic	sanitary landfill
chemical sensitizer	secure landfill
chronic toxicity	solidification
component separation process	source segregation
	still bottoms
hazardous	teratogenic
hazardous waste	toxic
infectious	waste minimization
leachate	
multiple chemical sensitivity (MCS)	

Discussion Questions

1. What is the difference between toxic substances and hazardous substances? How can these substances enter the environment?

2. In what ways is hazardous waste a social problem? In what ways is it an international social problem?

3. What is the difference between a sanitary landfill and a secure landfill? Are hazardous wastes a problem in both kinds of landfills? Why or why not?

4. Define bioremediation. How is it different from other methods of treating waste?

5. What is waste minimization? What strategies can companies or individuals use to minimize waste?

6. Imagine that a contaminated hazardous waste site has been identified. Over the years many companies have dumped their waste there or paid someone else to dump it for them. What legislation would apply to the situation, and what organizations might be involved in the cleanup?

Unrealized Resources: Waste Minimization and Resource Recovery

The biggest challenge we will face is to recognize that the conventional wisdom about garbage is often wrong.

William Rathje

From an environmental point of view, preventing waste or the creation of pollutants in the first place is best. When some waste is inevitable, it should be minimized. Next, waste or by-products should be reused or recycled, at the factory level or at the consumer level. Only after all such measures have been exploited should waste treatment and disposal—the traditional end-of-the-pipe approach—be considered.

World Resources Institute

Learning Objectives

When you finish reading this chapter, you should be able to:

1. Explain what is meant by unrealized resources and give some examples of these resources.

2. Describe how the generation and management of solid wastes vary worldwide.

3. Identify the major waste disposal methods in the United States and describe each.

4. Define waste minimization and resource recovery and explain why they are preferable to waste disposal.

Recently, a new museum opened in Lyndhurst, New Jersey. The event would scarcely be cause for notice, except that this museum is like no other. Paintings and sculptures do not grace its walls, nor do models of dinosaurs or prehistoric creatures prowl its galleries. Instead, the museum, located at the Hackensack Meadowlands Environment Center, displays common artifacts of contemporary American life—trash. Visitors view a first-rate display of old tins, empty bottles, rusting bicycles, crumpled chicken wire, squashed milk cartons, and long-emptied cereal boxes. As they do, they come face to face with one of our society's most visible gifts to future generations, our refuse. It is an unusual but appropriate museum for a culture that has been called the "throwaway" society. But is our society really so different from the many civilizations that have preceded it? At base, the answer is no. Humans have been throwing away their leftovers and unwanted artifacts for tens of thousands of years. What distinguishes contemporary society from previous cultures is the amount and the type of our throwaways. Because of soaring population, the volume of wastes is much greater. Our technological and consumer culture gives us relatively more products and more elaborate and excessive packaging, both made of more synthetic materials and much of which we use once and throw away. And many of these products are hazardous or toxic substances.

In this chapter we learn about materials that are usually thrown away, but could be used to benefit both human and natural systems. Managing wastes in an environmentally sound manner requires a thorough understanding of the physical, biological, and social boundaries of the resource. How much waste is produced, what happens to it, and what resources can be recovered are the physical boundaries of these "unrealized resources." In biological boundaries, we look at how natural systems manage waste products through the process of biodegradation and how humans can mimic natural systems. Since "garbage" is strictly a human concept, it is especially important to understand the social boundaries of solid waste resources. The most important of these are how the production and management of wastes differ among various countries, the environmental problems associated with waste disposal, and the misconceptions that obscure a true understanding of this nation's solid waste problem.

After describing the resource, we look at how solid wastes are managed, focusing on landfills and incineration. Exploring other management options, such as resource recovery and source reduction, can help us to eliminate or minimize many of these problems. As we will see, recovering valuable resources from refuse and reducing the amount of wastes produced initially can serve as important guiding principles in developing a sustainable and sustaining society.

Describing Unrealized Resources: Physical Boundaries

What Is Solid Waste?

A **solid waste** is any of a variety of materials that are rejected or discarded as being spent, useless, worthless, or in excess. The term **refuse** refers to things rejected as worthless or of insignificant value and is a common synonym for solid waste. **Garbage** refers strictly to animal or vegetable wastes resulting from the handling, storage, preparation, or consumption of food. It decomposes rapidly, often creating offensive odors. **Trash** is useless or worthless matter that is unsightly but does not contain odor-producing food wastes.

As the definition implies, a solid waste is the result of a conscious decision to throw something away, rather than any quality (or lack thereof) inherent in the object itself. In other words, what one individual may deem worthless and fit only for the trash can—a shirt that has been outgrown or that is out of style, for example—another individual may find valuable. Whenever an object is thrown away, regardless of its actual or potential value, it becomes a solid waste.

The **solid waste stream** refers to the collective and continual production of all refuse; it is the sum of all solid wastes from all sources. An estimated 5 billion tons (4.55 billion metric tons) of solid waste are produced each year in the United States alone (Figure 20-1). The two largest sources of solid wastes are agriculture (animal manure, crop residues, and other agricultural by-products) and mining (waste rock, dirt, sand, and slag, the material separated from metals during the smelting process). About 10 percent of the total solid waste stream is generated by industrial activities (scrap metal, slag, plastics,

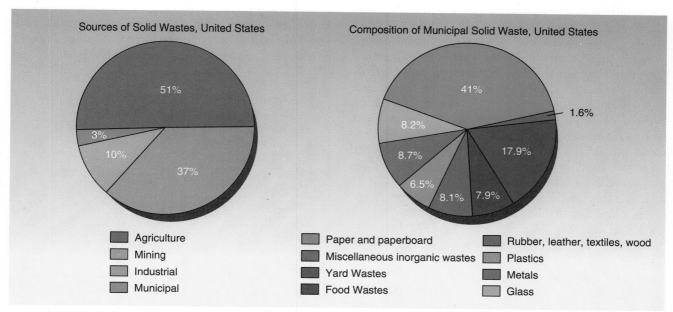

FIGURE 20-1: Sources of solid waste and composition of municipal solid waste. Agriculture and mining are the largest sources of solid waste. While most agricultural wastes are used to enrich the soil, the proportion that is not composted, about 10 percent, is a significant source of nonpoint water pollution. Mining wastes contribute to both water and air pollution, as do industrial and municipal wastes.

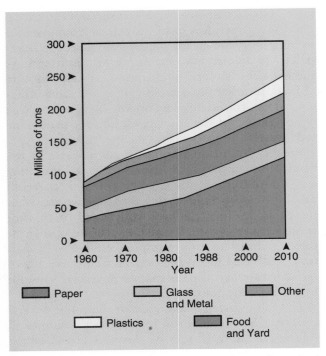

FIGURE 20-2: Municipal solid waste generation in the United States, 1960–2010. Per capita generation of municipal solid waste has increased over the past 30 years, a trend that is expected to continue. In 1960 each U.S. citizen generated an average of 2.6 pounds of refuse daily. By the year 2000 that figure will rise to about 4.5 pounds.

paper, fly ash, and sludge from treatment plants). Many industrial facilities have on-site waste disposal capabilities, either landfills or incinerators, to handle their refuse.

About 3 percent of the nation's solid waste stream is made up of **municipal solid waste** (MSW), refuse generated by households, businesses, and institutions. Paper and cardboard account for the largest percentage (41 percent) of refuse materials by weight of MSW. Yard wastes are the next most abundant material, accounting for almost 18 percent. Metals and glass each make up over 8 percent of MSW, food wastes just under 8 percent, and plastics about 6.5 percent.

Estimates of the U.S. solid waste stream range from about 3 to 8 pounds per person per day (1.35 to 3.6 kilograms). According to the EPA, 180 million tons (163.8 million metric tons) of MSW were generated in the United States in 1988, equivalent to about 4 pounds per person per day (1.8 kilograms). By the year 2000 the EPA estimates that waste generation in the United States will rise to 216 million tons (196.56 million metric tons) annually, almost 4.5 pounds per person per day (2.03 kilograms) (Figure 20-2).

What Materials Can Be Recovered from the Solid Waste Stream?

Solid wastes, like hazardous wastes, are misplaced or **unrealized resources.** They are wastes only because we have not had the foresight to recognize their value or the political will to stop their misuse. Routinely, Americans throw away potentially useful resources—newspapers, plastic bottles, glass jars, rubber tires, grass clippings, coffee grounds, apple peels, and aluminum and bimetal cans. Let us look at some of the uses for these unrealized resources.

Aluminum. Aluminum, the most abundant metal on earth, is never found in a free state; to be useful, it must be separated from aluminum ore. The refining process uses large amounts of energy and water and is a significant source of air and water pollution. Reusing aluminum metal rather than mining for new aluminum results in tremendous energy savings and reduces water and air pollution (Table 20-1). By far, the largest source of aluminum for reuse is cans; more than 50 percent of all aluminum cans are recycled. Nearly all of these are recycled into new aluminum cans. Other sources of aluminum are house siding, roof gutters, storm window and storm door frames, and lawn furniture.

Paper. One of the most frequently recycled materials, paper can be made into a variety of products, including newsprint, paper bags, record jackets, game boards, egg cartons, gift boxes, jigsaw puzzles, paper matches, game and show tickets, and insulation.

To make paper, tiny fibers (from various sources, including wood, corn stalks, and other organic matter) are bonded together to form a continuous sheet. In general, paper made from long fibers is stronger

 Table 20-1
Environmental Benefits Derived from Substituting Secondary Materials for Virgin Resources

Savings in or Reduction of	Aluminum (%)	Steel (%)	Paper (%)	Glass (%)
Energy use	90–97	47–74	23–74	4–32
Air pollution	95	85	74	20
Water pollution	97	76	35	—
Mining wastes	—	97	—	80
Water use	—	40	58	50

Source: Adapted from Robert Cowles Letcher and Mary T. Schell, "Source Separation and Citizen Recycling," in William D. Robinson, ed., *The Solid Waste Handbook* (New York: Wiley, 1986); Cynthia Pollock, *Mining Urban Wastes: The Potential for Recycling*, Worldwatch Paper 76 (Washington, D.C.: Worldwatch Institute, April 1987), p. 22.

than that made from short fibers; grocery bags, for example, which require strength, are made from long fibers. Making new paper products from recycled paper saves both trees and energy and thus is environmentally and commercially cost-effective. Typically, recycled paper is combined with virgin paper to create products, because 100 percent recycled paper is not of sufficient quality for many purposes.

Of the paper recycled, most fits into three basic categories: newsprint, mixed office waste, and waste paper from paper-converting plants, plants that make products such as stationery, envelopes, and business forms from paper. Each type of paper can be recycled into different products; consequently, the market for and the economic return on each varies widely.

About one-third of all newsprint in the United States is recovered for recycling. Recycling newspapers is important since this item constitutes the largest single component by volume of most landfills (14 percent). The major uses for recycled newsprint are cereal boxes, paperboard, and home insulation. The automotive industry, another important market for recycled newsprint, uses it in the insides of automobiles. The average car contains about 60 pounds of recycled newspaper. There is also a growing trend among U.S. papermakers to use old newsprint to produce new newsprint. This use should increase as technologies for de-inking newsprint and removing impurities improve.

Mixed office waste consists of computer paper, mimeograph paper, office stationery, and other white paper. Mixed office waste, if sorted, can be recycled into high grade paper. If unsorted, it can be made into roofing felt or as part of the filler ply in paperboard. High-grade waste paper from paper-converting plants consists of machine trimmings and clippings from envelopes, business forms, books, and catalogues. If it is sorted, it can go back into the same product line. If it is unsorted, high grade waste paper can be used in filler ply in paperboard.

Cardboard. About 50 percent of corrugated cardboard containers, which are generated in bulk from retail stores and factories, are recycled. Large handlers of corrugated cardboard, such as supermarkets, typically bale the cardboard and sell it directly to recycling mills. Factories and retail outlets, which handle smaller amounts of the material, often depend upon individuals who collect the cardboard and sell it to waste paper processors.

Glass. Recycled glass containers are most commonly used to produce new containers. Nearly 90 percent of recycled bottles are made into new bottles. Recycled glass is also used to manufacture fiberglass and

to produce glasphalt (a material used in street paving), bricks, tiles, and reflective paints on road signs.

Sand, soda ash, and limestone are the raw materials of glass. However, according to the Glass Packaging Institute, about 25 percent of any new glass container is made with recycled glass. The recycled glass (bottles and jars) is crushed into small pieces, which are then passed over a magnetic belt to remove any ferrous metals. The end product, known as cullet, is mixed with sand, soda ash, and limestone and melted to produce molten glass. Using cullet yields several important economic and environmental benefits. It conserves the raw materials of glass. It saves energy; since cullet melts at a lower temperature than the other materials, the furnace temperature can be reduced 10 °C for every 10 percent of cullet used in the mixture. It extends the life of the furnace. It reduces particulate emissions to the atmosphere. Recycled glass can account for as much as 83 percent of molten glass without adversely affecting its quality. However, the quality of the final product depends on using cullet that is free of contaminants. Aluminum, lead, stone, dirt, and ceramic materials—contaminants not removed by the magnetic conveyor belt—may cause weak spots or flaws in the glass or damage equipment.

Plastics. By the year 2000 plastics are expected to make up about 10 percent of the municipal solid waste stream. This volume and the durability of plastics make them one of the most potentially valuable of all unrealized resources. At present, only a few of the many types of plastics (each of which has a different chemical composition) are being recycled in significant quantities. These are polyethylene terephthalate (PET) soft drink containers, high-density polyethylene (HDPE), and polystyrene.

The PET soft drink container is the most commonly recycled plastic container. About 20 percent of all PET containers are recycled into various products such as carpet backing, fiber filling for sleeping bags and outerwear, bathtubs, shower stalls, corrugated awnings, swimming pools, floor tiles, paint, paintbrushes, scouring pads, automotive parts, and containers for nonfood products.

HDPE is the plastic used in milk jugs and the base cups of PET containers. Recycled HDPE is used to make new base cups for PET containers, trash cans, flowerpots, piping, and traffic cones. A small amount of HDPE is also being used to create "plastic lumber" for railroad ties, decking for boat piers and docks, and fencing. Plastic lumber has several advantages over wood: it does not rot, splinter, or chip, nor does it require painting.

Polystyrene is used in items such as fast food carryout containers ("clam shells"), cups, and plates. It can be cleaned, converted into pellets, and used to make building insulation, packaging material, and plastic lumber for walkways, benches, and other items.

Iron and Steel. Iron and steel products can be melted and reformed into new products. Scrap iron, such as in discarded automobiles and large appliances, has long been melted and reused in other products. Until recently, just a small portion of steel (primarily "tin" and bimetal cans) was reused, but this situation is changing. For example, most "tin" cans—for such food items as soups, canned vegetables, and pork and beans—are actually steel cans. The Steel Can Recycling Institute reported that 51 million tons (46.4 million metric tons) of iron and steel were recycled in 1987, an amount more than twice as much as all other recyclable materials combined.

Tires. Tires pose a mounting problem for solid waste officials. Many service stations now charge a small fee to take back old tires. Because in most places they cannot be landfilled unless they are shredded, they are often stockpiled or dumped illegally. That is one reason old tires are a common sight in ravines and creekbeds, on beaches, and at other illegal dumpsites. Tires, however, are a potentially valuable resource. Many can be retreaded, prolonging the life of the tire and conserving rubber. Retreading uses only about one-third of the energy needed to produce new tires, and a retreaded tire provides about three-quarters of the mileage of a new tire. Tires that cannot be retreaded can be used in a number of ways. Strung together, they act as effective controls against landslides and beach erosion, and they can be used as artificial reefs, providing habitat for marine life. Some power plants (primarily in Oregon, Washington, and California) use old tires for fuel; combustion of the tires produces enough energy to drive steam turbines which generate electricity. Considering the amount of old tires disposed of—Los Angeles County, for example, disposes of 250 tons (227.5 metric tons) of old tires daily—the benefits of recycling can be significant. A word of caution, however: burning tires generates a significant amount of air pollution unless the plant is equipped with pollution control devices.

Used Oil. The Department of Energy estimates that about 1.2 billion gallons (4.55 billion liters) of used oil are generated annually, a figure equivalent to about 78,000 barrels per day. Motor oil accounts for about 60 percent of used oils; the remaining 40 percent consists of industrial substances such as hydraulic, metal working, and cooling oils. About two-thirds of the used oil generated in the United States each year is recycled, usually as a fuel. The remaining one-third, some 400 million gallons (1.5 billion liters), is disposed of in landfills or is simply dumped on the ground or down storm sewers, where it poses a threat to local surface water and groundwater.

White Goods. Old appliances, known as white goods, can be salvaged from the waste stream, repaired, and reused. One concern with recycling these items is that they may contain polychlorinated biphenyls (PCBs), a toxin that can cause skin lesions, liver ailments, and certain kinds of cancer. PCBs, which are contained within the appliance's capacitor, cause a problem only when a leak in the capacitor allows the PCBs to escape and enter the environment. Appliances manufactured after 1981 do not contain PCBs.

Food and Yard Wastes. Grass clippings, leaves, brush, food scraps, and other organic matter are the raw materials of fertile soil. As these materials biodegrade, humus is formed and their constituent minerals are released. Humus is vitally important to soil productivity because it helps to retain water and maintain a high nutrient content.

Biological Boundaries

How Do Living Systems Manage Waste Products?

Recycling is as old as life itself, for in nature there is no such thing as an unusable waste. Living organisms use materials from their environment and return those materials, in a different form, to the environment. Thus, materials continually cycle through the ecosystem.

The decomposition of organic materials into their constituent compounds, which makes them available for reuse by other organisms, is one of the most fundamental biological processes. Soil microbes such as bacteria and fungi secrete enzymes that break down organic materials; the decomposed materials provide the organisms with energy. **Aerobic decomposition** occurs in the presence of oxygen by oxygen-requiring organisms. Some organisms do not require oxygen; when these organisms are present, **anaerobic decomposition** occurs.

Most decomposition in a landfill is anaerobic, since oxygen is present in limited amounts and is quickly depleted by aerobic organisms. The complex interactions of three classes of soil-dwelling microbes are responsible for most of the decomposition that occurs in landfills. Cellulolytic bacteria separate out the cellulose (a complex sugar) in plant wastes like wood and paper. Decomposition continues as acidogens act on the cellulose, fermenting the sugar into weak acids. The process is completed when methanogens convert the acids into carbon dioxide and methane.

A material that can be decomposed by natural systems is said to be biodegradable. Organic compounds, those containing carbon, can be readily degraded by decomposers when sufficient oxygen is available. Unfortunately, much of our waste is synthetic and not biodegradable. Plastics are perhaps the best example of a durable, nonbiodegradable substance. Because of their chemical structure, plastics cannot be decomposed by bacteria and other organisms. Marketing claims that certain plastic bags and other items are biodegradable are misleading. Biodegradable substances, such as cornstarch, are integrated into the matrix of these items and it is the cornstarch, not the plastic, that decomposes. As the cornstarch is digested by microbial organisms, the bag breaks up into smaller and smaller pieces of plastic, but the total volume of plastic remains the same. In fact, such a "biodegradable" item typically contains more plastic than its "nonbiodegradable" counterparts. Because the cornstarch tends to lower a material's tensile strength, the manufacturer must increase the total amount of plastic to make the item.

How Can Humans Mimic the Action of Living Systems?

Natural systems are cyclic: decomposers break down organic materials (both wastes and dead plants and animals) into their constituent compounds, which are then used by other organisms, which give off wastes and eventually die; the wastes and dead matter are again recycled by decomposers. In contrast, human systems tend to be linear: manufacturers produce objects from raw materials, consumers use the objects, consumers dispose of objects. But as the law of the conservation of matter indicates, materials can neither be created nor destroyed. Therefore, when we "dispose" of our wastes, we do not get rid of them; we simply move them from one place (our homes, schools, or businesses) to another (typically, a landfill or dump).

Were it not for the law of the conservation of matter, there would be no solid waste crisis. If we could destroy matter (and create more when we needed it), landfills would never become full, toxic and hazardous wastes would pose no problem, and the environmental and economic costs of waste disposal would be minimal. But since materials cannot be destroyed, society faces problems associated with solid wastes. The most promising solution to these problems is to mimic natural systems by reusing and recycling materials as much as possible. For materi-

als that cannot be recycled, such as some hazardous and toxic substances, the solution lies in reducing consumption and substituting nonhazardous alternatives for them. By reusing and recovering materials and reducing the volume of wastes generated, we can lessen the impact of human activities on the environment.

Social Boundaries

How Do Waste Production and Management Vary Worldwide?

Both the consumption and disposal of goods vary from country to country. The United States, with just 5 percent of the world's population, generates about 40 percent of the world's waste. The United States, Canada, and Australia generate about twice as much refuse per person as do Japan, Sweden, and China (Figure 20-3). In 1990 the EPA reported that about 73 percent of U.S. municipal solid wastes are buried in landfills, another 14 percent are burned or incinerated (usually to generate energy), and the remaining 13 percent are recycled. Other industrial

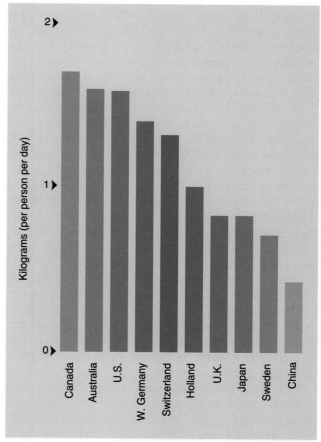

FIGURE 20-3: Per person waste generation, selected countries, 1989.

▶ Table 20-2
Recovery Rates (percentage) for Aluminum,
Paper, and Glass, Selected Countries[a]

Country	Aluminum	Paper[b]	Glass
Netherlands	40	46	53
Italy	36	30[c]	25
West Germany	34	40	39
Japan	32	48[c]	—
United States	28	27	10
France	25	34	26
United Kingdom	23	29	12
Austria	22	44	38
Switzerland	21	43	46
Sweden	18	42	—

[a]*Includes industrial recycling.* [b]*Data for 1984.* [c]*Data for 1983.*
Sources: Aluminum Association, Inc., *Aluminum Statistical
Review for 1985* (Washington, D.C.: 1986); U.N. Food and
Agriculture Organization, *Waste Paper Data, 1982–84* (Rome:
1985); *Glass Gazette* (Brussels), October 1986; U.S. glass data
from U.S. Department of Commerce, "Current Industrial
Reports: Glass Containers." May 1986, and Bill Clow, Owens,
Ill., private communication, August 28, 1986.

nations recycle a much larger percentage of their
refuse (Table 20-2).

Ours is a throwaway society for many reasons
(Figure 20-4). Americans as a whole are not used to
reusing and recycling materials (even though we
have embraced recycling in times of need). We have
become accustomed to disposable products, everything from paper towels to pens to cameras. Relatively cheap energy (compared to other industrial
nations) has encouraged waste in production processes. Low costs for land disposal have made it economically painless for consumers and industry alike
to throw away valuable resources. The federal government has not promoted recycling and the reduction of wastes as actively as have the governments of
other industrialized nations. For example, in the late
1980s the German government asked the nation's
beverage industry to undertake binding commitments to package a certain percentage of its products in reusable bottles and containers by 1991. The
figures are high: 90 percent for beers and mineral
waters, 80 percent for carbonated beverages, 35 percent for fruit drinks, and 50 percent for wines. If the
industry failed to meet these targets, the government vowed to introduce legislation to ensure that
recycling and waste reduction reach a level considered appropriate and practical.

The waste stream of less-developed countries is
typically much smaller than that of more-developed
countries for several reasons, and the reasons are
related to affluence (Table 20-3). Low per capita
incomes and widespread poverty mean that people
in LDCs cannot afford the goods and products common in MDCs. The waste stream in LDCs is also

smaller because items are reused and recycled far
more often. People in Mexico, China, the Philippines, and other countries recycle out of economic
necessity (Figure 20-5). It is not unusual to find
communities built around open dumps in large
urban areas in the developing world (Figure 20-6).
The poor who live in these communities scavenge in
the dumps for a living, digging for food and any
items that can be recycled and sold. Conditions are
wretched: the odor is foul and constant, insects and
rodents proliferate, a reliable source of clean water is
usually nonexistent, and disease is common. But
government officials are reluctant to close the
dumps and landfill the refuse. To do so would be an
expense few LDCs can afford. In addition, landfilling would make these resources unavailable; without
the dumps, the poor would have no means by which
to earn even their present meager living.

What Misconceptions Are Associated with Solid Waste Disposal?

One of the reasons that waste disposal has become
such a politically and socially volatile issue is that
debate on the topic is clouded by widely held but
erroneous beliefs. Many of the misconceptions associated with the issue of solid waste have to do with

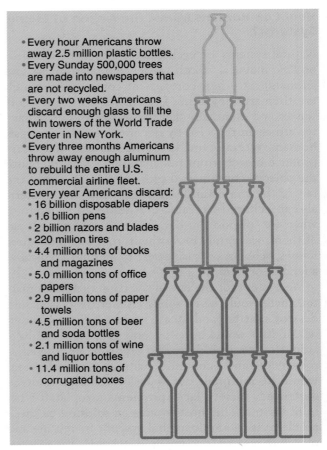

- Every hour Americans throw away 2.5 million plastic bottles.
- Every Sunday 500,000 trees are made into newspapers that are not recycled.
- Every two weeks Americans discard enough glass to fill the twin towers of the World Trade Center in New York.
- Every three months Americans throw away enough aluminum to rebuild the entire U.S. commercial airline fleet.
- Every year Americans discard:
  - 16 billion disposable diapers
  - 1.6 billion pens
  - 2 billion razors and blades
  - 220 million tires
  - 4.4 million tons of books and magazines
  - 5.0 million tons of office papers
  - 2.9 million tons of paper towels
  - 4.5 million tons of beer and soda bottles
  - 2.1 million tons of wine and liquor bottles
  - 11.4 million tons of corrugated boxes

FIGURE 20-4: Our throwaway society.

▶ Table 20-3
Per Capita Production of Trash, per Day,
Selected Cities

City	Per Capita Waste Generation Rate lbs per day (kg per day)
CITIES, MORE DEVELOPED COUNTRIES	
Los Angeles, United States	6.6 lbs (3.0 kg)
New York, United States	3.96 lbs (1.80 kg)
Tokyo, Japan	3.04 lbs (1.38 kg)
Paris, France	2.42 lbs (1.10 kg)
Singapore	1.91 lbs (0.87 kg)
Hong Kong	1.87 lbs (0.85 kg)
Hamburg, West Germany	1.87 lbs (0.85 kg)
Rome, Italy	1.52 lbs (0.69 kg)
CITIES, LESS DEVELOPED COUNTRIES	
Lahore, Pakistan	1.32 lbs (0.60 kg)
Tunis, Tunisia	1.23 lbs (0.56 kg)
Bandung, Indonesia	1.21 lbs (0.55 kg)
Medellin, Columbia	1.19 lbs (0.54 kg)
Calcutta, India	1.12 lbs (0.51 kg)
Manila, Philippines	1.1 lbs (0.50 kg)
Kano, Nigeria	1.01 lbs (0.46 kg)

Sources: Sandra J. Cointreau, *Environmental Management of Urban Solid Wastes in Developing Countries* (Washington, DC: World Bank, 1982); Clean Japan Center, "Recycling '86 Turning Wastes into RESOURCES," Tokyo, 1986; Jean-Michel Abt, National Agency for the Recovery and Elimination of Wastes (ANRED), Angers, France, private communication, November 4, 1986.

FIGURE 20-5: A child enjoying a ride alongside a cardboard recycler in China.

what exactly it is that we throw away, that is, the typical contents of landfills. In the 1980s researchers with the Garbage Project, headed by "garbologist" William Rathje, an anthropologist at the University of Arizona, set out to study scientifically solid waste disposal in the United States. They exhumed a total of 16,000 pounds (7,200 kilograms) of garbage from seven landfills across the country (two outside Chicago, two in the San Francisco Bay area, two in Tucson, and one in Phoenix). The items were weighed and sorted into 27 basic categories and 162 subcategories. Through their efforts, the Garbage Project researchers have debunked some common misconceptions about what kinds of things are found in landfills and what happens to these items once they are buried.

One widely held belief is that plastics are a major culprit in the filling up of landfills. The Garbage Project found that plastics of all kinds account for less than 5 percent of the average landfill's contents by weight and only 12 percent by volume. Disposable diapers, a highly visible and often cited symbol of our throwaway society, make up only 1 to 2 percent by weight, and fast-food packaging, another

highly visible item, accounts for only about 0.1 percent of the contents by weight. Whether by weight or volume, diapers, fast-food packaging, and other plastic items are far less a problem than are newspapers, old telephone books, and yard wastes—items that can easily be recycled.

Researchers with the Garbage Project are quick to dispute some other mistaken ideas about plastics in landfills. They argue that while it is easy to blame the shortage of landfill space on packaging and plastics, both offer important advantages. Adequate (not excessive) and effective packaging reduces food spoilage and thus waste. Packaging of prepared foods reduces individual generation of waste. For example, individuals who make orange juice from scratch instead of using packaged concentrate have to dispose of the rinds, which most people simply throw away. In contrast, the producers of orange juice concentrate sell the leftover rinds as animal feed.

FIGURE 20-6: Pepinadores scavenge the Mexico City dump near the Nezahualcoyotl slums on the northern edge of the city.

Because plastics do not degrade, they do not release hazardous or toxic substances into the environment. And finally, plastics occupy little volume in a landfill because they are usually compacted, first by the truck mechanism during collection and later by the pressure of layers of garbage.

Other popular misconceptions about landfills concern what happens to wastes once they are buried. Many people believe that because paper is biodegradable, it does not present a problem in a landfill. If the papers degrade, the volume of space they occupy will shrink, creating more room in the fill for additional wastes. But owing to the design of modern landfills, little air is available once wastes are buried, and available air is used up quickly. In the absence of air the process of biodegradation is slowed significantly. Organics like food scraps and yard waste, which should decompose rapidly, degrade only some 20 to 50 percent over 10 to 15 years. The remainder becomes mummified and retains its original weight, volume, and form. Newspapers as much as 30 or 40 years old have been unearthed in landfills, virtually intact and readable. (Actually, it may be a good thing that newspapers do not decompose readily in a landfill, for if they did they would release a significant amount of ink, which might then contaminate local surface water and groundwater.)

History of Management of Unrealized Resources

As long as humans have roamed the planet, they have left behind unwanted items and wastes. When populations were small, refuse could simply be discarded anywhere; the family or tribe would eventually move on to other areas and the wastes would gradually decompose. As the human population grew and settlements were formed, this disposal method became less attractive. As cities developed, residents began to dump their refuse on the edge of town, away from their homes, although there were exceptions—Parisians, for example, continued to dump their day's garbage into the streets outside their homes as late as the seventeenth century. Most civilizations have managed their refuse in one of four ways: by dumping it, burning it, converting it into something that could be used again, or reducing the amount of wastes produced (Figure 20-7). These same methods are used by societies today.

What Is Ocean Dumping?

Historically, coastal communities have dumped sewage and industrial wastes into the ocean. Because the ocean seems so vast, little thought is given to the

(a)

(b)

FIGURE 20-7: World's most scenic dump, Moab, Utah. (a) In the mid-1980s, the local chamber of commerce initiated a publicity campaign touting Moab as site of the world's most scenic dump. (b) The dump had actually been cleaned up several years earlier, part of an effort to encourage tourism and boost the local economy.

effects of ocean dumping on the marine ecosystem. But as we have learned, the volume and toxicity of wastes adversely affect marine life. Increased awareness of these adverse effects has resulted in a ban in the United States against dumping municipal sewage sludge and industrial wastes in near-shore coastal waters. However, some communities have not complied with the ban. Moreover, dumping in deep ocean waters is still allowed. The dumping of plastics from private cruise vessels, pleasure craft, and fishing boats also continues, despite its ban under an international treaty. Beach "cleanups" provide some idea of the extent of ocean dumping. In 1991 the annual effort organized by the Center for Marine Conservation included cleanups in 26 states, the

District of Columbia, three U.S. territories, Canada, Mexico, Guatemala, and Japan. More than 108,700 volunteers collected 1,300 tons (1,183 metric tons) of debris from over 3,600 miles (5,796 kilometers) of coastline. Plastics made up about 67 percent of all debris collected, far more than any other type. Metal, glass, and paper each accounted for about 10 percent of the total. The remaining waste included wood, rubber, and cloth.

Environmental Problems Associated with Ocean Dumping.

Ocean dumping poses a serious threat to the marine environment. Because plastics do not disintegrate, they are particularly dangerous to seals, sea turtles, sea birds, and other marine animals. Six-pack yokes and fishing wire can entangle an animal, causing it to drown or strangle. These plastic nooses can also cause deep wounds, and a fatal infection may result. Even apparently innocuous items, like plastic foam cups, balloons, and sandwich bags, can prove lethal in the marine environment. Turtles may mistake plastic bags for jellyfish, a favored morsel, and seabirds may mistake spherical resin pellets, the raw material used to make plastic, for fish eggs. If plastics are ingested, they can become lodged in an animal's intestinal system, resulting in intestinal blockage and ulceration.

Many communities and states have enacted laws to combat the problems caused by ocean dumping, both intentional and unintentional. Baltimore, Maryland; Louisville, Kentucky; and the state of Florida prohibit the intentional release of balloons into the atmosphere. Several states, including Massachusetts, have passed laws mandating that six-pack plastic yokes be manufactured from biodegradable plastic (a measure that will not prevent the ingestion of small particles of plastic). In 1989 Maine became the first state to ban the use of six-pack plastic yokes altogether.

Because the oceans are a global common, efforts to eliminate dumping must be international in scope. MARPOL Annex V is an international treaty that regulates ocean garbage disposal and prohibits the dumping of plastic at sea. By 1989 the United States and 39 other countries had signed the treaty.

What Is Landfilling?

Sometime early in the twenty-first century, as the world ushers in the new millennium, a monument to contemporary American society, currently under construction on New York's Staten Island, will be completed. Similar in size and shape to Egypt's Great Pyramid of Cheops, the Fresh Kills Landfill will be 200 feet higher than the Statue of Liberty, laying claim to the title of the tallest "mountain" on the U.S. Atlantic Coast south of Maine (Figure 20-8).

Modern landfills like Fresh Kills represent a contemporary twist on humankind's oldest form of waste management: refuse is dumped and then buried beneath layers of earth and additional

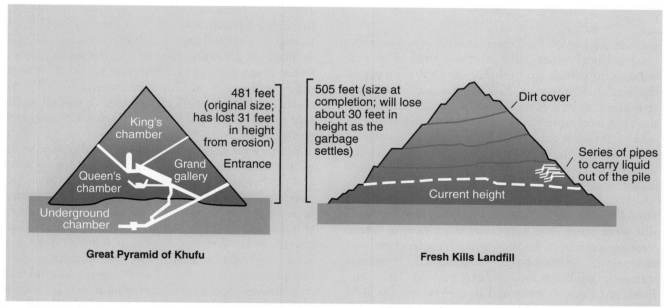

FIGURE 20-8: Final resting places compared. The Great Pyramid of Cheops at Giza was built by Cheops (also known as Khufu), a twenty-sixth century B.C. king of Egypt. The pyramid is made of 2.3 million blocks of sandstone and weighs about 5.8 million tons. Herodotus said it took 100,000 men 20 years to build it. The Fresh Kills garbage pyramid on Staten Island is being built by New York City and should be completed in 2005. Its 79.6 million cubic feet of garbage will weigh almost 50 million tons.

FIGURE 20-9: Researchers with the Garbage Project sort refuse at the Fresh Kills landfill site on Staten Island, New York, in 1990. Using a special drilling rig, researchers bring up trash from the bottom of the fill in order to determine the composition of New Yorkers' refuse.

garbage (Figure 20-9). While landfills are more sanitary and less esthetically offensive than litter and open dumps, they are probably the best example of the artificial "one-way" pathway for material goods that characterizes many human societies. Ironically, the safer and more environmentally sound a landfill is, the more it exacerbates the one-way path of materials and the less it functions like a cyclic natural system. To understand why this is so, let's take a closer look at how landfills operate.

How Landfills Operate. Until the mid-1970s, landfills were often little more than open dumps. There were no guidelines for their siting, construction, and operation, and there were no restrictions against dumping of hazardous substances. In 1976 the Resource Conservation and Recovery Act (RCRA) established classifications for various types of landfills and developed operating guidelines for each.

Secure landfills, designed to contain hazardous wastes, have a heavy plastic liner and a clay cap. Wastes are typically stored in barrels, and different

types of wastes are segregated to prevent any mixing of wastes should the containers leak. (Chapter 19 describes secure fills and their management in greater detail.)

All new landfills that are designed to receive non-hazardous residential, commercial, and industrial wastes must meet EPA standards for **sanitary landfills.** The bottom and sides are lined with thick layers of clay or plastic so that precipitation that percolates through the fill is prevented from entering underlying groundwater. Leachate moves through a network of drains throughout the fill to recovery points at which it is collected for treatment. The local groundwater is monitored for signs of contamination; monitoring continues for up to 30 years after the fill is closed (Figure 20-10).

EPA regulations also mandate that operators control the production of methane gas within the landfill. As garbage decomposes, gases are emitted; methane accounts for about 55 percent of these gases, carbon dioxide 45 percent, water vapor 4 percent, and trace gases the remainder. Because methane is explosive, its buildup is a potential danger. Moreover, unlike natural gas, methane does not lie in a reservoir; rather, it is emitted continuously from the surface of the landfill and migrates outward underground, where it may seep into basements of nearby homes and buildings, posing a safety threat.

Recovery systems can help to eliminate the potential danger caused by the production of methane. Collection pipes sunk vertically into the landfill tap the decomposition gas and bring it to the surface. There, it is dissipated to the air, burned, or collected through a vacuum system. The gas may be cleaned to extract the methane, which can then be sold as is or used to generate electricity. For example, 67 wells in the Rumpke landfill in Cincinnati, Ohio, produce about 3 million cubic feet (81,000 cubic meters) of clean methane gas per day. The gas is sold to the Cincinnati Gas and Electric Company for resale to residential customers.

Sanitary landfills are usually divided into a series of individual cells. To minimize exposure to wind and rain, work occurs in only one cell at a time. The portion of the site being filled is called the working face. Each day the collected garbage is spread out in thin layers, bulldozed to compact the volume, and covered with a layer of dirt about 6 inches deep or a layer of plastic or both. Covering the garbage each day prevents the garbage from blowing around and discourages animals from foraging in the fill, but it also prevents oxygen from reaching the garbage and thereby inhibits decomposition; in effect, garbage becomes "entombed" in the landfill, even those materials that would otherwise readily degrade. As each cell is filled, it is capped with a layer of clay

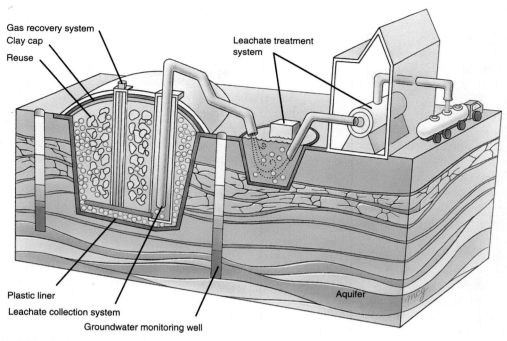

FIGURE 20-10: Cross-section of a modern sanitary landfill.

and earth. Native grasses planted atop the capped cell hold the soil in place, minimizing erosion. When the entire site is filled, it is capped with a final layer of clay and earth and landscaped. Capped landfills have been used as a variety of recreational sites such as parks, golf courses, and ski slopes.

The siting and construction of sanitary landfills are carefully regulated by law. Numerous factors are taken into consideration when a potential site is evaluated. Several of the most important factors are sources and direction of groundwater flow, soil composition, and site engineering.

Environmental, Economic, and Social Problems Associated with Landfills. In December 1984 two major landfills in New Jersey closed, leaving Philadelphia and southern New Jersey with no sites for waste disposal. The nearest landfills were hours away, and collection was halted as communities tried to find a way to deal with their mounting wastes. Refuse was left piled at curbside for several weeks in some areas. The public furor over the region's refuse problems focused national attention on solid waste issues and some of the environmental, economic, and social problems associated with landfilling: closing of existing landfills, lack of appropriate sites for new landfills, rising economic costs, community opposition to the siting of new landfills, and leachate contamination.

Closing of Existing Landfills. The lifespan of most landfills is about 10 years. The EPA estimates that about 50 percent of the nation's roughly 9,000 land-fills will close by the year 2000; as many as 80 percent may close by 2009. The closing of landfills in itself is not a problem. The problem is that old fills are often not replaced by new ones.

Lack of Appropriate Sites for New Landfills. A clay bed is considered the best substrate for a landfill, but clay sites are not evenly distributed throughout the country. High water tables or permeable underlying geologic strata (which allow water to percolate, leaching pollutants from the fill and carrying them into groundwater) prohibit the siting of new fills. In much of Florida, for example, the water table is simply too high to facilitate landfilling. Certain areas, particularly in the Northeast, are simply too congested.

Rising Economic Costs. Lack of suitable sites for landfills is a serious problem in the densely populated Northeast. As landfill space becomes increasingly scarce, the cost of dumping waste rises. Between 1982 and 1988 the National Solid Waste Association found that the average cost to dump wastes more than doubled, from about $11 per ton of garbage to almost $27. In some areas along the East Coast costs rose as high as $50 per ton by 1990. These costs, of course, are passed on to the consumer.

In those areas where landfill sites are not readily available, shipping or transporting wastes to out-of-state landfills significantly increases the cost of waste disposal. Even so, in many areas this option is currently less expensive than incineration or recy-cling. Thus, private haulers, who contract with communities in northeastern states, routinely ship wastes westward to areas where disposal costs, called **tipping fees,** are lower. Tipping fees at Ohio landfills, for example, are just one-fourth the cost of disposal on the East Coast. In 1989, 15 million tons (13.65 million metric tons) of waste were transported across state borders; 53 percent of that refuse originated in New Jersey (5.5 million tons or 5 million metric tons) and New York (2.4 million tons or 2.18 million metric tons). Compared to other states, New York and New Jersey ship their wastes the farthest (New York's refuse is shipped as far away as Louisiana, Missouri, and New Mexico) and to the most locations (both send refuse to 11 other states).

Just as some states have become trash exporters, others have become trash importers. Ohio accepts about 1 million tons (0.91 million metric tons) of out-of-state trash each year, 10 percent of the amount generated in state. Michigan, Indiana, Illinois, South Carolina, Georgia, West Virginia, and Virginia also accept out-of-state trash, although mounting public concern over limited landfill space and health and environmental effects is causing many states to reconsider that policy.

Community Opposition to the Siting of New Landfills. Even when a site can be found that is geologically sound, it may be in or near a community where there is a strong NIMBY attitude ("Not in my backyard"); the people acknowledge that a landfill is necessary and even desirable and will support it as long as it is not sited in their immediate community. One reason opposition to landfills runs so high is the fear that property values will decline if a landfill is located nearby; others are a fear of contaminated groundwater, the presence of odors, heavy truck traffic, and other undesirable living conditions. Consequently, landfills have traditionally been sited in rural areas or poorer neighborhoods where there is likely to be less organized resistance.

Leachate Contamination. Even when landfills can be sited, operators face numerous problems. Rainwater percolating through the fill can leach heavy metals, organic compounds, and other chemicals from the refuse and carry these substances with it as it continues to seep downward. A 1988 study by the EPA found that landfills can contain any of more than 200 chemicals, including such carcinogens as arsenic, methylene chloride, and carbon tetrachloride. If leachate reaches an underground aquifer, it may contaminate the water. Because about half of the U.S. population relies at least in part on groundwater for its drinking water, the problem is potentially serious.

An estimated one-fourth of all disposal sites in the United States violate federal groundwater protection standards. Many older landfills, including hundreds that have already closed, may leave a toxic legacy. Built during a time when refuse was less toxic or when protection standards were nonexistent, many of these fills are located in areas once considered "undesirable," such as wetlands or along riverfronts. Often, the idea was to fill the wetland in order to make it productive, that is, suitable for housing, development, or agriculture. Unfortunately, wetlands and river frontage are two areas where leachate is most likely to occur, since the underlying geologic strata make it easy for water to filter downward and migrate to the nearest source of water (an underground aquifer or nearby stream or river).

What Is Incineration?

As space in landfills decreases and disposal costs rise, interest in incineration increases. Incinerators offer some attractive advantages. When functioning properly, they reduce the volume of waste by about 80 to 90 percent. **Waste-to-energy incinerators** burn garbage to produce heat and steam, which can then be used to produce electricity. These plants can be profitable *if* landfill tipping fees are high enough. The municipality is charged for disposing of the waste and, under a federal law requiring utilities to "buy" power from waste-to-energy plants, the local utility is charged for the power generated. In 1990, 128 waste-to-energy plants incinerated 16 percent of the nation's solid waste, providing enough electricity to supply 1.1 million homes. By the year 2000 the EPA estimates that 350 incinerators will be in operation nationwide.

There are several kinds of waste-to-energy plants. In **mass burn incinerators** all incoming waste is fed into a furnace where it falls onto moving grates (Figure 20-11). The temperature inside the furnace may reach 2,400 °F (1,315 °C). As the waste burns, it heats water. The resultant steam is used to drive a turbine to generate electricity. Particulates are removed by electrostatic precipitators; acid gases are removed by scrubbers. In some, but not all mass burn incinerators, metals are recovered from the ash.

In **refuse-derived fuel incinerators,** incoming waste is separated before burning to remove noncombustible items. Glass and metals are separated for recycling or landfilling. The combustible wastes are then shredded into smaller, more uniform particles prior to incineration. The particles may be compacted into fuel pellets. The refuse derived fuel (RDF) is usually mixed with crushed coal and burned in a standard boiler, although specially designed boilers can burn RDF alone. While costs for sorting and processing wastes are higher than in a mass burn incinerator, an RDF plant offers environmental benefits: the removal of potentially harm-

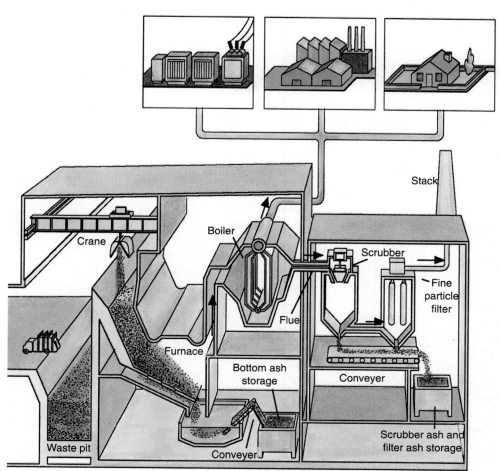

FIGURE 20-11: Mass burn incinerator. Collected solid waste is deposited in the waste pit. A crane transfers the waste to the furnace, where it is burned at high temperatures. The furnace heats a boiler that produces steam for electrical generation, the heating of buildings, and industrial processes. A conveyor removes ash that collects at the bottom of the furnace; the bottom ash, which may contain heavy metals and other toxic substances, must be landfilled. Smoke flows through a flue at the top of the furnace and into a scrubber, which sprays a wet or dry calcium compound into the smoke. The calcium compound reacts with heavy metals and toxic organic compounds and regulates them. A fine-particle filter removes tiny ash particles by passing them through either a porous bag or an electric field that attracts charged particles. The scrubber ash and filter ash are collected for landfilling. Illustration by Ian Worpole from MANAGING SOLID WASTE by Philip R. O'Leary, Patrick W. Walsh, and Robert K. Ham. Copyright © 1988 by Scientific American, Inc. All rights reserved.

ful materials before burning and the recovery of useful resources.

Environmental and Economic Problems Associated with Incineration. Incineration has significant environmental and economic drawbacks. Burning refuse poses threats to human and environmental health. Even those incinerators that are equipped with the best pollution control devices can release into the atmosphere small amounts of metal particulates such as mercury, acid gases such as hydrogen chloride, and classes of chemicals known as dioxins and furans. Both dioxins and furans are toxic and have been linked to birth defects and certain kinds of cancer. Ash may contain high concentrations of toxic heavy metals such as cadmium and lead and must be handled with care. Moreover, because ash must be landfilled, incineration does not entirely do away with the need to find acceptable and secure fill

sites. As of 1990 the EPA had not established specific regulations for handling ash; it is managed as regular refuse. Though some states have enacted their own laws for managing potentially toxic incinerator ash, these vary widely.

Incinerators are expensive to build, and equipping the incinerators with adequate pollution control devices increases the cost considerably. A large plant may cost as much as $400 million.

Another drawback is that incinerators burn paper and other unrealized resources that could be recycled and reused.

What Is Waste Minimization?

Given the myriad environmental and economic drawbacks associated with ocean dumping, landfills, and incinerators, municipalities and solid waste experts are searching for alternatives to waste disposal. The most promising options are waste minimization and resource recovery.

Waste minimization, also called **source reduction,** includes minimizing the volume of products, minimizing packaging, extending the useful life of products, and minimizing the amount of toxic substances in products. For example, the use of absorbent gel packs substantially lowered the volume of disposable diapers. In Germany a law went into effect in December 1991 that requires retailers, wholesalers, and manufacturers to take back packaging returned by their customers. The law's net effect is to make the businesses that are responsible for creating waste accept the responsibility for and the cost of disposing of it. It is expected that the law will reduce the production of unnecessary waste.

Reducing the amount of wastes produced makes good sense economically and environmentally because it can reduce disposal costs, extend the life of landfills, reduce emissions from incinerators, and reduce the amount of toxic wastes produced. For industry minimizing or eliminating the production of toxic wastes can relieve it from the burden of future liability for environmental and health hazards caused by those wastes. (Chapter 19 examines some of the ways that industries are modifying production processes in order to create less waste, particularly hazardous wastes.)

Consumers can also practice waste minimization. **Precycling** is the conscious effort to purchase merchandise that has a minimal adverse effect on the environment. What You Can Do: To Precycle suggests some ways that individuals can reduce the amount of wastes produced in the home, at school, the office, and in the marketplace.

What is Resource Recovery?

Resource recovery is an umbrella term that refers to the taking of useful materials or energy out of the waste stream at any stage before ultimate disposal. Recycling and composting, which remove useful materials from the waste stream in order to yield new products, are discussed below. Waste-to-energy incinerators, described earlier, are the focus of Environmental Science in Action in Chapter 13.

Recycling. Recycling is the collection, processing, and marketing of waste material for use in new products. One of the problems with the way we currently manage our wastes is that all materials are simply mixed together. Mixing wastes greatly reduces the value of recyclable materials. Paper, for instance, can be easily soiled if mixed with food or other organic wastes. Not surprisingly, then, efficient and effective collection and sorting are critical to a successful recycling effort.

Source separation is the act of separating recyclable items from other wastes at the point of generation prior to collection. Separating recyclables at their source keeps the items clean and thus preserves their value. Once they are separated, recyclables can be taken to a drop-off or buy-back center. At drop-off centers the individual is not paid for the recyclables; the recycler hauls the materials away and collects the profit from them. At buy-back centers people are paid for the recyclables they bring in. Single material buy-back centers collect and recycle just one resource, usually aluminum, glass, or paper; multimaterial centers collect many.

Unless sufficiently motivated, whether by economic or moral reasons, many people find drop-off and buy-back centers inconvenient and simply do not bother to make the effort to recycle. Consequently, many municipalities are looking to **curbside collection.** Recyclable materials are set out at the curb and picked up weekly, often with the regular refuse. Some communities require residents to sort recyclables and place each type (paper, glass, aluminum, plastic) in a separate container. Others allow **co-mingled recycling,** the collection of different types of recyclables in a single container. Co-mingling is more convenient for the homeowner and can make collection simpler and faster, but it increases the cost of the recycling program since workers must be hired to separate incoming wastes. The goals of curbside recycling programs are to recycle resources, lower disposal costs, and extend the life of landfills. The local government may assume responsibility for curbside collection, or it may contract with a private enterprise to operate the recycling effort.

Seattle claims one of the nation's most effective waste reduction and recycling efforts. About 34 percent of the waste stream is diverted from landfills; the target goal is 60 percent by 1995. The program serves all single-family homes and multifamily residences up to four units. A rate structure determines the fee for refuse disposal based on factors such as family size, age, backyard or curbside service, extra waste rate, yard waste rate, and size of containers. Basically, the more refuse a household generates, the more it must pay to have those wastes removed (Figure 20-12). Such volume-based pricing promotes waste minimization.

The chief advantage of curbside collection programs is that they usually increase the number of residents participating in a recycling effort. However, they can be expensive. Curbside collection requires major capital expenditures for collection vehicles and involves significant operating costs (for labor, maintenance, and administrative and transportation expenses), particularly if separate trips are made to collect regular refuse and recyclables. A high participation rate can offset costs, since rev-

- Steer clear of excessive packaging and buy sensibly packaged products. Avoid buying items that come in double packages, such as bottles packaged in boxes.
- Avoid plastic foam packaging altogether.
- Buy products in containers that are reusable or recyclable. Look for the recycled and recyclable logos.
- Write to manufacturers and ask them to use less packaging and to mark it as to its recycled and recyclable content.
- Buy products in concentrated form, in large-size containers, or in bulk. Avoid buying single serving packages, which have a high ratio of packaging to product.
- Buy durable products. Avoid

buying disposable products such as pens, razors, cameras, and plastic or paper plates and cups. Using it once is not enough!
- Bring along your own reusable bag to carry purchases, so that you don't need to take a bag (paper or plastic) from the store.
- To avoid kitchen wastes, plan meals carefully.
- Use jars or reusable plastic containers for left-over food. If aluminum foil is not badly soiled, rinse it and reuse.
- Reuse boxes for mailing, storage, and moving.
- Buy products that are recyclable.
- Recycle at home and in the office.
- If your community does not have a recycling program, tell your elected officials that you want one—and be ready to help

get the program started and operating.
- At work or school use both sides of paper, route material instead of copying, use electronic messages.
- Get off mailing lists; share magazines with friends and neighbors.
- Try to get your friends and neighbors interested in precycling and recycling. Do it gently; set a good example for others, and let your interest and enthusiasm sway them.
- Substitute nonhazardous products for household hazardous and toxic substances whenever possible. (See What You Can Do, pages 414–415 for many things you can do to detoxify your home.)

enue generated by the fees imposed for collecting the recyclables increases with each participant. Other revenue is generated by the sale of curbside materials.

Municipal Solid Waste Composting. Refuse composting operations convert organic wastes into useful material. Some municipalities pick up leaves and yard waste and compost them in rows on fields. The

FIGURE 20-12: Recycling bins at curbside, Seattle, Washington. Seattle boasts one of the nation's most effective waste reduction and recycling programs. The city charges $13.75 per month for weekly pick-up of a 30-gallon residential waste can; extra cans are $9 each.

material is turned periodically to encourage biodegradation. When the waste has composted, it is sold or given back to members of the community for use as a soil conditioner.

In compostable refuse operations refuse arriving at a composting facility is sorted to remove recyclable materials, which are sent to recycling centers. The remaining waste is reduced in volume by shredding or some other procedure and passed through a metals separator. A second screening step removes other noncompostable materials such as textiles and leather. The compostable wastes (soiled paper, food residues, yard wastes) are moved to a chamber specially designed to accelerate the composting process; temperature, moisture, and oxygen content are closely monitored in order to maximize the rate of decomposition. After decomposition, a final screening removes any remaining traces of noncompostable materials. The compost is set out to air, or cure. This compost is usually given away or sold for only a nominal fee because it is difficult to guarantee a consistent product.

In **co-composting** operations yard waste or compostable refuse is mixed with sewage sludge. In one method the refuse is mixed with sludge in digesters, vats that are 6 feet (1.8 meters) deep and about 100 feet (30 meters) in diameter. The digesters stir the mixture for 5 to 7 days. The temperature within the vats may rise to 150 °F (65.5 °C) as the bacteria decompose the refuse and sludge. After the mix-

FIGURE 20-13: A community composting effort Austin, Texas.

ture has been sufficiently decomposed, it is moved from the vats and set in mounds to cure. The co-composting operation can take anywhere from 30 to 180 days.

Composting can also be done on a smaller scale (Figure 20-13). Homeowners can compost yard waste to produce a soil conditioner for gardens and lawns. Environmental Science in Action: The New Alchemy Institute (pages 439–441) examines research on the role of composting and other recycling techniques for residential uses and small-scale commercial operations.

What Economic Factors Affect Recycling?

Economic factors can hinder or encourage recycling. During the economic boom of the 1950s and 1960s, there was little interest in recycling. Glass soda bottles were one exception, but the elimination of the deposit for return and the introduction and widespread use of "throwaway" steel, aluminum, and plastic containers soon reduced the reuse of glass. The environmental movement of the late 1960s and early 1970s spawned interest in recycling, and about 300 drop-off centers were started nationwide. The energy crisis of the 1970s provided some impetus for recyclers, but low landfill costs and a lack of markets for recycled materials hampered recycling efforts. The economic recession in the mid-1970s forced many recycling centers to close.

Recent years have witnessed the emergence of a number of economic factors that encourage recycling. As landfill capacity dwindles, particularly in the Northeast, tipping fees have become prohibitive for many communities. Every ton of recycled material saves a municipality the cost of its disposal. According to the Worldwatch Institute, these savings may be as high as $100 per ton in parts of New York, New Jersey, and Pennsylvania. Recycling also prolongs the life of existing landfills; depending on their extent and success, recycling efforts may significantly delay the process of siting a new landfill.

The most effective way to encourage recycling is to begin to charge consumers for the real cost of waste disposal. If households were charged for waste disposal in proportion to the amount of waste generated, more people would see the economic benefit of reusing and recycling materials. In evaluating the economic potential of a recycling program, municipalities should consider not only the revenue from the sale of recyclable items, but also the landfill and incineration costs that are avoided.

By 1990 several states, including Pennsylvania and New Jersey, had enacted mandatory recycling laws. By requiring consumers to recycle goods and materials, the laws ensure a supply of recyclables. They do not, however, ensure a demand for recyclables, also known as **secondary materials.** As long as demand remains low, recyclables will not be recycled. Secondary materials will simply pile up: collected from homes and businesses, but not given new useful life. By 1991, for example, solid waste officials in some communities had to pay waste haulers to cart collected newsprint away. In New York City, where approximately 55,000 tons (50,050 metric tons) of newsprint are collected annually, officials devised a temporary solution to the dilemma of what to do with it all by compressing it into 1-ton bales and exporting it in containerized cargo ships.

Demand for secondary materials remains low for several reasons. Conservationists point out that tax deductions and subsidies that benefit mining and logging interests keep the costs of many raw materials artificially low. Manufacturers are unwilling to refurbish their plants and machinery in order to use secondary materials when they can buy virgin materials cheaply. Moreover, the environmental and economic costs of procuring and using virgin materials are not built into the price of the goods produced. These costs include the energy needed to mine or secure raw materials and water and air pollution caused by the processing of mineral ores. As Table 20-1 illustrates, substituting secondary materials for virgin resources yields substantial environmental benefits. These benefits go largely unrealized, however, as long as the market price of raw materials does not reflect the real costs of using them.

While state governments and many nations are taking innovative steps to minimize waste and ensure markets for secondary materials, the U.S. federal government has been slow to act (Table 20-4). The 1976 Resource Conservation and Recovery Act instructed the government to purchase "items composed of the highest percentage of recovered materials practicable, consistent with maintaining a

Table 20-4
Innovative Waste Minimization and
Resource Recovery Programs, Selected
States and Nations

STATE PROGRAMS IN THE UNITED STATES
By 1992 twelve states, led by Oregon, had passed laws assessing a fee on the sale of all replacement tires, thus encouraging retreading of old tires.
Oregon, North Carolina, and California offer tax credits to companies that produce goods made with recycled materials.
New Jersey offers tax credits, loans, grants, and tax exemptions to companies that produce goods made with recycled materials.
Florida has established a unique means of encouraging source reduction. Beginning in 1992 an advance disposal fee of 1 cent per item is charged on retail store sales of every container, regardless of the material. The fee is refunded when consumers return containers to a recycling center.

PROGRAMS WORLDWIDE
Denmark requires companies to use *refillable*, not recycled or recyclable, bottles. These are cleaned and reused, not crushed or melted down, thus saving energy and materials.
More than 3,000 of Japan's 3,255 municipalities have substantial recycling programs. Neighborhood drop-off points are set up to receive four types of discards: glass, metal, and paper; hazardous materials and batteries; burnables, such as kitchen waste, soiled paper, and plastic; and nonrecyclables such as construction debris and other plastics.
Zurich, Switzerland, has approximately 500 community compost piles each shared by 200 to 300 families.

satisfactory level of competition," but guidelines have been published for only two materials thus far (recycled paper products and fly ash for use in making cement). Provisions that call for reducing waste at the source and encouraging recycling have not been enforced. A 1990 Senate bill that would have expanded markets for recycled products by requiring federal agencies to procure them in preference to virgin materials failed to pass Congress. Nationwide standards for products with recycled content have not been established, making it difficult for suppliers of recycled and recyclable materials, since it means that their markets often have varying requirements for what constitutes a recycled product.

What Is a Bottle Bill?

One way to establish and maintain a market for glass, aluminum, and bimetal containers is through container deposit legislation, or bottle bills. A bottle bill requires consumers to pay a deposit on beverage containers. The deposit is refunded when the container is returned to a grocery store, food market, or recycling center. Because bottle bills encourage peo-

ple to return empty containers, they create a steady supply of materials for recycling industries and thus boost recycling efforts. The first bottle bill was enacted by the state of Oregon in 1972; between 1972 and 1983 eight other states followed suit. Such legislation can be very effective. Since enacting a bottle bill, New York has seen the volume of containers sent to landfills drop by over 70 percent.

Bottle bills place direct responsibility for recycling on the companies that put the product in the marketplace. Users of the product pay for it; abusers— those who don't recycle—pay more. Objections to bottle bills include the arguments that consumers' money is tied up in the deposit, retailers must make space to accommodate returned containers, and bottle bills represent yet another instance of government restriction of personal freedoms. While mandatory deposits may impose some inconvenience on the consumer and retailer, they do tend to reduce litter and conserve natural resources and energy.

What Is an Integrated Solid Waste System?

How will we manage our unrealized resources? The EPA hopes to achieve the following target rates for managing the nation's municipal solid waste by the mid-1990s: landfilling 55 percent, recycling 25 percent, and incineration 20 percent. Many experts contend that the EPA's recycling goal is not ambitious enough; they maintain that anywhere from 60 to 80 percent of the typical municipal waste stream could be diverted from landfills or incinerators through an efficient and comprehensive recycling and composting program. A number of communities throughout the nation are approaching these rates. Most recycling programs, however, do not target all the potentially recyclable materials; generally, they focus on paper, aluminum and bimetal cans, and glass, although there is increasing interest in plastics. Municipal composting programs typically ignore food scraps, even though they can be easily composted along with yard wastes.

A growing number of solid waste experts maintain that the emphasis in solid waste management should be on recovering as many resources from the solid waste stream as possible; they contend that landfills and incinerators should be deemphasized as waste management options. Indeed, for many people the future of solid waste management lies in integrated systems, which use many, varied approaches to solid waste management. **Integrated solid waste systems** include waste minimization, reuse of materials, recycling, composting, energy recovery, incineration of combustible materials, and landfilling. With current technology, at least, landfills cannot be eliminated as a management option. They are needed to dispose of goods that cannot be recycled as well as ash from incinerators and waste-to-energy plants.

What Is Green Marketing?

American business is increasingly aware that many consumers are seriously concerned about environmental issues. Accordingly, many companies are becoming more environmentally conscious, and **green marketing**—the practice of promoting products based on the claims that they help or are benign to the environment—is becoming more prevalent. A 1990 survey of about 50 business executives by the Wirthlin Group, an opinion research organization, found that reducing the refuse created by their products and packaging is the most pressing environmental concern facing American companies, more pressing even than the greenhouse effect, ozone depletion, water pollution, and toxins.

In developing its environmental conscience, corporate America is essentially following consumers' lead. Two surveys conducted by the Opinion Research Corporation found that the buying public is using its pocketbook to express its concern over environmental issues. According to the surveys, 80 percent of consumers do some kind of recycling at home, 27 percent have boycotted a product on environmental grounds, and 71 percent have switched brands to purchase items that are biodegradable, benign in their effect on the ozone layer, or made from recycled products.

In order to ensure that products are truly environmentally friendly, many consumer advocates have called for a uniform national "green" label. Several other nations have established such standards. In 1978 West Germany adopted a "blue angel" label for environmentally preferable products. The German label uses a single criterion—use of recycled materials, for instance, or use of materials that do not harm the ozone layer—to judge a product. Canada and Japan have developed labels that use multiple criteria. These countries perform a cradle-to-grave review of an item to ensure that it is environmentally friendly during its production, use, and disposal. This approach seems most likely to protect human health and the environment.

Future Management of Unrealized Resources

One of the most pressing issues of the 1990s is what to do with the millions of tons of "stuff" Americans throw away each year. In rethinking the ways we manage our solid wastes, it is helpful to focus on measures that prevent resource overuse through conservation, protect human health and the environment, and preserve living systems.

Prevent Overuse Through Conservation

Precycling and waste minimization, both in industry and at home, can limit the use of materials which will only be thrown away. Reducing the production of wastes conserves resources (including energy used in production) and reduces the need for disposal. Recycling can also help to conserve natural resources. Changing our business and consumption patterns can help to conserve valuable natural resources and slow imports of foreign raw materials, particularly oil. Accordingly, recycling and composting should be the first options of any waste management system. Recovering all possible resources, both organic and inorganic, from the waste stream minimizes the amount that must be disposed of, whether by incineration or landfilling. Finally, legislation should be enacted to require the federal government to purchase recycled materials whenever possible. Passing such legislation could be particularly significant since the government represents the largest single "buyer" for recycled goods.

Protect Human Health and the Environment

Any form of waste disposal poses a potential threat to human health or the environment. The less waste produced and disposed of, the less is the impact on both human health and the environment. Thus, although society is increasingly turning to source reduction and resource recovery as economic necessities, these activities can also help to minimize environmental degradation and to cut costs associated with the mitigation of environmental damage. Wastes should be diverted to a landfill or incinerator only if they cannot be recycled or composted. Incinerators that burn refuse-derived fuel, generating steam or electric energy, are the most environmentally sound option *if* certain materials are minimized *and* we pay for pollution control. If, as a society, we decide on incineration to manage our wastes, then we should shift from plastic to paper products whenever possible. Landfills should be utilized only after all other options—source reduction, reuse, recycling, composting, and incineration—have been exhausted.

Preserve Living Systems

How we manage solid wastes affects other resources—air, soil, water, wilderness, and biological diversity, to name but a few. Mimicking living systems through the recycling of wastes and the consequent conservation of materials and energy can help alleviate many environmental problems and help to preserve living systems. In a direct way, resource recovery affords us the opportunity to build a sustainable and sustaining society.

Environmental Science in Action: The New Alchemy Institute

Frank Andrew Jones

Although most people realize that the glass, paper, and aluminum they throw away can be recycled, they often overlook another valuable resource disguised as waste: organic material, such as grass clippings, animal manure, and food scraps. The New Alchemy Institute (NAI), a nonprofit research and education center located on Cape Cod, Massachusetts, is a leader in developing innovative methods of nutrient recycling and composting. Like the medieval alchemists who sought to transform worthless metals into gold, the New Alchemists seek to transform material normally discarded or ignored into resources for producing food, energy, and shelter. NAI has designed simple and affordable technologies, such as a composting greenhouse, that make reducing and recycling human, animal, and vegetable waste a realizable goal.

Describing the Resource

Biological and Physical Boundaries

The New Alchemy Institute occupies 12 acres which were formerly a dairy farm in East Falmouth, Massachusetts. The grounds encompass a variety of experiments and demonstrations that promote self-sufficiency and ecological awareness. These energy- and cost-efficient projects are designed to meet the needs of homeowners, small-scale farmers, and educational institutions.

The NAI gardens, grown without any synthetic chemicals, include a vegetable garden and a large collection of herbs. The Cape Cod ark is an early bioshelter, or solar greenhouse, designed to sustain fish, plants, and other life forms with no fossil fuels. The pillow dome is a greenhouse designed to demonstrate a type of super insulation—triangles of light plastic filled with inert gas—that helps to retain heat in the wintertime. The superinsulated auditorium, a converted dairy barn, demonstrates advanced techniques for conserving energy. The visitors' center and store offer information on NAI's history and ongoing projects and sell books, garden supplies, and energy-saving devices. In the rest of this section, we will look in depth at the composting greenhouse, which recycles organic materials.

Theory Behind the Composting Greenhouse. Researchers at NAI became interested in developing a sustainable, affordable greenhouse in the early 1980s. Greenhouses built before the 1970s were designed to be heated during the winter. When the cost of oil rose markedly after the Organization of Petroleum Exporting Countries oil embargo of 1973, these greenhouses became expensive to operate. Solar greenhouses, designed for use without supplemental heat, are expensive to build because they must be heavily insulated to withstand the winter months, and insulation blocks sunlight and adds to cost. Various storage systems that "save" heat from the sun are an additional cost.

In 1983 NAI built the composting greenhouse with the help of the Biothermal Energy Center. The greenhouse was an attempt to reduce the need for extra insulation and heat storage by taking advantage of the heat generated when bacteria decompose organic material in compost. Using the heat from compost to supplement the sun's rays cuts costs without relying on nonrenewable energy and promotes better management of manure, which can easily pollute water if improperly stored. Composting reduces the amount of waste that must be disposed of and produces a rich fertilizer. Organic fertilizer can help farmers and gardeners decrease reliance on synthetic fertilizers, which are made from fossil fuels.

How the Composting Greenhouse Works. The basic structure of the greenhouse is a 12-foot by 48-foot frame (576 square feet) fashioned from electrical conduit. A double-layered plastic film, which is commercially available, makes up the shell of the greenhouse. The air space between the two layers of film provides extra insulation. The greenhouse is situated on an east-west axis, with the 25-cubic-yard compost chamber on the north side. The compost chamber has removable, insulated panels so it can be loaded from the outside conveniently.

The upper-level growing beds are on top of the compost chamber, and the lower-level beds are in front of it. Blowers periodically pull air from the greenhouse up through the compost. This ventilation system eliminates the labor-intensive practice of turning the compost every other day in order to provide oxygen to the aerobic bacteria which decompose the organic waste. Blowing through the compost, the air picks up heat, moisture, carbon dioxide, and ammonia produced by the bacteria and other organisms and carries it into perforated pipes running beneath the lower growing beds. Heat is bound up in the water molecules and when the moisture condenses in the pipes, it releases the stored energy into the soil in the form of heat. The water is also absorbed by the soil and eventually by the plants. The carbon dioxide in the pipes filters up through the soil and becomes available to the underside of the leaves of the plants. The added carbon dioxide in the greenhouse has been shown to produce 20 to 30 percent net increases in yields of vegetable crops. Nitrogen is a limiting nutrient in most greenhouses, and a short supply of it retards the growth of plants. Thus, the increased nitrogen from ammonia in the greenhouse also helps plant growth significantly.

Unlike other compost by-products, ammonia can damage plants even in low concentrations. The soil in the growing beds, however, acts as a biofilter: bacteria living in the soil transform the ammonia into useful nitrate, which is readily absorbed by plant root hairs. The texture and health of the biofilter soil are enhanced by populations of earthworms and manure worms which create vertical tunnels through the soil as they digest organic matter present in the topsoil.

Compost can be composed of a variety of organic materials, including sewage sludge, human waste (nightsoil), woodchips and sawdust, animal manure, garbage and yard wastes, garden and vegetable wastes, and even preshredded municipal solid waste mixed with sludge. Each type of material decomposes at a different rate and gives off different amounts of carbon dioxide and heat. To produce the appropriate amounts of heat and carbon dioxide as well as a high-quality compost, operators must often mix several materials. The operators must evaluate individual plant needs and seasonal differences to determine what materials will work best.

Overall, the self-sustaining greenhouse has proven successful: it produced over 100 tons (91 metric tons) of compost and tens of thousands of seedlings in its first full year. NAI researchers estimate that for every cubic yard of compost materials placed in the compost chamber, they save approximately $5.50 to $11.00 they would have had to spend on supplemental carbon dioxide and $3 to $6 that would have gone for heat.

Social Boundaries

New Alchemy is a philosophy of living as well as a research institute. The ideas that lie behind the institute grew out of concern over the increasing technologization and specialization of our world. The research at NAI is meant to address these concerns by enabling people to coexist in a mutually supportive and beneficial way with nature. The environmental and social problems we have created through chemical and mechanical technology cannot be solved by more of the same, and so NAI concentrates on providing biological analogues to chemical and mechanical methods of producing food, shelter, and energy.

NAI also believes we must change as a society in order to achieve self-reliant, ecologically harmonious life-styles. Its projects are designed to enable individuals to fulfill more of their own basic needs without technology or dependence on centralized services. For instance, the

composting greenhouse is a simple, affordable way to produce food and compost (which enhances soil fertility) while simultaneously disposing of organic waste. NAI's ultimate vision is a sustainable society composed of self-reliant communities capable of growing all of their own food and generating their own power from renewable resources.

Looking Back

Since its founding almost 20 years ago, NAI has been on the forefront of a new "biotechnology": appropriate technology that is life-encouraging and ultimately safer and simpler than a technology based on chemical and mechanical methods. By developing naturally based methods to increase self-reliance and by educating the public, it is trying to provide answers today to the problems of tomorrow.

Origins of New Alchemy Institute

The founders of the New Alchemy Institute, John Todd, Nancy Jack Todd, and Bill McLarney, originally lived in southern California. The idea for NAI started in the late 1960s, when John was teaching biology classes at San Diego State. Ecology and environmental degradation were just beginning to gain widespread public awareness, so John and Bill began to hold informal seminars on aspects of the overall problem. These seminars often lasted late into the night and focused on the questions "Can anything be done?" and if so, "What?"

John had also been exploring the practical side of ecological awareness by taking his students on field trips to the hills east of San Diego. His original purpose was to examine the area and discuss how to farm the land without degrading it. However, John soon realized that most students had no sense of the land and how to use its resources. He began a second phase of the project in which the students learned to group species of plants and animals taxonomically and to classify soil

types. By examining each species in its environment, they began to recognize certain patterns in the land. For instance, the existence of a plant that grows only near water in a seemingly dry area provides a clue that there is probably a hidden spring nearby. The ability to "read" a landscape in this way is essential for those who want to farm in harmony with the land with a minimum of dependence on chemical and mechanical devices.

These experiences inspired the philosophy that drives New Alchemy today. In 1970 John Todd expressed the essence of the New Alchemy ideal in a paper titled "A Modest Proposal." This paper served as the first New Alchemy newsletter and has since been translated into many languages. In "A Modest Proposal" John presented his vision of a biologically harmonious plan for the future of all people. He felt that

a plan for the future should create alternatives and help counter the trend toward uniformity. It should provide immediately applicable solutions for small farmers, homesteaders, native peoples everywhere, and the young seeking ecologically sane lives, enabling them to extend their uniqueness and vitality. Our ideas could also have a beneficial impact on a wider scale if some of the concepts were incorporated into society-at-large. Perhaps they could save millions of lives during crisis periods in the highly developed states.

Research and Education at NAI

As the first new alchemists began developing their ideas into projects, the group looked for funding and a site on which to conduct the research. They found the site on Cape Cod and were settled in by May 1972. Projects concentrated on human support systems essential for human survival, including wind energy, solar energy, and organic and greenhouse agriculture.

The creation of the Cape Cod ark bioshelter was the cornerstone of the early work at NAI. This solar greenhouse meets the energy, food, and shelter demands not only of humans, but also of bacteria, soil, plants, trees, flowers, insects, and fish. It is a semicontained ecosystem that conserves energy and recycles all materials. This innovative use of natural cycles and processes to meet human needs led to the diverse projects going on at NAI today.

In addition to designing systems for meeting basic human needs, NAI has also established a research station in Costa Rica. This station concentrates specifically on investigating solutions to the destruction caused by deforestation and offers Costa Rica help in avoiding the environmentally and economically disastrous mistakes that many developing nations make in their quest to modernize.

Besides researching and developing various biotechnical devices and methods, New Alchemy staff concentrate on educating people about other aspects of an ecologically sound, self-reliant life-style. They offer a catalogue of publications and devices to assist individuals in achieving this way of life. Their quarterly newsletter reports on NAI research and activities, along with updates on advances in biotechnology made by other organizations and individuals. NAI staff give tours, workshops, and programs for local schoolchildren, and the institute offers internships for college students. The staff also produces technical reports evaluating various ecologically sound methods of producing food, energy, and shelter, such as nontoxic pest management and solar greenhouses.

Looking Ahead

In 1980 John and Nancy Todd left the New Alchemy Institute to form two new corporations: Ocean Arks International, a nonprofit research organization, and the Four Elements Corporation, a profit enterprise set up to prove that the Todds' ideas were economically feasible. The New Alchemy Institute continues to serve as a leader in establishing nonharmful technologies and biological analogues to chemical and mechanical technologies. Through efforts like the composting greenhouse, it has demonstrated that it is possible to work with sunlight and living organisms, as does nature, to recycle wastes efficiently and safely. Education is one of the institute's most important functions. By spreading its ideas and products among as many people as possible, NAI could set the stage for a widespread movement toward environmentally sound means of providing food, shelter, and energy.

Summary

A solid waste is any material that is rejected or discarded as being spent, useless, worthless, or in excess. Refuse consists of things rejected as worthless or of insignificant value; the term is a common synonym for solid waste. Garbage is animal or vegetable waste resulting from the handling, storage, preparation, or consumption of food. Trash is useless or worthless matter that is unsightly but does not contain odor-producing food wastes. The solid waste stream is the collective and continual production of all refuse. The two largest sources of solid wastes are agriculture and mining. Municipal solid waste (MSW) is refuse generated by households, businesses, and institutions. Solid wastes, like hazardous wastes, are misplaced or unrealized resources.

Aerobic decomposition occurs in the presence of oxygen by oxygen-requiring organisms. Anaerobic decomposition occurs in the presence of organisms that do not require oxygen. In a landfill most decomposition is anaerobic. A biodegradable material can be decomposed by natural systems.

Natural systems are cyclic; in contrast, human systems tend to be linear. As the law of the conservation of matter indicates, materials can be neither created nor destroyed. The most promising solution to the solid waste problem is to reuse and recycle wastes as much as possible. For materials that cannot be recycled, the solution is to reduce consumption or to substitute nonhazardous alternatives for them.

Both the consumption of goods and the management of waste products vary from country to country. The United States generates about 40 percent of the world's waste. The waste stream of less-developed countries is typically much smaller than that of more-developed countries.

When populations were small, refuse could be discarded anywhere and would gradually decompose. As cities developed, residents began to dump their refuse on the edge of town. Historically, coastal communities and industries have dumped sewage wastes and sludge into the ocean. Many communities and states have enacted laws to combat the serious threats to the marine environment caused by ocean dumping, both intentional and unintentional.

Secure landfills are designed to contain hazardous wastes. Sanitary landfills are designed to receive nonhazardous wastes. The siting and construction of sanitary landfills are carefully regulated by law. Several of the most important factors considered when choosing a site are sources, direction of groundwater flow, soil composition, and site engineering.

The lifespan of most landfills is about 10 years. The EPA estimates that about 50 percent of the nation's landfills will close by the year 2000. Old fills are often not replaced by new ones, and suitable sites are not evenly distributed throughout the country. In those areas where landfill sites are not readily available, wastes are transported to out-of-state landfills. Disposal costs are called tipping fees.

Incinerators reduce the volume of waste by 80 to 90 percent. Waste-to-energy incinerators burn waste to produce electricity. In mass burn incinerators, all incoming waste is fed into a furnace. In refuse-derived fuel incinerators, incoming waste is separated before burning to remove noncombustible items.

Waste minimization includes minimizing the volume of products, minimizing packaging, extending the useful life of products, and minimizing the amount of toxic substances in products. Precycling is the conscious effort to purchase merchandise that has a minimal adverse effect on the environment. Resource recovery refers to taking of useful materials or energy out of solid waste at any stage before ultimate disposal. Resource recovery methods include recycling and composting. Integrated solid waste systems include waste minimization, reuse of materials, recycling, composting, energy recovery, incineration, and landfilling. Green marketing is the practice of promoting products based on the claims that they help or are benign to the environment.

Key Terms

aerobic decomposition
anaerobic decomposition
co-composting
co-mingled recycling
curbside collection
garbage
green marketing
integrated solid waste system
mass burn incinerator
municipal solid waste
precycling
recycling
refuse

refuse-derived fuel incinerator
resource recovery
sanitary landfill
secondary material
secure landfill
solid waste
solid waste stream
source reduction
source separation
tipping fee
trash
unrealized resources
waste minimization
waste-to-energy incinerator

Discussion Questions

1. What kinds of unrealized resources do you use and discard in your home and your school or office? Which of these could be reused or recycled? Identify resource recovery centers in your area and find out what kinds of resources they accept.

2. Compare the waste generation patterns of the United States, Canada, and Australia with those of other industrial nations. What factors might be responsible for the differences?

3. Some people argue that the United States has ample space for land-based waste disposal, particularly in the West and Midwest, and need not resort to resource recovery systems. Do you agree or disagree? Why or why not?

4. Discuss various methods proposed to encourage waste minimization and resource recovery. What are the advantages and disadvantages of each?

5. Your community's landfill is slated to close in two years. You have been asked to be a member of the problem-solving team charged with devising a solid waste management plan for the community. What steps would you take? What information would you seek? What role would landfilling, incineration, waste minimization, and resource recovery play in your plan?

6. Refer to What You Can Do on page 435. List at least three additional steps individuals could take to minimize waste and to recover resources from the waste stream.

7. Using specific examples to illustrate your answer, explain how the technologies and projects of the New Alchemy Institute attempt to mimic natural systems.

An Environmental Heritage

Preserving Threatened Resources

The Public Lands

Harmony with land is like harmony with a friend; you cannot cherish his right hand and chop off his left. That is to say, you cannot love game and hate predators; you cannot conserve the waters and waste the ranges; you cannot build the forest and mine the farm. The land is an organism.

Aldo Leopold

The federal lands have always provided the arena in which we Americans have struggled to fulfill our dreams. What we do with them tells a great deal about what we are—what we care for—and what is to become of us as a nation.

D. Michael Harvey

Learning Objectives

When you finish reading this chapter, you should be able to:

1. Discuss the biological and physical significance of the federal lands.

2. Identify the four major federal land managing agencies and describe how their missions differ.

3. Briefly explain the historical development of the system of federal lands, including the establishment of the four major federal land managing agencies.

4. Describe the major environmental problems and management issues associated with the National Park System, National Wildlife Refuge System, National Forest System, and National Resource Lands.

In 1872 the United States Congress set aside a vast stretch of land in Wyoming known for its natural beauty and scenic wonders as "a public park or pleasuring ground for the benefit and enjoyment of the people." In designating Yellowstone as a "nation's park," Congress established an American tradition of parks and preserves that would serve as a model for the world. In this chapter we discuss the public lands managed by the federal government, the most well-known of which are the national parks. We see just how vast and diverse these lands are, why they are biologically significant, what minerals and other resources they contain, and the diverse ways in which we use them. The historical aspects of federal land management are also addressed. An understanding of the management philosophies and traditions of the agencies that manage our federal lands can improve our appreciation of current environmental problems and issues. Finally, we discuss management strategies that can help ensure the preservation of the public lands for future generations.

Describing the Public Lands Resource: Biological Boundaries

What Are the Federal Public Lands?

The United States boasts one of the largest total acreages of publicly owned lands in the world—over 700 million acres (283 million hectares). Owned jointly by U.S. citizens, they are managed by various federal, regional, state, and local authorities. The federal government manages most of the public lands, an area equivalent to approximately one-third of the entire United States. Known as the **public domain,** the land holdings of the federal government are so vast that fully one-half have never been surveyed. Ninety percent of federally owned lands lie in the western states and Alaska (Figure 21-1).

The federal public lands are managed by numerous agencies. About 25 million acres (10 million hectares), set aside for military installations, air bases, training facilities, and corps water projects,

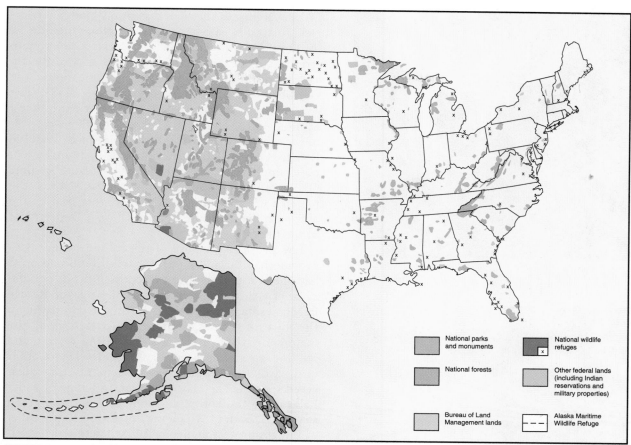

FIGURE 21-1: Federal lands in the United States—the public domain.

are managed by the Department of Defense. The Bureau of Reclamation manages 6 million acres (2.43 million hectares) for water development projects in the West, primarily irrigation. Another 6 million acres or so are managed by the Bureau of Indian Affairs, the Department of Energy, or the Tennessee Valley Authority. As important as these lands are, they are dwarfed by the acreage managed by the National Park Service, the Fish and Wildlife Service, the Forest Service, and the Bureau of Land Management. These four agencies manage more than 660 million acres (267 million hectares) of the public domain. Their holdings are the subject of this chapter. Because we consider wilderness to be a distinct and unique resource, it is discussed separately in Chapter 22, even though areas designated as wilderness are found in the lands managed by the four major agencies.

National Park System. The National Park System (NPS) is a network of 357 units, including 50 national parks, totaling over 80 million acres (32.4 million hectares). It includes areas designated as national parks, monuments, preserves, lakeshores, rivers, seashores, historic sites, memorials, military parks, battlefields, scenic or historic trails, recre-

ation areas, parkways, and others. These units vary greatly in size, from the 8.3 million acres (3.4 million hectares) of Wrangell–St. Elias National Park in Alaska to the one-third of an acre (one-twelfth of a hectare) of Ford's Theatre National Historic Site in Washington, D.C.

The National Park System is managed by the National Park Service of the Department of the Interior. In creating the service, Congress decreed that it should manage the parks in order to conserve their scenery, natural and historic objects, and wildlife and to provide for the enjoyment and use of the same by the public. The national parks enjoy a greater degree of protection than other federal lands. Logging, for example, is prohibited in the parks; mining is also forbidden, except in those cases where a claim to the mineral rights had been filed prior to the area's designation as a park.

The national parks are the country's best-known and most popular public lands (Figure 21-2). Many of the units within the system commemorate people and events that shaped our country. But the image of the park system is dominated by the large western parks—Yellowstone, Grand Canyon, Yosemite, and Glacier. Because of their natural grandeur and scenic wonders, they are called the "crown jewels" of the

(a)

(b)

FIGURE 21-2: National parks. (a) Thomas Jefferson's home at Monticello is part of the National Park System. (b) Mt. Goat soars over Hidden Lake in Montana's Glacier National Park, designated in 1910. The first areas to be named national parks were grand in both size and scenic beauty.

park system. They were and are the pride of our country, our counterpart to the cathedrals and castles of Europe and a part of our shared heritage.

National Wildlife Refuge System. The National Wildlife Refuge System (NWRS) consists of 456 units, totaling approximately 91 million acres (37 million hectares) of land and water, in 49 states (West Virginia is the lone exception) and 5 territories. Just a few refuges in Alaska alone include 12 million acres (5 million hectares). The system stretches from the north shore of Alaska to the tip of Florida and beyond to tropical locales in the Caribbean and South Pacific (Figure 21-3). The smallest refuge, less than 1 acre, is at Mille Lacs, Minnesota; the largest is Alaska's 20-million-acre (8-million-hectare) Yukon Delta National Wildlife Refuge. Over 1,000 species of birds, mammals, reptiles, amphibians, fish, and plants—including some 60 endangered species—are found within the nation's wildlife refuges, which are managed by the U.S. Fish and Wildlife Service (FWS), Department of the Interior.

The primary purpose of the refuge system is to provide habitat and haven for wildlife. Some units have been established to protect significant populations of a single species, such as the Agassiz Refuge in Minnesota, site of the largest moose population in the lower 48 states. Others are located along the major north-south flyways, providing resting and feeding areas for migratory waterfowl and other birds. Still other units serve as sanctuaries for endangered species. The Hawaiian Islands Refuge provides habitat for the Hawaiian monk seal and the green sea turtle, both of which are endangered. A number of refuges protect archeological artifacts and historically significant areas. For example, items recovered from the steamship *Bertrand*, which sank in the Missouri River in 1865, are preserved in an exhibit at the DeSoto Refuge in Iowa.

In addition to providing habitat and haven for wildlife, most refuges also permit secondary uses. Nonconsumptive uses such as wildlife observation,

FIGURE 21-3: The Great Dismal Swamp National Wildlife Refuge. The refuge covers over 100,000 acres in North Carolina and Virginia.

photography, nature study, hiking, and boating are permitted in most refuges; some refuges also allow consumptive uses such as hunting, trapping, and fishing. The development of oil, gas, and mineral resources is permitted in certain refuges, at the discretion of the secretary of the interior. In refuges that have been established on land managed by another federal agency, such as the Army Corps of Engineers, the Bureau of Reclamation, or the Department of Defense, the Fish and Wildlife Service has only secondary authority. It is responsible for managing the wildlife in the area, but has no authority to prohibit various land uses, such as military aircraft overflights, even when those uses may pose a threat to resident animals or plants.

National Forest System. The National Forest System is composed of 156 national forests and 19 national grasslands. These units total almost 191 million acres (77.3 million hectares) in 44 states, Puerto Rico, and the Virgin Islands. Most of the national forests are located in the West, but some are also found in the Northeast, South, and Midwest. The 156 national forest units cover 8.5 percent of the United States, an area as large as Wisconsin, Michigan, Illinois, Indiana, Ohio, and Kentucky combined (Figure 21-4).

The Forest Service, Department of Agriculture, manages the nation's forests and grasslands to accommodate a wide array of uses from commercial to recreational. Each year Americans use approximately 80 million tons (72.8 million metric tons) of paper products—about 600 pounds (270 kilograms) per person—and throw away 45 million tons (41 million metric tons). While most paper products come from private pine and poplar plantations in the southern United States, the national forests help to meet the country's substantial appetite for lumber and plywood for construction and wood furniture and other wood products. The national forests yield enough timber for about 1 million houses annually. But forests are not just timber stands; they are living ecosystems, intricate and unique associations of plants, animals, microorganisms, and abiotic components. The national forests are home to over 3,000 vertebrate species and 129 endangered or threatened species. As we learned in Chapter 6, forests perform many invaluable ecosystem functions, such as cleansing the atmosphere, protecting watersheds, maintaining soil and preventing erosion, and cycling nutrients.

The national forests provide special recreational opportunities. Visitors enjoy sight-seeing, driving, hiking, horseback riding, camping, canoeing, off-road-vehicle use, picnicking, snowmobiling, and cross-country skiing. Lands are leased to ski resorts, youth camps, and cabin owners. The national forests also

FIGURE 21-4: A rushing stream in Soleduck Valley, Olympic National Forest. This unit of the National Forest System blankets almost 650,000 acres in Washington.

hold much of the designated wilderness areas in the lower 48 states.

National Resource Lands. Officially known as the national resource lands, these areas, managed by the Bureau of Land Management are more commonly called the BLM lands (Figure 21-5). A 1987 Bureau of Land Management publication described them this way: "These lands in the lower 48 states are those which no one wanted; they were bypassed in favor of more promising areas during the course of settlement. The lands are the mountain tops, steep canyons, vacant valleys, and sandy deserts." They have been described as "stretches of picturesque poverty," "the lands nobody wanted," and "the lands no one knows." The BLM lands are those areas of the public domain that were not set aside as parks, forests, or wildlife refuges. Many units occur in a checkerboard fashion, interspersed with privately owned land, and many are unmarked. By the early 1990s the bureau managed approximately 300 million acres (121 million hectares), roughly 47 percent of federal lands, making it the nation's largest land-managing agency. The lands are located primarily in the arid and semiarid western states and Alaska. They are managed to provide multiple uses—economic, recreational, and wilderness. However, com-

FIGURE 21-5: Gilbert Badlands, Upper Blue Hills, Utah, a unit managed by the Bureau of Land Management.

pared to other federal lands, they host few visitors each year, in part because of a general lack of public awareness about them.

Why Are the Public Lands Biologically Significant?

The federal lands are biologically significant because they encompass an astounding diversity of living ecosystems—alpine meadows and prairies and coral reefs and mangrove swamps. They include the wind-swept Alaskan tundra and the sun-drenched beaches of St. Johns, U.S. Virgin Islands; the slick-rock canyons of Utah and the rocky shoreline of Maine; the bone-dry expanses of Death Valley and the tropical lushness of the Caribbean National Forest, Puerto Rico; the jagged, snow-covered peaks of the Rockies and the smooth, rounded slopes of the Great Smoky Mountains; the sagebrush expanses of Nevada and the sawgrass wetlands of the Everglades; the old-growth forests of the Pacific Northwest and the geologically young Great Lakes.

The public domain is biologically significant for other reasons as well. The diverse ecosystems represented by the nation's public lands perform a myriad of ecological functions (see Chapters 5 and 6). Forests and grasslands, for example, hold soil in place, minimizing erosion and helping to keep streams and rivers clean and clear. Wetlands slow and absorb floodwaters, helping to protect adjacent land areas, and provide critical habitat for many species of fish, shellfish, birds, and other animals.

Because the public lands encompass a wide array of ecosystems, they also protect a diverse complement of living organisms. Preserving species diversity, as well as the genetic diversity within species, is prudent: medicine, agriculture, and industry all ben-

efit from the genetic characteristics possessed by wild species. Many people want to preserve biological diversity also out of esthetic considerations; for them, sharing the planet with the greatest possible diversity of living organisms is a key ingredient in a high quality of life. Others argue that all species have a right to exist independent of their economic or esthetic value to humans; for them, preserving biological diversity is an ethical or moral mandate. (Chapter 23, Biological Resources, includes a complete discussion of these and other arguments in favor of preserving biological diversity.)

Physical Boundaries

The federal lands contain substantial mineral wealth. They hold one-quarter of the nation's coal, four-fifths of its huge oil shale deposits, one-half of its uranium deposits, and one-half of its estimated oil and gas reserves (including outer continental shelf deposits). Significant reserves of other strategic minerals needed for sophisticated technology and weaponry are also thought to lie on federal lands (Figure 21-6).

In addition to their mineral wealth, the federal lands are rich in renewable resources, particularly timber. Alaska and many of the western states have extensive forested areas which, if managed to yield a sustainable harvest, can provide numerous and varied benefits now and in the future. Grasslands are another renewable resource found on the public domain. Properly managed and maintained, the public ranges can support grazing by livestock. Care must be taken, however, to limit herd sizes so that they do not exceed the carrying capacity of an area. Semi-arid ranges, for example, can support fewer animals than can grasslands that receive more abundant rainfall.

The federal lands also hold one-half of the nation's naturally occurring steam and hot water pools, which can be used to generate electric power.

Alaska and the western states enjoy certain benefits from federal ownership. The government turns over 50 percent of its mineral leasing revenues to states and counties and 40 percent to federal reclamation funds that finance western water projects. The Forest Service gives the states 25 percent of its national forest revenues. Of the grazing fees collected by the Interior Department, 12.5 percent are given to state and local governments and 50 percent is spent on range improvement. The western states also collect large revenues by imposing mineral severance taxes on coal, oil and gas, and other resources.

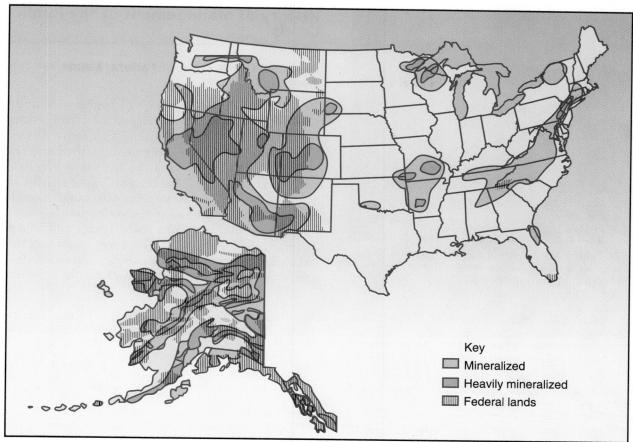

FIGURE 21-6: Mineral deposits on federal public lands. Oil, oil shale, coal, uranium, copper, cobalt, nickel, zinc, titanium, platinum, and molybdenum are some of the minerals found on the public domain.

Social Boundaries

We use the federal lands for many purposes. Each use presents a variety of environmental problems, from direct disruption of a natural ecosystem (as in logging) to indirect disruption (as in road building to permit access to, and activity in, an area).

The forests and BLM lands receive especially heavy commercial use, primarily logging, livestock grazing, and mining. These are **consumptive uses:** resources are consumed in the process. Commercial enterprises are also found in the national parks, primarily in the form of recreational concession contracts. Food, lodging, gift shops, and activities such as helicopter rides and river rafting trips are provided by private concerns that lease the privilege to operate within park boundaries. These concerns have a monopoly for the duration of their lease. Although some of the concessions are well managed, others are less than ideal. Complaints include expensive facilities and services and the poor condition of buildings and facilities. Also, some people feel that the concessions are too commercial, cheapening the park experience.

In addition to commercial uses, the federal lands offer many recreational, scientific, and wilderness opportunities. Recreational uses vary in their impact on the ecosystem. Some, including hunting and fishing, are consumptive. Others, such as bird-watching, photography, and nature study, are **nonconsumptive uses,** meaning that they do not result in the consumption or depletion of resources. Indeed, many people go to the federal lands solely for inspiration and solitude, seeking a respite from the urban environment and the pressures of everyday life. But even nonconsumptive uses can degrade an area if there are many users or if the users are not careful about how they disturb the land.

On all federal lands, conflicts arise among users. Recreational users vie for space and resources and argue over the correct use of the federal lands with ranchers, loggers, and miners. Conflicts even arise among recreational users, for example, hikers versus mountain bikers and off-road vehicle enthusiasts (Figure 21-7). At risk in all these conflicts is the integrity of the ecosystem. Once degraded, the "experience" and the benefits the ecosystem affords are altered. To understand the conflicts that arise, let us look at how the public domain has been and is being managed.

FIGURE 21-7: Mountain biker near Boulder, Colorado. Recreational activities like this have become a matter of heated debate among users of the public domain.

History of Management of the Public Lands

How Did the System of Federal Lands Develop?

By 1800 the United States had begun to acquire territories outside of the original 13 states, either through treaties (usually broken) with native American tribes or through agreements with European powers. For example, with the Louisiana Purchase of 1803 the United States bought France's rights to much of the territory west of the Mississippi, significantly expanding the size of the young nation.

Many people, including such figures as Thomas Jefferson, believed that the best way to protect the expanding nation against its enemies was to encourage settlement and development of the land, partic-

▶ Table 21-1
Important Dates in Federal Land Management

1832	Congress sets aside Hot Springs, Arkansas, as the first "national reservation."
1872	Congress designates Yellowstone in Wyoming, Montana, and Idaho, the country's first national park, as a "park and pleasuring ground."
1891	The first forest reserves are established and placed under the control of the Department of the Interior.
1892	John Muir founds the Sierra Club to promote the establishment of national parks and to protect existing parks.
1897	The Organic Act of 1897 gives the government the authority to protect and manage the forests.
1898	Gifford Pinchot is named the first chief forester of the Bureau of Forestry.
1905	Pinchot has the forest reserves transferred from Department of the Interior to the Department of Agriculture and reorganizes the bureau into the Forest Service, Department of Agriculture.
1903	President Theodore Roosevelt invokes the 1891 Forest Reserve Act to establish Pelican Island, Florida, as the first national wildlife refuge.
1906	The Antiquities Act gives the president the authority to set aside areas as national monuments.
1911	The Weeks Act grants the Forest Service the authority to acquire private timberlands.
1913	Congress authorizes the construction of a dam in Yosemite, California, which destroys the Hetch Hetchy Valley, an area similar in grandeur to the famed Yosemite Valley.
1916	The Organic Act of 1916 creates the National Park Service and National Park System.
1919	The National Park Association (later renamed the National Park and Conservation Association), a citizens' group, is established to promote the parks and their protection.
1933	President Franklin D. Roosevelt issues an executive order transferring historic military areas from the War Department and national monuments from the Agriculture Department to the National Park Service. The Civilian Conservation Corps (CCC) is established.
1934	The Taylor Grazing Act closes to indiscriminate settlement and use all unreserved and unappropriated public domain land in nine western states and the territory of Alaska. It also divides 142 million acres of rangeland into grazing districts and creates the Grazing Service to administer them.
1935	The Historic Sites Act establishes the nation's first comprehensive policy on historic preservation and sets up the designation "national historic sites." The Wilderness Society, a private, nonprofit organization, is founded by Robert Marshall, Benton MacKaye, Robert Sterling Yard, Aldo Leopold, and others to act as an advocate for wilderness preservation.
1937	Cape Hatteras National Seashore is established by the National Park Service; it would become the model for protected areas along the coasts and Great Lakes.
1946	The Reorganization Act merges the Grazing Service with the General Land Office to form the Bureau of Land Management (BLM) in the Department of the Interior.
1956	Mission 66, a $1-billion, 10-year project, is undertaken by the Park Service to construct and improve park facilities in order to accommodate increased visitation.
1958	Congress establishes the Outdoor Recreation Resources Review Commission (ORRRC) to conduct a comprehensive study of the nation's outdoor recreation needs.
1960	The Multiple Use and Sustained Yield Act redefines the purpose of the national forests to include recreation, soil, range, timber, watershed, wildlife, fishing, hunting, and mining on a "most judicious use" basis.

ularly the frontier. Consequently, the history of the federal lands is largely one of land disposal to private interests. The first patent conveying title to public lands was granted in 1788, and less than a quarter century later the General Land Office (GLO) was established to oversee the disposition of federal lands. Land disposal—to private citizens, businesses, and state and local governments—dominated federal policy for almost 150 years. The Homestead Act of 1862 granted free title to 160 acres (65 hectares) to anyone who settled the land and cultivated it for five years. Between 1860 and 1930 Congress gave more than 94 million acres (38 million hectares) to the railroads alone. Other lands were given away to build schools and roadways.

In 1907 vacant public lands amounted to approximately 400 million acres (162 million hectares); by 1932 that figure had been cut in half. In the mid-1930s President Franklin D. Roosevelt issued an executive order withdrawing all public lands within the lower 48 states from homesteading. He thereby effectively closed the public domain (outside Alaska) to major disposals, signaling a shift in policy from land disposal to retention of the remaining public lands. By this time the government had either given away or had sold 1.1 billion acres (445 million hectares).

Although the concept of land disposal had long dominated federal policy, the government had begun to set aside land units for special purposes as early as 1832. Over the next 150 years, numerous events, persons, and laws influenced the development of a system of federal lands (Table 21-1). How an area is managed and the uses permitted in the area depend on its designation as park, refuge, forest, or resource land.

1964	The Wilderness Act of 1964 provides a legal framework for wilderness preservation on federal lands.
	Congress enacts the Land and Water Conservation Fund Act, following a recommendation by the ORRRC.
1966	The National Wildlife Refuge System Administration Act brings all types of wildlife areas and game ranges under one authority, but the act fails to define the basic purpose of the system or to establish clear measures for protection of wildlife.
1968	Congress establishes the National Scenic Trails System and the National Wild and Scenic Rivers System.
1974	Congress adds 16 designated areas in the East to the National Wilderness Preservation System.
1976	The National Forest Management Act directs the Forest Service to develop 50-year unit-by-unit management plans for all Forest Service lands. Planning is to include economic, wildlife, wilderness, and recreational uses.
	The Federal Land Policy and Management Act establishes public land policy guidelines for administering, protecting, and developing all national resource lands managed by the Bureau of Land Management.
1978	Omnibus parks legislation authorizes 21 new National Park Service units, including the Santa Monica Mountains National Recreation Area (California) and Pinelands National Reserve, New Jersey.
	President Jimmy Carter doubles the size of the National Park Service by proclaiming 10 national monuments in Alaska and enlarging three existing units.
1979	Angry westerners demand that the government turn over most of the western lands to the states in which they lie. Known as the "Sagebrush Rebellion," the uproar did not last long, but it was a reminder of the ongoing tension between westerners and the federal government.
1980	The Alaska National Interest Lands Conservation Act (ANILCA) alters boundaries of the units proclaimed by Carter in 1978 and redesignates most units as national parks or national preserves.
	The Park Service issues *State of the Parks, 1980*, describing threats to park resources.
	The National Historic Preservation Act of 1966 is amended to allow leasing and reuse of historic structures managed by the Park Service.
1981	James Watt, a westerner and opponent of federal lands conservation programs, is appointed secretary of the interior by President Ronald Reagan. (As an attorney with the Mountain States Legal Foundation, based in Denver, Colorado, from 1977 to 1980, he challenged federal land policies in court, opposing stricter strip mining controls and suing to speed up applications for oil and mineral exploration on public lands.) Watt would resign under pressure in October 1983.
1982	Draft report by the Fish and Wildlife Service identifies 7,717 threats facing the National Wildlife Refuge System.
1984	Renovation of the historic sites, the Statue of Liberty and Ellis Island, begins with over $2 million in private funds.
1990	The Fish and Wildlife Service grants threatened status to the northern spotted owl, denizen and indicator species of the old growth forests of the Pacific Northwest.
1991	Despite President George Bush's campaign pledge of "no net loss" of wetlands, the administration redefines the degree of standing water necessary for such designation, potentially opening up 10 percent of 1 billion acres for development.

How Did the National Park System Develop?

While the concept of the "nation's parks" was immediately embraced by the American public, the early years of park management were plagued by problems. To begin with, there was no established park system, nor was there a standard method for adding units; parks were designated in haphazard fashion. The Antiquities Act of 1906 gave the president the authority to preserve as "national monuments" objects of cultural or historic significance, such as native American ruins, on public lands. President Theodore Roosevelt broadened the intent of the act in order to set aside spectacular natural areas such as Devil's Tower, Wyoming; the Petrified Forest and Montezuma's Castle, Arizona; and El Morro (Inscription Rock), New Mexico.

In the early twentieth century, the parks also suffered because of a lack of unified, organized management. Finally, the 1916 Organic Act established the National Park Service and the National Park System to protect and manage the nation's growing park areas. According to the act, the purpose of the service was to "conserve the scenery and the natural and historic objects and wildlife therein and provide for the enjoyment of the same in such manner and by such means as will leave them unimpaired for the enjoyment of future generations." In the years since the service's creation, many controversies have arisen as it struggles to make the parks accessible to citizens and at the same time preserve their natural character.

The National Park System's first director, Stephen Mather, convinced Congress to add numerous units to the system. Reasoning that the best way to protect the parks was to garner and solidify strong public support, he undertook several campaigns to encourage tourism. While Mather's strategy was successful, public enthusiasm for the growing park system had its drawbacks, and by the 1950s many units suffered the effects of too many visitors and too little funding to maintain roads and facilities. In 1955 the National Park Service instituted a 10-year expansion plan to improve roads, visitor centers, and overnight accommodations in order to increase public access and bring in more park visitors. It worked. Between 1955 and 1974 visitors to national parks and monuments more than tripled.

To meet the growing demand for outdoor recreation opportunities and at the same time to ease the pressure on existing parks and monuments, the Park Service began to establish national seashores and national recreation areas in the 1960s and 1970s. The service chose sites near urban areas to make it easy and convenient for people to use and enjoy the public lands. Battlefields, military sites, and other historic units were also added to the system, as were numerous monuments and open spaces in Washington, D.C. The seashores and recreation areas enabled people who had little hope of ever visiting the traditional parks to experience the public lands first-hand.

With the passage of the Alaska National Interest Lands Conservation Act (ANILCA) in 1980, more than 43 million acres (17.4 million hectares) were added to the park system.

What Environmental Problems Face the National Park System?

A 1980 National Park Service report entitled *State of the Parks* identified 4,345 threats to the parks. Some originate within park boundaries; others are external (Table 21-2).

Internal Threats to the National Park System. The ecological integrity of the national parks is threatened by various internal factors, including overuse, insufficient funding and park operations, threats to wildlife, commercial recreational operations, and energy and minerals development.

Overuse. Many national parks are being "loved to death" by the crowds who visit them annually. Crowded conditions exist in many of the parks, especially those close to urban areas, such as Shenandoah, Virginia, and the most famous parks, such as Yellowstone and Yosemite. The National Parks and Conservation Association (NPCA), a citizens' group and self-appointed watchdog of the National Park Service, reported that throughout the mid- to late-1980s park visits increased about 4 percent annually, roughly a few million more visits each year. Urban units, such as Golden Gate National Recreation Area near San Francisco, are under even greater pressure. Visits to these units are increasing at approximately 10 percent every year—more when gasoline prices are high. In 1990, there were an estimated 252 million recreational visits to park system units; by 2010, the number of visits is expected to rise to 500 million.

With the crowds come problems of overuse: heavily traveled trails become eroded, vegetation is trampled, and wildlife is crowded into an ever-shrinking space. The visitors also suffer. Too often crowded campsites, litter, noise, traffic jams, water pollution, and smog caused by automobile exhaust greet the visitor who comes to the park precisely to escape those conditions! Esthetic and intangible qualities—serenity, solitude for reflection, wilderness characteristics—are significantly diminished. Safety has also become a concern in recent years, as crime (especially vandalism) and drug use are becoming significant problems in some park units, particularly

▶ Table 21-2
Most Endangered National Parks

Everglades National Park, Florida. This park is famous for its special wildlife such as the American crocodile, Florida panther, and colorful bird species. Now the wildlife habitat and its vital water supply are threatened by urban development, water pollution, and military use.

Glacier National Park, Montana. Known for its nearly 50 glaciers, 200 lakes, numerous peaks and waterfalls, grizzly bear, bighorn sheep, moose, and at least 210 species of birds, this park is threatened by residential and commercial development on its borders, a wolf-kill program, oil development, and logging.

Grand Canyon National Park, Arizona. One of our most well-known parks, the Grand Canyon is threatened by 50,000 overflights each year by helicopters and small planes, the operation of the Glen Canyon Dam, which erodes beaches and harms fish, and uranium mining (test drilling has occurred within 200 yards of the park's boundary).

Great Smoky Mountains National Park, North Carolina and Tennessee. Our most visited national park, Great Smoky boasts 70 miles of the Appalachian Trail as well as the black bear, wild turkey, bobcat, 130 species of trees, and over 70 species of fish. Over the past 10 years, logging in the 6 national forests surrounding the park has more than doubled and logging roads have grown by 1,400 miles (2,254 kilometers). Further logging and road building threaten to destroy wildlife habitat and recreation areas.

Olympic National Park, Washington. Olympic contains glaciers, mountains, alpine meadowlands, and seacoast. It is home to the cougar, Roosevelt elk, black bear, Olympic marmot, sea lion, and whale. Mount Olympus has been designated a biosphere reserve and a world heritage site. The park is threatened by oil drilling off its coast and by logging and road building in the national forests that surround it.

Rocky Mountain National Park, Colorado. Located on the front range of the Rocky Mountains, this park features important areas of high alpine tundra and is home to the golden eagle, river otter, deer, and bighorn sheep. It is threatened by residential development (condominiums have been built within a few feet of the park's wilderness area), an expanding logging program, and overcrowding.

Santa Monica Mountains National Recreation Area, California. Stretching along 46 miles (74 kilometers) of Pacific coastline, this popular recreation area is threatened by the development of blocks of privately owned land. Congress has authorized acquisition of this land, but the administration has not yet purchased it.

Yellowstone National Park, Wyoming, Montana, and Idaho. Long treasured for its natural geysers, three mountain ranges, and diverse wildlife, including the grizzly, bald eagle, and trumpeter swan, Yellowstone is threatened by a planned 20 percent increase in logging in the national forests on its borders. Geothermal development and oil and gas drilling planned for the lands surrounding Yellowstone could even threaten its renowned geyser system.

Yosemite National Park, California. Famed for its granite peaks and domes, giant sequoia, and the country's highest waterfall, Yosemite is threatened by congestion within the park and by logging and vacation home development outside its borders.

Source: Adapted from The Wilderness Society.

those near urban areas. (Environmental Science in Action: Yosemite National Park, pages 468–471, explores more fully the problems caused by overuse.)

Insufficient Funding and Park Operations. Like many government agencies, the National Park Service often seems to operate on too little money for the job it is required to do. Combating problems caused by overuse, for example, is expensive. According to a 1986 report by the government's Council on Environmental Quality, the Park Service reported a conservative estimate for litter cleanup of $15 million, roughly 3 percent of its park operations budget. Park and recreation agencies spend over $500 million annually to prevent and repair the damage caused by vandals. The ramifications of inadequate funding became all too apparent during the Reagan administration (1981–1988) when budget constraints resulted in reductions in permanent park personnel at

many sites and deteriorating conditions at nearly all the facilities. In 1990, the NPCA released a report entitled *A Race Against Time* in which it identified the backlog of repair, maintenance, preservation, and safety projects as a major threat to the park system. According to the report, more than $2 billion is needed to undertake needed projects throughout the park system.

Insufficient funding has other, less obvious adverse effects. As fewer park rangers must handle increasing visitor numbers and criminal activity, rangers in some units spend much of their time on crowd control and law enforcement. Many people fear that the old time ranger—part naturalist, part teacher, part outdoorsperson—is giving way to a new breed of ranger, a sort of park police officer. To its credit, the Park Service has recognized this problem and has developed a training program in natural resources management for employees. Twenty trainees from throughout the park system partici-

Focus On:
The Greater Yellowstone Ecosystem Fires of 1988

Beginning in 1972 the Park Service adopted a policy to allow some naturally occurring fires (caused by lightning) to burn themselves out. Each natural wildfire is assessed separately and a decision is made as to whether or not to let it burn; those allowed to run their course are continuously monitored. The policy does not apply to human-caused fires and natural fires that threaten adjacent public or private lands, gateway communities, villages within the park, and significant resources within the park.

This type of fire management policy is an acknowledgment of the natural role of fire in maintaining the variety of habitats, vegetation types, and species typical of a healthy wilderness. In the Yellowstone forests, by reducing the amount of lodgepole pine, the dominant species, fire encourages new growth such as wildflowers, grasses, and shrubs and is thus beneficial to the forest ecosystem.

To many observers, however, the fires that spread through the Greater Yellowstone Ecosystem (GYE) during the summer of 1988 appeared to be anything but beneficial. After a wet spring, with rainfall well above normal for April and May, the summer was exceptionally dry, with virtually no precipitation during June, July, or August. Many natural wildfires started early in the summer, and about 20 were allowed to burn; of these, 11 burned themselves out. Those that survived into the dry weeks of mid-summer, however, met dramatically changed fire conditions. A series of extremely high winds associated with dry fronts spread the fires quickly. The low moisture content in leaves and grasses enabled many more lightning-caused fires to start. Because the unusual weather conditions increased the possibility that

fires would burn out of control, after July 21 the Park Service no longer allowed any natural fires to burn and fought all fires to the best of its ability. The various agencies within the GYE (Park Service, Forest Service, Fish and Wildlife Service, and others) joined together in their efforts to suppress fires in the region. Their task was difficult for two reasons: the large amount of downed trees and underbrush that had accumulated during the years when the Park Service fought all fires, and the unusual weather conditions. Typically, higher nighttime humidity, coupled with dying winds, encourages wildfires to "lie down" at night. In this condition, they can be controlled more easily. But low humidity and strong winds made the fires dangerous to tackle even after dusk. Consequently, fire-fighting efforts were directed at controlling the flanks of fires and protecting lives and property in the advancing paths of the fires.

The Yellowstone fires became a media event. Graphic photos of smoke rising from burning forest and often inaccurate reporting incited a public outcry. Three inaccuracies were especially unfortunate: (1) all fires in the Yellowstone ecosystem area were attributed to the Park Service's natural fire program in the park; (2) all fires in the ecosystem area (which includes more than 10 million acres of public land) were frequently referred to as the "Yellowstone *Park* fires"; and (3) burn acreages were oversimplified and exaggerated. The figures usually reported were perimeter figures—that is, the total area in which fires occurred—and did not reflect the amount of acreage within the perimeter that had actually burned.

Some of the most damaging fires began after the Park Service sus-

pended its natural fire program in mid-July, a fact that underscores what is probably the most serious misconception about firefighting efforts: that humans can always control fire if they so choose. Moreover, several major fires started outside the park and moved in, accounting for over half of the total burn in the ecosystem area. The north fork fire, which threatened Old Faithful, West Yellowstone, Mammoth Hot Springs, and Tower-Roosevelt Lodge, was probably started by a woodcutter's cigarette in Targhee National Forest. The north fork fire was fought from the day it began, but with little success.

Undoubtedly, the Greater Yellowstone Ecosystem has changed considerably as a result of the fires, but those changes are not necessarily bad. Habitat and species diversity has increased; spring wildflowers, for example, now thrive in sunny areas once darkened by stands of mature lodgepole pines. Elk, bison, and deer should benefit from improved forage quantity and quality in the next few years. If the Greater Yellowstone fires of 1988 serve as a reminder that humans cannot control nature at will, perhaps they will also eventually show the wisdom in working with and accepting natural occurrences. A park publication stated the case quite clearly: "the fires did nothing to Yellowstone that has not been done many times in the past. . . . One of the greatest challenges offered by national parks is a conceptual one: they compel us to take the long view, and consider nature's direction rather than our own. We are not protecting the parks merely for ourselves, but for many later generations, who will witness the revegetation of Yellowstone with an interest and excitement hard to appreciate through the smoke of 1988."

pate in the year-long program; they are exposed to a wide array of resource problems and issues, including air and water quality, integrated pest management, vegetation management, wildlife, minerals and mining, cultural resources management, natural resources law, and fire ecology. They also receive

hands-on training in various park units and complete a research project on a specific management issue.

Because the national parks are the nation's most highly visible public lands, they tend to be the focus of intense scrutiny. Management strategies adopted by the Park Service, even when they are environmentally sound, can prove controversial. No situation better illustrates this than the 1988 fires that raged through the Yellowstone ecosystem, which encompasses some 13 million acres or 5.25 million hectares (see Focus On: The Greater Yellowstone Ecosystem Fires of 1988).

Threats to Wildlife. The wildlife within the national parks is threatened by diminishing or insufficient habitat, the elimination of predators, and poaching. Some of the parks are too small to maintain viable populations of large mammals. A study conducted in the mid-1980s found that of 14 western parks, all but one had lost populations of such species as grizzly bear, wolf, lynx, gray fox, bighorn sheep, jackrabbit, otter, mink, and pronghorn antelope. The affected parks included Bryce Canyon, Sequoia–Kings Canyon, Yosemite, Zion, and Teton-Yellowstone.

The elimination of predators in national parks can wreak havoc within the ecosystem. Without their natural enemies, game such as elk, deer, bison, and moose become too plentiful and overbrowse the vegetation of the area, which can no longer support the increasing population. Many of the animals weaken and starve. By the early 1980s the Park Service, at the urging of many wildlife biologists and conservation groups, drew up a plan for reintroducing gray wolves to Yellowstone. Local ranchers, however, opposed the plan, fearing that the wolves would prey on their livestock. Throughout the 1980s, the two sides remained at a stalemate. In 1991 Congress approved a bill supporting wolf reintroduction to Yellowstone and several other areas in the lower 48 states and provided the U.S. Fish and Wildlife Service with funds to begin planning for the animal's restoration in the nation's oldest park.

With valuable game species and rare plants rapidly disappearing from private lands, poachers have set their sights on the national parks. Park rangers in Alaska consider trophy hunting to be the chief threat to wildlife in those units, and poaching is cited by managers at many parks nationwide as a serious and growing threat. The motive is money. A mounted, record-size bighorn sheep head sells for about $50,000. Prime elk horn, valued as an aphrodisiac, can fetch $400 an ounce.

It is impossible to determine how much poaching goes on, but officials believe that it is quite extensive, particularly in some parks and for some species. Targets of poachers include black bear in the Great Smoky Mountains National Park; grizzly bear, elk, and golden eagle in Yellowstone; brown bear, wolf, and moose in Alaska's Katmai and Denali; wild ginseng, a medicinal herb, in Shenandoah; and cactus, rare snakes, and rare lizards in Saguaro and Organ Pipe National Monuments (Figure 21-8). The

FIGURE 21-8: Conservation officers arrest a man for poaching white-tailed deer in Adirondack State Park, New York. Poachers pose a serious threat to wildlife in the United States, from cacti to black bear, ginseng to wolves.

species most vulnerable to poaching are those, like the grizzly bear, whose populations are already small because of habitat destruction and other causes. But heavy poaching can threaten even stable populations, since poachers take the best and fittest specimens (rather than the young, old, or infirm), leaving younger and weaker individuals to propagate the species.

Concessions System. In recent years, the concessions system has increasingly come under fire. Concessionaires are private businesses licensed to sell goods and services such as food and hotel space in the national parks. Facilities in some park units are not well maintained, and park visitors often complain about the expense of housing, food, gift items, and recreational opportunities. Another common complaint is the overcommercialization prevalent in many parks. The circus atmosphere and tacky goods offered by some concessionaires diminish the esthetic quality of the area. Finally, critics maintain that the concessionaires are not required to return a high enough proportion of their profits to the federal government. Under the present system, concessionaires return an average of just 2.5 percent of their gross receipts to the government. The money that is returned is put into a general fund for use in other programs; it is not earmarked specifically for park maintenance and restoration. Concessions contracts sometimes run for as long as 30 years and allow little room for renegotiation or competition. The result, critics argue, is that concessionaires enjoy a virtual monopoly in the national parks, realizing tremendous profits but failing to share in the burden of maintaining and preserving these areas.

In some park units commercial recreational operations are a serious threat to the area's natural and cultural features and the quality of the visitor experience. For example, helicopter flights in the Grand Canyon have become so numerous that they pose safety problems. A tragic mid-air crash in 1986 killed 25 people and resulted in tighter controls on overflights. Still, many feel that the restrictions are not tight enough. Some hikers complain that the noise and the presence of the hovering craft are disruptive intrusions on the serenity and solitude of the park.

Energy and Minerals Development. Some units of the National Park Service face another threat: the development of subsurface oil and mineral deposits. Although the Mining in the Parks Act, passed in 1976, closed most national parks and monuments to new mineral claims, it upheld claims that were filed prior to that time, as long as the claim is developed in accordance with federal regulations. The rights to subsurface oil and mineral deposits within approximately 100 units are owned or claimed by parties other than the Park Service. (A situation in which the government, or Park Service, owns the land, but private citizens or companies own the minerals beneath its surface is known as a split estate.) For example, approximately 247 mining claims exist within Great Basin National Park, Nevada. Although none is currently being developed due to low market prices, any one of these could be developed pending approval of a plan of operations.

External Threats to the National Park System. Not all threats to the national parks originate within park boundaries. Some of the most serious threats are external, including long-distance atmospheric pollution and activities on adjacent lands.

Atmospheric Pollution. Poor air quality is a serious problem in some national parks (Figure 21-9). Shenandoah, just 65 miles (105 kilometers) west of Washington, D.C., has the worst air quality of any national park, ranking first in terms of sulfates and second in terms of ozone. Smog obscures the spectacular views that built the park's fame, and acidic deposition is gradually poisoning streams and vegetation.

Few parks are immune from the ravages of air pollution; winds carry pollutants from urban areas, smelters, factories, and coal-fired power plants to even the most remote parks. The Golden Circle of the Southwest is a region encompassing five national parks (Grand Canyon, Bryce Canyon, Canyonlands, Capitol Reef, and Arches) and two national recreational areas (Lake Mead and Glen Canyon). A decline in visibility within the region from 60 to 40 miles (97 to 65 kilometers) is due mainly to the number of power plants in the area, including the

FIGURE 21-9: National parks threatened by air pollution.

Navajo Generating Station near Page, Arizona, located just 12 miles (19 kilometers) from Grand Canyon National Park, and the Four Corners Power Plant at Farmington, New Mexico. The Four Corners plant spews about 80,000 tons (72,800 metric tons) of sulfur dioxide into the Colorado plateau area each year. Relief may be in sight for the region, however, particularly for the Grand Canyon. In 1991 an agreement was reached between environmental groups and the owners of the Navajo Generating Station. According to the terms of the agreement, sulfur dioxide emissions from the coal-fired power plant will be cut by 90 percent before August 1999. Beginning in 1997 smokestack scrubbing equipment will be installed at the plant at an estimated cost of $1.8 billion. This landmark agreement was reached under a provision of the 1990 Clean Air Act designed to improve air quality in national parks.

Activities on Adjacent Lands. Mining, lumbering, and other activities conducted on federal and private adjacent lands can adversely affect the ecological integrity of park lands. In Yellowstone exploration and drilling for oil and gas threaten to disrupt the thermal activity of the geysers and hot springs; offshore drilling threatens Washington's Olympic National Park. Road building in the national forests, which surround many parks, threatens to disrupt the wildlife that cross the park and forest borders. Residential and commercial development is encroaching on historic areas like the Antietam National Battlefield in Maryland and the Manassas National Battlefield in Virginia, the sites of significant Civil War battles (Figure 21-10).

How Did the National Wildlife Refuge System Develop?

The National Wildlife Refuge System had a rather rocky start. In 1903 President Theodore Roosevelt invoked the 1891 Forest Reserve Act to establish

FIGURE 21-10: The cannons now lie silent in Manassas National Battlefield Park, Virginia. In the late 1980s a developer announced plans to construct a major shopping mall adjacent to the Manassas National Battlefield in Virginia. The developer's plans would have destroyed the site of General Lee's battle headquarters, then outside park boundaries. The public reacted with outrage at the plans. Finally, after months of controversy, President Reagan signed legislation in the autumn of 1988 to block construction of the mall.

Pelican Island, Florida, as the first national wildlife refuge. Roosevelt contended that the island's mangrove thickets qualified as forest. Later, he used provisions of the 1906 American Antiquities Act to protect other areas as wildlife refuges, including the National Bison Range in Montana. By the time he left office in 1908, Roosevelt had designated 51 national wildlife refuges.

Although the Bureau of Biological Survey had been tapped to care for the refuges, the sanctuaries received no congressional funding. Consequently, Audubon Society members often acted as refuge caretakers. Also, although Congress had approved the acquisition of land for new refuges under the Migratory Bird Conservation Act of 1929, it had failed to provide funding. In 1934 President Franklin Roosevelt named Aldo Leopold, J. N. (Ding) Darling, and other prominent conservationists to a committee to seek funding for the refuges. The committee introduced the concept of the duck stamp, a $1 stamp hunters were required to purchase each year. Stamp proceeds were used to buy land for new refuges. In 1935, 55 new refuges were created, including the Upper Souris Refuge in North Dakota, a favorite nesting place for the great blue heron and the black-crowned night heron.

In the 1940s F.D.R. merged the Bureau of Biological Survey (Department of Agriculture) with the Bureau of Fisheries (Department of Commerce) to form the Fish and Wildlife Service under the auspices of the Department of the Interior. The number of refuges grew steadily over the next several decades, and in 1966 Congress formally established the National Wildlife Refuge System. The Fish and Wildlife Service's mission was to "provide, preserve, restore, and manage a natural network of lands and waters sufficient in size, diversity, and location to meet society's needs for areas where the widest spectrum of benefits associated with wildlife and wild lands is enhanced and made available."

What Environmental Problems Face the National Wildlife Refuge System?

A 1982 draft report by the Fish and Wildlife Service identified 7,717 threats to the National Wildlife Refuge System (Table 21-3). As with the National Park System, these threats originate both internally and externally.

Internal Threats to the National Wildlife Refuge System. The management structure of the Fish and Wildlife Service and secondary uses allowed on refuges are the most serious internal threats to the National Wildlife Refuge System.

Management Structure. The refuge system is a victim of problems arising from the internal structure of the agency responsible for its management. The Fish and Wildlife Service is plagued by too many competing responsibilities. For instance, in addition to the care and maintenance of the refuges, it is responsible for administering the Endangered Species Act. It is also responsible for operating the nation's fish hatcheries and other tasks geared to commercial interests. The various functions all require significant funding, and the service departments that carry them out must compete for scarce money. Also, economic interests sometimes seem to take preference over measures designed to manage the refuges as sanctuaries for all wildlife, plants as well as animals and nongame as well as game species.

Secondary Uses. The primary purpose of the refuge system—to provide habitat and a haven for wildlife—is often undermined by secondary uses. According to the Refuge Recreation Act of 1962, any secondary or recreational use proposed for a refuge is supposed to be compatible with the system's primary purpose. The secretary of the interior has the authority to decide whether or not a use is compatible. Many conservationists maintain that the Interior Department is too willing to permit harmful secondary activities. A handful of refuges and controversial secondary uses are: Chincoteague, Virginia, where increasing numbers of swimmers and bathers stress nesting populations of waterfowl and shorebirds, such as the endangered piping plover; National Key Deer, Florida, where cars kill as many as 20 percent of this endangered species within

▶ Table 21-3
Most Endangered National Wildlife Refuges

Kesterson National Wildlife Refuge, California. Selenium poisoning from irrigation runoff has killed thousands of ducks and other waterfowl.

National Key Deer Wildlife Refuge, Florida. As many as 20 percent of the deer in this western Florida Keys refuge are killed by cars each year.

Loxahatchee National Wildlife Refuge, Florida. Nearby development has drastically changed water flow patterns in this Everglades region, bringing periods of both excessive and insufficient flow. The water is contaminated by heavy fertilizer use.

Stillwater National Wildlife Refuge, Nevada. It has become a chemical dumping ground for the Newlands irrigation project. Mercury in fish has been found at up to four times the amount permitted for human consumption.

Chincoteague National Wildlife Refuge, Virginia. A growing visitor population has put pressure on wildlife nesting.

Yazoo National Wildlife Refuge, Mississippi. Agricultural runoff is carrying high concentrations of pesticides into the refuge.

Arctic National Wildlife Refuge, Alaska. Oil companies and the Department of Interior want to explore for oil near the coast of the Beaufort Sea, which Congress has not permitted.

Lower Rio Grande National Wildlife Refuge, Texas. Nearby pesticide spraying and runoff, wastewater discharges, and planned real estate development are the major threats.

Upper Mississippi River National Wildlife Refuge, Wisconsin, Iowa, Minnesota, and Illinois. Soil erosion is damaging habitat and reproduction. Wastewater and pesticide-loaded agricultural runoff is polluting refuge waters, and heavy recreational use is hurting habitat.

Great Swamp National Wildlife Refuge, New Jersey. Population growth 26 miles west of Manhattan is straining two sewage treatment plants whose outflow eventually reaches the refuge. Road salt, fertilizer, and pesticides are reaching the refuge. Old toxic dumps threaten to contaminate the water. Water flow is increasingly variable.

Note: According to a report by the Fish and Wildlife Service, threats to the refuge system "will continue to degrade certain fish and wildlife resources until such time as mitigation measures are implemented. In some cases, this degradation or loss of resources is irreversible. It represents a sacrifice by a public that, for the most part, is unaware that such a price is being paid."
Source: Adapted from The Wilderness Society.

refuge boundaries each year; D'Arbonne, Louisiana, where the development of a natural gas field beneath the refuge is contaminating soil and water resources and may destroy foraging habitat for the endangered red-cockaded woodpecker; Cabeza Prieta, Arizona, where military air exercises conducted at the Barry M. Goldwater Air Force Range (which overlies much of the refuge) pose a threat to the endangered Sonoran pronghorn antelope.

External Threats to the National Wildlife Refuge System. Severe threats to the integrity of the wildlife refuges also arise from external sources. Chief among these are activities on adjacent lands and political pressures.

Activities on Adjacent Lands. Irrigation runoff from nearby farmlands, encroaching development, wastewater discharges, soil erosion, and contamination from old toxic waste dumps are just a few of the problems with which refuge managers must contend (Figure 21-11).

Political Pressures. Like all federal lands, the nation's wildlife refuges are often held hostage to political

pressures. In the 1980s conflicts between preservationists and developers arose on many units of the refuge system. The most hotly debated battle of recent years concerns the fate of the coastal plain of the Arctic National Wildlife Refuge, known as ANWR. (Chapter 11, Energy: Issues, contains a discussion of the political battle over oil development in the refuge.)

How Did the National Forest System Develop?

The first forest reserves were established in 1891 and placed under the control of the Department of the Interior. The Organic Act of 1897 gave the government the authority to protect and manage the forests. One year later President Theodore Roosevelt named Gifford Pinchot the first chief forester of the Bureau of Forestry. In 1905 Pinchot managed to have the forest reserves transferred from Interior to Agriculture. He then reorganized the bureau into the Forest Service.

Roosevelt was the first president to show a serious commitment to conservation. His friendship with both Gifford Pinchot and the preservationist John

FIGURE 21-11: Devil's Kitchen Lake, Crabtree Orchard National Wildlife Refuge, Illinois. The Crabtree Refuge has been placed on the EPA's Superfund list of toxic waste sites scheduled for cleanup.

Muir influenced his thinking. It was Pinchot's philosophy of **utilitarianism,** or conservation for economic reasons, that dominated official policy. Trained in Europe, Pinchot was committed to the scientific management of forests and the goal of sustained harvests. He set out the conservation principles, particularly the principle of multiple use, that would guide the Forest Service for decades. **Multiple use** is the management of land to accommodate a number of uses, from commercial to recreational. Pinchot also imposed grazing regulations and fees on ranchers, who had previously used the land free of charge.

Westerners were hostile to the conservation concerns of the Roosevelt administration, and they vented their hostility through their elected officials. In 1907 Congress passed legislation to prohibit new forest reserves in six western states without congressional approval. Unwilling to veto the bill, Roosevelt directed Pinchot to quickly draw up plans for the addition of new western forests. The Forest Service responded with plans to add 16 million acres (6.4 million hectares) in new reserves, enlarging the National Forest System to about 140 million acres (57 million hectares). Roosevelt approved the additions and then signed the bill which effectively prohibited, from then on, any president from doing what he had just done!

Easterners soon began to agitate for forests to be set aside in their region largely to reverse deforestation and to protect against flash floods. Little public domain remained east of the Mississippi, but with the passage of the Weeks Act of 1911, Congress granted the Forest Service the authority and the funds to acquire private timberlands. New Hampshire's White Mountain National Forest became the first national forest in the eastern United States. In the 1920s and 1930s many more eastern forests were added to the system, as landowners hurt by the Depression were anxious to sell their lands for much needed cash. The Forest Service also acquired grasslands from homesteaders who abandoned the damaged land during the Dust Bowl days.

The Forest Service's difficulties with western economic interests again came to the forefront after World War II. In 1946 a handful of ranchers launched a serious attempt to dismantle the Forest System. They met in Salt Lake City, Utah, and devised a plan to have the states take possession of the national forests as well as 142 million acres (57.5 million hectares) of grazing lands under the control of the Forest Service. According to the plan, supported by western politicians, a select group of ranchers would eventually buy the lands at 9 cents per acre. When the historian and social critic Bernard DeVoto learned of the plan and described it in *Harper's Magazine*, the ensuing public outcry spoiled the ranchers' plan.

About the same time that western forces were attempting to dismantle the forest system, the management philosophy of the Forest Service underwent a dramatic change. At its inception and until World War II, the Forest Service was basically a caretaker agency. Government foresters acted as custodians of the national forests and taught wise management practices to private foresters. The postwar boom in the U.S. housing industry and the need to rebuild war-torn Europe created a skyrocketing demand for lumber. When private forests could no longer meet that demand, timber executives cast a covetous eye on the national forests, and the Forest Service began to seek out the "business" of the timber industry. Harvesting in the nation's forests increased.

What Environmental Problems Face the National Forest System?

The rapid and dramatic change in Forest Service management philosophy after World War II is directly related to the factors (all of them internal) that pose the most serious threats to the integrity of the national forests today. These include an emphasis on logging over all other uses, clear-cutting, below-cost timber sales, and wilderness designation.

Emphasis on Logging Over Other Uses. By 1950 the Forest Service had begun to place less emphasis on its custodial role and greater emphasis on encouraging economic uses, particularly logging. The biggest obstacle to the Forest Service's plans to increase logging in the national forests was inadequate access—the lack of roads and the poor condition of existing roads. After much persuasion, the agency finally won congressional appropriations for road building. Subsequently, the timber cut in the national forests more than doubled between 1950 and 1969.

By the end of the 1950s it was clear that the Forest Service mandate of "multiple use" had been virtually ignored in the flurry of road building and logging. Congress passed the 1960 Multiple Use and Sustained Yield Act to define multiple use in the national forests. For the first time five land uses were designated as equally important: outdoor recreation (including wilderness), range, timber, watershed, and wildlife and fish habitat. On the surface the act appeared to have ensured the principle of true multiple use. Unfortunately, it did not work that way in many of the national forests. Industry and foresters, eager to continue in their quest to "mine" the forests, did not give up easily. Ralph Hodges, spokesperson for the lumber industry, complained that "by making all uses equal in priority, the forest manager will probably act on the basis of public pressure. This doesn't give protection to the lumber industry." Hodges's fears did not materialize, and the act did not halt the agency's bias toward timber and logging. As recently as 1986 the Forest Service's proposed annual budget called for $600 million for resource development and exploitation (including timber sales, minerals, and grazing) and a mere $170 million for stewardship programs (soil and water, fish and wildlife, recreation, trails, and land acquisition).

One budget item in particular underscores the agency's emphasis on timber production. In 1989 the Forest Service committed about $175 million to road construction, despite the fact that its road system, at 365,000 miles (587,650 kilometers), is already 8 times longer than the nation's interstate highway system. Roads benefit only logging; they do not benefit other forest uses such as recreation, watershed, and fish and wildlife habitat. Roads fragment habitat and accelerate erosion, particularly on steep slopes where the water is concentrated by the roads so that it carves gullies into the hillside and may cause landslides.

Clear-cutting. As the Forest Service became intent on increasing the harvest from the national forests, it began to change its methods of operation. Clear-cutting, once shunned by the agency, became a favored harvesting technique. **Clear-cutting** is simply cutting everything in an area regardless of size, age, and species. The consequences of clear-cutting are many and severe. It accelerates erosion and water runoff, which saps nutrients from the soil. Without the shading provided by trees growing on the banks,

water temperatures in streams rise. Salmon, trout, and other temperature-sensitive species are adversely affected by the warmer temperatures, as is the sportfishing industry, which depends on healthy fish populations. Streams suffer siltation, particularly after heavy rains or during spring thaws, as soil runs unimpeded off the land. Silt in the streams makes it more difficult for downstream municipal water treatment plants to purify drinking water; silt in reservoirs means they must be dredged more frequently. Clear-cutting diminishes the esthetic appeal of an area and potential recreational use and so has an undeniable adverse effect on local tourism.

In 1969 clear-cutting gleaned 61 percent of harvests from western forests and 50 percent of those from eastern forests (Figure 21-12). Why did the Forest Service come to rely so heavily on a technique it once decried as ecologically unsound? Clear-cutting is faster and cheaper than selective cutting. It makes it easier to contain damage in stands that have been infested with insects. After an area is clear-cut, it is easier to raise trees of commercial value, such as the Douglas fir, which require broad, open light as seedlings. Finally, clear-cutting prepares the ground for even-aged stands of trees, which can later be harvested in another clear cut. These stands—little more than tree farms—are a poor substitute for a forest.

The 1976 National Forest Management Act limited the application of clear-cutting and instructed the Forest Service to maintain species diversity, rather than simply to maximize the growth of commercially valuable species. It also reiterated the need for sustained-yield harvests, mandating that the sale of timber from each forest had to be limited to a

FIGURE 21-12: Clear cut in Olympic National Forest, Washington. Some of the old-growth forest of the Pacific Northwest is protected in national parks and wilderness areas, but much of it is found in 12 national forests where it is subject to logging.

quantity equal to or less than that which the forest could replace on a sustained-yield basis, provided that all multiple use objectives were met.

In recent years clear-cutting policies in the eastern national forests have been modified to reduce the environmental impacts of this technique; esthetic concerns still remain, however. Clear-cutting is especially controversial in the old-growth forests of the Pacific Northwest. These forests, consisting mainly of cedar, Douglas fir, Sitka spruce, and hemlocks 200 to many hundreds of years old, are one of our country's last substantial stretches of virgin forest (Figure 21-13). To the dismay of environmentalists, scientists, and many residents of Oregon and Washington, the logging industry is clear-cutting more and more old-growth in those states every day. Old-growth forests are unique ecosystems; many of the species that live there could not survive elsewhere, including in forests composed of younger trees. If the old-growth forests disappear, an untold wealth of diversity and information will die with them.

Below-cost Timber Sales. Since its inception, the Forest Service had always lost money on timber sales. Until the middle of the century the amount of money lost was minor, for several reasons. First, the service didn't sell much timber; just about 2 percent of the annual harvest came from the national forests. Second, the service simply didn't offer timber sales that were economically or ecologically unsound. Third, because the service was not trying to push timber and to boost sales, it did not spend much money on road construction and improvement; in other words, its costs were relatively low. But after World War II the service began to actively seek the business of timber companies. By 1992, 13 percent of the annual timber harvest in the United States came from the national forests, a six-fold increase over pre–World War II figures. The costs of doing business mounted with the increase in logging. Building and maintaining roads in areas which were often mountainous was costly. Consequently, the Forest Service began to offer **below-cost timber** sales, that is, the price paid by the loggers for the timber is lower than the costs incurred by the service to make the timber available. According to the Wilderness Society, below-cost timber sales resulted in average annual losses by the Forest Service (and ultimately, by taxpayers) of $327 million between 1979 and 1990. Ninety-eight of 120 forests lost money in 1990 alone.

No forest better illustrates the disastrous effects and costliness of below-cost timber sales than the Tongass National Forest in Alaska (Figure 21-14). The 16.8-million-acre (6.8-million-hectare) Tongass, our largest national forest, blankets most of Alaska's

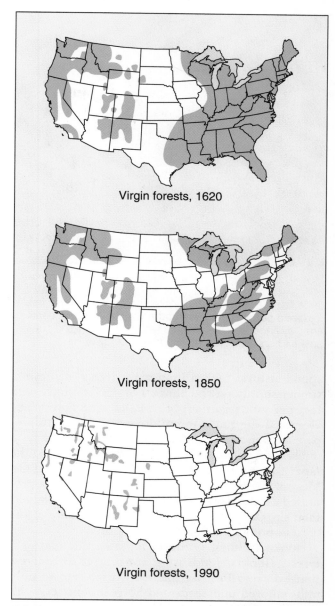

FIGURE 21-13: Virgin forests in the contiguous United States, 1620, 1850, 1990. Approximately 95 percent of the virgin forests have been cleared since 1620. According to the Wilderness Society, there may be as little as 2 million acres of old growth left in Oregon and Washington, just 10 percent of the virgin forests that once covered these two states. The Forest Service and the logging companies argue that opponents of clear-cutting old growth substantially underestimate the area of virgin forest left standing.

southeastern panhandle. Old-growth forest (300- to 800-year-old trees, 150 to 200 feet tall, and 5 to 10 feet in diameter), the largest concentration of grizzly bears and bald eagles on the planet, and five species of Pacific salmon who come to spawn, to name just a few, all find refuge in the Tongass. Beginning in 1980, however, their refuge was threatened. When the 1980 Alaska National Interest Lands Conservation Act (ANILCA) was passed, a provision of the bill required the government to

FIGURE 21-14: Tongass National Forest, Alaska. Receiving over 8 feet of rain annually, the Tongass contains the last large tracts of North America's temperate coastal rain forest. Icy island peaks and frozen tundra rise above valleys darkened by ancient stands of Sitka spruce and hemlock. The area harbors a spectacular and diverse array of life.

spend at least $40 million annually to maintain a timber supply on the Tongass of 450 million board feet per year *regardless of demand*. The Forest Service used the funds to build new roads and to prepare timber sales for two southeastern Alaska pulp mills, the Louisiana Pacific Ketchikan (LPK) and the Japanese-owned Alaska Pulp Corporation (APC). Most of the timber taken from the Tongass was processed into dissolving pulp, a wood-fiber product, then shipped to Pacific rim countries, where it was used to produce cellophane and rayon.

Unlike funding requests for all other national forests, Tongass funds were not subject to congressional review. To make matters worse, the two pulp mills enjoyed unprecedented 50-year contracts with the Forest Service. Under the terms of those contracts, the service had to sell two-thirds of the Tongass's annual timber supply to the mills at bargain-basement prices. For example, 1,000 feet of Sitka spruce cost the pulp mills about $2.55. Independent timber operators, who had to engage in competitive bidding, might spend as much as $175 for the same timber! Between 1982 and 1986 the government spent $287 million on Tongass timbering, but collected just $32 million in timber sales. The average annual loss was $50 million.

Throughout the 1980s the Forest Service's management of the Tongass was opposed by the region's major commercial fishing groups, subsistence users, hunting groups, tour operators, 16 of the 25 communities in southeast Alaska, local conservation groups, and diverse national groups, including the Audubon Society, Wilderness Society, National Wildlife Federation, Sierra Club, American Forestry Association, and the National Taxpayers Union. Finally, in 1990, despite opposition from Alaska's congressional delegation, the Tongass Timber Reform Act was passed. The act eliminated the $40 million annual appropriation and the mandated 450 million board feet supply and required that the contracts with the two pulp mills be renegotiated to eliminate the sale of timber at below-market value. It also added key fish and wildlife areas to existing Tongass wilderness.

Wilderness Designation. With the Wilderness Act of 1964, Congress instructed the Forest Service to review its lands and recommend any of those suitable for inclusion in the National Wilderness Preservation System. Official wilderness areas receive greater protection than other federal lands: commercial activities, road building, and construction of permanent structures are prohibited. Since the 1960s the Forest Service has conducted several inventories of its holdings and designated numerous wilderness areas. Critics, however, point out that by 1990 some 50 million acres (20 million hectares) of roadless areas within national forests in the lower 48 states lacked legal protection as wilderness. They argue that the agency's bias toward logging threatens the pristine character of these areas; logging roads would effectively exclude an area from inclusion in the National Wilderness Preservation System. (The intricacies of the Wilderness Preservation Act and the designation process are discussed in Chapter 22.)

How Did the Network of National Resource Lands Develop?

By the 1930s the majority of the federal lands remained unreserved; they did not belong to the National Park System, the National Wildlife Refuge System, or the National Forest System. They were designated the public domain. These lands were severely degraded by overgrazing and other deleterious practices.

Only after the Dust Bowl and its devastating effect on ranchers' livelihood did they recognize the need to limit grazing on the federal lands. In 1934 the Taylor Grazing Act authorized Interior to set up grazing districts on 142 million acres (53 million hectares) of public range. It also established a federal grazing service to lease rangelands to established ranchers. In return for this authority, Interior Secretary Ickes promised the livestock industry that grazing fees would be kept low and that bureaucracy within the administrative agency would be kept to a minimum. The result of this bargaining was a weak and underfunded Grazing Service with district boards dominated by industry. In the late 1940s the

Grazing Service was merged with the General Land Office to form the Bureau of Land Management (BLM). For the next several decades the effectiveness of the BLM continued to suffer because of underfunding and the excessive influence of the livestock and mining industries.

In 1976 the Federal Land Policy and Management Act (FLPMA) mandated that the BLM lands be managed to provide "the greatest good for the greatest number of people for as long as possible." Essentially a multiple-use planning mandate, FLPMA spelled out goals and procedures, but left BLM personnel with the authority to make resource decisions. FLPMA did not prioritize uses for BLM lands, just as the National Forest Management Act had not prioritized uses in the national forests. Congress's reluctance to set policy priorities and its deference to agencies has resulted in tremendous pressure on the BLM and Forest Service by various user groups.

What Environmental Problems Face the National Resource Lands?

The Bureau of Land Management has often seemed to accommodate commercial interests at the expense of the land and wildlife; its policies have consistently favored grazing and mining over recreation, wilderness, wildlife habitat, and ecological functions. Those who oppose BLM policies sometimes refer to the agency as the Bureau of Livestock and Mining or the Bureau of Livestock Management.

To be fair, several factors have made the BLM less effective than it must be in order to manage and protect the rich heritage of national resource lands. BLM lands are undoubtedly the most underappreciated of all federal land. Consequently, the BLM is probably the least well-known land-managing agency. And, despite the fact that it is in charge of far more land than the Park, Forest, and Fish and Wildlife services and that in addition to the land it manages (some 300 million acres or 121 million hectares) the BLM leases mineral rights beneath another 200 million acres (81 million hectares) in national forests and private lands, its budget is just one-third that of the Forest Service! It is little wonder, then, that a variety of environmental problems and issues are related to BLM lands. These include illegal harvesting and disruptive recreational use, commercial exploitation, and wilderness designation.

Illegal Harvesting and Disruptive Recreational Use. The vast acreage managed by the BLM makes it difficult to protect these lands against illegal harvesting of plants and animals and destruction of historic sites. The looting of archeological sites and the disruption of wildlife (including poaching and the theft of rare plants, especially cacti) are significant problems. Off-road vehicles (ORV) damage vegetation, are hazardous to wildlife, and damage the soil, causing erosion. Barbaric human behavior is responsible for declines in many species' populations. For instance, researchers documented a 30 to 60 percent decline in the tortoise population of the western Mojave Desert in California from about 1980 to 1986. According to the Audubon Society, 20 percent of the animals found dead were shot—ostensibly for target practice.

Commercial Exploitation. Despite the significant problems related to recreational use, commercial use poses the greatest threat to the integrity of BLM lands. Overgrazing, in particular, has severely damaged the vegetation in many areas (Figure 21-15). Too many livestock also mean that waterholes are depleted, streams are degraded by animal wastes, and riparian, or stream-bank, ecosystems are damaged or destroyed. As biologist Paul Ehrlich describes the problem:

Most of the West is arid, and lush vegetation is largely concentrated within a few yards of waterways. Naturally, cattle congregate there, trampling the vegetation, beating down stream banks, and defecating and urinating into the streams. The resultant load of eroded soils and wastes, combined with increased stream temperatures because of lack of shading, reduces or exterminates trout populations and makes water less pure for thirsty townships. The water-retaining capacity of riparian (stream-bank) ecosystems is reduced, making both droughts and floods more likely. Aesthetic and recreational values are also compromised, as camp sites become wall-to-wall cow pads and fish and game populations are reduced.

FIGURE 21-15: Cattle drive on BLM land, Harquahala Valley, Eagle Tail Mountains, Arizona. Grazing is permitted on 89 percent of BLM lands and 69 percent of Forest Service lands (excluding Alaska).

The issue of grazing on public lands is complicated and controversial. Overgrazing occurs because the BLM charges much lower grazing fees than do private landowners—about $1.87 per animal unit month (AUM) compared to $10 to $15. Some people argue that the public ranges are being damaged to subsidize a minority of ranchers who contribute only a fraction to the beef market. Others, however, point out that the lack of parity with private grazing fees is due to the lack of corresponding amenities. Private grazing fees typically include, for example, guaranteed weight gains, total supervision of livestock, and maintenance and improvement of pastures. Essentially, ranchers drop off their stock and pick them up later, fat and happy. In contrast, ranchers who use the public lands are responsible for controlling their herds and other tasks and are not reimbursed or credited for low weight gains or the loss of animals.

Wilderness Designation. Wilderness designation for BLM lands has become particularly controversial in recent years. The battle lines are familiar: commercial interests versus preservationists. So are the battle cries. Mining companies, ranchers, and oil and gas developers maintain that designating some areas as wilderness spells economic doom for communities. Preservationists warn that unless the BLM reverses its policies and begins to designate more of its most valuable holdings as wilderness, the little-known beauty and unique character of many national resource lands will be lost.

What Initiatives Are Being Taken to Protect the Federal Lands?

While the increasing public use of federal lands has resulted in adverse impacts on many parks, forests, refuges, and BLM lands, it has also produced a number of promising initiatives aimed at protecting and preserving the federal lands. In the early 1980s Ronald Reagan formed the President's Commission on Americans Outdoors to assess the country's remaining open, nondeveloped spaces and to determine what needs were currently unmet by the present system of public lands (including federal, state, and local holdings). The commission found that Americans have a deep and abiding love for the nation's public lands and are particularly concerned over the state of facilities and conditions in the national parks. (See Focus On: Americans Outdoors—The Legacy, the Challenge.) The commission made several recommendations to increase government support for the public lands. The most promising was the establishment of a $1 billion trust fund to provide for land acquisition by the federal government and to maintain lands already in the system.

National Park System. Another promising initiative for the federal lands also began in the early 1980s, when the National Park System Protection and Resources Management Act (commonly called the Park Protection Act) was introduced to Congress. The major thrust of the bill was twofold: to establish ways in which park administrators can track the condition of the natural and cultural resources within park boundaries and to give the Park Service a stronger voice in determining how other federal agencies use or affect park lands. For example, in the early 1980s the Department of Energy announced that one of the prime candidates for its nuclear waste dump was a site just 4,000 feet (1,200 meters) from the boundary of Canyonlands National Park, Utah. Public pressure eventually forced the department to reconsider the Canyonlands site. By 1992 the House of Representatives, in two different Congresses, had passed the Park Protection Act, but the Senate had not. Many park advocates warn that if steps are not soon taken to provide adequate protection, the parks face continued degradation and irrevocable change, change that will diminish or destroy the very qualities that make these areas worthy of national park status. Given the increased environmental awareness of the general public and their enthusiastic support of the national parks, the bill will once again be debated before Congress, perhaps with a more favorable outcome.

Two other promising pieces of legislation have also been introduced to Congress and have garnered significant support. The California Desert Protection Act is hailed as the most significant piece of park and wilderness legislation in a decade. It would create Mojave National Park, expand Death Valley and Joshua Tree national monuments and redesignate them as national parks, and designate 4.4 million acres (1.8 million hectares) of BLM wilderness. The Act passed the House of Representatives in late 1991 and in 1992 was under consideration in the Senate. The Concessions Policy Reform Act, introduced by Senator Dale Bumpers of Arkansas, would explicitly make clear that concessions should be restricted to necessary services that cannot be provided in surrounding communities. It would require that facilities be compatible with each park, both environmentally and esthetically. The bill would also alter the way concessions contracts are awarded, allowing the Park Service to select the bid that is best, in terms of both cost and service to the visitor, rather than simply the lowest bid. All contracts would be subject to competition, and minimum fees for each contract would be determined on a case-by-case basis to ensure that the government receives a fair return for its investment. Finally, the bill would return concessions revenues to the parks rather than to the general treasury.

Focus On:
Americans Outdoors—The Legacy, the Challenge
Gilbert Grosvenor, president, National Geographic Society, and vice-chairman, President's Commission on Americans Outdoors

When Lamar Alexander, then governor of Tennessee, was named chairman of the President's Commission on Americans Outdoors (PCAO), he described the mandate of the commission in these terms: "To find out what the American people will want to be able to do in the outdoors into the twenty-first century and how we can make sure they have appropriate places to do those things." We started from there.

Our task was undertaken in one of the most open processes of extensive public involvement that I have ever seen in government. There were many who were surprised by what they considered the conservation or resource-preservation emphasis of the commission's final recommendations, which called for increased investment in outdoor-recreation assets and strengthened protection of our natural environment. I have a simple explanation for this outcome: We listened to the people. We heard what they had to say, and we learned from them.

We heard from over 1,000 witnesses in 18 public hearings across the country. We visited recreation sites from senior citizen centers in Austin to hunting lodges in Alaska. Three hundred technical experts were organized into study teams, more than 100 researchers conducted a literature review, 800 citizens submitted letters and concept papers. Other contributions to the effort included a nationwide opinion poll of 2,000 American adults, special workshops, special conferences, and the involvement of 50 states.

We learned that Americans place a high value on easy access to quality outdoor experiences in their daily lives and they want the outdoor legacy passed on to their grandchildren to equal what they received from their forefathers. We also learned of serious threats to recreation opportunities and to the quality of the outdoor estate, including problems of funding, legal liability, environmental degradation, open-space loss, and information needs.

In retrospect, I feel the commission process epitomized the ideal in an open, free, democratic society. Anyone with an opinion to express was allowed to participate, and no one (except Hawaiians) had to travel more than 500 miles to testify. We came together as a group of 15 people with diverse economic and social backgrounds, but we were sympathetic to each other's interests and concerns. Private citizens as well as members of Congress and local and state officials were included among commission members. I feel our report is more valuable for the breadth of ideas represented, both by those from whom we heard and in the makeup of the commission itself. It is a consensus document, which I hope captures the best of the creative vision and energy of the American people.

The work of the commission offers the potential for a turning point for America, a time to reexamine the role of the outdoors and outdoor recreation in our lives. The single clearest lesson I learned, as I listened to Americans across this country, was that our people care deeply about our outdoor heritage, and many of them feel that government policy and actions, and sometimes individual behavior as well, do not reflect the importance which we as a people attach to the outdoors.

I hope as well that our efforts may inspire others to action. Our report is aimed primarily at citizens and communities, not at federal officials. Since the commission's report was released, we have seen a great surge in programs and activities by private organizations and local and state governments, particularly in the areas of creation of greenway networks, development of an ethic of respect for the outdoors, shaping and managing growth and development, and increased funding for outdoor programs. I believe that this renewed excitement about the importance of the outdoors is only the beginning.

I hope that the work of the commission may help to usher in a new era of caring and involvement by local citizens. We all care deeply about the quality of life in the places where we live, work, and play. We now know that natural assets need not be sacrificed for but can complement economic growth and development. We must look for new ways to engage a broader range of players at the table when development and preservation decisions are made. Our communities, and our society, will be the stronger for it. And it is our obligation to future generations to do our best at maintaining the magnificent outdoor legacy that is their birthright.

National Wildlife Refuge System. Recently, several initiatives have been taken to help restore the integrity of the national wildlife refuges. In 1990 Gerry Studds, representative from Massachusetts, introduced to Congress a bill that defined a clear and unequivocal purpose for the refuges. By legally mandating that the system's primary purpose is to preserve wildlife and habitat, the National Wildlife Refuge System Act would, in the words of Congressman Studds, "restore integrity to the wildlife refuge system." The act would grant the Fish and Wildlife Service the authority to prohibit any activity that damages refuge habitats or wildlife. Although the bill was not passed when first introduced, it was reintroduced in 1992 with greater support from conservationists, environmental groups, and others.

Another promising development is the preparation of a management plan for the refuge system. The Fish and Wildlife Service has identified 17 major refuge management issues that will be addressed in "Refuge 2003—A Plan for the Future." Among them are protection of biological diversity, compatibility of secondary uses, and endangered species management. With the cooperation of more than 40 national interest groups and conservation organizations, the service had developed a preliminary list of options to address these issues. In a series of public meetings held throughout the country in the spring of 1991, the service invited the public to comment on the issues and propose additional options.

National Forest System. Public participation may also prove to be the saving grace for the nation's forests. The 1976 National Forest Management Act mandated that all uses of each forest were to be laid out in 50-year plans to be completed by 1985, which were to be subject to public review. Public review was an important qualification and, for the Forest Service, a contentious one. The service's tradition of independence, forged in the early days by Pinchot's leadership, did not allow it to accept public review of its plans easily. Public scrutiny of the management plans for the national forests was intense, with critics arguing that the plans were clearly biased toward economic uses. Thanks to public participation, the plans for some forests were revised, leading to more balanced management plans for the George Washington National Forest, Virginia and West Virginia; the Gallatin National Forest, Montana; and the Payette National Forest, Idaho, among others. Many hope that this experience will lead to greater cooperation between the Forest Service and other groups in the coming years.

Perhaps the most promising sign for the national forests, however, has nothing to do with laws or management plans. It has to do with people. There is a growing movement within Forest Service ranks to promote responsible, balanced use of the land. One sign of this movement was the formation of a group known as the Association of Forest Service Employees for Environmental Ethics (AFSEEE) by Jeff DeBonis, a U.S. Forest Service timber planner in Oregon's Willamette National Forest. DeBonis and other foresters are speaking out against what they see as unsustainable target cuts on national forestland. These and other developments are indications that the Forest Service may yet reassume its past role as steward of the nation's forests.

National Resource Lands. In 1991 the House of Representatives approved an amendment to the Interior Department's appropriations bill that would phase out controversial below-market grazing fees on BLM rangelands. Fees would be raised incrementally until 1995. The legislation also directs the BLM and the Forest Service (which manages some 102 million acres or 41 million hectares of rangeland) to use the monies collected from grazing fees to restore riparian areas and fish and wildlife habitat, to cover the costs of administering the grazing program, and to abolish the BLM grazing boards, which traditionally have been dominated by ranchers. These measures have yet to be embraced by the Senate and enacted by the full Congress, but conservationists point out that their introduction is a welcome and long-overdue development and a hopeful sign for the nation's western ranges (Figure 21-16).

Future Management of the Public Lands

From the stark grandeur of Mount Denali to the other-worldly wonder of Mammoth Caves, from the now-quiet Civil War battlefields to the crowded monuments of Washington D.C., the federal lands encompass all that is best, most painful, and most important about our nation. Although these lands "belong" to us collectively, we hold them only for a short time before passing them on to our children and our children's children. The measures outlined below can help to ensure that we bequeath to future generations of Americans a healthy and balanced system of public lands.

FIGURE 21-16: Livestock can seriously damage streams in riparian areas. Revegetating and protecting riparian habitat are critical to restoring the western ranges.

What You Can Do:
To Protect Public Lands

- Venture out into the public lands; pick an area that interests you and get to know its natural and cultural history.
- When in a park, forest, refuge, or BLM area, practice low-impact camping and hiking. "Take only pictures; leave only footprints."
- Conserve paper products, energy, and mineral resources in order to lessen the need for logging and the development of oil, coal, and other mineral deposits on public lands.

- Use recycled lumber where possible; recycle your scrap wood, and reuse wood around your house.
- Design your building projects carefully to avoid wasting wood. And consider other materials, such as brick, adobe, and concrete block, when building a home or office.
- Adopt a public land unit in your region. Help the managers to maintain it in its natural state.
- Support legislation promoting the wise and sustainable use of

public lands. Let your elected officials know where you stand on issues pertaining to public lands.
- Join a conservation organization to support its efforts to protect public lands.

Prevent Overuse Through Conservation

Several measures should be taken to conserve the public domain. First, all logging operations on the federal lands should be both ecologically and economically sound; all costs, including the cost of new roads, should be considered when determining the profitability of a proposed operation. Recreational and wilderness uses should be given parity with economic uses; a restatement of Forest Service and Bureau of Land Management policy is in order to officially acknowledge recreation and wilderness as uses of equal importance on the national forests and BLM lands. Raising grazing fees and restricting herd sizes would greatly help to prevent overuse of rangelands. The fees collected should be used to restore and maintain degraded areas.

Protecting and conserving natural and cultural resources on federal lands are not inexpensive. Increased funding for the four major land-managing agencies is necessary in order to provide them with the resources they need to function effectively.

Finally, every one of us must make an effort to act as stewards, wisely using and preserving the federal lands. What You Can Do: To Protect Public Lands suggests ways that individuals can help to conserve the invaluable public domain.

Protect Human Health and the Environment

The protection of human and environmental health can be accomplished by taking several key steps. Strict enforcement of existing environmental protection laws, which cover air, water, wildlife, and cultural resources, would safeguard the public domain and its wealth of natural and cultural resources.

To protect national parks and their irreplaceable biological resources, an ecosystem approach to management should be adopted. One important facet of ecosystem management is the creation of buffer zones or greenbelts around park boundaries. Greenbelts buffer the parks from potentially harmful activities on adjacent lands, and they afford increased habitat for wildlife. By enacting the Park Protection Act, Congress could make ecosystem management of the nation's parks a reality.

The acquisition of additional lands is needed to provide adequate haven and habitat for wildlife. Priority should be given to acquiring wetlands and habitat critical for migratory birds and threatened and endangered species. (Presently, most refuge acreage is in Alaska.)

Preserve Living Systems

Currently, the nation's land-managing agencies are divided between the Department of the Interior and the Department of Agriculture, a system that is not always conducive to cooperation and effective management. Establishing a cabinet-level Department of Natural Resources and bringing all the land-managing agencies under its jurisdiction would promote interagency cooperation and coordination of land-managing efforts. Interagency cooperation is essential to facilitate ecosystem management, protect habitat, and manage animal populations in those areas in which federal lands under different jurisdictions lie adjacent to one another.

Effective management requires having as complete and thorough a knowledge base as possible. A biological inventory of the nation's public domain should be conducted, through the coordination of the various land-managing agencies, in order to

(Text continues on page 471)

Environmental Science in Action:
Yosemite National Park
April Rowan

Yosemite National Park is home to a host of natural wonders: thundering waterfalls, massive granite domes, and towering sequoias. Its unusual beauty was prized by the native Americans who once lived in the area, as well as by the European settlers who supplanted them. Today, Yosemite is one of our country's most popular national parks, attracting about 2.5 million visitors each year. Unfortunately, the activities of these visitors, combined with the facilities designed to attract their business, are threatening to rob Yosemite of its wild beauty and sublime grandeur.

Describing the Resource

Physical Boundaries

Yosemite National Park covers 1,189 square miles (3,082 square kilometers) on the western edge of the Sierra Nevada (Figure 21-17). Approximately 25 million years ago the earth shifted, uplifting the granite ridges that had lain below its surface and forming the mountain range. Two or three million years ago massive glaciers carved out valleys and lakes and sculpted granite peaks. One of the most striking legacies of the glaciers are the patches of polished rock found in the upper Yosemite region; they are patches of granite so compacted and polished by the pressure of tons of ice that they shine like mirrors in the sun. Even today, small glaciers exist in the upper reaches of the Sierra Nevada.

The park's best-known area is the Yosemite Valley. Roughly 7 miles long (11.3 kilometers) and 1 mile (1.6 kilometers) wide, it lies between 900-foot (300-meter) high granite walls which appear to rise almost perpendicularly from the valley floor. The valley features some of the park's most impressive granite peaks, including El Capitan, the Cathedral Rocks, and the Three Brothers. Their massive size and elegant, sculpted lines give the valley the air of a cathedral. Flowing through the valley is the Merced River, which forms some of the park's spectacular cataracts. For

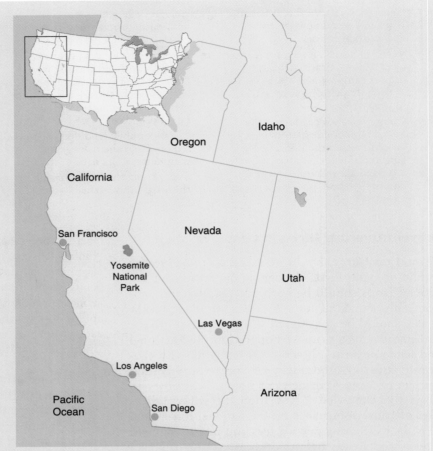

FIGURE 21-17: Yosemite National Park, California.

example, Yosemite Falls, North America's highest waterfall, thunders down 2,610 feet (783 meters) in three stages. Bridal Veil Falls, covered with the gauzy mist that gives it its name, often shimmers with rainbows (Figure 21-18).

Biological Boundaries

Yosemite National Park is home to a multitude of diverse species. Over 70 species of mammals, including elk, mule deer, and black bear, live in the park. It also contains at least 1,200 species of flowering plants and 37 tree species. Of the plants, 8 are considered endangered under the Endangered Species Act and 37 are designated rare by the state or the park.

Vegetation types range from those commonly found in the California deserts and lowlands to those native to glacial regions in Canada and Alaska. The foothill belt, generally dry and warm, extends up to an average altitude of 3,000 feet (900 meters); vegetation is largely shrubs and grass. The yellow pine belt extends between 3,000 and 6,200 feet (1,860 meters) and features a great variety of conifers, including yellow pine, incense cedar, and a variety of willows along the streams. It is also home to three major groves of the famed sequoia, some of which are approximately 250 feet tall, 30 feet in diameter, and 3,500 years old. Species of boreal origin, such as silver pine and snow bush, thrive in the upper coniferous belt, between 6,200 feet and the timberline. Only plants that can withstand harsh winters and little warmth, such as Arctic willow and alpine sorrel, are found above the timberline up to 13,090 feet (3,927 meters).

The many streams, rivers, and lakes in Yosemite are home to 11 species of fish, 5 of which occur naturally in the region. The others are stocked for sport fishing. In 1972, as part of an effort to restore the park to a more natural condition, managers planned to stop stocking nonnative fish. However, the California Department of Fish and Game protested, and limited stocking continues.

Over 350 black bears live in the park, and problems sometimes arise when they take food and supplies from campers. The bears have an uncanny ability to open any sort of container and retrieve backpacks from branches high above the ground. The park is trying to solve these problems through visitor education and rules prohibiting feeding of the bears, since feeding them acclimates the bears to humans and can lead the bears to raid campsites.

Although wolves once lived in Yosemite, there are none today, largely due to the efforts of the early park service to eliminate animals that might frighten visitors. At that time predator control was widely accepted as an appropriate management technique because the ecological effects of eliminating predators were unknown. Unfortunately, those effects are clearly visible in Yosemite today. The wolf's absence has allowed populations of its natural prey, such as elk, to grow to such proportions that the animals starve during the winter, forcing the park service to kill off portions of these populations to prevent mass starvation.

The bighorn sheep, another of Yosemite's original inhabitants, was eliminated from the park by 1914 because of disease, hunting, and destruction of parts of their habitat outside the park. The animal was restored to the park in March 1986, when the California Department of Fish and Game, the National Forest Service, and the National Park Service introduced a herd of 27 bighorns to Yosemite.

Yosemite is also home to several pairs of the endangered peregrine falcon. This species is threatened by the amount of the pesticide DDT that has accumulated in the food chain in and around Yosemite. Although DDT has been banned in the United States for over 20 years, use of DDT by Latin American countries and residual DDT in California continues to pollute the Yosemite area. As the falcons ingest more and more DDT, the shells of their eggs become thinner and thinner. These thin-shelled eggs have little chance of hatching, and so the adults produce few offspring by themselves. To remedy this situation, the Park Service replaces thin-shelled eggs with artificial eggs, and then, just before the eggs would have hatched, it replaces the artificial eggs with two or three newly hatched falcons raised in captivity. The parent falcons then raise the babies as their own.

Social Boundaries

European settlers first saw Yosemite in 1851, when an army battalion chasing a band of native Americans (who had been raiding mining camps) followed them into the valley. At that time, there were 22 native American villages in the area that now makes up the park. Within a few years the army had "conquered" the area, killing or displacing all the natives. The only reminder of their presence was the name of their tribe, Yosemite. The park's Indian Cultural Museum sponsors weekly demonstrations of native American traditions and provides other cultural and historical background on the area.

The Yosemite valley today would be almost unrecognizable to the native Americans who once lived there. By 1974 the valley contained restaurants, gift shops, grocery stores, service stations, liquor stores, swimming pools, tennis courts, kennels, horse and mule stalls, a bank, a skating rink, and a miniature golf course. At first, hotels and golf courses and other "improvements" were thought to enhance the park's appeal, but the exponential growth in visitors they caused has brought many typically urban problems to the supposedly natural area. Campfire smoke and automobile exhaust pollute the air. Park rangers spend a good portion of their time dealing with speeders, drunk drivers, and thieves. During the popular summer months overcrowding, noise, and litter all detract from the peaceful atmosphere people hope to find in a national park.

Looking Back

The history of Yosemite has been one of a continual struggle between those who have wanted to develop the park for maximum public enjoyment and those who have tried to preserve the beauty and natural quality or character of the environment.

How Public Use Grew from a Goal to a Problem

The Europeans who settled Yosemite gave little thought to their effect on the environment. Herds of cattle, sheep, and horses grazed meadows down to bare earth, and heavy logging also contributed to

FIGURE 21-18: Bridal Veil Falls, Yosemite National Park, California.

erosion. Then, in 1864 President Lincoln granted the Yosemite valley and the nearby Mariposa Big Tree Grove of sequoias to the state of California, with the instruction that these areas be preserved specifically for public use and recreation.

One of Yosemite's earliest and most ardent admirers was John Muir. He spent years exploring the wonders of the area, which he described in *The Yosemite*. Muir exulted in the powerful natural forces of Yosemite: he rushed outside to experience earthquakes, climbed tall trees to watch thunderstorms, and scaled precipitous heights to obtain the best view of waterfalls. Muir was part of the campaign to make Yosemite a national park, a goal realized in 1890. Sadly, this triumph was soon followed by a bitter defeat. The park's second major river, the Tuolumne, originally flowed through the Hetch Hetchy valley, similar to Yosemite in its grandeur. When the city of San Francisco proposed damming Hetch Hetchy to create a reservoir for the city, Muir and others fought the idea. The city prevailed, and today the waterfalls and domes of Hetch Hetchy are buried under tons of water.

The park was operated by the U.S. Army until 1917, when the newly created National Park Service assumed that duty. The early managers of the park did not share Muir's appreciation for the wilder side of nature. They tried to eliminate factors that might prevent visitors from enjoying the park in peace and comfort. Crews built roads and paths to areas of natural beauty. Naturally occurring fires were suppressed. Animals that might discourage visitors, such as wolves, were exterminated, allowing animals that visitors enjoyed, such as deer, to flourish. From the vantage point of the 1990s fire suppression and predator control seem to many people to be short-sighted and anthropomorphic, but, to be fair, these management techniques were once widely thought to be appropriate and sound. It was only relatively recently that ecologists, wildlife biologists, and resource managers recognized the importance of fire and predators to a healthy ecosystem.

With the popularity of the automobile, increased leisure time, and the dwindling amount of land remaining in a natural state, Yosemite and other parks became increasingly popular for vacations and weekend trips. A few hours' drive from the metropolitan areas of Los Angeles and San Francisco, Yosemite experienced heavy influxes of visitors. The overuse problem came to a head on the weekend of July 4, 1970. Over 76,000 people visited the park that weekend—a record crowd. Several thousand young people gathered in a meadow, playing loud music and smoking marijuana. Park rangers asked them to leave, to no avail. Finally, a dozen rangers, armed with clubs and Mace, dispersed the crowd and arrested 186 people. Because of this and similar incidents and other problems throughout the park system, the Park Service began to train rangers in law enforcement.

How the Park Service Tried to Protect the Park

Soon after the July 4th incident, the Park Service began limiting visitors and auto traffic in an attempt to protect the park from environmental degradation and to preserve the wilderness experience for park visitors. Meanwhile, they worked on a master plan aimed at finding a solution to the park's problems. The proposed plan included the complete elimination of private automobiles in the park, the construction of a mass transit system, and the removal of all Park Service and concession buildings from the valley to the town of El Portal, at the western entrance to the valley.

A storm of controversy soon raged around the plan. In Yosemite, as in most other national parks, all of the entertainment, restaurant, and lodging facilities had become consolidated under the management of a single concessioner. In Yosemite's case, the concessioner was the Yosemite Park and Curry Company, which in 1973 was acquired by MCA, a large entertainment company. MCA opposed any action that might limit its profits in any way. In the late 1980s MCA was purchased by a Japanese firm. The revelation that the concession contract for one of the nation's best-loved parks was held by a foreign company brought a public outcry that, in part, led MCA-Matsushita to sell its concession rights in Yosemite (more on that below).

Conservationists charged that MCA had influenced the Park Service's final master plan in favor of development rather than conservation and persuaded the assistant secretary of the interior to examine the plan. When the secretary deemed the plan inadequate, the Park Service developed a more comprehensive general management plan, which was approved in 1980. The plan's main goals, to be attained in 10 or 15 years, included designating 90 percent of the park as wilderness, forever free from development; removing substandard Park Service and concession staff housing and other facilities from Yosemite valley to El Portal; reducing concessioner-operated lodging facilities by 10 percent and reducing overnight facilities in the valley by 17 percent; reducing the use of private vehicles in the valley, with a long-range goal of eliminating them entirely; identifying and enforcing carrying capacities; and improving and expanding information, interpretation, and reservation services.

Identification and enforcement of carrying capacities, especially for overnight visitors, were important steps in preserving the park. Unfortunately, the conservationist's goal—enhancing the wilderness experience for everyone by restricting the number of people who could enjoy it at any one time—was misinterpreted by many people, who saw the restrictions as a blow to their individual freedom to enjoy Yosemite rather than as a step necessary to preserve this national treasure. Opponents of the plan, pointing to the removal of nonessential amenities from park grounds, accused the conservationists and the Park Service of limiting the enjoyment of Yosemite to the young and able-bodied.

In addition to changing visitor policies, the Park Service altered the way it managed the vegetation and the wildlife of the park to produce a more natural ecosystem. One of the first steps was to end the artificial suppression of fires. To prevent natural fires from running rampant, the Park Service first had to eliminate an unnatural buildup of fuel, the legacy of decades of the no-burn policy. Sequoia groves, which would naturally be open, were clogged with an understory dominated by white pine. The Park Service began a series of carefully controlled, purposefully set fires.

The Park Service also began a program to eliminate exotic plants brought into the park from other areas, using biological controls whenever possible. It revegetated sites stripped or altered by human activity and reduced the threat of further degradation by regulating the grazing of horses, mules, and burros used for recreational trips or field work. It monitored water quality to safeguard against the pollution of lakes, streams, and rivers. As part of its attempts to "re-naturalize" the park, it set up breeding programs to preserve peregrine falcons.

Looking Ahead

Although the Park Service has temporarily safeguarded Yosemite from the effects of crowds of people—especially noise, pollution, and development—the park's future remains in question because the nation cannot resolve the controversy over use versus preservation.

When the National Park Service was established in 1917, Congress defined its primary purposes as providing for the enjoyment of the public and managing the scenery, natural and historical objects, and wildlife within the parks in ways that preserve them unimpaired for the enjoyment of future generations. In other words, there are limits on how the public may use "our" national parks. We may not develop or alter them in ways that will prevent others from experiencing the same peaceful environment and breathtaking sights we enjoy today. This philosophy underlies the limits on visitors and the removal of concessions from the valley.

In 1990 MCA-Matsushita sold its concession rights in Yosemite to the National Park Foundation, a non-profit organization chartered by Congress to channel private donations to the parks. Under the terms of the agreement, the foundation

will purchase the assets of the Yosemite Park and Curry Company, (the MCA subsidiary) when that company's concession contract expires in 1993. The foundation will turn over the company's assets to the National Park Service; revenues gained by the next concessioner for the park will be used to repay the loan. This is a promising development, because it means that there can be real competition for the concession contract in 1993 and that the Park Service can choose to limit commercialization within Yosemite. Unfortunately, it does not guarantee that the general management plan will be implemented. Many of the goals, which were to be achieved between 1990 and 1995, have not been attained.

In a 1912 argument against damming the Hetch Hetchy valley, John Muir wrote that "no holier temple has ever been consecrated by the heart of man." Preservation, which includes limits on use, must be the top priority in order to safeguard the natural temples of Yosemite and other national parks for future generations.

assess the biological resources (at both the ecosystem and species levels) of the nation's parks, refuges, forests, and BLM lands.

Congress should formulate and enact an Organic Act for the National Wildlife Refuge System in order to clarify the role and importance of the refuge system. Such a law should unequivocally mandate that the primary purpose of the system is to protect and preserve ecosystems in order to provide safe haven for species; it should clearly prohibit incompatible or harmful secondary uses.

Finally, to preserve the nation's living systems, Congress should establish a $1 billion trust fund (after the recommendations of the President's Commission on Americans Outdoors) to enable the four land-managing agencies to acquire additional lands and to maintain lands already in the system. An emphasis should be placed on acquiring ecosystems

not currently represented in the nation's public domain and on habitats for rare (threatened or endangered) plants and animals.

Summary

The history of the federal lands is largely one of land disposal to private interests. This tradition dates back to the early part of the nineteenth century when land disposal was seen as a way of encouraging western expansion and settlement and thus securing the nation's frontier. By the mid-1930s the government had either given away or sold 1.1 billion acres.

Despite its tradition of land disposal, the government had begun to set aside land units for special purposes as early as 1832. Forty years later Yellowstone was designated a "nation's park." The 1891 Forest Reserve Act facilitated the establishment of the national forests; this law was also invoked to establish the first national wildlife refuge, Pelican Island, Florida, in 1903. By the 1930s the majority of the federal lands remained unreserved, not belonging to the National Park System, the National Wildlife Refuge System, or the National Forest System. The public domain, as it was called, was severely degraded due to overgrazing and other deleterious practices. Finally, in the late 1940s the Bureau of Land Management (BLM) was formed to oversee the public domain.

The National Park System is a network of 343 diverse units totaling almost 80 million acres. The Park Service, Department of the Interior, manages the system. According to congressional mandate, the Park Service is to manage the parks in order to conserve their scenery, natural and historic objects, and wildlife and to provide for the enjoyment and use of the same by the public. The most serious threats to national parks are crowding and overuse, insufficient funding, threats to wildlife (diminishing habitat, elimination of predators, and poaching), commercial recreational opportunities, energy and minerals development, atmospheric pollution, and activities on adjacent lands.

The National Wildlife Refuge System, consisting of 456 units, totals about 91 million acres of land and water in 49 states and 5 territories. The refuges are managed by the Fish and Wildlife Service, Department of the Interior. Their primary purpose is to provide habitat and haven for wildlife. Threats to the refuges include the internal management structure of the Fish and Wildlife Service, which is plagued by too many competing responsibilities, harmful or incompatible secondary uses on many refuges, activities on adjacent lands, and political pressures.

The National Forest System, encompassing almost 191 million acres in 44 states, Puerto Rico, and the Virgin Islands, is managed by the Forest Service, Department of Agriculture. The nation's forests and grasslands are managed to accommodate a wide array of commercial and recreational uses. The major threats to the national forests are an emphasis on logging over other uses, clear-cutting, below-cost timber sales, and lack of wilderness designation.

The National Resource Lands are also managed to accommodate multiple uses. Administered by the Bureau of Land Management, Department of the Interior, the National Resource Lands, or BLM lands, are located primarily in the arid and semiarid western states and Alaska. BLM lands equal approximately 300 million acres, roughly 47 percent of federal land. Threats to the BLM lands include illegal harvesting and disruptive recreational use, commercial exploitation (especially grazing), and lack of wilderness designation.

The federal lands are biologically significant for several reasons. First, they encompass an astounding diversity of living ecosystems. Second, these diverse ecosystems perform a myriad of ecological functions, such as watershed protection. Finally, because the public lands encompass a wide array of ecosystems, they protect a diverse complement of living organisms. In addition to their biological wealth, the federal lands contain one-quarter of the nation's coal, four-fifths of its huge oil-shale deposits, one-half of its uranium deposits, one-half of its naturally occurring steam and hot water pools, one-half of its estimated oil and gas reserves, and significant reserves of strategic minerals.

The federal lands are used for many purposes. Consumptive uses result in the depletion of a resource. Commercial consumptive uses include logging, mining, and oil development. The federal lands also offer a variety of recreational, scientific, and wilderness opportunities, many of which are nonconsumptive (they do not result in the depletion of a resource). However, even nonconsumptive uses can degrade an area if there are many users or if they are not careful about how they use the land.

A number of promising initiatives have been undertaken recently to protect and preserve federal lands. Legislation has been proposed to protect and improve the management of the national parks and refuges. The Fish and Wildlife Service is preparing a management plan for the refuge system and is inviting public participation in the preparation of the plan. Public participation, as required by the 1976 National Forest Management Act, has led to more balanced management plans for several of the nation's forests. BLM lands should benefit from a proposed plan to phase out controversial below-market grazing fees. The plan would also direct the BLM and Forest Service to use monies collected from grazing fees to restore riparian areas and fish and wildlife habitat.

Key Terms

below-cost timber sales
clear-cutting
consumptive use
multiple use
nonconsumptive use
public domain
utilitarianism

Discussion Questions

1. Describe the federal public lands: their extent, location, and biological, physical, and social significance.

2. Identify the four major federal land-managing agencies and describe their missions. How do their missions differ and how do those differences affect the way the land is managed?

3. What are the most serious problems facing the national parks, national forests, national wildlife refuges, and national resource lands? What actions do you think should be taken to alleviate these problems?

4. Briefly discuss the history of land acquisition and disposal that ultimately gave rise to the public domain.

5. Identify and describe the national park, national forest, national wildlife refuge, or BLM land nearest you. (If there is no federal land close to you, choose the site that you would most like to visit.) What ecosystems does it harbor? What rare species are found there? What environmental problems are of greatest concern? What management strategies are of greatest concern?

6. Do you agree with the goals of the general management plan devised for Yosemite National Park? What are its strengths and weaknesses? Should private automobiles be eliminated entirely from Yosemite (as called for in the long-range plans)?

Wilderness

Of what avail are forty freedoms without a blank spot on the map?

Aldo Leopold

When we establish a wilderness reserve or national park we say, in effect, thus far and no farther to development. We establish a limit. For Americans, self-limitation does not come easily. Growth has been our national religion. But to maintain an area as wilderness is to put other considerations before material growth. It is to respect the rights of non-human life to habitat. It is to challenge the wisdom and moral legitimacy of man's conquest and transformation of the entire earth. This acceptance of restraint is fundamental if people are to live within the limits of the earth.

Roderick Nash

Learning Objectives

When you finish reading this chapter, you should be able to:

1. Define wilderness and identify the single characteristic common to all definitions of wilderness.

2. Discuss the reasons most often given for preserving wilderness.

3. Summarize the cultural roots of the wilderness preservation movement and the development of the Wilderness Preservation Act of 1964.

4. Explain the process for designating an area as wilderness.

5. Briefly relate how wilderness areas are managed.

6. Identify the major threats to wilderness areas.

Wilderness. The word conjures up a kaleidoscope of sensory images. Roaring waters, the soulful howl of a lone wolf, the whisper of the wind in a cotton-wood stand, the murmur of insects unseen in sage-brush. Breathtaking vistas and distant horizons, unobscured by human artifacts. Wild rivers, now rushing and roiling through a cataract, now spreading out quietly in a deep, wide pool. These are among the sights and sounds of wilderness. But the word also brings to mind intangible ideas: awe and fear at the power of natural forces and also solitude, space, and serenity.

Wild lands are elemental. No cars, no houses, no gas, no electricity. The wilderness reminds us that water does not come from a tap nor light from an incandescent bulb. Your alarm clock is the sun, if you are a late sleeper, or the birds and other animals that stir in the half-light before dawn, if you are not. And for entertainment, there is the ever-changing landscape. Wilderness is the original theater in the round, a 360° panorama through which the seasons cycle, spring into summer into autumn into winter, against a variegated backdrop of rock and sand, forest and stream.

In this chapter we examine the wilderness resource, beginning with a discussion of what wilderness is and why it is biologically significant. The chapter focuses on those wilderness areas in the United States that lie in national parks, forests, wildlife refuges, and resource lands. We limit our discussion to federal wildernesses because, while a significant number of other wild areas are protected by state and local governments and private organizations such as the Nature Conservancy, these lands are typically not managed as wilderness nor are they usually large enough to encompass and protect entire ecosystems or aggregates of ecosystems. We look at the history of the wilderness preservation movement, how an area is officially designated as wilderness, how that designation affects its preservation and management, what problems face wilderness areas, what promising initiatives have been proposed for wilderness preservation, and what our future management options are.

Describing the Wilderness Resource: Biological Boundaries

What Is Wilderness?

The philosopher, the poet, the scientist, the bureaucrat—each defines wilderness differently. Wilderness, defined in philosophical, psychological, or esthetic terms, is not so much a place as a state of mind. "It is the feeling of being far removed from civilization, from those parts of the environment that man and his technology have modified and controlled," says the historian Roderick Nash. For some people, the concept of wilderness is unsettling precisely because it is not controlled or dominated by humans. Most contemporary societies are characterized by a profusion of technologies that separate humans from natural forces and processes. Consequently, many of us have "forgotten" how to live and survive without the use of our technologies, and we may feel bewildered or even frightened by the dominance of natural forces and processes in the wilderness.

Long ago, native Americans realized that the Europeans who had arrived to "settle" North America felt uncomfortable with living in the natural world. Luther Standing Bear, of the Oglala band of Sioux, described the difference in the European and native American perceptions of nature. "We did not think of the great open plains, the beautiful rolling hills, and winding streams with tangled growth, as 'wild.' Only to the white man was nature a 'wilderness' and only to him was the land infested with 'wild' animals and 'savage' people. To us it was tame." The native Americans had no specific word for wilderness. In most matters, from everyday diet to spiritual beliefs, the lives of the Plains Indians were integrated with the natural world around them, and the land was simply home.

Clearly, values are important when one tries to define wilderness. Not surprisingly then, there is no purely scientific definition of wilderness, since science attempts to define things in concrete, value-free terms. Nevertheless, ecologists generally agree that a **wilderness ecosystem** is one in which both the biotic and abiotic components are minimally disturbed by humans. The stipulation of minimal disturbance is common to all definitions of wilderness. In fact, it is the single most important characteristic of wilderness as defined by the 1964 Wilderness Preservation Act. According to the act, wilderness is "an area where the earth and community of life are untrammeled by man, where man himself is a visitor who does not remain." The act established the country's National Wilderness Preservation System, the first of its kind worldwide (Figure 22-1).

FIGURE 22-1: The Linville Gorge Wilderness in Pisgah National Forest, North Carolina, was one of the original components of the National Wilderness Preservation System. The wilderness area was enlarged in 1984 and now consists of 10,975 acres.

How Can Wilderness Areas Preserve Biological Diversity?

Wilderness areas serve as reservoirs of biological diversity, or biodiversity. There are several levels of diversity in living things. Diversity within a species, known as genetic diversity, is the variation among different populations and individuals of a particular species. Genetic diversity is the mechanism by which a species adapts to changes in its environment. Variation among different species of plants, animals, and microorganisms is species diversity. Variation in the distinct assemblages of species found in different physical settings is known as community or ecosystem diversity. Genetic diversity, species diversity, and ecosystem diversity are inextricably linked.

Clearly, our best chance to preserve maximum diversity lies in the preservation of ecosystem diversity. We cannot hope to preserve the greatest possible number of species unless we also preserve the habitats upon which they depend. Unfortunately, our National Wilderness Preservation System falls woefully short in this respect. Scientists have identified 233 distinct ecosystem types in the United States, but only 81 are represented in the wilderness system. Of those not represented, 50 do not occur on federal lands and so cannot be officially protected. The remainder—some 100 ecosystem types—do occur on federal lands, but are not now protected as wilderness areas. How much longer they can retain their integrity without official protection is unknown.

Physical Boundaries

When Congress passed the Wilderness Preservation Act of 1964, it designated 9.1 million acres (3.7 million hectares) of federal land as protected wilderness. These were the original constituents of the National Wilderness Preservation System (NWPS). By the late 1980s the system had grown nearly tenfold, to over 89 million acres (36 million hectares) in 445 areas (Figure 22-2). Most of the wilderness areas lie within national parks; the rest are found in national forests, national wildlife refuges, and national resource lands.

Most protected wildernesses are high alpine or tundra ecosystems, and most are in the western states or Alaska. Fifty-six million acres (23 million hectares), well over half of all designated wilderness acreage, lie in Alaska alone. But two-thirds of all recreational use in the wilderness system occurs on just 10 percent of the areas in the lower 48 states, especially in California, North Carolina, and Minnesota. In the contiguous United States, where the demand for wilderness recreational use is the greatest, protected wilderness areas comprise less than 2 percent of the total land area. By conservative estimates, another 5 percent qualifies for wilderness designation. Even if this land were added to the National Wilderness Preservation System (an unlikely event) less than 8 percent of the contiguous United States would be designated wilderness.

Some people argue that wilderness designation "locks up" valuable resources that could be used to spur economic development and that are essential for national security. But wilderness designation does not close down an area to ongoing economic activities; it only prevents the initiation of those activities within an area. The Wilderness Act contains special provisions that permit activities such as mining and grazing to continue if they have occurred before an area's designation as wilderness. Moreover, conservationists maintain that most of the lands that are rich in economic resources—both renewable resources such as timber and nonrenewable resources such as oil, gas, and strategic minerals—have already been designated as multiple-use areas.

Opponents of wilderness designation also claim that it is undemocratic, benefiting only a minority who enjoy backpacking and primitive camping. In *A Sand County Almanac*, Aldo Leopold countered such criticisms: "The basic error in such argument is that it applies the philosophy of mass production to what is intended to counteract mass production. The value of recreation is not a matter of ciphers. Recreation is valuable in proportion to the intensity of its experiences, and to the degree to which it *differs from* and *contrasts with* workaday life. By these

criteria, mechanized outings are at best a milk-and-water affair."

Wilderness designation does not deny recreational opportunities to anyone. For those who must rely on mechanical means of transportation, such as some handicapped and elderly persons, and for those who prefer it, the majority of the public lands offer access to natural areas. What designation does do is offer roadless areas to those who are able and who desire a nonmechanized outing. Our public lands provide a broad spectrum of recreational opportunities for people of all tastes and capabilities.

Social Boundaries

Why Is It Important to Preserve Wilderness?

Any conscious decision to preserve something—a species, an ecosystem, a cultural resource—implies that the object or entity has value. The dominant values shared by the people in a particular society determine what that society will preserve. Consequently, the arguments offered for preserving wilderness are also social or cultural in nature. These include arguments based on personal benefits, economics, ecological benefits, and cultural heritage.

Personal Benefits. For many people, wilderness is a refuge. It offers solitude and an escape from the pressures and complexities of contemporary life. The beauty and expansive grandeur of wild lands inspire awe and renew the human spirit, thus enhancing mental and emotional health. Accordingly, the psychological justification is one of the most important arguments for wilderness preservation. In addition, the ruggedness of wild lands can be used to enhance physical health.

Venturing into the wilderness enables us to put into perspective our everyday concerns and to focus, instead, on the basic necessities and challenges of survival. In that way, wilderness can also help us to develop and exercise self-reliance. It allows us to shed the trappings of our civilization and the conventions and conditions of our culture; we can experience, even if only briefly, the condition of being as near to nature as is possible. Hence, the personal benefits of wilderness extend to both the mind and body.

Economics. Economic arguments also support the concept of wilderness preservation. A growing segment of the population enjoys and actively seeks wilderness experiences. As the popularity of camping, hiking, and nature study increases, the economic benefits derived from these pursuits will increase as well. Wilderness areas benefit the local tourist

The National Wilde

FIGURE 22-2: The National Wilderness Preservation System, 1964–1989.

Preservation System
9

end

tional Park Wilderness

tional Forest Wilderness

tional Wildlife Refuge Wilderness

reau of Land Management Wilderness

eas less than 20,000 acres in size.

Wilderness

industry (accommodations, food) and the businesses involved in the manufacture and sale of outdoor gear and equipment.

Ecological Benefits. Ecological arguments are probably the most critical factors in support of wilderness preservation. Wild lands, as long as they are of sufficient size, are able to preserve ecosystem diversity, species diversity, and genetic diversity. The Greater Yellowstone Ecosystem illustrates the principles and applications of preserving biological diversity within an area of sufficient size. The area consists of the park, nearby Grand Teton National Park, and seven adjacent national forests. It is the last, best refuge in the contiguous United States for *Ursus horribilis*, the grizzly bear (Figure 22-3). Each male grizzly requires a range of approximately 200 square miles (518 square kilometers). The bears cross many types of habitats as they range, making the grizzly an important indicator species of the health of the entire ecosystem. If a healthy population of grizzly bears, which depend on so many different habitats, can survive, then those habitats are probably in good shape as well. A decline in the grizzly population may be an indication that a problem exists in one or more of the habitats within the bears' range.

Wilderness offers other ecological benefits. Healthy, unaltered ecosystems protect airsheds and watersheds and perform numerous other important functions. Additionally, wilderness areas are valuable for myriad research purposes. Scientists can study these undisturbed areas to determine how human actions have altered similar, but unprotected ecosystems. Using the wilderness ecosystem as a reference, they can gauge the effect of cultural actions and the extent of resulting changes in the ecosys-

tem. Further, wilderness areas may provide scientists with the clues and understanding required to restore the ecological health and integrity of altered ecosystems. Nature itself is likely to be the best teacher we have concerning how to heal ecosystems and restore stability. In addition, undisturbed areas are a natural laboratory in which evolutionary and ecological change can proceed apace without human influence or disruption.

Cultural Heritage. For our not-so-distant ancestors, wilderness was a formidable opponent against which they struggled, and sometimes lost, as they carved this nation out of the North American continent. For many people today, the connection of wilderness to our past and all it represents is reason enough to preserve wilderness. Destroying wilderness means destroying an integral component of our national heritage. If we relegate wilderness to the history books and the far reaches of our collective memory, we will have obliterated a part of our national identity. If, on the other hand, we preserve wilderness, we keep alive a fundamental shaper of these United States. Not everyone will be drawn to wilderness as a source of strength and renewal, but it will be there just the same, healthy and self-renewing, able to renew us and our children and our children's children.

History of Management of Wilderness

How Did the Wilderness Preservation Movement Develop?

The philosophical justification for wilderness preservation is embedded in the writings of early American philosophers and naturalists. To quote two of them, Henry Thoreau wrote, "In wildness is the preservation of the world," and John Muir said, "You lose consciousness of your own separate existence; you blend with the landscape and become part and parcel of nature." But it was not until the twentieth century that scientists and naturalists began to realize the ecological importance of protecting wild lands and began to work for the establishment of a formal system of protected wilderness (Figure 22-4).

Arthur Carhart and Aldo Leopold were both Forest Service employees working in the West in the 1920s. Though it had long since seen its wildest days, the land was still fairly unsettled and open. Carhart, surveying the Trapper's Lake area of Colorado for possible resort development, and Leopold, working the ranges around the Gila River of New Mexico, both began to urge the Forest Service to preserve certain roadless areas in their natural,

FIGURE 22-3: *Ursus horribilis*, grizzly bear. Grizzlies are omnivorous, feeding on berries as well as prey (especially elk and deer) and carrion.

(a)

(b)

FIGURE 22-4: (a) John Muir, right, with President Theodore Roosevelt at Glacier Point, Yosemite Valley, in 1906. A proponent of the modern preservationist movement, Muir spoke eloquently of the ability of wild lands to restore the human spirit. He was instrumental in having Yosemite declared a national park in 1890. (b) Robert Marshall on a canoe trip through the Quetico-Superior area of Minnesota in 1937. An avid outdoorsman, seasoned hiker, and employee of the Forest Service, Marshall spurred the preservation movement in the middle of the twentieth century.

undisturbed state, devoid of human structures and influence. At their prodding, the service abandoned its plans for a resort at Trapper's Lake and in 1924 set aside a portion of the Gila National Forest as the first "primitive area" within the national forests.

Another Forest Service employee, Robert Marshall, emerged as the leading advocate for wilderness preservation. Soon after joining the service in 1930, he wrote an article for *Scientific Monthly* which later became known as the Magna Carta of wilderness. In it, he defined wilderness as a region containing no permanent inhabitants and "no possibility of conveyance by mechanical means and . . . sufficiently spacious that a person in crossing it must have the experience of sleeping out." He believed that large size was an essential characteristic. According to Marshall, the two dominant attributes of such a region are, first, it would require any individual who exists in it to depend exclusively on personal efforts for survival, and second, it would preserve as nearly as possible the primitive environment, thus barring all roads, power transportation, and settlement.

As the Director of Forestry for the Bureau of Indian Affairs, Department of the Interior, Marshall created 16 new wilderness areas within the reservations. While head of the Division of Recreation and Lands, Forest Service, Marshall personally financed out-of-pocket expenses for a survey of remaining roadless areas greater than 300,000 acres (121,460 hectares) within the National Forest System. He conducted the survey himself over a span of two years, identifying 46 areas, 32 of which were in the western forests. He wished eventually to preserve 45 million acres (18.2 million hectares), approximately 9 percent of the National Forest System.

Marshall, like Carhart, Leopold, and a handful of others, ran up against several powerful obstacles to wilderness preservation. The logging and mining interests opposed making any area off-limits to development. They were supported by western politicians and others who sought the immediate economic benefits of development. Even organizations not directly involved with issues of preservation sometimes presented a problem for the movement. The Civilian Conservation Corps, initiated during the Depression to give work to unemployed young men, often created work to keep its men busy, and road building was especially popular. Unfortunately, building roads in areas where they were not needed destroyed that area's wilderness quality, and roads literally paved the way for other disruptive activities, particularly logging.

In 1935, concerned by the threats to wilderness preservation, especially the reluctance of land-managing agencies like the Forest Service and Park Service to preserve wilderness in the face of opposition from powerful economic interests, Marshall, Benton MacKaye (developer of the Appalachian Trail), Robert Sterling Yard (a journalist and former publicity chief for National Park Service Director Stephen Mather), Leopold, and a few others formed the Wilderness Society. At his death just three years later, Marshall bequeathed a large sum of money to the fledgling organization, enabling it to grow and achieve many conservation successes. More than 50 years after its inception, the Wilderness Society continues to act specifically as an advocate for wilderness preservation.

How Were Wilderness Areas Managed Prior to the Wilderness Preservation Act?

Prior to the signing of the Wilderness Preservation Act in 1964, there was no standard procedure for identifying, designating, and managing wilderness areas nationwide. The Forest Service did make several unsuccessful attempts at establishing such procedures, but by 1957 the Forest Service regulations had added a paltry 350,000 acres (141,700 hectares) to the wilderness administration system, just a fraction of the 14.2 million Marshall had personally protected while with the Bureau of Indian Affairs.

With the onset of World War II, the utilitarian philosophy within the Forest Service—which emphasized economic uses over recreation and wilderness preservation—became even stronger. Officials advocated the extraction of all natural resources within the national forests. Postwar economic growth exacerbated this trend. In the 1950s logging in the national forests became far more intense. But utilitarianism was not limited to the Forest Service or the Department of Agriculture. The secretary of the Department of the Interior supported a plan to build two major dams within the confines of Utah's Dinosaur National Monument, a unit of the National Park System. After a lengthy and hard-fought battle, opponents managed to squelch the plans and preserve the monument.

Though the battle over the Dinosaur was an important one, conservationists did not revel in their victory for long. They realized that development pressures would not be easily thwarted. Without national legislation designed to designate, protect, and govern the management of wilderness areas, the future held a continuing series of battles between developers and conservationists over the fate of wild areas throughout the nation's public lands. In such a future the developers, with greater economic resources on their side, would almost certainly be in a better position, and valuable wilderness lands would surely be lost. To prevent that scenario, conservationists began to mount a campaign to secure legal protection for wild lands throughout the United States.

How Was the Wilderness Act of 1964 Enacted?

The threat to Dinosaur National Monument spurred Howard Zahniser, president of the Wilderness Society, to begin drafting a bill to give legal protection to wilderness. Hubert H. Humphrey, the Democratic senator from Minnesota, introduced the bill in 1957. The opposition, consisting mainly of commercial interests (logging, mining, and ranching), was well organized and well financed and had significant political clout. Ironically, they were joined by the Park Service and the Forest Service. The former felt that the Organic Act, which created the Park Service, gave it the mandate to preserve and protect wilderness. Park Service officials considered the bill an indictment of their inability to manage wilderness. Forest Service officials felt that their agency alone could make such land use decisions. They argued that agency guidelines already afforded a vehicle for wilderness preservation.

Ultimately, the debate divided the country along geographic lines, with the South and Northeast in favor of wilderness legislation and the West opposed. It was a long battle. Zahniser rewrote the bill 66 times, choosing his words painstakingly: "a wilderness, in contrast with those areas where man and his own works dominate the landscape, is . . . recognized as an area where the earth and its community of life are untrammeled by man, where man himself is a visitor who does not remain."

Zahniser died one week after giving his last congressional testimony in support of wilderness preservation. Less than five months later, on September 3, 1964, the legislation for which he had worked so long and tirelessly was signed into law by President Lyndon Johnson. Wilderness preservation, long a concept in the American mind, had finally become fact.

The Wilderness Preservation Act of 1964 directed the Park Service, Forest Service, and Fish and Wildlife Service to review their holdings and recommend areas that qualified for protection. (Not until

1976 was the BLM also required to review its holdings for possible wilderness designation.) Congress was given the responsibility to approve the recommendations and to designate areas for inclusion in the National Wilderness Preservation System (NWPS). The act also enabled Congress to act independently to designate new wilderness areas. This provision slowed the designation process somewhat, but it invited wider public participation, something preservationists felt necessary.

To secure congressional support for the wilderness bill, preservationists were forced to compromise on several points. Perhaps the most important compromise concerned the exploration and development of mining, prospecting, and oil and gas drilling in wilderness areas. These activities were allowed to continue at the discretion of the secretary of the interior until December 31, 1983. Any claims filed by that date could be developed at any time in the future.

How Is an Area Designated as Wilderness?

An area being considered for inclusion in the National Wilderness Preservation System is known as a wilderness study area (WSA). The National Park Service, National Forest Service, Fish and Wildlife Service, and Bureau of Land Management are required to conduct thorough studies of all WSAs under their jurisdiction and recommend for wilderness designation those areas they believe meet congressional qualifications (Table 22-1). Congress then decides which areas and how much acreage to include. Specific bills are written to designate wilderness areas in various states. Those WSAs that pass congressional muster and the vagaries of the political process are then officially designated as wilderness and are included in the National Wilderness Preservation System.

According to the Wilderness Act, protected areas must be managed to meet the following objectives:

▶ Table 22-1
Qualifications for Wilderness Designation by Congress

A wilderness area...
Is affected primarily by the forces of nature, where humans are visitors who do not remain
Possesses outstanding opportunities for solitude or a primitive and unconfined type of recreation
Is undeveloped, federally owned, and generally over 5,000 acres (2,020 hectares) in size
Is able to be protected and managed so as to allow natural ecological processes to operate freely
May contain ecologic, geologic, or other features of scientific, educational, scenic, or historic value

to perpetuate for present and future generations a long-lasting system of high-quality wilderness that represents natural ecosystems; to provide opportunities for public use and enjoyment of the wilderness resource; to allow plants and animals indigenous to the area to develop through natural processes; to maintain watersheds and airsheds in a healthy condition; and to maintain the primitive character of wilderness as a benchmark for ecological studies.

How Did the Wilderness Act Affect the Preservation and Management of the Federal Lands?

The Wilderness Act immediately designated 9.1 million acres (3.7 million hectares) of land in 54 sites within the federal lands. Most wilderness areas were in the national forests, which were generally considered to hold the greatest potential for new additions to the National Wilderness Preservation System. Over the next 15 years additional wilderness areas were included in the system. In 1980, with the inclusion of more than 56 million acres (22.7 million hectares) of Alaskan wilderness, the system more than doubled in size.

Though the Forest Service did develop regulations for the protection and management of areas already classified as wilderness, the agency failed to recommend any additions to the wilderness system between 1964 and 1973. The Forest Service viewed wilderness designation as antithetical to the principle of multiple use and tried to buck the law on both sides of the Mississippi. In a landmark decision in 1969, Judge W. E. Doyle of the U.S. District Court for the District of Colorado ruled that Forest Service actions in the West thwarted "the purpose and spirit" of the Wilderness Act because they prevented congressional or presidential involvement in the designation process. In the East, meanwhile, the Forest Service was casting a very narrow interpretation on wilderness. Any evidence whatever of past alteration, even an abandoned mine shaft or overgrown track, eliminated an area from consideration for wilderness designation. Consequently, the agency maintained that it was virtually impossible to identify wilderness areas east of the front range of the Rockies. Of 256 areas surveyed in the East, only three were considered to be potential official wildernesses.

To correct the situation that was fast developing, Congress passed the Wilderness Act of 1973. The act immediately added 16 areas, covering some 207,000 acres (83,806 hectares) in 13 states. By its action, Congress made plain its opinion concerning wilderness designation: If an area had sufficiently recovered from human abuses and was on its way back to a pristine condition, it could be included in

(a)

(b)

FIGURE 22-5: (a) Most wilderness areas, including Idaho's Frank Church-River of No Return Wilderness, are in the western United States. (b) Additional wilderness areas are needed in the more populous eastern half of the country. So far, attempts to have part of the Great Smoky Mountains designated as wilderness have failed. The photo shows sunrise in Oconoluftie Valley, Great Smoky Mountain National Park, North Carolina.

the National Wilderness Preservation System (Figure 22-5).

The Park Service had its own problems with the Wilderness Act. The service wanted to put a road right through the center of the Great Smoky Mountains National Park, 400,000 acres (162,000 hectares) of which were being considered for wilderness designation. Under pressure from conservationists, the service eventually changed its plans and instead built a scenic loop around the perimeter of the park. Even so, opponents of wilderness managed to prevent the area from achieving official designation, and by 1990 the area was not yet included in the wilderness system.

What Is RARE?

In 1972 public pressure forced the Forest Service to launch RARE, roadless area review and evaluation. The results of the survey were disappointing. Only 12.3 million acres (5 million hectares) were recommended for wilderness designation, just 19 percent of the more than 56 million acres (22.7 million hectares) studied. The reaction by the conservation community was loud and forceful; critics maintained that the Forest Service had capitulated to timber and other economic interests. The Sierra Club brought a suit against the service, and to appease their detractors, Forest Service officials quickly agreed to prepare environmental impact statements, in accordance with the National Environmental Protection Act of 1969 (NEPA), before developing any roadless areas.

The Forest Service initiated RARE II, a far more comprehensive survey, in 1977, but again, the results were disappointing, with the agency recom-

mending only 15 million acres (6 million hectares) for inclusion in the wilderness system. Pressure from the conservation community prompted Congress to protect additional acreage based on the findings of the RARE II study. In 1984, for example, Congress passed more wilderness legislation for the lower 48 states than at any time in the previous 20 years. More than 8 million acres (3.24 million hectares) in 21 states, primarily in national forests, were officially designated as wilderness areas.

What Is the Federal Land Policy and Management Act?

The 1976 Federal Land Policy and Management Act directed the Bureau of Land Management (BLM) to review its holdings outside Alaska, identify all national resource lands that might qualify for inclusion in the National Wilderness Preservation System, place those areas under study, and, by 1991, recommend to the president areas that should receive official wilderness designation. The president would then have two years in which to convey those recommendations, or a variation thereof, to Congress for enactment.

Like the other land-managing agencies, the BLM did not enthusiastically embrace the concept of designated wilderness. In fact, many conservationists felt that the bureau was grossly negligent in its duty with respect to protecting and preserving wild lands. Although the Federal Land Policy and Management Act did not give Interior and the BLM the authority to decide which holdings should be initially reviewed for designation as a wilderness study area, the BLM assumed that authority. It identified 861 study areas, containing just under 23 million acres

(9.3 million hectares), and by 1988 recommended that 9.6 million acres (3.9 million hectares) in 704 of the study areas were suitable for inclusion into the wilderness system. The study areas recommended for wilderness designation by the BLM amounted to less than 4 percent of the agency's land holdings. Many conservationists strongly disagreed with the BLM's recommendations, maintaining that at least another 25 million acres (10.12 million hectares) should have been tagged as wilderness study areas.

How Are Wilderness Areas Currently Managed?

Designation alone cannot guarantee that a wilderness area will be preserved in perpetuity. Paradoxically, human intervention in the guise of management is needed to preserve an area in a state that is primarily characterized by its lack of human influence!

In the Wilderness Act of 1964 Congress instructed that wilderness areas be preserved in their essentially virgin state "for the use and enjoyment of the American people in such a manner as will leave them unimpaired for future use and enjoyment as wilderness." Accordingly, wilderness areas are open only for recreational activities such as hiking, sportfishing, camping, and nonmotorized boating. In some areas hunting and horseback riding are also allowed. Roads, structures, and logging are prohibited, as are livestock grazing and mining, unless the activity occurred before the area was designated as wilderness. Motorized vehicles, boats, and equipment are prohibited unless they are required in an emergency, such as fire control or a rescue attempt. Mineral exploration and identification are permitted (assuming a claim was filed before December 31, 1983) as long as motorized vehicles and equipment are not used.

To minimize the impact of visitors on wilderness ecosystems and thus preserve the unspoiled nature of the areas, wilderness managers have adopted a variety of tactics. To protect soil, water, and biological resources, officials may issue permits for designated campsites, limit the number of people hiking or camping at any one time, and develop trails. Additionally, wilderness managers may emphasize public education on wilderness values and appropriate use of these areas.

What Environmental Problems Face Wilderness Areas?

It is important to remember that only lands included in the National Wilderness Preservation System are officially managed as protected wilderness. Other wild lands remain undesignated and thus unprotected. Both those areas that enjoy official designation, "uppercase Wilderness" as the writer John G. Mitchell has called it, and those that do not, **de facto** (or, "in fact," if not in law) **wilderness,** face numerous threats to their ecological integrity.

To be sure, de facto wilderness is under greater threat, for if we do not officially protect it, there is little chance that it will remain wild for long. Too many competing interests are eagerly waiting to mine the rock, graze the ranges, plumb the soil, dam the waters, cut the timber, or roar across the hills, sand dunes, and dried salt beds in all-terrain vehicles, shattering the silence of wilderness as they unravel the delicate fabric of life it harbors.

It is also important to keep in mind that the National Wilderness Preservation System is only an administrative classification placed on lands under the jurisdiction of the federal land-managing agencies. Areas within the national parks, forests, refuges, and resource lands are designated as wilderness through a subjective process that engenders controversy more often than preservation. The agencies charged with recommending areas for inclusion in the wilderness system must balance wilderness preservation against the myriad other uses of federal lands. At best, wilderness review is a difficult and challenging process. At worst, it can become a farce, a review in name only, flawed by bureaucratic antipathy and intense lobbying by powerful development interests (Figure 22-6).

For de facto wilderness, the greatest threats are the loopholes of the designation process itself. Utah,

FIGURE 22-6: Balsam root and lupine in the Jarbridge Mountains Wilderness, Humboldt National Forest, Nevada. For twenty-five years, Nevada had only one designated wilderness area, the 64,000-acre Jarbridge, an original component of the National Wilderness Preservation System. Nevada's congressional delegation failed to protect any of the state's remaining wilderness until a bill passed in late 1989 designated an additional 13 wilderness areas, totaling over 700,000 acres.

with approximately 1.6 million acres (647,500 hectares) of designated wilderness, contains the lowest amount of protected acreage among the western states. Much of the state's vast federal lands are managed by the BLM. Many are among the wildest, most beautiful, and most interesting canyonlands our nation has to offer. Yet they have remained unprotected.

The BLM's wilderness review and recommendation procedures in Utah illustrate some of the flaws in the designation process. Millions of acres of de facto wilderness disappeared from the study process after district managers arbitrarily decided that the land did not offer "outstanding opportunities" for solitude or "a primitive and unconfined type of recreation." (According to the Wilderness Act, a roadless area must have one of these two characteristics in order to qualify for inclusion in the NWPS.) Conservationists point out that the BLM was not given the authority to make such discretionary judgments. Moreover, they claim that the real reason the lands were left out of the study process was economics: the land was reserved for exploitation by grazing and mining interests and developers (especially hydropower developers).

Historically, cattle ranchers in particular have exerted considerable influence over the BLM. In many potential wilderness study areas the BLM has undertaken "range improvement" projects that involve such drastic changes in the landscape that they effectively eliminate the areas from inclusion in the wilderness system. Perhaps the worst of these is the process of **chaining,** a technique that has scarred Tarantula Mesa and other areas in the Henry Mountains of southern Utah. Two bulldozers with an anchor chain strung between move slowly along, tearing down the pinyon-juniper forest, which has expanded in the absence of natural fires or prescribed burning. Grass seed is then applied to convert the denuded land to pasture. The purpose, according to the BLM, is to increase the mountains' carrying capacity for sheep and cattle. Sixty percent of the cost for "reclaiming" the land, as the agency calls its work, is borne by taxpayers; the ranchers' share is 9 percent, with the state contributing the balance.

Utah is not the only state in which conflicts have arisen between proponents and opponents of wilderness. In almost every western state there are significant conflicts over designation. In Colorado, for instance, the issue of federal reserved water rights in designated wilderness areas is an ongoing source of conflict. Reserved water rights ensure that upstream projects, such as dams and diversions for irrigation, will not deplete the water supply in an area. Under the Reagan administration the federal government failed to press for federal reserved water rights.

Arguing that water was not essential to the integrity of wilderness, the government effectively allowed special interest groups the right to drain rivers and streams before they reached protected wilderness areas. Economic self-interest compels some ranchers, farmers, and others to oppose reserved water rights, with the result that wilderness designation for many wilderness study areas in Colorado has been delayed.

Those areas designated as wilderness also face numerous threats. Chief among these are the problems caused by increasing recreational use: crowds, litter, damaged vegetation, soil erosion from heavily used trails and campsites, and water pollution. As more people use wilderness areas, the very qualities that define the areas are degraded. This is especially true in the eastern states, specifically New England. Some areas may soon require a permit system to limit use and protect the resource, a step already taken in the Boundary Waters Canoe Area Wilderness, the focus of Environmental Science in Action (pages 486–489).

What Initiatives Are Being Taken to Preserve Wilderness Areas?

In recent years a number of measures have been proposed to preserve wilderness areas more effectively. Of these, the most promising is a Memorandum of Understanding (MOU) developed in March 1990 between the Bureau of Land Management and the Nature Conservancy, a private, national, nonprofit organization whose primary interest is the preservation of biological diversity. According to the terms of the memorandum, the parties agree to "cooperate in identifying, evaluating, and protecting public lands that have exceptional natural resource values, and to place in public ownership [through land exchanges with private landowners] . . . lands identified as having such value." The memorandum does not in any way limit or affect the bureau's delegated authority, but if the bureau sincerely embraces the agreement, it could significantly improve its wilderness designation process.

The memorandum seems to represent an important change in the bureau's attitude toward the preservation of wilderness areas. Changing attitudes—not only of the federal land-managing agencies, but also of ordinary citizens, development interests, and wilderness users—are critical to the preservation of the nation's remaining wild lands. (See What You Can Do: To Preserve Wilderness.) To help effect that change in public attitudes and prevent the overuse and subsequent degradation of wild lands, the historian Roderick Nash has proposed that we divide wilderness areas into three categories.

The first would include the most popular and accessible areas. These would be intensively managed, with hiking trails, bridges, outhouses, assigned campsites, and extensive ranger patrols. The second category would include large and relatively remote wilderness areas. Left essentially unmanaged, these areas would be accessible to those who qualify for a special license by demonstrating wilderness skills. The third category would include large and unique wilderness areas that are off-limits to all visitors. These areas would be left undisturbed to act as a gene pool of biological resources and a natural setting for ecological and evolutionary processes.

Future Management of Wilderness

The establishment of the National Wilderness Preservation System was an important achievement, signaling that we as a society recognize the value of preserving wilderness. However, designated wilderness areas remain vulnerable to overuse and to activities on adjacent lands, such as logging and mining. In addition, many unique and valuable stretches of de facto wilderness remain unprotected by wilderness designation.

Prevent Overuse Through Conservation

Several measures can be undertaken to prevent the overuse of wilderness areas. A modest increase in grazing and mining fees, in those wilderness areas in which these activities are permitted, could generate additional funds for the upkeep of wilderness lands. Environmentally destructive range improvement techniques, specifically chaining, should be eliminated. Instead, prescribed burning could be used to control the spread of pinyon-juniper communities into grasslands and help to restore and maintain biological diversity.

Protect Human Health and the Environment

Protecting human and environmental health means safeguarding critical wilderness areas throughout the country. It is imperative that wilderness areas be designated and maintained in the populous eastern half of the United States. Several areas of high priority include Michigan's McCormick Tract and the Southern Appalachian Highlands (northern Georgia, eastern Tennessee, western North Carolina).

For many people, wilderness areas are essential for psychological and physical well-being. Sizable wilderness areas also provide essential environmental services: they help to maintain the gaseous composition of the atmosphere, generate and maintain soils, cycle nutrients, and help to preserve biological diversity. They also help to cleanse the atmosphere, hydrosphere, and lithosphere of pollutants.

Preserve Living Systems

The federal government should attempt to purchase lands representing the 50 ecosystems in the United States which are no longer on federal public lands. A minimum of three to five representatives of each ecosystem type is needed to protect against the catastrophic loss of an ecosystem and to preserve the greatest biological diversity.

A Wilderness Service that would have equal footing with the Park Service, Forest Service, Fish and Wildlife Service, and Bureau of Land Management should be established to ensure that an agency within the federal government will act as an advocate for wilderness preservation.

Preserving living systems can perhaps best be accomplished by making selected large and unique wilderness areas off-limits to all visitors. Protection would ensure that these areas remain undisturbed, thus safeguarding the biological diversity they harbor and allowing ecological and evolutionary processes to continue unimpeded.

485

Environmental Science in Action:
The Boundary Waters Canoe Area Wilderness
Greg Moody

The Boundary Waters Canoe Area Wilderness (BWCA) stretches for 120 miles (193 kilometers) along the United States–Canada border, providing a unique wilderness experience for the canoeists and campers who explore its vast network of lakes, rivers, and forests (Figure 22-7). Although the United States recognized the value of preserving this area in its natural state as early as 1902, the BWCA's history demonstrates the complexity of wilderness issues. Throughout the twentieth century conservationists have battled over the fate of the area with those eager to develop it for mining, logging, or recreation. Today, although the area is federally protected as wilderness, problems such as overuse, acid precipitation, and mining on adjacent lands threaten to degrade the wilderness environment and compromise its natural beauty.

Describing the Resource
Physical Boundaries
Spreading over approximately 1,030,000 acres (416,835 hectares) along the northern reaches of the Superior National Forest, the Boundary Waters Canoe Area is the largest wilderness area in the eastern United States and the nation's only canoe wilderness. Its most outstanding feature is the intricate tangle of more than 1,000 lakes linked by hundreds of miles of streams and short portages. The Ojibwa and Chippewa and, later, explorers and fur traders once navigated the canoe country. Bordering the area to the north is Canada's Quetico National Forest. Together, these two protected areas create a canoeist's paradise: over 3,000 square miles (7,776 square kilometers) of undisturbed woodland, islands, rivers, and lakes.

The BWCA owes much of its unusual nature to its geologic formation. The wilderness occupies the southern edge of the Canadian shield, a Precambrian formation extending from central and eastern Canada to the upper Midwest and New York. This formation was cre-

ated over 3 billion years ago, when volcanic lava flows solidified and then were uplifted and eroded. Ice age glaciers carved out the vast network of lakes and rivers we see today.

Biological Boundaries
Because of its size and proximity to relatively untrammeled areas in Canada, the BWCA is home to diverse species, including the timber wolf, largely exterminated in other areas of the United States, moose, deer, beaver, black bear, snowshoe hare, and porcupine. Bird watchers can find thrushes, partridges, and loons, and sportfishers can enjoy such fish as bass, northern pike, walleye, and lake trout. Despite extensive logging, the BWCA still contains 540,000 acres (218,620 hectares) of virgin forests, by far the largest area of forest untouched by logging in the eastern United States. These forests include stands of tall red and white pine, redolent cedar, and silvery birch.

Social Boundaries
The BWCA was one of the original areas included in the National Wilderness Preservation System, despite the fact that it was not "untrammeled" by human activity, as stipulated by the 1964 Wilderness Preservation Act. The area had endured decades of use and abuse by trappers, loggers, commercial fishers, and resort owners. However, the efforts of those committed to preserving the area's wilderness qualities, combined with its unique beauty and vast stretches of still untouched wild lands, earned the BWCA its wilderness designation.

Since that time, the National Forest Service, which manages the BWCA, has accomplished a transformation: many areas previously scarred by humans have now returned to almost pristine conditions, such as logged areas replanted with indigenous trees. The federal government also bought up private holdings in the wilderness, eliminated resorts and private cabins,

banned the initiation of mining, logging, and other commercial activities, and banned the use of snowmobiles, chainsaws, and machines in general. Even airplanes and outboard motors are restricted.

Today, the BWCA is the most heavily visited unit of the entire wilderness system. In 1987, the area had 180,000 visitors, a 10 percent increase from the previous year. In addition to problems caused by heavy use in general, the area has some problems caused by uneven concentration of use. Overnight visitors cluster in the 2,200 designated campsites. Over 50 percent of visitors enter through 7 entry points, while the other 80 entrances go relatively unused. The most popular portages are subject to erosion and crowding, thus damaging the environment as well as detracting from the wilderness experience.

The BWCA is more than a haven for canoeists and campers; it also serves as a valuable educational and scientific resource. The area's rich ecosystem has been the subject of research in wildlife behavior, acid deposition, forest ecology, nutrient cycles, lake systems, and vegetation history.

Looking Back

Since the early 1900s, this land has been the subject of controversy: should its resources be open to private industry and individual ownership, or should the unique area be preserved in its wild state for public enjoyment and as a standard of true wilderness?

Managing the BWCA Before the Wilderness Act

Growing public concern over watershed protection and similar conservation issues led the U.S. Land Office in 1902 to set aside 500,000 acres (202,420 hectares) in the area now known as the BWCA. This land was intended as a forest reservation, protected from settlement and development. The Land Office withdrew an additional 659,000

486

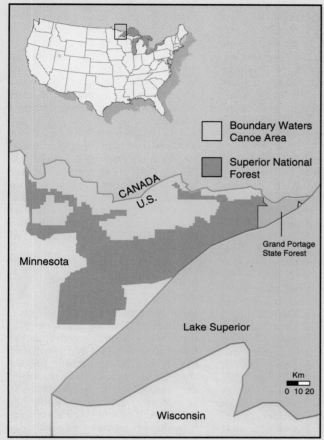

FIGURE 22-7: Boundary Waters Canoe Area Wilderness.

acres (266,800 hectares) in the same area by 1908.

In 1909 President Theodore Roosevelt designated 1.2 million acres (about 500,000 hectares) of the previously withdrawn area as the Superior National Forest. Quetico Forest, established the same year by the Canadian government, achieved park classification within four years.

Management of the area was confused and complicated from the beginning. Arthur Carhart visited Superior in 1919 and again in 1921 and recommended that it be kept as close to wilderness as possible. In 1926 Agriculture Secretary William Jardine halted plans by local governments to build roads to many of the lakes. He established a policy to protect the wilderness quality of Superior by prohibiting unnecessary road construction and preserving a 1,000-square-mile area with no roads whatsoever.

But other interests had their eyes on Superior. Proposed dams, requiring the construction of multiple roads, would have raised some lake levels 80 feet (24 meters), drowning scenic rapids and waterfalls. Conservationists pressured Congress to protect Superior from these ravages by law, and in 1930 the Shipstead-Newton-Nolan Act was passed. This act—the first in U.S. history specifically designed to protect wilderness—prohibited logging within 400 feet (120 meters) of lakeshores, forbade any alteration of natural water levels (such as those caused by hydropower dams), and withdrew public lands in the area from homesteading. Hydropower interests

tried to have the act repealed, but were unsuccessful.

Throughout the 1930s the federal government added to the protected acreage in Superior National Forest by buying tracts of land and repossessing tax-delinquent property. Unfortunately, the government frequently lacked funds to purchase pieces of land especially suited or crucial to the wilderness area.

After World War II resorts sprang up on privately owned land in Superior. Some of these resorts brought visitors in by floatplanes, which conservationists felt insulted and disturbed the wilderness environment. Although the wilderness area became known as the Superior Roadless Primitive Area in 1938, logging operations soon grew up at its perimeter. In addition, some logging companies pressured the Forest Service into allowing them to make incursions into the roadless area itself.

To help protect Superior from commercialization, concerned individuals and groups, such as the Izaak Walton League, gave or sold (at a loss) many key holdings to the Forest Service. For example, beginning in 1937 lawyer Frank Hubachek contributed hundreds of acres. The 1948 Thye-Blatnik Act gave the Forest Service authority to purchase private holdings scattered throughout the wilderness area. Congress provided $9 million to fund the purchase of private lands. In 1949 President Harry Truman joined the preservation effort, establishing an unprecedented 4,000-foot (1,200-meter) airspace reservation over the wilderness area that effectively prevented floatplanes from landing on wilderness lakes. Sportspeople claimed that the order was unconstitutional, but a district court of appeals backed Truman's policy.

The protection of the BWCA was fought by a consortium of powerful interests who wanted to develop the area: logging and mining companies, resort owners, and people who wanted to enjoy "nature" in mechanized comfort. These groups accused conservationists of wanting to create their own private wilderness or to increase the value of their own land holdings. Despite the controversy, in 1954 Congress extended the Thye-Blatnick Act to include almost all of the present BWCA and provided $2.5 million for land purchases. In 1958 the Superior Roadless Primitive Area officially became known as the Boundary Waters Canoe Area.

Effect of the Wilderness Act on the BWCA

Conservationists soon realized that despite the apparent support of Congress, the wilderness quality of Superior National Forest remained in danger. The Forest Service seemed unable to decide if logging or preservation was the higher priority, and confusion over what was and was not permitted in the BWCA abounded. With the passage of the 1964 Wilderness Act, the BWCA became a key unit in the new National Wilderness Preservation System. Unfortunately, the act did not end confusion over use of the BWCA because it allowed logging and motorized vehicles to continue in some sections of the area.

In a compromise effort to appease conservationists and developers, in 1965 the secretary of agriculture proposed a management plan dividing the BWCA into two zones. The interior zone contained over 618,000 acres (250,100 hectares) and 90 percent of the water area. Logging was strictly prohibited in this zone. Logging was allowed in the portal zone, the remaining 412,000 acres (167,000 hectares) of the area.

In 1972 the Minnesota Public Interest Research Group and the Sierra Club sued the Forest Service and several logging companies in an attempt to protect virgin forests in the portal zone. The federal district court that heard the case sided with the conservationists, but on August 30, 1976, a federal appeals court reversed the ruling to permit logging. The appeals court added that virgin timber cutting in the BWCA could be prevented only by substantive changes in policy made by Congress or the Forest Service.

Additional lawsuits sprang up over mining and motorized vehicles. Often, those who favored development and motorized recreation lived near the BWCA, while those who wanted to preserve the wilderness character came from St. Paul, Minneapolis, Chicago, and other distant cities. The conflict became regional and socioeconomic as well as use-oriented.

In 1978 Congress finally settled these management conflicts by adopting the BWCA Wilderness Act which ended logging and road building in any area, severely restricted mining, and provided funding to buy property from resort and lodge owners. It also cut motorboat use from 60 percent of the water area to 33 percent, with the eventual goal of phasing it out altogether. To protect and enhance the natural qualities of the area, the act called for restoring natural conditions to areas damaged or altered by human activity. Congress tried to compensate businesses and resorts for the removal of operations from the BWCA by providing more facilities outside its borders. The government also offered to assist the local merchants by purchasing resorts that became unprofitable as a result of the act.

Problem of Overuse in the BWCA

Although the BWCA Wilderness Act gave the area substantial protection from outside threats, it did not address an equally serious problem: the increased demand for authentic wilderness experiences (Figure 22-8). By the early 1970s the problem of overuse had grown so severe that managers worried that it would permanently degrade parts of the BWCA ecosystem.

In 1976 the Forest Service initiated a visitor distribution program to control overuse. It used travel simulator models to place daily limits on the number of visitors who could enter the wilderness. The computer models, which simulated lake-to-lake movements by campers, recommended optimum entry rates based on predicted occupancy levels of lakeside campsites. At first, the user

control program encountered opposition from commercial interests in the area, including canoe outfitters, resorts, and camps. Eventually, though, these groups were persuaded to support a program of gradual adjustment to recommended entry levels. The entry program went into effect in 1979.

All visitors must have entry permits between May 1 and September 30, the most popular season to visit the BWCA. These permits are free, but can be reserved ahead of time for a $5 charge. The number of people entering as a group is limited to 10 during the May–September season; larger parties cannot receive permits.

Current Environmental Threats to the BWCA

Many external and internal forces continue to threaten the integrity of the BWCA. Mining operations (for minerals such as copper-nickel-sulfide ores and precious metals) that occur outside the area's borders can affect the wilderness. Air and water pollution know no boundaries. Disrupting habitat outside the wilderness frequently disturbs animal populations that roam in and out of the protected area. Only by expanding the wilderness or establishing a buffer zone where damaging activities like mining are prohibited can we adequately protect the BWCA.

Although President Truman's airspace reservation saved the BWCA from the intrusion of floatplanes, a new aeronautical threat emerged in later years. A military airspace called the Snoopy Military Operations Area has been established over the BWCA, just above the protected airspace. Military jet flights, which increased in frequency by 10 times between 1983 and 1986, produce sonic booms and exhaust vapors that could disturb animal populations, inhibit plant growth, and generally detract from the wilderness experience. The military has not thoroughly studied how these flights affect the BWCA, although conservationists claim that studies of similar problems reveal a significant potential for environmental damage. In 1988 the Friends of the Boundary Waters Wilderness and other groups initiated legislation to prevent use of the airspace until adequate environmental reviews have been conducted, but the outcome of this action is still unknown.

Acid precipitation is another external threat. Because aquatic ecosystems are so sensitive to acid deposition, the BWCA's thousand lakes and related waterways make the area especially vulnerable, and the vast tracts of forest are at risk as well. A Minnesota study rated the BWCA as the area in the state most sensitive to acid deposition. To combat the problem, Minnesota passed the Acid Deposition Control Act in 1982, which directed the state to identify areas especially sensitive to acid deposition, set standards for those areas, and devise controls to maintain those standards. Unfortunately, since much of the acid deposition comes from outside the state, Minnesota acting on its own cannot eradicate this threat.

Other external threats include a proposed National Guard Training Facility near the BWCA's border. This facility would include areas for tank maneuvers, artillery ranges, and other activities likely to impinge upon and degrade the wilderness. Power plants, logging operations, and other commercial operations outside the BWCA can harm the wilderness as well. Fortunately, groups like the Friends of the Boundary Waters Wilderness sponsor citizen wilderness surveillance programs, in which concerned individuals uncover, report, and combat activities or developments that might harm the area.

Internal forces also pose a threat to the BWCA. Although the quota system has helped protect the area from overuse, even a modest number of visitors can destroy vital components of the area's ecosystem if the visitors do not understand and respect the wilderness environment. The Forest Service has found that well-publicized regulations, such as the prohibition of glass and metal containers, helps to solve this problem. Regulations are even more effective when explained by a uni-

FIGURE 22-8: Gunflint Lake, Boundary Waters Canoe Area Wilderness.

formed aide; a 1981 program in which an aide handed out and explained a brochure on tree damage at a campground reduced tree damage by 81 percent. The Forest Service must also repair what damage does occur. Besides replanting trees lost in logging operations, it is considering a program to reintroduce woodland caribou, a species which was driven out of the BWCA by human activity.

Looking Ahead

Conservation groups and other concerned citizens are vital to the health of the BWCA wilderness. The Friends of the Boundary Waters Wilderness played a major part in discontinuing the use of 8 small dams in the area, establishing pesticide regulations, and developing a program to convince the owners of mineral rights within the BWCA to give up those rights. Another group, the Boundary Waters Wilderness Foundation, has been instrumental in initiating acid precipitation studies and the proposal to reintroduce woodland caribou.

In order to encourage public involvement, the Forest Service conducts a variety of public infor-

mation programs. For instance, a slide and tape program instructs visitors on proper wilderness behavior when they pick up their entry permits. The Forest Service could expand educational efforts through activities such as school tours and workshops. It could also encourage universities in the area to offer courses in wilderness management and community education classes on wilderness issues.

Although public support is crucial to the maintenance of the BWCA, the area must also have an efficient and effective management structure if it is to endure as a lasting, protected wilderness. Currently, different aspects of the BWCA fall under the authority of several groups, including the Forest Service, the Fish and Wildlife Service, the Minnesota legislature, and Congress. An independent review board might be established to coordinate and evaluate the actions of all the agencies involved, or the agencies could be brought together into a compact specifically designed for managing the BWCA. One way or another, management must be coordinated if inaction and confusion among well-meaning parties are not to be the downfall of this hard-won wilderness area.

Summary

Wilderness can be defined in many different ways, but ecologists generally agree that a wilderness ecosystem is one in which the biotic and abiotic components are minimally disturbed by humans.

Opponents of wilderness preservation argue that it "locks up" valuable resources which could be used for economic development and that it benefits only a minority of recreational users. However, wilderness can provide many benefits to all people. Wilderness is a refuge from the pressures of contemporary life and can help people develop self-reliance. It can promote economic gain from recreation industries and tourism. It can preserve a diversity of ecosystems, individual species, and genetic information within species. Healthy ecosystems protect airsheds and watersheds and are valuable for research. Undisturbed areas enable evolutionary and ecological change to occur free from cultural influences.

Among the first advocates of wilderness preservation were Arthur Carhart, Aldo Leopold, and Robert Marshall. The Wilderness Society, formed in 1935, continues to act as an advocate for wilderness preservation.

After World War II economic uses of land were emphasized over recreation and wilderness preservation. To counteract this emphasis, the Wilderness Preservation Act, introduced in 1957 and passed in 1964, directed the Park, Forest, and Fish and Wildlife services to recommend areas for protection and also enabled Congress to designate wilderness. An area being considered for inclusion in the National Wilderness Preservation System is called a wilderness study area.

The Wilderness Preservation Act of 1964 designated 9.1 million acres (3.7 million hectares) of federal land as protected wilderness. These were the original constituents of the National Wilderness Preservation System (NWPS). By the late 1980s the wilderness system had grown nearly tenfold. Though the Forest Service developed regulations for the protection and management of areas already classified as wilderness, it did not recommend any additions to the system between 1964 and 1973. To correct the situation, Congress passed the Wilderness Act of 1973.

The 1976 Federal Land Policy and Management Act directed the Bureau of Land Management to review its holdings outside Alaska, identify areas that might qualify for the National Wilderness Preservation System, place those areas under study, and, by 1991, recommend to the president those areas which should receive official wilderness designation.

Only lands included in the National Wilderness Preservation System are officially managed as protected wilderness. Both officially designated areas and de facto wilderness areas face numerous threats to their ecological integrity. For de facto wilderness, the greatest threats are the loopholes of the designation process. Designated wilderness areas are threatened by overuse (litter, damaged vegetation, soil erosion from heavily used trails and campsites, and water pollution).

Key Terms

chaining
de facto wilderness
wilderness ecosystem

Discussion Questions

1. Define wilderness. What characteristics of wilderness areas distinguish them from other public lands?

2. Do you think that wilderness areas should be preserved or opened for development? Give arguments to support your answer.

3. Trace the development of the wilderness preservation movement in the United States from the time of Thoreau to the mid-1930s. Can you draw a parallel between the movement and the closing of the American frontier?

4. Discuss the major problems facing both designated and de facto wilderness areas and suggest ways in which these problems can be eliminated.

5. In what ways does the Boundary Waters Canoe Area typify the current pressures and problems faced by wilderness areas throughout the country? In what ways does it differ?

23

Biological Resources

For one species to mourn the death of another is a new thing under the sun. The Cro-Magnon who slew the last mammoth thought only of steaks. . . . The sailor who clubbed the last auk thought of nothing at all. But we, who have lost our last pigeons, mourn the loss.

Aldo Leopold

I heard on the news that there are only about 1,000 okapis left, only a few bantengs and about 1,000 small pandas in the whole world! I did not know that. With the water up to our necks, so to speak, it is high time to build another ark. When you look around in the countryside, you do see many such an "ark"—always with that gigantic eggcup next to it— but those arks are already chockfull of chickens and pigs . . . will we ever build a solid ark for the little panda?

Rien Poortvliet

Learning Objectives

When you finish reading this chapter, you should be able to:

1. Define biological diversity. Differentiate between genetic diversity, species diversity, and ecosystem diversity.

2. Identify the major threats to biological diversity.

3. Explain why it is important to preserve biological diversity.

4. Briefly relate how plant and animal species have been managed historically.

5. Describe how plant and animal species are managed today both within and outside of their natural habitats, and explain the advantages and disadvantages of on-site and off-site management.

6. Discuss the major legislation enacted in the United States and worldwide to protect and preserve wildlife.

Perhaps we could redefine environmental science as the study of building a solid ark for the panda—and the okapi and the banteng and the giant redwood and *Homo sapiens*—in short, for all species that share this living planet. Time is running short. For some species, like the California condor, we may be too late; their numbers are so low that their survival, at least in the wild, seems doubtful. Although the human species is not yet endangered, our fate is inextricably intertwined with that of all other species. We depend on them for survival: they provide us with food, clothing, shelter, and even breathable air. Wildlife is also a source of wonder for the human mind (Figure 23-1).

In this chapter we discover what biological resources are, the relationship between extinction and biological diversity, and the habitats where biological resources are found. We learn how humans benefit from biological resources, how our beliefs, attitudes, and activities affect them, and why we should make preserving them a priority. We look at how wild species have been managed in the past and how they are currently managed both outside of and

FIGURE 23-1: The majestic snow leopard, native of Asia.

within their natural habitats. We examine the major legislation governing management of wild species in the United States and worldwide, and we suggest measures to safeguard irreplaceable biological resources in the future.

Describing Biological Resources: Biological Boundaries

What Are Biological Resources?

All species, from the smallest microorganisms to the great blue whales, are **biological resources.** Many scientists classify living things into five kingdoms: Monera, Protista, Fungi, Plantae, and Animalia. (Figure 23-2). Some biological resources, including crop plants, livestock, and pets, are domesticated. We discussed domestic biological resources in Chapter 10, Food Resources, and mention them again here, but we focus primarily on wild biological resources, the earth's most important and perhaps most threatened resource.

Biological diversity, or **biodiversity,** refers to the variety of life forms that inhabit the earth. We often think of biodiversity at the level of **species diversity,** the millions of distinct species that share this planet with us. Scientists have formally identified approxi-

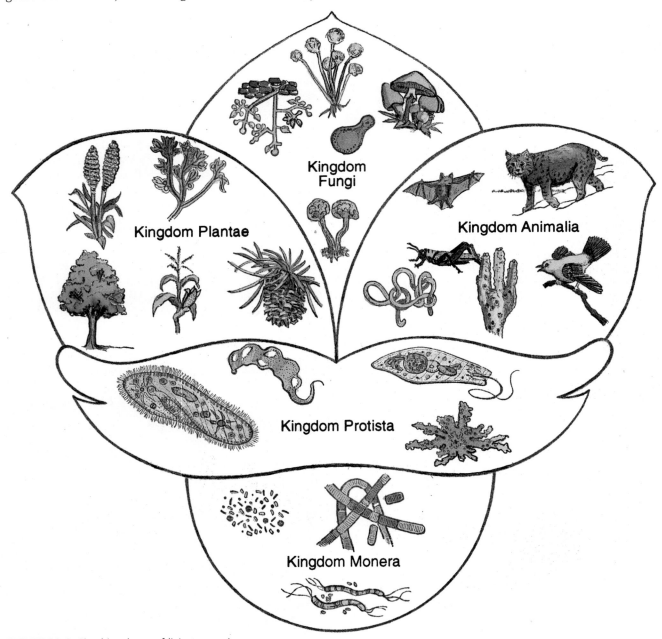

FIGURE 23-2: Five kingdoms of living organisms.

mately 1.4 million species. But this number represents only the species that they have named and categorized, and researchers have studied just 3 percent of these. (Many other species may be known and named by local peoples, but as indigenous cultures are increasingly fragmented and threatened by development pressures, the knowledge they possess may be lost as well.) The fact is that we do not even know how many species exist, *not even to the nearest order of magnitude.* Estimates of the total number of species range from 5 to 100 million, most of them microbes, insects, and tiny sea organisms. Only an estimated 1 percent of all species is larger than a bumblebee! Terry Erwin, a biologist with the Smithsonian Institution, conducted a now-famous study in the Panamanian tropical rain forest in the early 1980s. He discovered a tremendous number of previously unknown beetle species, so many that he estimates that the total number of insect species *alone* may be 30 million! In 1991 Edward O. Wilson and Paul Ehrlich, both highly-respected biologists and outspoken conservationists, estimated that there may be as many as 100 million species. Their estimate is based, in part, on Erwin's and others' work that indicates that the diversity of insects and other arthropods is probably much higher than had previously been estimated for the entire world flora and fauna. Wilson and Ehrlich point out that little study has been devoted to nematodes, fungi, mites, and bacteria, each of which is highly diverse, containing undescribed species that may total in the hundreds of thousands.

In addition to species diversity, we can measure biodiversity in terms of **ecosystem diversity,** the variety of habitats, biotic communities, and ecological processes in the biosphere. Some entire ecosystems, such as rain forests and coral reefs, are just as endangered as the individual species that inhabit them.

On the other end of the scale from ecosystem diversity is **genetic diversity,** which refers to the variation among the members of a single population of a species. Each member has a unique **genotype,** the individual's complement of **genes,** the fundamental physical units of heredity that transmit information from one cell to another and thus from one generation to another. The hereditary material of an organism is known as its **germ plasm.** The sum of all of the genes present in a population of organisms is known as a **gene pool.**

What Is the Relationship Between Extinction and Biological Diversity?

Whenever a species becomes extinct, biodiversity is diminished. That unique biological resource is lost

and cannot be replaced. Extinction is a natural process that has occurred throughout the earth's history, and many experts believe that only 1 to 2 percent of all the species who have ever lived on earth are alive now. But human activities have greatly accelerated the rate of extinction, perhaps by as much as 400 times (Figure 23-3).

Estimates of species extinctions range from 1 to 50 per day, a total of 365 to 18,250 each year. The International Union for the Conservation of Nature and Natural Resources (ICUN) estimates that 1 vertebrate species disappears annually. Worldwide, of the estimated 250,000 more complex plants, 40,000 to 50,000 are thought to be threatened; 200 native American plants are already extinct.

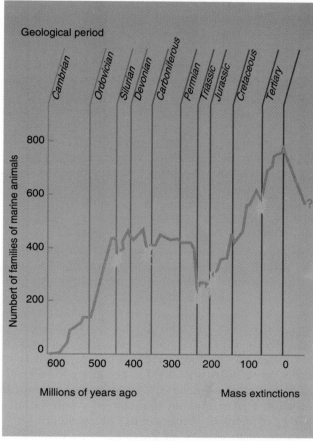

FIGURE 23-3: Extinctions over time, marine organisms. The diversity of the earth's life forms has gradually increased over time. Occasional mass extinctions—at the close of the Ordovician, Devonian, Permian, Triassic, and Cretaceous periods, when the number of families of marine organisms fell by 12, 14, 52, 12, and 11 percent, respectively—set back but did not do irreparable harm to biological diversity. However, human activity is now causing an unprecedented decline in biological diversity. If present trends continue, the 150 years from the beginning of the twentieth century to the middle of the twenty-first century will witness the extinction of more species than any other similar time span since life began.

If the total number of species on earth is much larger than was traditionally thought, as the work of Erwin and others suggests, chances are that the number of extinctions is also greater. Wilson and Ehrlich argue that as many as 100,000 species may become extinct each year. They point out that within the next 50 years we may lose one-quarter of all species now present on earth. Each area of rain forest or coral reef that is destroyed may be taking with it untold numbers of unidentified species. We simply do not know.

The disappearance of an individual species often has far-reaching effects. Each species plays a distinct role in its ecosystem, and when it is gone, the organisms that depended on it may be threatened as well. For example, in the Amazonian rain forest, female euglossine bees pollinate the Brazil nut tree, but the males pollinate a particular species of orchid during their courtship displays. If the orchid becomes extinct, the bees may also die out. The Brazil nut trees might go unpollinated, rendering them unable to set seed and produce nuts. Extinction also disrupts food chains. As the links between species are ruptured, the entire ecosystem becomes more vulnerable to stresses. Thus, diminishing species diversity threatens ecosystem diversity, just as the degradation of an ecosystem threatens the chances of survival of its inhabitants and thus threatens both species and genetic diversity.

Decreasing genetic diversity in turn threatens species diversity by making species more vulnerable to extinction. If the size of a population of a species decreases, as individuals die or are removed, the gene pool decreases. If no new individuals from other populations join the group, the population will be less genetically diverse than the original population, even if it regains its former size through new births. Further, the genetic variability among individuals will decrease with each successive generation because of inbreeding, unless individuals from other populations are added to the group. The resultant loss of genetic variability is known as **genetic erosion,** a phenomenon that lessens a population's ability to adapt to changes in its environment.

Genetic erosion is of great concern to wildlife biologists, zoo curators, managers of botanical gardens, and plant and animal breeders. The **minimum viable population size** is the fewest number of individuals necessary to maintain a viable breeding group without the short-term loss of genetic variability. This figure has not been determined for most species. Extensive, time-consuming, and expensive research is needed to determine it for a single species. It is unlikely that we will be able to conduct the necessary research for some species before some of their populations fall below the minimum viable population size.

Physical Boundaries

Where Are Species Located?

The planet harbors life everywhere, from the ocean floor to the rocky mountain summit, from high deserts to old-growth forests, from tropical pools to the icy waters of the Antarctic. All ecosystems host unique associations of interdependent organisms. Yet some areas are especially biologically rich, and some have an inordinate number of **endemic species**—species that are unique to the area and occur nowhere else. About 80 percent of Madagascar's plants, 50 percent of Papua New Guinea's birds, and 50 percent of the Philippines' mammals are endemic to those countries (Figure 23-4).

Rain forests are the most biologically rich of all ecosystems. Many of the megadiversity countries, so called because of the large numbers of species they contain, owe their biological wealth to rain forests. These countries include Brazil, which probably has more species than any other nation, Colombia, Indonesia, Peru, Malaysia, Ecuador, Zaire, and Madagascar. Second in terms of diversity are the coral reefs of the oceans, which have been called the rain forests of the sea. After the tropical rain forests, the land areas with the greatest diversity are those with a mild climate. The Mediterranean basin, coastal California, and the southern part of western Australia have a large number of endemic species. Several of the planet's biologically rich areas are the subject of Focus On: Where Diversity Reigns: Tropical Rain Forests, Coral Reefs, and Wetlands.

Social Boundaries

How Do Beliefs and Attitudes Affect the Management of Biological Resources?

Cultural attitudes and beliefs affect how people value, use, and manage biological resources. For example, different people may value an animal species for its meat and hide, for its beauty or symbolism, or for the sport it provides. (Chapter 25 discusses how beliefs and attitudes affect how we value and interact with nature.)

Some species that have become highly valued by some groups of humans have been called the **charismatic megafauna.** These are usually endangered mammals or birds, such as the California condor, bald eagle, panda, harp seal, and humpback whale, and their plight typically receives great media attention. The management of these species is well funded, by both private groups and governmental agencies. Although the focus of preservation is shifting more and more to saving habitat and entire ecosystems, the appeal of these species remains high

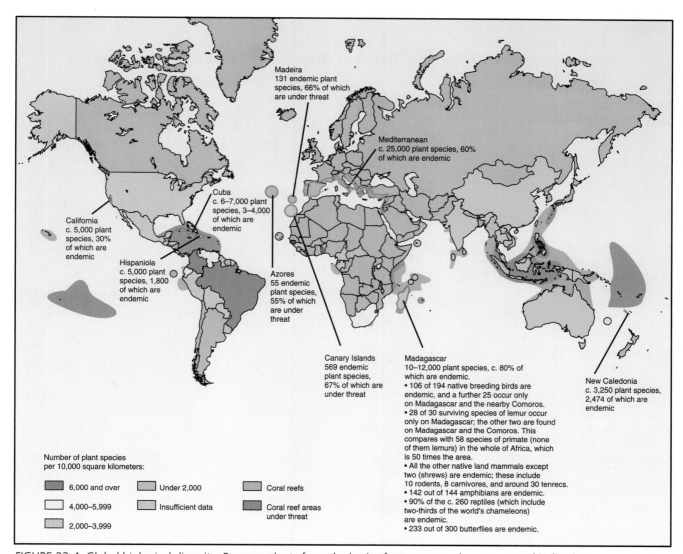

Madeira
131 endemic plant
species, 66% of which
are under threat

Mediterranean
c. 25,000 plant species, 60%
of which are endemic

Cuba
c. 6–7,000 plant
species, 3–4,000
of which are
endemic

California
c. 5,000 plant
species, 30%
of which are
endemic

Hispaniola
c. 5,000 plant
species, 1,800
of which are
endemic

Azores
55 endemic
plant species,
55% of which
are under
threat

Canary Islands
569 endemic
plant species,
67% of which are
under threat

Madagascar
10–12,000 plant species, c. 80% of
which are endemic.
• 106 of 194 native breeding birds are
  endemic, and a further 25 occur only
  on Madagascar and the nearby Comoros.
• 28 of 30 surviving species of lemur occur
  only on Madagascar; the other two are found
  on Madagascar and the Comoros. This
  compares with 58 species of primate (none
  of them lemurs) in the whole of Africa, which
  is 50 times the area.
• All the other native land mammals except
  two (shrews) are endemic; these include
  10 rodents, 8 carnivores, and around 30 tenrecs.
• 142 out of 144 amphibians are endemic.
• 90% of the c. 260 reptiles (which include
  two-thirds of the world's chameleons)
  are endemic.
• 233 out of 300 butterflies are endemic.

New Caledonia
c. 3,250 plant species,
2,474 of which are
endemic

Number of plant species
per 10,000 square kilometers:

⬛ 6,000 and over	◼ Under 2,000	◼ Coral reefs
◻ 4,000–5,999	◻ Insufficient data	◼ Coral reef areas under threat
◻ 2,000–3,999		

FIGURE 23-4: Global biological diversity. Because plants form the basis of ecosystems, they are a good indication of the biological diversity of a region. The map indicates the number of plant species per 10,000 square kilometers of land area. The number of endemic plants are indicated for selected areas. For the biologically rich island of Madagascar, figures are given for endemic plants and animals. The map also shows the location of coral reefs, the most diverse of aquatic habitats and second only to tropical rain forests in terms of biodiversity.

and they serve a useful function by alerting people to the need for preservation of all biological resources.

There are also some species that some people have historically learned to hate, from the wolf to the dandelion. We label undesirable animals **vermin** and undesirable plants **weeds.** Although we have spent a large amount of time and money trying to eliminate these misunderstood species, public attitudes toward some of them are beginning to change. The wolf, for example, is enjoying a surge in popularity in some areas of the United States, spurred in part by research and understanding of the animal and in part by books and films, such as *Of Wolves and Men* and *Never Cry Wolf*, that show the wolf in a sympathetic light. For many people the wolf, once considered vermin, has come to symbolize the

wilderness, and the wolf is likely to become more highly valued as people begin to value our shrinking wilderness more highly.

Species hunted for sport, such as Canada geese, are called **game animals.** They are highly valued by many different groups: people who hunt for recreation, manufacturers of hunting equipment, and people who regard the animals' presence and migration as a sign of continuity and stability. The management of game animals, like that of the charismatic megafauna, is well funded, with much of the financial support coming from hunting and fishing license fees, taxes on hunting and fishing equipment, permits, and the sale of duck stamps.

Nongame animals are species not hunted for sport. They include songbirds, meadow voles, turtles, and monarch butterflies. They are often the

(*Text continues on page 508.*)

Focus On:
Where Diversity Reigns: Tropical Rain Forests, Coral Reefs, and Wetlands

In the tropical rain forests, diversity reigns supreme. These forests are unparalleled in terms of number of species. Covering just 7 percent of the earth's surface, they are home to nearly half of all known species (Figure 23-5).

How species-rich are the tropical rain forests? According to the U.S. National Academy of Sciences, a typical patch of rain forest about 4 miles square (10 square kilometers) contains as many as 1,500 species of flowering plant, 750 tree species, 400 bird species, 150 butterfly species, 100 reptile species, 60 amphibian species, and an unknown number of insect species, (Figure 23-6). A single Brazilian river harbors more species of fish than all the rivers in the United States, and a single Peruvian wildlife preserve contains more bird species than does the United States. In one study, about 300 tree species were found in single-hectare plots in Peru; all of North America has just 700 native tree species. In that same study, just one Peruvian tree was found to host 43 species of ants, roughly equal to the total ant fauna of the entire British Isles. In the tropical forest of New Guinea as many as four species of pigeons, differing primarily in weight, rely on a particular fruit tree as their sole food source. The different species have adapted to feed on the fruit that grows on different-sized branches.

Tropical rain forests perform invaluable ecosystem services. They help to moderate both global and regional climates. Because carbon is stored in woody vegetation, tropical forests (and all forests) are an important global sink for carbon dioxide; intact and undisturbed, large stretches of forest may help to offset global warming. At the regional level, tropical forests play an important role in the cycle of precipitation, evaporation, and transpiration. Most rainfall is absorbed by plants and transpired, or is evaporated from the surface of vegetation and soil. Consequently, most of the precipitation that falls in the tropical rain forests is returned to the atmosphere, where it condenses

FIGURE 23-5: Tropical rain forest. The earth's most diverse habitats, tropical rain forests contain an estimated 30 percent of all bird species, 45 percent of all plant and animal species, and 96 percent of all arthropod species.

FIGURE 23-6: Tropical beetle, *Chrysochroa fulminasus*, Malaysia. Insects comprise the majority of species found in the tropics (as they do worldwide).

and again falls to the ground, beginning the cycle anew. Forests also hold the soil in place, minimizing erosion (Figure 23-7).

People in the developed world have benefited greatly from the products of the tropical rain forests: pharmaceuticals, woods, fibers, fruits, nuts, vegetables and other foods, spices, gums, resins, and oils. About 25 percent of all prescription drugs currently used in the United States were derived from tropical plants, yet only one percent of tropical plants have been studied for their medicinal properties. Unfortunately, these resources are typically ignored in economic assessments of forest use despite the fact that their value often outweighs that of timber.

The tropical rain forests have been reduced by almost one-half their original area. Between 1981 and 1990 an estimated 9 percent of the world's tropical forest was lost. In 1990, 42 million acres (17 million hectares) were deforested, an area equal to the size of Washington State. Some tropical forests have faced especially intense pressure; 98 percent of the tropical dry forest along the Pacific Coast of Central America is gone. Massive deforestation continues in Brazil, which has more tropical forest than any other nation. However, by national and international standards, the Brazilian Amazon is relatively untouched; nearly 90 percent of its forests are still standing. Unfortunately, logging and agricultural expansion have destroyed over 95 percent of Brazil's once extensive Atlantic coastal rain forests and the coniferous Araucaria forests in the south. Many other nations have essentially lost all their rain forests, including Benin, Côte d'Ivoire, El Salvador, Ghana, Haiti, Nigeria, and Togo. Where forests remain, they often occur in small fragments isolated by developed or degraded land. Fragmenting the forests interferes with vital ecological processes, such as nutrient cycling, and renders them less able to sustain viable populations of wildlife.

Tropical deforestation is linked to several causes. The first is economic

(a)

(b)

FIGURE 23-7: Costa Rica, with its diverse geography, boasts a wide variety of tropical forests including dry tropical rain forests and cloud forests. (a) Dry forests, such as the Palo Verde National Park, have a dry season and thus a different mix of species than that found in wetter areas. (b) Cloud forests, such as the Monteverde Cloud Forest Reserve, are found at higher altitudes than rain forests and dry tropical forests. Here, temperatures are cooler and much of the precipitation appears as mist or fog.

pressures. Exploiting the forest's timber resources or transforming it into pastureland for cattle ranching yields short-term benefits for a wealthy minority, but they are not sustainable and do not contribute to real, lasting economic development. Much of the timber logged from tropical forests is used to make cheap, disposable goods. Eighty percent of logs imported by Japan are used to make cheap plywood, much of which is used to make frames for molding concrete and scaffolding and is discarded after being used once or twice; tropical wood is also used to make disposable chopsticks and paper. In the United States and Europe, tropical wood is mostly used to make door and window frames, furniture, plywood, block-board, and veneer sheets, which could be manufactured from native hardwoods and softwoods harvested on a sustainable basis. Damage from logging extends beyond the actual process of cutting desirable trees. Many trees are injured or killed in order to remove target species; in Sarawak, Malaysia, 33 trees are damaged for every 26 trees removed. Logging also often opens up previously inaccessible areas to landless peasants. The extensive network of roads and skid trails needed to remove logged timber destroys even more forest. Erosion from roads and unprotected land chokes rivers with silt (Figure 23-8).

Tropical forests tend to contain large quantities of minerals and oil, and plans for exploiting these resources are increasingly contributing to deforestation. The Grande Carajas program in Brazil will open up one-sixth of eastern Amazonia to industry—an area the size of Great Britain and France. Some industries will mine for coal and bauxite, while others will be devoted to producing cash crops for export and sugar cane for the production of synthetic fuels. In the northwest Brazilian Amazon, thousand of "wildcat" miners have invaded, seeking gold, diamonds, uranium, titanium, and tin. These activities threaten the remaining homeland of the Yanomami Indians, who have already lost more than one-third of their territory to mining.

Forests are also being cleared for farming and ranching, typically agribusiness ventures conducted by the wealthy. The island of Negros in the Philippines, which once was almost completely forested, is now little more than a large sugar plantation. More than 50,000 acres

FIGURE 23-8: Newly constructed roads transect a recently burned tract of rain forest in Rondonia, Brazil.

(20,000 hectares) per year of the forested uplands of the island are being cleared every year as those who previously farmed the lower lands are driven into these marginal lands to grow their crops. In Thailand, much forest land in the east and northeast of the country has been cleared in order to increase the production of cassava, which is mainly exported to feed livestock in Europe. Peasants displaced by these cassava plantations are forced to clear more forest to grow the crops they need to survive. In Latin America at least 7,7000 square miles (20,000 square kilometers) of forest per year are lost to cattle ranching. Since 1950 two-thirds of the lowland tropical forests in Central America have been cleared, mostly for pastureland, and cattle now outnumber people in several countries. Most of this beef is exported, with 80 to 90 percent going to North America. This cheap imported beef has reduced the price of hamburgers by a few cents in North America, but the cattle industry does little to help the poor in the exporting countries, most of whom cannot afford to eat beef.

Large amounts of land in Brazil are also cleared for cattle ranches, but for a different reason: land speculation. Foot-and-mouth disease, which is endemic in Amazonia, prevents much beef from being exported, and even with government subsidies, only 3 out of 100 large ranches make a profit from livestock. However, under Brazilian law, anyone who clears an acre of forest may claim the land—and the mineral rights below it—and cattle ranching enables large amounts of land to be claimed with minimal labor. Not surprisingly, the areas most frequently cleared are close to gold strikes. Gold, not beef, makes the owners rich; at least half of the large ranches in Amazonia have never even sent a cow to market.

Another leading cause of deforestation in the tropics is slash-and-burn clearing for subsistence agriculture, which is directly related to population pressures and the inequitable distribution of agricultural land. In Central and South America, land ownership is concentrated in the hands of the wealthy elite. Less than 10 percent of the population typically owns nearly half the land; in non-Amazonian Brazil, just 4.5 percent of landowners control over 80 percent of the land. Their large holdings are often underused; land lies idle that could be more intensively cultivated to provide food crops and employment opportunities for the hungry and landless. The growing ranks of landless peasants, many of whom have been forcibly dispossessed of their own land to make way for development projects, must search elsewhere for the land and resources they need to survive; for many, that search has led to the tropical rain forests. Governments such as that of Brazil have tried to resettle large numbers of the urban poor in forest regions. These settlers use slash-and-burn techniques to clear small plots of land. Unfortunately, the soils underlying most of the tropical rain forests are not very fertile, since the nutrients are locked up in the vegetation. Moreover, the soils are easily degraded. Thus after only a few years the soil in cleared areas is exhausted and the peasants must move on to clear a new area, where the cycle is repeated.

Deforestation has many victims (Figure 23-9). The land itself is seriously and perhaps permanently degraded. The vast amounts of nutrients once bound up in the natural vegetation disappear, and the thin topsoil is susceptible to erosion. Nearby streams and rivers become choked with silt, killing many aquatic organisms. On the Philippine island of Palawan, erosion due to deforestation has almost destroyed coastal fisheries. Two-thirds of the rivers in Sarawak, Malaysia, are officially classified as "polluted" by soil erosion, and fish catches, which provide a major part of the native peoples' diet, have been drastically reduced. Flooding also becomes a a serious problem. Deforestation in India's watersheds more than doubled the amount of land classified as flood prone between 1971 and 1980. By 1984, the number had tripled. Peasants who try unsuccessfully to eke out a living from crops planted in a cleared area must move every few years to a new, more productive

FIGURE 23-9: Three-toed femal sloth, *Bradypus infuscatus*, sleeping in a tree, Panama. Deforestation means loss of habitat for an unknown number of species; loss of habitat is the single greatest threat to biological diversity in the tropics and worldwide.

area. Many Javanese peasants who were moved by the Indonesian government to the outer islands of the Indonesian archipelago in the mid-1970s and 1980s have been unable to survive in their new environment. Colonists were forced to clear more and more forests as previously cleared fields were degraded. In some areas, the soil is so poor that some colonists have starved, while others have returned to Java. The entire population of the region suffers if precipitation patterns are altered markedly. And finally, when wealthy landowners "develop" a tract of forest for timber or pastureland or when settlers clear an area to establish small farms, they displace the native peoples who originally lived there. These native peoples depend on the forest for food, shelter, medicines, and clothing, and have deep cultural and spiritual ties with the forest that gives meaning to their lives (Figure 23-10). Destruction of the forests in which they live causes physical and cultural destruction.

Slowing and preventing deforestation will require efforts in a number of areas. The first is to reform natural-resource accounting methods so that nontimber goods and ecosys-

FIGURE 23-10: Basketweaver from an indigenous tribe, Borneo. Among the endangered resources of the world's tropical rain forests are the cultures of the indigenous peoples who dwell in these dwindling habitats.

tem services are included in economic assessments of forest use (Figure 23-11). For example, in order to calculate the value of a forest, one should consider not just the economic value of the timber, but also the value of the forest as a carbon sink, habitat for wildlife and indigenous peoples, and moderator of the climate. The second is to change tenure laws in order to grant title or legal-use rights to forest inhabitants. Agrarian reform also can ease pressure on the rain forests; more equitable distribution of land would enable peasants to make a living on land that is suited

FIGURE 23-12: Great Barrier Reef, Queensland, Australia. Feather star crinoids, *Oxycomathus bennetti*, sitting atop a coral bommie.

FIGURE 23-11: In the Brazilian state of Para, a rubber tapper collects balata latex from *Manikara bidentata*. Rubber tapping is a sustainable extractive industry, providing natives with a reliable source of income while not damaging the forest.

for agriculture. Given a real opportunity to raise their standard of living, they would thus have a stake in the future.

International actions that can be taken to slow deforestation include reducing the demand for tropical timber, using debt reduction to finance conservation, and expanding and reforming development assistance (see Chapter 26, "Economics and Politics," for more on this subject).

Coral Reefs

Coral reefs have been called the rain forests of the oceans because of the abundance and diversity of life found there. Australia's Great Barrier Reef, which at 1,250 miles (2,000 kilometers) is the world's largest coral reef ecosystem, contains more than 3,000 animal species. The richest, most diverse coral reefs in the world are found in the Indo-West Pacific (Figure 23-12). These reefs contain more than 2,000 species of fish, 5,000 species of molluscs, 700 species of corals, and countless species of crabs, sea urchins, brittle stars, sea cucum-

bers, and worms. The diversity of coral reefs is largely due to great numbers of fish species, many as yet unidentified. Yet it is the hermatypic or reef-building coral animals themselves that make this spectacular ecosystem possible.

Reefs consist of the limestone skeletons of dead corals cemented together by the action of single-celled algae, called dinoflagellates, that live in association with the coral animals or polyps. The reef surface, which lies just below water level at low tide, is covered by a thin layer of live polyps. These small cup-shaped creatures, related to the sea anemones of temperate waters, have a ring of tentacles around a central mouth. Each individual secretes its own coral skeleton; these skeletons can be cemented together to form massive coral blocks. Most corals are small, and most live in colonies, but there are exceptions to both of these generalities (Figure 23-13).

Coral reefs are an aggregate of many coral colonies. The reefs grow rapidly; individual colonies grow at an estimated annual rate of 0.3 to 1 inch (8 to 25 millimeters). The

FIGURE 23-13: Coral polyps at night, Red Sea.

water between 72° and 82° F (22° to 28° C); they grow best in warm, shallow waters where light and oxygen are plentiful. They are rarely found at depths greater than 230 feet (70 meters). Hermatypic corals are not found in cool waters, such as those far north or south of the equator, and also do not occur on western continental coastlines, where zones of upwelling bring colder water to the surface. Coral reefs are also absent from areas where rivers transport fresh water and sediments to the ocean. Sediments smother the delicate corals and prevent light from reaching their surface, thus interfering with photosynthesis by the dinoflagellates.

There are three major classes of coral reef structures: fringing reefs, barrier reefs, and atolls. Fringing reefs grow as platforms from the shores of continents or islands. Barrier reefs are found offshore; deep lagoons separate the barrier reef from the coast. Atolls are very different from fringing reefs and barrier reefs. Found in the open ocean, atolls are circular reef structures that were once fringing-reef structures around volcanic islands (Figure 23-14). Rising sea levels or the sinking of the land surface caused

symbiotic dinoflagellates that live inside the polyps accelerate the rate at which the coral skeletons are formed, and they use carbon dioxide produced in the polyp's tissues during photosynthesis. In turn, the algae excrete nitrogenous materials that the polyp requires for growth. The shape of any one particular coral colony is a product of the water temperature, amount of wave action, and nutrient and light levels. These factors vary in different parts of a reef system.

Coral reefs develop in areas where the water is clear and the temperature does not fall below 68° F (20° C). These conditions typically occur in the coastal waters in the tropics. Hermatypic corals thrive in

FIGURE 23-14: Society Islands, French Polynesia.

the island to disappear below the water's surface. The reefs continued to grow upward, however, until only the coral remained as a circular or oval structure. Sometimes, the reef is capped by a sandy isle known as a cay in the Caribbean and a motu in the Pacific.

The waters in which coral reefs grow are nutrient poor, a fact that long puzzled scientists. We are beginning to understand, however, that the water is deficient in nutrients because they are locked in the biota. Nutrients are quickly removed from the water and cycled and recycled through the food webs of the reef, much as the nutrients in the tropical rain forests are quickly removed from the organic litter and returned to the vegetation, thus leaving the soil nutrient poor.

Coral reefs are extremely important ecosystems, performing many vital functions. They provide habitats for many organisms that cannot live elsewhere (Figure 23-15). Much of the primary production on reefs is carried out by the dinoflagellates living on the coral itself, forming the basis for a very complex food chain capable of supporting many different animals with many different life-styles. Many reefs support commercially valuable fisheries; according to one estimate,

the potential yield of fish and shellfish from coral reefs is 9 million tons per year. Coral reefs also protect coastal areas from being battered by waves. The quieter waters between the reef and the coast, called lagoons, provide homes for organisms too delicate to withstand direct battering by waves. This sheltering effect also helps to protect coastal soil from erosion. On the Indonesian island of Bali, the destruction of offshore coral has exposed the beaches to erosion. Storms have carried away soil, toppling palms growing near the shore. Coral reefs may also help regulate the amount of mineral salts in the world's oceans. Every year, runoff from land deposits massive amounts of mineral salts in the sea. The lagoons and closed seas formed by coral reefs serve as natural evaporation basins that eventually dry out, leaving behind huge salt deposits. Thus, large quantities of salt are removed from the sea. Finally, coral plays a major role in lowering the temperature of the earth's surface, a fact that has implications for global warming. Coral polyps remove carbon dioxide, an important greenhouse gas, from the atmosphere to build limestone skeletons.

These unique ecosystems are threatened by a variety of human activities. According to a recent survey, reefs in 93 out of the 109 countries with significant coral reefs are damaged. Coral reefs are especially vulnerable to any disturbance that stirs up sediment, which smothers coral polyps and interferes with photosynthesis by the dinoflagellates. Deforestation, especially of coastal areas and mangroves, contributes heavily to the sediment load of coastal waters. Logging in the watershed of Bascuit Bay in the Philippines has destroyed 5 percent of the coral reefs in the bay. Sewage and the runoff of fertilizers and pesticides have caused destructive algal blooms or poisoning of some reefs. Some people suspect that the proliferation of the crown-of-thorns starfish, *Acanthaster planci*, a major predator of coral polyps, is partly due to human activities. The crown-of-thorns starfish has ravaged many reefs in the Red Sea and the Indian and Pacific oceans, and it is feared that it may enter the Caribbean through the Panama Canal. It is not known if the infestations are a result of human-caused change, but over-fished reefs seem particularly vulnerable to invasion.

Many reefs have been mined for their limestone, which is used for construction. Coral miners dislodge coral using crowbars or explosive charges. Dynamite fishing is also extremely destructive, not only killing large numbers of fish and other animals, but also blowing apart the coral infrastructure. Dynamite fishing has destroyed some of the finest reefs in Kenya, Tanzania, and Mauritius.

Reefs also fall victim to the tourist industry. Coral and shells are collected in large quantities for sale, while tourists may also hack out pieces straight from the reef to take home as souvenirs. Boat anchors and chains also damage coral.

Nuclear testing is perhaps the most violent of human activities that affect coral. France has detonated about 100 nuclear devices on the Polynesian atoll of Mururoa, which is gradually sinking into the sea. The French are considering moving their testing site to another Pacific island, Fangataufa.

Coral reefs are also threatened by massive bleaching. When corals are

FIGURE 23-15: Queen angelfish, *Holacanthus ciliaris*, Florida Keys.

subjected to high water temperatures, even two degrees above normal, the colorful algae with which they live in symbiosis die. The disappearance of the algae exposes the white skeletons of the corals. If the temperature stress persists, the corals eventually die. Bleaching, now considered the greatest threat to reefs worldwide, occurred without warning in 1980, 1983, 1987, and 1990, affecting the Caribbean reefs most seriously. Although the causes of bleaching are unknown, some scientists believe it is a harbinger of global warming.

Wetlands

Wetlands are transitional zones between land and water. Accordingly, they exhibit an edge effect and tend to be relatively high in species diversity (recall the discussion of ecotones in Chapter 5). They are among the most biologically interesting and productive but least understood habitats, ranging in diversity from prairie potholes in the Dakotas of the United States to papyrus swamps in Uganda. Many wetlands have formed along riverbanks or near river deltas, including the Mississippi in the United States, the Nile in Egypt, the Okavanga in Botswana, the Rhône in France, and the Brahmaputra in India.

The presence of water is the distinguishing characteristic of wetlands. It may vary from standing water several feet deep to waterlogged soil without standing water. The soil in wetlands differs from that of adjacent uplands; it is often saturated long enough to become anaerobic. Wetland vegetation must be able to tolerate both flooding and the lack of oxygen in the soil.

There are two major categories of wetlands: inland and coastal. Inland wetlands are freshwater ecosystems and include marshes, swamps, riverine wetlands, and bogs. Coastal wetlands may be either fresh or salt water and are affected by tides. Examples include tidal salt marshes, tidal freshwater marshes, and mangroves.

Freshwater marshes are areas of incredible diversity. Dominated by

FIGURE 23-16: Cypress trees with Spanish moss, Florida.

emergent grasses and sedges, including cattail, wild rice, and bullrush, they have a high pH, high levels of nutrients, high productivity, and high rates of decomposition. Marshes usually are recharged by both precipitation and lake or river flooding, and it is the flooding that serves as a source of nutrients. Some marshes remain wet year-round while others are wet only seasonally. Marshes are very important habitats for wildlife, particularly migratory birds and waterfowl. The marshes at Point Pelee, a provincial park in Ontario, Canada, offer a wide spectrum of waterfowl and other wildlife. Over half of the waterfowl harvested in North America were produced in the marshes of Canada.

"Swamp"—for many people, the name evokes feelings of repulsion or dread. Popular culture is at least partially responsible for the poor

image these areas have. *Creature from the Swamp, Swamp Thing,* and other B-grade movies portray swamps as places of evil and terror. In reality, swamps are hauntingly beautiful. With standing water up to 5 or 6 feet deep throughout most or all of the year, there is little or no understory or emergent vegetation. Instead, swamps are dominated by woody vegetation, especially cypress, water tupelo, and black gum, trees well adapted to their watery environment. To obtain oxygen from the air (since it is typically in short supply in the often stagnant water), cypress and other trees send up pneumatophores, knobby aerial roots or "knees," through which air is funneled to the submerged roots (Figure 23-16). Reptiles, amphibians, and fish are common denizens of swamps. Among the best-known swamps in the United States are the bayous of

Louisiana and the Okefenokee Swamp in Georgia. Cienaga Grande, the "Great Swamp" region of Colombia, is one of the richest fishing grounds in that country.

Riverine or riparian wetlands, commonly called floodplains, are lands adjacent to rivers that undergo periodic flooding. During flooding, they receive a high input of nutrients from the surrounding area. Since they are found alongside rivers, floodplains have a linear form. A floodplain system includes the river channel, its floodplain, and all oxbows. An oxbow is a river channel formed by the natural meandering pattern of a river that becomes cut off from the main channel (Figure 23-17). Severe flooding, for example, may carve a new and straighter channel through an area, isolating a bend or meander

of a river. Although the oxbow becomes separated from the main river, it is recharged through precipitation and groundwater. Weston Lake in South Carolina is an oxbow lake formed from the Congaree River. The Congaree Swamp is a national monument, and includes the river, the floodplain, and the oxbow lake. Designated a Biosphere Reserve, it contains 50 tree species, including several 300-year-old Loblolly pines, the largest of which measures 27 feet in circumference.

Riverine wetlands vary from areas that are almost always flooded to areas that are rarely flooded. The frequency and duration of flooding are a product of climate (precipitation, spring thaw), floodplain levels (higher plains are flooded less frequently than low-lying areas), drainage area (the larger the area

drained, the longer the duration of the flooding), and the soil's ability to hold water. Hardwoods are the dominant vegetation, with the species composition depending on the frequency of flooding. Cypress, tupelo, and buttonbush are found in wetlands that are almost always flooded, while areas subject to progressively less frequent flooding host water locust, persimmon, sweetgum, box elder, sycamore, hickory, beech, and white oak. The soils of riverine wetlands are anaerobic part of the year, again depending on the frequency of flooding. One of the best-known riverine wetlands in the United States is the wetland formed by the Mississippi River in Louisiana. The Pantanal, a seasonal wetland in Brazil, is an important food source for birds. Every year, rains cause the rivers

FIGURE 23-17: Mara River, Kenya. An oxbow lake has formed (bottom right) where a meander of the river was cut off from the main channel. Other oxbow lakes may very well form along the river's twisting, winding path.

flowing through the Pantanal to flood the surrounding flat ground. As the Pantanal dries out, many fish are trapped and forced into smaller and smaller pools, where they become easy prey for birds.

Although floodplains are generally unsuitable for building, they have historically been important in agriculture. For thousands of years, seasonal flooding of the Nile River in Egypt was used to enhance the productivity of agricultural fields. In late summer, as the river began to rise, water was channeled into basins and released onto fields, where it was allowed to stand for 40 days or more. Today, however, seasonal flooding is prevented by a series of dams, and productivity in agricultural fields in the Nile Delta must be boosted by expensive artificial fertilizers. Tribal pastoralists in southern Sudan have also developed a migratory way of life that revolves around vast swamps known as the Sudd. At the end of the rainy season, the people burn the grasslands surrounding the floodplains, encouraging the growth of new, rich grass. As the floodwaters recede, the people bring their herds of cattle in to graze on the rich *toich* grass. When the rains return, the people move to higher ground, where they practice agriculture.

Bogs are generally found in cooler regions such as the northeastern United States, Canada, and the British Isles, but some tropical countries, such as Indonesia, possess rather extensive bogs. Although the water table in a bog is typically high, precipitation, not underground aquifers, is responsible for most water inflow. There is no outflow; the water is static and the land is permanently waterlogged. The oxygen supply in standing water soon becomes depleted; coupled with the cool climate, the static water contributes to slow rates of decomposition. Deposits then accumulate of peat, a brown, acidic material made up of the compressed remains of plants that have not decomposed. Nutrients thus remain locked up in plant and animal matter and are not quickly recycled through the ecosystem. Consequently, the productivity of a bog is low compared to that of the surrounding terrestrial system. Not surprisingly, bogs are colonized by plants that are acid tolerant, such as sphagnum moss, cranberry, pine, spruce, and tamarack. Eventually, enough peat may accumulate that it raises the level of the land making it less wet and suitable for other vegetation. Dried peat burns well, and for centuries has been used for fuel; until recently, almost every household in western Ireland relied on peat for heating and cooking. Countries in which fuelwood is scarce, such as China and Indonesia, are being encouraged to exploit their peat resources. Peat is also harvested for use as humus to improve the soil, or for use in potting compost. The destruction of peat bogs may have implications for global warming. Active peat bogs "fix" carbon in the soil, acting as sinks. However, when peat is dug up, carbon is released into the atmosphere, where it contributes to global warming.

Like inland wetlands, coastal wetlands exhibit great variety. Tidal salt marshes occur along coastlines throughout the world in the mid-to high latitudes, being replaced in the tropics and subtropics by mangroves. Salt marshes form wherever the accumulation of sediments is equal to or greater than the rate of land subsidence, as long as the area is protected from the destructive power of waves and storms. They are found near river mouths, in bays, along protected coastal plains, and around protected lagoons. The lower salt marsh is that area found in close proximity to the ocean; its water has the salinity of the adjacent seawater. The upper marsh is that area which is further removed from the ocean; its salinity is variable due to exposure and dilution by fresh water. Salt-tolerant grasses are the dominant vegetation in salt marshes. These wetlands are very productive ecosystems, but much of the primary productivity is washed out with the tides and is utilized by aquatic communities. Extensive salt marshes are located behind barrier beaches in Long Island, New Jersey, and North Carolina in the United States. Salt marshes along the Waddenzee of Germany and the Nether-lands are extremely important habitats for waterfowl and many other plants and animals. Also, almost the entire population of North Sea herring is dependent on these marshes at some point in their life cycle. Unfortunately, the Waddenzee is now seriously threatened by pollution.

Tidal freshwater marshes combine features of tidal saltwater marshes and inland freshwater marshes. The tide influences water level, as in a tidal salt marsh, but species diversity is very high, as in an inland marsh, because of the absence of the salt stress. Tidal freshwater marshes are found where a river enters the ocean or where precipitation is high. Plant diversity is high; vegetation varies with elevation, with submergent aquatic vegetation dominating in low-lying areas and emergent vegetation dominating at higher elevations. Tidal freshwater marshes are used by more birds than any other type of marsh. For example, the Camargue, in France, is a large wetland formed where the River Rhône flows into the Mediterranean (Figure 23-18). Much of the Camargue, which contains very fertile soil, has been drained for agriculture, but the remaining areas, which are breeding grounds for 15,00 to 20,000 greater flamingos and are visited by more than 300 species of migratory birds each year, have been designated a Biosphere Reserve. Approximately 280 species of birds are found in U.S. tidal freshwater marshes. They also serve as vital nurseries for many species of fishes. Unfortunately, tidal freshwater marshes are often found near urban areas, since people tend to settle along river mouths, and are therefore subjected to significant stresses, especially pollution and draining for development.

The mangrove swamp is the dominant type of coastal wetland in tropical and subtropical regions. It is a forested wetland, an association of salt-tolerant trees, shrubs, and other plants growing in brackish to saline tidal waters along tropical and subtropical coastlines. Mangroves cover approximately 39 million acres (15.8 million hectares)

FIGURE 23-18: Flamingos, *Phoenicopterus ruber*, feeding at dawn in the Camargue, a tidal freshwater marsh in France.

worldwide, with the largest concentrations in tropical Asia. The largest stretch of mangroves in the world is the Sundarbans, along the Lower Ganges Delta in Bangladesh.

Mangrove is a general term referring to any tree that can survive partly submerged in the relatively salty environment of coastal swamps. These trees do not require a saline environment and may in fact grow better when the salt content is low, but under such conditions, they tend to be outcompeted by other tree species. They have survived because they are able to adapt to a saline environment. Some mangrove species excrete excess salt through the action of special glands. Others isolate the salt in inactive tissues within the plant. Pneumatophores enable the tree to transport oxygen to its roots. Because they colonize shifting silt beds, mangroves must have some means of maintaining their stability. They do this through stilt roots, which act as buttresses (Figure 23-19). The mangrove roots also trap silt, and over time, mud banks accumulate. As the mud banks rise above the water level, other plants are able to invade the area. Mangroves may be totally inundated or they may have no standing water. They are important exporters of detritus for nearby estuaries. Half of the mangrove's primary productivity may be exported.

Wetlands have traditionally been misunderstood and underappreciated, but we are learning that these varied ecosystems provide essential services. Wetlands have been called the "kidneys" of the earth. They intercept and store runoff, helping to prevent flooding. The stored water can help to recharge shallow underground aquifers. As runoff flows through a wetland, suspended solids are precipitated out, reducing sediment loads to nearby waterways. Anaerobic and aerobic processes, such as denitrification, remove excess nutrients and other chemicals from the water, thus purifying it before it reaches the lake or ocean. For example, in a study of a Florida swamp, water hyacinth were found to remove 49 percent of the nitrogen and 11 percent of the phosphorus from the water.

FIGURE 23-19: Mangrove trees, *Rhizophora mangle*, on Sanibel Island, Florida. The roots of the mangrove help anchor the tree to shifting silt beds and also trap mud, detritus, and organic matter, thus enabling mud banks to accumulate. Other plants are then able to invade and colonize the mud banks.

Coastal wetlands, particularly mangrove swamps, protect shorelines from the erosive force of ocean tides and minimize storm damage. In addition, mangroves protect coral reefs and other offshore areas from land-based pollution.

Both coastal and inland wetlands are valuable reservoirs of biological diversity. They are important nesting and feeding habitats for birds and waterfowl and nurseries for fish and shellfish. Draining wetlands for questionable short-term economic gain might prove to be one of the biggest ecological disasters humans have foisted on the biosphere. Further, it is a questionable practice on economic grounds. Eugene Odum, an ecologist and wetlands researcher, estimates that an acre of Georgia coastal wetland is worth $20,000 to $50,000 per acre if measured in terms of the commer-

cial value of fish and shellfish produced and/or nurtured there. That same acreage converted and sold as farmland would be worth just $4,000 per acre!

In the past, wetlands have been drained and destroyed, without a thought to their importance, and converted to agricultural land or commercial development. Channelization of rivers in order to achieve better navigable depths or to prevent flooding has also had disastrous effects on riparian wetlands. Without protective wetlands, the straightened rivers are more prone to flooding.

Wetlands throughout the world are threatened (Figure 23-20). Australia and New Zealand have lost over 90 percent of their wetlands. In the United States, over half of the wetlands that existed before the advent of European settlement in

the seventeenth century have been lost. In some states and regions, the loss has been exceedingly high. California has the dubious distinction of leading the nation in this category, with over 90 percent of its wetlands destroyed. Many of the marshes and swamps that once ringed the Great Lakes, especially Lakes Erie and Michigan, were drained for conversion to farmland. Mangrove swamps have been significantly reduced in size in Africa, Latin America, and western Asia. Logging to produce pulpwood and charcoal and conversion to aquaculture ponds are the chief threats. For example, the Indonesian state of Kalimantan plans to clear 95 percent of all mangroves in order to accommodate pulpwood production, despite the fact that the fisheries nursed by those mangroves earn about seven times as much in

FIGURE 23-20: Endangered prairie potholes, Alberta, Canada. Formed over time by the action of receding glaciers, waterfowl-breeding potholes in agricultural land are being rapidly lost to the plow.

export revenue as all wood and charcoal production combined.

A more immediate threat to wetlands is their definition under national laws. While new advances in ecology have shown that wetlands are critical ecosystems, changes in the legal definition of "wetlands" can seriously hamper efforts to protect them. Currently, the Environmental Protection Agency (EPA), Army Corps of Engineers, U.S. Fish and Wildlife Service, and USDA Soil Conservation Service define "wetlands" similarly. Each organization requires positive indicators of hydrology,

hydrophytic vegetation, and hydric or saturated soils. The Fish and Wildlife Service also includes non-vegetated areas such as mudflats, gravel beaches, rocky shores, and sandbars. However, this definition does not always cover drier or seasonally saturated wetlands, which are just as valuable as wetter wetlands. The belief that wetter wetlands are more valuable has resulted in pressure to exempt wetlands—at least drier ones—from protection under the Clean Water Act. In 1991 the USEPA, Army Corps of Engineers, Department of Agriculture Soil, Conservation Service, and

USFWS proposed changes to federal criteria for identifying wetlands. These changes, proposed in response to criticisms from the Bush administration and some sectors of the public, would limit federal jurisdiction under Section 404 of the Clean Water Act to permanently inundated or flooded wetlands or wetlands that are saturated to the soil surface for prolonged periods and would reduce or eliminate protection for many seasonal wetlands. Such an action could prove disastrous for these already dwindling ecosystems in the United States.

victims of indifference; they may not suffer the same problems that face misunderstood or disliked species, but they do not arouse the strong sentiment for their preservation that charismatic megafauna and game animals arouse. Some states have check-off programs that allow residents to donate a portion

of their income tax refund to management programs for nongame species, but unfortunately, most such programs are not well funded. The value of these species cannot be neatly determined; they are not worth X amount of dollars annually to anyone and they have little emotional appeal. Even when a species' numbers are greatly reduced and its existence is threatened, as in the case of desert tortoises and certain warblers, there is no group particularly interested in saving it, so little money is allocated toward effective management.

Like nongame species, most plants, even those listed as rare, threatened, or endangered, do not generate much public support. In fact, most living things—plants, animals, and microorganisms—fall into this category. If they succumb to extinction, it will not necessarily be because of human greed, but because of our failure to recognize their true worth.

How Do Human Activities Affect the Management of Biological Resources?

Many human activities pose a significant threat to biological resources and biodiversity. Activities that directly affect biological resources include habitat degradation and destruction, overharvesting and illegal trade, and selective breeding. (Chapter 6 provides a discussion of how ecosystem degradation directly and indirectly affects wildlife.)

Habitat Degradation and Destruction. Without a doubt, the single greatest threat facing wildlife is degradation or loss of habitat. The type of habitat being destroyed most quickly is often home to the greatest variety of species. The coral reefs are being destroyed by mining, fishing with dynamite, siltation from development projects, and coral harvesting. An estimated 23,000 to 29,000 square miles (59,618 to 75,171 square kilometers) of tropical rain forests, an area larger than West Virginia and equal to the entire country of Costa Rica, are cleared annually. Such a loss affects not only exotic, tropical species, but also species more familiar to us.

Many songbirds that summer in the United States and Canada are threatened by the deforestation that is reducing or eliminating their tropical wintering grounds. At the same time, development and urban encroachment are reducing their summer breeding grounds in the north (Figure 23-21). Other habitats threatened in the United States include wetlands and grasslands, both of which are transformed for agriculture and development. Without immediate action these habitats and their native inhabitants may go the way of the tallgrass prairie. Prior to the 1850s tallgrass prairies extended over a quarter of a billion acres (101 million hectares) in central North America. Today, less than 1 percent of those prairies remain; the rest have been replaced by acre upon acre of corn and wheat.

In addition to out and out destruction, habitat can be "lost" to native species when it is rendered uninhabitable as a result of chemical contamination from hazardous waste dumps, industrial and municipal

FIGURE 23-21: Adult blue-winged warbler feeding insect larva to its young. According to the World Resources Institute, migrant songbirds whose populations decreased in the period from 1978 to 1987 include the olive-sided flycatcher (–5.7 percent), white-eyed vireo (–1.2 percent), blue-winged warbler (–1 percent), northern parula (–2.1 percent), chestnut-sided warbler (–3.8 percent), cerulean warbler (–0.9 percent), Canada warbler (–2.7 percent), scarlet tanager (–1.2 percent), and rose-breasted grosbeak (–4.1 percent). Between 1966 and 1978, the populations of all of these species except the cerulean and Canada warblers had increased anywhere from 0.3 to over 6 percent.

wastewater discharges, agricultural and urban runoff, and airborne toxins. For example, in 1983 resource managers at the Kesterson National Wildlife Refuge in California found toxic levels of selenium in the water, which they traced to irrigation runoff from nearby farms. Because high levels of selenium can be fatal to bird species, managers now must work to prevent waterfowl from settling in the area.

Introducing exotic species into an area can also wreak havoc on indigenous wildlife. A species brought to an area in which it is not native rarely has natural predators that keep its population under control. Consequently, introduced species may compete more successfully than native species that occupy the same niche and may eventually displace the native species. The economic costs associated with introduced species, like the environmental costs, are often high. The zebra mussel, discussed in Chapter 3, illustrates the devastating impact of introduced species. A native of the Caspian Sea, the zebra mussel was brought to the Great Lakes in the bilge water of an ocean-going ship sometime in the early 1980s. Its population soared throughout the decade, and by 1991, it had spread to the Ohio and Mississippi rivers. Millions of dollars have been spent in mitigating the damage to boats, docks, and pipes caused by the mussel and in attempting, thus far unsuccessfully, to control its population.

Overharvesting and Illegal Trade. Butterflies, parrots, lizard skins, rhinoceros horns, tortoise shells, orchids, elephant tusks, corals, and cacti: all are part of the huge global trade in wildlife and wildlife products worth an estimated $5 billion annually. Most of the wildlife trade is legal, governed by national laws and international treaties, the most notable of which is the Convention on International Trade in Endangered Species (CITES). However, about a quarter to a third of the wildlife trade is illegal. Worth an estimated $1.5 billion annually, the illegal trade in rare species is a serious drain on wild populations and a significant factor in driving species toward extinction. The World Wildlife Fund estimates that over 600 species of animals and plants worldwide face extinction as a result of the wildlife trade, while another 2,300 animals and 24,000 plants are endangered.

The major importers and consumers of wildlife and wildlife products are the United States, Japan, and Europe (Figure 23-22). Primates, birds, tropical fish, and reptile skins account for most of the U.S. imports. The leading European imports are exotic birds, reptile skins, monkeys, and small cats. The legal wildlife trade in the United States, worth about $250 million annually, is overshadowed by the illegal trade, worth about $300 million annually. Worldwide, the chief market for illegal wildlife products is Japan, where a clouded leopard coat may sell for over $100,000.

Social and cultural factors are responsible for the enormous wildlife trade. Fashion, custom, and tradition can endanger biological resources. (Focus On: The Black Rhino and the Species Survival Plan, page 514, examines the cultural factors that caused the demand for rhino horn to escalate in the 1970s, thus severely endangering rhino populations.)

Selective Breeding. In many areas of the world modern crop hybrids have replaced naturally occurring strains and semidomesticated or traditional strains, those bred by farmers over centuries. Inarguably, hybrid strains have yielded many benefits, especially in terms of productivity. But the reduction in the number and frequency of wild and traditional strains has come at a high price—the loss of rare and potentially valuable genes that carry traits such as resistance to pests or disease.

The genetic erosion of crop plants and the widespread use of genetically similar hybrid strains are probably the most serious threats facing modern agriculture. The genes of wild and semidomesticated plants are our only defense against this threat. The corn blight of 1974 in the United States dramatically illustrated both the danger of homogeneous strains and the value of traditional and wild strains. That summer, a blight swept through the country, threatening 80 percent of the corn crop. It eventually destroyed 15 percent of the crop; in some areas, the loss was as high as 50 percent. The blight was finally controlled by introducing a resistant strain from Mexico, the ancestral home of maize.

Wild biological resources constitute a genetic library of invaluable and irreplaceable characteristics. Genetic engineers are continually developing their ability to draw upon these characteristics to improve the quality of human life in myriad ways, including increasing crop yields and strengthening livestock strains. Unfortunately, the safety of this genetic library is in question. With every extinction—of both known and unknown species—we lose yet another valuable book (Figure 23-23). Most seed companies contribute to the problem by offering only a limited selection of modern hybrids, but Focus On: Seeds Blüm, page 516, discusses an innovative approach to preserving and marketing a wealth of crop species.

Loss of genetic variability is a problem in domesticated animals as well as plants; in the United States nearly half of the existing breeds of livestock are in danger of extinction. Breeds with economically desirable characteristics, such as high productivity (the most milk, the most meat per animal), are becoming more and more popular at the expense of other breeds. However, each of those other breeds also has valuable characteristics, such as disease resistance, adaptability to regional environments, low maintenance, special qualities as food (taste, fat content). We never know when these characteristics may become vitally important. For example, commercial cattle had all but replaced the native breeds in Africa's Chad Valley in the early 1970s. However, the advent of the area's periodic seven-year drought wiped out the imported stock, leaving farmers to rebuild herds with the few remaining local cattle, who had adapted to the drought cycles over the centuries. In the United States the American Minor Breeds Conservancy works to preserve genetic diversity in livestock.

Why Should We Preserve Biological Resources and Biological Diversity?

The arguments in support of preserving biological resources and maintaining maximum biological diversity fall into five categories: ecosystem services; benefits to agriculture, medicine, and industry; esthetics; ethical considerations; and evolutionary

FIGURE 23-22: Global wildlife trade. The regions that are the largest suppliers of the wildlife trade are South America (especially Bolivia, Argentina, Brazil, Peru, and Guyana), Africa (Senegal, Tanzania, Congo Republic, the Sudan, and South Africa), east Asia (Philippines, Taiwan, Thailand, Indonesia, Singapore, and Japan), and the United States, which exports cat skins, reptile skins, bear products, cacti, and ginseng.

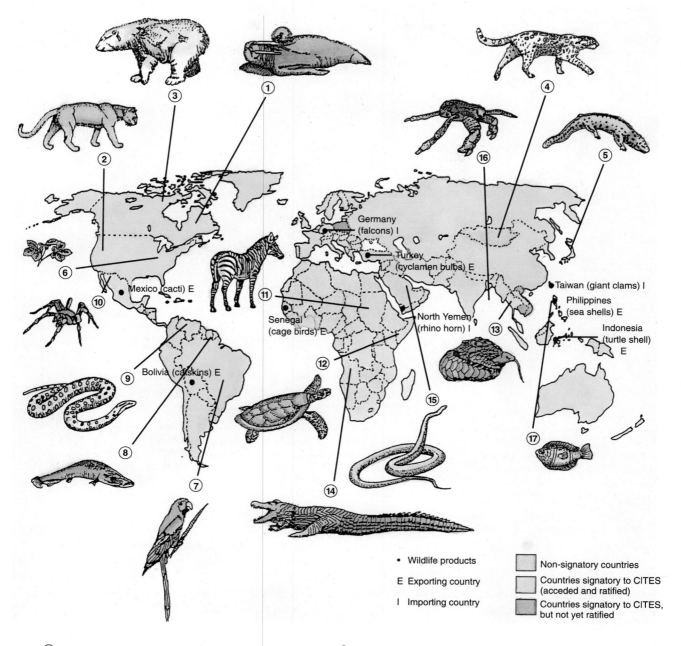

Germany
(falcons) I

Turkey
(cyclamen bulbs) E

Mexico (cacti) E

Senegal
(cage birds) E

North Yemen
(rhino horn) I

Taiwan (giant clams) I

Philippines
(sea shells) E

Indonesia
(turtle shell)
E

Bolivia (catskins) E

• Wildlife products

E Exporting country

I Importing country

Non-signatory countries

Countries signatory to CITES
(acceded and ratified)

Countries signatory to CITES,
but not yet ratified

① **Walrus**–A demand for walrus ivory threatens this species.

② **Cougar**–Habitat loss, overhunting.

③ **Polar bear**–Habitat loss, skin trade.

④ **Snow leopard**–Loss of prey.

⑤ **Giant salamander**–The Giant salamander is, at 180 centimeters, the largest living amphibian. The two species live in mountain streams in China and Japan where they are threatened by overhunting.

⑥ **American ginseng**–Exported to the Far East for its alleged medicinal properties

⑦ **Hyacinth macaw**–Smuggled into Bolivia for export.

⑧ **Arapaima**–This very large, scaly fresh water fish from the Amazon River basin is highly prized gastronomically. Its status is vulnerable due to overfishing (even by dynamiting), and export of smaller individuals for the aquarium trade.

⑨ **Anaconda**–Skin trade.

⑩ **Red-kneed tarantula**–Over collection for pets.

⑪ **Mountain zebra**–The mountain zebra now survives only in reserves, where it is still threatened by hunting. There is also a danger that the small herds could be wiped out by drought or anthrax.

⑫ **Loggerhead turtle**–The Loggerhead turtle of the oceans is a circumpolar species feeding mainly on shellfish and crustaceans. The Loggerhead turtle is threatened throughout its range by exploitation as food or for curios, and entanglement in nets. Its breeding beaches are also being destroyed, or disturbed by lights and spectators.

⑬ **Pangolin**–Its skin is being traded illegally to supply the U.S. boot industry.

⑭ **Nile crocodile**–Trade in wild populations now prohibited.

⑮ **Python**–Skin trade.

⑯ **Robber crab**–The robber or coconut crab is probably the largest terrestrial arthropod in the world. Frequently weighing over 3 kilograms and measuring 1 meter across the legs, this hermit crab lives in burrows on many islands in the Indo-Pacific region. Its larger size makes it attractive to humans both as a source of food and as curios for the tourist trade. However, intensive hunting has reduced its numbers so severely that in some places its collection is banned or regulated.

⑰ **Clown fish**–Large numbers exported from the Philippines by the aquarium trade.

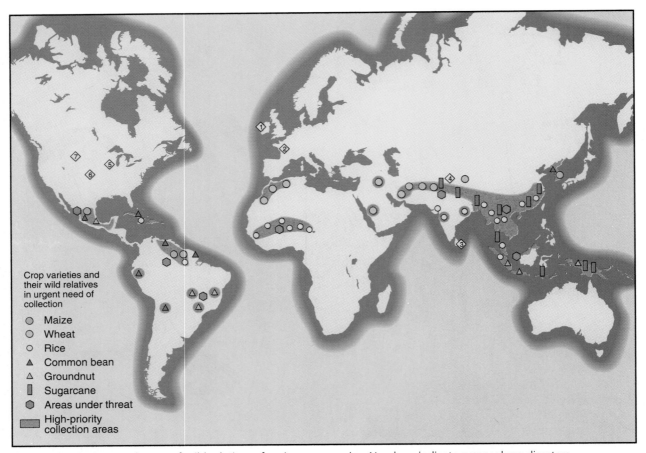

FIGURE 23-23: Ancestral areas of wild relatives of major crop species. Numbers indicate areas where disasters have occurred due to lack of genetic diversity: (1) 1840s, blight devastated Irish potato crops; 2 million people died; (2) 1860s, vine disease crippled Europe's wine industry; (3) 1870-1890, coffee rust destroyed Ceylon's coffee harvest; (4) 1942, rice crop in Bangladesh destroyed; millions died; (5) 1946, a fungus epidemic devastated U.S. oat crop; (6) 1950s, stem rust devastated U.S. wheat harvest; (7) 1970, maize fungus threatened 80 percent of U.S. corn acreage.

esthetics; ethical considerations; and evolutionary potential.

Ecosystem Services. Ecosystems perform many services, such as chemical and nutrient cycling and soil generation, which are beneficial to humans. For example, each year earthworms carry in their casings from 2.2 to 69 tons (2 to 63 metric tons) of soil to the surface of each hectare of land, effectively aerating the soil. Eliminating living resources may hinder the ability of ecosystems to perform vital functions. (Ecosystems and ecosystem services are discussed throughout this book, especially in Unit II, How the Biosphere Works.)

Benefits to Agriculture, Medicine, and Industry. Biological resources benefit agriculture in several ways. The genes of wild crop relatives may hold valuable characteristics, which, when introduced into crop species, can dramatically increase yields. Wild species are also the source of "new" food crops. Of the 80,000 edible plant species on earth, only a small percentage have been investigated for use as a

food crop, and just 8 supply 75 percent of the human diet. Similarly, most of the protein from livestock is derived from just 9 domesticated animals. Thus, the global human diet is precariously dependent on a handful of species. Agriculture also benefits from pollinating organisms such as bees and other insects. Without them, seed production would be impossible, the harvest of many vegetables would be reduced, and there would be no fruit.

Many important and familiar pharmaceuticals originated from biological sources, including digitalis, a heart stimulant derived from purple foxglove, and quinine, used to treat malaria. Worldwide, almost half of all pharmaceuticals are derived from biological resources; the annual monetary value of these products is approximately $40 billion. More significantly, biological resources are valuable as sources of medicines and drugs because they save lives. For example, two alkaloids derived from the rosy periwinkle, a tropical rain forest plant, have proven effective in treating leukemia and Hodgkin's disease (Figure 23-24). Animals also provide valuable medicinal commodities, including hormones,

FIGURE 23-24: Rosy periwinkle. A child suffering from leukemia in 1960 had only a one-in-five chance of long-term survival, but today would have a four-in-five chance. One important difference was discovery of two alkaloids derived from the rosy periwinkle, a tropical forest plant. These alkaloids have also proven effective in treating Hodgkin's disease, another form of cancer.

thyroid extracts, and estrogens. The slow-clotting blood of the Florida manatee has aided researchers in the study and treatment of hemophilia.

Industry—tourism, recreation, and manufacturing—realizes significant economic benefits from biological resources. Parks, reserves, and wildlife refuges, in the United States and abroad, are becoming increasingly popular as places to view and enjoy wild species. According to recent studies, Americans spend approximately $30 billion annually on recreational activities associated with wildlife: $16 billion on fishing, $7 billion on hunting, and $7 billion on nonconsumptive uses (such as whale watching, birding, and nature photography). Visitors to national and state parks spend additional millions each year.

Manufacturing industries also owe a tremendous economic debt to biological resources. For example, jojoba, once considered a desert weed, yields a wax which retails to Japan for over $3,000 a barrel. The wax is used as an industrial lubricant in place of sperm whale oil.

Esthetics. Many of us feel a need for natural areas and the diversity of species they harbor. Biologically rich areas such as the tropical rain forests are often considered to be among the most awe-inspiring regions of the planet. Many people feel that wild species imbue the earth with much of its beauty, grandeur, and mystery. Most of us do not revel in the physical, chemical, and biological conditions that enable earth to support life, although we recognize and appreciate those conditions. Instead, we revel in life itself, clover and condor, forest and field mouse (Figure 23-25).

Ethical Considerations. Many people argue that biological resources have inherent worth and so have a right to exist for their own sake, independent of their worth to humankind. Proponents argue that moral and ethical grounds alone are reason enough for humans to preserve biological resources and protect diversity.

In *A Sand County Almanac*, Aldo Leopold asks, "If the biota, in the course of aeons, had built something we like but do not understand, then who but a fool would discard seemingly useless parts? To keep every cog and wheel is the first precaution of intelligent tinkering." Eliminate a species, and the health of the ecosystem may suffer. For example, botanists estimate that approximately 25,000 plant species are currently threatened with extinction. They believe that between 20 and 40 animal species depend, for their survival, upon each plant species. Thus, for every plant species faced with extinction, many more animal species may also disappear. Leopold maintains that a land ethic is necessary to "change the role of *Homo sapiens* from conqueror of the land community to plain member and citizen of it. It implies respect for his fellow members and also respect for the community as such."

Homo sapiens has always been a tinkerer. Long ago, our ancestors domesticated certain animal species and practiced selective breeding for particular plant strains. We, too, are tinkerers, but our greater numbers and powerful technologies render our activities far more dangerous and potentially destructive. We must become intelligent tinkerers, protecting what we do not understand rather than

FIGURE 23-25: Desert habitat. In all its varied and unique forms, life intrigues and captivates humankind.

Focus On:
The Black Rhino and the Species Survival Plan
Edward Maruska, Director, Cincinnati Zoo and Botanical Garden

The Cincinnati Zoo and Botanical Garden has the dubious distinction of being the final living repository for two endemic species of birds. The most famous is Martha, the passenger pigeon who died here in September 1914 with much fanfare. She was the last living representative of a species of bird that during its zenith in the 1800s was the perhaps most abundant of all land birds. Migrating flocks are reported to have darkened the skies, their wing beats drowning out all other sound. Incas, a male Carolina parakeet, met his untimely end here four years after Martha's death. The Carolina parakeet was the only member of the Psittacine (parrot) family to nest exclusively within the United States. The zoo had searched frantically for mates for both birds, but to no avail.

Ironically, both species bred well in captivity, and with some foresight captive populations could have been bred and managed in a number of institutions. With new migratory bird laws and the decline of the millinery trade at the end of the century, the Carolina parakeet could probably have been reestablished in the wild. The passenger pigeon might have been more difficult to reintroduce because of its complex breeding and migratory patterns. However, we will never know for certain because with the passing of these species, we have lost all of our options as well.

FIGURE 23-26: The black rhino, native of Africa, at home in the Ngorongoro Crater, Tanzania.

Having witnessed extinction, we at the Cincinnati Zoo have vowed never to let that scenario repeat itself. And we believe we speak for all zoos. The mission for all zoological parks worldwide has expanded to include the propagation of endangered species. This does not mean we look forward to being the last repositories for endangered species, but we are forced into that position by situations that are occurring in the wild. Through educational programs, zoos widely encourage species preservation in their natural environments and preservation of those habitats. However, zoos may provide the most positive outlook for some endangered species that tend to do well in captivity because zoos are absolutely safe sanctuaries where the species' needs are met through carefully managed husbandry and breeding programs.

Zoos now approach the challenge of captive management of endangered species through the American Association of Zoological Parks and Aquariums (AAZPA) Species Survival Plan (SSP), a program that officially began in 1980. The goal of

eliminating, in our ignorance, that which may be invaluable to us. We must remember that what we lose—or more to the point, allow or cause to disappear—is irreplaceable.

Evolutionary Potential. In *Conservation Biology: An Evolutionary-Ecological Perspective*, biologists Michael E. Soulé and Bruce A. Wilcox write, "Perhaps even more shocking than the unprecedented wave of extinction is the cessation of significant evolution of new species of large plants and animals. Death is one thing—an end to birth is something else." All of the previous arguments—ecosystem services, benefits to agriculture, medicine, and industry, esthetics, and ethical considerations—are strong and convincing reasons to preserve diversity now and in the foreseeable future. But what if we were to take on a more universal, timeless perspective, more like the perspective the earth itself might have?

The time scale of the earth is far different from our own. None of us, for instance, will ever see a new species arise from an ancestor. Seeing things from the planet's perspective means thinking of time in terms of millions, rather than tens or hundreds of years. According to that time scale, the preservation of diversity is paramount because it alone can produce further diversity through adaptation and specialization.

the program is to maximize the potential of captive management by coordinating the breeding programs that house important species. Zoo geneticists have determined that from populations of a minimum of 50 animals (more would be desirable), we can maintain a healthy breeding population for up to 200 years. Loss of genetic variability (inbreeding) can be minimized when animals are collectively managed under a species specific program. To be included in the survival plan, a species must meet three basic criteria: (1) A breeding nucleus of species or subspecies must be available for captive management. (2) The continued existence of the species or subspecies in the wild must be in some degree of peril, as defined by the International Union for the Conservation of Nature and Natural Resources (IUCN), the U.S. Fish and Wildlife Service, or reliable field reports. (3) There must be available an organized group of captive propagation professionals with sufficient support to carry the species or subspecies program to captive preservation status.

Each SSP species (at the time of this writing there are 37) is managed by a propagation group consisting of member institutions that have a signed commitment to the SSP programs and numerous captive propagation professionals. They are led by a coordinator for the species. I serve as coordinator for the black rhinoceros *Diceros bicornis* (Figure 23-26). This species certainly meets the above criteria. In 1970 the estimated population of black rhino throughout its range south of the Sahara was 60,000, making it the most numerous of the five species of rhino. During the 1970s the value of rhino horn, which had been constant for decades, suddenly and dramatically escalated because of increased market demands in the Far East. The additional demand for medicinal purposes throughout the Orient and as an aphrodisiac in parts of Asia continues to threaten all rhino species. The heavy trade in rhino horn was further complicated by the North Yeminis' increased use of rhino horn for high-quality handles for jambias, their traditional daggers. From 1972 to 1978 the North Yeminis were importing 40 percent of all rhino horn which entered international trade. Much of the horn came from the black rhino. This demand, in turn, created an upsurge in poaching all over Africa.

The number of black rhinos in the wild has plummeted to fewer than 3,500, and wildlife biologists predict a continuing decline. The best hope for this species may lie in confining large numbers in fenced and heavily protected areas in the African countries in which it still remains. The only other hope for its survival as a species will depend on how successfully we manage it in captivity. Presently, in North America, we have a carrying capacity (zoos and private landowners) of about 100 black rhinos. We are managing 63 black rhinos of two subspecies. Our goal is to increase our present North American population to 100 specimens—50 of each subspecies. This increase will be realized through captive breeding and through the addition to the program of specimens from the wild.

With controlled breeding management of the SSP and its European counterparts, this magnificent species, present for over a million years, may yet have a bright future. When world conditions stabilize and the threat for black rhinos lessens, perhaps it may be repatriated over its former range from populations perpetuated in captivity. As David Western, IUCN African elephant and rhino specialist group member, has eloquently stated, "The black rhino is a flagship species for conservation. Its sheer size and bulk make it difficult to ignore. If we cannot save this species, there may be less hope for other endangered species."

History of Management of Biological Resources

How Have Biological Resources Been Managed Historically?

One of the earliest known gardens of nonnative, or exotic, species was a collection of medicinal plants established by the Chinese Emperor Shen Nung about 2800 B.C. In the tenth and eleventh centuries the Islamic Moors of the Iberian peninsula created gardens of plants from as far away as India for the purposes of pleasure, inspiration, and study. During the Middle Ages in Europe and the Middle East, collecting and cultivating fruit trees and ornamentals (both natives and exotics) was seen as a way of recreating Eden. In the seventeenth and eighteenth centuries the study of medicinal plants became increasingly popular throughout Europe and eventually led to the establishment of botanical gardens in England, France, Scotland, and Russia. In the United States privately owned gardens were established in the early eighteenth century in Pennsylvania, but public collections did not appear until the second half of the nineteenth century.

Like wild plant collections, collections of exotic animals have a long and storied history. Ancient Egyptian priests maintained temple menageries

Jan Blüm runs Seeds Blüm, a small seed company in Boise, Idaho. Seeds Blüm is devoted to preserving and increasing the use of heirloom plants, varieties that have been handed down over generations.

Seeds Blüm grew out of several of my interests which gradually came together. In 1976 I started my first garden and bumped headlong into my ignorance of plants and how they work. As a result of my own garden's obvious shortcomings, I started paying close attention to my neighbor, whose garden was successful. He took the opportunity to fill my head with how the old varieties were disappearing and how hybrids were making up more and more of our food supply. He also shared with me the descendants of some seeds he had been given as a wedding present in the 1920s. After experiencing these plants in my own garden and kitchen, I saw for myself how wonderful these heirloom varieties are!

At this time I was teaching vegetarian cooking classes and I had begun trading seeds to find new varieties of grains and beans to use as examples of the diversity of foods available to vegetarians. Eventually, I was growing 700 or 800 different kinds of beans, along with nearly 400 kinds of tomatoes. I couldn't understand why I wasn't seeing these varieties in seed catalogs or in the grocery stores. When a very tasty and unusually large sauce

tomato is available, why are we limited to the small varieties which don't even taste as good in sauce?

Then I began reading about genetic erosion and all the politics that contribute to it. I started thinking, "I've got to do something," but I felt pretty overwhelmed by the entrenched bias toward hybrids and the largeness of the task. Everyone I wrote to asking how I could best help replied that the crying need was for a seed company that would reintroduce these varieties to gardeners on a consistent basis. I agreed but felt I was not personally prepared to shoulder this task. However, the idea persisted. Finally, to assuage the demons in my head, I decided to write a seed catalog—the way I would *if* I were to have a seed company. Midway through the creation of my "fantasy" catalog, I realized that this was more than a trial run—it was the actual catalog for the actual seed company that I finally felt committed to making a reality. It united so many of my passions in life: writing, cooking, gardening, drawing, and making a positive contribution for future generations.

I sent out the first Seeds Blüm catalog in December 1981. Gardeners responded with enthusiasm. Garden writers and magazines did stories about our work and the wonderful old varieties that are our reason for existence. People from all over the country mailed us samples of seeds, sometimes accompanied

by actual squashes, tomatoes, and other produce. Other gardeners volunteered to help keep the varieties alive. We began to realize that Seeds Blüm is more than a seed company.

Our work has several facets. First and foremost, we collect and multiply for sale old varieties of flowers and vegetables (Figure 23-27). We sell these seeds through our mail order catalog. We multiply the seeds by working with volunteer seed multipliers who take small samples and multiply the seed amount so that a grower can produce a larger quantity for sale. Small-scale seed growers produce the heirloom varieties so that we can offer the seed to gardeners through the catalog. Seed guardians are a backup; they essentially adopt a variety and renew the seed each year, keeping it alive and genetically pure.

Another important part of our work is research and education. Often an heirloom variety has been grown and saved in one part of the country. We don't know where else it will grow well. To discover how these varieties do in various growing zones, we work with many of our customers who volunteer to be trial gardeners. They grow, eat, and preserve (can, freeze, root cellar) these varieties and report their experiences and impressions. I am also trying to learn more about the history of each variety, stories from different families and cultures that have saved and used the various dif-

including baboons, ibises, lions, dogs, and cats. In the twelfth century B.C. the Chinese Emperor Wen wang built Ling yu, the Garden of Intelligence, to house his animal collection. The Aztec ruler Montezuma maintained a vast menagerie in his palace at Tenochtitlán (Mexico City). When the explorer Hernán Cortés reached the palace in 1519, he found hawks, falcons, eagles, quetzals, jaguars, pumas, snakes, llamas, and vicunas.

The hunting preserves of sixteenth- and seventeenth-century Europe were stocked with such game animals as deer, wild boar, and birds, a practice that gave rise to the first zoo. In 1752 Francis I, Holy Roman emperor and husband of Maria Theresa

of Austria, commissioned a menagerie to be built on the palace grounds at Schonbrunn. The Schonbrunn menagerie and the early zoos that followed it were mainly the province of royalty and wealthy landowners. The public was seldom given the opportunity to view the animals. Gradually, however, zoos became more numerous and accessible to the general public. The first zoo in the United States opened in Philadelphia on July 1, 1874.

The chief function of the early zoos was entertainment. The overriding philosophy—"More is better"—led zookeepers to gather as large a collection of exotics as possible, even if each species was represented only by one or two members. As zoos

FIGURE 23-27: A sampling of heirloom vegetables.

ferent strains and varieties. The art of saving seeds of high quality has nearly been lost among gardeners. Part of our purpose here at Seeds Blüm is to rekindle interest and teach people how to save their own seeds.

The work is both fascinating and momentous. We are working to save our food chain by preserving genetic information. Plant breeders need this information to create new open pollinated and hybrid varieties to be planted on thousands of acres and then shipped to nationwide grocery stores, canning companies, processing plants, bakeries, cereal companies, and even cattle owners. There

has been a tragic inattention by our national government and industry giants toward management of our genetic heritage. Since we can manipulate genes but cannot create new ones, we must preserve what we have. Loss is irreversible—and our lives are directly dependent upon food.

The facts are sobering. Seeds Blüm is a small company and the task is large. However, the task is also immensely enjoyable. Garrison Wilkes (botanist at the University of Massachusetts) once said the secret of genetic preservation is that "people save what they value." We as humans tend to do what we enjoy

doing. For example, when I see Ragged Jack kale (an outstanding heirloom variety) being served in restaurant salads, I know that Ragged Jack kale will be grown each year (an act of preservation) because now it is valued for its flavor and beauty. One of our jobs in relating to gardeners is to translate preservation into enjoyment: to find the traits in old varieties that make them useful and valuable in our fast-paced indulgent culture.

People want to learn more. Besides strengthening what we are now doing, I would like to create both a vegetable botanical garden and a museum presenting the *true* history of agriculture unaffected by economically driven special interest groups. Arboretums and botanical gardens concentrate on ornamentals, which is lovely but incomplete. I want to present the public with gardens of vegetable varieties to observe and select for their own gardens. The museum would tell the story of how plants have come through human history and how they, like language and clothing, have changed according to the fashion of the day. The story will include the effect of politics and the seed industry itself.

Every region and many families have special heirloom varieties that are a part of their history. These plants must be carefully saved and passed along if they are to continue to exist other than as memories of extinct species.

became larger, more numerous, and more popular, they became a considerable drain on wild populations. When an animal died, a replacement was sought from the wild, but to obtain a baby chimpanzee, for example, five or six adults were often killed. Once an animal was captured, there was no guarantee that it would survive the trauma of transport. And, those that lived through the journey to a foreign land often succumbed to diseases and pests against which they had no defenses.

Unfortunately, the move to a new land was only the beginning of the animals' misery. They were usually kept in small, dark, poorly vented, and heavily barred cages. Little or no attention was paid to

the animals' psychological needs for comfortable and interesting surroundings, stimulation, companionship, and, perhaps most critically, the opportunity to pursue normal behaviors (such as the need of wolves to scent-mark trees).

How Are Biological Resources Currently Managed?

For the most part, current efforts to manage biological resources concentrate on preserving plant and animal species or their genetic material (germ plasm). Both species and germ plasm may be preserved off site (*ex situ*) or on site (*in situ*). Although efforts to manage and protect biological resources by preserving habitats are increasing, management efforts have traditionally focused on off-site management (especially through aquariums and zoological and botanical gardens).

Off-Site Management of Plants. Approximately 1,000 institutes worldwide maintain plant collections. They hold the primary responsibility for managing wild plant species outside their natural habitat. In the United States important collections are housed at the Missouri Botanical Gardens, Boston's Arnold Arboretum, and the New York Botanical Gardens. The Royal Botanic Gardens at Kew, England, established in 1759, is probably the leading botanical garden in the world, housing over 50,000 species from almost every country.

In addition to their recreational and educational functions, many contemporary botanical gardens sponsor research to add to our store of botanical knowledge. Many of these institutions also maintain collections of seeds or plant tissues. The collections and the research are often used by breeders in their quest to improve plant strains.

The Center for Plant Conservation, a network of 19 botanical gardens and research facilities in the United States, was established in 1984 to develop a "national collection" of all rare and endangered indigenous plants. It is an ambitious goal, considering that there are at least 3,000 such species, many located on private lands and many having only one or two known populations. By enlisting institutions in different biogeographic regions, the center has approximated the diversity of climates needed to compile and maintain a national collection. Each institution specializes in collecting and propagating species native to its region (Figure 23-28). The Fairchild Tropical Garden in Coral Gables, Florida, for example, cultivates species from the citrus and cycad families that are otherwise nearly extinct in the United States. For safety reasons, seeds are also stored at the Department of Agriculture's seed storage facilities. The cultivation of plants at participating institutions offers the species a secure environment while facilitating research into the plants' specific growing requirements and the general ecological niches they fill.

The international arena has no direct parallel with the United States' Center for Plant Conservation. The Royal Botanic Gardens at Kew, England, focuses on gathering information about Europe's rare and endangered species rather than on collecting and cultivating them. Limited space is a problem faced by Kew and other institutions. By the mid-1980s some institutions had begun efforts to preserve certain species. For example, the Botanical Gardens at the University of Tokyo undertook a study to propa-

FIGURE 23-28: Alan Boefer and Chris Dietrich, members of the horticulture staff, work with *Trifolium stoloniferum*, running buffalo clover, a federally endangered species, at the Missouri Botanical Garden, headquarters of the Center for Plant Conservation.

gate *Melastoma tetramerum*, an endangered species from the Bonin Islands, located some 620 miles (1,000 kilometers) south of Tokyo. Researchers propagated the species off site, conducting many tests to gather information about the plant. They then reintroduced it to its former habitat, studying the effect of various locations and environmental conditions on the success of the reintroduction. By conducting basic biological research, including systematic biology, ecology, and population genetics on exotic, little-understood species, institutions can contribute to conservation.

Despite these efforts, plant species often receive far less attention from the public and government than do animal species, especially mammals. Perhaps it is the close kinship that we share with animals that makes it easier for us to appreciate them. Even so, from a pragmatic standpoint, plant conservation is far more important to human survival than is animal conservation. Plants form the base of all food chains; without them we would have neither food to eat nor oxygen to breathe.

Off-Site Management of Animals. Private collections, wildlife refuges, sperm banks, and zoological parks are all used to preserve animals outside their natural habitat. Since zoos are by far the best known of these means and since they incorporate elements of wildlife refuges and sperm banks, we will look more closely at how zoos work to preserve animal species.

Unlike old-style zoos, most modern institutions focus on caring well for small numbers of species. In this way, they are able to house more individuals of a species. Having adopted a philosophy in which the zoo is seen as an "ark" for the world's vanishing wildlife, they are dedicated to four interrelated objectives: education, recreation, conservation, and research. (Environmental Science in Action: The St. Louis Zoo, pages 526–528, looks at how one modern zoo strives to accomplish these objectives.)

The best zoos strive to meet the physical and psychological needs of their animals in a manner that is both educational and entertaining. They do this primarily by exhibiting animals in settings that approximate the animals' natural habitats. Such naturalistic settings provide animals with more stimulation than traditional exhibits (bare cages) and allow them to pursue their normal behaviors—a wolf scent-marking a post or an otter floating on its back in a pool, breaking open clams against a rock perched on its belly. For example, many zoos equip their chimpanzee enclosures with an artificial termite mound. In the wild chimps use long sticks or straw to penetrate holes in termite mounds and draw out the tasty inhabitants. Because termites are extremely difficult to maintain, the artificial mounds usually contain honey, yet the chimps still enjoy fishing for the sweet treat. Visitors enjoy observing the animals' normal behaviors, and they learn much about the species and the intricate relationships between it, other species, and the environment.

Even the ways that zoos group animals can be educational. In the past animals typically were grouped taxonomically. All the members of the cat family might be grouped in one general area, primates in another, and the horse family in yet another. Taxonomic groupings teach people to recognize genus types and the evolutionary relationships among species. While taxonomic grouping is common, more and more zoos are beginning to group their animals zoogeographically, on the basis of where they live. This arrangement mimics how various species occur and interact in the wild and emphasizes ecology and habitat. The Metropolitan Toronto Zoo was the first to be arranged zoogeographically. Covering 704 acres (285 hectares), it is divided into four pavilions: The Americas, Australasia, Africa, and Indo-Malaya. Numerous paddocks within each pavilion house species particular to different vegetation types within those broad geographic areas.

The latest concept in zoo design is in the works at the San Diego Zoo, arguably this country's best. In 1985 officials announced plans to reorganize the whole zoo around ten bioclimatic zones: tropical rain forest, tropical dry forest, savanna, desert, grassland, temperate forest, taiga, tundra, montane, and islands. The idea is to immerse visitors in the world of specific bioclimatic regions, including plant and animal life. San Diego will be the first zoo totally organized around the concept. However, other zoos, including Chicago's Brookfield Zoo and New York's Bronx Zoo, have developed bioclimatic exhibits.

Zoos can be a strong participant in global preservation efforts. They promise real hope as reservoirs of species and genetic diversity. However, to fulfill that potential, curators must maintain viable breeding stocks of species to supply their own and other zoos rather than adding to the pressure on wild stocks. In addition, captive breeding populations hold out the hope that zoos will someday be able to return animals which have disappeared in the wild to their native habitat—if and when there is any wild habitat to which to return them.

Zoos have experienced both success and failure in their efforts to breed species. By 1962 the Arabian oryx had been virtually exterminated in the wild, the victim of hunters armed with Land Rovers and automatic weapons. That year, the last few wild animals were captured and sent to the Phoenix Zoo, where a breeding program was instituted using the wild oryx as well as some already in captivity. By 1980 the "world herd" at the Phoenix Zoo numbered more than 320 individuals. In 1982, 14 oryx were released into the wild in southern Oman, where they have flourished under the official protection of the Sultan of Oman. Breeding programs have also saved the golden lion tamarin, Pere David's deer, and Przewalski's horse, the original Mongolian wild horse, from extinction (Figure 23-29).

Captive breeding is not simply a matter of producing as many young as possible. Rather, programs

FIGURE 23-29: Golden lion tamarin in Brazilian reserve.

attempt to meet two objectives: to maximize the contribution of unrelated animals in order to reduce the effects of inbreeding and to attain a good age distribution in order to ensure a consistent number of individuals of breeding age. The International Species Inventory System (ISIS), a computer-based information system for wild animal species in captivity, was developed to overcome the problems caused by inbreeding, which include diminished fertility, reduced resistance to disease, and lessened competitive ability. ISIS includes over 2,500 registered species, housed in more than 285 institutions in 25 countries; it records basic biological information on each individual animal: age, sex, parentage, place of birth, and circumstance of death. This information is used to compile various reports and to analyze the status of captive populations. More importantly, it allows zoos to cooperate in the genetic and demographic management of their animals.

Using ISIS data, the American Association of Zoological Parks and Aquariums has developed species survival plans plans for approximately 40 threatened or endangered species. These plans indicate which animals should mate to preserve the maximum genetic diversity of the entire captive population across zoos. They also indicate how many animals should mate and when, in order to create a good-sized population with an optimal age distribution.

Despite the success of traditional captive breeding programs, they are plagued by several problems. Some animals, like the giant panda, simply do not breed well in captivity. Another problem is logistics. For example, in order for two black rhinos from different zoos to mate, one of them—an animal weighing thousands of pounds—must be moved to the other's institution. Doing so is costly and cumbersome and poses some risk to the animal. Breeding programs are also made more difficult by monomorphic species, species whose male and female look virtually identical. Many bird and reptile species are monomorphic. For a long time the only reliable method zoos had to determine the sex of a monomorphic animal was to open up the animal and look at the gonads, or reproductive organs, directly. But surgery, even with today's improved surgical techniques and highly trained personnel, carries risk. Fecal steroid analysis, pioneered at the London and San Diego zoos, eliminates the need for surgery. The animal's feces are examined to determine the ratio of estrogen, the female hormone, to testosterone, the male hormone. Although each sex produces both male and female hormones, the female produces a greater amount of estrogen, the male a greater amount of testosterone. Fecal steroid analysis can also be used to track an animal's reproductive cycle in order to tell when the animal is bio-logically ready to mate. Unfortunately, it requires sensitive and costly equipment, time, and a technician, and many zoos do not have these resources.

Researchers have solved some of the problems that complicate breeding programs by adapting alternative assisted reproduction methods that have been used for some time in the domestic livestock industry (Table 23-1). These methods include artificial insemination, embryo transplants, and cryopreservation (Figure 23-30).

On-Site Management of Plants and Animals. National parks, protected wilderness areas, and biosphere reserves offer the best hope for preserving most species and for preserving maximum biological

▶ Table 23-1
Assisted Reproduction Techniques

Artificial insemination (AI) of a female with semen acquired from a donor is fairly routine, if not always successful. Artificial insemination overcomes the logistical problems of breeding, since sperm samples can easily be transferred between institutions. Potentially, captive females could be fertilized with sperm obtained from wild animals, thus rejuvenating the genetic variability of the captive population without disrupting the wild population. Similarly, AI might someday be used to increase the genetic variability in a small and dwindling wild population by fertilizing wild females with semen from captive males.

Embryo transplants (ETs) can be used to help slow-breeding animals multiply more quickly. Given the right hormone, an animal can be made to produce several eggs rather than one, and the multiple eggs may all then be fertilized during mating. The fertilized eggs can be carefully flushed from the mother and injected into the uterus of a surrogate mother to complete their gestation period. Research is underway to perfect techniques for retrieving the fertilized eggs. Embryo transplants have been successfully undertaken in closely related species such as cow and gaur, horse and zebra, domestic cat and small exotic cat. Our knowledge of the basic biology of these domestic species is fairly complete. Although ETs hold great promise to increase the breeding capacity of zoos, much basic research must be completed before the technique can be used routinely with exotic species.

Cryopreservation holds great promise in the struggle to prevent the extinction of species. Unfertilized eggs, sperm, and embryos arrested at an early stage of development are frozen in liquid nitrogen at -196° C and held for long-term storage. Research is underway to determine the appropriate storage medium for germ plasm, since the storage medium varies from species to species and is unknown for most exotics. The frozen samples represent slices of the genetic diversity of a species. Cryopreserved germ plasm can be thawed and used in the future to infuse blood lines that have become genetically impoverished. A population can be sampled and embryos obtained and frozen which represent the genetic diversity of that population at a given time. In the future, if inbreeding threatens the population's health, the embryos can be used to rejuvenate the breed.

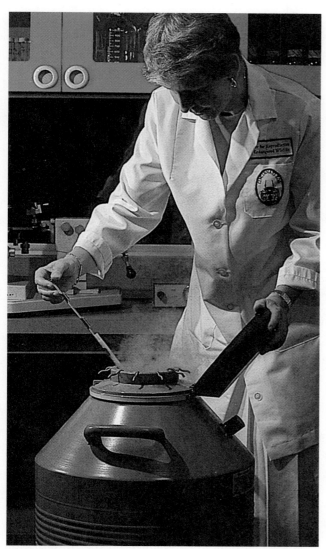

FIGURE 23-30: Dr. Betsy Dresser, Director, Cincinnati Zoo and Botanical Garden Center for the Reproduction of Endangered Wildlife. The use of cryopreservation to hold the germ plasm of rare plants and animals in long-term storage has earned the germ plasm bank the moniker the "Frozen Zoo and Garden."

diversity. Perhaps more importantly, they are the only way in which the relationships between organisms can be preserved. Many scientists contend that it is the relationships between organisms and between organisms and the nonliving environment that are of critical importance to the survival of the planet and the human race.

The amount of protected land throughout the world has grown rapidly in the past few decades, yet it accounts for only about 3 percent of the earth's land surface, far short of the 10 percent deemed necessary by the World National Parks Congress in 1982 (Figure 23-31). Moreover, existing parks and reserves do not represent the full complement of ecosystem types. Of the nearly 200 biogeographical regions in the world, one in eight is not represented by a single park; many others are represented by only one or two protected areas. Half of the global total of protected lands are found in North America, and most of these are in the boreal forest and semifrozen areas of Greenland and Canada. Many parks and reserves are subject to intensive agriculture and other intrusive human activities; they are "protected" on paper only. Here we will look at international efforts and private efforts within the United States to protect habitat (Chapters 21 and 22 examine protected public lands in the United States).

Biosphere Reserves. In the past thirty years many countries have established national parks or protected lands systems. Costa Rica, for example, has one of the finest systems of protected lands in the world. In addition to an extensive network of parks and nature reserves, Costa Rica protects by law a certain portion of its remaining forests, many of which lie on private lands. Another nation which is attempting to balance economic development and environmental preservation is Bhutan, a small nation in the eastern Himalayas that was for many years isolated from the outside world. Bordered by Tibet to the north and India to the east, west, and south, Bhutan is roughly the size of Switzerland. It has set aside approximately 20 percent of its land area in wildlife sanctuaries, forest reserves, nature and wildlife reserves, and one national park. While financial constraints and a lack of personnel have hampered the ability of the Department of Forestry to effectively manage these areas, Bhutan nonetheless has shown a remarkable commitment to protect its biological wealth. For example, in 1986 the country rejected a World Bank project, the proposed Manas-Sankosh Dam on the Manas River in the south, that would have flooded one of Bhutan's most diverse and significant wildlife areas. The area lies in the center of what is now Royal Manas National Park. The World Wide Fund for Nature, the international affiliate of the World Wildlife Fund, is assisting the Bhutan government in its attempt to establish a truly effective system of protected lands.

On the international front, a concerted preservation effort began in 1976, when the United Nations Educational, Social, and Cultural Organization (UNESCO) instituted the Man and the Biosphere (MAB) program to promote integrated ecological research and international cooperation in the field of environmental science. To address the problem of conserving biological resources, MAB has established a global system of biosphere reserves, which encompass concentric areas zoned for different uses. These reserves constitute representative examples of the earth's major ecosystems and serve as laboratories for research and on-site monitoring (refer again to Figure 23-31).

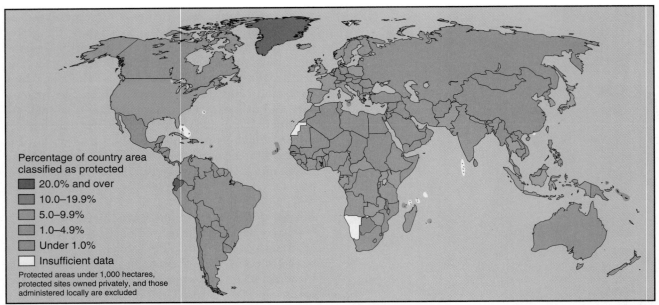

FIGURE 23-31: Protected lands worldwide. According to the World Resources Institute, between 1970 and 1989, 2,098 areas were added to the roster of protected lands, nearly as many as had been added in all previous years. The total acreage added during the past two decades far exceeded the amount protected before 1970, a reflection of our greater awareness of the need to set aside large natural areas to prevent genetic erosion and to ensure the long-term survival of as many species as possible.

Before we can establish an effective global network of biosphere reserves, several major obstacles must be overcome. Debt-burdened developing nations, many of which lie in the biologically rich tropical regions, are often desperate for the short-term cash promised by the immediate exploitation of land. They may feel it is simply impossible to set aside large tracts of land. A promising solution to this problem may be debt-for-nature swaps in which a conservation organization "buys" the debt of a nation at a discount from the bank(s) which made the original loan, in exchange for a promise by the debtor nation to establish and protect a nature reserve (see Chapter 26, Economics and Politics).

A second obstacle to the development of protected areas is the difficulty of designing effective reserves. Often, reserves become isolated islands of life surrounded by developed or agricultural lands (Figure 23-32). Genetic erosion eventually takes its toll on populations. Research is underway to determine the necessary size and shape that will enable reserves to protect and maintain populations of species. A single large reserve is better than several smaller reserves which, cumulatively, protect an equal amount of land. Similarly, reserves closer together are better than those located farther apart, and reserves with a corridor between them, allowing species to move between reserves, are preferable to those that do not have corridors.

Private Efforts in the United States: The Nature Conservancy. The Nature Conservancy, a private, non-profit organization, has created the largest private

system of natural sanctuaries in the world. Founded in 1951, the conservancy acquires land which represents endangered ecosystems or is home to threatened species. Other environmental groups sometimes criticize the Nature Conservancy because it refrains from lobbying on issues and accepts donations from corporations which may be involved in environmentally destructive activities, but the organization has established an impressive record. As of 1988 it had bought or engineered the preservation of 3.5 million acres (1.4 million hectares) in the United States alone. To attain its goals, the conservancy may use creative strategies, such as buying the controlling interest in a Mississippi lumber company to save a tract of bottomland hardwoods.

In recognition of the global need for preservation, the Nature Conservancy established an international program in the late 1970s. Aimed chiefly at Latin America, the program aided local conservation groups financially and with training in negotiation and acquisition. In 1987 some staff of the international program broke away to form their own group, Conservation International.

Pros and Cons of On-Site and Off-Site Preservation. Both off-site and on-site preservation have advantages and disadvantages. For many species, off-site preservation in a zoo or botanical garden offers the best or only hope for long-term survival. In some cases species may be preserved off site until a time when they can be reintroduced to the wild. Off-site institutions also play a vital educational role. Studies conducted in the early 1980s at Yale

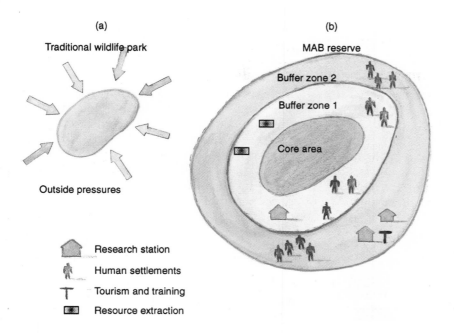

(a) Traditional wildlife park

Outside pressures

(b) MAB reserve

Buffer zone 2

Buffer zone 1

Core area

🏠 Research station

🧍 Human settlements

T Tourism and training

▦ Resource extraction

FIGURE 23-32: Biosphere reserves. (a) In traditional reserves the boundary is well defined: Within the boundary is "nature," and outside the boundary are people. Humans exert significant pressure on the natural area, particularly at the boundary. (b) Man and the Biosphere (MAB) reserves are established differently. A core area is strictly managed for wildlife; human activities are prohibited, although existing settlements of indigenous peoples are allowed. In a buffer zone just outside the core area, research and some human settlements are permitted. Also permitted is light resource extraction, such as rubber tapping, collection of nuts, or selective logging. In a second buffer zone, located outside the first, increased human settlements, tourism, and training are also allowed.

University and elsewhere found that both adults and children learn about wildlife mainly through zoos.

The most obvious disadvantage of off-site preservation is that it is an alternative only for those species known to us. Thousands of unidentified species are not represented in zoos and botanical gardens, and if their wild populations are lost, the species will be lost forever. Off-site preservation is also expensive. For example, in 1985 it cost more than $50,000 to house a family of gorillas. Moreover, zoos, botanical gardens, and gene banks can maintain only limited numbers of species. U.S. zoos can probably preserve self-sustaining populations of only 100 mammal species out of over 4,000 known mammal species.

In their natural habitat species generally adapt to environmental changes. When we preserve species off site, however, they do not have the opportunity to change along with their natural environment. Moreover, captive species undergo evolutionary changes in response to their new environment in a matter of relatively few generations. For example, by the fifth to eighth generation in captivity, Przewalski's horse foals were being born outside of the species' traditional, sharply defined foaling season. In the wild those foals would not survive. It seems probable that sustained preservation off site reduces the likelihood of a species' successful reintroduction to the wild.

In addition, we lack the basic biological knowledge essential for effective management of many exotic species, particularly in relation to captive breeding efforts. Finally, preserving species off site ignores the other components of their ecosystems. Many conservationists argue that we should focus on preserving ecosystems and habitats in order to preserve organisms of all five kingdoms, not just individual species of the plant and animal kingdoms.

The primary advantage of on-site preservation is that it protects the entire ecosystem, both known and unknown species, and the relationships between them. On-site preservation also allows a species to evolve with its environment. It is generally less expensive than off-site preservation, although there are costs involved in law enforcement and protection of the designated preserve against poachers and development pressures. If on-site preservation is to be successful, however, effective legislation is clearly needed to protect the habitats upon which species depend.

What Legislation Governs the Management of Wild Species in the United States?

In the United States, most legislation and management efforts have been aimed at game species. However, as populations of individual wild species declined, various laws were enacted to protect them.

For example, in 1913 the Wilson Tariff Act banned the import of wild bird plumes for women's hats. The Lacey Act, passed in 1900, was an exception to the rule of species-specific legislation. It prohibited the interstate traffic of any birds or mammals taken illegally. The Lacey Act was later amended to prohibit the import of wildlife killed, captured, or imported illegally from another country.

Historically, most wildlife legislation dealt with the trade or transport of wild animals or animal products. Such legislation failed to protect living animals and their habitats and ignored threatened plant species. In recognition of this fact, conservationists began to work for the protection of species and habitats several decades ago. Their efforts resulted in the Endangered Species Act of 1973.

The Endangered Species Act (ESA) mandated that a species or subspecies "endangered throughout all or a portion of its range" be listed with the secretary of the interior. Marine species were to be listed with the secretary of commerce. The Fish and Wildlife Service and the National Marine Fisheries Service (NMFS) were responsible for issuing regulations to protect endangered or threatened (likely to become endangered) species.

The ESA mandated that all federal agencies must consult with the Fish and Wildlife or Marine Fisheries Service before starting any project. The act prohibits killing, capturing, importing, exporting, or selling any endangered or threatened species, including plants. It also provides for citizens' suits to stop a project on the grounds that it violates the ESA if that project is begun without modifications necessary to protect endangered species. A seven-member panel, formally the Endangered Species Committee and informally known as the "God Squad," can grant an exemption to the Act under one of three conditions: (1) no reasonable and prudent alternatives exist; (2) the project is of national or regional significance; (3) the benefits clearly outweigh those of the alternatives.

Though the ESA is a fairly strong and potentially powerful law, it has its weaknesses. First, simply listing endangered species is difficult and expensive. Within two years the boundaries of the habitat of a species must be described, an economic impact study performed, and public hearings held. There are inadequate funds available to finance the study and listing of all species which are probably threatened with extinction. By 1992, over 600 species had been listed but hundreds more remained unstudied and unlisted. Also, protection is more likely to be given to mammals and other highly visible animals than to plants or invertebrates such as insects. Finally, politics can interfere with the protection of endangered species. In May 1992, the God Squad voted to allow timber sales on 13 of 44 tracts of fed-eral land controlled by the Bureau of Land Management. The tracts in question are prime habitat for the northern spotted owl, a threatened species that lives only in the old-growth forests of the Pacific Northwest. The conflict between timber interests and environmentalists might have been settled without the intervention of the Endangered Species Committee, as hundreds of others have been since the creation of the committee in 1978. Environmentalists accused Manuel Lujan, Secretary of the Department of Interior and chair of the committee, of convening the God Squad to force a showdown between timber interests and the northern spotted owl. Lujan has previously complained in public that the ESA is "too tough."

What Legislation Governs the Management of Wild Species Internationally?

Throughout the twentieth century countries have formally adopted various treaties and conventions designed to protect migratory species (Figure 23-33). For example, six bilateral treaties protecting migratory birds are currently in effect. By the 1970s the global community recognized the need for broad international action to save species. That recognition resulted in four conventions which are limited to neither a few species nor to geographic regions: the

FIGURE 23-33: Young green turtle. Effective management of migratory species, such as this green turtle, is especially difficult because it requires the cooperation and dedication of various states and governments.

What You Can Do: To Preserve Species

At Your Home, School, or Office

- Plant shrubs or trees around your home, school, or office to provide food and shelter for birds and other creatures.
- Put up bird houses and bird baths.
- Pull weeds instead of using herbicides.
- Don't use a lawn care service. If you must use one, find out what chemicals it uses and make sure it applies them correctly.
- Learn about gardening organically and try your hand at growing heirloom or traditional varieties suited to your region.
- Landscape with plants that aren't prone to insect and fungus problems.
- Let bird and insect predators take care of native garden pests.
- Use beer traps for slugs instead of baiting with poisons.
- If you live in an arid region, landscape with plants that do not require a great deal of water.
- Watch out for wildlife—give consideration to all living things you see crossing the road.

- Don't buy products that come from endangered animals or plants.
- Don't keep exotic pets or rare plants.
- Take advantage of the nongame wildlife checkoff on your state tax form.
- Vote for candidates who share your opinions on conservation and the preservation of biological diversity. If you do not know a candidate's opinion on a conservation issue, ask.
- Read books and articles on wildlife and environmental issues and watch nature programs on television.
- Teach children to respect nature and the environment. Take them on a hike, help them plant a tree or build a bird house, buy them a nature book or a subscription to a wildlife magazine.
- Volunteer your time to conservation projects designed to increase or protect wildlife habitat.
- Check your lifestyle—think about the effects of your daily actions on the environment.

- Join an organization that works exclusively for the protection of biological resources and habitat. Some suggestions: The Nature Conservancy, the Wilderness Society, Conservation International, World Wildlife Fund, and Cousteau Society.

Traveling

- Don't pick flowers or collect wild creatures for pets; leave them where you find them. If you want a remembrance, take pictures!
- Don't buy souvenirs or products made from wild animals or plants.
- When hiking, stay on the trail.
- Consider a vacation centered around wildlife observation and study. Choose an area or country with native wildlife in which you are particularly interested.
- Carefully research tourist businesses, such as safari tours, to make certain that you deal with a reputable firm.

Convention on Migratory Species of Wild Animals, the World Heritage Convention, the Convention on Wetlands of International Importance Especially as Waterfowl Habitat, and the Convention on International Trade in Endangered Species of Wild Flora and Fauna (CITES). Of these, the most important and effective is CITES.

CITES prohibits international trade in the approximately 600 most endangered species and their products and requires export licenses for an additional 200 threatened species. Over 100 countries have ratified CITES. Member states are required to submit reports and trade records to CITES. Permits issued by CITES are the only legal, recognized permits for international trade in wild animals, plants, or wildlife products.

Though it can be effective, CITES does have weaknesses. Only nations that have ratified the convention are bound to its restrictions. Many countries that have not joined the convention, such as China,

are significant participants in the wildlife trade. Moreover, a participating country can legally continue to trade in even the most highly endangered species simply by informing CITES of its intention to do so, an action known as taking out a reservation on a species. And as with all international conventions, enforcement is left to the countries involved. Consequently, there is wide variance in how well CITES is enforced in various nations. For example, Japan, although a participant in the convention, continues to import large quantities of prohibited items, including tortoise shell, rhino horn, and ivory. Exporting nations, which are usually the less-developed countries, need the foreign money brought in by the wildlife trade. Importing nations, on the other hand, are usually the more-developed countries. The primary enforcement problem faced by these nations is a lack of personnel qualified to detect and identify endangered species and wildlife products.

Environmental Science in Action: The St. Louis Zoo
Dana Wilson

The St. Louis Zoo illustrates how modern zoos are blending the conservation-oriented activities of breeding, research, and education with their traditional purpose of entertainment.

Describing the Resource
Physical Boundaries

The zoo occupies 83 acres (33.6 hectares) in Forest Park, a recreational and cultural area near downtown St. Louis. Visitors approaching the zoo see the impressive walk-through flight cage originally constructed for the 1904 World's Fair, an event that helped encourage public support for the creation, in 1914, of the zoo itself. They enter the zoo free of charge and, once inside, enjoy various displays and activities, including 15 major animal exhibits, a children's zoo, and an innovative education center.

Biological Boundaries

The St. Louis Zoo houses over 50 endangered species and more than 2,300 animals. The zoo makes a substantial contribution to preserving endangered species through breeding and research. In recent years its staff has made important advances in breeding many types of animals, including the rare lesser kudu (a type of antelope) and the black rhinoceros. Perhaps the zoo's greatest success has been its unsurpassed record for breeding the endangered black lemur, a small primate from Madagascar.

Social Boundaries

The St. Louis Zoo has always enjoyed strong community support, a vital ingredient in the success of any zoo. In 1913 the city set aside 77 acres (31 hectares) in Forest Park to establish a zoological park. By 1916 the citizens of St. Louis had approved a tax to finance the construction of the zoo, making St. Louis the first city to support its zoo with a tax. Eight decades later the tradition of community support continues.

The zoo also owes its success to the more than 300 volunteers who supplement the work of approximately 400 employees. Some volunteers answer visitors' questions, guide school groups around the exhibits, and conduct programs outside the zoo. Others help with clerical work, gift shops, information booths, and special events.

Looking Back

Like most zoos, the St. Louis Zoo's initial mission was entertainment. Gradually, its focus changed to conserving endangered species, and in the last two decades it has succeeded in strengthening three major areas that support this goal: naturalistic exhibits, breeding and research, and education.

Developing Naturalistic Exhibits

The St. Louis Zoo boasts one of the earliest attempts at a naturalistic habitat. The bear pits, built in 1921, had pools, trees, and walls made to look like rock—features that resembled the bears' natural environment and gave them more to do and more privacy than a standard cage exhibit.

Recent additions and renovations emphasize natural surroundings which make the animals feel at home, stimulate them, and give them privacy. The exhibit called Big Cat Country houses lions, tigers, jaguars, pumas, and leopards. The cats' yards are landscaped with rock formations, trees and shrubs, grassy hills, waterfalls, and even pools of water that allow them to swim. In 1977 the zoo completed similar renovations on the Primate House, home to groups of monkeys and lemurs. Primate curators have been careful to allow the animals to retain natural family groupings, thus encouraging normal behaviors and social organization. Complete with trees, rock outcrops, grassy floors, and swinging vines, each exhibit resembles the animals' natural habitat. The exhibits are enclosed with safety glass, protecting the animals from human germs, but enabling visitors to see into each area. Safety glass also separates admiring viewers from residents of the Herpetarium, home to the zoo's amphibians and reptiles (including its oldest resident, a 59-year-old crocodile). Renovated in 1978, it features indoor-outdoor exhibits for some of the larger animals and provides underground views of some of the aquatic displays.

In the Bird House, birds are grouped into enclosures that simulate their natural zoogeographical regions, ranging from rain forest to ocean shore. Over 8,300 tension wires run from floor to ceiling in front of the enclosures. Spaced closely enough to prevent birds from escaping, the wires are so fine that the visitor can focus them out when observing the birds. The wire barrier provides good ventilation and allows visitors to hear more sounds than would a glass barrier.

In 1981, with the assistance of the Smithsonian Institution, the zoo added the Living Coral Reef to the aquatic house. An example of a Caribbean coral reef, the display simulates a complete ocean floor ecosystem, from tiny microorganisms to large fish. The aquatic house also contains penguin displays and 21 aquariums exhibiting tropical and native fishes.

One of the zoo's most spectacular habitats, Jungle of the Apes, opened in 1987. Gorillas wander through large trees and mounds of rock, and orangutans live mainly among the tops of trees planted in their enclosure. Visitors are brought eye-to-eye with the orangutans by an elevated jungle bridge outside the glassed enclosure. The chimpanzees often take a break from climbing among rocks and trees to fish for honey in their termite mound. With the completion of the Jungle of the Apes, the zoo achieved an important goal: no animal was housed exclusively behind bars.

Aiding Conservation Through Breeding and Research

An expert staff of animal curators, veterinarians, and a reproductive physiologist help to develop successful breeding programs for a variety of rare species. The zoo participates actively in the International Species Inventory System (ISIS) and the Species Survival Plan (SSP). They even act as international record keeper for the black lemur, an endangered species for which the zoo holds the country's best breeding record: over 75 births in the past 20 years.

The zoo is also home to 5 tuatara, the only members of this endangered species from New Zealand to reside in a U.S. zoo. The exotic, lizardlike reptile has never before reproduced in captivity, and the zoo is concentrating on breeding it. Consequently, visitors cannot even view the tuatara, which dwell in a special enclosure, built to resemble their native habitat, in the basement of the Herpetarium. Research has led zoo staff to suspect that the female tuatara becomes fertile only once every six years, a disappointing but important discovery.

Other significant efforts to breed endangered species have focused on the Speke's gazelle, Malayan tapir, Bataleur eagles from sub-Saharan Africa, and Humboldt penguins, which are threatened by the destruction of their habitat (on the west coast of South America), oil slicks, and fishing nets.

In some cases the zoo enhances its breeding efforts through nontraditional methods. For instance, in 1980 one of the zoo's Speke's gazelles produced a fawn through artificial insemination. An interspecies embryo transfer successfully resulted in the 1984 birth of a zebra calf to a quarter horse mare. In the same year the zoo produced an eland calf from a frozen embryo.

Other research efforts are designed to help animals remain healthy after birth. One such effort began when one of the zoo's black rhinos mysteriously died in 1981. She had suffered from hemolytic anemia, a condition in which the red blood cells seem to self-destruct. After consulting with other zoos, St. Louis found that this problem affects a disproportionate number of black rhinos. It then formed a research team to determine the causes of the condition and means of preventing it.

The Cheetah Survival Center, established in 1974, makes breeding this notoriously hard-to-manage cat a priority while still exhibiting the cats and their offspring. Though the staff has had to overcome various problems (for example, the continuous presence of male cheetahs irritates females and can even prevent them from becoming fertile), the center has successfully produced 17 cubs so far.

Education and Entertainment

The St. Louis Zoo takes advantage of every chance to make learning fun. Visitors must understand and support the zoo's goals in order to provide it with the financial and moral support it needs to succeed. For example, visitors to the Jungle of the Apes learn about the natural habitats of gorillas, orangutans, and chimpanzees, which include rain forests in western Africa and southeastern Asia. They discover that destruction of the apes' natural habitats for farmland or development is the greatest threat to the primates' survival. Zoo officials hope visitors leave the exhibit with a new awareness of the urgent need to save the apes' natural home as well as to preserve the species in captivity.

Besides educating visitors through well-marked, naturalistic exhibits, the zoo runs a variety of programs designed to appeal to all ages. The Charles H. Yalem Children's Zoo targets younger visitors, although many adults also enjoy its displays of smaller animals, baby animals, and species native to Missouri. Other exhibits instruct visitors about animals' life-styles, including a child-size spider web and a tree-height walkway displaying tree-dwelling animals.

The Children's Zoo shares animals with a much bigger educational center—the Living World, a $17 million facility opened in 1989. The Living World uses live animals in conjunction with state-of-the-art technology to educate visitors about animal diversity, ecology, and conservation. The center contains two major exhibit halls, the Hall of Animals and the Ecology Hall.

Visitors to the Hall of Animals are greeted by a robot of Charles Darwin, who introduces his theory of evolution and the "great family of life" represented in the hall. The exhibit consists of a tour of the animal kingdom from one-celled organisms to mammals. Computer displays, videos, and live animals help bring the tour to life. The Ecology Hall focuses on the relationship of animals, including humans, to the environment (Figure 23-34). It features videos, weather satellite stations, computer question-and-answer programs, and interactive videos. One interactive video simulates the results of altering a particular environment, such as cutting down a rain forest.

For more traditional education, the Living World contains four classrooms for conservation-oriented courses for all ages. It also includes a teacher resource center and an extensive library. Annually, over 100,000 children attend courses at the center, which drew over 700,000 visitors in its first nine months.

Looking Ahead

The St. Louis Zoo has made great strides in becoming a center for conservation education and the breeding of endangered species, but several plans remain unrealized, such as an elephant house and a small mammal house. The zoo would also like to establish a preservation reserve designed to breed hoofed animals, such as the Speke's gazelle, and small mammals. The preserve would occupy approximately 10 acres (4 hectares) near the zoo.

Although the St. Louis Zoo's plans will expand its ability to breed and exhibit species, the zoo community agrees that the ultimate purpose of the zoo in today's society must be to protect the diversity of the animal kingdom from the effects of human activities—and no one zoo can accomplish this goal. Therefore, the St. Louis Zoo has already taken steps to reach out to other zoos in an attempt to establish some international ties. For example, Bruce Read, the zoo's curator of large mammals, traveled to Malaysia in 1989 to teach Malaysian zookeepers techniques for animal conservation and husbandry. He also developed a breeding program for the Malaysian (Malayan) tapir populations in three of the zoos. Read traveled to Vietnam as well, where he helped to build a breeding structure for the endangered kouprey, a rare species of wild cattle.

The St. Louis Zoo has also established an international alliance with the Beijing Zoo in China. The directors of the two zoos have discussed permanently exchanging certain species, such as the panda, the golden monkey, and certain types of North and South American reptiles and primates.

Because endangered species exist in developing and developed nations

FIGURE 23-34: Exhibit in the Ecology Hall of the St. Louis Zoo, Missouri. The 60-foot-long model of a Missouri Ozarks stream contains native fish, amphibians, and reptiles.

alike, zoos and breeding programs all over the world must share knowledge and animals in order to preserve our planet's biodiversity. Such cooperation, coupled with efforts to preserve natural habitats and ecosystems, can help to ensure that the earth of the twenty-first century possesses the beauty and variety that we enjoy today.

Future Management of Biological Resources

How we manage other species and ourselves in the coming years will indicate the true nature of *Homo sapiens:* conqueror and destroyer or steward and preserver. Environmentally sound management requires preventing overuse through conservation, protecting human and environmental health, and preserving living systems.

Prevent Overuse Through Conservation

The Endangered Species Act is considered by many to be the most significant environmental legislation ever written because it gives legal standing to wildlife. In order to conserve biological resources, the act should be reauthorized and strengthened when it comes up for renewal by Congress in 1992. Representative Gerry Studds of Massachusetts and 30 cosponsors have introduced a bill, known as the Endangered Species Act Amendments of 1992, which would give federal officials a deadline for developing and implementing recovery plans for ecosystems that contain threatened, endangered, and candidate species; encourage officials to develop conservation plans for whole ecosystems rather than for single species; expand citizens' rights to file lawsuits against violators of the ESA in emergency situations; increase funding levels for chronically underfunded federal endangered species programs; and establish a fund to help communities create plans to balance local development with habitat protection for endangered species. Reauthorization of the Act, which is opposed by real estate developers, the timber industry, and the Farm Bureau, is expected to be a lengthy and heated process.

To prevent the illegal slaughter of protected animals and overharvesting of plants, we must institute much stricter enforcement of the Lacey Act. An important step is the development of training programs for customs officials and others who are links in the effort to stop illegal trade in wildlife. Finally, the global community must put pressure on countries to ratify CITES and on countries that have already ratified it to enforce it properly.

Within the United States we can encourage Congress to fund the Fish and Wildlife Conservation Act to initiate programs for nongame species. In addition, Congress should establish an excise tax on outdoor goods (tents, cameras, canoes, backpacks, birdseed, and off-road vehicles) so that nonconsumptive users of wildlife share the cost of preserving biological resources. (What You Can Do: To Preserve Species, page 525, gives some suggestions for individual action.)

Protect Human and Environmental Health

To provide habitat for a variety of species in urban environments, we must establish undisturbed areas, or greenways, in and around cities. Buffer zones around national parks and preserves protect the integrity of the parks and help provide adequate habitat for species. In agricultural settings relatively undisturbed woodlots, hedgerows, and streambanks should be maintained around fields to preserve the flora and fauna of those areas and to provide a rich and varied landscape. Prohibiting the indiscriminate clearing of rights-of-way along railroads, highways, and telephone lines (especially via herbicides) also preserves wildlife habitat.

Preserve Living Systems

Within the United States, we should undertake a national survey to inventory the nation's biological resources and to provide a basis for monitoring and managing them. It would be relatively inexpensive to map (computerize) the country's flora and major elements of its fauna. In fact, the Nature Conservancy's Heritage Program has already begun such a task. Complete mapping would enable biologists to suggest areas where new reserves are needed to protect valuable biological resources. Thus, we would be able to put our limited funds to better use in preserving those areas with greatest diversity.

We should encourage similar surveys in other nations so they can identify their areas of greatest biotic diversity. In addition, we should support continued research into plant uses, both medicinal and otherwise, through international agencies or programs. Research is particularly pressing in the tropics because of the rate of deforestation. Increased ethnobotanical research (the study of the ways in which traditional or preliterate societies use plants) and ethnopharmacological research (the study of the medicinal use of plants by traditional societies) are imperative.

Finally, we must work to diminish habitat destruction and degradation in the United States and around the world. Although some legislation exists to protect endangered habitats like wetlands and old-growth forests in the United States, it must be strengthened and enforced. Internationally, we should carefully consider the consequences of our activities—such as loans used for development—on wildlife.

Summary

All species are biological resources. Biological diversity, or biodiversity, refers to the variety of life forms which inhabit the earth. Biodiversity is measured in terms of species diversity, the number of different species present in an area; ecosystem diversity, the variety of habitats, biotic communities, and ecological processes in the biosphere; and genetic diversity, the variation among the members of a single population of a species. Each member has a unique genotype, the individual's complement of genes. The hereditary material of an organism is its germ plasm. The sum of all of the genes present in a population is a gene pool.

Whenever a species becomes extinct, biodiversity is diminished. Extinction is a natural process which has occurred throughout earth's history. However, human activities have greatly accelerated the rate of extinction. The disappearance of an individual species often has far-reaching effects. Diminishing biological diversity threatens ecosystem diversity. Decreasing genetic diversity threatens species diversity by making species more vulnerable to extinction. The loss of genetic variability is genetic erosion. It lessens a population's ability to adapt to changes in its environment. The minimum viable population size is the fewest number of individuals necessary to maintain a viable breeding group without the short-term loss of genetic variability. This size has not been determined for most species.

All ecosystems host unique associations of interdependent organisms. Endemic species are unique to their particular ecosystem and occur nowhere else on earth. Two ecosystems unequaled in terms of biodiversity are the tropical rain forests and the coral reefs.

Our cultural attitudes and beliefs affect how we value, use, and manage biological resources. A group of species, such as endangered mammals or birds, which are highly valued by humans have been called the charismatic megafauna. We label undesirable animals vermin and undesirable plants weeds. Species hunted for sport are called game animals. Nongame animals are species not hunted for sport.

Cultural attitudes and beliefs give rise to many activities that pose a significant threat to biological resources and biodiversity. The single greatest threat facing wildlife is loss or degradation of habitat. Biological resources are also endangered by overharvesting and illegal trade and by selective breeding.

The arguments for preserving biological resources and maintaining maximum biological diversity fall into five categories: ecosystem services; benefits to agriculture, medicine, and industry; esthetics; ethical considerations; and evolutionary potential.

Wild plant collections and collections of exotic animals have a long history. The chief function of early zoos was entertainment. For the most part, current efforts to manage biological resources concentrate on preserving plant and animal species or their genetic material. Both species and germ plasm may be preserved off site or on site. Private collections, wildlife refuges, sperm banks, and zoological parks are all means used to preserve animals outside their natural habitat. Captive breeding programs attempt to meet two objectives: to maximize the contribution of unrelated animals in order to reduce the effects of inbreeding and to attain a good age distribution in order to ensure a consistent number of individuals of breeding age.

National parks, protected wilderness areas, and biosphere reserves offer the best hope for preserving most species and for preserving maximum biological diversity. They are the only means by which the relationships between organisms can be preserved.

In the United States, most legislation and management efforts have been aimed at game species. However, as populations of individual wild species declined, laws were enacted to protect them. Historically, most wildlife legislation dealt with the trade or transport of wild animals or animal products. Such legislation failed to protect living animals and their habitats and ignored threatened plant species. In recognition of the need to protect a diversity of species (plant and animal) as well as habitats, the Endangered Species Act of 1973 was passed.

Throughout the twentieth century countries have adopted various international treaties and conventions to protect migratory species; most of these are agreements that involve several countries in a specific region. Of the conventions developed to protect species on a global basis, the most significant is CITES, the Convention on Trade in Endangered Species of Wild Flora and Fauna.

Key Terms

biodiversity	genetic erosion
biological diversity	genotype
biological resources	germ plasm
charismatic megafauna	minimum viable
ecosystem diversity	population size
endemic species	nongame animal
game animal	species diversity
gene	vermin
gene pool	weed
genetic diversity	

Discussion Questions

1. Explain the difference between genetic, species, and ecosystem diversity and explain why it is important to preserve each of these components of biological diversity.

2. How do human activities threaten wildlife? How do they affect domesticated species? Give at least two examples of each.

3. What reasons are usually given in favor of preserving biological resources and maintaining maximum biological diversity? Briefly explain each.

4. Discuss how the role of zoos has changed over time. What are the implications for preserving genetic and species diversity?

5. How are national parks, protected wilderness areas, and biosphere reserves different from private collections, zoos, and sperm banks? Explain the difference in terms of species, genetic, and ecosystem diversity. Which do you think is more effective, off-site or on-site preservation, and why?

6. Describe the provisions of both the Endangered Species Act and the Convention on Trade in Endangered Species of Wild Flora and Fauna. What are the strengths and weaknesses of each?

7. In what ways does the St. Louis Zoo exemplify the commitment of contemporary zoos to education, research, and conservation?

Cultural Resources

Wilderness was never a homogeneous raw material. It was very diverse, and the resulting artifacts are very diverse. These differences in the end product are known as cultures. The rich diversity of the world's cultures reflects a corresponding diversity in the wilds that gave them birth.

Aldo Leopold

Culture is perhaps the only means by which individuals and nations can communicate: no common language is needed to feel the same fear of death, the same emotion at the sight of beauty, or the same anxiety at an uncertain future.

United Nations Educational, Scientific, and Cultural Organization

Learning Objectives

When you finish reading this chapter, you should be able to:

1. Describe why it is important to preserve cultural resources.

2. Support the relevance of studying environmental history.

3. Provide examples of environmental problems that hinder the preservation of cultural resources.

4. Provide examples of human actions that hinder the preservation of cultural resources.

5. Discuss at least three national and three international efforts to preserve cultural resources.

Imagine for a moment that representatives from every nation and people on earth come together to create an immense tapestry. Some sit close to their neighbors, speaking and exchanging ideas with one

another. Others are farther removed, and so they speak only among themselves. Each group uses materials brought from its homeland in order to fashion and decorate the cloth. The symbols and the designs which the groups weave are based upon their cultures' unique beliefs, values, and accumulated knowledge. Imagine further that each succeeding generation carries on the work of the tapestry making. Although there may be subtle changes in the materials used or the design wrought, all are based on materials and designs passed to the groups by their ancestors. The resulting tapestry is thus a kaleidoscope of color, texture, and design, a richly varied creation. What a different tapestry would result if all the groups had been given the same materials and instructed to weave the same design!

It is simplistic, perhaps, to equate cultural diversity to that varied tapestry. Yet, our global cultural heritage abounds with diverse and unique resources: Independence Hall, where the Declaration of Independence was signed in Philadelphia; the special knowledge and traditions of tribal peoples in Australia, Africa, and South America; the temples and buildings of the Nepalese in the Kathmandu Valley; the creation beliefs of native Americans; and the pyramids of ancient Egypt (Figure 24-1). Only by recognizing and preserving the world's cultural resources can we bequeath to our children the splendid cultural tapestry we enjoy. Recognition depends on our willingness to expand our definition of what constitutes a "resource." Preservation depends on our willingness and ability to preserve the environments that have given rise to these cultural resources.

In this chapter we discuss what cultural resources are, where they are found, why they are significant, and how they are threatened. After looking at how cultural resources have been managed historically, we suggest management strategies to preserve cultural resources for the future.

FIGURE 24-1: Guardian lion at the gate to the Taleju Temple, Kathmandu, Nepal.

Describing Cultural Resources: Physical and Biological Boundaries

What Are Cultural Resources?

A **cultural resource** is anything that represents a part of the culture (such as history, art, architecture, or archeology) of a specific people. All cultural resources arise from human thought or action. **Material culture** includes all tangible objects, such as tools, furnishings, buildings, sculptures, and paintings, which humans create to make living in the physical world easier or more enjoyable. **Nonmaterial culture** includes intangible resources such as language, customs, traditions, folklore, and mythology (Figure 24-2).

Cultural resources reflect or are part of the environment. Indigenous or tribal peoples typically possess a broad and deep knowledge of the natural system in which they live. For instance, they are usually familiar with the rhythms of the seasons and with the native plants and animals. Because this knowledge arises from their environment, it is specific to that place. In other instances, cultural

resources may actually be part of the environment. Temples, statues, and monuments are part of the cultural or built environment, and as such they are important features of the landscape.

Historic preservation is the field concerned with preserving material cultural resources, and historically, material cultural resources have been the focus of preservation organizations and programs. For example, the National Trust for Historic Preservation, the United States' largest preservation organization, defines a cultural resource as "a building, structure, site, object, or document that is of significance in [American] history, architecture, archeology or culture."

Culture, however, includes all the behavior patterns, arts, beliefs, institutions, and other products of human work and thought. Thus, the traditional bias toward equating cultural resources with material culture ignores many nonphysical aspects of culture. Since the 1970s the definition of cultural resources has been expanded to include living peoples and their differing cultural heritages. Slowly, we are realizing that the folklore, traditions, creation beliefs, and skills of a people are also resources. In 1984 the United Nations Educational, Scientific, and Cultural Organization (UNESCO) instituted a formal program to preserve these and other components of our nonmaterial or **living culture**. Unlike historic preservation, **cultural resources management** is concerned with preserving both material and nonmaterial culture.

Where Are Cultural Resources Found?

Cultural resources can be found everywhere. Museums specializing in art, archeology, or natural history are filled with paintings, statues, and other artifacts of many cultures and ages. Other artifacts, such as prehistoric cave paintings, can be viewed on their original sites. Monuments, churches, and other buildings provide a dramatic record of the beliefs, values, and history of those who built them. Written records of different cultures can be found in literature, historical accounts, diaries, and official documents preserved in libraries and archives.

Living history parks and exhibits bring the nonmaterial aspects of cultures and history alive for visitors (Figure 24-3). For example, extras from the 1990 movie *Glory*, which chronicled the Fifty-fourth Massachusetts Volunteer Infantry, an all-African-American regiment in the Union army during the Civil War, went on to create a re-enactment of a recruitment camp. As the actors performed and answered questions, visitors learned how African-American soldiers—some free, others runaway slaves—dealt with both the pressures of war and the prejudice of Caucasian Union soldiers.

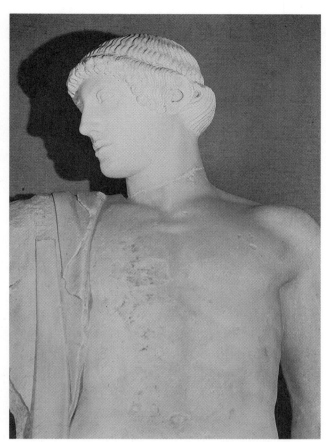

FIGURE 24-2: Statue of the Greek god Apollo. The ancient Greeks worshipped a number of anthropomorphic gods and goddesses. These deities played a variety of roles in Greek life, one of which was to explain natural phenomena. For example, the apparent movement of the sun, from sunrise to sunset, is described in the mythology as Apollo, the sun god, driving a chariot across the heavens.

Social Boundaries

Why Are Cultural Resources Significant?

The reasons cultural resources are significant can be grouped into six broad categories. Cultural resources provide us with a historical record of societies and their environments; are symbols of our heritage; are integral to cultural identity; are a valuable storehouse of environmental knowledge; provide economic benefits; and are essential for cultural diversity.

Historical Record of Societies and Their Environments. About A.D. 550, a group of native Americans settled in the high plateau country of what is now southwestern Colorado. These people, who lived in the area for 700 years, are called the Anasazi, a name derived from a Navajo word meaning "the ancient ones." Their dwellings, elaborate cities of sandstone perched in the sides of cliffs, are a tangible record of a segment of human history on

the North American continent. By studying the ancient dwellings and their surroundings, anthropologists have increased our knowledge about this land, the lives of the Anasazi, and how they modified their environment (Figure 24-4). This information may shed light on why the Anasazi culture disappeared. Studying the environmental effect of past cultures can also enable us to understand better how our own societies modify the environment.

History can be found in your local environment, and you need not be a historian or an archeologist to study and enjoy the resources around you. For instance, if you find an arrowhead, food grinder, hide scraper, or pottery shard in a creekbed, you might wonder, how old is it? Who made it? Did its maker leave it here, or did it belong to someone else? You might go to the library and look for answers in a book about the native Americans who lived in your area. Or, your search might take you to a specialist in that subject. In any case, finding the artifact has caused you to wonder—and learn—about the past of the area you call "home."

By preserving cultural resources, we maintain a record of cultural and environmental history. As long as a cultural resource is preserved, it continues to provoke both wonder and curiosity and remains an object of study and enjoyment for succeeding generations.

FIGURE 24-3: A Civil War reenactment, Port Hudson Battleground, Baton Rouge, Louisiana.

FIGURE 24-4: Balcony House ruins, Mesa Verde National Park. Because of their complex living structures and networks of roads linking population centers, the Anasazi civilization is considered to be one of the most advanced cultures to have developed north of Mexico.

Symbols of Our Heritage. A visitor to the museum and displays located at the Statue of Liberty and Ellis Island might expect to learn about the throngs of immigrants who entered the United States during the first half of the twentieth century. But visitors often come away with more than facts and figures. Their visit gives them a sense of the immigrant experience and a respect for the courage and determination of the immigrants. This heightened awareness expands their understanding and appreciation of the diversity that characterizes the United States. Because cultural resources enable us to experience personally our heritage, we often come to a better understanding of that legacy. Even those who never visit a particular resource can nevertheless cherish it as a symbol of their heritage. (This chapter's Environmental Science in Action: The Statue of Liberty, pages 544–546, discusses this important cultural resource and what it symbolizes to U.S. citizens and people throughout the world.)

Cultural resources often link us with the past and our heritage in a personal way. Imagine that you decide to rehabilitate a home in an old section of the city. Interested in the building's past and in the history of the entire neighborhood, you might decide to learn what ethnic groups settled the neighborhood, who lived in your building, and what life was like for the former residents and their neighbors. In doing so, you uncover the heritage which is now yours. Knowing something of the past enables us to see ourselves as part of a continuing, shared experience.

Cultural Identity. Cultural resources are a critical part of the psychological and social structure of the people to whom they belong. For instance, many immigrants to the United States retain a strong sense of cultural identity, living near and worshipping with others from the same background. By keeping alive aspects of their original culture such as language, food, or customs, they retain a sense of their unique heritage, including attitudes and practices that can help them succeed in their new environment.

The process by which one culture adapts or is modified through contact with another is called **acculturation.** Acculturation can be unsettling to the people whose life-style is altered, especially if they do not choose to come in contact with the new culture, as did the European immigrants to North America, but instead have a new life-style forced upon them by a dominant culture, as did the native Americans. As their way of life changes, including how and where they live and what occupations they pursue, acculturated people may give up many of the practices of their native culture. But the outward features of a culture (such as housing and employment choices) can be abandoned more easily than can the knowledge and the values that gave rise to those features. Thus, a people who have always lived in close association with the land may find urban life unsettling and disruptive. Problems with acculturation are believed to contribute to a high incidence of alcoholism and suicide among some native American groups.

Storehouse of Environmental Knowledge. The accumulated wisdom of a specific people cannot be replaced. Consider, for example, the peoples of the rain forests. They hold the secrets of living and flourishing in what to us is a mysterious and hostile environment; for example, they know which plants to use for food and medicines and which to avoid. This highly developed awareness of and knowledge about the environment is characteristic of many so-called primitive groups (Figure 24-5). The more we learn about the indigenous groups who live outside the industrialized world, the more we realize we can learn from them. We would be wise to encourage preservation of their knowledge so that it can be used by both native and nonnative inhabitants.

The beliefs and technologies of a culture might also be adapted or applied in another culture. The *Foxfire* books, for instance, are compilations of skills and techniques from the Appalachian region. Some of these techniques, for example, some of the old agricultural methods, are becoming more attractive as we become more aware of the environmental consequences and increasing cost of high-input, energy-intensive methods. In the first book of the series, editor Eliot Wigginton dedicates the work "to the people of these mountains in the hope that, through it, some portion of their wisdom, ingenuity and individuality will remain long after them to touch us all."

Economic Benefits. Maintaining, rehabilitating, or restoring material cultural resources is economically prudent. **Rehabilitation** is the act or process of returning a property to a state of utility through repair or alteration which makes possible an efficient contemporary use while preserving those features or portions of the property that are historically, architecturally, or culturally significant. **Restoration** is the process of accurately recovering the form and details of a property and its setting as it appeared at a specific period of time by means of the removal of later work or the replacement of missing earlier work.

In the long run it costs less to maintain buildings than it does to repair damages caused by neglect. By identifying the causes of property damage we can take steps to eliminate or control them and avoid expensive repairs. Moreover, it is sometimes cheaper to repair a building than to tear it down and replace it with another. Unlike new construction, restoration is labor-intensive and is thus not as influenced by rising materials costs.

Even more importantly, rehabilitation benefits an area economically by creating new jobs, both during construction and later (in new offices, shops, restaurants, and tourism activities), increasing property values in revitalized areas, increasing property tax revenues as rehabilitated buildings are returned to the tax rolls, and attracting new businesses, tourists, and visitors. (See Focus On: Rehabbing in Covington, Kentucky, page 536.)

FIGURE 24-5: The Inuit, Northwest Territories, Canada. Tremendously sensitive to their environment, the Inuit are able to track animals with an accuracy and efficiency that outsiders regard as uncanny.

Cultural Diversity. Cultural diversity broadens our view. As we open ourselves to other cultures, we benefit from the mental and emotional stimulation other beliefs, customs, and practices offer us.

Preserving cultural diversity increases our chances for long-term survival. Our species will certainly face changing environmental conditions in the future, changes that may be dramatic. Cultural diversity can provide new options and new solutions, enabling us to tackle perplexing problems and to adapt to environmental change.

What Are the Threats to Cultural Resources?

Threats to our cultural heritage may be intrinsic or extrinsic. Intrinsic factors pertain only to material resources such as buildings, monuments, and other structures. The location of a structure, the nature of the ground on which it stands, the soil type, faulty materials, and building defects may cause it to deteriorate more rapidly than expected.

Both material and nonmaterial cultural resources are threatened by extrinsic factors, including both long-term natural causes and the action of humans. Objects of material culture are subject to long-term natural deterioration as a result of interacting environmental factors. These can be physical, chemical, or biological. Physical factors include temperature fluctuations, frost, heat, rain, humidity, and winds. Chemical factors usually act through the atmosphere and water; for example, oxidation and salt spray can deteriorate structures. Biological factors include microorganisms, termites, rodents (particularly rats), bird droppings, and vining vegetation.

Human activities also pose a threat to material culture. The use of wells and the construction of tunnels can alter water tables, possibly damaging structures. Mechanical vibrations from heavy road traffic, railways, subways, industrial plants, and supersonic flight can cause structural damage.

Air pollution, particularly in the form of acid precipitation, can severely damage many materials used in monuments and buildings (see Figure 24-7, page 538). Acids weaken limestone and marble, and can corrode exposed materials such as metal culverts, roofs, bridges, and expressway support beams, unless they are painted and maintained. Acid precipitation also damages brick by dissolving the glassy fabric that holds the silica grains of brick together. The brick gradually weakens until it is little more than a "silica sponge."

Ultraviolet light combined with acid precipitation destroys wood. The acids weaken the cellulose fibers in wood, and the ultraviolet light decomposes lignin, the substance that holds the cellulose fibers together. Paint offers some protection.

Focus On:
Rehabbing in Covington, Kentucky
Stephen Oberjohn

During the 1950s and 1960s many cities measured progress by the number of old buildings torn down in the name of urban renewal. The buildings were replaced by an expressway architecture of high-rise buildings and fast-food restaurants. Few voices were heard in support of preserving old buildings and the way of life that accompanied them. This essay looks at one city that, for the most part, chose an alternative solution—restoration. It wasn't the most obvious, easiest, or most popular choice, especially when first proposed, but restoration has worked.

Covington, Kentucky, is located just across the Ohio River from Cincinnati, Ohio. It has a rich German and Irish heritage. The city played an important role in the early development of the west and was a strategic city during the Civil War. As a manufacturing and warehouse center on transportation routes provided by the Ohio River and several railroads, Covington attracted people from many areas. One of its significant populations is from Appalachia.

Covington has many wonderful treasures: the Roebling Suspension Bridge across the Ohio River, originally the longest suspension bridge in the world and the prototype for the Brooklyn Bridge; stately homes along the riverfront, one of which has a stone tunnel that may have been part of the underground railroad; and wonderful churches built by European immigrants during the

mid-1800s. But in 1980 the city also had major troubles. The population had declined from a high of 65,000 in 1930 to less than 50,000. As long-time residents moved to the suburbs, the buildings they left behind began to crumble. The wrecking ball was beginning to gather support as a way to demonstrate to the surrounding communities that Covington was "on the move."

Not everyone agreed with this approach. In particular, two areas of the city, Riverside and Mutter Gottes, resisted the wrecking ball and, more importantly, helped change attitudes about the value of restoring rather than destroying neighborhoods (Figure 24-6). In 1968 a prominent architect in Northern Kentucky wanted to raze the entire Riverside neighborhood district of historic homes to make way for a massive urban renewal project. The area had a rich local and regional history and culture, including architecturally significant homes that represent the best building designs of the nineteenth century.

First individually, then as a group, residents fought the plan. They became politically active, circulated petitions, and attended city commission meetings (which tended to get very heated). Without waiting for city government, indeed, despite city government, they began restoring their homes and the neighborhood. The debate raged for quite some time, but eventually the

FIGURE 24-6: The Oberjohn home in the Mutter Gottes section of Covington, Kentucky.

city learned a great deal about the power of an active and vocal citizens' group. The city killed the plan for urban renewal and later began serious planning for the restoration of the entire neighborhood.

Today professors bring their architecture students to the neighborhood to study design, and community celebrations along the Ohio River often take place in the area. Recognizing the history and culture

Changing fashions are also a potential threat to material resources. When artistic and architectural styles become unfashionable, works in those styles may be neglected or deliberately altered or destroyed. In the 1850s, workers cleaning the walls of the Bardi chapel in the Church of Santa Croce in Florence uncovered frescoes painted by the fourteenth-century master Giotto di Bondone; the frescoes had been whitewashed a hundred years before. The frescoes were restored and have since added much to our knowledge of the work and influence of the Italian artist.

War takes a terrible toll on material culture. During World War II many cathedrals and public buildings throughout Europe suffered significant damage from bombing raids. More recently, during Croatia's struggle for independence from Yugoslavia in 1991, many historic structures within the beautiful, centuries-old city of Dubrovnik suffered irreparable damage (Figure 24-8).

Development is another human activity that threatens material culture. For example, a series of 22 dams are being built along the Tigris and Euphrates rivers in Turkey. The modern needs that

of the area was critical to the future of the entire city, not just the Riverside area. Property values in this neighborhood have increased dramatically over the last few years, and today's developments west of the Riverside neighborhood complement the history of the city rather than replace it.

Of course, not all cities have an area of such obvious historic and cultural value. Another neighborhood that fought for preservation was an old working-class area of Covington. Though not as dramatic as Riverside, the restoration of this neighborhood has been no less critical to the revitalization of the city. This neighborhood is called Mutter Gottes, or "Mother of God." In 1975 the city was planning to condemn the modest homes along Covington Avenue in Mutter Gottes to make way for a highway. The preservation movement was led by one man who formed a partnership with other people concerned with the imminent destruction of more homes in Covington. By working with a bank that showed courage and a willingness to develop new methods of financing, each individual in the partnership bought one of the nine houses for sale on the street for $6,300 each. By investing as much sweat as money, these folks brought the street back to life.

Taking a lesson from the Riverside group, the Covington Avenue group became involved in city government and city politics, fighting for the election of officials who

were actively involved in rehabbing. By participating in stormy city commission meetings, the group was able to get the city to agree to several proposals, including closing the street to through traffic and eliminating on-street parking. When the city was ready to repave the deteriorated street with blacktop, the residents proposed that instead it restore the stone curbs and pave the street with bricks. As a sign of how far the city had come, it abandoned its precept that the city knows best and adopted the residents' plan.

Fifteen years later most of the original members of the Covington Avenue partnership still live in the homes they rehabbed. The group has formed other nonprofit partnerships as a way to buy old homes and hold on to them until a buyer interested in rehabbing can be found. They continue to remind the city government of the importance of preserving the old neighborhoods. Group activities are not restricted to hard work! Every year on the first Saturday of May, the neighbors gather in one backyard to celebrate the running of the Kentucky Derby.

The city gradually became more interested in preserving local history and culture until it even hired a historian and preservationist. As the momentum for rehabbing grew, the government embraced the concept that preservation and rehabbing are often the preferred alternative. The Housing Department, Urban Design Review Board, Historic Preservation staff, elected officials, and residents

of the city found ways to work together. Such cooperation was a far cry from the early 1960s.

The rehabbing movement spread beyond Riverside and Mutter Gottes. Today, many vibrant neighborhoods contribute to Covington's economy. Buildings that were once condemned are now part of the city's tax base. As residents return to abandoned areas, small businesses are revived and new ones appear. Large corporations are beginning to return to the city as well. Furthermore, the city is in a position to try new ideas as the tax base increases. One plan is to sell abandoned but structurally sound buildings for $1 each to a nonprofit organization that will rehab the buildings to provide shelter for the homeless.

Despite the success of rehabbing, much work remains. There are thousands of buildings that could be restored—or destroyed. Covington has progressed from being first on the federal government's list of most distressed cities in the country in the late 1970s to being a city with a solid future. This came about in great part because of the respect some innovative and stubborn citizens had for Covington's history and culture. By restoring first a building, then a neighborhood, rehabbers are offering a viable alternative to the wrecking ball.

the system will serve are in conflict with the need to preserve cultural resources. The dam system will provide electricity and irrigation water to farms and villages, but the reservoirs created by the dams will inundate an untold number of archeologically rich and virtually unexplored sites. Records indicate that one site, the ancient city of Hasankeyf, dates from the first millennium B.C. and served as a major center on an international trade route into the seventeenth century. Although full excavation of the Hasankeyf site would take about a hundred years, archeologists have only until 1999 before the area is

flooded by the completion of the Ilisu Dam. This particular dam will serve only to provide electricity (it has no irrigation systems), and engineers estimate that silt buildup will put an end to its use in only 70 years.

Vandalism also degrades or destroys cultural resources. Senseless damage, such as graffiti on monuments, endangers cultural resources throughout the world. In 1991 vandals struck Naj Tunich, a cave in Guatemala containing 90 drawings which represent the only known large body of Mayan inscriptions and sketches; 23 of the drawings were

FIGURE 24-7: Renaissance sculpture on Cloth Hall, Krakow, Poland. The sculpture's porous limestone has been damaged by airborne sulfuric acid from the smokestacks of nearby Nowa Huta Steel Mill. Poland has some of the worst air pollution in the world.

smeared with mud, scratched, or struck with hard objects (Figure 24-9).

Pillaging for profit is perhaps an even more serious problem. Artifact hunters equipped with metal detectors have pillaged Civil War battlefields in Virginia, Pennsylvania, and Maryland, searching for shell fragments, uniform buttons, and other historic treasures. The most serious cases of looting in the United States occur at native American sites. Since the early 1970s the value of native American artifacts has skyrocketed. By one estimate, 80 to 90 percent of the sites held by the Bureau of Land Management in Utah have been destroyed by treasure hunters. This problem is not confined to the United States; sites all over the world suffer the ravages of collectors, professional treasure hunters, and black-market traders.

Pillaging of a site is detrimental in several ways. First, important cultural resources—which are national property—are stolen. Second, pillaging disrupts a site to such an extent that the archeological record is ruined. The soil matrix, containing stores of information about how the inhabitants lived, what they ate, and their environment, becomes mixed and randomized. Finally, the desecration of a site that had religious significance for its society degrades the spiritual quality of the site (Figure 24-10).

Looting and vandalism are difficult to control. Although the National Park Service and Bureau of Land Management have attempted to protect cultural resources in the United States, these activities still continue. The federal lands are simply too vast to patrol adequately, and existing laws are not strong enough to halt the trade in illegally gotten artifacts.

The Archeological Resources Protection Act (ARPA), passed by Congress in 1979, levies stiff fines on anyone who removes archeological resources from public land or participates in their sale, purchase, transport, or receipt. However, the stipulation that law-enforcement officials prove that the artifact was taken from public property makes it difficult to prosecute cases, since defendants can simply assert that they obtained artifacts from private land.

Perhaps the greatest threat to nonmaterial cultural resources is acculturation. Whenever a group comes in contact with a foreign culture, it faces the difficult task of balancing the necessity to function in the new culture with the desire to retain its own distinct heritage. The dominant society can make that task easier—or far more difficult. Most often, minority cultures feel substantial pressure to conform to the practices of the dominant culture.

In 1988 Dr. Daryl Posey, an ethnobiologist from the United States working in Brazil, and two members of the Kayapo, a native Brazilian tribe, traveled to Florida to attend an international conference on the environmental threat posed by an Amazonian dam project. Upon their return to Brazil, the trio was arrested under the Brazilian Foreign Sedition Act, a little-used law that prohibits *foreigners* from interfering with the country's internal affairs. When the Kayapo leader arrived in court to make a preliminary statement, the judge refused to admit him because he was not "acculturated," that is, he was wearing his traditional garb rather than a suit. The judge also ordered psychiatric examinations to determine the extent of the Kayapo's resistance to

FIGURE 24-8: A woman sweeps up rubble after a December 1991 attack on Dubrovnik. Many historic buildings, structures, and sites in this medieval city, recognized as a World Heritage Site by the United Nations Educational, Scientific, and Cultural Organization, have been devastated by the civil war raging between Croatia and Serbia-dominated Yugoslavia.

FIGURE 24-9: Hieroglyphics carved in relief at Naj Tunich, a treasure of Mayan culture and a target of vandals in 1991.

acculturation. A lawsuit filed by lawyers for the Kayapo charged the Brazilian government with unconstitutional and racist treatment of its native population.

History of Management of Cultural Resources

How Have Cultural Resources Been Preserved in the United States?

Until fairly recently the private sector was the force behind the preservation movement in the United States. In the late nineteenth and early twentieth centuries preservation was at best a partnership of private groups and government; government action often came only at the insistence and prompting of citizens' groups. For instance, neither the Virginia legislature nor the federal government acted to preserve George Washington's Mount Vernon home. In 1853 private citizens rallied to the cause. The Mount Vernon Ladies Association of the Union, chartered in 1856, succeeded in raising the funds needed to save the estate, and in 1859 restoration began.

Despite its initial lack of involvement, the government became increasingly aware of the importance of cultural resources during the twentieth century and took on more and more responsibility.

Early Government Involvement in Preservation of Cultural Resources. The federal government did not formally acquire a historic property until 1864, when it purchased General Robert E. Lee's Virginia home, Arlington House. But the purchase was motivated by political considerations, not by the desire to

preserve a national historic and architectural treasure. The mansion and its grounds were considered spoils of war and were thus the responsibility of the federal government. In 1883 the Lees were properly reimbursed, after the Supreme Court ruled that they were the rightful owners. By this time, the surrounding area had already been marked as the site of the Arlington National Cemetery.

The government's concern for natural resources set a precedent that it eventually used to preserve cultural resources as well. The breathtaking natural beauty of the Yellowstone area led Congress to make it the first "nation's park" in 1872. Seventeen years later Congress used its power to protect an area's cultural resources; Casa Grande in Arizona was the first park tract to be designated a national monument solely on the basis of its historic value. The area's native American ruins were threatened by continuing vandalism, and concerned private organizations and government groups prompted the 1889 designation.

The Antiquities Act of 1906 was the country's first major piece of legislation designed to safeguard cultural resources. Intended to preserve prehistoric sites, it gave the president the power to designate national monuments and establish regulations to protect archeological sites on public lands. The law restricted archeological research on federal land to accredited educational or scientific institutions and stipulated that research findings must be made accessible in public museums. Although an important first step, the Antiquities Act failed to consolidate jurisdiction for protected sites in one agency. Instead, the responsibility for each site was left to

FIGURE 24-10: Theft of a native American grave. This Hopi gravesite in Arizona was disturbed by looters looking for artifacts. Such a disturbance is an affront and a violation to those whose ancestors are buried on the site; and the emotional and physical loss they experience is just as real and just as destructive as the physical desecration itself.

the department which had first dealt with it, resulting in duplicated efforts at some sites and no effort at others.

From the turn of the century until the 1930s preservation efforts by local and state authorities were sporadic, with the focus primarily on single homes or structures. However, progress was made in research and preservation studies throughout the country, and preservation legislation was passed at the local and state levels. For instance, a 1931 law establishing a historic district in Charleston, South Carolina, was the first local legislation of this type. Soon after, a number of states became involved with historic preservation, and by the late 1930s Indiana, California, Pennsylvania, Illinois, Ohio, and North Carolina were among those with designated historic sites.

Effects of World War II and Postwar Economic Growth on Preservation of Cultural Resources. In August 1935 Congress passed the National Historic Sites Act. This act authorized the National Park Service to acquire national historic sites and to designate national historic landmarks. The service was directed to identify, maintain, manage, and interpret historic sites for the public benefit and to establish public education programs. Many of the nation's most important cultural resources are sited within units of the National Park System and so are managed by the Park Service. In addition, the act provided for the continuation of the Historic American Buildings Survey (HABS). The survey had been initiated in 1933 by the Park Service, the American Institute of Architects, and the Library of Congress in order to inventory the nation's historic resources.

The U.S. government's commitment to historic preservation waned with the start of World War II. Of necessity, the private sector once again shouldered the responsibility of preserving the national heritage. These private efforts became increasingly important during the postwar period, when a national building boom threatened to destroy countless significant structures.

Realizing the danger to historic resources posed by the tremendous growth of the period, representatives from interested organizations nationwide met in 1947 to form a national preservation organization. The result was the National Council for Historic Sites and Buildings, and its intent was to enlist members and to generate and support interest in historic preservation nationally. The council also sought a congressional charter to establish a national trust, a legal entity through which it could acquire and operate historic properties. The charter was granted in October 1949, and the National Trust for Historic Preservation became a reality. Five years later, the National Council for Historic Sites and

Buildings merged into the National Trust for Historic Preservation. This enlarged private, nonprofit organization continued to encourage public participation in the preservation of buildings, sites, and objects significant in American history and culture.

A new threat to material culture emerged during the late 1950s and the 1960s with the widespread popularity of urban renewal projects. **Urban renewal** consisted of demolishing older buildings, often specific sections of a city neighborhood, to accommodate new and generally homogeneous structures. The often indiscriminate demolition fueled public sentiment in favor of historic preservation.

Major Efforts to Preserve Cultural Resources Since 1960. By the end of the 1950s preservationists began to enjoy support at all levels of government. In 1960 the Park Service initiated a listing of national historic landmarks which had "national significance in the historic development of the United States." In 1966 Congress passed the National Historic Preservation Act. The act directed the secretary of the interior to develop and maintain the National Register of Historic Places, including districts, sites, buildings, structures, and objects important in American history, architecture, archeology, engineering, and culture. The register was to include listings of local, state, and regional significance.

The National Historic Preservation Act also provided grants to the states and the National Trust. These grants were to be used to conduct statewide surveys of historic resources, to formulate the state historic preservation plans required by the act, and to assist in preserving the individual properties listed in the National Register. Finally, the act also created the Advisory Council on Historic Preservation, an independent agency to mediate between federal agencies and coordinate construction and preservation interests.

The National Environmental Policy Act of 1969 (NEPA), a wide-ranging piece of legislation that affected many segments of the environment, also affected the historic preservation movement. NEPA stressed the government's responsibility toward preservation and established the Council on Environmental Quality within the Office of the President. Among other duties the council was charged with seeing that all federal activities took environmental considerations into account. NEPA, which required each federal agency to prepare a detailed environmental impact statement before pursuing any activity that would significantly affect environmental quality, also required a similar social impact statement for any activity affecting historic and archeological sites.

Two years after the passage of NEPA, President Nixon issued Executive Order 11593, the Protection

FIGURE 24-11: View from Gateway Arch, St. Louis, Missouri. Urban revitalization has restored once-declining neighborhoods in St. Louis.

and Enhancement of the Cultural Environment. The order directed each federal agency to inventory and nominate to the National Register all federal properties of historic value which it administered. While the National Historic Preservation Act of 1966 had granted protection from federal projects to those structures and sites listed in the National Register, Executive Order 11593 granted similar protection to federal properties which were not yet listed in the register. In doing so, the order formally recognized historic preservation as a national priority.

The popularity of historic preservation increased with the Economic Recovery Tax Act of 1981, which provided significant new investment tax credits for rehabilitation of older or historic buildings. This economic incentive resulted in the revitaliza-

tion of many neighborhoods throughout the nation as individuals, professional developers, and corporations rehabilitated older buildings, preserving or restoring their significant cultural features, for private and commercial use. **Urban revival,** as this process is known, spurred economic development in these neighborhoods or historic districts, resulting in increased tax revenues at the federal, state, and local levels (Figure 24-11). Many different approaches, varying according to the degree in which they change a structure or site, are used to prolong the useful life of historic resources (Table 24-1).

How Are Cultural Resources Preserved Worldwide?

The chief international agency working to protect the global cultural heritage is the United Nations Educational, Scientific, and Cultural Organization (UNESCO), organized in 1945. A 1985 UNESCO publication, *Culture and the Future*, describes the organization's mission in this way:

Only forty years ago, a fair part of our planet was emerging from a war that had cost humanity fifty million lives. A few enlightened and cultivated spirits decided to take up the old Socratic notion of eliminating ignorance and incomprehension, so as to succeed, at last, in preventing men from killing one another at regular intervals. Thus was UNESCO born, and one of its aims was and still is to offer culture to all so that they will love life. This has not yet been achieved, but it is not for want of trying.

UNESCO member nations agree to a binding legal framework that obliges them to protect the global cultural heritage and establishes international con-

Table 24-1
Four Major Approaches to Renewing Older or Historic Structures

Approach	Description
Preservation	Preservation is the maintenance of a historic resource in its original state. A structure is cared for to maintain its existing form, integrity, and material and a site is cared for to maintain its existing form and vegetative cover. Maintenance is less expensive than other approaches and retains the original character of a structure to the greatest extent.
Stabilization	Stabilization is the use of methods designed to reestablish a weather-resistant enclosure or the structural stability of unsafe or deteriorated properties in order to prevent further damage. It involves little actual change to the state of a structure as it is when renewal begins. Stabilization was the primary method used on the Anasazi cliff dwellings.
Rehabilitation	Rehabilitation includes repairs and alterations that allow an efficient contemporary use of a structure yet also preserve features of historic, architectural, and cultural significance. Many rehabilitated buildings are converted to accommodate an adaptive use, that is, a use other than that which the building was originally designed to serve. For instance, a residential structure may be converted to a restaurant or specialty shop. Rehabilitation requires the use of new materials and the replacement of parts of the structure.
Restoration	Restoration may require significant changes from the current state of a structure. Its goal is to restore the form and details of a property and its setting to the way they were at a particular period of time. Missing features and details may be replaced and later alterations removed.

ventions concerning the protection of cultural treasures. Members are free not to ratify a convention, but if they do ratify it, they are bound to comply with its stipulations. UNESCO assists members in adapting their national legislation and administration to comply with convention requirements.

To date, UNESCO has established three conventions. The first concerns the protection of cultural property in the event of armed conflict. The second outlines means to prohibit and prevent the illegal import, export, and transfer of ownership of cultural property. This convention is intended to help put an end to the ages-old practice of pillaging works of art and archeological treasures. Many countries have been deprived of important pieces of their cultural heritage through these practices (Figure 24-12).

In 1972 UNESCO established its third convention, a document aimed at safeguarding property of outstanding universal value. It recognizes that the "deterioration or disappearance of any item of the

cultural or natural heritage constitutes a harmful impoverishment of the heritage of all the nations of the world." Each nation conducts an inventory of the cultural and natural property within its borders and composes a list of properties it deems to be of universal value. The lists are studied by the World Heritage Committee, a committee of representatives from 21 member nations. The committee decides which properties are to be included on the World Heritage List and assists the nations in protecting these properties. As of 1990, the World Heritage List included 322 sites in 69 countries (Figure 24-13).

Inclusion on the World Heritage List alone cannot guarantee that these treasures will be protected. Because many nations do not have the financial means to preserve their resources, the convention established a World Heritage Fund to aid members. The fund provides technical assistance, equipment, and supplies; finances studies to determine or prevent the causes of deterioration; and finances the planning of preservation measures and the training of local specialists in preservation and renovation techniques.

Future Management of Cultural Resources

Effectively managing our cultural resources can enable us to bequeath to our descendants a rich and varied global cultural heritage. Improving the management of cultural resources falls into three categories: preventing abuse of cultural resources, safeguarding cultural resources worldwide, and preserving cultural resources as part of living systems.

Prevent Abuse of Cultural Resources

The most direct way to prevent abuse of cultural resources is to strengthen and enforce laws that protect them. Although the reauthorized Clean Air Act will help reduce damage caused by acid precipitation, other forms of air pollution need to be investigated and regulated to control their effects on the materials that make up cultural resources.

The Archeological Resources Protection Act should be amended to shift the burden of proof to the defendant rather than the government; in other words, a person in possession of an artifact should have to be able to prove that the artifact was taken from private, not public, lands. A registry for artifacts found on private lands should be created to track the discovery and transfer of such objects. The owner of an artifact would supply a provenience, written by an archeologist. The provenience would document the object's origin, giving a complete

FIGURE 24-12: Rosetta Stone. A French officer in Napoleon's engineering corps discovered the Rosetta Stone in the mud of the Nile River. The text on the stone is repeated three times: in hieroglyphics, in demotic script (a simplified form of hieratic writing; the common, idiomatic form of the Egyptian language at the time of the stone); and in ancient Greek. By translating the Greek portion, French scholar Jean François Champollion deciphered the meaning of the hieroglyphics, thus enabling scholars to read the literature of ancient Egypt. The Stone, which was the key to our knowledge of ancient Egyptian language and culture, was removed from Egypt, taken to Paris, and later moved to Great Britain, where it now resides in the British Museum.

What You Can Do:
To Appreciate Cultural Resources

Cultivate an interest in other cultures.
- Learn a foreign language.
- Read literature, newspapers, and other documents in that language.
- Learn some of the customs of other cultures.
- Travel in countries with cultures different from your own.

Learn about your national cultural heritage.
- Visit museums, living history exhibits, and historical societies in places you visit.

Learn about your local cultural heritage.
- Visit natural history museums, living history exhibits, and historical societies in your area. Better yet, take an active part in reenactments your local groups stage. Contact your local groups for more information.
- Study your natural environment and learn how it influenced the cultures that developed there. (For example, the culture of the Plains Indians was very different from the culture of the native Americans who lived in the wooded Ohio Valley.)

Help preserve cultural resources in your area.
- Join a citizens' group concerned with historic preservation.
- Become involved in a rehabbing effort.

Learn about your personal history and culture.
- Research your family tree.
- Interview older relatives about their life stories.
- Study family customs or traditions; discover where those traditions come from, what they mean, and why they were important to your ancestors.

Spread your enthusiasm for cultural resources.
- Involve friends and family in your interest in cultural resources.
- Introduce children to the heritages of other cultures as well as their own.

Support legislation promoting the protection of cultural resources nationally and globally.

description of the artifact, a copy of the excavation report, a certificate of excavation by a licensed archeologist, and a certified appraisal. After registering the object, the owner would receive a nontransferrable title and an artifact documentation card.

In addition, we need to provide greater economic support to agencies, specifically, the National Park Service, the Bureau of Land Management, and the Forest Service, that fight pillaging and vandalism of cultural resources. Increased funding will enable these agencies to locate and protect material cultural resources on all public lands.

Another way to prevent abuse is to educate people, especially those visiting areas such as national parks with vulnerable resources, about the importance of cultural resources and how they should be treated. Education is a key tool for fostering respect for cultural resources and other cultures in general. (See What You Can Do: To Appreciate Cultural Resources.)

(a)

(b)

FIGURE 24-13: World heritage sites. World heritage sites may be designated natural sites, cultural sites, or combined natural and cultural sites. Two combined sites are (a) the Tanzania, home of diverse species of wildlife, including these greater and lesser flamingos, and (b) the Tansmania, Australia.

Environmental Science in Action: The Statue of Liberty

Karla Armbruster

Approximately 40 percent of the U.S. population, roughly 100 million Americans, trace their heritage to an immigrant whose first sight of this country was the Statue of Liberty, originally named *Liberty Enlightening the World*. Symbolizing freedom and opportunity, she has welcomed more than 17 million immigrants to these shores. Today, many immigrants do not see the statue, arriving overland from Mexico and Central America or by boat from Vietnam or Cuba. Nevertheless, the torch-bearing symbol of liberty is so powerful that even those who never see the statue are moved by the promise and hope it represents. Properly managing this important cultural resource will enable us to preserve it as a tangible reminder of that promise.

Describing the Resource

Physical and Biological Boundaries

The Statue of Liberty, rising 151 feet (45 meters) and weighing 225 tons (205 metric tons) stands on Liberty Island (formerly Bedloe's Island) in New York Harbor (Figure 24-14). A torch is held aloft in her right hand, and her left hand grasps a tablet inscribed with the date July 4, 1776, a commemoration of the Declaration of Independence. Unshackled chains lie at Liberty's feet. The seven rays on her crown, ranging up to 9 feet long and weighing 150 pounds each, symbolize the seven seas, the seven continents, and (at the time of her construction) the seven known planets. The monument rests on a granite pedestal 89 feet (27 meters) high which in turn rests on a foundation 65 feet (20 meters) high. Liberty towers 305 feet 1 inch (92 meters) from the foot of her base to the tip of her torch.

The statue's copper skin was created from 300 copper sheets, which were carefully hammered around a plaster model created by the statue's French designer, Frédéric Auguste Bartholdi. This sculptured skin, which measures 11,000 square feet (990 square meters) and is approximately the thickness of cloth drapery, serves a structural purpose in addition to its esthetic beauty. The hammered copper forms a rigid envelope, and the many folds of Liberty's robe distribute stress and minimize sagging.

An interior iron armature to support the statue was devised by Bartholdi's countryman, Alexandre Gustave Eiffel, who would later design Paris's Eiffel Tower. This "skeleton" was composed of over 1,800 2 inch by ⅝ inch iron straps which conformed to the inner surface of the copper skin. The straps were supple enough to allow the skin to move in response to winds and thermal stresses. Copper saddles (pieces of U-shaped metal) and copper rivets secured the iron straps to the skin. Flat iron bars connected the straps to a stronger interior structure, also made of iron. This structure was connected directly to the statue's interior iron pylon. Four legs, which ran through the pedestal and into the ground, anchored the pylon and provided a secure base for the statue.

Eiffel and Bartholdi knew that copper and iron react if they contact each other in the presence of an electrolyte such as water. They tried to prevent this reaction by inserting shields of asbestos cloth soaked in shellac between the two metals at every junction. But the shellac's ability to prevent the movement of water through the cloth was short-lived and asbestos alone could not prevent this movement. As saline water entered the statue's interior through various openings, including rivet holes and the joints of the copper sheets, it acted as an electrolyte. The water caused the iron bars of the statue's armature to corrode everywhere they contacted the copper skin and saddles.

The rust damaged the structure in two ways. As it ate into the bars, it progressively weakened the iron. As the rust accumulated, it occupied a greater volume under the saddles than had the original iron bars. The slow but powerful expansion of the rust pulled the rivets,

FIGURE 24-14: The Statue of Liberty, after restoration work was completed in 1986.

which secured the saddles, right through the copper skin, detaching many of the saddles and causing holes in the skin, which admitted more water, hastening the deterioration. By the 1980s a substantial number of the bars had lost as much as half their original thickness of iron.

Another problem contributing to the statue's degradation arose from alterations made to the flame of the torch. Bartholdi had originally fabricated the torch flame from solid sheets of copper, but prior to the statue's dedication in 1886, two rows of portholes were cut into the lower half of the flame so that it could be illuminated from inside. Because this illumination was dim, in 1892 the upper row of portholes was replaced by a continuous band of glass and a pyramid-shaped skylight was inserted at the top of the flame. In 1916 the entire surface of the flame was altered into panes of amber glass in a copper framework.

An elaborate glazing system, devised to prevent water from entering the interior through the mosaic-like flame, failed. The torch became the statue's single worst source of water penetration, and the leakage and resulting corrosion severely damaged both the torch and the upraised right arm.

Caretakers tried to combat the leakage from the torch and from the stretching seams and rivets, but to no avail. In 1911 they applied a bituminous paint (coal tar) to the interior of the statue in an attempt to isolate the iron and copper. As the paint peeled and blistered, it trapped moisture inside, further complicating the problem.

Social Boundaries

The Statue of Liberty has always been a potent symbol of values important to the United States and to the people of France, who gave us the statue. It represents liberty and freedom from oppression, the aspirations of humanity, justice and equality before the law, the struggle for American independence, American democracy, the American tradition of hospitality and welcome, opportunity, and hope for the future.

The importance of the statue to Americans and tourists from other countries is demonstrated by the millions of visitors it receives each year. Although the number of visitors declined after an all-time high of 3.1 million in 1987 (immediately following the statue's restoration), 2.6 million people still came to view the statue in 1989. The statue's universal value was confirmed in 1986, when it was designated a world heritage site.

Looking Back

Although the statue has always been an important monument in this country, its history demonstrates that we must always be vigilant in protecting our cultural resources from the damage caused by human use and environmental processes.

How the Statue Came to the United States

In 1865 the United States was emerging from the Civil War, the blackest period in the young nation's history. However, the abolition of slavery represented another step toward the ideal of freedom for all peoples, and, although President Abraham Lincoln had fallen victim to an assassin's bullet, the union was preserved.

The events in the United States were widely discussed in France, which was then laboring under the oppressive regime of the Emperor Napoleon III. At a dinner party in 1865, Edouard de Laboulaye, a French historian and legal scholar, proposed that France give the United States some sign of the long friendship between them—a monument dedicated to their shared ideals of freedom. Bartholdi, a guest at the dinner, envisioned the monument as a woman holding aloft a torch. Five years passed before the collapse of Napoleon's empire, but when the French reestablished their democracy in 1873, Laboulaye and Bartholdi were finally able to launch their project.

Although Congress had accepted the gift and approved the Bedloe's Island site chosen by Bartholdi, it would not appropriate any money for a pedestal for the statue. When only half the necessary funds had been raised by 1883, the project was in danger of failing. Joseph Pulitzer, an immigrant and owner of several newspapers, took up the cause and stimulated support for the statue. He embarrassed the rich for not contributing to the project and stressed that the statue was a symbol for all Americans. He published the names of all contributors—over 121,000 in all.

Finally, on July 4, 1884, the monument was presented to the American minister in France. It was then dismantled and shipped to the United States, where it was reassembled. *Liberty Enlightening the World* was formally dedicated by President Grover Cleveland on October 28, 1886.

How the Statue Was Assembled and Maintained

Unfortunately, U.S. workers who reconstructed the statue sometimes failed to match French rivet holes or plans for the framework. As a result, the head was assembled approximately 2 feet off its correct alignment, resulting in unexpected stresses on the monument and causing one spike of the crown to touch the statue's upraised arm and eventually puncture the copper skin. Over the years attempts were undertaken to maintain the statue, but its large size and waterbound location made these efforts sporadic. Meanwhile, the marine environment, combined with the well-meaning "improvements" to the design of the torch and the many leaks, caused the statue to deteriorate.

In 1924 the Statue of Liberty was declared a national monument; nine years later, it was placed under the jurisdiction of the National Park Service. Still, management efforts were largely "brushfire tactics," dealing with the worst of the deterioration in a haphazard manner. For instance, at least seven coats of paint were applied to the Statue's interior throughout this century in an attempt to prevent the copper-iron reaction. In 1937–1938, the most severely corroded iron bars were replaced.

How the Statue was Restored

Finally, in 1982 the Park Service realized the urgency of restoring the statue in preparation for its upcoming centennial celebration in 1986. Budget deficits forced the service to appeal to the private sector to raise the necessary funds. Then Secretary of the Interior James Watt appointed a Statue of Liberty–Ellis Island Centennial Commission to advise the government on restoration activities. Lee Iacocca, head of the Chrysler Corporation and himself the son of immigrants, was named chairman of the commission. Subsequently, a nonprofit private corporation, the Statue of Liberty–Ellis Island Foundation, was formed. The foundation assumed the responsibility for all fund-raising activities connected with the restoration and also assumed the power to make contractual decisions regarding the restoration work and to oversee the work in progress.

Restoration first required a careful analysis of the condition of the copper skin and the iron armature. The copper's green patina, produced by a complex reaction between copper and atmospheric gases in the presence of water, provides a natural shield for the skin, slowing the process of corrosion. Tests demonstrated that the skin was in surprisingly good shape; only 4 percent of its original thickness had corroded in 100 years.

The interior of the statue required far more attention. Architects Richard Hayden and Thierry Despont were in charge of this complex operation. More than 1,300 armature bars were replaced; only four bars in any one area could be removed and restored at one time, and only four areas far apart from one another could be worked on simultaneously. The replacement bars were made of stainless steel. Its properties of thermal expansion and elasticity are similar to the type of iron in the original bars, but it is stronger and reacts only minimally with copper. As an extra precaution, Teflon was placed as a buffer between the bars and any copper surfaces.

Another important improvement was the removal of the unattractive, damaging layers of paint on the statue's interior. Liquid nitrogen removed the outside layers, and the more abrasive bicarbonate of soda was chosen for the bottom layer of coal tar. It stripped the tar without scarring the copper, but it did have one unexpected effect. When the bicarbonate of soda leaked outside the statue, it reacted with the patina to streak the statue's surface a

bright turquoise blue! Fortunately, a water bath and time have almost completely erased the scars from this mishap. A new, simplified stairway built to the top of the statue reveals the freshly cleaned, cavernous interior and impressive skeleton for the admiration of visitors.

The torch and flame had become so severely damaged that they could not be repaired. French craftspeople replicated the original torch and flame, based on precise computer-generated models, and covered the flame with gold leaf. Special lights shining from the balcony around the base of the flame illuminate it beautifully.

Finally, the architects added an environmental control system to reduce the heat and humidity that had contributed to the statue's corrosion. By Liberty's centennial in July 1986, she was fully restored, ready to hold up her torch of freedom for future generations of Americans and immigrants.

Looking Ahead

The massive restoration effort saved the statue from destructive environmental and human forces, but it is still subject to renewed and new threats as time goes on. Therefore, the Park Service must vigilantly monitor the statue's condition in order to maintain the symbol of which our country is so proud.

While experts predict that the statue's skin could last for another 1,000 years if environmental conditions remain stable, increased acid deposition may result in color changes and an increased corrosion rate. Because acid precipitation is a mounting problem in our atmosphere, the Park Service plans to measure the skin's thickness at selected points from time to time. These spot checks will track how

quickly the statue's skin is corroding. If the rate begins to increase, the Park Service will be able to take steps to protect the statue.

Other factors the Park Service should monitor are other forms of air pollution, the corrosive marine environment, and vibrations from low-flying aircraft which could weaken the statue's structure. Admiring visitors also pose problems for the statue. Because the monument attracts over 14,000 visitors on some days, overcrowding and the noise, vandalism, and structural strain that accompany it could harm the statue and detract from the experience of visiting it. In 1989 the Park Service conducted a visitor study to help them better manage visits to the monument and satisfy visitors in the future. Better management will help to preserve the statue so that it may continue to inspire future generations. As Lee Iacocca said:

We didn't spend $230 million just so the statue won't fall in the harbor and become a hazard to navigation. We aren't fixing up Ellis Island so people will have a nice place to go on Sunday afternoon. We're doing it because we want to remember, and to honor, and to save the basic values that made America great. Values like hard work, dignified by decent pay. Like the courage to risk everything and start over. Like the wisdom to adapt to change. And maybe most of all—self-confidence. To believe in ourselves. Nothing is more important than that.

We're not preserving a statue here; we're preserving all that she stands for. And if that's not worth remembering, and honoring, and saving . . . if that's not worth passing on to our kids . . . then let me ask you—what the hell is?

Safeguard Cultural Resources Worldwide

Although the National Park Service prepares management plans for the properties under its jurisdictions, even the best-designed plan will be ineffective

without funds to implement it. Some cultural resources, such as the Statue of Liberty, enjoy such widespread public support that money can be raised privately to preserve them. However, many resources are not as well known, and government funding is essential for their preservation. Just as funds are needed to identify and protect cultural resources, monies are also needed to conduct studies and inventories, enable park personnel to undertake preservation measures, and staff our parks.

In addition, increased support should go to the formal educational programs found at many national parks. Park Service guidelines state that "intangible cultural traits and behaviors (songs, ceremonies, traditional uses, etc.) shall be accurately represented and interpreted" whenever they are associated with the material cultural resources within a national park. Formal educational efforts include visitor programs, displays, and classes.

The United States withdrew from UNESCO in 1984, citing poor management, a bloated bureaucracy, and UNESCO support for the Palestinian Liberation Organization. Its withdrawal marked the end of the organization's educational programs in this country and U.S. economic support for world heritage sites and other UNESCO programs. As of 1990, the United States had not rejoined UNESCO despite a change in management personnel and efforts to reform the organization's bureaucracy. The United States's rejoining would be an important step in safeguarding cultural resources worldwide.

Undoubtedly, the organization needs U.S. financial support and technical assistance, but perhaps more importantly, the United States can be an example to other countries by its willingness to set aside political differences in order to achieve the common goal of protecting the world heritage. As a member nation, we can and should encourage all nations to adopt the UNESCO international conventions. Our support of UNESCO programs to return cultural treasures to their countries of origin might influence other nations, particularly other western industrialized countries, to participate in such programs. Finally, as one of the world's wealthiest nations, we have a responsibility to provide economic support and technical assistance to needy countries who are striving to preserve their cultural resources. We must remember that their national cultural heritage is our human heritage.

Preserve Cultural Resources as Part of Living Systems

The widespread preservation of cultural resources will depend upon the development of a stewardship ethic. Preserving and effectively managing our global cultural heritage depend on an ethic that values all human life and respects natural and cultural diversity. In addition, perhaps the most important step we can take to preserve cultural resources is to recognize that they are integral parts of living systems, systems that also include our everyday lives and the natural environment. In other words, we need to incorporate the history and protection of the cultural environment into our efforts to preserve the natural environment.

Summary

A cultural resource is anything that embodies or represents a part of the culture of a specific people. All cultural resources arise from human thought or action. Material culture includes all tangible objects humans create to make living in the physical world easier or more enjoyable. Nonmaterial culture includes intangible resources such as language, customs, and beliefs. Cultural resources often reflect or are part of the environment. Historic preservation is concerned with preserving material cultural resources. Cultural resources management is concerned with preserving both material nonliving and living culture.

Cultural resources are significant because they provide us with a record of societies and their environments, are symbols of our heritage, are integral to cultural identity, are a valuable storehouse of environmental knowledge, provide economic benefits, and are essential for cultural diversity.

Intrinsic threats to our cultural heritage pertain only to material resources and include the position of the structure, building materials, and design. Material and nonmaterial cultural resources are threatened by extrinsic factors, including long-term natural causes and the action of humans, particularly war, development, and acculturation.

Until fairly recently the private sector was the force behind the preservation movement in the United States. The Antiquities Act of 1906 was the country's first major piece of legislation to safeguard cultural resources. Intended to preserve prehistoric sites, it gave the president the power to designate national monuments and establish regulations to protect archeological sites on public lands. In 1935 Congress passed the National Historic Sites Act, which authorized the National Park Service to acquire national historic sites and to designate national historic landmarks. In 1947 representatives from private organizations nationwide formed the National Council for Historic Sites and Buildings.

In 1960 the Park Service initiated a listing of national historic landmarks. In 1966 Congress passed the National Historic Preservation Act, which directed the secretary of the interior to develop and maintain a National Register of Historic Places. The National Environmental Policy Act of 1969 (NEPA) also affected the historic preservation movement because it required a social impact statement for any activity receiving federal funds that might affect historic and archeological sites. In 1971 President Nixon issued Executive Order 11593, which directed each federal agency to inventory and nominate to the National Register all federal properties of historic value which it administered. The Economic Recovery Tax Act of 1981 provided investment tax credits for rehabilitation of older or historic buildings.

The chief international agency working to protect the global cultural heritage is the United Nations Educational, Scientific, and Cultural Organization (UNESCO), organized in 1945. To date, UNESCO has established three conventions. The first concerns the protection of cultural property in the event of armed conflict. The second outlines means to prohibit and prevent the illegal import, export, and transfer of ownership of cultural property. The third convention is aimed at protecting the world cultural and natural heritage by safeguarding property of outstanding universal value. This convention established a World Heritage List of properties of universal value.

Key Terms

acculturation

cultural resource

cultural resources management

historic preservation

living culture

material culture

nonmaterial culture

rehabilitation

restoration

urban renewal

urban revival

Discussion Questions

1. Why is it important to preserve nonmaterial cultural resources?

2. How would you convince your town council or board of commissioners to preserve the birthplace and early homestead of Rachel Carson? (Note: Rachel Carson was born in Springdale, Pennsylvania.)

3. What are the biological, physical, and chemical conditions that do the most damage to material culture?

4. What is the relationship between the National Environmental Policy Act and cultural preservation?

5. What can we learn by studying environmental history that would help us to solve current environmental problems?

6. Describe the cultural heritage of your area. Which native-American tribe(s) originally lived there? What lifestyle(s) did they have? When did nonnative settlers arrive in your area? What country or countries did the majority of settlers arrive from? What traditions, customs, celebrations, and holidays observed by your community are a reminder of those people(s)?

UNIT VII

An Environmental Legacy
Shaping Human Impacts on the Biosphere

Religion and Ethics

Protecting the world's environment is part of the natural order, and those who damage it are showing contempt for the divine nature of created things.

Pope John Paul II

In short, a land ethic changes the role of Homo sapiens *from conqueror of the land community to plain member and citizen of it. It implies respect for his fellow-members, and also respect for the community as such.*

Aldo Leopold

How wonderful, O Lord, are the works of your hands! The heavens declare your glory, the arch of sky displays your handiwork. . . . A divine voice sings through all creation.

Jewish prayer

Learning Objectives

When you finish reading this chapter, you should be able to:

1. Explain how religious beliefs affect interactions with the environment.

2. Distinguish between anthropocentric and biocentric worldviews.

3. Give some examples of how religions are promoting environmentally sound management.

4. Define what is meant by an environmental ethic and explain how it differs from a frontier ethic.

5. Identify at least two current situations that pose an ethical dilemma and explain them in terms of environmental consequences.

The human race is at a turning point. In the coming decades our personal and societal attitudes, beliefs, and values will determine the ways in which we manage our threatened resources and work to solve the complex environmental problems that beset us. Will we make the hard choices necessary to take immediate action on such problems as acid precipitation, ozone depletion, and global warming? Or will we take a more moderate, easier course and hope that technology and luck will solve our difficulties?

Personal and societal ethics are the lines in the road that keep us from colliding into others and causing chaos as we move toward our individual destinies. We all follow some self-defined personal ethic. We make decisions according to our personal ethic, and these decisions, in part, determine the health and wholeness of our world. With knowledge and understanding, our decisions can intelligently control population and industrial growth and manage our planet's irreplaceable natural and cultural resources.

For many people, religious upbringing largely shapes their personal ethic. For others, secular values and concerns are the dominant forces. In this chapter we look at the roles of religion and ethics in forming the values and attitudes that so profoundly affect our behaviors. We see how religion and ethics relate to environmental problem solving and environmentally sound resource management. Finally, we examine several situations that illustrate the importance of religion and ethics in environmental concerns.

What Is Religion?

Religion can be defined as the expression of human belief in and reverence for a superhuman power. That power may be recognized as a creator and governor of the universe, a supernatural realm, or an ultimate meaning. In many religions rites and rituals are formal expressions of the followers' belief in a supreme being or a spiritual realm beyond physical existence. For example, burial rites may serve to usher the deceased person's soul from the earthly world to an eternal, spiritual life (Figure 25-1).

Religions are characterized by a core of beliefs or teachings that answer basic questions about the universe, existence, the world, and humankind: Who is God? How was the world created? What is the essential quality or characteristic of humans? Two important questions concern humanity's place in the world. Are humans part of or apart from "nature"? Is it humanity's role to act as lord and master of creation or as its steward?

Native American religious beliefs see humankind as part of nature. Many Eastern religions, particularly Taoism-Confucianism and Buddhism, also see humans as part of the natural world and emphasize that we can and should live in harmony with nature. The followers of Buddhism are directed to revere all living creatures. The Dalai Lama of Tibet, in a 1991 article, "A Universal Task," points out that according to Buddhist teaching, war, poverty, pollution, and suffering are the result of ignorance, selfish actions, and failure to see the essential interconnections among all life on Earth.

As we learned in Chapter 1, there are several interpretations of the Judeo-Christian view of humans and nature. For many centuries the concept of humans as lord and master, with dominion over all living things, dominated Judeo-Christian thinking. In recent years an alternative view—humans as the stewards of nature—has gained acceptance. Many Christians and Jews argue that God never intended humans to abuse or exploit the natural world or other species. They maintain, as the theologian Richard A. Baer, Jr., points out, that "though Biblical writers demythologized nature, they always treated it with deep respect as the glorious creation of a Supreme Architect." The Jewish prayer that follows illustrates this concept of stewardship: "And God saw everything that He had made, and found it very good. And He said: This is a beautiful world that I have given you. Take good care of it; do not ruin it. It is said: Before the world was created, the Holy One kept creating worlds and destroying them. Finally He created this one, and was satisfied. He said to Adam: This is the last world I shall make. I place it in your hands: hold it in trust."

Factors other than the tenets of a religion—particularly personal interpretations of religious teachings and how strictly individuals follow their professed beliefs—can play a significant role in shaping people's attitude and behavior toward the environment. The tenets of a particular religion may not necessarily determine how its followers relate to the natural world. For example, even countries where the dominant religions stress the harmony of humans in nature suffer from environmental degradation and

FIGURE 25-1: Muslim cremation ceremony, Bali, Indonesia.

resource abuse. Still, because religion is such a powerful force in many people's lives, it can be a primary influence on the human-nature relationship (Table 25-1). The values, beliefs, and attitudes inspired by religions affect how the faithful respond to specific environmental concerns. Let us look more closely at how religious beliefs affect environmental problem solving and resource management.

How Does Religion Affect Environmental Problem Solving and Management?

Religion can affect environmental problem solving and management in several ways. A religion that espouses the belief that God will ultimately solve all earthly problems can inspire in its followers a sense of security, causing them to delay taking action on a specific environmental problem or to take only moderate action. An article entitled "The Ozone Layer: A Different Perspective" appeared in the February 29, 1988, edition of *The Christian Science Monitor*. The author argues that "one portion of [God's] creation would not be capable of damaging

▶ Table 25-1
Estimated Religious Population of the World, mid-1990

	Number of Members	Percentage
Christians	1,758,778,000	33.3
Roman Catholics	995,780,000	18.8
Protestants	363,290,000	6.9
Orthodox	166,942,000	3.2
Anglicans	72,980,000	1.4
Others	159,786,000	3.0
Muslims	935,000,000	17.7
Nonreligious	866,000,000	16.4
Hindus	705,000,000	13.3
Buddhists	303,000,000	5.7
Atheists	233,000,000	4.4
Chinese folk religionists	180,000,000	3.4
New religionists	138,000,000	2.6
Tribal religionists	92,012,000	1.7
Jews	17,400,000	0.3
Sikhs	18,100,000	0.3
Shamanists	10,100,000	0.2
Confucians	5,800,000	0.1
Baha'is	5,300,000	0.1
Jains	3,650,000	0.1
Shintoists	3,100,000	0.1
Others	17,938,000	0.3

Source: Adapted from *The World Almanac, 1992.*

Religions affect environmental issues in other ways. They may limit the actions or options available to followers who are struggling with a specific environmental problem. For example, Catholicism is the major religion in many of the developing countries of Central and South America. Most of these countries, especially Mexico and El Salvador, face severe problems because of rapidly increasing populations (Figure 25-2). The Catholics in these countries tend to adhere to the church's teachings. As long as the church forbids the use of contraceptives as a means of birth control, family planning and other programs are likely to have little effect in curbing population growth.

Perhaps most importantly, religious beliefs affect environmental problem solving and resource management by their influence on a particular people's worldview. As we learned in Unit 1, a worldview is a powerful shaper of values, attitudes, beliefs, and, ultimately, behaviors. An **anthropocentric worldview,** with its belief in the preeminent position of humans in the world, may cause some people to abuse or overexploit natural resources, may encourage some people to disregard the needs of other species for habitat and food, and may cause some people to consider pollution and resource depletion

or destroying another portion of it. Nor could any aspect of His universe deteriorate." The author goes on to cite the founder of Christian Science, Mary Baker Eddy. In *Science and Health with Key to the Scriptures*, Eddy provides biblical examples of humanity's deliverance from harm: "The divine love, which made harmless the poisonous viper, which delivered men from the boiling oil, from the fiery furnace, from the jaws of the lion, can heal the sick in every age and triumph over sin and death."

In providing a Christian Science perspective on the ozone layer issue, the author implies that God will continue to deliver humanity from any and all harm, including a thinning ozone shield. While the author does not dispute the fact that the ozone layer is threatened, he insists that "a prayerful realization of God's nature and care for man, and for all creation, really does make a decisive difference on the human scene. Such prayer can even undergird and stimulate intelligent discussion and action on what to do about the ozone layer or the CFCs that are said to be depleting that layer." Further, prayer "can open a window in thought to perceive better the kind of creative and innovative thinking needed for solutions to this challenge. And most important, it can help all of us see more of man's actual, indestructible nature and of his safety in God's care."

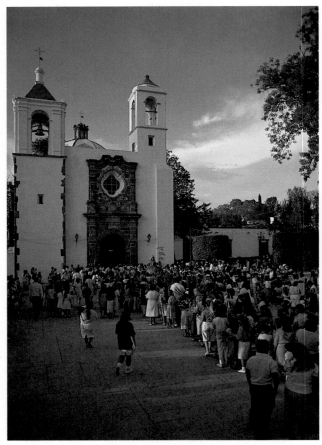

FIGURE 25-2: Worshipers participate in a Palm Sunday parade as part of a Catholic service in the highlands of Mexico.

as simply the necessary and natural consequences of economic progress.

An anthropocentric view may also affect the choice of solutions to environmental problems. For example, if the promise of cheap power generated by a proposed dam might attract new industry to an economically depressed area, the residents may well decide in favor of the dam even though it will destroy the habitat of a particular endangered species of fish. Unless other industries or ventures, such as tourism or sportfishing, can promise similarly quick and lucrative economic gains, they may be ignored, despite the fact that such industries can be developed without the loss of wildlife habitat. Anthropocentric decisions often produce solutions that work in the short term but are not sustainable.

In contrast, a **biocentric worldview** tends to foster a sense of care and concern for the natural world. Faced with solving an environmental problem, those with a biocentric worldview consider the needs of other species and the health and balance of the natural world, as well as the needs of the humans involved. Because it incorporates concern for ecosystem and biospheric balance, biocentrism tends to produce solutions that are long range in scope.

Anthropocentric and biocentric views do not necessarily stem from religious beliefs. A person with no religious beliefs or attachments could, for instance, hold either of these views. We focus on religion because different religions tend to provide a formal, traditional framework through which large numbers of people view their relationship with the environment.

How Are Religions Promoting Environmentally Sound Management?

In *A Sand County Almanac*, Aldo Leopold bemoaned the fact that "no important change in ethics was ever accomplished without an internal change in our intellectual emphasis, loyalties, affections, and convictions. The proof that conservation has not yet touched these foundations of conduct lies in the fact that philosophy and religion have not yet heard of it. In our attempt to make conservation easy, we have made it trivial."

At the time Leopold wrote those words, organized religions—at least the dominant western religion, Christianity—had done little to promote environmentally sound resource management. The past few decades, however, have witnessed a shift in the thinking and philosophy of some organized religions. Concern for the earth and the belief that humans are to act as stewards of God's creation are increasingly popular ideas among many religious groups.

The following are just a few examples that illustrate this shift toward environmental responsibility among various religions and religious groups:

- The North American Conference on Christianity and Ecology (NACCE) was founded in 1986 by an ecumenical group of Christians concerned about the environment. Its goals are to articulate the ecological dimension already present in Christianity, to help each church become a witness to Christian ecological understanding and action, and to help every Christian gain an ecological understanding. The NACCE is based in San Francisco.

- An interfaith Christian organization, the Eleventh Commandment Fellowship, has proposed an eleventh commandment, based on scripture, to define humanity's proper relationship with nature: "The earth is the Lord's and the fullness thereof; Thou shalt not despoil the earth nor destroy the life thereon." The fellowship's stated purpose is to "promote the eleventh commandment as a benevolent ethic of the environment, to teach its implications, and to address the roots of environmental problems, which are in people, not technology."

- Various initiatives within the Catholic church demonstrate a trend toward environmental awareness and action. These include the opening in 1988 of the Franciscan Center for Environmental Studies in Rome, an institute dedicated to the study of environmental issues; the papal encyclical, *"Sollicitudo Rei Socialis"* ("On Social Concerns"), which contained strong language on environmental concerns, including a warning that the Biblical directive which granted humankind dominion over the natural world has biological and moral limits that cannot be violated in the name of development; and a growing feeling that environmental studies should be an integral part of missionary training, in order to prepare missionaries to deal effectively with the environmental problems they encounter daily.

- In September 1986 representatives of five of the world's major religions (Judaism, Buddhism, Christianity, Hinduism, and Islam) met in Assisi, Italy, to participate in an interfaith service and to pledge the support of their faiths for the conservation of the earth and its resources. Known as the Pilgrimage to Assisi, the meeting was held in conjunction with the celebration of the twenty-fifth anniversary of the World Wildlife Fund. The international president of the fund, Prince Philip of the United Kingdom, declared that "a new alliance has been forged between the forces of religion and the forces of conservation."

- The popularity of earth-based spiritualities or traditions, which celebrate the cycle of life—birth, growth, decay, death, and regeneration—is indicative of the increasing interest in the philosophical and historical roots of the environmental crisis and of alternative worldviews to the dominant Judeo-Christian worldview. According to earth-based traditions, the cycle of life is embodied in the seasons of the year, the phases of the moon, and human, animal, and plant life. The spiritual traditions of native Americans, Africans, Asians, pre-Christian Europeans, and other tribal peoples are earth-based; the sacred is seen as present throughout the universe; all things are endowed with the spiritual (Figure 25-3). In general, earth-based traditions have no official dogma, no authorized texts, and no institutional, authoritative bodies.

Central to all earth-based spiritualities are the concepts of immanence and interconnection. Immanence is the belief that the Great Spirit, Goddess, or God is embodied in the living world and in all its components (human, animal, plant, and mineral) as well as in the interrelationships among them. Immanence implies that all components of the living world have inherent value: people of color as well as people of European descent, plants and stone as well as animals.

Interconnection, the relatedness of all things, is a familiar concept for many; it is the basis of a biocentric worldview. Humans are seen as part of, not separate from, nature; our fate is linked to the fate of the natural world. In "Feminist, Earth-based Spirituality and Ecofeminism," Starhawk, a feminist, peace activist, and author of *The Spiral Dance: A Rebirth of the Ancient Religion of The Great Goddess*, elaborates on the meaning of interconnections. "To ally ourselves with nature *against* human beings, as some environmentalists do, does nothing to challenge the essential split in our thinking. In the worldview of the earth religion, we *are* nature, and our human capacities of loyalty and love, rage and humor, lust, intuition, intellect, and compassion are as much a part of nature as the lizards and redwood forests."

Saint Francis of Assisi

Assisi, Italy, was chosen as the locale for the meeting of religious representatives and celebration of the World Wildlife Fund's twenty-fifth anniversary because it is the birthplace of Saint Francis, the Catholic monk whose life-style exemplified that of a steward of nature. Saint Francis has been proposed by some as the Patron Saint of Ecology. Born into wealth in A.D. 1182, the son of a successful Italian cloth merchant, Francis enjoyed a youth of revelry and popularity. He was always followed by a crowd of friends who were willing to join in any adventure he planned, no matter what the risks. And he was always forgiven for his pranks because people could not stay angry at him.

Francis had a life-long love of the outdoors. At the age of 24 he felt that he was called by God to spend his life developing his ethical sensitivity to both the beauty and the pain that he saw all around him in medieval Italy. He gave all of his possessions to his father and went to live in a leper colony. During the next 20 years Francis identified the four cornerstones of the order of monks that would become known as the Franciscan order: humility, simplicity, prayer, and poverty.

Saint Francis's devotion to and tenderness for all living things was founded on the concepts that all things are made by God, anything created by God should remind us of God because it belongs to Him, and certain things are worthy of our special love and reverence because they symbolize the character and activity of God.

At the end of his life, blind and weak, Francis wrote "The Canticle of Brother Sun" (Figure 25-4), a prayer that celebrates the universal brotherhood of all creatures—humans, animals, plants, flowers, earth, air, fire, water, sun, moon, and stars—and unites that brotherhood in ceaseless praise for the creator. Like Francis's own life, the canticle expresses the belief that the meaning of human life is found not in creating riches, but in fraternity; it is supported not by having, but by being one with and compassionate toward all creation.

FIGURE 25-3: Spirit house, Chiang Mai, Thailand. Buddhists construct spirit houses in order to provide spirits, whom they believe are present everywhere, with a place to reside.

The Canticle of Brother Sun

Most high, omnipotent, good Lord
to you alone belong praise and glory
honor and blessing.
No man is worthy to breathe your name.

Be praised, my Lord, for all your creatures,
In the first place, for the blessed Brother Sun
Who gives us the day and enlightens us through you.
He is beautiful and radiant with great splendor,
giving witness of you, Most Omnipotent One.

Be praised, my Lord, for Sister Moon and the stars,
formed by you so bright, precious, and beautiful.

Be praised, my Lord, for Brother Wind
and the airy skies, so cloudy and serene;
for every weather, be praised, for it is life-giving.

Be praised, my Lord, for Sister Water
so necessary, yet so humble, precious, and chaste.

Be praised, my Lord, for Brother Fire,
who lights up the night.
He is beautiful and carefree, robust and fierce.

Be praised, my Lord, for our sister, Mother Earth,
who nourishes and watches us
while bringing forth abundant fruits with colored flowers and herbs.

Be praised, my Lord, for those who pardon through your love
and bear weakness and trial.
Blessed are those who endure in peace,
for they will be crowned by you, Most High.

Be praised, my Lord, for our sister, Bodily Death,
whom no living man can escape.
Woe to those who die in sin.
Blessed are those who discover thy holy will.
The second death will do them no harm.

Praise and bless the Lord.
Render him thanks.
Serve him with great humility.
Amen.

FIGURE 25-4: *Saint Francis Preaching to the Birds,* a fresco by the Italian master Giotto.

The Amish Communities

The Amish, a Protestant group, exemplify environmentally sound living. They have established agrarian communities mostly in the eastern United States, especially in Ohio and Pennsylvania. Today's Amish people live much as most rural Americans did in the 1930s and 1940s. The Amish practice labor-intensive, low-input farming (Figure 25-5). They usually work their family farms with animal labor and organic farming techniques.

The Amish way of life is firmly bound to principles of Christian life such as moderation, simplicity of living, frugality, interdependence (neighborliness), and a prudent practice of ecology. The Amish also value hard work, self-reliant living, private property, the sanctity of marriage and family, the pursuit of practical skills, and caring for the land. Amid the pressures of mainstream American culture, it is hard for individuals to maintain these values in isolation, but the Amish people have formed tightly knit communities where these values can flourish.

The Amish do not reject new technologies out of hand, but instead wait as a community to see if new technologies will stand the test of time and will prove compatible with their beliefs. For example, most Amish sects do not reject all electricity; electricity generated on the farm is acceptable for use in business and farm work. But they do reject electricity brought into the home from the outside world, because such electricity reduces their self-reliance

FIGURE 25-5: An Amish farmer does his spring plowing, Lancaster County, Pennsylvania.

and could tie them to worldly distractions. Although many use only animal labor, some have accepted tractors.

The Biblically based traditions of frugality, physical labor, and ecological awareness have enabled Amish farms to prosper for years, even as high-tech, mainstream farms struggle to stay afloat. The Amish tradition of caring for the land may also help ensure that their lands remain fertile for years to come.

What Are Ethics?

Saint Francis lived according to principles of respect and care for the natural world, principles that were largely the result of his religious beliefs. Saint Francis adopted and followed what we now call an **environmental ethic.**

Ethics is a branch of philosophy concerned with standards of conduct and moral judgment. An **ethic** is a system or code of morals that governs or shapes attitudes and behavior. **Morals** are principles that help us to distinguish between right and wrong.

There is no universal human ethic. What is considered acceptable behavior in one society may be considered immoral or unethical in another. Religions too foster different ethics. In 1989 the religious leader of Iran, the Ayatollah Khomeni, issued a death sentence for the British author Salman Rushdie on the grounds that Rushdie had blasphemed the prophet Mohammed and the Islamic faith in his fictional work *The Satanic Verses* and therefore deserved to die. People around the world were horrified at the Ayatollah's action, for they considered it immoral. Yet millions of Muslims remained steadfast in their belief that Rushdie, not the Ayatollah, had committed the immoral act.

Clearly, it is difficult to define or to prescribe ethical behavior, particularly on a societal level. Ethics cannot be legislated. A change in societal ethics will only occur as the result of a change in the individual ethics of many people within the society.

In "The Ethical Quality of Life," Richard Baer emphasizes that the first area in which we can realistically hope to effect change is within ourselves. We must avoid adopting a "villain" theory to explain the environmental crisis. One of the most fruitless and counterproductive things we can do is to spend time pointing the finger at such "culprits" as General Motors and Standard Oil. In addition, we must avoid focusing exclusively on what is wrong with the environment—on the ugliness, the deterioration, the pollution—and instead reflect on what a quality environment might look like. We must have a goal in mind that we can work toward.

How Do Ethics Affect Environmental Problem Solving and Management?

The first European settlers in North America found a vast land of varied and abundant resources. Their Christian culture not only condoned the use of resources, but regarded it as a Biblical mandate. The land and its natural resources were exploited, generally without hesitation. When the resources of an area were depleted, the settlers simply moved westward, until they reached the Pacific (Figure 25-6). Their attitude and behavior constitute the **frontier ethic.**

Little, if any, frontier remains on the North American continent—or anywhere else in the world. Yet our society still operates with a frontier mentality. We seem to believe that the resources of the land can regenerate themselves over and over again or that technology will make available new, previously untapped resources. On the coastal plain of Alaska, one of the last remaining pristine places in our nation, the oil developers stand poised, ready to mine the Arctic National Wildlife Refuge if and when Congress grants its permission, even though there is just a 20 percent chance that they will recover enough oil to meet the nation's demand for several weeks.

Why does the frontier ethic continue to guide our use of natural resources, when the evidence clearly shows that resources are not unlimited and that the land cannot absorb unrestricted assaults on its ecological integrity? Most people now agree that we face serious environmental problems, problems caused (to one degree or another) by our life-styles and by the consequences of our technological society. Moreover, most people agree that these problems should be corrected and our precious resources

FIGURE 25-6: Frederic Remington's *An Episode in the Opening up of a Cattle Country.* The frontier has become an American legend, popularized in paintings, stories, and song.

protected. But these goals can be achieved only if the majority of people adopt and foster an **environmental ethic**. Such an ethic defines the role of humans as part of the natural community and places limits on human activities that adversely affect the natural world. There can be many different types of environmental ethics, but they share certain qualities. Robert Cahn, a former member of the Council on Environmental Quality and author of *Footprints on the Planet: A Search for an Environmental Ethic*, contends that "The main ingredients of an environmental ethic are caring about the planet and all of its inhabitants, allowing unselfishness to control the immediate self interest that harms others, and living each day so as to leave the lightest possible footprints on the planet."

The **stewardship ethic** is one such environmental ethic. It places humans in the role of stewards of nature. People are to act as caretakers and nurturers of the natural world. According to the resource management model, a stewardship ethic is a necessary component of environmentally sound resource management. It may be based on either a biocentric worldview or an anthropocentric worldview that recognizes a human responsibility to care for and preserve the natural world.

The stewardship ethic and all other formal environmental ethics share a common ancestry: the **land ethic,** developed half a century ago by the ecologist Aldo Leopold (Figure 25-7). The land ethic was the first formally developed statement of an environmental ethic, and it laid the groundwork for the environmental movement of the 1970s. Leopold recognized the need for an ethic to define humanity's relationship to the environment. He wrote:

An ethic, ecologically, is a limitation on freedom of action in the struggle for existence. An ethic, philosophically, is a differentiation of social from anti-social conduct. . . . All ethics so far evolved rest upon a single premise: the individual is a member of a community of interdependent parts. His instincts prompt him to compete for his place in that community, but his ethics prompt him also to cooperate. . . . The land ethic simply enlarges the boundaries of the community to include soils, waters, plants, and animals, or collectively: the land.

Leopold realized that a land ethic would not prevent the alteration or abuse of resources, but he maintained that it would affirm "their right to continued existence, and, at least in spots, their continued existence in a natural state." Our society has progressed to the point where many people do embrace the land ethic—in theory if not in practice. But with serious environmental problems increasing daily and threats to our precious resources multiplying, we

FIGURE 25-7: Aldo Leopold, former U.S. Forest Service employee, professor of wildlife management at the University of Wisconsin, and one of the founders of The Wilderness Society.

must begin to practice an environmental ethic in our personal lives and in society at large. (Focus On: Earth Spirit, page 558, discusses how we as individuals and as a society can achieve that ethic.)

The Eleventh Commandment Fellowship

In an essay entitled "Toward an Ethic of Ecology" Vincent Rossi observes that healing the environment must begin in each individual's personal life. We must examine our choices and actions—what we eat, what we wear, what we buy, our chosen career, our pastimes and hobbies—to determine if those choices are destructive to the environment. If our habits are destructive, it is our responsibility to change them. Rossi maintains that "nothing is too small to be overlooked. Everything we do *counts*, as far as its effect on the environment is concerned." He offers a simple plan to help individuals integrate an environmental ethic into their daily lives.

- Make the Eleventh Commandment the foundation of your personal environmental ethic.

Focus On:
Earth Spirit
Steven Van Matre

This is about giving up old ways of seeing, about loving the earth as a whole, about tapping into the universal flow of life, and about simplifying, yet dreaming. But most of all, it is about our needs as an unusual species of life, needs that must be met if we are to respond to the earth's cries for help.

First, we need a new kind of love to temper our passions. We must give up our childish ways of seeing things. Our love has been *ego*centric; it must become *eco*centric. For many of us, our approach to love has been like loving a toe, an earlobe, or an elbow. It's been an unhealthy captivation with the parts rather than the whole. We must learn to love ourselves less and the earth more. This will not be an easy task for we live in the age of nonsense. Bombarded with thousands of messages each day that proclaim how to be loved instead of how to be loving, we have learned to love objects instead of processes.

Second, we need new rituals and parables to remind us of our earthbound roles. A man in India explained to this writer that his mother taught him as a child to pat the earth each morning when he awoke, apologizing for the need of walking upon it. This is the kind of reminder we need today, a way of focusing our attention on what is truly important here. In the Hindu parable of Indra's net, the primeval god Indra casts his net of life into the voids of space. At the junction of each pair of threads in his net of life there is a crystal bead, and each crystal bead is a living thing, shining forth with its own glow, its own radiance into space. And the glow of every crystal bead in the net of life reflects the glow of every other bead.

This is the way life works on earth. Each living thing is a spark of sunlight energy, a crystal bead in the net of life. As humans, like other forms of life, we are only here for a few moments, a mere glistening in time on the film of life covering the planet. When we die, the sunlight energy holding the building materials together flickers out, and those materials that make us up are eventually taken up through the threads of the net by other living things to be used again. Life on the earth represents a continual process of birth and death, decay and rebirth as the building materials are used over and over again by all living things. You see, the earth is not like our mother; it is our mother. The sky is not like our father; it is our father. The union of earth and sky begets all living things in this oasis of space.

Third, we need to reach out and embrace life anew. Yes, to hug a tree, to play with the wind, to wear a new costume, to whisper secrets to a flower, to seek beauty in life's becomings. We must not be afraid to act joyously, to display our curiosity, to seek adventure. When asked how he remained so youthful at age 72, Ashley Montague replied, "The trick is to die young as late as possible." In this context, we should all strive to die so young.

According to the story of the origins of Zen, the founder of Buddhism had been asked to present a talk on truth to his followers. However, instead of talking, he merely took a flower from a nearby vase and held it up, gazing at it. Everyone in the group was puzzled by his unusual behavior, but suddenly one of them smiled knowingly. This disciple became the first teacher of Zen. In a sudden flash he had recognized the point the Buddha was making. Words are just that, words, and nothing more. Reality lies in doing, not thinking.

We have become so full of our own importance that we can no longer see much of our role playing and posturing and verbalizing for what they are, as Alan Watts suggests, merely the attention-getting antics of a different species. To get out of our own light, our own static a bit, we must take unusual steps, perhaps something as simple as going out at night without a light, or something as complicated as Thoreau's retreat to Walden Pond.

Fourth, we need new watchmen to call our attention to the dangers

- Learn everything you can about the ecological and environmental crisis in order to make informed decisions.

- Become familiar with ecologically appropriate techniques, practices, and devices, and find ways to use them in your own life. For example, if you garden, use organic gardening techniques. Install energy-saving devices on appliances in your home. Or start a compost heap in your backyard or neighborhood.

- Examine your life. One by one, eliminate habits or activities that are destructive, no matter how slight, to the environment. Incorporate in your life activities and practices that are beneficial to the environment. For example, walk or ride a bike for short distances instead of driving a car. Stop using products that you know are harmful to the environment, such as sprays containing chlorofluorocarbon. Recycle as much material as you can.

- Realize that what is healthful for the environment is healthful for you. What is not healthful for the environment, no matter what the short-term gains may be, will ultimately threaten your health and the health of your children.

ahead. On ships at sea men stood watch throughout the night, prepared to alert the passengers to unseen dangers. As we sail the cosmic seas, our life vessel may have entered its darkest hours, but the dangers here are all on board. We need a new generation to sound the alarm, a new crew of men and women to help set the course for a safe passage. There is no doubt that turbulent days lie ahead. Will we eventually become a death ship, an eternal reminder in the universe of the dangers inherent in the development of higher forms of life? Or as someone else has expressed it, will our epitaph read: "'Next time,' God said, 'no brains'"?

Fifth, we need new visions toward which to dedicate our lives. After all, ideas and their symbols change the world. There is an Arabian tale about a prince who inherited a city, only to find his inheritance in chaos. The traders from the caravans had preyed upon the townspeople and, instead of resisting, the townspeople had fallen victim to the evils of corruption and thievery. In an effort to change the situation, the prince heeded the urgings of his ministers and proclaimed a new and tougher code of law. However, instead of getting better, things got worse. The disputes, the robberies, the violence increased, and in the end, the caravans began bypassing the impover-

ished city altogether. At last, in desperation the prince commanded that the best craftsmen in the kingdom should be brought to him. When they had gathered, the prince directed them to build, in great secrecy, a model that he had designed of the most perfect city imaginable. When the artisans had completed their task, the prince had the model installed behind a special screen in the main mosque. Next, he ordered that all newcomers arriving at the city gates should be taken to see the model city, but instructed to tell no one of what they had seen, and no one else was to be allowed behind the screen.

The newcomers, overwhelmed by their vision of an incredibly beautiful city, left the mosque radiant from the experience. The townspeople asked the newcomers over and over what they had seen, but to no avail. Day after day the appearance of the newcomers continued to arouse the townspeople's curiosity, until at last, they went to the palace and demanded to see what was hidden behind the screen in the mosque. The prince agreed; the people were also transformed by the experience, and the prince's dream of a model city became a reality.

Paraphrasing George Bernard Shaw, Robert Kennedy said, "Some men see things as they are and say why; I dream things that never were

and say why not." Let's dream new dreams together. Why not a world of millions instead of billions of human passengers? Why not a world where the threat of nuclear annihilation is only a vague, unpleasant memory? Why not a world of small towns spaced miles apart? Why not a world where all of us, not just some of us, pursue more physical, labor-intensive lifestyles closer to the land? Why not a world where there are more tigers in the wild than in zoos?

Let's cage ourselves and let the animals run free.
Let's tear down our egocentric structures and systems and build anew.
Let's find new starts and new songs to follow.
And finally, as Thoreau suggests, let's build some foundations under our dreams.
For if we have the prowess to destroy the earth, then we can surely save it.

- Study the lives and works of great naturalists, such as Saint Francis of Assisi, Henry David Thoreau, John Muir, Rachel Carson, and Aldo Leopold to begin to appreciate the joy and spiritual fulfillment that can be found in attuning one's self with nature.
- Form environmental action groups in order to raise the collective consciousness about the environment and to promote positive environmental action based upon a spiritual foundation. For example, start a recycling project, a watchdog group to monitor environmental hazards in your area, or a bird-watching society.

How Can We Apply a Personal Ethic to Societal Problems?

It matters little to the effectiveness of environmental efforts whether our personal ethic rests on a religious or secular foundation. What counts is how we apply that ethic, in our private lives and in our society. In Focus On: The Conscience of the Conqueror, page 562, the late Edward Abbey, a writer noted for his fierce love and defense of wilderness, discusses how an environmental ethic might transform American society.

It is difficult to apply a personal ethical standard in societal situations. Since people's ideas of what is ethical or unethical behavior vary, it is almost impossible to get a group to agree on an approach to a specific issue. Moreover, in many societal issues, especially environmental ones, there is no clear-cut right and wrong.

The following sections present three environmental subjects that have received a great deal of publicity in recent years: the capture of the world's remaining California condors for the purposes of captive breeding; the discovery of taxol, a cancer-fighting substance, in the Pacific yew tree; and genetic engineering. The publicity is due, in large measure, to the fact that each raises serious ethical issues. As you read about these subjects, try to apply your own environmental ethic. What would you do if you were involved in the captive breeding program for the condor? If you suffered from or knew someone who suffered from ovarian cancer? If you worked in a genetic engineering lab that was designing a new organism for agricultural use?

The California Condor

As the rate of species extinctions accelerates, we are repeatedly confronted with issues involved in attempting to control another species' destiny. For example, do humans have the right to remove an entire species from the wild in the name of preservation? Such a situation occurred in the late 1980s. The imposing California condor, *Gymnogyps californianus*—with a body length of 46 to 55 inches (116.8 to 139.7 centimeters) and a wing span of 9.5 feet (2.85 meters), the largest of any North American bird—faced imminent extinction in the wild. Shooting, environmental contaminants (such as DDT and lead in bullets), environmental hazards (especially power lines), loss of foraging habitat, and reduction in food supply (condors feed on large, dead animals in open spaces), contributed to the decline in condor numbers. The situation was aggravated by the bird's low reproduction rate. Condors nest once every other year, with just one egg per nest.

By 1985, only five California condors were known to survive in the wild. Fearing that free birds would not survive in the wild much longer, in December 1985 the Fish and Wildlife Service recommended that all remaining wild California condors be trapped and transplanted to California zoos as part of the Condor Recovery Program (Figure 25-8).

This recommendation was opposed by the Audubon Society in a lawsuit lasting until mid-1986. While it acknowledged that the wild birds were in danger of extinction, the society maintained that if condors were kept in zoos, they would grow

FIGURE 25-8: The endangered California condor. The California condor's removal from the wild in 1987 spurred a controversy among members of the conservation community.

accustomed to humans and might not survive when they were reintroduced into their natural habitat. However, the trapping program was approved, and the last wild California condor was captured on Easter in 1987. The captive birds were housed at the Los Angeles Zoo and the San Diego Wild Animal Park. Successful breeding increased the captive population to 52 by 1990. The condor recovery team released two California condors in the winter of 1991–1992.

Despite the success of breeding efforts, there is no guarantee that the recovery program will be effective. Captive-bred condors may not be able to survive in the wild. They will have to be continuously fed uncontaminated meat by human caretakers. In addition, the ultimate success of the program will depend upon the availability of suitable habitat; the pressure of developers who want to build in the birds' remaining unoccupied natural habitats is a constant threat. Finally, researchers have no way of knowing for about 20 years whether the tiny condor population has the genetic diversity necessary to sustain the species over time.

The Pacific Yew

During the 1980s a new environmental dilemma surfaced which may pit the survival of a species directly against human life. The species is the Pacific (or western) yew, which lives in old-growth forests from northern California to Alaska (Figure 25-9). In 1979 the Pacific yew was discovered to contain taxol, a chemical that has shown promise in fighting ovarian cancer, which kills about 15,000 American women each year. In tests conducted by Johns Hopkins University doctors, taxol reduced the size of ovarian tumors by half in 40 percent of 40 patients for whom traditional treatments had failed. Taxol has also produced a 30 percent remission rate in a group of women who had not responded to other treatments. It is also being tested for its effectiveness against other forms of cancer.

Obtaining this remarkable chemical unfortunately presents several environmental problems. First, the Pacific yew is a slow-growing tree, requiring 200 years to mature. Most of the taxol in the Pacific yew is concentrated in the bark, and when the bark is stripped, the tree dies. Unfortunately, huge quantities of bark are required to produce taxol. The National Cancer Institute has contracted for 60,000 pounds (27,000 kilograms) of yew bark, which will yield only 2.5 pounds (1.13 kilograms) of taxol. To obtain this much bark, 12,000 trees will have to be cut down. If demand increases for this drug, the slow-growing species could be seriously depleted. Widespread felling of Pacific yews could also endanger other residents of the old-growth forests, including the already threatened northern spotted owl, which nests in the Pacific yew.

Taxol can also be extracted from limbs and needles of the yew, but chemical companies prefer not to use these sources because they contain less taxol. All efforts to synthesize the chemical have so far failed. Therefore, at present, the situation pits the existence of a species against human lives. The resolution of this conflict will depend directly on the religious and ethical values of decision makers. Does human life take precedence over living vegetation? Does human life take precedence over the existence of one (possibly more) species? Is the greatest value of the Pacific yew its cancer-fighting properties? The moral and ethical dilemma is especially bitter for ecofeminists and other female conservationists, because taxol is one of few drugs successful in treating ovarian cancer.

FIGURE 25-9: Pacific yew tree, California. The Pacific yew tree, once considered an undesirable species by loggers, has been found to contain taxol, a cancer-fighting substance. The yew is now highly sought after, and its management is a matter of intense debate.

Genetic Engineering

Genetic engineering, one of the most hotly disputed topics in science today, potentially has both appropriate and inappropriate applications. Our challenge as responsible, ethical members of society is to differentiate between the appropriate and inappropriate uses of this technology.

Genetic engineering is the process in which genes, the fundamental physical units of heredity that transmit information from one cell to another and hence one generation to another, are transferred from one organism to another. This is known as recombinant DNA technology. The genes are "cut" from the genetic material (deoxyribonucleic acid, or DNA) of one organism and "pasted" into the DNA of another. The organism that is most commonly used to receive the genes is the bacterium *Escherichia coli*. *E. coli* is used because its DNA structure is very simple compared to the DNA structure in plants or animals, and scientists understand this organism better than any other organism in the world.

If the gene that has been inserted into the bacterium's DNA tells the bacterium's cell machinery to produce insulin, then it will produce insulin. Because bacteria grow and reproduce rapidly, large quantities of insulin will be produced. Genetically engineered bacteria are grown under controlled laboratory conditions because biologists believe the bacteria are unable to live outside the laboratory.

Biologists are now learning how to transfer genes into the DNA of plant cells. Unlike bacteria, which

Focus On:
The Conscience of the Conqueror

If, as some believe, the evolution of humankind is the means by which the earth has become conscious of itself, then it may follow that the conservationist awakening is the late-flowering conscience of that world mind. A vainglorious exaggeration? Not at all, if in conservation we can see a logical extension of the traditional Christian ethic—and that of the other world religions—beyond narrowly human concerns to include the other living creatures with whom we share this planet. Not only those obviously beneficial to us, but even those that might appear to be competitors, even enemies. The broadening of the ethic cannot stop at this point; once we become generous enough in spirit to share goodwill with living things, we can advance to the nonliving, the inorganic, to the springs, streams, lakes, rivers, and oceans, to the winds and clouds, even the rocks that form the foundation of our little planet.

All is one, say the mystics. Well, maybe. Who knows? Some of us might prefer to stress the unique, the individual, the diversity of things. But it now seems well proven that all things, animate and inanimate, living and (as we say) nonliving, are clearly interdependent. Each form of life needs the others. . . . What the conscience of our race—environmentalism—is trying to tell us is that we must offer to all forms of life and to the planet itself the same generosity

and tolerance we require from our fellow humans. Not out of charity alone—though that is reason enough—but for the sake of our own survival as free men and women. Certainly the exact limits of what we can take and what we must give are hard to determine; few things can be more difficult than attempting to measure our needs, to find that optimum point of human population, human development, human industry beyond which the returns begin to diminish. Very difficult; but the chief difference between humankind and the other animals is the ability to observe, think, reason, experiment, to communicate with another through language; the mind is our proudest distinction, the finest achievement of our human evolution. I think we may safely assume that we are meant to use it. . . .

America offers what may be our final opportunity to save a useful sample of the original land. It is not a question merely of preserving forests and rivers, wildlife and wilderness, but also of keeping alive a certain way of human life, a wholesome and reasonable balance between industrialism and agrarianism, between cities and small towns, between private property and public property. Here it is still possible to enjoy the advantages of contemporary technological culture without having to endure the overcrowding and stress characteristic of this culture in less fortunate

regions. If we can draw the line against the industrial machine in America, and make it hold, then perhaps in the decades to come we can gradually force industrialism underground, where it belongs, and restore to all citizens of our nation their rightful heritage of breathable air, drinkable water, open space, family-farm agriculture, a truly democratic political economy. Why settle for anything less? And why give up our wilderness? What good is a Bill of Rights that does not include the right to play, to wander, to explore, the right to stillness and solitude, to discovery and physical freedom?

Dreams. We live, as Dr. Johnson said, from hope to hope. Our hope is for a new beginning. A new beginning based not on the destruction of the old but on its reevaluation. It will be the job of another generation of thinkers and doers to keep that hope alive and bring it closer to reality. If lucky, we may succeed in making America not the master of the earth (a trivial goal), but rather an example to other nations of what is possible and beautiful. Was that not, after all, the whole point and purpose of the American adventure?

Adapted from Edward Abbey's essay "The Conscience of the Conqueror," which appeared in *Abbey's Road*, E.P. Dutton, New York, copyright 1979.

are one-celled organisms, plants are composed of millions of cells which perform specific functions, such as photosynthesis or new growth or reproduction. The DNA in plants is much more complicated than the DNA in bacteria. Because isolating a specific gene in a plant that codes for the production of a specific protein is difficult, scientists are using genetically altered bacteria to carry selected genes into plants.

Beneficial Uses of Biotechnology. Currently, products of genetic engineering are being developed that may prove beneficial to human health, agriculture, and other areas. Products that could benefit human

health include insulin, used to treat diabetes; a human growth hormone used to treat dwarfism; factor VIII, a blood plasma protein that promotes clotting and is missing in the blood of hemophiliacs; tissue plasminogen activator, a blood protein that activates an enzyme that dissolves clots and may be useful in preventing strokes or heart attacks; interferon, an experimental antiviral, anticancer drug that was previously available in quantities too small for extensive clinical study; and vaccines, used to prevent currently incurable diseases such as herpes and rabies.

The practical applications of genetic engineering to agriculture include crops with increased resistance to salt, heat, cold, disease, drought, and other environmental conditions; plants resistant to commercial herbicides; plants that produce their own herbicide or insecticide; and food crops with enhanced nutritional value.

Genetically engineered bacteria have also been developed that feed on oil slicks. Other research is underway to develop organisms that degrade dioxins and PCBs; degrade insecticides and herbicides that persist in soils and streams; convert organic wastes into sugar, alcohol, and methane; and convert waste wood directly into grain alcohol.

Potentially Destructive Uses of Biotechnology. In addition to its beneficial uses, genetic engineering can have unforeseen destructive effects and might be used to produce intentional destruction.

What if we release into the environment a genetically altered organism before we have fully studied its potential impact on the environment? While genetically altered bacteria generally cannot survive outside of the laboratory, bacteria that inject their DNA into plant cells can manifest their genetically engineered traits in the environment. As this technology moves out of the laboratory and into the environment, it could force us to change our fundamental view of nature. We can now create viable organisms outside of the normal course of evolution. What will be the effect on native plant and animal species if a genetically altered organism with no natural competitors is released in the environment? Could genetically engineered bacteria that feed on oil slicks enter the holds of tankers and destroy the cargo? Will plants that have been engineered to produce their own herbicides and pesticides be so successful at warding off weeds and insects that they become like weeds themselves, growing where they are not wanted?

One organism that could soon be used outside of the laboratory is Ice Minus, produced by Advanced Genetic Sciences, Inc. This organism, which induces frost to form on plants at temperatures just above the freezing point, could prevent crop loss because of freezing. However, there is growing concern that large-scale spreading of this organism could significantly affect weather patterns.

There is also a danger that genetically engineered organisms might be used to produce biological or chemical weapons, such as plant viruses that might destroy all of a country's wheat or corn crops or deadly human diseases caused by altered bacteria or viruses.

In the 1970s the National Institutes of Health (NIH) issued a set of guidelines for all federally funded recombinant DNA research and established the Recombinant DNA Advisory Committee to review and approve all genetic engineering experiments. The NIH guidelines, while voluntary, were thought to be adequate for safeguarding the environment and the public as long as genetic engineering experiments were confined to the laboratory. But now that the research projects are being tested in the field and in clinics, more people are calling for a moratorium on this type of research until clear, effective federal regulations are developed. Robert Cooke, author of *Improving on Nature—The Brave New World of Genetic Engineering*, has expressed such concerns:

It's been a long, difficult road, a long journey up through the mud and mire of evolution, up through 750,000 generations of mostly ordinary people, of too many villains and not enough saints, since the "birth" of man. But we made it, and today we stand, here on the palace threshold, arrogantly planning a subtle biological coup d'etat which, once and for all, may topple Mother Nature from her throne. Do we know what we are doing?

Summary

Religion is the expression of human belief in and reverence for a superhuman power recognized as a supreme being, a supernatural realm, or an ultimate meaning. Religions are characterized by a unique core of beliefs or teachings that answer basic questions about the universe, existence, the world, and humankind. The values, beliefs, and attitudes inspired by religions affect, to varying degrees, how the faithful respond to specific environmental concerns.

Religion can affect environmental problem solving and management in several ways. A religion that espouses the belief that God will ultimately solve all earthly problems can cause followers to delay taking action on a specific environmental problem or to take only moderate action. Religion may also limit the actions or options available to followers who are struggling with a specific environmental problem.

Religion is an important component of a people's worldview. An anthropocentric worldview holds that humans have a preeminent position on earth. A biocentric worldview fosters a sense of caretaking and concern for the natural world. Anthropocentric and biocentric views do not always stem from religious beliefs.

The past few decades have seen a shift in the thinking of some organized religions. Concern for the earth and the belief that humans are to act as stewards of creation are increasingly widespread ideas among many religious groups.

Ethics is a branch of philosophy concerned with standards of conduct and moral judgment. An ethic is a system or code of morals that governs or shapes behavior. Morals are principles that help us to distinguish between right and wrong. There is no universal human ethic.

The first European settlers in North America followed a frontier ethic, in which they exploited the land and its natural resources. Our society still operates under a frontier mentality. The stewardship ethic places humans in the role of caretakers and nurturers of nature. The stewardship ethic and all other formal environmental ethics share a common ancestry: the land ethic developed half a century ago by Aldo Leopold.

It is difficult to apply our ethical standards in societal situations because many people have widely different ideas about what is ethical behavior and on many societal issues there is no clear-cut right and wrong.

Key Terms

anthropocentric worldview	environmental ethic
biocentric worldview	frontier ethic
ethic	religion
ethics	stewardship ethic
morals	land ethic

Discussion Questions

1. What are ethics? What are morals? How are they related to religion?
2. How do religious beliefs and ethical codes affect the environment?
3. What is the difference between a frontier ethic and a stewardship ethic?
4. Pick one of the three ethical controversies described in this chapter (the California condor, the Pacific yew, genetic engineering). Present the dilemmas involved in these controversies in terms of the religion or ethical code most familiar to you.
5. From your own knowledge of history, identify at least three situations in which differing worldviews, based upon different religious beliefs, led to conflicts over the use or ownership of natural resources.

Economics and Politics

The "key-log" which must be moved to release the evolutionary process for an ethic is simply this: quit thinking about decent land-use as solely an economic problem. Examine each question in terms of what is ethically and esthetically right, as well as what is economically expedient. A thing is right when it tends to preserve the integrity, stability, and beauty of the biotic community. It is wrong when it tends otherwise.

Aldo Leopold

The Biosphere in some form would survive the self-destruction of humanity; but humanity could not survive the destruction of the Biosphere. If a policy of Biospheric protection is idealistic, it is also pragmatically realistic. Without the implementation of a science-informed morality through enlightened political action, those forces that have made the modern world by discounting the future may well bring its history to a close.

Lynton Keith Caldwell

Growth for the sake of growth is the ideology of the cancer cell.

Edward Abbey

Learning Objectives

When you finish reading this chapter, you should be able to:

1. Explain the difference between value and price as it applies to resources.

2. Describe the relationship between environmentally sound resource management and economics.

3. Discuss the world debt crisis and explain how it relates to environmental problems.

4. Explain the relationship between politics and solutions to environmental problems, giving both national and international examples.

The resources to which we have access are controlled, in large measure, by economic and political factors. Nations and individuals alike fight for access to the world's resources, and these struggles are often played out in political or economic arenas. History and the daily newspaper provide countless examples: the revolutions against colonial rule that led to the independent nations of the United States, India, numerous African countries, and others; the Great Depression of the 1930s, when the U.S. economy was faltering and world politics were on the brink of chaos; the OPEC oil embargo of 1973, which popularized the phrase "energy crisis" and focused world attention on the oil-rich and suddenly powerful countries of the Middle East. Italy entered World War II, in part, because its burgeoning population was straining at its borders. The struggle over water rights in the western United States has led to political maneuvering and legal wrangling. The 1980s drought in Ethiopia, a widely publicized and graphic illustration of an environmental disaster, caused tremendous suffering and hardship for citizens. Their suffering was greatly exacerbated by a civil war that prevented famine relief supplies from reaching many of those who were starving to death.

As the world's human population increases, the demand for limited resources will continue to rise and the potential for conflict over access to resources will continue to grow. It will be a global challenge to determine which resource uses deserve priority.

The ecologist Garrett Hardin used early English grazing practices to illustrate the dilemmas involved in the management of shared resources. The village "common" was the open grazing land shared by the villagers. The common often became severely overgrazed and everyone suffered, but each villager had little incentive to restrict the number of grazing cattle because others would increase their herds even if that villager did not.

Much like individual villagers, various states and multinational corporations of the world act selfishly regarding the "global common"—the oceans and atmosphere. The adverse effects of some of these actions are quite obvious. For example, we use the oceans as a "sink" for our wastes, some of which wash up on our own beaches, while others carried by currents and tides wash up on foreign shores. The oceans have also suffered from overharvesting; both shellfish and finfish stocks have been depleted in many areas, and overharvesting has permanently reduced the catch of many valuable species. Some nations, specifically the United States, have refused to reduce emissions of greenhouse gases despite commitments by Canada, Europe, and Japan to reduce or stabilize carbon dioxide emissions by 2005. The United States, the world's largest producer of greenhouse gases, insists that more study on global warming is needed. Many economic and political actions are less obviously related to environmental degradation, but are nonetheless harmful. For example, fast-food restaurants that buy beef raised in Central America implicitly condone the destruction and conversion of rain forests to grazing land. Homeowners in arid regions who demonstrate an "oasis mentality" condone wasting a valuable resource (Figure 26-1).

In this chapter we define both economics and politics and see how each discipline affects environmental problem solving and resource management.

FIGURE 26-1: Oasis mentality: wasteful use of water in arid region. The Santa Catalina Mountains provide a backdrop of stark contrasts to this resort in Tucson, Arizona.

We examine several environmental issues that illustrate the impact and importance of economics and politics. Finally, we see how political and economic factors can be used to transform the United States into a true "conservator" society.

What Is the Study of Economics?

Ecology and economics have the same Greek root word, *oikos*, meaning "household." **Economics** is derived from the word *oikonomia*, meaning "the management of the house." Ecology is the study of the household, and economics is the study of its management.

The link between economics and ecology is not merely an exercise in etymology. If the earth is home to the human species, then economics, in its broadest sense, is the tool we use to manage our home and its many resources. Unfortunately, we have lost sight of its broader meaning and have instead come to equate economics with the study of money; rather than focusing on the value of the earth's resources, we focus on their price.

There is a great difference between **value** and **price.** The value of a forest, for example, is not just the price of the lumber that can be harvested. It includes the ecosystem services provided by the forest (cleansing of air and water and reduction of soil erosion), the biological diversity it harbors, and its esthetic and recreational aspects. Many resources are invaluable, that is, they have value beyond estimation, they are priceless. Can we put a price on fresh air, clean water, and fertile soil? How much is a wilderness worth in dollars? Obviously, these questions cannot be answered. Our failure to determine the true value of resources and to account for the environmental damage caused by our activities and products illustrates the inadequacy of our economic system. As long as economics considers price, or monetary cost, as the sole determinant of value, we will always have difficulty in attempting to solve environmental problems and manage resources. In effect, we will be trying to fit a round peg (a resource) into the square hole of economics.

If, as a society, we come to realize the true value of our resources, we could avoid many environmental problems. For example, farmers who value the soil use techniques, such as conservation tillage and crop rotation, to protect the soil's fertility. These farmers know that good crops can be produced only from good soil, and good (healthy) soil can be maintained only by good farming practices. They know that healthy soil is a renewable resource, capable of sustaining itself over many years, and is the vital element required to sustain them and their way of life.

Wendell Berry, farmer, poet, and environmental philosopher, writes that the ecological value of soil is "inestimable; we must value it, beyond whatever price we put on it, by respecting it." What we really need is a new economic system that determines the value, and not simply the price (or cost), of resources. How would such a system work? We can look at the Amish economy for some clues.

In the 1980s the nation's farmers faced a desperate economic future. Many were forced into bankruptcy or barely avoided it. Banking institutions foreclosed on hundreds of family farms. Yet through this difficult period, Amish farmers continued to prosper, in large measure because of their efficient use of resources and their economic system.

The Amish practice traditional, small-scale farming. Amish culture forbids or limits the use in the home of many modern conveniences, including electricity, telephones, and automobiles, but they have adapted modern farming technology to meet their religious restrictions. For example, Amish dairy farmers use diesel motors to generate their own electricity for the milk room, cooler, and milk machines. That option is cheaper than buying electricity and protects them from power outages. Most do not use tractors in their fields, because to do so might tempt them to expand the acreage they tend; acquiring additional acreage might lead to debt and would almost certainly drive other Amish off the land.

A look at the accounting ledger of an Amish farmer and the ledger of a typical "English" counterpart (the Amish term for non-Amish mainstream American agribusinesspeople) reveals some interesting differences. Obviously, Amish farmers' machinery costs are far below those of most farmers. The Amish also spend far less on herbicides, pesticides, and fertilizer. They use the manure from their livestock to fertilize the soil. The organic matter in the manure helps the soil to use nutrients more efficiently and soak up rain more easily, thereby reducing erosion. Perhaps the most telling difference, however, is in the way that the Amish account for land ownership and labor costs. The Amish hire no labor and consider their own labor as part of their profit, not a cost. Moreover, whereas agribusinesses figure the "cost of ownership" of the land as a fixed cost, the Amish consider land ownership a reward, not a cost. Unlike the dominant western economies, the Amish economy clearly considers more than the price of an item or service; it more closely accounts for the value of a given item. The differences between these two economies reveal why Amish farmers are thriving when so many practitioners of modern agribusiness have fallen victim to economic woes.

How Does Economics Affect Environmental Problem Solving and Management?

Too often, through our economic programs and actions, we show little respect for nature and resources. In this section we see how economics affects environmental problem solving and resource management.

Environmental Problem Solving

Economic considerations are an integral part of problem solving. Alternative proposed solutions are often compared to determine which is more cost-effective, that is, which solution achieves the desired environmental result with the lower monetary expenditure. Unfortunately, economic cost is often viewed as the sole or primary determinant in the selection of a solution. Problem solvers must be careful to weigh other factors, particularly ecological and esthetic effects and ethical considerations, before choosing a specific course of action.

Natural resource economists are important members of a problem-solving team. They organize and analyze information about natural resources. Using the tools of economics and the economic process, they advise problem solvers, resource managers, and others on policy matters.

Environmental Management

Economic analysis is a criterion for environmentally sound management. A simple example illustrates the economic importance of natural resources and underscores the need for sound and sustainable management. According to a report by ICF Technology, Inc., the net economic value associated with recreational salmon fisheries in the state of Washington was $43.7 million annually between 1982 and 1985. Properly managed, fisheries are a renewable resource that can yield sustained economic benefits (Figure 26-2).

In contrast, nonrenewable resources, such as fossil fuels, are ultimately limited both in their availability and useful quality. As the stores of easily obtainable fuels are consumed, developers must expend more energy at higher costs to extract fuels from remote places. As they do so, the market price paid by consumers rises. Moreover, as long as the demand for fuels remains constant or rises, the cost to consumers will continue to rise as the available stores of fuels dwindle. Thus, the ways we use and manage our energy and other nonrenewable resources become increasingly important. Conservation measures and energy-efficient technologies, for example,

(a)

(b)

FIGURE 26-2: Economic considerations can play an important role in improving the management of both resources and ecosystems. (a) Salmon in the Pacific Northwest are a renewable resource when managed within ecological limits. They support a fishery whose monetary value can easily be calculated. (b) Mangrove wetlands, such as this thicket in the Florida Everglades, buffer coastlines from storms, serve as fish and shellfish nurseries, and accumulate soil and other organic matter, thus building up the soil for other plants. Although difficult to calculate, the economic benefits of the services provided by ecosystems such as mangrove wetlands are very important.

become far more attractive options, both economically and ecologically, as energy supplies dwindle.

For most of the history of the United States, the economic system has stressed resource utilization over conservation. As long as resources were both cheap and plentiful, the faster we used them, the faster we built our economy and increased general economic well-being. Government incentives designed to increase economic activity have frequently resulted in environmental damage. Examples of destructive or wasteful programs include subsidies for pesticides and road building, biased utility regulation, below-cost timber sales, and underpriced irrigation services. In recent years, however, the rising costs of extracting and processing raw materials have changed our economic fortunes. Resource scarcity means that we can no longer look to increased resource exploitation in order to bolster our economy and improve economic well-being.

If we are serious about improving our nation's economy, we can begin by using our resources more efficiently. The United States lags far behind many countries, particularly Japan and Germany, in this area. Resources are no longer inexpensive and abundant, yet we continue to live as if they were. Wasteful practices in both our businesses and our homes are draining our limited resources. Let us look again at the example of energy resources. When oil and other fossil fuels were inexpensive, it did not seem to matter if we left lights on which were not being used or set the thermostat at a toasty 75° F. We used the car to make many trips throughout the day

and rarely, if ever, thought to combine our trips in order to conserve gasoline. The 1973 OPEC oil embargo and subsequent shortages prompted many people to realize that energy supplies should not be taken for granted. We began to practice energy conservation and to develop energy-efficient technologies. But our efforts waned, and we once again became complacent about energy availability, when oil prices fell. Government and industry began to talk about developing reserves to maintain the supply—and our complacency.

How Are Economic Programs Contributing to Environmentally Sound Management?

People are becoming more aware of the important link between sound environmental policies and sound economics. As a result, many economic programs, measures, and policies have been developed which promote and enhance environmental protection and resource management. The following is a sampling of both national and international initiatives.

• In 1988 Senators Timothy Wirth (Democrat from Colorado) and John Heinz (Republican from Pennsylvania) sponsored Project 88, a bipartisan effort to identify innovative solutions based on economic incentives to major environmental and resource management problems. The senators formed a 50-member team composed of academicians, industrialists, environmentalists, and public servants. The Project 88 Report, *Harnessing Market Forces to Protect Our Environment: Initiatives for the New President*, describes various techniques that could be implemented to address 13 major environmental and resource management issues, including the greenhouse effect, stratospheric ozone depletion, local air pollution, acid rain, indoor radon pollution, water supplies, wetland resources, toxic substances, and public land management. These techniques are designed to use the forces of the marketplace and entrepreneurial ingenuity to help reduce pollution and resource abuse. For example, the report recommends reducing government subsidies on federal lands, notably below-cost timber sales in remote, roadless areas. Such a move would foster environmental protection, decrease federal expenditures, and increase net revenues.

- In 1990, after years of study, Minnesota instituted a groundwater protection program which sets fees on chemical use, well construction, and other activities that endanger groundwater.

- Many states have tax checkoff programs that allow citizens to donate all or part of their state income tax refund to benefit wildlife and natural areas. Since its inception in 1984, Ohio's natural areas checkoff program has provided funds to acquire and protect lands, develop visitor facilities, produce public information and educational materials, and complete numerous special projects, such as evaluating Ohio's rivers for possible additions to the state's scenic river system.

- In most states, utility regulations mean that profits rise as electricity sales rise. These regulations discourage conservation measures: Utilities have little incentive to save energy (through customer-service programs to install efficient lighting, low-flow showerheads, and insulation) when they stand to make more money through increased demand. Programs instituted by California, New York, Oregon, and five New England states, however, are decoupling profits from electricity sales. The result is that utilities suddenly find it in their best interest to conserve electricity. In 1990 California's three largest utilities jointly proposed a plan whereby earnings will be tied to energy savings. If the utilities meet their conservation targets, one of them will be allowed electricity rates that yield an annual return of 14.6 percent on the funds invested in its conservation measures. In contrast, the utility would realize only a 10.7 percent return if it invested those same funds in the construction of a new power plant. The other two utilities will realize profits equal to 15 to 17 percent of the value of the energy efficiency measures they undertake for customers. Estimated costs of the utilities' efficiency programs is $500 million over two years. But they are expected to save more than twice that thanks to reduced demand for power.

- In 1988, José Sarney, then president of Brazil, suspended most tax credits that encouraged the conversion of the country's tropical rain forests to pasture, cash crops, and other land uses that promised only short-term economic gain. These credits were further reduced under the administration of President Fernando Collor de Mello. Deforestation in the Amazon, which peaked in 1987 at an estimated 19.8 million acres (8 million hectares), has since fallen, dropping to 4.5 million acres (1.8 million hectares) in 1990. The drop was attributed in part to a sluggish economy and a rainy dry season in 1989, which hindered the burning of forest tracts. Even so, Brazil's Secretary of State for Science and Technology, José Goldemberg, points to the government's stepped-up enforcement against illegal burning and the elimination of tax credits to farming and cattle-ranching projects as significant factors in the reduction in deforestation.

- Environmental or so-called green taxes can be a powerful tool for environmental protection. When the United Kingdom instituted a substantially higher tax on leaded gasoline, the market share of unleaded gasoline rose from 4 percent in April 1989 to 30 percent in March 1990. Fourteen members of the Organisation for Economic Co-operation and Development, a development group composed of the wealthier industrialized nations, have established green taxes on air and water pollution, waste, and noise, as well as various product charges such as fees on batteries. Unfortunately, the fees are usually too low to result in real changes in behaviors and activities. Nevertheless, in many nations, there is growing support for a comprehensive green tax code that would achieve numerous environmental objectives. Examples include fees on carbon emissions from the combustion of coal, oil, and natural gas to help slow global warming; taxes on virgin materials to encourage reuse and recycling; fees on the generation of toxic waste to encourage waste reduction and the development of nonhazardous alternatives; levies on the overpumping of groundwater to foster conservation and more efficient water use; and fees on emissions of air pollutants to reduce acid precipitation.

In developing a green tax code, care would have to be taken to determine tax levels that reduce threats to human and environmental health without disrupting the economy. Phasing in environmental taxes over a 5 or 10 year period would ease the impact on the economy.

Because most people will resist tax increases, creative ways must be found to institute new programs. Income taxes might be reduced to offset the introduction of green taxes; a balance could be struck between green taxes, which reflect the real cost of resources and of economic activities, and income taxes, which are designed to ensure that the wealthy pay a proportionately higher share than the poor. A tax code consisting solely of environmental charges would be undesirable for several reasons. First, living costs would likely rise under a green tax code (a charge on carbon emissions, for example, would result in higher heating oil prices), thus imposing greater hardship on the poor as they strive to obtain essential items. To compensate, income taxes for low-income persons might have to be lowered more than they are for others to ensure that the poor are not unfairly burdened by the green taxes. Moreover, because revenues from green taxes would decrease as production and consumption patterns adjust in response to the environmental fees, a mix of income and green taxes would help to maintain continuity.

- Development projects in less-developed countries too often have wreaked havoc on the local or regional environment and have not realized economic expectations. The United Nations Environment Programme (UNEP) was designed to combat this trend. UNEP has developed the five following criteria by which to judge environmentally sound (economic) development projects:

  1. Minimal waste in the use of natural resources
  2. Maximum productive use of all residues and leftover materials
  3. Respect for the integrity of ecosystems and careful evaluation of all impacts that could result from implementation of a project
  4. Minimal degradation of the environment and abatement of degradation that is unavoidable
  5. Emphasis on how environmental improvement and socioeconomic development can be complementary

- In 1990 the international community agreed to establish an environmental fund, known as the Global Environmental Facility, to ensure that resources will be available to poor countries to help combat global problems such as ozone depletion and climate change. The fund is managed by the World Bank in cooperation with the United Nations Development Programme and the United Nations Environment Programme. Global Environmental Facility monies are to be provided to LDCs for environmental investments that will yield benefits in four target areas: protection of the ozone layer, limitation of greenhouse gas emissions, preservation of biological diversity, and protection of international water resources. A fund of $1.4 billion was established for the first three years.

In the past it was believed that ventures with deleterious environmental effects could still produce economic benefits. Only recently have we begun to realize that the opposite is true: what is harmful to the environment will ultimately be harmful to cultural systems, including economies. As Lester Brown, president of the Worldwatch Institute, warns, "The earth's physical condition is deteriorating on several major fronts, threatening the long-term viability of the global economy."

The World Debt Crisis

The following section describes the world debt crisis and presents some solutions currently being developed to ease this crisis. As you read, carefully evaluate the ways in which environmental degradation and economic chaos are related.

International lending institutions have loaned billions of dollars to poor countries for use in development projects intended to alleviate poverty and increase these countries' ability to compete in the world market. Today, as these loans come due, the borrowing nations find that they do not have enough money to pay even the interest on the loans, let alone the principal. This debt crisis is not just a problem for LDCs; as long as developing nations are burdened by excessive debt, they will not be able to institute strong social improvement and environmental protection programs needed to improve their people's standard of living. This debt crisis is a problem for the world, and it warrants serious attention.

The Problem

According to the Worldwatch Institute, LDC debt had soared to $1.2 trillion by 1990, 44 percent of the collective gross national product of developing countries. The cost of servicing that debt was $140 billion in 1990. How did these countries get into such serious economic trouble, and what have they got to show for all of the money they have borrowed? How can they hope to repay these debts, when the interest alone often equals a country's entire export earnings? And what effect is this enor-

FIGURE 26-3: The Sardar Sarovar Dam, under construction, Gujerat, India, January 1991. Construction of the approximately 4,000-foot long Sardar Sarovar Dam, funded by the World Bank, requires the resettlement of 100,000 people.

mous debt having on the social and environmental well-being of LDCs?

Over the past 40 years billions of dollars have been spent on efforts to improve the economies of LDCs. The World Bank, the largest single lender worldwide, lends about $14 billion to approximately 100 developing countries each year. These loans have been based largely on the theory that successful economic development requires investment in large-scale, industrial projects (Figure 26-3). Yet the standard of living of millions of people in developing countries has fallen. Large-scale investments, whose benefits were expected to "trickle down" to the masses, have failed to eradicate hunger or stem the tide of migration from destitute rural villages to urban slums. And economic woes intensify environmental degradation, especially deforestation in the tropics and desertification in northern Africa.

Economic adversity and environmental problems are intertwined. Debtor countries often try to increase productivity to realize higher yields, and overexploitation is generally the result. People in poorer nations are compelled to use all of the resources of their scantily productive lands. They cut all trees that can be used or sold as fuel or fencing. They convert woodlands to pasturelands. Under such treatment, the quality of natural resources is degraded.

Governments facing huge debt burdens feel that they cannot afford to be concerned with environmental issues. LDC governments are scrimping on development projects, cutting government spending, and devaluing their currency to pay off debts to lending institutions. Social and environmental issues take a back seat to economic survival. Unfortunately, countries with the largest debt burdens also have most of the world's biodiversity; debt places their wealth of biological resources at an especially high risk. If the debt burden is not eased, the degradation of LDC environments is likely to continue.

Many LDCs invested billions of borrowed dollars in new infrastructure projects (roads and dams) that were intended to attract foreign capital. They hoped to turn their nations into industrial powers capable of competing with the West in major export markets. But most of the developing countries did not attract enough foreign capital to offset the cost of the infrastructure projects and the tax breaks they made for multinational corporations. Population pressure, use of inappropriate industrial technologies, low environmental consciousness of ruling classes, and exploitation by multinational corporations all contribute to the debt crisis and, consequently, to environmental devastation.

Mounting political pressures in the United States and Europe and the deteriorating environmental situation in LDCs have forced both lenders and borrowers to grapple with such questions as how to reduce the adverse environmental impacts of development projects and how to provide for the management of renewable natural resources to ensure that projects are sustainable over the long term.

Some Solutions

The following potential solutions to these intertwined economic and environmental problems are described below: require that development banks become more environmentally responsible; encourage sustainable development; increase and rechannel MDC development assistance to LDCs; encourage grassroots loans; and swap debt for nature.

Environmentally Responsible Development Banks. Development banks have made loans for environmentally destructive and socially disruptive projects. These loans are blocking economic progress in some poor countries because the projects they support often help destroy natural systems, such as forests, farmland, and watersheds that are essential for sustainable development. Large-scale, capital-intensive projects such as the trans-Amazon highway in Brazil and huge hydroelectric projects in Africa often displace local populations and can destroy their culture. International lenders must ensure that the projects they invest in are environmentally sound.

It is especially important that a sense of environmental responsibility be incorporated into the policies and practices of the four multilateral development banks: the World Bank and three regional institutions, the Inter-American Development Bank, the Asian Development Bank, and the African Development Bank. Collectively, these institutions provided $28 billion in loans to LDCs in 1989. Moreover, their priorities have a significant influence on both the lending decisions of commercial banks and the investment decisions of poor nations. The World Bank has tremendous potential to provide leadership in the area of environmentally sound economic development. Unfortunately, the Bank's tradition of funding large-scale projects such as road building, mining operations, dam construction, and irrigation schemes has made it a partner in the destruction of vast tracts of rain forest, the strip-mining of large areas of land, the harnessing and eventual siltation of once-wild rivers, and the pollution of waterways. In addition, it has often made loans to governments that have resulted in environmentally destructive policies, such as encouraging increased use of pesticides in an attempt to reduce insect damage and thus raise crop yields.

In its defense, the World Bank has long asserted that it is first and foremost a lending institution, and by the standards commonly used to judge such institutions, it is highly efficient; no country has ever defaulted on a World Bank loan. Even so, in the early 1980s, in response to pressure from international environmental groups, the World Bank began to institute reforms in its policies and practices. In 1987 the Bank established a central Environment Department and four regional environmental divisions to assess the potential impacts of proposed projects, evaluate environmental threats in the 30 most vulnerable LDCs, promote a program to slow deforestation and desertification in Africa, and contribute to a global program to promote the conservation of tropical forests. While there have been some notable achievements—such as a $237 million loan to improve sewerage, drainage, and water supply facilities in several Indonesian cities—many other projects are suspect, including a "sustainable forestry" project in Côte d'Ivorie that is expected to actually *increase* the pace of logging and deforestation.

Critics of the World Bank reforms maintain that, for the most part, the institution's nascent commitment to environmentally sound development projects is in name only; policy positions have not yet been put into practice. For example, improved energy efficiency is a stated priority for World Bank projects, yet coal plants and large hydropower facilities constitute the largest area of Bank lending, receiving 16 to 18 percent of loans in recent years. In contrast, less than 3 percent of the Bank's lending to energy and industry are for energy efficiency projects. Moreover, the Bank's process of environmental assessment is undermined by the fact that the borrowing countries themselves are responsible for the assessment. Eager to obtain funding, and lacking the necessary skilled personnel to prepare a valid assessment, these countries often give the green light to environmentally destructive, and ultimately economically unsound, development projects.

To effect real change at the World Bank, a necessary first step is to mandate that the environmental impacts of proposed projects be assessed by in-house staff. Another important step is to strengthen and enlarge the Environment Department. With a staff of just 54, out of a total World Bank professional staff of over 4,000, the department is sorely understaffed for the work it is attempting to do. Perhaps the most significant step the World Bank can take is to develop a coherent vision of environmentally sound development. Such a vision would enable the Bank to restructure its lending programs, supporting smaller, more labor-intensive projects appropriate to the needs and resources of LDCs, such as community woodlots and nurseries, agricultural cooperatives, integrated pest management programs for farmers, rural cookstove industries, urban bicycle factories, and job training programs (Figure 26-4). By shifting more funds to policy lending, which currently accounts for about 20 to 30 percent of its loan portfolio, the World Bank could encourage

FIGURE 26-4: Steel plant, Tanzania. All steel at this small plant is made from recycled metal.

governments to institute policies that foster both environmental and economic reforms. Examples include levying pollution taxes and eliminating pesticide subsidies. Because policy lending incurs lower overhead costs than project loans, increasing the share of policy loans would mean that project loans could be made at lower interest rates. Monies could thus be made available to grassroots organizations and other groups involved in small-scale development projects.

Sustainable Development. According to some LDC observers, environmental degradation has actually caused much of the LDC debt. Some nations have become financially indebted by abusing and depleting natural resources so they can no longer export products such as fish and wood. The reverse is also true: In trying to service foreign debts, some nations are forced to degrade their environments by using renewable resources faster than they can be replaced. Development projects that degrade the environment cannot be sustained over time; eventually, the resource base will be depleted or resources will be rendered unfit for consumption.

In 1991 the World Conservation Union (formally known as the International Union for the Conservation of Nature and Natural Resources or IUCN), the

United Nations Environment Programme, and the World Wide Fund for Nature jointly published *Caring for the Earth: A Strategy for Sustainable Living*. In it, they defined **sustainable development** as improving the quality of human life while living within the carrying capacity of supporting ecosystems. Sustainable development seeks to maximize human resource potential as well as the wealth provided by natural resources by managing all resources—natural, human, financial and physical—so that they can be used to serve the common good. Development is sustainable only when it meets the needs of present generations without compromising the ability of future generations to meet their own needs.

Sustainable development includes the **sustainable use** of renewable resources, that is, using these resources at rates that do not exceed their capacity for renewal. By definition, sustainable use does not apply to nonrenewable resources; because the supply of nonrenewable resources is finite, they cannot be used sustainably. At best, the life of nonrenewable resources can only be extended through recycling, conservation, and substitution measures.

Sustainable development leads to a **sustainable economy,** one which maintains its natural resource base. A sustainable economy persists and continues to develop by adapting to change and through improvements in knowledge, organization, technical efficiency, and wisdom. A sustainable economy is critical to the development of a **sustainable society,** a society that works with, not against, nature, to ensure its survival in the long-term. Box 26-1 contains a discussion of the evolution of thought concerning sustainable development and presents nine principles of a sustainable society.

Increased and Rechanneled Development Assistance by MDCs to LDCs. Sustainable living and sustainable societies will be impossible to achieve unless the global economy becomes more balanced. The gap between the rich and poor nations widens daily; action must be taken quickly to close this gap. Like development banks, the wealthy nations have an important role to play; by reforming their assistance or aid programs to LDCs, they can encourage ecologically sound development.

In 1989 the developed world provided $41 billion in nonmilitary aid to less-developed nations. Japan was the world's largest donor, contributing $8.95 billion, surpassing the United States, which contributed $7.66 billion, and France, which contributed $7.45 billion. In terms of share of gross national product (GNP), however, Norway was the world's most generous donor in 1989, providing aid to LDCs equal to 1.04 percent of its GNP. The U.S. aid level was just 0.15 percent of its GNP, while Japan's giving equalled 0.32 percent of GNP. The Organisation for Economic Co-operation and Development is trying to encourage its members to raise annual aid levels to 0.7 percent of GNP, a move that would double assistance to over $80 billion each year. In 1989, only three nations, Norway, the Netherlands, and France, reached that level, and in many donor nations aid levels are falling.

In addition to its comparatively low aid level, the United States channels most of its assistance to a few nations that are deemed strategically important. Almost 40 percent of the United States's nonmilitary aid goes to Israel, Egypt, and El Salvador, which collectively have just 2 percent of the world's population. Some countries tie their aid to the purchase of goods and services, a form of export promotion that ignores the real needs of poor countries. Exacerbating these problems is the fact that most aid is not targeted for sustainable development priorities such as reforestation, appropriate technologies, sustainable agriculture, energy efficiency, and family planning. For example, only about 7 percent of bilateral funds are earmarked for population and health programs each year.

To reform their assistance programs, the wealthy developed nations must focus on sustainable projects and distribute funds more equitably among needy nations. The MDCs can look to Norway as the global leader in development assistance. In addition to providing more aid as a share of its GNP than any other country, Norway emphasizes sustainable development. Nineteen percent of its development assistance is targeted for agriculture and fisheries projects; education receives 8 percent. The Norwegian government has reached out to the neediest of the LDCs; Bangladesh, India, and Tanzania are among the leading beneficiaries of the country's development aid. While most MDCs have been slow to follow Norway's lead, there are signs that some countries are moving in that direction. In the United States, Congress has proposed legislation that would mandate environmentally sustainable development as a major aim of U.S. foreign aid.

Grassroots Loans. By moving away from capital-intensive development projects such as large dams and investing in local communities and citizens, nations may be able to strengthen their most important resource, their people, and relieve their poverty. This trend would in turn encourage sustainable development and reduce environmental degradation. A key advantage to grassroots loan programs is that the funds recirculate in the borrower's community and are available to others (Figure 26-5). One successful program is sponsored by the Grameen ("rural") Bank of Bangladesh, which provides up to $10 million per month in loans that average about $70. Over 90 percent of the Bank's loans are made to women; it boasts a 98 percent repayment record. Another promising initiative is the "Partners in

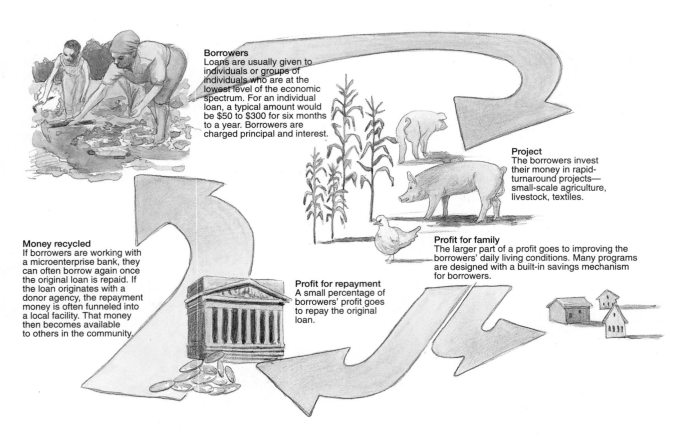

Borrowers
Loans are usually given to individuals or groups of individuals who are at the lowest level of the economic spectrum. For an individual loan, a typical amount would be $50 to $300 for six months to a year. Borrowers are charged principal and interest.

Project
The borrowers invest their money in rapid-turnaround projects—small-scale agriculture, livestock, textiles.

Money recycled
If borrowers are working with a microenterprise bank, they can often borrow again once the original loan is repaid. If the loan originates with a donor agency, the repayment money is often funneled into a local facility. That money then becomes available to others in the community.

Profit for repayment
A small percentage of borrowers' profit goes to repay the original loan.

Profit for family
The larger part of a profit goes to improving the borrowers' daily living conditions. Many programs are designed with a built-in savings mechanism for borrowers.

FIGURE 26-5: Grass-roots loans. Investments in people rather than in large projects can help people to help themselves out of poverty, and at the same time, protect the environment from the impact of major projects.

Development Programme," established in 1990 by the U.N. Development Programme. The "Partners in Development Programme" currently operates in 60 countries, making small grants ($25,000 per country) to nongovernmental organizations (NGOs) and grassroots organizations to support innovative projects. An NGO in Rwanda, for example, received $13,500 for a "cattle bank" that provides farmers with training and animals for start-up livestock projects.

Debt-for-Nature Swaps. In a **debt-for nature swap**, conservation organizations "buy" a portion of the debt of a nation at a discount from the bank that made the original loan. In exchange, the debtor nation agrees to establish and protect a nature reserve or implement a conservation program. The banks accept the discounted offer because the debt situation is so severe that many institutions fear debtor nations will not be able to pay them at all. The debt-for-nature swap allows them to recoup a portion of their loss. The debtor nation does not lose sovereignty, because the swap does not result in a foreign stake in a local corporation. Thus, the swap is advantageous for all participants.

The first debt-for-nature swap occurred in 1987 when Conservation International bought $650,000 of Bolivia's $4 billion external debt at a discounted price of $100,000. The Bolivian government set aside 3.7 million acres (1.5 million hectares) of Amazon River country around the existing Beni Biosphere Reserve. The Bolivians continue to own and manage the land. Since 1987 debt-for-nature swaps have generated over $60 million for conservation efforts while relieving almost $100 million in foreign debt. By 1991, 18 swaps had been arranged for countries including Costa Rica, the Dominican Republic, Ecuador, Poland, Madagascar, the Philippines, and Zambia. Costa Rica has made the largest volume of debt-for-nature swaps (Table 26-1).

Expanding on the basic concept underlying debt-for-nature swaps, some nations are writing off large sums of LDC debt in exchange for the adoption of sustainable development programs. In 1989 the Dutch government purchased $33 million worth of Costa Rica's debt in exchange for $10 million in local investments in reforestation, watershed management, and soil conservation. In 1991 the United States announced that it would exchange 10 percent of Poland's U.S. debts for an equivalent amount of

Box 26-1:
A Global Effort: Forging Sustainable Societies

In 1980 The World Conservation Union, the U.N. Environment Programme, and the World Wide Fund for Nature published the *World Conservation Strategy*, a document that asserted that humanity's future was in peril unless the peoples of the world began to conserve nature and natural resources. The *World Conservation Strategy* was among the first to point out that conservation is *not* the opposite of development. In fact, the authors maintained that conservation was impossible without development to alleviate the poverty and misery of hundreds of millions of the world's poor. In the decade since its publication, over 50 countries have prepared national and regional conservation strategies. The link between economics and the environment was further elucidated in *Our Common Future*, a 1987 report by the World Commission on Environment and Development, which increased recognition of the need for sustainable development and equity among all nations.

In 1991 the conservation groups published a second world conservation strategy, *Caring for the Earth: A Strategy for Sustainable Living*. Its goal is nothing short of significant change—a change in values, economies, and societies—in order to secure a better quality of life for humankind. The authors of *Caring for the Earth* intend it to be used by

all those who shape policy and make decisions that affect development strategies and environmental conditions: politicians and executives in the public and private sectors at both the national and international levels, of course, but also community leaders, business people, and citizens everywhere who are concerned for the future of Planet Earth and its human and nonhuman inhabitants.

Caring for the Earth is based on three points. The first is that the world's people want not just to survive but to secure a satisfactory life for all peoples and for their descendants. To do so we must learn to live differently, which requires a new, sustainable form of development. The second point is that because we depend on the Earth and its resources to meet our needs, the diminishment or degradation of those resources means that our needs, and those of our descendants, may go unmet, putting human survival at risk. After decades of unsustainable living, that risk is now perilously high. The third point is that all is not lost; we can choose to alter our behaviors and chart a new course for humanity. By creating sustainable means of development and living, distributing the benefits of development equitably among nations, and learning to care for the earth, we can ensure the survival of civilizations.

Caring for the Earth outlines nine principles needed to establish a sustainable society. They are as follows:

- Respect and care for the community of life
- Improve the quality of human life
- Conserve the earth's vitality and diversity (which requires conserving life-support systems and biological diversity and ensuring that renewable resources are used in a sustainable manner)
- Minimize the depletion of non-renewable resources
- Keep within the earth's carrying capacity
- Change personal attitudes and practices
- Enable communities to care for their own environments
- Provide a national framework for integrating development and conservation
- Create a global alliance

Caring for the Earth provides suggestions for achieving and maintaining the goals inherent in these principles. The document also outlines other actions that can be taken to foster sustainable living and provides guidelines for helping users to adapt the strategy to their particular needs and capabilities.

---

local investments in programs designed to restore the country's ravaged environment. The exchange is expected to make an estimated $350 million available for spending by a newly created foundation for the environment. The Worldwatch Institute and other environmental groups maintain that such debt-for-environment swaps can reduce debt burdens while promoting real, lasting change in the development paths followed by poor nations.

What Is Politics?

The debt burden of the world's developing countries illustrates the interdependence of nations' economies; it is difficult, if not impossible, for nations to

remain economically isolated. The debt crisis also underscores the relationship between economics and politics and between politics and the environment. Poverty breeds social discontent and environmental destruction. In the following sections we look more closely at politics and its relationship to environmental issues.

In its broadest meaning, **politics** encompasses the principles, policies, and programs of government. Different political systems, such as democratic republics (the United States) and communist states (the People's Republic of China), are based on different principles and consequently adopt different policies and programs in order to govern their societies. Even within a particular nation, all citizens do not agree on the best means of government. For

▶ Table 26-1
Debt-for-Nature Swaps (through March 1991)

Date	Country	Purchaser	Cost (dollars)	Face value of debt (dollars)	Conservation funds generated (dollars)
1/91	Costa Rica	RFA, MCL, TNC	360,000	600,000	540,000
8/90	Madagascar	WWF	445,891	919,363	919,363
8/90	Philippines	WWF	438,750	900,000	900,000
3/90	Costa Rica	Sweden, WWF, TNC	1,953,473	10,753,631	9,602,904
3/90	Dominican Republic	PRCT, TNC	116,400	582,000	582,000
1/90	Poland	WWF	11,500	50,000	50,000
8/89	Zambia	WWF	454,000	2,270,000	2,270,000
7/89	Madagascar	WWF	950,000	2,111,112	2,111,112
4/89	Ecuador	WWF, TNC, MBG	1,068,750	9,000,000	9,000,000
4/89	Costa Rica	Sweden	3,500,000	24,500,000	17,100,000
1/89	Costa Rica	TNC	784,000	5,600,000	1,680,000
1/89	Philippines	WWF	200,000	390,000	390,000
7/88	Costa Rica	Holland	5,000,000	33,000,000	9,900,000
2/88	Costa Rica	FPN	918,000	5,400,000	4,050,000
12/87	Ecuador	WWF	354,000	1,000,000	1,000,000
8/87	Bolivia	CI	100,000	650,000	250,000

WWF = World Wildlife Fund
TNC = The Nature Conservancy
PRCT = Puerto Rican Conservation Trust
MBG = Missouri Botanical Gardens
RFA = Rain Forest Alliance

FPN = National Parks Foundation of Costa Rica
CI = Conservation International
MCL = Monteverde Conservation League
CABEI = Central American Bank for Economic Integration

Source: World Wildlife Fund, Conservation Foundation, as reported in *The Information Please Environmental Almanac, 1992* (World Resources Institute, 1992).

example, the United States's two major political parties, the Democrats and the Republicans, favor different policies and programs.

During the presidential campaign of 1988 in the United States the environment became a major issue. Both of the major parties claimed to be the environmental party. The Republicans, under George Bush, lambasted Michael Dukakis, the Democratic candidate and governor of Massachusetts, for his failure to clean up Boston Harbor. The Democrats pointed to the numerous environmental setbacks of the Reagan era and warned that a Bush presidency promised more of the same. However, history has shown that concern for the environment is not the sole domain of either party. Our nation's two most environmentally aware and concerned presidents were Theodore Roosevelt and Jimmy Carter, one a Republican, the other a Democrat. Clearly, environmentalism is a philosophy that transcends party lines.

Environmentalism is conserving and, hence, conservative in the true sense of the word. Political decisions about environmental and natural resource issues should err on the side of caution—on the side of conservation and preservation.

How Does Politics Affect Environmental Problem Solving and Management?

Environmental Problem Solving

It is not a lack of knowledge that prevents us from alleviating or eliminating many of the serious environmental problems we face today. For instance, we possess the technology to reduce acid precipitation, and we have substitutes for the chlorofluorocarbons that are destroying the earth's protective ozone shield. Why then, if we know how to slow or reverse many of the threats to the biosphere, do we not use our knowledge? The answer is, in large measure, a lack of political will. During President Reagan's first term in office, he resisted calls for action on the acid precipitation issue, claiming that further study was needed. Against the best advice of the nation's top scientists, Reagan instituted a 10-year study of the problem. His unwillingness to act was evidence of a lack of political will to tackle and solve an environmental problem. Environmentalists accused President Bush of similar stonewalling when his administration refused, in 1992, to support reductions in carbon emissions. Just one month before the 1992 World Conference on Environment and Development, at which negotiators had hoped to sign a U.N.-brokered global warming treaty that would impose limits on carbon emissions, the United States convinced other industrialized nations to support a weakened version of the agreement. The U.S.-supported treaty calls on nations to assess their carbon emissions but does not require them to stabilize or reduce those emissions.

The political will to take action on issues usually stems from sufficient public demand. When enough people make enough noise, politicians begin to find ways to satisfy their constituents' demands. In the mid-1980s the media began to publicize reports of leaks and contamination at several of the nation's nuclear weapons manufacturing plants, including the Fernald Feed Materials Processing Plant in southwestern Ohio. Over the next several years the reports continued, revealing a decades-long history of radioactive releases and leaks. The local populace, concerned about the possibility of significant health threats, finally turned to local and state politicians. The pressure applied by Representative Thomas Luken and Senator John Glenn was instrumental in forcing the Department of Energy to release information about radioactive releases and environmental safety at Fernald.

Politics is an important component of international environmental issues as well. Consider, for example, the Nagymaros Dam, currently under construction on the Danube River at the Danube Bend, one of the most scenic spots along the river's 1,600-mile (2,560-kilometer) course (Figure 26-6). The bend is about 40 miles north of Budapest, Hungary. In 1977 Hungary and Czechoslovakia agreed to develop the river for hydropower generation and to improve navigation. The Czech government immediately began construction of Gabcikovo, a huge hydropower complex now nearly complete. The Hungarian government did not begin work immediately because of financial constraints and environmental concerns, including fears that the dam will drastically alter the river's hydrology, render some sections of the Danube biologically dead, and concentrate industrial and municipal pollution.

In 1985 the Austrian government persuaded Hungary to begin work on the Nagymaros, offering to pay 70 percent of the project's $3 billion cost in return for 20 years of 1.2 billion kilowatts of electricity annually. Austria was unable to build a dam across the Danube at Hainburg, Austria, because of pressure by the Greens, a political party of environmental activists. Austria stands to benefit the most from the power generated by Nagymaros; critics claim that the dam will provide Hungary with only

FIGURE 26-6: Nagymoros Dam on the Danube River in Hungary. Construction of the dam has led to political conflict among Hungary, the former Czechoslovakia, and Austria.

enough electricity to meet 2 percent of its annual needs. Any adverse environmental effects, however, will belong primarily to Hungary.

Nagymaros has become a major political issue in Hungary. Members of parliament, under attack for allowing the plan to proceed, blame the scientific community for not taking a firm stand on the dam's environmental effects. In response, the scientists claim that they were not asked to comment on the proposed dam in the first place, and they refuse to accept the blame for the project.

Several unexpected and positive developments have resulted from the struggle over Nagymaros. The first is the galvanization of Hungary's emerging environmental movement. "The Blues," as they are known by the western media, consist of approximately 30 environmental groups. Another positive development is that environmental issues are now being discussed publicly in the media, something that would not have happened even a few years ago.

Environmental Management

Political decisions affect how resources are managed. For example, Table 26-2 shows some of the important social and environmental programs that could have been pursued with the money that the U.S. Congress and other governments allocated to military programs. Politics also affects resource management when a particular issue becomes a political football. The struggle over the 1990 Tongass Reform Bill is a case in point. The bill was designed, in part, to reverse a portion of the Alaska Native Interests Lands Claim Act (ANILCA) of 1980. ANILCA provided an annual appropriation of $40 million for the

Table 26-2
Military Versus Social and Environmental Priorities

Military Priority	Cost (dollars)	Social and Environmental Priority
Trident II submarine and F-16 jet fighter programs	100,000,000,000	Estimated cleanup cost for the 3,000 worst hazardous waste dumps in the United States
Stealth bomber program	68,000,000,000	Two-thirds of estimated costs to meet U.S. clean water goals by 2000
Requested SDI funding, 1988–1992	39,000,000,000	Disposal of highly radioactive waste in the United States (nonmilitary)
2 weeks of global military spending	30,000,000,000	Annual cost of the proposed UN Water and Sanitation Decade
German outlays for military procurement and R&D, 1985	10,750,000,000	Estimated cleanup costs for West German sector of the North Sea
4 days of global military spending	8,000,000,000	UN Action Plan over 5 years to save the world's tropical forests
Development cost for Midgetman ICBM	6,000,000,000	Annual cost to cut U.S. sulfur dioxide emissions by 8–12 million tons per year to combat acid rain
2 days of global military spending	4,800,000,000	Annual cost of proposed UN Action Plan to halt desertification over 20 years in developing world
6 months of U.S. outlays for nuclear warheads, 1986	4,000,000,000	U.S. government outlays for energy efficiency, 1980–1987
SDI research, 1987	3,700,000,000	Cost to build a solar power system serving a city of 200,000
3 weeks of military spending of countries with literacy rates of 50 percent or less	2,400,000,000	Additional UNESCO budget needs, over a decade, to eliminate illiteracy worldwide
10 days of European Economic Community military spending	2,000,000,000	Annual cost to clean up hazardous waste sites in 10 European Economic Community countries by the year 2000
1 Trident submarine	1,400,000,000	5-year child immunization program against 6 deadly diseases, preventing 1 million deaths a year
3 B-1B bombers	680,000,000	U.S. government spending on renewable energy, 1983–1985
2 months of Ethiopian military spending	50,000,000	Annual cost of proposed UN Antidesertification Plan for Ethiopia
1 nuclear weapon test	12,000,000	Installation of 80,000 hand pumps to give villages in the developing world access to safe water
Operating cost of B-1B bomber for 1 hour	21,000	Community-based maternal health care in 10 African villages to reduce maternal deaths by half in one decade

Source: Adapted from Worldwatch Institute, based on various sources.

Tongass in order to promote lumbering and boost the local economy. The Tongass was the only national forest whose budget was not determined by Congress; its $40 million was guaranteed. Two companies were given exclusive contracts to buy Tongass timber. (See Chapter 21 for further discussion of forestry practices on the Tongass.)

After ANILCA was enacted, the Forest Service received approximately one cent for every dollar it spent (primarily on road construction) to make timber available to those two lumber companies. They were able to buy centuries-old trees, worth hundreds of dollars, for several dollars. The lumber companies and the timber industry did not bring economic well-being to the area, as promised,

because of reduced demand for timber. In the meantime, logging caused numerous adverse environmental effects, including erosion and subsequent siltation of waterways, and marred the beauty of the area.

The Tongass debacle persisted for almost a decade, thanks largely to the stonewalling techniques of Alaska's senators, who insisted, contrary to all logic, that the arrangement was good for the Tongass and good for Alaska. By 1989, however, public pressure and opinion were solidly against the management practices on the Tongass, and the reform bill was passed.

As the Tongass struggle illustrates, conservationists must possess a thorough knowledge of the political system in order to ensure that resources are managed wisely. They must know how bills are drafted and introduced, how to pressure industry, and how to lobby effectively.

Grassroots pressure can influence natural resource management decisions. Throughout the developing world, countless grassroots initiatives are achieving economic and environmental goals. Examples include reforestation efforts in Haiti, Pakistan, and Niger; the construction of small dams in the village of Sukhomairi, India; an NGO-sponsored project to promote the dissemination of improved woodstoves in Kenya; water development projects for villages in Kenya and Senegal; and the construction of a milk processing plant by a dairy farmers' cooperative in Durazno, Uruguay. Increasingly, the world's women are taking a leading role in grassroots environmental action. Focus On: Women and Conservation describes some achievements of this vital emerging force.

How Are Political Programs Contributing to Environmentally Sound Management?

Just as economics can contribute to environmentally sound resource management, politics, too, holds the potential for much environmental good. The following examples are a few of the numerous political developments of recent years that are having beneficial effects on the national and global environment.

- In 1972 the United States and the former Soviet Union signed a comprehensive Agreement on Cooperation in the Field of Environmental Protection. Scientists from the two countries cooperated on projects as diverse as joint symposia on aquatic toxicology, exchange of wild plant seeds, distribution of bands for snow geese, zoo exchanges, development of a joint monograph on comparative environmental law, and research into atmospheric effects and climate change.

- The Montreal Protocol to Control Substances that Deplete the Ozone Layer, agreed upon in 1987, called for halving the use of ozone-depleting chlorofluorocarbons by the end of the 1990s. This protocol, which included the major CFC-producing nations, was strengthened in June 1990 at the London Conference. An agreement was reached on financial arrangements, and some developing nations which had not originally participated in the accord decided to join, bringing the total participants to 66. Eleven more nations indicated they were seriously considering joining.

- The Basel Convention, signed in 1989, controls the international transport and disposal of haz-

ardous waste. It was signed on the spot by 34 countries, with 71 more nations tentatively agreeing to it. While the convention does not halt the commerce of toxic substances, it signals an international resolve to eliminate the dangers such commerce poses to the global environment.

- The Declaration of the Hague, signed by 17 countries in 1989, has been hailed as a revolutionary agreement because it calls for the creation of a new or newly strengthened institution within the United Nations that would be given the authority to make and enforce decisions on global environmental problems, *even in the absence of unanimous agreement among nations.* Nations historically have avoided any treaties or agreements that might infringe on their sovereignty, but the parties to the Hague Declaration, which now number 30, acknowledge two important facts that other nations either have yet to realize or to act on. First, one nation's environment cannot be separated from the environments of others; the nations of the world are inextricably linked with one another and environmental problems in one nation often affect another nation. Some problems, too, are global in nature, and so one country's activities can help or hinder the efforts of other nations to remedy these common problems. Second, the Hague Declaration signatories recognize that protection of the global environment is so critical that nations' sovereign rights must not be allowed to impede progress in slowing global warming, preserving the ozone layer, and so forth.

  Neither the United States nor the former Soviet Union signed the Hague Declaration. By 1992 it was for the most part a forgotten proposal. The Declaration is significant, however, because it underscores the growing need for new political mechanisms to tackle global environmental problems.

- In October 1989 members of the Convention on International Trade in Endangered Species (CITES) voted 76–11 to ban all trade in ivory in an attempt to save the African elephant from extinction. The ban was instituted after a decade of unrestrained poaching had caused a serious decline in elephant numbers in many African nations, including Kenya. At a CITES meeting in 1992 some African nations whose herds are not threatened attempted unsuccessfully to have the ban repealed.

- The 1992 World Conference on Environment and Development, known as the Earth Summit, took place in June in Rio de Janeiro, Brazil (Figure 26-8). Approximately 10,000 official delegates from 150 countries were expected to attend the Earth Summit, the largest U.N. conference ever held. As many as 20,000 citizens were expected to attend a Global Forum, a parallel conference held in conjunction with the U.N. meeting.

Focus On:
Women and Conservation

Long ignored by international lending institutions, national and provisional governments, and local power brokers, women of all nations are now taking part in efforts to shape environmentally sound government policy, foster sustainable development, halt environmental degradation, and preserve ecosystems and wildlife. Especially in developing countries, native women's groups and groups with high female participation are proving to be very effective. The following are but a few examples of efforts undertaken by women worldwide.

- Petra Kelly, the leader of the influential Green Party in Germany, was a charismatic leader who had achieved popular appeal for her comparatively new political party. In Germany's 1983 election the Green Party captured 5.6 percent of the votes and put 27 representatives into the Bundestag, Germany's legislative body. Kelly was killed in late 1992 under uncertain circumstances.

- The Green Belt movement in Kenya was founded in 1977 by a biologist named Wangari Maathai, the first Kenyan woman to earn a Ph.D. and the first to head a department at the University of Nairobi. In order to help villagers suffering from malnutrition and lack of firewood for fuel, she began a tree-planting program that is proving remarkably successful. The Green Belt movement encourages natives, especially women, to plant tree seedlings in public areas near their homes and villages to form protective tree belts. Foresters teach the people how to find the seeds of native trees, how to bed them in village nurseries, and how to transplant and care for them. The local people are paid a small amount for each *surviving* tree. Approximately 50,000 Kenyan women have joined the Green Belt movement. By 1989 these women had established 670 community tree nurseries and had planted 10 million trees in 1,000 tree belts (Figure 26-7).

FIGURE 26-7: Wangari Maathai, leader of the Greenbelt Movement, Kenya.

- The Chipko Movement of India also focuses on tree conservation. This movement has its roots in a late fifteenth-century Hindu sect that believed that the protection of trees and wild animals is a religious duty. By the early 1960s indiscriminate logging had resulted in severe

The organizers of the Earth Summit established an ambitious agenda. One of the most important topics of discussion was a treaty on global warming. Many countries have agreed to stabilize or reduce emissions of greenhouse gases in order to avert or minimize climate change, but several influential nations, principally the United States, have resisted such measures. Another agenda item concerned the Hague Declaration. Some nations would like to see the Declaration revived in order to deal more effectively and efficiently with global environmental problems. Those attending the conference were also expected to discuss proposed United Nations reforms, particularly with respect to the U.N. Environment Programme and the U.N. Development Programme. The aim of some of the proposed reforms is to more closely link the U.N.'s environment and development institutions to highlight the interrelatedness of environmental and economic goals. Reformers maintain that linking the environment and development institutions is the best way to ensure that these goals are achieved.

In the coming decade, no topic is likely to be of greater import than efforts to restore balance to the global economy by strengthening the economies of LDCs. For that reason, fostering and funding sustainable development was expected to be a highlight of the Earth Summit. Conference participants planned to consider strengthening the Global Environmental Facility,

FIGURE 26-8: Demonstrators at the 1992 Earth Summit protest U.S. environmental policy.

deforestation. Floods and land-slides became common, and women were forced to travel far from their homes in search of firewood and water. In 1973, when a sports company attempt-ed to cut trees near the village of Gopeshwar, a community leader urged the villagers to "hold fast," or "chipko," to the trees. When the loggers arrived, local women saved the trees by flinging their arms around them. The Chipko Movement has since spread among the hill women in north-ern India.

- Another group of Indian women guarded their forests using more contemporary methods. The women of Khirakot, a small vil-lage in the northern Indian state of Uttar Pradesh, collect fuel-wood from the surrounding community forests. They care-fully manage the forests in order to maintain the fuelwood sup-ply, thus practicing sustainable use of this renewable resource. When a contractor obtained a lease to mine soapstone in the hills near Khirakot, the women

organized to prevent the mining. They realized that mining activi-ties would prevent them from gaining access to the forests, and they feared that mining debris might destroy the forests. The women filed a formal court protest, and the decision went in their favor. The mines were closed, the forests preserved.

- In the mid-1980s the Boiteko Agricultural Management Asso-ciation (AMA) received a $35,000 grant from the African Development Foundation (ADF), a fund established by the U.S. Congress in 1980 to sup-port grassroots development in Africa, to conduct a horticultur-al and poultry project in Serowe, a village in eastern Botswana. With the assistance of a techni-cal manager and a financial con-sultant, the project is run by nine women and one man from Serowe. The project provides rural people with employment and a reliable source of income and at the same time ensures a stable source of vegetables and eggs for the village. AMA is able

to pay its members about $45 per month from the proceeds generated by the project. The original ADF grant enabled the project participants to build a chicken house and purchase wire fencing to enclose the vegetable- and egg-production operations. The project's success (in addi-tion to its vegetable and poultry operations, the project now includes a citrus orchard) has allowed the AMA to secure grants from other organizations in order to purchase a windmill, netting for a vegetable nursery, and a mechanical tiller.

- The United Nations Develop-ment Fund for Women (UNIFEM) was established after a Mexico City conference in 1975 to mark International Women's Year. The Conference also served as a kick-off for the United Nations Decade for Women (1976–1985). By the mid–1980s, UNIFEM had signif-icantly increased its work with NGOs, providing badly needed support to women's groups throughout the developing world.

the fund established to help poor nations imple-ment actions designed to combat global problems such as ozone depletion. Other funding mecha-nisms for sustainable development were also on the Earth Summit agenda.

A Piece of the Antarctic Pie—Exploration Now, Exploitation Later?

Encouraging signs in the political arena indicate that environmental issues may finally receive the national and international attention they deserve. As we approach the twenty-first century, one of the most important issues to be decided by politicians is the fate of the planet's last pristine continent, Antarcti-ca. The following section describes why Antarctica's resources are in danger of being exploited and how nations and concerned environmental groups are reacting.

The Antarctic continent is a land mass of 5.5 mil-lion square miles (14.3 million square kilometers), an area bigger than the United States and Mexico,

much of it covered with ice averaging 1 mile thick. The dominant form of continental vegetation is a lichen less than 1 inch high, chiefly inhabited by mites. The surrounding ocean is home to penguins, seals, whales, fish, and krill. Antarctica is the cold-est continent, and, although covered with ice, it is a desert, with annual precipitation of less than 3 inches.

The Antarctic landscape is deceiving, for its rich-ness is hidden. Its snow and ice hold large reserves of fresh water. Dry nations dream of hauling icebergs to their shores to supply fresh water for domestic use and irrigation. Others speculate about the min-eral wealth to be had from oil (tens of billions of barrels), ethane, methane, ethylene, copper, silver, and nickel. (See Chapter 17, "Mineral Resources.") However, these riches are locked under the mile-thick ice sheets and at the bottom of frozen seas. Even if needed resources are found in Antarctica, the costs of extracting them using today's technolo-gy would be staggering. So far, the resources exploited in Antarctica have been biological (whales, penguins, seals, and krill).

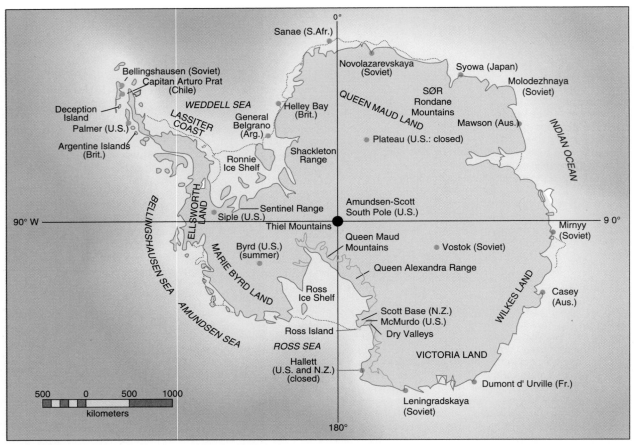

FIGURE 26-9: Territorial claims and scientific research stations in Antarctica.

At this time many nations, particularly LDCs, are arguing for their rights to claim whatever natural resources are discovered in Antarctica. These nations feel that a share of the continent's unplumbed resources will bestow economic and political clout on them.

Research expeditions are the only current "industry" on the continent. The pursuit of knowledge is protected by the Antarctic Treaty, signed by 12 countries (known as the treaty or consultative parties) on December 1, 1959. The two stated purposes of the treaty are to maintain Antarctica for peaceful uses, prohibiting all military activities, weapons testing, nuclear explosions, and disposal of radioactive wastes and to promote freedom of scientific investigation and international cooperation. Sections of Antarctica are claimed by seven nations—Australia, New Zealand, South Africa, Norway, Argentina, Chile, and the United Kingdom (Figure 26-9). Neither the United States nor the former Soviet Union claims Antarctic territory or honors the claims of others, but both consider the continent politically important. Basic provisions of the treaty and the 12 original signatories are shown in Table 26-3.

Several safeguards have been incorporated into the treaty to ensure its fairness and equitability. Because each party to the treaty must concur in all decisions, no activity or management practice that is prejudicial to any party can be enacted. Moreover, individuals designated by each party to the treaty have the right to inspect all stations, ports, and cargos of all nations in order to enforce the treaty's terms.

Some groups feel that the nations responsible for the management of Antarctica are short-sighted and selfish. Even though the parties to the treaty agree to set aside all territorial claims to the continent, they would have a clear advantage over other nations in any race to exploit resources. Realizing this, a group of less-developed countries led by Malaysia in 1985 attacked the treaty in the United Nations as a club set up to prevent poor nations from benefiting from the continent's wealth.

Environmentalists are also concerned about the ability of the treaty parties to protect the Antarctic environment. The consultative parties developed a framework for protecting the Antarctic environment from mineral or oil exploitation at a time when there was no substantial pressure for commercial development. Such pressure, which is bound to

▶ Table 26-3
Summary of Basic Provisions and
Signatories of the Antarctica Treaty

BASIC PROVISIONS

Article I. Antarctica shall be used for peaceful purposes only. All military measures, including
weapons testing, are prohibited. Military personnel and equipment may be used, however, for scientific purposes.

Article II. Freedom of scientific investigation and
cooperation shall continue.

Article III. Scientific program plans, personnel,
observations, and results shall be freely exchanged.

Article IV. The treaty does not recognize, dispute,
or establish territorial claims. No new claims shall
be asserted while the treaty is in force.

Article V. Nuclear explosions and disposal of radioactive wastes are prohibited.

Article VI. All land and ice shelves below 60° south
latitude are included, but high seas are covered
under international law.

Article VII. Treaty-state observers have free access,
including aerial observation, to any area and may
inspect all stations, installations, and equipment.
Advance notice of all activities and of the introduction of military personnel must be given.

Article VIII. Observers under Article VII and scientific personnel under Article III are under the
jurisdiction of their own states.

Article IX. Treaty states shall meet periodically to
exchange information and take measures to further treaty objectives, including the preservation
and conservation of living resources. These consultative meetings shall be open to contracting parties that conduct substantial scientific research in
the area.

Article X. Treaty states will discourage activities by
any country in Antarctica that are contrary to the
treaty.

Article XI. Disputes are to be settled peacefully by
the parties concerned or, ultimately, by the International Court of Justice.

Article XII. After the expiration of 30 years from
the date the treaty enters into force, any member
state may request a conference to review the operation of the treaty.

Article XIII. The treaty is subject to ratification by
signatory states and is open for accession by any
state that is a member of the UN or is invited by
all the member states.

Article XIV. The United States is the repository of
the treaty and is responsible for providing certified
copies to signatories and acceding states.

SIGNATORIES (CONSULTATIVE PARTIES)

Argentina	Japan
Australia	New Zealand
Belgium	Norway
Brazil	Poland
Chile	South Africa
Federal Republic of Germany	Soviet Union
	United Kingdom
France	United States
India	

Source: Deborah Shapley, *The Seventh Continent*

increase because of economic expansion in developed and developing nations, worries conservation
and environmental organizations. Members of the
international environmental community have
become more adamant during the last decade in
expressing concern that growth in scientific research
programs and facilities, logistics and supply operations, tourism, fishing, and prospective minerals
development will undermine Antarctic's value as a
pristine scientific laboratory and wilderness area and
will alter its role in the formation of the world climate. Many have called for a moratorium on all oil
and mineral exploration and drilling in Antarctica.
Others want to limit excessive krill and finfish harvests and abolish the seal and penguin industries.

In response to growing concern about the Antarctic environment and mounting distrust of the
consultative parties' motives, Greenpeace, an international conservation group, has established the first
private, nongovernmental scientific base in Antarctica to dramatize its call to make Antarctica a world
park off limits to war and commerce. Greenpeace
hopes that its presence will secure it a voice in
future treaty reviews, thus enabling it to take part in
the political process that decides Antarctica's fate.

Summary

Ecology and economics have the same Greek root word,
oikos, meaning "household." Ecology is the study of the
household, and economics is the study of its management.

There is a great difference between a resource's value
and its price. If, as a society, we were to realize the true
value of our resources, we could avoid many environmental problems.

Economic considerations are an integral part of problem
solving. Proposed alternative solutions are often compared
to determine which solution is more cost-effective. Unfortunately, economic cost is often viewed as the sole or primary determinant in the selection of a response to an
environmental problem.

Economic analysis is a criterion for environmentally
sound management. Renewable resources can yield sustained economic benefits. In contrast, nonrenewable
resources are ultimately limited both in their availability
and useful quality. For most of the history of the United
States, our economic system has stressed resource utilization over conservation. Increasingly, people are realizing
the important link between sound environmental policies
and sound economics.

Many less-developed countries have gone into debt from
large-scale development projects. Many are unable to pay
even the interest on these loans. However, the loans have
not helped improve economic conditions for citizens, and
economic woes intensify environmental degradation. Five
potential solutions to the economic and environmental
problems of debtor countries are the following: more environmentally responsible development banks, sustainable

development, increased and rechanneled development aid from MDCs, grassroots loans, and debt-for-nature swaps.

In its broadest meaning, politics encompasses the principles, policies, and programs of government. Different political systems are based on different principles, and they adopt different policies and programs in order to govern their societies. Even within a particular nation, all citizens do not agree on the best means of government. History has shown that concern for the environment is not the sole domain of any major national political party.

Lack of political will prevents us from alleviating or eliminating many serious environmental problems. The political will to take action on issues usually stems from sufficient public demand. Politics is an important component of international environmental issues as well as national ones.

Political decisions often affect how resources are managed. Just as economics can contribute to environmentally sound resource management, politics, too, holds the potential for much environmental good.

Many nations, particularly LDCs, are arguing for the right to claim the vast natural resources of Antarctica, hoping to increase their economic and political clout. So far, however, research is the only industry on the continent. The Antarctic Treaty of 1959 was signed to maintain Antarctica for peaceful uses and to promote freedom of scientific research and international cooperation.

Key Terms

debt-for-nature swap	sustainable economy
economics	sustainable society
politics	sustainable use
price	value
sustainable development	

Discussion Questions

1. What is the link between ecology and economics? How could this link be used to foster sound environmental policies?
2. What is the difference between value and price? How does this difference affect natural resource management?
3. How is the world debt crisis adversely affecting the environment?
4. How can politics contribute to environmental problems? How can politics contribute to solutions to these problems?
5. Why is Antarctica strategically important? What problems might arise from development in Antarctica?

Chapter

27

Law and Dispute Resolution

We are in this together. We all create the opportunities that result in the world's highest standard of living, and we create the problems that challenge us all. Together we must consider the options that most of us want and that make sense, and see if, together, we can continue to progress. This requires give and take.

E. Bruce Harrison

An issue may be so clear in outline, so inevitable in logic, so imperative in need, and so universal in importance as to command immediate support from any reasonable person. Yet that collective person, the public, may take a decade to see the argument and another to acquiesce in an effective program.

Aldo Leopold

Learning Objectives

When you finish reading this chapter, you should be able to:

1. Define environmental law or legislation.
2. Identify four reasons why environmental laws are not always enforced effectively.
3. Define environmental dispute resolution.
4. Explain how legislation and dispute resolution contribute to environmental problem solving and environmentally sound resource management.

Environmental disputes arise when parties disagree over alternative uses of available resources. Some people see the deserts of the Southwest as the perfect place to site a nuclear waste repository, but oth-ers see these "wastelands" as fragile ecosystems rich in native American heritage and beautiful vistas. People in southern California want to divert the water that runs out of the Sierra Nevada in the Sacramento and San Joaquin rivers to Los Angeles, but people near San Francisco Bay need that fresh-water runoff to prevent saltwater intrusion into the delta areas (Figure 27-1). In the Midwest chemical and steel companies located their operations along-side major waterways which served as a cheap source of fresh water and a convenient means of waste disposal. The Midwest's plentiful water sup-ply was the reason for the region's economic pros-perity, yet this prosperity came at the cost of the degradation and eventual elimination of recreational and wildlife uses of the waters. A local chemical industry claims that it has a right to discharge its wastes into streams; local fishers claim that they have a right to catch fish that are not poisonous to eat; the community claims that it has the right to a safe source of drinking water. But, the community's economic health is based in part on the continued successful operation of the industry. Who owns the river? Whose use should take priority? Clearly, opposing economic, cultural, and recreational inter-ests are the root of most disputes over natural resources.

Environmental disputes can be resolved through litigation, or they can be resolved through dispute resolution techniques such as negotiation and medi-ation. Both litigation and dispute resolution are valuable tools for settling individual disputes as they arise. But the value of these tools may be even greater over the long term as through them a body of environmentally sound decisions are reached to help establish and nurture a sustainable society.

In this chapter we define the fields of environ-mental law and dispute resolution and explore how both relate to resource management and environ-mental problem solving.

FIGURE 27-1: Tidal wetlands, San Leandro Bay Regional Park, Oakland, California. Conflicts arise when competing uses for water are proposed. Should water from northern California be diverted to Los Angeles or be left alone to continue to maintain wetlands near San Francisco Bay?

What Is Environmental Law?

Environmental law is that aspect of the legal system that governs the activities of persons, corporations, government agencies, and other public and private groups in order to regulate the impact of the activities on the environment and on natural resources. It includes pollution control, the allocation of scarce or important resources, and the protection and preservation of resources. It involves both of the two types of laws that are the basis of our whole legal system, common laws and statutory laws.

Common Law

Common law is a large body of written and unwritten principles and rules based on thousands of past legal decisions dating back to the beginning of the English legal system. Common law is built on **precedent,** a legal decision or case that may serve as an example, reason, or justification for a later one. When no applicable statute exists for a current case, the trial judge will review past legal decisions and attempt to find similarities between past cases and the current one. Based on these findings, the judge will make a decision that will then set precedent for the resolution of disputes in the future.

Common law attempts to balance competing interests in society. For instance, if the operation of a factory creates noise at levels that disturb nearby residents, a court may require that the factory reduce its noise levels during certain hours, thereby balancing the needs of the residents (rest, health, and enjoyment of property) with the utility of the factory (doing business, making a profit, and providing employment and taxes for the community).

Cases involving common law are based on three legal concepts: nuisance, trespass, and negligence. Known as **torts,** these are causes of action, that is, they denote wrongful acts for which a civil suit can be brought by an injured plaintiff against a defendant accused of committing the act. In order to file a lawsuit, a plaintiff must have standing (legal rights) and be an injured party. In the United States, two types of redress can be obtained through a civil suit: injunction and compensation. An **injunction** is a court order to do or refrain from doing a specified act. Abatement is an injunction that requires the defendant to stop or restrict the nuisance. **Compensation** is a monetary award for damages.

Nuisance, the most common cause of action in the field of environmental law, is a class of wrongs that arise from the unreasonable, unwarrantable, or unlawful use of a person's own property that produces annoyance, inconvenience, or material injury to another. Individuals may use their property or land in any way that they see fit, as long as they do so in a "reasonable" manner that does not injure or annoy others. Nuisance suits are generally settled by compensation or abatement (Figure 27-2).

Trespass is the unwarranted or uninvited entry upon another's property by a person, the person's agent, or an object that he or she caused to be deposited there. The property can be land, material possessions, or even one's body itself. The trespass may be intentional, or it may be unintentional but reckless, negligent, or abnormally dangerous. In the court case *Martin* v. *Reynolds Metal Company* the court decided that the deposition of airborne micro-

FIGURE 27-2: A field burning in the Willamette Valley, Oregon. Burning may be used by a field owner to clear fields of weeds and the previous season's crop residues, but the smoke from the burning fields may be a nuisance to nearby residents.

scopic fluoride compounds, originating from Reynolds Metal, onto Mr. Martin's property, constituted a trespass.

Negligence is the failure to exercise the care that "a prudent person" usually takes resulting in an action or inaction that causes personal or property damage. Statutes and regulations help the courts determine if a certain action is legal or not, but an action does not have to be illegal to be found negligent. To substantiate a negligence claim, the plaintiff must show that the action or lack of action was the "proximate" cause of an injury (that is, it occurred immediately or shortly before the effect) and that the consequences of the defendant's action were "reasonably foreseeable" by an "ordinary, prudent person."

Once harm has been found, the matter of liability must be determined. Of particular relevance to environmental law is the concept of **strict liability.** In reference to products, strict liability means that if harm results from a product, the maker of that product is liable for the harm done, no matter how careful the maker was and even if the results can be considered unforeseeable. The same principle applies to activities. For example, if a gas transmission line blows and harm is done, the gas company is still liable for the harm done even if the company is in no way at fault (it took proper care, using the best materials, techniques, design, and maintenance, and it turns out that a beaver bit into the pipe). Whether a party is to be held strictly liable for particular actions will be stated in statutes and regulations or can be derived from precedent (Figure 27-3).

FIGURE 27-3: Plane spraying pesticide on sweet-corn crop, California. Crop dusters may be liable for damage if the chemicals they use contaminate adjacent properties and homes.

Statutory Law

Statutory law is the body of acts passed by a local legislature or Congress. Statutes in the area of environmental law govern activities that affect the environment and human health and the management of resources. Statutes generally state the broad intentions of their specific provisions, such as to protect health and the environment by reducing air pollution or to use resources wisely by mandating conservation measures and standards.

Congress has enacted a number of important environmental and resource protection laws. These laws concern widely disparate aspects of the environment, from endangered species to hazardous waste sites to forests and agricultural pesticides. Whatever the difference in their content, they all attempt to protect or improve the environment in one of the following ways:

- By setting pollution level standards or limiting emissions of effluents for various classes of pollutants (e.g., Safe Drinking Water Act, Federal Water Pollution Control Act, and Clean Air Act)
- By screening new substances to determine their safety before they are widely used (e.g., Toxic Substances Control Act)
- By comprehensively evaluating the environmental impacts of an activity before it is undertaken (e.g., National Environmental Policy Act)
- By setting aside or protecting ecosystems, resources, or species from harm (e.g., Wilderness Preservation Act, Endangered Species Act, National Historic Preservation Act)
- By encouraging resource conservation (e.g., Farm Act of 1985)
- By cleaning up existing environmentally unsound hazardous waste disposal sites (e.g., Comprehensive Environmental Response, Compensation, and Liability Act)
- By preventing the creation of new unauthorized hazardous sites (e.g., Resource Conservation and Recovery Act)

Table 27-1 describes some of the most widely known and significant federal environmental laws. Most states have passed similar laws; some states, such as California and Minnesota, have enacted even stronger ones.

How Are Environmental Laws Enforced?

Generally, each environmental law specifies how it is to be enforced and who is to enforce it (Table 27-2). The Environmental Protection Agency (EPA), perhaps the best known of the federal agencies

Table 27-1
Major Provisions, Selected Federal Environmental Laws

Federal Clean Air Act (1970, 1977)

Establishes National Ambient Air Quality Standards (NAAQS) to control emissions from stationary and mobile sources of air pollution and protect the quality of the air nationally

Establishes National Emission Standards for Hazardous Air Pollutants (NESHAPS) to control hazardous air pollution from new stationary sources not already covered by NAAQS

Establishes an offset policy to prevent further degradation in areas not attaining NAAQS and to make reasonable progress toward compliance with NAAQS. If, for example, a new factory would emit 100 tons of sulfur dioxide per year, existing facilities in the area are required to decrease their emissions of sulfur dioxide by at least 100 tons per year, so that the level of the pollutant in the area will not increase because of the operation of the new facility.

Federal Water Pollution Control Act (1972), Clean Water Act (1987)

Defines pollution limits, requires water quality monitoring, provides grants for treatment, and studies wastewater treatment options

Establishes a system of standards, permits, and enforcements to eliminate discharges of pollutants into navigable waters, with an interim goal of attaining high enough water quality so people can swim and fish in these waters

Sets effluent limitations that restrict the amounts of pollutants that can be discharged to water and requires that, to meet limitations, certain equipment be installed at wastewater outlets to reduce the emissions of hazardous materials

National Environmental Policy Act (1969)

Imposes environmental responsibilities on all agencies of the federal government, specifying that they prepare a detailed statement of environmental impact (EIS) for each major federal project that may significantly affect the quality of the human environment. An EIS details what the project is, why it is needed, probable environmental impact of the proposed action and of all potential alternatives, adverse environmental effects that could not be avoided if the proposal were to be implemented, ways environmental impacts can be minimized, relationships between the likely short- and long-term impacts of the proposal on environmental quality, irreversible commitments of resources that would be made were the proposal to be implemented, objections voiced by reviewers to a draft EIS, names and qualifications of those persons responsible for developing the EIS, and references to document all statements and conclusions included in the report

Establishes in the executive office a three-member Council on Environmental Quality (CEQ) to determine the condition of the national environment, develop and recommend to the president new environmental policies and programs, evaluate and coordinate federal environmental programs and activities, counsel the president on environmental issues, problems and solutions, and establish guidelines for the preparation of environmental impact statements

Coastal Zone Management Act (1972)

Establishes a coastal zone management program with grants to participating states for coastal planning and management. Each participating state defines its coastal zone, identifies legal authorities for controlling shorelands, and shows how national goals will be achieved.

Wilderness Preservation Act (1964)

Establishes the National Wilderness Preservation System, which designates certain lands as wilderness areas and determines the desirability of adding other areas to the system

Defines a wilderness as having four characteristics: unnoticeable human impact; outstanding opportunities for solitude or primitive recreation; an area of at least 5,000 acres or of sufficient size to make preservation practicable; and ecological, geologic, or other value

Restricts the use of wilderness land. The agency administering any wilderness area is "responsible for preserving the wilderness character of the area" and for ensuring that these areas are devoted to "the public purposes of recreational, scenic, scientific, educational, conservational, and historic use."

charged with enforcing environmental policies, is the federal government's primary "environmental watchdog." A 1970 executive order by President Richard Nixon created the EPA to manage many of the environmental protection laws that issue from Congress.

Each state also has its own agencies that are responsible for air pollution, water pollution, solid and hazardous waste, state parks and forests, fish and game, public drinking water, and the coordinated review of public projects such as the construction of highways.

Unfortunately, environmental laws are not always enforced effectively. In some cases the allocation of authority is unclear, especially when either state or federal authorities could be in charge. At times the agency charged with enforcing a particular law is unwilling to do so because it disagrees with the principle, intent, or certain stipulations of the legislation. For example, many conservationists feel that the Forest Service and Bureau of Land Management have been remiss in nominating lands under their jurisdiction for wilderness designation (Figure 27-4). They contend that the agencies have allowed activi-

Major Provisions, Selected Federal Environmental Laws (*Continued*)

Federal Safe Drinking Water Act (1974, 1977)

Protects water supplies that serve the public by setting drinking water standards, which limit the amount of contamination permissible in water sources

Regulates underground injection of substances that might endanger drinking water sources

Comprehensive Environmental Response, Compensation, and Liability Act (1980)

Authorizes the president to require cleanup of toxic material releases

Makes facility owners liable for governmental cleanup costs and for destruction of natural resources owned by the government

Establishes Superfund, financed by a tax on the production of toxic chemicals, to be used to cover costs of restoring natural resources that have been damaged by toxic chemicals

Toxic Substances Control Act (1976)

Empowers the EPA to require chemical manufacturers to test toxic substances if insufficient data are available concerning the substance and if the substance may "present an unreasonable risk," "enter the environment in substantial quantities," or present the likelihood of "substantial human exposure"

Requires chemical manufacturers to notify the EPA before manufacturing a new chemical substance

Allows the EPA to apply restrictions on the use of any chemical if "there is a reasonable basis to conclude that the substance presents or will present an unreasonable risk of injury to health or the environment"

Resource Conservation and Recovery Act (1976)

Establishes cradle-to-grave regulation of wastes through licensing and notification requirements for those who generate, store, treat, or dispose of hazardous wastes

Authorizes the EPA to establish criteria for identifying hazardous wastes, specific requirements for hazardous waste containers and labels, record keeping requirements, and the system of manifests to accompany wastes when they are transported, documenting their nature, origin, routing, and destination.

Food Security Act (1985, 1990)

Popularly known as the Farm Bill, the Act includes four major provisions: swampbuster, sodbuster, conservation compliance, and conservation reserve program (CRP).

Swampbuster and sodbuster provisions deny federal price supports, crop insurance, and federal loans to farmers who cultivate environmentally sensitive lands.

Conservation compliance, required after 1990, mandates that farmers who cultivate highly erodible land implement conservation practices to minimize erosion. Failure to do so can result in a farmer becoming ineligible for all farm programs on their entire acreage.

Through the CRP, a voluntary program, farmers receive annual "rent" payments of about $50 per acre to withdraw highly erodible land from production. The land is planted in trees, grass, or a legume cover for 10 years, with the costs of the planting shared by the federal government.

ties such as mineral exploration and off-road vehicle use in wilderness study areas which are under review for inclusion into the National Wilderness Preservation System. Such disruptive activities effectively degrade the wilderness quality of an area and disqualify it for inclusion in the wilderness system. If the conservationists' claims are valid, the agencies' actions are violations of the Wilderness Preservation Act of 1964.

Environmental laws are also ineffective when the administration or Congress fails to provide the funds needed for enforcement. For example, the Endangered Species Act established procedures for the identification, study, and protection of threatened or endangered plant and animal species, but Congress has not appropriated sufficient funds to enable the Fish and Wildlife Service to study all those species which are threatened. Hence, despite the presence of a legal framework for preserving species, populations of many of our native flora and fauna continue to decline.

Problems can also occur when a government agency has conflicting priorities. For example, in the early years of nuclear power, a single agency, the

Table 27-2
Federal Agencies and Their Responsibilities for Human Health and the Environment

Department of Agriculture

Forest Service — Manages national forests, including road construction, timber harvesting and mining, and national grasslands

Soil Conservation Service — Gives technical and financial assistance to reduce soil erosion, prevent floods, and conserve water; conducts research on soil conservation methods and new strains of plants

Department of Commerce

National Oceanic and Atmospheric Administration — Conducts research on and monitors the oceans and atmosphere; provides ecological baseline and models to predict the impacts of air and water pollution; makes nautical charts and coastal maps; predicts weather

National Marine Fishery Service — Conducts research on marine fish and mammals; administers the Marine Mammal Protection Act and the Fishery Conservation and Management Act

Department of Defense

Army Corps of Engineers — Plans and builds dams, reservoirs, levees, harbors, and locks; enforces Section 404 of the Clean Water Act regarding the management of wetlands

Department of Energy

Office of Conservation and Renewable Energy — Directs energy conservation programs; conducts research on solar and renewable energy resources

Federal Energy Regulatory Commission — Licenses construction of nonfederal dams and other hydroelectric projects. Although it is within the DOE, it is a separate and independent agency.

Department of Health and Human Services

Occupational Safety and Health Administration — Conducts research on worker safety and health; enforces worker safety and health laws

Food and Drug Administration — Conducts research on safety of foods, drugs, and cosmetics; enforces health and safety laws pertaining to these substances

National Institutes of Health — Studies cancer through the National Cancer Institute; studies radiation and other environmentally related diseases through the National Institute of Environmental Health Services

Centers for Disease Control — Conducts epidemiological studies of disease

Department of the Interior

Fish and Wildlife Service — Manages national wildlife refuges and national fish hatcheries; enforces Endangered Species Act and Migratory Birds Treaty Act

National Park Service — Manages national parks, national monuments, Wild and Scenic Rivers System, and National Trail System

Bureau of Land Management — Manages approximately 300 million acres of public lands for multiple uses, including recreation, wildlife habitat, grazing, timber, and watershed protection

Office of Surface Mining — Administers the 1977 Surface Mining Control and Reclamation Act

Bureau of Reclamation — Manages water projects such as irrigation, flood control, and hydroelectric power

Department of Transportation

Coast Guard — Enforces maritime law, including oil spill liability section of Clean Water Act

Federal Highway Administration — Constructs and maintains federal highways; designs and builds roads in national parks; administers highway beautification program; regulates movement of hazardous cargo on the nation's highways

Environmental Protection Agency — Conducts research on causes, effects, and controls of environmental problems; enforces most environmental laws

Nuclear Regulatory Commission — Regulates and licenses all activities related to civilian nuclear materials and facilities

FIGURE 27-4: Shale hills on BLM land, Utah. Many BLM lands in Utah have wilderness potential but have not been designated and protected as such.

Atomic Energy Commission (AEC), was created to promote and to regulate atomic power. In this conflict of interest, regulation usually lost to promotion. The Energy Reorganization Act of 1974 helped solve the problem by abolishing the AEC and dividing its functions between the Energy Research and Development Administration (ERDA), which became the Department of Energy in 1977, and the Nuclear Regulatory Commission (NRC).

Many times the burden of administrative duties hampers agencies' ability to enforce laws. The EPA is often caught in a crossfire between the public, which seeks tighter controls, and the businesses it regulates, which commonly complain that regulations are too stringent and costly. As a result, the good intentions of an environmental law may take years to materialize.

Among the laws enforced by the EPA is the Comprehensive Environmental Response, Compensation, and Liability Act, also known as Superfund. The Superfund law was passed in 1980 and reauthorized in 1986 and 1990. By 1991, only 34 hazardous waste sites had been cleaned and removed from the National Priorities List, an inventory of the nation's most dangerous hazardous waste sites. Another 65 had been cleaned and were in the process of being de-listed. The NPL currently lists over 1,200 sites, many of them in the Northeast and Great Lakes regions. An additional 31,000 sites are under investigation as possible Superfund sites. The average cost of a federal cleanup of one site is estimated at $21 to $30 million.

The EPA is not entirely to blame for the slow progress made on cleaning up hazardous waste sites. The EPA cannot look to other agencies or private firms to reap the benefits of others' experiences.

The agency has had to "invent" cleanup techniques. Moreover, its efforts have been hampered by inadequate funding. Also, the failure of Congress to reauthorize Superfund on schedule in 1985 severely hampered and delayed cleanup operations. The reauthorization of the act in 1990 increased the funding level and mandated cleanup schedules for sites on the Superfund list. Even so, it remains questionable as to whether or not real, significant progress will be made in cleaning up the nation's uncontrolled hazardous waste sites.

How Do Environmentally Sound Laws Contribute to Environmental Problem Solving and Management?

Environmental Problem Solving

Legal specialists are necessary members of problem solving teams because they provide information about the legal system: what laws are relevant to a specific problem and the legal ramifications of various alternatives. Such factors shape the alternative solutions available to the problem solving team. For example, a small community faced with the challenge of upgrading its sewage treatment plant must first determine the acceptable amounts of various substances in effluent. To do so, they would have to look to the Federal Water Protection Act and Clean Water Act. Environmental laws provide guidelines for implementation (both means and deadlines) and monitoring of the upgraded sewage treatment facility.

Environmentally Sound Resource Management

The relationship between environmental law and resource management is perhaps not as clear-cut as the relationship between law and environmental problem solving. Nevertheless, the legal system contributes to sound resource management in important, if sometimes subtle, ways. Let us look briefly at the criteria of environmentally sound management and see how each is embodied in various environmental laws.

Stewardship Ethic. The Endangered Species Act gives to humans the role of caretaker of those species whose existence is threatened (often because of human activities) (Figure 27-5).

Biocentric Worldview. In attempting to preserve some areas in a natural state, free of human influence, the Wilderness Act of 1964 implicitly acknowledges the rights of nonhuman life.

FIGURE 27-5: American bald eagle. The bald eagle is endangered in all states except Alaska, Washington, Oregon, Minnesota, Wisconsin, and Michigan, where it is threatened.

Natural System Knowledge. The Food Security Act of 1990, with its Swampbuster and Sodbuster provisions, illustrates how a knowledge of natural systems (in this case, wetlands and the process of erosion) can result in legislation which is environmentally sound. The Farm Act prohibits farmers who drain and cultivate wetlands from receiving any federal assistance.

Political System Knowledge. A knowledge of the political system is vital to drafting and garnering support for environmental legislation. Many, if not most, environmental laws, for example, the Wilderness Act of 1964, the Federal Lands Policy and Management Act, and the Superfund Amendments Reauthorization Act, were enacted only after extensive lobbying and political compromise.

Sociocultural Considerations. Sociocultural considerations are perhaps best exemplified by the National Historic Preservation Act, which provides for the protection of our material cultural heritage. Other laws also consider sociocultural factors. For example, some states, such as Wisconsin and Alaska, have enacted hunting laws that place limitations (seasons, weapons, amount) on hunters, but make exceptions for native Americans. The exceptions recognize that subsistence hunting is important in the native American culture (Figure 27-6).

Natural and Social System Research. According to the National Environmental Policy Act, any pro-

posed project involving federal funds must undergo a comprehensive review to assess its environmental impact. Natural and social system data are essential to ensure a thorough and accurate assessment.

Economic Analysis. Environmental laws have attempted to put a monetary price on the value of resources such as clean air, clean water, and soil. Many environmental laws include provisions that require the economic analysis of various alternative solutions or economic penalties for failure to protect resources. For example, the Comprehensive Environmental Response, Compensation, and Liability Act (CERCLA) established a fund, financed by a tax on the production of toxic chemicals, to cover the costs of restoring natural resources that have been damaged by toxins. As we learned in Chapter 26, however, economic cost is not equivalent to the true value of a resource, and thus the tax imposed on the producer does not equal the value of the damaged resources or the ultimate costs to society. Clearly, our laws have not been able to account for the real value of natural resources and the costs to society of restoring resources and ecosystems damaged by pollution.

Public Participation. Several important environmental laws mandate public participation. CERCLA requires that community relations specialists be available to answer public concerns about the cleanup process and keep the public informed on its progress. NEPA requires that a draft of an environmental impact statement for a major government project be submitted to the public for review and comment and then be revised accordingly. The National Forest Management Act directed the Forest

FIGURE 27-6: Nez Percé fisherman on a tributary of the Salmon River, Idaho. Subsistence fishing and hunting are an integral part of the culture of many native American groups.

Service to draw up 50-year management plans for the national forests and submit them for public review. The Forest Service, citing delays in the implementation of the management plans, sought a court order to end the public review process. Their efforts failed, however, as the courts recognized the value of public input.

Environmental Education. The Resource Conservation and Recovery Act mandates the development and implementation of educational programs to promote citizen understanding of the need for environmentally sound solid waste management practices. The 1990 National Environmental Education Act (NEEA) supports a variety of educational activities. In fiscal year 1992, Congress appropriated $6.5 million for NEEA activities, 38 percent of which was designated for a grants program.

How Is the Field of Environmental Law Changing?

Proponents of environmental law have accomplished a great deal since the 1960s, when federal environmental protection laws were first enacted. Open burning, careless dumping of trash, unregulated industrial discharges to air and water, and uncontrolled draining of wetlands are far less common than they were three decades ago (Figure 27-7). One reason is the work of public interest legal organizations. Major public interest groups formed to do legal work associated with environmental issues include the Natural Resources Defense Council, the Environmental Defense Fund, the Sierra Club's Legal Defense Fund, and the Center for Science in the Public Interest.

The Natural Resources Defense Council (NRDC) has been called the most effective lobbying and litigating group on U.S. environmental issues. The NRDC has become a "shadow EPA" to ensure that comprehensive environmental protection laws are adopted and enforced. The NRDC was launched in 1970 by the Ford Foundation, which brought together a group of Yale Law School graduates and established lawyers who wanted to set up a public interest environmental law firm. The NRDC, now with 55,000 supporting members, has built its reputation by winning lawsuits and forcing agencies to write and rewrite regulations and to take enforcement action. Table 27-3 presents highlights from the NRDC's 20-year history.

In the 1970s environmental protection was chiefly directed at reducing smog, cleaning up the nation's water, and preserving wilderness areas. Some measures of air quality have improved significantly in many locales. Water quality improvements

FIGURE 27-7: Cars junked along a riverbank in the Great Smoky Mountains of North Carolina. Sights such as this one have become less common since the passage of environmental laws in the late 1960s and early 1970s.

have been modest overall, but spectacular in some places.

While air, water, and wilderness remain high priorities, a new area of concern has emerged. We are becoming increasingly aware of the threats to human and environmental health posed by toxic substances. The regulation of toxic substances is one of the fastest growing areas of environmental law. Environmental law and enforcement need to be innovative to meet the regulatory demands created by toxic pollutants. The following section illustrates a creative judicial solution to one problem caused by the release of a toxic pollutant.

Virginia Environmental Endowment: Translating National Concerns into Local Action

In 1977 the Allied Chemical Corporation was fined $13.2 million for polluting the James River in Virginia with 1.5 million gallons of the pesticide Kepone. Rather than automatically requiring Allied to pay this fine in the usual manner, Judge Robert R. Merhige, Jr., encouraged the company to develop a way for the fine to be used to benefit the people of Virginia.

Allied proposed that it make a voluntary contribution of $8 million to start an environmental fund for Virginia. Judge Merhige responded favorably to this suggestion and reduced Allied's fine by that amount. Thus, for the first time ever, the judicial system was instrumental in turning a routine matter of a fine into a constructive way to help the environment.

Since the Virginia Environmental Endowment (VEE) was established in 1977, it has provided $5,370,992 in grants for projects that improve the

Table 27-3
Highlights of Activities of the Natural
Resources Defense Council

1973	Compels the EPA to establish regulations restricting lead additives in gasoline
1974	Fights and wins a court order requiring the EPA to issue water pollution control regulations for all of the nation's industries
1975	Forces the Nuclear Regulatory Commission to adopt tougher regulations controlling the storage and disposal of uranium mill tailings
1976	Leads and wins the fight to ban the use of CFCs in aerosol products Stops three logging companies from clear-cutting slopes adjacent to Redwood National Forest
1983	Forces the National Steel Company to comply with air pollution control laws and pay $2.5 million in back penalties Spearheads a successful campaign to protect 40 million acres of fragile coastal areas in California, Massachusetts, and Florida from offshore oil leasing program
1984	Compels oil refineries to tighten pollution controls and reduce toxic discharges by 400,000 pounds per year Wins court action that will lead to strict controls on sulfur dioxide emissions from power plant smokestacks Wins Supreme Court victory giving the public the right to obtain chemical industry data on the health effects of pesticides
1985	Wins "citizen suit" that finds Bethlehem Steel Corporation liable for 350 pollution violations at its Chesapeake Bay plant Leads a coalition of citizens' groups in successful negotiations with the chemical industry to strengthen safety provisions of the federal pesticide law

	Helps to bring about congressional hearings and a Defense Department study on the severe environmental consequences of nuclear war
1986	Launches an agreement with the Soviet Academy of Sciences to enable scientists to monitor nuclear test sites in both countries Leads negotiations with the oil industry resulting in an agreement to protect 240 million environmentally sensitive acres in Alaska's Bering Sea Stops Norfolk Southern Corporation from building a coal transfer operation in Delaware Bay, an important ecological and recreational waterway
1987	Launches a campaign focused on daminozide (Alar), a plant growth regulator used on apples, to increase public awareness about the dangers, particularly to children, of pesticides used on fruits and vegetables
1991	Receives the Energy Conservation Award of the National Energy Resources Organization for its role in negotiating the Energy Efficiency Blueprint for California with utilities, state agencies, and other public interest groups Compels California to test and treat poor children for lead poisoning, as required by federal law. Because few states have as yet complied with the law requiring that children be protected from this known and severe health threat, the case is likely to set a crucial precedent and force compliance by reluctant state governments.

quality of Virginia's environment. The priorities of the endowment are:

- To implement the recommendations on water, land use, waste management, and the Chesapeake Bay made by the Governor's Commission on Virginia's Future
- To promote practical, cooperative strategies for managing toxic and hazardous substances in ways that protect public health, water quality, and the environment
- To enable citizens to identify local environmental problems, develop creative solutions, and share the results with other communities
- To support neutral mediation services for the resolution of environmental disputes
- To support legal research and education on the policies and laws that affect Virginia's environment

In 1981 the U.S. attorney in Philadelphia selected the VEE to administer a $1 million fund paid as a fine by the FMC Corporation as part of a court settlement. Similar in spirit to the fund that created the VEE, the $1 million is used to support projects that study water quality and the effects of water pollution on human health in the Ohio and Kanawha valleys.

The VEE funds programs that actively involve community members in developing solutions to environmental needs. In providing grants and loans for the development and implementation of innovative programs, it encourages new ideas that enable citizens, government, and business to work together to improve the environment (Table 27-4).

Many of the projects that the VEE has supported could easily be adapted for use in other parts of the country. For example, the toxic substances educa-

Table 27-4
Selected Programs of the Virginia
Environmental Endowment

TOXIC SUBSTANCES

MCV Division of Clinical Toxicology and Environmental Medicine. After several years of supporting Kepone research at the Medical College of Virginia (MCV), VEE awarded a challenge grant in 1983 to establish MCV's Division of Clinical Toxicology and Environmental Medicine.

Environmental Defense Fund. To bring national scientific and legal expertise to Virginia, VEE granted funds to the Environmental Defense Fund to establish a Virginia office in 1980. The Environmental Defense Fund provides technical assistance and monitors the work of state and federal agencies to protect Virginia's environment.

LOCAL ENVIRONMENTAL IMPROVEMENT

Virginia Water Project, Inc. The Virginia Water Project published and distributed to local government officials the first handbook of its kind describing steps to take in managing a drinking water emergency.

Student Environmental Health Project. The project provides year-round laboratory services for soil and water analysis and places qualified students in communities requesting technical assistance in documenting and resolving environmental health problems.

ENVIRONMENTAL MEDIATION

Virginia Toxics Roundtable. This group works to avoid conflict and build consensus on environmental issues among industrial, environmental, and government interests. Members develop solutions to the problems posed by toxic substances. A significant achievement was working with legislators to secure passage of the Virginia Hazardous Waste Facilities Siting Act of 1984.

Institute for Environmental Negotiation at the University of Virginia. Founded in 1981, the institute has resolved more than 50 cases, ranging from land use to low-level radioactive waste management.

ENVIRONMENTAL LAW AND PUBLIC POLICY

Virginia Journal of Natural Resources Law. Published by the University of Virginia Law School, this journal is one of a few in the nation that is devoted exclusively to environmental and natural resources law.

University of Virginia. The VEE awarded the University a two-year grant in 1990 to support a visiting professor in the T. C. Williams School of Law and to develop a foundation course in environmental law.

WATER QUALITY PROGRAM

Ohio River Basin Research and Education Consortium. The consortium of academic institutions and corporations promotes research and education on water resources of the Ohio River basin.

Kentucky Governmental Accountability Project. Created to monitor the enforcement of water quality statutes in Kentucky, this project provides technical assistance and legal information to community organizations, landowners, farmers, and environmentalists.

Friends of the Rappahannock. In 1990, this Fredericksburg, Virginia organization was awarded a two-year grant to support water quality, conservation, and land use programs.

tional curriculum developed by Virginia Commonwealth University could be used by middle- and high-school students nationwide. The water contamination notification handbook detailing the steps to take to manage an emergency if water supplies are contaminated could be used in any community.

What Is Environmental Dispute Resolution?

Many people have become dissatisfied with litigation because lawsuits are costly, it often takes years to reach a decision, decisions can be appealed, and decisions may or may not be satisfactory to all or any of the parties involved. Dissatisfaction with litigation is leading many people to seek negotiated reso-

lutions to environmental disputes. Interestingly, however, negotiation often relies on litigation; an actual or threatened lawsuit may be necessary before a party is willing to come to the table. **Dispute resolution** is the process of negotiation and compromise by which disputing parties reach a mutually acceptable solution to a problem

A neutral third party, called a **mediator,** is called in to facilitate negotiations—to help the parties see their common interests, define the problem, and find a solution. The mediator has no authority to impose a settlement, but can only assist parties in settling their differences. The mediated dispute is settled when the parties themselves reach what they consider to be a workable solution. The mediator is likely to spend most of his or her time in meetings with the parties alone, helping them to explore positions and formulate alternatives, advising on the mediation process, carrying messages among the parties, and "trying out" offers.

Dispute resolution becomes a feasible option under certain conditions. The conflict must have reached the point at which the issues are clearly defined. The parties to the dispute must feel that a balance of power exists between them and that they cannot achieve their specific objectives without negotiations. The parties must volunteer to participate in negotiations. There should be a formally agreed-upon agenda for discussion focusing on specific resource and pollution issues; disputes over values cannot be solved by negotiation. Both sides must be willing to explore new ideas and possibilities and to negotiate honestly with each other.

What Are the Advantages and Disadvantages of Dispute Resolution?

Compared to litigation, dispute resolution ensures a better chance that all parties will realize their objectives, hastens resolution, and normally costs less. Mediation tends to promote a more accurate view of problems than litigation. The parties are encouraged to discuss the uncertainties of their position, to look for points of agreement, and to build a better understanding of the opposing position. Lawsuits, in contrast, encourage each party to bring up the evidence that favors its goal and to ignore or dismiss unfavorable information that might weaken its stand. Lawyers sometimes increase the distance between parties; mediators assist parties to find the common ground necessary to reach a solution.

Negotiation has some drawbacks. First, funding for negotiation services is inadequate. Second, some parties fear that they will lose power and status if they enter negotiations, because doing so implies that they are willing to compromise. Third, some people lack faith in the outcome of negotiations, because, unlike court orders, resolutions cannot be imposed and legally enforced. Even so, dispute resolution is a viable alternative to litigation. In the future we can look forward to increased confidence in the negotiation process.

How Does Dispute Resolution Contribute to Environmental Problem Solving and Management?

Environmental Problem Solving

Dispute resolution brings parties together to find mutually acceptable solutions to environmental conflicts. Usually, those conflicts involve the manage-

ment of a particular resource. Because dispute resolution contributes to environmentally sound resource management, it may help to avoid or minimize environmental problems.

Environmentally Sound Resource Management

Dispute resolution techniques can be effective in cases in which the management of resources involves several different managing agencies and different groups of users. Coordinated resource management planning (CRMP) is one dispute resolution strategy used to develop resource management plans. It encourages communication, compromise, and commitment. CRMP brings together a planning team of managers and users who work, from beginning to end, to develop the resource management plan (Figure 27-8). Through open communication, the participants gain a greater understanding of the needs and desires of others, while informing the others of their own needs and desires. Once they have reached a basic level of understanding, they find it easier to reach a compromise or consensus. If all parties feel that they had an equal hand in reaching the consensus, it is more likely they will remain committed to the plan.

The value of CRMP is that it can avoid or resolve conflicts among various users, among managers, and between users and managers, replacing animosity with a willingness to compromise and seek a consensus. In this respect, it differs markedly from other planning strategies which often deteriorate into vocal dissension, power struggles, or lawsuits and countersuits.

CRMP is not a magic process. Its success depends on hard work, the willingness of all parties to reach a compromise, and, often, the skill of the mediator.

FIGURE 27-8: Slash and burn area, Cascade Mountains. Environmental conflicts, such as disputes over the best use of wilderness areas, can sometimes be worked out through compromise and negotiation.

Summary

Environmental law is that aspect of the legal system that governs the activities of persons, corporations, government agencies, and other public and private groups in order to regulate the impact of the activities on the environment and on natural resources. Environmental protection resides in common and statutory law. Common law is a body of written and unwritten rules based on precedent. Cases involving common law are based on four legal axioms: nuisance, trespass, negligence, and strict liability. Statutory law is the body of specific statutes governing actions and procedures passed by the legislative branch of a local or national government.

Congress has enacted several important environmental and resource protection laws. These laws set pollution level standards, screen new substances for safety before they are widely used, evaluate the environmental impacts of activities before they are undertaken, protect various ecosystems, resources, or species, clean up environmentally unsound hazardous waste disposal sites, and prevent new unauthorized hazardous sites from forming. Each environmental law specifies how it is to be enforced and who is to enforce it. The Environmental Protection Agency (EPA) is perhaps the best known of the federal agencies charged with enforcing environmental policies. Each state also has its own agencies.

Environmental laws are not always enforced effectively. Allocation of authority may be unclear. The agency charged with enforcing a law may disagree with the law. The Administration or Congress may not provide the funds needed for enforcement. An agency may have a conflict of interest that keeps it from enforcing a law. Finally, the burden of administrative duties may hamper an agency's ability to enforce laws.

Environmentally sound laws are important to environmental problem solving and management. The legal system comes into play at each step of the problem-solving process. Certain environmental laws illustrate and support the stewardship ethic, a biocentric world view, knowledge of natural systems, knowledge of political systems, sociocultural considerations, natural and social system research, economic analysis, public participation, and environmental education.

Environmental law has advanced significantly since the 1960s, when the first such federal laws were enacted. The Natural Resources Defense Council (NRDC) has been called the most effective lobbying and litigating group on U.S. environmental issues. One of the most innovative legal solutions to an environmental problem is the Virginia Environmental Endowment, funded by the Allied Chemical Corporation as part of a legal settlement.

Dispute resolution is the process of negotiation and compromise by which disputing parties reach a mutually acceptable solution to a problem. A neutral third party, a mediator, facilitates negotiations. The parties in dispute must volunteer to participate in negotiations. The mediator has no authority to impose or enforce a settlement. Coordinated resource management planning is one dispute resolution strategy used to develop environmentally sound resource management plans.

Key Terms

common law	nuisance
compensation	precedent
dispute resolution	statutory law
environmental law	strict liability
injunction	tort
mediator	trespass
negligence	

Discussion Questions

1. What is environmental law, and how is it formed?

2. Why are environmental laws not always enforced effectively?

3. Imagine that you work for an environmental consulting firm. How does environmental law affect the way you approach environmental problem solving?

4. How do environmental laws reflect society's priorities for environmental management? Give at least three examples.

5. What is environmental dispute resolution? If you were a mediator, what steps might you take to help opposing parties resolve a dispute without litigation?

Environmental Education

The objective is to teach the student to see the land, to understand what he sees, and to enjoy what he understands.

Aldo Leopold

Do not try to satisfy your vanity by teaching a great many things. Awaken people's curiosity. It is enough to open minds; do not overload them. Put there just a spark. If there is some good flammable stuff, it will catch fire.

Anatole France

And the world cannot be discovered by a journey of miles, no matter how long, but only by a spiritual journey, a journey of one inch, very arduous and humbling and joyful, by which we arrive at the ground at our feet, and learn to be at home.

Wendell Berry

Learning Objectives

When you have finished reading this chapter, you should be able to:

1. Discuss what is included in environmental education.

2. Explain the difference between formal and informal environmental education.

3. Describe two approaches to integrating environmental science into school curricula.

4. Explain how environmental education affects environmental problem solving and management.

In a sense this entire textbook, not just this chapter, has been about environmental education. All of us who teach or study environmental science are taking an active role in educating ourselves and others about the world in which we live. In this chapter we look specifically at the discipline of environmental education—what it is, how it is taught, and how it contributes to environmental problem solving and environmentally sound management.

What Is Environmental Education?

Along with the spread of environmental awareness and activism into a broader spectrum of society in the 1960s came the emergence of environmental education as a discrete entity (Figure 28-1). Many people attempted to define this emerging discipline.

FIGURE 28-1: Environmental rally, Washington, D.C., Earth Day 1970. The environment movement of the late 1960s and early 1970s stimulated environmental education.

One definition, developed under the leadership of William Stapp, past president of the North American Association for Environmental Education, appeared in the first issue of *Environmental Education* in 1969: "Environmental education is aimed at producing a citizenry that is knowledgeable concerning the biophysical environment and its associated problems, aware of how to help solve these problems, and motivated to work toward their solution." This definition called for appreciably more teaching and learning about the environment and the education of all citizens, in and out of school, children as well as adults. It gave as the goal of education the development of skills needed to solve environmental problems and the motivation of people to be actively involved in environmental problem solving.

In the 1970s the concept of values became part of the definition of environmental education. Environmental education seeks to awaken and explore people's values. While we may intuitively value the natural world and the services it provides, the structure of contemporary society often clouds our intuition. As a result, we lose sight of our real relationship to nature and of our dependence upon the natural world. Our technocratic society separates, even isolates, us from the natural world. But it is an isolation that we can break through. Sigurd Olson, a wilderness ecologist, proponent of wilderness preservation, and author of numerous popular and scientific articles, once observed that "Awareness is becoming acquainted with the environment, no matter where one happens to be. Man does not suddenly become aware of or infused with wonder; it is something we are born with. No child needs to be told its secret; he keeps it until the influence of gadgetry and the indifference of teen-age satiation extinguish its intuitive joy." This inherent joy and awareness must be acknowledged and nourished (Figure 28-2). These feelings are the most basic form of an ecological consciousness, and they are the first planks laid down in the construction of an environmentally literate citizen.

In the United States, kindergarten through twelfth grade programs provide the opportunity for fostering environmental literacy. Educators generally agree that environmental education should be a multidisciplinary program incorporated throughout the school years. Unfortunately, there is no comprehensive, uniform base for curriculum development and implementation. Often, programs are loosely organized and lack a clear sense of direction.

A widely accepted goal for environmental education might provide this needed sense of direction for curriculum development and implementation. Such a goal was formulated at the World Intergovernmental Conference on Environmental Education in

FIGURE 28-2: A group with the Peffer-Western Environmental Education Program, Miami University, Oxford, Ohio. "If I had influence with the good fairy who is supposed to preside over the christening of all children," wrote biologist Rachel Carson, "I should ask that her gift to each child in the world be a sense of wonder so indestructible that it would last throughout life, as an unfailing antidote against the boredom and disenchantments of later years, the sterile preoccupation with things that are artificial, the alienation from the sources of our strength."

1977, held in Tbilisi, Georgia, USSR: "To develop a citizenry that is aware of, and concerned about, the total environment and its associated problems, and which has the knowledge, attitudes, motivations, commitment, and skills to work individually and collectively toward solutions of current problems and the prevention of new ones."

In order to develop an environmentally aware and concerned citizenry, educators must convey an appreciation for the natural world, an understanding of the structure and function of nature, and a mastery of problem-solving skills. Objectives must be directed toward these higher levels of cognitive thinking. Research has shown that the majority of environmental education programs focus on fostering environmental awareness rather than developing problem-solving skills. Educators have responded to this situation by developing the following broad goals for curriculum development: ecological foundations (programs based on natural science); conceptual awareness (issues and values); issue investigation and evaluation; and environmental action skills (training and application).

These goals represent a comprehensive effort to standardize environmental education and foster problem-solving, action-oriented skills, but awareness, appreciation, and a basic understanding of ecological processes remain important components of environmental education (Figure 28-3). All components—from simple appreciation of natural process-

FIGURE 28-3: An elementary school teacher discovers some of the wonders to be found in a stream during a hands-on environmental science workshop for teachers. Such activities provide teachers with the knowledge and techniques they need to help students experience and understand the natural world.

es to sophisticated problem-solving skills—are inter-related and compatible.

Implementing the goals of environmental education in the classroom remains a challenge. But the gap between theory and practice narrows as more public schools incorporate environmental education into their curricula. One prominent source of environmental education materials is the Institute for Earth Education, an international, nonprofit, volunteer organization based in Warrenville, Illinois. The institute's programs focus primarily on understanding basic ecological systems and understanding how these systems affect people's daily lives. They are designed to help students understand their place in the environment and learn to live in harmony with the earth. Programs for upper elementary students include Sunship Earth™ and Earthkeepers™. Sunship Earth™, first published in 1979, is a complete 5-day outdoor experience with pre- and post-classroom activities. Its goals are to foster planetary awareness, sensory awareness, environmental consciousness and ecological understanding of seven key concepts: energy flow, cycles, diversity, communities, interrelationships, change, and adaptation. Earthkeepers™, a 2 1/2 day program, requires fewer leaders and props than Sunship Earth™. It stresses the key concepts of energy flow, cycles, interrelationships, and change.

The New Alchemy Institute in East Falmouth, Massachusetts has created a more agriculturally oriented program called the Green Classroom, which has been integrated into the science curriculum of nearby elementary schools. The Green Classroom is an organic gardening program combining classroom instruction and hands-on activities. Students select and order vegetable seeds, tend gardens, build compost heaps, and chart temperature and growth. Lessons in social studies, mathematics, art, and language are integrated into the program. In 1989, all 16 fourth grade classes in Falmouth Public Schools participated in the Green Classroom program; an on-site garden was located at each of the district's four elementary schools.

The School for Field Studies (SFS), based in Beverly, Massachusetts, offers environmental education programs at sites around the world (Figure 28-4). SFS operates five centers where students participate in semester or month-long summer courses. They are the Center for Rainforest Studies in Australia; the Center for Marine Resource Studies in the Caribbean; the Center for Wildlife Management Studies in Kenya; the Center for Marine Mammals Study in Baja, Mexico; and the Center for Sustainable Development Studies in Costa Rica. Eleven additional satellite sites located at environmental "hot spots" around the world enable students to examine specific issues, such as diminishing biological diversity, within the framework of a traditional academic discipline, such as ecology. In fact all SFS programs investigate one or more environmental problems in the field from the perspectives of biological sciences and social sciences, such as anthropology, political science, and economics. Open to students 16 years and older, SFS programs combine theoretical training and hands-on experience. Students receive college credit for their field experience.

FIGURE 28-4: Rain forest study, School for Field Studies, Queensland, Australia.

The interdisciplinary approach followed by SFS is also used by the Institute of Environmental Sciences (IES) at Miami University in Oxford, Ohio. This graduate program offers courses in environmental policy and law as well as courses in ecology. The entire program stresses the importance and value of using other disciplines besides science to understand and solve environmental problems. Its introductory course, Principles and Applications of Environmental Science, is open to undergraduates as well as graduate students.

Another college-level program is the Au Sable Institute of Environmental Studies in Michigan. The Au Sable Institute approaches environmental science from a Christian perspective, emphasizing stewardship. Graduate and undergraduate students from both Christian and secular colleges and universities are eligible for admission. The Institute offers certificates for naturalists, land resources analysts, water resources analysts, and environmental analysts, although students may choose not to study for one of these certificates.

How Is Environmental Education Taught Formally?

There are at least two approaches to adding environmental education to the school curriculum. Some teachers advocate adding courses called "environmental education." The advantages of this approach are that schools can hire teachers for such classes who are trained in presenting environmental information to students and that teachers of other subjects would be unaffected by this addition to the school's curriculum. At Wisconsin's Kaukauna High School, students in the Environmental Action course must first have a prerequisite class in ecology in order to obtain the basic scientific knowledge required to understand, and act on, environmental issues. Once enrolled in the Environmental Action course, they apply their ecological knowledge to real-life situations. In the past, students have analyzed the recyclable material content of the school's trash to help bring Kaukauna High into compliance with a new state recycling law. Another project entailed creating an ecologically appropriate land-use design for the grounds of a nearby dog racing track.

A major disadvantage of establishing separate courses for teaching environmental principles is that students might not see how important these ideas are to them. For example, will children raised in the city, with fenced concrete playgrounds and window-boxed flowers, appreciate the importance of rain forests, when those forests seem so far removed from their lives? People care about places and things that matter to them. If they do not have some background in general science, geography, or genetics, can they understand the immediate importance to them of the rain forest? If they have never experienced the silence of a forest or the power in the roar of the sea, are they likely to change the way they live to protect these wild places? Education must touch students—through poetry, music, physics experiments, and experience. To be most effective, education should be taken out of the classroom and into the world so that people see the effects of their action, or inaction, on their environment. It is important that teachers accept environmental education as a critical part of the school curriculum so that students experience the joy of being healthy and young in a healthy environment and become creative in resolving environmental problems (Figure 28-5). The earth can be helped only by people who have the will, self-esteem, and knowledge to make a difference. Michael J. Cohen, author of *Our Classroom Is Wild America*, *Across the Running Tide*, and *Prejudice Against Nature: A Guidebook for the Liberation of Self and Planet*, has said, "It makes sense to address the wholeness of a person (their life processes, sensations, feelings, and intellect) if a person is to be fully educated. Active experiences that expand one's consciousness of life educate most honestly and reliably. They are the best teachers, making possible the discovery and practice of new foundations for survival of ourselves and the planet."

FIGURE 28-5: A group of teachers in hands-on a science workshop. The success of environmental education depends on motivated teachers who are interested in and knowledgeable about the natural world. Effective environmental educators have retained their childlike sense of wonder and are able to share it with their students.

Instead of establishing separate environmental education courses, some teachers advocate making environmental issues an integral part of all courses in the curriculum. Environmental topics would be studied as they are appropriate to each class. The Rachel Carson Project in the Corvallis, Oregon, school system incorporates environmental topics into standard high school courses:

- English. Read essays and poems from *Audubon* magazine, satirical pollution poems by comedian Henry Gibson, and Edward Abbey's *Desert Solitaire*.
- Drivers' education. Become familiar with Detroit's efforts to reduce automobile pollution and alternatives to internal combustion engines.
- Typing. Use script from various environmental publications.
- Physics. Apply the laws of thermodynamics to the problem of producing energy in the future.

For a truly comprehensive environmental education program, environmental topics should be introduced in all grades and in all courses of a school curriculum. The elements of any environmental curriculum should be integrated into the existing system so skillfully that the elements are not recognizable as separate units. Environmental education should be a daily process of increasing awareness of the environment. Noel McInnis, a leading educator, feels that a root of our environmental problems is that we have not learned to think ecologically. If teachers in all disciplines regularly used environmental studies as their vehicles of instruction in working toward their normal course objectives, students would automatically become aware of the importance of the environment.

To assure that the nation's students receive a sound environmental education, educators point out that schools must stress the importance of the discipline and help teachers incorporate environmental issues into their specific curriculum. Unfortunately, only a few states—Wisconsin, Arizona, and Maryland—mandate environmental education in schools; a handful of others encourage, but do not require, it. Wisconsin is recognized as a leader in the field of environmental education. To be certified, new elementary teachers must have completed a college-level environmental program and have passed competency tests in environmental science. Every Wisconsin school system must have in place plans that incorporate environmental science across the curricula. Currently, the state is evaluating the districts' plans. Many other states have looked to Wisconsin for help in preparing curricula or in developing environmental education legislation. One promising sign on the environmental education front was the resurrection in 1990 of the Office of Environmental Education under the auspices of the federal Environmental Protection Agency. A victim of neglect, the office had languished and died nearly a decade earlier during the Reagan administration. Its responsibilities include awarding education grants, creating federal internships, and recognizing outstanding environmental achievement through an awards program.

Environmental Education at Oak Park–River Forest High School

The following shows how one group of students took the responsibility for environmental education into their own hands. Imagine a group of high school students establishing their own environmental information service to increase environmental awareness among their classmates, younger students, and members of the community. Students in the community of Oak Park, Illinois, a suburb of Chicago, got together in 1970 to plan a conservation workshop to commemorate Earth Day. For this workshop, classes were rescheduled so that students could hear speakers in all disciplines discuss environmental concerns. An attorney spoke to history classes about environmental legislation; a member of the President's Environmental Board spoke to science classes about the technical aspects of water pollution; a member of a conservation group talked to English classes about ecological concepts and personal life-styles. The students were shown that environmental education does not belong only in the biology lab.

The conservation workshop had a permanent impact on the high school. Today, the high school course offerings include environmental science and field biology courses, earth science and physical science courses, and advanced placement science courses that stress environmental topics. In addition, the students run a pollution control center, where they provide information about environmental issues, sending free pamphlets, periodicals, and books on almost any environmental topic to other students. They also share their knowledge of environmental issues by giving lectures at local elementary schools that do not have environmental courses. Students promote environmental awareness in the community through newspaper articles, a newsletter, displays, posters, photographs, and television and radio coverage. Many students speak to local civic and service groups about pertinent environmental issues and participate in national and international environmental conferences. Supported by grants from local organizations and corporations, these students are taking an active role in their education and sharing their knowledge with others.

How Is Environmental Education Taught Informally?

Many target groups for environmental education are outside of the formal school system: local elected officials, civic groups, industrial and commercial leaders, general adult audiences, and the media. It is important that these groups become environmentally aware because their action or lack of action influences local environmental quality.

To reach target groups outside of the formal school system, informal educational programs are becoming increasingly important. Zoos, nature centers, museums, aquariums, and wildlife refuges are just some examples of the diverse forums in which informal environmental education programs are offered. In addition to reaching those outside the formal educational system, such programs can supplement school programs for children. For children and adults alike, experiential activities available at zoos, museums, and parks engage the imagination and stimulate learning. Many educators are turning toward issues-related programs in informal forums in the hope that these will be a positive force in changing societal attitudes and behaviors.

Significant informal educational programs are also offered by national and international environmental organizations. For example, Earthwatch, a nonprofit institution founded in 1971, sponsors scientists, artists, and teachers undertaking research designed to improve human understanding of the planet, the diversity of its inhabitants, and the quality of life on earth. Based in Boston, Earthwatch also has offices in Los Angeles; Oxford, England; and Sydney, Australia. Earthwatch members have the opportunity to volunteer for the research expeditions sponsored by the organization. Volunteers of all ages and from all walks of life join expeditions such as the efforts to save leatherback turtles in the Caribbean, study the rain forest canopy in Australia, and unearth Inca villages in the Andes of Peru. Groups such as the Sierra Club, the National Audubon Society, the National Wildlife Federation, Conservation International, Worldwatch Institute, and the Cousteau Society can act as invaluable vehicles for environmental education. The Audubon Society, for example, offers several ecology camps and workshops every year for participants ranging from elementary school students to adults. Participants can go on one- or two-week expeditions or stay at special camps to learn about animals, plants, ecology, and/or nature photography. The National Wildlife Federation sponsors a program called COOL IT! for college-level students. This program offers resources such as directories, campus visits, and lobbying guides to member groups as well as information on careers, fundraising, cultural diversity, and many specific environmental issues. Members of COOL IT! can use these resources to educate other students and faculty on their campuses about environmental issues.

Children have begun forming environmental organizations of their own as well, many of which include education as a vital element. Kids for Saving Earth (KSE) was started by 11-year-old Clinton Hill of Plymouth, Minnesota, who died of cancer in 1989. About 3,600 KSEs have been established in the United States and in 13 other countries. The objectives of KSE are to increase awareness of environmental issues, practice environmentally friendly techniques, and educate both adults and children. Club activities include discussions about current environmental topics, recycling projects, and letter-writing campaigns to public officials, including the president. Members also work to educate their families and friends about the three R's—reduce, reuse, and recycle.

Nonprofit corporations are also contributing to informal environmental education efforts. For example, the San Francisco Bay–Delta Aquatic Habitat Institute (AHI) is an independent, nonprofit corporation that researches and monitors pollution in the San Francisco Bay–Delta area. In addition to its research functions, AHI makes information on the area available to all interested individuals and organizations. AHI and the National Oceanic and Atmospheric Administration have developed a portable computer-based public education display, *Exploring the Estuary*, which provides information on both the ecology of the area and on complex management issues.

Environmental education is more than a passing movement. While still in the developmental stage, it has the potential to direct the environmental consciousness of society. But more effort is needed to weave environmental education throughout the social fabric. Only then will we achieve a sustainable society that embraces and promotes an environmental ethic.

How Does Environmental Education Contribute to Environmental Problem Solving and Management?

The Tbilisi declaration resulted from the world's first intergovernmental conference on environmental education. The declaration established goals for environmental education. These goals illustrate how the discipline can and does contribute to environmental problem solving and management.

- Consider the environment as a whole; show how biological and physical phenomena affect social,

Focus On:
Some Overall Imperatives of the Environmental Education Movement

William B. Stapp, Professor of Environmental Education, School of Natural Resources, University of Michigan

Environmental problems exist in all countries of the world, at every stage of economic development and in all political ideologies. It is evident that there can be no hope of a workable solution to environmental problems unless education is suitably modified to enable people from all walks of life to comprehend the fundamental interaction between humans and their environment, problems in this interaction, and potential changes to avoid or at least reduce these problems.

Within the attitudes and actions of our human population and its organizations lie the behavioral roots of such ills as pollution, wasted energy, and destruction of the environment. There is a general lack as yet of any global ethic effectively encompassing the world environment—an ethic that espouses attitudes and behaviors on the part of individuals and societies and that should be consonant with humanity's place and critical role in the biosphere.

In order to move toward a global ethic, I would like to outline a few environmental education imperatives for consideration. They are representative of the thinking of many environmental educators in different regions of the world and have been expressed in thoughtful papers, meaningful discussions, meetings, conference proceedings, and publications. These imperatives are not all-inclusive, but represent some important thoughts that could be modified and expanded upon.

1. Environmental education should design programs to reach three major audiences:

   a. The world's general public, from early childhood through adulthood and beyond, in both the formal and the informal education sectors

   b. Specific professional and social groups whose actions have an influence on the environment (engineers, architects, economists, planners, policy makers, manual workers, industrialists, and the like)

   c. Scientists, technologists, and other professionals who deal directly with environmental problems (agriculturists, horticulturists, foresters, biologists, hydrologists, ecologists, and others).

2. Environmental education should be a comprehensive, continual, life-long process that is always responsive to changes in a rapidly changing world. It should prepare the individual for life through an understanding of the major issues of the contemporary world and the acquisition of skills and attitudes needed to play a productive role in improving life and protecting humanity's and nature's environment with due regard given to ethical values.

3. Environmental education should forge ever closer links between educational processes and real life, building activities around the environmental issues that are faced by particular communities and nations and using an interdisciplinary, comprehensive approach to analyze those issues.

4. Environmental education should show quite clearly the ecological, economic, political, social, and educational interdependence of the modern world, in which decisions and actions by the different countries can have international repercussions. It should help

economic, political, technological, cultural, historical, moral, and esthetic disciplines.

- Integrate knowledge from the disciplines across the natural sciences, social sciences, and humanities.

- Examine the scope and complexity of environmental problems and thus the need to develop critical thinking and problem-solving skills and the ability to synthesize data from many fields.

- Develop awareness and understanding of global problems, issues, and interdependence; help people think globally and act locally.

- Consider both short- and long-term effects of solutions to problems of local, regional, national, and international importance.

- Relate environmental knowledge, problem solving, values, and sensitivity at every level.

- Emphasize the role of values, morality, and ethics in shaping attitudes and actions affecting the environment.

- Stress the need for active citizen participation in solving environmental problems and preventing new ones.

- Enable learners to play a role in planning their learning experiences and provide an opportunity for making decisions and accepting their consequences.

- Make environmental education a life-long process, including formal methods from preschool to post-secondary level and informal methods to reach all ages and education levels.

Focus On: Some Overall Imperatives of the Environmental Education Movement, by William Stapp

to develop a sense of responsibility and solidarity among nations and regions as a foundation for a new international order which will guarantee conservation and improvement of humanity's and nature's environment.

5. Environmental education should be an integral part of every nation's educational process. It should be centered on practical environmental issues, be interdisciplinary, aim at building up a sense of values in people, contribute to the public well-being, and be concerned with the survival of the human species. Its focus should reside mainly in the enlightened initiative of the teachers and the receptiveness of the learners, whose involvement and action should be guided by both immediate and future concern on the local, national, and global levels.

6. Environmental education should recognize that public involvement and participation in planning and decision making are valuable means of testing and integrating economic, social, and ecological objectives. Further, public participation heightens consciousness-raising, fosters critical reflection on information collected and reviewed, facilitates group processing and conflict resolution, and promotes empowerment of participants for committed actions by people (through collective planning action) to exercise improved control over the decisions, resources, and institutions that affect their lives.

7. Environmental education should make effective use of the mass media—radio, television, and the press—both inside and outside the school system, involving the stimulation and production of appropriate programs, the training of mass media personnel in environmental matters, and the coordination of the ministries and other agencies and institutions involved, both governmental and nongovernmental.

8. Environmental education should work toward the effective development and interlinking of centers to help promote pertinent research, programs, and policies. These centers should work with existing public and private institutions and with all facilities available to society for the education of the population—including the formal education system, different forms of informal education, and the mass media.

9. Environmental education should adopt the goals established by the Tbilisi Conference, which were affirmed by the North American Association for Environmental Education (NAAEE) in 1983.

10. Environmental education should strive to integrate the above imperatives into international and national policies and plans for the development of education of the general public, of appropriate social and professional groups, and of specialists in environmental fields, while promoting pertinent educational reform and concomitant innovation throughout the world.

suggests ways that environmental education can be structured to help improve the quality of our human environment.

Environmental Education: Where Do We Go from Here?

Becoming environmentally educated does not mean that you must always fight problems and never find joy in the natural world. It does mean having the tools to fight for what you love and loving the world that you live in.

It is easy to focus on what is wrong with the environment rather than on what is right with it. Too much bad news can be overwhelming. Because there are so many problems and so many are very complex, we may feel powerless to do anything about any of them.

We need to view nature, not as so many overwhelming problems that only specialists can understand, but as a whole, living, nurturing entity that can benefit immeasurably from loving concern and attention. The biologist Rachel Carson merged the divergent specialties of modern science into a comprehensive whole. Carson's *Silent Spring* revealed that, by disregarding the effects that our activities have on natural systems, we are destroying their delicate balance. We need to step out of our laboratories, offices, classrooms, and homes and develop anew our sense of the earth (Figure 28-6). Try the exercise presented in Box 28-1: Developing a Sense

Box 28-1:
Developing a Sense of Place

1. Trace the water you drink from precipitation to tap.

2. How many days until the moon is full (plus or minus a couple of days)? (From your birthday this year.)

3. Describe the soil around your home.

4. What were the primary subsistence techniques of the culture(s) that lived in your area before you?

5. Name five native edible plants in your bioregion and their season(s) of availability.

6. From what direction do winter storms generally come in your region?

7. Where does your garbage go?

8. How long is the growing season where you live?

9. On what day of the year are the shadows the shortest where you live?

10. Name five trees in your area. Are any of them native? If you can't name names, describe them.

11. Name five resident and five migratory birds in your area.

12. What is the land-use history by humans in your bioregion during the past century?

13. What primary geological event/process influenced the land form where you live? (Explain.)

14. What species have become extinct in your area? (Minimum of three.)

15. What are the major plant associations in your region?

16. Trace your wastewater from toilet to precipitation

17. What spring wildflower is consistently among the first to bloom where you live?

18. What kinds of rocks and minerals are found in your bioregion?

19. List 5 adaptations for winter survival that plants and animals exhibit in your area.

20. Name some beings (nonhuman) which share your place.

21. Do you celebrate the turning of the summer and winter solstice? If so, how do you celebrate?

22. How many people live next door to you? What are their names?

23. How much gasoline do you use a week, on the average?

24. What energy costs you the

most money? What kind of energy is it?

25. What developed and potential energy resources are in your area?

26. What plans are there for massive development of energy or mineral resources in your bioregion?

27. What is the largest or closest wilderness area in or near your bioregion?

28. Interview a conservation, environmental, or ecology professional in your bioregion for their sense of place.

29. Interview an environmental activist from your region for their sense of place.

30. How long has it been since you've walked beyond where the sidewalk ends (why so long)? Take a walk in a special place on a special day with a special person (optional). Describe your experience.

Adapted from a list found in *Deep Ecology: Living as if Nature Mattered*, Devall and Sessions, Gibbs, Smith, Inc. Salt Lake City, Utah

FIGURE 28-6: Where the sidewalk ends, the journey of discovery begins.

of Place, to determine how finely tuned your sense of the earth is.

Environmental education is a powerful tool. It can shape a knowledgeable, active citizenry prepared with the skills and desire to solve environmental problems—a citizenry that understands, accepts, and values its connection with the earth and therefore works to maintain a relationship with the planet that benefits both the earth and humankind.

Because the world is ever changing, the process of environmental education must be life-long. (Figure 28-7). "The waves echo behind me," wrote Anne Morrow Lindbergh. "Patience—Faith—Openness is what the sea has to teach. Simplicity—Solitude—Intermittency . . . But there are other beaches to explore. There are more shells to find. This is only a beginning."

WHERE THE SIDEWALK ENDS

Shel Silverstein

There is a place where the sidewalk ends
And before the street begins,
And there the grass grows soft and white,
And there the sun burns crimson bright,
And there the moon-bird rests from his flight
To cool in the peppermint wind.

Let us leave this place where the smoke blows black
And the dark street winds and bends.
Past the pits where the asphalt flowers grow
We shall walk with a walk that is measured and slow,
And watch where the chalk-white arrows go
To the place where the sidewalk ends.

Yes, we'll walk with a walk that is measured and slow,
And we'll go where the chalk-white arrows go,
For the children, they mark, and the children, they know
The place where the sidewalk ends.

FIGURE 28-7: The life-long process of learning from nature. One of the benefits of an environmental education is that it allows one "To see a world in a grain of sand / And a heaven in a wild flower / Hold infinity in the palm of your hand / And eternity in an hour" (William Blake).

Summary

Environmental education seeks to awaken and explore a person's values. It attempts to educate all citizens about the environment and its problems and to motivate them to solve these problems. Educators generally agree that environmental education should be a multidisciplinary program incorporated into the curriculum in all grades.

In order to develop an environmentally aware and concerned citizenry, educators must convey an appreciation for the natural world, an understanding of the structure and function of nature, and a mastery of problem-solving skills. Most current environmental education programs focus on fostering environmental awareness rather than developing problem-solving skills. Educators have responded to this situation by developing the following broad goals for curriculum development: ecological foundations (programs based on natural science), conceptual awareness (issues and values), issue investigation and evaluation, and environmental action skills (training and application). Some teachers advocate adding a separate environmental field to the curriculum. Others advocate making environmental issues an integral part of all courses in the curriculum.

Many groups outside of the formal school system, such as elected officials, business leaders, and members of the media, are important targets for environmental education because their actions or inaction can influence local environmental quality.

Informal programs encompass a broad array of educational approaches at diverse sites. As with formal educational programs, informal programs are increasingly developed around problem-oriented objectives.

The Tbilisi Declaration resulted from the world's first intergovernmental conference on environmental education. The declaration's goals illustrate how education can and does contribute to environmental problem solving and management.

Discussion Questions

1. What is environmental education? How do modern definitions differ from older definitions?

2. Why are values an important part of environmental education?

3. What are two approaches to incorporating environmental education into school curricula? What are the advantages and disadvantages of each?

4. Imagine you are an elementary teacher who has been asked to design an environmental science curriculum for your school. What will be the major features of your program, and how will it be integrated into the existing curriculum?

5. What can you do to become environmentally educated outside of the formal classroom?

Environmental Science in Action Contributors

Karla Armbruster

Education
B.A. in English Literature, Miami University, 1985
M.A. in English Literature, Ohio State University, 1989
Candidate for Ph.D. in English Literature, Ohio State University

Work experience: Karla has worked at a pharmaceutical publishing company in Cincinnati, Ohio, and as a science writer on this textbook. Currently, she is a teaching associate and writing tutor at the Ohio State University in Columbus, Ohio. She is helping Dr. Kaufman develop a book of expanded case studies based on many of the case studies in this book.

Comments on **Biosphere 2000**: "Working on this project has changed the way I look at the world, although in some ways it has also affirmed the intuitive belief I've always had that human beings are not the center and purpose of the universe."

Lisa Bentley

Education
B.S. in Management, *magna cum laude*, Wright State University, 1987
Lisa plans to pursue a master's degree in environmental management and "pursue a career in the ever-expanding field of environmental cleanup in the Air Force."

Work experience: Lisa works for Wright Patterson Air Force Base in Dayton as a contract negotiator.

Comments on **Biosphere 2000**: "My initial paper was on exploring the history of Lake Erie, and that was an incredible opportunity. I think this paper was unique because I got to see the whole spectrum of environmental management. Lake Erie went from almost total devastation to a pretty remarkable comeback, and I thought it illustrated that while people have the ability to harm the environment, we also have the capability to coexist with our environment. We just have to make the effort. That realization that what one individual does to the environment makes a difference stuck with me, and has been an influence on how I live and how I've raised my son."

Doug Davidson

Education
B.A. in Public Administration, Miami University, 1987
Currently pursuing a master's degree in interdisciplinary studies, with an emphasis on economic development, at the Ohio State University.

Work experience: Doug is currently Director of Public Relations with the Chamber of Commerce in Columbus, Ohio.

Comments on **Biosphere 2000**: "Part of my professional theory on the field includes ideas that came to life with my involvement in the original Resource Management class."

Steve Douglas

Education
B.A. in Psychology, Miami University, 1987
M.A. in Experimental Psychology, Miami University, 1992

Work experience: Steve is currently interning in IBM's Human Factors Department in Florida.

Comments on **Biosphere 2000**: "Seven or eight years ago, I felt that there was merit in this undertaking, in that we were forming a textbook of ecological information. In 1992, I feel that the years since have demonstrated vividly the need for a guidebook to secure a future for this planet. Anyone not disturbed by the rapid deterioration of our world and the loss of its beauty needn't bother trying to help save it. Just keep your head buried in the sand and enjoy your view."

Michael J. Fath

Education
B.A. in Microbiology and Chemistry, Miami University, 1987
M.A. in Microbiology and Molecular Genetics, Harvard University, 1990
Candidate for Ph.D. in Microbiology and Molecular Genetics, Harvard University, expected 1993.

Work experience: Mike has worked as a biochemistry teaching fellow at Harvard Medical School and a biology teaching fellow at Harvard College.

Comments on **Biosphere 2000**: "I fondly remember back in 1983 the honors seminar ZOO 180, Resource Management. When Don suggested we turn our final case studies into a book, it seemed like the natural thing to do. But I never imagined that it would grow and mature into such an extraordinary project! I can't wait to see the finished product! May *Biosphere 2000* serve as a useful guide to students for years to come."

Amy J. Franz

Education
B.A. in Zoology, Miami University, 1988
Shoal's Marine Lab, New Hampshire

Work experience: Amy worked as a seacoast naturalist at Odiorne Point State Park in Rye, New Hampshire. She is currently a naturalist with the Hamilton County Park District in southwest Ohio and, with Donald Kaufman, directs an environmental education project for elementary teachers in the Cincinnati area, sponsored by the Ohio Environmental Protection Agency.

Comments on **Biosphere 2000:** "Education is the key to changing society's values, and this book is the tool. It has been a long project, but the result is worth it."

Pam M. Gates

Education
B.A. in Political Science, Miami University, 1986
J.D., University of Texas

Work experience: Pam is an attorney with Frost & Jacobs.

Comments on **Biosphere 2000:** At the beginning of this project, Pam recalls thinking, "I can't believe my name is going to be published!" Now her thoughts are, "Is it really done?!"

Molly A. Grannan

Education
B.S. in English Education, *magna cum laude*, University Honors, Miami University, 1987
Studied Linguistics in Education at the Ohio State University, Summer 1989

Work experience: For the past five years, Molly has been a teacher/newspaper advisor/drama technical director at St. Francis DeSales High School in Columbus, Ohio.

Comments on **Biosphere 2000:** "Looking back, I can't believe we actually had the nerve (or naivete!) to undertake this project. Maybe it was better to be young and not know. I still use this project as an example for my students when I'm teaching the writing process. It's a great way to show the importance of revising."

James E. Johnson II

Education
B.A. in Linguistics, Miami University, 1990
John plans to attend the Cincinnati College of Law in the fall of 1992, possibly majoring in Environmental Law.

Work experience: From August 1990–May 1991, John was a foreign trainee at International Meeting Point in Turku, Finland. He is currently employed by Elder-Beerman Department Stores.

Comments on **Biosphere 2000:** "It was wonderful to be a part of this project at the time, and I'm very excited that the book is finally going to be published."

Frank Andrew Jones

Education
B.S. in Zoology, Miami University, 1990
Candidate for master's degree in botany, Miami University

Work experience: Andy is currently working with Dr. Kaufman and Cecilia Franz on a new textbook on biodiversity.

Comments on **Biosphere 2000:** "A fulfilling and worthwhile experience second only to my first attempt at the north face of Everest."

Lindsay Koehler

Education
B.A. in English-Journalism

Work experience: Lindsay works at Manning Selvage & Lee Public Relations, and helps run Beauty World, a family business. She also acts as a public relations consultant for Tangles Salon and White Gallery.

Comments on **Biosphere 2000:** " 'The Book,' as we called *Biosphere 2000* back then, was a tremendous project to be a part of. We were actually treated as adults, given both the respect and the responsibility we sought. I'll admit we were all pretty impressed with ourselves. We felt we had gone beyond the basic college experience—we were doing something for the *real world*."

Tracy J. Linerode

Education
B.A.s in Political Science and French, Miami University, 1990
Currently attending Vanderbilt Law School

Work experience: Tracy was a Congressional Intern with the U.S. House of Representatives from 1990–1991. She was a consultant for the Kettering Foundation in Dayton, Ohio, and is the author of a nationally published National Issues Forum Guidebook. Tracy has also been a law clerk at the firm of Seiller & Handmaker in Louisville, Kentucky, and for Justice Martha Daughtrey of the Tennessee Supreme Court in Nashville. She is currently a law clerk at the firm of Brouse & McDowell in Akron, Ohio.

Comments on **Biosphere 2000:** "I can't believe this book is finally being published! I thought my 50-page paper that I submitted for the book was long until I got to law school!"

Deborah A. Lokai

Education
B.A.s in International Studies and German, Miami University, 1987
M.A. in German, Wayne State University
Candidate for Ph.D. in Modern Languages, Wayne State University

Work experience: Debbie was a part-time Teaching Assistant in German at Wayne State University and is now a full-time faculty member.

Comments on **Biosphere 2000:** "I thought the project was an exciting and motivating way to learn. The thought of spreading our concern for the environment by first educating ourselves in the issues and then educating others was innovative at the undergraduate level. I've traveled extensively throughout the world since this project began, and everywhere I go I find myself examining the environment, questioning people on environmental issues—I have gained a special awareness through this project, and I hope that many students in the future will gain such an awareness by reading this text."

Pamela A. Marks (Marsh)

Education
B.S.s in Decision Sciences and Economics, Miami University, 1988
M.S. in Applied Statistics, Bowling Green State University, Bowling Green, Ohio, 1991

Work experience: Pam interned as a data analyst at Weyerhauser and at Moritz Market Research. She also has been a statistical consultant for Bowling Green State University and a statistics instructor at Miami University.

Comments on Biosphere 2000: "I was very excited about the project from the beginning—I wondered if we could actually do it. Now I am very glad it has been accomplished and am proud that my name will be in the book."

Greg Moody

Education
B.S. in Economics, Minor in Political Science, Miami University, 1989

Work experience: Greg has been a Senior Legislative Assistant to Congressman David Hobson from January 1991 to the present. From December 1989 to January 1991 he was Legislative Assistant to then State Senator David Hobson.

Comments on Biosphere 2000: "It was a great academic experience and gave me a better way to look at the world."

Tom Pedroni

Education
B.A. in Philosophy (Critical Social Theory) through the School of Interdisciplinary Studies, Miami University, 1990
Tom plans eventually to attend graduate school at the University of Wisconsin–Madison, majoring in Critical Curriculum and Instruction.

Work experience: Tom has been Field Manager for the Public Interest Research Group in Michigan and Wisconsin, and a trilingual tour guide at Sequoia National Park in California. He has also been a volunteer coordinator and fund-raising coordinator for the *Madison Insurgent*, a progressive/radical movement newspaper.

Cheryl Puterbaugh

Education
B.A. in English, Miami University, 1987
Master of Technical and Scientific Communication, Miami University, 1988

Work experience: Cheryl has worked for three years in the management systems division of Procter & Gamble, Cincinnati, Ohio, as a technical communicator.

Comments on Biosphere 2000: "I'll admit I wasn't a believer at first, but not long after that first summer when we wrote the prospectus, I was a true convert. Still involved in the project now after (what is it? more than

eight years?) all this time, anyway. I've made friends that will last a lifetime, and helped to protect the environment at the same time."

April Rowan

Education
B.A.s in Speech Communication and Philosophy, Miami University, 1989
Candidate for master's degree in Speech Communication, Miami University, autumn 1992

Work experience: April has worked in the environmental health department of the British government, and has been *au pair* in Italy. She currently teaches communication courses at Miami University.

Marsha Shook

Education
B.S. in Social Science, and Secondary Education Certification, University of Rio Grande, 1988

Work experience: For the last four years, Marsha has taught American History and American Government at Licking County Joint Vocational School in Newark, Ohio. She is Chair of the Social Studies Department.

Comments on Biosphere 2000: "The project was the best experience I had during my two years at Miami and throughout my college career. As I teach young people, I realize the huge task of communicating a 'land ethic.' It is much more important for my students to gain a respect and obligation for the land than to learn the structure of the federal court system! I hope I can strive to live and teach a land ethic."

Cynthia R. Tufts

Education
B.A. in Zoology, Miami University, 1989
Candidate for D.D.S. at the Ohio State University, expected 1993

Comments on Biosphere 2000: "I feel kind of unique— lots of dentists contribute to books, but I wonder how many of them have been a part of a book on environmental science. . . ."

Dana I. Wilson

Education
B.A. in Anthropology, *summa cum laude*, Miami University, 1986
M.A. in Anthropology, Northwestern University
J.D., University of Michigan, Ann Arbor

Work experience: Dana is an associate at Strasburger & Price, L.L.P., Dallas, Texas, specializing in intellectual property law.

Comments on Biosphere 2000: " 'The Book' as it was colloquially known to those of us working on it, began as a project for a class but became a project for the environment. In these days of heightened environmental awareness, the interests we developed for our topics can now raise the interest levels of others. I still find that the project affects my way of thinking—I can't visit any zoological park without finding myself analyzing the habitats in detail. Hopefully, others will do the same."

Organizations and Publications

Environmental Organizations

Accord Associates
5500 Central Avenue
Boulder, CO 80301
(303) 444-5080
Mediates environmental disputes.
Publication: *Accord*.

The Acid Rain Foundation, Inc.
1410 Varsity Drive
Raleigh, NC 27606
(919) 828-9443
Supports research and supplies educational resources on global atmospheric issues.

African Wildlife Foundation
1717 Massachusetts Avenue, NW
Washington, DC 20036
(202) 265-8393
Involved in wildlife conservation projects in Africa, including wildlife management, research, and education. Publication: *African Wildlife News*.

Air Pollution Control Association
P.O. Box 2861
Pittsburgh, PA 15230
(412) 232-3444
Collects and disseminates information on air pollution control and hazardous waste management.

The Alaska Wildlife Alliance
P.O. Box 202022
Anchorage, AK 99520
(907) 277-0897
Dedicated to preserving Alaskan wildlife.

Alliance for Environmental Education
10751 Ambassador Drive, Suite 201
Manassas, VA 22110
(703) 335-1025
Professional organization for formal and informal educators. Publication: *Alliance Exchange Newsletter*.

American Association of Botanical Gardens and Arboreta, Inc.
P.O. Box 206
Swarthmore, PA 19081
(215) 328-9145
Dedicated to research and education related to botanical gardens and arboreta of North America. Publication: *The Public Garden*.

American Association of Zoological Parks and Aquariums
7970-D Old Georgetown Road
Bethesda, MD 20814
(301) 907-7777
Promotes welfare of zoological parks and aquaria as educational and conservation facilities. Publications: *AAZPA Newsletter*, *Directory—Zoological Parks and Aquariums in the Americas*.

American Cetacean Society
P.O. Box 2639
San Pedro, CA 90731-0943
(213) 548-6279
Dedicated to conservation, education, and research of marine species. Publications: *Whalewatcher, Journal of the American Cetacean Society*, *Whale News, ACS National Newsletter*.

American Committee for International Conservation, Inc.
c/o Mike McCloskey
Sierra Club
330 Pennsylvania Avenue, SE
Washington, DC 20003
(202) 547-1144
Promotes and assists international research and conservation activities.

American Conservation Association, Inc.
30 Rockefeller Plaza, Room 5402
New York, NY 10112
(212) 649-5600
Promotes understanding of conservation and the preservation and development of natural resources for public use.

American Forest Council
1250 Connecticut Avenue, NW
Washington, DC 20036
(202) 463-2459
Promotes the development and productive management of commercial forest lands by government, industry, and private landowners. Publication: *American Tree Farmer*.

American Geographical Society
156 Fifth Avenue, Suite 600
New York, NY 10010-7002
(212) 242-0214
Sponsors research and publishes scientific and popular books, periodicals, and maps.

American Institute of Biological Sciences, Inc.
730 11th Street, NW
Washington, DC 20001-4584
(202) 628-1500
Professional organization for biologists. Publication: *BioScience*.

American Nature Study Society
5881 Cold Brook Road
Homer, NY 13077
(607) 749-3655
Promotes environmental education and avocation. Publications: *ANSS Newsletter*, *Nature Study, A Journal of Environmental Education and Interpretation*.

American Rivers
801 Pennsylvania Avenue, SE
Suite 303
Washington, DC 20003
(202) 547-6900
Dedicated to preservation, management, and protection of American rivers.

American Society for Environmental History
Department of History
Oregon State University
Corvallis, OR 97331
(503) 754-3421
Seeks understanding of human ecology through the perspectives of history and humanities. Publication: *Environmental Review*.

American Solar Energy Society
2400 Central Avenue
Boulder, CO 80301
Disseminates information on solar energy.

American Water Resources Association
5410 Grosvenor Lane, Suite 220
Bethesda, MD 20814
(301) 493-8600
Promotes water resources research and disseminates information on water resources science and technology.

Audubon Naturalist Society of the Central Atlantic States, Inc.
8940 Jones Mill Road
Chevy Chase, MD 20815
(301) 652-9188
Active in public education, conservation issues, and natural science studies. Publications: *Naturalist News, Atlantic Naturalist, Naturalist Review*.

Bio-Integral Resource Center
P.O. Box 7414
Berkeley, CA 94707
(415) 524-2567
Provides information on least-toxic pest control.

Bioregional Project (North American Bioregional Congress)
Turtle Island Office
1333 Overhulse Rd., NE
Olympia, WA 98502
Disseminates information about restoring bioregions in the United States.

Bounty Information Service
Stephens College Post Office
Columbia, MO 65215
(314) 876-7186
Promotes the removal of bounties in North America.

The Camp Fire Club of America
230 Camp Fire Road
Chappaqua, NY 10514
(914) 941-0199
Works to preserve forests, woodlands, and wildlife for future generations.

Center for Environmental Information
46 Prince Road

Rochester, NY 14607
(716) 271-3550
Dedicated to conserving endangered and threatened species and their habitats. Publication: *The CEE Report*.

Center for Holistic Resource Management
5820 4th Street, NW
Albuquerque, NM 87107
(505) 344-3445
Promotes community development and resource management.

Center for Marine Conservation
1725 DeSales Street, NW, Suite 500
Washington, DC 20036
(202) 429-5609
Dedicated to conserving endangered and threatened marine species and their habitats.

Center for Plant Conservation
P.O. Box 299
St. Louis, MO 63166
(314) 577-9450
Network of botanical gardens and arboreta dedicated to conservation and study of rare and endangered U.S. plants.

Center for Science Information
4252 20th Street
San Francisco, CA 94114
(415) 553-8772
Educates decision makers and journalists about the environmental applications of biotechnology.

Center for Science in the Public Interest
1501 16th Street, NW
Washington, DC 20036
(202) 332-9110
Consumer advocacy organization on health and nutrition issues.

Children of the Green Earth
Box 95219
Seattle, WA 98145
Helps young people worldwide plant and care for trees and forests. Publications: *Newsletter, Tree Song*.

Clean Water Action Project
317 Pennsylvania Avenue, SE
Washington, DC 20003
(202) 547-1286
Citizen's organization lobbying for water pollution control and safe drinking water.

Clean Water Fund
National Office
317 Pennsylvania Avenue, SE
3rd Floor
Washington, DC 20005
(202) 547-2312

Advances environmental and consumer safety through grass-roots participation.

Climatic Institute
316 Pennsylvania Avenue, SE
Suite 403
Washington, DC 20003
Sponsors conferences on science and policy related to global climate.

The Coastal Society
5410 Grosvenor Lane, Suite 110
Bethesda, MD 20814
(301) 897-8616
Promotes the understanding and wise use of coastal environments.

Concern, Inc.
1794 Columbia Road, NW
Washington, DC 20009
(202) 328-8160
Disseminates environmental information.

Conservation Foundation
1250 42nd Street, NW
Washington, DC 20076
Concerned with conservation, environmental impact of foreign aid, and environmental education. Publication: *Conservation Foundation Letter*.

The Conservation Fund
1800 North Kent Street, Suite 1120
Arlington, VA 22209
(703) 525-6300
Advances land and water conservation.

Conservation International
1015 18th Street, NW, Suite 1000
Washington, DC 20036
(202) 429-5660
Dedicated to preserving tropical ecosystems and wildlife.

Consumer Federation of America
1424 18th Street, NW
Washington, DC 20009
Dedicated to consumer and environmental safety.

COOL IT! Program
National Wildlife Federation
1400 16th Street, NW
Washington, DC 20036-2266
Provides information and resources on a variety of environmental issues to college students.

The Cousteau Society, Inc.
930 West 21st Street
Norfolk, VA 23517
(804) 627-1144
Dedicated to environmental education. Publications: *Calypso Log, Dolphin Log*.

Defenders of Wildlife
1244 19th Street, NW

Washington, DC 20036
(202) 659-9510
    Dedicated to preserving the
    diversity of the world's wildlife.

Earth First!
P.O. Box 2358
Lewiston, ME 04241
    Actively promotes preservation of
    wildlife and natural systems.

Earth Island Institute
300 Broadway, Suite 28
San Francisco, CA 94133
(415) 788-3666
    Dedicated to conserving,
    preserving, and restoring the
    global environment.

Earthwatch
P.O. Box 403N
680 Mt. Auburn Street
Watertown, MA 02272
(617) 962-8200
    Promotes and assists research on
    various environmental issues
    worldwide.

The Ecological Society of America
Dr. Hazel R. Delcourt, Secretary
Department of Botany
University of Tennessee
Knoxville, TN 37996
    Professional organization.

Educational Communications, Inc.
P.O. Box 35473
Los Angeles, CA 90035
(213) 559-9160
    Promotes environmental and
    science education for public
    welfare.

Environmental Action Foundation, Inc.
1525 New Hampshire Avenue, NW
Washington, DC 20036
(202) 745-4870
    Promotes research and educational
    programs on complex
    environmental issues. Publication:
    Power Line.

Environmental Action, Inc.
1525 New Hampshire Avenue, NW
Washington, DC 20036
(202) 745-4870
    Lobbies for political and social
    change on many environmental
    issues. Publication: *Environmental
    Action.*

Environmental and Energy Study Institute
122 C Street, NW
Washington, DC 20001
(202) 628-1400
    Dedicated to public policy
    research, analysis, and education.

Environmental Data Research Institute
797 Elmwood Avenue

Rochester, NY 14620
(716) 473-3090
    Provides information on
    organizations, publications, and
    funding.

Environmental Defense Fund, Inc.
257 Park Avenue South
New York, NY 10010
(212) 505-2100
    Organization of lawyers,
    scientists, and economists
    dedicated to improving
    environmental quality and public
    health. Publication: *EDF Letter.*

Environmental Education Coalition
Pocono Environmental
Education Center
Box 1010
Dingmans Ferry, PA 18328
(717) 828-2319
    Enhances the role of member
    organizations in improving
    environmental education.

The Environmental Law Institute
1616 P Street, NW, Suite 200
Washington, DC 20036
(202) 328-5150
    Dedicated to research and
    education on environmental law
    and policy.

Environmental Policy Institute
218 D Street, SE
Washington, DC 20003
(202) 544-2600
    A public-interest advocacy group
    influencing public policies
    through research, education,
    lobbying, litigation, and organizing
    coalitions.

Farralones Institute
15920 Coleman Valley Road
Occidental, CA 95465
    Disseminates information on
    ecologically sound living and
    agriculture in urban areas.

Fauna & Flora Preservation Society
1 Kensington Gore
London SW7 2AR
United Kingdom

Felicidades Wildlife Foundation, Inc.
Box 490
Waynesville, NC 28786
(704) 926-0192
    Dedicated to the care and
    rehabilitation of native wildlife
    and the environmental education
    of youth.

Fish and Wildlife Reference Service
5430 Grosvenor Lane, Suite 110
Bethesda, MD 20814
(301) 492-6403
    Information clearinghouse for fish
    and wildlife management
    research reports.

Food and Agriculture Organization of the United Nations
Via delle Terme di Caracalla
Rome 00100
Italy
    Dedicated to improving nutrition,
    agriculture, and living conditions
    of hungry people.

Food First
1885 Mission Street
San Francisco, CA 94103
    Conducts research and education
    on agriculture, especially in
    developing countries.

Forest Trust
P.O. Box 9238
Santa Fe, NM 87504-9238
(505) 983-8992
    Dedicated to protecting and
    improving forest ecosystems and
    resources.

Freshwater Foundation
Box 90
2500 Shadywood Road
Navarre, MN 55392
(612) 471-8407
    Supports research and education
    about proper use and management
    of surface water and groundwater.

Friends of the Earth
218 D Street, SE
Washington, DC 20003
(202) 544-2600
    Dedicated to the preservation,
    restoration, and rational use of the
    earth.

Friends of the Earth Foundation, Inc.
1045 Sansome Street
San Francisco, CA 94111
(415) 433-7373
    Conducts scientific, educational,
    and literary programs for the
    preservation, restoration, and
    rational use of the earth.

Friends of Trees
P.O. Box 1466
Chelan, WA 98816
    Dedicated to conserving forests
    and wildlife.

The Fund for Animals, Inc.
200 West 57th Street
New York, NY 10019
(212) 246-2096
    Dedicated to preserving wildlife,
    saving endangered species, and
    promoting humane treatment for
    all animals.

Gaia Institute
Cathedral of St. John the Divine
1047 Amsterdam Avenue at 112th
Street
New York, NY 10025

Global Greenhouse Network
1130 17th Street, NW, Suite 530
Washington, DC 20036
Facilitates international cooperation on global warming at both governmental and grass-roots levels.

The Global Tomorrow Coalition
1325 G Street, NW, Suite 915
Washington, DC 20005-3104
(202) 628-4016
Dedicated to public environmental education.

Green Party
831 Commercial Drive
Vancouver, British Columbia
Political party based on environmental issues.

Greenpeace (CANADA)
427 Bloor Street
West Toronto, Ontario M5S 1X7
Conducts lobbying, organizing, and direct action on specific environmental issues.

Greenpeace USA, Inc.,
1436 U Street, NW
Washington, DC 20009
(202) 462-1177
Conducts lobbying, organizing, and direct action on specific environmental issues. Publication: *Greenpeace Magazine.*

Household Hazardous Waste Project
Box 87
901 South National Avenue
Springfield, MO 65804
Provides information on household toxics and safe alternatives.

The Human Ecology Action League, Inc.
P.O. Box 66637
Chicago, IL 60666
(312) 665-6575
Information clearinghouse on environmental conditions hazardous to human health.

Human Environment Center
1001 Connecticut Avenue, NW
Suite 829
Washington, DC 20036
(202) 331-8387
Provides education and services to encourage integration of environmental organizations and joint activities.

INFORM
381 Park Avenue South
New York, NY 10016
(212) 689-4040
Dedicated to public education on practical actions for the conservation and preservation of natural resources.

Institute for Alternative Agriculture
9200 Edmonston Road, Suite 117
Greenbelt, MD 20770
Researches and promotes methods of sustainable agriculture.

Institute for Conservation Leadership
2000 P Street, NW, Suite 413
Washington, DC 20036
(202) 466-3330
Conducts training and research to increase the number and effectiveness of volunteer conservation leaders and organizations.

Institute for Earth Education
P.O. Box 288
Warrenville, IL 60555
(509) 395-2299
Develops and disseminates focused environmental education programs.

Institute for Local Self-Reliance
1717 18th Street, NW
Washington, DC 20009
Dedicated to helping communities become more self-reliant in resources and economic development.

International Association of Fish and Wildlife Agencies
444 North Capitol Street, NW
Suite 534
Washington, DC 20001
Dedicated to conserving, protecting, and managing wildlife and related natural resources.

International Council for Outdoor Education
P.O. Box 17255
Pittsburgh, PA 15235
(412) 372-5992
Develops and implements education programs in conservation and outdoor recreation.

International Fund for Animal Welfare
P.O. Box 193
Yarmouth Port, MA 02675
(508) 362-4944
Dedicated to the protection of wild and domestic animals.

International Oceanographic Foundation
3979 Rickenbacker Causeway
Virginia Key
Miami, FL 33149
(305) 361-5786
Encourages public education and study of the oceans.

International Union for Conservation of Nature and Natural Resources (IUCN)
Avenue du Mont-Blanc
CH-1196
Gland, Switzerland
Promotes scientifically based action for the conservation of wildlife.

Island Resources Foundation
Red Hook Center, Box 33
St. Thomas, VI 00802
(809) 775-6225
Dedicated to the study of island systems and improved resources management.

The Izaak Walton League of America
1401 Wilson Boulevard, Level B
Arlington, VA 22209
(703) 528-1818
Promotes environmental education and protection of natural resources.

John Muir Institute for Environmental Studies, Inc.
743 Wilson Street
Napa, CA 94559
(707) 252-8333
Conducts research and education on various environmental problems.

Keep America Beautiful, Inc.
9 West Broad Street
Stamford, CT 06902
(203) 323-8987
Dedicated to improving waste-handling practices in communities in the United States through public education.

Land Trust Exchange
1017 Duke Street
Alexandria, VA 22314
(703) 683-7778
Dedicated to assisting local and regional land conservation groups.

League of Conservation Voters
2000 L Street, NW, Suite 804
Washington, DC 20036
(202) 785-8683
Political action committee promoting the election of public officials who will work for a healthy environment.

League of Women Voters of the U.S.
1730 M Street, NW
Washington, DC 20036
(202) 429-1965
Lobbies on a wide range of environmental issues and seeks to educate citizenry and encourage political responsibility.

Marine Technology Society
2000 Florida Avenue, NW, Suite 500
Washington, DC 20009
(202) 462-7557
Professional organization promoting the exchange of information on marine-related subjects.

Monitor
1506 19th Street, NW
Washington, DC 20036
(202) 234-6576
　　Coordinating center and
　　information clearinghouse on
　　endangered species and marine
　　mammals.

National Arbor Day Foundation
211 North 12th Street, Suite 501
Lincoln, NE 68508
(402) 474-5655
　　Sponsors National Arbor Day,
　　Tree City USA, Conservation
　　Trees educational programs, and
　　other educational programs.

**National Association of Biology
Teachers**
11250 Roger Bacon Drive, Room 19
Reston, VA 22090
(703) 471-1134
　　Assists teachers to improve the
　　teaching of biology. Publications:
　　*The American Biology Teacher,
　　News and Views.*

National Audubon Society
950 Third Avenue
New York, NY 10022
(212) 832-3200
　　Conducts research, education, and
　　action in protecting and managing
　　natural resources. Publications:
　　*Audubon, American Birds,
　　Audubon Wildlife Report,* and
　　others.

**National Center for Policy
Alternatives**
2000 Florida Avenue, NW
Washington, DC 20003
　　Promotes environmentally and
　　socially responsible public policy.

**National Center for Urban
Environmental Studies**
516 North Charles Street, Suite 501
Baltimore, MD 21201
(301) 727-6212
　　Conducts research, training,
　　technical assistance, and
　　information dissemination on
　　developing various urban
　　programs.

National Clean Air Coalition
530 7th Street, SE
Washington, DC 20003
(202) 523-8200
　　Active in attaining and
　　maintaining air quality.

**National Coalition Against the Misuse
of Pesticides**
530 7th Street, SE
Washington, DC 20003
(202) 543-5450
　　Dedicated to improving safety in
　　pesticide use.

**National Coalition for Marine
Conservation**
P.O. Box 23298
Savannah, GA 31403
(912) 234-8062
　　Devoted to the conservation of
　　ocean fish and the protection of
　　the marine environment.

**National Environmental Health
Association**
720 S. Colorado Boulevard
South Tower, Room 970
Denver, CO 80222
(303) 756-9090
　　Professional organization of
　　environmental health
　　practitioners. Publications:
　　*Journal of Environmental Health,
　　Environmental Health Trends
　　Report,* and others.

National Geographic Society
17th and M Streets, NW
Washington, DC 20036
(202) 857-7000
　　Disseminates and encourages
　　dissemination of information
　　about geography, environmental
　　problems, and natural resource
　　conservation. Publication:
　　National Geographic.

**National Outdoor Leadership School
(NOLS)**
Post Office Box AA
Lander, WY 82520
(307) 332-6972
　　Teaches how to enjoy and
　　conserve the wilderness and how
　　to lead and teach others.

National Park Foundation
P.O. Box 57473
Washington, DC 20037
(202) 785-4500
　　Provides private-sector support for
　　the enhancement and
　　improvement of the nation's park
　　system.

**National Parks and Conservation
Association**
1015 31st Street, NW
Washington, DC 20007
(202) 944-8530
　　Dedicated to preserving,
　　promoting, and improving the
　　nation's park system. Publication:
　　National Parks Magazine.

**National Recreation and Park
Association**
3101 Park Center Drive, 12th Floor
Alexandria, VA 22302
(703) 820-4940
　　Dedicated to improving park and
　　recreation leadership, programs,
　　and facilities.

National Science Teachers Association
1742 Connecticut Avenue, NW
Washington, DC 20009
(202) 328-5800
　　Dedicated to improving the
　　teaching of science at all levels.
　　Publications: *The Science Teacher,
　　Journal of College Science
　　Teaching,* and others.

**National Solid Waste Management
Association**
1730 Rhode Island Avenue, NW
Suite 100
Washington, DC 20036
　　Gathers statistics and disseminates
　　information on waste management
　　and resource recovery.

National Water Resources Association
955 L'Enfant Plaza North, SW
Room 1202
Washington, DC 20024
(202) 488-0610
　　Promotes development,
　　conservation, and management of
　　the water resources of 17 western
　　state associations.

National Wetlands Technical Council
1616 P Street, NW, Suite 200
Washington, DC 20036
(202) 328-5150
　　Council of scientists organized to
　　provide scientific assistance in
　　national wetlands conservation.

National Wildlife Federation
1400 16th Street, NW
Washington, DC 20036
(202) 797-6646
　　Dedicated to creating and
　　encouraging awareness of the
　　need for wise use and proper
　　management of natural resources.

Natural Resources Council of America
1015 31st Street, NW
Washington, DC 20007
(202) 333-8498, -8595
　　Concerned with sound
　　management of natural resources
　　in the public interest.

Natural Resources Defense Council
40 West 20th Street
New York, NY 10011
(212) 727-2700
　　Dedicated to research,
　　organization, and litigation on
　　various environmental issues.
　　Publications: *Newsline, The
　　Amicus Journal.*

The Nature Conservancy
1815 North Lynn Street
Arlington, VA 22209
(703) 841-5300
　　Preserves ecologically significant
　　areas by purchasing them and
　　managing them as sanctuaries.

New Alchemy Institute
237 Hatchville Road
East Falmouth, MA 02536
Conducts research and educational programs on self-sufficient agriculture, aquaculture, solar and wind energy, and energy conservation.

New York Zoological Society
The Zoological Park
Bronx, NY 10460
(212) 220-5100
Operates several zoological parks and research facilities and promotes zoological research and public education throughout the world.

North American Association for Environmental Education
P.O. Box 400
Troy, OH 45373
(513) 698-6493
Supports individuals and groups involved in environmental education, research, and service. Publication: *Environmental Communicator.*

North American Lake Management Society
1000 Connecticut Avenue, NW
Suite 202
Washington, DC 20036
(202) 833-3382
Dedicated to protecting, restoring, and managing lakes and reservoirs and their watersheds.

The Oceanic Society
1536 16th Street, NW
Washington, DC 20036
(202) 328-0098
Works to protect and preserve the marine environment for people and wildlife.

Ontario Environment Network
P.O. Box 125, Station P
Toronto, Ontario M5S 2Z7
Conducts research and education on environmental issues.

Ontario Public Interest Research Group
229 College Street
Toronto, Ontario K1P 5C5
Conducts research and education on environmental issues.

Permaculture Association
P.O. Box 202
Orange, MA 01364
Conducts research and education on sustainable agriculture.

Permaculture Institute of North America
6488 South Maxwelton Road
Clinton, WA 98236
Provides information on sustainable agriculture.

Physicians for Social Responsibility
639 Massachusetts Avenue
Cambridge, MA 02139
Conducts research and education on the health effects of nuclear war.

Planned Parenthood Federation of America, Inc.
810 Seventh Avenue
New York, NY 10019
(212) 541-7800
Provides family-planning services and information.

Pollution Probe
12 Madison Avenue
Toronto, Ontario M5R 2S1
Conducts research and education on environmental issues.

Population Crisis Committee
1120 19th Street, NW, Suite 550
Washington, DC 20036
(202) 659-1833
Provides public education, policy analysis, and liaison to support internation population and family planning.

Population Reference Bureau, Inc.
777 14th Street, NW, Suite 800
Washington, DC 20005
(202) 639-8040
Gathers and publishes information on the implications of U.S. and international population dynamics. Publication: *Population Bulletin.*

Probe International
100 College Street
Toronto, Ontario M5G 1L5
Conducts research and education on environmental and resource issues.

Public Environment Center, Inc.
One Milligan Place
New York, NY 10011
(212) 691-4877
Conducts research and information programs on regional and city planning, urban design, and environmental policies and issues.

Rachel Carson Council
8940 Jones Mill Road
Chevy Chase, MD 20815
(301) 652-1877
International clearinghouse for information on chemical contamination, especially pesticides.

RAIN
2270 Northwest Irving
Portland, OR 97210
Conducts research and education on appropriate technology and community self-sufficiency.

Rainforest Action Network
300 Broadway, Suite 29A
San Francisco, CA 94133
(415) 398-4404
Cooperates with other organizations on direct-action campaigns to protect rain forests.

Rainforest Alliance
270 Lafeyette Street, Room 512
New York, NY 10012
(212) 941-1900
Develops and promotes alternatives to activities that cause tropical deforestation.

Rainforest Information Centre
P.O. Box 368
Lismore, New South Wales 2480
Australia
Provides information on rain forest issues.

Renew America
1400 16th Street, NW, Suite 710
Washington, DC 20036
(202) 232-2252
Provides information and makes recommendations to policymakers to encourage development of a safe and sustainable environment.

Resources for the Future
1616 P Street, NW
Washington, DC 20036
(202) 328-5000
Conducts research and education on environmental problems and resource conservation.

Rocky Mountain Institute
1739 Snowmass Creek Road
Snowmass, CO 81654
Conducts research and education on environmental and resource conservation issues, especially energy.

Save the Whales, Inc.
P.O. Box 2397
1426 Main Street, Unit E
Venice, CA 90291
(213) 392-6226
Educates the public about marine mammals, their environment, and their preservation.

The School for Field Studies
16 Broadway
Beverly, MA 01915-4499
(508) 927-7777
Offers study programs on ecological and environmental issues worldwide.

Scientists' Institute for Public Information
355 Lexington Avenue
New York, NY 10017
(212) 661-9110
Conducts multidisciplinary public-information programs on science, technology, and environmental issues.

Sea Shepherd Conservation Society
P.O. Box 700-S
Redondo Beach, CA 90277
(213) 373-6979
    Actively promotes protection of
    marine animals and habitat.

Sierra Club
730 Polk Street
San Francisco, CA 94109
(415) 776-2211
    Lobbies and educates on many
    environmental issues. Publication:
    Sierra.

Smithsonian Institution
1000 Jefferson Drive, SW
Washington, DC 20560
(202) 357-1300
    Conducts and supports
    environmental education
    programs and research and
    develops natural history and
    anthropological collections.
    Publication: *Smithsonian
    Magazine*.

Society for Conservation Biology
Biology Department
Montana State University
Bozeman, MT 59717
(406) 994-4548
    Professional organization
    dedicated to protecting biological
    diversity.

Society for Range Management
1839 York Street
Denver, CO 80206
(303) 355-7070
    Professional organization which
    promotes understanding of
    rangeland ecosystems and their
    management and use.

Soil and Water Conservation Society
7515 Northeast Ankeny Road
Ankeny, IA 50021-9764
(515) 289-2331
    Professional organization which
    promotes the science and art of
    good land and water use
    worldwide.

Student Conservation Association, Inc.
P.O. Box 550
Charlestown, NH 03603
(603) 826-4301
    Promotes work and conservation
    education for students from high
    school to graduate school.

Tomorrow Foundation
10511 Saskatchwan Drive
Edmonton, Alberta T6E 4S1
    Conducts research and education
    on environmental issues.

Union of Concerned Scientists
26 Church Street
Cambridge, MA 02238
(617) 547-5552

Conducts policy and technical
research, public education, and
legislative advocacy on issues
concerning advanced technologies.

**University Research Expeditions
Program**
University of California
Berkeley, CA 94720
(415) 642-6586
    Promotes public involvement in
    ongoing worldwide scientific
    research and education in the
    environmental, natural, and social
    sciences.

Water Pollution Control Federation
601 Wythe Street
Alexandria, VA 22314-1994
(703) 684-2400
    Promotes the development and
    dissemination of technical
    information on water quality and
    water resources.

The Wilderness Society
900 17th Street NW
Washington, DC 20006
(202) 833-2300
    Devoted to preserving wilderness
    and wildlife and fostering a land
    ethic. Publication: *Wilderness*.

Wilderness Watch
P.O. Box 782
Sturgeon Bay, WI 54235
(414) 743-1238
    Dedicated to the sustained use of
    America's forests and waters, placing
    ecological considerations first.

Wildlife Conservation International
Division of the New York Zoological
Society
185th Street and South Boulevard,
Building A
Bronx, NY 10460
(212) 220-5155
    Conducts research on endangered
    species and ecosystems, with an
    emphasis on conservation.

Wildlife Management Institute
1101 14th Street, NW, Suite 725
Washington, DC 20005
(202) 371-1808
    Promotes use of natural resources
    for the welfare of the nation.

**Wildlife Preservation Trust
International, Inc.**
34th Street and Girard Avenue
Philadelphia, PA 19104
(215) 222-3636
    Awards grants internationally in
    captive propagation of endangered
    species, research, education, and
    fieldwork.

The Wildlife Society
5410 Grosvenor Lane
Bethesda, MD 20814
(301) 897-9770

Professional and student
organization involved in wildlife
research, management, education,
and administration.

Woods Hole Research Center
P.O. Box 296
Woods Hole, MA 02543
    Conducts research on
    environmental issues, with
    emphasis on oceans and global
    warming.

World Resources Institute
1735 New York Avenue, NW
Washington, DC 20006
(202) 638-6300
    Analyzes and disseminates
    information on environmental and
    natural resource issues, with
    emphasis on national and
    international policy.

**World Society for the Protection of
Animals**
P.O. Box 190
29 Perkins Street
Boston, MA 02130
(617) 522-7077
    Promotes conservation and
    protection of both domestic and
    wild animals.

Worldwatch Institute
1776 Massachusetts Avenue, NW
Washington, DC 20036
(202) 452-1999
    Identifies, analyzes, and publicizes
    emerging global problems and
    trends.

Worldwide
1250 24th Street, NW, Suite 500
Washington, DC 20037
(202) 331-9863
    Dedicated to strengthening the
    role of women in developing and
    implementing sound
    environmental and natural
    resource policies.

World Wildlife Fund (Canada)
60 St. Clair Avenue, Suite 201
Toronto, Ontario M4T 1N5
    Dedicated to protecting
    endangered wildlife and wildlands,
    especially in tropical forests.

World Wildlife Fund (United States)
1250 24th Street, NW
Washington, DC 20037
(202) 293-4800
    Dedicated to protecting
    endangered wildlife and
    wildlands, especially in tropical
    forests. Publication: *Focus*.

The Xerces Society
10 Southwest Ash Street
Portland, OR 97204
(503) 222-2788

Dedicated to protecting rare and endangered invertebrates and promoting positive attitudes toward insects.

Zero Population Growth, Inc.
1400 16th Street, NW, Suite 320
Washington, DC 20036
(202) 332-2200
Promotes balance among people, resources, and environment by advocating population stabilization worldwide.

Organizations Dealing with Hunger in the United States

For more information on hunger in the United States and how you can help, contact the following organizations:

Bread for the World
802 Rhode Island Avenue, NE
Washington, DC 20018
(202) 269-0220

Center for Budget and Policy Priorities
236 Massachusetts Avenue, NE
Suite 305
Washington, DC 20002
(202) 544-0591

Children's Defense Fund
122 C Street, NW
Washington, DC 20001
(202) 628-8787

Ecumenical Child Care Network
Child Advocacy Office
475 Riverside Drive, Room 572
New York, NY 10115
(212) 870-3342

Food Research and Action Center
1319 F Street, NW, Room 500
Washington, DC 20004
(202) 393-5060

Interfaith Action for Economic Justice
110 Maryland Avenue, NE
Washington, DC 20002-5694
(202) 543-2800

National Council of Churches, Washington Office
110 Maryland Avenue, NE
Washington, DC 20002-5694
(202) 544-2350

Physician Task Force on Hunger in America
Harvard University
School of Public Health
677 Huntington Avenue
Boston, MA 02115
(617) 732-1000

Second Harvest
National Food Bank Network
343 South Dearborn Street

Chicago, IL 60604
(312) 341-1303

Environmental Publications

Acid Precipitation Digest
Acid Rain Information Clearinghouse
33 South Washington Street
Rochester, NY 14608

African Wildlife News
African Wildlife Foundation
1717 Massachusetts Avenue, NW
Washington, DC 20036

Alternate Sources of Energy
Alternate Sources of Energy, Inc.
107 South Central Avenue
Milaca, MN 56353

The American Biology Teacher
Journal of the National Association of Biology Teachers
11250 Roger Bacon Drive, Room 319
Reston, VA 22090

American Demographics
American Demographics, Inc.
P.O. Box 68
Ithaca, NY 14851

American Forests
American Forestry Association
1516 P Street, NW
Washington, DC 20005

The Amicus Journal
Natural Resources Defense Council
40 West 20th Street
New York, NY 10011

Annual Review of Energy
Department of Energy
Forrestal Building,
1000 Independence Avenue, SW
Washington, DC 20585

Audubon
National Audubon Society
950 Third Avenue
New York, NY 10022

Audubon Wildlife Report
National Audubon Society
950 Third Avenue
New York, NY 10022

BioScience
American Institute of Biological Sciences
730 11th Street, NW
Washington, DC 20001

Buzzworm: The Environmental Journal
P.O. Box 6853
Syracuse, NY 13217

Ceres
Food and Agriculture Organization of the United Nations (FAO)

UNIPUB, Inc.
P.O. Box 433
New York, NY 10016

Conservation Biology
Blackwell Scientific Publications, Inc.
52 Beacon Street
Boston, MA 02108

Conservation Directory
National Wildlife Federation
1400 Sixteenth Street, NW
Washington, DC 20036-2266

Demographic Yearbook
Department of International Economic and Social Affairs
Statistical Office
United Nations Publishing Service
United Nations
New York, NY 10017

Earth Island Journal
Earth Island Institute
300 Broadway, Suite 28
San Francisco, CA 94133

The Ecologist
Ecosystems Ltd.
73 Molesworth Street, Wadebridge
Cornway PL27 7DS
United Kingdom

Ecology
Ecological Society of America
Center for Environmental Studies
Arizona State University
Tempe, AZ 85281

Ecology Law Quarterly
University of California
Boalt Hall School of Law
Berkeley, CA 94720

E: The Environmental Magazine
Earth Action Network, Inc.
Subscription Department
P.O. Box 6667
Syracuse, NY 13217

Endangered Species Update
School of Natural Resources
University of Michigan
Ann Arbor, MI 48109

Environment
Heldref Publications
4000 Albemarle Street, NW
Washington, DC 20016

Environment Abstracts
Bowker Publishing
245 West 17th Street
New York, NY 10011

Environmental Action
Environmental Action Foundation, Inc.
1525 New Hampshire Avenue, NW
Washington, DC 20036

Environmental Defense Letter
Environmental Defense Fund
257 Park Avenue South
New York, NY 10010

Environmental Ethics
Department of Philosophy
University of Georgia
Athens, GA 30602

The Environmental Professional
National Association of Environmental
Professionals
Subscription Department
P.O. Box 15210
Alexandria, VA 22309-0210

Environmental Quality
Council on Environmental Quality
722 Jackson Place, NW
Washington, DC 20006

Environmental Science & Technology
American Chemical Society
1155 16th Street, NW
Washington, DC 20036

EPA Journal
Environmental Protection Agency
Government Printing Office
Washington, DC 20402

Everyone's Backyard
Citizens' Clearinghouse on Hazardous
Waste
P.O. Box 926
Arlington, VA 22216

Family Planning Perspectives
Planned Parenthood–World Population
Editorial Offices
666 Fifth Avenue
New York, NY 10019

FDA Consumer
U.S. Department of Health and
Human Services
Public Health Service
5600 Fishers Lane
Rockville, MD 20857

FOCUS
World Wildlife Fund
1250 24th Street, NW
Washington, DC 20037
(202) 293-4800

The Futurist
World Future Society
P.O. Box 19285, Twentieth Street
Station
Washington, DC 20036

Gaia Quarterly
Cathedral of St. John the Divine
1047 Amsterdam Avenue at 112th
Street
New York, NY 10025

**Garbage: The Practical Journal for the
Environment**
P.O. Box 51647
Boulder, CO 80321-1647
1-800-888-9070

Greenpeace Magazine
Greenpeace USA
1436 U Street, NW
Washington, DC 20009

Harrowsmith
Ferry Road
Charlotte, VT 05445

**In Business: The Magazine for
Environmental Entrepreneuring**
Subscription Department
419 State Avenue
Emmaus, PA 18049
(215) 967-4135

INFORM Reports
INFORM
381 Park Avenue South
New York, NY 10016

International Environmental Affairs
University Press of New England
17½ Lebannon Street
Hanover, NH 03755

**International Journal of Air Pollution
Control and Hazardous Waste
Management (JAPCA)**
Air Pollution Control Association
P.O. Box 2861
Pittsburgh, PA 15230

International Wildlife
International Wildlife Federation
Membership Services
National Wildlife Federation
8925 Leesburg Pike
Vienna, VA 22184
1-800-432-6564

Issues in Science and Technology
National Academy of Sciences
2101 Constitution Avenue, NW
Washington, DC 20077-5576

**Journal of the American Public Health
Association**
1015 18th Street, NW
Washington, DC 20036

Journal of Environmental Education
In association with the North
American Association for
Environmental Education
The Helen Dwight Reid Educational
Foundation
4000 Albemarle Street, NW
Suite 504
Washington, DC 20016

Journal of Environmental Health
National Environmental Health
Association
720 South Colorado Boulevard
South Tower 970
Denver, CO 80222

Journal of Forestry
Society of American Foresters
5400 Grosvenor Lane
Bethesda, MD 20814

Journal of Range Management
Society for Range Management
1839 York Street
Denver, CO 80206

**Journal of Soil and Water
Conservation**
Soil and Water Conservation Society
7515 Northeast Ankeny Road
Ankeny, IA 50021-9764

**Journal of the Water Pollution Control
Federation**
601 Wythe Street
Alexandria, VA 22314-1994

Journal of Wildlife Management
Wildlife Society
5410 Grosvenor Lane
Bethesda, MD 20814

Mother Jones
P.O. Box 50032
Boulder, CO 80322-0032

National Geographic
National Geographic Society
P.O. Box 2895
Washington, DC 20077-9960

**National Parks and Conservation
Magazine**
National Parks and Conservation
Association
1015 31st Street, NW
Washington, DC 20007

National Wildlife
Membership Services
National Wildlife Federation
8925 Leesburg Pike
Vienna, VA 22184
1-800-432-6564

Natural History
American Museum of Natural History
Central Park West at 79th Street
New York, NY 10024

Natural Resources Journal
University of New Mexico
School of Law
1117 Stanford, NE
Albuquerque, NM 87131

Natural Resources News
School of Natural Resources
University of Michigan
Ann Arbor, MI 48109-1115

Nature
711 National Press Building
Washington, DC 20045

The Nature Conservancy News
1815 North Lynn Street
Arlington, VA 22209

New Scientist
128 Long Acre
London, WC2
England

Newsline
Natural Resources Defense Council
122 East 42nd Street
New York, NY 10168

Not Man Apart
Friends of the Earth
530 7th Street, SE
Washington, DC 20003

Ocean Watch
The Oceanic Society
1536 16th Street, NW
Washington, DC 20036

**Organic Gardening & Farming
Magazine**
Rodale Press, Inc.
33 East Minor Street
Emmaus, PA 18049

Orion
Subscription Department
136 East 64th Street
New York, NY 10021

Oryx
Journal Subscriptions Department
Marston Book Services
P.O. Box 87
Oxford, England

Pollution Abstracts
Cambridge Scientific Abstracts
7200 Wisconsin Avenue
Bethesda, MD 20814

Population and Vital Statistics Report
United Nations Environment
Programme
New York North American Office
Publications Sales Section
United Nations
New York, NY 10017

Population Bulletin
Population Reference Bureau
777 14th Street, NW, Suite 800
Washington, DC 20005

Renewable Energy News
Solar Vision, Inc.
7 Church Hill
Harrisville, NH 03450

Rocky Mountain Institute Newsletter
1739 Snowmass Creek Road
Snowmass, CO 81654

Science
American Association for the
Advancement of Science
1333 H Street, NW
Washington, DC 20005

Science News
Science Service, Inc.
1719 N Street, NW
Washington, DC 20036

Scientific American
415 Madison Avenue
New York, NY 10017

Sierra
Sierra Club
730 Polk Street
San Francisco, CA 94109

**Smithsonian Institution Research
Reports**
Office of Public Affairs
1000 Jefferson Drive, SW
Washington, DC 20560

Smithsonian Magazine
Smithsonian Institution Press
955 L'Enfant Plaza, Suite 2100
Washington, DC 20560

Solar Age
Solar Vision, Inc.
7 Church Hill
Harrisville, NH 03450

The State of the Environment
OECD Publications and Information
Center
1750 Pennsylvania Avenue, NW
Suite 1207
Washington, DC 20006-4582
or
OECD Publications Office
2, rue Andre-Pascal
75775 PARIS CEDEX 16

State of the States
Renew America
1001 Connecticut Avenue, NW
Suite 719
Washington, DC 20036

State of the World
Worldwatch Institute
1776 Massachusetts Avenue, NW
Washington, DC 20036

Statistical Yearbook
Department of International Economic
and Social Affairs
Statistical Office
United Nations Publishing Service
United Nations
New York, NY 10017

Technology Review
Room E219-430
Massachusetts Institute of Technology
Cambridge, MA 02139

Transition
Department of Geography
University of Cincinnati
Cincinnati, OH 45221

The Trumpeter Journal of Ecosophy
P.O. Box 5883, Station B
Victoria, British Columbia
Canada, V8R 6S8

Wilderness
Wilderness Society
1400 I Street, NW
Washington, DC 20005

Wildlife Conservation
New York Zoological Park
Bronx, NY 10460

World Development Report
World Bank, Publications Department
1818 H Street, NW
Washington, DC 20433

World Magazine
PMG Inc.
303 Linwood Avenue
Fairfield, CT 06430

World Rainforest Report
Rainforest Action Network
300 Broadway, Suite 298
San Francisco, CA 94133

World Resources
World Resources Institute
1735 New York Avenue, NW
Washington, DC 20006

World Watch
Worldwatch Institute
1776 Massachusetts Avenue, NW
Washington, DC 20036

Worldwatch Papers
Worldwatch Institute
1776 Massachusetts Avenue, NW
Washington, DC 20036

Yearbook of World Energy Statistics
Department of International Economic
and Social Affairs
Statistical Office
United Nations Publishing Service
United Nations
New York, NY 10017

Glossary

Abiota   Nonliving component of the environment.

Abortion   An elective surgical procedure in which the lining of the uterus is removed along with the developing embryo.

Abundant metal   Any metal that constitutes greater than 1 percent of the earth's continental crust, for example, iron, aluminum, manganese, titanium, and magnesium. Compare *Scarce metal.*

Abyssal zone   The bottom waters of the ocean beyond the continental shelf, usually below a depth of 1,000 meters.

Acculturation   The process by which one culture is modified through contact with another.

Acid mine drainage   The seepage of sulfuric acid solutions from mines and mining wastes dumped at the surface.

Acid precipitation   Rain, snow, or fog that contains higher than normal levels of sulfuric or nitric acid, which may damage forests, aquatic ecosystems, and cultural landmarks.

Acid rain   See *Acid precipitation.*

Acid surge   A period of short, intense acid deposition in lakes and streams as a result of the release (by rainfall or spring snowmelts) of acids stored in soil or snow.

Active solar system   System for temperature control in buildings that utilize devices such as fans and pumps to enhance the distribution of collected and stored solar heat. Compare *Passive solar system.*

Actual population growth rate   The growth rate of a country that includes the natural growth rate and the effects of migration. See *Natural population growth rate.*

Acute effect   An effect that appears shortly after exposure to a toxic substance or disease-causing organism.

Acute toxicity   The ability of a substance to trigger an adverse reaction immediately after an organism is exposed to it. Compare *Chronic toxicity.*

Adaptation, genetic   A genetically determined characteristic that improves an organism's ability to survive and transmit its genes to the next generation.

Adaptation, physiological   A change in an organism's physiological response to a stimulus or substance that occurs after prolonged exposure to that stimulus or substance.

Advanced material   Any material developed since the 1950s that exhibits greater strength, greater hardness, or better thermal, electrical, or chemical properties than a traditional material.

Aerobic   Requiring oxygen.

Aerobic decomposition   The degradation of organic material by living organisms in the presence of oxygen.

Age distribution   Percentage of the population or number of people of each sex at each age level in a population.

Agent Orange   A chemical vegetation killer that was used by the United States during the Vietnam War and that had harmful effects on humans.

Age-specific fertility rate   The number of live births per 1,000 women of a specific age group per year.

Agreement on Cooperation in the Field of Environmental Protection, 1972   An agreement between the former Soviet Union and the United States to have scientists from both countries cooperate on diverse projects in the field of environmental protection.

Agribusiness   Large-scale farming conducted by corporations.

Agrichemicals   Fertilizers, herbicides, and pesticides that are applied to plant crops to increase yield or reduce pest damage.

Agriculture   The intentional tending of a particular plant species for human use.

Agroecology   The study of agroecosystems and the long-term management of agricultural land.

Agroecosystem   A crop-field ecosystem.

Airborne toxins   Hazardous substances that have been released into the atmosphere and are carried by air currents.

Air pollutant   Any substance present in or released to the atmosphere that adversely affects human health or the environment.

Air pollution alert   Level of air pollution dangerous to humans defined as an air quality index reading of 100 or greater.

Air quality index   Index showing the frequency and degree of air pollution in a particular region, usually determined by the pollutant with the highest measured level.

Air quality standards   Standards limiting air pollutants that are harmful to human health and well-being.

Albedo   Reflectivity, or the fraction of incident light that is reflected by a surface.

Algae   A group of relatively simple, mostly aquatic plants that occur in fresh and salt water, and also in damp terrestrial areas.

Algal bloom   An extensive growth of algae in a body of water, sometimes as a result of an increase in the phosphate and nitrate content of the water caused by fertilizer and detergent runoff.

Alkaline   Having the properties of or containing a soluble base. See *pH.*

Alpha particle Slow-moving particle consisting of two protons and two neutrons which cannot penetrate paper or skin but is dangerous when inhaled or ingested.

Alternative agriculture Approach to agriculture to halt the long-term reduction of soil fertility caused by conventional, high-input agriculture; the use of techniques such as polyculture, organic fertilization, and biological control of pests to maintain croplands.

Alternative energy resource Resource that is available but not widely used; resource other than conventional fossil fuels.

Ambient Surrounding, prevailing.

Ammonia A colorless gas composed of nitrogen and hydrogen (NH_3); the main form in which nitrogen is available to living cells.

Anaerobic Not requiring oxygen. Compare *Aerobic*.

Anaerobic decomposition The degradation of organic material by living organisms in the absence of oxygen. Compare *Aerobic decomposition*.

Anemia A condition in which the number of red blood cells, amount of hemoglobin, or total volume of blood is deficient that may be caused by various diseases, poisoning, or malnutrition; causes weakness and fatigue.

Anthracite coal A rare coal with low sulfur content and high heating value. Compare *Bituminous coal*, *Subbituminous coal*, *Lignite coal*.

Anthropocentric worldview The belief that humans are not part of nature and that nature is that part of the world devoid of human influence. Compare *Biocentric worldview*.

Antinatalist policy A policy that discourages natality or births.

Antiquities Act of 1906 The first major piece of legislation in the United States designed to safeguard cultural resources by giving the president the power to designate national monuments and establish regulations to protect archeological sites on public lands.

Applied ecology The study of the ecological consequences of human activities and of ways to limit damage to or restore natural ecosystems.

Aquaculture The production of aquatic plants or animals in a controlled environment.

Aquatic Pertaining to marine and freshwater environments. See *Terrestrial*.

Aquifer Underground, water-saturated zone which may extend from a few square miles to several thousand square miles.

Arable land Region that is able to sustain agriculture on an annual basis.

Area strip mining A method of mineral extraction in which the entire deposit is removed from the surface down.

Asbestos A substance formerly widely used for thermal insulation that is made of magnesium silicate minerals, minute particles of which can be inhaled and cause respiratory problems.

Associated gas Natural gas that is found with petroleum and can be extracted only when the petroleum is brought to the earth's surface and refined. Compare *Nonassociated gas*.

Atmosphere A 500-kilometer thick layer of colorless, odorless gases known as air that surrounds the earth and is composed of nitrogen, oxygen, argon, carbon dioxide, and other gases in trace amounts.

Autotroph An organism that can produce complex chemicals to satisfy its nutritional needs from simple chemicals and a source of energy, such as sunlight. See *Phototroph*, *Chemotroph*. Compare *Heterotroph*, *Saprotroph*.

Bacteria Unicellular, or rarely multicellular, prokaryote organisms.

Barrier contraception A type of birth control that relies on a physical device, such as a condom or diaphragm, to prevent sperm from reaching the egg.

Basel Convention of 1989 An agreement signed by 34 countries and tentatively approved by 71 to control the international transport and disposal of hazardous wastes.

Below-cost timber sales The sale of timber to lumber companies at prices lower than the actual cost incurred by the Forest Service to make that timber available.

Benthic community Community of organisms, primarily invertebrate animals, which lives at the bottom of a body of water.

Beta particle Fast-moving particle, similar to an electron but containing more energy, which can penetrate the outer layers of skin and cause damage to the body.

Big bang theory The theory that the universe—all matter, energy, and space—arose from an infinitely dense, infinitely hot point called a singularity which exploded and began to expand 13 to 20 billion years ago.

Binding force The attractive force that holds together the constituents of an atom, atomic nucleus, or molecule.

Bioaccumulation The storage of chemicals in an organism in higher concentrations than are normally found in the environment.

Biocentric worldview The belief that humans are a part of nature and that humans are subject to all natural laws. Compare *Anthropocentric worldview*.

Biochemical conversion Process of converting biomass through fermentation or digestion to produce energy in the form of biogas or alcohol.

Biodegradable Capable of being broken down by natural systems.

Biodiversity The variety of life forms that inhabit the earth; biodiversity includes the genetic diversity among members of a population or species as well as the diversity of species and ecosystems. See *Biological diversity*.

Biogas The methane-rich gas produced from the fermentation of animal dung, human excreta, or crop residues.

Biogeochemical cycle A series of biological, chemical, and geological processes by which materials cycle through ecosystems.

Biointensive agriculture A form of organic agriculture that relies on the use of biological controls to maintain croplands and strives to put all available land to use.

Biological diversity The variety of different species found within a particular ecosystem or community and the genetic variability found within a species.

Biological oxygen demand (BOD) A measure of the amount of oxygen needed to decompose the organic matter in water.

Biological resources All plant and animal species, domesticated and wild.

Biomagnification The accumulation of chemicals in organisms in increasingly higher concentrations at successive trophic levels.

Biomass The sum of all living material in a given environment.

Biome A major ecological community of organisms, both plant and animal, usually characterized by the dominant vegetation type, for example, a tundra biome and a tropical rain forest biome.

Bioremediation The use of living organisms (bacteria) to clean up hazardous substances in the environment.

Biosphere The part of the earth's surface and its immediate atmosphere that is inhabited by living organisms.

Biota Living component of the environment.

Biotechnology The industrial use of living microorganisms, such as bacteria and other biological agents, to perform chemical processing or to produce materials such as animal food. See also *Genetic engineering*.

Biotic potential The ability of species to reproduce regardless of the level that an environment can support. See *Carrying capacity*.

Bituminous coal An abundant coal with medium to high sulfur content and high heating value. Compare *Anthracite coal*, *Subbituminous coal*, *Lignite coal*.

Bottle bill Legislation to encourage recycling by requiring consumers to pay a deposit on bottle or can beverage containers, which is refunded when the containers are returned to a food market or recycling center.

Breeder reactor A type of fission reactor which eventually produces more fuel (plutonium 239) than it consumes.

Cancer A general term for a disease process in which the growth control mechanism of cells becomes altered and the cells reproduce with abnormal rapidity.

Capitalism An economic system based on free-market activities in the pursuit of profit.

Carbon cycle The natural circulation of carbon in the biosphere.

Carbon dioxide A colorless gas (CO_2) formed as a product of animal respiration and decay and combustion of animal and plant matter. A primary pollutant originating chiefly from fossil fuel combustion which contributes to the greenhouse effect.

Carbon monoxide A colorless, odorless toxic gas (CO) formed as a product of incomplete combustion of carbon. A primary pollutant originating chiefly from imperfect combustion in fuel engines.

Carcinogenic Cancer causing.

Carnivore An organism that consumes only animals. Compare *Detrivore*, *Herbivore*, *Omnivore*.

Carrying capacity The maximum number of a particular species that an environment can support without degradation to the environment.

Cellulolytic bacteria Bacteria that feed on the cellulose in plants and products, such as wood and paper.

Center of diversity Area of great biological diversity which was a site of the earliest domestications of plants.

CERCLA See *Superfund*.

CFC See *Chlorofluorocarbon*.

Chaining A forest management technique in which a chain is strung between two bulldozers and then pulled along, tearing down any trees in its path.

Chain reaction A series of reactions in which each reaction causes additional and increasingly larger reactions.

Charismatic megafauna A group of species, typically large mammals, which are highly valued by humans.

Chemical sensitizer A substance that causes an individual to become extremely sensitive to low levels of that particular substance or to experience reactions to other substances.

Chemosynthesis The process of producing energy from inorganic materials through simple chemical reactions.

Chemotroph A producer organism that converts inorganic chemical compounds into energy.

Childhood mortality rate The number of children between the ages of 1 and 5 who die per 1,000 births per year.

Chlorofluorocarbon A chemical used as a coolant, solvent, or propellant which is a catalyst in the destruction of atmospheric ozone.

Chronic toxicity The ability of a substance to cause an adverse reaction in an organism after it accumulates in the organism through repeated exposures. Compare *Acute toxicity*.

Chronic undernutrition The consumption of too few calories and too little protein over an extended period of time.

Clean Air Act Legislation originally passed in 1963 that calls for states to develop air quality standards and implementation plans to curb atmospheric pollutants.

Clear-cutting A forest management technique in which an entire stand of trees is felled and removed.

Climate The long-term weather pattern of a particular region.

Climax A plant community that has reached stability and is in equilibrium with the climatic conditions prevailing at the time.

Clockwork universe The view of the nature of the universe, proposed by Francis Bacon in the seventeenth century, that God established the laws of nature and that science could discover God's patterns and arrive at truth.

Cloning See *Tissue culture*.

Coal gasification The conversion of solid coal to a gas that can be burned as fuel.

Co-composting A form of composting in which organic wastes are mixed with sewage sludge to yield a high-quality, nutrient-rich material.

Cogeneration The use of waste heat from one process to power a second process, for example, the use of steam generated in the production of electricity to heat buildings.

Coliform A bacteria normally inhabiting the colon, for example, *Escherichia coli* and *Enterobacteria aerogenes*, high levels of which in drinking water or swimming water may indicate the presence of disease-causing organisms in the water.

Collection center Facility for the collection of recyclable materials.

Combined sewer system Sewer system in which pipes that carry sewage are coupled with pipes that remove rainwater from streets after storms.

Co-mingled recycling The collection of different types of recyclable items in a single container.

Commensalism An interaction between two species that benefits one species, and neither harms nor benefits the other. Compare *Mutualism*, *Parasitism*, *Symbiosis*.

Common law A system of law, originating in England, based on custom and past court decisions rather than on government or ecclesiastical statutes.

Community All of the populations of organisms that interact in a given area at a given time.

Competition An interaction between two or more organisms that are striving to attain the same limited resource, such as food, water, space, or mates.

Competitive exclusion principle The principle that competition between two species with similar requirements will result in the exclusion of one of the species.

Component separation A waste-treatment method in which particular hazardous materials are removed from a waste stream without the use of chemicals.

Composite An advanced material that is a matrix of one material reinforced with fibers or dispersions of another.

Composting A waste-disposal method in which bacteria decompose biodegradable wastes such as paper, food scraps, and yard wastes to form compost or humus.

Comprehensive Environmental Response, Compensation, and Liability Act (CERCLA) See *Superfund.*

Computer model In environmental science, a sophisticated mathematical equation that helps applied ecologists understand how ecosystems respond to stress and to predict the effects of management strategies on a particular environment.

Conservation A management philosophy that encourages rational use of the environment to provide a long-term quality life for humans without sacrificing other constituents of the biotic community.

Conservation ecology Field of science that applies ecological principles and knowledge in order to preserve or conserve species and communities. Also known as preservation ecology.

Conservation Revolution Environmental movement in the 1970s in the United States which increased awareness of the country's energy problems and led to environmental legislation.

Conservation tillage The farming practice of disturbing the soil as little as possible in order to reduce soil erosion, lower labor costs, and save energy.

Consumer An organism that cannot produce its own food and must get it by eating or decomposing other organisms. In economics, one who uses goods and services.

Consumerism Wasteful consumption of resources to satisfy wants rather than needs.

Consumptive Tending to consume, as by use, decay, or destruction.

Containment vessel An enclosure surrounding a nuclear reactor designed to prevent the accidental release of radiation.

Continental shelf The gently sloping sea floor around the earth's land masses.

Continuous cropping The farming practice of growing the same crop year after year without allowing the land to lie fallow periodically to restore itself.

Contour plowing The farming practice of tilling the soil along the natural contour of the land in order to protect against erosion.

Contour strip mining A method of mineral extraction in which a series of shelves or terraces are cut on the side of a hill or mountain to remove a mineral from a deposit near the earth's surface.

Contour terracing The farming practice of constructing ridges along the natural contour of the land in order to reduce soil erosion from the unimpeded runoff of rainwater.

Conventional agriculture The practice of maintaining croplands through the use of synthetic fertilizers, pesticides and herbicides, and heavy machinery.

Conventional fuel An energy source, such as fossil fuels, that is currently widely used.

Convention on International Trade in Endangered Species (CITES) International legislation enacted in 1975 which prohibits international trade in approximately 600 endangered species and requires export licenses for an additional 200 threatened species.

Cooperation An interaction between members of the same species or an association of two dissimilar species that aids in the survival of one or both parties.

Coral reefs A limestone ridge near the surface of the sea formed from the calcareous skeletons of reef-building coral and other marine organisms.

Cornucopian view of population growth The view that population growth is a positive influence on future economic growth and that any problems arising from excess population will be resolved by human inventiveness.

Cover crop A crop planted to cover the ground after a main crop is harvested in order to reduce erosion and maintain soil fertility.

Creationism A theory of the origin of life that postulates that living organisms were created by an all-powerful being (God) 6,000 years ago, according to the record set forth in the Bible.

Critical mass The amount of a fissionable material necessary to sustain a chain reaction.

Critical mineral A mineral considered essential to a country's economic activity.

Crop rotation The farming practice of changing the crop planted in a particular area in successive years to avoid soil depletion and maintain soil fertility.

Cross-media pollutant A pollutant that moves through various media (soil, water, air) from one environment to another.

Crude birth rate The number of live births per 1,000 people, per year.

Crude death rate The number of deaths per 1,000 people, per year.

Cryopreservation A method of preserving living material by freezing and storing in liquid nitrogen at very low temperatures.

Cultural carrying capacity The maximum number of people a given amount of land can support at a given standard of living.

Cultural diversity The variety of values, attitudes, beliefs, and actions within a population and in different populations.

Cultural eutrophication The addition of organic nutrients to bodies of water through human activities.

Cultural resource Anything that embodies, symbolizes, or represents a part of the culture of a specific people.

Cultural resources management A field of interest concerned with preserving both material culture and living culture.

Culture The beliefs, attitudes, values, behaviors, and practices of a group or society.

Curbside collection Method of source separation in which households remove recyclables from other garbage and place them in separate containers to be picked up with the garbage.

Darwinism The theory of evolution formulated by Charles Darwin that different species of plants and animals arise by a process of slow and gradual changes over successive generations, brought about by natural selection.

DDT Dichlorodiphenyltrichloro-ethane, a chlorinated hydrocarbon that was at one time widely used as a pesticide.

Debt-for-nature swap An agreement between a debtor nation and a bank in which the bank agrees to absolve a loan if significant amounts of undisturbed land are set aside for conservation.

Decommissioning The process of dismantling and decontaminating a nuclear reactor once it has reached the end of its useful life.

Decomposer See *Detrivore*.

Deep ecology A belief in the unity of humans, plants, animals, and the earth.

Deep well injection A waste disposal method in which liquid wastes are pumped deep underground into rock formations that geologists believe will contain them permanently.

De facto wilderness A wilderness area not protected by official designation.

Defoliant A herbicide that causes premature removal of leaves from plants.

Deforestation The cutting down and clearing away of forests.

Demographic transition The movement of a nation from high population growth to low growth.

Demographic trap The inability of a country to pass the second phase of demographic transition, which is characterized by high birth rates and low death rates.

Demography The statistical study of characteristics of human populations, such as size, distribution, age, income.

Density-dependent factor A biotic factor in an environment that limits population growth and is more pronounced when population density is high.

Density-independent factor An abiotic factor in an environment that sets upper limits on the population, and is unrelated to population density.

Dependency load The number of dependents (those under age 15 or over 65) in a population.

Desalination The process by which salt and suspended minerals in seawater are removed to make drinkable water.

Desert An area of land in which evaporation exceeds precipitation, usually where annual precipitation is less than 250 millimeters per year.

Desertification The creation of a desert due to overgrazing, overcultivation, deforestation, poor irrigation procedures, or climate changes.

Detritus food web A food web that includes several levels of consumers that derive energy from decomposing plant and animal material or animal waste products.

Detrivore An organism that consumes dead or decaying tissues or wastes. Also known as detritus feeder or scavenger. Compare *Carnivore*, *Herbivore*, *Omnivore*.

Direct radiation exposure Exposure to a source of radioactivity.

Dispute resolution The process of negotiation and compromise in which two disputing parties meet face to face and reach a mutually acceptable solution to the dispute.

Disruption An acute disturbance in a natural system that can be traced directly to a specific human activity. Compare *Disturbance*.

Disturbance A variation in some factor in an ecosystem beyond the normal range of variation resulting in a change in the ecosystem.

Disturbance ecology A branch of ecology concerned with assessing the impact of stress on organisms, populations, and ecosystems.

DNA Deoxyribonucleic acid, a complex molecule found in the cells and chromosomes of almost all organisms that acts as the primary genetic material.

Domestication An evolutionary process in which genes useful for survival in captivity prevail over genes necessary for survival in the wild.

Doubling effect The hypothesis that a doubling in atmospheric carbon dioxide will increase worldwide temperature by 3° C (9° F), causing a number of diverse climatic changes.

Doubling time The number of years it takes for a population to double assuming current growth rates remain the same.

Dry steam reservoir A rare and highly preferred geothermal energy source in which underground water deposits have already been vaporized.

Dynamic equilibrium An ecosystem's ability to react to constant changes thereby maintaining relative stability.

Ecological toxicology A branch of ecology concerned with determining the effects of toxic materials on individual organisms.

Ecology The study of the structure, function, and behavior of the natural systems that comprise the biosphere.

Economics The study of the production, consumption, and distribution of goods and services.

Ecosystem A system of interdependent and interacting living organisms and their immediate physical, chemical, and biological environments.

Ecosystem development The maturation process of an ecosystem.

Ecosystem diversity The variety of plants and animals in a specific ecosystem.

Ecotone The transitional zone of intense competition for resources and space between two communities.

Ectoparasite A parasite that lives on the outside of the host's body. Compare *Endoparasite*.

Effluent Any contaminating substance, usually a liquid, that enters the environment through an industrial, agricultural, or sewage plant outlet.

Emission standard The maximum amount of a specific pollutant permitted to be discharged from a particular source in a given environment.

Emissivity The relative power of a surface to reradiate solar radiation back into space in the form of heat, or long-wave infrared radiation.

Endangered species A plant or animal species that no longer can be relied on to reproduce in numbers sufficient to ensure its survival.

Endangered Species Act, 1973 Legislation that mandates that a species or subspecies "endangered throughout all or a portion of its range" be listed with the Secretary of the Interior (marine species with the Secretary of Commerce); makes the U.S. Fish and Wildlife Service and the National Marine Fisheries responsible for issuing regulations to protect endangered and threatened species; and prohibits the killing, capturing, importing, exporting, and selling of plants or animals listed as endangered or threatened.

Endemic species A plant or animal species confined to or exclusive to a specific area.

Endoparasite A parasite that lives inside of the host's body cavity, organs, or blood. Compare *Ectoparasite*.

End product substitution A method of minimizing hazardous substances in the environment in which consumers choose products that will generate a minimum volume of hazardous resources.

Energy The ability to do work, to move matter from place to place, or to change matter from one form to another.

Energy efficiency The ratio of production or output to the amount of energy input.

Entropy   The measure of the degree of molecular disorder within an organism or the degree of disorder within a system.

Environment   The system of interdependent living and nonliving components in a given area over a given period of time, including all physical, chemical, and biological interactions.

Environmental activist   A person who works for the protection of organisms, land, water, air, and other natural resources from pollution and other harmful effects of human activities.

Environmental degradation   A change in the earth's life support systems that interferes with the earth's capacity to maintain a maximum range of tolerances for life.

Environmental education   Information and training to awaken and explore a person's intuitive value of the natural world.

Environmental ethics   Principles of respect and care for the natural world.

Environmental impact statement   An analysis of a proposed project to predict its likely repercussions on the social and physical environment of the surrounding area.

Environmentalism   Concern for the environment above and beyond personal concerns.

Environmental law   That part of the legal system that deals with natural resource protection, pollution standards, and liability related to the state and protection of the environment.

Environmentally sound management   Methods of managing any resource, natural or cultural, which minimize or prevent environmental degradation.

Environmental problem solving   A five-step interdisciplinary and goal-directed model to develop workable solutions, which can be applied to almost any situation.

Environmental Protection Agency   The U.S. government agency responsible for federal efforts to control pollution of air, land, and water by human activities and to sponsor research and education on the impact of humans on the environment.

Environmental refugee   An individual who is forced to abandon his or her home because the land can no longer support him or her.

Environmental resistance   Controlling influence of the environment which determines a successful level of numbers for a given population.

Environmental science   The interdisciplinary study of the complex interactions between natural and cultural systems, including the fields of physical geography, ecology, geology, and aspects of economics, politics, and sociology.

Environmental toxicology   The study of the effects of poisons and pollutants on ecosystems and the environment as a whole.

Epidemiology   The study of the distribution and causes of health disorders.

Epilimnion   Upper layer of a lake, which is heated by the sun and tends to be warmer than the underlying water. Compare *Hypolimnion*.

Escherichia coli   See *Coliform*.

Estuary   A semienclosed, nutrient-rich, coastal body of water usually composed of fresh and salt water.

Ethic   A system of morals which governs or shapes behavior.

Euphotic zone   The area in a body of water where light penetrates enough to support photosynthesis and populations of plankton. Compare *Neritic zone, Pelagic zone*.

Eutrophication   A mix of physical, chemical, and biological processes, occuring either naturally or through human activities, by which a body of water becomes rich in nutrients.

Eutrophic lake   A lake that is warm and shallow, with a low oxygen content and a relatively high amount of nutrients, phytoplankton, and other organisms. Compare *Oligotrophic lake*.

Evaporation   The process of energy absorption that enables a liquid to change state and become a gas.

Evolution   An explanation of the way in which present-day organisms developed, involving changes in the genetic makeup of populations passed on to subsequent generations. See *Creationism, Darwinism, Natural selection*.

Evolutionary potential   The potential for a new species to arise from an ancestor through adaptation and specialization over millions of years.

Exclusionary land use   See *Restricted land use*.

Exclusive economic zone   Waters extending 200 miles from a country's shoreline within which it has exclusive rights to exploit mineral and fish resources established by the U.N.; Law of the Sea Convention in 1977.

Exponential growth   A rate of growth that follows a geometric rate of increase, in which each increase differs from the preceding one by a constant ratio, creating a J-curve when plotted on a graph.

Extinction   The disappearance of a species from the biota.

Facilitation model of succession   A theory that proposes that one community prepares the ecosystem for a subsequent community.

Family planning   Measures that enable parents to control the number of children they have (if they so desire), and the spacing of their children's births.

Famine   The widespread scarcity of food, with subsequent suffering and starvation in the population.

FAO   The Food and Agriculture Organization, a specialized agency of the United Nations, founded in 1945 to coordinate programs of food, agriculture, forestry, and fisheries development in order to improve the standards of living of rural populations and combat malnutrition and hunger.

Fauna   The total animal population that inhabits an area.

Federal Water Pollution Control Act of 1972   Legislation that stipulated how industry must treat pollutants and classified pollutants into three categories: toxic, conventional, and unconventional. Also known as the Clean Water Act.

Fertility   A measurement of a population's potential for future population growth.

Fertilizer   Material added to soil to supply essential nutrients for crop growth.

Field gene bank   Storage facility that preserves genetic plant material through annual germinations.

First law of thermodynamics   See *Law of the Conservation of Energy*.

Fishery   Concentrations of particular aquatic species in a given aquatic area suitable for commercial exploitation.

Fission products   The radioactive by-products of a fission reaction.

Fission reaction   Reaction that occurs when neutrons hit the unstable nucleus of a radioactive atom, splitting it into two smaller fragments while discharging more neutrons and large amounts of energy.

Flora   The total plant population that inhabits an area.

Fluidized bed combustion   Process for burning coal more efficiently, cleanly, and cheaply, in which a mixture of powdered coal and limestone is suspended by a stream of hot air during combustion.

Fly ash   Extremely fine particles of ash produced when coal is burned in modern forced-draft furnaces, particularly those associated with electric power plants.

Food chain   Successive levels within an ecosystem illustrating the energy transfers between organisms as a result of consumption.

Food security   The ability of a nation to feed itself on an ongoing basis.

Food Security Act of 1985   More commonly known as the 1985 Farm Bill, legislation that implemented several significant conservation programs aimed at reducing the deterioration of U.S. farmland.

Food web   A complex system of various interdependent food chains in a given ecosystem.

Forestry   The practice of planting, tending, and managing forests primarily for the exploitation of timber for commercial or local subsistence needs.

Fossil fuel   Any naturally occurring carbon or hydrocarbon fuel derived from anaerobic decomposition of organic material in the earth's crust, including natural gas, oil, coal, oil shale, and tar sands.

Fossil groundwater   Groundwater sources that were formed thousands of years ago when the earth's rainfall pattern was much greater.

Frontier ethic   Belief by American settlers, which is still popular today, that natural resources are inexhaustible or will regenerate and that exploration will discover new, untapped resources.

Fungicide   A chemical used to kill fungi.

Fungus   A saprotrophic or parasitic organism that may be unicellular or made up of tubular filaments and which lacks chlorophyll.

Fusion reaction   Reaction that occurs under extremely high temperatures in which individual nuclei of a particular element fuse to form another element.

Gaia hypothesis   James Lovelock's worldview that compares the earth to a self-regulating, living organism.

Game animal   Any animal hunted chiefly for sport.

Gamma ray   Short, intense burst of electromagnetic energy given off by a radioactive substance, capable of penetrating lead or concrete 1 meter thick and of causing tissue damage in living organisms.

Gaseous cycle   The circulation of a gas through the environment, primarily in the atmosphere.

Gene   The basic unit of heredity.

Gene bank   Storage facility to preserve genetic material through various freezing and drying methods.

Gene pool   The sum of all the genes present in a population of organisms.

General fertility rate   The number of live births per 1,000 women of childbearing age per year.

Generalist   A species that can survive in many different habitats. See *Specialist*.

Genetic diversity   Variation among the members of a single population of species.

Genetic engineering   The human manipulation and transfer of genes from one organism to another to improve the productivity or survivability of economically important organisms.

Genetic erosion   The reduction of genetic diversity in a gene pool due to decreasing population size.

Genotype   The genetic makeup of an organism or group of organisms.

Geopressurized reservoir   A geothermal energy source in which underground water and methane gas are subjected to extremely high temperatures and pressures.

Geothermal energy   Energy generated by the natural heat and pressure occurring beneath the earth's surface.

Germ plasm   The genetic material of an organism.

Giardiasis   The most common waterborne disease in the United States today arising from ingestion of the protozoan *Giardia*.

Global warming   The increase in global temperature predicted to arise from increased levels of carbon dioxide, methane, and other gases in the atmosphere.

Grassed waterway   A trench planted in grass to slow the course of runoff and absorb excess water and eroding soil.

Grassroot loan   Small-scale loans to individuals in a community rather than to capital-intensive development projects.

Great hunger belt   Equatorial region in which most of the world's hungry live, spanning parts of Southeast Asia, the Indian subcontinent, the Middle East, Africa, and Latin America.

Greenhouse effect   The prevention of the reradiation of heat waves to space by carbon dioxide, methane, and other gases in the atmosphere.

Greenhouse gas   A gas that contributes to the greenhouse effect, such as carbon dioxide, chlorofluorocarbons, ozone, methane, and nitrous oxide.

Green manure   A cover crop, such as alfalfa, that is planted and then plowed under to improve soil structure and fertility.

Green marketing   The promotion of products based on claims that they are benign to the environment.

Green revolution   A group of measures to improve agricultural productivity in less-developed countries including the development of high-yield cereal varieties, their adoption and diffusion in the third world, the increased use of fertilizers, pesticides, and irrigation.

Gross national product (GNP)   Total market value in current dollars of all final goods and services produced by an economy during a year.

Gross primary productivity   The total amount of energy produced by autotrophs over a given period of time.

Groundwater   Water found in underground hollows and aquifers, recharged by rainfall and infiltration from surface waters and wetlands.

Habitat   The specific environment or geographic region in which a species is found.

Half-life   The period of time it takes for half of the atoms of a radioactive material to decay into the next element of the decay process, ranging from minutes for some low-level radioactive materials to billions of years for some highly radioactive materials.

Hazardous substance   A toxic substance that has the potential to threaten human health and the environment if poorly handled or stored. Compare *Toxic substance*.

Herbicide   Any chemical substance, usually synthetic, that injures or kills plant life by disrupting normal hormonal functions.

Herbivore   An organism that consumes only plants. Compare *Carnivore, Detrivore, Omnivore*.

Heterotroph   A consumer organism that cannot produce its own food and must rely on eating other organisms to satisfy its nutritional needs. Compare *Autotroph, Saprotroph*.

High-density polyethylene (HDPE)   A modern plastic used in packaging that can be recycled.

High-input agriculture See *Conventional agriculture.*

High-level nuclear waste Nuclear reactor fuel or fission products that will remain radioactive for tens of thousands of years and must be handled or stored with extreme care.

High-pressure cell Atmospheric condition in which air sinks toward the ground.

High-yield variety (HYV) Special genetic strain of a hybrid crop developed by scientists to permit substantial gains in crop yield.

Historic preservation The preservation of material culture representative of events in a society's history.

Homeostasis The maintenance by an organism of a constant internal environment, such as regulation of blood sugar levels by insulin; a process which involves self-adjusting mechanisms in which the maintenance of a particular level is initiated by the substance to be regulated.

Homo sapiens The Latin name of the human species, literally "wise man."

Hormonal contraception A type of birth control that alters a female's hormone levels to inhibit production of the egg.

Hot dry rock reservoir A geothermal energy source in which subsurface rock is heated to high temperatures as a result of the intrusion of molten rock from the earth's center into the earth's crust.

Humus Partially decomposed plant and animal matter that is the organic component of soil and gives it the ability to retain water and maintain a high nutrient content.

Hybridization The crossing of one or more varieties of a species to produce an offspring with particular desired qualities.

Hydrocarbon An organic compound containing hydrogen and carbon and often occurring in fossil fuels. A primary pollutant originating chiefly from fossil fuel combustion.

Hydroelectric energy A means of producing electricity which exploits the energy present in falling water. Water at the top of a fall or dam is in a high state of gravitational potential energy; as the water drops, its potential energy is converted to kinetic energy. As the water strikes the blades of a turbine, spinning the turbine shaft, the kinetic energy is converted to mechanical energy, which can be used to drive a generator to produce an electric current.

Hydrological cycle The circulation of water through bodies of water, the atmosphere, and land.

Hydroponics The growing of plants in a nutrient-rich solution.

Hydrosphere The total mass of free water in solid or liquid state on the earth's surface.

Hypolimnion Lower layer of a lake, which receives little heat from the sun and tends to be colder than overlying water. Compare *Epilimnion.*

Igneous rock Rock formed by the cooling and crystallization of magma.

Illegal harvesting The taking of plants or animals prohibited by law in terms of season, age, size, or species.

Indicated reserve A fossil fuel deposit that is thought to exist and to be likely to be discovered and available for use in the future. Also known as inferred reserve. Compare *Proven reserve, Subeconomic reserve.*

Indirect radiation exposure Exposure to radiation removed from its original source, for example, exposure to radioactive substances transmitted through the food chain.

Inertia The tendency of a natural system to resist change.

Infant mortality rate The number of infants who die before age 1 per 1,000 births per year.

Inhibition model of succession The theory that certain species are able to inhibit or prevent succession within an ecosystem.

Insecticide Any chemical substance, usually synthetic, used to deter, destroy, or repel insects.

Integrated pest management The combined use of several methods, biological and chemical, to control insect pests.

Integrated solid waste management The combined use of various methods of waste disposal, including recycling, composting, incineration, and landfills, to minimize the negative impact of the waste stream on the environment.

Intensive agriculture A method of farming in which high inputs of capital and labor lead to a high output per unit of area.

International Seabed Authority A group appointed by the United Nations to regulate the seabed resources of the high seas.

International Species Inventory System (ISIS) A computer-based information system for wild animals in captivity developed to allow zoos to cooperate in the genetic and demographic management of their animals.

International Union for the Conservation of Nature (IUCN) An independent, international organization founded in 1948 whose main purpose is to promote and initiate the conservation of habitats and natural resources throughout the world.

Interspecific competition Competition between members of different species for limited resources such as food, water, or space. Compare *Interspecific competition.*

Interspecific cooperation Cooperation between members of different species. Compare *Intraspecific cooperation.*

Intraspecific competition Competition between members of the same species for limited resources such as food, water, or space. Compare *Interspecific competition.*

Intraspecific cooperation Cooperation between members of the same species. Compare *Intraspecific cooperation.*

Intrauterine device (IUD) A birth control device implanted in a woman's uterus that prevents the attachment of the zygote to the wall of the uterus.

In vitro preservation A method of preserving plant tissue for future propagation by storing it under regulated temperature and light conditions.

Ionizing radiation Energy released from a fission reaction which is powerful enough to pull electrons away from the atoms of materials that it strikes, producing positively charged ions; can be extremely damaging to living tissue.

Isotope One of two or more forms of the atom of an element that differs from the other forms in the number of neutrons.

Keystone species A species whose activities determine the structure of the biotic community.

Kinetic energy Energy associated with the motion of matter.

Kwashiorkor Childhood disease arising from protein insufficiency.

Lake A body of standing water that occupies a depression on the earth's surface and is completely surrounded by land.

Land ethic The principle developed by ecologist Aldo Leopold of cooperation between humans and other biospheric components.

Land farming A waste disposal method that relies on the biological processes of microorganisms within the soil to decompose any organic chemicals present in the waste.

Landfilling The disposal of waste in excavated sites or in sites such as former quarries, abandoned mine workings, gravel pits, and clay pits.

Land race A variety of a species adapted to specific local conditions such as climate and soil type.

Landscape ecology A field of study of the structure, function, and dynamics of ecosystems by looking at specific geographic areas in a holistic way.

Land subsidence The gradual sinking of the ground surface to a lower level.

Land use The way in which a particular parcel of land is used.

Land use planning The practice of studying information on a specific parcel of land and its possible uses before choosing a course of action.

Law of the conservation of energy Law that states that during a physical or chemical change energy is neither created nor destroyed, but it may be changed in form and moved from place to place.

Law of the conservation of matter Law that states that during a physical or chemical change matter is neither created nor destroyed, but it may be changed in form and moved from place to place.

Law of the minimum Law that states that survivability is primarily determined by the minimum amounts of limiting factors in an ecosystem.

Law of the Sea, 1982 An international agreement that coastal nations exercise total sovereignty over a territorial sea extending 12 nautical miles offshore and have an *exclusive economic zone* extending 200 miles.

LD_{50} test Test to determine the lethal dose of a substance, defined as the dose after which 50 percent of test subjects die.

Leachate The solution formed when water percolates through a solid, such as rainwater draining through soil.

Lead pollution The accumulation of lead in organic tissue, primarily in plants, animals, and humans, which may produce behavioral changes, blindness, and ultimately death.

Less-developed country (LDC) A country that has low to moderate industrialization and low to moderate average GNP per person. Compare *More-developed country, Third world.*

Lifeboat ethics A guide to action based on the belief that each nation is like a lifeboat, with limited capacity (resources) and that people in the lifeboats with enough resources to support their passengers easily cannot rescue everyone from the overcrowded lifeboats without jeopardizing their own safety.

Life expectancy The average number of years a newborn can be expected to live.

Lignite coal An abundant coal with low sulfur content and low heating values. Compare *Anthracite coal, Bituminous coal, Subbituminous coal.*

Limiting factor An abiotic factor such as light or temperature that determines the habitability of an ecosystem for a given organism.

Limnetic zone Area of a body of fresh water where light penetrates enough to support photosynthesis and populations of plankton. Compare *Littoral zone, Profundal zone.*

LISA See *Low-input sustainable agriculture.*

Lithosphere The rigid outer layers of the earth's crust and mantle.

Littoral zone Shallow area of a body of fresh water where light penetrates to the bottom and rooted vegetation dominates. Compare *Limnetic zone, Profundal zone.*

Living culture A group of people and its cultural heritage, including folklore, traditions, creation beliefs, skills, and technologies.

Low-input sustainable agriculture The practice of maintaining croplands through limited use of synthetic fertilizers, pesticides, and herbicides.

Low-level nuclear waste Nuclear reactor fuel or fission products that will remain dangerous for a few hundred years or less.

Low-till planting The practice of tilling the soil just once in the fall or spring, leaving half or more of previous crop residue on the ground's surface.

MAB program A UNESCO project of the 1970s designed to promote the long-term conservation of representative ecosystems worldwide based on sound scientific grounds.

Macroconsumer An organism that feeds by ingesting or engulfing part or entire bodies of other organisms, living or dead.

Macronutrients A chemical that is a major constituent of the complex organic compounds found in all living organisms and which is required for the construction of proteins, fats, and carbohydrates.

Magma Melted rock found deep within the earth's crust where it is exposed to tremendous pressure and high temperatures.

Malabsorptive hunger Hunger that results when the body loses its ability to absorb nutrients from food consumed.

Malnutrition The consumption of inadequate levels of specific nutrients essential for good health.

Malthusian view of population growth Viewpoint developed by eighteenth-century English economist Thomas Malthus that population increases geometrically, while food increases arithmetically, thus causing population to exceed sustainable levels.

Marasmus Childhood disease arising from insufficient amounts of protein and calories.

Marsh An area of spongy, waterlogged ground dominated by grasses with large numbers of surface water pools.

Mass burn incinerator Plant that burns all incoming waste, regardless of its composition, at temperatures as high as 2400 °F.

Material culture Tangible objects, such as tools, furnishings, buildings, and art works, that humans create.

Matter That which constitutes the substance of physical forms, has mass, occupies space, and can be quantified.

Meltdown The overheating of a nuclear reactor core so that fuel rods melt, burn through the containment vessel, and bore toward the center of the earth releasing large amounts of radiation to the environment.

Memorandum of Understanding (MOU) 1990 An agreement developed by the Bureau of Land Management and the Nature Conservancy to cooperate in identifying, evaluating, and protecting public lands that have exceptional natural resource value and placing them in public ownership.

Metamorphic rock A type of rock that forms when rocks lying deep below the earth's surface are heated to such a degree that their original crystal structure is lost. As the rock cools, a new crystalline structure is formed.

Microconsumer An organism that feeds on waste products of living organisms or the tissue of dead organisms.

Micronutrient A chemical that is required by organisms in small amounts for the construction of proteins, fats, and carbohydrates.

Migration   Movement from one geographic area to another.

Mineral   Any naturally occurring solid inorganic substance of definite chemical composition and crystalline structure.

Mineral resource   A naturally occurring concentration of an economically valuable and workable mineral.

Minerals policy   Laws, regulations, and agreements established by a nation to govern mineral production, use, and commerce.

Minimum critical diet   The minimum amount of food needed to remain healthy and maintain body weight, assuming little physical activity.

Minimum viable population   The minimum number of individuals required to maintain a viable breeding group without suffering short-term loss of genetic variability.

Mining Law of 1872   Legislation that stated that all valuable mineral deposits in the land belonging to the United States shall be free and open to exploration and purchase under regulations prescribed by law.

Monoculture   The extensive cultivation of one crop or one variety of a crop plant.

Monomorphic species   A species in which the male and female look almost identical.

Montreal Protocol   An agreement reached by 45 nations in 1987 to reduce by 50 percent the production of chlorofluorocarbons by 1998 and to freeze the production of halons at present levels.

Morals   Standards of right and wrong behavior.

More-developed country (MDC)   A country that has a high degree of industrialization and moderate to high average GNP per person. Compare *Less-developed country*.

Multiple chemical sensitivity   An illness characterized by an intolerance of one or more classes of chemicals.

Multiple land use   Using land for several different purposes, for example forestry, recreation, and wildlife habitat. Compare *Restricted land use*.

Municipal solid waste   Waste generated by households and small businesses.

Mutagenic   Tending to increase the frequency or extent of genetic mutations.

Mutation   A change in the genetic material of an organism.

Mutualism   A mutually beneficial interaction between two species.

National Ambient Air Quality Standards   Standards established by the Environmental Protection Agency that limit the quantities of air pollutants permitted to be emitted by specific sources such as power plants and factories.

National Environmental Policy Act (NEPA)   Legislation passed in 1969 that charges federal agencies with restoring and maintaining environmental quality throughout the country and requires federal agencies to prepare environmental impact statements for any major project.

National Forest System   Designated forests and grasslands managed by the Forest Service to accommodate a wide array of uses from commercial interest to recreation.

National Park Service   A division of the Department of the Interior responsible for managing national parks in order to conserve their scenery, natural and historic objects, and wildlife and to provide for the enjoyment and use of the parks by the public.

National Resources Defense Council   A public-interest environmental law firm established in 1970 to lobby national agencies for environmental protection.

National Wilderness Preservation System   Designated wilderness areas within the National Forest System, National Parks System, Natural Wildlife Refuge System, and Natural Resource Lands, designated to preserve diversity through setting aside wide expanses of roadless area.

National Wildlife Refuge System   Designated areas managed by the Fish and Wildlife Service to provide a habitat and haven for wildlife.

Natural gas   A fossil fuel composed of several different types of gases of which methane, ethane, and propane are most prevalent.

Natural population growth rate   Difference between the crude death rate and the crude birth rate expressed as a percentage. Compare *Actual population growth rate*.

Natural selection   The mechanism of gradual evolutionary change proposed by Charles Darwin by which organisms that are best adapted to the environment in which they live

produce more viable young, increasing their proportion in the population and ensuring the survival of desirable genetic traits (in relation to the environment) and the elimination of undesirable ones.

Nature   The sum of all living organisms interacting with the earth's physical and chemical components as a complete system.

Nature Conservancy   A private, national, nonprofit organization whose primary interest is the preservation of biological diversity and pristine habitats.

Negative feedback   Situation in which a change in a system in one direction provides information that causes the system to change in the opposite direction. Compare *Positive feedback*.

Nekton   Free-swimming aquatic organisms. Compare *Plankton*.

Neritic zone   The part of the euphotic zone that overlies the continental shelf or surrounds islands and supports large fish populations. Compare *Euphotic zone*, *Pelagic zone*.

Net energy efficiency   The total amount of energy produced in a process after energy lost is taken into account.

Net primary productivity   The total amount of energy produced each year at the producer level minus what producers need for their own life processes.

Niche   A position occupied by a particular species within its community that determines its activities and relationships with other species; can be occupied by only one species at a time.

Nitrogen cycle   The natural circulation of nitrogen through the environment.

Nitrogen fixation   The conversion of atmospheric nitrogen into organic nitrogen compounds.

Nitrogen oxide   A *primary pollutant* originating chiefly from the fossil fuel combustion.

Nonassociated gas   Free-flowing natural gas that is found apart from petroleum reserves. Compare *Associated gas*.

Nonbiodegradeable   A substance that cannot be broken down by natural systems, and which is thus unusable to organisms.

Non-ionizing radiation   Energy such as radio waves, heat, and light released from a fusion reaction which is not powerful enough to create ions and is not generally considered a health risk.

Nonmaterial culture Intangible resources, such as language, customs, traditions, folklore, and mythology.

Non-point source pollution Pollution that cannot be traced to a specific source. Compare *Point source pollution*.

Nonrenewable resource A resource, such as fossil fuels, that exists in finite supply or is consumed at a rate faster than the rate at which it can be renewed. Compare *Renewable energy resources*.

No-till planting The practice of sowing seeds without turning over the soil, thus allowing plants to grow amid the stubble of the previous year's crop.

Not in my backyard syndrome (NIMBY) Opposition to having waste and pollutants processed near one's home or community.

Nuclear energy Electricity derived from fission reactions.

Nuclear Regulatory Commission The agency of the U.S. government that licenses and regulates commercial nuclear facilities.

Nuclear winter theory Theory that a global nuclear war would pump such large quantities of smoke, soot, and debris into the atmosphere that sunlight reaching the earth would be severely reduced causing the earth's temperature to drop substantially.

Ocean power Energy that can be derived from the seas through means such as harnessing tides and thermal currents.

Oil See *Petroleum*.

Oil shales A minor fossil fuel source composed of fine-grained, compacted sedimentary rocks that contain varying amounts of a solid, combustible, organic matter called kerogen.

Old-growth forest Uncut, virgin forest which may contain massive trees hundreds of years old.

Oligotrophic lake A lake that is cold and deep, with a high oxygen content and a relatively low amount of dissolved solids, nutrients, and phytoplankton. Compare *Eutrophic lake*.

Omnivore An organism that consumes both plants and animals. Compare *Carnivore*, *Detrivore*, *Herbivore*.

Open pit surface mining A method of mineral extraction in which a large pit is dug and the exposed ore removed.

Opportunistic species A species that takes advantage of the weaknesses of other species or its own ability to exploit temporary habitats or conditions.

Optimum population size The number of individuals an environment or habitat can best support.

Organic agriculture The practice of maintaining cropland without the use synthetic fertilizers, pesticides, or herbicides.

Organic compounds A compound that contains carbon.

Overburden Overlying vegetation, soil, and rock layers that are removed to expose ore deposits for surface mining.

Overdraft The withdrawal of water from an aquifer faster than the aquifer can naturally replenish itself.

Overgrazing Consumption of vegetation on rangeland by grazing animals to the point that the vegetation cannot be renewed or is renewed at a rate slower than it is consumed.

Overpopulation A situation in which the number of people in an area cannot be supported adequately by the available resources, leading to declining standards of living and failure to realize human potential fully.

Oxygen cycle The natural circulation of oxygen through the environment.

Ozone An atmospheric gas (O_3) that when present in the stratosphere helps protect the earth from ultraviolet rays, but when present near the earth's surface is a primary component of urban smog and has detrimental effects on both vegetation and human respiratory systems.

Ozone layer A layer of concentrated ozone in the stratosphere about 20 to 30 miles (32 to 48 kilometers) above the earth's surface.

Parasitism An interaction between two species that benefits one species, the parasite, and harms the other species, the host. Compare *Commensalism*, *Mutualism*.

Parent material The raw mineral material from which soil is eventually formed.

Particulates Tiny particles of solids or liquid aerosols that are light enough to be transported in the air and may cause respiratory problems in humans and atmospheric haze in the environment.

Passive solar system System for temperature control in buildings that utilizes only the natural forces of conduction, convection, and radiation to distribute collected and stored solar heat. Compare *Active solar system*.

Pasture Grassland used for livestock production.

PCB Polychlorinated biphenyl, a group of at least 50 widely used compounds which accumulate in food chains and may produce harmful effects in organisms.

Pelagic zone The area in an ocean so deep that light cannot penetrate and organisms must rely on nutrients that filter down from above. Compare *Euphotic zone, Neritic zone*.

Pellagra A disease associated with a niacin-deficient diet that can cause weakness, spinal pain, convulsions, and idiocy.

Permafrost A layer of permanently frozen ground beneath the earth's surface found in frigid regions.

Perpetual fuels Energy source that is inexhaustable, such as solar, wind, ocean, and geothermal fuels.

Persistent pollutant Nonbiodegradable pollutant which accumulates in natural systems over time.

Petroleum A liquid fossil fuel composed primarily of hydrocarbon compounds with small amounts of oxygen, sulfur, and nitrogen compounds.

pH Numeric value that indicates the relative acidity or alkalinity of a substance on a scale of 0 to 14, with acid solutions below 7, neutral solutions at 7, and basic solutions above 7.

Phase separation process A physical treatment, such as filtration, centrifugation, sedimentation, and flotation, which separates wastes into liquid and solid components.

Phosphorus cycle The natural circulation of phosphorus through the environment.

Photochemical reaction A reaction induced by the presence of light.

Photochemical smog An atmospheric haze that occurs above industrial sites and urban areas resulting from reactions between pollutants produced in high temperature and pressurized combustion processes, such as the combustion of fuel in a motor vehicle.

Photosynthesis The process of using radiant energy from sunlight to produce chemical compounds in chlorophyll-containing plants.

Phototroph An organism that produces complex chemicals through photosynthesis.

Photovoltaic solar system System that transforms sunlight directly into electricity utilizing light-sensitive material to absorb the solar energy.

Pioneer organism The initial species that colonizes an ecosystem, breaking down weathered rock and minerals to form soil, or the initial species in a habitat after a major disturbance.

Plankton Microscopic plants (phytoplankton) or animals (zooplankton) that float passively or swim weakly in a body of water. Compare *Nekton*.

Poaching Trespassing on a preserve or private property to hunt or the killing of animals that are legally protected.

Point source pollution Pollution that can be traced to an identifiable source. Compare *Non-point source pollution*.

Pollutant A substance that adversely alters the physical, chemical, or biological quality of the earth's living systems or that accumulates in the cells or tissues of living organisms in amounts that threaten the health or survival of that organism.

Polyculture The cultivation of a variety of crops.

Polyethylene terephthalate (PET) A recyclable type of plastic.

Pool An area of relatively low oxygen content in running-water habitats that houses the consumers and decomposers of biomass. Compare *Riffle*.

Population The size, density, demographic distribution, and growth rate of any definable nation, region, or continent. Individuals of a particular species with definable group characteristics.

Population density The number of individuals per unit of space.

Population momentum Tendency of a population to continue to increase in absolute numbers despite declines in fertility rate due to a large base of childbearing women.

Population policy A government's planned course of action designed to influence and regulate its constituents' choices or decisions on fertility or migration.

Population profile A graphical representation of the age distribution of a population.

Positive feedback A situation in which a change in a system in one direction provides information that causes the system to change further in the same direction. Compare *Negative feedback*.

Potential energy Any energy that can be released to do work.

Power The rate at which work is performed or the rate at which energy is expended to do work.

Predator An organism, typically an animal, that acquires food by killing and consuming other organisms.

Predator–prey interaction Relationship between two organisms of different species in which one organism, the predator, feeds on another organism, the prey.

Prescribed burn Deliberately set fire in a forest to prevent more destructive fires or to kill off unwanted plants that compete with a desirable species.

Preservation A strict form of conservation in which use of a resource or area is limited to nonconsumptive activities. Also, the process of maintaining the existing condition of a historical site. Compare *Restoration, Rehabilitation*.

Pressurized hot water reservoir A geothermal energy source in which underground water, heated under intense pressure, produces a mixture of steam, scalding water, and dissolved materials.

Prey An organism that is killed and consumed by another organism.

Primary consumer In a food chain, an organism that consumes producers (green plants). Compare *Secondary consumer, Tertiary consumer*.

Primary pollutant A pollutant that is emitted directly into the atmosphere, the six primary pollutants being carbon dioxide, carbon monoxide, sulfur oxides, nitrogen oxides, hydrocarbons, and particulates. Compare *Secondary pollutant*.

Primary productivity The rate at which autotrophs store energy over a given period of time. Compare *Gross primary productivity, Net primary productivity*.

Primary succession The development of a biotic community in an area previously devoid of organisms.

Primordial soup The mixture in sea water of organic molecules, such as proteins and amino acids, from which it is theorized that life arose.

Producer An autotroph capable of synthesizing organic material, thus forming the basis of the food web.

Profundal zone Deep area of a body of fresh water where light cannot penetrate and organisms must rely on nutrients that filter down from above. Compare *Littoral zone, Limnetic zone*.

Pronatalist policy A policy that encourages natality or births.

Proven reserve A mineral concentration or fossil fuel deposit from which the resource can be extracted profitably with current technology. Also known as an economic resource. Compare *Indicated reserve, Subeconomic reserve*.

Public domain Public land.

Pyramid of biomass Conceptual tool used to illustrate that total biomass tends to decrease at each subsequent trophic level and the size of each individual organism tends to increase.

Pyramid of energy Conceptual tool used to illustrate the inefficiency of energy transfers from one trophic level to another.

Pyramid of numbers Conceptual tool used to illustrate the tendency toward large population on lower trophic levels and small population on higher trophic levels.

Quality of life A complex set of indicators that provides a definition of the general condition of a human population in a given area.

Rad Unit used to measure the amount of radiation absorbed per gram of tissue.

Radiation Energy released by an atom; radiation takes two basic forms, ionizing and nonionizing radiation.

Radiational cooling The reradiation of heat from the ground after sunset faster than from the air masses above it, keeping the cooler, denser air near the earth's surface.

Radioactive decay The process by which a radioactive atom seeks stability by emitting particles and energy so that the number of neutrons will eventually equal the number of protons in the nucleus.

Radioactive fallout Dirt and debris contaminated with radiation which is produced by atomic tests and spread throughout the environment by winds and rain.

Radioisotope See *Isotope*.

Radon 222 A naturally occurring radioactive gas arising from the decay of uranium 238 which may be harmful to human health in high concentrations.

Rain shadow effect The phenomenon that occurs as a result of the movement of air masses over a mountain range. As an air mass rises to clear a mountain, the air cools and precipitation forms. Often, both the precipitation and the pollutant load carried by the air mass will be dropped on the windward side of the mountain. The air mass is then devoid of most of its moisture; consequently, the lee side of the mountain receives little or no precipitation and is said to lie in the rain shadow of the mountain range.

Range The extent of the distribution of a species.

Range of tolerances Range of abiotic factors within which an organism can survive from the minimum amount of a limiting factor that the organism requires to the maximum amount that it can withstand.

Rare species A species at risk of local or general (worldwide) extinction because of its small total population.

Recycling The recovery and reuse of materials from wastes.

Reforestation The planting of trees on land previously covered by forest that was removed by natural or human agency.

Refuse-derived fuel incinerator Plant that separates incoming waste before burning to remove noncombustible items and recyclable materials.

Rehabilitation The process of repairing and altering a historical site while maintaining features of the site with significant historical, architectural, and cultural value. Compare *Preservation, Restoration.*

Relaxation The tendency for biological diversity to decrease within a wildlife reserve as the reserve's carrying capacity is approached.

Rem Measure of the damage potential caused by a specific dose of radiation. The Nuclear Regulatory Commission has set 0.17 rem annually as the maximum safe exposure for the general public.

Renewable resourc A resource, such as biomass, that exists in virtually infinite supply or is consumed at a rate lower than that at which it can be renewed. Compare *Nonrenewable energy resource.*

Replacement fertility rate The fertility rate needed to ensure that the population is just "replaced" by its offspring, ranging from 2.1 to 2.5 depending on the mortality rate of the population.

Resiliency The ability of a natural system to return to a state of equilibrium after a disturbance.

Resource Something that serves a need, is useful, and is available at a particular cost.

Resource base An estimate of the total amount of a fossil fuel or mineral contained in the earth's crust based on proven, subeconomic, and indicated reserves.

Resource Conservation and Recovery Act (RCRA) Legislation passed in 1976 and amended in 1984 that establishes a legal definition for hazardous wastes and guidelines for managing, storing, and disposing of hazardous resources in an environmentally sound manner.

Resource recovery An umbrella term referring to the taking of useful materials or energy out of the solid waste stream at any stage before ultimate disposal, for example, recycling, composting.

Respiration The process by which organisms produce energy by capturing the chemical energy stored in food.

Restoration The process of returning a historical site to its original condition by making such changes as removing later alterations and adding missing features. Compare *Preservation, Rehabilitation.*

Restoration ecology A branch of applied ecology that has three major goals: repairing, restoring, or replacing native biotic communities; maintaining the present diversity of species and ecosystems by finding ways to preserve biotic communities or to protect them from human disturbances so they can evolve naturally; and increasing knowledge of biotic communities.

Restricted land use Using land for only one or two purposes. Also known as exclusionary land use. Compare *Multiple land use.*

Rhizobium A rod-shaped bacterium in the genus *Rhizobium,* capable of fixing nitrogen in the root nodules of beans, clover, and other legumes.

Rhythm method Form of birth control based on the natural cycle of ovulation in the female.

Ridge tilling The practice of planting a crop on top of raised ridges.

Riffle An area of high oxygen content in running-water habitats that houses the producers of biomass. Compare *Pool.*

Riparian Relating to the banks of a natural waterway, such as a river.

Roadless Area Review and Evaluation (RARE) A study performed by the Forest Service for the first time in 1972 under public pressure to increase the number of wilderness areas included in the National Wilderness Preservation System.

Rule of 70 Rule for finding the amount of time required for a population to double in numbers: doubling time is equal to 70 divided by the annual growth rate.

Runoff Water that flows over the land surface from higher elevations to lower elevations by following the contours of the land.

Safe Drinking Water Act of 1974 Legislation that set national drinking water standards, called maximum contaminant levels, for pollutants, and established standards to protect groundwater from hazardous wastes injected into the soil.

Salinity A measure of the concentration of dissolved salts in water.

Salinization The deposition on farmland of salty minerals during the evaporation of irrigation water.

Sanitary landfill Landfill designed to receive only nonhazardous and nontoxic waste. Compare *Secure landfill.*

Saprotroph An organism that uses enzymes to feed on waste products of living organisms or tissues of dead organisms. Compare *Autotroph, Heterotroph.*

Scarce metal Any metal that constitutes less than 1 percent of the earth's continental crust. Compare *Abundant metal.*

Scavenger A heterotroph that consumes dead organic material.

Seasonal hunger The period of time between the new harvest and the point at which reserves from the previous harvest run out.

Secondary consumer In a food chain, an organism (usually animal) that consumes primary consumers. Compare *Primary consumer, tertiary consumer.*

Secondary pollutant A pollutant formed from the interaction of primary pollutants with other primary pollutants or with atmospheric compounds such as water vapor. Compare *Primary pollutant.*

Secondary succession The development of a biotic community in an area following a serious disturbance of the original community in that area.

Second law of energy Law that states that with each change in form, some energy is degraded to a less useful form and given off to the surroundings, usually as low-quality heat.

Second Law of Thermodynamics See *Second Law of Energy.*

Secure landfill A landfill designed to receive hazardous and toxic wastes with specially engineered systems to prevent the escape of these substances into the environment and reduce the production and release of leachate. Compare *Sanitary landfill.*

Sedimentary rock Rock formed by the breakdown of igneous rock through the action of wind and rain erosion and chemical reactions.

Selective cutting The cutting of intermediate-aged, mature, or diseased trees in an uneven-aged forest stand either singly or in small groups.

Self-sufficiency A way of life in which basic needs of food, clothing, and shelter are met and sustained by the efforts of the immediate community without recourse to goods and services produced by others.

Sense of the earth An intimate knowledge of the local environment— location and use of plant species, habits and movements of animals, and seasonal weather patterns. Often best exhibited by early hunter-gatherer societies.

Septic tank An underground concrete tank large enough to accommodate the wastewater flow from a building in which bacteria liquify solids and the liquid then passes out to a drainage field.

Siltation The obstruction of streams and lakes by soil deposited as a result of erosion.

Singularity The infinitely hot and dense dot from which, according to the big bang theory, the entire universe arose.

Slash-and-burn A method of land clearing in which vegetation is cut, allowed to dry, and then set on fire prior to the cultivation of the soil and the planting of crops.

Sludge A viscous, semisolid mixture of bacteria- and virus-laden organic matter, toxic metals, synthetic organic chemicals, and settled solids removed from domestic and industrial wastewater at sewage treatment plants.

Smog A dense, discolored haze containing large quantities of soot, ash, and gaseous pollutants such as sulfur dioxide and carbon dioxide.

Soil A naturally occurring mixture of inorganic chemicals, air, water, decaying organic material, and living organisms.

Soil degradation A deterioration of the quality and capacity of a soil's life-supporting processes.

Soil erosion The accelerated removal of soil through various processes at a greater rate than soil is formed.

Soil fertility Measure of the soil's mineral and organic content; the ability of a soil to supply the required type and amount of nutrients for optimum growth of a particular crop when all the other growing factors are favorable.

Soil horizon A horizontal layer formed as a soil develops and distinct in color, texture, structure, and composition.

Soil loss tolerance level The amount of soil that can be lost through erosion without a subsequent decline in fertility. Also known as a soil's T-value or replacement level.

Soil productivity The ability of the soil to sustain life, especially vegetation.

Soil profile A vertical profile of soil horizons.

Soil structure The tendency of the particles of a soil to form larger aggregates (crumbs, chunks, or lumps) largely depending on the amount of clay and organic material it contains. Also known as soil tilth.

Soil texture The coarseness of a soil, depending largely on the proportion of sand, silt, and clay it contains.

Solar energy Radiant energy originating from the sun.

Solar ponds A method of generating electricity using salt water and fresh-water ponds heated by the sun.

Sole source aquifer An aquifer that serves as the principal drinking water source for a community.

Solidification A waste disposal method in which liquid, semisolid, or solid hazardous waste is transformed into a more benign solid product by entombing the waste in a concrete block.

Solid waste stream The sum of all waste—industrial, agricultural, residential—produced in a year.

Source reduction Reducing the amount of waste initially entering the waste stream by product changes such as minimizing excessive packaging and extending the useful life of products.

Source segregation Keeping used hazardous resources apart from nonhazardous resources in order to reduce the overall volume of material that must be treated as hazardous.

Source separation The removal of recyclables from the waste stream and the sorting of recyclables to maintain the purity and value of the recyclable material.

Spaceship Earth Term coined by environmentalists of the 1960s which likens the closed system of a spaceship to the planet Earth, with humans as the pilots of the spaceship, responsible for its well-being.

Specialist A species that requires one particular type of habitat for survival. See *Generalist.*

Speciation The process of new species formation in which populations become genetically isolated most often because of geographic separation for a period of time during which heritable variations occur in the populations.

Species A group of individuals or populations potentially able to interbreed and unable to produce fertile offspring by breeding with other sorts of animals and plants.

Species diversity The variety of plants and animals in a given area.

Species survival plan Plan developed using the International Species Inventory System indicating which animals of a species in zoos should mate to preserve the maximum genetic diversity of the entire captive population.

Split estate Situation in which the federal government owns a piece of land, but private citizens or companies own the minerals beneath the surface.

Spontaneous radioactivity The release of mass in the form of particles, energy in the form of rays, or both, by unstable atoms of a particular element, such as uranium.

Standard of living Measurement of the quality of life which a given population accepts as desirable.

Staple A main element of a diet.

Starvation The consumption of insufficient calories to sustain life.

Statute A law passed by a legislature.

Sterilization Nonreversible method of birth control in which the sperm tubes or the oviducts are altered to prevent the occurrence of fertilization.

Stewardship ethic A guide for behavior based on the belief that humans are caretakers and nurturers of the natural world.

Still bottom The concentrated, highly toxic mixture of metals and solvents that results from resource recovery distillation techniques.

Strategic mineral A mineral that is essential to national defense but which exists in relatively short supply.

Stratosphere An atmospheric layer extending from 6 or 7 miles to 30 miles above the earth's surface.

Strip cropping The practice of alternating rows of grain with low-growing leaf crops or sod.

Subbituminous coal An abundant coal with low sulfur content and low heating values. Compare *Anthracite coal, Bituminous coal, Lignite coal.*

Subeconomic reserve A mineral concentration or fossil fuel deposit that has been discovered, but from which the resource cannot be extracted at a profit at current prices or with current technologies. Compare *Proven reserve, Indicated reserve.*

Subsistence farming The production of food and other necessities to satisfy the needs of the farm household.

Subsoil Relatively infertile lower layer of soil in which dissolved minerals accumulate.

Subsurface mining A method of mineral extraction in which a shaft is dug down to the level of the deposit and the ore is then extracted and hauled to the surface.

Succession The gradual, sequential, and somewhat predictable changes in the composition of an ecosystem's communities from an initial colonization of an area by pioneer organisms to the eventual development of the climax community.

Sulfur cycle The natural circulation of sulfur through the environment.

Sulfur oxide A primary pollutant originating chiefly from the combustion of high-sulfur coals.

Superfund A fund originally established by the U.S. government in 1980 to clean up hazardous waste sites and to respond to uncontrolled releases of hazardous substances.

Surface mining A method of mineral extraction in which the overlying vegetation, soil, and rock layers are removed to expose an ore deposit.

Surface water Water found on the surface of the earth, recharged by precipitation and runoff. Compare *Groundwater.*

Sustainable In harmony with natural systems and acting to maintain the health and integrity of the environment.

Sustainable agriculture Farming methods that sustain and protect both fertility and soil productivity.

Sustainable development Managing the economy and renewable resources of an area for the common good of the entire community and the environment.

Sustained yield The amount of harvestable material that can be removed from an ecosystem over a long period of time with no apparent deleterious effects on the system.

Sustaining Nurturing and supporting a rich diversity of life.

Symbiosis Any intimate association of two dissimilar species regardless of the benefits or harm derived from it.

Synergism An interaction between two substances that produces a greater effect than the effect of either one alone. An interaction between two relatively harmless components in the environment to form a more potent pollutant.

Tailings The rock surrounding a valuable mineral in an ore deposit which is discarded.

Tar sands A minor fossil fuel deposit composed of sandstones which contain bitumen, a thick, high-sulfur, tarlike liquid that may be purified and upgraded to synthetic crude oil.

Ten percent rule Rule stating that, in general, 90 percent of available energy is lost as low-quality heat when members of one trophic level are consumed by members of another.

Teratogenic Causing malformations in fetuses.

Terrestrial Of or pertaining to land environments.

Territorial waters Any area of water over which an adjacent country claims jurisdiction; an area within which a country has the sole right to exploit mineral and fish resources.

Tertiary consumer In a food chain, an organism at the top that consumes other organisms. Compare *Primary consumer, Secondary consumer.*

Thermal pollution An undesired rise in temperature in an environment above that which occurs through natural solar radiation; caused by the release of warm substances into the environment.

Thermocline The area of sharp temperature gradient that exists between the epilimnion and the hypolimnion.

Third world Those countries located mainly in Africa, Asia, and Latin America that are neither industrial market economies (first world) nor centrally planned economies (second world).

Threatened species A species that is severely exploited at present or inhabits an area of major environmental disturbance, is unlikely to adapt to those changes, and will most likely become endangered.

Throwaway society A society in which objects are manufactured to be short-lived, disposable, and nonrepairable and people habitually discard objects rather than repairing, reusing, or recycling them.

Tidal energy Energy originating from fluctuations in the ocean's water level.

Tilth Soil structure.

Tipping fees Disposal cost for dumping garbage in a landfill.

Tissue culture The production of plants from individual plant cells rather than plant seeds. Also known as cloning.

Tolerance model of succession Theory that subsequent communities in an ecosystem are not determined by present communities, but by an increased level of tolerance to environmental changes.

Topography The shape and contour of land formations.

Topsoil Highly fertile layer of soil found immediately below the surface and composed primarily of humus, living organisms, and minerals.

Tort A claim for which a civil suit can be brought by an injured plaintiff.

Total fertility rate The average number of children a woman will bear during her life, based on the current age-specific fertility rate.

Toxicant A chemical that can cause serious illness or death. Also called toxin.

Toxicology The study of poisonous materials, their effects, and their antidotes.

Toxic substance A chemical substance that adversely affects human health and the environment. Compare *Hazardous substance.*

Transpiration The loss of water vapor through the pores of a plant.

Transuranic nuclear waste Human-made radioactive elements such as plutonium, americium, and neptunium, that have atomic numbers higher than that of uranium.

Trash conversion The burning of municipal and industrial refuse and waste to produce electrical energy.

Trickle drip irrigation An irrigation technique in which water is delivered slowly to the base of plants through perforated or permeable pipes or tubes.

Trophic level A group of organisms with the same relative position in the food chain.

Trophy animal An animal valued for its meat, hide, beauty, or symbolism.

Troposphere The atmospheric layer that extends from the earth's surface to 6 or 7 miles above the surface.

Turnover The mixing of the upper layer and lower layer of a lake, which most often occurs in the spring and fall, due to dramatic changes in surface water temperature.

T-value See *Soil loss tolerance level.*

Ultimately recoverable resource An estimate of the total amount of fossil fuel that will eventually be recovered based on discovery rates, future costs, demand and market values, and future technological developments.

Undernutrition The consumption of inadequate levels of protein and calories which over an extended period of time gradually weakens an individual's capacity to function properly and to ward off disease.

United Nations Education, Scientific, and Cultural Organization (UNESCO) An agency established to reduce international tension by encouraging exchange of scientific and cultural ideas and improving education.

Unrealized resource Material that is usually thrown away, but could be used to benefit human and natural systems.

Upwelling An upward movement of ocean water masses that brings nutrients to the surface and creates a region of high productivity.

Urbanization An increase in the number and size of cities.

Urban revival Revitalization of neighborhoods for private or commercial use.

Urban sprawl The spread of urban areas which substantially reduces farmland.

Vermin Animal species that are regarded as pests.

Vital statistics Information about the essential needs and makeup of a population.

Waste minimization A variety of strategies to reduce the amount of hazardous substances used and hazardous wastes that must be disposed of.

Waste-to-energy incinerator Special plant engineered to burn garbage to produce heat and steam that is then used to produce electricity.

Water Quality Act of 1987 Legislation that reauthorized the original Safe Drinking Water Act and provided $20 billion to curb water pollution, primarily through the construction of wastewater treatment plants.

Watershed The area from which a body of water receives runoff.

Water table The upper surface of the zone of permanent groundwater saturation.

Wave power The stored kinetic energy contained within ocean waves, which can be used to generate electricity.

Weather The day-to-day pattern of precipitation, temperature, wind, barometric pressure, and humidity.

Weed A plant species that is regarded as a pest.

Wetland Vegetated land area that is occasionally or permanently covered by water and acts as a natural boundary between land areas and bodies of water.

Wilderness The domain of nature, a region undisturbed by human artifacts and activity. An area in which both the biotic and abiotic communities are minimally disturbed by humans.

Wilderness Act of 1964 Law that created the National Wilderness Preservation System to protect certain areas from motorized vehicles, roads, and structures.

Wilderness study area An area being considered for inclusion in the National Wilderness Preservation System.

Wildlife management Field of study that seeks to sustain populations of wildlife species.

Wind energy Energy that originates from air currents and can be collected using windmills or wind turbines.

Wind farms A large utility project consisting of many wind turbines that produce electricity for the surrounding population.

Wind turbine A device for producing electricity in which large blades or rotors are driven by wind.

Work The product of the distance that an object is moved times the force used to move it.

World Resources Institute A policy research center in Washington, D.C., established to help governments, international organizations, and the private sector meet human needs and foster economic growth without undermining the natural resources and environmental diversity of the biosphere.

Worldview A way of perceiving reality, the earth, and humanity's place on the earth and the attitudes, values, and beliefs based on that perception.

X-ray An ionizing radiation that is produced by bombarding a metallic target with fast electrons in a vacuum, is capable of penetrating various thicknesses of solids, and is a powerful mutagen.

Zero population growth The growth rate at which births are equal to deaths.

Zone of leaching The layer of soil between topsoil and subsoil through which water and dissolved minerals pass.

Bibliography

Chapter 1

Abbey, Edward. *Desert Solitaire*. New York: Ballantine Books, 1968.

American Society of Zoologists. *Science as a Way of Knowing: Human Ecology*. Thousand Oaks, Calif.: American Society of Zoologists, 1985.

Berry, Wendell. *A Continuous Harmony: Cultural and Agricultural Essays*. New York: Harcourt Brace Jovanovich, 1972.

Brown, Lester R., Christopher Flavin, and Sandra Postel. *Saving the Planet: How to Shape an Environmentally Sustainable Global Economy*. New York: W.W. Norton and Company, 1991.

Brown, Lester, et al. *State of the World 1992*. New York: Norton, 1992.

Brown, Lester R., and Sandra Postel. "Thresholds of Change." In *Environment 89/90*. Guilford, Conn.: Dushkin, 1989.

Callicot, J. Baird. *Companion to a Sand County Almanac: Interpretive and Critical Essays*. Madison: University of Wisconsin Press, 1987.

Carson, Rachel. *The Sense of Wonder*. New York: Harper & Row, 1956.

Carson, Rachel. *Silent Spring*. Boston: Houghton Mifflin, 1962.

Chown, Marcus. "The Big Bang." *New Scientist* (October 22, 1987).

Clark, William C. *Managing Planet Earth*. Washington, D.C.: Scientific American, 1990.

Commoner, Barry. *The Closing Circle: Nature, Man, and Technology*. New York: Knopf, 1972.

Council on Environmental Quality and U.S. Department of State. *Global 2000 Report to the President: Entering the 21st Century*, vol. 1-3. Gerald O. Barney, Study Director. Washington, D.C.: Government Printing Office, 1980–1981.

Daniel, Joseph E., ed., et al. *1992 Earth Journal*. Boulder, Colo.: Buzzworm Books, 1991.

Davis, Kingsley, et al., eds. *Population Resources in a Changing World. Current Readings*. Stanford, Calif.: Morrison Institute for Population and Resource Studies, 1989.

Devall, Bill, and George Sessions. *Deep Ecology: Living As If Nature Mattered*. Salt Lake City, Utah: Smith, 1985.

Eckholm, Erik P. *Down to Earth: Environment and Human Needs*. New York: Norton, 1982.

Ehrlich, Anne H., and Paul R. Ehrlich. *Earth*. New York: Franklin and Watts, 1987.

Ehrlich, Paul R. "Human Ecology for Introductory Biology Courses: An Overview." *American Zoologist* 25 (1985): 379–394.

Fladers, Susan. *Thinking Like a Mountain: Aldo Leopold and the Evolution of an Ecological Attitude Toward Deer, Wolves, and Forests*. Columbia: University of Wisconsin Press, 1974.

Goldsmith, Edward, and Nicholas Hilyard, eds. *The Earth Report: The Essential Guide to Global Ecological Issues*. Los Angeles: Price Stern Sloan, 1988.

Hardin, Garret. "The Tragedy of the Commons." *Science* 162 (1968): 1243–1248.

Leopold, Aldo. *A Sand County Almanac and Sketches Here and There*. New York: Oxford University Press, 1949.

Lovelock, James. *The Ages of Gaia*. New York: Norton, 1988.

Lovelock, James. *Gaia: A New Look at Life on Earth*. New York: Oxford University Press, 1979.

Margulis, Lynn, and Lorraine Olendzenski. *Environmental Evolution*. Cambridge, Mass.: MIT Press, 1992.

Marsh, George Perkins. *The Earth as Modified by Humans*. New York: Arno, 1874.

Marsh, George Perkins. *Man and Nature*. New York: Scribner, 1864.

McKibben, Bill. *The End of Nature*. New York: Random House, 1989.

McLuhan, T. C. *Touch the Earth. A Self-Portrait of Indian Existence*. New York: Dutton, 1971.

Meadows, Donella H. *Limits to Growth: A Report for the Club of Rome's Projection on the Predicament of Mankind*. New York: Universe Books, 1974.

Myers, Norman, ed. *Gaia: An Atlas of Planet Management*. Garden City, N.Y.: Doubleday, 1984.

Nash, Roderick. *The American Environment: Readings in the History of Conservation*. Reading, Mass.: Addison-Wesley, 1968.

Nash, Roderick. *The Rights of Nature: A History of Environmental Ethics*. Madison: University of Wisconsin Press, 1988.

Nash, Roderick. *Wilderness and the American Mind*. New Haven, Conn.: Yale University Press, 1967.

Rifken, Jeremy. *Declaration of a Heretic*. Boston: Routledge & Kegan Paul, 1985.

Schumacher, Ernest. *Small Is Beautiful: Economics As If People Mattered*. New York: Harper & Row, 1973.

Silk, Joseph. *The Big Bang. The Creation and Evolution of the Universe*. San Francisco: Freeman, 1980.

Speth, Timothy C. "The Ecological Lessons of the Past: An Anthropology of Environmental Decline." *The Ecologist* 19, no. 3 (1989).

Time Magazine. "Planet of the Year: Endangered Earth." January 2, 1989, 24–63.

Udall, Stewart. *The Quiet Crisis*. New York: Holt, Rinehart and Winston, 1963.

Wilson, Edward O. *Biodiversity*. Washington, D.C.: National Academy Press, 1988.

World Resources Institute. *World Resources 1990–1991: A Guide to the Global Environment*. New York: Oxford University Press, 1990.

World Resources Institute. *World Resources 1992–1993: A Guide to the Global Environment*. New York: Oxford University Press, 1992.

World Watch Institute. *State of the World 1989: A World Watch Institute Report on Progress Toward a Sustainable Society*. New York: Norton, 1989.

Chapter 2

Allen, J. L. *Environment 92/93*. Guilford, Conn.: Dushkin, 1992.

Barrett, Gary. "A Problem Solving Approach to Resource Management." *Bioscience* 35 (1985): 432–437.

Botkin, Daniel B. *Discordant Harmonies: A New Ecology for the Twenty-First Century.* New York: Oxford University Press, 1990.

Brower, Kenneth. *One Earth.* San Francisco: Collins, 1990.

Brown, Lester. *Building a Sustainable Society.* New York: Norton, 1981.

Cahn, Robert, ed. *An Environmental Agenda for the Future.* Washington, D.C.: Island Press, 1985.

Commoner, Barry. *Making Peace with the Planet.* New York: Pantheon Books, 1990.

Council on Environmental Quality and U.S. Department of State. *Global 2000 Report to the President: Entering the 21st Century*, vol. 1-3. Gerald O. Barney, Study Director. Washington, D.C.: Government Printing Office, 1980–1981.

Devall, Bill. *Simple in Means, Rich in Ends: Practicing Deep Ecology.* Salt Lake City, Utah: Peregrine Smith Books, 1988.

Joseph, Lawrence E. *Gaia: The Growth of an Idea.* New York: St. Martin's Press, 1990.

Leopold, Aldo. *Aldo Leopold's Wilderness: Selected Early Writings by the Author of A Sand County Almanac.* Harrisburg, Penn.: Stackpole, 1990.

Lewin, Roger. "Case Studies in Ecology." *Science* 232 (1986): 25.

Managing Planet Earth: Readings from Scientific American.. New York: Freeman, 1990.

Miller, Alan. "Environmental Problem Solving, Psychosocial Factors." *Environmental Management* 6, no. 6 (1982): 535–541.

Miller, Alan. "The Influence of Personal Biases on Environmental Problem Solving." *Journal of Environmental Problem Solving* 17, no. 2 (1983): 133–143.

Mills, Stephanie. *In Praise of Nature.* Washington, D.C.: Island Press, 1990.

Orians, Gordon H. "The Place of Science in Environmental Problem Solving." *Environment* 28 (1986): 12–21.

Park, Chris C. *Ecology and Environmental Management: A Geographical Perspective.* Boulder, Colo.: Westview Press, 1980.

Rambler, Mitchell, Lynn Margulis, Rene Fester, eds. *Global Ecology: Towards a Science of the Biosphere.* San Diego, Calif.: Academic Press, 1989.

Rees, William E. "The Ecology of Sustainable Development." *The Ecologist* 20, no. 1 (1990): 18–24.

Ridker, Ronald G., and William D. Watson. *To Choose a Future: Resource and Environmental Consequences of Alternative Growth Paths.* Baltimore: Johns Hopkins University Press, 1980.

Rothkrug, Paul, and Robert L. Olson. *Mending the Earth: A World for Our Grandchildren.* Berkeley, Calif.: North Atlantic Books, 1991.

Ruckelshaus, William D. "Toward a Sustainable World." *Managing Planet Earth: Readings from Scientific American.* New York: W. H. Freeman and Company, 1990.

Sagan, Dorion. *Biospheres: Metamorphosis of Planet Earth.* New York: McGraw-Hill, 1990.

Sahtouris, Elisabet. *Gaia: The Human Journey from Chaos to Cosmos.* New York: Pocket Books, 1989.

U.S. Department of Agriculture. *Using Our Natural Resources, Yearbook of Agriculture.* Washington, D.C.: Government Printing Office, 1983.

Ward, Barbra. *Progress for a Small Planet.* New York: Norton, 1979.

Wattenberg, Ben J. *The Good News Is the Bad News Is Wrong.* New York: Simon & Schuster, 1984.

World Commission on Environment and Development. *Our Common Future.* New York: Oxford University Press, 1990.

Worster, Donald, ed. *The Ends of the Earth: Perspectives on Modern Environmental History.* New York: Cambridge University Press, 1988.

Young, Oran. *Resource Management at the International Level.* New York: Nichols, 1977.

Chapter 3

Barrett, Gary W., and Claire A. Puchy. "Environmental Science: A New Direction in Environmental Studies." *International Journal of Environmental Studies* 10 (1977): 157–160.

Bertelman, Thomas, et al. *Resources, Society, and the Future.* Oxford: Pergamon Press, 1980.

Carpenter, Steven R., and James F. Kitchell. "Consumer Control of Lake Productivity." *Bioscience* 38 (1988): 764–769.

Evans, F. C. "Ecosystem and the Basic Unit in Ecology." *Science* 123 (1956): 1127–1128.

Gittleman, John L. *Carnivore Behavior, Ecology, and Evolution.* Ithaca, N.Y.: Comstock, 1989.

Naiman, Robert J., Jerry M. Meillo, and John E. Hobbie. 1986. "Ecosystem Alteration of Boreal Forest Streams by Beaver." *Ecology* 67 (1986): 1254–1269.

Odum, Eugene. "The Emergence of Ecology as a New Integrative Discipline." *Science* 195 (1977): 1289–1292.

Perry, John S. "Managing the World Environment." *Environment* 28, no. 1 (1986): 10–40.

Risser, Paul G. "Toward a Holistic Management Perspective." *Bioscience* 35, no. 7 (1985): 414–418.

Royce-Malgren, Carl H., and Winsor H. Watson III. "Modification of Olfactory Related Behavior in Juvenile Atlantic Salmon by Changes in pH." *Journal of Chemical Ecology* 13, no. 3 (1987): 533–546.

Scientific American. The Biosphere. San Francisco: Freeman, 1970.

Simon, Herbert A. *The Colonization of Complex Systems in Hierarchy Theory.* New York: Patee, 1973.

Smith, Robert Leo. *Ecology and Field Biology.* New York: Harper & Row, 1980.

Worster, David. *Nature's Economy: The Roots of Ecology.* San Francisco: Sierra Club Books, 1977.

Chapter 4

Bolin, Bert. "The Carbon Cycle." *Scientific American* 223, no. 3 (1970): 124–132.

Borman, F. H., and Gene E. Likens. "The Nutrient Cycles of an Ecosystem." *Scientific American* 223, no. 3 (1970): 92–101.

Cook, E. K., and R. A. Berner. *The Global Water Cycle.* Englewood Cliffs, N.J.: Prentice-Hall, 1987.

Delwiche, C. C. "The Nitrogen Cycle." *Scientific American* 223, no. 3 (1970): 136–146.

Devey, Edward S. "Mineral Cycles." *Scientific American* 223, no. 3 (1970): 148–158.

Erlich, Paul R., and Harold A. Mooney. "Extinction, Substitution and Ecosystem Services." *Bioscience* 33 (1983): 248–254.

Hobbie, John, Jon Cole, Jennifer Dugan, R. A. Houghton, and Bruce Peterson. "Role of Biota in Global CO_2 Balance: The Controversy." *Bioscience* 34 (1984): 492–498.

Jordan, Carl F. *Nutrient Cycling in Tropical Forest Ecosystems: Principles and Their Application in Management and Conservation.* New York: Wiley, 1982.

Kellog, W. W., R. D. Cadle, E. R. Allen, A. L. Lazarus, and E. A. Martell. "The Sulfur Cycle." *Science* 175 (1972): 587–596.

Miller, G. Tyler, Jr. *Energy, Kinetics, and Life.* Belmont, Calif.: Wadsworth, 1971.

National Academy of Science. *Productivity of World Ecosystems.* Washington, D.C.: National Academy of Science, 1975.

Phillipson, J. *Ecological Energetics.* New York: St. Martin's Press, 1966.

Pimm, Stuart L. *Food Webs.* New York: Chapman and Hall, 1982.

Prigogine, Ilya, Gregoire Nicoles, and Agnes Babloyantz. "Thermodynamics and Evolution." *Physics Today* 25, no. 11 (1972): 23–28; 25, no. 12 (1972): 138–141.

Schindler, David W. "Evolution of Phosphorus Limitation in Lakes." *Science* 195 (1977): 260–262.

Svensson, Bo H., and R. Soderlund, eds. "Nitrogen, Phosphorus and Sulfur, Global Cycles." *Ecological Bulletins*, no. 22. Stockholm: Royal Swedish Academy of Sciences; 1976.

Chapter 5

Allee, W. C. "Cooperation Between Species." *American Scientist* 60 (1951): 348–357.

Boucher, D. H., S. James, and K. H. Keeler. "The Ecology of Mutualism." *Annual Review of Ecological Systems* 13 (1982): 315–347.

Cheng, T. E. *Aspects of the Biology of Symbiosis*. Baltimore: University Park Press, 1971.

Clements, Frederic E. "Plant Succession: An Analysis of the Development of Vegetation." Publication no. 242. Carnegie Institution of Washington, 1916. (Reprints in book form. New York: Wilson, 1928.)

Cornell, Joseph H., and Ralph G. Slayter. "Mechanisms of Succession in Natural Communities and Their Role in Community Stability and Organization." *American Naturalist* 111 (1977): 1119–1114.

den Boer, P. J. "The Present Status of the Competition Exclusion Principle." *Trends in Ecological Evolution* 1 (1986): 25–28.

Egerton, Frank N. "Changing Concepts of the Balance of Nature." *Quarterly Review of Biology* 48 (1973): 322–350.

Erlich, Paul R., and Peter H. Raven. "Butterflies and Plants: A Study of Coevolution." *Evolution* 18 (1965): 586–608.

Ewald, P. W. "Host-Parasite Relations, Vectors, and the Evolution of Disease Severity." In *Annual Review of Ecology And Systematics* 14 (1984): 365–485.

Gause, G. F. "Ecology of Populations." *Quarterly Review of Biology* 7 (1932): 27–46.

Hardin, Garrett. "The Competitive Exclusion Principle." *Science* (1960): 1292–1297. Vol. 131.

Head, Suzanne, and Robert Heinzman, eds. *Lessons of the Rainforest*. San Francisco: Sierra Club Books, 1991.

Keddy, Paul. *Competition*. New York: Chapman and Hall, 1989.

Kerr, Richard A. "No Longer Willful, Gaia Becomes Respectable." *Science* 240 (1988): 293–296.

Kussler, Jon A. *Our National Wetland Heritage: A Protection Guidebook*. Washington, D.C.: Environmental Law Institute, 1983.

Lovelock, James E. *Gaia: A New Look at Life on Earth*. New York: Oxford University Press, 1979.

Lyman, Francesca "What Gaia Hath Wrought: The Story of a Scientific Controversy." *Technology Review* 92 (1989): 24–31.

McIntosh, Robert P. *The Background of Ecology: Concept and Theory*. New York: Cambridge University Press, 1985.

Moffett, Mark W. "Life in a Nutshell." *National Geographic* 175, no. 6 (1989): 782–796.

Odum, Eugene P. "Population Regulation and Genetic Feedback." *Science* 159 (1969): 1432–1437.

Patten, Bernard C., and Eugene P. Odum. "The Cybernetic Nature of Ecosystems." *American Naturalist* 118 (1981): 886–895.

Pontin, A. J. *Competition and Coexistence of Species*. Boston: Pitman Advanced Publishing Program, 1982.

Shugart, Herman H. *A Theory of Forest Dynamics: The Ecological Implications of Forest Succession Models*. New York: Springer-Verlag, 1984.

Tiner, Ralph W. *Wetlands of the United States: Current Status and Present Trends*. Washington, D.C.: Government Printing Office, 1984.

Whitmore, Timothy C. *An Introduction to Tropical Rain Forests*. New York: Oxford University Press, 1990.

"A world in the shallows: the American Wetlands." *Wilderness Magazine*, 1985. Vol. 49, no. 171.

Chapter 6

Blaikie, Piers M., and Harold Brookfield. *Land Degradation and Society*. New York: Methuen, 1987.

Collins, Mark. *The Last Rain Forests: A World Conservation Atlas*. New York: Oxford University Press, 1990.

Davidson, Art. *In the Wake of the Exxon Valdez: The Devastating Impact of Alaska's Oil Spill*. San Francisco: Sierra Club Books, 1990.

Ellis, William S., and Steve McCurry. "Africa's Sahel: The Stricken Land." *National Geographic* 172 (1987): 140–179.

Freedman, Bill. *Environmental Ecology: Impacts of Pollution and Other Stresses on Ecosystem Structure and Function*. San Diego, Calif.: Academic Press, 1989.

Gradwohl, Judith, and Russell Greenberg. *Saving the Tropical Forests*. Washington, D.C.: Island Press, 1988.

Hecht, Susanna B., and Alexander Cockburn. *The Fate of the Forest: Developers, Destroyers, and Defenders of the Amazon*. New York: Verso, 1989.

Myers, Norman. *The Primary Source: Tropical Forests and Our Future*. New York: Norton, 1984.

Schlesinger, W. H., et al. "Biological Feedbacks in Global Desertification." *Science* 247 (1990): 1043–1049.

Sears, Paul B. *Deserts on the March*. Washington, D.C.: Island Press, 1988.

Thompson, Jon. "East Europe's Dark Dawn." *National Geographic* 179, no. 6 (1991): 36–69.

Vesilind, P. Aarne. *Environmental Pollution and Control*. Stoneham, Mass.: Butterworth, 1983.

Waid, John S. *PCB's and the Environment*. Boca Raton, Fla.: CRC Press, 1987.

Wolf, Edward C. "Survival of the Rarest." *World Watch* 4, no. 2 (199): 12–20.

World Resources Institute. *World Resources 1990–91*. New York, Oxford: Oxford University Press, 1990.

World Resources Institute. *World Resources 1992–1993*. Oxford University Press, 1992.

World Watch Institute. *State of the World 1990*. New York: Norton, 1990.

World Watch Institute. *State of the World 1992*. New York: 1992.

Chapter 7

Aber, John D., and William R. Jordan III. "Restoration Ecology: An Environmental Middle Ground." *BioScience* 35 (1985): 399–400.

Altieri, Miguel A. *Agroecology: The Scientific Basis of Alternative Agriculture*. Boulder, Colo.: Westview Press, 1987.

Altieri, Miguel A., and Susanna B. Hecht. *Agroecology and Small Farm Development*. Boca Raton, Fla.: CRC Press;

Berger, John J. *Environmental Restoration: Science and Strategies for Restoring the Earth*. Washington, D.C.: Island Press, 1984.

Bioscience 40, no. 7 (July–August 1990). Special issue on long-term ecological research.

Bonnicksen, Thomas M. "Restoration Ecology: Philosophy, Goals and Ethics." *The Environmental Professional* 10 (1988): 25–35.

Carrol, C. Ronald, John H. Vandermeer, and Peter M. Rosset. *Agroecology*. New York: McGraw-Hill, 1990.

Clements, William H., John H. Van Hassel, Donald S. Cherry, and John Cairns Jr. "Colonization, Variability, and the Use of Substratum-Filled Trays for Biomonitoring Benthic Communities." *Hydrobiologia* 173, (1989): 45–53.

Hammer, Donald A. *Constructed Wetlands for Wastewater Treatment: Municipal, Industrial, and Agricultural*. Chelsea, Mich.: Lewis, 1989.

Hunter, Christopher J. *Better Trout Habitat: A Guide to Stream Restoration and Management*. Washington, D.C.: Island Press, 1990.

Knight, Dennis H., and Linda L. Wallace. "The Yellowstone Fires: Issues in Landscape Ecology." *BioScience* 39 (1989): 700–706.

Kustler, John A., and Mary E. Kentula, eds. *Wetland Creation and Restoration: The Status of the Science*. Washington, D.C.: Island Press, 1990.

Levine, M. Beth, A. Tilghman Hall, Mary W. Barrett, and Douglas H. Tayler. "Heavy-Metal Concentrations During Ten Years of Sludge Treatment to an Old-Field Community." *Journal of Environmental Quality* 18, no. 4 (1989): 411–418.

Lugo, Ariel E. "The Future of the Forest: Ecosystem Rehabilitation in the Tropics." *Environment* 30 (1988): 16–25.

Matthews, Samuel W. "Is Our World Warming?" *National Geographic* 178, no. 4 (1990): 66–99.

Reid, Walter V., and Kenton Miller. *Keeping Options Alive: The Scientific Basis for the Conservation of Biodiversity*. New York: World Resources Institute, 1989.

Risser, Paul G., James R. Karr, and Richard T. T. Forman. "Landscape Ecology: Directions and Approaches." *Illinois Natural History Survey*. Special Publication no. 2. Champaign Ill.: Illinois Natural History Society.

Rodiek, Jon. 1990. *Wildlife and Habitat in Managed Landscapes.* Washington, D.C.: Island Press, 1990.

Sheehan, Patrick J., Donald R. Miller, Gordon C. Butler, and Philippe Bourdeau, eds. *Effects of Pollutants at the Ecosystem Level.* New York: Wiley, 1984.

Slobodkin, Lawrence B. "The Intellectual Problems of Applied Ecology." *BioScience* 38 (1988): 156–162.

Soulé, Michael E., and Kathryn A. Kohm. *Research Priorities for Conservation Biology.* Washington, D.C.: Island Press, 1989.

Urban, Dean, Robert V. O'Neill, and Herman H. Shugart. "Landscape Ecology." *BioScience* 37 (1986): 119–128.

Western, David, and Mary C. Pearl, eds. *Conservation for the Twenty-First Century.* New York: Oxford University Press, 1989.

Winner, Robert, Heather A. Owen, and Marianne V. Moore. "Seasonal Variability in the Sensitivity of Freshwater Lentic Communities to a Chronic Copper Stress." *Aquatic Toxicology* 17 (1990): 75–92.

Zonneveld, Isaak S., and Richard T. Forman, eds. 1990. *Changing Landscapes: An Ecological Perspective.* New York: Springer-Verlag, 1990.

Chapter 8

Banister, Judith. *China's Changing Population.* Stanford, Calif.: Stanford University Press, 1987.

Brown, Becky J., Mark G. Hanson, Diana M. Liverman, and Robert W. Meredith, Jr. "Global Sustainability: Toward Definition." *Environmental Management* 11, no. 6 (1987): 713–719.

Brown, Lester R. "Feeding Six Billion." *World Watch Journal* 2, no. 5 (1989): 32–40.

Brown, Lester R., and Jodi Jacobson. "Our Demographically Divided World." World Watch Paper 74. Washington, D.C.: World Watch Institute, 1986.

Brown, Lester R., and Jodi L. Jacobson. "The Future of Urbanization: Facing the Ecological and Economic Constraints." *World Watch Paper 77.* Washington, D.C.: World Watch Institute, 1987.

Brown, Lester R., and Edward C. Wolf. "Reversing Africa's Decline." World Watch Paper 65. Washington, D.C.: World Watch Institute, 1985.

Chandler, William U. "Investing in Children." World Watch Paper 64. Washington, D.C.: World Watch Institute, 1985.

Coale, Ansley J., and Edgar M. Hoover. *Population Growth and Economic Development in Low-Income Countries.* Princeton, N.J.: Princeton University Press, 1958.

Cook, James. "More People Are a Good Thing: An Interview with Julian Simon." *Forbes,* December 21, 1981, 70–72.

Cutler, M. Rupert. "Human Population: The Ultimate Wildlife Threat." *Vital Speeches* 53 (1987): 691–697.

Daly, Herman E. *The Steady State Economy: Alternative to Growthmania.* Washington, D.C.: Population-Environmental Balance, 1987.

Durning, Alan B. "Action at the Grassroots: Fighting Poverty and Environmental Decline." World Watch Paper 88, Washington, D.C.: World Watch Institute, 1989.

Ehrlich, Anne. "Critical Masses: World Population 1984." *Sierra,* July–August 1984, 36–40.

Ehrlich, Paul, and Anne Ehrlich. "The Population Explosion: Why Isn't Everyone as Scared As We Are?" *The Amicus Journal* 12, no. 1 (1990): 22–29.

Jacobson, Jodi L. "Environmental Refugees: A Yardstick of Habitability." World Watch Paper 86. Washington, D.C.: World Watch Institute, 1988.

Jacobson, Jodi L. "Planning the Global Family." World Watch Paper 80. Washington, D.C.: World Watch Institute, 1987.

Keyfitz, Nathan. "The Growing Human Population." *Scientific American* 261 (1989): 151–169.

Lappé, Frances Moore, and Rachel Schurman. *Taking Population Seriously.* San Francisco: Institute for Food and Development Policy, 1988.

Liu, Cheng. *China's Population: Problems and Prospects.* Beijing: New World Press, 1981.

Martin, Linda G. "The Graying of Japan." *Population Bulletin* 44, no. 2 (1989).

Meadows, Donella H., Dennis Meadows, Jorgen Randers, and William Behrens. *The Limits to Growth.* New York: The New American Library, Inc., 1974.

Misch, Ann. "*Purdah* and Overpopulation in the Middle East." *World Watch* 3, no. 6 (1990): 10–11.

O'Hare, William P., and Judy C. Felt. "Asian Americans: America's Fastest Growing Minority Group." *Population Trends and Public Policy* 19 (1991).

Population Crisis Committee. *Population Briefing Paper,* no. 20 (1988).

Population Information Program. *Population Reports* 13, no. 5 (1985).

Population Reference Bureau. *Asian Americans: America's Fastest Growing Minority Group.* William P. O'Hare and Judy C. Felt. Population Trends and Public Policy Number 19, February 1991.

Preston, Samuel H. "Population Growth and Economic Development." *Environment* 28, no. 2 (1986): 6–9, 32–33.

Reining, Priscilla, and Irene Tinker, eds. *Population: Dynamics, Ethics and Policy.* Washington, D.C.: American Association for the Advancement of Science, 1975.

Repetto, Robert. "Population, Resources, Environment: An Uncertain Future." *Population Bulletin* 42, no. 2 (1987).

Thurow, Lester R. "Why the Ultimate Size of the World's Population Doesn't Matter." *Technology Review* 89 (1986): 22–29.

Ulph, Owen. "On the Limits to Growth, the Human Condition, and the Meaning of Overshoot and Collapse."

Valdivieso, Rafael, and Cary Davis. "U.S. Hispanics: Challenging Issues for the 1990's." *Population Reference Bureau* 17 (1988).

Weber, Susan, ed. *USA by Numbers: A Statistical Portrait of the United States.* Washington, D.C.: Zero Population Growth, 1988.

World Bank. *Development Report 1990.* New York: Oxford University Press, 1990.

World Resources Institute. "Human Settlements." In *World Resources 1990–1991.* Oxford: Oxford University Press, 1990.

World Resources Institute. "Population and Health." In *World Resources 1990-1991.* Oxford: Oxford University Press, 1990.

World Resources Institute. "Population and Human Development." *World Resources 1992–1993: A Guide to the Global Environment.* New York: Oxford University Press, 1992.

Zero Population Growth. *ZPG's Urban Stress Test.* Washington, D.C.: Zero Population Growth, 1988.

Chapter 9

Boserup, Ester. *Economic and Demographic Relationships in Development.* Baltimore: Johns Hopkins University Press, 1990.

Brown, Lester R., and Jodi L. Jacobson. "Our Demographically Divided World." *Worldwatch Paper 74.* Washington, D.C.: Worldwatch Institute, 1986.

Butts, Yolanda. *International Amazonia: Its Human Side.* Chicago: Social Development Center, 1989.

Che-Alford, Janet. *Population Profile of China.* Toronto: Thompson Educational Publishers, 1990.

Cordell, Dennis D., and Joel W. Gregory. *African Population and Capitalism: Historical Perspectives.* Boulder, Colo.: Westview Press, 1987.

Eekelaar, John M. *An Aging World: Dilemmas and Challenges for Law and Social Policy.* Oxford: Clarendon Press, 1989.

Ehrlich, Paul R., and Anne H. Ehrlich. *The Population Explosion.* New York: Simon & Schuster, 1990.

Fornos, Werner. *Gaining People, Losing Ground: A Blueprint for Stabilizing World Population.* Washington, D.C.: Population Institute, 1987.

Grant, Lindsey. *Cornucopian Fallacies.* Washington, D.C.: The Environmental Fund, 1982.

Harris, Marvin. *Cannibals and Kings: The Origins of Culture.* New York: Random House, 1977.

Hinrichsen, Don. "The Decisive Decade: What We Can Do About Population." *The Amicus Journal* 12, no. 1 (1990): 30–32.

Jacobson, Jodi L. "Abortion in a New Light." *World Watch* 3, no. 2 (1990): 31–38.

Johnson, Stanley. *World Population and the United Nations: Challenge and Response.* Cambridge: Cambridge University Press, 1987.

Kasun, Jacqueline R. *The War Against Population: The Economics and Ideology of World Population Control.* San Francisco: Ignatius Press, 1988.

Kirkby, R. J. R. *Urbanization in China: Town and Country in a Developing Economy, 1949–2000.* New York: Columbia University Press, 1985.

Lancaster, Henry O. *Expectations of Life: A Study in the Demography, Statistics, and History of World Mortality.* New York: Springer-Verlag, 1990.

Ornstein, Robert E., and Paul Ehrlich. *New World, New Mind: Moving Toward Conscious Evolution.* Garden City, N.Y.: Doubleday, 1989.

Population Crisis Committee. "Access to Birth Control: A World Assessment." Population Briefing Paper no. 19. Washington, D.C., 1987.

Rhoades, Robert E. "The World's Food Supply at Risk." *National Geographic* 179, no. 4 (1991): 74–105.

Ross, John A. *Family Planning and Child Survival: 100 Developing Countries.* New York: Center for Population and Family Health, Columbia University, 1988.

Russell, Thornton. *American Indian Holocaust and Survival: Population History Since 1492.* Norman: University of Oklahoma Press, 1987.

Santos, Miguel A. *Managing Planet Earth: Perspectives on Population Ecology and the Law.* New York: Bergin & Garvey, 1990.

Simmons, Ozzie G. *Perspectives on Development and Population Growth in the Third World.* New York: Plenum, 1990.

Simon, Julian Lincoln. *Population Matters: People, Resources, Environment and Immigration.* New Brunswick, N.J.: Transaction, 1990.

UNESCO. *Population Education in Asia: A Source Book.* Bangkok: UNESCO, Regional Office for Education in Asia, 1975.

Wattenberg, Ben J. *The Birth Dearth.* New York: Pharos Books, 1989.

Wattenberg, Ben J. *The First Universal Nation: Leading Indicators and Ideas About the Surge of America in the 1990's.* New York: Free Press. 1991.

Chapter 10

Brown, Lester R. "The Changing World Food Prospect: The Nineties and Beyond." *World Watch Paper 85.* Washington, D.C.: Worldwatch Institute, 1988.

Brown, Lester R., and Gail W. Finsterbusch. *Man and His Environment: Food.* New York: Harper & Row, 1972.

Brown, Lester R., and John E. Young. "Growing Food in a Warmer World." *World Watch* 1, no. 6 (1988): 31–35.

Byron, William. *The Causes of World Hunger.* New York: Paulist Press, 1982.

Crosson, Pierre R., and Norman J. Rosenberg. "Strategies for Agriculture." In *Managing Planet Earth: Readings from Scientific America.* New York: W. H. Freeman and Company, 1990.

Danhoff, Clarence H. *Change in Agriculture: The Northern United States, 1820–1870.* Cambridge, Mass.: Harvard University Press, 1969.

Durning, Alan B. "Action of the Grassroots: Fighting Poverty and Environmental Decline." *World Watch Paper 88.* Washington, D.C.: World Watch Institute, 1989.

Durning, Alan B. "Life on the Brink." *World Watch* 3, no. 3 (1990): 22–30.

Durning, Alan B. "Poverty and the Environment: Reversing the Downward Spiral." *World Watch Paper 92.* Washington, D.C.: World Watch Institute, 1989.

Eckholm, Erik P., and Frank Record. *The Two Faces of Malnutrition.* Washington, D.C.: World Watch Institute, 1976.

Ehrlich, Paul R., and Anne H. Ehrlich. "Population, Plenty, and Poverty." *National Geographic* 174 (1988): 914–946.

Fox, Robert W. "The World's Population Problem." *National Geographic* 166 (1984): 176–186.

Galbraith, John Kenneth. *The Nature of Mass Poverty.* Cambridge, Mass.: Harvard University Press, 1979.

Gupte, Pranay. *The Crowded Earth: People, Politics and the Population.* New York: Norton, 1984.

Hulse, Joseph H. "Food Science and Nutrition: The Gulf Between the Rich and the Poor." *Science* 216 (1982): 1291–1295.

The Hunger Project. *Ending of Hunger: An Idea Whose Time Has Come.* New York: Praeger, 1985.

Jackson, Wes. *New Roots for Agriculture.* San Francisco: Friends of the Earth, 1980.

Kaplan, Robert D. *Surrender or Starve: The Wars Behind the Famine.* Boulder, Colo.: Westview Press, 1988.

Kelsey, Darwin P. *Farming in the New Nation: Interpreting American Agriculture.* Washington, D.C.: The Agricultural History Society, 1972.

Knowles-Nobel, Loretta. *Starving in the Shadow of Plenty.* New York: Putnam, 1981.

Lappé, Frances Moore. *Diet for a Small Planet, 10th Anniversary Revised Edition.* New York: Ballantine Books, 1982.

Lappé, Frances Moore, and Joseph Collins. *World Hunger: Twelve Myths. A Food First Book.* New York: Grove Press, 1986.

Lenihan, John, and William W. Fletcher. *Food, Agriculture and the Environment.* San Diego, Calif.: Academic Press, 1976.

Levi, Carolyn. "Growing Fish Salad: An Experiment in Integrated Aquaculture." *Nor'Easter* 3, no. 1 (1991): 15–17.

Mellor, John. "The Entwining of Environmental Problems and Poverty." *Environment,* November 1988, 8.

Mengisteab, Kidane. *Ethiopia: Failure of Land Reform and Agricultural Crisis.* New York: Greenwood Press, 1990.

Office of Technology Assessment. *Enhancing Agriculture in Africa: A Role for U.S. Development Assistance.* Washington, D.C.: Government Printing Office, 1988.

Paulino, Leonardo. *Food in the Third World: Past Trends and Projections into 2000.* Washington, D.C.: International Food Policy Research Institute, 1986.

Pimentel, David, and Carl W. Hall. *Food and Natural Resources.* San Diego, Calif.: Academic Press, 1989.

Power, J. F., and R. F. Follet. "Monoculture." *Scientific American* 256 (1987): 78–87.

Shepard, Jack. *The Politics of Starvation.* New York: Carnegie Endowment for International Peace, 1975.

Tangley, Laura. "Beyond the Green Revolution." *BioScience* 37 (1987): 176–181.

United Nations Children's Fund (UNICEF). *The State of the World's Children 1989.* New York: Oxford University Press, 1989.

U.S. Department of Commerce. *Statistical Abstract of the United States 1991.* Washington, D.C.: Bureau of the Census, 1991.

Wennergren, E. Boyd, et al. *Solving World Hunger: The U.S. Stake.* Published for the Consortium for International Cooperation in Higher Education. Cabin John, Md.: Seven Locks Press, 1986.

Wilkes, Hilbert G., et al. "Plant Genetic Resources: Why Privatize a Public Good?" *BioScience* 37 (1987): 647–652.

Witwer, Sylvan, et al. *Feeding a Billion— Frontiers of Chinese Agriculture.* East Lansing: Michigan University Press, 1987.

Wolf, Edward C. "Beyond the Green Revolution: New Approaches for World Agriculture." World Watch Paper 73. Washington, D.C.: World Watch Institute, 1986.

World Resources Institute. *World Resources 1990-1991.* Oxford: Oxford University Press, 1990.

World Resources Institute. *World Resources 1992-1993: A Guide to the Global Environment.* New York: Oxford University Press, 1992.

Chapter 11

"Alaska After Exxon." *Newsweek.* September 18, 1989, 50–60.

Commoner, Barry. *The Poverty of Power: Energy and the Economic Crisis.* New York: Bantam Books, 1977.

Deudeny, Daniel, and Christopher Flavin. *Renewable Energy: The Power to Choose.* New York: Norton, 1983.

Energy and Environment: Readings from Scientific American. New York: Freeman, 1980.

Gibbons, John H., Peter D. Blair, and Holly L. Gwin. "Strategies for Energy Use." In *Managing Planet Earth: Readings from*

Scientific America. New York: W. H. Freeman and Company, 1990.

Holing, Dwight. *Coastal Alert: Energy, Ecosystems, and Offshore Drilling.* Washington, D.C.: Island Press, 1990.

Kumar, Ashok. *Environmental Challenges in Energy Utilization During the 1990's.* Air and Waste Management Association, 1989.

Long, Robert Emmet. *Energy and Conservation.* New York: Wilson, 1989.

Office of Technology Assessment. *Energy Use and the U.S. Economy.* Washington, D.C.: Government Printing Office, 1990.

Skinner, Samuel K., and William K. Reilly. *The* Exxon Valdez *Oil Spill: A Report to the President.* Prepared by the National Response Team May 1989. Washington, D.C.: 1989.

Soussan, John. *Primary Resources and Energy in the Third World.* New York: Routledge, 1988.

World Resources Institute. "Energy." *World Resources 1992–1993: A Guide to the Global Environment.* New York: Oxford University Press, 1992.

Yergin, Daniel. "Energy Security in the 1990's." *Foreign Affairs,* Fall 1988, 110–132.

Chapter 12

Alm, Alvin L. *Coal Myths and Environmental Realities: Industrial Fuel Use in a Time of Change.* Boulder, Colo.: Westview Press, 1984.

Brown, Lester R., Christopher Flavin, and Sandra Postel. "The Efficiency Revolution." In *Saving Planet Earth: How to Shape an Environmentally Sustainable Global Economy.* New York: W.W. Norton and Company, 1991.

Canby, Thomas Y. "After the Storm." *National Geographic* 179, no. 8 (1991): 2–35.

Energy Information Administration. *Energy Facts.* Washington, D.C.: U.S. Department of Energy, 1985.

National Coal Association. *1983 Facts About Coal.* Washington, D.C.: National Coal Association, 1983.

World Resources Institute. "Energy." In *World Resources 1990–1991.* Oxford: Oxford University Press, 1990.

World Resources Institute. "Energy." *World Resources 1992–1993: A Guide to the Global Environment.* New York: Oxford University Press, 1992.

Chapter 13

The Audubon Energy Plan, Summary of the Second Edition. New York: The Audubon Society, 1984.

Brower, Michael. *Cool Energy; the Renewable Solution to Global Warming: A Report by the Union of Concerned Scientists.* Cambridge, Mass.: Union of Concerned Scientists, 1990.

Brown, Lester R., Christopher Flavin, and Sandra Postel. "Building a Solar Economy." *Saving the Planet: How to Shape an Environmentally Sustainable Global Economy.* New York: W.W. Norton and Company, 1991.

Burnett, W. M. "Changing Prospects for Natural Gas in the United States." *Science* 244 (1989): 305–311.

Flavin, Christopher, and Alan B. Durning. *Building on Success: The Age of Energy Efficiency.* Washington, D.C.: World Watch Institute, 1988.

Flavin, Christopher, and Nicholas Lenssen. *Beyond the Petroleum Age: Designing a Solar Economy.* Washington, D.C.: World Watch Institute, 1990.

Flavin, Christopher, and Cynthia Pollock. "Replacing Fossil Fuels with Renewable Energy That Is Environmentally Benign." *Natural History* 94 (1985): 80–84.

Gibbons, John H., et al. "Strategies for Energy Use." *Scientific American,* September 1989, Vol. 261, no. 3, 136–143.

Goldemberg, José, et al. *Energy for a Sustainable World.* Washington, D.C.: World Resources Institute, 1987.

Hamburg, Robert. "Assessing the Benefits of Biogas." *Environment* 30, no. 10 (1988).

Hershkowitz, Allen. "Burning Trash: How It Could Work." *Technology Review* 90 (1987): 44–52.

Horton, Tom. "Paradise Lost." *Rolling Stone,* December 14, 1989, 150–182; December 28, 1989, 241–246.

Ivey, Mark, and Ronald Grover. "Alcohol Fuels Move Off the Back Burner." *Business Week,* June 29, 1987, 100–101.

Lenssen, Nicholas, and Jon E. Young. "Filling Up in the Future." *World Watch* 3, no. 3 (1990): 18–26.

Lowe, Marcia D. "The Bicycle: Vehicle for a Small Planet." World Watch Paper 90. Washington, D.C.: World Watch Institute, 1989.

Mackenzie, James J. *Breathing Easier: Taking Action on Climate Change, Air Pollution, and Energy Insecurity.* Washington, D.C.: World Resources Institute, 1988.

Office of Technology Assessment. *Replacing Gasoline: Alternative Fuels for Light-Duty Vehicles.* Washington, D.C.: Government Printing Office, 1990.

Quillen, Ed. "At Home in a High-Altitude Think Tank." *Country Journal,* July 1986, 42–51.

San Martin, Robert. "Renewable Energy Power for Tomorrow." *Futurist* 23 (1989): 37–41.

Stanfield, Rochelle L. "Less Burning, No Tears." *National Journal,* August 13, 1988, 2095–2098.

World Resources Institute. "Energy." In *World Resources 1990-1991.* Oxford: Oxford University Press, 1990.

Chapter 14

Abrahamson, Dean Edwin. *The Challenge of Global Warming.* Washington, D.C.: Island Press, 1989.

"Acidification." *Ambio* 18, no. 3 (1988). Special issue on acid deposition and its effects.

Bates, Albert K. *Climate in Crisis: The Greenhouse Effect and What You Can Do.* Summertown, Tenn.: Book, 1990.

Fisher, David E. *Fire and Ice: The Greenhouse Effect, Ozone Depletion, and Nuclear Winter.* New York: HarperCollins, 1990.

Flavin, Christopher. *Slowing Global Warming: A World Wide Strategy.* Washington, D.C.: World Watch Institute, 1989.

French, Hilary F. "You Are What You Breathe." *World Watch* 3, no. 3 (1990): 27–34.

Goldsmith, Edward, and Nicholas Hildyard, ed. *The Earth Report: The Essential Guide to Global Issues.* Los Angeles, Calif.: Price Stern Sloan, Inc., 1988.

Graedel, Thomas E., and Paul J. Crutzen. "The Changing Atmosphere." *Scientific American,* Vol. 261, no. 3: 58–68.

Gribbin, John R. *Hothouse Earth: The Greenhouse Effect and Gaia.* New York: Grove Weidenfeld, 1990.

Hall, Bob, and Mary Lee Kerr. *1991–1992 Green Index.* Washington, D.C.: Island Press, 1991.

Houghton, Richard A., and George M. Woodwell. "Global Climatic Change." *Scientific American* 260 (4): 116–119, 126–132.

Lean, Geoffrey, Don Hinrichsen, and Adam Markham. *Atlas of the Environment.* New York: Prentice Hall Press, 1990.

Lyman, Francesca. *The Greenhouse Trap: What We're Doing to the Atmosphere and How We Can Slow Global Warming.* Boston: Beacon Press, 1990.

MacKenzie, James J., and Mohamed T. El-Ashry. *Air Pollution's Toll on Forests and Crops.* New Haven, Conn.: Yale University Press, 1989.

MacKenzie, James J., and Mohamed T. El-Ashry. *Winds: Airborne Pollution's Toll on Trees and Crops.* Washington, D.C.: World Resources Institute, 1988.

Mathews, Samuel W. "Is Our World Warming?" *National Geographic* 178, no. 4 (1990): 66–99.

National Research Council. *Biologic Markers of Air-Pollution Stress and Damage in Forests.* Washington, D.C.: National Academy Press, 1989.

Regens, James L., and Robert W. Rycroft. *The Acid Rain Controversy.* Pittsburgh, Pa.: University of Pittsburgh Press, 1988.

Rifkin, Jeremy. *Entropy: Into the Greenhouse World.* New York: Bantam Books, 1989.

Schneider, Stephen H. "The Changing Climate." *Scientific American,* September: Vol. 261, no. 3, 70–79.

Schneider, Stephen H. *Global Warming: Are We Entering the Greenhouse Century?* San Francisco: Sierra Club Books, 1989.

Smith, Joel B., and Dennis A. Tirpak. *The Potential Effects of Global Climate Change on the United States of America.* New York: Hemisphere, 1990.

"Special Report on the Greenhouse Effect." *Newsweek,* July 11, 1988, 16–23.

Sun, Marjorie. "Academy Dispels Doubt on Acid Rain." *Science* 231 (1986): 1500.

Treshow, Michael, and Franklin K. Anderson. *Plant Stress from Air Pollution.* New York: Wiley, 1990.

U.S. Environmental Protection Agency. *National Air Quality and Emissions Trends Report, 1990.* Washington, D.C.: Environmental Protection Agency, 1991.

Vatavuk, William. *Estimating Costs of Air Pollution Control.* Chelsea, Mich.: Lewis, 1990.

Wellburn, Alan. *Air Pollution and Acid Rain: The Biological Impact.* New York: Wiley, 1988.

World Resources Institute. "Atmosphere." In *World Resources 1990–1991.* Oxford: Oxford University Press, 1990.

World Resources Institute. "Climate Change: A Global Concern." In *World Resources 1990–1991.* Oxford: Oxford University Press, 1990.

World Resources Institute. "Global Systems and Cycles." In *World Resources 1990–1991.* Oxford: Oxford University Press, 1990.

World Resources Institute. "Atmosphere and Climate." *World Resources 1992–1993: A Guide to the Global Environment.* New York: Oxford University Press, 1992.

Yost, Nicholas, et al. "Wetlands: Through Murky Waters." *NAEP Newsletter* 17, no. 2 (1992).

Chapter 15

Boyle, Robert H., and R. Alexander Boyle. *Acid Rain.* New York: Schocken Books, 1983.

Cairns, John, Jr., and Arthur C. Burkema, Jr. *Restoration of Habitats Impacted by Oil Spills.* Boston: Butterworth, 1984.

Carrier, Jim. "Water and the West: the Colorado River." *National Geographic* 179, no. 6 (1991): 2–35.

Carson, Rachel. *The Edge of the Sea.* Boston: Houghton Mifflin, 1955.

Cousteau, Jacques-Yves. *The Cousteau Almanac: An Inventory of Life on Our Water Planet.* Garden City, N.Y.: Doubleday, 1981.

Cummins, Joseph E. "The PCB Threat to Marine Mammals." *The Ecologist* 18, no. 6 (1988): 193–195.

Danson, James. *Superspill: The Future of Ocean Pollution.* London: Jane's, 1980.

"The Dirty Seas." *Time,* August 1, 1988, 44–50.

"Don't Go Near the Water." *Newsweek,* August 1, 1988, 42–48.

Duplaix, Nicole. "South Florida Water: Paying the Price." *National Geographic* 178, no. 1 (1990): 89–113.

Edgerton, Lynne T. *The Rising Tide.* Washington, D.C.: Island Press, 1990.

Environmental Law Institute. *Clean Water Deskbook.* Washington, D.C.: Environmental Law Institute, 1988.

Furgurson, Ernest. "Lake Tahoe—Playing for High Stakes." *National Geographic* 180, no. 3, (1992): 113–132.

Hall, Bob, and Mary Lee Kerr. *1991–1992 Green Index.* Washington, D.C.: Island Press, 1991.

Horan, N. J. *Biological Wastewater Treatment Systems: Theory and Operation.* New York: Wiley, 1990.

Jones, E. Bruce, and Timothy J. Ward. *Watershed Management in the Eighties.* New York: American Society of Engineers, 1985.

LaRiviere, J. W. Maurits. "Threats to the World's Water." *Scientific American* 261, no. 3, (1989): 80–94.

Lean, Geoffrey, Don Hinrichsen, and Adam Markham. *Atlas of the Environment.* New York: Prentice Hall Press, 1990.

Moore, James W. *Balancing the Needs of Water Use.* New York: Springer-Verlag, 1989.

National Research Council. *Managing Coastal Resources.* Washington, D.C.: National Academy Press, 1990.

Postel, Sandra. "Conserving Water: The Untapped Alternative." World Watch Paper 67. Washington, D.C.: World Watch Institute, 1985.

Postel, Sandra. "Water: Rethinking Management in an Age of Scarcity." World Watch Paper 62. Washington, D.C.: World Watch Institute, 1984.

Reisner, Marc. *Cadillac Desert: The American West and Its Disappearing Water.* New York: Penguin Books, 1986.

Reisner, Marc, and Sarah F. Bates. *Overtapped Oasis: Reform or Revolution for Western Water.* Washington, D.C.: Island Press, 1989.

Scudder, Thayer. "Conservation vs. Development: River Basin Projects in Africa." *Environment* 31, no. 2 (1989): 4–9.

Sedeen, Margaret, ed. *Great Rivers of the World.* Washington, D.C.: National Geographic Society, 1984.

Sierra Club Legal Defense Fund. *The Poisoned Well: New Strategies for Groundwater Protection.* Washington, D.C.: Island Press, 1989.

Speidel, David H. *Perspectives on Water: Uses and Abuses.* New York: Oxford University Press, 1988.

Waggoner, Paul E., ed. *Climate Change and U.S. Water Resources.* New York: Wiley, 1990.

Weber, Michael, and Richard Tinney. *A Nation of Oceans.* Washington, D.C.: Center for Environmental Education, 1986.

White, Gilbert F. *Strategies of American Water Management.* Ann Arbor: University of Michigan Press, 1971.

Williams, W. D., and N. V. Aladin. "The Aral Sea: Recent Limnological Changes and Their Conservation Significance." *Aquatic Conservation: Marine and Freshwater Ecosystems* 1 (1991): 3–23.

World Resources Institute. "Freshwater." In *World Resources 1990–1991.* Oxford: Oxford University Press, 1990.

World Resources Institute. "Freshwater." In *World Resources 1992–1993: A Guide to the Global Environment.* New York: Oxford University Press, 1992.

World Resources Institute. "Oceans and Coasts." In *World Resources 1990–1991.* Oxford: Oxford University Press, 1990.

World Resources Institute. "Oceans and Coasts." In *World Resources 1992–1993: A Guide to the Global Environment.* New York: Oxford University Press, 1992.

Chapter 16

Berry, Wendell. *A Continuous Harmony.* San Diego: Harcourt Brace Jovanovich, 1970.

Berry, Wendell. *The Unsettling of America: Culture & Agriculture.* San Francisco: Sierra Club Books, 1977.

Bouwman, A. F., ed. *Soils and the Greenhouse Effect.* New York: Wiley, 1990.

Brown, Lester R. "The Growing Grain Gap." *World Watch* 1, no. 5 (1988): 8–18.

Brown, Lester R. "The Vulnerability of Oil-Based Farming." *World Watch* 1, no. 2 (1988): 24–29.

Crosson, Pierre R., and Norman J. Rosenberg. "Strategies for Agriculture." In *Managing Planet Earth: Readings from Scientific America.* New York: W. H. Freeman and Company, 1990.

Dover, Michael J., and Lee M. Talbot. *To Feed the Earth: Agro-ecology for Sustainable Development.* Washington, D.C.: World Resources Institute, 1987.

Ebeling, Walter. *The Fruited Plain: The Story of American Agriculture.* Berkeley: University of California Press, 1979.

Eisenberg, Evan. "Back to Eden." *Atlantic* 264 (1989): 57–77.

Ellis, William S. "California's Harvest of Change." *National Geographic* 179, no. 2 (1991): 48–74.

Faulkner, Edward H. *Plowman's Folly and a Second Look.* Washington, D.C.: Island Press, 1987.

Francis, Charles A., et al. *Sustainable Agriculture in Temperate Zones.* New York: Wiley, 1990.

Hall, Bob, and Mary Lee Kerr. *1991–1992 Green Index.* Washington, D.C.: Island Press, 1991.

Journal of Soil and Water Conservation 45, no. 1 (1990).

Lal, Rattan. *Soil Erosion in the Tropics: Principles and Management.* New York: McGraw Hill, 1990.

Leon, Geoffrey, Don Hinrichsen, and Adam Markham. *Atlas of the Environment.* New York: Prentice Hall Press, 1990.

Little, Charles E. *Green Fields Forever: The Conservation Tillage Revolution in America.* Washington, D.C.: Island Press, 1987.

Logan, William Bryant, ed. "Living Soil." *Orion* 11, no. 2 (1992).

Mollison, Bill. *Permaculture: A Practical Guide for a Sustainable Future.* Washington, D.C.: Island Press, 1990.

Paddock, Joe, Nancy Paddock, and Carol Bly. *Soil and Survival: Land Stewardship and the Future of American Agriculture.* San Francisco: Sierra Club Books, 1986.

Steiner, Frederick. *The Living Landscape: An Ecological Approach to Landscape Planning.* New York: McGraw-Hill, 1991.

Weber, Peter. "U.S. Farmers Cut Soil Erosion by One-Third." *World Watch* 3, no. 4 (1990): 5–6.

World Resources Institute. "Forests and Rangelands." *World Resources 1992–1993: A Guide to the Global Environment.* New York: Oxford University Press, 1992.

Chapter 17

Abelson, Philip H., ed. "Advanced Technology Materials," *Science* 208 (1980).

Abelson, Philip H., and Allen L. Hammond, eds. *Materials: Renewable and Nonrenewable Resources.* Washington, D.C.: American Association for the Advancement of Science, 1976.

Bates, Robert L., and Julia A. Jackson. *Our Modern Stone Age.* Los Altos, Calif.: Kaufmann, 1982.

Branson, Branley Allan. "Is There Life After Strip Mining?" *Natural History,* August 1986, 31–36.

Brown, Lester, et. al. "Mining the Earth." In *State of the World, 1992.* New York: Norton, 1992.

Cameron, Eugene N. *At the Crossroads— The Mineral Problems of the United States.* New York: Wiley, 1986.

Chandler, William U. "Minerals Recycling: The Virtue of Necessity." World Watch Paper 56. Washington, D.C.: World Watch Institute, 1983.

Cook, Earl. "Limits to Exploitation of Nonrenewable Resources." *Science* 191 (1976): 677–682.

Door, Ann. *Minerals Foundations of Society.* Alexandria, Va.: American Geological Institute, 1987.

Frosch, Robert A., and Nicholas E. Gallopoulos. "Strategies for Manufacturing." *Managing Planet Earth: Readings from* Scientific American. New York: W. H. Freeman, 1990.

Goeller, H. E., and Alvin M. Weinberg. "The Age of Substitutability: What Do We Do When the Mercury Runs Out?" *Science* 191 (1976): 683–689.

Mikesell, Raymond F. *Non-Fuel Minerals: Foreign Dependence and National Security.* Ann Arbor: University of Michigan Press, 1987.

Mineral Commodity Summaries 1988. Bureau of Mines, United States Department of the Interior, 1988.

Park, Charles F., Jr. *Earthbound: Minerals, Energy, and Man's Future.* San Francisco: Freeman, Cooper, 1975.

Wolfe, John A. *Mineral Resources: A World Review.* New York: Chapman and Hall, 1984.

Chapter 18

Beckman, Peter R., et al. *The Nuclear Predicament: An Introduction.* Englewood Cliffs, N.J.: Prentice-Hall, 1989.

Burns, Michael E. *Low-Level Radioactive Waste Regulation: Science, Politics, and Fear.* Chelsea, Mich.: Lewis, 1988.

Colglazier, E. William, Jr. *The Politics of Nuclear Waste.* New York: Pergamon Press, 1982.

Flavin, Christopher. *Reassessing Nuclear Power: The Fallout from Chernobyl.* Washington, D.C.: World Watch Institute, 1987.

Gale, Robert Peter, and Thomas Hauser. *Final Warning: The Legacy of Chernobyl.* New York: Warner Books, 1988.

Gershey, Edward L., et al. *Low-Level Radioactive Waste: From Cradle to Grave.* New York: Van Nostrand Reinhold, 1990.

Jasper, James M. *Nuclear Politics: Energy and the State in the United States, Sweden, and France.* Princeton, N.J.: Princeton University Press, 1990.

Krieger, David, and Frank Kelly, eds. *Waging Peace in the Nuclear Age: Ideas for Action.* Santa Barbara, Calif.: Capra Press, 1988.

Kruschke, Earl R., and Byron M. Jackson. *Nuclear Energy Policy: A Reference Handbook.* Snata Barbara, Calif.: ABC-CLIO, 1990.

Leone, Bruno, and Judy Smith. *The Energy Crisis: Opposing Viewpoints.* St. Paul, Minn.: Greenhaven Press, 1981.

Lifton, Robert Jay. *The Future of Immortality and Other Essays for a Nuclear Age.* New York: Basic Books, 1987.

Lovins, Amory, and L. Hunter Lovins. *Energy/War, Breaking the Nuclear Link.* San Francisco: Friends of the Earth, 1980.

Megaw, James. *How Safe? Three Mile Island, Chernobyl and Beyond.* Toronto: Stoddart, 1987.

Miller, Peter. "A Comeback for Nuclear Power?" *National Geographic* 179, no. 8 (1991): 60–89.

Murray, Raymond L., and Judith A. Powell. *Understanding Radioactive Waste.* Columbus, Ohio: Battelle Press, 1988.

Nader, Ralph, and John Abbots. *The Menace of Atomic Energy.* New York: Norton, 1977.

Nealy, Stanley M. *Nuclear Power Development: Prospects in the 1990's.* Columbus, Ohio: Battelle Press, 1989.

Porro, Jeffrey, Paul Doty, Carl Kaysen, and Jack Ruina et al. *The Nuclear Age Reader.* New York: Knopf, 1989.

Rifkin, Jeremy. *Declaration of a Heretic.* Boston: Routledge & Kegan Paul, 1985.

Schneider, Stephen. *The Oil Price Revolution.* Baltimore: Johns Hopkins University Press, 1983.

Thomas, Steve D. *The Realities of Nuclear Power: International Economic and Regulatory Experience.* New York: Cambridge University Press, 1988.

Wagner, Henry N., Jr., and Linda E. Ketchum. *Living with Radiation: The Risk, the Promise.* Baltimore: Johns Hopkins University Press, 1989.

Chapter 19

Cohen, Gary, and John O'Connor. *Fighting Toxics: A Manual for Protecting Your Family, Community, and Workplace.* Washington, D.C.: Island Press, 1990.

Epstein, Samuel S., Lester Brown, and Carl Pope. *Hazardous Waste in America.* San Francisco: Sierra Club Books, 1982.

Piasecki, Bruce. *Beyond Dumping: New Strategies for Controlling Toxic Contamination.* Westport, Conn.: Quorum Books, 1984.

Russell, Dick. "Dances with Waste." *The Amicus Journal* 13, no. 4 (1991): 28–29.

Schwartz, Seymour I., and Wendy B. Pratt. *Hazardous Waste from Small Quantity Generators.* Washington, D.C.: Island Press, 1990.

Wentz, Charles A., Jr. *Hazardous Waste Management.* New York: McGraw-Hill, 1989.

Chapter 20

Bagcji, Amalendu. *Design, Construction, and Monitoring of a Sanitary Landfill.* New York: Wiley, 1990.

Blumberg, Louis, and Robert Gottlieb. *War on Waste: Can America Win Its Battle with Garbage?* Washington, D.C.: Island Press, 1989.

Bonomo, Luca, and A. E. Higginson. *International Overview on Solid Waste Management.* San Diego, Calif.: Academic Press, 1988.

Brown, Lester R., Christopher Flavin, and Sandra Postel. "Reusing and Recycling Materials." In *Saving the Planet: How to Shape an Environmentally Sustainable Global Economy.* New York: Norton, 1991.

Carra, Joseph S., ed. *International Perspectives on Municipal Solid Wastes and Sanitary Landfilling.* San Diego, Calif.: Academic Press.

Denison, Richard A., and John Ruston. *Recycling and Incineration.* Washington, D.C.: Island Press, 1990.

Frosch, Robert A., and Nicholas E. Gallopoulos. "Strategies for Manufacturing." *Managing Planet Earth: Readings from* Scientific American. New York: W. H. Freeman and Company, 1990.

Gould, Robert N. *1992–1993 Materials Recovery and Recycling Yearbook.* New York: Governmental Advisory Associates, 1992.

Institute for Local Self-Reliance. *Beyond 40 Percent: Record-Setting Recycling and Composting Programs.* Washington, D.C.: Island Press, 1990.

Kharbanda, O. P., and Ernest A. Stallworthy. *Waste Management: Toward a*

Sustainable Society. New York: Auburn House, 1990.

Lazare, Daniel. "Recycled But Not Used." *The Amicus Journal* 13, no. 4 (1991): 20–27.

Morris, David. "As If Materials Mattered." *The Amicus Journal* 13, no. 4 (1991): 17–20.

Newsday. Rush to Burn: Solving America's Garbage Crisis. Washington, D.C.: Island Press, 1989.

Rathje, William L. "Once and Future Landfills." *National Geographic* 179, no. 5 (1991): 116–134.

Reed, Sherwood C., E. Joe Middlebrooks, and Ronald W. Cities. *Natural Systems for Waste Management and Treatment.* New York: McGraw-Hill, 1988.

Robinson, William D. *The Solid Waste Handbook: A Practical Guide.* New York: Wiley, 1986.

Selke, Susan E. *Packaging and the Environment: Alternatives, Trends, and Solutions.* Lancaster, Penn: Technomic, 1990.

Senior, Eric. *Microbiology of Landfill Sites.* Boca Raton, Fla.: CRC Press, 1990.

Wolf, Nancy, and Ellen Feldman. *Plastics: America's Packaging Dilemma.* Washington, D.C.: Island Press, 1990.

Chapter 21

Fishbein, Seymour L., ed. *Wilderness U.S.A.* Washington, D.C.: National Geographic Society, 1973.

Fisher, Ron, ed. *Our Threatened Inheritance: Natural Treasures of the United States.* Washington, D.C.: National Geographic Society, 1984.

Fox, Stephen. *The American Conservation Movement.* Madison: The University of Wisconsin Press, 1981.

Grove, Noel. "Greenways: Paths of the Future." *National Geographic* 177, no. 6 (1990): 77–99.

Hunter, Malcom L. *Wildlife, Forests and Forestry: Principles of Managing Forests for Biological Diversity.* Englewood Cliffs, N.J.: Prentice-Hall, 1990.

Junkin, Elizabeth Darby. *Lands of Brighter Destiny: The Public Lands of the American West.* Golden, Colo.: Fulcrum, 1986.

Muir, John. *The Yosemite.* New York: Century, 1912.

O'Toole, Randal. *Reforming the Forest Service.* Washington, D.C.: Island Press, 1988.

Pritchard, Paul C. "The Best Idea America Ever Had." *National Geographic* 179, no. 8 (1991): 36–59.

Reisner, Marc. "The Sting." *The Amicus Journal.* 10, no. 2 (1988).

The Report of the President's Commission: Americans Outdoors. Washington, D.C.: Island Press, 1987.

Wellman, J. Douglas. *Wildland Recreation Policy: An Introduction.* New York: Wiley, 1987.

Zaslowsky, Dyan, and the Wilderness Society. *These American Lands.* New York: Holt, Rinehart and Winston, 1986.

Zuckerman, Seth, and The Wilderness Society. *Saving Our Ancient Forests.* Venice, Calif.: Living Planet Press, 1991.

Chapter 22

Adams, Ansel. *The American Wilderness.* Boston: Little, Brown, 1990.

Agee, James K., and Darryll R. Johnson. *Ecosystem Management for Parks and Wilderness.* Seattle: University of Washington Press, 1988.

Borrelli, Peter. "Timber!" *The Amicus Journal* 12, no. 1 (1990): 41.

Cahill, Tim. "The Splendors of Lechuguilla Cave." *National Geographic* 179, no. 3 (1991): 34–59.

Findley, Rowe. "Along the Santa Fe Trail." *National Geographic* 179, no. 3 (1991): 98–123.

Findley, Rowe. "Will We Save Our Own?" *National Geographic* 178, no. 3 (1990): 106–136.

Fox, Stephen. *The American Conservation Movement.* Madison: The University of Wisconsin Press, 1981.

Gladden, James N. *Boundary Waters Canoe Area: Wilderness Values and Motorized Recreation.* Ames: University of Iowa Press, 1990.

Hodgson, Brian. "Alaska's Big Spill: Can the Wilderness Heal?" *National Geographic* 177, no. 1 (1990): 5–43.

Lanting, Franz. "Botswana: A Gathering of Waters and Wildlife." *National Geographic* 178, no. 6 (1990): 5–37.

Lee, Douglas B., and Franz Lanting. "Okavango Delta: Old Africa's Last Refuge." *National Geographic* 178, no. 6 (1990): 38–69.

Lopez, Barry. *Arctic Dreams: Imagination and Desire in a Northern Landscape.* New York: Scribner, 1986.

Lopez, Barry. *Crossing Open Ground.* New York: Scribner, 1988.

Lopez, Barry. *Desert Notes: Reflections in the Eye of a Raven.* New York: Avon Books, 1990.

Lopez, Barry. *Of Wolves and Men.* New York: Scribner, 1978.

Lopez, Barry. *River Notes: The Dance of the Herons.* Kansas City, Kan.: Andrews and McMeel, 1979.

McPhee, John. *Encounters with the Archdruid.* New York: Farrar, Straus & Giroux, 1971.

Norse, Elliot. "What Good Are Ancient Forests?" *The Amicus Journal* 12, no. 1 (1990): 42–45.

Soderberg, K. A., and Jackie DuRette. *People of the Tongass: Alaska Forestry Under Attack.* Bellvue, Wash.: Free Enterprise Press, 1988.

Stiak, Jim. "Old Growth!" *The Amicus Journal* 12, no. 1 (1990): 34–40.

Chapter 23

Ackerman, Diane. "Last Refuge of the Monk Seal." *National Geographic* 180, no. 1 (1992): 128–144.

Balog, James. "A Personal Vision of Vanishing Wildlife." *National Geographic* 177, no. 4 (1990): 84–103.

Bertness, Mark D. "The Ecology of a New England Salt Marsh." *American Scientist* 80 (1992): 260–268.

Brower, Kenneth. "The Destruction of Dolphins." *Atlantic* 263 (1989): 35–55.

Brown, Lester, Christopher Flavin, and Sandra Postel. "Protecting the Biological Base." In *Saving the Planet: How to Shape an Environmentally Sustainable Global Economy.* New York: Norton, 1991.

Chadwick, Douglas. "Elephants—Out of Time, Out of Space." *National Geographic* 179, no. 5 (1991): 2–49.

Di Silvestor, R., ed. *Audubon Wildlife Report 1986.* New York: National Audubon Society, 1986.

Durrell, Lee. *State of the Ark.* Garden City, N.Y.: Doubleday, 1986.

Eckholm, Erik P. *Down to Earth: Environment and Human Needs.* New York: Norton, 1982.

Ehrlich, Paul R., and Anne H. Ehrlich. *Extinction: The Causes and Consequences of the Disappearance of Species.* New York: Ballantine Books, 1983.

Ehrlich, Paul R., and E. O. Wilson. "Biodiversity Studies: Science and Policy." *Science* 253 (1991): 758–762.

Franklin, William L. "Patagonia Puma: Lord of Land's End." *National Geographic* 179, no. 1 (1991): 102–113.

Hallé, Francis. "A Raft Atop the Rain Forest." *National Geographic* 178, no. 4 (1990): 129–138.

Lean, Geoffrey, Don Hinrichsen, and Adam Markham. *Atlas of the Environment.* New York: Prentice Hall Press, 1990.

Mann, Charles C., and Mark L. Plummer. "The Butterfly Problem." *Atlantic* 269, no. 1 (1992): 47–73.

Matthiesson, Peter. *Wildlife in America.* New York: Viking Press, 1987.

McNeely, Jeffrey, et. al. *Conserving the World's Biological Diversity.* Gland, Switzerland: International Union for Conservation of Nature and Natural Resources, 1990.

Mitsch, William J., and James G. Gosselink. *Wetlands.* New York: Van Nostrand Reinhold Co., 1986.

Myers, Norman. *The Sinking Ark: A New Look at the Problem of Disappearing Species.* Oxford: Pergamon Press, 1979.

Myers, Norman. *A Wealth of Wild Species: Storehouse for Human Welfare.* Boulder, Colo.: Westview Press, 1983.

National Geographic Society. "Tropical Rainforests." *National Geographic* 163, no. 1 (1983): 2–65.

Norton, Bryan G. *The Preservation of Species: The Value of Biological Diversity.*

Princeton, N.J.: Princeton University Press, 1986.

Office of Technology Assesment. *Technologies to Maintain Biological Diversity*. Washington, D.C.: Government Printing Office, 1987.

Poten, Constance J. "America's Illegal Wildlife Trade." *National Geographic* 179, no. 9 (1991): 106–132.

Reisner, Marc. "The Sting." *The Amicus Journal* 10, no. 2 (1988): 40–47.

Risley, Dave. "Reflicts of the Past: Wetlands on Private Lands." *Wild Ohio* 2, no. 1 (1992).

Rowell, Galen. "Falcon Rescue." *National Geographic* 179, no. 4 (1991): 106–115.

Ryan, John C. "Life Support: Conserving Biological Diversity." *Worldwatch Paper 108*. Washington, D.C.: Worldwatch Institute, 1992.

Thorne-Miller, Bruce, and John Catena. *The Living Ocean: Understanding and Protecting Marine Biodiversity*. Washington, D.C.: Island Press, 1990.

Tuttle, Merlin D. "Bats—The Cactus Connection." *National Geographic* 179, no. 6 (1991): 131–140.

U.S. Fish and Wildlife Service. *Wetlands: Meeting the President's Challenge*. Washington, D.C.: U.S.F.W.S., 1990.

Van Dyk, Jere. "Long Journey of the Pacific Salmon." *National Geographic* 178, no. 1 (1990): 3–37.

Ward, Fred. "Florida's Coral Reefs Are Imperiled." *National Geographic* 178, no. 1 (1990): 89–113.

Wilson, Edward O. *Biodiversity*. Washington, D.C.: National Academy Press, 1988.

Wilson, Edward O. "Rain Forest Canopy: The High Frontier." *National Geographic* 179, no. 12 (1991): 78–107.

Wilson, Edward O. "Threats to Biodiversity." *Scientific American* 261, no. 3 (1989): 108–16.

Winckler, Suzanne. "Stopgap Measures." *Atlantic* 269, no. 1 (1992): 74–82.

World Resources Institute. "Forests and Rangelands." *World Resources 1992–1993: A Guide to the Global Environment*. New York: Oxford University Press, 1992.

World Resources Institute. "Wildlife and Habitat." In *World Resources 1990–1991*. Oxford: Oxford University Press, 1990.

World Resources Institute. "Wildlife and Habitat." *World Resources 1992–1993: A Guide to the Global Environment*. New York: Oxford University Press, 1992.

Yost, Nicholas, et al. "Wetlands: Through Murky Waters." *NAEP Newsletter* 17, no. 2 (1992).

Ziffer, Karen A., J. Martin Goebel, and Susanna Mudge. "Ecotourism." *Orion* 9, no. 2 (1990): 42–45.

Chapter 24

Baldwin, James. *Whole Earth Ecolog: The Best of Environmental Tools & Ideas*. New York: Harmony Books, 1990.

Beckwith, Carol, and Angela Fisher. "The Eloquent Surma of Ethiopia." *National Geographic* 179, no. 2 (1991): 77–102.

Budianky, Stephen. "The Trees Fell—and so did the People." *U.S. News & World Report* February 9, (1987).

Craig, Bruce. "Bones of Contention: The Controversy over Digging Up Human Remains in Parks." *National Parks* July/August (1990): 16–17.

Cultural Survival, Inc. "Deforestation: The Human Costs." *Quarterly* 6, no. 2 (1982).

Culture and the Future. Paris, France: United Nations Educational, Scientific and Cultural Organization, 1985.

Daniel, John. "Stealing Time." *Wilderness* 53, no. 188 (1990): 18–38.

Denslow, Julie S., and Christie Padoch. *People of the Tropical Rainforest*. Berkeley: University of California Press, 1988.

Errahmani, Abdelkader Brahim. "The World Heritage Convention: A New Idea Takes Shape." *Courier*. Paris, France: UNESCO, 1990.

Hunter, Robert. *Warriors of the Rainbow: A Chronicle of the Greenpeace Movement*. New York: Holt, Rinehart and Winston, 1979.

Ki-Zerbo, Joseph. "Oral Tradition as a Historical Source." *Courier*. Paris, France: UNESCO, 1990.

National Trust for Historic Preservation. *The Brown Book*. Washington, D.C.: The Preservation Press, 1983.

Ofenstein, Sharon. "Restoring the Statue of Liberty." *CRM Bulletin* 9, no. 2 (1986).

Peterson, Ivars. "A Material Loss." *Science News* 128 (1985): 154–156.

Robbins, Jim. "Violating History." *National Parks*, July/August (1987): 26–31.

Sadek, Hind. "Treasures of Ancient Egypt: Preserving Civilization's Legacy in World Cultural Parks." *National Parks*, May/June (1990): 16–17.

Smith, Brent W., and Lynn E. Dwyer. "Unresolved Issues in an Emerging Profession: Some Problems in Cultural Resource Management." *Environmental Professional* 4 (1982): 177–180.

Spirn, Anne Whiston. "From Uluru to Cooper's Place: Patterns in Cultural Landscape." *Orion* 9, no. 2 (1990): 32–39.

Sugarman, Aaron. "The Treasures of America . . . Looted!" *Condé Nast Traveler*, July (1992).

Ward, Diane Raines. "In Anatolia, a Massive Dam Project Drowns Traces of an Ancient Past." *Smithsonian* 21, no. 5 (1990): 29–40.

Wigginton, Eliot, ed. *The Foxfire Book*. Garden City, N.Y.: Doubleday, 1972.

Chapter 25

Attfield, Robin. *The Ethics of Environmental Concern*. Oxford: Basil Blackwell, 1983.

Austin, Richard Cartwright. *Environmental Theology*. Atlanta, Ga.: J. Knox Press, 1988.

Berry, Thomas, and Thomas Clark, S. J. *Befriending the Earth*. Mystic, Conn.: Twenty-Third Publications, 1991.

Birch, Charles, and John B. Cobb, Jr. *The Liberation of Life: From the Cell to the Community*. New York: Cambridge University Press, 1981.

Borrelli, Peter. "Epiphany." *The Amicus Journal*. Winter 1986.

Cunningham, Lawrence. *Saint Francis of Assisi*. Boston Mass.: Twayne Publishers, 1976.

Dalai Lama of Tibet. "A Universal Task." *EPA Journal* 17, no. 4 (1991).

Davis, Donald Edward. *Ecophilosophy: A Field Guide to the Literature*. San Pedro. Calif.: Miles, 1989.

Dower, Nigel. *Ethics and the Environmental Responsibility*. Brookfield, Vt.: Avebury, 1989.

Dudley, William. *Genetic Engineering: Opposing Viewpoints*. San Diego, Calif.: Greenhaven Press, 1990.

EarthKeeping News 1, no. 1 (1991).

Engel, J. Ronald, and Joan Gibb Engel. *Ethics of Environment and Development*. Tucson: University of Arizona Press, 1990.

Hargrove, Eugene C. *Foundations of Environmental Ethics*. Englewood Cliffs, N.J.: Prentice-Hall, 1989.

Lappé, Marc. *Broken Code: The Exploitation of DNA*. San Francisco: Sierra Club Books, 1984.

Leopold, Aldo. *A Sand County Almanac, and Sketches Here and There*. New York: Oxford University Press, 1949.

Meeker, Joseph W. "The Assisi Connection." *Wilderness* 51 (1988): 61–64.

Nash, Roderick. *The Rights of Nature*. Madison: University of Wisconsin Press, 1989.

Nossal, G. J. V. *Reshaping Life: Key Issues in Genetic Engineering*. New York: Cambridge University Press, 1989.

"The Ozone Layer: A Different Perspective." *The Christian Science Monitor*. Monday, February 29 (1988).

Plant, Judith, ed. *Healing the Wounds*. Philadelphia: New Society Publishers, 1989.

Rifkin, Jeremy. *Declaration of a Heretic*. Boston: Routledge & Kegan Paul, 1985.

Roberts, Elizabeth, and Elias Amidon. *Earth Prayers from Around the World: 365 Prayers, Poems, and Invocations for Honoring the Earth*. New York: Harper-Collins, 1991.

Rolston, Holmes. *Environmental Ethics, Duties to and Values in the Natural World*. Philadelphia: Temple University Press, 1988.

Rolston, Holmes. *Philosophy Gone Wild: Essays in Environmental Ethics*. Buffalo, N.Y.: Prometheus Books, 1986.

Rossi, Vincent. "The Eleventh Commandment: Toward an Ethic of Ecology." *Epiphany Journal*.

Santos, Michael. *Genetics and Man's Future: Legal, Social, and Moral Implications of Genetic Engineering.* Springfield, Ill.: C. C Thomas, 1981.

Shrader-Frechette, K.S. *Environmental Ethics.* Pacific Grove, Calif.: Boxwood Press, 1981.

Swimme, Brian. *The Universe is a Green Dragon.* Santa Fe, N.M.: Bear and Company, 1984.

Weiss, Edith Brown. *In Fairness to Future Generations: International Law, Common Patrimony, and Intergenerational Equity.* Ardsley-on-Hudson, N.Y.: Transactional Publishers, 1989.

Witt, Steven C. *Biotechnology, Microbes and the Environment.* San Francisco: Center for Science and Information, 1990.

Chapter 26

Allin, Craig W. *The Politics of Wilderness Preservation.* Westport, Conn.: Greenwood Press, 1982.

Baumol, William J., and Wallace Oates. *The Theory of Environmental Policy.* New York: Cambridge University Press, 1988.

Boo, Elizabeth. *Ecotourism: The Potentials and Pitfalls,* vol. 1, 2. Washington, D.C.: World Wildlife Fund, 1990.

Brown, Lester R., Christopher Flavin, and Sandra Postel. "Banking on the Environment." In *Saving the Planet: How to Shape an Environmentally Sustainable Global Economy.* New York: Norton, 1991.

Brown, Lester R., Christopher Flavin, and Sandra Postel. "Green Taxes." In *Saving the Planet: How to Shape an Environmentally Sustainable Global Economy.* New York: Norton, 1991.

Brown, Lester R., Christopher Flavin, and Sandra Postel. "Reshaping Government Incentives." In *Saving the Planet: How to Shape an Environmentally Sustainable Global Economy.* New York: Norton, 1991.

Capra, Fritjof, and Charlene Spretnak. *Green Politics.* New York: Dutton, 1984.

Caring for the Earth: A Strategy for Sustainable Living. Gland, Switzerland: The World Conservation Union, 1991.

Daly, Herman E., and John B. Cobb, Jr. *For the Common Good: Redirecting the Economy Toward Community, the Environment, and a Sustainable Future.* Boston: Beacon Press, 1989.

deBeer, Jenne H., and Melanie J. McDermott. *The Economic Value of Non-timber Forest Products in Southeast Asia.* Gland, Switzerland: International Union for the Conservation of Nature and Natural Resources, 1989.

Decker, Daniel J., and Gary R. Goff, eds. *Valuing Wildlife: Economic and Social Perspectives.* Boulder, Colo.: Westview Press, 1987.

Dixon, John A., and Paul B. Sherman. *Economics of Protected Areas: A New Look at Benefits and Costs.* Washington, D.C.: Island Press, 1990.

Fisher, Anthony C. *Resource and Environmental Economics.* New York: Cambridge University Press, 1981.

French, Hilary F. "Strengthening Global Environmental Governance." In *State of the World, 1992.* New York: Norton, 1992.

Goodland, Robert, ed. *Race to Save the Tropics: Ecology and Economics for a Sustainable Future.* Washington, D.C.: Island Press, 1990.

Grassroots Development: The African Development Foundation. Washington, D.C.: Office of Technology Assessment, Congress of the U.S., 1988.

Ingram, Helen. *Water, Politics, Community, and Change.* Albuquerque: University of New Mexico, 1990.

Ledec, George, and Robert Goodland. *Wildlands: Their Protection and Management in Economic Development.* Washington, D.C.: World Bank, 1988.

Norse, Elliot A. *Ancient Forests of the Pacific Northwest: Sustaining Biological Diversity and Timber Production in a Changing World.* Washington, D.C.: Island Press, 1990.

Oppenheimer, Michael. *Dead Heat: The Race Against the Greenhouse Effect.* New York: Basic Books, 1990.

Raphael, Ray. *Tree Talk: The People and Politics of Timber.* Washington, D.C.: Island Press, 1981.

Repetto, Robert, et al. *Wasting Assets: Natural Resources in the National Income Accounts.* Washington, D.C.: World Resources Institute, 1989.

Repetto, Robert, and Malcolm Gillis. *Public Policies and the Misuse of Forest Resources.* New York: Cambridge University Press, 1988.

Sweezy, Paul. "Socialism and Ecology." *Monthly Review* 41 (1989): 1–9.

Tobin, Richard J. *The Expendable Future: U.S. Politics and the Protection of Biological Diversity.* Durham, N.C.: Duke University Press, 1990.

World Resources Institute. "Policies and Institutions: Natural Resources Accounting." In *World Resources 1990–1991.* Oxford: Oxford University Press, 1990.

World Resources Institute. "Forests and Rangelands." *World Resources 1992–1993: A Guide to the Global Environment.* New York: Oxford University Press, 1992.

World Resources Institute. "Policies and Institutions: Nongovernmental Organizations." *World Resources 1992–1993: A Guide to the Global Environment.* New York: Oxford University Press, 1992.

World Resources Institute. "Sustainable Development." *World Resources 1992–1993: A Guide to the Global Environment.* New York: Oxford University Press, 1992.

Ziffer, Karen A., J. Martin Goebel, and Susanna Mudge. "Ecotourism." *Orion* 9, no. 2 (1990): 42–45.

Chapter 27

Benneman, Russell L., and Sarah M. Bates. *Land-Saving Action.* Washington, D.C.: Island Press, 1984.

Bingham, Gail. *Resolving Environmental Disputes: A Decade of of Experience.* Washington, D.C.: The Conservation Foundation, 1986.

BNA Editorial Staff, eds. *U.S. Environmental Laws.* Washington, D.C.: Bureau of National Affairs, 1988.

Crowfoot, James E., and Julia M. Wondolleck. *Environmental Disputes: Community Involvement in Conflict Resolution.* Washington, D.C.: Island Press, 1990.

Environmental Law Institute. *Environmental Law Deskbook.* Washington, D.C.: Environmental Law Institute, 1989.

Gaskins, Richard H. *Environmental Accidents: Personal Injury and Public Responsibility.* Philadelphia: Temple University Press, 1989.

Sagoff, Mark. *The Economy of the Earth: Philosophy, Law, and the Environment.* New York: Cambridge University Press, 1988.

Santos, Miguel A. *Managing Planet Earth: Perspectives on Population, Ecology, and the Law.* New York: Bergin & Garvey, 1990.

Simon, John J. *Our Common Lands: Defending the National Parks.* Washington, D.C.: Island Press, 1988.

Stone, Christopher D. *Should Trees Have Standing? Toward Legal Rights for Natural Objects.* Los Altos, Calif.: Kaufman, 1973.

Turner, Tom. *Wild by Law: The Sierra Club Legal Defense Fund and the Places It Has Saved.* San Francisco: Sierra Club Books, 1990.

Westing, Arthur H. *Global Resources and International Conflict: Environmental Factors in Strategic Policy and Action.* Oxford: Oxford University Press, 1986.

Chapter 28

Berkmuller, Klaus. *Environmental Education About the Rainforest.* Ann Arbor: Wildland Management Center, University of Michigan, 1984.

Brough, Holly. "Nature's Classroom." *World Watch* 4, no. 3 (1991).

Cohen, Michael J. *Across the Running Tide.* Freeport, Me.: Cobblestone, 1979.

Cohen, Michael J., Ed. D. *Prejudice Against Nature: A Guidebook for the Liberation of Self and Planet.* Freeport, Me.: Cobblestone, 1984.

Eagan, David J., and David W. Orr. *The Campus and Environmental Responsibility.* San Francisco: Jossey-Bass Publishers, no. 77 (Spring 1992).

The Earth Works Group. *50 Simple Things You Can Do to Save the Earth.* Berkeley, Calif.: Earth Works Press, 1989.

Elkington, John, et al. *Going Green: A Kid's Handbook to Saving the Planet.* New York: Viking Press, 1990.

Ferguson, Marilyn, et. al. *Who is Who in Service to the Earth and 41 Visions of a Positive Future*. North Carolina: VisionLink Education Foundation, 1991.

Gibbons, Boyd. "Missouri's Garden of Consequence." *National Geographic* 178, no. 2 (1990): 124–140.

Gray, David B. *Ecological Beliefs and Behaviors: Assessment and Change*. Westport, Conn.: Greenwood Press, 1985.

Head, Suzanne, and Robert Heinzman. *Lessons of the Rainforest*. San Francisco: Sierra Club Books, 1990.

Iozzi, Louis A. "What Research Says to the Educator—Part Two: Environmental Education and the Affective Domain." *Journal of Environmental Education* 20, no. 4 (1989): 6–14.

Monroe, Martha C., and Stephen Kaplan. "When Words Speak Louder Than A.ctions: Environmental Problem Solving in the Classroom." *Journal of Environmental Education* 19, no. 3 (1988): 38–42.

Student Environmental Action Coalition. *The Student Environmental Action Guide: 25 Simple Things We Can Do*. Berkeley, Calif.: Earth Works, Group, HarperCollins Publishers, 1991.

Van Marte, Steve. *Earth Education: A New Beginning*. Warrenville, Ill.: Institute for Earth Education, 1990.

Van Marte, Steve, and Bill Weiller. *The Earth Speaks*. Warrenville, Ill.: Institute for Earth Education, 1983.

Wilson, *Inside Outward Bound*. Charlotte, N.C.: East Woods Press, 1981.

Credits

Photographs

Illustrations, Tables, and Boxes

Figure 5-7: Susan Friedmann.

Figure 5-13 (a–c): Figure from ECOLOGY AND OUR ENDANGERED LIFE-SUPPORT SYSTEMS by Eugene P. Odum. Copyright © 1989 by Sinauer Associates, Inc., Publishers. Reprinted by permission of Sinauer Associates, Inc., Publishers.

Figure 5-17: Map of Everglades National Park: courtesy Laura Taylor.

Figure 5-21: Table from FUNDAMENTALS OF ECOLOGY, Third Edition by Eugene P. Odum. Copyright © 1971 by Saunders College Publishers, reproduced by permission of the publisher.

Table 5-2: Table from FUNDAMENTALS OF ECOLOGY, Third Edition by Eugene P. Odum. Copyright © 1971 by Saunders College Publishers, reproduced by permission of the publisher.

Focus on: Fields of Gold, p. 78: As adapted from Richard B. Fischer's essay entitled "Goldenrods: An Ecological Goldmine," which first appeared in THE AMERICAN BIOLOGY TEACHER, Volume 47, Number 7, October 1985.

Figure 6-6: "World Desertification Hazards" from STATE OF THE ENVIRONMENT 1985. Copyright © OECD, 1985. Reprinted by permission.

Figure 6-13: Susan Friedmann.

Figure 7-12: Map of Fermi Lab Area: courtesy Laura Taylor.

Figure 8-3: Susan Friedmann.

Figure 8-4: "The Growing Human Population," by Nathan Keyfitz. Copyright © 1989 by Scientific American, Inc. All rights reserved.

Figure 8-9: Source: CONNECTIONS, LINKING POPULATION AND THE ENVIRONMENT, Population Reference Bureau, 1991.

Figure 8-15: Source: IMR data from Population Reference Bureau, 1991 World Population Data Sheet; data on clean water from STATE OF THE WORLD'S CHILDREN, 1990.

Figure 8-16: OUR OWN FREEDOM, Buchi Emeta; taken from Parents Plan International. Reprinted with permission of Childreach.

Table 8-1: Source: Data from 1991 WORLD POPULATION DATA SHEET OF THE POPULATION REFERENCE BUREAU, INC.

Table 8-2: Source: Data from 1991 WORLD POPULATION DATA SHEET OF THE POPULATION REFERENCE BUREAU, INC.

Table 8-3: Source: Data from 1991 WORLD POPULATION DATA SHEET OF THE POPULATION REFERENCE BUREAU, INC.

Table 8-4: Excerpt from table "Gross National Product per capita, top ten and bottom ten countries, 1991" from THE 1992 INFORMATION PLEASE ENVIRONMENTAL ALMANAC, compiled by World Resources Institute. Copyright © 1991 by World Resources Institute. Reprinted by permission of Houghton Mifflin Company.

Table 8-6: "Women's Status in the Highest and Lowest Ranked Countries" from POPULATION CRISIS COMMITTEE. Reprinted by permission of Population Crisis Committee.

Box 9-1: Information adapted from a series on personal ecology by Monte Paulsen, editor and publisher of Casco Bay Weekly, in Portland, Maine. Reprinted by

permission of Monte Paulsen.

Figure 10-3: DIET FOR A SMALL PLANET by Frances Moore Lappé (Ballantine Books, 1991). Copyright © 1971, 1975, 1982, 1991 by Frances Moore Lappé. Reprinted by permission of the author and her agents, Raines & Raines, 71 Park Avenue, New York, NY 10016.

Figure 10-4: Reprinted with permission from DIET FOR A NEW AMERICA by John Robbins (Stillpoint Publishing, Walpole, NH 03608: 800/847-4014.)

Table 10-2: Table from STATE OF THE WORLD 1990 by Lester R. Brown et al. Copyright © 1990 by Worldwatch Institute. Reprinted by permission of W.W. Norton & Company.

Table 10-4: from NEW BOOK OF WORLD RANKINGS, Third Edition by George Thomas Kurian. Copyright © 1991 by Facts On File. Reprinted with permission of Facts On File Inc., New York.

Table 10-5: from NEW BOOK OF WORLD RANKINGS, Third Edition by George Thomas Kurian. Copyright © 1991 by Facts On File. Reprinted with permission of Facts On File Inc., New York.

Table 10-6: Sources: October 1985 food prices were extracted from the BULLETIN OF LABOUR STATISTICS 1986, published by the ILO. Prices from the foreign currencies were converted to U.S. dollars at the exchange rates of October 1985. For GNP per capita per day, we divided UNICEF's 1985 GNP per capita figures by 365. Percent of income spent on food is from SOLVING WORLD HUNGER: THE U.S. STAKE, by the Consortium for International Cooperation in Higher Education, 1986. Reprinted with permission of Childreach.

Table 10-7: from NEW BOOK OF WORLD RANKINGS, Third Edition by George Thomas Kurian. Copyright © 1991 by Facts On File. Reprinted with permission of Facts On File Inc., New York.

Table 10-8: Table from STATE OF THE WORLD 1990 by Lester R. Brown et al. Copyright © 1990 by Worldwatch Institute. Reprinted by permission of W.W. Norton & Company.

Figure 11-7: "Trans-Alaskan Pipeline" reprinted by permission of Exxon Corporation, courtesy of American Petroleum Institute.

Figure 12-2: Susan Friedmann.

Figure 12-3: Line drawing of a coal-fired power plant courtesy of Cincinnati Gas and Electric Company. Reprinted by permission.

Figure 12-4: From the book ATLAS OF THE ENVIRONMENT by Geoffrey Lean, Don Hinrichsen, and Adam Markham. Copyright © 1990 by Banson Marketing Ltd. Used by permission of the publisher Prentice Hall Press, a Division of Simon & Schuster, New York.

Figure 12-10(a): Graph, "World Price of Oil, 1970–87" from a Worldwatch Paper. Reprinted by permission of Worldwatch Institute.

Table 13-1: "Cost of renewable electricity" from WORLDWATCH Magazine, Volume 4, Number 5, September/October 1991. Copyright © 1991 by Worldwatch Institute. Reprinted by permission.

Figure 14-16: "A Brief History of Lung Cancer" from BIOLOGY: THE SCIENCE OF LIFE, Third Edition by Robert A. Wallace et al. Copyright © 1991 by HarperCollins Publishers, Inc.

Figure 15-4: Map from 1991–1992 GREEN INDEX: A STATE-BY-STATE GUIDE TO THE NATION'S ENVIRONMENTAL HEALTH by Bob Hall and Mary Lee Kerr. Copyright © 1991 Institute for Southern Studies. Reprinted by permission of Island Press, a Division of The Center for Resource Economics.

Table 15-2: Table showing Water Scarcity, Selected Countries and Regions from WATER FOR AGRICULTURE: FACING THE LIMITS by Sandra Postel, Worldwatch Paper 93, December 1989. Reprinted by permission of Worldwatch Institute.

Box 15-1: From "Tips to Stem the Flow Of Water Waste" from CHEMECOLOGY, October 1991. Based on "50 Ways to Save a Drop," published by the Southwest Florida Water Management District. Reprinted by permission.

Figure 16-4: Susan Friedmann.

Figure 16-12: "Ten-Year Crop Rotation" by Alexandra Schulz as appeared in HARROWSMITH Magazine, September/October, 1989.

Table 16-4: Table "Annual Off-Site Damage from Soil Erosion in the United States" from THE 1992 INFORMATION PLEASE ENVIRONMENTAL ALMANAC, compiled by World Resources Institute. Copyright © 1991 by World Resources Institute. Reprinted by permission of Houghton Mifflin Company.

Figure 17-3: Data adapted from THE EARTH REPORT: THE ESSENTIAL GUIDE TO GLOBAL ECOLOGICAL ISSUES edited by Edward Goldsmith and Nicholas Hildyard. Copyright © Mitchell Beazley Publishers, 1988. Reprinted by permission of Price Stern Sloan, Inc.

Figure 17-11: "For 1 Jet-Fighter Engine" from U.S. NEWS & WORLD REPORT, February 8, 1982. Copyright © February 8, 1982 by U.S. News & World Report. Reprinted by permission.

Table 17-3: Adapted from illustration (modified) by Ed Bell from page 146 of STRATEGIES FOR MANUFACTURING by Robert A. Frosch and Nicholas Gallopoulos. Copyright © 1989 by Scientific American, Inc. All rights reserved.

Figure 18-2: "Radiation" from RADIATION: DOSES, EFFECTS, RISKS by United Nations Environment Programme. Copyright © 1985 by United Nations Environment Programme. Reprinted with permission.

Figure 18-3: "Radiation" from RADIATION: DOSES, EFFECTS, RISKS by United Nations Environment Programme. Copyright © 1985 by United Nations Environment Programme. Reprinted with permission.

Figure 18-4: Reprinted with permission from RADIOACTIVE WASTES AT THE HANFORD RESERVATION: A TECHNICAL REVIEW. Published by the National Academy of Sciences, Washington, D.C., 1978.

Figure 18-9: Line drawing of The World's Nuclear Firepower by James Geier from PROMOTING ENDURING PEACE. Reprinted by permission of Promoting Enduring Peace.

Figure 18-10: Map of nuclear weapons production sites in the United States from Physicians for Social Responsibility. Reprinted by permission.

Figure 18-11: "The Radioactive Cloud: A Calculation of Its Reach" from THE NEW YORK TIMES, May 16, 1986. Copyright © 1986 by The New York Times Company.

Index